THE MAMMALS OF THE INDOMALAYAN REGION

THE MAMMALS OF THE INDOMALAYAN REGION: A SYSTEMATIC REVIEW

by
G.B. Corbet and J.E. Hill

NATURAL HISTORY MUSEUM PUBLICATIONS

OXFORD UNIVERSITY PRESS
1992

Oxford University Press, Walton Street, Oxford OX2 6DP
Oxford New York Toronto
Delhi Bombay Calcutta Madras Karachi
Petaling Java Singapore Hong Kong Tokyo
Nairobi Dar es Salaam Cape Town
Melbourne Auckland

and associated companies in
Berlin & Ibadan

Oxford is a trade mark of the Oxford University Press

Published in the United States
by Oxford University Press, New York

A catalogue record for this book is available from the British Library

Library of Congress Cataloging-in-Publication Data

Corbet, G.B. (Gordon Barclay)
The mammals of the Indomalayan Region: a systematic review/by
G.B. Corbet & J.E. Hill.
p. cm.
Includes bibliographical references (39 pp.) and index.
ISBN 0-19-854693-9: £60.00 ($90.00 U.S.)
1. Mammals—Asia, Southeastern—Classification.
2. Mammals—South Asia—Classification.
3. Mammals—Asia, Southeastern—Geographical distribution.
4. Mammals—South Asia—Geographical distribution.
I. Hill, John Edwards. II. Title.
QL729.A785C67 1992 92-299
599.0954—dc20 CIP

ISBN 0-19-854693-9

Typeset in times by Spottiswoode Ballantyne
Printed by Bookcraft

CONTENTS

PREFACE

The Indomalayan (or Oriental) Region has been recognized for over a century as one of the major zoogeographical regions of the world, with about a thousand species of mammals, over 20 per cent of the world's mammalian fauna. However, there appears to have been no previous attempt to produce a comprehensive, integrated account of its mammal fauna. Hitherto, four separate nomenclatorial lists have together covered the region, all of them now very out of date but nevertheless providing a valuable foundation for the present work. These are:

Ellerman & Morrison-Scott (1951, 1965): *Checklist of Palaearctic and Indian mammals 1758–1946*, which covers the region north of 10°N, i.e. north of the Isthmus of Kra in peninsular Thailand.

Chasen (1940a): *A handlist of Malaysian mammals*, covering the Malayan peninsula and the Greater Sunda Islands except Palawan.

Laurie & Hill (1953): *List of land mammals of New Guinea, Celebes and adjacent islands, 1758–1952*, covering all the Indonesian islands not dealt with by Chasen.

Taylor (1934): *Philippine land mammals.*

Except for Taylor's work on the Philippines these did not purport to provide descriptions nor, with some exceptions, keys to identification, but they did provide a valuable stimulus for further taxonomic work by postulating the limits of species, by pinpointing gaps and by erecting hypotheses of relationships that could be subjected to testing as new information became available. However, the research that has been conducted during the past 50 years has for the most part been highly fragmented and localized. Some of the results have been summarized in a few important regional works but there are many gaps.

Our objective in compiling this work has been to bring together in one place an account of every species of mammal in the region, documenting the names and synonyms in full, and placing them within a taxonomic framework that takes account of all relevant published research. The order Chiroptera (bats) has been dealt with by JE Hill, based largely upon research undertaken while on the staff of The Natural History Museum, London. The remaining orders and the Introduction have been dealt with by GB Corbet, based primarily upon a critical review of the existing literature, but the results have been tested and many gaps filled by use of the extensive collections of The Natural History Museum, supplemented by visits to many other museums with important collections of Indomalayan mammals. We have scrutinized and criticized each other's text but the taxonomic opinions expressed are those of the individual author.

About 7000 scientific names are documented (including c.340 based upon fossils described in Recent genera), with type localities in the case of species-group names. These are backed up by a bibliography of about 3000 items covering the original citation of all names as well as relevant, recent research.

Ideally, this kind of synthesis should be based upon more detailed syntheses of the component countries of the region and upon more detailed revisions of particular taxa, without artificial geographical limitations. Very few modern sources of this kind exist and for many groups we have had to depend upon a mixture of primary, and more or less obsolete and incomplete, secondary sources.

There is nothing definitive about this volume—it is a presentation of the current state of our knowledge, using our personal judgement and experience to choose the 'most probable' solution where alternative views have been expressed or where, as is so often the case, there are many alternative ways of combining the scattered data into an integrated whole. We hope that it will serve a useful purpose for reference and also act as a stimulus to further research that will show where we have gone wrong.

ACKNOWLEDGEMENTS

This project has been made possible by a generous grant to The Natural History Museum, London, from the trustees of the Moorgate Trust Funds, founded by the late Sir John Ellerman. Our sincere thanks go to them, and to the Earl of Cranbrook and the late Sir Terence Morrison-Scott for their valuable support and encouragement. We are grateful to the trustees and staff of The Natural History Museum for providing facilities and much cooperation. Especial thanks are due to Iain Bishop, Juliet Jewell, Jean Ingles, Paula Jenkins, Daphne Hills and Martin Sheldrick of the Mammal Section (Department of Zoology) and to Ann Datta, Carol Gokce and Paul Cooper of the Zoological Library.

The authorities and staff of the other museums and institutes listed below kindly provided GB Corbet with access and facilities for study of their collections and/or gave generously of their time to discuss problems:

Bangkok, Thailand. Thailand Institute of Scientific and Technological Research: Dr Yenbutra Songsakdi.

Beijing, China. Institute of Zoology, Academia Sinica: Prof. Ma Yong, Dr Wang Sung, Dr Gao Yaoting, Dr Feng Zuo-jian.

Bogor, Indonesia. Museum Zoologi: Mr B Boeadi, Mr A Suyanto; WWF: Dr C Santiapillai.

Bombay Natural History Society, India: Mr J C Daniel, Mr N Chaturvedi, Mr M Muni, Mr I Kehemicar, Dr V Krishnamurthy, Mr N Haque.

Boston, USA. Museum of Comparative Zoology, Cambridge: Ms Maria Rutzmoser.

Calcutta, India. Zoological Survey of India: Dr V C Agrawal, Dr S Chakraborty, Dr P K Das.

Chicago, USA. Field Museum of Natural History: Dr L R Heaney, Dr B Patterson, Mr W Stanley.

Kota Kinabalu, Sabah, Malaysia. Sabah Museum: Mr Raymond Goh; WWF: Dr J Payne.

Kuala Lumpur, Malaysia. Institute of Medical Research: Dr I Singh; University of Malaya: Dr D Wells.

Kuching, Sarawak, Malaysia. Sarawak Museum: Mr Charles Leh, Mr Yakup Nawi.

Kunming, Yunnan, China. Institute of Zoology: Prof. Wang Ying-xiang, Mr Jiang Xuelong.

Leiden, Netherlands. Nationaal Natuurhistorisch Museum (formerly Rijksmuseum van Natuurlijke Historie): Dr C Smeenk.

New York, USA. American Museum of Natural History: Dr G G Musser, Dr K F Koopman, Ms Marie Lawrence.

Singapore. Zoological Reference Collection, National University of Singapore: Mrs C M Yang.

Washington, DC, USA. National Museum of Natural History, Smithsonian Institution: Dr M Carleton, Dr R S Hoffmann, Dr W C Wozencraft.

Study by GB Corbet in museums in China, Singapore and Borneo was supported by an exchange scholarship and a travel grant from the Royal Society which are gratefully acknowledged.

The following persons kindly read parts of the text and enabled many useful corrections and additions to be made: Dr P J J Bates (Indian Insectivora); Dr J E C Flux (Lagomorpha); Dr P Grubb (ungulates); Dr L R Heaney (especially Soricidae, Sciuridae and Philippine Chiroptera); Ms Paula Jenkins (Soricidae, Tupaiidae, Primates); Dr K F Koopman (Chiroptera); Dr G G Musser (Murinae); Dr I Muul (Sciuridae, Pteromyidae); Mr M C Sheldrick and Dr C Smeenk (Cetacea); Mr R Wirth (Sciuridae); Dr W C Wozencraft (Carnivora).

Dr N van Strien kindly gave us access to his detailed unpublished documentation on the nomenclature of mammals of the Australasian Archipelago.

The un-numbered drawings have been prepared by Ray Burrows and the numbered figures by Sally Alexander. The opportunity has been taken to include a few hitherto unpublished drawings of rodent skulls and teeth prepared by A J E Terzi for Sir John Ellerman, now in the Library of The Natural History Museum, as well as some repeated from Ellerman's *Families and Genera of Living Rodents*.

INTRODUCTION

1. The Indomalayan Region

The first attempt to identify the major, natural zoogeographical regions of the earth that has stood the test of time was that of Sclater (1858), based mainly upon the distribution of birds. He recognized six regions, including a 'Regio Indica' which has been accepted by most subsequent authorities with relatively minor changes of boundary. Wallace (1876) changed the name to 'Oriental Region' and defined it more precisely, taking the distribution of mammals into account. Since the name 'Oriental Region' has always been somewhat misleading in implying inclusion of northern China and Japan, the name 'Indomalayan Region' has been proposed (Udvardy, 1975; Clark *et al.*, 1988) and appears to be gaining acceptance. Whereas the western boundary (the western edge of the Indus Basin) and the western part of the northern boundary (the Himalayas) are moderately well defined, there has been considerable controversy as to where best to draw the line in the southeast, in relation to the Australian Region, and in the northeast, in China, in relation to the Palaearctic Region.

Alfred Russel Wallace travelled extensively amongst the islands between Asia and Australia (Wallace, 1869) where he was most influenced by the birds in drawing his 'Wallace's line' such as to keep the Philippines, Borneo and Bali in the Oriental Region, and the Moluccas, Sulawesi and Lombok in the Australian Region. This line represented the western limit of certain characteristically Australian groups such as the cockatoos and the marsupials. The opposite view was expressed by Lydekker (1896) who drew his line between the Moluccas and New Guinea, being in effect the eastern limit of many Asian taxa.

The area between these lines is clearly a transition zone, consisting essentially of islands that have never been in contact with either the Asiatic or Australian continental shelves (at least not in sufficiently recent times to acquire a continental fauna). Some authors, e. g. Heilprin (1887), have recognized this as a transition zone without allocating it to either region; others, e. g. Weber (1902), have drawn a compromise line, between Sulawesi and the Moluccas. In this area the species of non-flying mammals derived from the west (c. 80 species, c. 60 of them endemic) greatly outnumber those from the east (c. 17 species, only 10 of them endemic); we therefore use Lydekker's line for the purpose of this work, thereby ensuring more complete treatment of the truly Indomalayan taxa at the expense of including a few with Australian affinities (Map 1).

In the case of the bats the situation is similar, but mingling of the two faunas has clearly been greater and any line is more arbitrary than in the non-flying mammals. For example there are 63 species of bats on Sulawesi and neighbouring islands and 54 on the Lesser Sunda Islands and Timor, of which only 11 and 3 respectively are endemic. If the Moluccas are added then there are 108 species in this transition zone. Of these, 15 are wide-ranging, distributed both to the west and east of its boundaries, 43 occur also only to the west and 21 only in the Australian Region, while 29 are endemic to the zone. Tate (1946) studied the distribution of bats in the Australasian archipelago, and Koopman (1989) reviewed distributional patterns in Indomalayan bats, but did not include the Moluccan fauna.

The islands of Weigeo and Misool and the Aru archipelago are excluded since they are on the New Guinea shelf and carry several additional species characteristic of New Guinea, e. g. marsupials of the genera *Thylogale* and *Dorcopsis*. More detailed reviews of this boundary zone can be found in Raven (1935), Brongersma (1936a), Mayr (1944), Scrivenor *et al.* (1943), Groves (1976), Cranbrook (1981) and Whitmore (1987).

The situation in China is very different since here there are fewer physical barriers to dispersal and there is therefore a much greater degree of intermingling of two continental faunas, Indomalayan and Palaearctic. Definition of a boundary in China is further complicated by the disturbance caused by ancient and extensive cultivation, leaving little natural habitat throughout large areas in the lowlands of eastern China. The arbitrary boundary used here, at 35°N, and indeed the entire Indomalayan-Palaearctic boundary, is the one used in *The mammals of the Palaearctic Region* (Corbet, 1978). No species with Indomalayan affinities occur entirely north of this line on the mainland, although a few widespread Indomalayan species like *Panthera tigris* do extend further north; these are included here in their entirety if they extend only marginally beyond the boundary. The mammal fauna of the region on either side of the lower Yangtze river was analyzed by Huang (1985) who concluded that the lower Yangtze did indeed separate areas in which Palaearctic and Indomalayan species predominated. However in mapping the zoogeographic regions within China Zhang & Zhao (1978) placed the corresponding median line a little north of the Yangtze.

Except in China and Japan the boundary with the Palaearctic Region used here defines the greatest extent of the Indomalayan fauna rather than the median line through the transition zones. It can be described in more detail as follows, beginning in the west (Map 1). It lies between Iran and Pakistan but crosses Afghanistan so as to include only the eastern parts draining to the Indus. From the eastern end of the Hindu Kush to central China it follows the tree-line at about 3000 metres so as to exclude the alpine fauna with Palaearctic affinities but to include the Himalayan/Chinese montane forest fauna comprising endemic genera such as *Soriculus, Ailurus* and *Budorcas* as well as the more widespread Indomalayan elements. Tibet (Xizang) is excluded except for small parts of the Brahmaputra, Salween and Mekong valleys. Nepal, Bhutan, Yunnan and Sichuan are included except for the high alpine zones, but Qinghai is excluded. Use of 35°N as the northern boundary in China means that the southern parts of Gansu and Shaanxi are included along with most of Honan and the whole of Anhui and Jiangxu, while Shanxi, Hebei and Shandong are excluded.

In Japan the boundary is drawn at 30°N so that Yakushima, Tanegashima and all the main islands are excluded while the

Map 1. Boundary of the Indomalayan Region.

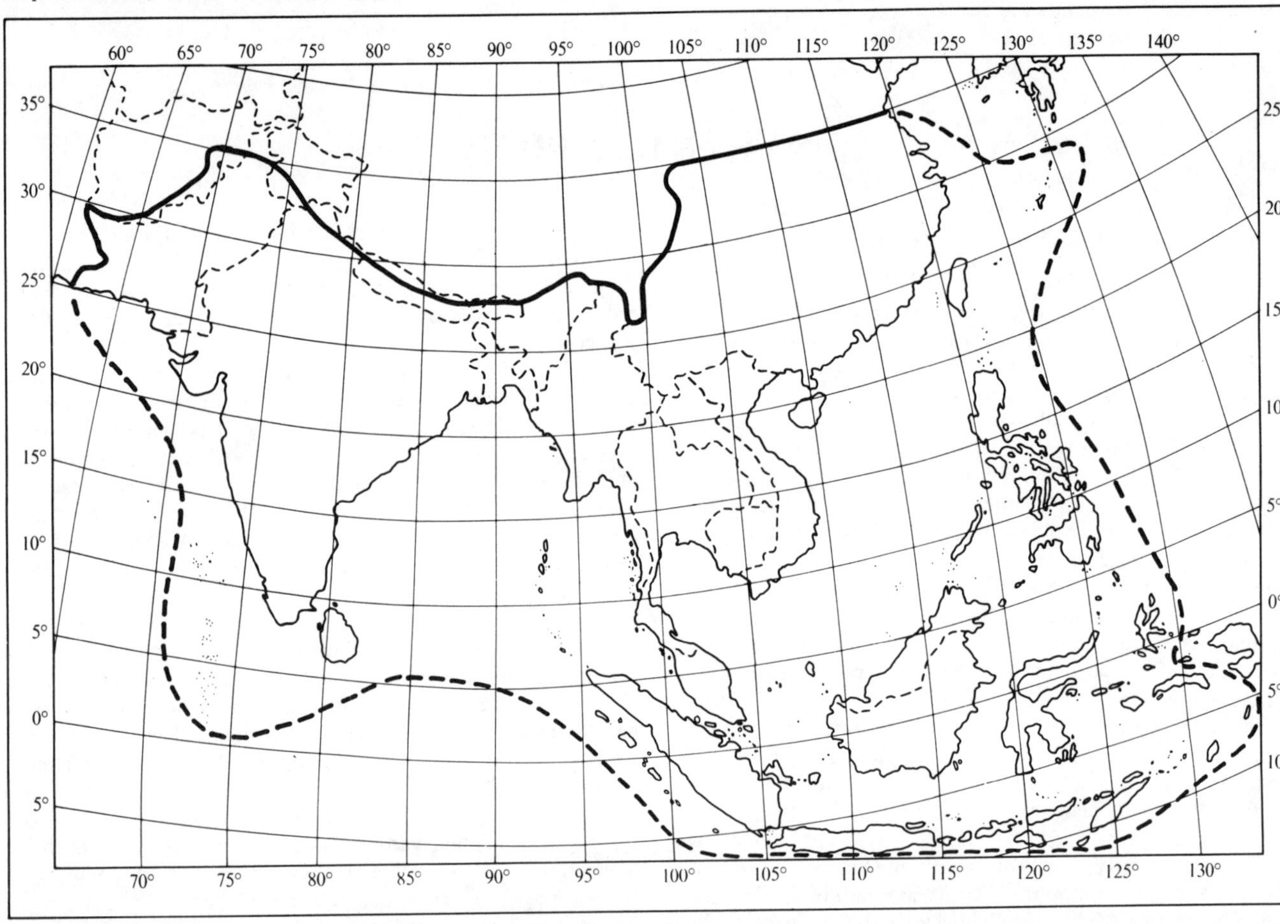

Ryukyu Islands are included. A few species with Indomalayan affinities are found only on the main islands of Japan and are mentioned in the appropriate place in the text, without full treatment (*Chimarrogale platycephala, Macaca fuscata, Petaurista leucogenys, Naemorhedus crispus*).

In the Indian Ocean the Laccadive, Maldive, Andaman, Nicobar and Mentawai islands are included, along with Christmas Island, but the more isolated oceanic islands are excluded. In the Pacific all oceanic islands, i. e. east of the Ryukyu Islands and the Philippines, are excluded.

2. The Indomalayan mammal fauna

The composition of the mammal fauna of the Indomalayan Region as defined here is summarized in Table 1. The additional species in the second column fall into the following categories (the categorization of each species is indicated in Appendix 1):

* Palaearctic species that occur marginally in the Indomalayan Region as defined here, e. g. *Hemiechinus auritus, Crocidura pergrisea, Rhinolophus ferrumequinum, Mustela nivalis, Tamias sibiricus, Acomys cahirinus, Meriones libycus*.
* Australo-Papuan species that occur marginally in the Indomalayan Region as defined here, e. g. *Phalanger orientalis, Spilocuscus maculatus, Petaurus breviceps, Melomys fulgens*.
* Cetaceans of temperate oceans that occur marginally or sporadically in the tropics.
* Domesticated species.
* Introduced species (perhaps only *Ondatra zibethicus, Rattus norvegicus* and, more doubtfully, *Mus musculus*; it is possible that some other marginal species have been introduced by man, e. g. amongst the marsupials on the Moluccas).

Table 1. Composition of the Recent Indomalayan mammal fauna.

	Number of extant, indigenous genera/ species with predominantly or extensively Indomalayan distributions	Total number of Recent genera/ species included in this work
Peramelina	0/0	2/2
Diprotodontia	3/5	5/8
Pholidota	1/3	1/3
Insectivora	17/69	20/77
Scandentia	5/16	5/16
Dermoptera	1/2	1/2
Chiroptera	35/243	66/307
Primates	11/61	11/61
Carnivora	35/75	37/83
Cetacea	22/30	24/32
Sirenia	1/1	1/1
Proboscidea	1/1	1/1
Perissodactyla	3/4	4/7
Artiodactyla	19/50	23/61
Rodentia	95/333	115/369
Lagomorpha	4/7	6/11
TOTAL	253/897	322/1041

The only orders of mammals that are confined to the Indomalayan Region are the Scandentia (16 species of tree shrews) and Dermoptera (2 species of flying lemurs). Endemic families or subfamilies are the Galericinae (6 species of shrew-hedgehogs), Craseonycteridae (1 species, the hog-nosed bat), Harpyionycterinae (2 species of fruit bats), Tarsiidae (5 species of tarsiers), Hylobatidae (11 species of gibbons), Ailuridae (1 species, the red panda), Platacanthomyinae (2 species of rodents). About 44 of the 896 species of predominantly Indomalayan mammals fall into these endemic orders, families and subfamilies.

The remaining families and distinctive subfamilies represented in the Indomalayan Region are shared with other zoogeographical regions as follows (ignoring species that spill over marginally into other regions):

Shared only with the Afrotropical Region: Manidae (3 species of scaly anteaters), Rhinopomatidae (3 species of mouse-tailed bats), Nycteridae (2 species of slit-faced bats), Loridae (4 species of lorises), Cercopithecidae (39 species of monkeys), Ponginae (1 species of ape), Viverridae (20 species of civets etc.), Herpestidae (8 species of mongooses), Rhinocerotidae (3 species of rhinos), Tragulidae (3 species of mouse-deer), Elephantidae (1 species of elephant), Hystricidae (8 species of porcupines), Rhizomyinae (4 species of bamboo rats).

Shared only with the Palaearctic Region: Moschidae (1 species of musk deer).

Shared only with the Neotropical Region: Tapiridae (1 species of tapir).

Shared with both Afrotropical and Palaearctic regions: Erinaceinae (6 species of hedgehogs), Hyaenidae (1 species of hyaena), Suidae (7 species of pigs).

Shared with both Afrotropical and Australian regions: Pteropodidae (74 species of fruit bats), Megadermatidae (2 species of false vampire bats), Hipposideridae (40 species of leaf-nosed bats).

Shared with Afrotropical, Palaearctic and Australian regions: Rhinolophidae (42 species of horse-shoe bats), Murinae (213 species of rats and mice).

Shared with Afrotropical, Neotropical and Australian regions: Emballonuridae (11 species of sheath-tailed bats), Molossidae (13 species of free-tailed bats).

Shared with both Palaearctic and Nearctic regions: Talpidae (10 species of moles), Pteromyidae (38 species of flying squirrels), Zapodidae (2 species of jumping mice).

Shared with Afrotropical, Palaearctic and Nearctic regions: Bovidae (28 species of cattle etc.).

Shared with Palaearctic, Nearctic and Neotropical regions: Ursidae (4 species of bears), Cervidae (20 species of deer).

Shared with Afrotropical, Palaearctic, Nearctic and Neotropical regions: Soricidae (55 species of shrews), Vespertilionidae (119 species of bats), Canidae (9 species of dogs), Mustelidae (22 species of weasels etc.), Felidae (16 species of cats), Sciuridae (62 species of non-flying squirrels).

3. The subdivisions of the Indomalayan Region

Most attempts to recognize natural subregions within the Indomalayan Region have been based primarily upon vegetation rather than the distribution of animals. Udvardy (1975) for example recognized 27 divisions based mainly upon the dominant vegetation. Consideration of the distribution of mammals indicates some important boundaries that need to be superimposed upon the vegetation maps, in particular: 1. Many species with Chinese or Indochinese distributions extend a considerable way westwards along the Himalayan forests without extending south into peninsular India, giving the Himalayan montane forests a very characteristic fauna albeit sharing many species with peninsular India. 2. The Isthmus of Kra in peninsular Thailand, although corresponding to a boundary between rain forest in the south and more seasonal forest in the north, appears to be a faunal boundary out of proportion to the vegetational and climatic differences. 3. The Mentawai Islands, although showing many affinities with Sumatra and the other Greater Sunda Islands, have a high degree of endemism that justifies treating them on a par with each of the Greater Sunda Islands. 4. The mammal fauna of Palawan clearly reflects its position on the Sunda shelf and distances it from the other Philippine islands which have a much more insular fauna.

These, and other considerations, are reflected in Map 2 which suggests the recognition of six major subregions, as detailed below.

A. Indian Subregion

The endemic genera in this subregion, e.g. *Boselaphus, Tetracerus, Antilope, Funambulus, Golunda*, mostly have their nearest relatives in the west and north, although they coexist with much larger numbers of more widespread Indomalayan species.

A1. Indus Division

This area is transitional to the Palaearctic Region, with many characteristic arid-zone genera of the central and southwestern Palaearctic such as *Hemiechinus, Capra, Ovis, Acomys, Gerbillus, Meriones* and *Jaculus* that do not extend further into the Indomalayan Region. However they overlap with several truly Indomalayan taxa that reach their western limits here, e. g. *Viverricula, Millardia*.

A2. Peninsular Indian Division

Besides the endemic groups like *Boselaphus, Tetracerus* and *Golunda*, other western taxa penetrate far into the peninsula without going further east, e. g. *Paraechinus, Hyaena, Tatera*. These are characteristic mammals of the drier habitats. The wetter forests, e. g. in the southwest of the peninsula, have more species with eastern affinities, e. g. squirrels of the genera *Ratufa* and *Petinomys* and the endemic bat genus *Latidens*. Other more widespread eastern elements are found amongst the tree-shrews (*Anathana*) and the viverrids such as *Paradoxurus*.

A3. Sri Lanka

Sri Lanka has a number of endemics in the montane forests of the central highlands, e. g. the shrews *Feroculus* and *Solisorex* and the rat *Srilankamys*, as well as endemic species of *Crocidura, Suncus, Mus* and *Rattus*. It also shares many species or species- pairs with the southwestern part of peninsular India, e.g. *Semnopithecus vetulus/johni*, *Petinomys fuscocapillus*, *Ratufa macroura*. In other respects the fauna is a somewhat impoverished version of that of peninsular India.

A4. Maldive and Laccadive Islands

These are truly oceanic islands with a fauna of introduced shrews (*Suncus*), rats, mice and rabbits, and two bats (*Pteropus giganteus* and *P. hypomelanus*).

B. Himalayan Subregion

This subregion is transitional to the Palaearctic Region in that several Palaearctic taxa descend some way into the forest zone, e. g. species of *Apodemus*, *Microtus* and *Ochotona*. However there are also several endemic (or nearly endemic) genera: *Uropsilus, Nectogale, Soriculus, Sphaerias, Ailuropoda, Ailurus, Budorcas, Eupetaurus*, as well as representatives of more widespread southeastern groups, many of which extend along the Himalayan forest zone but not into peninsular India, e. g. *Dremomys* and *Niviventer*.

C. Indochinese Subregion

This subregion has rather few endemic genera—*Craseonycteris, Eudiscopus, Ia, Pygathrix, Lipotes, Vernaya, Chiromyscus, Hadromys*. However it is characterized by many species that reach their limits either about the River Brahmaputra or at the Isthmus of Kra.

Map 2. The subdivisions of the Indomalayan Region. The shaded areas are transitional to the Palaearctic Region; subregion F is transitional to the Australian Region. See text for identification of the subdivisions.

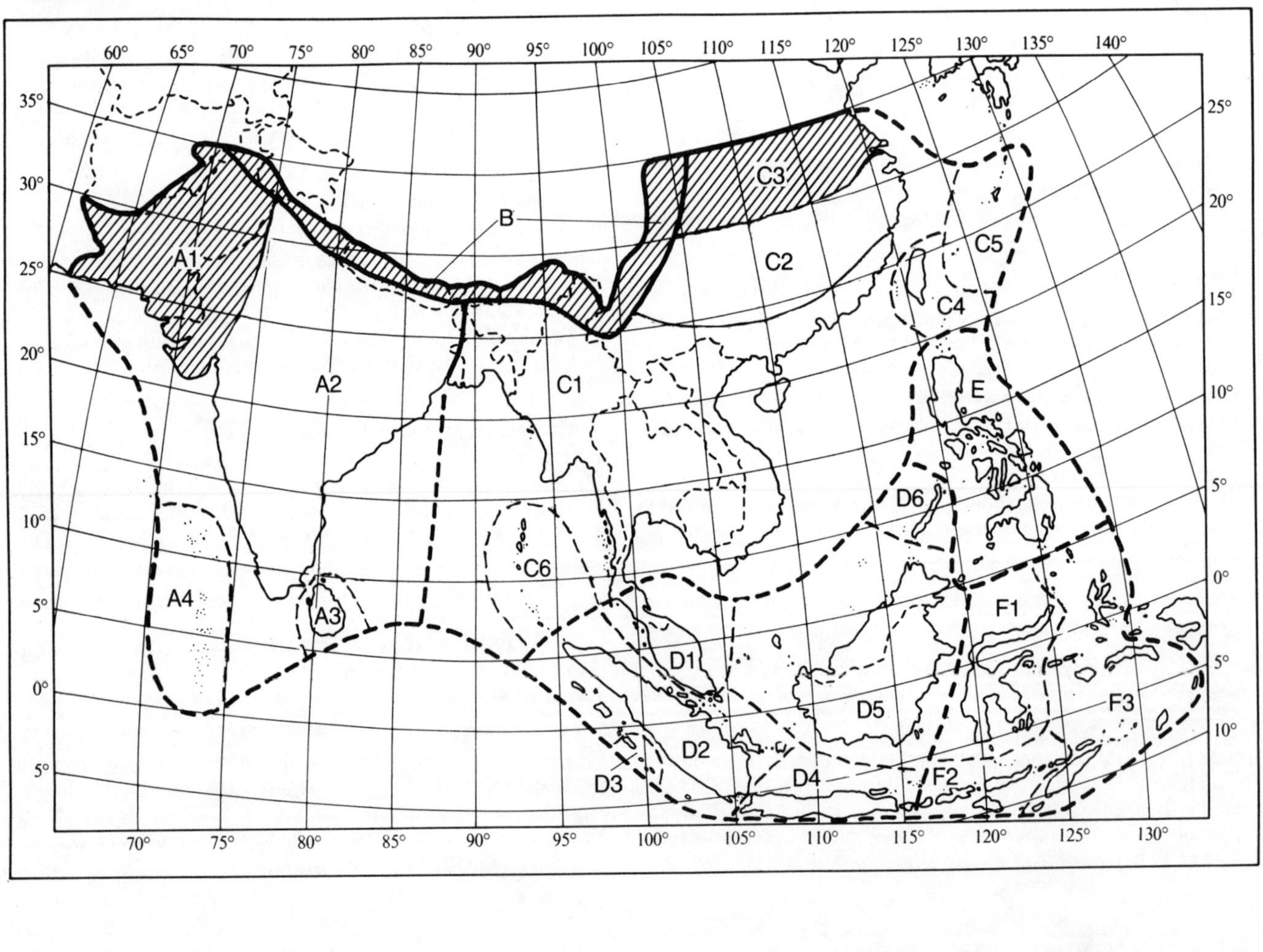

C1. Indochinese Division

This division includes many tropical forest genera that do not extend further west into India or north into temperate China, e. g. *Tupaia, Cynocephalus, Arctictis, Berylmys, Hapalomys*, although these all extend south into Malaya and to some extent to the Greater Sunda Islands. However there are many species boundaries at the Isthmus of Kra defining the southern boundary of the subregion, which also lacks many Sunda taxa that do not extend north of the isthmus, e. g. *Echinosorex, Sundasciurus*.
There is one endemic (but monospecific) bat family, Craseonycteridae, and two endemic bat genera *Eudiscopus* and *Ia*.

C2. Southern Chinese Division

Here Palaearctic taxa form a significant minority, e. g. *Meles, Apodemus*, although Indomalayan taxa predominate, e. g. *Melogale, Viverra, Muntiacus, Vandeleuria*, many at their northern limit.

C3. Central Chinese Division

This is transitional to the Palaearctic Region, containing a predominence of Palaearctic groups such as *Erinaceus, Vespertilio, Meles, Apodemus, Myospalax* as well as the northern outliers of Indomalayan taxa, e. g. *Arctonyx, Paguma, Niviventer*.

C4. Taiwan

The island of Taiwan is on the continental shelf of China but is much more isolated than Hainan Island which has a fauna very similar to that of the mainland. There are no endemic genera on Taiwan but several endemic species, some belonging to Palaearctic genera, especially in the montane zone, e. g. *Macaca cyclopis, Microtus kikuchii*, and many others on the borderline between species and distinctive subspecies, e. g. *Petaurista alborufus lena*, all closely related to forms on the mainland of south China.

C5. Ryukyu Islands
These islands are oceanic, with a small fauna but several distinctive endemics, especially the monospecific genera *Tokudaia, Diplothrix* and *Pentalagus*.

C6. Andaman and Nicobar Islands
These have a highly impoverished, 'oceanic' fauna consisting largely of species that are likely to have been introduced by human agency, but there are also a number of intriguing endemics suggesting some earlier, but highly selective, colonization from the eastern mainland. Endemics are *Tupaia nicobarica, Crocidura andamanensis, C. hispida, Rhinolophus cognatus, Rattus palmarum* and *R. stoicus*.

D. Sundaic Subregion
This subregion comprises the Malayan peninsula, the Mentawai Islands and the Greater Sunda Islands, which lie on the Asiatic continental shelf and have all been connected to the Asiatic mainland at some time or times as recently as the Pleistocene. Successive contact and separation of islands has led to extensive speciation and the subregion also contains many endemic genera that do not extend north of the Isthmus of Kra.

D1. Malayan Division
The fauna here is dominated by genera that occur also on the Greater Sunda Islands, but also includes a few that are southern outliers of northern taxa that do not extend to the islands, e. g. *Talpa, Tamiops, Hapalomys*.

D2. Sumatran Division
The fauna of Sumatra is very similar to that of peninsular Malaya and Borneo. *Nesolagus* is the only endemic genus, but several endemic species occur and it shares with peninsular Malaya a number of genera and species that do not extend to the other Sunda islands, e. g. *Naemorhedus, Rhizomys*.

D3. Mentawai Division
The Mentawai Islands of Siberut, Sipura, N Pagai and S Pagai have a very distinctive fauna which shows many similarities with, but also many significant differences from that of mainland Sumatra (and from the other islands of the Mentawai chain—Batu Is, Pini, Nias and Simeulue—which have a fauna much closer in character to that of mainland Sumatra). There are no endemic genera (unless the genus *Simias* is recognized for *Nasalis concolor*), but there are many distinctive species, e. g. *Presbytis potenziani, Hylobates klossi, Callosciurus melanogaster, Petinomys lugens, Maxomys pagensis*, and many other forms that are on the borderline between distinct species and well-marked subspecies. Many of the endemic species and subspecies are characterized by melanic pelage compared with their vicariants on the Sumatran mainland, e. g. *Callosciurus melanogaster, Iomys sipora, Hylobates klossi* and *Pteropus melanotus* (but the last also extending to the Nicobar and Andaman Is and to Christmas I).

D4. Javan Division
The fauna of Java is somewhat impoverished compared with that of Sumatra and Borneo. However many of the absences are probably due to extinctions caused by loss of habitat with or without direct exploitation, since Java has been more heavily populated and intensively cultivated than the other islands over at least 2000 years. Some of the absences are shown to be secondary by the presence of Pleistocene or Holocene fossils, e. g. of *Pongo, Elephas* and *Tapirus*. Other absences are more difficult to explain, e. g. *Echinosorex, Petaurillus* and *Aeromys*. On the other hand there is one endemic genus, *Kadarsanomys*, and *Pithecheir* is shared only with peninsular Malaya.

D5. Bornean Division
The fauna of Borneo is very rich, mainly by virtue of a large number of endemic species. There are no endemic genera but *Haeromys* is one that occurs elsewhere only on Palawan and Sulawesi. On the other hand many widespread Sunda species occur that show very little if any differentiation amongst the Sunda Islands or Malaya.

D6. Palawan Division
The fauna of Palawan is less well known than that of the other Sunda Islands but nevertheless appears to be significantly less species-rich than that of Borneo. On the other hand it shows a very much closer affinity with Borneo than with the other Philippine islands. Typical Sunda genera occurring on Palawan but not elsewhere in the Philippines are *Manis, Tupaia, Herpestes, Mydaus, Chiropodomys, Sundamys* and *Hystrix*. *Palawanomys* is endemic, but none of the many other endemic Philippine genera occur.

E. Philippine Subregion
The Philippines (except Palawan) have probably never been broadly connected to the Asiatic mainland but the fauna shows evidence of a number of colonizations from Asia and subsequent diversification on the islands. There are also many recent introductions. Endemic genera are *Podogymnura, Urogale, Ptenochirus, Haplonycteris, Alionycteris, Otopteropus* and many murid rodents, e.g. *Apomys, Batomys, Bullimus, Phloeomys, Rhynchomys*.

F. Wallacean Subregion
The islands constituting this subregion have probably never had sufficient connection with either the Asiatic or New Guinea continental shelves to receive overland colonization. All but Sulawesi have impoverished faunas.

F1. Sulawesi Division
Sulawesi has many endemic genera of rodents—*Hyosciurus, Prosciurillus, Rubrisciurus, Bunomys, Taeromys, Eropeplus, Melasmothrix, Margaretamys, Paruromys* —with predominantly Asiatic affinities. Other endemic genera are *Neopteryx, Styloctenium, Macrogalidea* and *Babyrousa*. In addition there are endemic species of *Crocidura, Tarsius, Macaca, Bubalus* and *Rattus*. Eastern colonists are represented by the two endemic but monospecific genera of marsupials, *Stigocuscus* and *Ailurops*, and possibly by some of the rodents.

F2. Lesser Sunda Division
The islands from Lombok eastwards have very impoverished faunas. Endemic rodents are the genera *Komodomys* and *Papagomys* and *Bunomys naso*; there are endemic species of the bat genera *Pteropus, Acerodon* and *Nyctimene*, and doubtfully endemic species of *Crocidura* and *Suncus*. Most other species are likely to have been introduced from the west, e. g. in the genera *Macaca, Rattus* and *Hystrix*.

F3. Moluccan Division
All these islands have very impoverished faunas. There are some endemic species of *Rattus* and *Melomys* (otherwise in New Guinea). The only endemic genus, the marsupial *Rhynchomeles* on Seram, doubtfully justifies generic separation from New Guinea *Echymipera*. The fauna of many of the islands, especially Timor and others in the southeast, is very poorly known.

4. The time-scale

This work is primarily concerned to document those species of mammals that are currently living in the Indomalayan Region or that have occurred there within historic time. However an attempt has been made to record other specific and subspecific names that have been used in Recent genera and have their type localities in the region, irrespective of geological age. No attempt has been made to include fossil forms recorded in the region but named from localities elsewhere, e. g. the many fossil taxa described from northern China but recorded further south.

It is hoped that this will help with the integration of palaeontological and neontological studies. In particular it might help to avoid duplication of nomenclature and will be relevant to the study of historical zoogeography. However it should be noted that such names are simply listed; in most cases no critical taxonomic assessment has been made of extinct forms. It is likely that more detailed study will show many nominal Pleistocene and Holocene species to fall within the range of variation of Recent species.

The literature on fossil mammals is even more scattered than that on Recent species and no recent checklists are available. Major sources have been Lydekker (1884b), Trouessart (1897-1905), Teilhard de Chardin & Leroy (1942) (for China), Badam (1979) (for India) and Savage & Russell (1983) as well as the more detailed literature.

5. The scientific discovery and description of Indomalayan mammals: a historical perspective and recent literature

Although many accounts of mammals occur in early literary works throughout those parts of the region with a literary tradition, their integration into a coordinated body of knowledge was only made possible by the development in 18th century Europe of an efficient, international system of nomenclature, epitomized by the 10th edition of Linnaeus' *Systema naturae* (1758) which is still used as the base-line for zoological nomenclature. Linnaeus (1758) listed about 34 species of mammals from the Indomalayan Region (plus a few others clearly recorded in error). This number rose rapidly during the late 18th and early 19th century, especially as a result of the development of European trade routes.

During the 19th century new species were described from the Indian Subregion at an ever-increasing rate, beginning with the work of T Hardwicke, E Blyth and T Horsfield in India and B H Hodgson in Nepal, augmented by J E Gray at the British Museum in London. In the latter half of the century J Anderson, first curator of the Indian Museum at Calcutta, and W T Blanford, latterly editor of the Fauna of British India series, continued the work in India, while Oldfield Thomas began his prolific output at the British Museum.

A major initiative began in 1912 with the launch of an ambitious Mammal Survey of India by the Bombay Natural History Society. Professional collectors were employed beginning with C A Crump and the project continued until 1923, the resulting material being described in London by Oldfield Thomas, R C Wroughton, T B Fry and others.

A second edition of the mammal part of the *Fauna of British India* series began to appear in 1939 and 1941 with Pocock's volumes on the primates and carnivores, followed by Ellerman's rodents (written in the 1940s but not published until 1961), but no further parts have appeared so that Blanford's volumes of 1888 and 1891 remain the latest comprehensive descriptive work on the mammals of India.

Description of Chinese mammals was dominated by Alphonse Milne-Edwards in Paris, based especially upon material collected by Père Armand David, but significant additions were made by Robert Swinhoe, a British consular official stationed in various parts of China including Taiwan from 1854 to 1875. Another French missionary, P Heude, based at Shanghai towards the end of the 19th century was responsible for a great proliferation of species names, especially of ungulates, although only a few have stood the test of time as valid species.

In the Indonesian region some early collecting of mammals was undertaken by French voyages around the end of the 18th century and later by the American Dr Thomas Horsfield and the British administrator in Java, Sir Stamford Raffles. However a major initiative came in the 1820s with the dispatch of the first of a succession of collectors by the Rijksmuseum van Natuurlijke Historie newly founded by C J Temminck in Leiden (Netherlands). The resultant novelties were described especially by Salomon Müller and Herman Schlegel as well as by Temminck. Subsequently Dutch work was undertaken especially by F A Jentink based in Leiden and H J V Sody in Java. This was supplemented by the work of H C Robinson, C B Kloss and F N Chasen, operating from Malaya and Singapore.

Much of the early descriptive work was done with little regard to comparison with material from other parts of the region, resulting in a proliferation of species names that have proven to be synonymous. Oldfield Thomas and Bonhote appear to have been the first, around 1890-1905, to follow the example of the ornithologists and some American mammalogists in recognizing larger 'polytypic' species and using subspecific names. Although subspecific names were increasingly used over the next few decades, the proliferation of species continued, especially in the Sunda subregion with the work of Miller and Lyon in Washington, based upon the extensive collections of Dr W L Abbott.

Comprehensive syntheses of data on parts of the region during the period up to 1940 were few. Blanford's *Mammalia* in the *Fauna of British India* (1888,1891) was an important synthesis that has only been superceded in parts. Taylor's (1934) work on the mammals of the Philippines also stands as an important landmark although it used a very split classification by modern standards and made few comparisons with mammals from elsewhere. Likewise for China G M Allen's survey (1938,1940) still stands as the only comprehensive work on that country although the first volume of a new mammal fauna of China has appeared recently (Gao *et al.*, 1987).

In the Sunda subregion Chasen's checklist of 1940 was a landmark in adopting a bold approach in ordering the vast number of nominal species and subspecies from that area.

From the mid-19th century regional mammal faunas began

to be published. However in spite of the recognition of the 'Oriental Region' as a natural zoogeographic region no previous work appears to have been devoted to the mammals of the entire region. The principal regional works are listed below along with modern successors, country by country. Many of these references have been used, but not separately quoted, in compiling the accounts of the ranges of individual species and in preparing the maps.

The most recent of these regional works are usually the best means of identification of mammals and the best sources of detailed distributional data provided the more recent data given here are taken into account. They are annotated using the following abbreviations:

C: comprehensive list covering all species recorded from the area (but not including the following features unless these are separately indicated);
Id: keys or diagnoses designed for identification;
MI: monochrome illustrations of most species;
CI: coloured illustrations of most species;
D: verbal accounts of distribution;
M: maps for most species.
Parentheses, e. g. (M), indicate that, in this case, maps are included for only some of the species dealt with.

Afghanistan
Hassinger (1973): C.

Pakistan
Roberts (1977): C, Id, MI, D, M; Ahmad & Ghalib (1979): C, D.

India
Blanford (1888b, 1891): C, Id, MI, D; Wroughton (1918b, 1919a): summary of the results of the Indian Mammal Survey (including Pakistan, Sri Lanka, and Burma): Id, D; Pocock (1939): Primates, Carnivora (part); (1941): Carnivora; Ellerman (1961): Rodentia: Id, D; Kinnear (1952): a history of mammalogy in India; Kurup (1966, 1968): Assam, C, D; Agrawal & Bhattacharyya (1977a): Tripura.
Also many papers in the Journal of the Bombay Natural History Society.

Andaman and Nicobar Is.
Tikader & Das (1985): C, D.

Sri Lanka
Kelaart (1851): C; Eisenberg & McKay (1970): C; Hill (1980a), McKay (1984): C; Phillips (1980,1981,1984): C.

Maldive Islands
Hill (1958): C.

Bangladesh
Khan (1984); Sarkar & Sarkar (1984): C; Khan (1985): C, M.

Kashmir
Chakraborty (1983).

Nepal
Hodgson (1841b): C; Gruber (1969), Weigel (1969), Frick (1969); Mitchell (1975): C (except bats); Abe (1982).

Burma
Anthony (1941); U Tun Yin (1967).

Thailand
Lekagul & McNeely (1977): C, Id, MI, M.

Vietnam
Van Peenen *et al.* (1969): C, MI, M; Dao (1985): C.

Laos
Deuve & Deuve (1963).

China: general
Allen (1938b, 1940): C, Id, (M); Shou (1964):(M).

China: Hainan
Xu *et al.* (1983): C.

China: Sichuan
Hu & Wang (1984): C.

China: Xizang (Tibet)
Feng *et al.* (1986): C, Id, M.

China: Yunnan
Kao *et al.* (1962): C (except Muridae); Wang, Y-x. & Yang, G-h. (1986): C; Wang (in prep.): C, Id, M.

China: Zhejiang
Zhuge (1982): C.

Hong Kong
Marshall, P (1967): C.

Taiwan
Kuroda (1952a): history and zoogeography; Jones *et al.* (1971): C, Id.

Ryukyu Islands
Imaizumi (1960): C, Id; Imaizumi (1970): Insectivora, Chiroptera, Primates, Lagomorpha, Id, MI, D, M.

Malaya, Sumatra, Java, Borneo
Chasen (1940a): C (nomenclatorial checklist).

Malaya and Singapore
Harrison (1966): C; Medway (1969,1978,1983): C, Id, (CI), D.

Singapore
Yang *et al.* (1990), Sigurdsson & Yang (1990): C.

Java
Sody (1989): C.

Borneo
Banks (1931):C, Id; Medway (1965b, 1977): C, Id; Payne *et al.* (1985): C, Id, CI, D.

Sunda Islands to Moluccas
van der Zon (1979): C, D (Indonesia); van Strien (1986): C, D (also includes New Guinea); Laurie & Hill (1954): C, D (Sulawesi to New Guinea); Musser (1987a): C, D (Sulawesi only).

Philippines
Taylor (1934): C, Id, D; Rabor (1977): D; Heaney *et al.* (1987): C, D.

6. Classification

An explanation of the way classification has been treated needs to be divided according to the four main steps in formulating any biological classification: (1) recognition and definition of the species; (2) grouping the species in genera; (3) grouping the genera in a succession of more-inclusive categories; and (4) recognition and description of the main groups *within* each species.

6.1. The species

In many groups of Indomalayan mammals much remains to be done before the morphological and geographical limits of the species can be satisfactorily defined. This requires the examination of large samples from throughout the range and with a few exceptions no attempt has been made to undertake comprehensive species-level revisions in the course of this work. Instead we have attempted to compare existing studies and where conflicting views exist to arrive at the 'most probable' conclusion on the basis of a review of the literature and examination of relevant and accessible museum collections.

Many compromises have been necessary, especially in trying to reconcile recent data derived from karyology and biochemical taxonomy with classical morphological taxonomy. These techniques frequently indicate the presence of pairs of very similar, 'sibling' species, but often on the basis of samples that are too small or localized to provide a generalization about the entire species. Much of this work has been performed using zoo animals of doubtful provenance, and too often no voucher specimens have been preserved to facilitate integration of the new data with morphological data.

Where detailed studies have demonstrated the presence of sympatric sibling species, these have been accepted irrespective of the degree (or absence) of morphological differences. However in the case of series of vicariant forms that are allopatric (e. g. on islands) or possibly parapatric, a wider concept of the species has been adopted pending the demonstration of discontinuities based upon adequate samples.

6.2. The genus

The degree of difference justifying the recognition of separate genera is much more subjective and arbitrary than in the case of species. Because the allocation of species to genera affects the naming of the species, instability of classification at the generic level is much more inconvenient than instability at higher levels of the hierarchy. Instability arises in many ways: personal preferences for small or large genera; studies of relationships between species that do not take account of all the species in the genera concerned; the use of diverse methods of classification, giving incongruent results.

Although recently developed techniques of cladistic analysis have introduced some badly needed objectivity into classification, the results are often less objective than one might hope, and especially at the level of genera and species many almost equally probable phylogenies can be produced by subtle differences in the technique employed and by the many subjective decisions that need to be made in choosing and scoring characters and assessing their polarities. Likewise morphometric analysis, while useful in helping to discriminate between species, can be a poor indicator of relationships between species and can easily lead to changes in generic classification that do not stand up to subsequent analysis by different techniques.

The use of fairly large genera is preferred here in order to minimize nomenclatural changes. The study of relationships within large genera can best be expressed in diagramatic form as dendrograms or cladograms and does not require an elaborate nomenclature designating every pair of sister taxa.

6.3. Higher classification

A regional compilation like this is not the place for originality in higher classification, which, to be effective, needs to be studied without geographical limitations. The classification used here is fairly conservative, adopting recent changes only if they have had time to gain acceptance by consensus. The interrelationships, and indeed the content, of the orders of mammals have been the subject of much recent study using biochemical techniques (reviewed by Benton, 1988a).

6.4. Subspecific variation

In theory, provided the species themselves can be clearly defined, the description of geographical variation within the species can be treated as a separate matter, to be ignored or taken into account according to its relevance to the subject in hand. This has been done here as far as is practicable, by using a separate heading 'Variation' within the account of each species. However this approach is somewhat complicated by the fact that in the International Code of Zoological Nomenclature the names of species and of subspecies are interchangeable, all being treated as names 'in the species group', as explained under 'Nomenclature' on p. 9.

Geographical variation within a species can take many forms:

(1) There may be discrete, spatially isolated (allopatric) populations whose members have been shown, on the basis of adequate samples, to be readily distinguishable from those of all other such populations, e. g. on islands. These can usefully be designated as formally named subspecies (provided at least 75% of individuals can be identified), and this is done here in the very few cases where convincing diagnoses are available.

(2) There may be populations that are distinguishable as in (1) except for a narrow zone where they are contiguous (parapatric) and show some degree of intergradation. These can also be designated as formally named subspecies and normally indicate zones of secondary contact, but the detection of such contact zones requires intensive geographical sampling and few examples in the Indomalayan Region have been studied in sufficient detail. (Similar parapatric populations without intergradation are best designated as separate species.)

(3) Populations may show geographical variation of a continuous nature, without sharp boundaries. If change in a characteristic, like size, is gradual and consistent over a considerable area, the variation can be described as clinal, but clines can involve several characters that do not necessarily vary in a congruent manner. Formal subspecific names are not useful in describing such variation, even to designate the ends of clines, especially since the type localities of existing names are rarely situated in convenient parts of the range for this purpose.

(4) Individual populations from widely separated parts of the species' range may be mutually distinguishable but if no information is available from intervening areas, or if the samples are too small to be sure that the differences are not due to age, sex, season etc., it may be impossible to predict whether the actual pattern of variation falls into categories 1, 2 or 3 or is random. A large proportion of formal subspecific names in the literature have been applied in this kind of situation. These are

simply listed here, sometimes with an indication of the kind of differences on which they were based—small size, dark pelage etc.—pending a comprehensive study to determine the actual pattern of variation in the species. It may be helpful to use such names in identifying individuals, but only if the specimen concerned does indeed show the characters upon which the name was based. It is more misleading than useful to use such names only on geographical grounds. Any such subspecies must be considered provisional as long as its geographical and phenetic limits have not been defined in relation to adjacent subspecies.

6.5. Domestic animals

Domestic forms of mammals are included provided that either they have been domesticated from wild ancestors within the region, e. g. *Bos frontalis* and probably *Bubalus bubalis*, or there is a long tradition of keeping them in the region. They are usually excluded from works of this kind but their inclusion seems desirable for a number of reasons. Firstly many domestic animals are free-ranging and there are many feral populations (i. e. derived from domestic stock but now living quite independently of human control) so that the domestic status of any individual or population is not always self-evident. Secondly such free-ranging or feral animals compete with wild species and therefore play a significant part in the ecology of the wild mammal fauna. Finally, scientific names given to domestic animals are potentially valid according to the Code of Zoological Nomenclature and therefore need to be taken into account in considering the nomenclature of both wild and domestic forms. Most of the more distinctive domestic animals are treated and named as separate 'domestic species' rather than including them in the wild ancestral species. This is justified on taxonomic grounds, since they form a discrete taxon that is not equivalent to a wild geographical subspecies, and it also reduces ambiguity in nomenclature (Corbet & Clutton-Brock, 1984). (See also under *Nomenclature* below.)

6.6. Summary of taxonomic changes

In the many places where previous classifications have been unstable or unsatisfactory the solution used here has been derived mainly by the choice of one out of several existing published alternatives. However there are several exceptions where an original, unpublished or unorthodox solution has been adopted and the more important are listed below:

Suncus montanus and *S. zeylanicus*. Recognized as species separate from *S. murinus*.
Crocidura baluensis and *C. orientalis*. Recognized as species separate from *C. fuliginosa*.
Paradoxurus lignicolor. Recognized as a species distinct from *P. hermaphroditus*.
Hemigalus. Includes *Chrotogale*.
Cynogale lowei. Treated as a species distinct from *C. bennettii*.
Callosciurus erythraeus. Includes *flavimanus*.
Callosciurus simus. *Glyphotes* is included in *Callosciurus*.
Dremomys gularis. Recognized as a species separate from *D. rufigenis*.
Trogopterus pearsonii. *Belomys* is included in *Trogopterus*.
Petaurista. The limits of *P. petaurista*, *P. philippensis* and *P. alborufus* differ from those in previous classifications; *P. caniceps* and (less certainly) *P. sybilla* are treated as separate from *P. elegans*.
Hylopetes. *H. phayrei* is held to include *Petinomys electilis*; *Iomys winstoni* is included in *Hylopetes*.
Petinomys. The Philippine form *mindanensis*, often included in *Hylopetes*, is considered to be certainly a *Petinomys* and is tentatively included in *P. crinitus*.
Vandeleuria. *V. nilagirica* is tentatively retained as a species separate from *V. oleracea*.
Chrotomys silaceus. *Celaenomys* is included in *Chrotomys*.
Bunomys. *Paulomys* is included.
Melasmothrix. *Tateomys* is included.

7. Nomenclature

An attempt has been made to comply as far as possible with the *International Code of Zoological Nomenclature* (3rd edition, 1985). The problems that arise in applying the Code differ according to the taxonomic rank concerned.

Species-group names. Below each species-heading the name is given as originally proposed (but ignoring capitalization of trivial names), spelling out the name of the genus only if it differs from the one used in the heading. This is followed by the original author, date, page and type locality. Synonyms are then given in order of date, listing only those that were proposed as species, or as subspecies of a different species. Generic names are shortened to the initial letter provided they are the same as the genus used in the heading or the same as the last generic name given in full in the synonymy. Generally only original citations of names are given, but some other combinations of generic and specific names are included, with author and date of the combination, if an unexpected generic name is involved. These are distinguished by preceding the author's name with a colon: '*Tana chrysura*:Lyon, 1908', or '*Tana chrysura* (Gunther, 1878):Lyon, 1908'. Subsequent mis-spellings and other minor variants of spelling are ignored unless they have been widely used and need to be corrected. Objective synonyms, i. e. those based upon the same type specimen, are placed immediately after the earlier synonymous name, irrespective of date, preceded by '='.

Species-group names that have been used only as subspecies of the species in which they are here placed are listed under *Variation*. However it should be remembered that all names of species are available for application to subspecies and *vice versa*.

To help with problems of priority, names of extinct species and subspecies described in Recent genera are given, usually in the introduction to the genus, if the type locality is in the Indomalayan Region. However even for genera now confined to the Indomalayan Region this does not necessarily amount to a comprehensive list of species-group names since there may be others based upon fossil material from elsewhere.

Names based upon domestic forms in the region are included, if they conform to the requirements of the Code, whether in 'domestic species' or within the synonymy of wild species. However since it has not been customary to include such names in nomenclatural checklists or taxonomic revisions devoted primarily to wild species, it is likely that many have been overlooked. The nomenclature of domestic animals and their wild ancestors has been very unstable, with many conflicting views. The solution adopted here, as expounded more fully by Corbet & Clutton-Brock (1984), is to treat the more distinctive domestic forms with long-standing scientific names as 'domestic species' named separately from their wild ancestral species; but to include in the wild species those, such as the Bali cattle in *Bos javanicus*, that are less distinct and have not traditionally been separately named in the scientific literature.

Nomina nuda, i. e. names that are not available because they are not accompanied by a description or equivalent 'indication', are ignored unless they have previously been listed as available names.

The spelling of species-group names has as far as possible been brought into line with the 3rd edition of the Code. In the case of names based upon personal names the Code advocates adherence to the original spelling while advising that new names should be spelled with a single "i", e. g. *smithi*. There is therefore no way of knowing whether *smithi* or *smithii* is correct without consulting the original citation of the name.

No attempt has been made to give data on type specimens. Many of the earlier names were published without designation of types, but for the Indian subregion Thomas (1918b) selected lectotypes for many such names.

It is normally quite unnecessary to quote author and date of a specific or subspecific name unless nomenclature is under discussion, especially if the nomenclature used follows a single source such as this.

Genus-group names. All synonyms based upon species named from the region are given following the original citation, even in the few cases where subgenera are used. Where subgenera are sufficiently clear-cut for *all* the species in the genus to be allocated to a subgenus unambiguously, they have been used in the case of larger genera, especially to enable the identification tables to be broken down to manageable units. However in most contexts there is no need to include the subgenus when naming a species. As with species-group names, objective synomyms, i. e. those based upon the same nominal type species, are preceded by '=' and immediately follow the earlier synonymous name.

Family-group names. Family-group names have only been subject to the same criteria of priority as other names since the first edition of the present Code in 1961, i. e. their priority date depends upon their first use as family-group names (including tribes, subfamilies and superfamilies, but allowing some flexibility of spelling). Previously they were often used according to the priority of the generic names on which they are based and this has led to some instability. Author, date and reference have been given for all family-group names, but there are several cases where application of the Code is unclear and early application of names difficult to find. In cases of doubt preference has been given to the most familiar name. No attempt has been made to list later synonyms unless the validity of the one used is in doubt.

Higher taxa. The Code does not apply to taxa above the family group and therefore no attempt has been made to give sources for these names. Most such names follow Simpson (1945).

8. Identification

Identification tables or keys are included at each level of the taxonomic hierarchy. Tables are generally used in preference to dichotomous keys and also serve as diagnoses of the taxa included. The characters chosen are those most useful for practical identification and do not necessarily include all those that have been used in defining and justifying the taxa although some more obscure or difficult characters, including chromosome data, are included if they are of particular significance. In some cases additional notes on identification are given in the species accounts.

In the case of genera and higher taxa that extend beyond the Indomalayan Region, character-states that are not applicable to the whole taxon, i. e. including non-Indomalayan species, are enclosed in parentheses in the tables. In some cases the more distinctive and useful character-states are shown in **bold**.

Besides acting as a guide to identification the tables help to justify the classification employed. Where possible these tables should be used in conjunction with recent local handbooks (as listed under section 5, p. 7), since these have the advantages of dealing with a much smaller group of species and, usually, of having descriptions and illustrations that can help to confirm identifications.

For convenience external measurements and other externally visible characters are given first in the tables, followed by internal ones. However in the case of some of the more difficult groups of small mammals it is necessary to resort to cranial or dental characters.

The recording and use of size differences in identification present some problems. The usual measure of overall size, length of head and body, is notoriously difficult to take in a standardized way and all such measurements should be treated as very approximate. In some cases where adulthood is difficult to define only upper limits are given. Although absolute tail-length is subject to less error it is often not clear what has been used as the origin (anus, pelvis) and of course when converted to a percentage of length of head and body it acquires all the uncertainty of that measurement. Length of hind foot is a particularly useful measure for inter-species comparisons since it reaches its maximum length at an early age. However precise use of collections or published data is often hindered by uncertainty as to whether claws are included (designated c. u. — *cum unguis*) or excluded (s. u. —*sine unguis*). All measurements quoted are 's.u.' unless otherwise stated.

The length of the skull is often the most useful precise measure of overall size, but again there are several ways of expressing it:

Greatest length of skull: from most posterior to most anterior point including teeth, irrespective of the bone or tooth concerned. (Sometimes, but not always the same as condylo-incisive length or occipito-nasal length, rarely condylo-canine length as in the Megadermatidae).

Condylobasal length: posterior surface of occipital condyles to anterior point of premaxillae.

Condylo-incisive length: posterior surface of condyles to anterior point of first incisors.

Condylo-canine length: posterior surface of condyles to anterior surface of canines.

Occipito-nasal length: posterior point of occipital bones (which may or may not be the condyles) to anterior point of nasals. (Used especially in lagomorphs and some rodents where the occiput extends well behind the condyles.)

The measure used in the tables is usually a compromise between the one most convenient to take and the one most quoted in the literature relating to the group concerned.

The length of the upper tooth-row is a particularly useful measure, especially in the case of small species with fragile skulls since in damaged skulls (including fossils and those from raptor pellets) the teeth often remain intact when the braincase (and therefore the condyles) are detached. It is also relatively independent of age provided the last molar has erupted. Depending upon the group the most convenient measure may be the entire tooth-row (e. g. in insectivores and carnivores), the canine to the last molar (e. g. in bats) or the post-diastema length, i. e. premolars, if present, and molars (e. g. in rodents, lagomorphs and most ungulates). Such measurements are taken from the most anterior point of the crown of the first tooth to the most posterior point of the last, not from the alveoli unless this is specifically stated.

The measurements given in the tables are approximate and are intended only as a guide to identification, not as a basis for further taxonomic research and analysis. They have been derived from many sources, published and unpublished. Where the range of a measurement is small it is likely to be based upon a small sample and specimens outside the range should be expected. Although the measurements generally exclude clearly juvenile individuals, the criteria of adulthood used is not always clear. The upper limit of a range is therefore generally more useful than the lower or even a median or mean. All measurements are in millimetres unless otherwise stated.

Since much of the taxonomy is very provisional, it is very important, when identifying animals that have been the subject of research, e. g. relating to physiology, reproduction, behaviour, chromosomes or parasitology, to record the locality from which they were collected. This is especially important when establishing captive colonies, whether in the laboratory or in zoos. If for example a squirrel is identified simply as *Callosciurus erythraeus* without further details, this can only be interpreted in the widest sense in which that name has been used (as in this volume). However if *C. erythraeus* in this sense subsequently proves to be an assemblage of smaller species, such an identification loses most of its value. With a locality, the identity of the specimen is more likely to be determinable, however the taxonomy changes. In theory the use of a subspecific name should serve the same purpose but in practice this is not always so unless the subspecies have been unusually well defined and the specimen is identified to subspecies on the basis of its characters, not just on geographical grounds. Whenever practicable, voucher specimens should be preserved if there is any doubt about the identification of the animals studied, although in some groups photographs can serve this purpose provided they show potentially diagnostic characters.

9. Distribution

In compiling the maps and describing the range, priority has been given to marginal records. Since it is far beyond the scope of this work to identify or list all locality records for most species, the amount of detail depends upon the secondary sources available. Where a detailed spot-map for the entire range of a species has been published this has been used, simplifying only to accommodate to the scale used here. For most species however detailed maps are only available for a few areas, e. g. Pakistan and to a lesser extent China. The list of sources in Section 5 indicates those containing maps.

In describing distribution in the text references to sources are given only where they supplement the regional works listed in Section 5. Small islands close to other parts of the range are not detailed in the case of synanthropic species, e. g. of shrews or rats, that have been extensively introduced to small islands, or in the case of the mobile bats. For most of the smaller mammals no attempt has been made to distinguish between recent and older records: the maps represent the accummulated data since scientific recording began. In the case of the larger mammals, and especially those whose ranges are known to have declined considerably, a variety of treatments has been necessary according to the data available. An indication is given in the caption of the map if the current range is known to be significantly different from that shown in the map and further explanation is given under 'Range' in the text.

In considering the significance of patterns of distribution it should be remembered that extensive sea-going trade has been conducted between many parts of the region for more than 2000 years, especially between the southeastern islands and both India and China (Hall, D. G. E., 1981). There must therefore have been many opportunities for the introduction of species beyond their original range, both deliberately and inadvertently. Such introductions are more likely to have been successful and permanent on islands where the pre-existing fauna was impoverished. Known introductions are mentioned in the text, and attention is drawn to the probability of populations having arisen thus when this seems especially likely. However the possibility of such artificial extensions of range needs to be always kept in mind, especially when analysing the patterns of distribution on islands.

In quoting type localities the original spelling is retained for particular localities but names of provinces and countries are translated to their modern forms whenever this can be done with confidence. Likewise altitudes are given in the original units. Under the heading **Range** modern spellings are used as far as possible and altitudes are given in metres. In China, the modern, Pinyin, names are used for the provinces except that the familiar 'Tibet' is retained instead of 'Xizang'. 'Malaya' is used for Peninsular Malaysia, and 'the Moluccas' rather than the Indonesian provincial name Maluku. Obsolete spellings and some of the more frequent but obscure type localities are included in the Gazetteer (Appendix 3).

See p. 429 for a list of some omissions from the maps.

10. Ecology

No attempt has been made to review all the literature on the ecology of Indomalayan mammals, but for some species a few words are given under this heading to indicate such aspects as habitat, diet, whether predominantly terrestrial or arboreal, diurnal or nocturnal etc., especially where these are known to be different from related species. Similar information is given in the introductory text to some genera and higher taxa if it applies to all species. Much information of this kind is derived from casual observation and anecdotal accounts. Where the results of more detailed ecological studies have been published a reference is given, especially to recent work that contains bibliographies of earlier work.

Ecological information is also summarized in some of the regional handbooks, e.g. Roberts (1977) for Pakistan, Lekagul & McNeely (1977) for Thailand, Medway (1969-83) for Malaya, Medway (1965b,1977a) and Payne *et al.* (1985) for Borneo. More general accounts of the ecology of parts of the Region are those of Whitten *et al.* (1987a) for Sulawesi, Whitten *et al.*(1987b) for Sumatra, Mani (1974) for India and McKay (1984) for Sri Lanka.

11. Conservation

Many species in the region are endangered by exploitation and/or by loss of habitat, often exacerbated by small original ranges. In the case of those species particularly controlled by the Convention on International Trade in Endangered Species (CITES) or listed in the *1990 IUCN red list of threatened animals* (World Conservation Monitoring Centre, 1990) the category given to the species is included under 'Remarks'.

The categories are as follows:

CITES. Appendix I. Endangered species, trade in which is normally prohibited.

Appendix II. Threatened species, trade in which is controlled by permits.

Appendix III. Additional species for which export permits are required by the countries named.

It should be noted that in order to implement CITES many countries require permits for other species as well.

IUCN. Ex. Extinct.
- E. Endangered.
- V. Vulnerable.
- R. Rare.
- I. Indeterminate (but considered to be E,V or R).
- K. Insufficiently known (but suspected to be E,V or R).
- T. Threatened (used for species in which different subspecies fall in diferent categories).

A letter in parentheses ( ) indicates that the categorization applies to only part of the species.

12. Sources

The nomenclatural data have been derived primarily by combining the four 'parent' checklists detailed in the Preface, but the original sources of most names have been checked for bibliographic accuracy. Other data have been derived from the regional sources listed on page 7, augmented by the periodical literature. The Zoological Record for the years 1938-1990/91 has been searched. It is likely that most literature appearing up to the end of 1990 has been seen and some publications that appeared up to July 1991 have also been used.

To avoid cluttering the text, references to the original publication of all scientific names are given in the terminal bibliography. Long titles have been abbreviated, leaving enough, hopefully, to facilitate location in a library and to show the scope of the work. A detailed bibliography of works on the mammals of SE Asia, covering the Indomalayan Region south and east of India and China, was published by Jones & Jones (1976).

13. Authorship

The section on the order Chiroptera is by J E Hill, the remainder by G B Corbet.

MAMMALS OF THE INDOMALAYAN REGION

Infraclass MARSUPIALIA

Marsupials

The marsupials constitute a clearly defined taxon of about 270 species in Australia, New Guinea and South America with a few species reaching C and N America and, in the Indomalayan Region, ten species in three families on the islands of E Indonesia, west as far as Sulawesi. This taxon has been given various ranks from a single order to a subclass containing up to six Recent orders (Aplin & Archer, 1987; Marshall *et al.*, 1990). The latter arrangement is more concordant with the ordinal classification of the placental mammals although there have been problems in implementing it because of the dubious monophyly of some of the orders. However the two orders included here appear to be generally recognized as monophyletic (Table 2). In addition to the possession of a marsupial pouch, both orders share the peculiarity of having the second and third digits of the hind feet fused (syndactylous).

ORDER PERAMELINA

Bandicoots

Includes the Australian family Thylacomyidae in addition to the Peramelidae represented here.

Family Peramelidae

Bandicoots

Peramelina Gray, 1825: 340; type genus *Perameles* Geoffroy.

A distinctive family of about 20 species in Australia and New Guinea, plus one genus and species endemic to the island of Seram, Indonesia. In addition one New Guinea species occurs on the Kai Islands. The most recent comprehensive revision is that of Tate (1948). Bandicoots are terrestrial and feed mainly upon invertebrates. See Table 3.

Table 3. The genera and species of Peramelidae.

	Echymipera rufescens	*Rhynchomeles prattorum*
Range in region	Kai Is	Seram
Pelage: texture	Spiny	Normal
colour	Agouti-banded (brown/yellow)	Uniform, dark chestnut-brown
Tail (mm)	c.80	90–130
Width of rostrum at C	8.4–9.2	5.1–6.3
M^4 (last molar) transverse width	4.0–4.5 ($\geq$ that of M^3)	2.0–2.5 ($<$ that of M^3)

ECHYMIPERA Spiny bandicoots

Echymipera Lesson, 1842: 192; type species *Perameles kalabu* Lesson.

New Guinea, Queensland; 3 species of which one reaches the Kai Islands.

Echymipera rufescens Rufous spiny bandicoot

Perameles rufescens Peters & Doria, 1875: 541; Kai Is, Moluccas.

Table 2. The orders and families of marsupials.

	PERAMELINA Peramelidae	DIPROTODONTIA Phalangeridae	Petauridae
Range in region	Moluccas	Sulawesi, Moluccas, Timor	N Moluccas
Genera/species in region	2/2	4/7	1/1
Habit	Terrestrial	Arboreal	Arboreal
Head and body length	To 320mm	To 650mm	To 155mm
Pelage	Sleek or spiny, short	Woolly or shaggy	Soft, silky
Gliding membranes on flanks	0	(0)	(+)
Functional digits on fore feet	3	5	5
Muzzle	(Very long & slender)	Short	Short
Tail	Thin, short-haired, < 1/2 h&b	Thick at base, distal half naked, = h&b	Bushy, > h&b
Dental formula	(4.1.3.4/ 3.1.3.4)	2–3.1.2–3.4/ 2.0.2–3.4	(3.1.3.4/ 2.0.3.4)

Echymipera rufescens

Peramelopsis welsianus Heude, 1897: 143; Kai Is.
Perameles keiensis Cohn, 1910: 727; Kai Is.
And other synonyms from the Aru Is and New Guinea.

Range (Map 3). Kai Is, Moluccas; and the Aru Is, New Guinea and N Queensland.

RHYNCHOMELES

Rhynchomeles Thomas, 1920a: 429; type species *R. prattorum* Thomas.

Map 3. *Phalanger, Stigocuscus, Rhynchomeles, Echymipera*

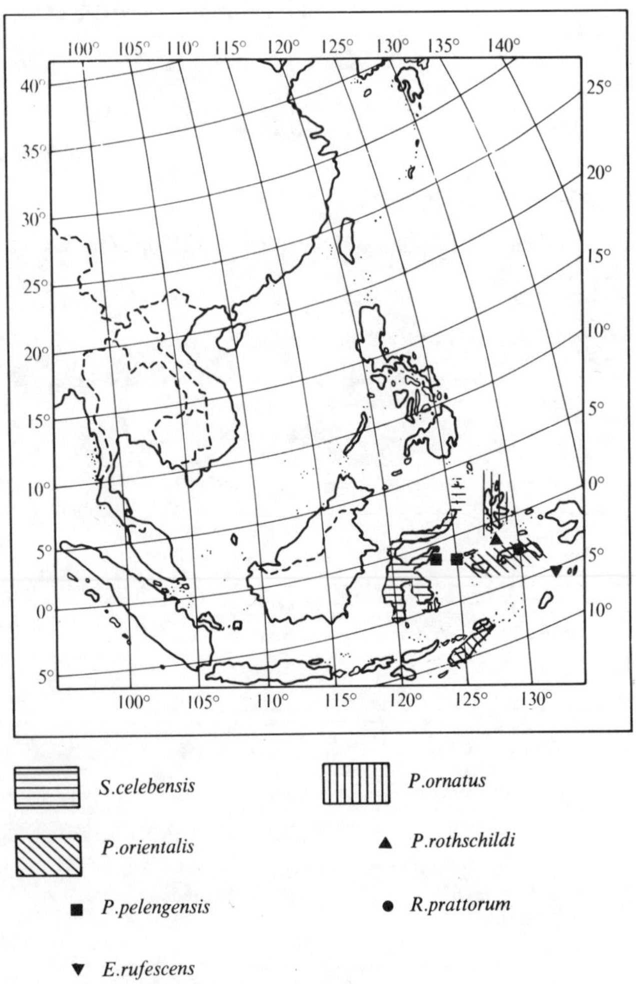

A monospecific genus, only marginally distinct from *Echymipera* of New Guinea, sharing with it four pairs of upper incisors (five in other genera of Peramelidae) and very long, slender rostrum, but differing in having soft, non-spiny pelage. A comprehensive revision of the family might well indicate that generic separation from *Echymipera* is not justified.

Rhynchomeles prattorum Seram bandicoot

R. prattorum Thomas, 1920a: 429; Mt Manusela, Seram, Indonesia, 2000m.

Range (Map 3). Known only from the island of Seram, Indonesia.

Variation. The pelage is uniform dark brown except for a variable amount of white below—usually a narrow streak on the thorax but in some individuals more extensive.

ORDER DIPROTODONTIA

Diprotodont marsupials

Includes the Australian families Macropodidae, Phascolarctidae, Vombatidae, Acrobatidae and Tarsipedidae as well as the Phalangeridae and Petauridae dealt with here (Table 2).

Family Phalangeridae

Phalangers, cuscuses

Phalangistadae Gray, 1821: 308; type genus *Phalangista* E Geoffroy.
= Phalangeridae Thomas, 1888a: 126 (*Phalanger* Storr). Maintained in accordance with Article 40(b)(i) of the *Code*.

Australia and New Guinea. Seven species extend westwards into the Moluccas and Sulawesi. These have most often been considered congeneric, in *Phalanger*, but a recent revision of the family, followed here, recognizes four genera amongst these, along with *Trichosurus* (including *Wyulda*) in Australia (Flannery *et al.*, 1987). A revision of cuscuses allied to *Phalanger orientalis* by Menzies & Pernetta (1986) agrees with the following in most respects but retains all species in *Phalanger*, as does a further review by Groves (1987).

Cuscuses are very palatable and commonly hunted; they are also known to be carried alive in canoes on long voyages as a source of meat (Brongersma, 1958: 11) and are therefore very likely to owe their origin on at least some of the islands to dispersal in this way. They are slow, mostly arboreal omnivores but feed predominantly upon leaves and fruit. See Table 4.

PHALANGER Typical cuscuses

Phalanger Storr, 1780: 33; type species *Didelphis orientalis* Pallas.
= *Phalangista* Geoffroy & Cuvier, 1795: 187 (*nomen nudum*).
= *Balantia* Illiger, 1811: 77.
= *Sipalus* Fischer, 1813: 581.
Coescoes Lacépède, 1799: 5 (*C. amboinensis* Lacépède).
= *Cuscus* Lesson, 1826: 150.

New Guinea, NE Queensland; c. 10 species, of which 4 extend west of New Guinea, as far as Timor and Sulawesi (George, 1987) (Table 5).

Table 4. The genera of Phalangeridae.

	Phalanger	*Stigocuscus*	*Spilocuscus*	*Ailurops*
Range in region	Moluccas, Timor	Sulawesi	Moluccas	Sulawesi
Species in region	4	1	1	1
Pelage of male spotted	+/0	0	+	0
Dorsal stripe	+	0 or slight	0	0
Pelage: texture	Soft, woolly	Soft, woolly	Soft, woolly	Harsh,long
Ears: inside surface	Naked	Naked	Furred	Furred
Frontal bones	Concave	Concave	Convex	Concave
I^3	$<<I^2$	$<I^2$	$<I^2$	$>I^2$
Unicuspid teeth between C and large P^3	2	1	1	1

Table 5. The species of *Phalanger*.

	orientalis	*ornatus*	*pelengensis*	*rothschildi*
Range in region	S Moluccas, Timor	N Moluccas	Peleng Is (Sulawesi)	Gt Obi I (Moluccas)
Head & body (mm)	>450	360–400	300–365	360–400
Male white-spotted	0	+	0	0
Dorsal colour (ad. male)	White	Brown	Gold or grey-brown	Grey-brown
Dorsal stripe	+	+	0	+
Ventral pelage	White-buff	White	Light orange-brown	White/cream

Phalanger orientalis Grey cuscus, Common cuscus

Didelphis orientalis Pallas, 1766: 59; Amboina.
D. indica Mueller, 1776: 61; Amboina.
D. molucca Gmelin, 1788: 106; 'Philippines and Moluccas'.
Coescoes amboinensis Lacépède, 1799: 5; Amboina.
Phalangista rufa Desmarest, 1818: 473; locality unknown.
Phalanger cavifrons Temminck, 1824: 17; Moluccas.
Also other, extralimital, synonyms.

Range (Map 3). The S Moluccas and Timor eastwards, including the islands of Buru, Seram, Ambon, Goram, Saparua, Sula Bessi, Banda, Timor, Wetar and Leti; probably also on Babar I (Menzies & Pernetta, 1986). Also in most of New Guinea and the surrounding islands east to the Solomons, and in N Queensland.

Variation. Two subspecies in the region:
P. o. orientalis. S Moluccas. P^4 and P_4 small, with 2 main cusps, remaining conical.
P. o. timorensis Menzies & Pernetta, 1986: 588; Tjamplong, Timor. Islands of Timor, Wetar and possibly Babar. P^4 and P_4 large, up to 5 cusps, wearing to a ridge.

P. o. orientalis was considered by Menzies & Pernetta (1986) to occur also in most of lowland New Guinea; George (1987) limited it to the Moluccas.

Remarks. This species has in the past been held to include the form *ornatus* (N Moluccas) which is here treated as a separate species following George (1987). The cuscuses on Timor and Wetar were recorded as *P. gymnotis* (type locality Aru Is, New Guinea) by Feiler (1978a) and tentatively by Laurie & Hill (1954) and van Strien (1986). These are here considered to be *P. orientalis* following George (1987). CITES: Appendix II.

Phalanger ornatus Moluccan cuscus

Cuscus ornatus Gray, 1860a: 374; Batjan I, Moluccas.

Range (Map 3). N Moluccan islands of Halmahera, Bacan, Ternate, Morotei, Tidore and Gebe.

Remarks. Included as a race of *P. orientalis* by most previous authors, but treated as a separate and distinctive species by George (1987), related most closely to *P. gymnotis* and *P. leucippus* (New Guinea).

Phalanger pelengensis Peleng Island cuscus

P. celebensis pelengensis Tate, 1945a: 3; Peleng I, Sulawesi.
P. c. mendeni Feiler, 1978a: 387; Taliabu I (Sula Is), 300m.

Range (Map 3). Peleng and Taliabu Is, E of Sulawesi.

Variation. The form *mendeni* (Taliabu I) was based on pelage colour—greyer brown with hands and ears tan.

Remarks. Treated as generically and specifically distinct from *Stigocuscus celebensis* following George (1987).

Phalanger rothschildi Rothschild's cuscus, Obi Island cuscus

P. rothschildi Thomas, 1898a: 433; Great Obi I, Moluccas.

Range (Map 3). Great Obi I, Moluccas.

Remarks. Included in *Stigocuscus celebensis* by Feiler (1978a, c) but here treated as generically and specifically distinct from it following George (1987).

STIGOCUSCUS

Stigocuscus Gray, 1862a: 319; type species *Cuscus celebensis* Gray.

A monospecific genus, previously included in *Phalanger* but separated by Flannery *et al.* (1987) and George (1987) on the grounds that it is more closely related to the Australian genus *Trichosurus* (including *Wyulda*) than to *Phalanger*.

Stigocuscus celebensis Small cuscus, Temboeng

Cuscus celebensis Gray, 1858: 105; Macassar (Ujung Pandang), S Sulawesi; original description composite—type selected by Thomas (1888a).
Phalanger sangirensis Meyer, 1897: 34; Sangihe I.

Range (Map 3). Sulawesi and the adjacent islands of Sangihe and Siau to the north and Muna in the southeast.

Variation. The following subspecific names have been proposed:
Phalanger c. callenfelsi Hooijer, 1950a: 28; subfossil, SW Sulawesi; likely to be less than 8000 years old (Musser, 1987).
P. c. feileri Groves, 1987: 571; Rurukan, Minahassa, N Sulawesi, 1000m.

Reviewed by Feiler (1978a, c), who however included other forms (*rothschildi, pelengensis, mendeni*) here excluded, and by Groves (1987). Three subspecies can be recognized:
S. c. celebensis. S Sulawesi. Grey above, no dorsal stripe, sharp demarcation with white underside. Includes the subfossil *callenfelsi*.
S. c. feileri. N Sulawesi. Fawn above, no dorsal stripe.
S. c. sangirensis. Sangihe Is. Pelage dark brown with yellowish sheen, slight dorsal stripe. Large. Rostrum shorter and broader.

Remarks. The forms *rothschildi, pelengensis* and *mendeni* have usually been included in this species but are here listed as *Phalanger rothschildi* and *P. pelengensis* following George (1987).

SPILOCUSCUS

Spilocuscus Gray, 1862a: 316; type species *Phalangista maculata* Desmarest.

Spilocuscus maculatus

Moluccas, New Guinea and NE Australia. Most often included in *Phalanger* but treated as a separate genus, with three species, by George (1987). Only one species in the region.

Spilocuscus maculatus Common spotted cuscus

Phalangista maculata E Geoffroy, 1803: 149; 'Moluccas', corrected and restricted by Tate (1945a) to Dorei (= Manokwari), NW New Guinea.
= *P. maculata* Desmarest, 1818: 472.
P. variegata Schinz, 1821: 258; locality unknown.
P. chrysorrhous Temminck, 1824: 12, pl. 1; 'Moluccas', restricted to Ambon by Husson (1955b: 293).
Also other, extralimital, synonyms.

Range (Map 4). Moluccas, including the islands of Buru, Seram, Ambon, Banda, Pandjang, Tioor and the Kai Is; and Selayar I (Sulawesi). Also in much of New Guinea and adjacent islands, and in NE Queensland.

Map **4**. *Spilocuscus, Ailurops, Petaurus*

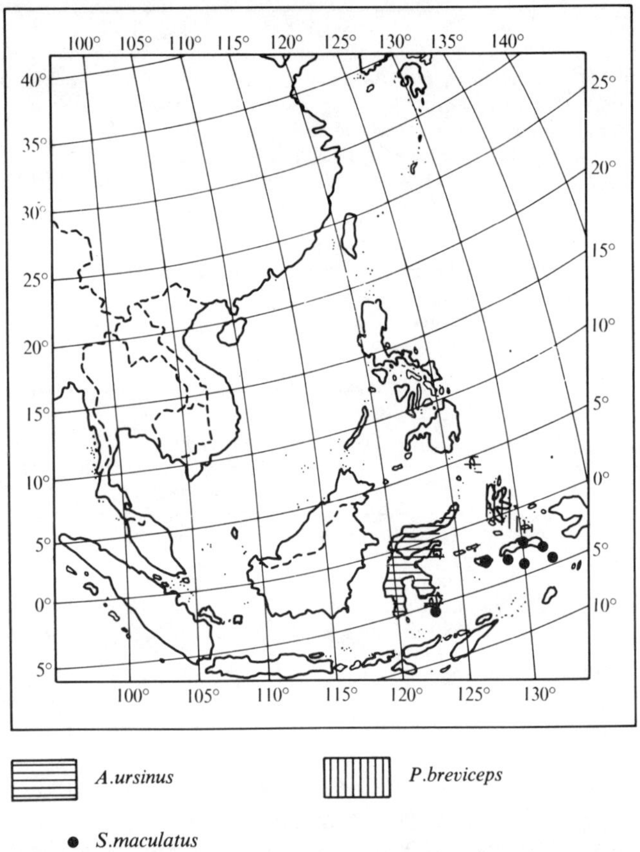

Variation. Reviewed by Feiler (1978b), including colour illustrations. The only subspecies in the region is *P. m. chrysorrhous* which also occurs on the Aru islands and in S New Guinea. It is characterized by the females being unspotted and the males having grey on the back of the head and dark brown to black spots.

Remarks. CITES: Appendix II.

AILUROPS

Ailurops Wagler, 1830: 26; type species *Phalangista ursina* Temminck. (*Ceonyx* Temminck, 1827 was proposed in synonymy and is therefore not available—Flannery *et al.*, 1987.)
= *Eucuscus* Gray, 1862a: 316.

A monospecific genus, previously included in *Phalanger* but separated as the sole member of a subfamily Ailuropinae by Flannery *et al.* (1987), supported by George (1987).

Ailurops ursinus Bear phalanger, Large Sulawesi cuscus, Bubutu

Phalangista ursina Temminck, 1824: 10, pl. 1; Menado, Minahassa, N Sulawesi.
Phalanger melanotis Thomas, 1898f: 2; Lirung, Talaud Is.
P. furvus Miller & Hollister, 1922: 115; Rano Rano, C Sulawesi, 1800m.

Range (Map 4). Sulawesi, adjacent islands of Muna, Butung, Malengi (Togian Is), Peleng and Lembeh; Lirung (Talaud Is).

Variation. The following subspecific names have been proposed:
Phalanger u. intermedius Hooijer, 1952: 241; Batoe Edjaja cave, near Bonthain, SW Sulawesi; subfossil, diagnosed only on size, being intermediate between *ursinus* and *furvus*.
P. u. togianus Tate, 1945a: 4; Malenge, Togian Is, Sulawesi.
P. u. flavissimus Feiler, 1977: 193. Peleng I.

Variation was reviewed by Hooijer (1952) and Feiler (1977) who recognized the following subspecies:
A. u. ursinus. Sulawesi mainland. Dorsal pelage dark brown, slightly flecked with yellow; I^3 un-notched.
A. u. togianus. Malenge I. I^3 bipartite.
A. u. melanotis. Lirung (Talaud Is). Pelage much lighter.
A. u. flavissimus. Peleng I. Yellowish above, especially at base of tail.

Remarks. The subfossil material described by Hooijer from SW Sulawesi is likely to be younger than 8000 BP (Musser, 1987a).

Family Petauridae

Gliding phalangers etc.

Petaurinae Gill, 1872: 25; type genus *Petaurus* Shaw & Nodder.

Australia and New Guinea; 5 genera, frequently included in the Phalangeridae but given separate family rank by Kirsch (1968) on the basis of serological comparisons. One wide-ranging species occurs in the Moluccas (Table 2).

PETAURUS Lesser gliding possums

Petaurus Shaw & Nodder, 1791: pl. 60; type species *P. australis* Shaw & Nodder.
= *Ptilotus* Fischer, 1814: 512.
Petaurula Matschie, 1916: 261 (*P. breviceps* Waterhouse).
Petaurella Matschie, 1916: 261 (*P. papuanus* Thomas).

Australia and New Guinea; 4 species of which one occurs in the Moluccas. All have gliding membranes.

Petaurus breviceps Sugar glider

P. breviceps Waterhouse, 1838e: 880 (and 1839b: 152); New South Wales, Australia.
P. b. papuanus Thomas, 1888a: 158; NE New Guinea.

Range (Map 4). Islands of Halmahera, Ternate and Bacan, N Moluccas (Tate, 1945b). Also in most of New Guinea and adjacent islands (including Misool), and in E and N Australia.

Variation. The Moluccan population is referable to *P. b. papuanus*.

Remarks. The form *papuanus* was given specific rank by Matschie (1916) but its inclusion in *P. breviceps* is given some support by their ability to interbreed in captivity (Fleay, 1947: 88). For a general account of the species see Smith (1973).

Infraclass EUTHERIA

Placental mammals

The fourteen orders of placental mammals in the Indomalayan Region are all very discrete taxa that are generally recognized as orders in recent classifications, except that the two suborders of Cetacea are sometimes given ordinal rank, and the same has been suggested for the two suborders of Chiroptera (Hill & Smith, 1984,88; Pettigrew, 1991), as discussed under the headings of these orders. The principal distinguishing features of five orders are so conspicuous as to require only brief mention:

Pholidota: body and long tail covered with large, overlapping horny scales; no teeth—scaly anteaters.

Chiroptera: membranous wings, between elongated digits 2-5 of fore limb and hind leg; capable of sustained flight—bats.

Cetacea: no external trace of hind limbs; tail-tip expanded into lateral, horizontal flukes; teeth absent, or simple and conical; totally aquatic (mainly marine) and predaceous—whales, dolphins and porpoises.

Sirenia: no external trace of hind limbs; tail-tip expanded and horizontally flattened; teeth 2.0.3.3/2.0.3.3 in the sole Indomalayan species, cheek-teeth flat-crowned; totally aquatic (mainly inshore, marine), herbivorous—sea cows.

Proboscidea: nose elongated to form prehensile trunk—elephant.

The most conspicuous characters of the remaining nine orders are detailed in Table 6.

ORDER PHOLIDOTA

Pangolins, Scaly anteaters

A small order of uncertain affinity confined to the Afrotropical and Indomalayan regions. Probably most closely related to the order Xenarthra containing the anteaters, sloths and armadillos in the Neotropical Region and with them forming a sister-group to all the other placental mammals (Novacek *et al.*,

Table 6. The more overt characters of nine orders of eutherian mammals.

	Insectivora	Scandentia	Dermoptera	Primates	Carnivora	Perissodactyla	Artiodactyla	Rodentia	Lagomorpha
No.of families in region	3	1	1	5	8	3	6	7	2
Head & body	35–300	100–320	330–420	90–2000	120–2000	1800–4200	400–3000	60–900	350–530
Tail as % of head & body	5–160%	70–135%	60–70%	0–160%	0–120%	3–15%	0–15%	10–200%	0–15%
Digits with claws (C), thin nails (N) or thickened hooves (H)	C	C	C	C/N	C	H	H	C	C
Long diastema in front of cheek-teeth	0	0	0	0	0	+	+(0)	+	+
Canines	+	+	+	+	+	+(0)	0(+)	0	0
Functional digits on fore/hind feet	5/5	5/5	5/5	5/5	4–5/4–5	1–4/1–3	2–4/2–4	4–5/4–5	5/4
Gliding membrane between fore and hind limbs	0	0	+	0	0	0	0	0(+)	0

Table 7. The species of *Manis*.

	pentadactyla	*crassicaudata*	*javanica*
Range	China-Nepal	India etc.	Sunda
Head & body	400–580	600–750	400–650
Tail	300–380	c.450	350–560
% of head & body	c.50–70%	c.70–80%	c.80–90%
Claws of hind feet:length cf those of fore feet	<1/2	<1/2	>1/2
Rows of scales round mid-body	15–18	11–13	17–19
Mid-dorsal scales (posterior margin)	Most truncate	Most convex	Most convex
No.of rows of scales on flanks with keel	3(4)	0–3	c.5
Scales along edge of tail	16–19	14–15	>20
Terminal scale under tail	Membranous	Hard	Membranous
Ear with protruding rim (5–6mm)	+	0	0
Post-anal sac in male	+	0	0
Nasal bones	Biconvex	Parallel-sided or biconcave	As last
Width:length	c.1/2	c.1/4–1/3	c.1/4–1/3

1988). The integument of large, horny scales is unique, and teeth are completely absent. They feed almost entirely upon ants and termites. There is only one family.

Family Manidae

Pangolins

Manidae Gray, 1821: 305; type genus *Manis* Linnaeus.

Manis javanica

Afrotropical and Indomalayan regions; 2 genera, one in the Indomalayan Region. Although all seven species are often placed in a single genus *Manis*, the African ones share a suite of apparently derived characters—elongate and bifurcate xiphisternum, median line of scales on the tail not reaching tip, absence of bristles between the scales, and absence of ear pinnae —that appear to justify their generic separation in *Phataginus* Rafinesque, 1821 leaving the Asian species in *Manis* (Pocock, 1924a; Patterson, 1979).

MANIS Asiatic pangolins

Manis Linnaeus, 1758: 36; type species *M. pentadactyla* L.
= *Pholidotus* Brisson, 1762: 18.
= *Pangolinus* Rafinesque, 1821: 214.
= *Pangolin* Gray, 1873b: 8.
Phatages Sundevall, 1843b: 258, 273 (*M. laticauda* Illiger, in *M. crassicaudata*).
Paramanis Pocock, 1924a: 722 (*M. javanica* Desmarest).

Most of the Indomalayan Region; 3 species (Table 7).

The following nominal fossil species have been described from the region:
M. sindiensis Lydekker, 1876: 82(=64); Sind, Pakistan; Manchhar beds, Siwaliks. Allocated to *Nestoritherium* (order Perissodactyla) by Colbert (1935).
M. palaeojavanica Dubois, 1907: 455; Kedoeng Broeboes, central Java; Pleistocene. A distinctive giant species, 2-2.5 times the size of *M. javanica*, recorded also in Borneo where it has been dated to about 42-47000 BP by indirect C^{14} dating (Hooijer, 1960a).
M. lydekkeri Dubois, 1908: 1268; Carnal Caves, India. (Based upon an isolated phalanx.)
Manis sp. Pickford, 1976: 21; Potwan Plateau, Pakistan; Upper Miocene. Based upon an isolated phalanx.

All species are found mainly in forest, living in burrows in the ground but foraging to some extent by climbing.

The name *M. pentadactyla* was in the past used for the Indian pangolin, *M. crassicaudata*. This led to an error in Lekagul & McNeely (1977) where the photograph of a skull captioned *M. pentadactyla* (p. 331) is actually an *M. crassicaudata* from Sri Lanka, showing the characteristic narrow nasals of that species. The skull captioned *M. javanica* (p. 329) is that of *M. pentadactyla*.

Manis pentadactyla Chinese pangolin

M. pentadactyla Linnaeus, 1758: 36; Taiwan.
M. brachyura Erxleben, 1777: 98.
M. auritus Hodgson, 1836b: 234; lower and central Nepal.
M. dalmanni Sundevall, 1843b: 256,278, pl. 4; near Canton, S China.
Pholidotus assamensis Fitzinger, 1872: 57.
Phatages bengalensis Fitzinger, 1872: 72.
Pholidotus kreyenbergi Matschie, 1907: 234; Nanking, Jiangsu, China.
M. pusilla J Allen 1906: 465, pl. 69; Hainan I, China.

Range (Map 5). Nepal to Yunnan and SE China, north to the Yangtze, and south to E Burma, N Thailand and N Vietnam; Hainan and Taiwan.

Remarks. The karyotype was described by Quan & Wang (1984): 2n=38. CITES: Appendix II.

Map 5. *Manis*

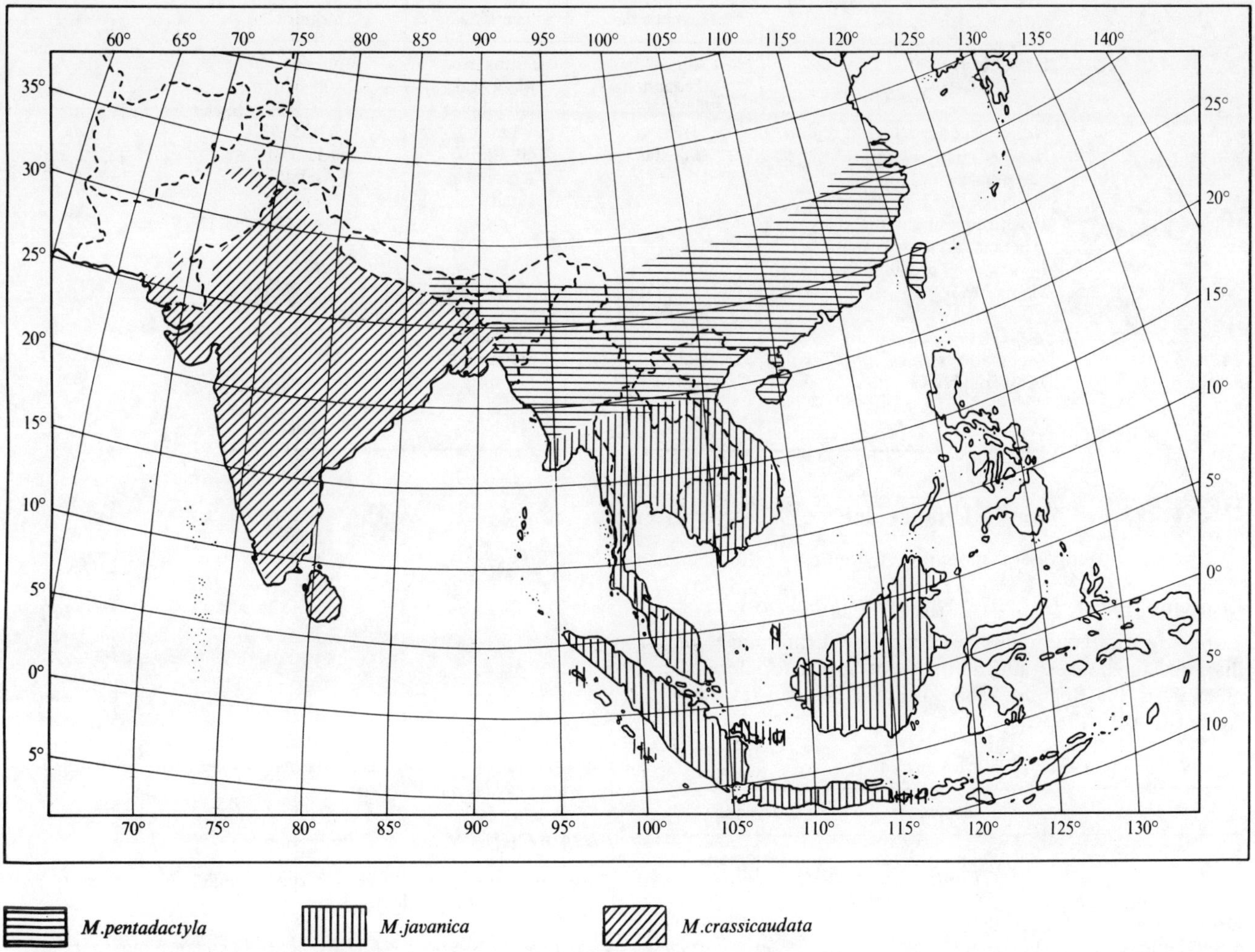

Manis crassicaudata Indian pangolin

M. laticauda Illiger, 1815: 90; *nomen nudum.*
M. crassicaudatus Gray, 1827a: 282; India.
Pholidotus indicus Gray, 1865b: 368.

Range (Map 5). Pakistan, peninsular India, NE to Bengal and Bangladesh; Sri Lanka.

Ecology. Found mainly in forest but extending into very arid thorn forest in Pakistan.

Remarks. CITES: Appendix II.

Manis javanica Sunda pangolin

M. javanica Desmarest, 1822: 377; Java.
M. leptura Blyth, 1842b: 454; locality unknown.
M. aspera Sundevall, 1843b: 253; Sumatra.
M. leucura Blyth, 1847b: 1274; Arakan, Burma.
M. guy Focillon, 1850: 513, pl. 10; 'Afrique', probably in error.
M. sumatrensis Ludeking, 1862: 40; Sumatra.
Pholidotus malaccensis Fitzinger, 1872: 50; Malacca, Malaya.
P. labuanensis Fitzinger, 1872: 56; Borneo.
= *M. labuanus* Fitzinger: Chasen, 1940a.
P. culionensis Elera, 1915: 274; Culion I, Calamian Is, Philippines.

Range (Map 5). S Burma, Thailand, Laos, S Vietnam, Malaya; Sumatra, Java, Borneo, Palawan; and on the small islands of Phuket (Thailand), Penang (Malaya), Singapore; Bintan, Batam, Bulan, Galang, Kundur (Riau Is); Lingga, Singkep, Sanglar (Lingga Is); Bangka, Belitung; Nias, Pagai (W Sumatra); Bunguran (N Natuna Is); Karimata; Bali, Lombok; Culion (Calamian Is). Although it has been suggested that it might occur on Luzon, Philippines (IUCN, 1986) this is unlikely to be the case.

Remarks. CITES: Appendix II.

ORDER INSECTIVORA

Insectivores

Contains six families of which three are represented in the Indomalayan Region (Table 8). The family Tupaiidae, which has been included in the past, is here placed in a separate order Scandentia. The classification of the Insectivora has been reviewed by Gill (1875), Cabrera (1925), Van Valen (1967), Gureev (1979) and Butler (1988).

Table 8. The families of Insectivora.

	Erinaceidae	Talpidae	Soricidae
Range in region	Almost throughout	China-Malaya & Himalayas	Throughout
No. of genera/species in region	6/12	4/12	10/55
Head & body	100–300	60–180	35–150
Ear pinnae	Large	Absent or small	Small or absent
I^1 with prominent 2nd cusp behind large 1st cusp	0	0	+
I_1 very long, procumbent	0	0	+
M^1 and M^2	Quadrate, low-cusped	Sharp cusps connected by W-shaped ridge	As Talpidae
Zygomatic arches	Robust	Slender	Absent
Tympanic bones	Annular	Forming bulla	Annular, loosely attached

Family Erinaceidae

Hedgehogs, moonrats, gymnures

Erinacini Fischer, 1817: 372; type genus *Erinaceus* L.

Two very discrete subfamilies: one, the Erinaceinae, only marginal in the west and north, the other, the Galericinae, endemic to the Region (Table 9). For a detailed review see Corbet (1988).

Table 9. The subfamilies of Erinaceidae.

	Erinaceinae	Galericinae
Range	Pakistan, W India, C China	S China-Sunda, Philippines
Genera/species in region	3/6	3/6
Dorsal pelage	Spiny	Hairy (or bristly)
Tail (% of head & body)	c.10%	10–90%
Lower incisors	2:2	3:3
P_3	Absent	Present

Subfamily Erinaceinae

Hedgehogs

Africa, Palaearctic; 4 genera of which 2 reach the arid zone of the Indomalayan Region and 2 are marginal in C China (Table 10). Hedgehogs have also been recorded in five localities in Sichuan as '*Erinaceus europaeus miodon*' (Hu & Wang, 1984) and in extreme NW Yunnan (Wang Yingxiang, pers. comm.), but it is uncertain whether these records refer to *Erinaceus* or to *Hemiechinus*.

Table 10. The genera of Erinaceinae

	Erinaceus	*Hemiechinus*	*Paraechinus*
Range	C China	NW India etc., C China	NW India etc.
Species in region	1	3	2
Medial naked zone on crown (Fig.1)	Narrow (c.4–6mm)	Absent	Wide (c.8–10mm)
Spines	Smooth	Varied	Papillate, grooved
Auditory region of skull	Normal	Slightly inflated	Strongly inflated

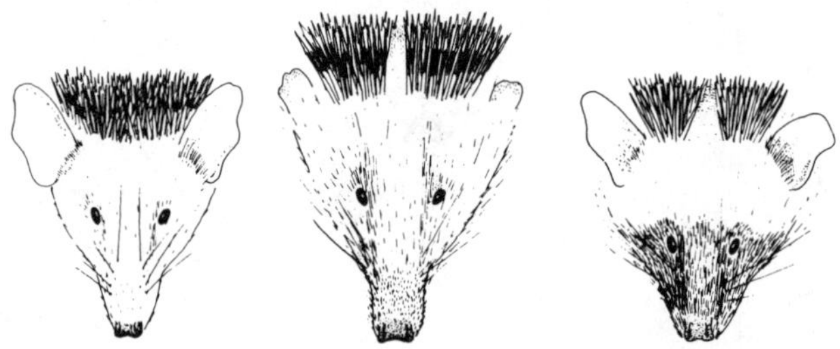

Fig. **1**. Heads of (a) *Hemiechinus auritus*; (b) *Erinaceus amurensis*; (c) *Paraechinus micropus*.

ERINACEUS Woodland hedgehogs

Erinaceus Linnaeus, 1758: 52; type species *E. europaeus* L. Conserved by Opinion 75 (1921).

W Europe-W Siberia; NE China etc. ; 3 species, one marginal in the Indomalayan Region.

The following nominal fossil species has been described from the region:
E. koloshanensis Young & Liu, 1951: 47; Koloshan, near Chungking, Sichuan, China, [29°31'N, 106°35'E]; Middle Pleistocene. On the evidence available it is not clear whether this is a form of *Erinaceus s.s.* or of *Hemiechinus*.

Erinaceus amurensis Amur hedgehog

Erinaceus amurensis

E. europaeus var. *amurensis* Schrenk, 1859:pl. iv; Gullsoja, near Aigum, R Amur, E Siberia.
E. dealbatus Swinhoe, 1870a: 450; Beijing.

E. kreyenbergi Matschie, 1907: 135; Shanghai.
E. hanensis Matschie, 1907: 138; Hankow, Hubei.
And other, extralimital synonyms.

Range (Map 6). Lowland China from c. 29°N (i. e. a little south of the Yangtze) north to the Amur Basin and Korea. Perhaps also in Sichuan (recorded as '*Erinaceus europaeus miodon*' by Hu & Wang, 1984).

Variation. The three apparently isolated segments appear to be distinguishable as follows:
E. a. amurensis. Manchuria etc. Spiny pelage dark; ventral pelage dark greyish brown with a variable amount of white on the throat and chest.
E. a. dealbatus. Beijing to Yangtze. Spiny pelage pale; ventral pelage white.
E. a. kreyenbergi. South of Yangtze. Spiny pelage dark; ventral pelage uniform brown.

Remarks. Has been included in *E. europaeus* (W Europe), e. g. by Ellerman & Morrison-Scott (1951), but not obviously more closely related to it than to *E. concolor* (E Europe etc.) which is undoubtedly distinct from *E. europaeus*.

HEMIECHINUS Steppe hedgehogs

Hemiechinus Fitzinger, 1866: 565; type species *Erinaceus platyotis* Sundevall, in *H. auritus*.
Mesechinus Ognev, 1951: 8 (*Erinaceus dauuricus* Sundevall).

N Africa to W China and NW India; 4 species, 3 in Indomalayan Region (Table 11). Most easily characterized by the complete absence of a medial naked zone on the crown although this is likely to be a primitive feature (Fig. 1).

Hemiechinus auritus Long-eared hedgehog

Erinaceus auritus Gmelin, 1770: 519; Astrakhan, SE Russia.
E. megalotis Blyth, 1845b: 353; Kandahar, Afghanistan.
And many other extralimital synonyms.

Range (Map 6). Subdesert and dry steppes of Afghanistan and W Pakistan; W to Libya, N to the Gobi Desert and S Russia.

Variation. The population in Afghanistan and Pakistan is characterized by very large size (head and body 260-300mm) and uniformly brown ventral pelage: *H. a. megalotis.*

Map **6**. *Erinaceus, Hemiechinus, Paraechinus*

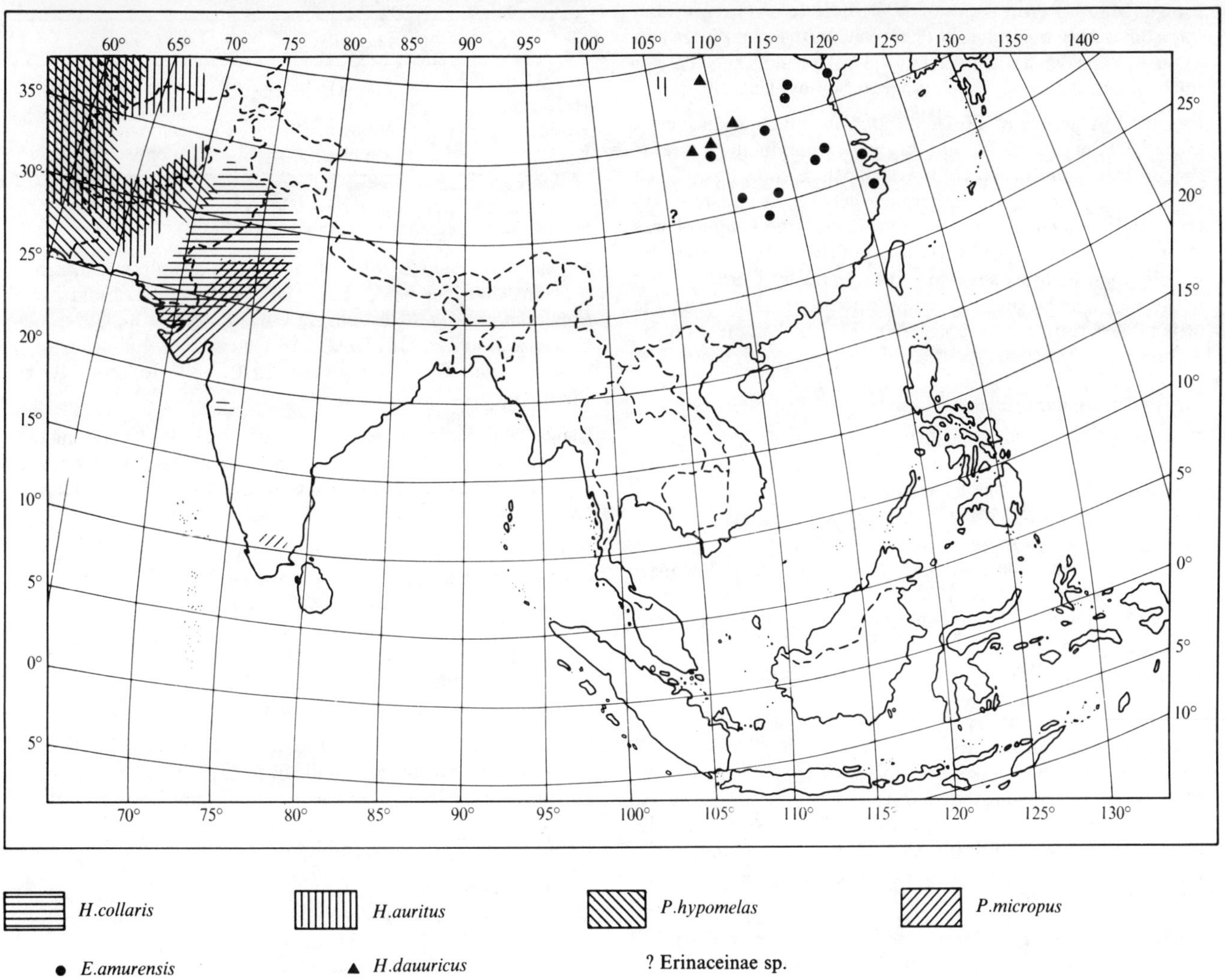

Table 11. The species of *Hemiechinus* .

	auritus	*collaris*	*dauuricus*
Range in region	Afghanistan, Pakistan	Pakistan, NW India	C China
Head & body	260–300	c.170	c.200
Ears	Much longer than spines	Much longer than spines	No longer than spines
Spines	Slightly papillate and grooved		Smooth
Pelage (including spines)	Medium brown	Blackish brown	Medium brown
P^2 (6th tooth)	2-rooted	1-rooted	2-rooted

Remarks. The form *megalotis* has been treated as a separate species, e. g. by Ellerman & Morrison-Scott (1951), but has been shown to intergrade with more normally sized *H. auritus* in SW Afghanistan (Niethammer, 1973). The form *collaris* has also been included, but is here given specific rank.

Hemiechinus collaris

Erinaceus collaris Gray, 1830(in 1830-35):pl. 8; Doab, between the rivers Jumna and Ganges, India.
E. grayi Bennett, 1832: 123; Himalayas.
E. spatangus Bennett, 1832: 124; Himalayas.
E. blanfordi Anderson, 1878a: 208; Sind, Pakistan.

Range (Map 6). Arid zones of NW India (east to Agra) and Pakistan west to the Indus (both banks), north to Jamnur, south as far as Pune (Maharashtra), but the last perhaps an isolate or introduction (specimen in Zoological Survey, Calcutta).

Remarks. A vicariant of *H. auritus*, in which it has been included (Corbet, 1978), but probably marginally warrants specific status on the basis of the dark pelage and single-rooted P^2, and the extreme size difference between it and the very large and adjacent *H. a. megalotis*. The name *blanfordi* was erroneously included in the synonymy of *Paraechinus hypomelas* by Ellerman & Morrison-Scott (1951) and by Corbet (1988) because of a mislabelled type specimen. It is clear from the original description that it is a *Hemiechinus*, as pointed out by Agrawal (1973).

Hemiechinus dauuricus

Erinaceus dauuricus Sundevall, 1842: 237; Dauuria, Transbaikalia, USSR. Conserved by Opinion 1243 (1983), with precedence over *E. sibiricus* Erxleben.
E. sibiricus Erxleben, 1777: 172; Siberia. Suppressed in relation to *E. dauuricus* by Opinion 1243 (1983).
E. hughi Thomas, 1908a: 44 (and 1909a: 966); Paohi, Shaanxi, China [= Baoji, 34°20'N, 107°10'E].
E. miodon Thomas, 1908a: 44 (and 1909a: 965); Yulinfu, Shaanxi, China, 4000ft.
H. sylvaticus Ma, 1964: 31,35; Quin-shui District, N slope of Li-Shan, Shanxi, China [c. 35°30'N, 112°E].

Range (Map 6). Qinling Mts, S Shaanxi and SW Shanxi, central China; also in semi-arid zone in N China from C Shaanxi to W Manchuria, NE Mongolia and the upper Amur Basin in USSR; perhaps also in Sichuan (records of '*Erinaceus europaeus miodon*' from five localities in Sichuan by Hu & Wang (1984) are on geographical grounds likely to be *H. dauuricus* or *Erinaceus amurensis*).

Variation. The southern form, *H. d. hughi* (synonym *sylvaticus*) is characterized by darker spines (due to longer terminal dark tips), and very pale ventral pelage. Apparently also by some cranial differences, e. g. more vertical occiput and more prominent frontal ridges, but only two skulls are available.

PARAECHINUS Desert hedgehogs

Paraechinus Trouessart, 1879a: 242; type species *Erinaceus micropus* Blyth.
Macroechinus Satunin, 1907: 189 (*E. hypomelas* Brandt).

S & W India to W Sahara, N to Aral Sea; mainly in desert; 3 species, 2 in Indomalayan Region (Table 12).

Table 12. The species of *Paraechinus*.

	micropus	*hypomelas*
Range	S & NW India	Indus Valley-Arabia
Head & body	c.180	c.225
Face	White mask	All dark
P^3	1 root	3 roots

Paraechinus micropus Indian hedgehog

Erinaceus micropus Blyth, 1846b: 170; Bahawalpur, Pakistan.
E. nudiventris Horsfield, 1851: 136; Madras dist., India.
Hemiechinus mentalis Fitzinger, 1867: 874; ? Himalayas.
E. pictus Stoliczka, 1872: 223; W Cutch, NW India.
P. intermedius Biswas & Ghose, 1970: 469; Jodhpur, Rajasthan, India.

Range (Map 6). In two segments: from the Indus (including the W bank at the delta) through Rajasthan and Gujarat, south to at least Bombay and Pune, east to Agra ; and in Tamil Nadu, S India.

Variation. The following subspecies has been described on the basis of slight differences in the pigmentation of the spines:
P. intermedius kutchicus Biswas & Ghose, 1970: 475; Bhuj, Kutch, Gujarat, 106m.

The isolated southern population is little known but may justify subspecific rank (*P. m. nudiventris*) on the basis of shorter ears, brown collar and very rudimentary P^3.

Remarks. The isolation and apparently limited distribution of the southern population is difficult to explain. The possibility of introduction by human agency should not be ruled out.

Paraechinus hypomelas Brandt's hedgehog

Erinaceus hypomelas Brandt, 1836: 32; N Iran.
E. jerdoni Anderson, 1878a: 209; Karachi, Pakistan.
P. amir Thomas, 1918a: 232; Kandahar, Afghanistan.
And other extralimital synonyms.

Range (Map 6). Indus Valley, E Afghanistan and Baluchistan; west to Arabia and north to Turkmenia.

Variation. Two subspecies can be recognized in Pakistan:
P. h. hypomelas (including *amir*). W Pakistan including Baluchistan, Afghanistan. Large: head & body 205-285mm.
P. h. jerdoni. Indus Valley. Small: head & body 150-205mm.

Remarks. The Indus subspecies has often been called *P. h. blanfordi* but it is clear that this name is a synonym of *Hemiechinus collaris* (Agrawal, 1973).

Subfamily Galericinae

Moonrats, gymnures

Galerices Pomel, 1848: 249; type genus *Galerix* Pomel, 1848 (extinct).
Gymnurinae Gill, 1872: 19 (*Gymnurus*).
Hylomyidae Anderson, 1879: 138 (*Hylomys*).
Echinosoricinae Cabrera, 1925 (*Echinosorex*).

Forest zone of SE Asia; 3 Recent genera (Table 13). Also 8 extinct genera in N America, Europe, E Africa and Pakistan; Oligocene to Miocene.

Table 13. The genera of Galericinae.

	Echinosorex	*Hylomys*	*Podogymnura*
Range	Malaya, Sumatra, Borneo	S China, Greater Sunda Is	Philippines
No. of species	1	3	2
Head & body	c.300	c.110–140	160–200
Tail (% of h&b)	80–90%	10–55%	30–40%
Sagittal dark stripe	0	+/0	0
Upper premolars	4:4	4:4 or 3:3	3:3
Lower premolars (Fig.2)	4:4	4:4 or 3:3	3:3

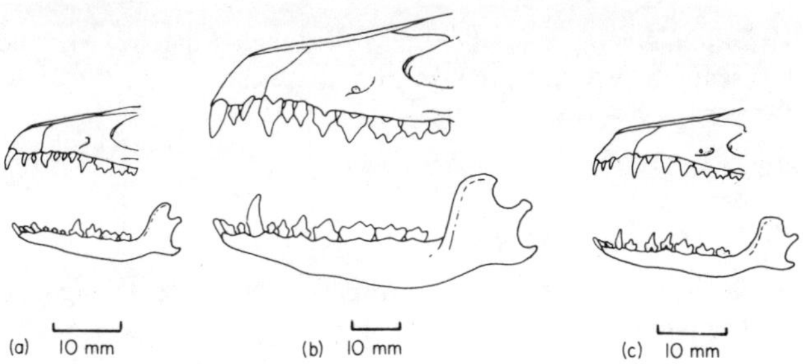

Fig. 2. Dentition of (a) *Hylomys suillus*; (b) *Echinosorex gymnurus*; (c) *Podogymnura truei*.

ECHINOSOREX

Gymnura Lesson, 1827a: 171; type species *Viverra gymnura* Raffles. Pre-occupied by *Gymnura* Kuhl, 1824, a fish.
= *Echinosorex* Blainville, 1838a: 742.

Contains one species. Fossils possibly referable to *Echinosorex* have been recorded from the Miocene of Pakistan (Munthe & West, 1980).

Echinosorex gymnurus Moonrat, gymnure

Viverra gymnura Raffles, 1821: 272; Bencoolen, Sumatra.
= *Gymnura rafflesii* Lesson, 1827a: 171.
G. alba Giebel, 1863: 277; Banjermasin, S Borneo.
G. borneotica Fitzinger, 1868a: 133; Borneo.
G. birmanica Trouessart, 1879a: 240; Bankachon, S Tenasserim, Burma.

Range (Map 7). Malayan peninsula S of c. 12°30'N, Sumatra and Borneo.

Map 7. *Hylomys, Echinosorex, Podogymnura*

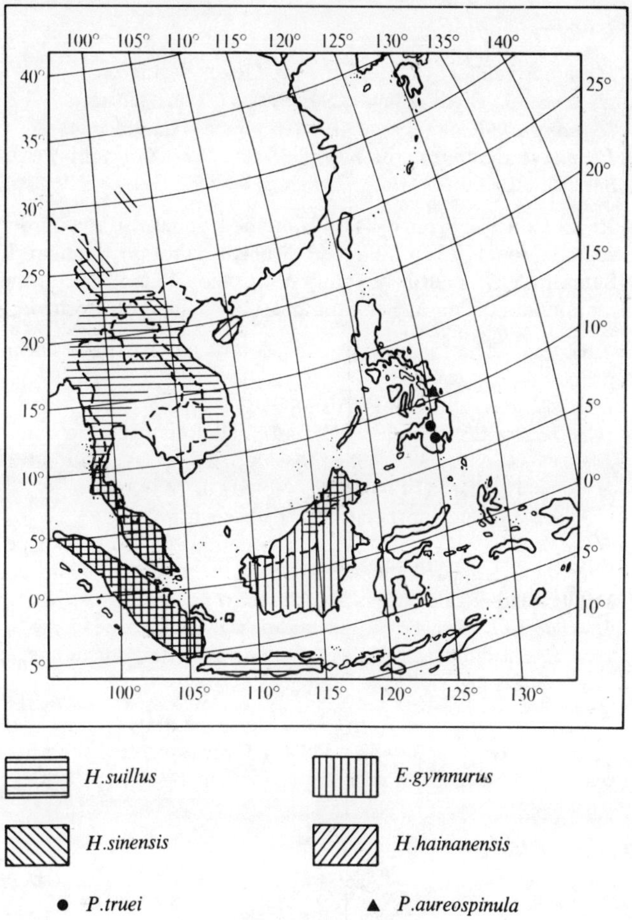

Variation. The following subspecific names have been proposed:
Gymnura rafflesi var. *candida* Günther, 1876: 425; Labuan and adjacent mainland of Borneo.
G. g. minor Lyon, 1909a: 453; Trang, S Thailand, 2000ft.
Two moderately discrete subspecies can be recognized:
E. g. gymnurus. Sumatra and Malaya. Smaller: condylobasal length of skull (cbl.) c. 75-85mm; body pelage dark.
E. g. albus. Borneo. Larger: cbl. c. 80-90mm; all white in most of range, some dark hair in pelage in west and on island of Labuan.

HYLOMYS

Hylomys Müller, 1841: 50; type species *H. suillus* Müller.
Neotetracus Trouessart, 1909: 389 (*N. sinensis* Trouessart).
Neohylomys Shaw & Wong, 1959: 422 (*N. hainanensis* Shaw & Wong).

Contains 3 species, although the genus has often been restricted to *H. suillus* (Table 14).

Table 14. The species of *Hylomys.*

	suillus	*sinensis*	*hainanensis*
Range	Burma-Borneo	SW China etc.	Hainan
Tail (% of h&b)	10–15%	c.55%	c.30%
Upper premolars	4:4*	3:3*	4:4
Lower premolars	4:4	3:3	3:3

* Not completely constant.

Hylomys suillus

H. suillus Müller, 1841: 25,50; Mt Gede, W Java.
H. peguensis Blyth, 1859a: 294; Pegu, Lower Burma.
H. siamensis Kloss, 1916a: 10; Hinlap, E Thailand, 900ft.
H. parvus Robinson & Kloss, 1916: 269; Korinchi Peak, Sumatra, 10000ft.

Range (Map 7). From c. 24°15'N on the Yunnan/Burma border south through Indochina and Malaya; and on Tioman I, Sumatra, Java and in N and NW Borneo. To c. 900m on the mainland, 3000m on Sumatra and 1100-3400m on Borneo.

Variation. The following subspecific names have been proposed:
H. s. dorsalis Thomas, 1888b: 407; Borneo.
H. s. microtinus Thomas, 1925: 497; Thai-nien, N Vietnam.
H. s. maxi Sody, 1933: 438; Giesting, Lampongs, S Sumatra.
H. s. tionis Chasen, 1940a: 12; Tioman I, Malaya.
Three discrete subspecies can be recognized:
H. s. suillus. Java, Sumatra, mainland. No dorsal stripe. Specimens from Korinchi Peak, Sumatra at 2200-3000m are small: head & body 90-105mm (*parvus*).
H. s. dorsalis. Borneo. Slight mid-dorsal dark stripe.
H. s. tionis. Tioman I, Malaya. No dorsal stripe; dorsal pelage more reddish.

Remarks. A specimen from a Late Pleistocene site (Niah Cave) in Borneo (carbon-dated to 38000 BP on associated material) was c. 15% larger than Recent specimens (Medway, 1963).

Hylomys suillus

Hylomys sinensis

Neotetracus sinensis Trouessart, 1909: 390; Tatsienlu, Sichuan, China, 2545m.

Range (Map 7). Montane areas of Sichuan, Yunnan and adjacent parts of Burma and N Vietnam, between 300 and 2700m.

Variation. The following four nominal subspecies (all described in *Neotetracus*) have been described but they are doubtfully discrete and boundaries have not been defined:
H. s. sinensis. Sichuan. Dorsal stripe faint or absent; ventral pelage grey or faintly buff.
H. s. fulvescens (Osgood, 1932: 239). Chapa, N Vietnam. Large; no dorsal stripe; ventral pelage fulvous.
H. s. cuttingi (Anthony, 1941: 58); Hpimaw, NE Burma, 9000ft. Dark above, buffy below.
H. s. hypolineatus (Wang & Li, 1982: 427,429); Leng-Shuihuo, Da Tang, Tengchong, W Yunnan, 2100m. C, W, SW Yunnan. Faint dorsal stripe on head and posterior back; colour lighter; skull with post-orbital processes.

Hylomys hainanensis

Neohylomys hainanensis Shaw & Wong, 1959: 422; Pai-sa Hsian, Hainan I, China.

Range (Map 7). Hainan Island, China.

PODOGYMNURA

Podogymnura Mearns, 1905: 436; type species *P. truei* Mearns.

Southern Philippines; 2 species (Table 15). Reviewed by Poduschka & Poduschka (1985).

Table 15. The species of *Podogymnura.*

	truei	*aureospinula*
Range	Mindanao	Dinagat I
Head & body	c.160	c.200
Dorsal pelage	Soft	Bristly
Sagittal and frontal crests	Obscure	Prominent

Podogymnura truei

P. truei Mearns, 1905: 436; Mt Apo, Mindanao.

Range (Map 7). Mindanao, Philippines (known from Mt Apo, Mt Katanglad and the western extremity).

Variation. Averages smaller on Mt Katanglad than on Mt Apo but scarcely justifying recognition as subspecies (*P. t. minima* Sanborn, 1954: 283).

Remarks. IUCN: V.

Podogymnura aureospinula

P. aureospinula Heaney & Morgan, 1982: 13; Dinagat I, Philippines.

Range (Map 7). Dinagat I, S Philippines.

Family Talpidae

Moles

Talpini Fischer, 1817: 372; type genus *Talpa* L.

Holarctic, and northern part of Indomalayan Region; three Recent subfamilies of which two reach this region (Table 16). The only comprehensive (but out-dated) revision is that of Stroganov (1948). Other reviews of classification within the family are by Hutchison (1974) for Eurasia and Yates (1982) for N America. An extensive fossil record in N America and the Palaearctic; the only fossil find in the Indomalayan Region appears to be a fragment identified as Talpidae sp. from the Miocene of Pakistan (Munthe & West, 1980).

Table 16. The subfamilies of Talpidae.

	Uropsilinae	Talpinae
Range in region	S China etc.	China, Himalayas -Malaya
Genera/species in region	1/3	3/9
Fore feet	Normal	Enlarged
Tail/head & body	c.80–90%	<60%
Ear pinnae	+	0
Dental formula	2.1.3–4.3/ 1–2.1.3–4.3	(3.1.3–4.3/ 3.0–1.3–4.3)

Subfamily Uropsilinae

Uropsilinae Thomas, 1912a: 397; type genus *Uropsilus* Milne-Edwards.

One genus, although three have been recognized in the past.

UROPSILUS Chinese shrew-moles

Uropsilus Milne-Edwards, 1872a: 92; type species *U. soricipes* Milne-Edwards.
Nasillus Thomas, 1911c: 49 (and 1912b: 129) (*N. gracilis* Thomas).
Rhynchonax Thomas, 1912b: 130 (*R. andersoni* Thomas).

S China and adjacent parts of Burma; 3 species, earlier regarded as conspecific, e. g. by Ellerman & Morrison-Scott (1951), on the grounds that the variability of dental formula was erratic, but in a detailed revision Hoffmann (1984) demonstrated that this is not the case and that three species could be recognized, with areas of sympatry and each with a characteristic dental formula (Table 17). In using Table 17 note that, when present, P^3, P_3 and I_2 can be very small. The species of *Uropsilus* are shrew-like in appearance and probably also in ecology, without the subterranean adaptations of the Talpini.

Table 17. The species of *Uropsilus*.

	soricipes	*gracilis*	*andersoni*
Range	C Sichuan	Sichuan, Yunnan, N Burma	C Sichuan
Tail/head & body (%)	82–85%	c.90–98%	c.93%
Dental formula	2.1.3.3/ 1.1.3.3	2.1.4.3/ 1.1.4.3	2.1.4.3/ 2.1.3.3

Uropsilus soricipes

U. soricipes Milne-Edwards, 1872a: 92; Moupin (=Baoxing), Sichuan, China.

Range (Map 8). A small area of Sichuan at altitudes of 1500-2700m. Precise localities were listed by Hoffmann (1984). Sympatric with *U. gracilis* in some areas.

Variation. Anomalies in dental formula (extra P^3, P_3 or I^2) were found in four out of 291 individuals (Hoffmann, 1984).

Uropsilus gracilis

Nasillus gracilis Thomas, 1911c: 49; Mt Chinfusan, near Nanchwan, Sichuan, China, 4000ft.
Nasillus investigator Thomas, 1922a: 393; Kiukiang-Salween divide at 28°N, Yunnan, 11000ft.
Rhynchonax andersoni atronates Allen, 1923b: 2; Mucheng-Salween drainage, SW Yunnan, 7000ft.
R. a. nivatus Allen, 1923b: 2; Ssu-shan, Likiang Range, W Yunnan, 12000ft.

Range (Map 8). S Shaanxi (Wu & Li, 1982), Sichuan, W Yunnan and N Burma, at altitudes of 1200-4250m. Sympatric with *U. soricipes* in parts of C Sichuan.

Map **8**. *Uropsilus, Scapanulus*

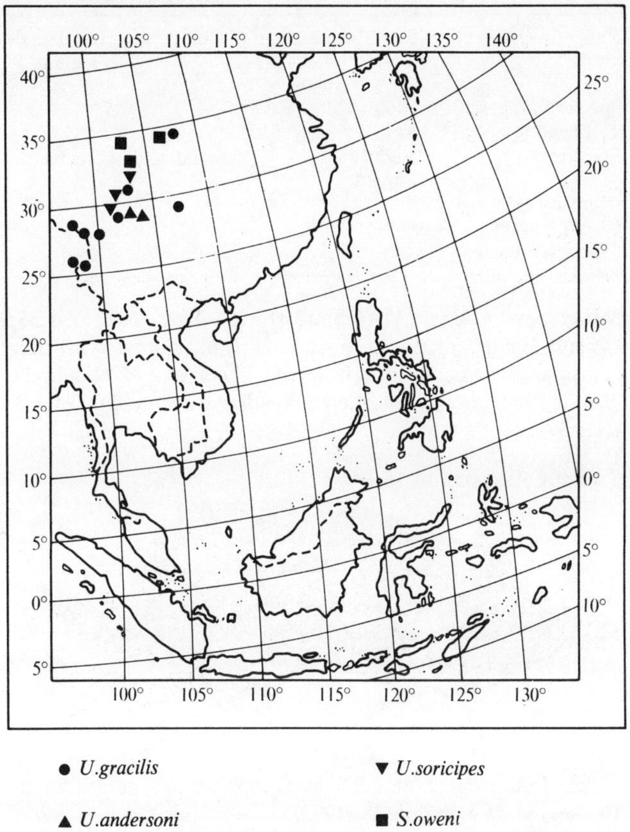

Variation. The forms listed in the synonymy were distinguished on size and colour but probably do not represent discrete subspecies (Hoffmann, 1984). Anomalies in dental formula (extra upper incisors, missing P_3) were seen in 3 out of 154 specimens (Hoffmann, 1984).

Uropsilus andersoni

Rhynchonax andersoni Thomas, 1911c: 49 (and 1912b: 130); Omisan (= Emei Shan), S Sichuan, China, 9500ft.

Range (Map 8). S Sichuan, China.

Variation. Two forms described as subspecies of *U. andersoni* are conspecific with *U. gracilis* (Hoffmann, 1984). No variation in dental formula was found in 15 individuals (Hoffmann, 1984).

Subfamily Talpinae

Moles

The true, fossorial moles, comprising about seven genera in the Holarctic Region, one of which (*Talpa*) extends substantially into the Indo-Malayan Region, and two localized genera in central and southern China. The Nearctic genera, along with the Asian *Scapanulus, Scaptonyx* and *Urotrichus*, are sometimes given subfamily rank as Scalopinae but are probably better treated as a tribe within the Talpinae. Three genera occur in this region (Table 18). In addition a new genus and species has recently been described from the Ryukyu Is:

Table 18. The genera of Talpinae.

	Scaptonyx	*Scapanulus*	*Talpa*
Range in region	SW China etc.	C China	Himalayas - S China - Malaya
Species in region	1	1	5
Tail/head & body	50%	35%	<25%
Fore feet	Slightly elongate	Circular	Circular
I^1 greatly enlarged	0	+	0
C^1 larger than adjacent teeth	+	0	+

Nesoscaptor uchidai Abe, Shiraishi & Arai, 1991: 48,53; Uotsuri-Jima, Senkaku Is, Ryukyu Is. It has not been possible to assess it critically but it seems likely that it should be included in *Talpa* as used here. The dental formula is 3.1.3.3/2.1.3.3.

TALPA Eurasian moles

Talpa

Talpa Linnaeus, 1758: 52; type species *T. europaeus* L. Conserved by Opinion 75 (1922).
Mogera Pomel, 1848: 246 (*T. wogura* Temminck).
Parascaptor Gill, 1875: 110 (*T. leucura* Blyth).
Euroscaptor Miller, 1940b: 443 (*T. klossi* Thomas).
Eoscalops Stroganov, 1941: 270 (*T. longirostris* Milne-Edwards).
Also other extralimital synonyms.

Eurasia SE to Malaya but excluding arid zones and India south of the Himalayas; as used here, in its widest sense, contains about 15 species of which 5 occur in the region (Table 19). It is frequently divided amongst up to six genera, but these are difficult to define satisfactorily without leaving a residue of doubtfully related species. In the interests of stability of nomenclature it seems best to recognize a large genus *Talpa*.

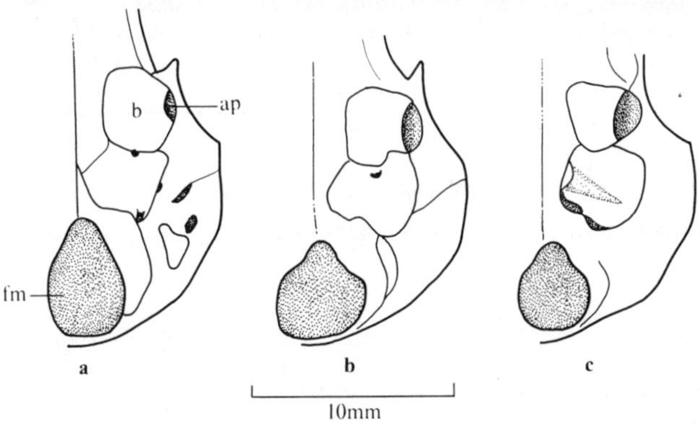

Fig. 3. *Talpa*. Ventral aspect of left auditory bulla of (a) *T.longirostris*; (b) *T.micrura* (holotype of *T.klossi*); (c) *T.leucura*. ap: aperture of bulla; b: bulla; fm: foramen magnum.

Talpa micrura Himalayan mole

T. micrura Hodgson, 1841b: 221; Nepal.
T. cryptura Blyth, 1843b: 177; Darjeeling, India.
T. klossi Thomas, 1929a: 206; Hue Nya Pla, 10 miles NW of Raheng, Thailand, 2500ft.
T. parvidens Miller, 1940a: 203; Blao, upper Donai R, Annam, Vietnam.

Range (Map 9). Himalayas west probably to E Nepal (but precise localities in Nepal appear to be lacking); also known from isolated localities in Assam, Thailand, Laos, Vietnam and Malaya. Specimens from Chapa, N Vietnam were recorded by Miller (1940b) as *T. longirostris*, but both *T. longirostris* and *T. micrura* appear to be represented by specimens from Chapa in BMNH. Recorded mainly at altitudes of 1000-2000m, but down to 100m in Bengal and Assam.

Variation. The following subspecific name has been proposed: *T. klossi malayana* Chasen, 1940a: 13; Kuala Terla Tea Estate, Cameron Highlands, Pahang, Malaya, 5000-5500ft; described solely on the basis of darker pelage.

Table 19. The species of *Talpa*.

	micrura	*grandis*	*longirostris*	*leucura*	*insularis*
Range	Himalayas Indochina Malaya	Sichuan Yunnan	Sichuan etc.	N Burma etc.	S China Taiwan
Tail/h&b	5–10%	5–10%	c.15%	c.10%	c.10%
Skull:greatest length	30–34	34–37	32–33	27–31	28–33
Upper premolars	4	4	4	3	4
(Upper teeth each side)	11	11	11	10	11
Lower incisiform teeth (preceding large caniniform P_1)	4	4	4	4	3
(Lower teeth each side)	11	11	11	11	10
Aperture of bulla (mm) (Fig.3)	Wide c.3	Wide c.3	Small c.1.5–2.0	Wide c.3	Wide c.3
Area behind bulla (+/− transverse groove) (Fig.3)	0	0	0	+	0
Molar width (external)/upper tooth-row	59–67%	60–61%	52–59%	60–67%	58–67%

Much confusion has been caused by the tendency for the pelage of preserved skins to fade, as clarified by Cranbrook (1962). Too few specimens are available to assess overall geographic variation. Tail length appears to be especially variable— minimal in *micrura s. s.*, longer in *klossi* and *parvidens*.

Remarks. Most of the talpine moles of eastern Asia were included in *T. micrura* by Schwarz (1948), followed by Ellerman & Morrison- Scott (1951). Only *T. longirostris, T. grandis* and the Japanese *T. mizura* are closely related to *T. micrura*, all the others being very distinctive.

Talpa grandis

Euroscaptor grandis Miller, 1940b: 444; Omei Shan, Sichuan, China, 5000ft.

Range (Map 9). Uncertain because of difficulty of identification: C Sichuan (Omei Shan and Chengdu); W Yunnan (specimens in Institute of Zoology, Kunming).

Remarks. This species was described as being close to '*T. klossi*' (in *T. micrura*), differing only in larger size and with a 'clavate' tail. *T. longirostris* is known from the same locality as the holotype of *T. grandis* (BMNH).

Talpa longirostris

T. longirostris Milne-Edwards, 1870a: 341; Moupin, Sichuan, China.

Range (Map 9). Sichuan, China and adjacent areas (S Shaanxi, E Chinghai and W Yunnan); and in N Vietnam (specimens from Chapa in BMNH and Field Museum, Chicago). Recorded altitudes range from 1800-2900m.

Remarks. The type species of *Eoscalops* Stroganov. This species has been much confused with *T. micrura* etc. to the south, but seems clearly definable on the basis of long, narrow rostrum and small aperture of the auditory bulla (Fig. 3), and probably also by longer tail.

Talpa leucura

T. leucura Blyth, 1850: 215; Cherrapunji, Khasi Hills, Assam.

Range (Map 9). Montane forest in Assam, Burma, Laos, Yunnan and SW Sichuan, at altitudes of 1000-2500m.

Remarks. A distinctive species in spite of previous inclusion in *T. micrura*, but doubtfully justifying separation in the genus *Parascaptor* of which it is the type species.

Map **9**. *Talpa, Scaptonyx*

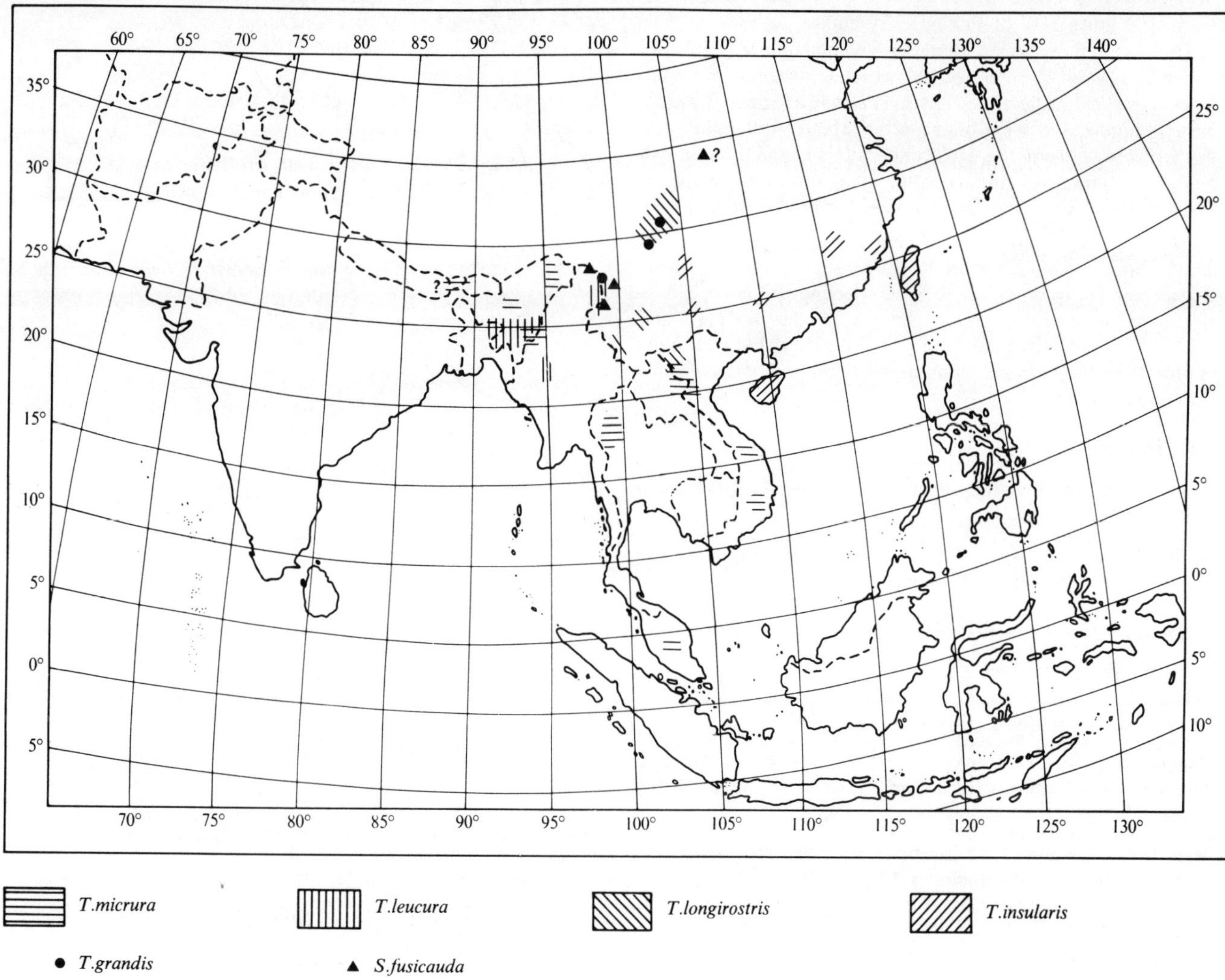

Talpa insularis

T. insularis Swinhoe, 1863: 356; Formosa.
Mogera latouchei Thomas, 1907a: 463; Kuatun, Fujian, SE China, 3500ft.
M. hainana Thomas, 1910a: 535; Mt Wuchi, Hainan I, China.

Range (Map 9). Islands of Taiwan and Hainan; mainland SE China, S of Yangtze.

Variation. The mainland form appears to be definable on the basis of small size, but it is doubtful it the two island forms are separable from each other:
T. i. insularis. Taiwan. Largest: upper tooth-row (utr) 11.6-14.2mm (n=20).
T. i. hainana. Hainan. Medium: utr 12.0-12.8mm (n=5).
T. i. latouchei. Mainland China. Smallest: utr 10.6-11.8mm (n=12).

Remarks. A member of the *Mogera* group, closely related to *T. wogura* and *T. robusta* of Japan, Korea etc. The three forms included here are separable from these northern species by the more slender form of P^1 and P^2, contrasting with the more elongate form of P^3 (in *T. wogura* all three are elongate, with two divergent roots); and by darker colour—black rather than brown.

[*Talpa moschata*]

Scaptochirus moschata Milne-Edwards, 1867a: 375; Swanhwafu, 100 miles NW of Peking, N China.

This distinctive species, characterized by reduced premolars above and below (3/3) and very short, wide rostrum, and often placed in a monospecific genus *Scaptochirus* Milne-Edwards, occurs at present in NE China south to about 36°N. It has also been recorded from the Holocene cave site of Shenxian in Jiangxi Province, S China (Zhang, 1985) and from the Lower Pleistocene of Sichuan (Teilhard de Chardin & Leroy, 1942).

[*Talpa, incertae sedis*]

T. macrura Hodgson, 1858: 176. 'Darjeeling, India'.
From examination of the holotype in the BMNH and information given by Anderson (1881) it seems that this name is almost certainly based upon a specimen of *Talpa europaea* imported into India.

SCAPTONYX

Scaptonyx Milne-Edwards, 1872a: 92; type species *S. fusicauda* Milne-Edwards.

A distinctive, monospecific genus with some affinity to the North American 'scalopine' moles.

Scaptonyx fusicauda Long-tailed mole

S. fusicauda Milne-Edwards, 1872a: 92; border of Kukunor and Sichuan, China.
= *S. fusicaudatus* Milne-Edwards, 1872b: 278.

Range (Map 9). S Shaanxi (Wu & Li, 1982), E Chinghai, Sichuan, Yunnan and N Burma; montane, at altitudes of 2100-4100m.

Variation. One subspecies has been described on the basis of larger C^1 and smaller molars:
S. f. affinis Thomas, 1912c: 514; 12 miles SE of Atunsi, NW Yunnan, 13500ft.

Variation was discussed by Hill (1962a).

SCAPANULUS

Scapanulus Thomas, 1912a: 396; type species *S. oweni* Thomas.

A distinctive, monospecific genus, related to the N American 'scalopine' moles with which it shares a greatly enlarged I^1 and small C^1.

Scapanulus oweni Kansu mole

S. oweni Thomas, 1912a: 397; 23 miles SE of Taochou, Gansu, China, 9000ft.

Range (Map 8). S Gansu, E Chinghai, SW Shaanxi and N Sichuan, China; montane, recorded between 2700 and 3000m.

Family Soricidae

Shrews

Soricini Fischer, 1817: 372; type genus *Sorex* L.

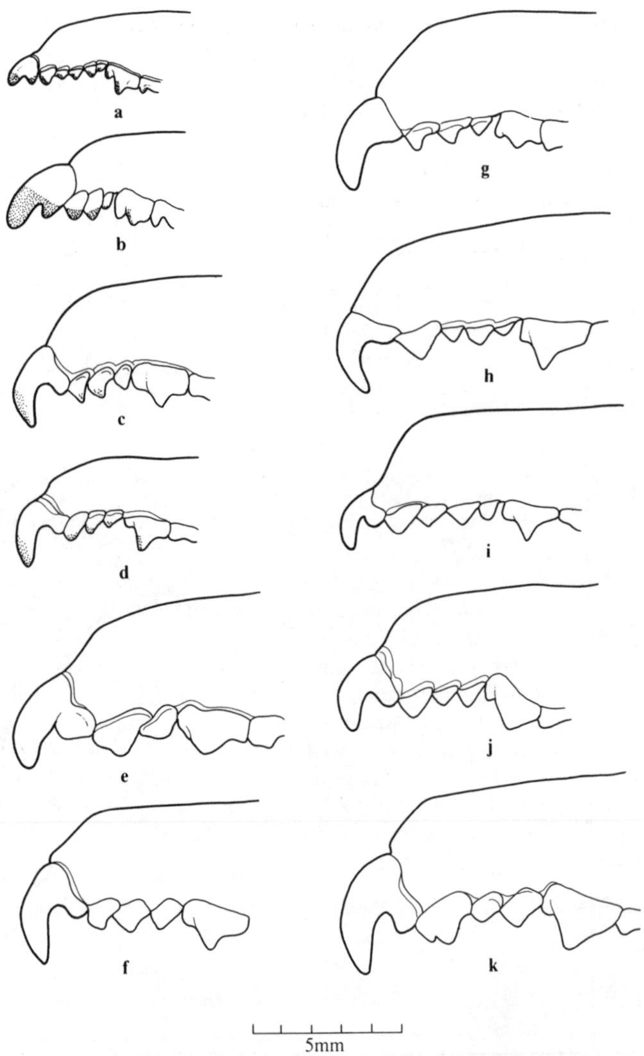

Fig. **4**. Soricidae. Left side of rostrum of: (a) *Sorex bedfordiae*; (b) *Blarinella quadraticauda*; (c) *Soriculus nigrescens* (a 4th unicuspid tooth is present but not visible in this aspect); (d) *Soriculus hypsibius*; (e) *Anourosorex squamipes*; (f) *Chimarrogale himalayica*; (g) *Nectogale elegans*; (h) *Suncus zeylanicus*; (i) *Feroculus feroculus*; (j) *Crocidura fuliginosa*; (k) *Solisorex pearsoni*.

Table 20. The genera of Soricidae.

	Sorex	*Blarinella*	*Soriculus*	*Anourosorex*	*Chimarrogale*	*Nectogale*	*Suncus*	*Feroculus*	*Crocidura*	*Solisorex*
Species in region	2	1	10	1	3	1	8	1	23	1
Range	Himalayas W China	SW China	Himalayas S China	S China- E Himalayas	China- Borneo	SW China etc.	Widespread	Sri Lanka	Widespread	Sri Lanka
Head & body	50–80	60–75	50–100	85–110	80–120	90–130	35–150	100–120	60–120	120–135
Tail with a few scattered, long, erect hairs	0	0	0	0	0	0	+[1]	+	+(0)	0
Tail with complex fringes of white hairs	0	0	0	0	0	+	0	0	0	0
Tail length (% h&b)	80–105%	**c.50%**	45–170%	**c.10%**	70–80%	c.100%	45–95%	c.60%	60–160%	c.50%
Ear pinnae	+	+	+	0	+[3]	0	+	+	+	+
Hind feet with prominent fringes of white hairs	0	0	0	0	+	+	0	0	0	0
Claws on fore feet considerably longer than on hind feet	0	+	0[2]	+	0	0	0	+	0	+
Teeth red-tipped when unworn	+	+	+[3]	0	0	0	0	0	0	0
No. of upper unicuspid teeth (Fig.4)	5	5[4]	3–4	2	3	3	4	4	3	3

Notes. 1: except *S.dayi*; 2: except *S.nigrescens*; 3: very small; 4: posterior two rudimentary.

Worldwide except for Australasia and most of S America; two subfamilies are generally recognized: Soricinae and Crocidurinae. The subfamilies and genera, including extinct forms, were revised by Repenning (1967). The genera *Anourosorex, Chimarrogale* and *Nectogale* were included in the subfamily Crocidurinae by Simpson (1945) but in the Soricinae by Repenning (1967) based upon dental and skeletal characters. In the case of *Chimarrogale* and *Nectogale* this was supported by Vogel & Besançon (1979) on the basis of hair structure. Accounts, in Russian, dealing with all species, living and extinct, were produced by Gureev (1971,1979). Much of the taxonomy is still very provisional. Species with predominantly Indomalayan distributions included here number 45, in 10 genera (Table 20).

The number and relative sizes of the upper unicuspid teeth are useful for identification (Fig. 4). However the posterior unicuspids are often very small and may be invisible when the toothrow is viewed from the outer side. These may be missing in a small proportion of individuals; the abnormal presence of extra unicuspids, especially symmetrically, is likely to be even less frequent. The permanent dentition is complete at an early age, usually on leaving the nest, and therefore the length of the toothrows is a particularly useful indicator of overall size. (However *Suncus murinus* appears to be exceptional in showing considerable increase in body size after the dentition is complete and even after reproductive maturity is reached.)

Subfamily Soricinae

SOREX

Sorex Linnaeus, 1758: 53; type species *S. araneus* L. Conserved by Opinion 91 (1926).
= *Corsira* Gray, 1838b: 123; and many other extralimital synonyms.

Throughout the Holarctic Region, with c. 60 species of which c. 7 reach the fringe of the Indomalayan Region in the Himalayas and W China. Of these, two are included here, since they are distinctive endemic species, extending below 3000m. The others, mainly above 3000m, are *S. sinalis* (S Shaanxi and W Sichuan), *S. excelsus* (W Yunnan and C Sichuan), *S. cansulus* (S Gansu), *S minutus* or a close relative (Nepal), *S. thibetanus*, probably including *bucharensis, kozlovi* and *planiceps* (Himalayas, W China) and *S. minutissimus* (SW Sichuan). All were reviewed by Hoffmann (1987). The two species detailed here (Table 21) are unique in the genus in having a narrow dark mid-dorsal stripe.

Table 21. The species of *Sorex*.

	cylindricauda	*bedfordiae*
Range	Sichuan	Gansu - Nepal
Size	Larger	Smaller
Head & body	70–77	50–70
Hind feet	15–16	11–15
Greatest length of skull	19.5–20.2	17.0–18.6
Upper tooth-row	8.2–8.6	7.0–7.8
Length of tail (% of h&b)	c.80%	c.85–105%

Sorex cylindricauda Greater stripe-backed shrew

S. cylindricauda Milne-Edwards, 1872a: 92; Moupin, W Sichuan, China.

Range (Map 10). Central Sichuan, China, at altitude of c. 3000m.

Remarks. The forms included here in *S. bedfordiae* have been included in this species, e. g. by Ellerman & Morrison-Scott (1951). It is now generally agreed that these are specifically distinct, being sympatric in central Sichuan. However, there is doubt as to whether *cylindricauda* is the correct name for this larger species (Hoffmann, 1987).

Sorex bedfordiae Lesser stripe-backed shrew

S. bedfordiae Thomas, 1911a: 3 (and 1911b: 164); Omisan, Sichuan, China, 9500ft.
S. wardi Thomas, 1911a: 3 (and 1911b: 165); Taochou, Gansu, 9000ft.

S. cylindricauda nepalensis Weigel, 1969: 167; Ringmo, NE Nepal, 3920m.

Range (Map 10). S Gansu, Sichuan, W Yunnan, N Burma and Nepal, at altitudes of 2100 to 4400m.

Variation. The following subspecific names have been proposed:
S. b. gomphus Allen, 1923b: 3; Mucheng, Salween drainage, W Yunnan, 7000ft.
S. wardi fumeolus Thomas, 1911c: 49 (and 1912b: 132); Weichoe, Siho R, W Sichuan, 6000-11000ft.

The named forms have been described on the basis of variation in size, tail-length and distinctness of the dorsal stripe but no comprehensive review has been made.

Remarks. Has been included in *S. cylindricauda* but now generally agreed to be specifically distinct.

BLARINELLA

Blarinella Thomas, 1911b: 166; type species *Sorex quadraticauda* Milne-Edwards.

Contains one species, closely related to *Sorex*.

Blarinella quadraticauda Sichuan short-tailed shrew

Sorex quadraticauda Milne-Edwards, 1872b: 261, pl. 38; Moupin, Sichuan, China.
B. griselda Thomas, 1912a: 400; 42 miles SE of Taochou, Gansu, 10000ft.
B. wardi Thomas, 1915a: 336; Hpimaw, NE Burma, 8000ft.

Range (Map 10). N Burma through Yunnan and Sichuan to Shaanxi and S Gansu, at altitudes of 2000-3500m.

SORICULUS

Soriculus Blyth, 1854: 733; type species *Corsira nigrescens* Gray.
Chodsigoa Kastschenko, 1907: 251 (*S. salenskii* Kastschenko).
Episoriculus Ellerman & Morrison-Scott, 1951: 56 (*Sorex caudatus* Horsfield).

China and Himalayas; 10 species (Table 22). (Plus many fossil species in Europe.) *Chodsigoa* and *Episoriculus* have been used as genera but are here treated as subgenera in *Soriculus* following a revision of the genus by Hoffmann (1986). *Chodsigoa* has been treated as a subgenus of the otherwise north American genus *Notiosorex* by Gureev (1971,1979). The following fossil species has been named from China:
Chodsigoa youngi Zheng, 1983; Hexian Man locality, Anhui, China, Middle Pleistocene (probably a *nomen nudum*).

Map **10**. *Sorex, Blarinella, Soriculus* (part)

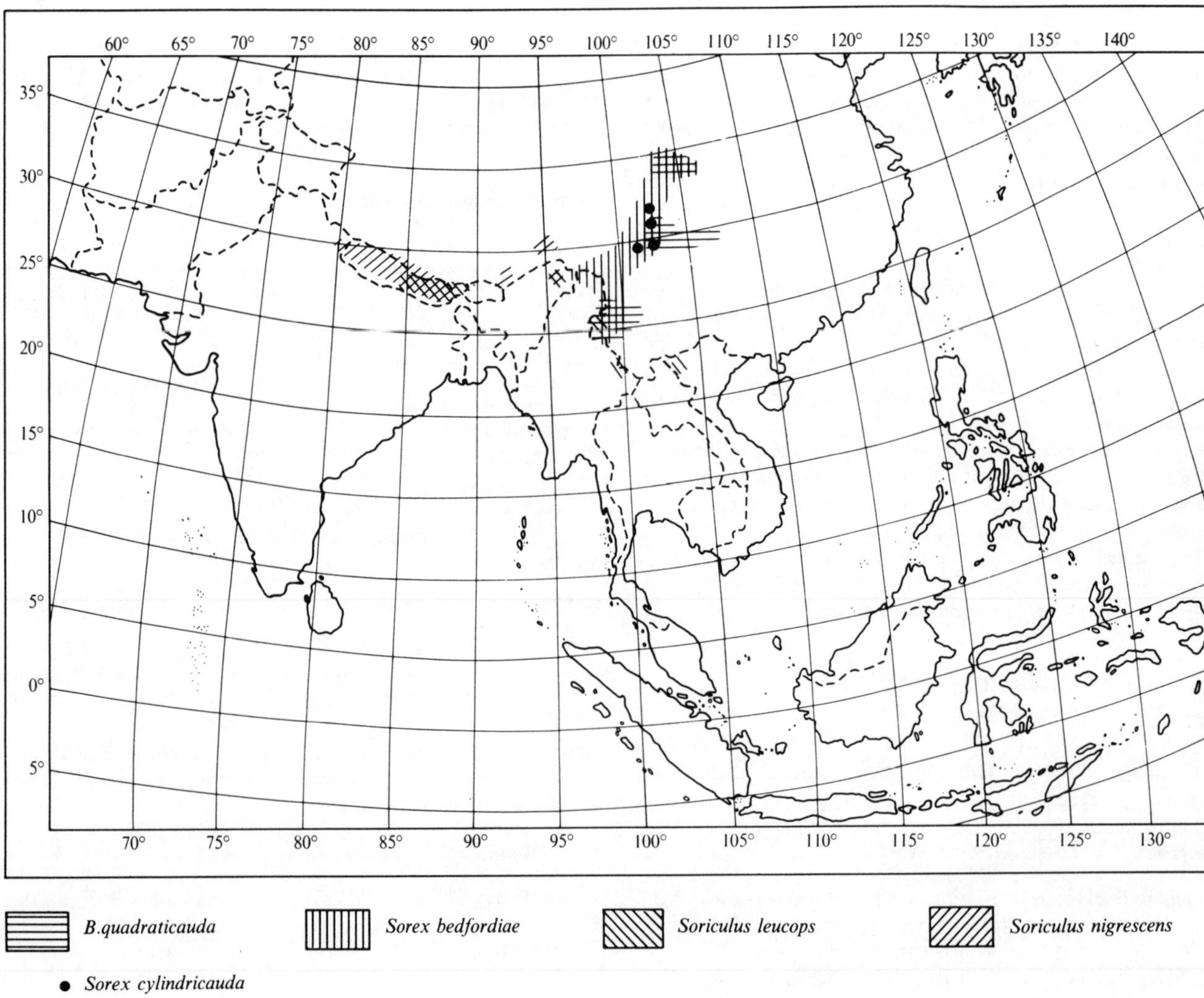

Subgenus *Soriculus*

Soriculus nigrescens Sikkim large-clawed shrew

Corsira nigrescens Gray, 1842: 261; Darjeeling, India.
Sorex aterrimus Blyth, 1854: 733; Darjeeling.
Sorex sikimensis Horsfield, 1855: 111; Darjeeling.
Soriculus radulus Thomas, 1922b: 429; Dreyi, Mishmi Hills, N Assam, 5140ft.

Table 22. The subgenera of *Soriculus*.

Subgenus	*Soriculus*	*Episoriculus*	*Chodsigoa*
No. of species	1	4	5
Normal no. of upper unicuspid teeth	4	4	3
Tail length (% of h&b)	45–60%	80–170%	80–140%
Fore claws	2.5–4.0mm	1–2mm	1–2mm

Range (Map 10). Himalayas from Kumaon east through Nepal and Bhutan to N Burma and SE Tibet, at altitudes of 1560-4300m.

Variation. The following subspecific names have been proposed:

S. n. pahari Hinton, 1922: 1053; Gnatong, Sikkim, 12300ft.
S. n. caurinus Hinton, 1922: 1054; Khati, Kumaon, India, 7600ft.
S. n. centralis Hinton, 1922: 1054; Bouzini, Nepal.

Two isolated specimens from Burma are smaller than those from the main Himalayan range and can tentatively be treated as a subspecies:

S. n. nigrescens. Himalayas. Larger: greatest skull length 20.6-23.8mm.
S. n. radulus. Burma. Smaller: skull 19.7-20.9mm.

Ecology. Fossorial, in broadleaved and coniferous forest.

Subgenus *Episoriculus* (Table 23).

Soriculus caudatus

Sorex caudatus Horsfield, 1851: 135; Darjeeling, India.
S. gracilicauda Anderson, 1877a: 282; Sikkim.
S. sacratus Thomas, 1911a: 4 (and 1911b: 165); Omei Shan, Sichuan, 6000ft.

Table 23. The species of *Soriculus* (*Episoriculus*).

	caudatus	*fumidus*	*leucops*	*macrurus*
Range	Kashmir-Sichuan	Taiwan	Nepal-Vietnam	Nepal-Vietnam
Head & body (max.)	–74	–71	–81	–73
Tail (cf h&b)	85–105%	c.75%	100–105%	130–170%
Skull length (condylo-incisive)	16.7–18.8	17.8–19.0	18.8–21.0	16.8–18.6
Upper tooth-row	7.3–8.0	7.4–8.0	8.1–9.1	7.1–7.7
Skull width (max.)	7.9–9.2	9.2–9.5	9.2–10.1	8.3–9.0

Map **11**. *Soriculus* (part)

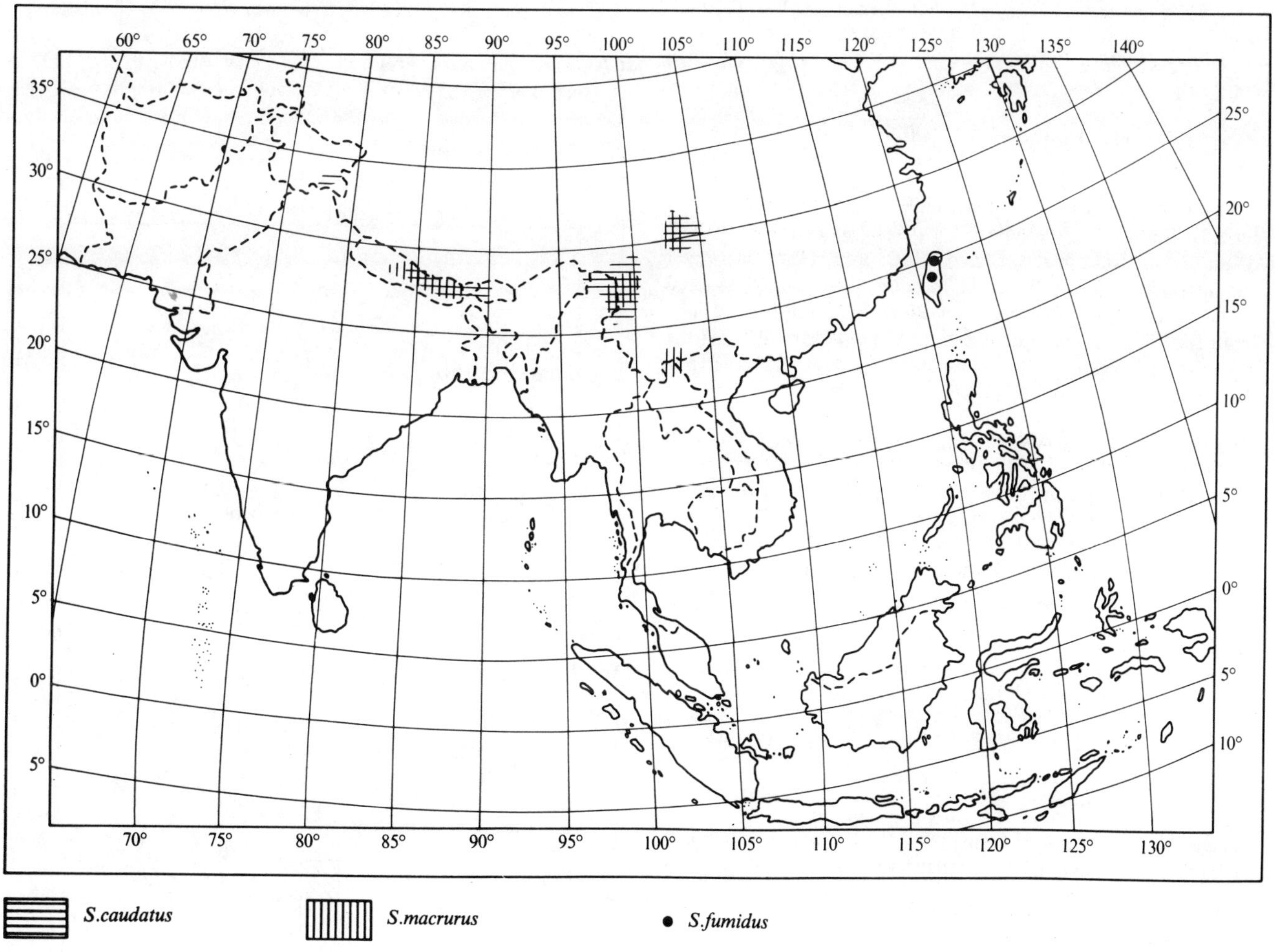

Table 24. The species of *Soriculus* (*Chodsigoa*).

	hypsibius	*lamula*	*salenskii*	*smithii*	*parca*
Range	Sichuan-Hebei	Yunnan-Gansu	Sichuan	Sichuan, Shaanxi	Sichuan-Vietnam
Head & body	73–99	54–75	c.80	72–96	66–84
Hind feet	15–18	11–16	c.22	16–19	18–20
Tail (% of h&b)	80–85%	80–85%	c.140%	c.120%	c.120%
Skull length (condylo-incisive)	19.6–22.1	18.0–18.8	c.25	21.0–23.3	19–20.5
Upper tooth-row	9.2–9.4	7.1–7.6	?	9.5–10.1	8.0–8.8

Range (Map 11). Kashmir; C Nepal to N Burma, Yunnan and Sichuan, at altitudes of 1800-3600m.

Variation. Geographical variation is slight. The following subspecific names have been proposed:
S. c. umbrinus Allen, 1923b: 5; Muching, Salween drainage, SW Yunnan, 7000ft. Small and dark.
S. c. soluensis Gruber, 1969: 309; Ringmo, Solu Prov., E Nepal. Small, long rostrum.

Remarks. The form *baileyi* was included by Ellerman & Morrison- Scott (1951) but was considered a race of *S. leucops* by Hoffmann (1986).

Soriculus fumidus

S. fumidus Thomas, 1913a: 216; Mt Arizan, central Taiwan, 8000ft.
? *Chodsigoa sodalis* Thomas, 1913a: 217; Mt Arizan, Taiwan, 8000ft.

Range (Map 11). Taiwan, from 1000 to 3200m, in forest and dwarf bamboo.

Variation. Two out of 19 skulls lack fourth upper unicuspid teeth (including the holotype of *Chodsigoa sodalis*, leading to its original allocation to *Chodsigoa*) (Hoffmann, 1986). However the inclusion of *sodalis* in *S. fumidus* must be considered tentative since the holotype of *sodalis* (a skull only) also differs from the 13 available specimens of *S. fumidus* (including the holotype) in other dental characters—narrow M^3 and less robust facies of all the teeth.

Remarks. Close to *S. caudatus* and tentatively included in that species by Ellerman & Morrison-Scott (1951). Further records of *Chodsigoa sodalis* by Kuroda (1935) referred to misidentified *Crocidura attenuata* and subsequent attempts to find *sodalis* have been unsuccessful (Jameson & Jones, 1977). Chromosomes: 2n=40, FN=52 (Harada & Takada, 1985).

Soriculus leucops

S. leucops Hodgson in Horsfield, 1855: 111; Nepal.
S. minor Dobson, 1890a:pl. xxiv; Manipur.
S. baileyi Thomas, 1914b: 683; Tsu R, Mishmi Hills, Assam, 7500ft.
S. gruberi Weigel, 1969: 170; S of Mt Everest, NE Nepal.

Range (Map 10). Central Nepal to N Burma, SW Yunnan and N Vietnam, at altitudes up to 2900m in Burma.

Remarks. This species has been confused with *S. macrurus* and has been called *S. baileyi*.

Soriculus macrurus

S. macrurus Blanford, 1888b: 231; Darjeeling, N India.
S. irene Thomas, 1911c: 49 (and 1912b: 132); Yuanchingsien, SW Sichuan, 5200ft.

Range (Map 11). Central Nepal to Sikkim; NW Burma and W and S Yunnan to Sichuan and N Vietnam.

Remarks. Included in *S. leucops* by Osgood (1932) and others, but very distinctive.

Subgenus *Chodsigoa* (Table 24).

Soriculus hypsibius

S. hypsibius de Winton, 1899: 574; Yangliupa, NW Sichuan, China.
S. (*Chodsigoa*) *berezowskii* Kastschenko, 1907: 252; Chodsigou, N Sichuan.
Chodsigoa larvarum Thomas, 1911c: 49 (and 1912b: 133); Eastern Tombs, 65 miles E of Peking, China, 1000ft.

Map **12**. *Soriculus* (part)

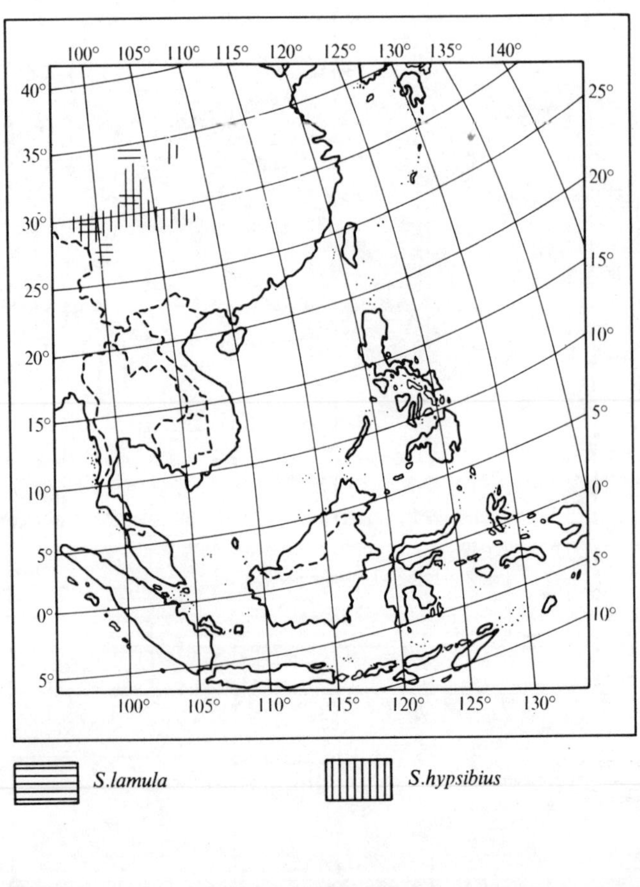

Range (Map 12). Two disjunct areas in China: Sichuan and S Shaanxi (Qinling Shan); and in Hebei.

Remarks. The forms *lamula* and *parva* have been included but are here separated as *S. lamula.*

Soriculus lamula

Chodsigoa lamula Thomas, 1912a: 399; 46 miles SE of Taochou, Gansu, China, 9500ft.
C. hypsibius parva Allen, 1923b: 5; Ssushancheng, Likiang Range, W Yunnan, 9000ft.

Range (Map 12). NW Yunnan, C Sichuan and S Gansu, China, at about 2000-3000m. Also recorded from a single specimen in Fujian (SE China) (Lehmann, 1955).

Variation. The Yunnan form, *parva*, is slightly smaller and darker than those in Sichuan and Gansu.

Remarks. Included in *S. hypsibius* by several authors, but the two are sympatric in several parts of Sichuan.

Soriculus salenskii

S. (*Chodsigoa*) *salenskii* Kastschenko, 1907: 253; Linganfu, N Sichuan, China.

Range (Map 13). Known only from the type locality in N Sichuan, China.

Remarks. This species was held to include also the forms *smithii, parva* and *furva* by Ellerman & Morrison-Scott (1951). Of these *smithii* might possibly be conspecific, but since *S. salenskii* is known only from the incomplete holotype which, according to the original measurements, appears to be very large, *S. smithii* is tentatively treated as distinct.

Map **13**. *Soriculus* (part)

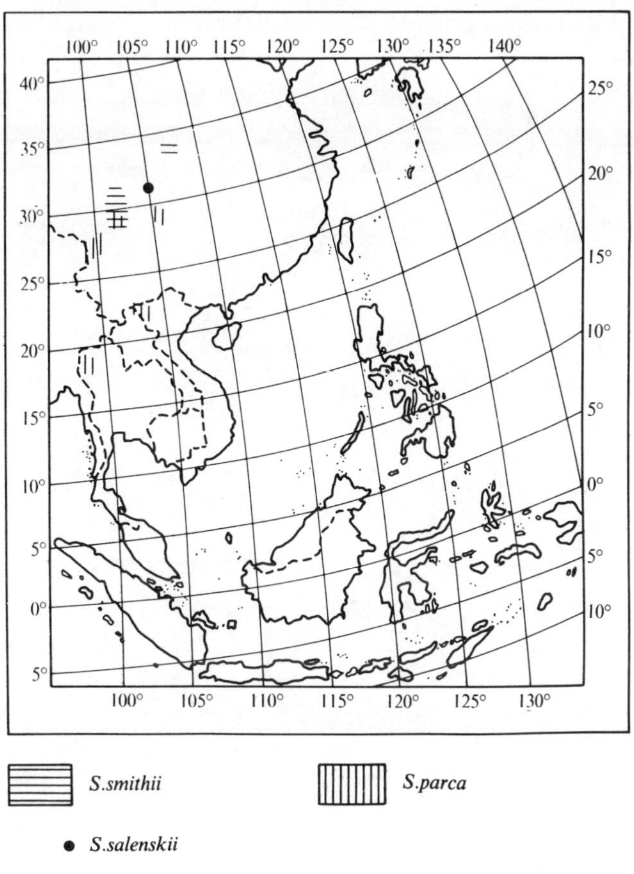

Soriculus smithii

Chodsigoa smithii Thomas, 1911b: 166; Tatsienlu, Sichuan, China, 9000ft.

Range (Map 13). Central Sichuan and Qinling Mts, S Shaanxi, China.

Remarks. Possibly conspecific with the little known *S. salenskii.* The forms *parca* and *furva*, although described as subspecies of *S. smithii*, are clearly distinct.

Soriculus parca

Chodsigoa smithii parca Allen, 1923b: 6; Homushu Pass, W Yunnan, 8000ft.
C. lowei Osgood, 1932: 249; Chapa, N Vietnam.
C. s. furva Anthony, 1941: 71; Mt Imaw Bum, N Burma, 9000ft.

Range (Map 13). S Sichuan to NW Yunnan, NE Burma, N Thailand and N Vietnam, at altitudes up to 2700m.

Variation. The Burmese form *furva* was described mainly on the basis of darker colour.

Remarks. Included in *S. salenskii* by Ellerman & Morrison-Scott (1951), but distinctly smaller than both *S. salenskii* and *S. smithii.*

ANOUROSOREX

Anourosorex Milne-Edwards, 1870a: 341; type species *A. squamipes* Milne-Edwards.
Pygmura Anderson, 1873: 229(footnote) (*A. assamensis* Anderson).

Contains a single Recent species. The following fossil species have been described from Pleistocene deposits in China; others have been described from Japan and Europe.
A. kui Young & Liu, 1951: 48; Koloshan, near Chungking, Sichuan.
A. edwardsi Zheng, 1985: 44; Guizhou Prov.
A. qianensis Zheng, 1985: 47; Guizhou Prov.

Anourosorex squamipes Mole-shrew

A. squamipes Milne-Edwards, 1872b: 264, pl. 38; ? Moupin, Sichuan, China.
A. assamensis Anderson, 1875: 282; Subsasugu, Assam.

Range (Map 14). W & C China from Shaanxi and Hubei to Yunnan (up to 3000m); N Burma, N Thailand, Assam, Bhutan, N Vietnam, probably Laos; and on Taiwan from 300-3000m.

Variation. The following subspecies have been described on the basis of small differences in size and colour but it is doubtful if any justify recognition as discrete subspecies:
A. s. capnias Allen, 1923b: 10; To-mu-lang, Chung-tien district, Yunnan, 7000ft. Small, pelage grey.
A. assamensis capito Allen, 1923b: 11; Mucheng, Salween drainage, Yunnan, 7000ft. Small, ventral pelage greenish or yellowish.
A. s. yamashinae Kuroda, 1935: 288; Taiheizan, Taihokusiu, N Taiwan. Tail short, palate short. Chromosomes: 2n=50, FN=96 (Herada & Takada, 1985).

A. s. schmidi Petter, 1963: 445; Bomdi La, NE Frontier Agency, Kameng Frontier Division, Assam, India, 2700m. Based upon a single large specimen (greatest length of skull 30.5mm); other specimens from Assam are also large (Mandahl & Das, 1969).

There is a cline of increasing size from south to north (Hoffmann, 1987).

Ecology. Mainly in forest, burrowing in leaf-litter and topsoil. Syntopic with *Soriculus fumidus* in Taiwan (Alexander *et al.*, 1987).

CHIMARROGALE

Chimarrogale Anderson, 1877a: 262; type species *Crossopus himalayicus* Gray.
Crossogale Thomas, 1921d: 243 (*C. sumatrana* Thomas).

E Asia from Borneo to the Himalayas, China and Japan; living in streams and rivers, mainly montane. Four species are recognized here, following Hoffmann (1987) (Table 25). One of these, *C. platycephala* Temminck, 1842 in Japan, is extralimital, but has been held (e. g. by Ellerman & Morrison-Scott, 1951) to include all the other named forms in the genus. It is similar to *C. himalayica* but is larger with relatively longer tail and broader skull.

Table 25. The species of *Chimarrogale*.

	himalayica	*phaeura*	*styani*
Range	Himalayas, China, Taiwan	Malaya, Sumatra, Borneo	SW China
Ventral pelage	Brown-grey	Brown-grey	Grey
Transition on flank	Gradual	Gradual	Sharp
Size	Larger	Larger	Smaller
Head & body (max.)	130mm	117mm	108mm
Max.skull length	(24.7*)25.3–28.1	25.1–29.0	23.6–24.8
Upper tooth-row	10.9–12.8	11.4–13.4	10.1–10.6
I^1 with accessory, medial cusp (Fig.5)	0	+	0

* Taiwan & E China

Chimarrogale himalayica Himalayan water shrew

Crossopus himalayicus Gray, 1842: 261; Chamba, NE Punjab, India.
C. leander Thomas, 1902b: 165; Kuatun, NW Fukien, China, 1200m.
C. varennei Thomas, 1927: 45; Dakto, Annam, Vietnam.

Map 14. *Anourosorex, Nectogale, Chimarrogale, Feroculus*

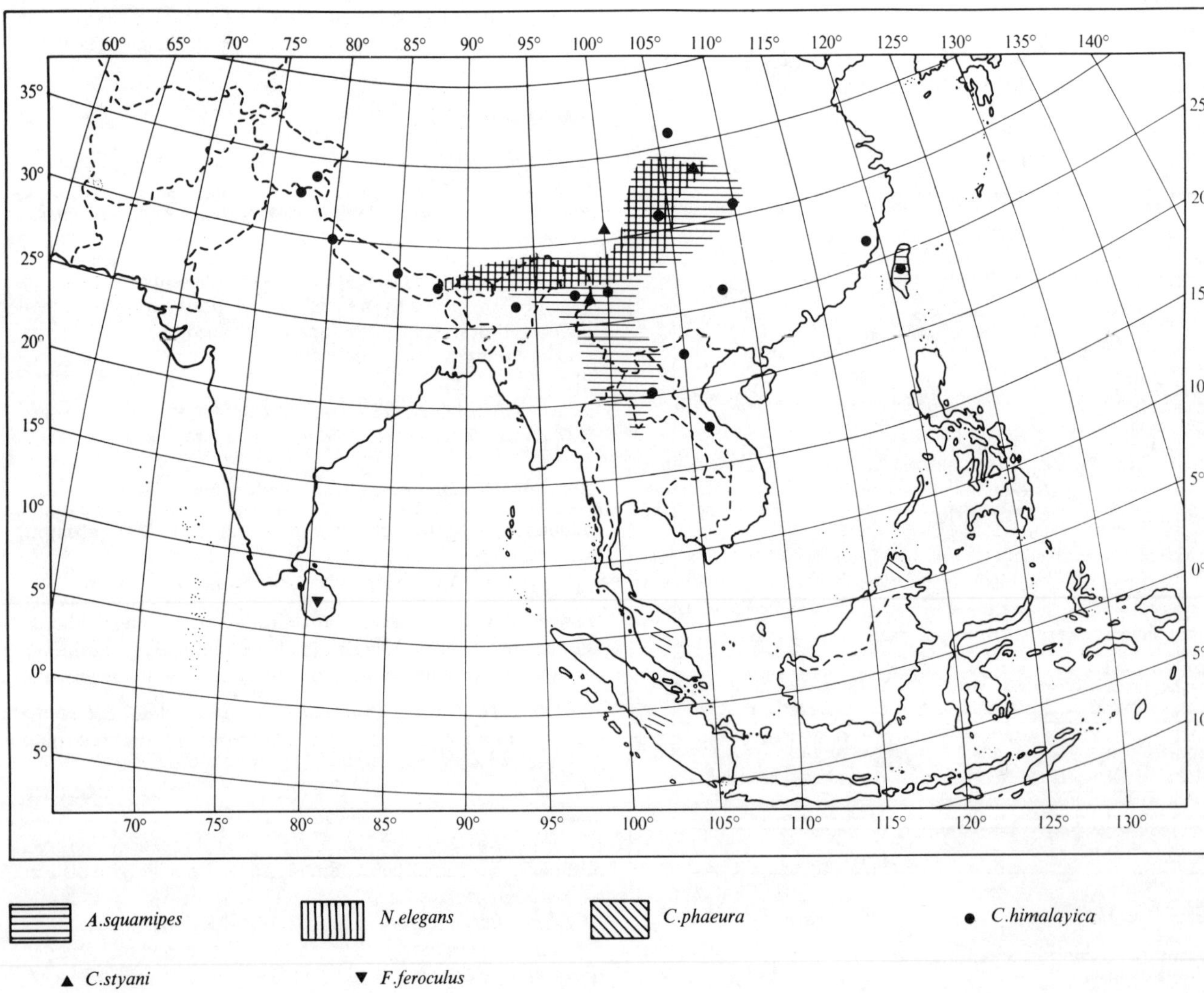

Range (Map 14). Most of China; the Himalayas west to Kashmir; N Burma, Laos, N Vietnam; Taiwan (Jones & Mumford, 1971). At altitudes of 800-1500m in the Himalayas. Mapped by Zheng & Wang (1985).

Variation. Size decreases from the Himalayas through SW China to SE China and Taiwan.

Ecology. In mountain streams.

Chimarrogale styani Styan's water shrew

C. styani de Winton, 1899: 574; Yangliupa, NW Yunnan, China.

Range (Map 14). SW China in Sichuan, S Shaanxi, SE Chinghai, E Tibet, W Yunnan; and in NE Burma; at altitudes of 1570-3100m in Burma. Mapped by Zheng & Wang (1985).

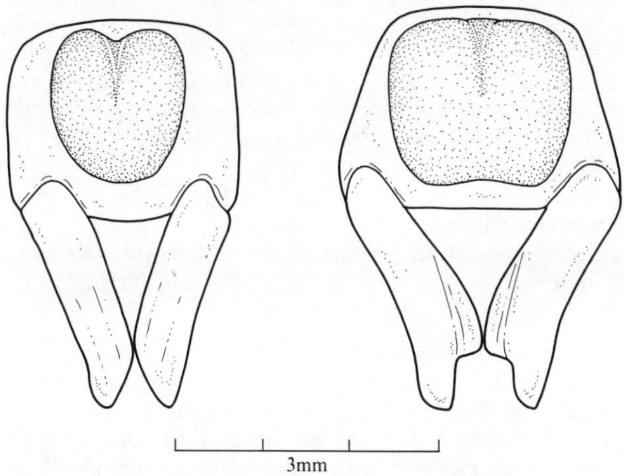

Fig. 5. *Chimarrogale*. First upper incisors (anterior view) of (left) *C.himalayica* and (right) *C.phaeura*.

Chimarrogale phaeura Sunda water shrew

C. phaeura Thomas, 1898b: 246; Mt Kinabalu, N Borneo.
Crossogale sumatrana Thomas, 1921d: 244; Padang Highlands, Sumatra.
Chimarrogale hantu Harrison, 1958: 282; Ulu Langat Forest Reserve, Selangor, 20km E of Kuala Lumpur, Malaya.

Range (Map 14). Malayan peninsula, Sumatra, N Borneo (only known from Mt Kinabalu and Mt Trus Madi); up to 1600m in Malaya, from 460-1700m in Borneo.

Variation. *C. hantu* (Malaya) was distinguished from *C. phaeura s.s.* (Borneo) on the basis of deeper braincase (7.7-8.0mm compared with 7.3-7.5mm in Borneo), but the difference is slight. The Sumatran form is known only by the damaged holotype which is larger than specimens from Malaya and Borneo (greatest length of skull 29.0mm; cf 26.0-27.2 in Malaya, n=5, and 25.3-27.6 in Borneo, n=4).

Remarks. This species has been included in *C. platycephala* (Ellerman & Morrison-Scott, 1951). However it seems distinctive, but not sufficiently so to justify generic or subgeneric separation in *Crossogale* of which *C. sumatrana* is the type species.

NECTOGALE

Nectogale Milne-Edwards, 1870a: 341; type species *N. elegans* Milne-Edwards.

Contains one, very distinctive species, strongly adapted to aquatic habitats. The multiple fringes of white hairs on the tail are unique in the family.

Nectogale elegans

N. elegans Milne-Edwards, 1870a: 341; Moupin, Sichuan, China.
N. sikhimensis de Winton, 1899: 573; Lathong, Sikkim, 10000ft.

Range (Map 14). Mountains of SW China (S Shaanxi, Sichuan, SE Tibet, Yunnan), N Burma and Himalayas west to Sikkim, perhaps to E Nepal (Frick, 1969); at altitudes of c. 900-2270m.

Variation. *N. sikhimensis* was based upon larger size and browner pelage, but no differences in size nor proportions between samples from Sikkim, Burma and Sichuan were found by Hoffmann (1987).

Ecology. Aquatic, in montane rivers and streams.

Subfamily Crocidurinae

Crocidurinae Milne-Edwards, 1872b: 256.

Suncus murinus

SUNCUS

Suncus Ehrenberg, in Hemprich & Ehrenberg, 1833:folio k; type species *S. sacer* Ehrenberg (in *S. murinus*).
Pachyura de Selys Longchamps, 1839: 32 (*S. etruscus* Savi).
Paradoxodon Wagner, 1855: 805 (*S. melanodon* Blyth, in *S. etruscus*).
Plerodus Schulze, 1897: 90 (*Crocidura suaveolens* Blasius, in *S. etruscus*).
Podihik Deraniyagala, 1958a: 5 (*P. kura* Deraniyagala, in *S. etruscus*).

S Asia, S Europe, Africa; c. 12 species of which c. 7 occur in the Indomalayan Region. Although distinguishable from the closely related *Crocidura* by the retention of four upper unicuspid teeth in each row (three in *Crocidura*), it is doubtful if the species of *Suncus* are monophyletic with respect to *Crocidura*.

There has been no revision, and the delimitation of species is very provisional (Table 26). The forms in Pakistan, India, Sri Lanka and Burma were reviewed by Lindsay (1929a).

Suncus murinus House shrew

Sorex murinus Linnaeus, 1766: 74; Java.
= *Sorex myosurus* Pallas, 1785: 337.
Sorex caerulaeus Kerr, 1792: 207; Java.
Sorex caerulescens Shaw, 1800: 533; India.
Sorex indicus Geoffroy, I.,1811: 183; Pondicherry, S India.

Table 26. The species of *Suncus*

	murinus	*montanus*	*zeylanicus*	*stoliczkanus*	*dayi*	*ater*	*mertensi*	*etruscus*
Range	Throughout	Sri Lanka S India	Sri Lanka	Pakistan India	S India	Borneo	Flores	Widespread
Head & body	100–160	80–105	110–120	60–80	c.70	75	62–68	35–55
Tail	45–90	45–62	c.100	37–50	63	57	54–62	24–32
As % of head & body	50–75%	60–90%	c.85%	c.60%	90%	75%	83–95%	48–75%
Hind feet	17–26	14–17	c.20	11–14	15	12	?	7.0–9.5
Dorsal pelage	Pale grey to dark brown	Dark brown	Dark brown	Light/medium grey-brown	Medium brown	Blackish brown	Dark reddish brown	Medium to dark brown
Vibrissae on tail	+	+	+	+	0	+	+	+
Skull: greatest length	27–37	24–28	c.28	20–23	19.4	21.3	20–21	13.7–14.7
Upper tooth-row	12.4–16.0	10.7–12.6	c.12.3	8.7–10.1	8.6	9.2	8.6–9.0	5.3–6.4
1st upper unicuspid with anterior cusp	0	0	0	0	0	+	0	+/0
I_1: no.of cusps on upper edge	0	0	0	0	2	?	1	0

Sorex sonneratii Geoffroy, I.,1827: 132; India.
Sorex giganteus Geoffroy, I., 1831: 117; Bengal.
Sorex serpentarius Geoffroy, I.,1831: 119; Pondicherry, S India.
Sorex nemorivagus Hodgson, 1845: 269; C Nepal.
Sorex soccatus Hodgson, 1845: 270; C Nepal.
Sorex griffithi Horsfield, 1851: 134; ?Assam—see Lindsay (1929).
Sorex kandianus Kelaart, 1852: 30; Sri Lanka.
Sorex heterodon Blyth, 1855a: 31; Cherrapunji, Khasi Hills, Assam.
Sorex kelaarti Blyth, 1855a: 32; Galle, Sri Lanka.
Sorex saturatior Hodgson in Horsfield, 1855: 110; Gangtok, Sikkim.
Sorex grayii Motley & Dillwyn, 1855: 1; Labuan, Borneo.
Sorex swinhoei Blyth, 1859a: 285; Amoy, S China.
Sorex viridescens Blyth, 1859a: 285; S Malabar, India.
Sorex tytleri Blyth, 1859a: 285; Dehra Dun, N India.
Crocidura microtus Peters, 1870: 589; Hong Kong.
C. (*Pachyura*) *waldemarii* Peters, 1870: 590; Bengal.
C. (*P.*) *ceylanica* Peters, 1870: 591; Paradenia, Sri Lanka.
C. (*P.*) *media* Peters, 1870: 592; Paradenia, Sri Lanka.
C. (*P.*) *sumatranus* Peters, 1870: 593; Palembang, Sumatra.
C. (*P.*) *fuscipes* Peters, 1870: 594; Singapore I.
C. (*P.*) *luzoniensis* Peters, 1870: 595; Daraga, S Luzon, Philippines.
C. (*P.*) *fulvocinerea* Anderson, 1877a: 263; Gauhati, Assam.
C. (*P.*) *blythii* Anderson, 1877a: 264; Assam.
C. (*P.*) *sindensis* Anderson, 1877a: 266; Karachi, Pakistan.
C. (*P.*) *pealana* Anderson, 1877a: 267; Sibsagar, Assam.
C. (*P.*) *blanfordii* Anderson, 1877a: 269; Khandalla, W Ghats, India, 2000ft
C. (*P.*) *rubicunda* Anderson, 1877a: 277; Paresnath Hill, E of Hazaribagh, Bihar, India.
C. andersoni Trouessart, 1879b: 253; Khasi Hills, Assam.
C. edwardsiana Trouessart, 1880a: 330; Jolo, Sulu Is, Philippines.
Sorex beddomei Anderson, 1881: 179; Kollegal Hills, Coimbatore dist., S India.
Pachyura mülleri Jentink, 1888b: 166; Timor.
P. unicolor Jentink, 1888b: 166; Ternate, Moluccas.
P. semmelinki Jentink, 1889b: 213; Bandanaira, Banda Is, Moluccas.
P. kroonii Kohlbrugge, 1896a: 197; Plihari, SE Borneo.
Crocidura (*Pachyura*) *kükenthali* Matschie, 1901a: 279; Tobelo, Halmahera, Moluccas.
Pachyura occultidens Hollister, 1913a: 303; Iloilo, Panay I, Philippines.
C. muschata Hatori, 1915; Taiwan.
Pachyura palawanensis Taylor, 1934: 78; Taytay, Palawan, Philippines.
Also several extralimital synonyms.

Range (Map 15). Throughout the Indomalayan Region except that it does not extend north of c. 25°N in inland China (to Shanghai at 31°N near the coast). Also through Iran and Arabia to Egypt, in East Africa, Madagascar, islands in the Indian Ocean (Reunion, Comores) and Pacific Ocean (Guam etc. —Barbehenn, 1974) and in S Japan. Perhaps established also in parts of New Guinea (Heim de Balsac & Heim de Balsac, 1956). Up to 2825m near Darjeeling, W Bengal (Ghose, 1976b) but only to 300m in Taiwan (Jameson & Jones, 1977).

Variation. The following additional names have been proposed as subspecies:
Pachyura murina celebensis Revilliod, 1911: 515; Palu, NC Sulawesi.
P. caerulea riukiuana Kuroda, 1924: 3; Okinawa, Ryukyu Is.
S. m. mentosus Dao, 1960a: 230; Ho-Xa, Vinh-Linh, Quang-tri Prov., central Vietnam.
= *Sunsus menaosus*; Dao, 1962: 725.

There is considerable variation in colour (from light bluish grey to dark brown or black) and size (head and body of adults from about 100 to 160mm). Between them the checklists of Ellerman & Morrison-Scott (1951), Chasen (1940a) and Laurie & Hill (1954) combined 48 nominal species into *S. murinus* and Hooijer (1950a) proposed that three nominal species in the Philippines, *luzoniensis, occultidens* and *palawanensis*, should also be included. The inclusion of *luzoniensis* was supported by a study of the karyotype (2n = 40) by Medina & Leonard (1977). Some of these forms are so different that they seem very unlikely to be conspecific and evidence of sympatry of two different forms has been found in Sri Lanka (Phillips, 1980-84).

Until a comprehensive revision is undertaken the best that can be done is to treat as separate species the most distinctive form (as *S. montanus*) and also *S. zeylanicus* in view of apparent sympatry with *S. murinus*.

The pattern of subspecific variation in *S. murinus s.s.* is complicated by the presumed frequent translocation and introduction through human trade. The following more distinctive subspecies can be recognized tentatively, although it is unlikely that there are any discrete boundaries between them on the mainland. No attempt is made to allocate all the other synonyms to these forms although a few suggestions are made.

S. m. murinus. The large, grey, thick-tailed form most often found in lowland synanthropic sites and the only one on many of the marginal islands. The largest ones, often called *S. m. caeruleus* or *S. m. caerulescens*, appear to occur in the lower Ganges area (head and body up to 160mm). Those on the Indonesian islands tend to be darker and smaller than the common urban form in India.

S. m. sindensis. A very pale grey form in the arid zone of Pakistan and Rajasthan.

S. m. nemorivagus. A smaller, darker form in the Himalayan foothills and Assam, possibly including *soccatus, griffithi, saturatior* and *tytleri*.

S. m. viridescens (perhaps the earlier names *indicus* or *serpentarius* apply). A smaller, but grey and thick-tailed form in S India, probably including *blanfordii*.

The chromosome complement shows Robertsonian polymorphism and also varies geographically (2n=30-40) (Yosida, 1985). Throughout most of the range 2n=40, but forms with lower numbers are found in S India, Sri Lanka and peninsular Malaya. In view of the possibility that more than one species might be involved it is important that any karyological studies be backed up with adequate voucher specimens so that the karyological and morphological data can be correlated.

This species appears to be unusual in the Soricidae in continuing to grow significantly after the full dentition has developed and reproduction has begun, resulting in an unusual degree of variability in size amongst 'adult' animals. There is however also considerable geographical variation in size.

Ecology. Most forms occur primarily in and around houses and other synanthropic habitats, but several of the montane forms live in natural habitats, including grassland, scrub and forest, and only occasionally enter houses.

Remarks. In Japan this species has been developed as a laboratory animal, and a variety of mutant strains have been established (Oda *et al.*, 1988). It has also been the subject of investigation as a carrier of leptospirosis (Kundin *et al.*, 1970) and arbovirus (Carey *et al.*, 1971).

Map **15**. *Suncus*

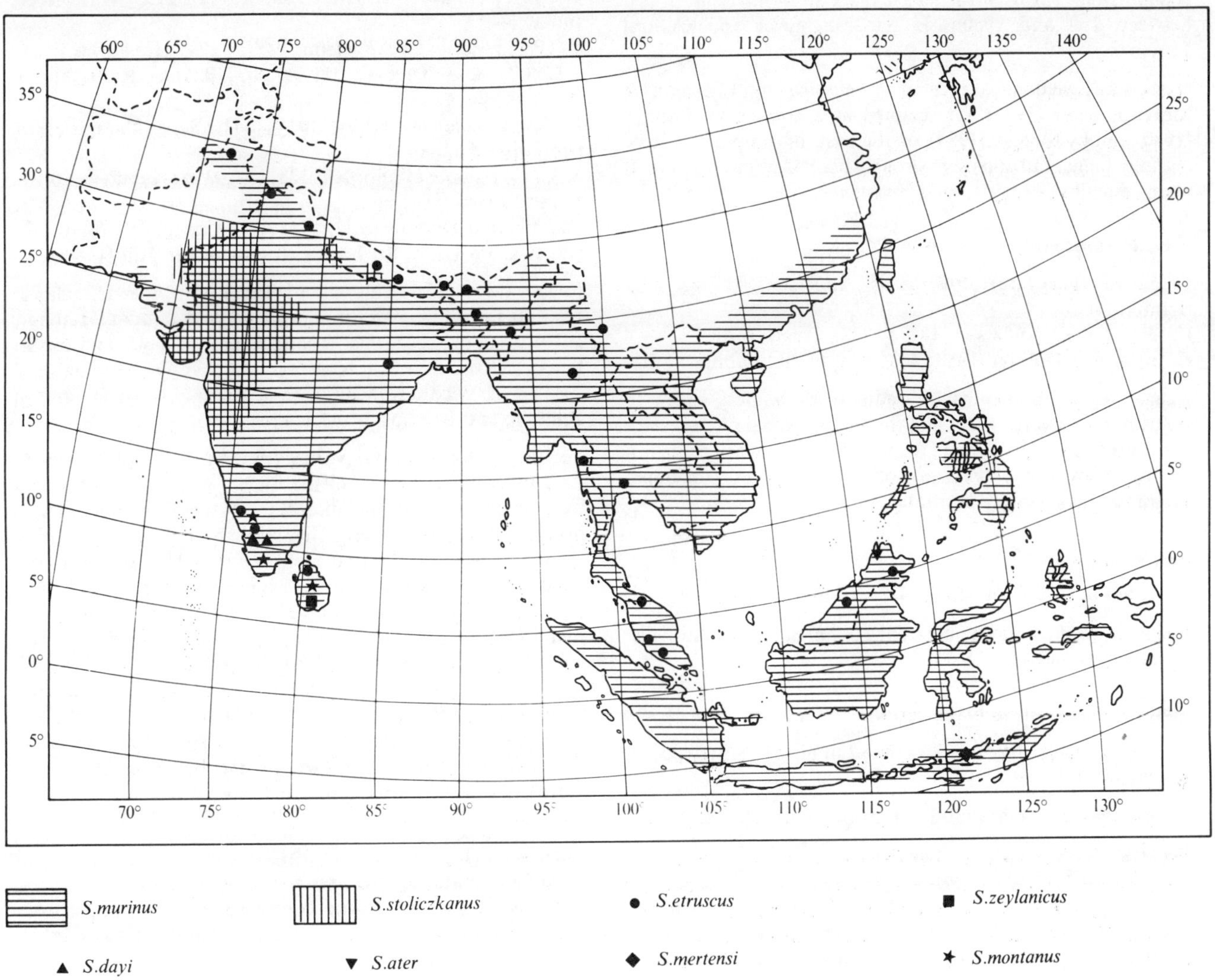

Suncus montanus

Sorex montanus Kelaart, 1850b: 211; Pidurutalagala, Nuwara Eliya, Sri Lanka.
Sorex ferruginea Kelaart, 1850b: 212; Dimbula, Nuwara Eliya, Sri Lanka.
? *Sorex niger* Horsfield, 1851: 135; Madras, India.

Range (Map 15). Central and S Sri Lanka; the Nilgiri and Palni Hills and perhaps other parts of S India, if *niger* is conspecific; mainly montane but down to 150m in the rain forest of S Sri Lanka (Phillips, 1980).

Variation. The following subspecific name can be tentatively included:
S. niger malabaricus Lindsay, 1929a: 334; Virajpet, S Coorg, India.

Ecology. In Sri Lanka inhabits humid forest; rarely enters buildings (Phillips, 1980).

Remarks. This form was included in *S. murinus* by Ellerman & Morrison-Scott (1951), but is very much smaller and darker.

Suncus zeylanicus

Suncus zeylanicus Phillips, 1928: 313; Gonogamma Estate, Kitulgala, Sri Lanka, 900ft.

Range (Map 15). Central and S Sri Lanka, in moist forest between 150 and 1000m in Sabaragamuwa and Central Provinces.

Remarks. Listed as a race of *S. murinus* by Ellerman & Morrison-Scott (1951) but retained as a species by Phillips (1980) and by McKay (1984), on the basis of sympatry with *S. murinus*, longer tail and more strongly hooked upper incisors. It seems equally distinct from *S. montanus*.

Suncus stoliczkanus Anderson's shrew

Crocidura (Pachyura) stoliczkanus Anderson, 1877a: 270; Bombay, India.
C. (P.) subfulva Anderson, 1877a: 278; Cutch, India.
C. leucogenys Dobson, 1888: 428; Ajmere, Rajputana, India.

Range (Map 15). Peninsular India from Madras north to Rajasthan (Prakash, 1972) and the Indus; Nepal (Abe, 1982). A record from SW Yunnan (Wang & Li, 1980) was erroneous, having been based upon a young *S. murinus* (specimen examined in Kunming Institute).

Suncus dayi

Crocidura dayi Dobson, 1888: 428; Trichur, Cochin, India.

Range (Map 15). Known only from Trichur and the Palni Hills, S peninsular India.

Suncus ater Bornean black shrew

S. ater Medway, 1965b: 38; Lumu-Lumu, Mt Kinabalu, N Borneo, 5500ft.

Range (Map 15). Known only from Mt Kinabalu, Borneo.

Remarks. Known from a single individual (in MCZ, Harvard) previously identified as *Crocidura foetida* (= *C. fuliginosa*), but distinguished from it by darker pelage, less robust teeth and anterior cusp on the first upper unicuspid tooth, as well as by the presence of the 4th unicuspid characteristic of *Suncus*.

Suncus mertensi Flores shrew

S. mertensi Kock, 1974: 198; Rana Mese, Flores, Indonesia.

Range (Map 15). Island of Flores, Lesser Sunda Is, Indonesia.

Remarks. A medium-sized, non-commensal species.

Suncus etruscus

Sorex etruscus Savi, 1822: 60; Pisa, Italy.
Sorex perrotteti Duvernoy, 1842: 29; Nilgiri Hills, S India.
Sorex pygmaeus Hodgson, 1845: 269. Pre-occupied by *S. pygmaeus* Laxmann, 1769, in *S. minutus*, a Palaearctic species.
Sorex melanodon Blyth, 1855a: 33. Pre-occupied by *S. melanodon* Wagler, 1832, in *S. araneus* L.
Sorex micronyx Blyth, 1855a: 33; Landour, Dehra Dun dist, United Provinces, N India.
Sorex hodgsoni Blyth, 1855a: 34; Darjeeling.
Sorex nudipes Blyth, 1855a: 34; Amherst, Tenasserim, Burma.
Sorex atratus Blyth, 1855a: 34; Cherrapunji, Khasi Hills, Assam.
Crocidura (Pachyura) macrotis Anderson, 1877a: 271; Tenasserim, Thailand.
C. (P.) nitidofulva Anderson, 1877a: 272; Lower Bengal, India.
C. (P.) nilagirica Anderson, 1877a: 274; Ootacamund, Nilgiri Hills, S India.
C. (P.) travancorensis Anderson, 1877a: 275; Travancore, India.
C. (P.) pygmaeoides Anderson, 1877a: 279; Himalayas.
? *C. (P.) hosei* Thomas, 1893a: 343; Bakong R, Baram, E Sarawak, Borneo.
Pachyura malayana Kloss, 1917: 282; Bang Nara, Patani, peninsular Thailand.
S. fellowes-gordoni Phillips, 1932a: 124; West Haputale, Ohiya, Central Prov., Sri Lanka.
Podihik kura Deraniyagala, 1958a: 5; Kandulu R, Madirigiriya, N-C Province, Sri Lanka; for status see Hill (1980a).

Range (Map 15). Throughout India and the Himalayan foothills from Pakistan through Nepal and Bhutan (Hutterer, 1979) to Yunnan, Burma, Thailand and Malaya, and on Sri Lanka; also on Borneo if *hosei* is conspecific. West through SW Asia to the Mediterranean zone of Europe, N Africa and in Ethiopia and perhaps W Africa.

Variation. Dorsal colour varies from light greyish brown to dark brown and the size of the fourth unicuspid teeth varies from less than half the diameter of the third and almost invisible in external lateral view, to only slightly smaller than the third and clearly visible from the side. However, samples available are too small to determine whether there is any regional pattern. The Bornean form can provisionally be treated as a subspecies, *S. e. hosei*, distinguished by short tail (c. 50% of head and body), less flattened skull and the presence of an accessory cusp on the anterior side of the first unicuspid tooth. Size is greatest in Borneo and Sri Lanka (upper tooth-row >5.9mm). Measurements of Malayan specimens given by Davison (1979) were confused by the inclusion of a large individual which later proved to be *Crocidura monticola* (Davison *et al.*, 1982).

Remarks. This is possibly the smallest of all terrestrial mammals, matched only by some close relatives, e. g. *S. infinitesimus*, in Africa. The chromosomes of a sample from Mysore, India were described by Aswathanarayana *et al.* (1987).

Table 27. The species of large *Crocidura* on the mainland, the Greater Sunda Islands and Timor (greatest length of skull >21mm, upper tooth-row >9.5mm, head & body >70mm).

	fuliginosa	*orientalis*	*baluensis*	*tenuis*	*lasiura*
Range	China-Borneo	Java	Borneo ? Sumatra	Timor	China
Head & body	70–95	85–90	80–100	? c.70	75–95
Tail	45–70	68–78	70–86	c.40–45	35–42
As % of h&b	c.55–80%	80–90%	75–100%	? c.65%	40–55%
Hind feet	14–16	15–16	15–17	c.14.5	13–16
Dorsal pelage: length (dark base/total)	3/5	3/5	5–6/8–9	–	–
Tail vibrissae: extent from base	10–50%	0	0–15%	0	50–60%
Skull: greatest length	22–24	23–24	23–25	?	21–23
Upper tooth-row length	9.6–10.7	9.5–11.5	10.9,11.0		9.2–10.5
2nd unicuspid tooth:width	0.8–0.9	0.60–0.75	0.8–0.9	0.9–1.0	0.7–0.8
Width of coronoid process (Fig.6, p. 42)	1.1–1.5	<1	1.2–1.5	c.1.7	1.2–1.3

FEROCULUS

Feroculus Kelaart, 1852: 31; type species *Sorex macropus* Blyth, in *S. feroculus* Kelaart.

Contains a single species, most closely related to the African *Sylvisorex* and to *Suncus*.

Feroculus feroculus

Sorex feroculus Kelaart, 1850b: 211; Nuwara Eliya, Central Prov., Sri Lanka, 6000ft.
S. macropus Blyth, 1851a: 163; Nuwara Eliya.
Corsira newera-ellia Kelaart, 1851: 340; Nuwara Eliya.
= *S. newera* Wagner, 1855: 564.
Crocidura macropus, Blanford, 1888: 237.

Range (Map 14). Sri Lanka. Blanford (1888) also included the Palni Hills, S India, but this awaits confirmation.

Ecology. Montane swamps and marshes around 2000m.

CROCIDURA

Crocidura Wagler, 1832: 275; type species *Sorex leucodon* Hermann. Conserved by Opinion 91 (1926) and Direction 24 (1955).

S Eurasia, Africa; c. 120 species, of which c. 26 in the Indomalayan Region, but taxonomy still very provisional. There has been no comprehensive revision of Indomalayan species but the relationship of some Indomalayan species with Palaearctic species was reviewed by Jenkins (1976) and those of Malaya and the Greater Sunda Islands by Jenkins (1982).

The following nominal species were based upon damaged or immature types, or inadequate descriptions and have not been allocated in confidence to better known species, although most are unlikely to represent independent species (Jenkins, 1982):
C. weberi Jentink, 1890a: 124; Singkarah, Sumatra. (Holotype badly damaged.)
C. vosmaeri Jentink, 1888b: 165; Banka I, Sumatra. (Holotype damaged and juvenile. Between *C. orientalis* and *C. neglecta* in size.)
C. beccarii Dobson, 1887: 556; Mt Singalang, Sumatra.
C. minuta Otten, 1917: 103; E Java. (No type preserved; best treated as unidentifiable.)
C. parvicauda Taylor, 1934: 83; Saub, Cotabato, Mindanao, Philippines. (Based upon a damaged, immature specimen: uniform greyish brown, condylobasal length 20mm, upper tooth-row 9.1mm.)

In tabulating the characters of the species they have been divided into four groups according to size (Tables 27,28,29,30). However there is a little overlap, even in the more precise cranial measurements, and any specimens of borderline size will need reference to two tables. Since some species are known from only a single, or very few, specimens, it is inevitable that some further specimens will fall outside the ranges of measurements given here. A great deal of further work will be required before the species can be adequately discriminated and diagnosed.

Crocidura fuliginosa

Sorex fuliginosa Blyth, 1856a: 362; Schwegyin, near Pegu, Burma.
? *C. foetida* Peters, 1870: 584; Borneo.
? *C. doreae* Peters, 1870: 584; Sarawak, Borneo.
C. brunnea Jentink, 1888b: 164; Java.
C. brevicauda Jentink, 1890a: 124; Tjibodas, W Java.
? *C. lepidura* Lyon, 1908b: 662; Kateman R, E Sumatra.
C. melanorhyncha Jentink, 1910a: 198; Mt Pangerango, Java, 3000ft.
? *C. malayana* Robinson & Kloss, 1911b: 243; Maxwell's Hill, Perak, Malaya, 3300ft.
C. major Kloss, 1911: 117; Great Redang I, Malaya. Preoccupied by *C. major* Wagler, 1832, in *C. russula* Hermann.
= *C. klossi* Robinson, 1912: 589.
C. aoris Robinson, 1912: 589; Aor I, Malaya.
C. dracula Thomas, 1912d: 686; ? near Mongtze (Mengtsz), S Yunnan, China.
C. negligens Robinson & Kloss, 1914a: 232; Samui I, peninsular Thailand.
C. maporensis Robinson & Kloss, in Robinson, 1916: 63; Mapor I, Riau Is.
C. aagardi Kloss, 1917a: 283; Bang Nara, Patani, peninsular Thailand.
C. tionis Kloss, 1917b: 127; Tioman I, Malaya.
C. gravida Kloss, 1917b: 127; Dayang Bunting I, Malaya.
C. praedax Thomas, 1923c: 656; Likiang Valley, C Yunnan, China, 9500ft.

Table 28. The species of large *Crocidura* on the Philippines, Sulawesi, Nicobar and Andaman islands (greatest length of skull >21mm, upper tooth-row >9.5mm, head & body >80mm).

	mindorus	*elongata*	*nigripes*	*rhoditis*	*nicobarica*	*hispida*	*andamanensis*	*jenkinsi*
Range	Philippines	Sulawesi	Sulawesi	Sulawesi	Nicobar I	Mid.Andaman	S Andaman	S Andaman
Head & body	90–95	90–94	80–82	83	107–120	85	114?	107
Tail	67,72	c.100–120	50–60	70	90,95	103	86	95
As % of head & body	70–80%	c.130%	65–75%	84%	75,90%	121%	75%	89%
Hind feet	16–18	20–22	14,15	17	24,26	25.5	25	26
Dorsal pelage	Medium brown	Medium brown	Very dark brown	Medium brown	Sooty brown	Grey with brown wash	Medium brown	Medium brown
texture	Soft	Soft	Soft	Soft	Bristly	Bristly	Soft	Bristly
length (grey bases/ total)	4/7	3/5	3/6	2/4	4		7	7
Ventral pelage	As dorsum	Buffy grey	As dorsum	Buffy grey			Grey	
Tail vibrissae: extent from base of tail	c.60%	10–20%	30–50%	c.50%	c.65%	c.50%	c.75%	c.65%
Skull: greatest length	22–24	24–25	21–23	22	27,28	28	26	28
Upper tooth-row	9.8–10.2	10–4–11.0	10.1–10.6	10.2	12.4,12.8	12.3	11.9	12.7
Molar width (M^2-M^2, external)	6.6	6.8–7.4	7.3–7.8	8.4	8.5	8.4	7.9	8.5
Unicuspid teeth - (3rd:2nd)	≤	>>	≤	>	>	>	>>	>

Table 29. The species of medium-sized *Crocidura* (skull 17–21mm, upper tooth-row 7–10mm, head & body c.60–80).

	attenuata	*orii*	*maxi*	*greyi*	*miya*	*pergrisea*	*zarudnyi*	*gueldenstaedti*	*paradoxura*
Range in region	China-Sumatra	Ryukyu Is	Sumatra Java etc.	Philippines	Sri Lanka	Kashmir	Pakistan Afghanistan	Himalayas SW China	Sumatra
Head & body	65–75	78,90	57–70	68–77	65,80	72	50–60	60–75	66
Tail	50–80	51	35–60	50–55	90,100	53	40–53	36–52	105
As % of h&b	70–110%	65%	60–90%	70–75%	110,155%		75–90%	60–70%	160%
Hind feet	12–15	15	12–13	13–15	15,16	13	11–13	12–14	16
Dorsal pelage	Dark grey-brown		Dark brown	Dark brown	Medium brown	Pale grey	Pale grey - brown	Medium brown	Dark brown
Ventral pelage	Very slightly lighter		Slightly lighter	As dorsum	Slightly lighter	Creamy grey	Very pale grey	Light brown - grey	
Tail vibrissae - extent from base	30–50%		30–50%	75%	30–50%	80%	75%	75%	0
Skull: greatest length	20–21	20.7	17–20	19–21	20	19–20	16–19	18–20	21
Upper tooth-row	7.8–9.8	9.1,10.0	7.3–8.5	8.8–9.3	8.7–8.8	8.4–8.6	7.1–8.3	7.6–8.8	8.8
Unicuspid teeth: 3rd:2nd	>		>	≥	=	<	≤	≥	

Table 30. The species of small *Crocidura* (skull <18mm, upper tooth-row <7.5mm, head & body <70mm).

	suaveolens	*horsfieldi*	*monticola*	*lea*	*levicula*
Range in region	China, Taiwan	Sri Lanka, India - Ryukyu Is	Malaya - Lesser Sunda Is	Sulawesi	Sulawesi
Head & body	60–65	60–75	55–65	c.60	c.64
Tail	30–50	40–45	30–55	51	35
As % of h&b	50–70%	65–75%	c.60%	85%	c.55%
Hind feet	10–12	c.12	10–11	14	11
Dorsal pelage (all soft)	Grey-brown	Medium brown	Dark brown	Very dark grey-brown	Medium brown
Tail vibrissae (extent from base)	c.100%	50%	30–50%	50%	65%
Skull: greatest length	16.0–17.5	16.5–18.0	15.5–18.0	18.0	c.16.5
Upper tooth-row	6.7–7.7	6.8–8.0	6.6–7.6	7.8	7.2
Unicuspid teeth-3rd:2nd	>	>	≤	>>	>

C. grisescens Howell, 1928a: 60; Kuatun, Fujian, SE China.
? *C. palawanensis* Taylor, 1934: 88; Sir J Brooke Point, Palawan, Philippines.
C. orientalis lawuana Sody, 1936a: 52; Mt Lawu, E Java, 2000m.

Range (Map 16). S China to Malaya, Sumatra, Java, Borneo and adjacent small islands of Hong Kong (Romer, 1963), Con Son (Vietnam); Samui (Thailand); Great Redang, Tioman, Aor, Dayang Bunting (Malaya); Mapor (Riau Is); Panjang (S Natuna Is). From the lowlands up to at least 1700m on Mt Kinabalu, Borneo.

Variation. The following have been described as subspecies:
C. brunnea pudjonica Sody, 1936a: 52; Poedjon, E Java, 1200m.
C. dracula mansumensis Carter, 1942: 1; Mansum, N Burma, 25°47'N, 96°16'E, 3200ft. (Darker than *dracula*.)

Most of the forms listed in the synonymy were described on the basis of small differences in colour, size and relative length of tail. A specimen from Palawan has the tail very long, about equal to head and body.

Remarks. The above account follows the reviews by Jenkins (1976, 1982) except that *C. orientalis* and *C. baluensis* have been excluded, but it must be considered very provisional. It is possible that *C. tenuis* from Timor is conspecific in which case that would be the valid name. However it is better kept separate until its relationships can be reassessed on more adequate information. It is also possible that some of the taxa from Sulawesi should be considered conspecific but these do not predate the name *fuliginosa*. The Sumatran form *lepidura* has more robust teeth than other *C. fuliginosa* and may not be conspecific. In peninsular Malaya two species, provisionally referred to *C. fuliginosa* and *C. malayana*, were recognized on the basis of distinctive karyotypes by Ruedi *et al.* (1990), but these cannot yet be separated morphologically. For discussion of the problems surrounding the use of the name *fuliginosa* see Medway (1965b: 40-42 or 1977a: 18-20).

Crocidura orientalis

C. orientalis Jentink, 1890a: 124; Tjibodas, W Java.

Range (Map 16). Known for certain only from the vicinity of the type locality at Cibodas, W Java, including Mt Pangrango and Mt Gede, at altitudes of 1800-2700m.

Recognition. This species is distinguished from the probably parapatric forms of *C. fuliginosa* (*brevicauda, melanorhyncha*) by longer tail, absence of caudal vibrissae, narrower, much less

Map **16**. *Crocidura* (part), *Solisorex*

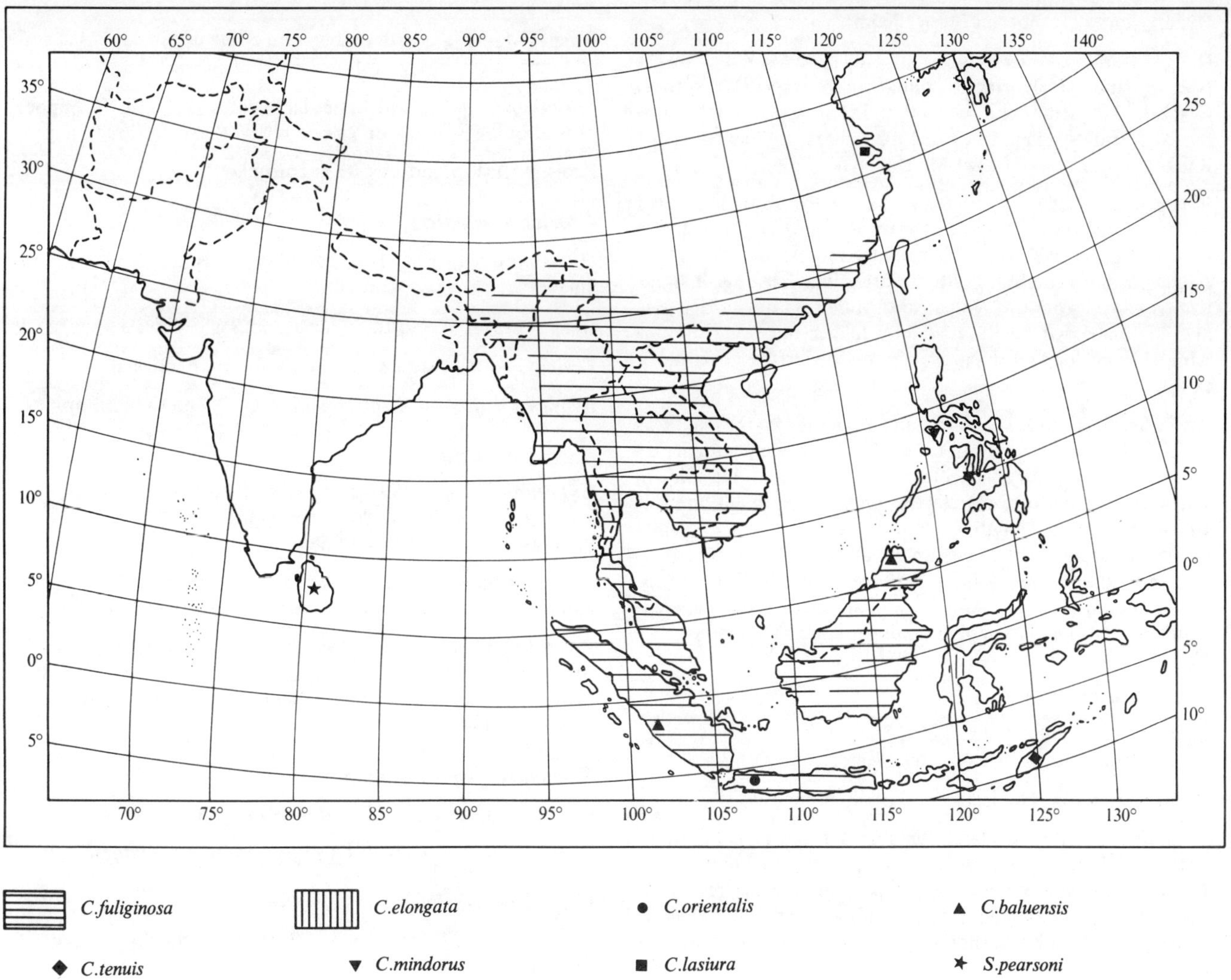

robust teeth, and more pointed coronoid and slender angular processes of the mandibles (Fig. 6). It is smaller on average but with a slight overlap.

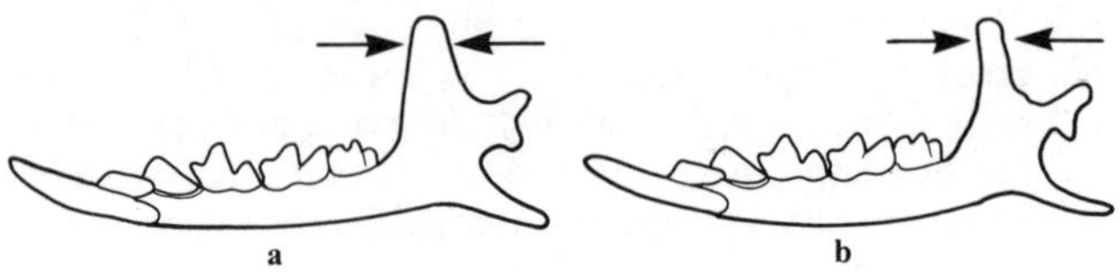

Fig. **6**. *Crocidura*. Mandibles of (a) *C.fuliginosa*; (b) *C.orientalis*, both from W Java.

Remarks. This form was included in *C. fuliginosa* by Jenkins (1976, 1982), but seems separable on a sufficient number of characters to justify recognition as a species. However the form *lawuana* from E Java, described as a subspecies of *C. orientalis* by Sody (1936a) appears to be a form of *C. fuliginosa* rather than of *C. orientalis*.

Crocidura baluensis

C. baluensis Thomas, 1898b: 247; Mt Kinabalu, N Borneo.
C. villosa Robinson & Kloss, 1918a: 21; Sungai Kumbang, Korinchi, W Sumatra, 4700ft.
? *C. fuliginosa kelabit* Medway, 1965b: 43; Kelabit Uplands, Baram, Sarawak, Borneo (over 1000m).

Range (Map 16). Known for certain only from Mt Kinabalu (N Borneo) and Mt Korinchi (Sumatra), but specimens from Kelabit, Mt Mulu and elsewhere in Sarawak might also represent this species. Recorded from 300m to 3600m on Mt Kinabalu and at 1400m on Mt Korinchi.

Recognition. From *C. fuliginosa* and *C. orientalis* distinguished by larger size (upper tooth-row over 10.5mm) and distinctive long pelage, equally dark but slightly more reddish than *C. fuliginosa* and sometimes with a very slight frosting of light-tipped hairs. From the sympatric *C. fuliginosa* on Borneo it is also distinguished by the distinctly longer tail (over 70mm) which has the long vibrissae usually confined to the proximal 5-10%.

Remarks. These forms were included in *C. fuliginosa* by Jenkins (1976,1982) and Medway (1965b, 1977a), but seem sufficiently distinctive to justify specific rank. On Mt Kinabalu in particular there is a strong correlation between the differences in length of pelage, length of tail, tail vibrissae and size, in spite of an apparently wide overlap in altitude. Chasen (1940a) suggested that *baluensis* and *vellosa* were close to *C. orientalis*, the montane species on Java, but there are sufficient differences to justify keeping them separate meanwhile.

Crocidura tenuis

Sorex tenuis Müller, 1840: 26,50; Timor.
C. macklotii Jentink, 1888b: 164; Timor.

Range (Map 16). The island of Timor, Indonesia.

Remarks. Possibly includes the forms here included in *C. fuliginosa*, but only known from two damaged holotypes in Leiden. It is therefore best treated as a separate species until a more accurate assessment is possible. It resembles *C. fuliginosa* in size, robust teeth and short tail, but lacks caudal vibrissae (as does *C. orientalis*).

Crocidura elongata

C. elongata Miller & Hollister, 1921b: 101; Temboan, SW of Tondano Lake, NE Sulawesi.

Range (Map 16). North and central Sulawesi.

Ecology. In lowland and montane forest; nocturnal.

Remarks. A distinctive, long-tailed species.

Crocidura mindorus

C. mindorus Miller, 1910: 392; Mt Halcon, Mindoro, Philippines, 1938m.
C. grandis Miller, 1910: 393; Grand Malindang Mts, Mindanao, 1910m.
? *C. negrina* Rabor, 1952: 6; Dayongan, Negros I, Philippines, 1300m.

Range (Map 16). Mindoro, Mindanao and Negros islands, Philippines.

Variation. *C. grandis* was described only on the basis of slightly larger size and *C. negrina* only on the basis of darker, less reddish colour.

Crocidura nigripes

C. nigripes Miller & Hollister, 1921b: 101; Temboan, SW of Tondano Lake, NE Sulawesi.

Range (Map 17). North and central Sulawesi, including island of Lembeh (NE Sulawesi).

Variation. *C. n. lipari* Miller & Hollister, 1921b: 101 (Gimpoe, central Sulawesi) was described on the basis of larger size.

Ecology. In lowland forest; nocturnal.

Crocidura rhoditis

C. rhoditis Miller & Hollister, 1921b: 102; Temboan, NE Sulawesi.

Range (Map 17). North, central and SW Sulawesi.

Ecology. In lowland and montane forest; nocturnal.

Remarks. Similar to *C. grayi* Dobson, 1890 on the Philippines.

Crocidura lasiura

C. lasiura Dobson, 1890c: 31; Ussuri R, Manchuria.
? *C. campuslincolnensis* Sowerby, 1945: 1; Shanghai.
Plus other extralimital synonyms.

Range (Map 16). Ussuri region of USSR and Manchuria, Korea; also Shanghai if *campuslincolnensis* is conspecific.

Remarks. *C. campuslincolnensis* was based upon a single specimen which, from its combination of large size and relatively short tail, does not seem referable to any Indomalayan species.

Crocidura hispida Andaman shrew

C. hispida Thomas, 1913c: 468; Middle Andaman I.

Range (Map 17). Known only from Middle Andaman Island.

Crocidura nicobarica Nicobar shrew

C. nicobarica Miller, 1902b: 776; Great Nicobar I, Bay of Bengal.

Range (Map 17). Great Nicobar I, Bay of Bengal.

Remarks. The skull and teeth of *nicobarica* were described as being exactly as in Kashmir *Suncus murinus* except for the absence of 4th unicuspids which might suggest that it is an *S. murinus* with aberrant dentition, but they are equally similar to those of other large *Crocidura*. Krumbiegel (1978) listed *nicobarica* in *Suncus* without explanation.

Crocidura andamanensis

C. andamanensis Miller, 1902b: 777; MacPherson Strait, S Andaman I.

Range (Map 17). South Andaman Island.

Remarks. Described on the basis of a single individual with long (c. 7 mm) soft pelage. The head and body length was recorded as 114mm, but examination of the dry skin suggests that about 100mm would be more likely.

Crocidura jenkinsi

C. jenkinsi Chakraborty, 1978a: 303; Wright Myo, S Andaman I.

Range (Map 17). S Andaman I, Indian Ocean.

Remarks. Known only by the holotype, distinguished by long, bristly pelage and tail shorter than head and body.

Crocidura attenuata Grey shrew

C. attenuata Milne-Edwards, 1872b: 231, pls 38B, 39A; Moupin, Sichuan, China.
C. rubricosa Anderson, 1877a: 280; Sibsagar, Assam.
C. kingiana Anderson, 1877a: 281; Sikkim.
? *C. fuliginosa trichura* Dobson, in Thomas, 1889a: 532; Christmas I, Indian Ocean.
? *C. aequicauda* Robinson & Kloss, 1918a: 22; Sungai Kring, Korinchi Peak, W Sumatra, 7200ft.
C. grisea Howell, 1926a: 137; 75 miles SW of Yenpingfu, Fujian, China, 500ft.
C. tanakae Kuroda, 1938: 81; Shohosha, Horigai, Taichusiu, C Taiwan.

Range (Map 17). S China from Sichuan, Hubei and Jiangsu south to Burma and Malaya (Davison, 1984) and the Himalayas west to Nepal, doubtfully Kashmir and Pakistan;

Map **17**. *Crocidura* (part)

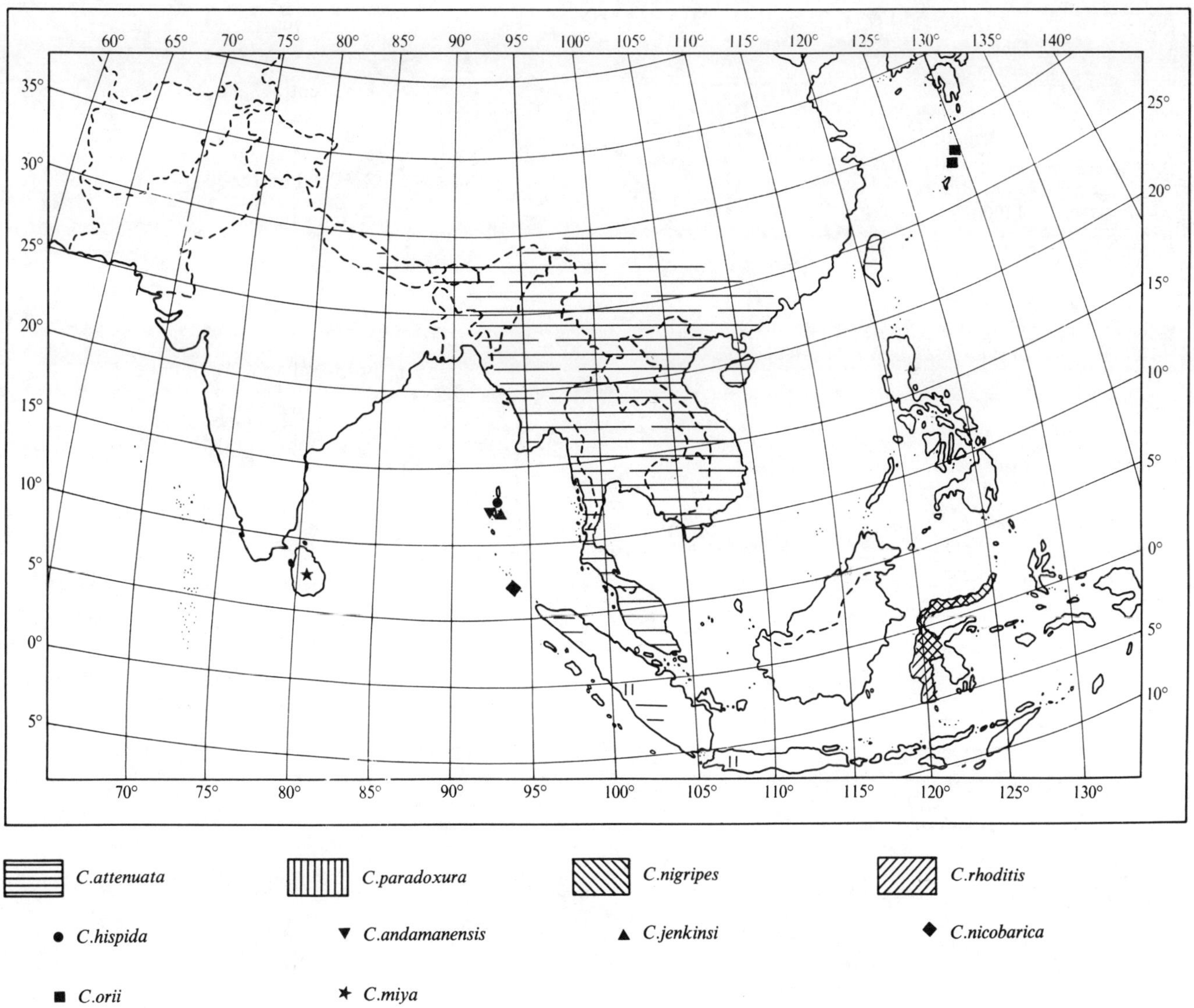

the islands of Taiwan and Hainan; perhaps formerly on Christmas I (Indian Ocean) if *trichura* is conspecific, but probably extinct there (C Tidemann, pers. comm.). Probably also in Sumatra (Jenkins, 1982).

Variation. There is considerable variation in tail length amongst the forms included here. In particular, specimens with the tail equal to or exceeding the head and body are found in N Burma and on Sumatra (holotype of *aequicauda*) and formerly on Christmas I (*trichura*), and it remains doubtful if these should be considered conspecific with *C. attenuata s.s.* from China and the Himalayas where the tail is generally substantially shorter than the head and body.

Crocidura orii Amami shrew

C. dsinezumi orii Kuroda, 1924: 3; Komi, Amami-Oshima, Ryukyu Is.

Range (Map 17). Known only from the islands of Amami-Oshima and Tokuoshima in the Ryukyu Islands where it is sympatric with the smaller *C. horsfieldi* (Imaizumi, 1974).

Remarks. Treated as a subspecies of *C. russula* by Ellerman & Morrison-Scott (1951), but given specific rank by Imaizumi (1961, 1970) who distinguished it from *C. dsinezumi* from the main islands of Japan by longer pelage (6.0-6.5mm), longer claws on fore feet (2.7mm), shorter braincase and posterior position of hind border of infraorbital foramen.

Crocidura neglecta

C. neglecta Jentink, 1888b: 165; Sumatra.
C. maxi Sody, 1936a: 53; E Besoeki, E Java.

Range (Map 18). Sumatra (perhaps including Mt Korinchi), E Java, Sumba, Flores and perhaps Ambon (Moluccas) (Jenkins, 1982). Sympatric with *C. monticola* on Java.

Remarks. Sumatran specimens are represented only by the damaged and faded holotype of *neglecta* (Leiden Museum) and possibly a specimen from Mt Korinchi (Singapore collection).

Crocidura grayi

C. grayi Dobson, 1890b: 494; Philippines [Luzon].
C. halconus Miller, 1910: 391; Mt Halcon, Mindoro, 1910m.
? *C. beatus* Miller, 1910: 392; Mt Bliss, Mindanao, 1460m.

Map 18. *Crocidura* (part)

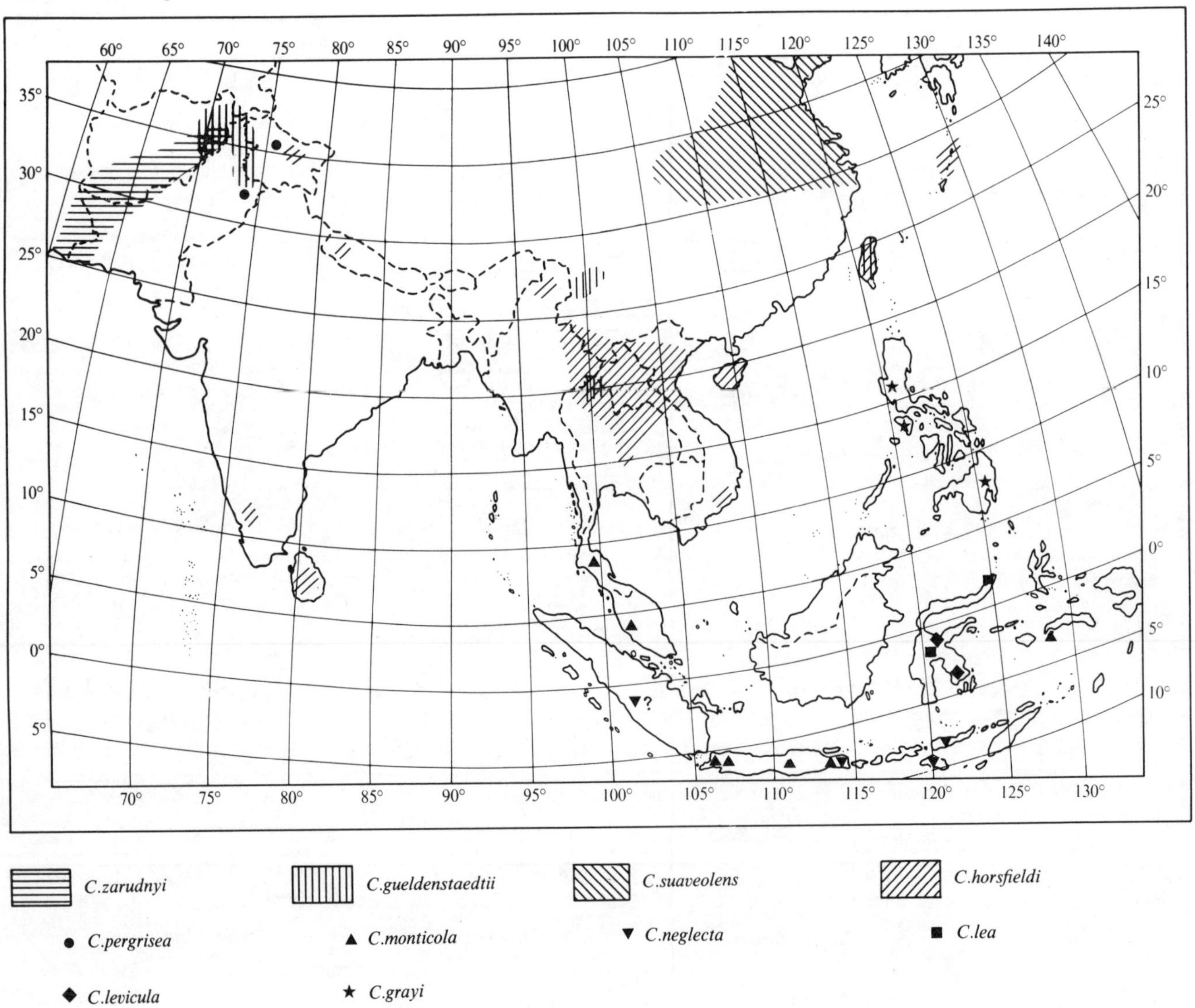

Range (Map 18). Luzon, Mindoro and Mindanao, Philippines.

Variation. The Mindanao form *beatus* was distinguished by darker colour; that on Mindoro by small differences in the relative sizes of the unicuspid teeth.

Crocidura miya Sri Lankan long-tailed shrew

C. miya Phillips, 1929a: 113; Moolgama, Nilambe dist, Kandyan Hills, Sri Lanka, c. 3000ft.

Range (Map 17). Sri Lanka, in the central highlands around 1000–2000m.

Crocidura pergrisea Pale grey shrew

C. pergrisea Miller, 1913a: 113; Skoro Loomba, Shigar, Baltistan, Kashmir, 9500ft.
Including several extralimital synonyms.

Range (Map 18). Kashmir, and west to the Kopet Dag and Asia Minor. In Kashmir known only from the type locality at 2900m.

Remarks. Has been held to include also the form *zarudnyi* Ognev but this is given specific rank following Hassinger (1973) and others.

Crocidura zarudnyi

C. tatianae Ognev, 1921: 338; E Iran, near Baluchistan border. Pre-occupied by *C. nyansae tatianae* Dollman, 1915 (Africa).
= *C. zarudnyi* Ognev, 1928: 341.

Range (Map 18). W Pakistan and Afghanistan; also in SE Iran; at altitudes from 100-2250m.

Variation. The following subspecies from E Afghanistan was described on the basis of larger size and greyer pelage:
C. z. streetorum Hassinger, 1970: 7; 30km NW of Ghazni, E Afghanistan, 33°43'N, 68°15'E.

Crocidura gueldenstaedtii

Sorex güldenstaedtii Pallas, 1811: 132, pl. 9; near Dushet, Georgia, USSR.
? *Sorex (C.) myoides* Blanford, 1875a: 106; Leh, Ladak (P. J. J. Bates, pers. comm.).
C. pullata Miller, 1911a: 241; Kotihar, Kashmir, 7000ft.
? *C. vorax* Allen, 1923b: 8; Ssu Shan, Likiang Range, Yunnan, China, 12000ft.
? *C. rapax* Allen, 1923b: 9; Yinpankai, Mekong R, Yunnan, China, 9000ft.
C. pamirensis Ognev, 1928: 366; L Drum, S slope of Pamir mts, USSR, 12000ft.
? *C. russula hosletti* Jameson & Jones, 1977: 465; 10 miles W of Taichungk, Taichung Hsien, Taiwan.
Also many extralimital synonyms.

Range (Map 18). Uncertain because of taxonomic uncertainties; from the Caucasus and E Mediterranean, reaching the fringes of the Indomalayan Region in E Afghanistan (1000-3000m), N Pakistan, Kashmir and perhaps at high altitudes in Yunnan (2000-2900m) if *vorax* and *rapax* are conspecific; doubtfully in Taiwan (if *hosletti* is conspecific) and Hainan (Xu & Liu, 1985).

Remarks. This form was included in the European *C. russula* by Ellerman & Morrison-Scott (1951), but is now generally agreed to be specifically distinct. However it is possible that some or all of these forms are conspecific with *C. suaveolens* Pallas (Catzeflis *et al.*, 1985). A further species, *C. serezkyensis* Laptev, 1929, occurs in the Pamirs and therefore might be found in Kashmir (Stogov, 1985). It is possible that the forms in Yunnan and Taiwan are related to the Japanese *C. dsinezumi* rather than to *C. gueldenstaedtii.*

[*Crocidura leucodon* Bicoloured white-toothed shrew]

Sorex leucodon Hermann, in Zimmermann, 1780: 382; Strasburg, France.
And several extralimital synonyms.

Range. ? Kashmir Valley; Europe to Caucasus and Elburz Mts, doubtfully further east.

Remarks. A single shrew from Daksum in the Kashmir Valley collected in 1974 was recorded as this species by Chakraborty (1983). Shrews from central Asia that were included in *C. leucodon* by Ellerman & Morrison-Scott (1951) are almost certainly specifically distinct, namely *C. sibirica* Dukelski, 1930, with clear differences in genitalia (Dolgov, 1974). The identity of the Kashmir form must be considered uncertain until it can be more critically assessed. *C. leucodon* and *C. sibirica* are distinguishable from *C. russula* and *C. gueldenstaedtii* by having a sharper line of demarcation between dorsal and ventral pelage.

Crocidura suaveolens

Sorex suaveolens Pallas, 1811: 133, pl. 9; Khersones, Crimea, USSR.
C. shantungensis Miller, 1901b: 158; Chimeh, Shandong, China.
C. ilensis phaeopus Allen, 1923b: 7; Wanhsien, Sichuan, China.
Also other, extralimital, synonyms.

Range (Map 18). A widespread species extending from W Europe to central Asia and extending marginally into the Indomalayan Region in China from Sichuan to Zhejiang and on Taiwan (Jameson & Jones, 1977).

Remarks. Inclusion of the above synonyms in *C. suaveolens* must be considered very provisional. Allen (1938b) used the name *C. ilensis* Miller, 1901 for these Chinese forms.

Crocidura horsfieldi Horsfield's shrew

Sorex horsfieldi Tomes, 1856: 23; Sri Lanka.
C. retusa Peters, 1870: 585; Paradenia, Sri Lanka.
? *C. indochinensis* Robinson & Kloss, 1922: 88; Dalat, Langbian Plateau, Annam, Vietnam, 5000ft.
C. watasei Kuroda, 1924: 1; Komi, Amami-Oshima, Ryukyu Is.
C. tadae Tokuda & Kano, 1936: 429; Koto-sho = Botel Tobago I, E of Taiwan.

Range (Map 18). Poorly known: Sri Lanka, S India (Mysore: Rao & Aswathanarayana, 1978), Kashmir, Nepal, N Burma, Thailand, Vietnam, S Yunnan, Hainan, Botel Tobago I (SE of Taiwan) and the Ryukyu Islands of Okinawa, Yoron, Iejima, Okinoerabu, Tokuno, Kikai and Amami.

Variation. The following insular subspecies have been described:
C. h. wuchihensis Wang, in Shaw *et al.*, 1966: 275; Shui-man, W slope of Wuchih Mt., Hainan, China. Small, pelage dark.
C. h. kurodai Jameson & Jones, 1977: 461; Linkou, Taipai Hsien, Taiwan. Does not appear distinguishable from other forms.

Remarks. Close to *C. suaveolens.*

Crocidura monticola

C. monticola Peters, 1870: 584; Mt Lawu, near Surakarta, Java, 3500ft.
C. bartelsii Jentink, 1910a;197; Mt Pangerango, W Java.

Range (Map 18). Peninsular Thailand (Davison, 1984), Malaya, Java and Borneo; on Ambon (Bogor Museum) and perhaps also on the islands of Sumbawa, Lombok, Sumba, Flores, Komodo, Obi and Timor, but all these need confirmation (Jenkins, 1982); specimens from Sumba and Flores in Bogor Museum appear to agree with *C. neglecta*.

Crocidura lea

C. lea Miller & Hollister, 1921b: 102; Temboan, NE Sulawesi.

Range (Map 18). North and central Sulawesi.

Ecology. In lowland and montane forest; nocturnal.

Crocidura levicula

C. levicula Miller & Hollister, 1921b: 103; Pinedapa, (near Mapane), C Sulawesi.

Range (Map 18). Central and SE Sulawesi.

Ecology. In lowland and montane forest; diurnal.

Remarks. Possibly conspecific with *C. lea*, both being considerably smaller than the other *Crocidura* species from Sulawesi.

Crocidura paradoxura

C. paradoxura Dobson, 1887: 566; Mt Singalang, Sumatra, 2000m.

Range (Map 17). Mt Singalang, Sumatra at 2000m and probably also in W Java (Mt Pangrango, 1450m).

Remarks. Known for certain only from the holotype, characterized by a relatively longer tail (159% of head and body) than in any other Indomalayan *Crocidura*. Even if the length of head and body has been underestimated, the absolute length of the tail (105mm) would make it difficult to allocate to any other species. A specimen from Cibereum, W Java at 1450m (Bogor Museum) may also be of this species: head and body recorded as 65mm (but appears more like 77mm on the dry skin) and tail 92mm, i. e. 142 or 120% of head and body. The skull length of 20.9mm is compatible with the larger body measurement.

SOLISOREX

Solisorex Thomas, 1924a: 94; type species *S. pearsoni* Thomas.

Contains a single, distinctive species, most closely related to *Crocidura*. The claws of the fore feet are enlarged (c. 5mm compared with c. 3mm in the hind feet) and the claws of the hind feet are also peculiar in having that of the 2nd digit much broader than the others.

Solisorex pearsoni Pearson's long-clawed shrew

Solisorex pearsoni Thomas, 1924a: 94,95; Hakgala, near Nuwara Eliya, central Sri Lanka.

Range (Map 16). The highlands of central Sri Lanka, between 1000 and 2000m.

ORDER SCANDENTIA

Treeshrews

Contains a single family, Tupaiidae, whose relationships have led to much uncertainty. It has been included in the orders Insectivora and Primates, but is now usually placed in a separate order Scandentia. The subject has been extensively reviewed by Campbell (1974) and the contributors in Luckett (1980a).

Family Tupaiidae

Treeshrews

Tupaiadae Bell, 1839: 994; type genus *Tupaia* Raffles.

Endemic to the Indomalayan Region. The only comprehensive revision is that of Lyon (1913) who used a very split classification. In spite of an enormous literature on the subject of the relationships of tupaiids—Elliott (1971) listed 1035 references on the family—there has been no subsequent review of the species-level classification. The generic classification has been reviewed by Luckett (1980b) and by Dene *et al.* (1980). Allocation of early Tertiary fossils to Tupaiidae is very uncertain but more definite evidence of former occurrence beyond the present range was provided by finds in the Siwalik formation of the W Himalayas (Miocene—Pliocene), which however were not allocated to a genus (Jacobs, 1980).

Most species are predominantly arboreal, diurnal and omnivorous, feeding on both fruit and invertebrates.

Ptilocercus is very distinctive and is generally treated as the sole member of a subfamily Ptilocercinae (Table 31).

Table 31. The subfamilies of Tupaiidae.

	Tupaiinae	Ptilocercinae
Genera/species in region	4/15	1/1
Pelage of tail	Uniformly haired	Naked with feathery 'flag' at tip
C^1	Large, caniniform	Small, as premolars
P^3 and P_3	Well developed	Reduced

Subfamily Tupaiinae

Of the four genera recognized here, *Urogale* and *Dendrogale* are most distinctive; *Anathana* is close to *Tupaia*. A further genus *Lyonogale* (replacing *Tana*) is often recognized. However recent studies have differed in the distribution of species between *Tupaia* and *Lyonogale* (cf. Luckett, 1880a and Dene *et al.*, 1980); they are therefore all included here in *Tupaia* (Table 32).

TUPAIA

Tupaia Raffles, 1821: 256; type species *T. ferruginea* Raffles, in *T. glis*.
= *Cladobates* Cuvier, F.,1824: 251, pl. 17.
= *Hylogale* Temminck, 1827:xix.
Sorex-glis Cuvier, in Geoffroy & Cuvier, 1821: 35,1 (*Sorex glis* Diard & Duvaucel).
= *Glisorex* Desmarest, 1822: 536.
= *Glisosorex* Giebel, 1855: 18.
= *Glirisorex* Scudder, 1882: 131.

Table 32. The genera of Tupaiinae.

	Tupaia	*Anathana*	*Dendrogale*	*Urogale*
Range	Throughout except S India & Mindanao	Peninsular India	Indochina, Burma	Mindanao etc.
No. of species	11	1	2	1
Tail	Bushy, distichous	Bushy, distichous	Short-haired, terete	Slightly bushy, terete
Posterior palatal vacuities	+	+	0	0
Zygomatic foramina	Large	Small	Very small	Very small
I^2	$\leq I^1$	$=I^1$	$=I^1$	$>I^1$
C^1: no. of roots	1	1	2	1
I_3	c.half I_1	c.half I_1	c.half I_1	Rudimentary
C_1,cf adjacent teeth	≥	=	=	>

Glipora Jentink, 1888a: 118 (*T. minor* Günther).
Tana Lyon, 1913: 134 (*T. tana* Raffles). Pre-occupied by *Tana* Reed, 1888 (Insecta).
= *Lyonogale* Conisbee, 1953: 46 .

Throughout the range of the family except for peninsular India and Mindanao; 12 species (Table 33). *T. tana* and *T. dorsalis* (and *T. minor*) have sometimes been separated in a genus or subgenus *Lyonogale* (formerly *Tana*).

Tupaia glis Common treeshrew

Tupaia glis

Sorex glis Diard, 1820: 478; Penang I, Malaya.
T. ferruginea Raffles, 1821: 256; Bencoolen, Sumatra (Lyon, 1913).
T. müllendorffi Matschie, 1898: 39; Culion I, Calamanian Is.
T. sordida Miller, 1900a: 231; Tioman I, Malaya.
T. chrysomalla Miller, 1900a: 232; Siantan I, Anamba Is.
T. phaeura Miller, 1902a: 157; Singkep I, Lingga Is.
T. castanea Miller, 1903a: 54; Bintang I, Riau Is.
T. pulonis Miller, 1903a: 56; Aor I, E of Malaya.
T. tephrura Miller, 1903a: 57; Tana Bala I, Batu Is, Sumatra.
T. chrysogaster Miller, 1903a: 58; N Pagi I, Mentawai Is.
T. discolor Lyon, 1906a: 602; Banka I.
T. siaca Lyon, 1908b: 661; Little Siak R, E Sumatra.
T. lacernata Thomas & Wroughton, 1909b: 535; Langkawi I.
T. cuyonis Miller, 1910: 393; Cuyo I, Philippines.
T. raviana Lyon, 1911: 167; Rawi I, Batung Is.
T. pemangilis Lyon, 1911: 168; Pemanggil I, E of Malaya.
T. obscura Kloss, 1911: 117; Great Redang I, E of Malaya .
T. riabus Lyon, 1913: 88; Riabu I, Anamba Is.
T. anambae Lyon, 1913: 89; Jimaja I, Anamba Is.

Range (Map 19). Malayan Peninsula, south of the Isthmus of Kra, Sumatra, Java, Borneo and Palawan; and on the following small islands—E Coast Malaya: Samui, Pennan, Perhentian, Redang, Tioman, Pemangil, Aor; W coast of Malaya: Panjang, Telibon, Langkawi, Terutau, Penang; S of Malaya: Singapore, Batam, Bintang, Mapor, Lingga, Sinkep, Banka; W of Sumatra: Tana Bala, Siberut, Sipura, N Pagai, S Pagai; W of Borneo: Siantan, Riabu, Jimaja; S Philippines: Balabac, Culion, Busuanga (Calamian Is), Cuyo. Up to 1120m in Borneo and 1420m in Malaya.

Variation. The following subspecific names have been proposed:

Mainland and major islands.
T. ferruginea longipes Thomas, 1893a: 343; Sarawak.
T. f. palawanensis Thomas, 1894a: 367; Palawan, Philippines. (Doubtfully conspecific with *T. glis.*)
T. f. hypochrysa Thomas, 1895d: 665; Java.
T. f. demissa Thomas, 1904a: 723; Tanjong Bringin, Lower Langkat, NE Sumatra. (Very dark, with pale, straw-coloured tail.)
T. f. wilkinsoni Robinson & Kloss, 1911a: 173; Ko-khaw, Trang, peninsular Thailand.
T. longipes salatana Lyon, 1913: 77; Pangkallahan R (15 miles from mouth), SE Borneo.
T. g. jacki Robinson & Kloss, 1918a: 15; Siolak Daras, Korinchi Valley, W Sumatra, 3100ft.
T. g. phoeniura Thomas, 1923a: 255; Deli, NE Sumatra.

Small islands.
T. f. batamana Lyon, 1907a: 656; Batam I, Riau Is.
T. f. longicauda Kloss, 1911: 117; E Perhentian I, E of Malaya.
T. f. penangensis Robinson & Kloss, 1911b: 242; Penang, Malaya.
T. f. operosa Robinson & Kloss, 1914a: 233; Koh Samui, E of Malaya.
T. f. ultima Robinson & Kloss 1914a: 234; Koh Pennan, E of Malaya.
T. castanea redacta Robinson, 1916b: 63; Mapar I, Riau Is.
T. g. siberu Chasen & Kloss, 1928c: 833; Siberut I, Mentawai Is.
T. lacernata kohtauensis Shamel, 1930: 71; Koh Tau, E of Malaya.
T. g. cognata Chasen, 1940a: 9; Panjang, W of peninsular Thailand.
T. g. umbratilis Chasen, 1940a: 9; Telibon I, W of peninsular Thailand.

Table 33. The species of *Tupaia.*

	glis	*belangeri*	*nicobarica*	*splendidula*	*montana*	*javanica*	*minor*	*gracilis*	*picta*	*tana*	*dorsalis*
Range:											
Malaya	+						+			+	
Sumatra	+					+	+			+	
Java	+					+					
Borneo	+			+	+		+	+	+	+	+
Other	Palawan	Thailand -S China	Nicobars								
Head & body (max.)	220	200	190	185	225	155	140	165	240	320	220
Hind feet	42–55	36–45	c.45	40–44	38–45	37–39	28–34	37–43	40–46	40–57	40–46
Tail (% of h & b)	c.100%	c.100%	c.110%	70–90%	75–85%	c.120%	115–125%	115–135%	65–90%	70–80%	70–80%
Mid-dorsal stripe	0	0	0	0	0	0	0	0	+	+	+*
Dorsal pelage reddish	+	0	0	+	+	0	0	0	+	+	(+)
Rhinarium: posterior edge straight (S) or V-shaped (V)	S	S	S	S	S	S	S	S	S	V	V
Light stripe on side of neck	+	+	0	+	+	+	+	+	+	+	+
Pairs of teats	2	3					2				
Rostrum: length	Medium	Medium	Medium	Medium	Medium	Medium	Very short	Very short	Medium	Very long	Long
I^1-I^2 (length of gap, mm)	1–2	1–2	1–2	1–2	1–2	1–2	<1	<1	0.5–1.5	2–2.5	1.5–2.5
I^2-C^1 (" " " ")	3–4	3–4	3–4	3.5–4.5	3–4	2–3	2–3	2.5–3.5	3.5–4.0	5.5–6.5	4.5–5.5
Chromosomes (2n, FN)	60, 76	62, 76			68, 74		66,			68, 72	

* On anterior part of back only.

T. palawanensis busuangae Sanborn, 1952: 112; Busuanga I, Calamian Is, S Philippines.

Many species and races (listed above) have been described on the basis of colour and size. Most forms were reviewed by Hill (1960). In Malaya dorsal colour becomes paler towards the north both on the mainland and on the islands.

Ecology. Predominantly terrestrial, feeding mainly on invertebrates but also fruit; territorial, diurnal, in plantations and gardens as well as forest (Langham, 1982; Payne *et al.*, 1985).

Remarks. The forms *belangeri* and *chinensis* etc. north of the Isthmus of Kra have been included in *T. glis*, e. g. by Ellerman & Morrison-Scott (1951). They are here excluded because of apparently incompatable karyotypes (Elliot *et al.* 1969; Arrighi *et al.*, 1969). However the morphological and geographical boundaries between these have not been adequately clarified. Some named forms from around the Isthmus of Kra may therefore be allocated to the wrong species.

The form on Palawan I, S Philippines was originally described as a subspecies (of *T. ferruginea*) but was given specific rank by Lyon (1913) diagnosed only by small differences in colour, especially the dark tail. This has been followed in subsequent listings but since it is no more distinct than many other forms normally included in *T. glis* that course is followed here. However the scanty data on karyology indicates that it might be specifically distinct (2n=52 compared with 60 in *T. glis* from both Malaya and Borneo—Arrighi *et al.*, 1969).

Tupaia belangeri Northern tree shrew

Cladobates belangeri Wagner, 1841a: 42; Siriam, near Rangoon, Burma.
T. peguanus Lesson, 1842: 93; ? Pegu, Burma.
Sciurus dissimilis Ellis, 1860: 71; Condore I, Vietnam.
T. chinensis Anderson, 1879: 129, pl. 7; Ponsee, Kakhyen Hills, 3185ft, and Muangla, Sanda Valley, 2400ft, W Yunnan, China.
T. modesta Allen, 1906: 481; Lei-mui-mon, Hainan I, China.
T. concolor Bonhote, 1907a: 2 (and 1707b: 7); Nhatrang, Annam, Vietnam.
T. clarissa Thomas, 1917a: 200; Bankachon, Tenasserim, Thailand.
T. glis cambodiana Kloss, 1919b: 357; Klong Yai, SE Thailand.
T. g. olivacea Kloss, 1919b: 358; Pak Bu, near Tachin, C Thailand.
T. g. cochinchinensis Robinson & Kloss, 1922: 87; Trangbom, 30 miles E of Saigon, Vietnam.

Range (Map 19). Thailand and Burma, south to the Isthmus of Kra; north through Indochina to Guangxi, Yunnan and S Sichuan in China; northwest to Assam, Bangladesh (Kock & Posamentier, 1986) and E Nepal (Johnson *et al.*, 1980); the island of Hainan. Up to 2300m in NE Burma and 2500m in Yunnan.

Variation. The following subspecies have been named:
T. b. yunalis Thomas, 1914a: 244; Mongtsze, S Yunnan.
T. b. siccata Thomas, 1914a: 243; Zibugang, Lower Chindwin, Burma.
T. b. laotum Thomas, 1914a: 244; Nan, Thailand, 290m.
T. concolor sinus Kloss, 1916b: 36; Koh Chang (I), SE Thailand.
T. b. tenaster Thomas, 1917a: 201; Tagoot, Great Tenasserim R, S Thailand.
T. b. assamensis Wroughton, 1921b: 599; Mokokchung, Naga Hills, Assam, 5000ft.

Map 19. *Tupaia* (part)

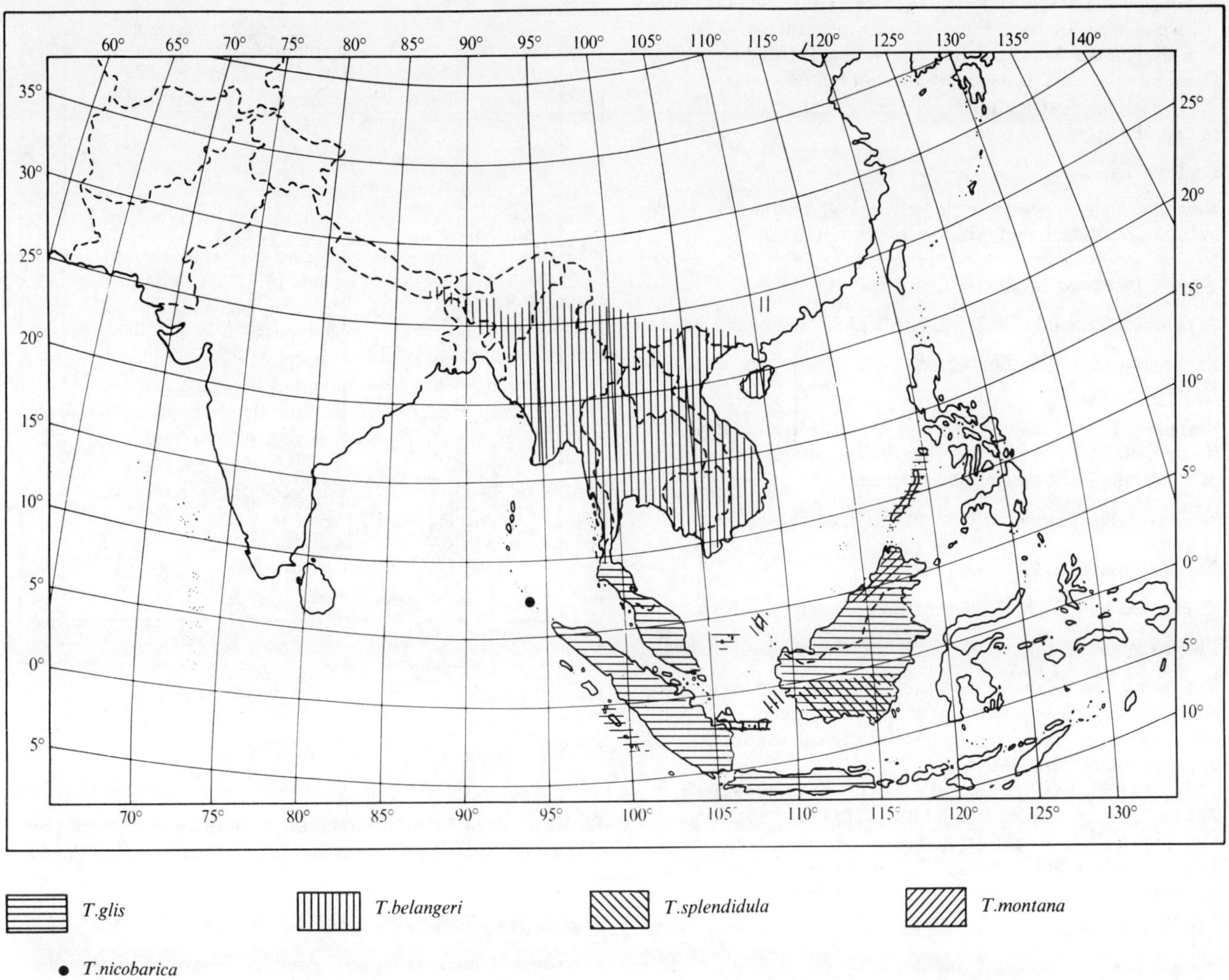

T. dissimilis annamensis Robinson & Kloss, 1922: 87; Daban, S Vietnam, 650ft.
T. b. versurae Thomas, 1922b: 428; Dening, Mishmi Hills, N Assam, 2250ft.
T. b. lepcha Thomas, 1922b: 428; Narbong, near Darjeeling, India, 2000ft.
T. b. brunetta Thomas, 1923b: 84; King I, Mergui Is, S Burma.
T. b. tonquinia Thomas, 1925: 497; Bao-ha, Tonkin, N Vietnam.
T. b. pingi Ho, 1936: 78; Bao-peng, Hainan I.
T. b. gaoligongensis Wang, 1987: 218,229; Hpimaw, Lushui Co., W Yunnan, 26°N, 98°38'E, 1900m. (Erroneously listed in the abstract of Wang (1987) and elsewhere as *T. b. gongshanensis* which is therefore a *nomen nudum*.)
T. b. yaoshanensis Wang, 1987: 225,230; Loxiang, Jinxiu Yao Aut. Co., NE Guanxi, China, 23°80'N, 110°25'E, 1000m.

Variation is considerable but there has been no comprehensive review. The Chinese forms were reviewed by Wang (1987), the Indian/Burmese forms by Agrawal (1975) and the Thai forms by Lekagul & McNeely (1977).

Remarks. Has been included in *T. glis*, e. g. by Ellerman & Morrison-Scott (1951), but here given specific rank following Elliot *et al.* (1969) and Arrighi *et al.* (1969) although the differences between the two species have not been clarified on more than a local level.

Tupaia nicobarica Nicobar treeshrew

Cladobates nicobarica Zelebor, 1869: 17, pls 1,2; Great Nicobar I.

Range (Map 19). Great and Little Nicobar Islands.

Variation. The form on Little Nicobar I has been described as a subspecies, *T. n. surda* Miller, 1902b: 774, on the basis of duller, less yellow colour.

Tupaia splendidula Ruddy treeshrew, Red-tailed treeshrew

T. splendidula Gray, 1865a: 322; Borneo.
= *T. ruficaudata* Mivart, 1867: 293.
T. mülleri Kohlbrugge, 1896a: 196; Bandjermasin, S Borneo.
T. natunae Lyon, 1911c: 168; Bunguran I, N Natuna Is.
T. carimatae Miller, 1906a: 61; Karimata I, Borneo.

Range (Map 19). Lowlands of S Borneo and adjacent islands of Bunguran and Laut (N Natuna Is) and Karimata (SW Borneo).

Variation. One subspecific name has been proposed:
T. s. lucida Thomas & Hartert, 1895: 490; Laut I, N Natuna Is.
Two subspecies can be provisionally recognized:
T. s. splendidula. Mainland Borneo, Natuna Is (Medway, 1961). Dorsal pelage dark reddish brown, tail lighter.
T. s. carimatae. Karimata I. Dorsal pelage light reddish brown, tail similar.

Ecology. Diurnal.

Remarks. Chasen (1940a) included *lucida* and *natunae* in *T. glis* but the above synonymy follows Medway (1961).

Tupaia montana Mountain treeshrew

T. montana Thomas, 1892a: 252; Mt Dulit, Sarawak, 5000ft.

Range (Map 19). NW Borneo, montane, at altitudes from 370-3170m.

Variation. *T. m. baluensis* Lyon, 1913: 95 (Mt Kinabalu, NE Borneo, 3000ft) was distinguished by the absence of a black patch on the back and by smaller size.

Ecology. Diurnal, mainly terrestrial (Payne *et al.*, 1985).

Tupaia javanica Javan treeshrew

T. javanica Horsfield, 1822; near Banjoewangie, E Java.

Range (Map 20). Bali, Java, W Sumatra and Island of Nias.

Variation. The following subspecific names have been proposed mainly on the basis of variation in colour. However colour is very variable even in one locality and cannot be used to define races (Hill, 1960).
T. j. balina Thomas, 1913b: 505; Mt Bratan, Bali, 4000ft.
T. j. occidentalis Robinson & Kloss, 1918a: 16; Sungei Penoh, Korinchi Valley, W Sumatra, 2600ft.
T. j. bogoriensis Sody, 1937: 213; Buitenzorg, W Java, 300m.
T. j. tjibuniensis Sody, 1937: 213; Tjiboeni, Bandoeng, W Java, 1350m.

Tupaia minor Pygmy treeshrew, Lesser treeshrew

T. minor Günther, 1876: 426; Borneo, opposite island of Labuan.
T. malaccana Anderson, 1879: 134, pl. 7; Malacca, Malaya.
T. sincipis Lyon, 1911: 169; Singkep I, Lingga Is.

Range (Map 20). Malay peninsula (including peninsular Thailand), Sumatra, Borneo and small islands of Singkep and Lingga (Lingga Is), Banggi, Balembangan and Laut. Up to 1700m in Borneo.

Variation. The following subspecific names have been proposed:
T. m. caedis Chasen & Kloss, 1932: 40; Balembangan I, N Borneo.
T. m. humeralis Robinson & Kloss, 1919a: 265; Rimbo Pengadeng, Bencoolen, W Sumatra, 1000m.
Four subspecies can be recognized:
T. m. minor. Borneo, including Banggi and Balembangan Is. Lower back russet, tail very dark.
T. m. humeralis. Sumatra. Intermediate between *T. m. minor* and *T. m. malaccana*.
T. m. malaccana. Malaya, Lingga Is. Less russet on back, tail less dark.
T. m. sincipis. Sinkep I. Back very russet, tail very dark, large (hind foot 34-35mm).

Map **20**. *Tupaia* (part)

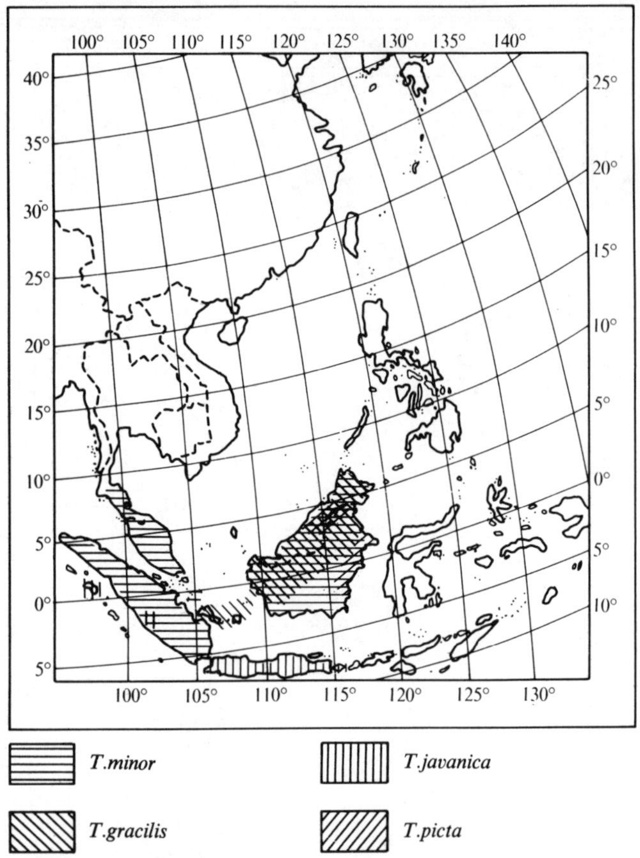

Ecology. Diurnal and arboreal, in plantations and gardens as well as forest; feed on insects and fruit (D'Souza, 1974; Payne *et al.*, 1985).

Tupaia gracilis Slender treeshrew

T. gracilis Thomas, 1893b: 53; Apoh R, base of Mt Batu Song, Baram, N Sarawak.
T. inflata Lyon, 1906a: 600; Tanjong Rengsam, Banka I.

Range (Map 20). Most of Lowland Borneo, except SE, up to 1200m; also on the islands of Banggi, Karimata, Banka and Billiton.

Variation. The following subspecific name has been proposed:
T. g. edarata Lyon, 1913: 118; Telok Edar, Karimata Is.
Three subspecies are marginally distinguishable:
T. g. gracilis. Mainland Borneo and Banggi. Large: hind foot >40mm; less russet above.
T. g. inflata. Banka and Billiton. Smaller: hind foot <40mm; more russet on posterior back.
T. g. edarata. Karimata I. Smaller: hind foot <40mm; dorsal pelage brownish.

Ecology. Diurnal and arboreal (Payne *et al.*, 1985).

Tupaia picta Painted treeshrew

T. picta Thomas, 1892a: 251; Apok, Baram, Sarawak, Borneo.

Range (Map 21). NW and E Borneo, below 1000m.

Variation. *T. p. fuscior* Medway, 1965b: 74 (Labuan Klambu, E Kalimantan, Borneo) was described on the basis of smaller size and duller colour.

Map **21**. *Tupaia* (part), *Anathana*, *Dendrogale*

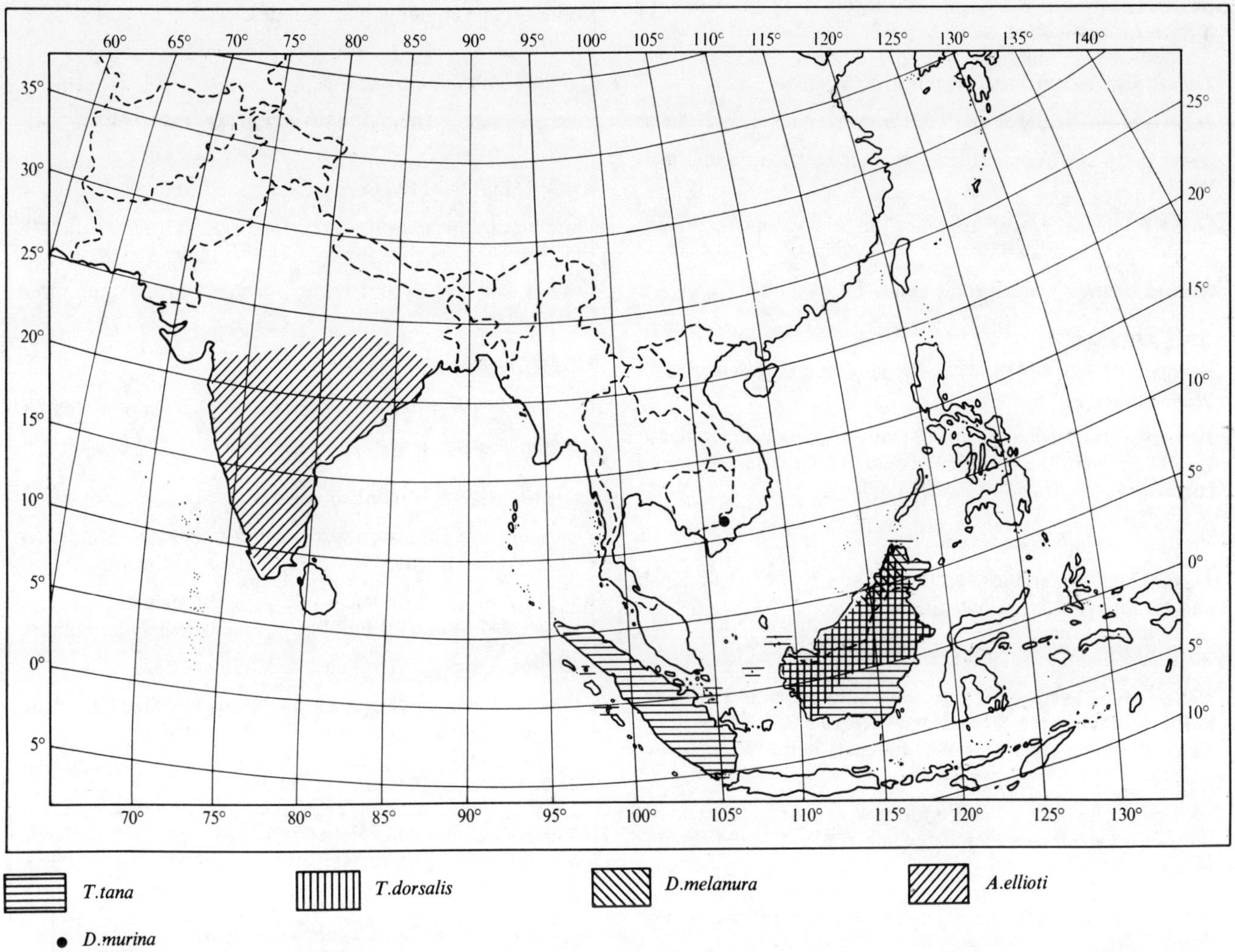

Ecology. Diurnal (Payne *et al.*, 1985).

Tupaia tana Large treeshrew, Terrestrial treeshrew

T. tana Raffles, 1821: 257; ? Bencoolen, Sumatra.
Cladobates speciosus Wagner, 1841a: 43; Borneo, restricted to Banjermasin, S Borneo (Chasen, 1940a).
T. bunoae Miller, 1900a: 229; Bunoa I, Tanbelan Is.
T. sirhassenensis Miller, 1901a: 133; Sirhassen I, S Natuna Is.
T. cervicalis Miller, 1903a: 59; Tana Bala I, Batu Is, W Sumatra.
Tana lingae Lyon, 1913: 145; Lingga I, E Sumatra.
Tana paitana Lyon, 1913: 150; Paitan R, N Borneo.

Range (Map 21). Sumatra and Borneo (up to 1500m) including islands of Tana Balu and Tana Mara (Batu Is); Tuangku (Banjak ls); Lingga, Banggi; Sirhassen (S Natuna Is); Big Tambelan and Bunoa (Tambelan Is).

Variation. The following subspecific names have been proposed:
Tupaia t. var. *chrysura* Günther, 1876: 427, pl. 36; Borneo, opposite Labuan I.
Tana t. besara Lyon, 1913: 141; opposite Jambu I, below Tynan, N Bank Kapuas R, W Borneo. Very large; dark and dull.
Tana t. utara Lyon, 1913: 141; Mt Dulit, Sarawak, 3000ft. Brightly coloured, sharply contrasting dorsal pattern.
Tana tana tuancus Lyon, 1913: 145; Tuangku I, Banjak I, W of Sumatra. Bright, reddish.
Tana cervicalis masae Lyon, 1913: 148; Tana Masa, Batu Is.
Tupaia t. banguei Chasen & Kloss, 1932: 44; Banguey I, N Borneo.
Tupaia t. nitida Chasen, 1933b: 197; Mt Poi, W Sarawak, 500ft.
Tana t. nainggolani Sody, 1936a: 54; Perlak, E Atjeh, N Sumatra, <100m.
Tana t. griswoldi Coolidge, 1938: 45; Kiau, Mt Kinabalu, N Borneo, 3300ft.
Tupaia t. kelabit Davies, 1958: 124; Bario, Kelabit Plateau, Fourth Division, Sarawak, 3700ft.
Tupaia t. kretami Davies, 1962: 46; Bukit Kretam, SE end of Dewhurst Bay, Kinabatangan Dist, N Borneo.

There has been no critical review since the very split classification of Lyon (1913). The more clear-cut island races appear to be as follows:
T. t. cervicalis. Batu Is. Bright ruddy pelage with well defined black area on lower back.
T. t. lingae. Lingga I. Small (hind foot 40mm, condylobasal length of skull 50.5mm). Rostrum slender.

Ecology. Diurnal, mainly terrestrial, in dense forest; feed on arthropods, earthworms and fruit (Payne *et al.*, 1985).

Remarks. Frequently separated from *Tupaia* in genus *Lyonogale* (formerly *Tana*) but this is difficult to define when all species are considered.

Tupaia dorsalis Striped treeshrew

Tupaja dorsalis Schlegel, 1857: 59; lower Kapuas R, W Borneo.

Range (Map 21). Borneo, almost throughout the lowlands, up to 1000m.

Ecology. Diurnal, mainly terrestrial, in primary and secondary forest (Payne *et al.*, 1985).

Remarks. Often placed in the genus *Lyonogale* (= *Tana*).

ANATHANA

Anathana Lyon, 1913: 120; type species *Tupaia ellioti* Waterhouse.

Peninsular India; one species, in many respects very similar to *Tupaia* although the dentition appears to share some derived characters with *Urogale* (Butler, 1980).

Anathana ellioti Madras treeshrew

Tupaia ellioti Waterhouse, 1850: 107; hills between Cuddapah and Nellore, Eastern Ghats, India.
A. wroughtoni Lyon, 1913: 123; Mandvi, near Bombay.
A. pallida Lyon, 1913: 124; Munbhum, Bihar, India.

Range (Map 21). Most of the forested parts of peninsular India, north to Bihar in the east and the Satpura Hills (SW Madhya Pradesh) in the west. To at least 1400m (Chorazyna & Kurup, 1975).

Variation. The three described forms were distinguished on the basis of colour, but they probably intergrade without definable boundaries:
A. e. ellioti. SE India. Upper parts reddish brown; feet buff.
A. e. wroughtoni. NW India. Upper parts dull grizzled brown; feet greyish.
A. e. pallida. NE India. Upper parts paler; feet buff.

Remarks. Anatomy has been described by Verma (1965).

DENDROGALE

Dendrogale Gray, 1848: 23; type species *Hylogalea murina* Schlegel & Müller.

Indochina and Borneo; two species (Table 34). A supposedly related fossil genus, *Prodendrogale*, has been described from the Upper Miocene of Yunnan, China (Qiu, 1987).

Table 34. The species of *Dendrogale*.

	murina	*melanura*
Range	S Indochina	Borneo
Ventral pelage	Pale cream or buff, sharply demarcated	Dark reddish or buffy grey, not sharply demarcated
Head & body (max.)	120mm	135mm

Dendrogale murina Northern smooth-tailed treeshrew

Hylogalea murina Schlegel & Müller, 1845: 167; 'Pontianak, W Borneo', probably an error.
Tupaia frenata Gray, 1860b: 217; Cambodia.

Range (Map 21). S Vietnam, Cambodia and E Thailand.

Remarks. The anatomy has been described in detail by Davis (1938).

Dendrogale melanura Bornean smooth-tailed treeshrew

Tupaia melanura Thomas, 1892a: 251; Mt Dulit, N Sarawak, 5000ft.

Range (Map 21). NW Borneo, montane, from 900 to 3300m.

Variation. *D. m. baluensis* Lyon, 1913: 132 (Mt Kinabalu, NE Borneo) was described on the basis of lighter colour.

Ecology. Diurnal, terrestrial and arboreal in montane forest (Payne *et al.*, 1985).

UROGALE

Urogale Mearns, 1905: 435; type species *U. cylindrura* Mearns.

A distinctive genus with one species.

Urogale everetti Mindanao treeshrew

Tupaia everetti Thomas, 1892a: 250; Zamboanga, Mindanao.
U. cylindrura Mearns, 1905: 435; Mt Apo, Mindanao.

Range (Map 22). Mindanao and adjacent smaller islands of Dinagat and Siargao, Philippines; mainly in montane forest.

Variation. Discussed by Angst & Mann (1971).

Ecology. Terrestrial as well as arboreal; feed on fruit, vertebrates and invertebrates (Wharton, 1950).

Map **22**. *Urogale, Ptilocercus, Cynocephalus*

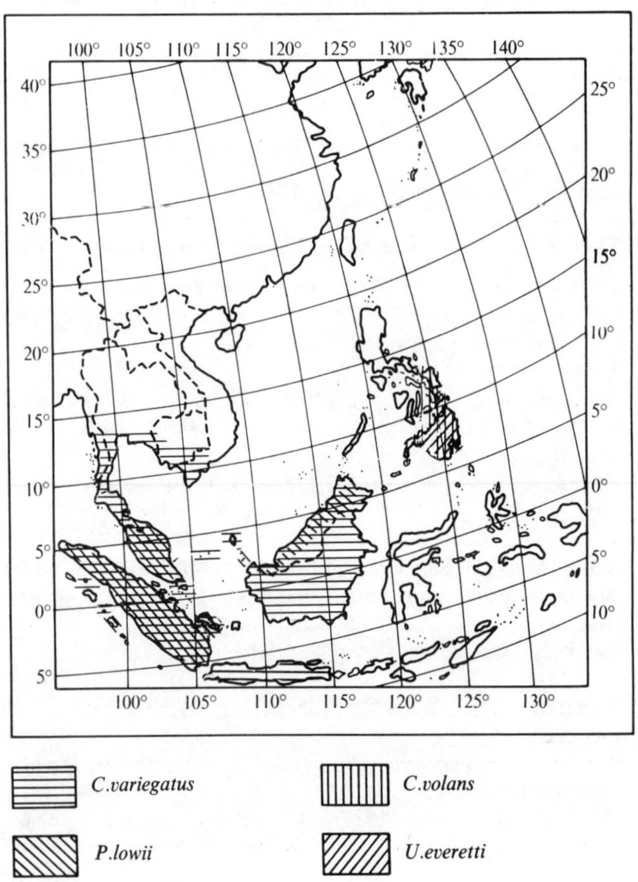

Subfamily Ptilocercinae

Ptilocercinae Lyon, 1913: 4; type genus *Ptilocercus* Gray.

Contains one genus and species, *Ptilocercus lowii.*

PTILOCERCUS

Ptilocercus Gray, 1848a: 167 (and 1848b: 23); type species *P. lowii* Gray.

Ptilocercus lowii Pen-tailed treeshrew

P. lowii Gray, 1848b: 24; Kuching, Sarawak.

Range (Map 22). N and NW Borneo, Sumatra, Malaya and peninsular Thailand; small islands of Labuan, Sirhassen, Banka, Pinie (Batu Is), Kariman (Riau Is) and Siberut (Mentawai Is).

Variation. *P. l. continentis* Thomas, 1910b: 426 (near Kuala Lumpur, Malaya) was described on the basis of greyer pelage, longer dark eye-stripe and narrower muzzle, but too few specimens are available to assess the validity of the distinctions.

Ecology. Nocturnal insectivores, foraging mainly in the canopy but also on the ground (Lim, 1967).

ORDER DERMOPTERA

Flying lemurs, colugos

Contains a single Recent family with two species, usually now considered congeneric. Formerly included in the order Insectivora. They are arboreal, nocturnal herbivores, capable of gliding flight.

Family Cynocephalidae

Galeopithecidae Gray, 1821: 300. Suppressed by Opinion 1077 (1977).
Cynocephalidae Simpson, 1945: 54; type genus *Cynocephalus* Boddaert.

Contains two species, both entirely Indomalayan, which have been placed in separate genera but are now usually considered congeneric.

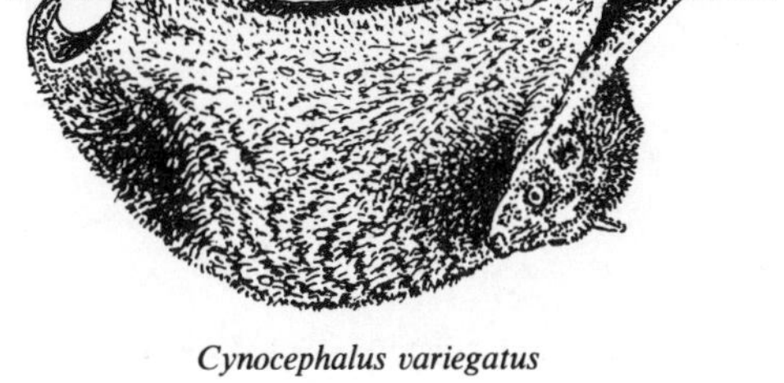

Cynocephalus variegatus

CYNOCEPHALUS

Cynocephalus Boddaert, 1768: 8 (footnote); type species *Lemur volans* L. Conserved by Opinion 1077 (1977).
= *Galeopithecus* Pallas, 1783: 208. Suppressed by Opinion 1077 (1977).
= *Geleopus* Rafinesque, 1815: 54.
= *Dermopterus* Burnett, 1829a: 268.
= *Pleuropterus* Burnett, 1829a: 268.
= *Colugo* Gray, 1871c: 98.
Galeopterus Thomas, 1908b: 254 (*Galeopithecus temminckii* Waterhouse, in *C. variegatus*).

Contains two clearly distinct species, although their distinction is confused by both sexual and non-sexual dichromatism (Table 35). Females are grey above, pale tawny brown below; males may be brownish grey above or entirely tawny.

Table 35. The species of *Cynocephalus* (Fig.7).

	variegatus	*volans*
Range	Indochina - Gt Sunda Is	Philippines
Sharp contrast on neck between tawny pelage below membrane and greyer above*	+	0
I^1 (ant.-post. length)	Larger, c.$\frac{1}{2}$ I^2	Smaller, c.$\frac{1}{4}$ I^2
I^2 and C^1 with accessory cusps in front of and behind main cusp	+	0
Nasals forming prominent ridge on top of rostrum	+	0
Width of rostrum (in front of zygomatic arch): width of braincase	=	>
Parietal ridges	Widely divergent	Closer

* In red-phase males there is a sharper contrast in tone in *C.variegatus* than in *C.volans.*

Cynocephalus variegatus Malayan flying lemur

Galeopithecus variegatus Audebert, 1799:sig. Rr; Java.
G. temminckii Waterhouse, 1839d: 119; Sumatra.
G. undatus Wagner, 1839a: 326; ? Java.
G. pumilus Miller, 1903a: 46; Adung I, Butung Is.
G. aoris Miller, 1903a: 47; Aor I, Malaya.
G. gracilis Miller, 1903a: 49; Sirhassen I, S Natuna Is.
G. natunae Miller, 1903a: 50; Bunguran I, N Natuna Is.
G. saturatus Miller, 1903a: 51; Tana Bala I, Batu Is, W Sumatra.
G. tuancus Miller, 1903a: 53; Tuangku I, Banyak Is, W Sumatra.
G. tellonis Lyon, 1908a: 139; Tello I, Batu Is, W Sumatra.
G. taylori Thomas, 1908k: 102; Tioman I, Malaya.
Galeopterus peninsulae Thomas, 1908e: 303; Semangko Pass, Malaya.
G. chombolis Lyon 1909b: 486; Chombol I, Riau Is.
G. borneanus Lyon, 1911a: 124; Chantung, near Klumpang Bay, SE Borneo.
G. lautensis Lyon, 1911a: 125; Laut I, SE Borneo.
G. abbotti Lyon, 1911a: 126; Panebangan I, Borneo.
G. lechei Gyldenstolpe, 1920: 17; Toembang Maroewe, E Borneo.
G. hantu Cabrera, 1924: 128; N Sarawak, Borneo.

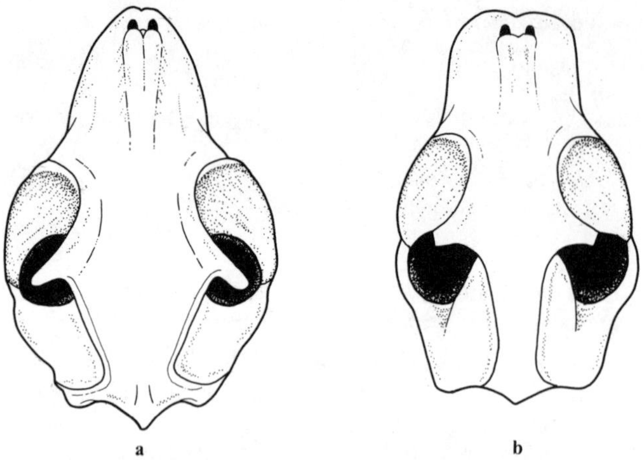

Fig. 7. *Cynocephalus*. Skulls of (a) *C.variegatus*; (b) *C.volans*.

Range (Map 22). From S Vietnam and S Thailand to Malaya, Sumatra, Java and Borneo, and many adjacent small islands, including Tioman, Aur, Singapore, Bangka and some in the Riau, Lingga, Natuna, Anamba, Batu and Banyak archipelagos.

Variation. Two forms have been described as subspecies: *Galeopterus v. terutaus* Chasen & Kloss, 1929a: 11; Terutau I, Straits of Malacca.
G. v. perhentianus Chasen & Kloss, 1929a: 11; E Perhentian I, off Trengganu, E coast Malaya.

There is considerable geographical variation in size and colour but there has been no comprehensive revision based upon adequate samples and comparisons are made difficult by the sexual dichromatism and individual variation. Variation was reviewed by Chasen & Kloss (1929b) who included in *C. variegatus* the nominal species listed in the synonymy above.

Ecology. Mainly nocturnal, in primary and secondary forest and plantations, feeding on foliage, buds and sap.

Cynocephalus volans Philippine flying lemur

Lemur volans Linnaeus, 1758: 30; Philippines. Conserved by Opinion 1077 (1977).
Galeopithecus rufus Desmarest, 1820: 108; Pelew I.
G. ternatensis Desmarest, 1820: 108; Ternate, probably in error.
G. philippinensis Waterhouse, 1839d: 119.

Range (Map 22). Southern Philippines: Mindanao, Basilan, Tongquil, Samar, Leyte, Bohol, Dinagat and Siargao. (Pampanga Prov., S Luzon has been given as the type locality of *volans* but there is no firm evidence that it has occurred on Luzon.)

ORDER CHIROPTERA

Bats

Bats are found throughout most of the world but the majority are tropical or subtropical. Second only to the ubiquitous Rodentia in number of species, the order includes some 188 modern genera and c. 977 modern species. It is divided into two suborders, one, the Megachiroptera or Old World fruit bats containing a single family, the other, the Microchiroptera, the predominantly insect-eating bats, including the remaining 17 families (Table 36). Both suborders and 10 families occur in the region, which is occupied by 65 genera and c. 307 species, so including approximately one third of the known bats.

Most bats are predominantly insectivorous but many are frugivorous or nectarivorous, eating flowers and flower products. Some are carnivorous, capturing small invertebrates, frogs, lizards, small rodents or even other bats, and a few catch small fish. Preferred roosts vary widely and include trees, foliage, tree hollows or holes, under bark or shale, caves, rock shelters, ruins, or other man-made structures such as mines, culverts and buildings. Some species are gregarious and strongly colonial, others form small groups, are semisolitary or rarely solitary.

It has been suggested on morphological and physiological grounds that bats are related to the Insectivora, Scandentia or Dermoptera, and more recently, that the Pteropodidae are related to Primates, leading to speculation (e. g. Hill & Smith,

Table 36. The suborders of Chiroptera.

	Megachiroptera	Microchiroptera
Margin of ear uninterrupted, forming a complete ring	+	0
Tragus in ear	0	+/0[1]
2nd digit usually with claw	+[2]	0
Tail present and well developed	0	+/0[3]
Mandible with angular process broad, low or almost absent (B); or long and narrow (L)	B	L
Molars simple, flat-crowned, or with simple, low, irregular cusps (S); or complex with regular pattern of cusps and ridges, forming a definite W or N surface outline (C) (Fig.10)	S	C

1. The tragus is absent in the Rhinolophidae and Hipposideridae, but a complex noseleaf is present in both families.
2. Absent in *Dobsonia, Neopteryx* and *Eonycteris.*
3. An external tail is lacking from the Craseonycteridae, and is rudimentary or absent in the Megadermatidae, but an extensive tail membrane is present in both families.

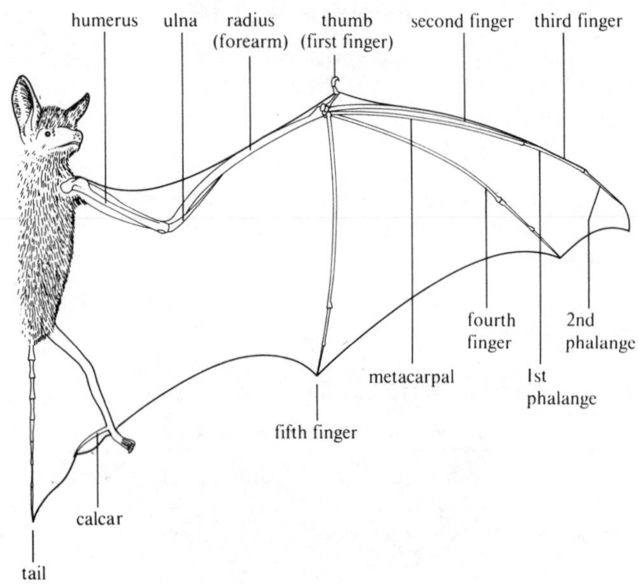

Fig. 8. The structure of a bat's wing.

1984,1985,1988) that the order may be biphyletic. This question has been discussed in detail by Pettigrew (1991) and Baker *et al.* (1991) who summarised existing evidence with new data and interpretations, but there seems as yet no agreed consensus of opinion.

The current classification of the order is based largely on Miller (1907) who provided a comprehensive review with diagnoses of the suborders and of the families and genera then known. Later treatments or reviews include Hill (1974a) who discussed the familial classification with especial emphasis on the features of the wing used by Miller, Van Valen (1979), Koopman (1982-85) and Koopman (1984c) who provided a detailed diagnostic survey of the families of bats. A semi-popular treatment can be found in Hill & Smith (1984,1985,1988). The distributional patterns of bats in the area bounded by northeastern India, Southern China, the Philippines, Sulawesi and the Lesser Sunda Islands were discussed in some detail by Koopman (1989).

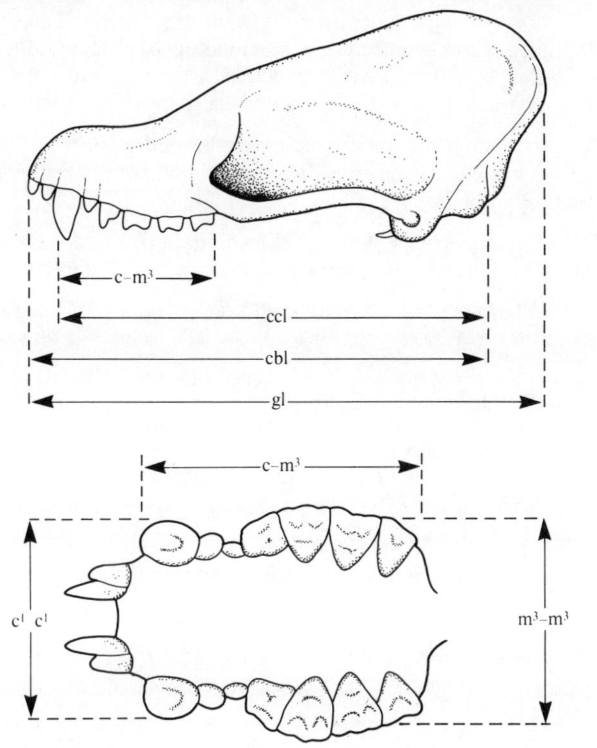

Fig. 9. The skull of a bat showing the measurements used in the text and tables. gl: greatest length; cbl: condylobasal length; ccl: condylocanine length; C-M^3: length of maxillary tooth-row; C^1-C^1: external width across upper canines; M^3-M^3: external width across last upper molars.

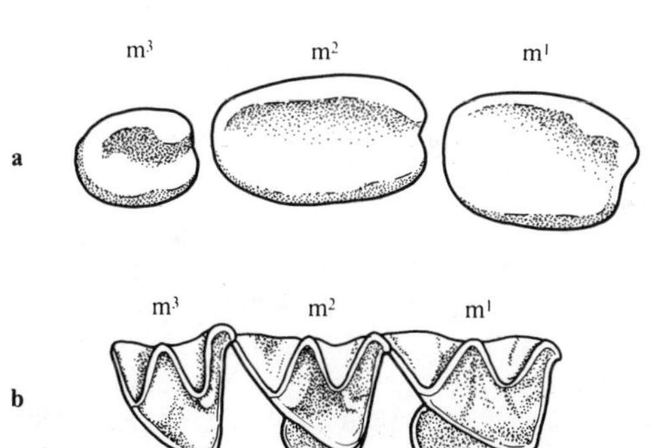

Fig. 10. Molar tooth structure of: (a) a megachiropteran; (b) a microchiropteran.

SUBORDER MEGACHIROPTERA

The Megachiroptera include one family, the Pteropodidae.

Family Pteropodidae

Old World fruit bats

Pteropidae Gray, 1821: 299; type genus *Pteropus* Brisson.
= Pteropodidae Bonaparte, 1838: 112.
For the correct formation of the family name see Handley (1980).

A family of c. 41 modern genera including c. 163 species distributed through the Old World tropics and subtropics from Africa to Australia and the islands of the W Pacific. 25 genera and c. 74 species occur in the region. Fossils are known from the Oligocene and Miocene of Europe, the Miocene of Africa and from the Pleistocene of Madagascar and the East Indies. Pteropodids range from relatively small to very large (forearm c. 40-220mm); they lack a tragus or noseleaf, the interfemoral membrane is usually narrow; and the tail generally short (the exception is *Notopteris* from Vanuatu, New Caledonia and Fiji) or absent. All except *Dobsonia, Neopteryx, Eonycteris, Melonycteris* (*Nesonycteris*) and *Notopteris* have a claw on the second finger: such a claw is lacking from all living microchiropterans. The majority have low, flat-crowned molar teeth adapted to their diet of fruit, sometimes with strong and massive jaws. In some the number and size of incisor and post-canine teeth are reduced: those that have adopted flower- or nectar-eating habits tend to have lighter jaws and smaller teeth. Rarely (*Harpyionycteris, Pteralopex*) the canines have strong supplementary cusps and the molars are also cuspidate, presumably to disintegrate hard, coarse or fibrous fruit pulp.

Pteropodids subsist on fruits, flowers, nectar and pollen and are active in the evening and at night, sometimes flying relatively long distances in search of food. Most roost in trees or foliage, some forming large colonies while others roost in smaller groups or are solitary. A number of species roost in the semi-lit areas of rock shelters or caves: one genus, *Rousettus*, is known to echolocate, using a system of clicks produced by the tongue, and may be found in the deeper, darker regions of caves or tombs.

The family is divided into four modern subfamilies (Table 37). The Pteropodinae includes the more generalized fruit- and flower- eating genera; the Harpyionycterinae the more aberrant genus *Harpyionycteris* from the Philippines and Sulawesi which is sometimes referred to the Pteropodinae. The subfamily Nyctimeninae contains the tube-nosed genera *Nyctimene* and *Paranyctimene* (New Guinea) which on occasion have also been subsumed into the Pteropodinae. The Macroglossinae contains those genera that are especially adapted for a diet of nectar and pollen.

Table 37. The subfamilies of Pteropodidae.

	Pteropodinae	Harpyionycterinae	Nyctimcninae	Macroglossinae
Nostrils strongly tubular	0	0	+	0
Tongue long, highly extensible, with papillate tip	0	0	0	+
Bony post-palate broad	0	0	+	0
Mesopterygoid fossa wide, deep	0	0	+	0
Lower incisors present	+	+	0	+
Anterior teeth strongly proclivous	+/0	+	0	0
Lower canines at tip of mandible	+/0	+	+	0
Canines and postcanines highly cuspidate	0	+	0	0

Subfamily Pteropodinae

Old World tropics and subtropics; 31 genera, of which 20 occur in the region. The subfamily is based on the generally less specialized tongue, skull and dentition: the tongue simple, not especially extensible, lacking filiform papilllae at its tip; the bony palate narrowing gradually behind the toothrows, with relatively narrow mesopterygoid fossa; and the canine and principal post-canine teeth lacking any extensive development of accessory cusps. The key below has been modified from Andersen (1912a).

1. Lower incisors 2-2 .... **2**

Lower incisors 1-1 .... **10**

2. Upper incisors 1-1 .... ***Rousettus***

Upper incisors 2-2 .... **3**

3. 5 upper postcanine teeth on each side, anterior tooth sometimes rudimentary or deciduous; facial axis of skull conspicuously deflected in relation to the braincase, a line projected through the maxillary alveoli passing through the middle or upper edge of the occipital condyle, or above the condyle **4**

3 or 4 upper postcanine teeth; facial axis of skull little deflected, alveolar line usually passing through or below lower edge of occipital condyle (except *Sphaerias*, in which the incisors are slanted forward and have triangularly pointed crowns) **6**

4. A short tail; occipital region of skull not elongated ***Rousettus***

No tail; occipital region elongate, sub-tubular **5**

5. Last upper premolar and 1st upper molar with well developed antero-internal tubercle; last lower premolar and 1st and 2nd lower molars with sharply defined inner basal ledge .... ***Acerodon***

Without these features .... ***Pteropus***

6. 3 upper postcanine teeth on each side .... ***Dyacopterus***

4 upper postcanine teeth on each side **7**

7. A short tail; postorbital foramen present, large .... ***Cynopterus***

Tail lacking or rudimentary; postorbital foramen absent **8**

8. Interfemoral membrane not extending beyond middle of tibia, calcar absent; incisors proclivous, their crowns triangularly pointed; postcanine teeth unusually narrow... ***Sphaerias***

Interfemoral membrane extending beyond middle of tibia, calcar present; incisors sub-vertical; postcanine teeth not unusually narrow **9**

9. Larger: forearm >70mm; last lower premolar and 1st lower molar broadened, with surface cusp .... ***Thoopterus***

Smaller: forearm <50mm; last lower premolar and 1st lower molar not broadened, lacking any surface cusp ..... ***Chironax***

10. 2nd finger lacking claw **11**

2nd finger with small claw **12**

11. Upper incisors 1-1; lower canines at tip of mandible; rostrum short .... ***Dobsonia***

Upper incisors 2-2; lower canines normally positioned; rostrum longer .... ***Neopteryx***

12. 5 upper postcanine teeth on each side **13**

4 upper postcanine teeth on each side **14**

13. Larger: forearm >90mm .... ***Styloctenium***

Smalller: forearm <50mm .... ***Balionycteris***

14. Upper incisors 2-2 .... **15**

Upper incisors 1-1 .... **18**

15. A short tail **16**

Tail lacking **17**

16. Larger: forearm >75mm; postorbital foramen present .... ***Ptenochirus***

Smaller: forearm <65mm; postorbital foramen absent .... ***Penthetor***

17. Pelage thick, woolly; hindlimbs strongly furred to base of claws; dorsal surface of rostrum much lower anteriorly than posteriorly; 2nd upper premolar with antero-internal basal cusp .... ***Aethalops***

Pelage not especially dense or woolly; hindlimbs not markedly furred; dorsal surface of rostrum almost horizontal; 2nd upper premolar lacking antero-internal cusp ***Megaerops***

18. Larger: forearm >60mm, greatest length of skull >33mm, C-M[1] >11mm .... ***Latidens***

Smaller: forearm <50, greatest length of skull <26, C-M[1] <9mm **19**

19. 5 lower postcanine teeth on each side.......... ***Alionycteris***

4 lower postcanine teeth on each side......................... **20**

20. Interorbital region not much constricted, least interorbital width exceeding the external width across the upper canines (at alveoli); a very shallow median interorbital depression; length of upper molar (M[1]) distinctly greater than its width, the tooth extending posteriorly beyond the lower anterior internal margin of the zygomatic arch .... ***Haplonycteris***

Interorbital region sharply constricted, least interorbital width less than width across canines; a deep median interorbital depression; M[1] little longer than wide, extending posteriorly only slightly beyond lower internal margin of zygomatic arch.. ***Otopteropus***

PTEROPUS Flying foxes

Pteropus

Pteropus Brisson, 1762: 153; type species *Vespertilio vampyrus niger* Kerr.
= *Spectrum* Lacépède, 1799: 15. Pre-occupied by *Spectrum* Scopoli, 1777.
= *Sericonycteris* Matschie, 1899: 3, 30.
Eunycteris Gray, 1866a: 64; (*P. phaiops* Temminck, in *P. melanopogon* Peters; see Andersen, 1912a).
Pselaphon Gray, 1871c: 110 (*P. pselaphon* Layard).
Desmalopex Miller, 1907: 60 (*P. leucopterus* Temminck).

Old World tropics and subtropics from Mafia and Pemba Is (off NE Tanzania, but never recorded from the African mainland), Madagascar to Indian Ocean islands; Pakistan to China, Philippines and Australia and islands of west Pacific. About 55 species, of which 22 in the region. This extensive genus is in need of revision, having been last examined in its entirety by Andersen (1912a); however, known material of some species has increased little or sometimes not at all since then. Little fundamental revisionary work has been carried out since then, although many of the taxa recognized as species by Andersen have since been reduced to subspecific rank in local studies. Andersen recognized 16 species groups, 12 occurring in the Indomalayan Region, although in several instances represented by only a single species (Table 38). These groups have been retained here, although no doubt their number might

Table 38. The species-groups of *Pteropus*.

	hypomelanus	*mariannus*	*caniceps*	*melanotus*	*melanopogon*	*chrysoproctus*	*lombocensis*	*pselaphon*	*personatus*	*vampyrus*	*alecto*	*conspicillatus*
Range in region	Throughout	Ryukyu Is	Sulawesi, Moluccas	Andaman Is W Sumatra Is	Moluccas	Moluccas	Lesser Sunda Is	Philippines	Moluccas	Throughout	Lesser Sunda Is	Moluccas
No. of species in region	5	1	2	1	1	1	1	1	2	4	1	2
Forearm	105–150	135–145	135–145	125–165	175–205	160–180	110–125	c.140	85–105	145–220	140–175	175–185
Ears long, pointed	0	0	+	0	0	0	0	0	0	+	0	0
Ears very small	0	0	0	0	0	0	0	+	0	0	0	0
Interfemoral membrane deep in centre	0	0	0	0	0	0	0	0	0	+	0	0
Tibia pilose dorsally	0*	0	0	0	0	+	+	+	+	0	0	0
Back blackish	0	+	0	+	0	0	0	0	0	+	+	+
Rostrum short	0	0	0	0	0	+	+	+	+	0	0	0
Large premolars with distinct basal ledges	+	+	+	+	+	+	+	+	+	0	0	0
Dentition unusually heavy	0	0	0	0	+	+	0	+	0	0	0	0
Dentition unusually weak	0	0	0	0	0	0	0	0	+	0	0	0

* Except in *P.dasymallus* which has the dorsal surface pilose proximally.

be reduced by further revision, the table being based chiefly on the definitions provided by Andersen.

Pteropus hypomelanus group (Table 39)

This group is primarily Indo-Australian in distribution, extending from the Maldive Is, Burma and the Nicobar Is to the Ryukyu and Philippine Is, Australia, the Solomon Is, New Caledonia and the Loyalty Is, with a single Mascarene species.

Pteropus hypomelanus Small flying fox

P. hypomelanus Temminck, 1853: 61; Ternate I, N Molucca Is.
? *P. vociferus* Peale, 1848: 19; Mangsi I, Balabac Strait, N of Borneo.
P. condorensis Peters, 1869a: 393; Condor I, Vietnam.
P. tricolor Gray, 1871c: 108; Ternate I, N Molucca Is.
P. macassaricus Heude, 1897: 177, pl. 5; Makassar, S Sulawesi.
P. lepidus Miller, 1900a: 237; Saddle I, Tambelan Is.
P. geminorum Miller, 1903a: 60; S Twin I, Mergui Is, Burma.
P. cagayanus Mearns, 1905: 433; Cagayan, Sulu I, Philippines.
P. enganus Miller, 1906e: 822; Dua I, near Enggano I, W Sumatra.
? *P. mearnsi* Hollister, 1913b: 112; Isabela, Basilan I, Philippines.

Range (Map 23). Maldive Is, Burma, Thailand, Vietnam; Luzon and Mindanao, Philippines; Sulawesi, some of the Moluccas and east to the Solomon Is. Including the following small islands. Maldive Is: Addu Atoll; Mergui Is, Burma: Malcolm, Gregory Group, Sir John Hayes (Kunthi Kyun), Barwell, S Twin; Thailand: Panjang, Muntia, Kuda, Mak, Rang, Som, Samui, Phangan; Vietnam: Con Son I; Malaya: Paya, Jarak, Berhala, Rumpia, Lallang (Sembilan Is), W Perhentian, Great Redang, Lantinga, Tioman, Pemanggil, Aor; W Sumatra: Tapah, Simalur, Siberut, Sipora, Dua, Enggano; Anamba Is: Jimaja; N Natuna Is: Pandak, Panjang, Laut; S Natuna Is: Serasan, Panjang, Subi; Tambelan Is: Saddle; Borneo: Labuan, Mangalum; Mantanani, Derawan, Bilang-bilangan, Balik Kukup, Ereban; Philippines: Maripipi (Heaney, 1991), Marinduque, Romblon, Cuyo, Panay, Leyte, Guimaras, Negros and Cebu (Klingener & Creighton, 1984; L R Heaney, pers. comm.), Dinagat, Camiguin (N of Mindanao), Mangsi, Cagayan Sulu, ? Basilan, Sibutu; Sangihe Is: Sangihe, Siao; Talaud Is: Lirung; Sulawesi: Peleng; Moluccas: Halmahera, Ternate.

Variation. The following subspecific names have been proposed:
P. h. var. *tomesi* Peters, 1869b: 626; Labuan I, off Sabah, Borneo.
P. h. canus Andersen, 1908: 361; Pandak I, N Natuna Is.
P. h. annectens Andersen, 1908: 361; Sirhassen I, S Natuna Is.
P. h. robinsoni Andersen, 1909c: 534; Rumbia (= Rumpia) I, Straits of Malacca.
P. h. fretensis Kloss, 1916f: 247; Jarak I, Straits of Malacca.
P. h. simalurus Thomas, 1923l: 592; Tapah I, near Simalur I, W Sumatra.
P. h. maris Allen, 1936: 343; Heratara, Addu Atoll, Maldive Is.

This species demonstrates a strong predilection for islands. Consequently numerous subspecies have been described, distinguished chiefly by small differences in size and colour. Hill (1975) discussed colour variation among specimens from the islands off S Thailand.

Two extralimital subspecies have been recognised: *P. h. vulcanius* Thomas, 1915g: 388 from Vulcan (= Manumudar) I, NE New Guinea; and *P. h. luteus* Andersen, 1908: 362 from Kiriwina I, Trobriand Is, E New Guinea, the latter extending to many of the east Papuan islands and to the Solomon Is: Koopman (1982) considered them synonymous.

No specimens of *mearnsi* from the southern Philippines have been examined but this may be a small subspecies of *P. hypomelanus*. However Heaney (pers. comm.) believes it to represent *P. griseus speciosus*. Bergmans & Rozendaal (1988) suggested that since specimens from the Sangihe and Talaud Is and from N Sulawesi may be larger than those from SW Sulawesi their taxonomic status should be reviewed when more extensive material is available.

Pteropus faunulus

P. faunulus Miller, 1902b: 785; Car Nicobar I, Nicobar Is.

Range (Map 23). Nicobar Is: Car Nicobar, Camorta, Nancowry.

Variation. Specimens from Camorta and Nancowry have a darker back than those from Car Nicobar, and the head is greyish rather than brownish, but the available material is very limited.

Pteropus griseus Grey flying fox

P. griseus Geoffroy, E.,1810a: 94, pl. 5; Timor I, Moluccas.
P. pallidus Temminck, 1825: 184, pl. 15; Banda I, Moluccas.
P. speciosus Andersen, 1908: 364; Malanipa I, off Zamboanga, Mindanao, Philippines.
P. mimus Andersen, 1908: 364; Makassar, S Sulawesi.

Range (Map 24). S Philippines, Sulawesi and the Moluccas; including the following islands. Philippines: Basilan (Mickleburgh *et al.*, in prep.), Mindanao, Caluya, Malanipa; observed on Semirara, Carabao, Borocay and Sibay (Alcala & Alviola, 1970); Sulu Is: Sanga Sanga, Sibutu; Talaud Is (Feiler, 1990); Java Sea: Solombo Besar, Mata Siri; Peleng (Koopman, pers. comm., ex Tate, 1942a, as *P. caniceps*); Salayer (Bergmans & Rozendaal, 1988); S Moluccas: Timor, Samao, Djampea, Bonerato; Banda.

Variation. Hitherto considered a distinct species, *speciosus* (Philippines) differs from *mimus* (Sulawesi) only in blackish

Table 39. The species of the *Pteropus hypomelanus* group.

	hypomelanus	*faunulus*	*griseus*	*pumilus*	*dasymallus*
Range in region	Burma-Moluccas	Nicobar Is	Philippines	Philippines	Ryukyu Is, Taiwan
Forearm	120–150	107–118	113–130	c.110	125–140
Skull: greatest length	61–69	53–54	54–63	45–56	59–62.5
C-M^2	22.8–26.2	19.3–19.8	20.3–23.2	16.5–19.2	21.4–24.4
Dorsal surface of tibia	Naked	Naked	Naked	Naked	Pilose proximally

Map 23. *Pteropus* (part)

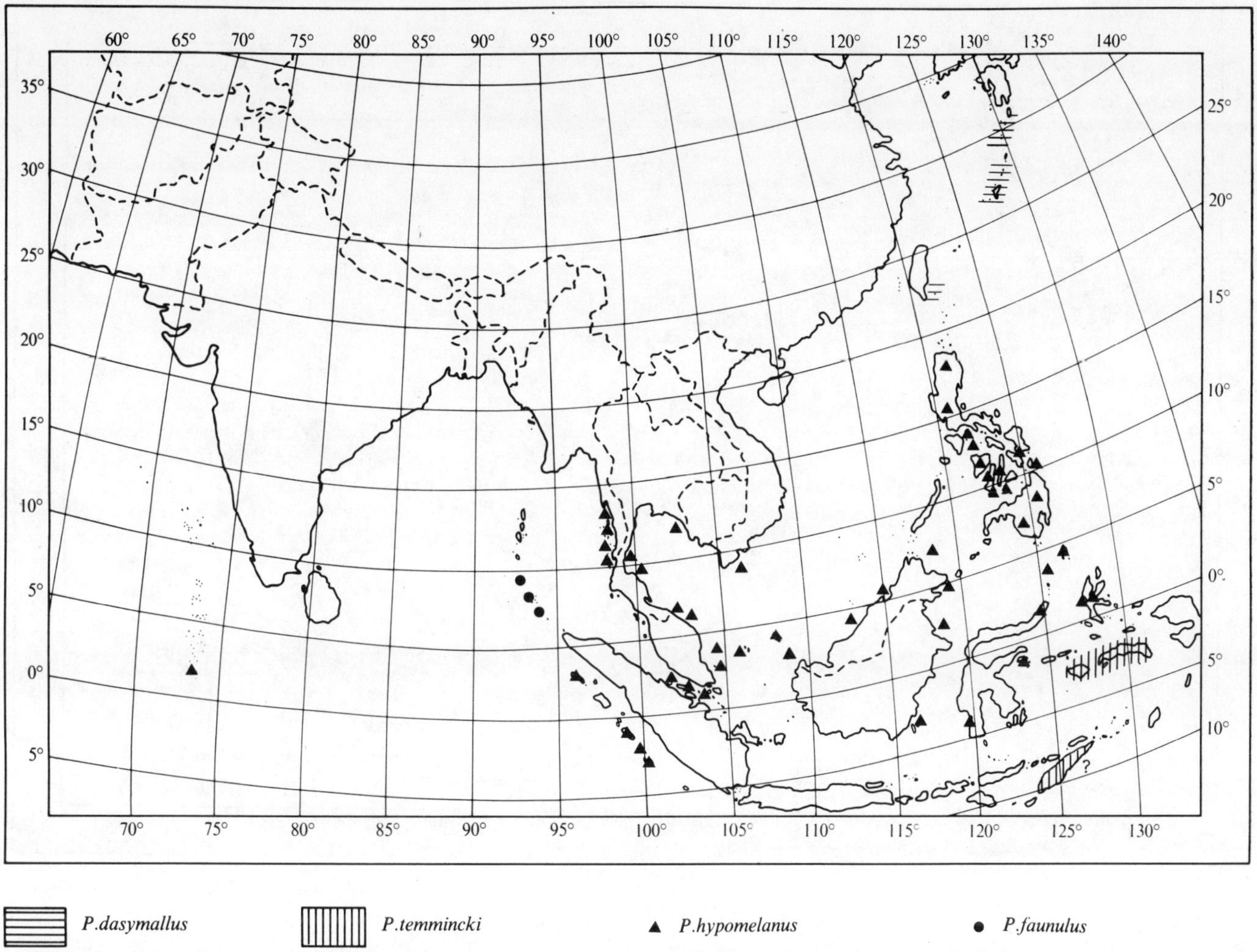

rather than brownish back and in darker underparts; *pallidus* (Banda I) and *griseus* (S Moluccas) are a little smaller and more grey brown, the last especially so.

Remarks. It seems that *speciosus* has been retained as distinct chiefly because Andersen (1912a) recorded an example from Luzon as *mimus*, so overlapping the distribution ascribed to *speciosus*. However, Taylor (1934) in listing *mimus* from the Philippines suggested that this, the only Philippine record, collected by A B Meyer, might be a Sulawesian example with an incorrect label. Moreover Chasen (1940a) in recording *speciosus* from Sibutu I, Solombo Besar I and Mata Siri I remarked 'A Malaysian outlier of a species occurring in Celebes and further east. There is, no doubt, an older name than *speciosus* available for the 'species', but I have no knowledge of the extralimital races'.

Specimens from Negros I and Mactan I, off Cebu I, recorded as *speciosus* by Sanborn (1952a) are apparently young examples of a larger species, probably *P. hypomelanus* (Klingener & Creighton, 1984). Goodwin (1979) discussed specimens from Timor and Semau.

Pteropus pumilus

P. pumilus Miller, 1910: 394; Palmas (= Miangas) I, SE of Mindanao, Philippines.

P. balutus Hollister, 1913b: 111; Balut I, Sarangani Is, S of Mindanao, Philippines.

P. tablasi Taylor, 1934: 169; near Odiongan, Tablas I, Philippines.

Range (Map 25). Philippine Is: Luzon, Mindanao, Maripipi (Heaney, 1991), Mindoro, Tablas, Leyte, Negros, Batu Batu, Camiguin (N of Mindanao), Balut (S of Mindanao); Sulu Is: Tawitawi, Bongao, Sanga Sanga; Talaud Is: Miangas.

Variation. Specimens on Camiguin differ from those on other islands in having steel-grey fur on the head, shoulders, the front of the mantle and on the anterior part of the chest (Klingener & Creighton, 1984).

Remarks. Klingener & Creighton (1984) examined small *Pteropus* from the Philippines and concluded that *balutus* and *tablasi* are synonyms of *pumilus* which, like Andersen (1912a), they considered a member of the *P. hypomelanus* group.

Pteropus dasymallus Ryukyu flying fox

P. rubricollis Siebold, 1824: 13; Ryukyu Is. Pre-occupied by *P. rubricollis* Geoffroy, E.,1810.

P. dasymallus Temminck, 1825: 180, pl. 10; Kuchinoerabu I, N Ryukyu Is (restricted by Kuroda, 1933a).

Map **24**. *Pteropus* (part)

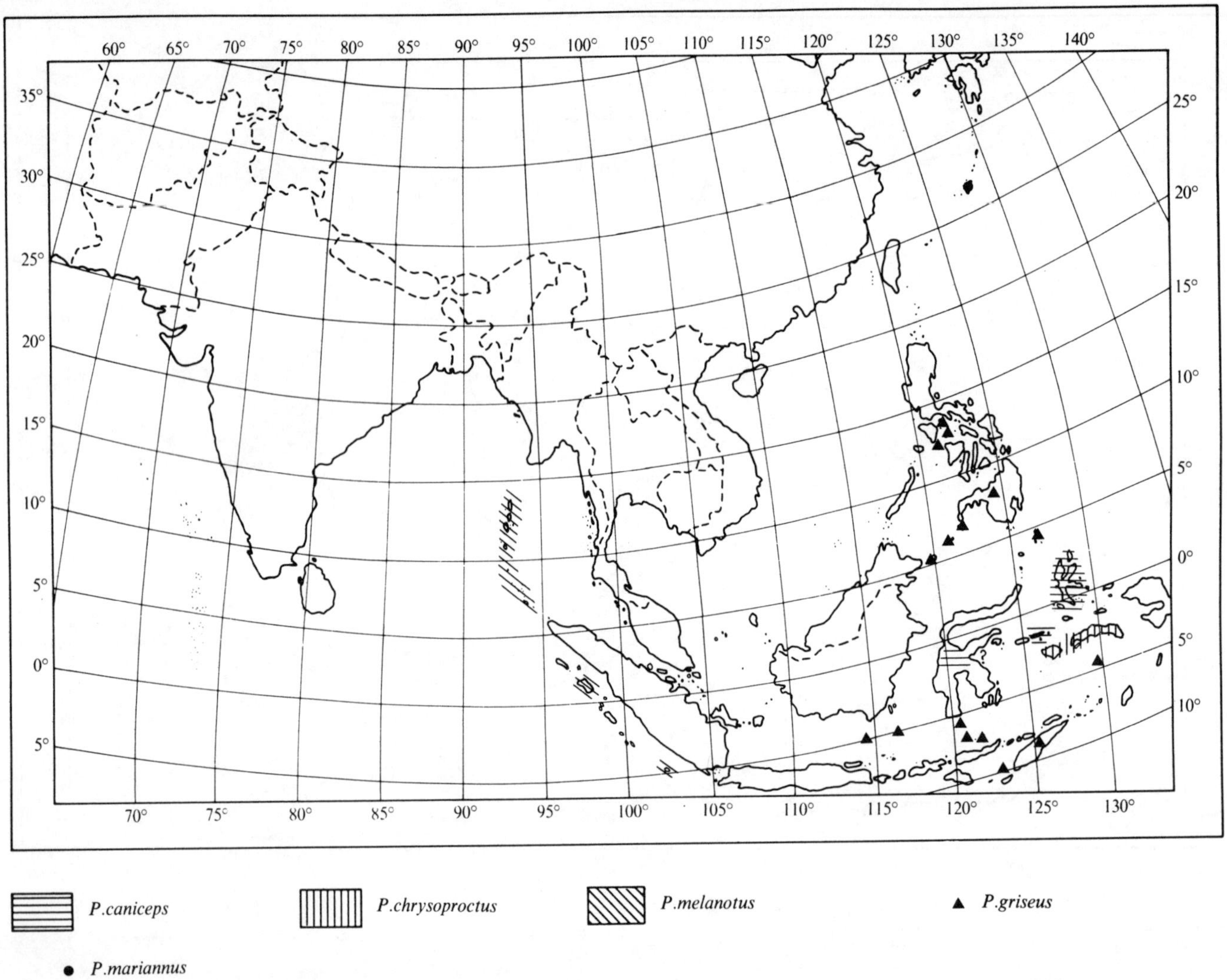

P. formosus Sclater, 1873: 193, pl. 22; Taku, Taiwan. Designated as Kashoto (= Lutao, Green) I, E coast of Taiwan by Kuroda (1933a).
P. daitoensis Kuroda, 1921: 210; 'Kita-Daitojima' (possibly an error for Minami Daitojima) I, Borodino (= Daito) Is, Ryukyu Is (see Kuroda, 1940; Yoshiyuki, 1989).
P. yamagatai Kishida, 1929: 125; Kuchinoerabu I, N Ryukyu Is.

Range (Map 23). Ryukyu islands of Kuchinoerabu, Takara, Okinawa, Mina, Ishigaki, Iriomote, Hateruma, Yonayuki and some smaller islands; Taiwan; Kashoto I, E coast Taiwan; Daito (= Borodino) Is: Kita-daito, Minami-daito.

Variation. The following subspecific names have been proposed:
P. d. inopinatus Kuroda, 1933a: 315; Okinawa I, Ryukyu Is.
P. d. yayeyamae Kuroda, 1933a: 315; Ishigaki I, Yayeyama Is, Ryukyu Is.

Yoshiyuki (1989) recognised four subspecies in the Ryukyu Is, distinguished chiefly by small differences in size and colour, the largest (*dasymallus*) in the north, the smallest (*yayeyamae*) in the south, the central islands occupied by intermediate forms (*inopinatus, daitoensis*).

Pteropus mariannus group

This group is Polynesian in distribution, extending to the Ryukyu Islands.

Pteropus mariannus Marianas flying fox

P. mariannus Desmarest, 1822: 547; Mariana Is, W Pacific.
P. loochoensis Gray, 1871c: 106; Ryukyu Is.
Pselaphon luchuensis Seitz, 1892: 364.
P. keraudreni var. *loochooensis* Fritze, 1894: 854; Okinawa I, Ryukyu Is.

Range (Map 24). Ryukyu Is: Okinawa; also to Mariana, Palau and Caroline Is.

Remarks. The subspecies on Okinawa I is *P. m. loochoensis*; there are other subspecies on islands of the W Pacific. Yoshiyuki (1989) considered *loochoensis* to be specifically distinct on account of its longer fur and smaller premolars. IUCN: E.

Pteropus caniceps group

This group is limited to Sulawesi, some associated islands and the N Moluccan Is.

Map **25**. *Pteropus* (part)

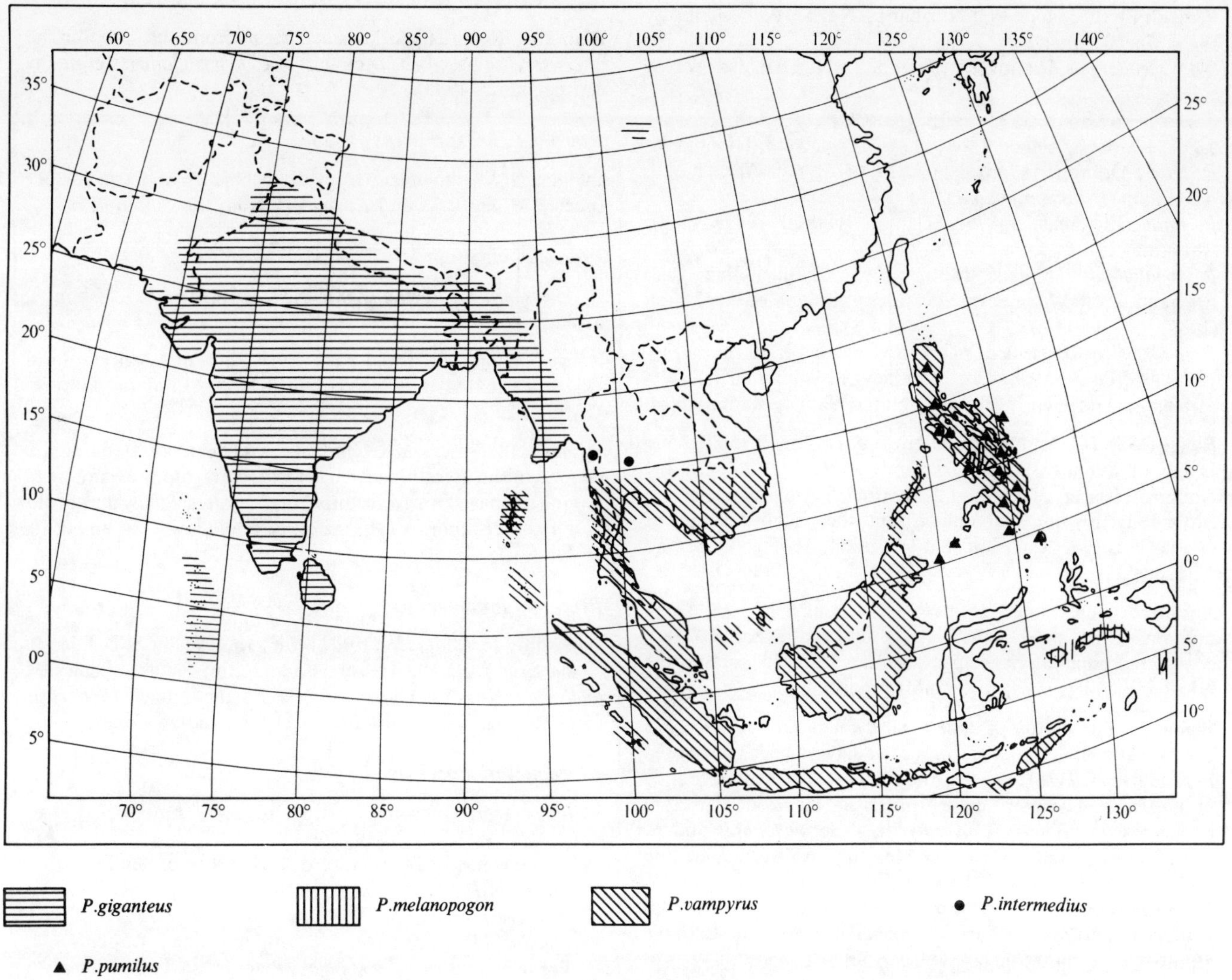

Pteropus caniceps Ashy-headed flying fox

P. caniceps Gray, 1871c: 107; Batchian (= Bacan) I, N Moluccas.
P. affinis Gray, 1871c: 108; Gilolo (= Halmahera) I, N Moluccas.
P. macklotii var. *batchiana* Gray, 1871c: 110; Batchian (= Bacan) I, N Moluccas.
P. fuscus Dobson, 1878: 59,552, pl. 4; Sulawesi. Pre-occupied by *P. fuscus* Geoffroy, 1803; Desmarest, 1803; and Blainville, 1840 which is *P. niger* etc.
= *P. dobsoni* Andersen, 1908: 370.

Range (Map 24). Sulawesi (but see Bergmans & Rozendaal, 1988); Sula Is: Sula Bessi; N Moluccas: Halmahera, Morotai, Ternate, Bacan.

Variation. Two subspecies can be recognised:
P. c. caniceps. N Moluccas, Sula Is. Slightly smaller, longer ears, slightly smaller teeth and paler underparts.
P. c. dobsoni. Sulawesi.

Remarks. The record of *P. caniceps* from Siao, Sangihe Is (Andersen, 1912a; Laurie & Hill, 1954; Van Strien, 1986) is based on specimens of *Acerodon celebensis* according to Bergmans & Rozendaal (1988) who also discussed the occurrence of *P. caniceps* on Sulawesi and Peleng I, regarding the identification of a series from the latter island by Tate (1942a) with some reservation. According to Koopman (pers. comm.) these specimens are in fact *P. griseus*.

[*Pteropus argentatus*]

P. argentatus Gray, 1843: 194. *Nomen nudum*.
= *P. argentatus* Gray, 1844: 30; ? Ambon I, Moluccas.

Range. ? Moluccas: Ambon.

Remarks. Andersen (1912a) placed *argentatus* with *caniceps* and *dobsoni* in a *caniceps* group. The only known material, the immature, much damaged holotype of uncertain provenance, is inadequate to determine its relationship to its supposed or any other associates. Tate (1942a) identified a series of specimens from N Sulawesi with *argentatus* but Musser, Koopman & Califia (1982) have demonstrated that these should be referred to *Acerodon celebensis*.

Pteropus melanotus group

This group includes only *P. melanotus*.

Pteropus melanotus

P. edulis Blyth, 1846c: 367; Nicobar Is. Not *P. edulis* Geoffroy, 1810a (in *P. vampyrus*).
? *P. nicobaricus* Fitzinger, 1861: 389; Nicobar Is. *Nomen nudum*.
P. melanotus Blyth, 1863a: 20; Nicobar Is.
? *P. nicobaricus* Zelebor, 1869: 11; Car Nicobar I, Nicobar Is.
P. tytleri Dobson, 1874d: 3 (and 1876: 189); S Andaman I, Andaman Is. *Nomen nudum*.
P. natalis Thomas, 1887c: 511, pl. 41 (and 1889a: 532); Christmas I, S of Java.
P. modiglianii Thomas, 1894d: 106; Kifa-juc, Bua Bua, Enggano I, off W Sumatra. Lectotype designated by Felten (1960).
P. niadicus Miller, 1906d: 64; Teliwaa, Nias I, W Sumatra.
P. tytleri Mason, 1908: 162; Rutland I, S Andaman Is.
P. satyrus Andersen, 1908: 362; Narcondam I, Andaman Is.

Range (Map 24). Andaman and Nicobar Is; Nias and Enggano Is (W of Sumatra); and Christmas I (S of Java). On the Andaman islands of Rutland, S Sentinel, S Andaman, Barren, Little Jolly Boy and Narcondam; and the Nicobar islands of Great Nicobar, Car Nicobar, Trinkut, Tillanchong and Nancowry.

Variation. The named forms were distinguished by small differences in size, colour and fur length. Saha (1980) considered populations from the Andaman Is (*tytleri*) and the Nicobar Is (*melanotus*) to be indistinguishable.

Remarks. The status of *tytleri, nicobaricus* and *satyrus* was discussed by Hill (1967,1971b), the last having been considered by Andersen (1912a) to be the Andamanese representative of *P. hypomelanus*. There are close similarities in colour variation between *P. m. tytleri, P. m. satyrus, P. m. melanotus* and *P. hypomelanus geminorum* from the Mergui Is in S Burma, leading Hill (1971b) to suggest that possibly *P. melanotus* and *P. hypomelanus* are more closely linked than is apparent from Andersen (1912a). In turn, *P. melanotus* appears to have affinities with the Malagasy *P. rufus* and its allies.

Pteropus melanopogon group

In the Indomalayan Region this group contains only *P. melanopogon* from the Molucca and Aru Islands but Andersen (1912a) included also *P. livingstonii* from the Comoro Is, off E Africa.

Pteropus melanopogon Black-bearded flying fox

P. phaiops Temminck, 1837: 65, pl. 35; Ambon I, Moluccas. Not of Temminck, 1825.
P. melanopogon Peters, 1867b: 330; Ambon I, Moluccas.
P. chrysargyrus Heude, 1897: 177, pl. 5; Little Kai I, Kai Is, Moluccas.

Range (Map 25). Moluccas: Buru, Seram, Ambon (and adjacent small islands of Boano, Saparua, Goram, Manovolka), Banda, Kai Is, Timor Laut; also on the Aru Is (New Guinea).

Variation. Three subspecies:
P. m. melanopogon. Moluccas, except Kai Is.
P. m. keyensis (*P. m.* var. B *keyensis* Peters, 1867b: 330). Kai Is. Slightly smaller; insertions of wing membranes on back further apart; more yellowish.
P. m. aruensis (*P. m.* var. A *aruensis* Peters, 1867b: 330). Aru Is. As *keyensis* but with silvery whitish back and rump.

Remarks. Records of *P. melanopogon* from the Sangihe Is, Sulawesi, are based on specimens of *P. hypomelanus* (Bergmans & Rozendaal, 1988).

Pteropus chrysoproctus group

Includes *P. chrysoproctus* from the Moluccas and two further species, *P. cognatus* and *P. rayneri*, from the Solomon Is.

Pteropus chrysoproctus

P. chrysoproctus Temminck, 1837: 67, pl. 35; Ambon I, Moluccas.

Range (Map 24). Moluccas: Buru, Seram, Ambon (and adjacent small islands of Arsilulu, Keffing, Panjang, Goram, Watubella).

Remarks. Bergmans & Rozendaal (1988) demonstrated that the records in the literature of *P. chrysoproctus* from the Sangihe Is are in fact based on an example of *P. hypomelanus*, and could find no other reports of the species from Sulawesi or any of its off-lying islands.

Pteropus lombocensis group

According to Andersen (1912a) this group includes besides *P. lombocensis* from the Lesser Sunda Is the further species *P. molossinus* from the Caroline Is (W Pacific) and *P. rodricensis* from Rodriguez I, in the Mascarene Is (Indian Ocean).

Pteropus lombocensis

P. lombocensis Dobson, 1878: 34, pl. 3,4; Lombok I, Lesser Sunda Is.
P. tricolor Heude, 1897: 177, pl. 5; Flores, Lesser Sunda Is. Preoccupied by *P. tricolor* Gray, 1871c (in *P. hypomelanus*).
= *P.* (*Sericonycteris*) *heudei* Matschie, 1899: 32.
P. temmincki Hartert, 1898: 456; Ombay, Alor I, Lesser Sunda Is. Pre-occupied by *P. temmincki* Peters, 1867b.
= *P. solitarius* Andersen, 1908: 367.

Range (Map 26). Lesser Sunda Is: Lombok, Flores, Alor.

Variation. The eastern subspecies, *P. l. solitarius* from Alor I, is smaller and paler than the nominate subspecies from the more westerly islands.

Pteropus pselaphon group

This group is primarily Micronesian in distribution, with one species, *P. leucopterus*, in the Philippines.

Pteropus leucopterus

P. leucopterus Temminck, 1853: 60; ? Philippines.
P. chinensis Gray, 1871c: 111; ? China, but more probably Luzon I, Philippines (see Andersen, 1912a).

Range (Map 26). Philippine islands of Luzon, Dinagat and Catanduanes (L Heaney, pers. comm.).

Pteropus personatus group

The *temmincki* group of Andersen (1912a), its two species occurring in the Moluccas, one extending to the Bismarck Archipelago (Table 40).

Map **26**. *Pteropus* (part)

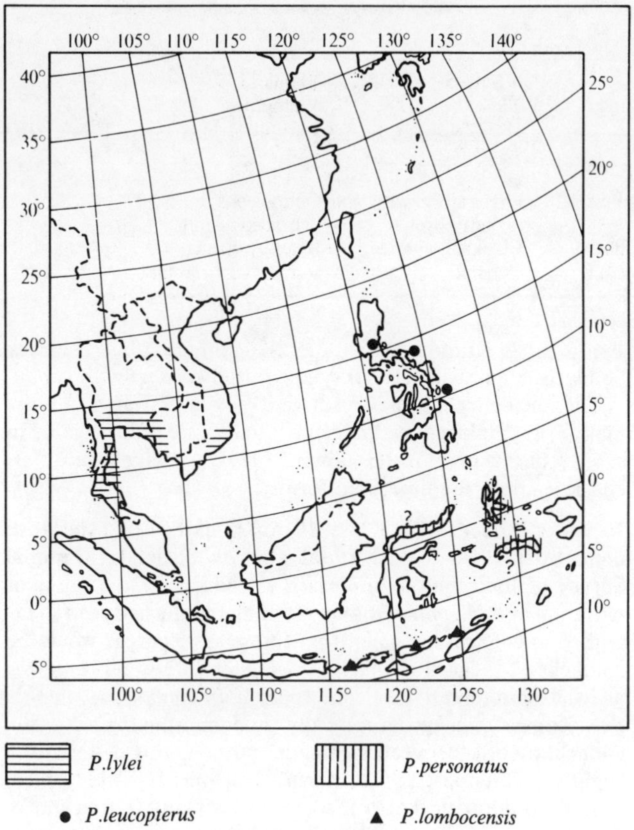

Table 40. The species of the *Pteropus personatus* group.

	personatus	*temmincki*
Range in region	N Moluccas	C,S Moluccas
Forearm	86–96	94.5–104
Brown facial stripe through and above eye	+	0
Teeth much reduced	+	0

Pteropus personatus Masked flying fox

P. personatus Temminck, 1825: 189; Ternate I, N Moluccas.

Range (Map 26). Moluccas: Halmahera, Morotai, Ternate, Bacan; ? Seram (Jentink, 1887,1888a; Andersen, 1912a); ? N Sulawesi.

Remarks. Bergmans & Rozendaal (1988) considered the provenance of the specimens on which the Sulawesian record is based to be questionable, these having perhaps originated from the N Moluccas.

Pteropus temmincki Temminck's flying fox

P. griseus Temminck, 1837: 81, pl. 36; Ambon I, Moluccas (see Andersen, 1912a). Pre-occupied by *P. griseus* Geoffroy, 1810a.
P. temmincki Peters, 1867b: 331; Ambon I, Moluccas. (See Andersen, 1912a).
= *P.* (*Sericonycteris*) *petersi* Matschie, 1899: 31.
P. liops Thomas, 1910c: 383; Buru I, Moluccas.

Range (Map 23). Moluccas: Buru, Seram, Ambon, ? Timor; also to islands off NE New Guinea and the Bismarck Archipelago.

Variation. *P. t. liops* from Buru I differs in longer, denser fur and browner back and underparts. A larger subspecies, *P. t. capistratus* Peters, 1876 (New Ireland) occurs in the Bismarck Archipelago and on some of the small offshore islands of NE New Guinea.

Remarks. Goodwin (1979) discussed the reputed occurrence of this species on Timor I, based in part on a historical confusion of this species with *P. griseus* and on two specimens reputedly from Timor in the collections of the BMNH. One of these, 7.1.1. 250, appears to be a juvenile of *P. griseus*: the other, 58.1.18.2, is correctly *P. temmincki* but was purchased from a dealer and its provenance may be questionable. Schwarz (1914) reported *temmincki* from Timor and Semau I but according to Goodwin one of his specimens (from Semau) is apparently *P. griseus*.

Pteropus vampyrus group (Table 41)

This group occupies almost exactly the area of this work, but does not extend to the Ryukyu Is, has not been recorded from Sulawesi, and apparently does not occur in the Moluccas.

Pteropus giganteus Indian flying fox

Vespertilio gigantea Brünnich, 1782: 43; Bengal, India.
P. medius Temminck, 1825: 176; Calcutta; Pondicherry, India.
P. edwardsi Geoffroy, I.,1828: 699; Bengal, India. Pre-occupied by *P. edwardsi* Geoffroy, 1810a.
P. leucocephalus Hodgson, 1835a: 700; central region of Nepal.
P. assamensis McClelland, 1839: 148; Assam, India.
P. kelaarti Gray, 1871c: 104; Sri Lanka.
P. ariel Allen, G. M.,1908: 28; Male Atoll, Maldive Is.

Range (Map 25). Pakistan, Nepal, India, Burma, Sri Lanka; Maldive Is: Male, Hulule, Gan; Andaman Is, where probably an occasional and rare vagrant (Mason, 1908; Hill, 1967); also a single record from Qinghai Prov., China (Wang, Z-y. & Wang, S, 1962).

Variation. *P. g. ariel* from the Maldive Is has a generally shorter forearm than the nominate subspecies, with a shorter, slightly narrower skull; *P. g. leucocephalus* from Nepal, Sikkim, Assam Manipur and possibly Burma differs in longer, softer fur. A further subspecies has been described from China:
P. g. chinghaiensis Wang, Z-y. & Wang, S.,1962: 494; Ming He County, Qinghai, China, [c. 36°30'N, 103°E]. Based upon a single, living individual, possibly vagrant.

Remarks. Possibly or even probably conspecific with *P. vampyrus* (see below). If this is accepted it should be noted that *vampyrus* is the prior name and would introduce a major change into the Indian literature, where *giganteus* has been established and widely used for many years.

Pteropus intermedius

P. intermedius Andersen, 1908: 368; Amherst, near Moulmein, S Burma.

Range (Map 25). S Burma and W Thailand.

Remarks. Kloss (1916a) and Lekagul & McNeely (1977) considered *intermedius* a subspecies of *P. vampyrus*, the latter with the reservation that it might prove a valid species; Honacki *et al.* (1982) included it in *P. vampyrus*. Hill (1975) reported two examples (not in fact seen but identified by Kitti Thonglongya) from Nakhon Sawan in W Thailand; the record

Table 41. The species of the *Pteropus vampyrus* group.

	giganteus	*intermedius*	*vampyrus*	*lylei*
Range	Pakistan-SW China	S Burma, W Thailand	S Burma-Timor	Thailand, Vietnam
Forearm	c.137–180	c.180	180–220	145–160
Skull: greatest length	62–76	72.5	72–91	61–67
Mantle	Generally yellowish	Yellowish buff	Generally tawny or buffy brown	Ochraceous or ochraceous buff
Underparts	Buffy or buffy brown, paler than back	Brownish black, similar to back	Blackish, similar to back	Blackish brown, sometimes brighter

from C Thailand by Kloss (1916a) was shown to be of *P. vampyrus malaccensis* by Kloss (1919b) who also dismissed the queried occurrence of *intermedius* in S Thailand listed by Andersen (1912a).

This species is of especial interest since both in size and coloration it seems to be intermediate between the Indian *P. giganteus* and the predominantly Malaysian *P. vampyrus*. The former extends eastward to Pegu in S Burma, the latter northward to Tenasserim, C Thailand and Indochina, *intermedius* being somewhat smaller than *malaccensis*, the subspecies of *P. vampyrus* in this northernmost part of its range. Lack of material prevents any real consideration of its status.

Pteropus vampyrus Large flying fox

Vespertilio vampyrus Linnaeus, 1758: 31; Java (designated by Andersen, 1912a).
= *V. caninus* Blumenbach, 1797: 73.
V. celaeno Hermann, 1804: 13; Batavia, Java.
? *V. nudus* Hermann, 1804: 15; type locality unknown.
P. edulis Geoffroy, E.,1810a: 90; Timor I.
P. javanicus Desmarest, 1820: 109; Java.
P. funereus Temminck, 1837: 63, pl. 35; Timor I (restricted by Andersen, 1912a).
P. pluton Temminck, 1853: 56; Bali (restricted by Kuroda, 1933b).
? *P. sumatrensis* Ludeking, 1862: 46; Agam, W Sumatra (*n. v.* —from Chasen, 1940a).
P. pteronotus Dobson, 1878: 48; Java.
P. lanensis Mearns, 1905: 432; Pantar, near L Lanao, Lanao Prov., Mindanao, Philippines.

Range (Map 25). S Burma, Vietnam, S Thailand, Malaya; Sumatra, Java, Borneo; Philippines; Lesser Sunda Is; including the following islands: Mergui Is: Tavoy, Kisseraing, Gregory, Malcolm, Barwell; Thailand: Kut, Mesan, Pra Noi; Vietnam: Phu Quoc; Riau Is: Bintang, Batam; Lingga Is; Mentawai Is: Sipora, Pagai Is; Sumatra: Bangka, Sebesi, Sibuku; Krakatau Is, Panaitan; Anamba Is: Siantan; N Natuna Is: Bunguran Panjang; Borneo: Banggi, Balambangan; Philippines: Luzon, Catanduanes, Mindoro, Bohol (Mickleburgh *et al.*, in prep.), Romblon, Samar, Panay, Leyte, Negros, Palawan, Cebu, Cabo, Canon, Dinagat, Mindanao; Bali, Lombok, Sumbawa, Savu, Timor. Perhaps vagrant to the Andaman Is; possibly a seasonal migrant to Nicobar Is (Mason, 1908; Hill, 1967).

Variation. The following subspecific names have been proposed:
P. v. malaccensis Andersen, 1908: 368; Kuala Tembeling, Pahang, Malaya, 200ft.
P. v. natunae Andersen, 1908: 369; Panjang I, N Natuna Is.
P. v. kopangi Kuroda, 1933b: 339; Kopang, Lombok I, Lesser Sunda Is. Van Strien (1986) synonymised *kopangi* with *pluton*.

The incidence of melanism and colour variation in the mantle, as employed by Andersen (1912a), are of little value in subspecific distinction (Goodwin, 1979). Average size differences separate the described forms.

Remarks. Andersen (in Kloss, 1916b) remarked that the work done by the Bombay Natural History Society's Mammal Survey of India had so increased knowledge of the range of variation of *P. giganteus* that everything that he had seen so far tended to confirm his belief that future systematists would be compelled to regard *giganteus, ariel, intermedius* and *vampyrus* as local representatives of one species, *P.* (*vampyrus*) *intermedius* being both geographically and morphologically the connecting link between *giganteus* and *vampyrus*, with *ariel* a slightly differentiated island form. However, available collections from the critical area of southern mainland Burma and W Thailand remain little changed since Andersen wrote, and it seems premature to take this step until variation in this region is more properly understood.

Pteropus lylei Lyle's flying fox

P. lylei Andersen, 1908: 317; Bangkok, Thailand.

Range (Map 26). Thailand and Vietnam.

Pteropus alecto group

This group has a single representative, *P. alecto*.

Pteropus alecto Black flying fox, Central flying fox

P. alecto Temminck, 1837: 75; Menado, N Sulawesi.
P. atterimus Temminck, 1846: 333; Bawean I, Java Sea. *Nomen nudum*.
? *P. nicobaricus* Heude, 1897: 176, pl. 5; Macassar, S Sulawesi. Not of Zelebor, 1869.
P. nicobaricus b. *aterrimus* Matschie, 1899: 17; Bawean I, Java Sea.
P. baveanus Miller, 1906d: 63; Bawean I, Java Sea.
P. morio Andersen, 1908: 369; Waingapo, Sumba I, Lesser Sunda Is.

Range (Map 27). Sulawesi; Saleyer I, Bawean I; Kangean Is; Lesser Sunda Is: Lombok, Sumba, Savu; also to Australia.

Variation. Three subspecies in region:
P. a. alecto. Sulawesi, Salayer I, Lombok. Largest.
P. a. aterrimus. Bawean I, Kangean Is. Smaller.
P. a. morio. Sumba I, Savu I. Smallest; longer fur.

P. a. gouldi Peters, 1867d from N Australia is similar to *P. a. alecto* but has longer fur, narrower rostrum and palate, and weaker dentition. Variation in *P. a. alecto* was discussed by Bergmans & Rozendaal (1988).

Pteropus conspicillatus group

The two species comprising this group occur in the Moluccas, one, *P. conspicillatus*, extending to NE Australia (Table 42).

Table 42. The species of the *Pteropus conspicillatus* group.

	conspicillatus	*ocularis*
Forearm	157–183	129–141
Skull: greatest length	72–80	62–62.5
C-M^2	27.5–31.2	24.0–25.7

Map **27**. *Pteropus* (part), *Acerodon* (part)

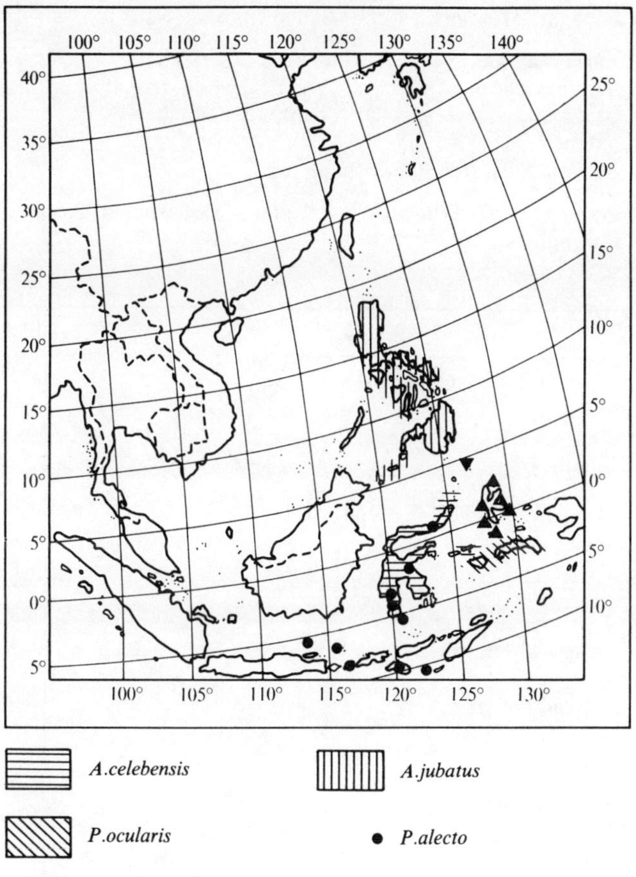

Pteropus conspicillatus

P. conspicillatus Gould, 1850: 109; Fitzroy I, Queensland, Australia.
P. chrysauchen Peters, 1862b: 576; Bacan I, N Moluccas.
P. mysolensis Gray, 1871c: 105; Mysol I, W New Guinea.

Range (Map 27). N Moluccas: Morotai, Halmahera (Peterson *et al*., 1990), Ternate, Bacan, Gebe, Obi; also to New Guinea, its N and E offshore islands and NE Queensland.

Variation. Subspecies in region is *P. c. chrysauchen*, extending to W New Guinea; nominate subspecies, with more pronounced paler ring round eye and slightly weaker dentition, extends therefrom to NE Australia.

Pteropus ocularis Seram flying fox

P. ocularis Peters, 1867b: 326; Seram I, Moluccas.
P. mysolensis var. *ceramensis* Gray, 1871c: 105; Seram I, Moluccas.

Range (Map 27). Moluccas: Buru and Seram.

ACERODON Wallacean flying foxes

Acerodon Jourdan, 1837a: 156; type species *Pteropus jubatus* Eschscholtz, 1831.

Philippines, Sulawesi and Lesser Sunda Is; 6 species (Table 43). Very closely related to *Pteropus* of which it might be regarded as a subgenus. Often found on small offshore islands; commonly in swampy forest.

Acerodon celebensis Sulawesi flying fox

Pteropus celebensis Peters, 1867b: 333; Sulawesi.
P. arquatus Miller & Hollister, 1921b: 100; Koelawi, C Sulawesi.

Range (Map 27). Sulawesi, and small islands of Siao (Sangihe Is: Bergmans & Rozendaal, 1988), Mangole (Sula Ia), Malenge (Togian Is); Salayar I.

Remarks. Musser, Koopman & Califia (1982) demonstrated that specimens from Sulawesi reported by Tate (1942a) as *Pteropus argentatus* are in fact referable to *A. celebensis*.

Acerodon macklotii Lesser Sunda flying fox

Pteropus macklotii Temminck, 1837: 69, pl. 35; Timor.
P. floresii Gray, 1871c: 106; Flores I, Lesser Sunda Is.
A. gilvus Andersen, 1909a: 24; Waingapo, Sumba I, Lesser Sunda Is.

Table 43. The species of *Acerodon*.

	celebensis	*macklotii*	*leucotis*	*humilis*	*jubatus*	*lucifer*
Range	Sulawesi	Lesser Sunda Is	Philippines	Talaud Is	Philippines	Panay I, Philippines
Size						
Forearm	125–142	135–160	137–142	c.140	180–205	165
Skull: greatest length	59–66	62–69	62–69		75–84	72
C-M^2	22.5–25.8	27.4–29.2	24.5–26.5	24.8	30.9–35.4	30.8
Ears sub-acutely pointed, longer than muzzle from front of eye to tip of nose	+	+	0	0	0	0
Pelage pale	+	+	+/0	0	0	0
Mantle buffy	+	+	+/0	0	0	0
P^3 with postero-internal cusp	0	0	+	+	+	+

Range (Map 28). Lesser Sunda Islands of Lombok, Sumbawa, Flores, Sumba, Alor and Timor.

Variation. The following subspecific names have been proposed:
A. m. alorensis Andersen, 1909a: 24; Alor I, Lesser Sunda Is.
A. m. prajae Sody, 1936a: 51; Praya, central Lombok I, Lesser Sunda Is.

The named forms are based on slight differences in size and colour and in some instances refer only to single specimens. Goodwin (1979) discussed the subspecies and suggested that it seems improbable that all are distinct.

Acerodon leucotis Palawan flying fox

Pteropus leucotis Sanborn, 1950: 189; Singay, Busuanga I, Calamianes Is, Philippines.

Range (Map 28). Philippine islands of Busuanga (Calamianes), Palawan and Balabac (Musser, Koopman & Califia, 1982).

Variation. One subspecies has been described:

Pteropus l. obscurus Sanborn, 1950: 190; Santiago, Ihahig, Palawan.

Remarks. Musser, Koopman & Califia (1982) have shown that *leucotis* should be transferred to *Acerodon* from *Pteropus*.

Acerodon humilis Talaud Islands flying fox

A. humilis Andersen, 1909a: 24; Lirong, Talaud Is.

Range (Map 27). Talaud Is, north of Sulawesi.

Acerodon jubatus Golden-crowned flying fox

Pteropus jubatus Eschscholtz, 1831: 1, pl. 16; Manila, Luzon I, Philippines.
P. pyrrhocephalus Meyen, 1833: 604, 606, pls 45, 46. Luzon.
P. auri-nuchalis Elliot, 1896: 77, pl. 12; Leyte I, Philippines.

Range (Map 27). Philippine islands of Luzon, Mindoro, Samar, Panay, Negros, Leyte, Biliran, Maripipi (both Heaney, 1991), Dinagat, Mindanao, Cebu, Banga, Basilan, Sulu Is.

Variation. One subspecies has been described:
A. j. mindanensis Andersen, 1909a: 24; Mindanao, Philippines. Based upon larger size but doubtfully distinct.

Acerodon lucifer Panay flying fox

Pteropus lucifer Elliot, 1896: 78, pl. 13; Concepcion, Panay I, Philippines.

Map **28**. *Acerodon* (part), *Rousettus* (part)

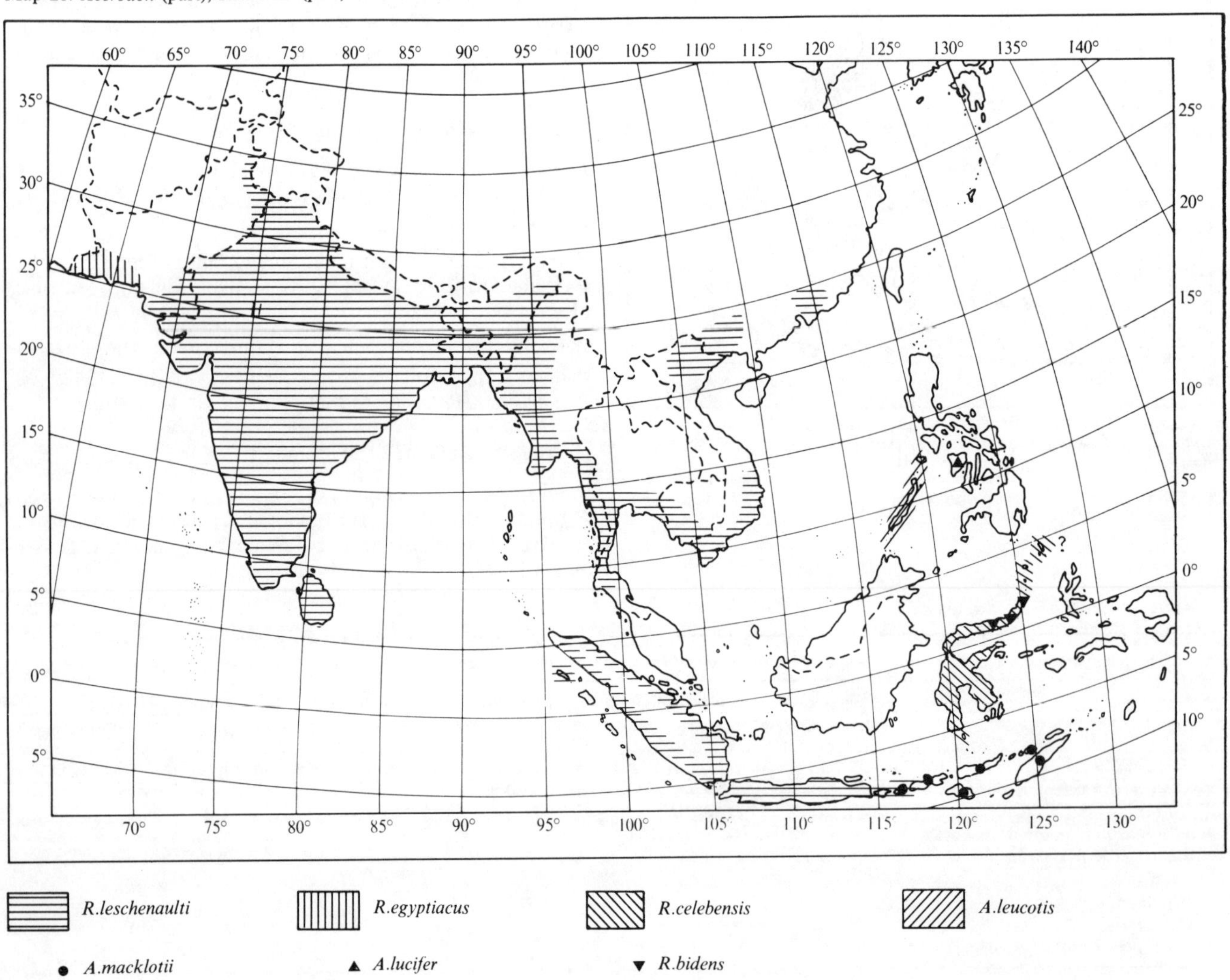

Range (Map 28). Known only from Panay I, Philippines where it has not been seen since 1888 and is possibly extinct (Heaney *et al.*, 1987). Details in Heaney & Heideman (1987).

Remarks. IUCN: Ex.

ROUSETTUS Rousettes

Rousettus Gray, 1821: 299; type species *Pteropus egyptiacus* Geoffroy.
Xantharpyia Gray, 1843a: 37 (*P. amplexicaudatus* Geoffroy).
Senonycteris Gray, 1871c: 115 (*P. seminudus* Kelaart).
Boneia Jentink, 1879c: 117 (*B. bidens* Jentink).

Pakistan through SE Asia to Philippines and Solomon Islands; also to SW Asia and Africa. 10 species; 6 species in 2 subgenera in the Indomalayan Region (Table 44).

Subgenus *Rousettus* (Table 45)

Rousettus egyptiacus

Pteropus egyptiacus Geoffroy, 1810a: 96; Great Pyramid, Giza, Egypt.
= *P. aegyptiacus* Geoffroy, E.,1818: 134, pl. 3 (unjustified emendation).
R. arabicus Anderson & de Winton, 1902: 86, 88, 89-90; Lahej, near Aden, S Yemen.
Also other extralimital synonyms.

Range (Map 28). S Pakistan and possibly adjoining parts of NW India (Das, 1986a); also to Israel, Lebanon, Cyprus and much of sub-Saharan Africa.

Variation. Subspecies in region is *R. e. arabicus*, extending westwards to S Iran, Arabian peninsula and E Ethiopia.

Table 44. The subgenera of *Rousettus*.

Subgenus	*Rousettus*	*Boneia*
Species in region	5	1
Forearm	<90	>90
Premaxillae separated	0	+
Upper canine (C^1) grooved anteriorly	0	+
Coronoid process	High	Low
Upper incisors	2:2	Usually 1:1
Basal antero-posterior diameter of C^1	$\frac{2}{3}$ P^4 length	= P^4 length
P3/3 and P4/4 higher, with surface ridges more evident (H), or low and flattened (L)	H	L
Crown area of I_2:I_1	1.5×	3×
I_1: height relative to I_2	Slightly less	Only to cingulum

Rousettus leschenaulti

Pteropus leschenaulti Desmarest, 1820: 110; Pondicherry, India.
P. pyrivorus Hodgson, 1835a: 700; Nepal.
Cynopterus marginatus Gray, 1843a: 38 (not of Geoffroy, 1810).
C. affinis Gray, 1843a: 39; Himalayas.
P. seminudus Kelaart, 1850c: 216; Mt Lavinia, Sri Lanka.
Eleutherura fuliginosa Gray, 1871c: 118; Laos Mts, Thailand.
E. fusca Gray, 1871c: 119; ? India.
Cynonycteris infuscata Peters, 1873: 487; Calcutta, India.
R. shortridgei Thomas & Wroughton, 1909c: 19 (and 1909d: 374); Kalipoetjang, Java.

Range (Map 28). NE and SE Pakistan (where sometimes a summer visitor), India, Sri Lanka, Burma; S China: S Tibet (Cai & Zhang, 1980b), Amoy, Guangxi, Fujian, doubtfully Hong Kong (Romer, 1974); Thailand, Laos, ? Cambodia, Vietnam; Sumatra, Simalur I (both Kitchener *et al.*, 1990); Java, Bali, Lombok.

Variation. Three subspecies are recognized:
R. l. leschenaulti. Pakistan to Vietnam. Darker, nape and shoulders pilose.
R. l. seminudus. Sri Lanka. Paler, nape and shoulders semi-naked.
R. l. shortridgei. Java. Larger, with broader rostrum and palate.

Rousettus amplexicaudatus Geoffroy's rousette

Pteropus amplexicaudatus Geoffroy, 1810a: 96, pl. 4; Timor I.
Eleutherura infumata Gray, 1871c: 118; Flores I.
? *E. philippinensis* Gray, 1871c: 119; Philippines.
Cynonycteris minor Dobson, 1873b: 203, pl. 14; Java.
C. bocagei Seabra, 1898: 160, pl. 1; Timor I.
R. stresemanni Stein, 1933: 91; Japen I, New Guinea.

Range (Map 29). S Burma, Thailand, Malaya, Sumatra, Java, Borneo, Philippines, Sulawesi, Lesser Sunda Is to Moluccas and extralimitally to Solomon Is. Including the following smaller islands: Sumatra: Mentawai Is, Enggano; Java: Krakatau Is; Philippines: virtually all islands including Sulu Is; Lesser Sundas: Bali, Noesa Penida, Lombok, Sumba, Flores, Alor, Roti, Ndai, Timor, Kisar, Ndao, Savu; Sulawesi: Batu Daka, Malenge (Togian Is: Owen *et al.*, 1987; Hill, 1991); Peleng, Lembeh, Talisai; Moluccas: Halmahera, Ternate, Ambon, Seram. Recorded from Tripura, NE India by Agrawal & Bhattacharyya (1977), but see Rookmaaker & Bergmans (1981) who considered the record to be more probably of *R. leschenaulti*.

Table 45. The species of *Rousettus* (*Rousettus*).

	egyptiacus	*leschenaulti*	*amplexicaudatus*	*spinalatus*	*celebensis*
Range in region	S Pakistan	Pakistan - Java	S Burma - Moluccas	Sumatra, Borneo	Sulawesi
Antitragal lobe reduced or obsolete	0	0	0	0	+
Wings inserted on spinal midline	0	0	0	+	0
Tibia haired	0	0	0	0	+
Pelage long, rather dense	0	0	0	0	+
Last premolars,molars elongate, narrow	0	0	0	0	+
Last lower molars elliptical, $1\frac{1}{2}$-2× as long as wide	0	+	0	0	0
Last lower molars subcircular	0	0	+	0	0

Variation. Three possible subspecies in region:
R. a. amplexicaudatus. Burma to Sulawesi, Moluccas and Timor.
R. a. infumatus. Sumatra, Java, Lesser Sunda Is except Timor.
R. a. stresemanni. ? Philippines.

Subspecific validities and boundaries are uncertain and were discussed by Hill (1983) and Bergmans & Rozendaal (1988).

Remarks. Reviewed in considerable detail by Rookmaaker & Bergmans (1981).

Map **29**. *Rousettus* (part), *Neopteryx*

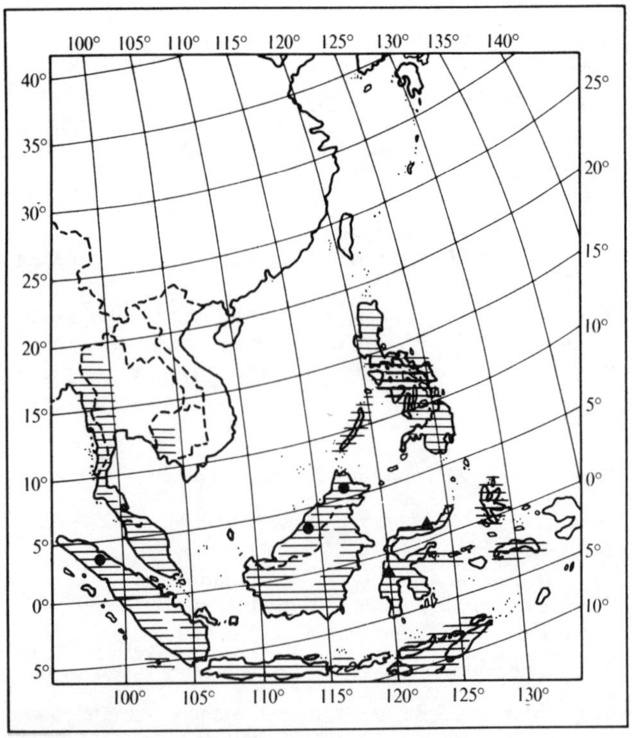

R.amplexicaudatus ● R.spinalatus

▲ N.frosti

Rousettus spinalatus

R. spinalatus Bergmans & Hill, 1980: 95; in or near Medan, or in or near Prapat, N Sumatra.

Range (Map 29). N Sumatra; Borneo: Sarawak, Sabah (Hill & Francis, 1984; Francis, 1989).

Rousettus celebensis Sulawesi rousette

R. celebensis Andersen, 1907a: 503,509; Mount Masarang, Sulawesi, 3500ft.

Range (Map 28). Sulawesi and adjacent islands—Togian Is: Batu Daka, Malenge, Talata Koh, Togian, Walea Kodi, Walea Bahi (Owen *et al.*, 1987; Hill, 1991); Sangihe Is: Sangihe, Siao, Tahulandang; ?Talaud Is.

Remarks. Reviewed by Rookmaaker & Bergmans (1981), Hill (1983) and Bergmans & Rozendaal (1988).

Subgenus *Boneia*

Rousettus bidens

Boneia bidens Jentink, 1879c: 117; Boné, N Sulawesi.
B. menadensis Thomas, 1896a: 242; Menado, N Sulawesi.

Range (Map 28). N Sulawesi; Lembeh I.

Remarks. Bergmans & Rozendaal (1988) synonymyzed *Boneia* with *Rousettus*, pointing out that the characters on which it was based are to be found in *Rousettus* species described after Andersen (1912a) reviewed these genera, or in other *Rousettus* species. However the features shown in Table 44 suggest that subgeneric rank is justified.

NEOPTERYX

Neopteryx Hayman, 1946: 569; type species *N. frosti* Hayman.

Sulawesi; one species.

Neopteryx frosti

N. frosti Hayman, 1946: 571; Tamalanti, W Sulawesi, 3300ft.

Range (Map 29). Sulawesi, in lowland primary forest at 225m in the north and at 1000m in the west.

Remarks. Known for many years only from the holotype. Bergmans & Rozendaal (1988) recorded three further examples (one obtained in the market place) from N Sulawesi, with additional descriptive details.

STYLOCTENIUM

Styloctenium Matschie, 1899: 33; type species *Pteropus wallacei* Gray.

Sulawesi etc.; one species.

Styloctenium wallacei Stripe-faced fruit bat

Pteropus wallacei Gray, 1866a: 65; Macassar, S Sulawesi.

Range (Map 30). Sulawesi; Malenge, Walea Kodi (Togian Is) (Owen *et al.*, 1987; Hill, 1991).

Remarks. Detailed measurements for a small series were provided by Bergmans & Rozendaal (1988).

DOBSONIA Naked-backed fruit bats

Dobsonia Palmer, 1898: 114; type species *Cephalotes peronii* E Geoffroy.
Pteronotus Rafinesque, 1815: 54; *nomen nudum*, apparently based upon *Pteropus paliatus* Geoffroy, ? in *C. peronii* Geoffroy.
= *Hypoderma* Geoffroy, E.,1828: 28. Pre-occupied by *Hypoderma* Latreille, 1825 (Insecta, Diptera).
? = *Tribonophorus* Burnett, 1829: 269 (*T. desmarestii* Burnett, ? in *C. peronii* Geoffroy).

Philippines to N Queensland; 11 species, 5 in the Indomalayan Region (Table 46).

Dobsonia minor Lesser naked-backed fruit bat

Cephalotes minor Dobson, 1879: 875; Amberbaki, NW New Guinea.

Table 46. The species of *Dobsonia*.

	minor	*exoleta*	*moluccense*	*peronii*	*viridis*
Range in region	Sulawesi	Philippines, Sulawesi	Moluccas	Lesser Sunda Is	Sulawesi, Moluccas
Forearm	<95	106–131	133–160	108–117	109–129
Post-canine teeth simple, lacking well-marked antero-internal and posterior basal ledges, without surface ridges	+	0	0	0	0
M^1 with well-developed antero-internal basal ledge	0	0	0	0	+
Antero-internal corner of M_1 usually differentiated into a distinct cusp or ledge	0	0	0	+	+

Range (Map 30). C Sulawesi (Boeadi & Bergmans, 1987; Bergmans & Rozendaal, 1988); also in New Guinea and some small adjacent islands.

Remarks. The sole specimen from Sulawesi differs slightly from examples from New Guinea.

Dobsonia exoleta Sulawesi naked-backed fruit bat

D. exoleta Andersen, 1909b: 531, 533; Tomohon, Minahassa, Sulawesi.
? *D. viridis chapmani* Rabor, 1952: 2; Negros I, Philippines.

Range (Map 31). Sulawesi; Togian Is: Malenge; Muna I; perhaps also the Philippine islands of Negros and Cebu if *chapmani* is conspecific.

Variation. Two possible subspecies:
D. e. exoleta. Range except Philippines.
? *D. e. chapmani.* Philippines. Usually a little larger.

Remarks. The form *chapmani* was initially described as a subspecies of *D. viridis* but Bergmans (1978) found that it lacked the distinguishing features of that species. It is usually slightly larger than specimens of *exoleta* from Sulawesi which it appears to represent in the Philippines where it is the only *Dobsonia* known to occur. L Heaney (pers. comm.) considers it specifically distinct. De Jong & Bergmans (1981) reviewed the specimens of *D. exoleta* then known from Sulawesi. Heaney & Heideman (1987) and Heaney *et al.* (1987) recorded *D. e. chapmani* on Cebu I and remarked that is was last seen on Negros in 1964 and is probably extinct through habitat destruction and hunting. IUCN: (Ex?)—*D. e. chapmani.*

Dobsonia moluccense Greater naked-backed fruit bat

Hypoderma moluccense Quoy & Gaimard, 1830: 86, Atlas, pl. 11; Ambon, Moluccas.

Range (Map 30). Molucca Is: Halmahera, Ambon, Buru, Seram; ? Semaoe, Lesser Sunda Is (see Bergmans, 1978); extralimitally in New Guinea.

Variation. The subspecies in the region is *D. m. moluccense*; a second subspecies *D. m. magna* Thomas, 1905d occurs in New Guinea and perhaps in Halmahera.

Dobsonia peronii Western naked-backed fruit bat

Cephalotes peronii Geoffroy, 1810a: 104, pl. 7; Timor.
? *Pteropus paliatus* Geoffroy, 1810a: 99; type locality unknown.
? *Tribonophorus desmarestii* Burnett, 1829a: 269; *nomen nudum.*
D. sumbana Andersen, 1909b: 531,533; Sumba I, Lesser Sunda Is.

Range (Map 30). Lesser Sunda Is: Babar, Wetar, Timor, Alor, Flores, Sumba, Komodo, Sumbawa, Lombok, Noesa Penida, Bali (Kitchener & Foley, 1985).

Variation. Three subspecies, discussed by Bergmans (1978).
D. p. grandis Bergmans, 1978: 6; Nbabi, S of Dompu, Sumbawa I. Western Lesser Sunda Is: Bali-Komodo, ? Flores.
D. p. sumbana. Sumba I.
D. p. peronii. Eastern Lesser Sunda Is: Timor, probably also Alor, Babar and Wetar.

Dobsonia viridis Greenish naked-backed fruit bat

Cephalotes viridis Heude, 1897: 176, footnote, pl. 5; Kai Is.
D. crenulata Andersen, 1909b: 532; Ternate I, Moluccas.

Range (Map 31). Sulawesi; Togian Is (Owen *et al.*, 1987; Hill, 1991); Peleng I; Sangihe Is; Molucca Is: Morotai, Halmahera, Ternate, Rau, Bacan, Buru, Ambon, Seram, Banda; Kai Is.

Variation. Two subspecies:
D. v. viridis (including *D. v. umbrosa* Thomas, 1910c: 384; Seram I). Sulawesi and S Moluccas.
D. v. crenulata. N Moluccas.

Remarks. Hill (1983) summarized the taxonomic history of *D. viridis* and referred *crenulata* to it as a subspecies, but Bergmans & Rozendaal (1988) retained it as a valid species.

CYNOPTERUS Dog-faced fruit bats, Short-nosed fruit bats

Cynopterus Cuvier, F.,1824: 248; type species *Vespertilio sphinx* Vahl.
Pachysoma Geoffroy, E.,1828: 26 (*Pteropus titthaecheilus* Temminck); pre-occupied by *Pachysoma* Macleay, 1821 (Insecta, Coleoptera).
Niadius Miller, 1906c: 83 (*Cynopterus princeps* Miller).

Most of Indomalayan Region; 4 species (Table 47). The account by Andersen (1912a) is contradictory in its treatment of *C. brachyotis* and *C. sphinx*, leading Hill & Thonglongya (1972) and Hill (1983) to further review these species so far as available material allowed. Further detailed study is needed, especially to resolve the difficulties surrounding these species and the many named forms usually associated with them; Table 47 is provisional and reflects these uncertainties. Kitchener & Maharadatunkamsi (1991) reviewed the *brachyotis* group on the basis of a comprehensive multivariate analysis.

Map **30**. *Styloctenium*, *Dobsonia* (part), *Cynopterus* (part)

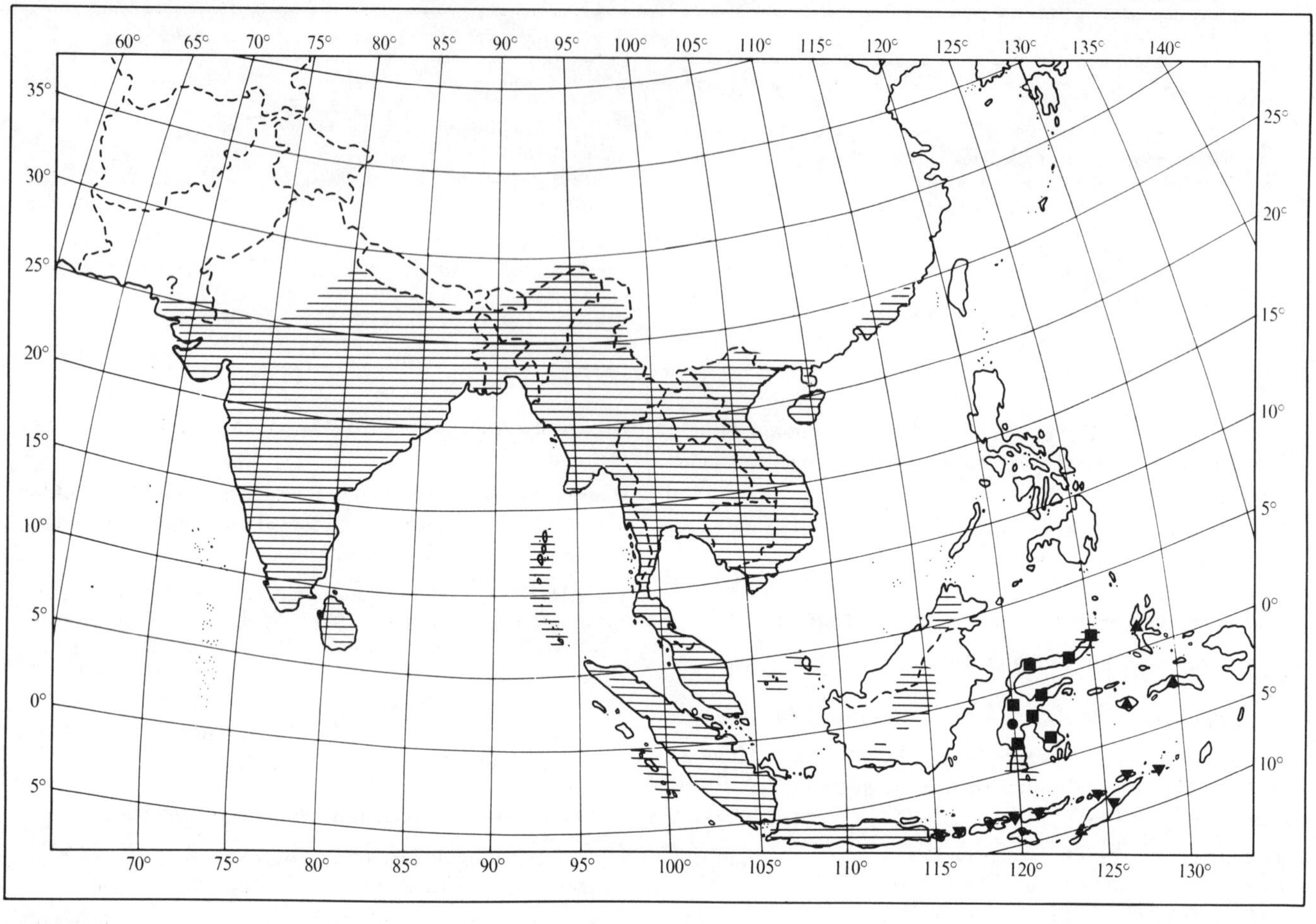

C.sphinx ● D.minor ▲ D.moluccense ▼ D.peronii

■ S.wallacei

Table 47. The species of *Cynopterus*.

	brachyotis	*sphinx*	*titthaecheilus*	*horsfieldii*
Forearm	<67	66–78	73–83	64–89.5
Condylobasal length	<29.5	29–35	34–37	29–37
Molars broad, sub-square or sub-rectangular (B) or narrower, more oval (N)	N	N	N	B
P_4 and M_1 with well-developed surface cusp	0	0	0	+
Rostrum stout and massive	0	0	+	+

Cynopterus brachyotis Lesser dog-faced fruit bat

Pachysoma brachyotis Müller, 1838: 146; Dewei R, Borneo.
? *Pachysoma duvaucelii* Geoffroy, E.,1828: 28; Sumatra.
? *P. brevicaudatum* Geoffroy, I.,1828: 705; Sumatra.
P. luzoniense Peters, 1861: 708; Iriga, S Camarines, Luzon, Philippines.
C. grandidieri Peters, 1869a: 394; 'Zanzibar'.
C. marginatus var. (*Pachysoma scherzeri*) Zelebor, 1869: 13; Car Nicobar.
C. marginatus var. *ceylonensis* Gray, 1871c: 122; Sri Lanka.
C. marginatus var. *philippensis* Gray, 1871c: 123; Philippines.
C. marginatus var. *cumingii* Gray, 1871c: 123; Philippines.
C. brachysoma Dobson, 1871a: 105 (and 1873a: 201, pl. 14); Andaman Is.
C. marginatus var. *andamanensis* Dobson, 1873b: 201, pl. 14); Andaman Is.
C. montanoi Robin, 1881a: 90; Kesang, near Malacca, Malaya.
C. minutus Miller, 1906d: 63; Nias I, W Sumatra.
C. minor Revilliod, 1911: 517; Lambuja, SE Sulawesi.
C. babi Lyon, 1916a: 438; Babi I, W Sumatra.
C. archipelagus Taylor, 1934: 182; Polillo, Polillo I, Philippines.
? *C. nusatenggara* Kitchener & Maharadatunkamsi, 1991: 312; Desu Belo, Jerewah, W Sumbawa I, Lesser Sunda Is.

Range (Map 32). Sri Lanka; SW India (Das, 1986a); ? NE India, Nepal; S China (Guangzhou, Hong Kong), Vietnam, Thailand, S Burma, Malaya; Sumatra, Java, Borneo, Sulawesi and throughout the Philippines; also on the Lesser Sunda Is of Lombok, Sumbawa, Sumba, Moyo, Komodo, Flores, Adonara and Lembata if *nusatenggara* is conspecific; also on the following smaller islands: Andamans, Great Nicobar, Car Nicobar, Trinkut; Thailand: Samui, Kra, Klum; Malaya: Penang, Langkawi, Pangkor, Redang, Pisang, Tioman, Tulai, Sri Buat, Tinggi; Ayer Merbau; Merambong; Singapore; Riau

Map **31**. *Dobsonia* (part), *Cynopterus* (part)

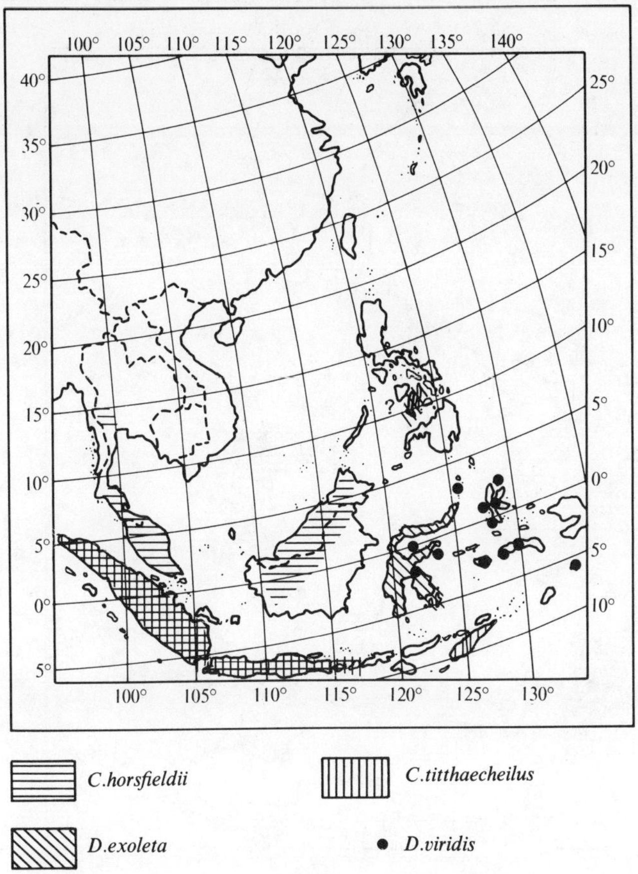

Is: Karimon, Kundur, Batam; Sumatra: Bangkalis, Babi, Nias, Pini, Batu Is, Enggano, Bangka, Belitung; Krakatau Is; Java: Dua, Madura, Bawean; Borneo: Balembangan; Bali; Noesa Penida; Setoko; Sugi; Bulan; Makapan; Panaitan; Kangean Is; Matasiri; Laut; Bunyu; Sibutu; Talaud Is; Sangihe Is (Bergmans & Rozendal, 1988); Sulawesi: Togian Is: Batu Daka, Walea Kodi (Owen *et al*., 1987; Hill, 1991); Peleng, Labobo, Potil Besar.

Variation. The following nominal subspecies have been described:
C. b. javanicus Andersen, 1910: 624; Buitenzorg, Java.
C. b. insularum Andersen, 1910: 624; Kangean Is.
C. b. concolor Sody, 1940a: 393; Meok, Enganno I.
C. b. hoffeti Bourret, 1944: 4; Cho-Bo, near Hanoi, Vietnam.
C. b. altitudinis Hill, 1961a: 631; Mount Brinchang, Cameron Highlands, Pahang, Malaya, 5500ft.

Geographical variation involves mainly size and colour, the montane *altitudinis* for instance being conspicuously darker than its lowland counterparts.

Remarks. Allocation of the name *scherzeri* (Nicobar Is) to this species is uncertain since *C. sphinx* also occurs there. The inclusion of *minor* (Sulawesi) was supported by Bergmans & Rozendaal (1988). *C. nusatenggara* from the Lesser Sunda Is seems likely to be a subspecies of *C. brachyotis*. Its authors considered *minutus* and *luzoniense* to be specifically distinct, the former occurring in Sumatra, Java, Borneo and Sulawesi, the latter in the Philippines and Sulawesi. They also thought *insularum* to be a subspecies of *C. sphinx* rather than of *C. brachyotis*.

Cynopterus sphinx Short-nosed fruit bat, Indian fruit bat

Vespertilio sphinx Vahl, 1797: 123; Tranquebar, Madras, India.
V. fibulatus Vahl, 1797: 124; Tranquebar, Madras, India.
Pteropus pusillus Geoffroy, 1803: 49; India.
P. marginatus Geoffroy, 1810a: 97; Bengal.
Pachysoma brevicaudatum Temminck, 1837: 92, pl. 35 (not *P. brevicaudatum* Geoffroy, I.,1828).
C. angulatus Miller, 1898: 316; Trong, S Thailand.
C. pagensis Miller, 1906d: 62; N Pagi I, Mentawai Is.

Range (Map 30). India, probably extreme SE Pakistan, Sri Lanka, Burma, Thailand, Cambodia, Vietnam, SE China: Fujian, Hong Kong, Hainan; Malaya, Sumatra and Borneo; Java and Sulawesi (both Kitchener & Maharadatunkamsi, 1991); and on the smaller islands of the Andamans, Car Nicobar, Nao-Tchao and Kwang-Taheou-Wan (S China); Chang, Phuket, Lontar, Muntia, Pennan, Samui (Thailand); Siberut, Sipora, N Pagai (Mentawai Is); Krakatau Is (Tidemann *et al*., 1990); Verlaten; Anamba Is; N Natuna Is; Serasan (S Natuna Is); also Bali, Sangeang (Lesser Sunda Is) and Salayer (all Kitchener & Maharadatunkamsi, 1991).

Variation. The following subspecific names have been proposed:
C. s. gangeticus Andersen, 1910: 623; Lucknow, Uttar Pradesh, India.
C. s. serasani Paradiso, 1971: 296; Serasan I, S Natuna Is.

Geographical variation is chiefly in size, the large Indian *gangeticus* and the slightly smaller *sphinx* merging in Burma and Thailand into the characteristically smaller *angulatus* that extends over most of the remaining range.

Remarks. The taxonomic history was reviewed by Hill & Thonglongya (1972) and Hill (1983).

Cynopterus titthaecheilus

Pteropus titthaecheilus Temminck, 1825: 198, pl. 15; Buitenzorg, Java (restricted by Andersen, 1912).
? *Pachysoma diardii* Geoffroy, E.,1828: 27; Sumatra.
? *C. major* Miller, 1906d: 62; Mojeia R, Nias I, W Sumatra.
? *C. sphinx terminus* Sody, 1940b: 404; Niki Niki, S Timor, 750m.

Range (Map 31). Sumatra, Java, Bali, Lombok and possibly Timor; and the small islands of Nias (if *major* is conspecific); Sertung, Panjang and Rakata (Krakatau Is); Sebesi.

Remarks. Sympatric or nearly sympatric *brachyotis, sphinx* (*angulatus*) and *titthaecheilus* from S Sumatra were compared in detail by Hill (1983) who recognized the last as a distinct species; it having been considered before then to be a subspecies of *C. sphinx*, after Andersen (1912a). *C. s. gangeticus* from India approaches or equals *C. t. titthaecheilus* in size, but the latter differs from *gangeticus* and *C. s. sphinx* in its heavier, more robust skull, with a broader, higher, more substantial rostrum. It differs similarly from *C. s. angulatus* but these features are more pronounced in comparison with this smaller form of *C. sphinx*.

The large size of *major* from Nias I (forearm 75.5-82.0, condylobasal length 32.7-34.4, C-M[1] 11.7-12.4) suggests that it may represent this species rather than *C. sphinx* to which Chasen (1940a) referred it. Similarly, *terminus* from Timor also has a long forearm (75.2-83) but has a relatively compact skull (cbl 30.3-32.2, C-M[1] 10.9-11.8), with a short, stout rostrum (Sody, 1940b; Goodwin, 1979). It seems likely that *terminus* represents *C. titthaecheilus* but its relatively small skull might justify its recognition as a distinct species.

Map 32. *Cynopterus* (part), *Megaerops* (part)

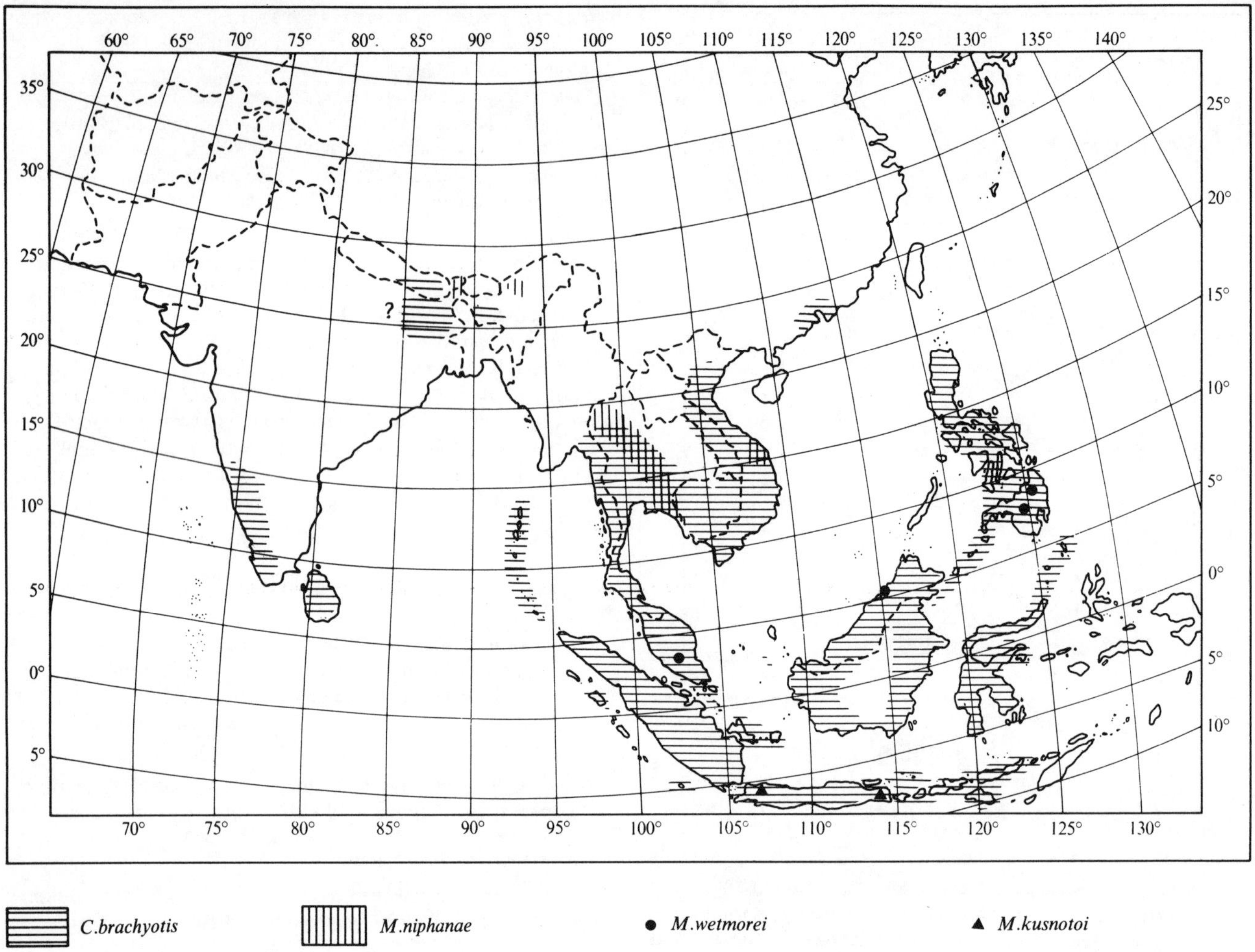

Cynopterus horsfieldii Horsfield's fruit bat

C. horsfieldii Gray, 1843a: 38; Java.
C. princeps Miller, 1906d: 61; Majeia R, Nias I, W Sumatra.
Niadias minor Lyon, 1908b: 665; Siak R, E Sumatra (preoccupied by *C.* (*Cynonycteris*) *minor* Trouessart, 1878.
= *C. h. lyoni* Andersen, 1912a: 827.
C. (*Niadius*) *harpax* Thomas & Wroughton, 1909a: 439; Semangko Pass, Malaya, 3000ft.
C. persimilis Andersen, 1912b: 640; Sarawak, Borneo.

Range (Map 31). S Thailand, Malaya, Sumatra, Java, Lombok and Borneo; and the small islands of Simeulue, Nias, Labuan; Sertung and Rakata (Krakatau).

Variation. Four subspecies are currently recognized:
C. h. harpax. Thailand, Malaya, Sumatra. Cheek-teeth narrowed.
C. h. princeps. Nias I. Largest.
C. h. persimilis. Borneo. Upperparts grey brown.
C. h. horsfieldii. Java, Lombok. Smallest.

MEGAEROPS

Megaera Temminck, 1841: 357; pre-occupied by *Megaera* Wagler, 1830 (Reptilia) and *Megaera* Robineau-Desvoidy, 1830 (Insecta).
= *Megaerops* Peters, 1865a: 256; type species *Pachysoma ecaudatum* Temminck.

NE India to Philippines, Borneo and Java. Four species (Table 48).

Megaerops ecaudatus Tail-less fruit bat

Pachysoma ecaudatum Temminck, 1837: 94 (and 1841: 274, footnote, pl. 69); Padang, W Sumatra.

Range (Map 33). C, S Thailand, ? Vietnam, Malaya, Sumatra, Borneo.

Remarks. Some of the earlier records of this species from India, Thailand and Vietnam (Hill, 1983, Van Peenen *et al.*, 1969) are referable to *M. niphanae.*

Megaerops kusnotoi

Megaerops kusnotoi Hill & Boeadi, 1978: 427; Hanjuang Ciletuh, Lengkong, S Sukabumi, W Java, 700m.

Range (Map 32). Java.

Remarks. IUCN: K.

Table 48. The species of *Megaerops.*

	ecaudatus	*kusnotoi*	*niphanae*	*wetmorei*
Range	Thailand-Borneo	Java	NE India-Indochina	Malaya, Borneo, Philippines
Forearm	50.5–57	49–53	52–60.5	46–52.5
Skull: greatest length	25.6–27.7	25.0–25.9	24.4–28.2	22.2–24.3
$C\text{-}M^1$	7.8–8.3	8.0–8.3	7.7–9.0	6.8–7.7
Upper lip deflected into lateral flange at side of mouth, terminating in vertical ridge	0	+	0	0
Rostrum in lateral profile parallel-sided (P) or sloping and trapezoidal (T)	P	P	T	T
$I^3 : I^2$	<	=	<	<

Map 33. *Megaerops* (part), *Ptenochirus*

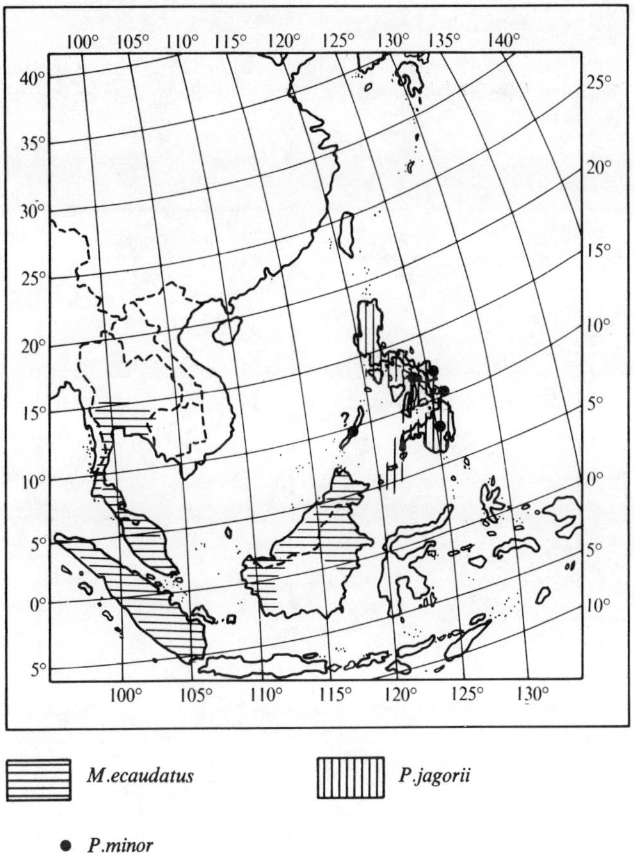

Megaerops niphanae

Megaerops niphanae Yenbutra & Felten, 1983: 2; Amphoe Pak Thong Chai, Nakon Ratchasima Prov., Thailand, 14°30'N, 101°56'E.

Range (Map 32). NE India (Pashok, Darjeeling; Tirap Dist., Arunachal Pradesh), Thailand, Vietnam. Sympatric with *M. ecaudatus* in much of Thailand.

Remarks. Specimens reported from NE India as *M.* (*?*) *ecaudatus* by Hill (1983) and as *M. ecaudatus* by Saha (1984) are more probably referable to this species.

Megaerops wetmorei

Megaerops wetmorei Taylor, 1934: 191; near Tatayan, Cotobato, Mindanao I, Philippines.

Range (Map 32). Malaya, Borneo (Brunei, perhaps Kalimantan), Mindanao (Philippines).

Variation. Two subspecies can be recognized:
M. w. wetmorei. Mindanao. Neck tufts absent or obscure.
M. w. albicollis Francis, 1989: 2878; Pasoh Forest Reserve, Malaya, 2°58'N, 102°17'E. Malaya, Borneo. Conspicuous white tufts on neck.

PTENOCHIRUS

Ptenochirus Peters, 1861: 707; type species *Pachysoma* (*Ptenochirus*) *jagorii* Peters.

Philippines. Two species (Table 49).

Table 49. The species of *Ptenochirus.*

	minor	*jagorii*
Size	Smaller	Larger
Forearm	62–77	76–86.5
Skull: greatest length	29.9–34.4	34.5–37.2
$M^1\text{-}M^1$ (external)	9.4–10.5	11.2–12.4

Ptenochirus jagorii

Pachysoma (*Ptenochirus*) *jagorii* Peters, 1861: 707; Daraga, Lalbay, Luzon, Philippines.

Range (Map 33). Philippine islands of Luzon, Camiguin (N of Luzon), Mindoro, Tablas, Leyte, Carabao, Biliran, Maripipi (both Heaney, 1991), Negros, Bohol, Dinagat, Siargao, Mindanao, Camiguin (N of Mindanao) and some smaller islands; Sulu Is: Bongao, Sanga Sanga.

Ptenochirus minor

Ptenochirus minor Yoshiyuki, 1979: 75; Baracatan, Mt Talomo, Davao City Prov., Mindanao, Philippines.

Range (Map 33). Philippine islands of Biliran, Leyte (both Heaney, 1991), Dinagat and Mindanao; very doubtfully on Palawan.

DYACOPTERUS

Dyacopterus Andersen, 1912a: 651; type species *Cynopterus spadiceus* Thomas.

? S Thailand, Malaya, Sumatra, Borneo and Philippines; two species (Table 50).

Dyacopterus spadiceus Dyak fruit bat

Cynopterus spadiceus Thomas, 1890a: 235; Baram, Sarawak, Borneo.

Range (Map 34). ? S Thailand, Malaya, NW Borneo.

Remarks. A specimen recorded as *D. spadiceus* from Luzon, Philippines (Kock, 1969a) is very large and likely to be *D. brooksi*. It is likely that the record from Mindanao (Heaney *et al.*, 1987) is also *D. brooksi*: it too is of a large specimen (L Heaney, pers. comm.). IUCN: K.

Dyacopterus brooksi

Dyacopterus brooksi Thomas, 1920f: 284; Lebang Tandai, upper Ketuan River, 100 miles N of Bencoolen, Sumatra.

Range (Map 34). Sumatra, ? Luzon, ? Mindanao (Philippines).

Remarks. Few specimens of *Dyacopterus* are known and the status of *brooksi* remains uncertain. It has been included in *D. spadiceus*, e. g. by Honacki *et al.* (1982) but the difference in size suggests that it is better treated as a separate species until further specimens clarify the situation. IUCN: K.

Table 50. The species of *Dyacopterus*.

	spadiceus	*brooksi*
Size	Smaller	Larger
Forearm	75.5–81.5	81–91.5
Skull: greatest length	36.0–37.4	39.6–42.5
C-M[1] (alveoli)	14.9–15.5	15.9–16.8

CHIRONAX

Chironax Andersen, 1912a: 658; type species *Pteropus melanocephalus* Temminck.

One species.

Chironax melanocephalus

Pteropus melanocephalus Temminck, 1825: 190; Gunung Karang, Bantam, W Java (restricted by Bergmans & Rozendaal, 1988).

Map **34**. *Dyacopterus, Latidens, Penthetor, Thoopterus, Sphaerias*

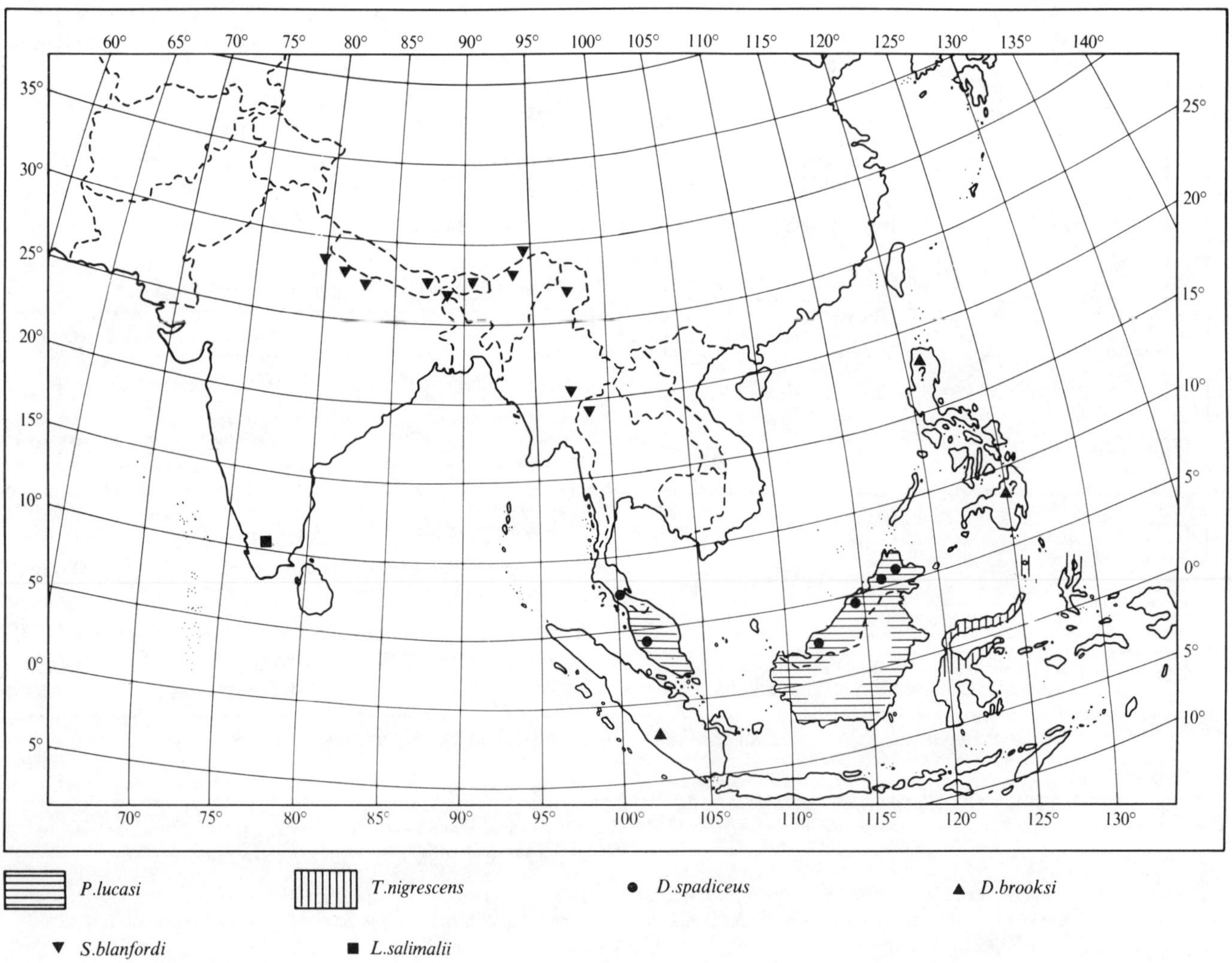

Range (Map 35). S Thailand, Malaya, Sumatra (including Nias I), Java, Borneo: Sabah, Brunei (Hill & Francis, 1984; Francis, 1989), C Kalimantan (Robinson, 1990); Sulawesi.

Variation. The following subspecies has been described from Sulawesi mainly on the basis of small differences between it and Javan specimens: absence of antero-external basal lobes on the second upper premolars, wider postorbital and interorbital regions of the skull and relatively longer, narrower rostrum.
C. m. tumulus Bergmans & Rozendaal, 1988: 46; Sungei Moinakom, Dumoga-Bone National Park, N Sulawesi, 625m.

LATIDENS

Latidens Thonglongya, 1972: 151; type species *L. salimalii* Thonglongya.

One species.

Latidens salimalii

L. salimalii Thonglongya, 1972: 153; High Wavy Mts, Madura Dist., S India, 2500ft.

Range (Map 34). S India.

Remarks. Known only from the holotype which was collected in 1948. IUCN: K.

PENTHETOR

Penthetor Andersen, 1912a: 665; type species *Cynopterus* (*Ptenochirus*) *lucasi* Dobson.

One species.

Penthetor lucasi Lucas's short-nosed fruit bat, Dusky fruit bat

Cynopterus (*Ptenochirus*) *lucasi* Dobson, 1880: 163; Sarawak, Borneo.

Range (Map 34). Malaya, Singapore I, Riau Archipelago, Borneo.

Variation. Specimens at 2000m on Mt Brinchang, Malaya are darker dorsally and greyer ventrally than lowland examples (Hill, 1961a).

THOOPTERUS

Thoopterus Matschie, 1899: 72; type species *Cynopterus marginatus* var. *nigrescens* Gray.

One species.

Thoopterus nigrescens

Cynopterus marginatus var. *nigrescens* Gray, 1871c: 123; Morty (= Morotai) I, N Moluccas.
= *C. latidens* Dobson, 1878: 86.

Range (Map 34). Sulawesi and small islands of Sangihe (N of Sulawesi) and Morotai (N Moluccas); very doubtfully on Luzon, Mindanao or other Philippine islands.

SPHAERIAS

Sphaerias Miller, 1906c: 83; type species *Cynopterus blanfordi* Thomas.

One species.

Map 35. *Chironax*, *Balionycteris*, *Haplonycteris*

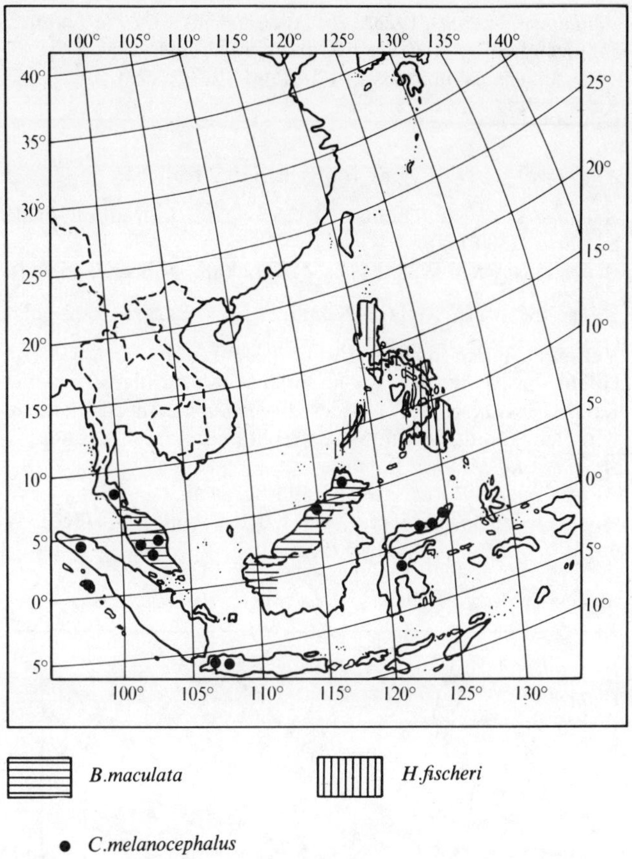

Sphaerias blanfordi Blanford's fruit bat, Mountain fruit bat

Cynopterus blanfordi Thomas, 1891a: 884; Leito, Cheba, Karin Hills, Burma, 1000m.

Range (Map 34). N India (Uttar Pradesh), E Nepal (Lekagul & McNeely, 1977), Bhutan, Burma, NW Thailand, S Tibet, S China.

Variation. The following subspecies was described on the basis of uniform blue-brown pelage, two greyish yellow round patches on underside of neck and lack of odontoid papillae on the inner sides of the lips:
S. b. motuoensis Cai & Zhang, 1980a: 443; Beibing, Motuo region, S Tibet.

BALIONYCTERIS

Balionycteris Matschie, 1899: 72; type species *Cynopterus maculatus* Thomas.

One species.

Balionycteris maculata Spotted-winged fruit bat

Cynopterus maculatus Thomas, 1893a: 341; Sarawak, Borneo.

Range (Map 35). S Thailand, Malaya, Borneo; and the small islands of Durian and Galang (Riau Is).

Variation. The following subspecific name has been proposed but is considered synonymous with *maculata*:
B. m. seimundi Kloss, 1921a: 229; junction of Tahan and Teku rivers, foot of Gunung Tahan, Pahang, Malaya.

AETHALOPS

Aethalodes Thomas, 1923a: 251; type species *A. alecto* Thomas. Pre-occupied by *Aethalodes* Gahan, 1888 (Insecta).
= *Aethalops* Thomas, 1923f: 14 (and 1923g: 178).

One species.

Aethalops alecto Pygmy fruit bat, Grey fruit bat

Aethalodes alecto Thomas, 1923a: 251; Indrapura Peak, Sumatra, 7300ft.
A. aequalis Allen, 1938a: 497; Mt Kinabalu, N Borneo, 5500ft.

Range (Map 36). Malaya, Sumatra, Java, Borneo; Lombok.

Variation. Three subspecies were recognized by Boeadi & Hill (1986) based mainly upon small dental differences, but Kitchener *et al.* (1990) considered *aequalis* specifically distinct, with some additional minor external and cranial features:
A. a. alecto. Malaya, Sumatra.
A. a. aequalis. Borneo: Sabah and Sarawak.
A. a. ocypete Boeadi & Hill, 1986: 264; Cibodas, Mt Gede, W Java, c. 6°50'S, 106°44'E, 1450m.

Map **36**. *Aethalops, Alionycteris, Harpyionycteris, Otopteropus*

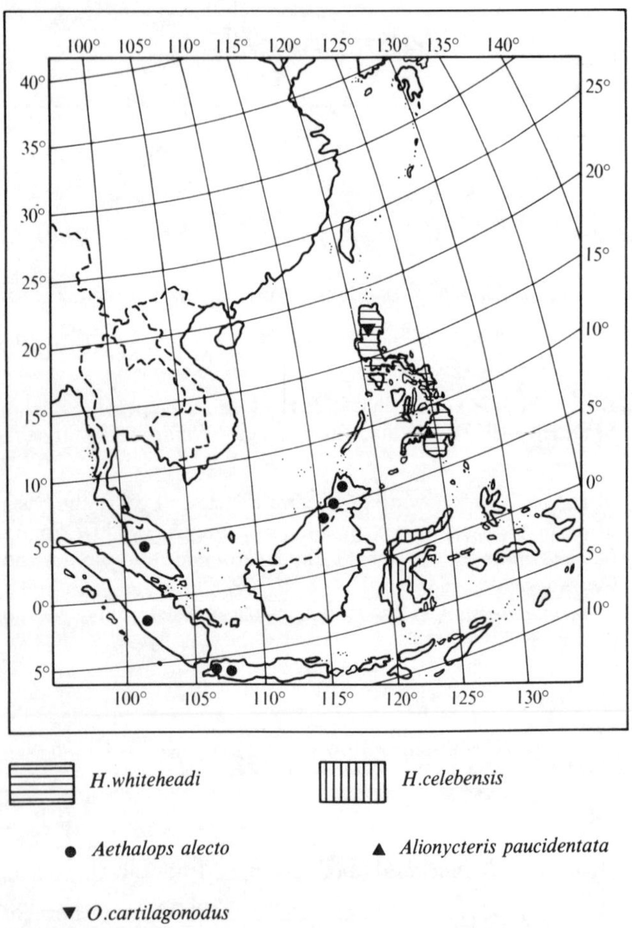

HAPLONYCTERIS

Haplonycteris Lawrence, 1939: 31; type species *H. fischeri* Lawrence.

One species.

Haplonycteris fischeri

H. fischeri Lawrence, 1939: 33; Bignay, Mt Halcon, Mindoro, Philippines.

Range (Map 35). Philippine islands of Luzon, Mindoro, ? Palawan, Leyte, Biliran (both Heaney, 1991), Negros, Dinagat and Mindanao.

ALIONYCTERIS

Alionycteris Kock, 1969b: 319; type species *A. paucidentata* Kock.

One species.

Alionycteris paucidentata

A. paucidentata Kock, 1969b: 322; Mt Katanglad, Bukidnon Prov., Mindanao, Philippines.

Range (Map 36). Known only from the type locality on Mindanao, Philippines.

Remarks. The dense fur, which extends to the toes and to the underside of the proximal part of the forearm, suggests that this species is likely to be found at relatively high elevations although altitude has not been recorded. IUCN: K.

OTOPTEROPUS

Otopteropus Kock, 1969c: 329; type species *O. cartilagonodus* Kock.

One species.

Otopteropus cartilagonodus

O. cartilagonodus Kock, 1969c; Sitio Pactil, Mountain Prov., N Luzon, Philippines.

Range (Map 36). Luzon, Philippines.

Remarks. Apparently adapted to relatively high altitudes, with long, thick fur that extends on to the tail membrane. IUCN: K.

Subfamily Harpyionycterinae

Harpyionycterinae Miller, 1907: 77; type genus *Harpyionycteris* Thomas.

Philippines and Sulawesi; one genus. Multicuspidate dentition suggests adaptation to a diet of tough-textured fruit. The aberrant genus *Harpyionycteris* is the sole representative of this group, which has been ranked from tribe (Koopman, 1982-5) to family (Schulz, 1970).

HARPYIONYCTERIS

Harpyionycteris Thomas, 1896a: 243; type species *H. whiteheadi* Thomas.

Philippines and Sulawesi; two, or perhaps only one, species (Table 51). Characterized by strongly proclivous anterior dentition, the upper incisors and the canines sloped and projected forward, relatively massive, strongly cusped i^2, and relatively large size of the anterior lower premolar and last molars. It is also noteworthy for its highly cuspidate canine and molariform teeth, once regarded as a virtually unique feature. However, a similar degree of cuspidation is approached or found in *Pteralopex* and appears to be an adaptation to

Table 51. The species of *Harpyionycteris*.

	whiteheadi	*celebensis*
Range	Philippines	Sulawesi
C^1 projected forward, tips almost level with tips of incisors	+	0
Posterior part of palate with moderate to deep median trough	0	+

maintain a maximum of occlusal area despite relatively small size (Hill & Beckon, 1977).

Harpyionycteris whiteheadi Harpy fruit bat

H. whiteheadi Thomas, 1896a: 244; Mindoro, Philippines, 5000ft.

Range (Map 36). Philippine islands of Luzon, Mindoro, Negros, Leyte, Biliran, Maripipi (both Heaney, 1991), Mindanao and Camiguin (N of Mindanao).

Variation. *H. w. negrosensis* Peterson & Fenton, 1970: 12 (Balanan, Siaton, Negros I, Philippines) was distinguished mainly by slightly narrower skull.

The lower incisors are rudimentary and variable in occurrence (Peterson & Fenton, 1970).

Harpyionycteris celebensis

H. celebensis Miller & Hollister, 1921b: 99; Gimpoe, C Sulawesi.

Range (Map 36). Sulawesi.

Variation. The lower incisors are rudimentary and variable. Of nine examples four had two small, slightly spatulate incisors tightly sandwiched between the anterior basal part of the canines; three had one lower incisor, in one instance laterally placed, suggesting that the other had been lost, and in two specimens lower incisors were lacking. Lower incisors were present in adults (in two adults both present, in two others one present) as well as in immature examples (Hill, 1983).

Remarks. Tate (1951a) suggested that *celebensis* was probably conspecific with *H. whiteheadi*, an opinion followed by Laurie & Hill (1954). However, Peterson & Fenton (1970) considered them to represent distinct species. The question was reviewed by Hill (1983) who pointed out on the basis of further specimens from Sulawesi that some of the characters upon which Peterson & Fenton relied were diagnostically valueless or unreliable. They are retained as two species, with some reservation, but their status must remain uncertain until more material becomes available, especially of the Philippine form. A similar view was adopted by Bergmans & Rozendaal (1988) who recorded additional specimens from Sulawesi.

Subfamily Nyctimeninae

? Harpyiinae Robin, 1881b: 3.
Nyctymeninae Miller, 1907: 75; type genus *Nyctimene* Borkhausen.

Australasian region, with two genera. One (*Nyctimene*) extends westwards to Timor, Sulawesi and the southern Philippines with 14 species, of which 4 occur in the area covered here; the other (*Paranyctimene*) is monospecific and occurs only in New Guinea. Sometimes subsumed into the Pteropinae, the subfamily is based on the strongly tubular nostrils; broad bony postpalate with wide, deep mesopterygoid fossa; the disposition of the lower canines at the extremity of the mandible, closely adpressed; absence of lower incisors; and presence of four circumvallate papillae towards the base of the tongue.

NYCTIMENE Tube-nosed fruit bats

Nyctimene Borkhausen, 1797: 86; type species *Vespertilio cephalotes* Pallas.
= *Cephalotes* Geoffroy, 1810: 104.
= *Harpyia* Illiger, 1811: 118. Pre-occupied by *Harpyia* Ochsenheimer, 1810 (Insecta, Lepidoptera).
= *Gelasinus* Temminck, 1837: 100. Pre-occupied by *Gelasinus* Van der Hoeven, 1827 (Crustacea).
Uronycteris Gray, 1863a: 262 (*Cynopterus* (*Uronycterus*) *albiventer* Gray).
Bdelygma Matschie, 1899: 82, 84 (*Harpyia major* Dobson, extralimital).

S Philippines through Sulawesi and New Guinea to Solomon Is and NE Australia; 14 species, 4 in this region (Table 52). A fifth, *N. celaeno*, has been erroneously reported from Halmahera (see below under *N. albiventer*).

Table 52. The species of *Nyctimene*.

	minutus	*rabori*	*cephalotes*	*albiventer*
Range in region	Sulawesi, Moluccas	Philippines	Sulawesi, Moluccas	Moluccas
Forearm	51–55	71–81	60–71	50–59
Dorsal pelage mottled	+	0	0	0
Condylobasal length	—	33–35	28–31	24–28
$C\text{-}M^1$	8.8–10.0	11.2–12.1	9.3–11.2	8.7–9.3
P^4 with inner cusp fused to outer cusp	+	0	0	0

Nyctimene minutus Lesser tube-nosed fruit bat

N. minutus Andersen, 1910: 622; Tondano, Minahassa, N Sulawesi.
N. varius Andersen, 1910: 622; Mount Mada, W Buru I, Moluccas.

Range (Map 37). Sulawesi; Moluccas: Obi (BMNH), Buru.

Variation. Two subspecies are recognized:
N. m. minutus. Sulawesi, Obi I.
N. m. varius. Buru I. Longer fur, heavier teeth.

Nyctimene rabori

N. rabori Heaney & Peterson, 1984: 3; Lake Balinsayao, 6 km W of Dumaguete City, Sibulan Municipality, Negros Oriental Prov., Negros I, Philippines, 9°21'N, 123°10'E, 835m.

Range (Map 37). Negros I, Philippines.

Remarks. Possibly conspecific with *N. cephalotes*, differing chiefly in larger size and in some cranial features mostly related to its greater size. IUCN: E.

Map 37. *Nyctimene*, *Eonycteris* (part)

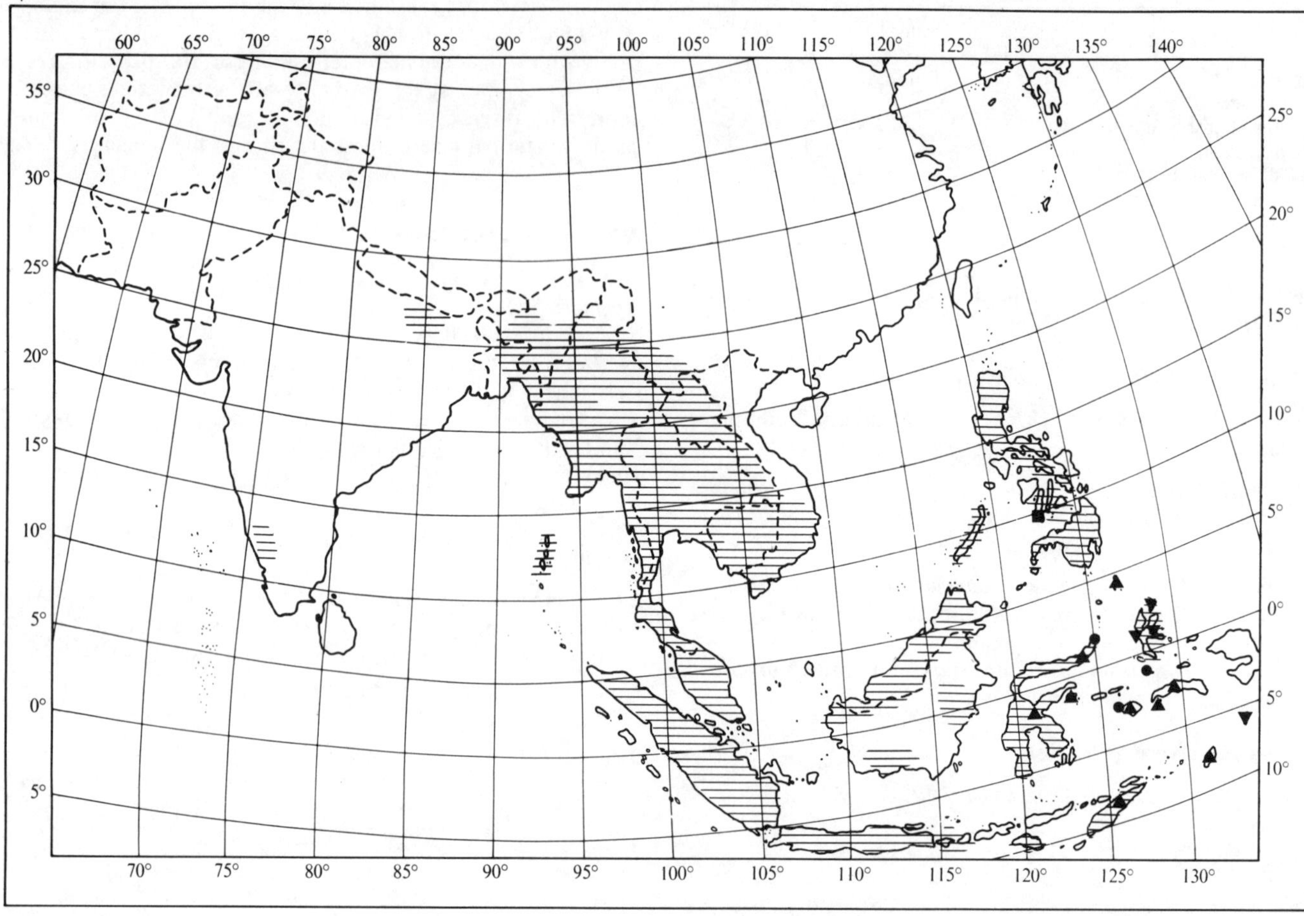

E.spelaea ● N.minutus ▲ N.cephalotes ▼ N.albiventer

■ N.rabori

Nyctimene cephalotes Pallas's tube-nosed bat

Vespertilio cephalotes Pallas, 1767: 10, pls 1,2; 'Moluccas', fixed on Ambon by Andersen (1912a).
= *V. c. melinus* Kerr, 1792:xviii, 98.
= *Cephalotes pallasi* Geoffroy, 1810b: 107.

Range (Map 37). Sulawesi; Peleng I; Talaud Is (Feiler, 1990); Molucca Is: Ambon, Buru, Seram; Timor; Timor Laut; Tanimbar I; extralimitally to New Guinea.

Remarks. Discussed by Smith & Hood (1983) and Heaney & Peterson (1984).

Nyctimene albiventer Common tube-nosed bat

Cynopterus (*Uronycteris*) *albiventer* Gray, 1863a: 262; Morty (= Morotai) I, N Moluccas.

Range (Map 37). N Molucca Is: Morotai, Halmahera, Ternate; Kai Is; extralimitally to New Guinea, Bismarck Is, Admiralty Is etc., perhaps to N Queensland.

Variation. Subspecies in region is *N. a. albiventer*; a second subspecies *N. a. papuensis* Andersen, 1910 in New Guinea etc.

Remarks. Dollman (1930) recorded *N. celaeno* Thomas, 1922d (Legare R, Geelvinck Bay, NW New Guinea) from Halmahera, but the specimen (BMNH 29.5.27.4) on which this was based is in fact an immature *N. albiventer*. Halmahera was listed in the distribution of *N. celaeno* by Laurie & Hill (1954) and van Strien (1986) on account of this incorrect record.

Subfamily Macroglossinae

Macroglossi Dobson, 1875b: 346; type genus *Macroglossus* Cuvier.
Carponycteriinae Lydekker, in Flower & Lydekker, 1891 (*Carponycteris* Lydekker).
= Kiodotinae Palmer, 1898: 111.

Indomalayan and Australian regions (5 genera) and Africa (1 genus). Two genera occur widely in the Indomalayan Region; one Australian genus reaches its western limit in the Moluccas (Table 53). Recognition as a subfamily is based solely upon the development of the tongue which is more extensible than in the Pteropodinae, narrower, tapering to a sharper point and with small differences in the structure and abundance of the lingual papillae. They feed upon nectar and pollen.

Table 53. The genera of Macroglossinae.

	Eonycteris	*Macroglossus*	*Syconycteris*
Range in region	N India Philippines, Timor	? N India Philippines, Moluccas	Moluccas
No. of species in region	2	2	2
Claw on 2nd digit	0	+	+
External tail evident	+	0	0
Interfemoral membrane	Narrow	Narrow	Rudimentary
Calcar evident	+	+	0
Space between inner incisors > that between inner and outer tooth	0	+	0
Upper incisors	Small, peg-like	Small, peg-like	Larger, crown chisel-like
Lower incisors	Subequal	Subequal	Outer c.2x inner

EONYCTERIS Dawn bats, Cave fruit bats

Eonycteris Dobson, 1873a:148 (and 1873b:204); type species *Macroglossus spelaeus* Dobson.
Callinycteris Jentink, 1889: 209 (*C. rosenbergii* Jentink).

N India to Philippines, Sulawesi and Timor; two species (Table 54). Gregarious, sometimes forming large colonies in caves or hollow trees. Reviewed by Tate (1942a) and Hill (1983).

Eonycteris spelaea Dawn bat, Cave fruit bat

Macroglossus spelaeus Dobson, 1871a: 105, 106;1871b: 261; Farm Caves, Moulmein, Tenasserim, Burma.
Callinycteris rosenbergii Jentink, 1889: 210, pl. 9; Tulabolu, Gorantalo, N Sulawesi (restricted by Rozendaal, 1984).
= *E. bernsteini* Tate, 1942a: 345 (lapsus for *rosenbergii*).

Range (Map 37). SW, N India (Sinha, 1990) and S China to Andaman Is, Malaya, Sumatra, Java, Borneo, Bali, Lombok, Sumba, Timor; Sulawesi and Muna I; Philippine islands of Luzon, Tablas, Negros, Cebu, Siquijor, Leyte, Biliran, Maripipi (last three Heaney, 1991), Mindanao, Siargao and Palawan; Sulu Is: Sanga Sanga; Moluccas: Halmahera (Peterson *et al.*, 1990).

Variation. Three subspecies can be tentatively recognized but their boundaries remain to be defined (Hill, 1983; Rozendaal, 1984; Bergmans & Rozendaal, 1988).
E. s. spelaea. Western part of range.
E. s. rosenbergii. N Sulawesi. Darker, M_3 small or absent.
E. s. glandifera Lawrence, 1939: 38; Montalban Caves, Rizal Prov., Luzon. Philippines to Java etc. Larger, rostrum less tapering.

Table 54. The species of *Eonycteris.*

	spelaea	*major*
Range	Widespread	Philippines, Borneo, Mentawai Is
Size	Smaller	Larger
Forearm	61–78	75–85
Skull: greatest length	32–37	35–40
C-M^2	11.1–13.3	13.2–14.4
1st upper premolar: I^3 (crown area)	About equal, sometimes absent	Distinctly larger

Eonycteris major

E. major Andersen, 1910: 625; Mt Dulit, Sarawak, Borneo.
E. robusta Miller, 1913c: 73; Montalban Caves, Rizal Prov., Luzon.
= *E. longicauda* Taylor, 1934: 131.

Range (Map 38). Philippine islands of Luzon, Lubang, Leyte, Biliran, Maripipi (both Heaney, 1991), Negros, Siargao and Mindanao; Borneo; ? N Pagai I (Mentawai Is) (Tate, 1942a). The last record was possibly based upon a large example of *E. spelaea* (L Heaney, pers. comm.).

Variation. Two subspecies:
E. m. major. Borneo; ? Mentawai Is.
E. m. robusta. Philippines. Cheekteeth narrower.

Remarks. Specimens from Luzon referred by Taylor (1934) to *E. robusta* are *E. spelaea*, his *E. longicauda* being a renaming of *E. robusta* Miller (Lawrence, 1939).

MACROGLOSSUS Long-tongued fruit bats

Macroglossus Cuvier, F., 1824: 248; type species *Pteropus minimus* Geoffroy.
= *Kiodotus* Blyth, 1840: 69, footnote (replacement for *Macroglossus* Cuvier, 1824, thought to be pre-occupied by *Macroglossum* Scopoli, 1777, an insect).
= *Rhynchocyon* Gistel, 1848: 9. Pre-occupied by *Rhynchocyon* Peters, 1847 (Macroscelidea).
= *Carponycteris* Lydekker, in Flower & Lydekker, 1891: 654.
Odontonycteris Jentink, 1902: 140 (*M. lagochilus* Matschie).

NE India and Burma through SE Asia to Philippines and N Australia. Two species (Table 55). Reviewed by Hill (1983).

Table 55. The species of *Macroglossus.*

	minimus	*sobrinus*
Size	Smaller	Larger
Forearm	36–44	44–50
Skull length (condylobasal)	22.0–26.0	23.8–28.6
C-M^2	8.0–9.2	9.0–10.0
Internarial groove extends to upper lip	+	0
Anterior extremity of mandible below incisors	Pointed	Squared, forming angular chin

Macroglossus minimus Common long-tongued fruit bat

Pteropus minimus Geoffroy, 1810a: 97; Java.
= *Macroglossa kiodotes* Lesson, 1827a: 115.
P. rostratus Horsfield, 1822:pt 3; Java.
= *Macroglossa horsfieldi* Lesson, 1827a: 115.
M. lagochilus Matschie, 1899: 96; Buru I, Moluccas.
Odontonycteris meyeri Jentink, 1902: 131,140; Tabukan, Gt Sangihe I, Sulawesi.
? *M. fructivorus* Taylor, 1934: 125; Tatayan, Cotabato, Mindanao, Philippines.

Range (Map 38). S Vietnam, S Thailand, Malaya, Borneo, Java, throughout the Philippines, Sulawesi, the Lesser Sunda Is, Moluccas. Including the following smaller islands: Vietnam: ? Con Son; Thailand: Samui; Malaya: Sri Buat, Tioman; Sumatra: Nias; Java: Madura, Kangean, Rakata (Krakatau); Natuna Is: Serasan, possibly Bunguran; Lombok, Timor; Sulawesi: Sangihe Is, Peleng; Moluccas: Halmahera, Bacan,

Map 38. *Eonycteris* (part), *Macroglossus*, *Syconycteris*

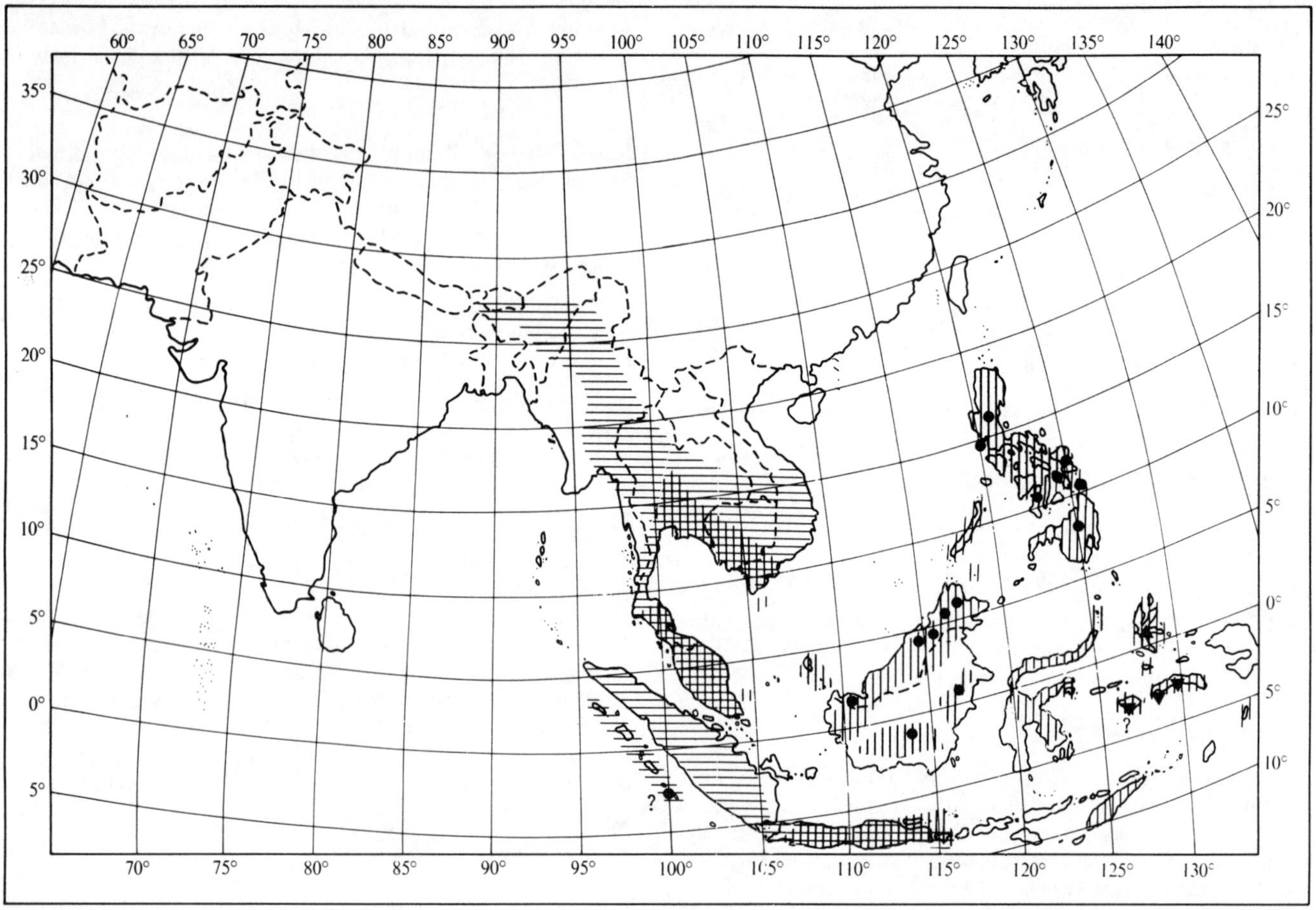

M.sobrinus — *M.minimus* — ● *E.major* — ▼ *S.australis* — ▲ *S.carolinae*

Buru, Seram, Ambon, Banda; Kai Is. Also east to New Guinea, the Solomons and Queensland. To altitude of 1000m in Borneo.

Variation. Three subspecies:
M. m. lagochilus. S Thailand, S Vietnam to Philippines and C, S Moluccas.
M. m. minimus Java, Madura I, Kangean I, Bali.
M. m. nanus Matschie, 1899: 97 (Lamellana, New Britain, Bismarck Archipelago). N Moluccas, Banda I. Small.

The distinction between *M. m. minimus* and *M. m. lagochilus* is small and these may prove synonymous.

Remarks. *M. fructivorus* from Mindanao was based upon a small, immature individual (forearm 35.2) (K Koopman, pers. comm.) and it seems likely to be synonymous with *minimus*.

Macroglossus sobrinus Hill long-tongued fruit bat.

M. minimus sobrinus Andersen, 1911: 641,642; Mt Igari, Perak, Malaya.
M. minimus fraternus Chasen & Kloss, 1928c: 836; Sipora I, Mentawai Is.

Range (Map 38). NE India (Saha, 1985, as *M. minimus*), Burma, Vietnam, Laos, Thailand (including Samui I), Malaya, Sumatra (including Simalur, Nias, Siberut, Sipora and N Pagai Is), Rakata (Krakatau), Java and Bali. To 2000m in Thailand.

Variation. The following subspecies are recognized:
M. s. sobrinus. Range except Mentawai Is.
M. s. fraternus. Mentawai Is.

Specimens from Bali are smaller in some respects than those from Java (Kitchener & Foley, 1985).

SYCONYCTERIS Blossom bats

Syconycteris Matschie, 1899: 94, 98; type species *Macroglossus minimus* var. *australis* Peters.

Moluccas to New Guinea and Australia. Three species, of which two occur in the Moluccas (Table 56).

Table 56. The species of *Syconycteris*

	australis	*carolinae*
Range	Central Moluccas	Halmahera
Size	Smaller	Larger
Forearm	<53	c.60
Skull: greatest length	<30	c.34

Syconycteris australis Common blossom bat (Australia)

Macroglossus minimus var. *australis* Peters, 1867a: 13, footnote; Rockhampton, Queensland, Australia.
S. crassa major Andersen, 1911: 643; Ambon, Moluccas.

Range (Map 38). Seram, Ambon, probably Buru and other small islands near Ambon; east to New Guinea and E Australia.

Variation. The Moluccan form, *S. a. major*, is larger than those further east.

Remarks. Reviewed by Hill (1983).

Syconycteris carolinae Halmaheran blossom bat

S. carolinae Rozendaal, 1984: 200; S base of Gunung Gamkunora, NW Halmahera, Indonesia, 1°20'N, 127°31'E, c. 180m.

Range (Map 38). Halmahera, Moluccas.

SUBORDER MICROCHIROPTERA

A suborder of 17 modern families, containing c. 147 genera and c. 814 species, represented in most parts of the world, its distribution that of the order; 40 genera and c. 232 species occur in the region. As might be expected in so large an assemblage there is considerable variation in form and structure: microchiropterans range in size from very small to moderately large (forearm 22.5-115mm); many species have nose leaves or other dermal outgrowths above the nostrils or on the lower lip; a tragus is usually present but is lacking in two families; the interfemoral membrane is generally well developed, and the tail is usually relatively long, rarely short or absent. The second finger lacks a claw. The molar teeth have a characteristic W-pattern of raised cusps and ridges, although this may be distorted in a few species, and in some frugivorous or nectarivorous species in the New World the cusps and ridges are much reduced or virtually absent.

The entire gamut of chiropteran dietary habits is found in the suborder, but there are no species in the Old World that subsist on flowers and flower products, this niche being occupied by the Megachiroptera, and the Old World has no sanguinivorous vampire bats. Microchiropterans likewise occupy the entire range of roosting sites, although the majority of species tend to roost in enclosed situations rather than in more open roosts such as those used by many megachiropterans. Many are strongly gregarious, forming large colonies in caves or similar shelters. All navigate and seek and capture food by means of a system of echolocation based on ultrasonic signals produced by the larynx.

1. Tail with its terminal portion protruding from the dorsal surface of the interfemoral membrane, or emerging from and projecting far beyond its edge **2**

Tail quite or almost completely enclosed in interfemoral membrane, with at most the terminal two vertebrae projecting from its edge, or external tail lacking **4**

2. Free terminal portion of tail protruding from middle of dorsal surface of interfemoral membrane; 2nd finger without phalanges .... **Emballonuridae**

Free terminal portion of tail emerging from edge of interfemoral membrane; 2nd finger with phalanges **3**

3. Tail long, projecting beyond edge of interfemoral membrane for more than one half its length, free portion slender, thread-like; low dermal ridge surmounting tip of muzzle; nostrils slit-like; 2nd finger with two phalanges .... **Rhinopomatidae**

Tail shorter, projecting beyond edge of interfemoral membrane for about one half its length, free portion generally rather stout; no dermal ridge surmounting tip of muzzle; nostrils circular or subcircular; 2nd finger with one rudimentary phalanx .... **Molossidae**

4. Muzzle strongly suiform, with nostrils set in flattened narial pad; tragus with oblate swelling in anterior margin; external tail and calcar lacking; narial branches of premaxillae fused above to form a tubular structure .... **Craseonycteridae**

Muzzle generally pointed, nostrils not set in flattened pad; tragus when present not swollen; external tail and calcar present; narial branches of premaxillae not fused above **5**

5. Muzzle lacking complex noseleaves or any marked modification, at most a low, horseshoe-shaped ridge above nostrils .... **Vespertilionidae**

Muzzle with complex narial foliations, or a deep median furrow **6**

6. Tragus present **7**

No tragus **8**

7. A long, erect, rather fleshy noseleaf; tragus bifid; upper incisors lacking .... **Megadermatidae**

Muzzle with deep median furrow, margined by fleshy dermal outgrowths; tragus undivided; upper incisors 2:2 .... **Nycteridae**

8. Posterior part of noseleaf usually triangular or subtriangular in outline, with erect, pointed tip; a vertical, strap-like sella lying above and behind nostrils, joined to posterior part of leaf by median connecting process (Fig. 11); toes II-IV with three phalanges .... **Rhinolophidae**

Posterior part of noseleaf rounded, elliptical or tridentate in outline; no sella, instead sometimes a strap-like structure lying in front of nostrils over anterior part of leaf (Fig. 12); toes II-IV with two phalanges .... **Hipposideridae**

Family Rhinopomatidae

Mouse-tailed bats

Rhinopomatidae Dobson, 1872c: 221; type genus *Rhinopoma* Geoffroy.
= Rhinopomidae Miller, 1907: 80.
For correct formation of family name see Handley (1980).

A distinctive family of one genus and 3 species, found chiefly in arid and semiarid regions from NW Africa to India, perhaps to Burma and Thailand, and to Sumatra. No fossils are known. Rhinopomatids are small to moderate in size and are characterized externally by a swollen muzzle with a distinct transverse dermal ridge above oblique, slit-like valvular nostrils; large conjoined ears with tragus; and by a long, mouse-like tail which projects considerably beyond the edge of a very narrow interfemoral membrane. They are insectivorous, roosting in old ruins, caves and rock clefts, sometimes in large colonies.

Miller (1907) considered them to be the most primitive of the microchiropteran families, largely on account of the presence of two phalanges in the second digit, the free premaxillae, and the structure of the shoulder joint. However, the family has numerous features that suggest that it is much less primitive than Miller believed (Hill, 1974a).

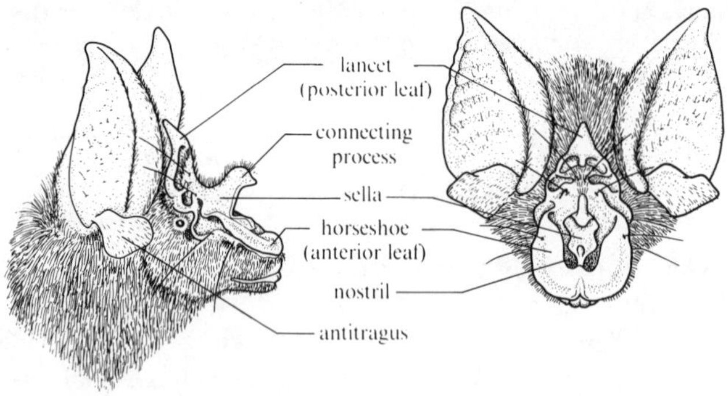

Fig. **11**. Noseleaf of *Rhinolophus*.

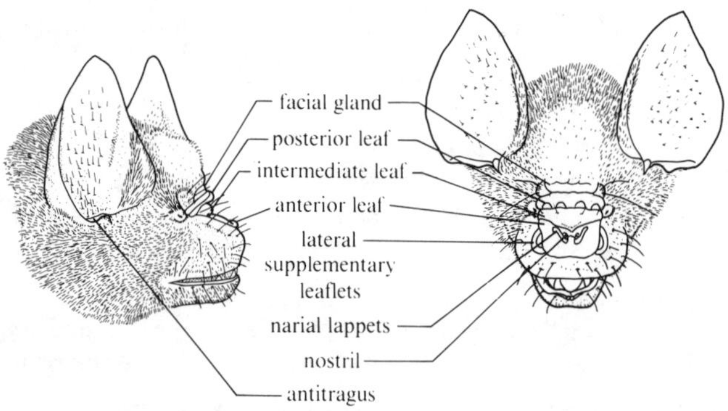

Fig. **12**. Noseleaf of *Hipposideros*.

RHINOPOMA Mouse-tailed bats

Rhinopoma Oken, 1816: 926. Not available (Opinion 417, 1956).
Rhinopoma Geoffroy, E.,1818: 113; type species *Vespertilio microphyllus* Brünnich.

Pakistan, India, Burma, Thailand; Sumatra; also to SW Asia and N Africa; mainly in arid and semi-arid areas; 3 species, all represented in the region (Table 57). Reviewed in detail by Hill (1977).

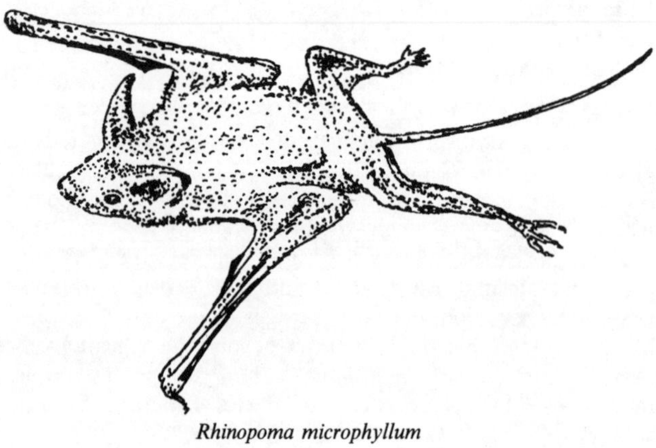

Rhinopoma microphyllum

Table 57. The species of *Rhinopoma*.

	microphyllum	*hardwickii*	*muscatellum*
Range in region	Pakistan, India, N Sumatra	Pakistan, India, ?Burma, Thailand	W Pakistan
Forearm	62–75	53–63.5	45–55
Skull: condylobasal length	17.6–21.2	15.9–17.8	14.2–16.2
Tail: forearm	Usually <	Usually >	Usually >
Sagittal crest	Pronounced	Low	Low
Supraorbital ridges	High	Low	Low
Narial swellings	Low, almost flat	Prominent, globose	Prominent, angular
Anterior projection of narial swellings	Not exceeding nasals	Slightly exceeding nasals	Considerably exceeding nasals
Foremost point of narial swellings	Behind or above rear of C^1	Above rear of C^1	Above front of C^1
Posterior margin of palate	Angular	Rounded	Angular

Rhinopoma microphyllum Greater mouse-tailed bat

Vespertilio microphyllus Brünnich, 1872: 50, pl. 6; Arabia and Egypt.
R. sumatrae Thomas, 1903a: 497; Balighe, near Lake Toba, N Sumatra.
R. kinneari Wroughton, 1912b: 767; Bhuj, Cutch, India.
Also many extralimital synonyms.

Range (Map 39). Afghanistan, Pakistan, India, ? Burma, Thailand; N Sumatra; also west to SW Asia, Egypt, Sudan, Mauritania, Senegal and Nigeria.

Variation. Indian and Sumatran specimens average a little larger than those from Pakistan.

Rhinopoma hardwickii Lesser mouse-tailed bat

R. hardwickii Gray, 1831a: 37; India.
Including many extralimital synonyms.

Range (Map 39). Afghanistan, Pakistan, India, ? Burma (Jerdon, 1867), ? Thailand (Cantor, 1846); also to SW Asia, N Africa and Kenya. Records from Burma and Thailand may refer to *R. microphyllum* (Hill, 1977); Lekagul & McNeely (1977) referred the Thai record by Cantor to *R. hardwickii* and suggested that it may be accidental.

Remarks. Reviewed by Qumsiyeh & Jones (1986).

Rhinopoma muscatellum

R. muscatellum Thomas, 1903a: 498; Wadi Bani Ruha, Muscat, Oman.
R. pusillum Thomas, 1920c: 25; Sib, SE Iran.

Range (Map 39). S Afghanistan, W Pakistan; also through Iran and Oman, probably to Somalia and Ethiopia.

Variation. The form in Pakistan and Afghanistan is referable to the following subspecies:
R. m. seianum Thomas, 1913d: 90; Seistan, Iran. Distinguished from the nominate form (which occurs in Iranian Baluchistan and may reach Pakistan) by larger size and proportionally larger bullae.

Remarks. Reviewed by Qumsiyeh & Jones (1986).

Map 39. *Rhinopoma*, *Emballonura* (part)

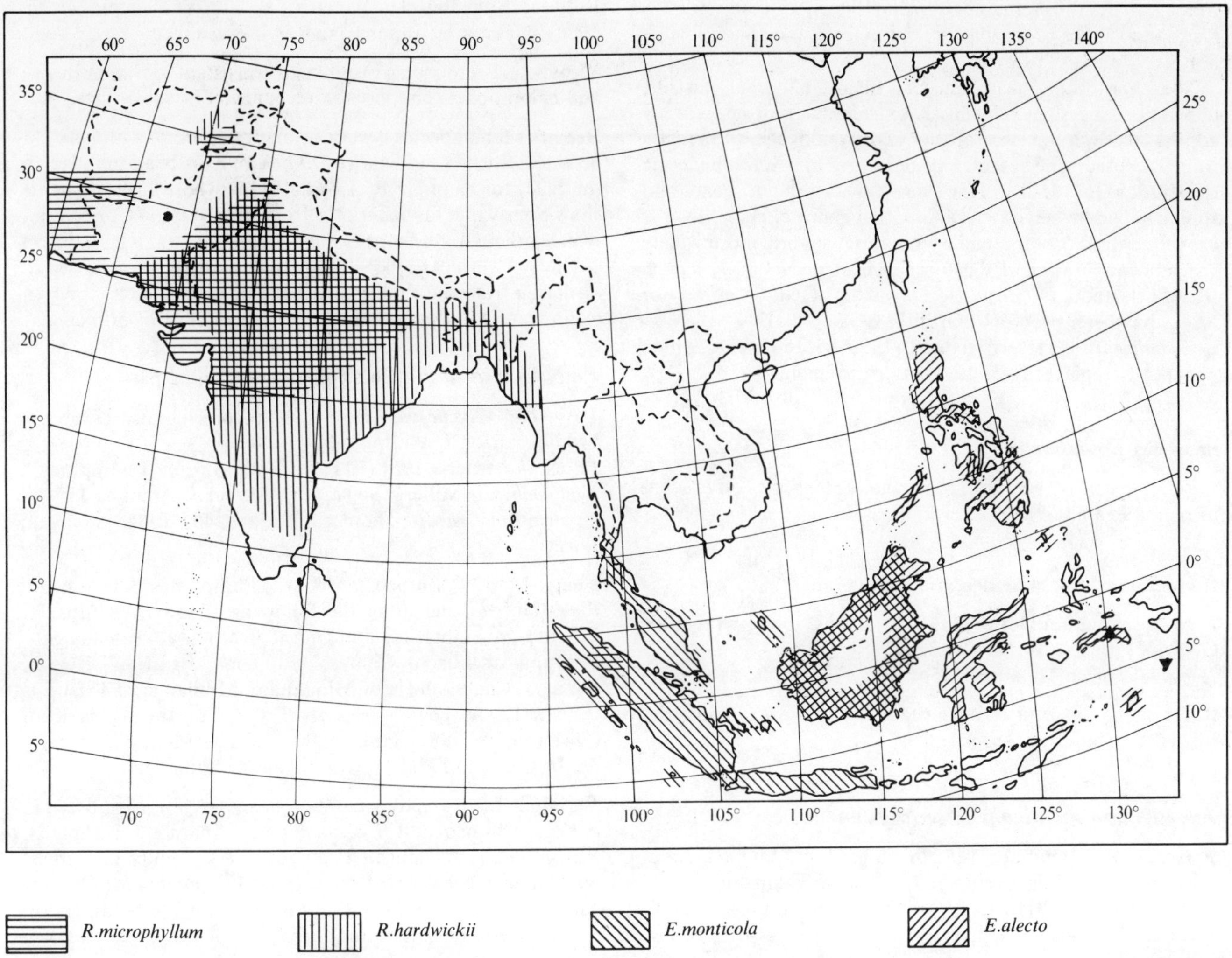

Family Emballonuridae

Sheath-tailed bats

Emballonuridae Dobson, 1875b: 349; type genus *Emballonura* Temminck.

A family of c. 12 modern genera and c. 48 species, distributed through the tropics and subtropics of the world, with 2 genera (Table 58) and 11 species in the Indomalayan Region. Emballonurids range from small to large (forearm c. 32-95mm), with variable ears, often joined anteriorly at the base over the forehead and with a small or moderate tragus; there is no noseleaf; a glandular pouch or area may be present on the throat; tail partly enclosed in moderate tail membrane, its terminal part projecting from the upper surface.

They are insectivorous and occupy a wide range of roosting sites, from caves, crevices, rock or root overhangs, hollow trees, or foliage to ruins or other man-made structures. Some are gregarious, forming large or small colonies, others solitary or nearly so.

Table 58. The genera of Emballonuridae.

	Emballonura	*Taphozous*
No.of species in region	5	6
Forearm	<50	>55
Skull: greatest length	<15	>20
Upper incisors	2:2 Small	1:1 Minute, sometimes absent
Lower incisors	3:3	2:2
Median division between basisphenoid pits	Usually low, ridge-like, sometimes lacking	A narrow, conspicuous plate

EMBALLONURA Old World sheath-tailed bats

Emballonura Temminck, 1838: 22; type species *E. monticola* Temminck.
Mosia Gray, 1843b: 117 (*M. nigrescens* Gray).

A genus of 10 species in two subgenera, one species occurring in Madagascar, the remainder variously distributed from S Burma to islands of the Pacific; 5 species in the region (Table 59).

The short blunt head, unusual tragus, elliptical nostrils, upper lip, long slender penis, hyoid musculature and closely adpressed lower incisors of *nigrescens* remove it sufficiently from the other species of *Emballonura* to justify subgeneric recognition for *Mosia*. However, a short, blunt head and similarly widely separated oblique but more circular nostrils are found in *E. beccarii*, and a similar but slightly more widely spaced configuration of the lower incisors occurs in *E. sulcata* (Pacific Is) and *E. atrata* (Madagascar). Generic status for *Mosia* has been proposed (Griffiths *et al.*, 1991).

Emballonura is closely related to the African *Coleura* which apparently replaces it in the Ethiopian region.

Subgenus *Emballonura*

Burma to islands of Pacific, Madagascar; 9 species of which 4 in region (Table 59).

Emballonura beccarii Beccari's sheath-tailed bat

E. beccarii Peters & Doria, 1881: 693; Ansus, Jobi (= Japen) I, Geelvinck Bay, NW New Guinea.
E. meeki clavium Thomas, 1915h: 137; Ara, Kai Is, Moluccas.

Range (Map 39). Kai Is, Moluccas; also to New Guinea, the Bismarck Is and the Trobriand Is.

Emballonura monticola Lesser sheath-tailed bat

E. monticola Temminck, 1838: 25, pl. 2; Mount Munara, Java.
E. peninsularis Miller, 1898: 323; Trang, S Thailand.
E. pusilla Lyon, 1911a: 132; Mankol, Kendawangan R, SW Kalimantan, Borneo.

Range (Map 39). S Burma and Thailand, Malaya, Sumatra, Java, Borneo, S Sulawesi (Tate & Archbold, 1939a), and the following associated small islands. Mergui Is: Lampi; Singapore; Malaya: Redang, Tioman, Aor, Ubin; Riau Is: Bintang, Bulan, Karimon Anak, Sanglar, Kundur; W Sumatra: Simalur, Nias, Batu, Babi, Sipora, Mentawai Is, Enggano; Bangka; Billiton; Krakatau Is: Panjang, Rakata (Tidemann *et al.*, 1990); Natuna Is; Butang Is; Rawi; Karimata.

Variation. There seems insufficient consistent variation in size and coloration to support the recognition of subspecies.

Remarks. This species is very similar to *E. alecto* with which it occurs in Borneo and Sulawesi. The two have been confused in the past, for example by Lyon (1911a) whose measurements show clearly that his '*monticola*' from Borneo is in fact *E. alecto*, which although similar to *E. monticola* in forearm length differs chiefly in its larger skull (see below). Dobson (1878) recorded *E. monticola* from the Philippines but this record is based on an example of *E. alecto*, as thus is his account of *monticola*.

Emballonura alecto Philippine sheath-tailed bat

Vespertilio (*Nycticeus*) *alecto* Eydoux & Gervais, 1836b: 7; Manila, Luzon I, Philippines.
E. discolor Peters, 1861: 711; Paracale, Luzon, Philippines.
E. anambensis Miller, 1900a: 236; Mobur I, Anamba Is.
E. monticola rivalis Thomas, 1915h: 140; Bida, Sarawak, Borneo.

Range (Map 39). Borneo, most of the Philippines, Sulawesi and the Moluccas; including the following islands: Anamba Is: Mobur; Moeratua Is; Bunguran; Karimata; Panebangan; Philippines: Luzon, Caluya, Guimaras, Palawan, Negros, Dinagat, Camiguin (N of Mindanao), Mindanao; ? Talaud Is (Feiler, 1990); Togian Is: Batu Daka, Togian, Walea Kodi (Owen *et al.*, 1987; Hill, 1991); Peleng; Moluccas: Seram, Ambon, Goram, Kai Is; Tanimbar Is: Yamdena.

Variation. Four weakly separable subspecies are recognised: *E. a. rivalis* (Borneo) and *E. a. anambensis* (Anamba I, Natuna Is, Karimata I, Panebangan I) average very slightly the largest while *E. a. palawanensis* Taylor, 1934: 197 (Iwahig R at base of Thumb Peak, Palawan, Philippines) apparently differs in smaller teeth and other minor features of the lower incisors.

Remarks. Similar in most respects to *E. monticola* but differing as a rule in its larger skull and the presence of a small diastema between the upper anterior and posterior premolars. The two are sympatric in Borneo and Sulawesi but here the larger skull of *E. alecto* distinguishes it from *E. monticola*.

Table 59. The species of *Emballonura*.

	nigrescens	*beccarii*	*monticola*	*alecto*	*raffrayana*
Range in region	Sulawesi, Moluccas	Kai Is	Burma - Sulawesi	Philippines - Moluccas	Seram
Forearm	30–37	36–39	42–48	43–49	39–41
Condylocanine length	9.9–11.0	11.0–11.3	11.8–12.8	12.7–13.8	12.8–13.5
C-M^3	4.0–4.5	4.3–4.5	4.8–5.4	5.3–5.9	5.1–5.4
Tragus narrow, tip directed posteriorly and outwardly	+	0	0	0	0
Nostrils elliptical	+	0	0	0	0
Upper lip projecting beyond lower lip	0	+	+	+	+
Basisphenoid pits shallow, together more or less circular	+	+	0	0	0
Basisphenoid pits moderate to deep, together more or less chordate	0	0	+	+	+
Basisphenoid pits deeply incised into basioccipital	0	0	0	0	+

Emballonura raffrayana

E. raffrayana Dobson, 1879: 876; 'Gilolo'. Correct type locality Mefor (= Noemfor) I, Geelvinck Bay, NW New Guinea (Thomas, 1914g).
E. stresemanni Thomas, 1914g: 442; Mt Lumutu (= Lamutai), W Seram, Moluccas, 400m.

Range (Map 39). Moluccas: Seram; also to New Guinea, Bismarck Is, Solomon Is and Vanuatu.

Remarks. IUCN: K.

Subgenus *Mosia*

Sulawesi to Solomon Is; one species.

Emballonura nigrescens

Mosia nigrescens Gray, 1843b: 117; 'South America', probably Ambon I, Moluccas.
E. papuana Thomas, 1914g: 443; Wakatimi, Mimika R, SW New Guinea.

Range (Map 40). Sulawesi and the Moluccas; also to New Guinea and the Solomon Is. Including the following islands: Togian (Owen *et al.*, 1987; Hill, 1991); Halmahera (Hill & Rozendaal, 1989), Ternate, Buru, Seram, Ambon, Kai Is.

Variation. Two subspecies in region:
E. n. nigrescens. S Moluccas and Kai Is. Braincase slightly less inflated, narrower.
E. n. papuana. Sulawesi, N Moluccas, New Guinea. Braincase more inflated, wider.

TAPHOZOUS Tomb bats, Pouched bats

Taphozous Geoffroy, E.,1818: 113; type species *T. perforatus* Geoffroy.
Saccolaimus Temminck, 1838: 6 (and 1841a: 277, 279) (*T. saccolaimus* Temminck).
= *Taphonycteris* Dobson, 1876b: 548.
Liponycteris Thomas, 1922g: 267 (*T. nudiventris* Cretzschmar).

Africa and the islands of the Indian Ocean through S Asia to the Philippines and Australia; c. 17 species in 3 subgenera (Table 60). All subgenera and 6 species in this region.

Saccolaimus has been variously considered to be a subgenus of *Taphozous* (e. g. by Tate, 1941d; Simpson, 1945; Ellerman & Morrison-Scott, 1951) or generically distinct (e. g. Chasen, 1940a; Barghoorn, 1977; Koopman, 1984b; Robbins & Sarich, 1988; Chimimba & Kitchener, 1991). It stands rather apart from *Taphozous* and *Liponycteris* by virtue of the structure of its auditory region and its large anterior upper premolar.

Map **40**. *Emballonura* (part), *Taphozous* (part)

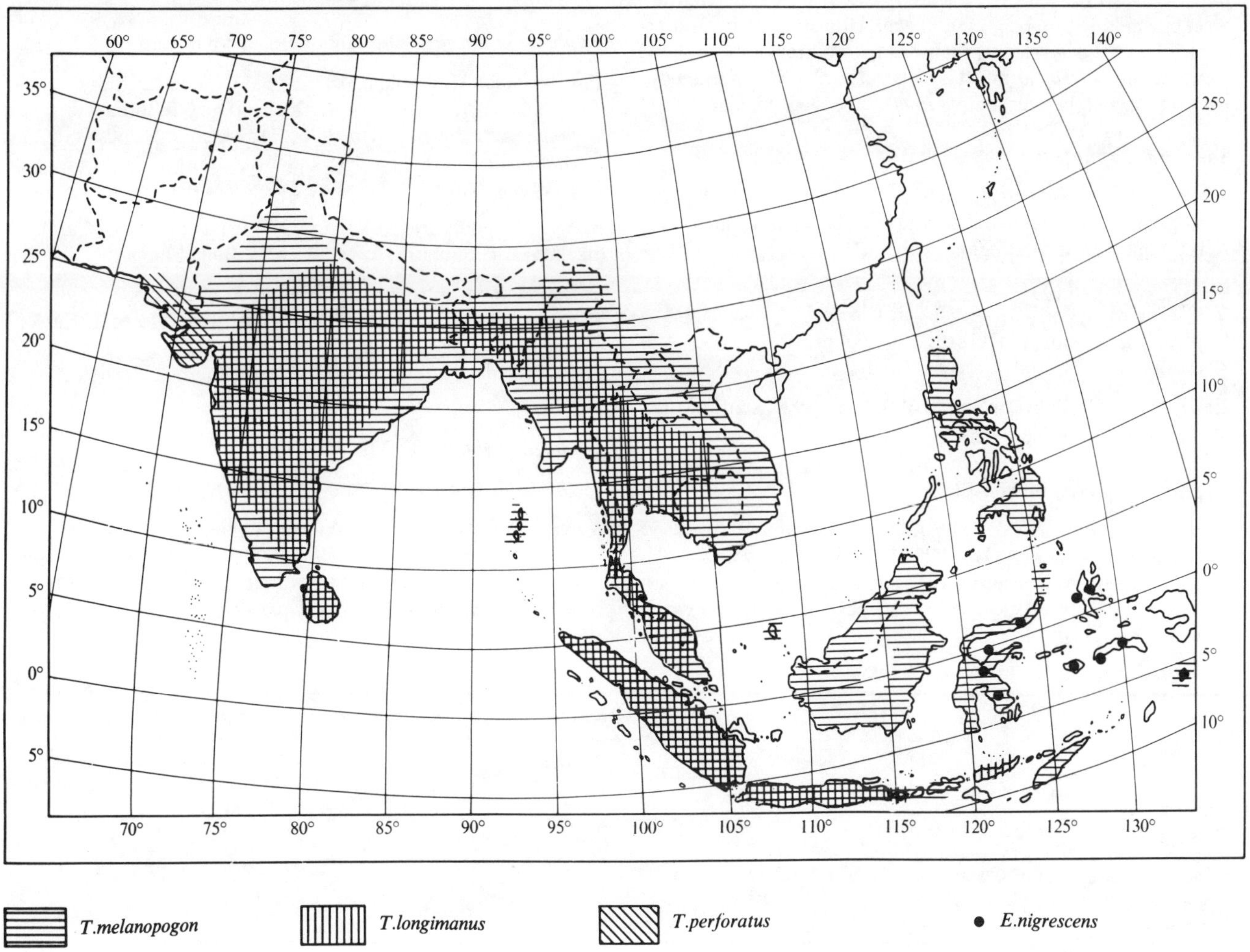

Table 60. The subgenera of *Taphozous*.

	Taphozous	*Liponycteris*	*Saccolaimus*
Species in region	4	1	1
Body partially naked posteriorly	0	+	0
Frontal region of skull concave	+	0	0
Well-developed sagittal crest, projecting behind occiput	0	+	+
Tympanic bullae complete, concealing much of cochleae	0	0	+
Posteroventral face of basisphenoid in contact with tympanic bulla and cochlea	0	0	+
Paroccipital process well developed, relatively long	0	+	+
Mandible concave beneath first premolar	+	0	0
Anterior upper premolar large, c.1/2 crown area of posterior premolar	0	0	+

Barghoorn (1977) and Chimimba & Kitchener (1991) (for Australian species) adduced other cranial features to support the distinction of *Saccolaimus* from *Taphozous*, including for the former a wider skull, especially across the zygoma, its flattened rostrum, and the posterior extension of the sagittal crest beyond the posterior cranial wall. However, such features occur also in *Liponycteris* which has the auditory morphology characteristic of *Taphozous*. The supraorbital foramen, thought by Barghoorn to be present in *Saccolaimus*, is not invariably so.

Subgenus *Taphozous*

Africa through S Asia to Australia; c. 11 species of which 2 are Ethiopian, c. 8 Indo-Australian and one occurs in Africa and S Asia; 4 species in this region (Table 61).

Taphozous perforatus Egyptian tomb bat

T. perforatus Geoffroy, E.,1818: 126; Egypt.
Also some extralimital subspecies and synonyms.

Range (Map 40). Pakistan and NW India (Kutch, Kathiawar); also to SW Arabia and Africa.

Taphozous melanopogon Black-bearded tomb bat

T. melanopogon Temminck, 1841a: 287; Bantam, W Java.
T. bicolor Temminck, 1841a: 290; E India.
T. philippinensis Waterhouse, 1845: 9; Philippines.
T. solifer Hollister, 1913c: 157; thought to be from Pekin, China, but Allen (1938b) considered that it came from some more tropical locality, perhaps the Philippines.
T. cavaticus Hollister, 1913c: 159; Muara Dua, near Emmahaven, S of Padang, W Sumatra.
T. achates Thomas, 1915n: 60; Savu I, near Timor I.

Range (Map 40). Sri Lanka, India, S China to Malaya, Sumatra, Java, Borneo, the Philippines, Lesser Sunda Islands, Sulawesi and east to Timor and the Kai Is. Including the following islands: Hainan; Andamans; Tarutau; Malaya: Langkawi, Penang, Tikus, Angsa, Pisang; Natuna Is: Bunguran; Berhala (N Borneo); Philippines: Luzon, Mindoro, Lubang, Mindanao; Sangihe Is (Feiler, 1990); Talaud Is (Musser, 1987a); Togian Is: Walea Bahi (Owen *et al.*, 1987; Hill, 1991); Bali, Nusa Penida, Lombok, Sumbawa, Savu; Timor.

Variation. One subspecific name has been proposed:
T. m. fretensis Thomas, 1916j: 5; Teratau I, Straits of Malacca.

Remarks. Tate (1941d) allocated *bicolor* to *T. longimanus*; Ellerman & Morrison-Scott (1951) considered it a synonym of *T. m. melanopogon*, which they listed as the Indian subspecies, but *fretensis*, generally paler than *melanopogon*, intervenes in Malaya. It is likely that *philippinensis* is a subspecies of *T. melanopogon* (Koopman, in Honacki *et al.*, 1982); the provenance of *solifer* is doubtful but apparently it is very close to *philippinensis*. Specimens from Timor are discussed by Goodwin (1979) and from Sulawesi by Hill (1983,1991).

Taphozous longimanus Long-winged tomb bat

T. longimanus Hardwicke, 1825b: 525; Calcutta, India.
T. fulvidus Blyth, 1841a: 975; Darjeeling, NE India.
T. brevicaudis Blyth, 1841a: 976; Travancore, S India.
T. cantori Blyth, 1842a: 784; Calcutta, India.
T. kampenii Jentink, 1907: 66; Batavia, Java.

Range (Map 40). Sri Lanka; India, Burma, Thailand, Cambodia, Malaya; Sumatra, Java, Borneo, Bali, Flores.

Variation. Two subspecific names have been proposed:
T. l. var. *leucopleurus* Dobson, 1875a: 385; Flores I, Lesser Sunda Is.
T. l. albipinnis Thomas, 1898b: 246; Labuan I, Borneo.

Taphozous theobaldi Theobald's bat

T. theobaldi Dobson, 1872b: 152; Tenasserim, Burma.

Range (Map 41). C India; Burma, Thailand, Vietnam; Java; Borneo (Kalimantan) and Sulawesi (Hill *et al.*, 1990). Chasen listed *T. theobaldi* from the Malay States, but this appears to be based on a misidentified *T. melanopogon* (Medway, 1978, ex

Table 61. The species of *Taphozous* (*Taphozous*).

	perforatus	*melanopogon*	*longimanus*	*theobaldi*
Range in region	Pakistan, NW India	India - Timor	India - Flores	India - Borneo
Forearm	<65	<70	<65	>70
Well developed gular pouch in male	0	0	+	0
Inferior ante-orbital process well developed	0	+	0	+
Anterior upper margin of zygoma curved (C) or flat (F)	F	F	C	F
Zygoma thin post-orbitally, lacking dorsal postorbital projection	+	0	0	0
Basisphenoid pits narrow, angular anteriorly	0	0	0	+

Hill, pers. comm.) in the BMNH received from the Selangor Museum.

Variation. A poorly represented subspecies *T. t. secatus* Thomas, 1915n: 60 has been described from Asirgarh, Nimar, Madhya Pradesh, India.

Subgenus *Liponycteris*

Savanna and semi-arid regions from W Africa to Tanzania, S Asia east to Burma; 2 species, one in region.

Taphozous nudiventris Naked-rumped tomb bat

T. nudiventris Cretzschmar, 1830-1831: 70; Egypt.
T. kachhensis Dobson, 1872c: 221; Kutch, India.
Also many extralimital synonyms.

Range (Map 41). Burma, India, Pakistan; also west to Israel and W Africa. Chasen (1940a) listed this species (as *T. kachhensis nudaster*) from the Malay States but this record appears to be based on a misidentified *T. melanopogon* in the BMNH (Medway, 1969; 1978).

Variation. One subspecific name has been proposed:
T. kachhensis nudaster Thomas, 1915n: 59; Pagan, near Mt Popa, Burma.

Subgenus *Saccolaimus*

Africa; India to Philippines and Australia; 4 species, one in this region.

Taphozous saccolaimus Pouch-bearing bat

T. saccolaimus Temminck, 1838: 14; Java. Lectotype designated by Tate (1941f).
T. crassus Blyth, 1844a: 491; Mirzapore, Allahabad, Uttar Pradesh, India.
T. pulcher Blyth, 1844a: 492; Madras, India.
Saccolaimus affinis Dobson, 1875a: 232; Labuan I, Borneo.
T. pluto Miller, 1910: 396; Mercedes, near Zamboanga, Mindanao I, Philippines.
Taphonycteris capito Hollister, 1913a: 308; Pandan, Catanduanes Is, Philippines.
Saccolaimus flavimaculatus Sody, 1931b: 355; Kutei, Kalimantan, Borneo.

Also a subspecies and synonym in New Guinea and Australia.

Range (Map 41). Sri Lanka, India, Great Nicobar I, ? Burma, S Thailand, Malaya; Sumatra, Java, Borneo; Philippines: Luzon, Catanduanes, Mindoro, Mindanao; Sulawesi (Feiler, 1980; Hill & Rozendaal, 1989); Talaud Is (Musser, 1987a); Timor; also to Solomon Is and NE Australia.

Map **41**. *Taphozous* (part), *Craseonycteris*

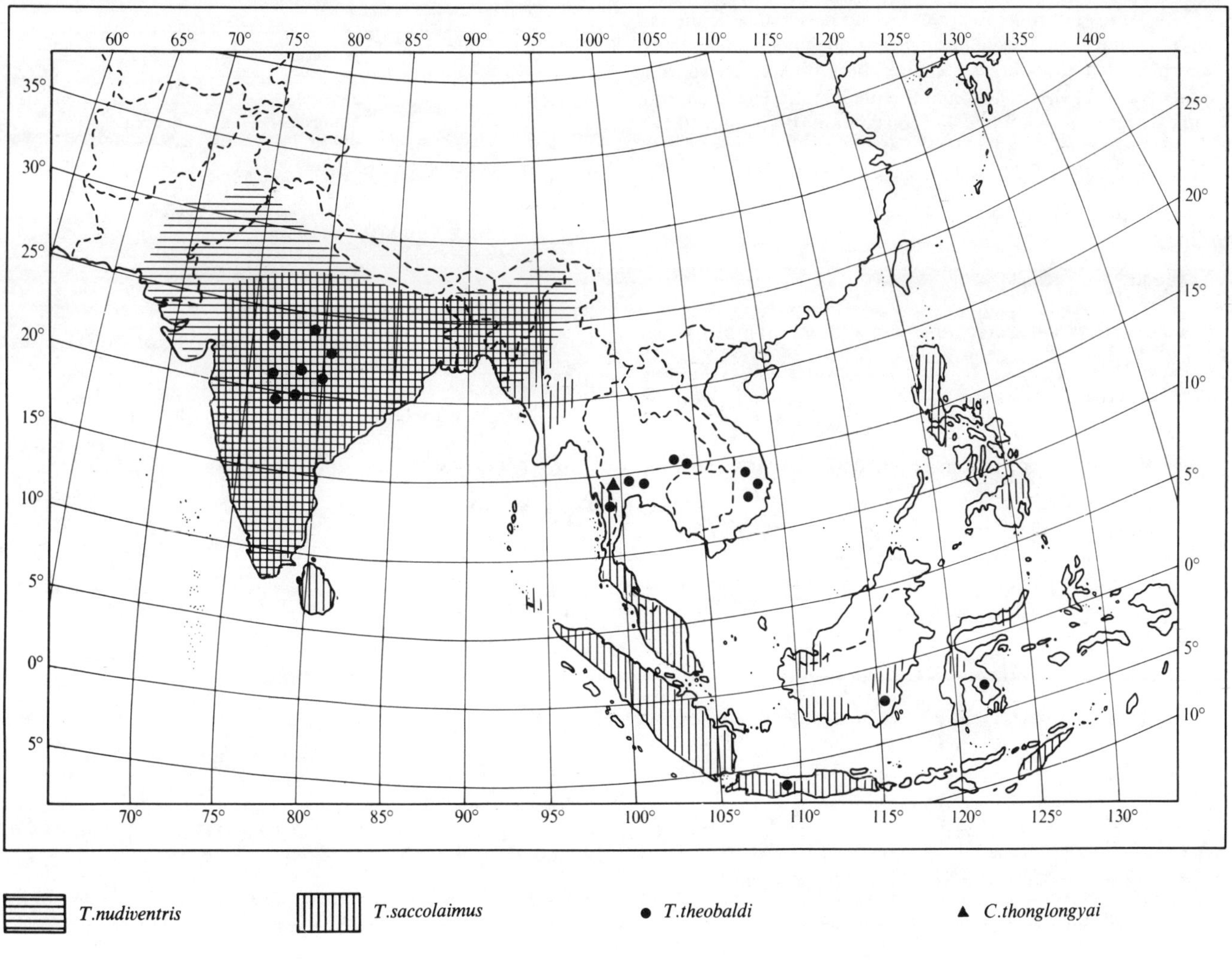

Variation. Subspecies are very arbitrary. Goodwin (1979) pointed out that it seemed that the species displayed as much individual variation in structure and colour in a particular island population as over a wide geographical range, and that until more complete samples became available, it seemed undesirable to attempt the recognition of subspecies.

Family Craseonycteridae

Hog-nosed bat; Bumblebee bat

Craseonycteridae Hill, 1974a: 303; type genus *Craseonycteris* Hill.
For formation of family name see Hill & Smith (1981).

A family of one genus and species so far limited to a small area of limestone caves in W Thailand. No fossils are known. The genus can be recognized by its small size, the sole species being the smallest of known bats (forearm 22.5-25.8mm, condylobasal length of skull 9.5-10.3mm) and amongst the smallest of known mammals; its rather suiform muzzle; large separate ears with tragus, which has a curious oblate thickening about halfway along its anterior part; the lack of any external tail; a moderate uropatagium lacking any calcarial support; and cranially by its free premaxillae, their palatal branches small and totally co-ossified, their narial branches exceptionally developed uniquely to fuse above the narial aperture, forming a subtubular flange around the narial opening.

This family has some features tending towards the Rhinopomatidae and the Emballonuridae but does not appear to be intermediate between these families, but rather to derive from a dichotomy of the rhinopomatid-emballonurid stock, sharing some features of each family but considerably derived in others.

CRASEONYCTERIS

Craseonycteris Hill, 1974a: 304; type species *C. thonglonglyai* Hill.

Thailand; one species.

Craseonycteris thonglongyai Hog-nosed bat, Bumblebee bat

C. thonglongyai Hill, 1974a: 305; cave near Forestry Station, Ban Sai Yoke (= Yok), Kanchanaburi, Thailand, 14°26'N, 98°51'E.

Range (Map 41). Known from a limited area within or near the Sai Yok National Park, W Thailand.

Remarks. Reviewed in detail by Hill & Smith (1981). A further account, with illustrations of the head and skull and a description of the habitat, is provided by Nabhitabhata *et al.* (1982). IUCN: R.

Family Nycteridae

Slit-faced bats, Hollow-faced bats, Hispid bats

Nycteridae Dobson, 1875b: 348; type genus *Nycteris* Cuvier & Geoffroy.

A well-marked family of one genus and 14 species, of which 2 species in the Indomalayan Region, distributed through much of Africa, in SW Asia, and in SE Asia from S Burma and Thailand to Borneo, Java and the Kangean Is.

Nycterids range from small to moderate in size (forearm 32-66mm); they are characterized by complex cutaneous facial outgrowths bordering each side of a deep median longitudinal groove; by long, large ears, joined anteriorly over the forehead by a low band of integument, with a small but well-developed tragus; and by the tail which is enclosed in the uropatagium and has uniquely a T-shaped cartilaginous tip embedded in the membrane edge. They feed on a variety of arthropods ranging from insects to scorpions, on occasion small vertebrates, and are sometimes colonial, roosting in caves, under rock outcrops, in man-made structures, hollow trees or vegetation, or even in the abandoned burrows of other animals. The family is generally considered to be most closely related to the Megadermatidae.

NYCTERIS

Nycteris Cuvier & Geoffroy, 1795: 186; type species *Vespertilio hispidus* Schreber, 1774; *nomen nudum*, validated by Opinion 111, 1929). Hall (1981) disregarded this and used *Nycteris* Borkhausen, 1797 for the Nearctic genus usually known as *Lasiurus* Gray, 1831.
Petalia Gray, 1838a: 494 (*N. javanica* Geoffroy).
= *Pelatia* Gray, 1866c: 83.

Africa, SW Asia, SE Asia to Java; 14 species of which 2 occur in the Indomalayan Region (Table 62).

Table 62. The species of *Nycteris*.

	javanica	*tragata*
Range	Java, Bali, Kangean Is	S Burma - Sumatra, Borneo
Forearm	43–50	47–55
Condylocanine length	Usually <19.0	Usually >19.2
C-M^3	Usually <8.0	Usually >8.0
Upper incisors	Generally tricuspid	Generally bicuspid

Nycteris javanica Javan slit-faced bat

N. javanica Geoffroy, E.,1813a: 20; Java.

Range (Map 42). Java; Nusa Penida I (Tate, 1941a); Kangean I. A possibly erroneous record (Dobson, 1878) from Timor, repeated by Tate (1944) who also listed it from Sulawesi, as did Carter *et al.* (1944), apparently wrongly.

Variation. Two subspecies:
N. j. javanica. Java.
N. j. bastiani Bergmans & van Bree, 1986: 332; Gua Ainkuan, Kangean I, Kangean Is; smaller, with more or less contrasting rufous on sides of head, neck and back.

Remarks. The following species may prove to be conspecific with *N. javanica*.

Nycteris tragata

Petalia tragata Andersen, 1912c: 546; Bidi Caves, Sarawak, Borneo.

Range (Map 42). S Burma, Thailand, Malaya, Sumatra and Borneo.

Remarks. The status of *tragata* needs appraisal. It is often thought to be a subspecies of *N. javanica*, after Chasen (1940a),

Map 42. *Nycteris, Megaderma*

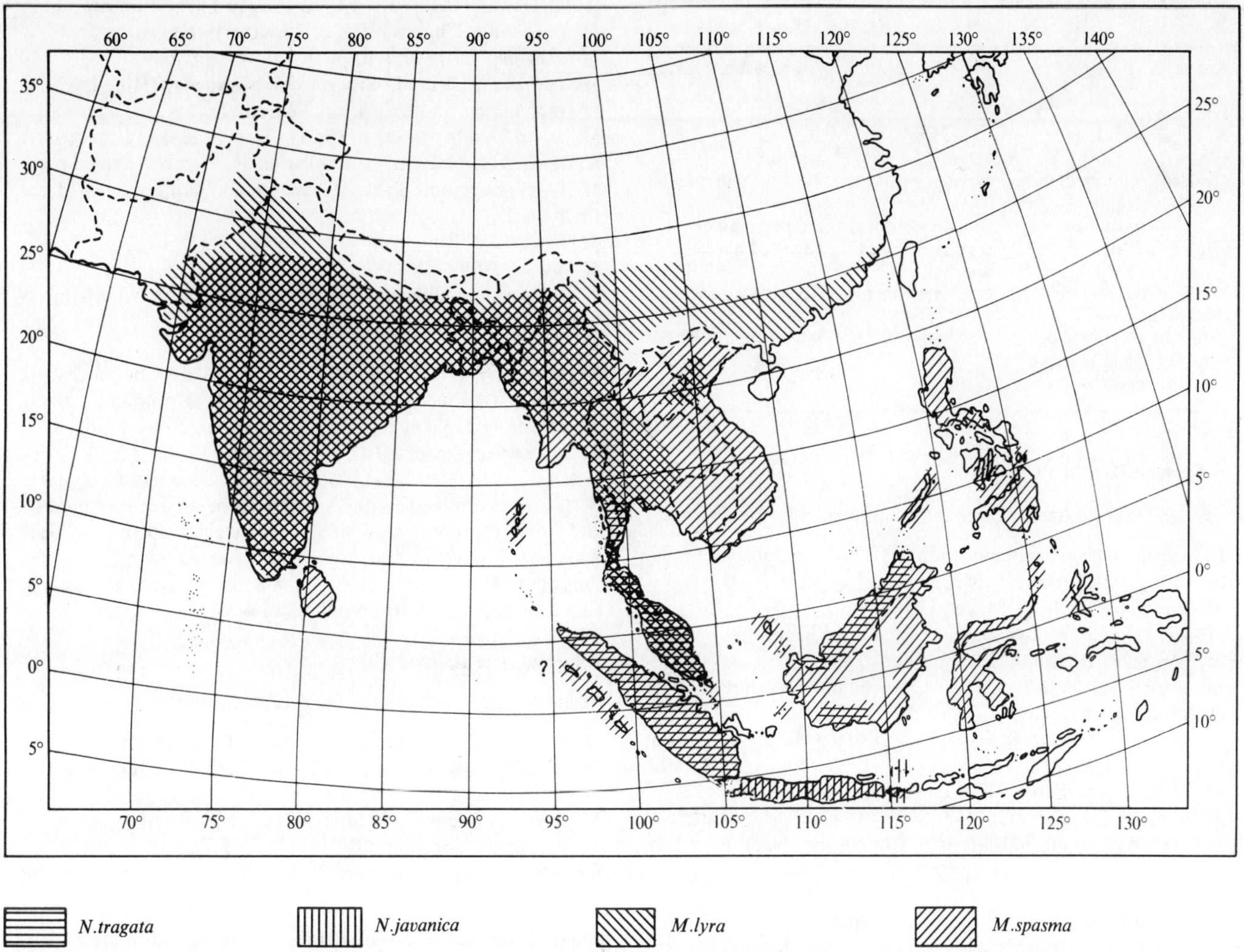

but Koopman (1975; in Honacki *et al.*, 1982) considered both distinct species. Ellerman & Morrison-Scott (1951) listed it in *N. javanica* but later (1955) remarked that *tragata*, with bicuspid upper incisors, could not be considered a race of *N. javanica* in which they averred the upper incisors are tricuspid. However Lekagul & McNeely (1977) stated that the incisors are variable in both. These authors treated *tragata* as a subspecies of *javanica*, with longer rostrum and palate, larger teeth, and slightly lighter pelage. Bergmans & van Bree (1986) reviewed published measurements of both forms and suggested that the status of *tragata* be re-examined. Certainly the differences between the two forms are small.

Family Megadermatidae

False vampire bats

Megadermatidae Allen, 1864:iii, 1; type genus *Megaderma* Geoffroy. See Handley (1980) for correct formation of family name.

A family of 4 genera and 5 species in C, E Africa, SE Asia and Australia, with one genus and 2 species, in 2 subgenera, in the Indomalayan Region. Moderate to large in size (forearm 50-115mm) and characterized by a long, erect noseleaf and by large ears united at the base by a high band of integument over the forehead, with bifurcated tragus.

They roost in caves, crevices, buildings, hollow trees, or in the branches of trees and bushes, singly or in small groups, occasionally in larger seasonal colonies. Insectivorous, taking a variety of arthropods; some distinctly carnivorous, capturing a variety of small vertebrates, including other bats.

MEGADERMA False vampire bats

Megaderma Geoffroy, 1810b: 190; type species *Vespertilio spasma* Linnaeus.
Eucheira Hodgson, 1847a: 891 (*M. schistacea* Hodgson, in *M. lyra* Geoffroy). Pre-occupied by *Eucheira* Westwood, 1838 (Insecta).
Spasma Gray, 1866c: 83 (*Vespertilio spasma* L.).
Lyroderma Peters, 1872a: 195 (*M. lyra* Geoffroy).

E Afghanistan to Philippines and Moluccas; 2 species (Table 63). The differences in the noseleaf and in the frontal region of the skull are considerable and *Lyroderma* has been recognized as a valid genus by some authors.

Table 63. The species of *Megaderma*.

	spasma	*lyra*
Range	India-Philippines, Moluccas	Afghanistan-Malaya
Forearm	52–63	60–71
Skull: greatest length	22.8–28.5	26.0–32.0
C-M^3	8.7–11.4	10.5–12.1
Noseleaf: intermediate leaf	Broad, chordate	Narrow, sub-circular
posterior leaf	Broadly elliptical, slightly rounded above	Subrectangular, flattened above
Narial emargination of skull	Relatively narrow	Relatively wide
Width of frontal expansion cf external width across toothrows	<	=

Subgenus *Megaderma*

Megaderma spasma Lesser false vampire bat

Vespertilio spasma Linnaeus, 1758: 32 (based on Seba, 1734: 90, pl. 56, fig. 1); Ternate I, Moluccas.
M. trifolium Geoffroy, 1810b: 193; Java.
M. philippinensis Waterhouse, 1843b: 69; Philippines.
M. horsfieldii Blyth, 1863a: 23; India.
M. carimatae Miller, 1906a: 63; Tanjong Karimata Tua, Karimata Is, Borneo.
M. natunae Andersen & Wroughton, 1907: 131,133; Bunguran I, N Natuna Is.
M. lasiae Lyon, 1916a: 439; Lasia I, off W Sumatra.
M. siumatis Lyon, 1916a: 439; Siumatis I, off W Sumatra.
M. niasense Lyon, 1916a: 440; Mojeia R, Nias I, off W Sumatra.

Range (Map 42). India, Sri Lanka, Burma, Thailand, Cambodia, Laos, Vietnam, Malaya, Sumatra, Java, Borneo, Philippines (Luzon, Negros, Cebu, Dinagat, Mindanao, Palawan), Sulawesi to N Moluccas. Including the smaller islands of S Andaman; Con Son (Vietnam); Tarutau (S Thailand); Aor (Malaya); Riau Is; Nias, Siumatis, Siberut, Lasia, Babi, ? Simalur (Sumatra); Rakata (Tidemann *et al.*, 1990), Panaitan (Krakatau Is); Kangean, Sepanjang (Java); S Natuna Is; Tambelan Is; Bunguran (N Natuna Is); Karimata (Borneo); Lombok; Togian I; Taliabu (Sula Is); Sangihe Is; Ternate (N Moluccas).

Variation. Sinha (1977) reviewed subspecific variation and recognized only the nominate subspecies from most of the range and *M. s. kinabalu* from N Borneo. However he examined no specimens from the many small islands from which subspecies have been named, only examples from the mainland, Sri Lanka, Sumatra, Java, Sulawesi and the Philippines, and Bergmans & van Bree (1986) did not think this conclusion justified. For the present the subspecies accepted by most authors are listed below.

M. s. horsfieldii. Peninsular India.
M. s. ceylonense Andersen, 1918: 384; Trincomalee, Sri Lanka.
M. s. majus Andersen, 1918: 383; Kin, lower Chindwin, Burma. Burma (part).
M. s. minus Andersen, 1918: 383; Cambodia. Thailand, Indochina.
M. s. medium Andersen, 1918: 383, 384; Singapore I. S Burma (Tenasserim), Tarutau I, Malaya, Riau Is, N Sumatra.
M. s. abditum Chasen, 1940a: 35; Aor I, Johore Is.
M. s. lasiae. Lasia and Babi Is, off W Sumatra.
M. s. niasense. Nias I; Siberut (Mentawai Is) (Hill, 1983).
M. s. siumatis. Siumatis I, off W Sumatra.
M. s. trifolium. Sumatra (part), Krakatau Is, Java (part), Kangean I, S Natuna Is, Tambelan Is, Borneo (part).
M. s. pangandarana Sody, 1936a: 46; Pangandaran, SC Java. Java (part).
M. s. natunae. Bunguran I, N Natuna Is.
M. s. carimatae. Karimata I.
M. s. kinabalu Chasen, 1940a: 35; Kiau, Mt Kinabalu, N Borneo, 3000ft.
M. s. philippinensis. Philippine Is.
M. s. celebensis Shamel, 1940b: 352; Likoepang, Sulawesi. Sulawesi, Togian I, Sangihe Is (Hill & Rozendaal, 1889), Sepanjang (Kangean Is) (Bergmans & van Bree, 1986).
M. s. spasma. Ternate I (N Moluccas).

A specimen from S Andaman I (Hill, 1967) has a very large skull when compared with those from the adjacent mainland, and may represent a distinct subspecies. Examples from Kangean I have in the past been referred to *trifolium* but Bergmans & van Bree (1986) found others from the nearby island of Sepanjang to be nearer to *celebensis.*

Subgenus *Lyroderma*

Megaderma lyra Greater false vampire bat

M. lyra Geoffroy, 1810b: 190; India (? east coast, Madras).
Vespertilio (*M.*) *carnatica* Elliot, 1839: 96; Dharwar, S Mahratta, India.
M. spectrum Wagner, 1844b: 569, pl. : Kashmir.
M. schistacea Hodgson, 1847a: 889; NE Bengal, India.
Eucheira sinensis Andersen & Wroughton, 1907: 135, 136; Amoy, Fukien, China.

Range (Map 42). E Afghanistan, Pakistan, India, Sri Lanka, Burma, S China (Sichuan, SW Hunan, Guangdong, Fujian), Vietnam (Topál, pers. comm.), Thailand, Malaya.

Variation. The following subspecific name has been proposed: *Eucheira l. caurina* Andersen & Wroughton, 1907: 135, 136; Surat Dist., India.

Ellerman & Morrison-Scott (1951) recognized two subspecies in India but Brosset (1962) considered that *caurina* could not be maintained as distinct. Two subspecies are generally accepted:
M. l. lyra. Afghanistan to Burma.
M. l. sinensis. China, Thailand, Malaya; slightly larger.

Family Rhinolophidae

Horseshoe bats

Rhinolophidae Bell, 1836: 599; type genus *Rhinolophus* Lacépède.

A family of one modern genus with c. 60 species throughout the tropics, subtropics and temperate zones of the Old World from Europe and Africa to Japan, New Guinea, the Bismarck Archipelago and E Australia. Fossils occur from the Eocene of Europe, the Miocene and Pliocene of Africa, and the Miocene of Australia. Rhinolophids range from small to moderate in size (forearm c. 30-75mm) and are characterized by a complex noseleaf consisting of a usually triangular or subtriangular posterior vertical cellular lancet, a generally wide anterior leaf (the horseshoe) that may on occasion almost completely cover the muzzle, and by a flattened median strap-like transverse sella surmounting the leaf and extending from a point above and behind the nostrils to the lower part of the lancet to which in the majority it is joined by a median longitudinal connecting process (Fig. 11, p. 82). The ears are moderate to large, and lack a tragus, and the tail is completely enclosed in the uropatagium.

Rhinolophids are insectivorous, catching their prey in flight or by gleaning, and occur in a wide variety of habitats, from forest to more open areas. Roosting in caves, buildings, hollow trees and foliage, some are solitary and others gregarious, roosting in moderate groups. They often hang freely in the roost, with the wings wrapped round the body.

RHINOLOPHUS Horseshoe bats

Rhinolophus Lacépède, 1799: 15; type species *Vespertilio ferrumequinum* Schreber. Conserved by Opinion 91 (1926) and Direction 24 (1955).
= *Rhinocrepis* Gervais, 1836: 617.
Phyllorhina Leach, 1816: 5 (*Vespertilio minutus* Montagu, in *V. hipposideros* Bechstein). See Hill (1963b).
Aquias Gray, 1847a: 15 (*R. luctus* Temminck; *R. trifoliatus* Temminck).
Phyllotis Gray, 1866c: 81 (*R. philippinensis* Waterhouse). Preoccupied by *Phyllotis* Waterhouse, 1837 (Rodentia).
Coelophyllus Peters, 1867e: 427 (*R. coelophyllus* Peters).
Euryalus Matschie, 1901b: 225 (*R. mehelyi* Matschie).
Rhinophyllotis Troughton, 1941: 342 (*R. megaphyllus* Gray).
Rhinomegalophus Bourret, 1951: 607 (*Rhinomegalophus paradoxolophus* Bourret).

Rhinolophus lepidus

Old World tropics, subtropics and temperate regions from Africa east to New Guinea and E Australia; c. 60 species in 6 groups, all groups and c. 42 species in the region (Table 64).

This extensive genus stands in need of modern revision. The arrangement adopted here is based largely on Andersen (1905d), as it has been modified by later authors. None since Andersen has examined the genus in its entirety.

Rhinolophus trifoliatus group (formerly the *philippinensis* group)

This group was examined and discussed by Andersen (1905a), Tate & Archbold (1939b) and Tate (1943). Andersen included *philippinensis, sedulus, trifoliatus, luctus, mitratus* and the African *maclaudi*, later (1905d) adding further taxa since reduced chiefly to subspecies; subsequently (1918) he renamed it the *luctus* group instead of the *philippinensis* group. Tate (1943) added *macrotis, coelophyllus* and *rex*, but *macrotis* and its allies, with *pearsonii*, were recognized as a distinct group by Andersen (1905b, 1905d). However, such a group does not appear to be a natural entity, as Tate (1943) recognized when he moved *pearsonii* into association with *luctus*, again a grouping that appears wrong. The aberrant noseleaf and short palate of *coelophyllus* also suggest that it has no affinity with the *trifoliatus* group. The group is represented in the Afrotropical Region by *R. maclaudi* of W and C Africa; otherwise it extends from Pakistan to NE Australia, and northward to S China and the Philippines.

1. Lateral margins of internarial expansion passing beneath base of sella to join base of connecting process **2**

Lateral margins of internarial expansion integral with margins of sella, not extending posteriorly beneath it **4**

2. Posterior leaf more or less triangular, rounded above; internarial expansion trapezoid with projecting lateral 'wings'; base of sella abruptly widened, = in width to internarial cup; smaller: forearm 44-47 .... ***R. marshalli***

Posterior leaf low, rounded; internarial expansion subcircular; sella not widened basally, its basal width less than width of internarial cup; larger: forearm >54 .... **3**

3. Larger: forearm 59-63 .... ***R. rex***

Smaller: forearm 54-55 .... ***R. paradoxolophus***

4. Internarial septum expanded into a cup-like structure, subcircular; sella lacking lateral basal lappets, not narrowed above **5**

Internarial septum usually linear, its lateral margins more or less parallel, sella with lateral projecting lappets at its base, narrowed above, more wedge-like... **7**

5. Connecting process high, its height exceeding that of the sella... ***R. macrotis***

Connecting process low, its height below that of the sella **6**

6. Three mental grooves in lower lip .... ***R. philippinensis***

[One mental groove in lower lip .... ***R. mitratus***]

7. Smaller: forearm 39-44; weak anterior sagittal crest.......... .... ***R. sedulus***

Larger: forearm 46.5-78; well developed anterior sagittal crest **8**

8. Smaller: forearm 46.5-54, C-M^3 8.0-9.4; upper incisors minute, widely separated .... ***R. trifoliatus***

Larger: forearm 54-78, C-M^3 9.5-12.3; upper incisors moderate, their tips convergent .... ***R. luctus***

Table 64. The species-groups of *Rhinolophus*.

	philippinensis group	*arcuatus* group	*fumigatus* group	*pusillus* group	*ferrumequinum* group	*hipposideros* group
Range in region	Pakistan - Moluccas	Burma - Moluccas	NE India - China	Widespread	Widespread	Kashmir
Ears very large	+	+	+	0	+/0	0
Antitragal lobe very large	+	+	+	0	0	0
Connecting process triangular, pointed	0	0	0	+	0	0
Sella	Large	Large	Moderate	Small	Small	Small
Lateral projections at base of sella	+/0	0	0	0	0	0
Internarial region expanded	+	+	0	0	0	0
Palate long, 1/3 or more length of maxillary toothrow	+	0	0	0	0	0
Cochleae very large	0	0	0	0	0	+
Basioccipital very narrow	0	0	0	0	0	+

Rhinolophus philippinensis Philippine horseshoe bat

R. philippinensis Waterhouse, 1843b: 68; Luzon, Philippines.
R. achilles Thomas, 1900c: 145; Kai Is.
R. maros Tate & Archbold, 1939b: 1, 9; Talassa, near Maros, S Sulawesi, 300m.

Range (Map 43). Philippine islands of Luzon, Mindoro, Negros, Mindanao; Borneo; Sulawesi; Timor I; Kai Is; also to NE Queensland, Australia.

Variation. The following subspecific names have been proposed:
R. p. alleni Lawrence, 1939: 46; lower slopes of Mt Halcon, Mindoro I, Philippines.
R. p. sanborni Chasen, 1940a: 39; Tapadong Caves, near Lahad Data, E coast of Sabah, Borneo.
R. p. montanus Goodwin, 1979: 112; Quoto Lou caves, near village of Lequi Mia, 5 miles (18km) S of Ermera, Timor, c. 1220m.

Subspecific variation was discussed by Goodwin (1979).

Rhinolophus sedulus Lesser woolly horseshoe bat

R. sedulus Andersen, 1905a: 244, 247; Sarawak.

Range (Map 43). Malaya; Borneo.

Rhinolophus trifoliatus Trefoil horseshoe bat

R. trifoliatus Temminck, 1834: 24, pl. 1 (and 1835: 27, pl. 31); Bantam, W Java.
R. solitarius Andersen, 1905a: 244, 250; Tanjong Pamuja, Bangka I.
? *R. edax* Andersen, 1918: 378; Singapore I.

Range (Map 43). NE India, S Burma, S Thailand, Malaya; Sumatra, Java, Borneo; plus small islands of Pintu Gedang (E Malaya), Bangka, Nias (Sumatra).

Variation. One subspecific name has been proposed:
R. t. niasensis Andersen, 1906b: 658; Nias I.

Remarks. The original description of *edax* is very brief and relates chiefly to its larger size when compared with *sedulus*. Chasen (1940a) listed it as a subspecies of *R. sedulus* and the name has been used since for the Malayan form of this species. However, the holotype of *edax* is large (forearm 50.0, C-M^3 8.0) and its skull has the characteristic narrow, deeply incised postnarial depression of *R. trifoliatus*, with a similar high frontal crest, longer infraorbital canal and more massive teeth. The transfer of *edax* to *R. trifoliatus* was anticipated by Tate (1943) who considered them conspecific.

Rhinolophus luctus Woolly horseshoe bat

R. luctus Temminck, 1834: 23, pl. 1 (and 1835: 24, pl. 30); Tapos, Java.
R. morio Gray, 1842: 257; Singapore.
R. perniger Hodgson, 1843a: 414; Nepal.
R. lanosus Andersen, 1905a: 248; Kuatun, NW Fujian, China.
R. beddomei Andersen, 1905a: 253; Wynaad, Madras, India.
R. geminus Andersen, 1905a: 244, 253; Kediri, E Java, 2000-3000ft.
R. formosae Sanborn, 1939: 41; Taiwan.

Range (Map 44). Sri Lanka; India, Nepal, Burma, SE China (Fujian), Vietnam (Topál, pers. comm.), Thailand, Malaya; Sumatra, Java, Borneo; Bali (Senckenberg Museum, examined); Hainan, Taiwan.

Variation. The following subspecific names have been proposed:
R. b[eddomei] sobrinus Andersen, 1918: 378; Kala Oya, North Central Prov., Sri Lanka.
R. m[orio] foetidus Andersen, 1918: 378, 379; Baram, Sarawak, Borneo.
R. lanosus spurcus Allen, 1928a: 3; Nodoa, Hainan I, China.

Rhinolophus rex

R. rex Allen, 1923a: 3; Wahnsien, Sichuan, China.

Range (Map 43). S China (Sichuan, Guizhou).

Remarks. Redescribed by Hill (1972b). This species has an unusually low, rounded, convex posterior leaf, unlike the triangular posterior leaf of most other *Rhinolophus*, excepting only the following, *R. paradoxolophus*.

Rhinolophus paradoxolophus

Rhinomegaphyllus paradoxolophus Bourret, 1951: 607; Rochepercée Cave, near Chapa, Lao Key Prov., Tonkin, Vietnam, 1700m.

Range (Map 43). NE Thailand; N Vietnam.

Map 43. *Rhinolophus* (part)

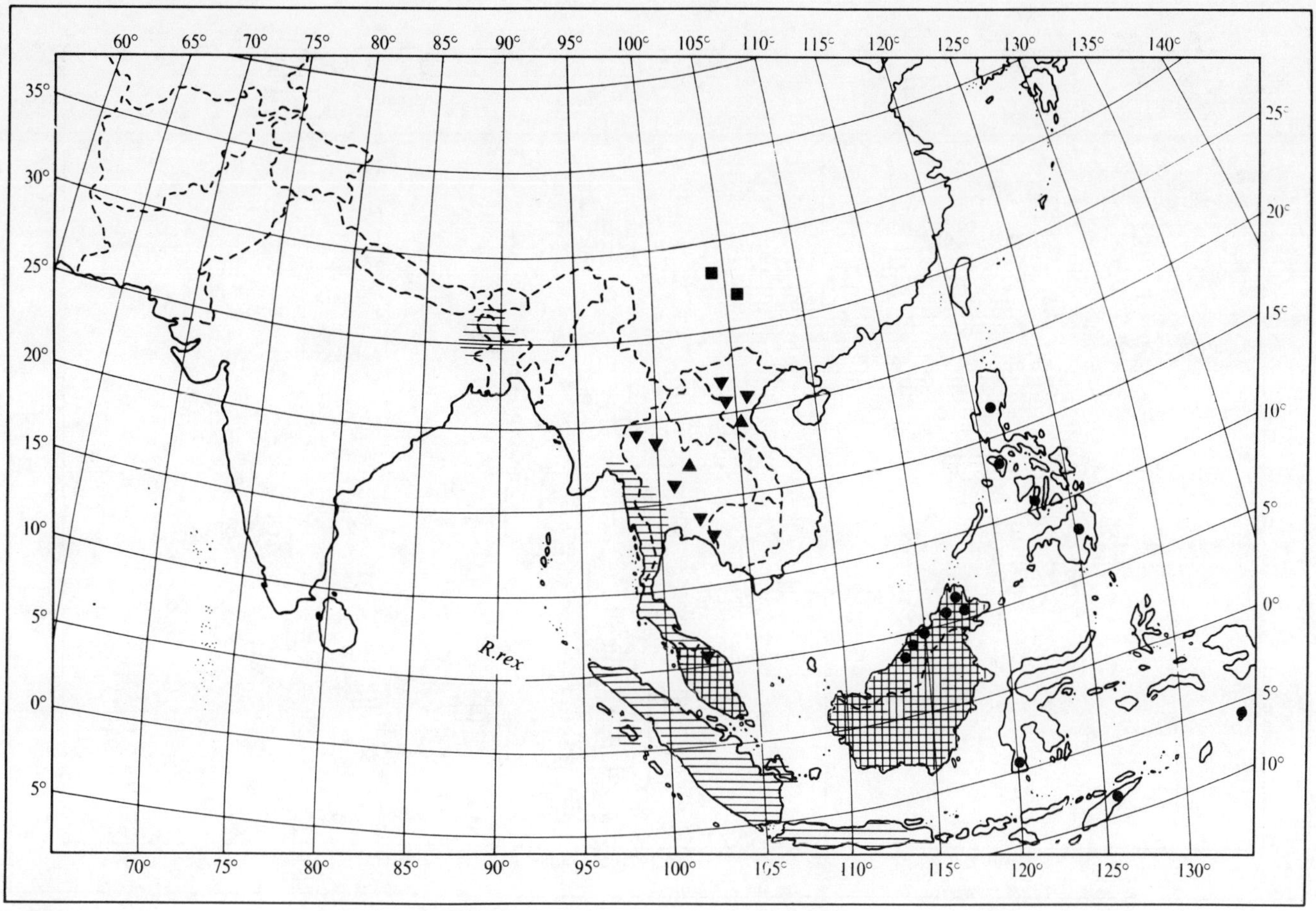

R.trifoliatus (horizontal hatching) R.sedulus (vertical hatching) ● R.philippinensis ▲ R.paradoxolophus

▼ R.marshalli ■

Remarks. Very similar to but smaller than *R. rex*, with which it may be conspecific. Dorst (1954) provided an account of some cranial features and the dentition of the holotype, which was redescribed by Hill (1972b). A further specimen was recorded from Thailand, with comments, by Thonglongya (1973a).

Rhinolophus marshalli Marshall's horseshoe bat

R. marshalli Thonglongya, 1973a: 590; Khao Soi Duo, Amphoe Pong Nam Ron, Chanthaburi, SE Thailand.

Range (Map 43). Thailand; Vietnam (Hill & Topál, 1990); N Malaya (Zubaid & Davison, 1988). Further Thai specimens were described and measured by Yoshiyuki (1990).

Remarks. The ears, noseleaves and skulls of *R. rex, R. paradoxolophus* and *R. marshalli* were illustrated by Thonglongya (1973a).

Rhinolophus macrotis Big-eared horseshoe bat

R. macrotis Blyth, 1844a: 485; Nepal.
R. hirsutus Andersen, 1905b: 289; Guimaras I, Philippines.
R. episcopus Allen, 1923a: 2; Wanhsien, Sichuan, China.

Range (Map 44). NW Pakistan (Harrison Zoological Museum, specimen examined), Nepal, Thailand, S China (Sichuan, SW Guangxi, Fujian), Vietnam, Malaya; Sumatra; Philippines: Guimaras, Negros (Heaney, pers. comm.). Widespread but uncommon in the Philippines according to Heaney *et al.* (1987)

Variation. The following subspecific names have been proposed:
R. m. dohrni Andersen, 1907b: 6, 29; Soekaranda, Deli, NW Sumatra.
R. m. siamensis Gyldenstolpe, 1917: 12; Doi Par Sakang, NW Thailand.
R. episcopus caldwelli Allen, 1923a: 3; Yuki, Fujian, China.

According to Osgood (1932) *caldwelli* and *siamensis* are apparently sympatric in N Vietnam. However, the specimens that he referred to *siamensis* are intermediate in size between *caldwelli* and very much smaller specimens from Thailand that evidently represent the former subspecies.

[*Rhinolophus mitratus*]

R. mitratus Blyth, 1844a: 483; Chaibassa, Orissa, India.

The status of *mitratus* is uncertain; it is known apparently only from the holotype in the collection of the Zoological Survey of India, Calcutta. Sinha (1973) provided an account of this specimen, with measurements.

Map **44**. *Rhinolophus* (part)

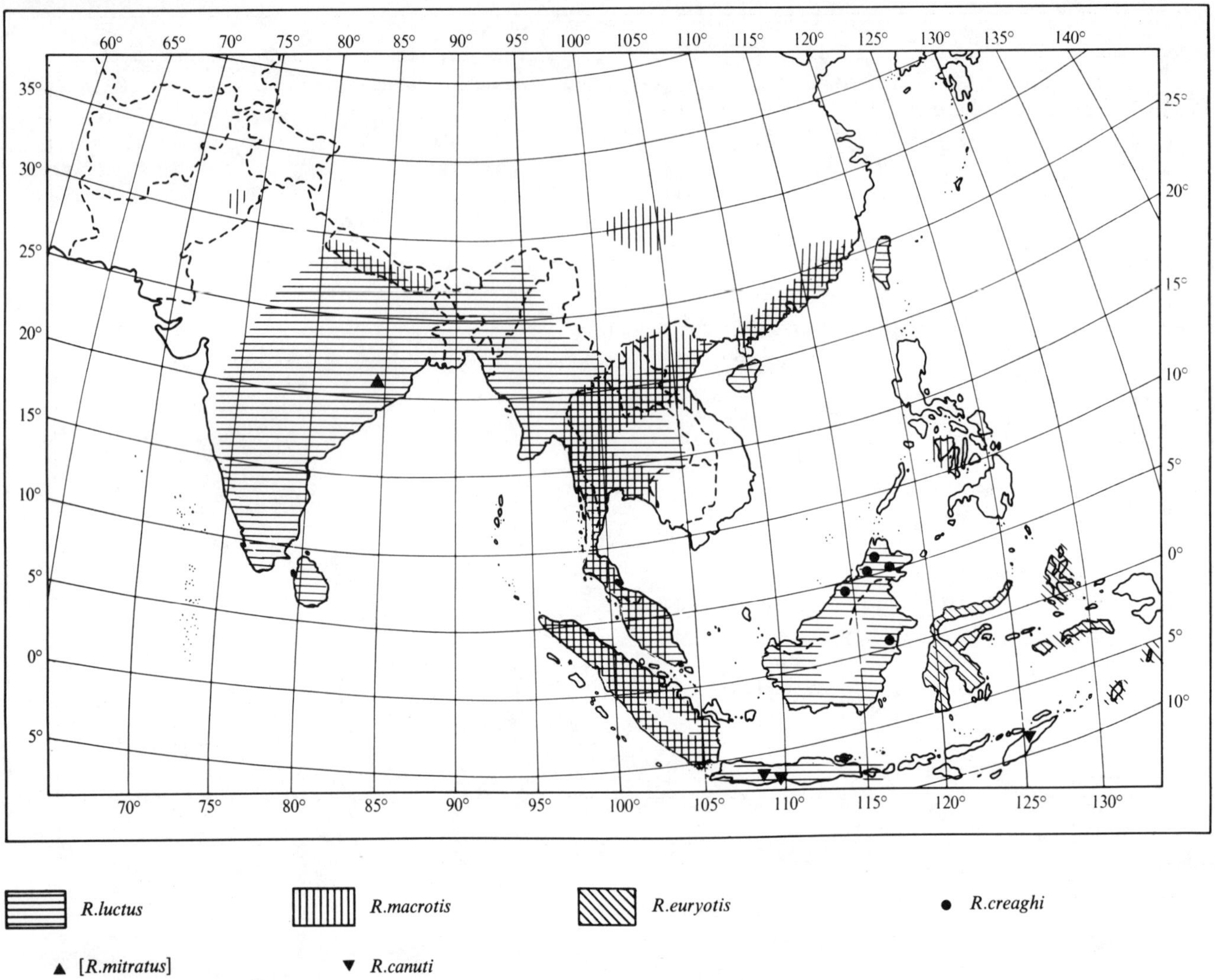

Tate & Archbold (1939b) listed it with *philippinensis*; Tate (1943) considered that the original description suggested an affinity with *trifoliatus* rather than with *luctus*, but Sinha (1973) thought *mitratus* to be closer to *R. pearsoni* in body size and noseleaf shape. However, features noted and illustrated by Dobson (1876a) contradict this view and support the view that it should be grouped with *philippinensis* and its allies.

Rhinolophus arcuatus group

The status and taxonomy of this group has changed little since it was first established by Andersen (1905e, d), apart from the addition of more recently described taxa or of others transferred into it. The group currently contains 9 species and is confined entirely to Australasia. It was briefly reviewed by Hill & Schlitter (1982).

1. Anterior leaf scarcely emarginated medially, but with two broad, parallel swollen ridges extending medially from edge of leaf to internarial region, enclosing a groove that widens posteriorly to terminate at a low median projection .... ***R. euryotis***

Anterior leaf narrowly emarginated medially, the emargination prolonged posteriorly as a narrow, linear groove on face of leaf, extending less than halfway to internarial region **2**

2. Connecting process low or obsolete; lower part of posterior leaf densely pilose **3**

Connecting process evident; lower part of posterior leaf at most sparsely haired **4**

3. Upper part of sella thickened posteriorly; connecting process obsolete; hairs at base of posterior leaf forming a dense, bristly sub-conical tuft .... ***R. creaghi***

Upper part of sella lacking any posterior thickening; connecting process low, rounded; hairs at base of posterior leaf long, dispersed .... ***R. canuti***

4. Posterior leaf thickened and folded to form a vertical fissure enclosing rear of connecting process **5**

Rear of connecting process attached directly to face of posterior leaf, not enclosed **6**

5. Postnarial rostral depression prominent, moderately deep, enclosed by broad, well-developed supra-orbital ridges..***R. coelophyllus***

Postnarial rostral depression shallow, little developed, supra-orbital ridges narrow .... ***R. shameli***

6. Upper part of sella forming a small, triangular pouch, its opening directed downwards................................... ***R. inops***

Upper part of sella unmodified **7**

7. Large: forearm 66-71 .... ***R. rufus***

Small to moderate: forearm <60.................................... **8**

8. Larger: forearm 53-57, C-M^3 9.5-10.4; internarial cup narrow; a prominent frontal depression............. ***R. subrufus***

Smaller: forearm 42-51, C-M^3 7.5-8.6; internarial cup expanded; shallow frontal depression .... ***R. arcuatus***

Rhinolophus euryotis Broad-eared horseshoe bat

R. euryotis Temminck, 1834: 24 (and 1835: 26, pl. 29, 32); Ambon I, Moluccas.
R. tatar Bergmans & Rozendaal, 1982: 169, 170; Moinakom R, Dumoga Nature Reserve, N Sulawesi, 0°41'N, 124°3'E, 525m.

Range (Map 44). Sulawesi; Togian Is (Owen *et al.*, 1987; Hill, 1991); Moluccas: Halmahera (Hill & Rozendaal, 1989; Peterson *et al.*, 1990), Bacan, Buru, Seram, Ambon, Timor Laut; Kai Is; also to New Guinea, Bismarck Is and the Aru Is (*R. e. aruensis* Andersen, 1907b).

Variation. The following subspecific names have been proposed:
R. e. timidus Andersen, 1905e: 285; Bacan I, Moluccas.
R. e. praestans Andersen, 1905e: 286; Kai Is.
R. e. burius Hinton, 1925: 256; Lehsoela, Buru I, Moluccas.
Subspecies were reviewed briefly by Hill (1983). Specimens from Sulawesi are generally smaller than those from the more eastern parts of the range.

Rhinolophus creaghi Creagh's horseshoe bat

R. creaghi Thomas, 1896a: 244; Sandakan, Sabah, Borneo.
R. pilosus Andersen, 1918: 379; Marengan, Soemenep, E Madura I, off NE Java.

Range (Map 44). Borneo; Madura I.

Variation. The subspecies *R. c. pilosus* from Madura I has a slightly larger anterior upper premolar (P^2) than *R. c. creaghi* from Borneo, and the second lower premolar (P_3) is not wholly extruded from the toothrow; in the nominate subspecies P_3 is minute, extruded, or sometimes lacking.

Rhinolophus canuti

R. canuti Thomas & Wroughton, 1909c: 18 (and 1909d: 378); Kalipoet-jang, Tji-Tangoei R, S Java.
R. creaghi timorensis Goodwin, 1979: 105; Lia Hoo Cave, near Fatu Maca Village, 7 miles (11km) S of Baucau, Timor I, c. 550m.

Range (Map 44). Java; Timor I.

Variation. The subspecies *R. c. timorensis* differs from *R. c. canuti* in more developed connecting process, a sparser covering of hairs at the base of the posterior leaf and a relatively narrower skull.

Remarks. Hill (1958b) considered *canuti* to be a subspecies of *R. creaghi*; subsequently Hill & Schlitter (1982) keyed and listed it as a distinct species, allying *timorensis* to it as a subspecies. The shape of the sella, the degree of development of the connecting process, and the diffuse pilosity of the lower part of the posterior leaf, the hairs not forming a distinct tuft as in *R. creaghi* and *pilosus*, suggest that specific rank is more appropriate.

Rhinolophus arcuatus

R. arcuatus Peters, 1871a: 305; Luzon I, Philippines.
R. toxopei Hinton, 1925: 256; Buru I, Moluccas, 1400m.

Range (Map 45). Sumatra, Borneo (Sarawak); Philippines: recorded from Luzon, Mindoro, Guimaras, Negros, Camiguin (N of Mindanao), Mindanao, probably widespread (Heaney *et al.*, 1987); Sulawesi (Hill, 1988); Lesser Sunda Is: Wetar; Moluccas: Buru; also to New Guinea.

Variation. The following subspecific names have been proposed:
R. a. exiguus Andersen, 1905e: 283; Zamboanga, Mindanao, Philippines.
R. a. beccarii Andersen, 1907c: 477; Bua Cave, Upper Padang, Sumatra.
R. a. angustifolius Sanborn, 1939: 43; Wetter (= Wetar) I, Moluccas.
R. a. proconsulis Hill, 1959b: 88; Bungoh Cave, Bau, 1st Division, Sarawak, Borneo.
A single example from Sulawesi is similar to *R. a. proconsulis* from Borneo (Hill, 1988); sufficient representation may show the Sulawesian population to be subspecifically distinct.

Rhinolophus coelophyllus Croslet horseshoe bat

R. coelophyllus Peters, 1867e: 426, pl. 35; Salween R, Burma.

Range (Map 45). Burma, Thailand, W Malaya; islands of Koh Lak (S Thailand), Langkawi (NW Malaya).

Variation. Two examples reported by Gyldenstolpe (1917) from Koh Lak are cranially smaller than mainland specimens and may represent an undescribed subspecies (Tate, 1943; Hill & Thonglongya, 1972). Yoshiyuki (1990) described and measured an extensive series from Thailand.

Rhinolophus shameli Shamel's horseshoe bat

R. coelophyllus shameli Tate, 1943: 3; Koh Chang, off SE Thailand.

Range (Map 45). Burma, Thailand (including Koh Chang I), Cambodia.

Remarks. For a discussion of *R. shameli* see Hill & Thonglongya (1972) who considered it a distinct species.

Rhinolophus inops

R. inops Andersen, 1905e: 284; Jodaya, Mount Apo, Davao, Mindanao, Philippines, 4000ft.

Range (Map 45). Philippines: Mindanao, Biliran, Leyte, Negros (L. Heaney, pers. comm.).

Map **45**. *Rhinolophus* (part)

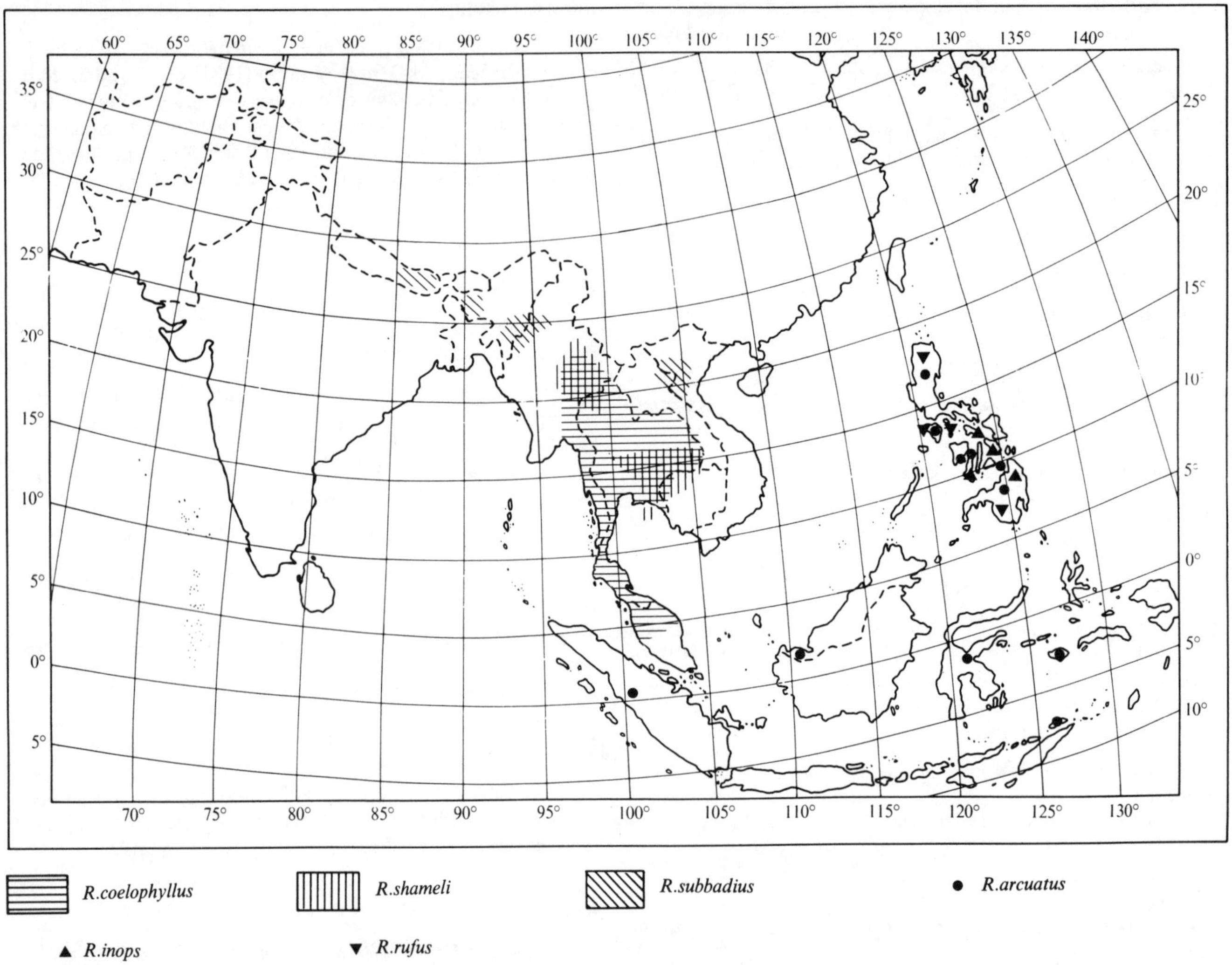

Rhinolophus rufus

R. luctus var. *rufus* Eydoux & Gervais, 1836a: 9; Manila, Luzon, Philippines.
? *Aquias Eudoxii* Fitzinger, 1870: 194; Manila, Luzon, Philippines.

Range (Map 45). Philippines: Luzon, Marinduque, Mindoro, Mindanao.

Remarks. Redescribed by Lawrence (1939).

Rhinolophus subrufus

R. rufus Peters, 1861: 710; Paracali, Luzon, Philippines. Preoccupied by *R. rufus* Eydoux & Gervais, 1836.
R. subrufus Andersen, 1905e: 283; Manila, Luzon, Philippines.
R. bunkeri Taylor, 1934: 214; Saub, Cotabato, Mindanao, Philippines.

Range (Map 46). Philippines: Luzon, Mindoro, Tablas, Negros, Mindanao, Camiguin (N of Mindanao).

Variation. The subspecies *R. s. bunkeri* from S Mindanao differs in slightly longer forearm, more massive skull lacking any postnarial depression, and larger molariform teeth. It was discussed by Lawrence (1939).

Rhinolophus fumigatus group

Andersen (1905b, d) placed *pearsonii* (including *yunanensis*) in a *macrotis* group (here subsumed into the *philippinensis* group), followed by Tate & Archbold (1939); later, Tate (1943) associated *pearsonii* and its probable synonym *chinensis* with *R. luctus*, but with some hesitation since to do so involved accepting two sympatric subspecies (*R. l. chinensis, R. l. lanosus*) of widely different forearm length in NW Fujian, China. However, although *pearsonii* and *yunanensis* have the large ears with well developed antitragal lobe characteristic of the *philippinensis* group, differences in noseleaf and skull suggest that these alliances may be wrong. Both differ from *R. macrotis* (and *R. philippinensis*) in relatively linear, only slightly expanded internarial septum, and in broadened zygomata, their total width exceeding and not less than the mastoid width. In this last respect they resemble *R. trifoliatus* or *R. luctus* but these species differ in the presence of well developed lateral lappets at the base of the sella. The combination of large ears and broad noseleaf, relatively simple internarial region and sella, and relatively broadened zygomata suggests that *pearsonii* and *yunanensis* may represent the otherwise Afrotropical *fumigatus* group, which they appear to link to the *philippinensis* group. As in the African species allocated to the *fumigatus* group they differ from each other chiefly only by size: they may

Map 46. *Rhinolophus* (part)

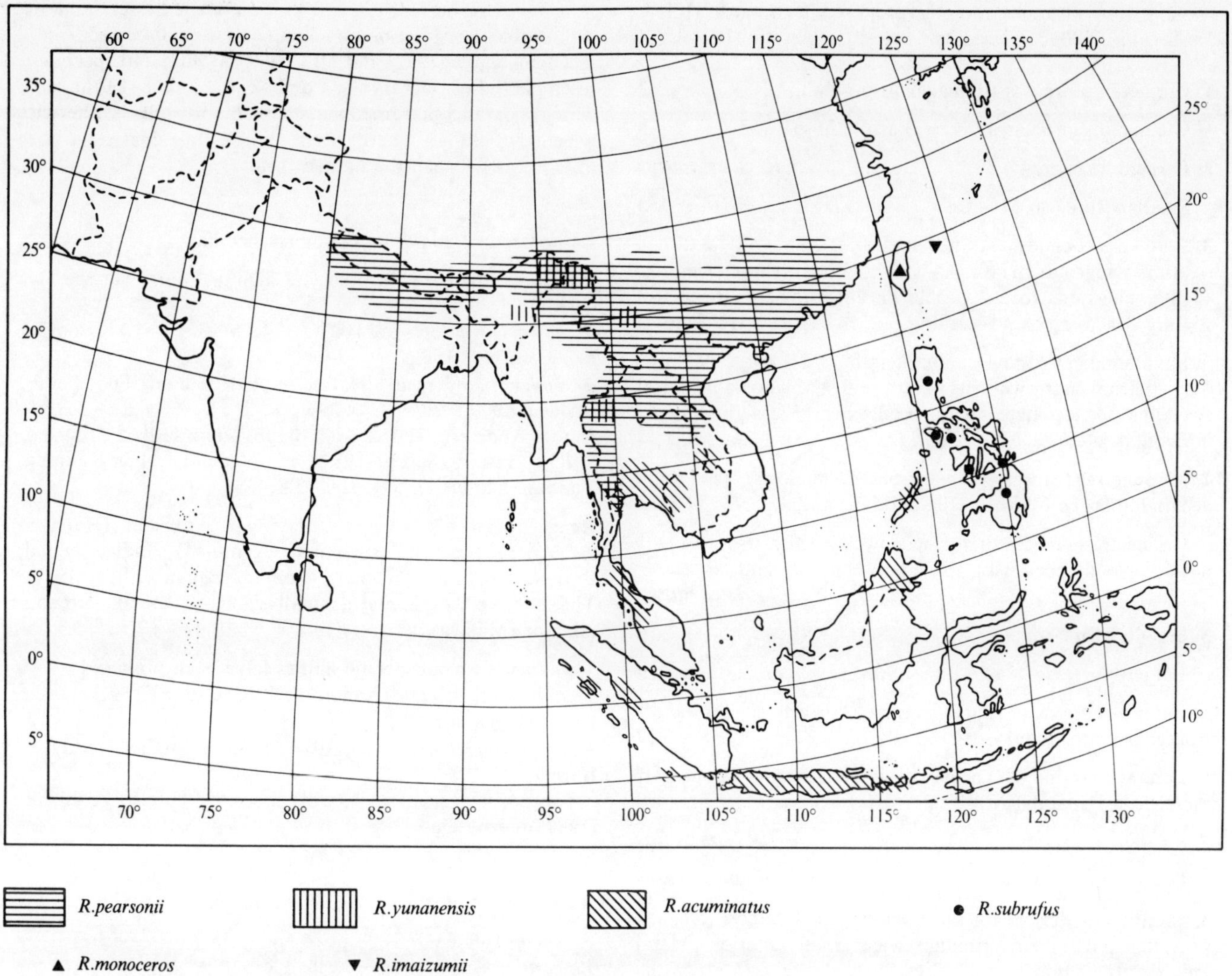

be distinguished from the African forms by the lack of pilosity of the sella, connecting process and lancet, and by a relatively large anterior upper premolar.

Hill (1986) reviewed *pearsonii* and *yunanensis* in detail, recognizing the latter as a distinct species, a view supported by Yoshiyuki (1990). See Table 65.

Table 65. The species of the *Rhinolophus fumigatus* group.

	pearsonii	*yunanensis*
Forearm	48–57	55–60
Condylocanine length	19.9–21.2	22.4–24.3
Zygomatic width	11.0–12.1	12.2–13.9
C-M^3	8.3–10.0	10.1–11.1
Length right mandibular ramus from condyle	15.0–17.0	17.7–19.3

Rhinolophus pearsonii Pearson's horseshoe bat

R. pearsonii Horsfield, 1851: 33; Darjeeling, West Bengal, NE India.

R. larvatus Milne-Edwards, 1872b: 248, pl. 37a; Moupin, Sichuan, China. Pre-occupied by *R. larvatus* Horsfield, 1823.

Range (Map 46). NE India, Nepal, Burma, China (S Tibet, Yunnan, Sichuan, Guangxi, Hunan, Guangdong, Fujian, Guangzhou, Anhui), Vietnam, Thailand; Malaya (Honacki *et al.*, 1982). Has been recorded up to 3300m.

Variation. One subspecies has been described but its validity seems doubtful:

R. p. chinensis Andersen, 1905b: 289; Kuatun, Fujian, SE China.

Rhinolophus yunanensis

R. yunanensis Dobson, 1872a: 336; Hotha, Yunnan, China.

Range (Map 46). NE India (Meghalaya), N Burma, SE China (Yunnan), Thailand.

Rhinolophus pusillus group

Primarily Oriental with 10 species in the region, extending westwards through S Europe and N Africa, with three further purely Afrotropical species, and eastwards to Japan and the Lesser Sunda Islands. Andersen (1905d, f;1918) provided the first detailed classification of the group. No complete treatment has since appeared: Tate & Archbold (1939b) summarized the classification of more recently proposed taxa, and Hill & Yoshiyuki (1980) reviewed and keyed the Asiatic species.

1. Connecting process triangular in profile 2

Connecting process an erect, narrow horn 5

2. Larger: forearm 46–53 .... ***R. acuminatus***

Smaller: forearm 31–42 3

3. Skull larger: condylocanine length 14.2–16.4, C-M^3 6.0–7.1; rostrum wider, its width 4.4–5.0; median rostral swellings slightly elevated, rostral profile curving upwards near tip, slightly concave behind swellings .... ***R. lepidus***

Skull smaller: condylocanine length 13.2–14.6, C-M^3 5.4–6.2; rostrum narrower, its width 3.8–4.5; median rostral swellings not elevated, rostral profile nearly straight, almost horizontal 4

4. Connecting process broadly based, its lower part fully occupying space between sella and lancet............. ***R. pusillus***

Connecting process slightly narrowed basally, its lower part not completely occupying space between sella and lancet...... .... ***R. cornutus***

5. Smaller: forearm <43 6

Larger: forearm >44 9

6. Lateral margins of lancet straight or nearly so, the lancet triangular with rounded tip... 7

Lateral margins of lancet slightly concave, its upper part more or less spatulate 8

7. Smaller: forearm 31.5–34.5, C-M^3 5.1–5.4 .... ***R. subbadius***

Larger: forearm 34–38.5, C-M^3 5.4–5.7 ...... ***R. monoceros***

8. Median anterior rostral swelling moderately inflated; posterior palatal emargination wider than anterior palatal emargination; upper canines relatively slender.. ***R. cognatus***

Median anterior rostral swellings low, only slightly inflated; posterior palatal emargination narrower than anterior; upper canines relatively massive .... ***R. imaizumii***

9. Sella narrowing abruptly, cuneate; connecting process pointing dorsally; 1st phalanx of 4th finger more than half as long as 2nd; anterior and posterior lower premolars subequal in crown area .... ***R. blasii***

Lateral margins of sella parallel; connecting process pointing forward; 1st phalanx of 4th finger less than half as long as 2nd; crown areas of anterior and posterior premolars very unequal .... ***R. mehelyi***

Rhinolophus acuminatus Acuminate horseshoe bat

R. acuminatus Peters, 1871a: 308; Gadok, Java.
R. sumatranus Andersen, 1905f: 133; Lower Langkat, Sumatra.
R. calypso Andersen, 1905f: 134, pl. 4; Kifa Juc, Bua Bua, Enggano I, off W Sumatra.
R. circe Andersen, 1906b: 657; Nias I, off W Sumatra.

Range (Map 46). Laos, Cambodia, S Thailand, Malaya; Sumatra (including Nias and Enggano Is), Java, Borneo, Palawan; Lesser Sunda Is: Bali, Lombok.

Variation. One subspecific name has been proposed:
R. a. audax Andersen, 1905f: 133; Lombok I, Lesser Sunda Is.

Subspecific distinctions are slight and were discussed by Hill & Thonglongya (1972) and Hill (1974b). Mainland specimens resemble those from Java or Lombok in the shape of the sella, which is parallel-margined at its base, but the sella is otherwise widened basally. There seem to be no consistent size differences between the named forms.

Rhinolophus lepidus Blyth's horseshoe bat

R. lepidus Blyth, 1844a: 486; ? Calcutta, India, but see Das (1986a).
R. monticola Andersen, 1905f: 124; Masuri (= Mussoorie), Kumaon, NW India.
R. refulgens Andersen, 1905f: 124, pl. 4; Gunong Igar, Perak, Malaya, 2000ft.
R. feae Andersen, 1907c: 474; Biapo, Karin Hills, SE Burma.
? *R. osgoodi* Sanborn, 1939: 40; Nguluku, N of Likiang, Yunnan, China, 27°5'N, 100°15'E.

Range (Map 47). Afghanistan, NW Pakistan (Harrison Zoological Museum, specimen examined), India, Nepal, Burma; S China: Sichuan, Yunnan, Fujian (Zhen, 1987); Thailand, Malaya; Sumatra; small islands of Tioman, Pemanggil, Aor (Malaya).

Variation. Two subspecific names have been proposed:
R. r[efulgens] cuneatus Andersen, 1918: 376, 377; Sukaranda, Deli, NE Sumatra.
R. l. shortridgei Andersen, 1918: 376; Pagan, Irrawaddy R, Burma.

Subspecific variation was discussed by Hill & Yoshiyuki (1980) and by Das (1986a).

Rhinolophus pusillus Least horseshoe bat

R. minor Horsfield, 1823:pt. 6, pl. ; Java. Pre-occupied by *Vespertilio ferrumequinum minor* Kerr, 1792.
R. pusillus Temminck, 1834: 29, pl. 1; Java.
R. minutus Miller, 1900a: 235; Siantan I, Anamba Is. Preoccupied by *Vespertilio minutus* Montagu, 1808.
= *R. minutillus* Miller, 1906f: 41.
R. gracilis Andersen, 1905f: 129, pl 4; Malabar coast, India.
R. blythi Andersen, 1918: 376; Almora, Kumaon, India, 5500ft.
R. pagi Tate & Archbold, 1939b: 7; N Pagi I, Mentawai Is.

Range (Map 47). India, Nepal, S China (Tibet, Sichuan, Guangxi, Anhui, Fujian, Hainan I, Hong Kong), Vietnam, Thailand, Malaya; Sumatra (Agronoff *et al.*, 1984; BMNH), Java, Borneo; and the small islands of Langkawi (in American Museum NH, Koopman, pers. comm.), Penang, Tioman, Aor (Malaya); N Pagai (Mentawai Is); Siantan (Anamba Is); Rakata (Krakatau Is); Madura (Java); Lombok (Lesser Sunda Is).

Variation. The following subspecific names have been proposed:
R. blythi szechuanus Andersen, 1918: 376,377; Chunking, Sichuan, China.
R. b. calidus Allen, 1923a: 1; Yenping, Fujian, SE China.
R. b. parcus Allen, 1928a: 2; Nodoa, Hainan I, China.
R. p. lakkhanae Yoshiyuki, 1990: 23; Tham Song Phi Nong, Chom Thong, Chiang Mai, Thailand, 18°35'N, 98°29'E.

Eight subspecies were listed by Hill & Yoshiyuki (1980).

Map **47**. *Rhinolophus* (part)

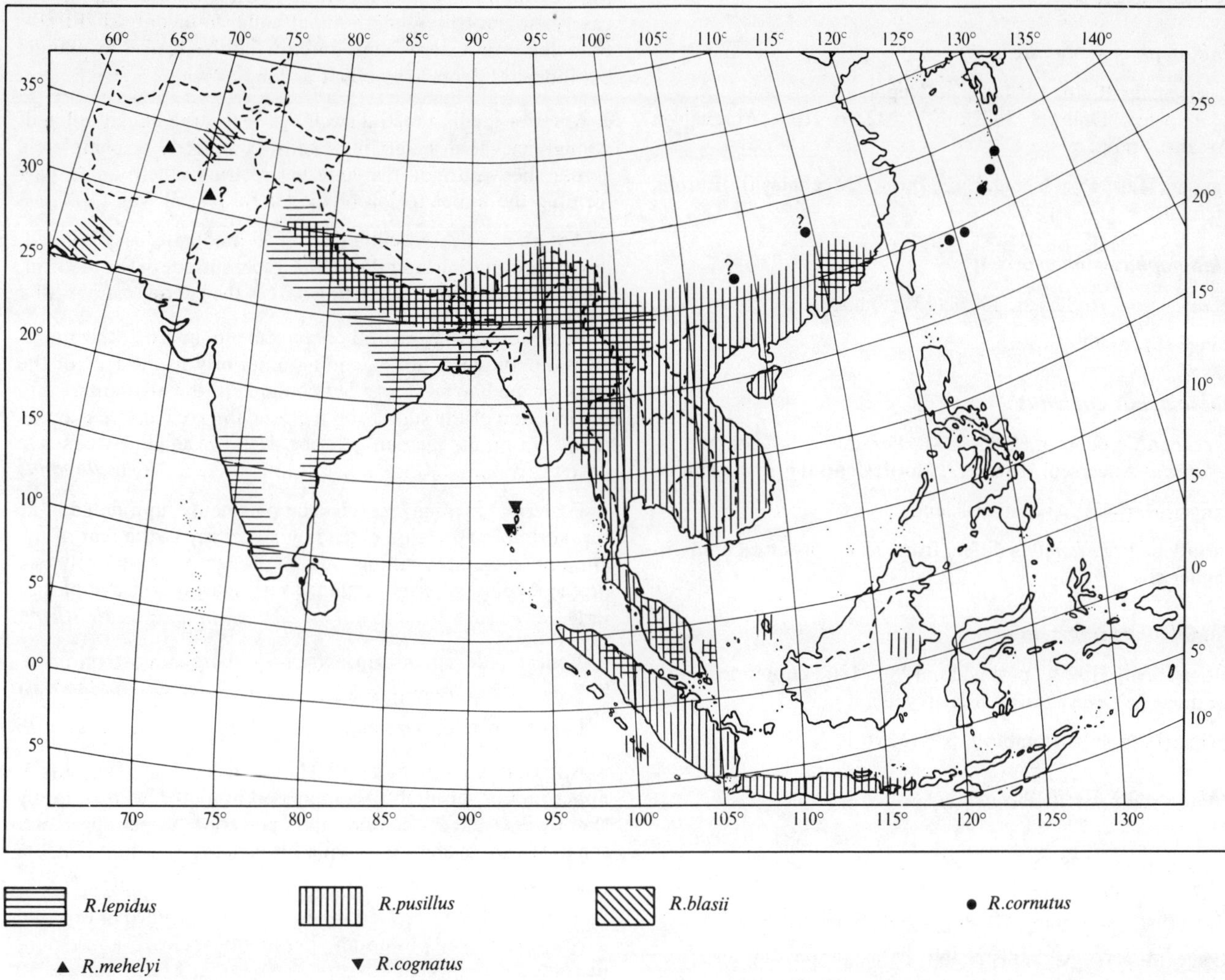

Remarks. Sinha (1973) considered *gracilis* a subspecies of *R. cornutus*, near to *blythi*, which he associated with this species. Examination of the holotype confirms that it is very similar to *blythi* and *szechuanus*.

Rhinolophus cornutus Little Japanese horseshoe bat

R. cornutus Temminck, 1834: 30 (and 1835: 37); Japan.
R. perditus Andersen, 1918: 376; Ishigaki I, S Ryukyu Is.
R. miyakonis Kuroda, 1924: 5; Nishisato, Miyakojima I, Ryukyu Is.

Range (Map 47). Ryukyu Is: Amami-oshima, Kakeroma, Tokunoshima, Okineorabu, Okinawa, Miyakojima, Ishigaki (Yaeyama Is); also to the main Japanese islands and possibly to China.

Variation. Two subspecific names have been proposed:
R. c. orii Kuroda, 1924: 4; San-Mura, Tokunishima, N Ryukyu Is, 300ft.
R. c. pumilus Andersen, 1905f: 127; Okinawa I, Ryukyu Is.

Hill & Yoshiyuki (1980) recognized five subspecies (one, *R. c. cornutus*, extralimital to the region). However, Yoshiyuki (1989) and Yoshiyuki *et al.* (1989) considered *pumilus* specifically distinct from *R. cornutus* on account of its darker membranes, wider zygomata and larger, more massive upper canines, with *miyakonis* as its subspecies. Yoshiyuki (1989) also thought *perditus* to be specifically distinct, differing similarly from *R. cornutus* and from *pumilus* in shorter tibia, generally longer skull, longer maxillary toothrow, and in the unextruded condition of the anterior upper premolar (P^2) and second lower premolar (P_3). No comparisons were made with specimens from the Asian mainland (see Remarks). Yoshiyuki *et al.* (1989) provided a further account of *R. c. orii.*

Remarks. This species may extend to E China. It is very probable that *pusillus* and *cornutus* are conspecific and they have been treated as such in the past (e. g. by Ellerman & Morrison-Scott, 1951). However, Mell (1922) mentioned a specimen of *pumilus* from the northern part of Guangdong and Wang *et al.* (1962) reported *R. c. pumilus* and *R. blythi* (= *R. pusillus*) from Guangxi. Specimens of *R. c. pumilus* reported by Thomas (1911b, 1912b) (some identified by Andersen) and consequently referred to *pumilus* by Allen (1938b), were in fact later referred to *R. pusillus szechuanus* by Andersen, and are so labelled by this worker. Probably they were re-identified in the course of preparing the diagnoses published (1918) on his behalf by Thomas. Moreover, the specimen from Fuzhou (Foochow), Fujian tentatively referred to *pumilus* by Andersen (1905f) has also been later allocated to *szechuanus* in the same way. Yoshiyuki (1990) however doubted the conspecificity of *pusillus* and *cornutus*.

Rhinolophus subbadius

R. subbadius Blyth, 1844a: 486; Nepal.
R. garoensis Dobson, 1872a: 337; Garo Hills, Meghalaya (Assam), India.

Range (Map 45). Nepal, NE India (Meghalaya), Burma, Vietnam.

Rhinolophus monoceros

R. monoceros Andersen, 1905f: 131; Baksa, Taiwan.

Range (Map 46). Taiwan.

Rhinolophus cognatus

R. cognatus Andersen, 1906c: 181; Port Blair, S Andaman Is.
R. famulus Andersen, 1918: 377; North Central I, Andaman Is.

Range (Map 47). Andaman Islands.

Variation. *R. c. famulus* differs from *R. c. cognatus* chiefly in smaller size.

Rhinolophus imaizumii

R. imaizumii Hill & Yoshiyuki, 1980: 180; Otomi-do Cave, Iriomote I, Yayeyama Is, S Ryukyu Is.

Range (Map 46). Iriomote I, S Ryukyu Is.

Rhinolophus blasii Blasius's horseshoe bat

R. clivosus Blasius, 1857: 33; Italy etc. Not *R. clivosus* Cretzschmar, 1828.
= *R. blasii* Peters, 1866c: 17.
Also other extralimital synomyms and subspecies.

Range (Map 47). Afghanistan, NE Pakistan; possibly also from SW Pakistan (known from Iranian Baluchistan); also to S Europe, N, E and S Africa.

Variation. The subspecies in Afghanistan and probably in Pakistan is *R. b. meyeroemi* Felten, 1977; Pashtunkot (Grotte Zarmast, 1295m), Farian Prov., Afghanistan.

Rhinolophus mehelyi Mehely's horseshoe bat

R. mehelyi Matschie, 1901b: 225; Bucharest, Roumania.

Range (Map 47). Extends to Afghanistan and may occur in Pakistan; otherwise to Portugal, Spain and N Africa.

Rhinolophus ferrumequinum group

Formerly the *simplex* group of Andersen (1905d, f) or the *megaphyllus* group of that author (1918), the *ferrumequinum* group is the most widely distributed in the genus, extending from W Europe and Africa to E Australia. It contains 21 species, of which 13 occur in the region.

1. Facial part of skull elongate, the supraorbital crests combining at a point behind the centre of the orbital cavity; supraorbital length (from junction of crests to nares) > width across anterior lateral rostral swellings; supraorbital depression longer than wide .... ***R. megaphyllus***

Facial part of skull shorter, the supraorbital crests combining at a point above or in front of the centre of the orbital cavity; supraorbital length about equal to or only slightly > than the width across the anterior lateral rostral swellings; supraorbital depression about as long as wide **2**

2. Anterior median rostral swellings very large, much inflated, strongly elevated, extending almost completely or completely across the width of the anterior rostrum, their outer part forming the upper region of the lateral rostral wall **3**

Anterior rostral swellings small to moderate, less inflated, lower, terminating laterally on the upper surface of the rostrum, their outer margins not forming part of the lateral rostral wall **4**

3. Posterior median rostral compartments inflated, the supraorbital depression not extending anteriorly to the rear of the anterior median swellings but confined to the area anterior to the junction of the supraorbital crests, the area anterior to this raised; in profile median anterior swellings shallowly concave posteriorly .... ***R. malayanus***

Posterior median rostral compartments uninflated, the supraorbital depression extending anteriorly to the rear of the anterior median swelling, although deeper posteriorly; in profile median anterior swellings very sharply concave posteriorly .... ***R. stheno***

4. Lancet reduced, its tip extremely short, almost rudimentary .... ***R. thomasi***

Lancet normal, unreduced **5**

5. Anterior upper premolar (P^2) larger, more than twice bulk of upper incisor, not tightly compressed in toothrow, only rarely slightly extruded; canine and posterior upper premolar separated by moderate or wide interspace **6**

P^2 smaller, sometimes absent, at most twice bulk of upper incisor, tightly compressed in row, slightly extruded or external to row, canine and posterior upper premolar closely adjacent or in contact **10**

6. Larger: forearm >46; lancet hastate, abruptly narrowed at centre, lateral margins strongly concave, tip well developed, slender .... ***R. rouxi***

Smaller: forearm <47; lancet triangular, not abruptly narrowed at centre, lateral margins more or less straight **7**

7. Greatest length of skull generally <17.9, C-M^3 6.0–6.8; median anterior rostral swellings narrower, their width 3.1–3.6; occipital region not especially swollen and not projecting posteriorly **8**

Greatest length of skull generally >17.8, C-M^3 6.9–7.6; median anterior rostral swellings wider, their width >3.7; occipital region rather bulbous, projecting posteriorly **9**

8. Noseleaf moderate, median anterior rostral swellings not reduced .... ***R. celebensis***

Noseleaf small, median anterior rostral swellings reduced. .... ***R. virgo***

9. 2nd phalanx of 3rd digit less than 1.5x length of 1st phalanx .... ***R. borneensis***

2nd phalanx of 3rd digit ⩾1.5x length of 1st phalanx .... ***R. nereis***
[*R. anderseni*]*

10. Anterior upper premolar (P^2) not extruded or only very slightly displaced.. ***R. affinis***

P^2 when present external to toothrow, canine and P^4 in contact.. **11**

11. Smaller: forearm 48–52, condylobasal length 16.2–17.4, C-M^3 6.6–7.2 .... ***R. bocharicus***

Larger: forearm >53, condylobasal length >19.0, C-M^3 >7.6 .... ***R. ferrumequinum***

* *R. anderseni* and its subspecies *aequalis* have not been examined but on descriptions should key to *R. nereis* in this arrangement.

Rhinolophus megaphyllus Smaller horseshoe bat

R. megaphyllus Gray, 1834a: 52; Murrumbidgee R, E Australia.

R. truncatus Peters, 1871a: 307; Bacan I, N Moluccas.

R. keyensis Peters, 1871a: 307; Kai Is. See also Andersen, 1906: 4.

R. simplex Andersen, 1905f: 76, pl. 3; Lombok I, Lesser Sunda Is.

R. nanus Andersen, 1905f: 82, pl. 3; Goram I, SE of Seram I, Moluccas.

R. robinsoni Andersen, 1918: 375; Khao Nawng, Bandon, S Thailand.

R. klossi Andersen, 1918: 375; Pemanggil I, E of Johore, Malaya.

Range (Map 48). Thailand, Malaya (including islands of Tioman, Pemanggil, Aor); Lesser Sunda Is: Lombok, Sumbawa, Komodo; Moluccas: Bacan, Ternate, Seram, Goram (Gorong) E of Seram, Wetar; Kai Is; also to New Guinea and adjacent islands, Bismarck Archipelago and E Australia.

Variation. The following subspecific names have been proposed:

R. keyensis annectens Sanborn, 1939: 37; Wetter (= Wetar) I, Moluccas.

R. robinsoni siamensis McFarlane & Blood, 1986: 220; Big House Cave, Aungkang Region, N Thailand, 19°50'N, 99°10'E. Pre-occupied by *R. macrotis siamensis* Gyldenstolpe, 1917.

= ***Rhinolophus megaphyllus thaianus*** Hill, *nom. nov.*

As understood here, *R. megaphyllus* includes besides *megaphyllus* and its commonly accepted subspecies from New

Map **48**. *Rhinolophus* (part)

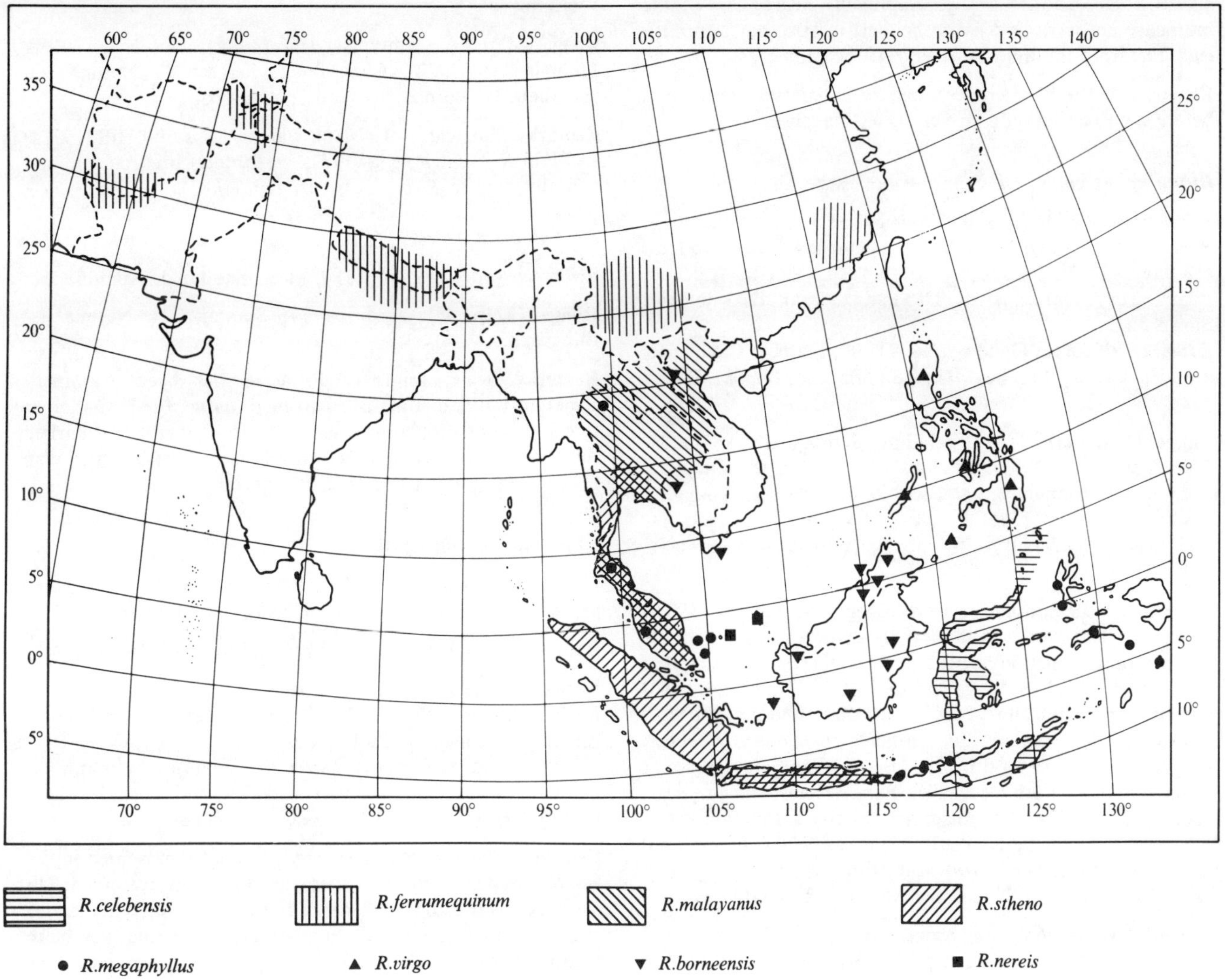

Guinea and Australia the more western *keyensis*, *simplex* and *robinsoni* which until now have been regarded as distinct species (Corbet & Hill, 1980,1986,1991; Honacki *et al.*, 1982). Although the available material is limited there seems little to support the concept of distinct species in the Moluccas (*keyensis*) or Lesser Sunda Is (*simplex*), or the similar separation of *robinsoni* (with *thaianus* and *klossi*) in Thailand and Malaya. Andersen (1905f) in an initial study of the eastern forms placed much weight on relatively small variations in the sella which scarcely seem to justify specific distinction, and Koopman (1982) suggested that those from the Moluccas and the Lesser Sunda Is might well be subspecies of *R. megaphyllus*.

The major variation lies in the width of the sella, especially basally, and in the degree of inflation of the median anterior rostral swellings. The forms from Thailand and Malaya and from New Guinea to E Australia have a wide sella and prominent, almost hemispherical narial swellings. In those from the Moluccas the sella is a little narrower, but the narial swellings remain well developed. In the Lesser Sunda Is *simplex* is the least developed in both respects with a relatively narrower sella and lower narial swellings, in these features approaching the allied *R. borneensis*, but retaining the long facial region of *R. megaphyllus*.

Three specimens from Seram I, Moluccas in the BMNH cannot be confidently allocated. In general the shape of the sella is like that of *keyensis*, and they are a little larger than *keyensis, annectens* or the geographically nearby *nanus*, the braincase not narrowed as in the last. In size they approach *truncatus* from the more northerly of the Moluccas.

Remarks. Hill (1972a) redescribed *robinsoni* and *klossi*, both before then resting only on very brief diagnoses.

Rhinolophus celebensis Sulawesi horseshoe bat

R. celebensis Andersen, 1905f: 83, pl 3; Makassar, S Sulawesi.
R. madurensis Andersen, 1918: 375; Soemenep, E Madura I, off NE Java.
R. javanicus Andersen, 1918: 375: Pangandaran, Dirk de Fries Bay, S Java.
R. borneensis parvus Goodwin, 1979: 102; Lia Hoo Cave, near Fatu Haca village, 7 miles (11km) S of Baucau, Timor I, Lesser Sunda Is.

Range (Map 48). Java (including Sertung, Krakatau Is: Tidemann *et al.*, 1990), Madura I; Kangean Is: Kangean, Sepanjang (Bergmans & van Bree, 1986); Sulawesi; Togian Is (Owen *et al.*, 1987; Hill, 1991); Sangihe I; Talaud Is: Karakelang (both Hill & Rozendaal, 1989); Lesser Sunda Is: Bali, Timor.

Variation. Four subspecies were recognized by Hill (1983). Goodwin (1979) drew attention to slight variations in the position of the junction of the supraorbital crests, the shape of the median narial swellings and the width of the interorbital constriction (postorbital constriction) that separate *celebensis, javanicus* and *parvus*, but did not include *madurensis*, which is very similar to *parvus*, differing only in slightly less globular narial inflations. Both are smaller than *celebensis* and *javanicus*, leading Hill (1983) to speculate that specific distinction (as *R. madurensis*) might be justified. Bergmans & van Bree (1986) later synonymized *parvus* with *madurensis*, which they regarded as a distinct species, but on the basis of a limited sample.

Remarks. When describing *parvus* Goodwin (1979) considered *celebensis* and *javanicus* conspecific with *R. borneensis* but remarked that this last differed more widely from these than they did from each other. Hill (1983) reviewed this grouping and recognized *R. borneensis* as a distinct species.

Rhinolophus virgo

R. virgo Andersen, 1905f: 88; Pasacao, S Camarines, Luzon, Philippines.

Range (Map 48). Philippines: Luzon, Palawan, Negros, Mindanao, Jolo (Sulu Is); widespread and common throughout the Philippines according to Heaney *et al.* (1987).

Remarks. Possibly conspecific with *R. celebensis* (Hill, 1983). A series from Palawan and Negros I measured by Sanborn (1952a) largely bridges the size differences between the very small specimens described by Andersen and *celebensis*, but *R. virgo* is retained as distinct by virtue of its apparently small noseleaf and small rostral inflations.

Rhinolophus borneensis Bornean horseshoe bat

R. borneensis Peters, 1861: 709; Labuan I, N Borneo.
R. spadix Miller, 1901a: 136; Sirhassen I, S Natuna Is.
R. importunus Chasen, 1939a: 188; Tjiawitali, near Wijnkoopsbaai (= Wynkoops Bay), S coast of W Java.
R. chaseni Sanborn, 1939: 38; Pulo Condore (= Con Son I), S Vietnam.

Range (Map 48). Cambodia, ? Vietnam (for both see Hill & Thonglongya, 1972); Con Son I; Serasan (S Natuna Is); Karimata Is; Borneo; Labuan I; Banggi I.

Remarks. The status of *chaseni* was discussed by Hill (1972a) and Hill & Thonglongya (1972): *importunus* was re-examined and allocated to *R. borneensis* by Hill (1983).

Rhinolophus nereis

R. nereis Andersen, 1905f: 90, pl. 3; Siantan I, Anamba Is.

Range (Map 48). Anamba Is: Siantan; N Natuna Is: Bunguran.

Remarks. No example of *R. nereis* has been examined: apparently it is distinguished from *R. borneensis* by its larger size and slightly different wing structure. Hill (1983) remarked that evidently it represented *R. borneensis* with which van Strien (1986) later listed it as conspecific.

Rhinolophus anderseni

R. anderseni Cabrera, 1909: 305; type locality unknown, probably Luzon I, Philippines (Taylor, 1934).

Range (Map 49). Philippines: probably Luzon; Palawan.

Variation. One subspecies has been described:
R. a. aequalis Allen, 1922: 2; Puerto Prusse (= Puerto Princesa), Palawan I, Philippines.

The metacarpals in *aequalis* are subequal in length and not graduated as in the nominate subspecies, and its skull is very slightly larger, but available measurements are few.

Remarks. This poorly known species is not represented in the BMNH and no definitive specimen has been examined. On descriptions it comes to *R. nereis* in the key given here, in size and to some extent on wing structure: in the nominate subspecies the length of the second phalanx of the third digit

Map **49**. *Rhinolophus* (part)

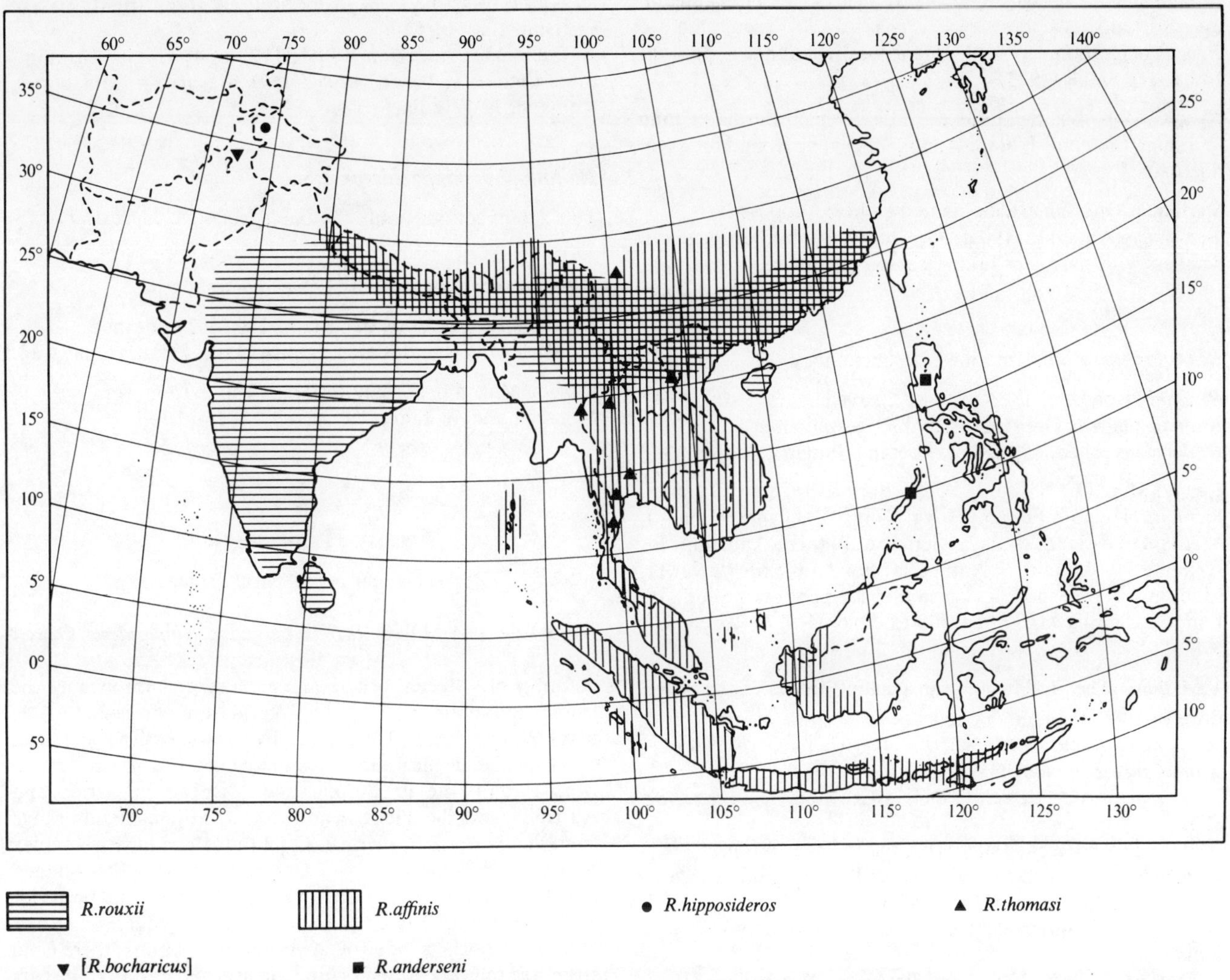

exceeds 1.5 times the length of the first phalanx, but in *aequalis* the length of the second phalanx is very slightly less.

Rhinolophus malayanus Malayan horseshoe bat

R. malayanus Bonhote, 1903a: 15; Biserat, Jalor, S Thailand.

Range (Map 48). Vietnam, Laos, Thailand, Malaya (including Langkawi I).

Rhinolophus stheno

R. stheno Andersen, 1905f: 91, pl. 3; Selangor, Malaya.

Range (Map 48). Thailand, Malaya; Sumatra; Java.

Remarks. On the basis of specimens of *malayanus* and *stheno* from Thailand, McFarlane & Blood (1986) suggested that the external features used by Lekagul & McNeely (1977) to separate these species are not diagnostically reliable. They further postulated that the two are more closely related than formerly supposed. However, the features of the cranial rostrum quite clearly establish the distinction between them.

Rhinolophus thomasi Thomas's horseshoe bat

R. thomasi Andersen, 1905f: 100, pl. 3: Karin Hills, SE Burma.

Range (Map 49). S China (Yunnan), Burma, Vietnam, Thailand.

Variation. Two subspecific names have been proposed:
R. t. septentrionalis Sanborn, 1939: 40; Nguluko, N of Likiang, Yunnan, China, 27°5'N, 100°15'E.
R. t. latifolius Sanborn, 1939: 39; Muong Moun, Tonkin, Vietnam.

According to Sanborn (1939) *septentrionalis* is larger than *thomasi*, with slightly extruded anterior upper premolars, and differs from *thomasi* and *latifolius* in a narrower, less rounded posterior edge to the palate; *latifolius* is similar in size to *thomasi*, but has a slightly longer ear and tail, and its sella is much wider. As in *thomasi* P^2 is in the toothrow.

Rhinolophus rouxii

R. rouxii Temminck, 1835: 30b; Pondicherry and Calcutta, India.
R. rubidus Kelaart, 1850b: 209; Kaduganava, Sri Lanka.
= *R. fulvidus* Blyth, 1851a: 182 (error for *rubidus*).

R. cinerascens Kelaart, 1852: 13; Fort Frederick, Sri Lanka.
R. rammanika Kelaart, 1852: 14; Anamapoora Hill, Kaduganava, Sri Lanka.
R. petersii Dobson, 1872a: 337; type locality unknown, perhaps India: see Sinha (1973).

Range (Map 49). Sri Lanka; India, Nepal, Burma, China (Yunnan, Sichuan, Fujian, Anhui, Zhejiang, Hong Kong), N Vietnam.

Variation. One subspecific name has been proposed:
R. r. sinicus Andersen, 1905f: 98; Chinteh, Anhui, China. Skull smaller, with narrower braincase and shorter toothrows.

Rhinolophus affinis Intermediate horseshoe bat

R. affinis Horsfield, 1823:pt 6, pl. ; Java.
R. andamanensis Dobson, 1872: 337; S Andaman I.
R. hainanus Allen, 1906: 482; Pouten, Hainan I, China.

Range (Map 49). N India, Nepal, Burma, S China (Yunnan, Sichuan, Hunan, Fujian, Hong Kong, Zhejiang, Hainan), Vietnam, Thailand, Malaya; Sumatra, Borneo; Andaman Is; Anamba Is: Siantan; N Natuna Is: Bunguran; Mentawai Is; Madura I; Kangean Is: Kangean, Sepanjang; Lesser Sunda Is: Lombok, Sumbawa, Sumba, Flores, Solor (Hill & Rozendaal, 1989).

Variation. The following subspecific names have been proposed:
R. a. himalayanus Andersen, 1905f: 103, pl. 3; Masuri (= Mussoorie), Kumaon, NW India.
R. a. macrurus Andersen, 1905f: 103; Taho, Karennee, SE Burma.
R. a. tener Andersen, 1905f: 103, pl. 3; Pegu, Burma.
R. a. superans Andersen, 1905f: 104; Pahang, Malaya.
R. a. nesites Andersen, 1905f: 104; Bunguran I, N Natuna Is.
R. a. princeps Andersen, 1905f: 106, pl. 3; Lombok, Lesser Sunda Is.

Andersen (1905f, 1907c) and Bergmans & van Bree (1988) discussed variation in this species. In general, Andersen found that the more southerly or southeasterly forms have longer ears, a broader horseshoe and a larger skull with broader narial swellings and longer toothrows, but specimens recorded from the Kangean Is by Bergmans & van Bree are relatively small.

[*Rhinolophus bocharicus*]

R. bocharicus Kastschenko & Akimov, 1917: 221; Murghab R, S Russian Turkestan.

Range (Map 49). Extralimital in Turkestan, NE Iran, Afghanistan, but may be found in N Pakistan.

Rhinolophus ferrumequinum Greater horseshoe bat

Vespertilio ferrumequinum Schreber, 1774:pl. 62; France.
R. nippon Temminck, 1835: 30a; Japan.
R. tragatus Hodgson, 1835a: 699; Nepal.
Also many extralimital synonyms.

Range (Map 48). Pakistan, Kashmir, N India (Kumaon), Nepal, Sikkim, China (Sichuan, Yunnan, Fujian); also to W Europe, N China and Japan.

Variation. Three subspecific names are relevant to the region:
R. f. irani Cheesman, 1921: 575; Shiraz, Iran, 5200ft. To SW Pakistan.
R. f. proximus Andersen, 1905f: 112, pl. 4; Gilgit, Kashmir.
R. f. regulus Andersen, 1905f: 112; Masuri (= Mussoorie), Kumaon, NW India.

Rhinolophus hipposideros group

This group contains only the one species, *R. hipposideros.*

Rhinolophus hipposideros Lesser horseshoe bat

Vespertilio hipposideros Bechstein, 1800: 629; France.
R. midas Andersen, 1905f: 138, pl. 4; Jask, Persian Gulf, Iran.

Range (Map 49). Kashmir (Gilgit); and from Iran west to Morocco and W Europe.

Family Hipposideridae

Old-World leaf-nosed bats, Trident bats

Hipposideridae Miller, 1907; type genus *Hipposideros* Gray.

A family of 9 Recent genera and c. 63 species throughout the tropics and subtropics of the Old World from Africa and S Asia to the Philippines, N Australia and Vanuatu, with 6 genera and c. 40 species in the Indomalayan Region. Fossils are known from the Eocene in Europe, the Miocene in Africa and Australia and the Pleistocene in Asia. Hipposiderids range from small to large (forearm c. 32–114mm), and have a complex noseleaf consisting usually of an anteriorly horseshoe-shaped structure, lacking any posterior vertical lancet, the posterior leaf instead with a more or less semicircular edge, sometimes with folds, pockets, or swollen vertical projections; there is no strap-like sella above and behind the nostrils (Fig. 13). The ears are moderate, lacking a tragus, and the tail is completely enclosed in the uropatagium, in some reduced or rudimentary.

They are insectivores, catching insects in flight or by gleaning; roosting in caves, rock crevices and hollow trees; gregarious or occurring in small groups, in pairs or singly.

The family is sometimes subsumed into the Rhinolophidae but differs sharply enough to justify familial distinction. Tate (1941a) suggested that the Hipposideridae (which he considered to have no more than a subfamilial relationship within the Rhinolophidae) might be divided into two subfamilies, with the majority of genera in the Hipposiderinae, but with *Coelops* given a subfamily of its own, the Coelopsinae. Since Tate wrote, Dorst (1947) has described *Paraceolops* from a heavily damaged example that remains unique, and although this author allocated it to the Coelopsinae many of its essential diagnostic features of noseleaf and skull remain unestablished. For the present this division has not been adopted but further study of *Paracoelops* may well show that with *Coelops* their considerable differences from the remaining hipposiderids should be recognised in this way.

1. Ears exceptionally long and broad, extending far beyond tip of muzzle when laid forward .... ***Paracoelops***

Ears small to large, at most extending just beyond muzzle when laid forward **2**

2. Anterior noseleaf lobate, lying above a wide, lobed anterior supplementary leaflet on each side, these projecting beyond margin of anterior leaf and united with the underside of its free edge; a small shield-shaped or chordate process at centre of upper margin of anterior leaf (Fig. 13e) .... ***Coelops***

Anterior leaf horseshoe-shaped, not deeply emarginated into two lobes; supplementary leaflets when present not projecting widely beyond anterior leaf, not united with its underside but instead arising from the muzzle; centre of upper margin of posterior leaf lacking any flattened process **3**

3. A flat, strap-like process extending anteriorly over the centre of the anterior leaf from the internarial region; posterior leaf strongly cellular (Fig. 13a) .... ***Triaenops***

Anterior leaf single, lacking any median strap-like process; posterior leaf not strongly cellular...................................... **4**

4. Upper margin of posterior leaf not lobulated, or if so upper part of leaf supported from rear by two narrow septa demarcating three posteriorly directed pockets (Fig. 12, p. 82)... ***Hipposideros***

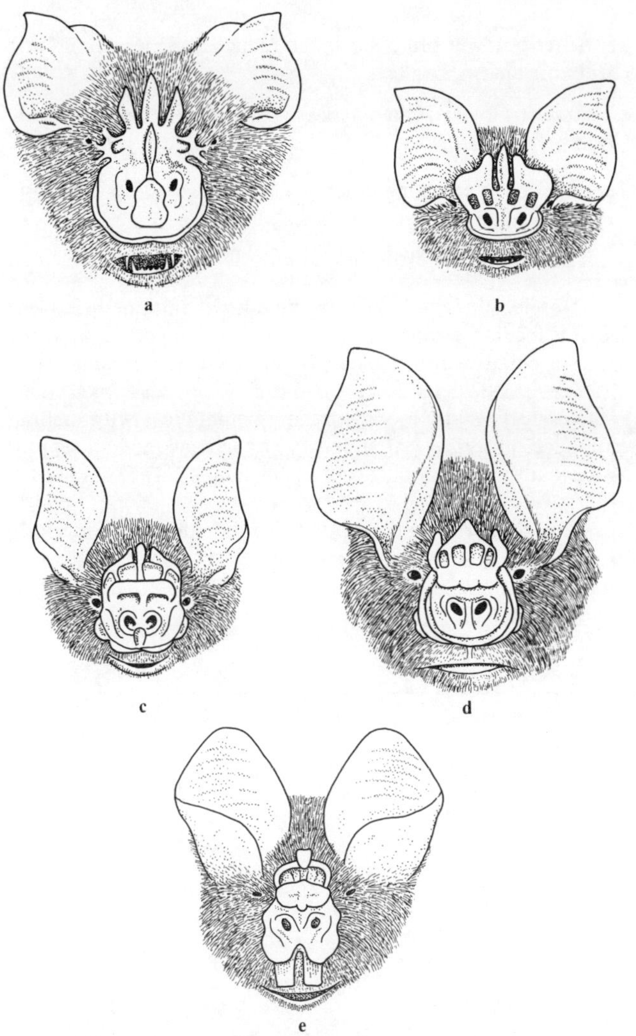

Fig. 13. Noseleaves of Hipposideridae: (a) *Triaenops persicus*; (b) *Aselliscus stoliczkanus*; (c) *Aselliscus tricuspidatus*; (d) *Asellia tridens*; (e) *Coelops frithii*.

Upper margin of posterior leaf with three fleshy projections lacking posterior support **5**

5. Central projection the largest, flanked by smaller lateral projections, their points widely separated (Fig. 13d); premolars 1/2 on each side .... ***Asellia***

Projections similar in size, closely adjacent (Fig. 13b); premolars 2/2 on each side .... ***Aselliscus***

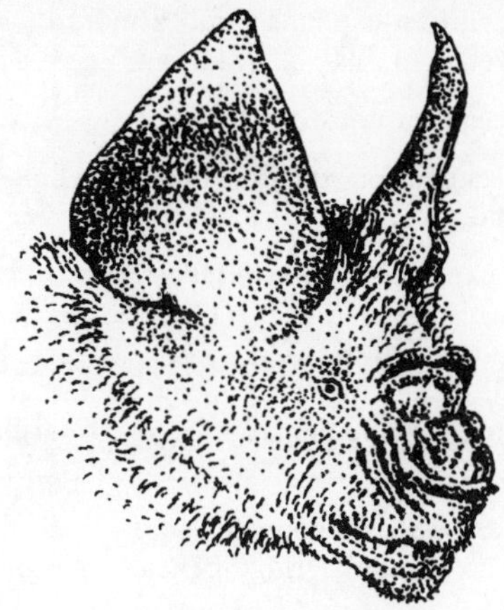

Hipposideros diadema

HIPPOSIDEROS

Hipposideros Gray, 1831a: 37; type species *Vespertilio speoris* Schneider.
Phyllorrhina Bonaparte, 1837: 3 (*Rhinolophus diadema* Geoffroy).
Gloionycteris Gray, 1866c: 82 (*R. armiger* Hodgson).
Rhinophylla Gray, 1866c: 82 (*Phyllorrhina labuanensis* Tomes).
Speorifera Gray, 1866c: 82 (*Rhinolophus vulgaris* Horsfield, in *R. larvatus* Horsfield).
Chrysonycteris Gray, 1866c: 82 (*H. fulvus* Gray).
Cyclorhina Peters, 1871a: 326 (*Phyllorrhina obscura* Peters and *P. doriae* Peters).
Thyreorhina Peters, 1871a: 327 (*P. coronata* Peters).
And other extralimital synonyms.

Tropics and subtropics of Old World, including much of Africa except Sahara and extreme south; SW Arabia; Afghanistan east to Philippines, N Australia and Vanuatu; c. 54 species of which 33 occur in the Indomalayan Region, in five of the species-groups recognized by Hill (1963b) in a review of the genus (Table 66).

Hipposideros bicolor group

Throughout the Region; 22 species, one of very doubtful validity.

1. Lateral supplementary leaflets none or one **2**

Lateral supplementary leaflets two or three **16**

2. Anterior upper premolar minute or absent, if present extruded, with 2nd premolar and canine in contact; anterior lower premolar 1/4 length and 1/3 or less height of 2nd **3**

Anterior upper premolar present, only minute exceptionally; anterior lower premolar 1/3 or more length and height of 2nd **5**

3. Posterior noseleaf supported by well defined septa, delimiting obvious pockets in the front face of its upper part; (Malaya, Borneo) .... ***H. dyacorum***

Posterior leaf lacking supporting septa **4**

4. Anterior leaf with well-defined anterior emargination; (Malaya, Sumatra, Borneo) .... ***H. sabanus***

Anterior leaf not emarginated; (Borneo)........... ***H. doriae***

5. Internarial septum expanded into a more or less disc-like structure between nostrils **6**

Internarial septum not disc-like **7**

6. Internarial expansion small, lobulated, ellipsoidal, thickened; (Thailand) .... ***H. halophyllus***

Internarial expansion large, circular, not especially thickened; (Malaya, Borneo) .... ***H. ridleyi***

7. Posterior leaf lacking supporting septa; (anterior leaf emarginated; a low glandular ridge beneath margin of anterior leaf) **8**

Posterior leaf with supporting septa; anterior leaf not emarginated; no glandular ridge beneath margin of anterior leaf **9**

8. Forearm 43–45mm; (Philippines) .... ***H. obscurus***

Forearm 47mm; (Philippines)........................ ***H. coronatus***
(not seen; characters in couplet 7 assumed)

9. Internarial septum thickened and/or bulbous; anterior half of zygomata slender **10**

Internarial septum not inflated; anterior half of zygomata massive **12**

10. Baculum strongly curved, almost the arc of a circle; (C India; forearm 36.0–37.5, condylobasal length 13.0–13.9, C-M^3 5.0–6.0; not seen) .... ***H. durgadasi***

Baculum not greatly curved **11**

11. Superior projection of zygoma lacking or low; anterior upper premolar not much less in crown area than upper incisor, usually not much extruded; (N India–Philippines, Borneo; slightly smaller: forearm 32–36.5, condylocanine length 12.6–13.9, C-M^3 4.8–5.5) .... ***H. cineraceus***

A low superior zygomatic projection; anterior upper premolar markedly less in crown area than upper incisor; usually extruded or partially extruded from row; (India–Philippines; slightly larger: forearm 34.5–42.5, condylocanine length 13.2–15.0, C-M^3 5.1–5.9) .... ***H. ater***

12. Posterior projecting portion of vomer thickened.......... **13**

Posterior projecting portion of vomer blade-like.......... **15**

13. Ears smaller, length 15.5–19.5mm; tympanic bullae relatively small: length 2.46–2.66, width 1.27–1.57, more widely separated, intercochlear distance 1.66–2.03; (S Thailand–Philippines) .... ***H. bicolor***

Ears larger, their length 19.0–23.6; tympanic bullae relatively large: length 2.76–3.20, width 1.46–1.98, less widely separated, intercochlear distance 1.14–1.50........................ **14**

14. Zygoma strong, with slight superior projection; well-defined sphenoidal depression; (S India–S China, N Malaya) .... ***H. pomona***

Zygoma slender, little or no superior projection; shallow sphenoidal depression; (Kangean Is, Sulawesi, Moluccas) .... ***H. macrobullatus***

15. Anterior lower premolar much reduced, 1/3 length of 2nd; (Afghanistan–Sri Lanka) .... ***H. fulvus***

Anterior lower premolar less reduced, 1/2 length of 2nd; (Malaya) .... ***H. nequam***

16. Anterior lateral supplementary leaflet extending anteriorly beneath leaf to median line **17**

This leaflet not extending anteriorly beneath leaf **18**

17. Noseleaf not excessively modified: intermediate leaf lacking median eminence; posterior leaf supported by three septa of equal width, not deeply pocketed; (Philippines) .... ***H. pygmaeus***

Noseleaf greatly modified: intermediate leaf with median eminence; posterior leaf supported by narrow median septum and broad lateral septa, deeply pocketed; (Borneo).. ***H. coxi***

18. Second lateral supplementary leaflet extending anteriorly beneath the leaf; often an incipient third leaflet; (N Moluccas) .... ***H. papua***

Table 66. The species-groups of *Hipposideros*.

	bicolor group	*pratti* group	*armiger* group	*speoris* group	*diadema* group
Species in Region	21	2	2	2	6
Forearm	32–53	73–89	63–100	53–67	73–101
Ears rounded or broadly triangular	+	+	0	0	0
Ears triangular, moderately or acutely pointed	0	0	+	+	+
Ear with small marginal projection at antitragal lobe	+/0	0	0	+	0
Fleshy outgrowths behind posterior leaf	0	+	+	0	0
Condylocanine length	12.6–17.1	24.9–27.9	23.0–29.8	15.8–29.8	25.0–34.7
C-M^3	4.8–7.5	10.5–12.0	8.7–12.8	6.7–9.5	11.0–15.2
Upper incisors	Weak	Strong	Strong	Strong	Strong
Crown area of I_2:I_1	=	>	>	>	>

This leaflet not extending anteriorly; no trace of a third leaflet **19**

19. Ears haired for 1/2 their length; tips of upper incisors strongly convergent; (Mentawai Is) .... ***H. breviceps***

Ears haired for 2/3 their length; tips of upper incisors not strongly convergent **20**

20. Antitragal lobe subangular, its height about 1/3 of ear length; no antitragal projection; intermediate leaf as wide as or wider than posterior leaf; (India, Sri Lanka–Borneo, Java) .... ***H. galeritus***

Antitragal lobe rounded, short, its height about 1/8 of ear length; an acutely pointed antitragal projection; intermediate leaf narrower than posterior leaf; (Malaya, Sumatra–Philippines) .... ***H. cervinus***

Hipposideros bicolor Bicoloured leaf-nosed bat

Rhinolophus bicolor Temminck, 1834: 19, pl. 1 (and 1835: 18, pls 29,32); Anjer coast, NW Java (lectotype designated and type locality restricted by Tate (1941c); see also Bergmans & van Bree (1986) who re-examined the syntype series).

H. g[entilis] atrox Andersen, 1918: 380, 381; Semangko Gap, Selangor, Malaya, 2800ft.

H. g[entilis] major Andersen, 1918: 380, 381; Bua Bua, Enggano I, W Sumatra.

H. javanicus Sody, 1937a: 215; Babakan, Kroja, Tjilatjap, C Java.

H. erigens Lawrence, 1939: 56; lower slopes of Mt Halcon, N side, near Calapan, Mindoro I, Philippines.

Range (Map 50). S Thailand, N Malaya (Perlis, Kedah), Sumatra, Java, Borneo, Philippine islands of Palawan, Mindoro and Luzon; Flores (Lesser Sunda Is); possibly Timor. Also on the small islands of Tarutau (Thailand), Tioman (Malaya), Nias, Enggano, Bangka (Sumatra), Sepanjang (Kangean Is); doubtfully on Taiwan.

Variation. Four subspecies are generally recognised:
H. b. atrox. S Thailand to Sumatra. Anterior lower premolar smaller.
H. b. major. Nias and Enggano Is, perhaps synonymous with *atrox*.
H. b. bicolor. Borneo, Java to Lesser Sunda Is. Anterior lower premolar larger.
H. b. erigens. Philippines. Not seen.

Map **50**. *Hipposideros* (part)

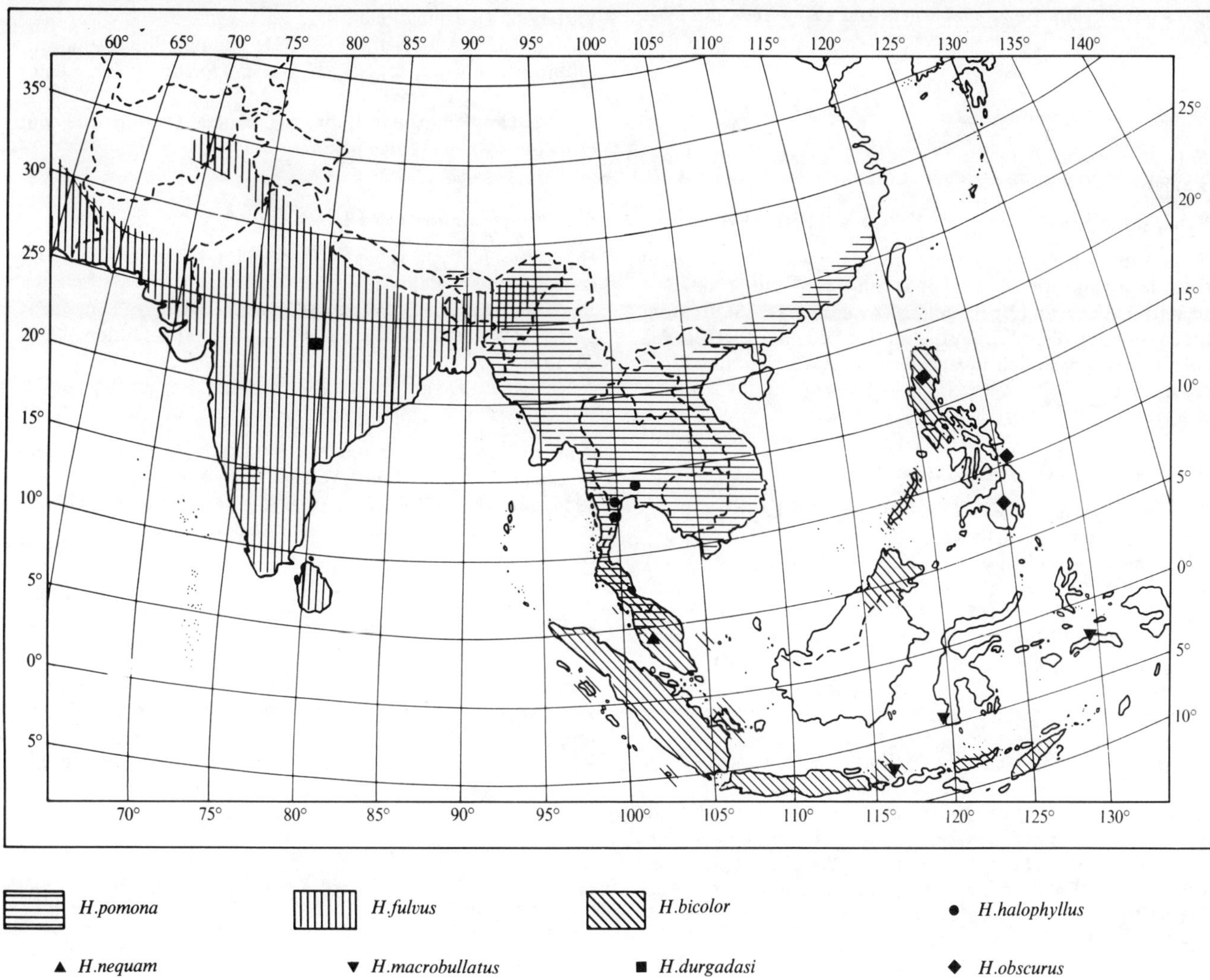

Hipposideros pomona

H. pomona Andersen, 1918: 380, 381; Haleri, N Coorg, India (a few miles N of Mercara, Coorg Dist., Karnataka).
H. gentilis Andersen, 1918: 380, 381; Thayetmyo, Burma.

Range (Map 50). S India; Sikkim, Assam, Burma; S China: Fujian, Hong Kong; Thailand, Laos, Vietnam, N Malaya (Perlis).

Variation. One subspecific name has been proposed:
H. g[entilis] sinensis Andersen, 1918: 380, 381; Foochow, Fujian, China.
Three subspecies are currently listed:
H. p. pomona. S India. Zygomata with low jugal projection; anterior lower premolar moderate.
H. p. gentilis. N India through Burma and Thailand (part) to N Malaya. No jugal projection; anterior lower premolar slightly larger.
H. p. sinensis. Thailand (part), S China, Indochina. Like *H. p. gentilis* but anterior lower premolar very large.

Remarks. Formerly thought conspecific with *H. bicolor* but recognised as a distinct species by Hill *et al.* (1986). The form *pomona* is known only from the holotype, its skull being in the BMNH, the alcoholic specimen apparently in the Calcutta collection. Cranially it agrees closely with *gentilis* and *sinensis*: examination of the alcoholic specimen by GBC showed that it lacks lateral supplementary leaflets and also agrees closely in size.

Hipposideros macrobullatus

H. bicolor macrobullatus Tate, 1941c: 357; Talassa, near Maros, S Sulawesi, 300m.

Range (Map 50). SW Sulawesi; Kangean I; Seram (Moluccas).

Remarks. Formerly considered conspecific with *H. bicolor* but listed as a valid species by van Strien (1986) and reported as such from Kangean I by Bergmans & van Bree (1986). Reasons for considering it specifically distinct are provided by Hill *et al.* (1986); it is very similar to *H. pomona* with which it may prove conspecific.

Hipposideros ater Dusky leaf-nosed bat

H. ater Templeton, 1848: 252; Colombo, Sri Lanka.
= *H. atratus* Kelaart, 1850b: 208.
Phyllorhina antricola Peters, 1861: 709; Paracali, Luzon I, Philippines.
P. amboinensis Peters, 1871a: 323; Ambon, Moluccas.
H. nicobarulae Miller, 1902b: 781; Little Nicobar I.
H. albanensis saevus Andersen, 1918: 380, 381; Kei Is.
H. gentilis toala Shamel, 1940b: 352; Toeare, Sulawesi.
Also other, extralimital synonyms.

Range (Map 51). Sri Lanka, India, Burma, S Thailand, Malaya, Sumatra, N Borneo, Java, Bali, Lombok; Philippine islands of Luzon, Mindoro, Palawan, Mindanao; Sulawesi and the Moluccas; also on the smaller islands of Little Nicobar, Car Nicobar, Great Nicobar, Camorta; Con Son (Vietnam); Marinduque, Balabac (Philippines); Togian Is (Owen *et al.*, 1987; Hill, 1991); Peleng, possibly also Sangihe and Talaud Is (Sulawesi); Buru, Seram, Ambon, Halmahera, Kai Is (Moluccas). Also in New Guinea and N Australia.

Variation. Five subspecies are recognised in the region: differences are small but in general the anterior lower premolar varies in length from one half or less of the second in western subspecies to almost equal to the second in those from the eastern part of the range, its height varying similarly from one half to two-thirds that of the second tooth.

H. a. ater. India, Sri Lanka.
H. a. nicobarulae. Nicobar Is.
H. a. saevus. Burma to Moluccas (including *toala*).
H. a. amboinensis. Ambon, Moluccas; may be synonymous with *saevus* or with the extralimital *H. a. aruensis* Gray, 1858 from the Aru Is and New Guinea.
H. a. antricola. Philippines.

Hipposideros fulvus Fulvous leaf-nosed bat

H. fulvus Gray, 1838a: 492; 'Madras', actually apparently Dharwar, India.
= *Rhinolophus fulgens* Elliot, 1839: 99.
H. murinus Gray, 1838: 492; Dharwar, India.
Phyllorhina aurita Tomes, 1859b: 76; India.
Phyllorrhina atra Fitzinger, 1870: 877; Sri Lanka.

Range (Map 50). Afghanistan, NE, S Pakistan, India, Sri Lanka.

Variation. Two subspecies:
H. f. pallidus Andersen, 1918: 381; Junaghadh, Kathiawar, India. Afghanistan to NE, N India. Dorsum paler, venter creamy with grey tinge.
H. f. fulvus. Peninsular India, Sri Lanka. Dorsum chestnut brown or brown, venter paler.

Hipposideros cineraceus Least leaf-nosed bat

H. cineraceus Blyth, 1853a: 410; near Pind Dadan Khan, Salt Range, Punjab, Pakistan.
Phyllorhina micropus Peters, 1872b: 256; Dehra Dun, near Simla, NW India (ascribed to Hutton; see also Hutton, 1872: 703).
? *H. wrighti* Taylor, 1934: 237; Baguio, Benguet, Luzon I, Philippines.

Range (Map 51). Extreme NE Pakistan; N India, Burma, Thailand, Laos, Vietnam, Malaya, Sumatra (Agronoff *et al.*, 1984; BMNH), Krakatau Is: Rakata (Tidemann *et al.*, 1990); Borneo, ? Luzon (Philippines). Also on the Riau Is and Kangean I.

Variation. Specimens from the eastern part of the range average slightly larger than those from more westerly locations (Hill & Francis, 1984).

Remarks. The taxonomic history of *H. wrighti* Taylor, 1934 was summarized by Lawrence (1939) who considered that Taylor might have redescribed *H. ater antricola*. However, Hill & Francis (1984) suggested that on the basis of the original description it more probably represents *H. cineraceus*.

Hipposideros durgadasi

H. cineraceus durgadasi Khajuria, 1970: 623; near Katungi village, Jabalpur Dist, Madhya Pradesh, India.

Range (Map 50). Central India.

Map **51**. *Hipposideros* (part)

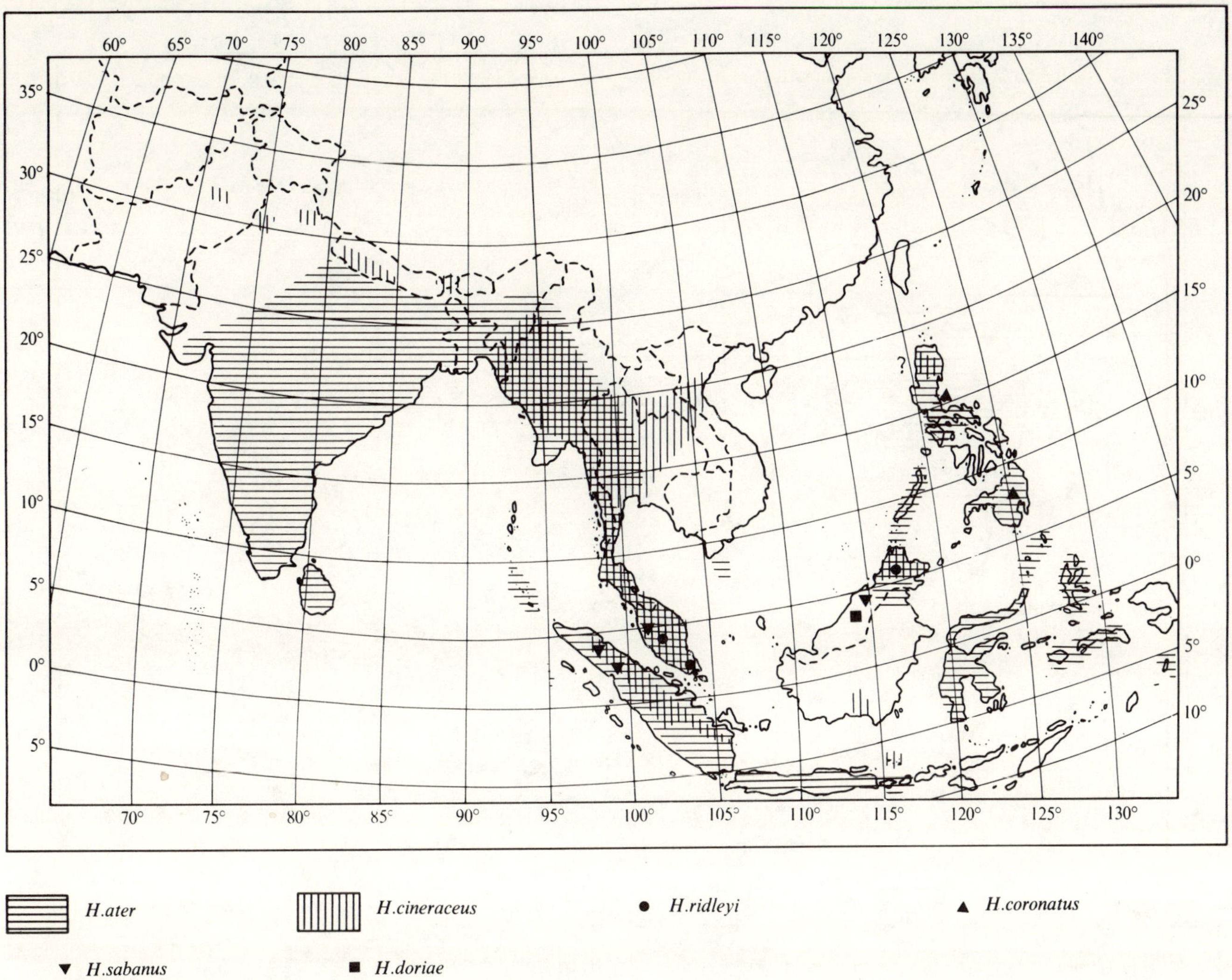

Remarks. A further account of this was given by Khajuria (1980). It was considered specifically distinct by Topál (1975) on account of its characteristic strongly curved baculum; Khajuria (1980) also treated it as a distinct species. No specimens have been seen.

Hipposideros halophyllus

H. halophyllus Hill & Yenbutra, 1984: 77; Khao Sa Moa Khon, Tha Woong, Lop Buri, Thailand.

Range (Map 50). Thailand: Lop Buri, Rat Buri, Phet Buri.

Hipposideros nequam Malay leaf-nosed bat

H. nequam Andersen, 1918: 380, 381; Klang, Selangor, Malaya.

Range (Map 50). Selangor, Malaya.

Remarks. Known only from the holotype (in BMNH), of which cranially only the rostrum and damaged mandible remain.

Hipposideros coronatus

Phyllorhina coronata Peters, 1871a: 327; Mainit, Surigao, NE Mindanao I, Philippines.

Range (Map 51). Philippine islands of Polillo (Cuy, 1980,1981) and Mindanao.

Hipposideros ridleyi Ridley's leaf-nosed bat

H. ridleyi Robinson & Kloss, 1911b: 241; Botanic Gardens, Singapore.

Range (Map 51). Malaya, Singapore I, N Borneo (Hill, 1983; Hill & Francis, 1984).

Remarks. IUCN: I.

Hipposideros dyacorum Dyak leaf-nosed bat

H. dyacorum Thomas, 1902: 271; Mt Mulu, Baram, Sarawak, Borneo.

Range (Map 52). Malaya: Perlis (Hill & Zubaid, 1989); Borneo.

Hipposideros sabanus

H. sabanus Thomas, 1898d: 243; Lawas, 'North Borneo', actually NE Sarawak, Borneo.

Range (Map 51). Malaya: Perak; Sumatra; Borneo: Sarawak.

Map **52**. *Hipposideros* (part)

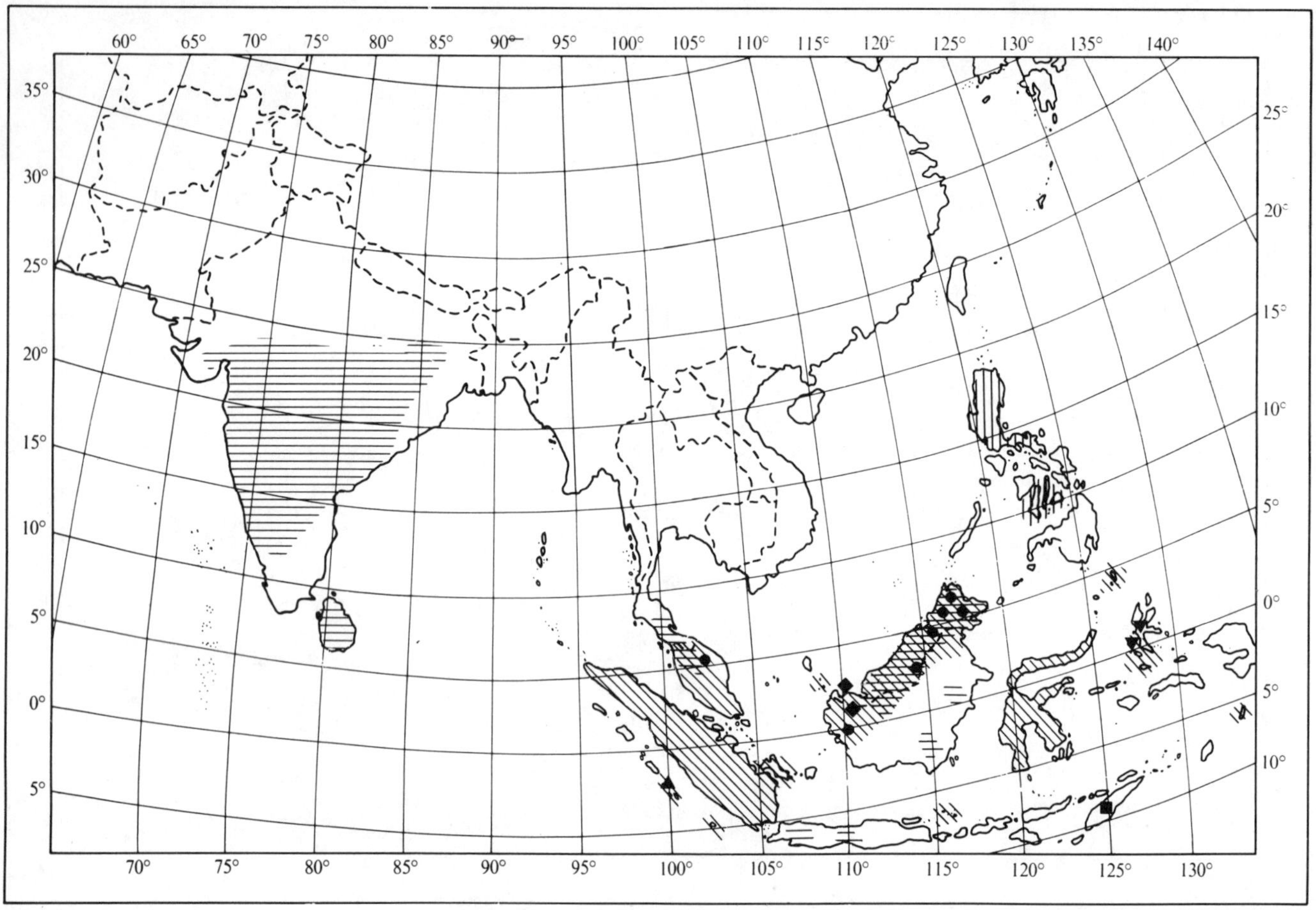

H. galeritus — *H. pygmaeus* — *H. cervinus* — ● *H. dyacorum*

▲ *H. breviceps* — ▼ *H. papua* — ■ [*H. crumeniferus*] — ◆ *H. coxi*

Hipposideros doriae

Phyllorhina doriae Peters, 1871a: 326; Sarawak, Borneo.

Range (Map 51). Borneo: Sarawak.

Remarks. No example has been examined, the holotype (presumably in the Genoa Museum) being apparently the only representative. Its features suggest that it is very similar to if not identical with *H. sabanus* as thought many years ago by Thomas (MS). A more detailed account was given by Hill (1963b).

Hipposideros obscurus

Phyllorhina obscura Peters, 1861: 709; Paracali, Luzon I, Philippines.

Range (Map 50). Philippine islands of Luzon, Dinagat and Mindanao.

Hipposideros pygmaeus

Rhinolophus pygmaeus Waterhouse, 1843b: 67; Philippines.

Range (Map 52). Philippine islands of Luzon, Negros and Bohol.

Hipposideros galeritus Cantor's leaf-nosed bat, Fawn leaf-nosed bat

H. galeritus Cantor, 1846: 183; Penang I, Malaya.
Phyllorhina longicauda Peters, 1861: 708; Surakarta, Java.
P. brachyota Dobson, 1874c: 237; central India.
H. insolens Lyon, 1911a: 129; near Lowatsi, upper Pasir R, SE Borneo.

Range (Map 52). India, Sri Lanka, Thailand, Malaya (including Penang I), Borneo, Java.

Variation. Four subspecies have been recognised:
H. g. brachyotus. India, Sri Lanka. Similar to nominate subspecies but available material limited.
H. g. galeritus. Thailand, Malaya. Relatively smaller.
H. g. insolens. Borneo. Larger, with more robust skull.
H. g. longicauda. Java. Relatively smaller, with small tympanic bullae.

Remarks. The taxonomic history of this species and its distinction from *H. cervinus* were reviewed by Jenkins & Hill (1981).

Hipposideros cervinus Gould's leaf-nosed bat

Rhinolophus cervinus Gould, 1854:pl. 34; Cape York and Albany I, Queensland, Australia.
Phyllorhina labuanensis Tomes, 1859a: 537; Labuan I, Borneo.
H. batchianus Matschie, 1901a: 273; Bacan I, Moluccas.
H. schneidersi Thomas, 1904a: 722; Upper Langkat, Sumatra (misprint for *schneideri*).
H. celebensis Sody, 1936a: 47; Mampoe Cave, 20 km N of Watoe Pone, S Sulawesi.

Range (Map 52). Malaya, Sumatra, Borneo, Mindanao, Sulawesi, Moluccas; and to N Queensland and Vanuatu. Also on the smaller islands of Kaban (Johore), Singapore, Karimon (Riau Is), Sipura, N Pagai, Enggano, Bangka (Sumatra); S Natuna Is; Togian Is (Owen *et al.*, 1987; Hill, 1991); Peleng; Karakelang (Talaud Is—Hill & Rozendaal, 1989); Kangean; Bacan, Kai Is (Moluccas).

Variation. Two, perhaps three subspecies in the region; the extralimital *Phyllorhina cervina* var. *misoriensis* Peters, 1906:pl. 5L, probably from Misor (= Schouten) I, New Guinea, has not been seen and is provisionally associated with this species by Jenkins & Hill (1981).

H. c. labuanensis. Malaya and Sumatra to Philippines and Borneo. Generally larger.
H. c. cervinus. Sulawesi and adjacent islands, to Queensland. Generally smaller. Probably including *batchianus.*

Remarks. On occasion considered conspecific with *H. galeritus*, the taxonomic history of *H. cervinus* and its recognition as a distinct species was reviewed in detail by Jenkins & Hill (1981).

[***Hipposideros crumeniferus***]

Rhinolophus crumeniferus Leseur & Petit, 1807:pl. 35; Timor I.

This early name is based upon an illustration and it seems possible that it represents *H. cervinus cervinus* or a closely related form, but pending the discovery of original specimens or the collection of topotypes the name must remain *incertae sedis*. See Hill (1963b) for a further explanation.

Hipposideros breviceps

H. breviceps Tate, 1941c: 358; N Pagi I, Mentawai Is.

Range (Map 52). Mentawai Is: North Pagai Island.

Hipposideros coxi Cox's leaf-nosed bat

H. coxi Shelford, 1901: 113; Mt Penrisen, Sarawak, Borneo 4200ft.

Range (Map 52). Borneo: SW Sarawak.

Hipposideros papua Geelvinck Bay leaf-nosed bat

Phyllorhina papua Thomas & Doria, 1886: 204; Korido, Misori (=Schouten) I, Geelvinck Bay, Irian Jaya, New Guinea.

Range (Map 52). Moluccas: Halmahera, Bacan (Hill & Rozendaal, 1989); also to W New Guinea.

Remarks. A third supplementary leaflet may be present on the nose in this species, but is always very small.

Hipposideros pratti group

Burma, SW China to Vietnam and Malaya; 2 species (Table 67).

Table 67. The species of the *Hipposideros pratti* group.

	pratti	*lylei*
Range	Vietnam	Burma - Malaya
Forearm	81–89	73–84
Median emargination of anterior leaf	Simple	Flanked by shallow lateral emarginations
Lateral margins of anterior and posterior leaves continuous	0	+
Condylocanine length	27.7–27.9	24.9–25.9
C - M^3	11.9–12.0	10.5–11.3
Rostrum with lateral pits	0	+
Step-like discontinuity between roofs of narial and mesopterygoid canals	+	0
Vomer projecting beyond posterior edge of palate	+	0

Hipposideros pratti Pratt's leaf-nosed bat

Hipposiderus pratti Thomas, 1891b: 527; Kiatingfu, Sichuan, China.

Range (Map 53). S China: Sichuan, Zhijiang, Fujian; Vietnam.

Hipposideros lylei Shield-faced leaf-nosed bat

H. lylei Thomas, 1913d: 88; Chiendao Cave, 50 miles N of Chiang Mai, Thailand, 350m.

Range (Map 53). Burma, Yunnan (Lu *et al.*, 1965), Thailand, Malaya.

Hipposideros armiger group

N India to Malaya; Ryukyu Is, Taiwan; 2 species (Table 68).

Table 68. The species of the *Hipposideros armiger* group.

	armiger	*turpis*
Range	N India-Taiwan, Malaya	S Thailand, Ryukyu Is
Forearm	79–100	63–80
Condylocanine length	26.4–29.8	23.0–23.6
C - M^3	11.1–12.8	9.7–10.2

Hipposideros armiger Himalayan leaf-nosed bat

Rhinolophus armiger Hodgson, 1835a: 699; Nepal.
Phyllorhina swinhoii Peters, 1871b: 616; Amoy, Fujian, China.
H. tranninhensis Bourret, 1942a: 20 (and 1942c: 12); Plaine des Jarres, Tran-Ninh, Vietnam.

Range (Map 53). Nepal, Assam, Burma; S China: Sichuan, Yunnan, Fujian and adjacent provinces in SE; Hong Kong, Hainan, Taiwan; Thailand, Vietnam, Malaya; Langkawi I.

Variation. Three subspecific names have been proposed:
H. a debilis Andersen, 1906a: 37; Province Wellesley, Malaya.
H. a. terasensis Kishida, 1924a: 42; Taiwan.
H. a. fujianensis Zhen, 1987: 238, 239; Shanghang, Zhangping and Zhaoan counties, Fujian, China.

Map **53**. *Hipposideros* (part)

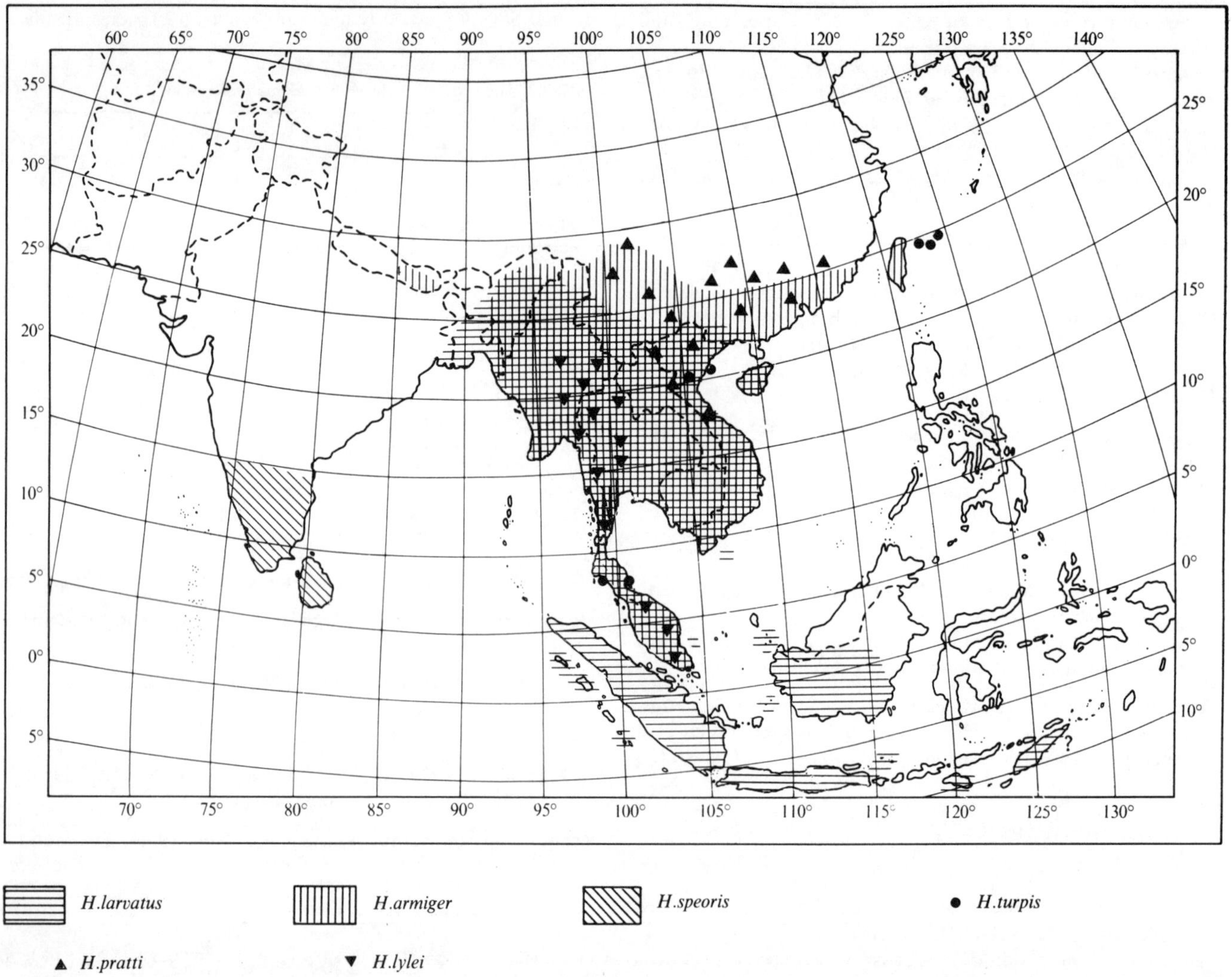

Four subspecies can be listed: *terasensis* from Taiwan is likely to be indistinguishable from *H. a. armiger* assording to Tate (1941c) while *fujianensis* from Fujian appears to be distinguished from the nominate form only by small details of the noseleaf and by a subcircular rather than transverse opening of the frontal gland, *tranninhensis* from Vietnam by its more trilobate posterior leaf.

H. a. armiger. Range of species except Taiwan and Vietnam.
H. a. tranninhensis. Vietnam.
H. a. terasensis. Taiwan.
H. a. fujianensis. Fujian (part).

Hipposideros turpis Lesser leaf-nosed bat

H. turpis Bangs, 1901: 561; Ishigaki (I), S Ryukyu Is.
H. pendleburyi Chasen, 1936: 133; near foot of Khao Ram, Nakon Sri Thamrat, Thailand.

Range (Map 53). S Thailand; Vietnam (Topál, pers. comm.); Ryukyu islands of Ishigaki, Iriomote, Yokayuni and Yaeyama (Yoshiyuki, 1989).

Variation. Two subspecies:
H. t. pendleburyi. S Thailand. Slightly larger.
H. t. turpis. S Ryukyu Is. Slightly smaller.

Hipposideros speoris group

W, C Africa; N India to Borneo, Java and Lesser Sunda Is; 3 species, 2 in region (Table 69).

Hipposideros larvatus

Rhinolophus larvatus Horsfield, 1823:pt 6, pl. ; Java.
R. vulgaris Horsfield, 1823:pt 6, pl. ; Java.
R. deformis Horsfield, 1823:pt 6, pl. ; Java.
R. insignis Horsfield, 1823:pt 6, pl. ; Java.
Phyllorhina leptophylla Dobson, 1874b: 234; Khasi Hills, Assam.
H. barbensis Miller, 1900a: 233; St Barbe I, S China Sea.
H. poutensis Allen, 1906: 483; Pouten, Hainan.

Table 69. The species of the *Hipposideros speoris* group.

	larvatus	*speoris*
Range	N India-Borneo, Lesser Sunda Is	C,S India, Sri Lanka
Forearm	52–67	45–55
Anterior noseleaf medianly emarginated	+	0
Condylocanine length	18.5–21.5	15.8–17.5
C - M^3	8.1–9.5	6.5–7.5
Prominent supraorbital ridges	0	+
Rostrum flattened	0	+
Vomer projecting beyond rear of palate	+	0
Upper canine antero-internally grooved, with small posterior cusp	0	+

Range (Map 53). Assam, Burma, Yunnan (Lu *et al.*, 1965), Thailand, Hainan, Vietnam, Malaya, Sumatra, Krakatau Is, Java, Borneo, Bali, Sumba, ? Timor. Also on the small islands of Con Son (Vietnam), Tioman, Aor (Malaya), Simeulue, Nias, N Pagai (Sumatra), St Barbe I (S China Sea); S Natuna Is, Butang, Serasan, Karimata (Borneo); Panebangan; Kangean, Sepanjang (Kangean Is).

Variation. The following subspecific names have been proposed:
H. l. grandis Allen, 1936a: 345; Akunti, upper Chindwin, Burma, 500ft.
H. l. neglectus Sody, 1936a: 46; Roema Manoeal, south foot of Mt Kenepai, C Kalimantan, Borneo.
H. l. alongensis Bourret, 1942a: 27; Bay d'Along, Vietnam.
H. l. sumbae Oei, 1960: 28; Sumba I, Lesser Sunda Is. Possibly also on Timor (Bergmans & van Bree, 1988).

Varies in size over its range, with large subspecies in Assam, Burma and Indochina, the smallest on St Barbe I, Java and the Lesser Sunda Is (Bergmans & van Bree, 1986). Specimens in Malaya and Borneo are intermediate in size.

Hipposideros speoris Schneider's leaf-nosed bat

Vespertilio speoris Schneider, 1800:pl. 59b; Tranquebar, India (Tate, 1941c; briefly reviewed by Oey & Feen, 1958).
Rhinolophus marsupialis Desmarest, 1820: 126.
R. dukhunensis Sykes, 1831: 99; Deccan, India.
H. penicillatus Gray, 1838a: 493; Madras, India.
H. templetonii Kelaart, 1850b: 208; Sri Lanka.
H. aureus Kelaart, 1853: 18; Sri Lanka.
H. blythi Kelaart, 1853: 20; Sri Lanka.

Range (Map 53). Central and S India, Sri Lanka.

Variation. Two very weakly defined subspecies have been recognised:
H. s. speoris. S India, Sri Lanka. Slightly larger on average.
H. s. pulchellus Andersen, 1918: 383; Vijaynagar, Bellary, India. SC India. Slightly smaller on average.

Hipposideros diadema group

Africa south of Sahara, Madagascar; India, Sri Lanka to Philippines, N Australia and Solomon Is; 7 species, 6 in the Indomalayan Region (Table 70).

Hipposideros lankadiva

H. lankadiva Kelaart, 1850c: 216; Kandy, Sri Lanka.
H. indus Andersen, 1918: 382; Gersoppa, Kanara, India.

Range (Map 54). Peninsular India, Sri Lanka.

Variation. Two subspecific names have been proposed:
H. i[ndus] mixtus Andersen, 1918: 382; Kolar, E Mysore, India.
H. i. unitus Andersen, 1918: 382; Mundra, Saugor, Central Provinces, India.

Available material is very limited; although four subspecies have been recognised there seems little between *indus* and *mixtus* and they are likely to prove synonymous. Mainland specimens average smaller than those from Sri Lanka, those from S India being slightly smaller than examples from more northerly locations.

Hipposideros schistaceus

H. schistaceus Andersen, 1918: 382; Vijayanagar, Bellary, India.

Range (Map 54). S India: Andhra Pradesh.

Hipposideros diadema Diadem leaf-nosed bat

Rhinolophus diadema Geoffroy, E.,1813b: 263, pls 5,6; Timor I.
R. nobilis Horsfield, 1823:pt 6, pl. ; Java.
Phyllorhina nicobarensis Dobson, 1871b: 262; Nicobar Is.
P. masoni Dobson, 1872a: 338; Moulmein, Burma.
R. griseus Meyen, 1883: 608, pl. 46; San Matheo Cave, Montalban, near Manila, Luzon, Philippines.
H. euotis Andersen, 1905c: 502; Bacan I, Moluccas.
Also several extralimital synonyms.

Range (Map 54). Burma, Thailand, Vietnam, Malaya, Sumatra, Java, Borneo, Philippines (Luzon, Mindoro, Mindanao etc.), Sulawesi, Lesser Sunda Is, Timor, Moluccas; to N Australia and Solomon Is. Also on the following smaller islands: Nicobars; Tinggi (Malaya); Nias, Enggano (Sumatra); Rakata (Krakatau Is) (Tiedmann *et al.*, 1990); Bunguran (N Natuna Is); Catanduanes, Leyte, Guimaras, Negros, Cebu, Dinagat (Philippines); Sepanjang (Kangean Is); Bali; Lombok, Sumbawa; Nusa Penida, Moyo, Sumba, Flores, Savu, Roti (all Kitchener, pers. comm.); Kalao, Sangihe, Karakelang (Talaud Is) (both Hill & Rozendaal, 1989); Togian Is (Owen *et al.*, 1987; Hill, 1991); Bacan, Buru, Seram, Yamdena (Tanimbar Is—Hill & Rozendaal, 1989), Kai Is (Moluccas).

Variation. The following subspecific names have been proposed:

H. d. vicarius Andersen, 1905c: 499; Sarawak, Borneo.
H. d. enganus Andersen, 1907b: 8; Kifa-juc, Bua Bua, Enggano I, Sumatra.
H. d. speculator Andersen, 1918: 381, 382; Kalao I, Flores Sea, Sulawesi.
H. d. custos Andersen, 1918: 381, 382; Kei Is.
H. d. anderseni Taylor, 1934: 246; Novaliches, Rizal Prov., Luzon, Philippines.
H. d. natunensis Chasen, 1940a: 43; Bunguran I, N Natuna Is.
H. d. ceramensis Laurie & Hill, 1954: 58; Teleoti Bay, Seram, Moluccas.

This species has been divided into numerous subspecies over its extensive range. Two broad groups can be discerned, one of larger subspecies on the Asiatic mainland and on Sumatra,

Table 70. The species of the *Hipposideros diadema* group.

	lankadiva	*schistaceus*	*diadema*	*lekaguli*	*dinops*	*inexpectatus*
Range in region	India, Sri Lanka	S India	Burma-Moluccas	Thailand, N Malaya	Sulawesi	Sulawesi
Forearm	76–93	73–82	74–92	71–80	93–97	96–101
Median ridge or projection on intermediate leaf	0	0	+	+	+	0
Upper part of posterior leaf heavily lobulated, with 3 posteriorly directed pockets separated by narrow septa	0	0	0	+	0	0
Condylocanine length	26.7–30.8	25.0–25.2	26.0–30.3	22.3–24.4	30.4–33.9	c.34.7
C-M^3	12.1–14.4	11.0–11.6	11.2–13.8	9.2–9.9	13.4–14.7	14.8–15.3
Zygomatic width	16.9–20.6	15.8–16.0	15.9–19.6	13.8–15.0	20.2–22.0	21.9–22.4
Frontal region convex or flattened	+	+	0	+	0	0
Frontal depression	0	0	+	0	+	0
Palation V-shaped	+	+	0	0	0	0
Well developed sphenoidal depression	+	+	0	0	0	0
Mandible exceptionally heavy	0	0	0	0	0	+

Map **54**. *Hipposideros* (part)

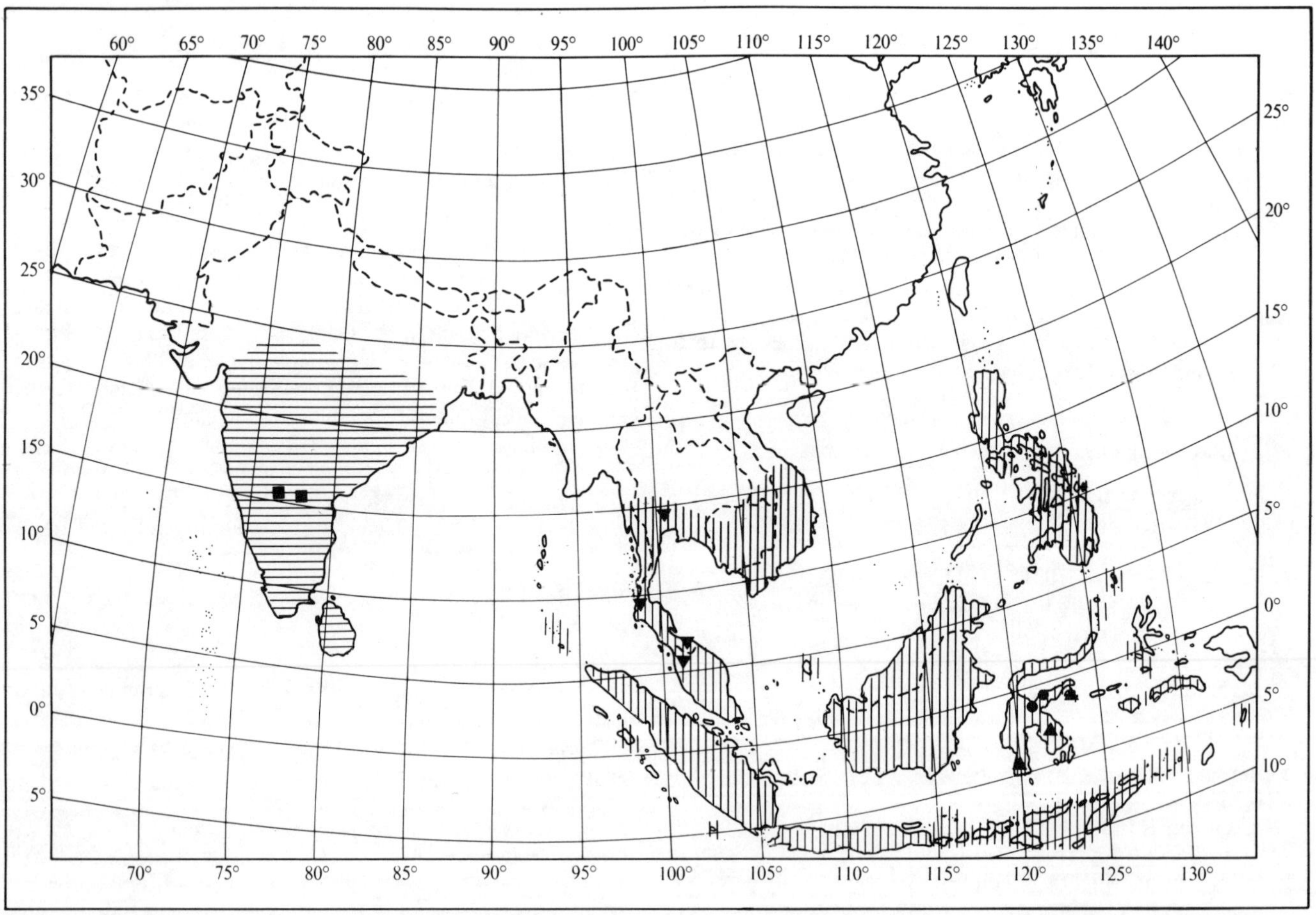

H.lankadiva
H.diadema
● H.inexpectatus
▲ H.dinops
▼ H.lekaguli
■ H.schistaceus

Java and Borneo and associated islands; and a group of smaller subspecies from the Philippines eastwards. There is overlap between the groups, with larger forms on Bacan and Seram in the Moluccas, and there is little evidence of a regular pattern of variation.

Hipposideros lekaguli

H. lekaguli Thonglongya & Hill, 1974: 285; Phu Nam Tok Tap Kwang, Kaeng Khoi, Suraburi, Thailand, c. 14°34'N, 101°9'E.

Range (Map 54). Thailand; N Malaya: Kedah (Hill *et al.*, 1985).

Hipposideros dinops

H. dinops Andersen, 1905c: 502; Rubiana I, Solomon Is.
H. pelingensis Shamel, 1940b: 353; Peleng I, E of Sulawesi.

Range (Map 54). Sulawesi; Peleng Is; extralimitally in Solomon Is.

Variation. *H. d. pelingensis* differs from the nominate subspecies in having the rostrum slightly narrower, the upper canines a little narrower at the base, and slightly shorter toothrows.

Remarks. See Hill (1971a, 1983) for a discussion of this species.

Hipposideros inexpectatus

H. inexpectatus Laurie & Hill, 1954: 60; Posso (= Poso), N Sulawesi.

Range (Map 54). N Sulawesi.

Remarks. Additional specimens from N Sulawesi were recorded by Feiler (1981).

TRIAENOPS

Triaenops Dobson, 1871c: 455; type species *T. persicus* Dobson. C, E Africa to Pakistan; Madagascar and Aldabra I; 2 species, one in the Indomalayan Region.

Triaenops persicus Persian trident bat

T. persicus Dobson, 1871c: 455, pl. 18; Shiraz, Iran, c. 4750ft. Also several extralimital synonyms.

Range (Map 55). S Pakistan (Kock, 1980); extralimitally as the range of the genus except Aldabra.

Map **55**. *Triaenops, Aselliscus, Asellia, Coelops, Paracoelops*

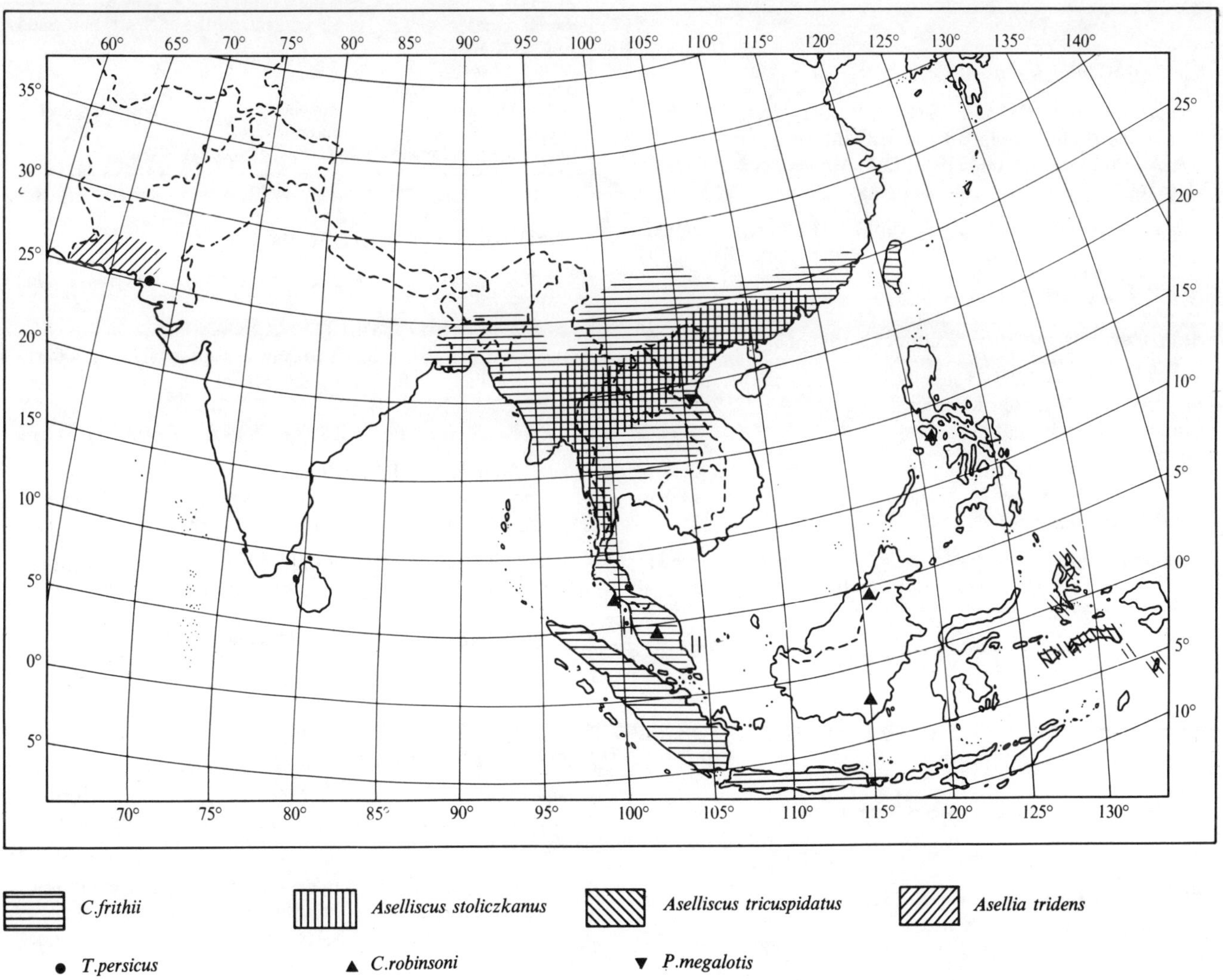

Variation. The subspecies in Pakistan is *T. p. persicus*, extending west to Iran.

Remarks. Reviewed by Hill (1982).

ASELLISCUS

Aselliscus Tate, 1941a: 2; type species *Rhinolophus tricuspidatus* Temminck.

Burma to S China and Malaya; Moluccas to Vanuatu; 2 species (Table 71).

Table 71. The species of *Aselliscus*.

	stoliczkanus	*tricuspidatus*
Range in region	Burma-S China, Malaya	Moluccas
Processes on upper margin of posterior noseleaf	Large, virtually in contact (Fig.13b)	Small, moderately separated (Fig.13c)
Ears	Broad, about as wide as long	Narrow, width less than length
Anterior rostrum	Level with postorbital region	Elevated above postorbital region
Maxillary toothrows	Convergent anteriorly	Nearly parallel
Anterior lower premolar	1/2 length of posterior lower premolar	About = length of posterior lower premolar

Aselliscus stoliczkanus Stoliczka's trident bat

Asellia stoliczkana Dobson, 1871a: 106; Penang I, Malaya.
Phyllorhina trifida Peters, 1871c: 513; Burma.
Triaenops wheeleri Osgood, 1932: 224; Muong Moun, Tonkin, Vietnam.

Range (Map 55). Burma; S China: Guizhou, Guangxi; Thailand, Laos, Vietnam; Malaya: Penang and Tioman Is (Zubaid, 1988).

Aselliscus tricuspidatus Temminck's trident bat

Rhinolophus tricuspidatus Temminck, 1834: 20, pl. 1 (also 1835: 20, pls 29,32); Ambon, Moluccas.
Also other extralimital synonyms.

Range (Map 55). Moluccas: Morotai, Bacan, Buru, Seram, Ambon, Goram, Kai Is; extralimitally to Solomon Is and Vanuatu.

Variation. Subspecies in region is *A. t. tricuspidatus*, confined to the Moluccas; other subspecies in eastern part of range; all reviewed by Schlitter *et al.* (1983).

ASELLIA

Asellia Gray, 1838a: 493; type species *Rhinolophus tridens* Geoffroy.

W Africa to Pakistan; 2 species, one in the region.

Asellia tridens Trident bat

Rhinolophus tridens Geoffroy, E.,1813b: 265; Egypt.
Including extralimital synonyms.

Range (Map 55). S Pakistan; and to Israel and W Africa.

Variation. Subspecies in region is *A. t. murraiana* (*Phyllorhina tridens* var. *murraiana* Anderson, 1881: 113; Karachi, Pakistan). This extends to Israel and to NW and W Africa, the nominate subspecies apparently being confined to N and NE Africa (Owen & Qumsiyeh, 1987).

COELOPS

Coelops Blyth, 1848: 251; type species *C. frithii* Blyth.
Chilophylla Miller, 1910: 395 (*Chilophylla hirsuta* Miller).

E India to S China, Philippines, Java, Bali, Taiwan; 2 species (Table 72).

Table 72. The species of *Coelops*.

	frithii (Fig.13e)	*robinsoni*
Range	E India-Java, Bali	Malaya, Borneo, Philippines
Forearm	33.5–36.5	38–44
Lappets below anterior noseleaf	Narrow, elongate	Wide, rounded
Narial region of leaf	Not depressed	Depressed
Prominent ridges from internarial process to lateral margin of leaf	0	+
Anterior rostral swellings	Little inflated, slightly convex	Inflated, strongly convex
Posterior dorsal rostral surface	Sloping anteriorly	Nearly horizontal
Interior pocket in maxillary root of zygoma	+	0
Upper toothrows	Convergent anteriorly	Nearly parallel
Posterior lower premolar	Narrow, elongate	Wide

Coelops frithii Tail-less leaf-nosed bat

C. frithii Blyth, 1848: 251; Sunderbans, Bengal, India/Bangladesh.
C. bernsteini Peters, 1862a: 117, footnote; Java.
C. inflata Miller, 1928: 85; Yenpingfu, Fujian, China, 2000ft.
C. sinicus Allen, 1928a: 4; 2 miles NE of Wanhsien, Sichuan, China.
C. formosanus Horikawa, 1928: 339; Kuraru, Koshun, Taiwan.

Range (Map 55). E India, Burma; China: Sichuan, Fujian; Thailand, Laos, Vietnam, Malaya, Sumatra (Bogor Museum—C M Francis, pers. comm.), Java, Bali, Taiwan.

Variation. The available material is insufficient to establish the diagnostic limits of subspecies or their boundaries.

Coelops robinsoni Malayan tail-less leaf-nosed bat

C. robinsoni Bonhote, 1908: 4; foot of Mt Tahan, Pahang, Malaya.
Chilophylla hirsuta Miller, 1910: 395, pl. 18; Alag River, Mindoro I, Philippines.

Range (Map 55). Tarutau I (Thailand); Malaya; Borneo: Sarawak, Kalimantan (Robinson, 1990); ? Philippines: Mindoro.

Variation. It seems likely that *hirsuta* from the Philippines is no more than subspecifically related to *robinsoni*. It is known apparently only from the holotype skin in the US National Museum.

Remarks. The status of *C. robinsoni* was discussed by Hill (1972a, 1983).

PARACOELOPS

Paracoelops Dorst, 1947: 436; type species *P. megalotis* Dorst.

Vietnam; one species.

Paracoelops megalotis

P. megalotis Dorst, 1947: 436; Vinh, Annam, Vietnam.

Range (Map 55). Vietnam: known only from the holotype in the Paris Museum.

Remarks. The holotype is badly damaged. Dorst (1947, footnote) noted that the noseleaf was generally damaged and deformed, and that the skull had been broken in skinning, only the rostrum and lower jaw remaining. Until further material can be studied many of its external and cranial features must remain unknown.

Family Vespertilionidae

Evening bats

Vespertilia Rafinesque, 1815: 54; type genus *Vespertilio* Linnaeus. Conserved by Direction 98 (1958).
= Vespertilionidae Gray, 1821: 299.

A family of c. 44 genera and c. 350 species in 6 subfamilies, distributed throughout the world except for the Polar and sub-Polar regions and some small oceanic islands; 5 subfamilies (Table 73), 24 genera and c. 119 species occur in the Indomalayan Region. It is the largest in content of the bat families; one genus, *Myotis*, is the most widely distributed of bat genera. Fossil vespertilionids occur from the Eocene in Europe, the Miocene in Africa and Asia, the Pleistocene in Australia and South America and the Oligocene in North America. Members of the family range from small to moderate in size (forearm 24–80mm); there is no noseleaf, at most only a low transverse dermal ridge above the nostrils, the ears vary from small to very large, the tragus may be short and blunt to long and pointed, and the tail is enclosed in an extensive membrane, its tip sometimes protruding beyond the edge. A few have strongly tubular nostrils, or adhesive pads on the wrist at the base of the thumb, on the sole of the foot or on both.

For the most part vespertilionids are insectivorous, but a few species catch and eat fish. Insects are captured in flight, from foliage, on the ground or skimmed from the surface of water. A wide variety of roosts is utilized, from caves and rock shelters to man-made structures such as wells or mine shafts, but some roost in hollow trees, among foliage, under loose bark, in the internodal spaces of bamboos, even in disused bird nests or in flowers. Some species regularly colonise the attics of houses or other occupied buildings. Many species are gregarious and colonial, a few semi-solitary or solitary.

Subfamily Vespertilioninae

Cosmopolitan within the range of the family; contains the majority of vespertilionids with c. 36 genera and c. 289 species, of which 18 genera and c. 88 species occur in the region. The least specialized of the subfamilies, chiefly characterized by the absence of any of the modifications that distinguish the other sub-familial groups. To some extent the classification of the subfamily is uncertain and the subject of varying opinion. In some instances the features on which genera are based are few or unstable. Classifications of the Vespertilioninae were reviewed by Hill & Harrison (1987) who examined bacular variation in the subfamily in an attempt to resolve some of these difficulties. Menu (1985, 1987) analysed tooth morphology and discussed its systematic and phylogenetic implications.

1. Inner margins of ears closely approximated, joined over forehead by a band of integument; nostrils opening on upper surface of nose **2**

Inner margins of ears widely separated on head, not joined; nostrils opening anteriorly or antero-laterally **3**

2. Ears very long, about as long as forearm, extending far beyond tip of muzzle when laid forward; lower premolars 3–3 .... ***Plecotus***

Ears short, when laid forward barely extending (<3mm) beyond tip of muzzle; lower premolars 2–2 .... ***Barbastella***

3. Tragus long, slender, bluntly pointed, more or less straight, its length along the anterior margin at least three times its greatest width (Fig. 14a); cheekteeth usually 6/6 on each side .... ***Myotis***

Tragus shorter, wider, generally rounded at tip, usually curved forwards, its length along the anterior margin usually less than three times its greatest width (Fig. 14b); cheekteeth fewer than 6/6 on each side **4**

Table 73. The subfamilies of Vespertilionidae.

	Vespertilioninae	Miniopterinae	Murininae	Kerivoulinae	Nyctophilinae
Genera in region	17	1	2	2	1
Nostrils strongly tubular	0	0	+	0	0
Muzzle surmounted by low dermal ridge	0	0	0	0	+
Ear funnel-shaped	0	0	0	+	0
2nd phalanx of 3rd finger very long, nearly 3x 1st phalanx	0	+	0	0	0
Premolars 3/3 (cheekteeth 6/6)	+*/0	0	0	+	0
Anterior upper premolar (P^2) large, c. = P^4 (Fig.16)	0	0	+	0	0
Sternum short and wide	0	0	0	+	0

* In region, only in *Myotis*.

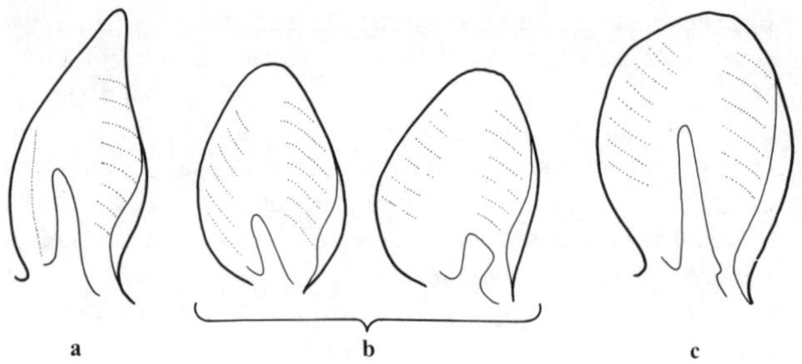

Fig. **14**. Ear and tragus of (a) *Myotis*; (b) *Pipistrellus*; (c) *Kerivoula*.

4. Upper premolars 2–2 .... **5**

Upper premolars 1–1 .... **10**

5. Skull greatly flattened; premolars 2/3 on each side, 2nd lower premolar much reduced, intruded from row .... ***Eudiscopus***

Skull not flattened; premolars 2/2 on each side.............. **6**

6. Outer upper incisor (I^3) minute, sometimes absent, when present not extending beyond cingulum of inner incisor (I^2) **7**

I^3 extending distinctly to or beyond cingulum of I^2 .... **8**

7. Small, forearm 32–38, greatest length of skull 13.5–14.8; outer upper incisor (I^3) when present with small pointed cusp... ***Scotozous***

Large, forearm 71–77, greatest length of skull 27.0–27.2; I^3 with flat crown .... ***Ia***

8. 5th finger only slightly longer than metacarpal of 3rd or 4th finger .... ***Nyctalus***

5th finger usually longer than metacarpal and 1st phalanx of 3rd or 4th finger **9**

9. No obvious thickened callosities or pads at base of thumb or on foot; outer upper incisor (I^3) postero-lateral to I^2, its concavity posteriorly directed ....***Pipistrellus***

Small thickened callosities or pads at base of thumb and on foot; I^3 lateral to I^2, its concavity directed laterally (Fig. 16f)...***Glischropus***

10. Upper incisors 2–2 .... **11**

Upper incisors 1–1 .... **15**

11. Distinct thickened callosities or pads at base of thumb and on foot; skull much flattened, the occipital depth < one third its greatest length...***Tylonycteris***

No callosities or pads at base of thumb or on foot; skull not noticeably flattened, occipital depth > one third its greatest length.. **12**

12. Upper canine with distinct posterior supplementary cusp, extending from one third to one half its length (Fig. 15f)...***Philetor***

Upper canine lacking any pronounced posterior supplementary cusp... **13**

13. Inner upper incisor (I^2) large, caniniform; I^3 low, usually large, generally crowded inward between I^2 and canine, or completely displaced to lie almost posteriorly to the inner tooth (Fig. 16h, i)...***Hesperoptenus***

I^2 not especially caniniform; I^3 small, not crowded inward, separated from canine... **14**

14. Rostrum noticeably concave laterally; narial emargination extending about halfway to interorbital constriction; anterior palatal emargination wider than long .... ***Vespertilio***

Rostrum evenly convex laterally; narial emargination not extending halfway to interorbital constriction; anterior palatal emargination longer than wide ....***Eptesicus***

15. Ears very large, extending beyond tip of muzzle when laid forward; auditory bullae very large, much wider than the intervening space ....***Otonycteris***

Ears not greatly enlarged; auditory bullae not expanded **16**

16. 1st and 2nd upper molars with reduced mesostyle (central labial cusp), the W-pattern of ridges and cusps distorted or nearly absent; 1st and 2nd lower molars with 2nd triangle smaller than 1st...***Scotophilus***

1st and 2nd upper molars with unreduced mesostyle and normal W-pattern of ridges and cusps; 1st and 2nd lower molars with 2nd triangle larger than 1st........................... **17**

17. Smaller: forearm 34–39, greatest length of skull 14.6–15.6; no dorsal pattern of white stripes or spots; anterior face of upper canine faintly grooved; 3rd upper molar not especially reduced, with well developed mesostyle, metacone (posterior medial cusp) and 3 ridges ....***Scotoecus***

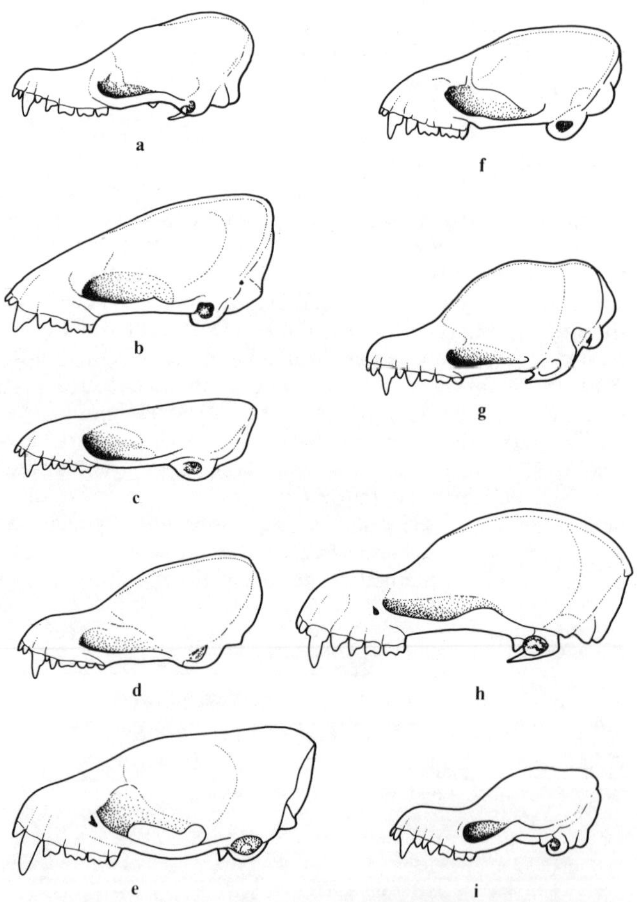

Fig. **15**. Skulls of representative species of Indomalayan Vespertilionidae: (a) *Myotis muricola*; (b) *Scotophilus kuhlii*; (c) *Tylonycteris robustula*; (d) *Pipistrellus javanicus*; (e) *Pipistrellus stenopterus*; (f) *Philetor brachypterus*; (g) *Miniopterus australis*; (h) *Murina cyclotis*; (i) *Kerivoula intermedia*. (After Payne *et al.*, 1985.)

Larger: forearm 50–62, greatest length of skull 19.1–22.0; usually a dorsal pattern of white stripes or spots; upper canine not grooved anteriorly; 3rd upper molar reduced with no more than 2 distinct ridges .... ***Scotomanes***

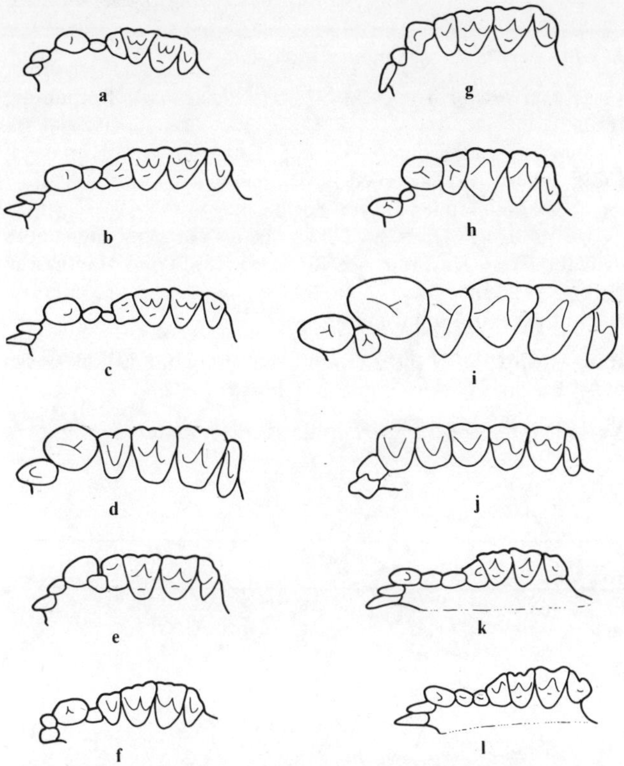

Fig. **16**. Upper left toothrows of: (a) *Myotis ridleyi*; (b) *Myotis hasseltii*; (c) *Myotis adversus*; (d) *Scotophilus kuhlii*; (e) *Pipistrellus javanicus*; (f) *Glischropus tylopus*; (g) *Philetor brachypterus*; (h) *Hesperoptenus doriae*; (i) *Hesperoptenus tomesi*; (j) *Murina cyclotis*; (k) *Kerivoula whiteheadi*; (l) *Kerivoula intermedia*. (After Payne *et al*., 1985.)

MYOTIS Mouse-eared bats

Myotis Kaup, 1829: 106; type species *Vespertilio myotis* Schreber, in *V. myotis* Borkhausen. Conserved by Opinion 98 (1958).
Leuconoë Boie, 1830: 256 (*Vespertilio daubentonii* Kuhl).
Selysius Bonaparte, 1841: 3 (*Vespertilio mystacinus* Kuhl).
= *Brachyotus* Kolenati, 1856: 131. Pre-occupied by *Brachyotus* Gould, 1837 (Aves).
Trilatitus Gray, 1842: 258. Included *Vespertilio hasselti* Temminck, *V. macellus* Temminck, and *V. blepotis* Temminck (a *Miniopterus*).
Pternopterus Peters, 1867d: 706 (*Vespertilio* (*P.*) *lobipes* Peters, in *V. muricola* Gray).
Exochurus Fitzinger, 1870: 75. Based upon *Vespertilio macrodactylus* Temminck, *V. horsfieldii* Temminck and *V. macrotarsus* Waterhouse.
Chrysopteron Jentink, 1910c: 74 (*Kerivoula weberi* Jentink, in *V. formosus* Hodgson).
Rickettia Bianchi, 1917:lxxvii (*Vespertilio* (*Leuconoe*) *ricketti* Thomas).
Megapipistrellus Bianchi, 1917:lxxvii (*Pipistrellus annectans* Dobson).
Dichromyotis Bianchi, 1917:lxxviii (*Vespertilio formosus* Hodgson).
Anamygdon Troughton, 1929: 87 (*A. solomonis* Troughton, in *Leuconoe moluccarum* Thomas).
And other synonyms based upon extralimital species.

Myotis

Almost cosmopolitan; c. 91 species in 8 or 9 subgenera, of which 5 subgenera and c. 27 species occur in the region (Table 74). It has probably the widest natural distribution of any genus of terrestrial mammal except man. The genus stands in need of detailed and overall revision, and its classification is provisional. Findley (1972) examined infrageneric relationships, while Tate (1941e) and more briefly Hill (1983) reviewed the Indomalayan species.

Subgenus *Myotis*

1. Larger: forearm >53, condylobasal length >18.6, C-M^3 >8.0 .... **2**

Smaller: forearm usually <53, condylobasal length <18.5, C-M^3 <8.0 .... **3**

Table 74. The subgenera of *Myotis*.

	Myotis	*Chrysopteron*	*Selysius*	*Leuconoe*	*Rickettia*
Species in region	6	2	10	8	1
Wing, forearm and interfemoral membrane conspicuously parti-coloured black, red, orange and yellow	0	+	0	0	0
Ear relatively small	0	0	+	+	+
Foot, including claw: tibia	<1/2	<1/2	<1/2	1/2–3/4	=
Braincase low, dorsal profile of skull gradually sloped	+	0	0	+/0	0
I_1 and I_2 4-lobed	+	+	0	0	0

2. Smaller: forearm 53–63, condylobasal length 18.6–22.0, C-M^3 8.0–9.5; outer upper incisor (I^3) not much exceeding I^2 in crown area; (N India).. ***M. blythii***

Larger: forearm 60–66, condylobasal length 22.0–23.1, C-M^3 9.0–10.8; I^3 about twice crown area of I^2; (E China, N Thailand).. ***M. chinensis***

3. Smaller: forearm <45, condylobasal length <15.3, C-M^3 <6.6 .... **4**

Larger: forearm >45, condylobasal length >15.5, C-M^3 >7.5 .... **5**

4. Outer margin of ear with conspicuous emargination; (? W Pakistan).. ***M. emarginatus***

Outer margin of ear lacking any conspicuous emargination; (S China, N Thailand).. ***M. altarium***

5. Pelage not especially short or velvety; dorsal surface brown, with a slight greyish tinge; hair bases little darker than remainder of hair; ventral surface buffy brown; (NE India) .... ***M. sicarius***

Pelage short, velvety; dorsal surface drab greyish brown, the hairs dark based, otherwise drab; ventral surface whitish grey, hairs slaty based; (S China).. ***M. pequinius***

Myotis blythii Lesser mouse-eared bat

Vespertilio blythii Tomes, 1857: 53; Nasirabad, Rajputana, India.
V. murinoides Dobson, 1873d: 205; Chamba, NW Himalaya, India, 3000ft. Pre-occupied by *V. murinoides* Lartet, 1851.
= *V. dobsoni* Trouessart, 1878: 248.
? *M. africanus* Dobson, 1875f: 260; 'Gaboon' but more probably from Kashmir, see Blanford (1891) and Hayman & Hill (1971: 35).
Also other extralimital names.

Range (Map 56). N India (Rajputana, Punjab, Kashmir); also to W Europe, N Africa and N China.

Variation. Subspecies in N India is *M. b. blythii.*

Map **56**. *Myotis* (part)

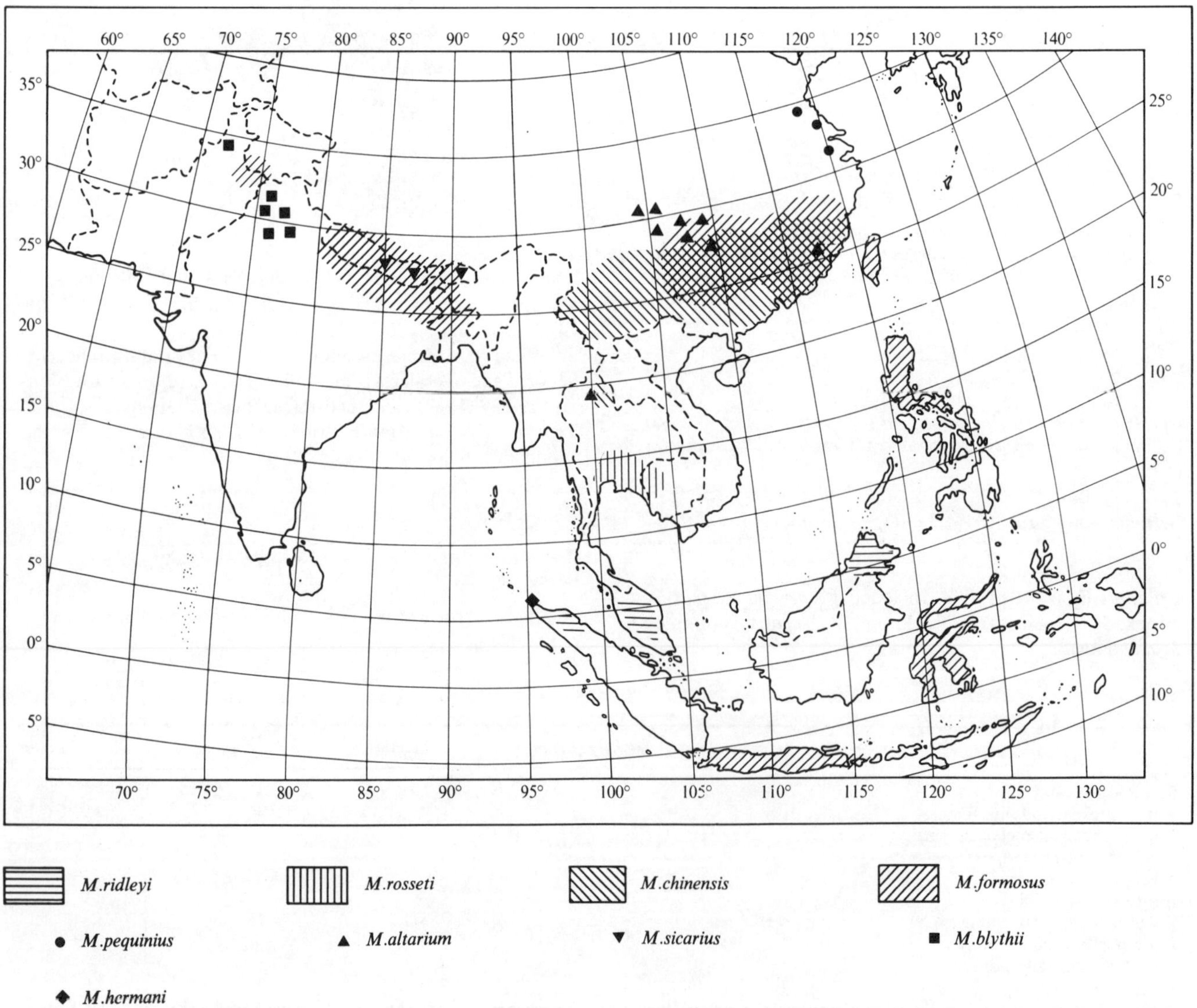

Myotis chinensis

Vespertilio chinensis Tomes, 1857: 52; S China.

Range (Map 56). China (Yunnan, Sichuan, Guangxi, Guizhou, Hunan, Zhejiang, Fujian, Hong Kong); N Thailand; also to N China.

Variation. A weakly differentiated subspecies *M. c. luctuosus* Allen, 1923a: 5 (Wahnsien, Sichuan) has been described, differing in having a more or less uniformly grey ventral surface lacking any black flank stripe.

Remarks. Unlikely to prove conspecific with either *M. myotis* or *M. blythii* (Corbet, 1978), although included in the otherwise Palaearctic *M. myotis* by Ellerman & Morrison-Scott (1951).

[*Myotis emarginatus* Geoffroy's bat]

Vespertilio emarginatus Geoffroy, 1806: 198; Charlemont, Givet, Ardennes, France.
V. desertorum Dobson, in Blanford, 1875b: 304; Jalk, E Iran.
And other extralimital names.

Range. Not so far recorded from the region but occurs close to the Iran-Pakistan border (Roberts, 1977) and it is likely to occur in Pakistan.

Myotis altarium

M. altarium Thomas, 1911a: 3 (and 1911b: 161); Omei Shan, Sichuan, China.

Range (Map 56). S China (Sichuan, Guizhou, Fujian); N Thailand (Blood & McFarlane, 1988).

Remarks. Blood & McFarlane (1988) re-described *M. altarium* on the basis of a specimen from Thailand, and concluded that it is best considered a member of the subgenus *Myotis*, rather than *Selysius* to which Tate (1941e) and Findley (1972) allocated it.

Myotis sicarius

M. sicarius Thomas, 1915k: 608; N Sikkim, India.

Range (Map 56). NE India: Sikkim; Nepal.

Myotis pequinius

M. (*Leuconoe*) *pequinius* Thomas, 1908j: 637; 30 miles W of Pekin, Chihli, China, 600ft.

Range (Map 56). S China (Jiangsu); also to N China (Henan, Shandong, Hebei).

Subgenus *Chrysopteron* (Table 75).

Table 75. The species of *Myotis* (*Chrysopteron*).

	formosus	*hermani*
Range in region	N India-Sulawesi	Sumatra
Forearm	45–51	c.60
Condylobasal length	16.3–17.8	20.0
C-M^3	6.8–7.3	8.6

Myotis formosus Hodgson's bat

Vespertilio formosa Hodgson, 1835a: 700; Nepal.
V. rufopictus Waterhouse, 1845: 3, 8; Philippines.
V. rufoniger Tomes, 1858a: 79, pl. 60; Shanghai, Jiangsu, China.
Kerivoula pallida Blyth, 1863a: 34; Chaibassa, Orissa, India.
V. auratus Dobson, 1871e: 186; Darjeeling, NE India.
V. dobsoni Anderson, 1881: 143; Purneah, Bengal, India. Preoccupied by *V. dobsoni* Trouessart, 1878.
= *V. andersoni* Trouessart, 1897: 129.
Kerivoula weberi Jentink, 1890a: 129, pl. 11; Loka, near Bontain, S Sulawesi.
Chrysopteron bartelsii Jentink, 1910c: 74; Mt Panerango, Java, 10000ft.
M. watasei Kishida, 1924a: 36,38; Terason, Taiwan.
? *M. flavus* Shamel, 1944: 191; Enri, Taiwan.

Range (Map 56). N India, S China (Guizhou, Guangxi, Fujian), Taiwan, Luzon (Philippines), Java, Bali, Sulawesi; also to E Afghanistan, Tibet, N China, Korea and Tsushima I, Japan.

Variation. Known specimens are quite inadequate for any study of infraspecific variation.

Remarks. IUCN: K (*M. bartelsii*).

Myotis hermani Herman's bat

M. hermani Thomas, 1923a: 252; Sabang, NW Sumatra.

Range (Map 56). Sumatra.

Remarks. Findley (1972) subsumed *hermani* into *M. formosus*; however, the holotype of *hermani* is very much larger than the largest of *formosus* as it is represented in the BMNH, but is a very old individual. IUCN: R.

Subgenus *Selysius*

The following key does not include *Myotis oreias* from Singapore which is poorly known.

1. Smaller: forearm 27–41.5, condylocanine length 11.0–14.1 **2**

Larger: forearm 38–47, condylocanine length 14.0–15.6 .. **9**

2. Braincase domed, rising abruptly from the rostrum **3**

Braincase flatter, not rising abruptly from the rostrum ... **4**

3. Larger: forearm 39–40; occiput elevated; rostrum uptilted; anteorbital foramen narrowly separated from anterior rim of orbit .... ***M. frater***

Smaller: forearm 29.5–35; occiput not elevated; rostrum horizontal; anteorbital foramen widely separated from anterior rim of orbit .... ***M. siligorensis***

4. Dorsal pelage overall cinnamon to cinnamon-rufous, hairs dark brown based; hairs on feet reddish-orange .... ***M. patriciae***

Dorsal pelage brownish or blackish-brown, lacking any reddish tinge; hairs on feet brownish **5**

5. Rostrum short and broad, the postorbital width subequal to the length orbit-gnathion; canines short, barely exceeding the posterior premolars in height; 2nd upper and lower premolars ($P^3/_3$) absent (Fig. 16a) **6**

Rostrum longer, narrower, the postorbital width much less than the length orbit-gnathion; canines not shortened; 2nd upper premolar (P^3) very rarely lacking, 2nd lower premolar (P_3) present 7

6. Base of thumb expanded to form a fleshy, wrinkled pad; sole of foot similarly expanded into a broad, concave pad .... ***M. rosseti***

No pad at base of thumb or on sole of foot..... ***M. ridleyi***

7. Larger: forearm 34.5–41.5, condylocanine length 12.5–14.1, C-M^3 5.4–5.9; 2nd upper premolar very small, occasionally absent, when present intruded into recess between anterior and posterior premolars, these in contact or nearly so; 2nd lower premolar slightly intruded or partially intruded, compressed, occasionally absent.......... ***M. ater***

Smaller: forearm 31–36.5, condylocanine length 11.2–12.6, C-M^3 4.7–5.4; 2nd upper premolar small, rarely very small, usually in row or only slightly intruded, separating anterior and posterior premolars; 2nd lower premolar usually in row, sometimes slightly intruded.......... 8

8. Dorsal pelage brownish, hairs black-based, tipped with shiny, paler ochraceous brown; ventrally whitish, hairs black-based, tipped with white or creamy white..... ***M. mystacinus***

Dorsal pelage darker, blackish, hairs similarly black-based but less profusely tipped with shiny brown; ventral pelage black-based but tipped with ochraceous brown .... ***M. muricola***

9. Anteorbital foramen widely separated from anterior rim of orbit.......... ***M. montivagus***

Anteorbital foramen narrowly separated from anterior rim of orbit.......... ***M. annectans***

Myotis frater

M. frater Allen, 1923a: 6; Yenping, Fujian, SE China. And extralimital synonyms.

Range (Map 57). SE China (Fujian); and from Afghanistan, Tibet and S USSR to Manchuria, Korea and Japan.

Myotis siligorensis Himalayan whiskered bat

Vespertilio siligorensis Horsfield, 1855: 102: Siligori, Nepal.
? *V. darjilingensis* Horsfield, 1855: 102.
M. sowerbyi Howell, 1926a: 138; Yenpingfu, Fujian, S China, 3000ft.

Map 57. *Myotis* (part)

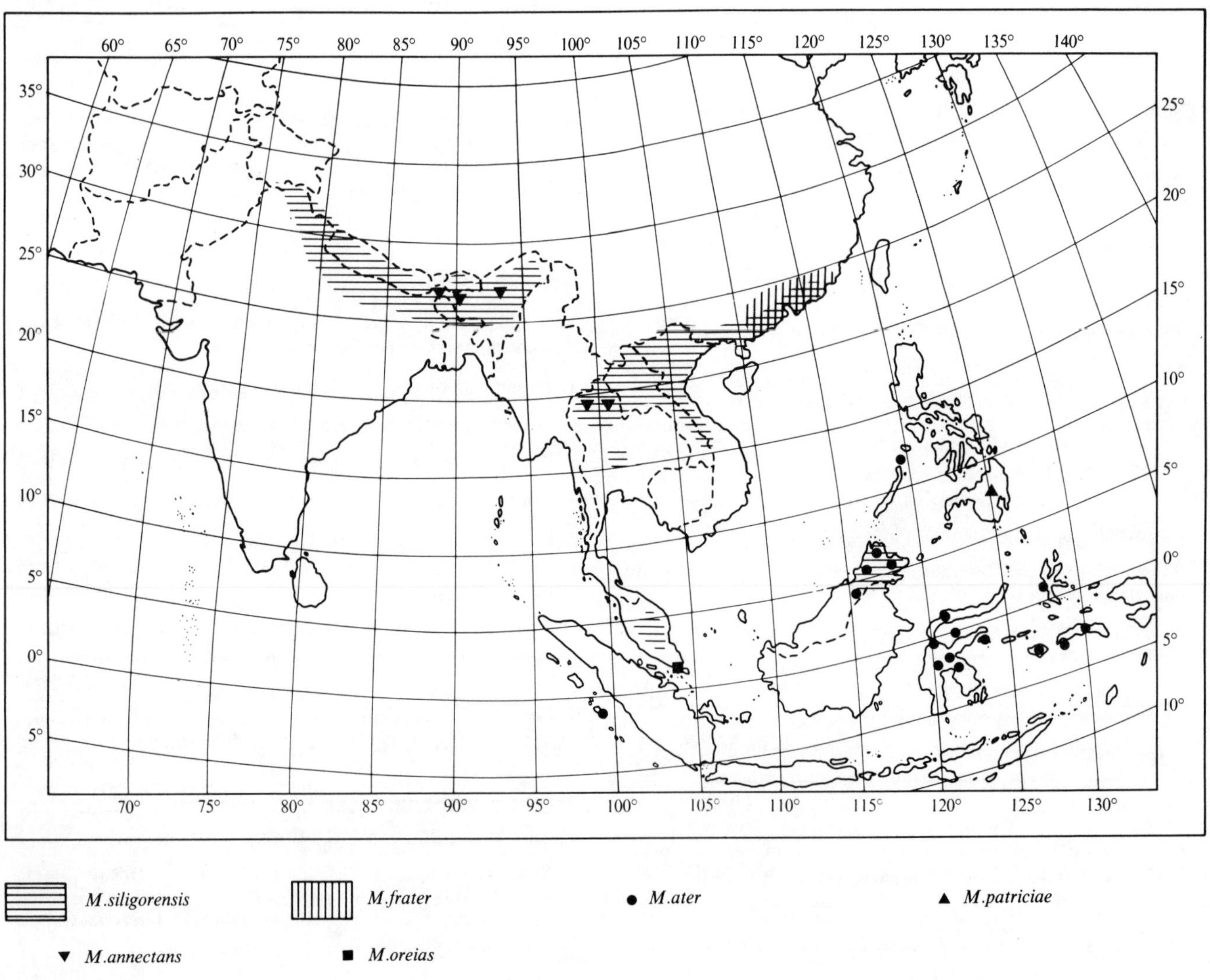

Range (Map 57). N and NE India, Thailand, SE China (Fujian), Vietnam; Malaya (Hill, 1972a); Borneo (Sabah—Kobayashi *et al.*, 1980).

Variation. The following subspecific names have been proposed:
M. s. alticraniatus Osgood, 1932: 232; Mouong Moun, Tonkin, Vietnam.
M. s. thaianus Shamel, 1942: 323; Chiengmai, N Thailand.

Myotis rosseti Thick-thumbed myotis

Glischropus rosseti Oey, 1951: 4; Cambodia.

Range (Map 56). Thailand (Bangkok) and Cambodia.

Remarks. This species has been reviewed by Hill (1969) and Hill & Topál (1973). Described initially in *Glischropus* on account of the pads on thumbs and feet, it was subsequently referred to *Pipistrellus* chiefly because of the absence of the second upper and lower premolars ($P^3/_3$) (Hill, 1969), and finally to *Myotis* by Hill & Topal (1973).

Myotis ridleyi Ridley's bat

Pipistrellus ridleyi Thomas, 1898e: 361; [Kepong], Selangor, Malaya.

Range (Map 56). Malaya (Perak, Pahang, Selangor); ? Sumatra (Ärnback-Christie-Linde, 1909); Borneo (Sabah–Hill & Francis, 1984).

Remarks. Accounts of this species were provided by Hill (1969) and Hill & Topál (1973). As with *M. rosseti*, its anomalous dentition, with the second upper and lower premolars lacking, led to its initial allocation to *Pipistrellus*.

Myotis ater

Vespertilio ater Peters, 1866c: 18; Ternate I, Moluccas.
V. adversus var. *amboinensis* Peters, 1866d: 400; Ambon I, Moluccas.
M. abbotti nugax Allen & Coolidge, 1940: 137; Bundutuan, Mt Kinabalu, Sabah, Borneo, 3500ft.

Range (Map 57). Mentawai Is, W Sumatra: Siberut (Hill, 1983); Borneo (Sarawak, Sabah); Philippines: Culion I; Sulawesi; Togian Is: Malenge (Hill & Rozendaal, 1989); Vagian I; Peleng I; Moluccas: Ternate, Buru, Seram, Ambon. Also to Papua New Guinea, possibly Australia (Hill, 1983).

Variation. The population in NW Borneo and on Culion I, Philippines may represent a distinct subspecies, *M. m. nugax*, but is poorly known (Hill & Francis, 1984).

Remarks. This species was reviewed in some detail by Hill (1983).

Myotis mystacinus Whiskered bat

Vespertilio mystacinus Kuhl, 1819: 202; Germany.
V. pallidiventris Hodgson, 1844b: 286. *Nomen nudum.*
V. nipalensis Dobson, 1871d: 214; Katmandu, Nepal.
M. meinertzhageni Thomas, 1926c: 609; junction of Nubra and Shyok rivers, Ladak, Kashmir.
And many other extralimital names.

Range (Map 58). ? N Pakistan; Kashmir; NE India (Bhutan Duars); Nepal; S China (Jiangxi, Fujian, ? Hainan). Also in the Palaearctic from W Europe and N Africa to Korea and Japan.

Variation. Available specimens are too few for any assessment but possibly two subspecies, *M. m. meinertzhageni*, paler, and *M. m. nipalensis*, somewhat darker, might be recognized in N India.

Myotis muricola

Vespertilio tralatitius Temminck, 1840: 228, pl. 57; Java. Preoccupied by *V. tralatitius* Horsfield, 1824.
V. muricola Gray, 1846b: 4; Nepal. (*V. muricola* Hodgson, 1841b: 212 and 1842b: 908 are *nomina nuda*).
V. caliginosus Tomes, 1859b: 73; India.
V. (*Pternopterus*) *lobipes* Peters, 1867d: 706; Akyab, Arakan, Burma.
V. blanfordi Dobson, 1871d: 214; Himalayas.
V. moupinensis Milne-Edwards, 1872b: 253, pl. 37a; Moupin, Sichuan, China.
M. niasensis Lyon, 1916a: 441; Siab Bay, Nias I, off W Sumatra.
M. latirostris Kishida, 1932: 153; C Taiwan.
M. browni Taylor, 1934: 288; near Saub, Cotabato, Mindanao I, Philippines.
M. herrei Taylor, 1934: 289, 290; cave at Montalban, Rizal Prov., Luzon I, Philippines.

Range (Map 58). From N Pakistan and Kashmir through N India, Nepal and Burma to S China (Yunnan, Sichuan, Fujian); Taiwan; south through Vietnam, Laos and Thailand to Malaya; Sumatra, Java, Borneo; Luzon, Mindanao (Philippines); Sulawesi; and on the smaller islands of Mergui Is, Nias, Simalur (W Sumatra), Sipura (Mentawai Is), Bunguran (Natuna Is), Kangean Is; Bali, Lombok (Kitchener *et al.*, 1990), Sumbawa, Sumba, Flores; Ambon (Moluccas). Also to Afghanistan and S Tibet.

Variation. The following subspecific name has been proposed:
M. m. orii Kuroda, 1935: 290; Taihezon, Taihokusiu, N Taiwan.

Remarks. This species and its separation from *M. mystacinus* in Pakistan and N India were discussed in some detail by Hill (1983), who amplified the view of Corbet (1978) that *mystacinus* and *muricola* should be considered specifically distinct.

Myotis patriciae

M. patriciae Taylor, 1934: 286; Agusan Prov., Mindanao I, Philippines.

Range (Map 57). Mindanao I, Philippines.

Remarks. This very small form (forearm 27mm) was based on a young adult. It is apparently close to *M. muricola*: according to Koopman (in Honacki *et al.*, 1982) it is almost certainly a synonym of *M. muricola browni*.

Myotis montivagus Burmese whiskered bat

Vespertilio montivagus Dobson, 1874c: 237; Hotha, Yunnan, China.
M. peytoni Wroughton & Ryley, 1913: 13; Gersoppa Falls, Kanara, S India, 1300ft (not Pashok, near Darjeeling, NE India as quoted erroneously by Hill, 1962a).

Map **58**. *Myotis* (part)

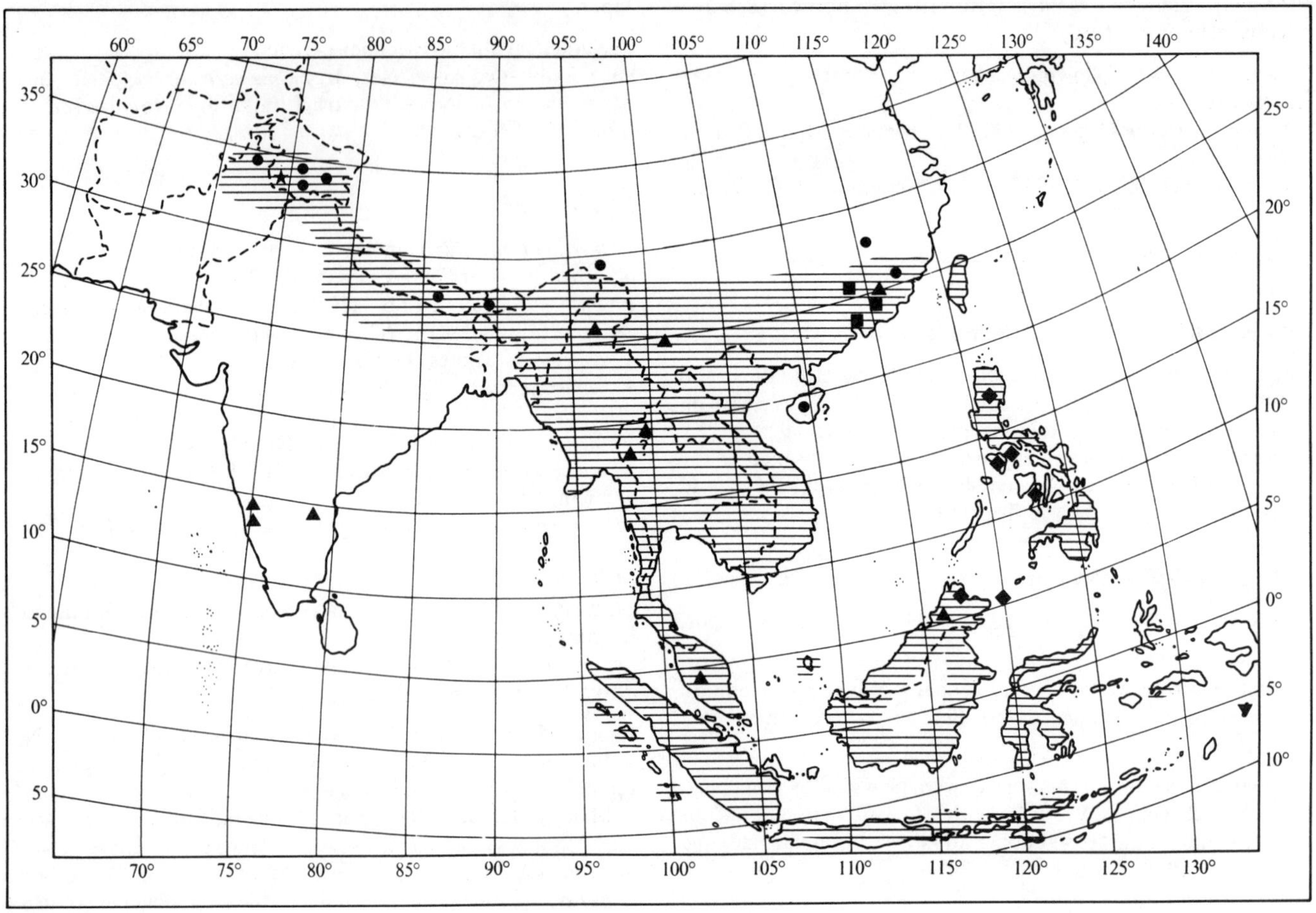

M.muricola ● *M.mystacinus* ▲ *M.montivagus* ▼ *M.stalkeri*

■ *M.fimbriatus* ◆ *M.macrotarsus* ✶ *M.longipes*

Range (Map 58). S India; N Burma, S China (Yunnan, Fujian); ? W, NE Thailand, Malaya; Borneo (Sabah).

Variation. Two subspecific names have been proposed:
M. peytoni federatus Thomas, 1916j: 3; Semangko Pass, Malaya, 2700ft.
M. m. borneoensis Hill & Francis, 1984: 309; Sepilok, Sabah, Borneo, 5°32'N, 117°56'E.

Four subspecies can be recognized; the Bornean *M. m. borneoensis* is cranially the largest, while *M. m. montivagus* from S China and N Burma is considerably smaller. In S India and Malaya *peytoni* and *federatus* are intermediate in size but tend towards *borneoensis*. Overall there is some variation in the displacement of the second lower premolar (P_3), which is in the toothrow in *montivagus*, sometimes partially intruded in *peytoni*, and almost wholly intruded in *federatus* and *borneoensis*.

Remarks. The species was discussed, with measurements, by Hill (1962a) and Hill & Francis (1984), while Das (1987) reviewed the type series of *V. montivagus*. Findley (1972) suggested that *montivagus* might be more distantly separated from *peytoni, federatus* and, by implication, *borneoensis* than in the arrangement adopted here, which is based on Hill (1962a).

Myotis annectans Hairy-faced bat

Pipistrellus annectans Dobson, 1871d: 213; Naga Hills, Assam, NE India.
M. primula Thomas, 1920g: 248; Pashok, near Darjeeling, NE India, 3500ft.

Range (Map 57). NE India; NE Thailand.

Remarks. Reviewed by Topál (1970a) who transfered the species to *Myotis* from *Pipistrellus*.

Myotis oreias

Vespertilio oreias Temminck, 1840: 270; Singapore I.

Range (Map 57). Known apparently only from the original specimen from Singapore.

Remarks. The status of *oreias* must remain uncertain. Tate (1941e) examined the holotype (Leiden Museum) and remarked 'Either a *Selysius* or one of the long-eared groups *Isotus* or *Paramyotis*', providing some measurements and noting that the skull is in fragments. According to Sullivan (in Honacki *et al.*, 1982) it may be a synonym of *M. mystacinus*, i. e. *M.*

muricola, but the toothrow is rather long (6.1mm) for this species.

Subgenus *Leuconoe*

Leuconoe was given generic rank by Menu (1987), with *Myotis* among others as one of its subgenera; however, the latter is by far the earliest name.

1. Larger: forearm >45, condylobasal length 16.9–17.6, C-M^3 6.9-7.4 .... **2**

Smaller: forearm <44, condylobasal length <15.8, C-M^3 <6.4 .... **3**

2. Skull generally smaller, condylobasal length c. 16.9; 2nd upper premolar (P^3) usually little intruded from toothrow ; (Philippines, Borneo) .... ***M. macrotarsus***

Skull generally larger, condylobasal length c. 17.6; P^3 intruded, adjacent premolars almost in contact; (Kai Is).......... ***M. stalkeri***

3. Outer margin of tibia with a fringe of moderate hairs **4**

Outer margin of tibia lacking any fringe of hairs **5**

4. Shallow lateral depressions on each side of rostrum above anterior margin of orbit .... ***M. longipes***

Moderate to pronounced lateral depressions on each side of rostrum .... ***M. fimbriatus***

5. 2nd upper premolar (P^3) in toothrow, separating adjacent premolars .... ***M. daubentonii***

P^3 intruded into recess between adjacent premolars, these sometimes in contact or nearly so **6**

6. Wing inserted on metatarsus .... ***M. horsfieldii***

Wing inserted at ankle or on tibia **7**

7. Fur short; post-palatal extension short, its median spicule lacking thin bony supporting laminae; 2nd upper premolar (P^3) minute, completely intruded, adjacent premolars in contact or nearly so; 2nd lower premolar (P_3) small, partially in row or completely intruded (Fig. 16b, p. 119) .... ***M. hasseltii***

Fur dense, woolly; post-palatal extension long, its median spicule supported by thin bony laminae; 2nd upper (P^3) and lower (P_3) premolars less reduced, upper not usually much intruded, lower usually in row or only slightly intruded (Fig. 16c, p. 119) .... ***M. adversus***

Myotis macrotarsus

Vespertilio macrotarsus Waterhouse, 1845: 3, 5; Philippines.

Range (Map 58). Philippine Is: Luzon, Marinduque, Mindoro, Guimaras, Mindanao, Tawitawi (Sulu Is); Borneo (Sabah). The record from Long Lama, Baram, Sarawak (Medway, 1977a) was based upon an example of *Kerivoula papillosa* in the BMNH.

Variation. One subspecies has been described:
M. m. saba Davis, 1962: 40; Semporna, Lahad Datu Dist., Sabah, Borneo.
Larger than *M. m. macrotarsus*, lacking the reddish element in its coloration.

Myotis stalkeri Kai myotis

M. (*Leuconoe*) *stalkeri* Thomas, 1910c: 384; Ara, Kai Is.

Range (Map 58). Kai Is, E Moluccas.

Remarks. Very probably conspecific with *M. macrotarsus* as Findley (1972) suggested, but very poorly known as yet.

Myotis fimbriatus

Vespertilio fimbriatus Peters, 1871b: 617; Amoy, Fujian, China.
M. hirsutus Howell, 1926a: 139; Yenpingfu, Fujian, China, 2000ft.

Range (Map 58). SE China (Fujian, Hong Kong).

Remarks. The status of *fimbriatus* is open to some question. Several authors (Tate, 1941e; Ellerman & Morrison-Scott, 1951; Wallin, 1969) have suggested that it is related to or perhaps conspecific with the Palaearctic *M. capaccinii*. However, Corbet (1978) remarked that a probable paratype of *fimbriatus* did not suggest conspecificity with *capaccinii*, an opinion confirmed by further study of the two apparent paratypes in the BMNH.

Myotis longipes

? *M. theobaldi* Blyth, 1856a: 363; caves near Matar Nag, N of Islamabad, Kashmir. Indeterminable: see Blanford (1891) and Thomas (1915k).
Vespertilio macropus Dobson, 1872d: 209; caves of Bhima Divi, Kashmir, 6000ft. Pre-occupied by *V. macropus* Gould, 1854.
= *V. longipes* Dobson, 1873e: 110.
? *V. megalopus* Dobson, 1875f: 261; 'Gaboon', but more probably from Kashmir: see Blanford (1891); Thomas (1915k); and Hayman & Hill (1971: 33).

Range (Map 58). Kashmir; ? Vietnam (Osgood, 1932); also west to Afghanistan.

Remarks. Ellerman & Morrison-Scott (1951) suggested that *longipes* might be allied to or represent *M. capaccinii*, probably following Tate (1941e). It was reviewed by Hanák & Gaisler (1969) who considered it a distinct species, a view supported by Corbet (1978) from the examination of a paratype.

Myotis daubentonii Daubenton's bat, water bat

Vespertilio daubentonii Kuhl, 1819: 195; Hanau, Hessen-Nassau, Germany.
? *V. laniger* Peters, 1871b: 617; Amoy, Fujian, China.
Also many extralimital names.

Range (Map 59). NE India, Vietnam (both Topál, pers. comm.), China (SE Tibet, Yunnan, Fujian), Hong Kong (BMNH), Hainan I; also widely distributed in the Palaearctic from W Europe to Japan.

Myotis horsfieldii

Vespertilio horsfieldii Temminck, 1840: 226; Mt Gede, Java.
M. dryas Andersen, 1907b: 33; Port Blair, S Andaman I.
Leuconoe peshwa Thomas, 1915k: 611; Poona, Bombay, India.
L. lepidus Thomas, 1915m: 171; Baram, Sarawak, Borneo.
? *M. jeannei* Taylor, 1934: 284; near Caldera, Zamboanga, Mindanao I, Philippines.
M. deignani Shamel, 1942: 324; Chiengmai, Thailand.

Map **59**. *Myotis* (part), *Plecotus*

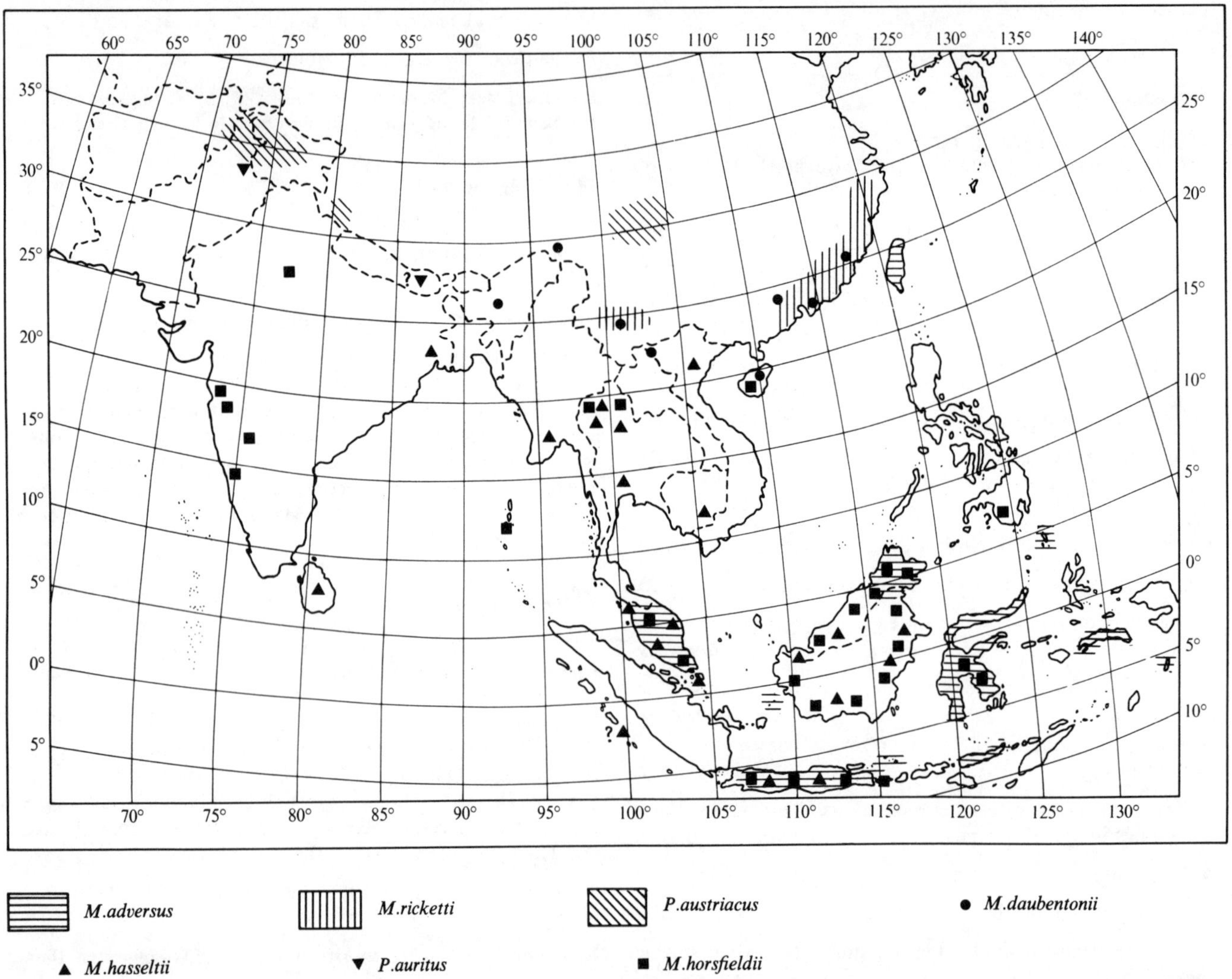

Range (Map 59). S, C India; Hainan I, ? Hong Kong (Findley, 1972); S Andaman I; Thailand, Malaya; Borneo; ? Mindanao (Philippines); Java, Bali, Sulawesi.

Variation. The species was briefly reviewed and its size variation tabulated by Hill (1983) who associated *deignani* (Thailand and S China) and provisionally *peshwa* (India), *dryas* (S Andaman I) and *jeannei* (Mindanao I) with *horsfieldii* (Malaya to Sulawesi) as subspecies. The range of variation is small and all may not prove valid.

Myotis hasseltii

Vespertilio hasseltii Temminck, 1840: 225; Bantam, Java.
V. macellus Temminck, 1840: 230; Borneo.
? *M. berdmorei* Blyth, 1863a: 35; 'Tenasserim provinces', Burma. Based on Blyth, 1859a: 293; according to Blanford (1891: 330) from Schwegyin, the type material lost.
? *M. abbotti* Lyon, 1916a: 441; N Pagi I, Mentawai Is, W Sumatra.
M. adversus continentis Shamel, 1942: 323; Bangkok, Thailand.

Range (Map 59). Sri Lanka; NE India, Vietnam (both Topál, pers. comm.); Burma, Thailand, Cambodia, Malaya; Sumatra, Java, Borneo; ? Sulawesi (Jentink, 1887,1888a); and the small islands of Langkawi (Malaya), Riau Is; ? N Pagai I (Mentawai Is).

Variation. Size variation was tabulated by Hill (1983) who recognized four subspecies, namely *continentis* (Burma to Cambodia), *hasseltii* (Malaya to Java), *macellus* (Borneo) and provisionally *abbotti* (N Pagai I). Specimens from Sri Lanka are slightly larger in some respects than *continentis* which otherwise they closely resemble. The brief description of *M. berdmorei* by Blyth (1859a: 293) suggests this species.

Myotis adversus

Vespertilio adversus Horsfield, 1824:part 8; Java.
M. carimatae Miller, 1906a: 62; Telok Edar, Karimata I, Borneo.
M. taiwanensis Ärnback-Christie-Linde, 1908: 235; Takao, Anping, Tainan, Taiwan.
Leuconoe moluccarum Thomas, 1915m: 170; Ara, Kai Is.
And extralimital names.

Range (Map 59). Taiwan; Malaya; Java, Borneo (Sabah, Karimata I), Sulawesi; Mendanau I; Kangean I (Bergmans & van Bree, 1986); Lesser Sunda Is: Flores (Hill & Rozendaal, 1989); Togian Is (Owen *et al.*, 1987; Hill, 1991); Peleng I;

Talaud Is: Karakelang (Hill & Rozendaal, 1989); Moluccas: Seram, Ambon, Kai Is; also east to Vanuatu and E Australia.

Variation. The species was briefly reviewed by Hill (1983) who tabulated size variation over most of the range. Four subspecies were recognized in the region: *taiwanensis* is poorly known, but the more southerly *adversus* (Java, Lesser Sunda Is, perhaps Malaya), *carimatae* (Borneo and adjacent islands) and *moluccarum* (Sulawesi to Solomon Is) differ in small points of size and cranial morphology.

Remarks. The classification of *M. horsfieldii, M. hasseltii* and *M. adversus* was summarized by Hill (1983).

Subgenus *Rickettia*

Allen (1936b, 1938b) considered that *Rickettia* justified generic rank, but it is clearly closely related to but more extreme than the subgenus *Leuconoe* into which it was incorporated by Findley (1972).

Myotis ricketti Rickett's big-footed bat

? *Vespertilio* (*Leuconoe*) *pilosa* Peters, 1869a: 403; 'Montevideo', Uruguay.
V. (*L.*) *ricketti* Thomas, 1894f: 300; Foochow, Fujian, China.

Range (Map 59). S and E China (Yunnan, Fujian), Hong Kong; also to NE China.

Remarks. It seems very probable that *pilosa* is a prior and correct name for this species. Its status was discussed by Miller & Allen (1928) and Allen (1936b), the latter author (1938b) adopting it as the valid name for this species, a lead not so far followed by subsequent authorities. As yet no critical examination has been made of the holotype, which apparently survives in the Paris Museum (Carter & Dolan, 1978).

PLECOTUS Long-eared bats

Plecotus Geoffroy, E.,1818: 112; type species *Vespertilio auritus* L.
And other extralimital synonyms.

Nearctic and Palaearctic Regions, marginal in the Indomalayan Region; 6 species in 2 subgenera, one (*Corynorhinus*) Nearctic, the other Palaearctic; 2 species reach the edge of the Indomalayan Region (Table 76); the recently described *P. taivanus* Yoshiyuki, 1991; 189 from Taiwan has not been assessed nor formally included (*Bull. natn. Sci Mus. Tokyo* A, 17: 189–95).

Plecotus auritus Brown long-eared bat

Vespertilio auritus Linnaeus, 1758: 32; Sweden.
? *P. homochrous* Hodgson, 1847h: 895; Nepal.
? *P. puck* Barrett-Hamilton, 1907: 521; Murree, Punjab, Pakistan, 7500ft.
And many other extralimital names.

Range (Map 59). N India and Pakistan; also through much of the Palaearctic from W Europe to Japan.

Remarks. Classification in E Asia uncertain; Corbet (1978) provisionally referred *homochrous* with synonym *puck* to *P. auritus*.

Table 76. The species of *Plecotus*.

	auritus	*austriacus*
Width of tragus	c.5	c.6
Length of thumb	Usually >6	Usually <6.1
Height of canine	Usually <1.7	Usually >1.7
Proximal zone of dorsal pelage	Usually brown or brownish grey	Dark or black
Proximal zone of ventral pelage	Usually dark brown or buff	Blackish
Paired basal limbs of baculum	Slender, longer than wide	Broad, about as long as wide

Plecotus austriacus Grey long-eared bat

Vespertilio auritus austriacus Fischer, 1829: 117; Vienna, Austria.
P. wardi Thomas, 1911e: 209; Leh, Ladak, Kashmir, NW India.
P. ariel Thomas, 1911a: 3 (and 1911b: 160); Tatsienlu, Sichuan, China, 8400ft.
And many extralimital names.

Range (Map 59). N Pakistan, Kashmir (Gaisler, 1970; Nath, 1987), SW China (Sichuan); also through much of the southern Palaearctic from SW Europe to Mongolia and W China.

BARBASTELLA

Barbastella Gray, 1821: 300; type species *Vespertilio barbastellus* Schreber.
= *Synotus* Keyserling & Blasius, 1839: 305.

A genus of two species occurring from Europe and N Africa to China and Japan; one species marginally in the region.

Barbastella leucomelas Eastern barbastelle

Vespertilio leucomelas Cretzschmar, 1830 or 1831: 73, pl. 18b; Arabia Petraea (= Sinai, Egypt).
Plecotus darjelingensis Hodgson, in Horsfield, 1855: 103; Darjeeling, NE India.
= *B. blanfordi* Bianchi, 1917:lxxv.
And other extralimital names.

Range (Map 60). Afghanistan, N Pakistan, N India (Rajputana, Bhutan Duars, Sikkim, Assam), Nepal, SW China (Yunnan, Sichuan, Gansu); possibly Indochina; also west to Turkey, Israel, perhaps Eritrea; north to Japan.

OTONYCTERIS

Otonycteris Peters, 1859: 223; type species *O. hemprichii* Peters.

An apparently monospecific genus distributed from North Africa to NW India.

Otonycteris hemprichii Hemprich's long-eared bat

O. hemprichii Peters, 1859: 223; locality unknown, restricted to Nile Valley between Assuan in Egypt and Northern Province, Sudan by Kock (1969d).
O. cinereus Satunin, 1909: 281,297; Nukendzaga, Ge Dist., NE Iran.
And other extralimital names.

Range (Map 60). N Afghanistan, N Pakistan, Kashmir (Nath, 1987); also west to Turkey, Israel, Algeria and Niger.

Variation. Subspecies in region is *O. h. cinerea*, extending to Iran.

Remarks. The status of *cinerea* was discussed by DeBlase (1980). Ognev (1928) regarded it as a valid species on account of darker colour, longer ears, and larger size. Ellerman & Morrison-Scott (1951) and Corbet (1978) listed it as a possible subspecies of *O. hemprichii.*

SCOTOMANES

Scotomanes Dobson, 1875c: 371; type species *Nycticejus ornatus* Blyth.
Scoteinus Dobson, 1875c: 371 (*Nycticejus emarginatus* Dobson). See Hill (1974c) for use of *Scotomanes*, rather than of *Scoteinus*.

NE India through S China and N Thailand to Vietnam; one species.

Scotomanes ornatus Harlequin bat

Nycticejus ornatus Blyth, 1851b: 517; Cherrapunji, Khasi Hills, Assam, India.
N. nivicolus Hodgson, in Horsfield, 1855: 104; N region of Sikkim Himalayas, India.
? *N. emarginatus* Dobson, 1871d: 211; ? India.

Range (Map 60). NE India, N Burma, S China (Yunnan, Sichuan, Fujian, Guangdong, Guangxi, Hunan, Hainan); N Thailand, Vietnam.

Variation. The following subspecific names have been proposed:
S. o. imbrensis Thomas, 1921l: 772; Khonsh-nong, Jaintia Hills, Assam, India, 3000ft.
S. o. sinensis Thomas, 1921l: 772; Kuatun, NW Fujian, China.
Specimens from the eastern part of the range are slightly smaller and darker, with richer coloration, than those from more westerly locations.

Remarks. Previously referred to *Scoteinus* or to *Nycticejus, emarginatus* was allocated to *Scotomanes* by Sinha & Chakraborty (1981) who remarked that apart from a shorter forearm and toothrow and the absence of white markings it agreed very closely with *S. ornatus.* However its forearm (55.2mm) falls within the range of variation (50.5–60.9) of a long series of *S. ornatus* in the BMNH and in some *ornatus* the white markings are faint and indistinct, especially in females. The holotype and only known specimen is female and it seems likely that

Map **60**. *Barbastella, Otonycteris, Scotomanes, Scotophilus*

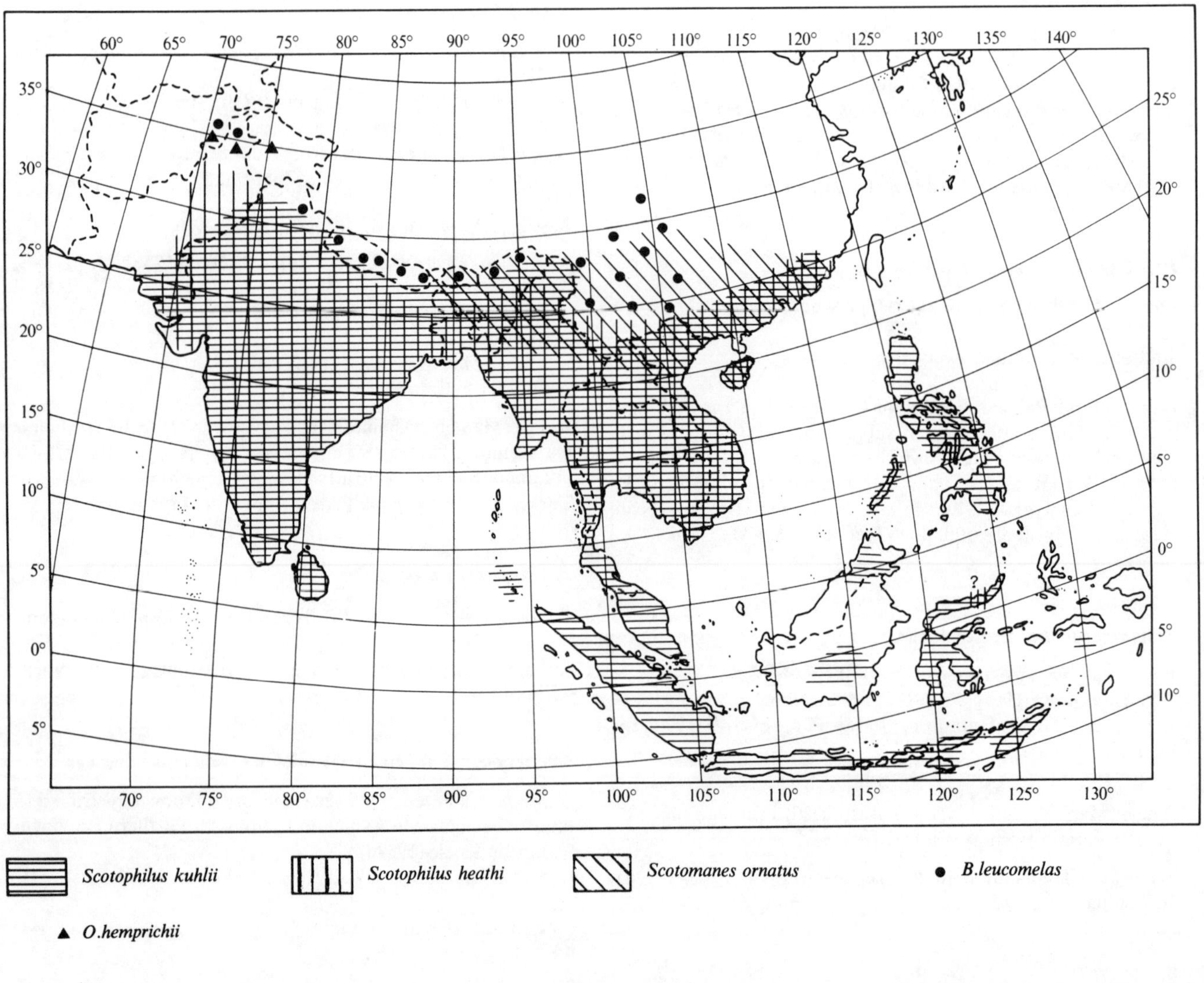

emarginatus and *ornatus* are synonymous. Differences in toothrow lengths are small (in *emarginatus* C-M^3, C-M_3 7.2, 8.0; in *ornatus* 7.4–8.5, 8.2–9.3).

SCOTOPHILUS Yellow bats

Scotophilus Leach, 1821: 69,71; type species *S. kuhlii* Leach.
= *Pachyotus* Gray, 1831a: 38.

Most of Africa and SW Arabia; and through most of the Indomalayan Region; 7–12 species, 2 in the region (Table 77). Although the largest of *S. kuhlii* overlap the smallest of *S. heathii*, such overlap is rare in areas where the two species are sympatric.

Table 77. The species of *Scotophilus*.

	kuhlii	*heathii*
Range in region	Pakistan-Timor	Afghanistan-Vietnam, ? Sulawesi
Forearm	45–59	55–65.5
Condylobasal length	16.7–19.2	18.5–21.3
C^1-C^1	8.0–9.2	9.0–10.5
C-M^3	6.0–7.2	7.1–8.3

Scotophilus kuhlii Asiatic lesser yellow house bat

S. kuhlii Leach, 1821: 72; locality unknown.
Vespertilio temminckii Horsfield, 1824:part 8; Java, restricted to W Java by Sody (1936a).
? *S. noctula* var. *sumatrana* Gray, 1838: 497; locality unknown, ? Sumatra.
S. castaneus Gray, 1838a: 498; Malacca, Malaya; *nomen nudum*: see Thonglongya (1973b).
S. fulvus Gray, 1843a: 31; Java, and Madras, India.
Nycticejus castaneus Horsfield, 1851: 38; Malacca, Malaya; see Thonglongya (1973b).
? *Nycticejus swinhoei* Blyth, 1860a: 88; Amoy, S China.
S. wroughtoni Thomas, 1897d: 275; Kim, Surat dist, W India.
S. gairdneri Kloss, 1917a: 284; Paknampo, C Thailand.

Range (Map 60). ? Afghanistan (Aellen, 1959), E, SE Pakistan, India, Sri Lanka, Nicobar Is, Burma, SE China (? Guangzhou, Fujian, Hong Kong, Hainan); Thailand, Cambodia, Vietnam, Malaya; Taiwan; Riau Is, ? Sumatra, Borneo; Philippines: Luzon, Catanduanes, Mindoro, Cuyo, Panay, Sibuyan, Palawan, Negros, Carabao (observed), Ticao, Romblon, Mindanao; Krakatau Is; Java, Bali; Lesser Sunda Is: Lombok (Kitchener *et al.*, 1990), Flores, Sumba, Savu; Timor, Sulawesi; Banda Is; also in Aru Is (New Guinea).

Variation. The following subspecific names have been proposed:
S. castaneus consobrinus Allen, 1906: 485; Rintoi, Hainan I.
Pachyotis temmincki panayensis Sody, 1928: 90; Panay I, Philippines.
S. temmincki collinus Sody, 1936a: 48; Djembrana, Bali I, c. 50m.
S. castaneus solutatus Sody, 1936a: 49; Tjandiroto (= Candirata), E Java.

A number of presumed subspecies could be listed but all are very provisional: although there are many available specimens from India in the BMNH relatively few are available from the more easterly part of the range. For the present *gairdneri* can be retained for specimens from Thailand and Cambodia on account of their slightly darker dorsal colour when compared with Indian examples; *castaneus* from Malaya is similar to *gairdneri* but is darker ventrally. The largest specimens occur apparently in E Java (*solutatus*). Possibly *collinus* and *solutatus* are synonyms of *S. k. temminckii* as Chasen (1940a) listed them. Tate (1942b) and Kitchener & Foley (1985) could find no difference in size between Balinese specimens and others from Java. Goodwin (1979) also found specimens from Timor to be similar in size to material from Java and Bali, while Kitchener *et al.* (1990) report rather large examples from Lombok.

Remarks. Hill (in Peterson, 1968) and Hill & Thonglongya (1972) provided justification for the use of *kuhlii* for this species, rather than *temminckii* by which name it had for many years been known.

Scotophilus heathii Asiatic greater yellow house bat

S. heathii Horsfield, 1831: 113; Madras, India.
Vespertilio belangeri Geoffroy, 1834: 87; towns near Pondicherry, Coromandel coast, India.
Nycticejus luteus Blyth, 1851a: 157; 'Bengal; Coromandel, India'.
S. flaveolus Horsfield, 1851: 37; 'many parts of continental India'.
S. kuhlii insularis Allen, 1906: 485; Rintoi, Hainan I, China.
S. solutatus watkinsi Sanborn, 1952: 4; Pak Nam Pho, Nakan Sawan Prov., C Thailand.
? *Pachyotis temmincki celebensis* Sody, 1928: 90; Toli-Toli, N Sulawesi.

Range (Map 60). Afghanistan, Pakistan, India, Burma, S China (Yunnan, ? Guangdong, Guangxi, Fujian, Guangzhou, Hainan); Thailand, Vietnam; Sri Lanka; ? Sulawesi.

Variation. There is considerable variation in colour as in *S. kuhlii*, but the recognition of subspecies is uncertain and provisional. No examples of *celebensis* have been examined, but Tate (1942b) considered it a subspecies of *S. heathii* with which it agrees in size. The smaller *S. kuhlii* was recorded from Sulawesi by Jones & Maa (1976).

EPTESICUS Serotines

Eptesicus Rafinesque, 1820: 2; type species *E. melanops* Rafinesque, in *Vespertilio fuscus* Palisot de Beauvois (N America).
Cnephaeus Kaup, 1829: 103 (*Vespertilio serotinus* Schreber).
= *Noctula* Bonaparte, 1837:fasc. xxi.
= *Cateorus* Kolenati, 1856: 131.
Amblyotus Kolenati, 1858: 252 (*A. atratus* Kolenati, in *Vespertilio nilssonii* Keyserling & Blasius).
Pachyomus Gray, 1866d: 90 (*Scotophilus pachyomus* Tomes, in *E. serotinus*).
Pareptesicus Bianchi, 1917:lxxvii (*Vesperugo pachyotis* Dobson).
Rhyneptesicus Bianchi, 1917:lxxvii (*Vesperugo nasutus* Dobson).
And other extralimital synonyms.

The Americas, Palaearctic and Africa, predominantly Palaearctic in the Old World, extending marginally into the northern part of the Indomalayan Region; c. 19 species in 2, possibly 3 subgenera. The definition and species content of *Eptesicus* has been discussed by Heller & Volleth (1984),

Horacek & Hanák (1985–86), Hill & Harrison (1987) and Menu (1987). Hill & Harrison provided detailed diagnoses of the subgenera, groups and subgroups within the genus, which is essentially similar to *Pipistrellus*, differing chiefly in features of the baculum and penis, and in the invariable absence of the anterior upper premolar (P^2).

Only the subgenus *Eptesicus* is represented in the region, with 5 species. A further species, *E. gobiensis* Bobrinskii, 1926, in the mountains of central Asia has been recorded in Kashmir and Nepal. The characters of the species groups are given in Table 78.

Eptesicus serotinus group

Distribution as for the genus; contains the majority of the species, with 4 in the region (Table 79).

Eptesicus serotinus Serotine

Vespertilio serotinus Schreber, 1774: 167, pl. 53; France.
Scotophilus pachyomus Tomes, 1857: 50; Rajputana, India.
Vesperus andersoni Dobson, 1871d: 211; Momein (Tengueh), Yunnan, S China.
? *E. horikawai* Kishida, 1924b: 127,139; Taiwan.
And many other extralimital names.

Range (Map 61). N Pakistan, N India, N Thailand, S China (Yunnan, Hunan, Fujian, Zhejiang), Taiwan; also through much of the Palaearctic from W Europe to Korea and in Africa.

Table 78. The species-groups of *Eptesicus*.

	serotinus group	*nilssonii* group	*nasutus* group
Species in region	4	1	1
Forearm	38–56	38–43	33–42
Braincase elongated	+	+	0
Rostrum short	0	0	+
Upper surface of rostrum transversely convex	0	+	0
Basioccipital pits present	+	+	0
Tympanic bullae large, covering cochleae	0	0	+
Inner upper incisor (I^2) unicuspid	+/0	0	+
I^3 large, = in crown area to I^2	+/0	+	0

Map **61**. *Eptesicus, Ia*

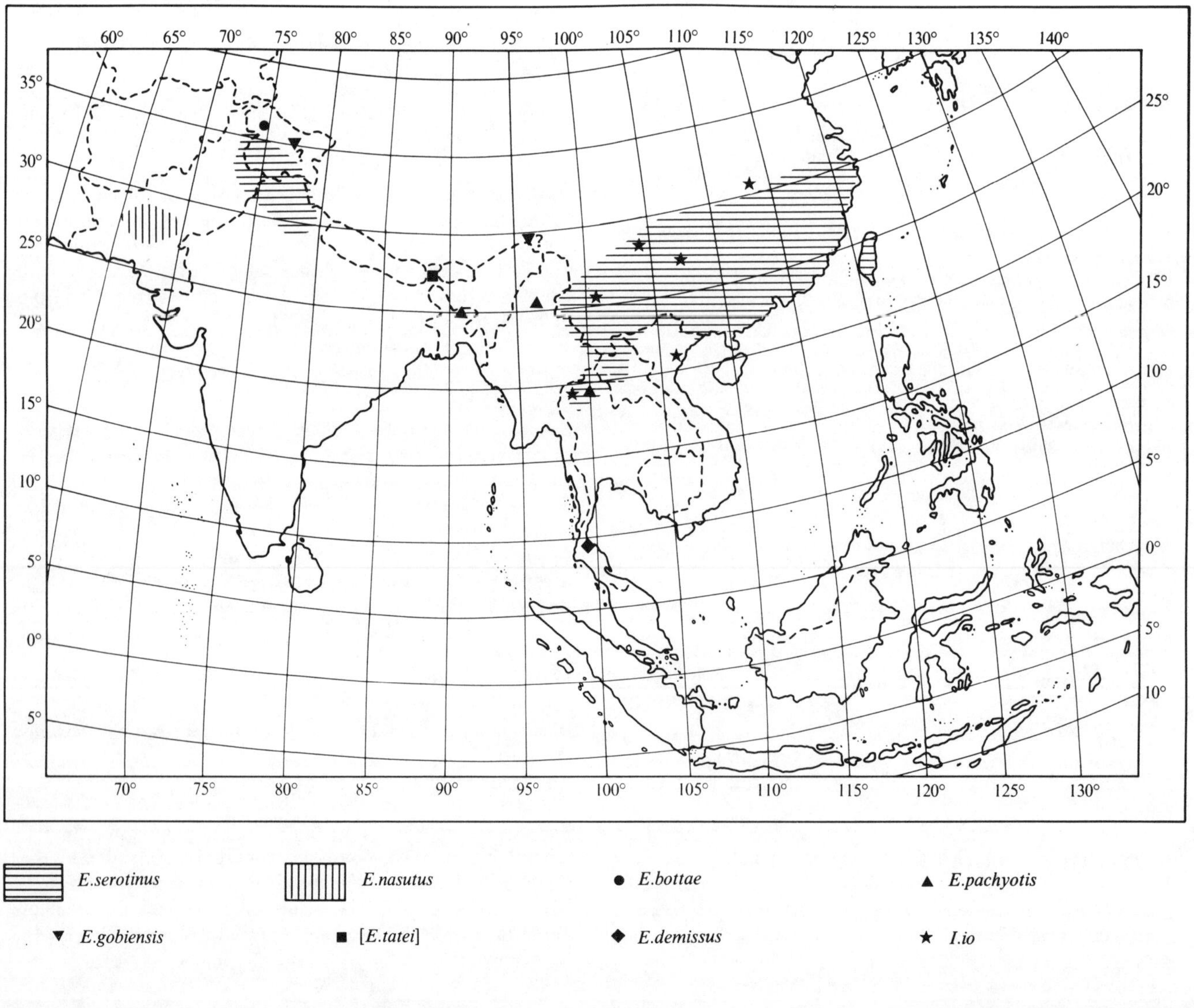

Variation. The following subspecific name has been proposed from the region:
Vesperugo s. var. *gabonensis* Trouessart, 1897: 108; based on *V. s.* var. *a* of Dobson (1878: 192), allegedly from Gabon, Africa, but almost certainly from Kashmir or N India (Hayman & Hill, 1971: 44; Ibánez & Valverde, 1985).

Colour and size variation in a number of the many subspecies was reviewed by Gaisler (1970). From N Pakistan and N India *E. s. pachyomus* is a little smaller in some respects than *E. s. andersoni* from N Thailand and S China. Possibly *horikawai* may be the appropriate subspecific name for the population of Taiwan, but from its description is very dark.

Table 79. The species of the *Eptesicus serotinus* group.

	serotinus	*bottae*	*pachyotis*	*demissus*
Range in region	Pakistan, NW India	Pakistan, ? N India	NE India-N Thailand	Thailand
Forearm	48-56	42–43.5	38–39	42
Proximal part of ear thick and fleshy	0	0	+	0
Prominent basi-occipital pits	0	0	0	+
Outer upper incisor (I^3) = I^2 in crown area	0	0	0	+

Eptesicus bottae Botta's serotine

Vesperus bottae Peters, 1869a: 406; Yemen.
E. ognevi Bobrinskii, 1918: 12; Bokhara district, Soviet Turkestan.
And other extralimital names.

Range (Map 61). N, perhaps also W, Pakistan; west to NE Egypt, Turkey and Soviet Turkestan.

Remarks. Roberts (1977) recorded a small, comparatively pale specimen (BMNH 8.12.12.1) from Shankagarh, Astor Dist., Gilgit, Kashmir as probably referable to *E. b. ognevi* which it appears to represent. Furthermore, the type locality of *E. b. taftanimontis* Roguin, 1988: 598 (Kasheh, W slope of Kuhe Taftan, Province Seistan and Baluchistan, Iran, 28°35'N, 61°E) is very close to the Iran-Pakistan border and this subspecies may be found in Pakistan.

Eptesicus pachyotis Thick-eared bat

Vesperugo (Vesperus) pachyotis Dobson, 1871d: 211; Khasi Hills, Assam, NE India.

Range (Map 61). NE India, N Burma and N Thailand.

Eptesicus demissus Surat serotine

E. demissus Thomas, 1916j: 1; Bandon, S Thailand.

Range (Map 61). S Thailand.

Eptesicus nilssonii group

Europe to E Siberia, possibly Japan; 3 species, one recorded from the fringe of the region.

Eptesicus gobiensis

E. nilssonii gobiensis Bobrinskii, 1926: 96; Burchasteitala, Gobi Altai Mts, Mongolia.
E. n. centralasiaticus Bobrinskii, 1926: 96; Ushchele Khatu, near Russk, Orin-Nor, Tibet.
E. n. kashgaricus Bobrinskii, 1926: 97; Khotun-Tagh, mountains of Russki, near Kashgar, Chinese Turkestan.

Range (Map 61). Recorded (as *E. nilssonii*) from Kashmir (Ellerman & Morrison-Scott, 1951; Siddiqui, 1969; DeBlase, 1980) and Nepal (Honacki *et al.*, 1982); also to N Afghanistan, S USSR, Mongolia and E Tibet.

Remarks. This species is sometimes considered conspecific with *E. nilssonii* but see Strelkov (1986). Felten (1971) identified a specimen from Afghanistan (formerly provisionally referred to *E. bottae* by Gaisler, 1971) as *E. g. kashgaricus* and this subspecies may occur in Pakistan. Elsewhere the subspecies in the region may be *E. g. centralasiaticus*.

Eptesicus nasutus group

Contains only *E. nasutus*.

Eptesicus nasutus

Vesperugo (Vesperus) nasutus Dobson, 1877: 311; Shikarpur, Sind, Pakistan.
And extralimital names.

Range (Map 61). Pakistan; and west to Saudi Arabia and Yemen. Erroneously recorded from Punjab, India, by Tate (1942b) (Neuhauser, 1970).

[*Eptesicus tatei*]

Nycticeius atratus Blyth, 1863a: 31; Darjeeling, NE India. Preoccupied by *Amblyotus atratus* Kolenati, 1858, in *Eptesicus nilssonii.*
= *E. tatei* Ellerman & Morrison-Scott, 1951: 158.

Range (Map 61). Known only from the holotype from Darjeeling, India.

Remarks. No example has been seen, and available information is inadequate to define or key the species. Its forearm length (c. 43mm) suggests the *E. nilssonii* group.

VESPERTILIO

Vespertilio Linnaeus, 1758: 31; type species *V. murinus* L.
Vesperugo Keyserling & Blasius, 1839: 312 (contained 13 species, including *V. murinus* L.).
Vesperus Keyserling & Blasius, 1839: 313. Pre-occupied by *Vesperus* Latreille, 1829.
Meteorus Kolenati, 1856: 131 (contained several species, one being *V. discolor* Natterer, in *V. murinus* L.
Aristippe Kolenati, 1863: 40 (in part; included *V. murinus* L.).
Marsipolaemus Peters, 1872b: 260 (*Vesperugo* (*M.*) *albigularis* Peters, in *V. murinus* L.).

Palaearctic, marginally on the northern fringe of the Indomalayan Region; 2 or 3 species (one at least strongly migratory), of which 2, perhaps 3, in the region (Table 80).

Table 80. The species of *Vespertilio*.

	murinus	*superans*	*orientalis*
Range in region	? N Pakistan, ? Kashmir	C China	C,S China, Taiwan
Forearm	<47	>45	>45
C-M^3	<6.1	>6.0	>6.0
Ventral pelage	Pale	Pale	Dark
Baculum	Almost rectangular	Triangular	Triangular
Pseudobaculum	+	0	+

Vespertilio murinus Particoloured bat

V. murinus Linnaeus, 1758: 32; Sweden.
And many extralimital synonyms.

Range (Map 62). Doubtfully in N Pakistan, and Kashmir (Ellerman & Morrison-Scott, 1951); through much of the Palaearctic Region from W Europe to E Siberia.

Remarks. Old records may refer to *Myotis myotis*, to which the name *Vespertilio murinus* was formerly applied.

Vespertilio superans

V. murinus superans Thomas, 1899: 770; Sesalin, Ichang, Hubei, China.

Range (Map 62). From the Yangtze north to N China, E Siberia and Japan.

Remarks. Possibly should include *V. orientalis*, below.

Vespertilio orientalis

V. orientalis Wallin, 1969: 307; Horobe, Ashiro, Ninoke Dist., Iwate Prefecture, Honshu, Japan.

Range (Map 62). E China (E Sichuan, Hubei, Fujian) and Taiwan; also in N China and Honshu, Japan.

Remarks. Possibly conspecific with *V. superans* Thomas, 1899 (Yoshiyuki, 1989; Yoon *et al.*, 1990).

IA

Ia Thomas, 1902b: 163; type species *I. io* Thomas.
Parascotomanes Bourret, 1942a: 23 (*Scotomanes* (*P.*) *beaulieui* Bourret).

Map **62**. *Vespertilio, Tylonycteris, Eudiscopus*

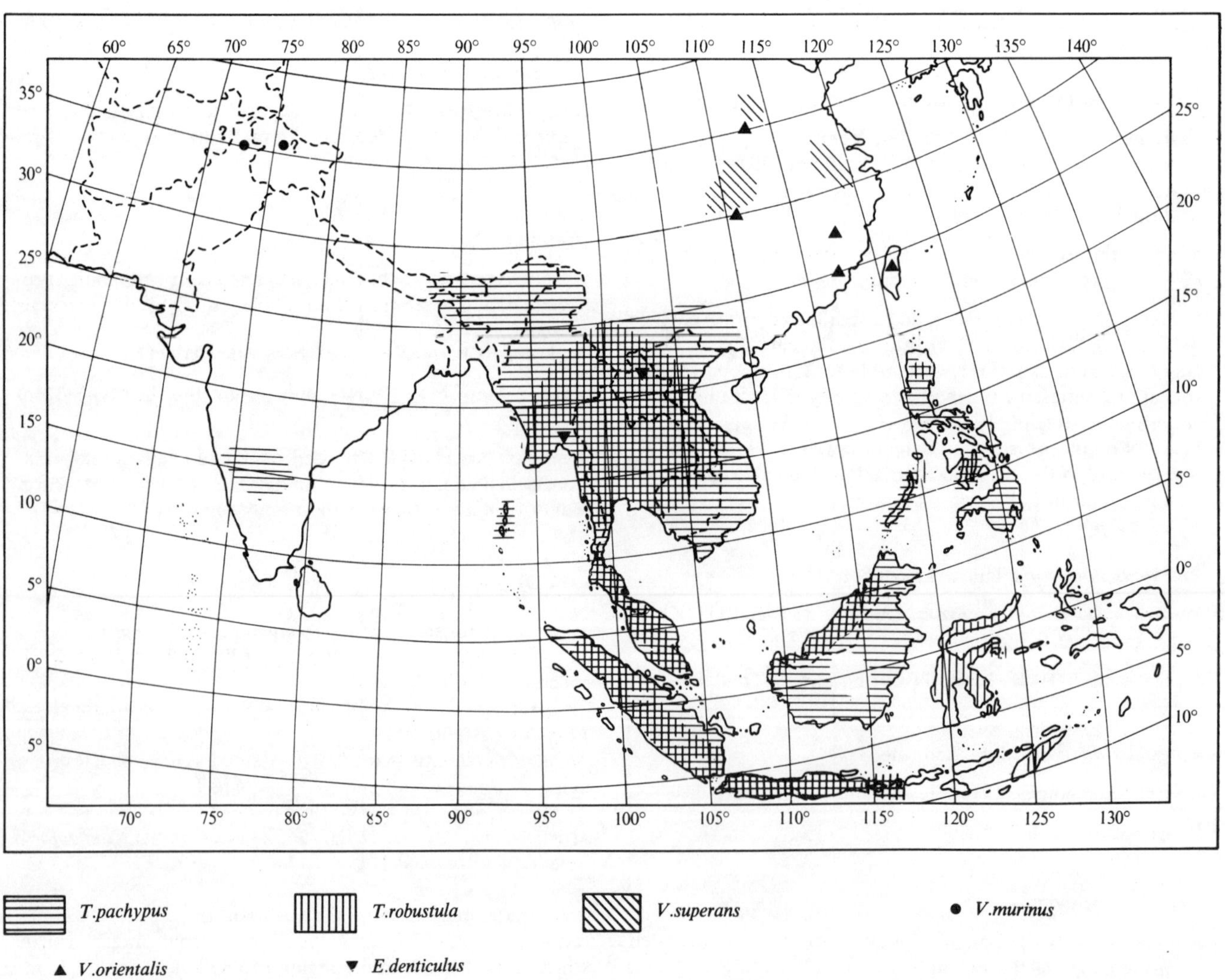

Contains one species. The status of this genus has varied over the recent past, Ellerman & Morrison-Scott (1951) for instance following Simpson (1945) in considering it a subgenus of *Pipistrellus*, perhaps adopting a suggestion by Tate (1942b) that it might represent one of the more specialized species groups in that genus. However, Topál (1970b) considered on bacular grounds that *Ia* was close to *Eptesicus*, and restored it to generic rank. Hill & Harrison (1987) took a similar view, but Menu (1987) referred it to the synonymy of *Eptesicus*.

Ia io Great evening bat

I. io Thomas, 1902b: 164; Chungyang, S Hubei, China.
Scotomanes (*Parascotomanes*) *beaulieui* Bourret, 1942a: 23; Tran-Ninh, Laos.
I. longimana Pen, in Pen *et al.*,1962: 113,130, pl. 1; Hueitung, Sichuan, China, 1700m.

Range (Map 61). NE India (Assam), S China (Sichuan, Hubei, Yunnan, Guizhou), N Thailand, Laos, N Vietnam.

Remarks. Reviewed by Topál (1970b).

TYLONYCTERIS

Tylonycteris Peters, 1872c: 703; type species *Vespertilio pachypus* Temminck.

Indomalayan Region; 2 species (Table 81).

Table 81. The species of *Tylonycteris*.

	pachypus	*robustula*
Forearm	22.1–26.7	24.8–28.6
Condylobasal length	10.0–11.4	11.2–13.1
Condylocanine length	9.7–11.2	11.3–12.8
Lachrymal width	4.7–5.5	5.4–6.5
Width across supraorbital tubercles	4.8–5.7	5.5–7.2
Braincase width	5.9–6.6	6.6–8.0
Mastoid width	6.4–7.3	7.0–8.5
Rostrum expanded	0	+
Supraorbital tubercles	Small	Well developed

Tylonycteris pachypus Bamboo bat, Flat-headed bat

Vespertilio pachypus Temminck, 1840: 217, pl. 54; Bantam, W Java.
Scotophilus fulvidus Blyth, 1859a: 293; Schwegyin, R Sittang, SE Burma.
= *T. rubidus* Thomas, 1915l: 227.
T. meyeri Peters, 1872c: 705; S Luzon, Philippines.
T. aurex Thomas, 1915l: 228; Astoli, Belgaum, S of Bombay, India.

Range (Map 62). S, NE India, Sikkim, Assam, Andaman Is, S China (Yunnan, Guangxi, ? Guangdong), Burma (including Tavoy I, Mergui Is), Thailand, Laos, Vietnam, Malaya; Sumatra, Borneo; Philippines (throughout); Java, Bali, Lombok.

Variation. The following subspecific name has been proposed: *T. p. bhaktii* Oei, 1960: 25; Sewela, E Lombok I, Lesser Sunda I.

Specimens from the northwestern part of the range, from India through Thailand to Vietnam, are generally slightly brighter in colour than those from Malaya and more easterly localities. The Philippine form *meyeri* is the smallest, Philippine examples being discusssed in detail by Heaney & Alcala (1986).

Tylonycteris robustula Greater flat-headed bat

T. robustula Thomas, 1915l: 227; Upper Sarawak, Borneo.
T. malayana Chasen, 1940a: 52; Jor, Batang Padang Dist., Perak, Malaya.

Range (Map 62). Burma, Yunnan, Thailand, Laos, Vietnam, Malaya; Sumatra, Borneo (Sarawak, Sabah); Philippines: Luzon, Calauit (Heaney & Alcala, 1986); Java, Bali; Lombok; Sulawesi, Peleng I; Timor I.

EUDISCOPUS

Discopus Osgood, 1932: 236; type species *D. denticulus* Osgood. Pre-occupied by *Discopus* Thomson, 1864 (Coleoptera) and by *Discopus* Zelinka, 1888 (Rotifera).
= *Eudiscopus* Conisbee, 1953: 30.

A monospecific genus known only from Burma and Laos; reviewed by Koopman (1972).

Eudiscopus denticulus Disc-footed bat

Discopus denticulus Osgood, 1932: 236; Phong Saly, Laos.

Range (Map 62). C Burma (Pegu, Yoma) and Laos.

PIPISTRELLUS Pipistrelles

Pipistrellus Kaup, 1829: 99; type species *Vespertilio pipistrellus* Schreber.
Romicia Gray, 1838a: 495 (*R. calcarata* Gray, in *P. kuhlii*).
Hypsugo Kolenati, 1856: 131 (*Vespertilio savii* Bonaparte, fixed by Wallin, 1969).
Nannugo Kolenati, 1856: 131 (included *Vespertilio nathusii* Keyserling & Blasius, *V. kuhlii* Kuhl and *V. pipistrellus* Schreber).
Falsistrellus Troughton, 1943: 349 (*Vespertilio tasmaniensis* Gould).
Arielulus Hill & Harrison, 1987: 250 (*Vespertilio circumdatus* Temminck).
Attelepharca Menu, 1987: 126 (no type species designated, therefore not available).
And other names based upon extralimital species.

A genus of c. 70 species in 7 subgenera, of which 4 subgenera and 26 species occur in the region (Table 82). Widely distributed in the Old World from Europe and Africa to Japan, the Solomon Is, and Australia, the genus also occurs in North America where two species are found. Its classification is open to considerable discussion and was last reviewed by Hill & Harrison (1987) who made extensive use of characters of the baculum in defining subgenera. Tate (1942b) examined the Indo-Australian species.

Subgenus *Pipistrellus*

Inner upper incisor (I^2) usually bicuspid; outer large, similar in crown area to inner and 1/2 or more its height; anterior upper premolar (P^2) similar in crown area to I^2.. ***pipistrellus*** group

Table 82. The subgenera of *Pipistrellus*.

	Pipistrellus	*Hypsugo*	*Falsistrellus*	*Arielulus*
Species in region	8	11	3	3
Dorsal pelage conspicuously tipped with copper, bronze or yellow	0	0	0	+
Braincase relatively high, rounded	+	0	0	+
Braincase noticeably long	0	0	+	0
Zygomata strong	0	+/0	+	0
Maxillary tooth-rows evidently convergent	0	0	+	0
Interdental palate noticeably longer than wide	+	0	+	0
Anterior upper premolar (P^2) much reduced, rarely absent	0	+	0	+
Baculum long, with cylindrical shaft	+	0	0	0
Baculum relatively short, shaft flattened	0	+	+	0
Baculum very small, Y-shaped	0	0	0	+

I^2 usually unicuspid; I^3 small, its crown area 1/2 or less crown area of I^2, its tip extending only slightly beyond the cingulum of that tooth; P^2 1/2 or less in crown area that of I^2 .... ***kuhlii*** group

Pipistrellus pipistrellus group

1. Anterior upper premolar (P^2) not much intruded from row, upper canine clearly separated from posterior premolar (P^4), anterior tooth visible laterally .... ***pipistrellus*** subgroup

P^2 intruded from row, usually to lie in a recess between canine and P^4, those teeth closely approximated or in contact, anterior premolar obscured laterally, often only its tip visible **2**

2. Large: forearm usually >35, condylobasal length usually >14.0, C-M^3 usually >5.4; cranial profile slightly convex, raised over frontal region....................... ***ceylonicus*** subgroup

Smaller: forearm <35, condylobasal length <13.0, C-M^3 <5.2; cranial profile almost straight from occiput to nares **3**

3. Supraorbital region distinctly broadened to produce abruptly incurving lateral margins to the anterior part of the postorbital region; rostrum broad, dorsally flattened; palate strongly domed .... ***javanicus*** subgroup

Supraorbital region not especially broadened; rostrum narrow, not dorsally flattened; palate weakly domed............ .... ***coromandra*** subgroup

pipistrellus subgroup

Pipistrellus pipistrellus Common pipistrelle

Vespertilio pipistrellus Schreber, 1774:pl. 54 (and p. 167); France.
P. aladdin Thomas, 1905b: 23 (and 1906b: 521); Derbent, 50 miles W of Isfahan, Iran, 6500ft.
And many extralimital names.

Range (Map 63). N Pakistan (Gilgit), Kashmir; and to W Europe, N Africa and Turkestan.

Variation. Subspecies in region is *P. p. aladdin.*

ceylonicus subgroup

Pipistrellus ceylonicus Kelaart's pipistrelle

Scotophilus ceylonicus Kelaart, 1852: 22; Trincomalee, Sri Lanka.
Vesperugo indicus Dobson, 1878: 222; Mangalore, Malabar coast, India.
P. chrysothrix Wroughton, 1899: 720; Mheskatri, Surat Dangs, India.
P. raptor Thomas, 1904c: 387; Tonkin, Vietnam.
P. shanorum Thomas, 1915o: 29; Pyaunggaung, N Shan States, Burma.

Range (Map 63). Pakistan, India, Sri Lanka, Burma, S China (Guangxi, Hainan), Vietnam; Borneo (Sabah).

Variation. The following subspecific names have been proposed:
P. c. subcanus Thomas, 1915o: 30; Yalala, Junagadh, Kathiawar, India.
P. c. borneanus Hill, 1963c: 534; Kiduk Aruk Camp, Sungei Kiduk Arok, c. 3 miles N of Mount Trus Madi, Sabah, Borneo, 4500ft.
P. c. tongfangensis Wang, in Shaw *et al.*, 1966: 264, 275; Tongfang, Hainan I, China.

Lal (1984) reviewed the Indian populations and concluded that all should be referred to a single subspecies, *P. c. indicus.*

Pipistrellus minahassae

Vesperugo minahassae Meyer, 1899: 14, pl. 4; Tomohon, Minahassa, N Sulawesi.

Range (Map 63). N Sulawesi.

Remarks. The status of *minahassae* was discussed by Tate (1942b) and by Hill & Harrison (1987). The skull and dentition of the holotype have never been described and Tate's remarks are based upon a referred specimen in the American Museum of Natural History. Hill & Harrison placed *minahassae* provisionally in the *ceylonicus* subgroup but its affinities can

Map **63**. *Pipistrellus* (part)

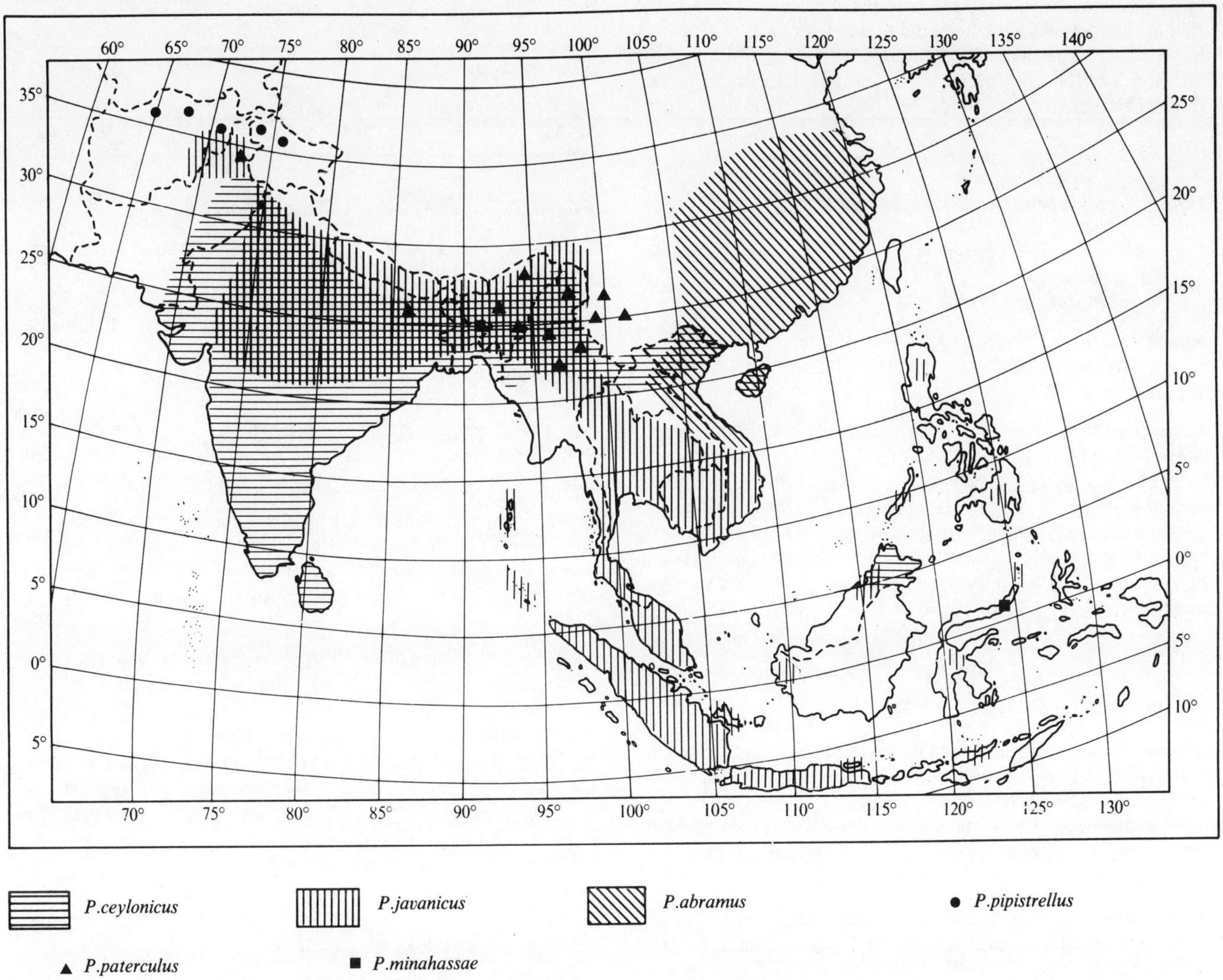

only be determined with certainty by further study of the original material. The features of the specimen described by Tate suggest that if this specimen represents *minahassae* the species should be referred to *P.* (*Pipistrellus*), but this placement is clearly speculative and no attempt has been made to incorporate it into any of the keys here.

javanicus subgroup

1. Skull small: condylobasal length 11.3–11.5, C-M^3 4.1–4.4; upper canine lacking posterior cusp, or with no more than a trace of such a cusp; 'horns' of bifid tip of baculum strongly deflected ventrally, forming a ring to some extent (baculum straight, very long, c. 9–12mm) .... ***P. paterculus***

Skull larger: condylobasal length >11.8, C-M^3 >4.4; upper canine with rudimentary posterior cusp; 'horns' of bifid tip of baculum not strongly deflected ventrally, more nearly in axial plane of shaft, not forming an incipient ring **2**

2. Braincase little inflated, only a slight convexity above the frontal region; rostrum slightly widened supraorbitally; palate rather narrow in relation to its length; upper molars not especially massive; baculum very long, c. 9–12, strongly curved, with a double sinuosity in the vertical plane............ ....***P. abramus***

Braincase slightly inflated, with a small but definite convexity above the frontal region; rostrum rather narrow supraorbitally; palate relatively wide in relation to its length; upper molars relatively massive; baculum shorter, c. 5–9, straight or nearly so .... ***P. javanicus***

Pipistrellus paterculus

P. paterculus Thomas, 1915o: 32; Mount Popa, N Burma.

Range (Map 63). N India, N Burma, S China (Yunnan).

Variation. The nominate subspecies occupies the Indian and Burmese part of the range: a slightly larger subspecies (*P. p. yunnanensis* Wang, 1982: 343; Luku, Lushui Xian) has been described from W Yunnan, China.

Remarks. Ellerman & Morrison-Scott (1951) listed *paterculus* as a possible subspecies of *P. abramus* and were followed in this by Hill (1962a). Soota & Chaturvedi (1980) remarked that the baculum in specimens ascribed to *paterculus* is doubly curved, as in *abramus*, contrary to the findings of Thomas (1915o) and Hill & Harrison (1987) who noted that it is straight or nearly so in *paterculus*. The smaller size and the distinctive bifid tip of the baculum (which is reminiscent of the baculum of *Scotoecus*) suggest that *paterculus* should be given specific rank.

Pipistrellus abramus Japanese pipistrelle

Vespertilio abramus Temminck, 1840: 232, pl. 58; Nagasaki, Kyushu, Japan.
And extralimital synonyms.

Range (Map 63). S China (Sichuan, Fujian, Hunan, Hong Kong, Hainan I); Taiwan; Vietnam; also to N China and Japan.

Remarks. Many recent treatments (e. g. Corbet, 1978; Corbet & Hill, 1980,1986; Honacki *et al.*, 1982) have included *abramus* in *P. javanicus* as a valid subspecies. However, it differs from *P. javanicus* in its long and sinuous baculum (Thomas, 1928c; Hill & Harrison, 1987): otherwise the two are very similar and can be separated only with difficulty if other conventional characters are relied upon. Both occur in Vietnam (Thomas, 1928c; Hill & Harrison, 1987).

Pipistrellus javanicus Javan pipistrelle

Scotophilus javanicus Gray, 1838a: 498; Java.
Vespertilio meyeni Waterhouse, 1845: 3, 7; Philippines.
P. camortae Miller, 1902b: 779; Camorta I, Nicobar Is.
P. babu Thomas, 1915o: 30; Murree, Punjab, Pakistan, 8000ft.
P. tralatitius bancanus Sody, 1937b: 233; Bangka I, Sumatra.
P. peguensis Sinha, 1969: 83; Pegu, Burma.

Range (Map 63). E Afghanistan, N Pakistan, N, C India, SE Tibet, Burma, Thailand, Vietnam, Malaya; Andaman Is (Das, 1990), Nicobar Is: Great Nicobar I, Camorta, Car Nicobar; Sumatra, Bangka I, Borneo; Philippines: Luzon (Taylor, 1934, as *P. abramus*), Mindanao (Cuy, 1981, as *P. abramus*), Palawan; Java, Madura I; Lesser Sunda Is: Flores (Hill & Rozendaal, 1989); Sulawesi; Talaud Is: Karakelang (Hill & Rozendaal, 1989); possibly to Aru Is, New Guinea (Hill, 1983).

Remarks. For many years *bàbu* has been considered a distinct species but it has much in common with *P. javanicus* and there seems no reason why it should not be considered conspecific. Cranially and dentally it agrees with *javanicus*. Soota & Chaturvedi (1980) discussed the status of *camortae* and considered it a distinct species, but made no direct comparison with *P. javanicus*; similarly *peguensis* was not compared with *javanicus* when described in 1969. Das (1990) reviewed the status of *camortae* and considered that with *peguensis* and *javanicus* it should be considered a distinct species, but made no direct comparisons. Both *camortae* and *peguensis* seem very close to *javanicus* with which they are here considered conspecific.

This species was for many years known as *P. tralatitius* (Horsfield, 1824), but Laurie & Hill (1954) regarded this name as indeterminable.

coromandra subgroup (Table 83).

Table 83. The species of the *Pipistrellus coromandra* subgroup.

	coromandra	*tenuis*
Range in region	Pakistan - Burma	Pakistan - Moluccas
Condylobasal length	11.2–12.5	10.1–11.6
M^3-M^3	5.2–5.8	4.7–5.4

Pipistrellus coromandra Indian pipistrelle

Scotophilus coromandra Gray, 1838: 498; Pondicherry, Coromandel coast, India.
Vespertilio coromandelicus Blyth, 1851a: 159.
Myotis parvipes Blyth, 1853b: 581; Masori (?= Mussoorie), Kumaon, India.
? *Vesperugo blythii* Wagner, 1855: 742; Sri Lanka.
? *V. nicobaricus* Fitzinger, 1861: 390; Nicobar Is.
Scotophilus coromandelianus Blyth, 1863a: 33.
? *Vesperugo micropus* Peters, 1872b: 259; Masuri, NW India (and Hutton, 1872: 708; Dehra Dun, near Simla, NW India).

Range (Map 64). NW Pakistan, India, Sri Lanka, Burma; N Thailand (Yenbutra & Felten, 1986); Nicobar Is: Great Nicobar, Car Nicobar (Bhattacharya, 1976), ? Camorta (Hill, 1967, as *P. javanicus camortae*); also to Afghanistan and SE Tibet.

Variation. Specimens from NW Pakistan in the BMNH agree with the description of Afghan examples by Gaisler (1970), who found these to be slightly larger and more greyish rather than brownish when compared with Indian specimens. Although it seems that he did not intend formally to recognise the Afghan population as a distinct subspecies, his proposal of *P. c. afghanus* (1970: 44, Afghanistan) appears to satisfy the requirements of the Code, and this name may prove applicable to these more westerly populations.

Pipistrellus tenuis Least pipistrelle

Vespertilio tenuis Temminck, 1840: 229; Sumatra (Tate, 1942b).
Scotophilus nitidus Tomes, 1859: 538; Labuan I, Borneo.
Vesperugo papuanus Peters & Doria, 1881: 696; Salawatti I, NW Irian Jaya, New Guinea.
P. mimus Wroughton, 1899: 722; Mheskatri, Dangs, Surat Dist., W India.
P. murrayi Andrews, 1900: 26; Christmas I, Indian Ocean.
? *P. subulidens* Miller, 1901a: 134; Sirhassen I, S Natuna Is.
P. portensis Allen, 1906: 487; Porten, Hainan I, China.
P. principulus Thomas, 1915l: 231; Gauhati, Assam, India.
P. coromandra tramatus Thomas, 1928c: 144; Thai-nien, Tonkin, Vietnam.
P. sewelanus Oei, 1960: 27; Sewela, Lombok I, Lesser Sunda Is.

Range (Map 64). Pakistan, India, Sri Lanka, Burma, Thailand, S China (? Yunnan, Fujian, Hainan), Vietnam, Cambodia, Malaya; Sumatra, ? Serasan I (S Natuna Is), Borneo, Labuan I; Philippines: Negros, Mindanao; Java, Bali, Lombok, Timor; Sulawesi; Moluccas: Seram, Ambon, Kai Is; Christmas I; also east to New Guinea, the Bismarck Is, and perhaps N Australia; and west to Afghanistan and Iran.

Variation. The following subspecific name has been proposed from the region:
P. mimus glaucillus Wroughton, 1912b: 769; Multan, Punjab, India.

Map **64**. *Pipistrellus* (part)

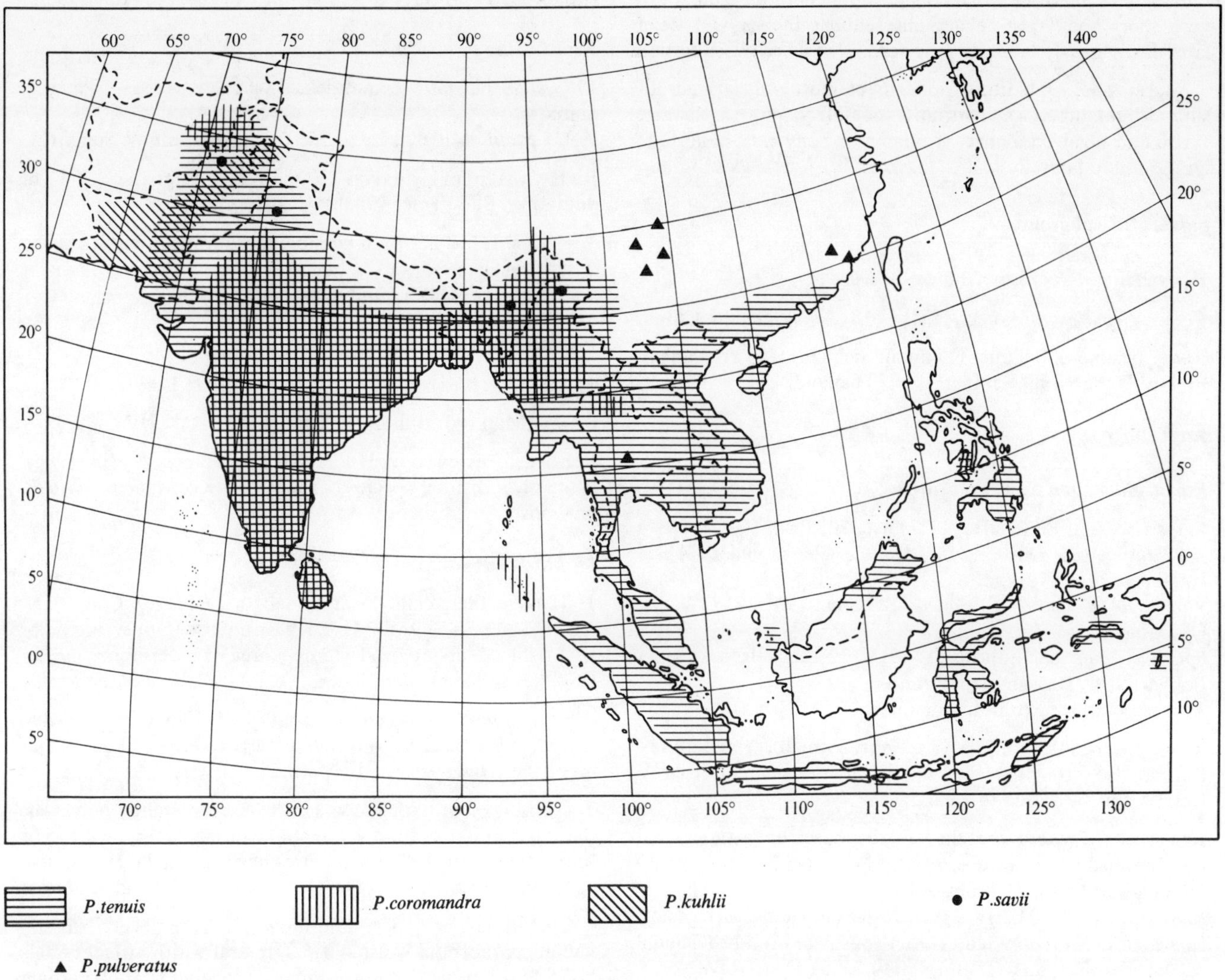

Remarks. Koopman (1973) considered *subulidens* conspecific with *P. tenuis* but from its description it has a wide rostrum and may represent *P. javanicus* as was thought by Chasen (1935b, 1940a). Kitchener *et al.* (1986) treated *papuanus* as a distinct species occurring west to the Moluccas and Timor, but its differences from *tenuis* and its allies are small.

Pipistrellus kuhlii group

Pipistrellus kuhlii Kuhl's pipistrelle

Vespertilio kuhlii Kuhl, 1819: 199; Trieste, Italy.
P. lepidus Blyth, 1845b: 340; Kandahar, Afghanistan.
And many extralimital names.

Range (Map 64). Pakistan and Kashmir; also to Turkestan, W Europe and much of Africa.

Subgenus *Hypsugo*

This is sometimes given generic rank, e. g. by Horácek & Hanák (1985–86) and Menu (1987), but it was retained as a subgenus of *Pipistrellus* by Hill & Harrison (1987).

Postorbital region, supraorbital region and rostrum not greatly widened; supraorbital tubercles if present small ....***savii*** group

Postorbital region very wide; supraorbital region much widened to include well-developed supraorbital tubercles; rostrum short and wide............................ ***stenopterus*** group

Pipistrellus savii group

1. Anterior upper premolar (P^2) relatively not greatly reduced, about equal or nearly equal to anterior upper incisor (I^3) in crown area ....***pulveratus*** subgroup

P^2 relatively reduced, its crown area not exceeding 2/3 that of I^3...**2**

2. Braincase rather low and flat, elongate; basal pits shallow or lacking ....***savii*** subgroup

Braincase inflated, globular, raised posteriorly; well-developed or deep basal pits **3**

3. Supraorbital area slightly widened with very small supraorbital tubercles; rostrum short; no jugal eminence on zygomata; a very short bony post-palate; outer upper incisor (I^3) lying laterally or nearly laterally to inner....... ***imbricatus*** subgroup

Supraorbital area little widened, at most only a trace of supraorbital tubercles; rostrum moderately long; a distinct zygomatic jugal eminence; a moderate bony post-palate; I^3 lying postero-laterally to I^2 .... ***lophurus*** subgroup

pulveratus subgroup

Pipistrellus pulveratus Chinese pipistrelle

Vesperugo pulveratus Peters, 1871b: 618; Amoy, Fujian, China.

Range (Map 64). S China (Yunnan, Sichuan, Hunan, Jiangxi, Fujian, Hong Kong, Hainan I); C Thailand.

savii subgroup

Pipistrellus savii Savi's pipistrelle

Vespertilio savii Bonaparte, 1837:fasc. 20; Pisa, Italy.
? *P. austenianus* Dobson, 1871d: 213; Cherrapunjee, Khasi Hills, Assam, NE India.
Vesperugo (*Vesperus*) *caucasicus* Satunin, 1901: 462; Tiflis, Caucasus.
Eptesicus tamerlani Bobrinskii, 1918: 12-16; Baisunski Bay, Bokhara dist, Russian Turkestan.
And many other extralimital names.

Range (Map 64). N India (Kashmir, Punjab), ? Assam, N Burma; also through much of the southern Palaearctic from W Europe and N Africa to Japan.

Remarks. The record from the Punjab is based upon a specimen misidentified as *Eptesicus nasutus* and reported by Tate (1942b) (Neuhauser, 1970). The subspecies in N India is probably *E. s. caucasicus*. Tate (1942b) assigned *austenianus* to the *savii* group; Ellerman & Morrison-Scott (1951) considered it a provisional subspecies of *P. savii*. If BMNH 14.7.8.63 from Maymyo, N Burma correctly represents *austenianus* then it is certainly close to *P. savii* (Hill, 1962a) and may represent a further subspecies.

imbricatus subgroup

1. Rostrum and palate relatively long; interdental palate decidedly longer than wide; anterior lower premolar (P_2) about 3/4 crown area of posterior premolar (P_4) .... ***P. imbricatus***

Rostrum and palate relatively short; interdental palate about as long as wide; P_2 about 1/2 or less crown area of P_4....... **2**

2. Skull generally larger: condylobasal length 11.6–12.0, condylocanine length 11.5–11.6, C^1-C^1 3.9–4.3, M^3-M^3 5.3–5.5, C- M^3 4.0–4.2 .... ***P. macrotis***

Skull generally smaller: condylobasal length c. 11.1, condylocanine length c. 10.8, M^3-M^3 c. 5.1, C^1-C^1 c. 4.0, C-M^3 c. 3.8 .... ***P. vordermanni***

Pipistrellus imbricatus

Vespertilio imbricatus Horsfield, 1824:part 8; Java.

Range (Map 65). Philippines: Luzon, Palawan, Negros; Borneo (Sarawak—Francis & Hill, 1986); Kangean Is; Java; Bali (Tate, 1942b); Lombok; S Sulawesi (Tate, 1942b).

Remarks. Records of *imbricatus* from peninsular Malaya (Medway, 1978) refer to *P. macrotis* (Francis & Hill, 1986).

Pipistrellus macrotis

Vespertilio macrotis Temminck, 1840: 218, pl. 54; Padang, Sumatra.
? *P. curtatus* Miller, 1911b: 25; Enggano I, off W Sumatra.

Range (Map 65). Malaya, Sumatra, ? Enggano I (W of Sumatra); Bali (Tate, 1942b).

Variation. It has not been possible to examine *curtatus* but the description leaves little doubt that it is very close to *P. macrotis* and may represent a valid subspecies.

Pipistrellus vordermanni

Vesperugo vordermanni Jentink, 1890c: 152; Billiton I.

Range (Map 65). Billiton I; Borneo (Sarawak–Hill, 1983).

Remarks. Reviewed by Hill (1983) and Francis & Hill (1986). Probably a distinct species but may prove conspecific with *P. macrotis*.

lophurus subgroup

1. Skull smaller: condylocanine length 12.5–12.9, C^1-C^1 4.5–4.6, M^3-M^3 5.8–5.9, C-M^3 4.5–4.8; anterior upper premolar (P^2) reduced, about 1/6–1/4 crown area of outer upper incisor (I^3) [NE India-Thailand].................................... ***P. cadornae***

Skull larger: condylocanine length 13.1–13.5, C^1-C^1 4.6–4.8, M^3-M^3 5.9–6.3, C-M^3 4.6–4.9; P^2 less reduced, about 1/2 or more the crown area of I^3 .... **2**

2. Overall colour dark brown; rostrum very slightly narrower, the supraorbital width c. 4.7, rostral width at lachrymals c. 5.0; braincase slightly inflated; anterior upper premolar (P^2) almost equal in crown area to outer upper incisor (I^3) .. ***P. lophurus***

Overall colour paler, mid-brown; rostrum very slightly wider, supraorbital width 4.8–5.2, rostral width at lachrymals 5.2–5.5; braincase not much inflated; P^2 about 1/2–2/3 crown area of I^3 .... ***P. kitcheneri***

Pipistrellus cadornae Cadorna's pipistrelle

P. cadornae Thomas, 1916c: 416; Pashok, Darjeeling, NE India, 3500ft.

Range (Map 65). NE India; N Burma (Hill, 1962a); N Thailand.

Pipistrellus lophurus

P. lophurus Thomas, 1915c: 413; Maliwun, Victoria Point, Tenasserim, S Burma.

Range (Map 65). S Burma.

Pipistrellus kitcheneri

P. kitcheneri Thomas, 1915d: 229; Buntok, Barito River, Kalimantan, Borneo.

Range (Map 65). Borneo (Kalimantan, ? Sabah—Francis & Hill, 1986).

Remarks. Revised by Francis & Hill (1986) who suggested the possible conspecificity of *kitcheneri* and *lophurus*, in which

Map **65**. *Pipistrellus* (part)

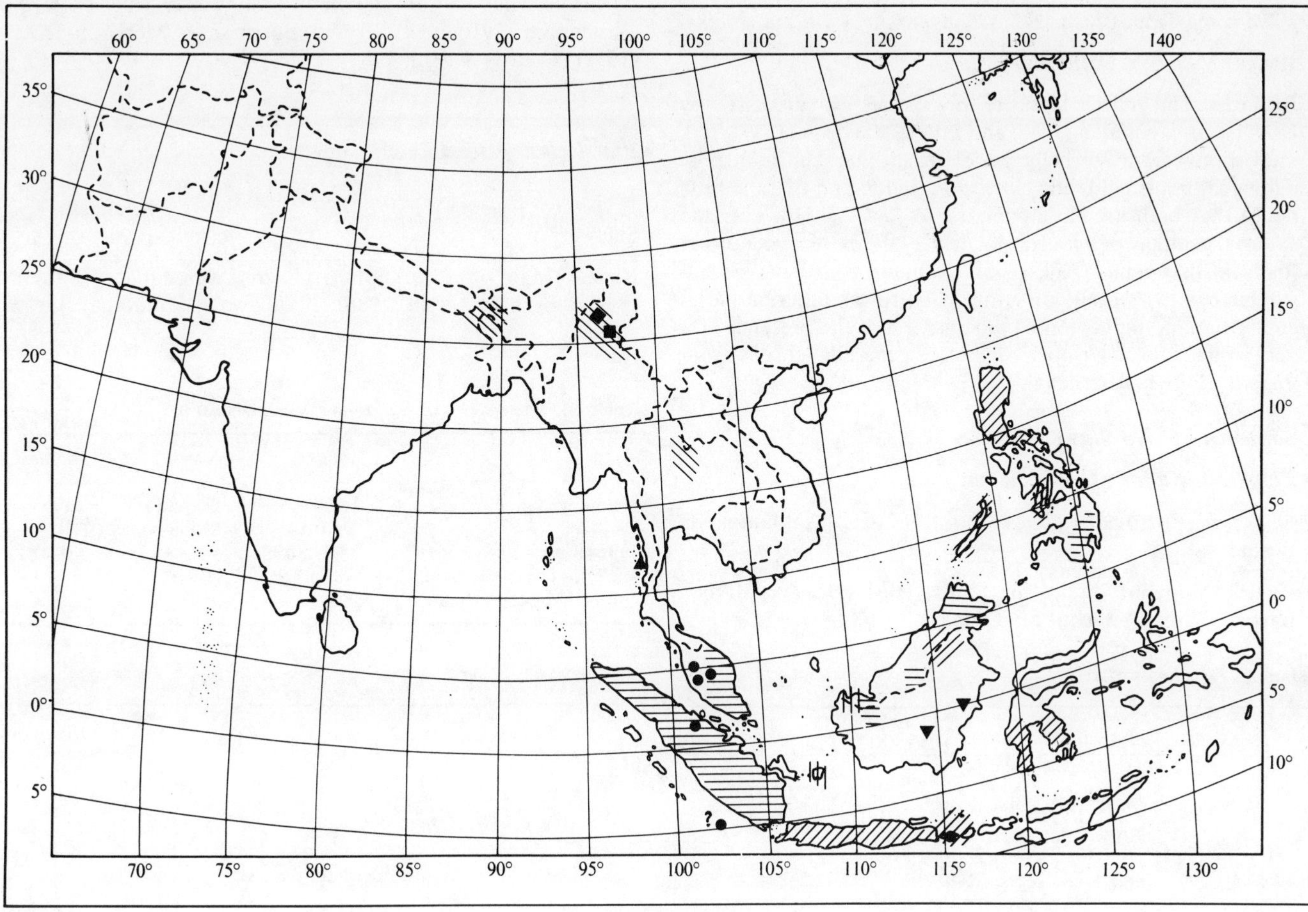

P.stenopterus — *P.vordermanni* — *P.cadornae* — *P.imbricatus*

● *P.macrotis* — ▲ *P.lophurus* — ▼ *P.kitcheneri* — ■ *P.joffrei*

◆ *P.anthonyi*

event *lophurus* (published 2 Feb. 1915) is apparently to be considered the prior name, *kitcheneri* being published only in February, 1915.

Pipistrellus stenopterus group

1. Skull larger: condylobasal length 15.1–15.8, C-M^3 5.5–5.9; supraorbital tubercles prominent but not protruding laterally; anterior upper premolar (P^2) large, equal to or exceeding outer upper incisor (I^3) in crown area; anterior lower premolar (P_2) slightly exceeding posterior lower premolar (P_4) in crown area .... ***P. stenopterus***

Skull smaller: condylobasal length 14.4–14.6, C-M^3 5.0–5.3; supraorbital tubercles greatly developed, protruding laterally; P^2 minute, about 1/4–1/3 crown area of I^3; crown area of P_2 about 1/2 that of P_4 .... **2**

2. Uniformly pale brown .... ***P. joffrei***

Uniformly dark brown .... ***P. anthonyi***

Pipistrellus stenopterus Narrow-winged pipistrelle

Vesperugo stenopterus Dobson, 1875e: 470; Sarawak, Borneo.

Range (Map 65). Malaya, Riau Is, Sumatra, Borneo (Sarawak, Sabah); Philippines: Mindanao.

Remarks. Variously referred to *Nyctalus* or to *Pipistrellus* in the past, this species was allocated without doubt to *Pipistrellus* on bacular grounds by Hill & Harrison (1987).

Pipistrellus joffrei Joffre's pipistrelle

Nyctalus joffrei Thomas, 1915l: 225; Kachin Hills, N Burma.

Range (Map 65). N Burma.

Remarks. On occasions allocated to *Nyctalus*, the generic status of *joffrei* was reviewed by Hill (1966) and Hill & Harrison (1987).

Pipistrellus anthonyi

P. anthonyi Tate, 1942b: 252; Changyinku, N Burma, 7000ft.

Range (Map 65). N Burma.

Remarks. The generic status of this species was discussed by Hill (1966), Koopman (1983) and Hill & Harrison (1987). It is known only from the holotype, of which the skull is heavily damaged, but is evidently close to *P. joffrei*, and there is little doubt that both are closely related to *Philetor*. However the external genitalia of *joffrei* do not display the extreme condition found in that genus (Hill, 1966). Cranially both have several similarities to *Philetor*, differing in narrower anterior narial emargination, high coronoid process, and in the presence of a very small anterior upper premolar, generally absent from *Philetor* (Hill & Rozendaal, 1989).

Subgenus *Falsistrellus* (Table 84).

Pipistrellus affinis Chocolate bat

Vesperugo (*Pipistrellus*) *affinis* Dobson, 1871d: 213; Bhamo, NE Burma.

Range (Map 66). India, Sri Lanka (BMNH), Nepal, NE Burma, China (Yunnan, SE Tibet).

Remarks. Indian examples in the BMNH were formerly labelled *mordax* and the species in India was listed as such by Wroughton (1916d, 1918b) and Ellerman & Morrison-Scott (1951) (Francis & Hill, 1986).

Pipistrellus petersi Peters' pipistrelle

Vesperugo petersi Meyer, 1899: 13, pl. 4; Minahassa, N Sulawesi.

Range (Map 66). Philippines: Luzon, Mindanao; Borneo (Sabah–Francis & Hill, 1986); N Sulawesi; Moluccas: Buru, Ambon.

Table 84. The species of *Pipistrellus* (*Falsistrellus*).

	affinis	*petersi*	*mordax*
Range in region	India - Yunnan	Borneo - Moluccas	Java
Forearm	38.5–40.5	40.0–41.8	40–42
Anterior upper premolar c. = I^3 in crown area	+	0	?

Map **66**. *Pipistrellus* (part)

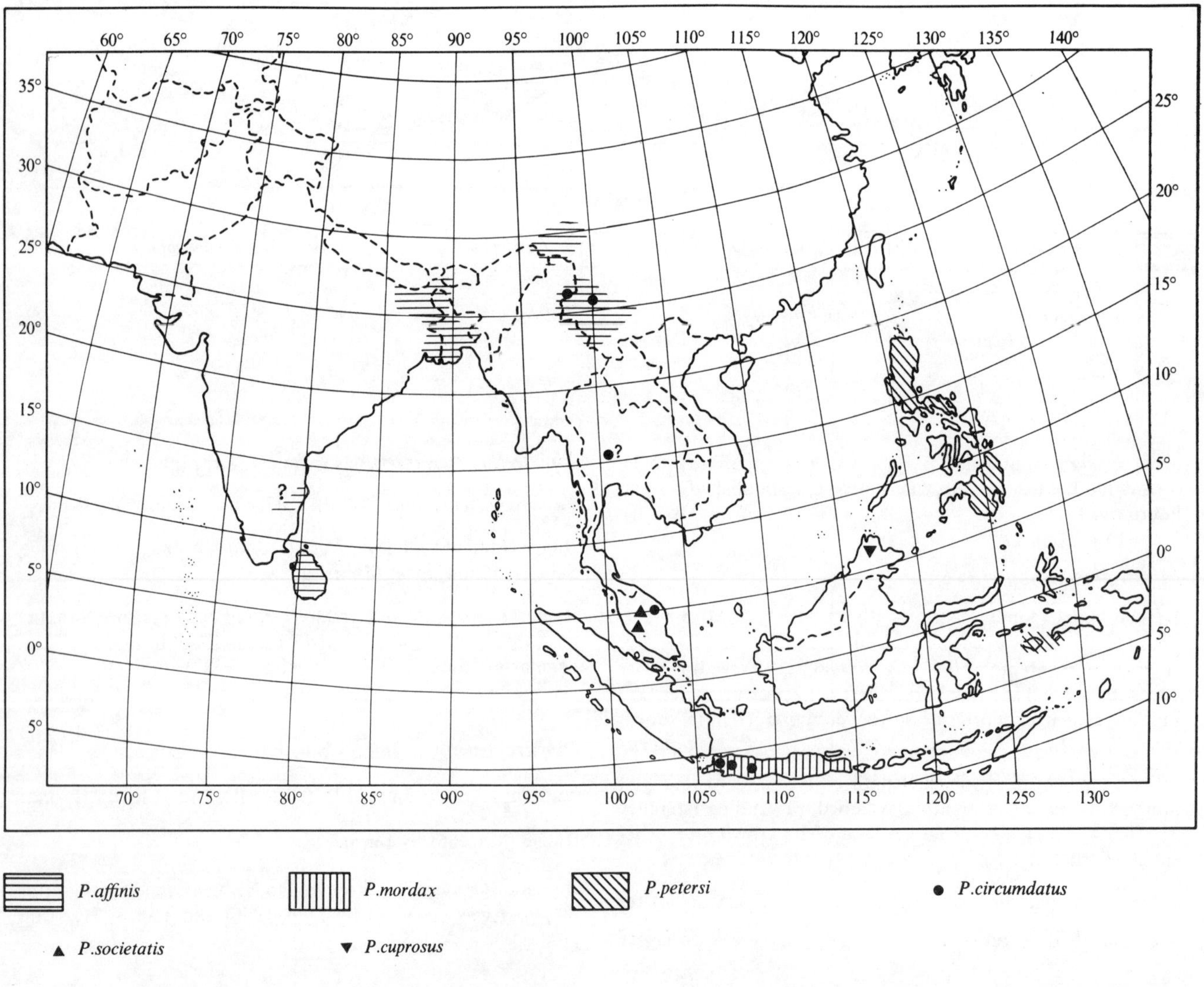

Remarks. As pointed out by Francis & Hill (1986) this species is similar in every respect to *P. affinis* although presently the two are geographically separated. It seems likely that ultimately they will prove to be conspecific.

Pipistrellus mordax

Vesperugo mordax Peters, 1866d: 402; Java.

Range (Map 66). Java; ? Philippines (Menu, 1987).

Remarks. Probably conspecific with *P. petersi*, and so, as pointed out, with *P. affinis*. It has not been possible to examine the holotypes of *affinis*, *petersi* or *mordax*, and no attributed examples of the last are available in London. Sody (1937a) recorded four specimens of *mordax* from Java with forearm lengths (40–42) in close agreement with Peters's holotype (40): Menu (1987) found one of Sody's specimens to correspond similarly with the measurements given by Peters. Moreover, Dr I Horácek (pers. comm.) has examined the holotype in Berlin and could find almost no differences in cranial and dental characters between it and *affinis*. It should be noted that *mordax* is the earliest name.

Subgenus *Arielulus* (Table 85).

Heller & Volleth (1984) considered *circumdatus* and *societatis* conspecific (but had examined only the latter) and transferred both to *Eptesicus* on bacular and karyological ground. However, Hill & Francis (1984) in describing the related *cuprosus* discussed this question and retained the three species in *Pipistrellus*, while Hill & Harrison (1987) considered *circumdatus* and its allies to represent a distinct subgenus in *Pipistrellus*, their bacula conforming more appropriately to that genus rather than to *Eptesicus* as they understood the two genera.

Table 85. The species of *Pipistrellus* (*Arielulus*).

	circumdatus	*societatis*	*cuprosus*
Range in region	Yunnan - Java	Malaya	Borneo
Forearm	40.5–44	36.5–41	34.5–36.5
Band of pale hair across muzzle, in front of eyes	0	+	+
Condylobasal length	14.9–16.0	13.9–14.8	12.9–13.0
Palatal length	>7.5	<7	<7
Length of bony post-palate	>1.4	<1.2	<1.2
C-M^3	5.9–6.4	5.2–5.6	4.8–4.9
Frontal profile straight	0	+	+
Distinct frontal depression	+	0	0

Pipistrellus circumdatus Black gilded pipistrelle

Vespertilio circumdatus Temminck, 1840: 214; Tapos, Java.

Range (Map 66). China (NW Yunnan), N Burma; ? SW Thailand; Malaya; Java.

Variation. A subspecies *P. c. drungicus* Wang, 1982: 345; Dilan, Drung Valley, has been described from NW Yunnan, China, differing chiefly in minor dental characters.

Remarks. Specimens from Burma, Malaya and Java were reviewed by Hill (1972a, 1974b).

Pipistrellus societatis

P. societatis Hill, 1972a: 34; Base Camp, Gunong Benom, Pahang, Malaya, 3°51'N, 102°11'E, 800ft.

Range (Map 66). Malaya.

Remarks. Further specimens were recorded by Heller & Volleth (1984,1989) who considered *societatis* to be a lowland subspecies of *P. circumdatus*. However, *societatis* (and *cuprosus*) differ sharply from *P. circumdatus* in shorter palate and much reduced bony post-palatal extension, as well as in other morphological details.

Pipistrellus cuprosus

P. cuprosus Hill & Francis, 1984: 312; Sepilok, Sabah, Borneo, 5°52'N, 117°56'E.

Range (Map 66). Borneo (Sabah).

GLISCHROPUS

Glischropus Dobson, 1875e: 472; type species *Vesperugo tylopus* Dobson.

Burma to the Moluccas; 2 species (Table 86). Menu (1987) considered *Glischropus* a synonym of *Pipistrellus*, averring that the dental features of *javanus* linked *tylopus* to this large genus and precluded subgeneric status.

Table 86. The species of *Glischropus*.

	tylopus	*javanus*
Range	Burma - Sumatra, Philippines, Moluccas	Java
Forearm	27.5–31	c.33
Lower leg (tibia, foot and claws)	18–20	21.5
Cranial profile	Nearly straight	Noticeably flexed
Braincase inflated	0	+

Glischropus tylopus Thick-thumbed pipistrelle

Vesperugo tylopus Dobson, 1875e: 473; North Borneo.
Vespertilio (*G.*) *batjanus* Matschie, 1901a: 277; Bacan I, N Moluccas.

Range (Map 67). Burma, Thailand, Malaya; Sumatra, Borneo, Palawan; Bacan (N Moluccas).

Glischropus javanus

G. javanus Chasen, 1939a: 187; Mt Pangrango, W Java.

Range (Map 67). W Java.

Remarks. Tate (1942b) thought that *javanus* might be a valid species: no specimen has been seen and it may prove to be a subspecies of *G. tylopus* as Chasen (1939a) suggested. However, Menu (1987) noted dental differences between *tylopus* and *javanus*. IUCN: K.

SCOTOZOUS

Scotozous Dobson, 1875c: 372; type species *S. dormeri* Dobson.

A monospecific genus limited chiefly to India. On occasion referred to *Pipistrellus* as a subgenus on account of an apparent or supposed resemblance to the African *P.* (*P.*) *rueppellii*, *Scotozous* was combined with *Scotoecus* (in which event

Map **67**. *Glischropus, Scotozous, Scotoecus, Philetor*

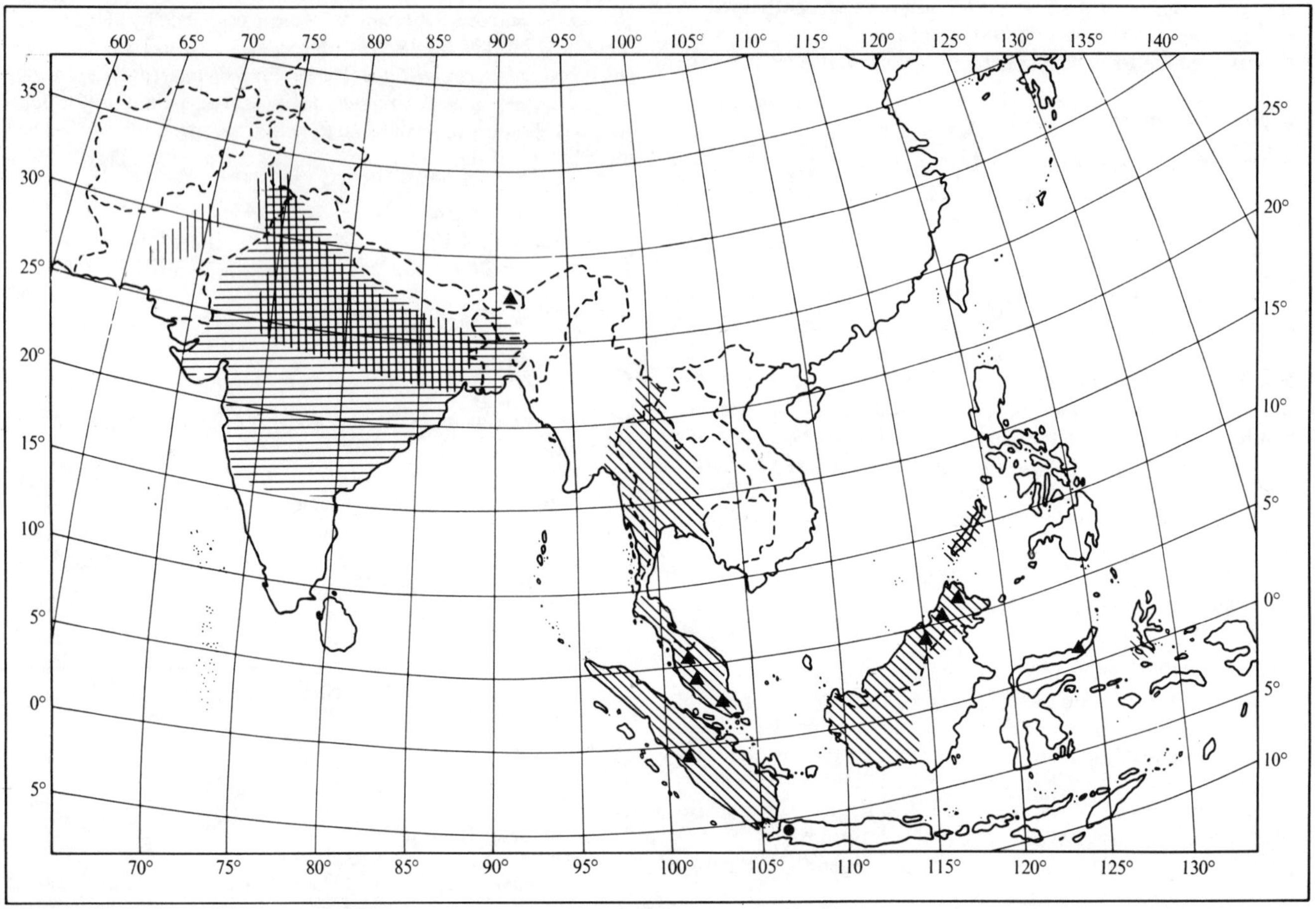

Scotozous dormeri *Scotoecus pallidus* *G.tylopus* ● *G.javanus*

▲ *P.brachypterus*

Scotozous is the prior name) by Menu (1987). However, *Scotozous* differs from *Pipistrellus inter alia* in its large, massive unicuspid inner upper incisor (I^2) which is in contact with the canine or nearly so, the outer upper incisor (I^3) when present no more than a minute, pointed spicule, reaching barely to the cingulum of I^2, and lying nearly laterally to this tooth, completely external to the toothrow. The genus differs from *Scotoecus* in more ligulate tragus, a narrower, uninflated rostrum, the narial and pre-palatal emarginations not deepened, in having the anterior face of the upper canine rounded, not flattened and grooved, and in the lack of any extreme development of the penis and baculum (Hill & Harrison, 1987).

Scotozous dormeri

S. dormeri Dobson, 1875c: 373; Bellary Hills, India.

Range (Map 67). NE Pakistan (Sialkot, perhaps Sind); India to West Bengal.

Variation. One subspecies has been described:
S. d. caurinus Thomas, 1915o: 33; Junagadh, Kathiawar, Gujarat, India, 400ft. The series available in the BMNH does not confirm that *caurinus* can be distinguished by larger skulls and teeth.

SCOTOECUS

Scotoecus Thomas, 1901b: 263; type species *S. albofuscus* Thomas.

Africa, Pakistan and India; 5 species, one in the region. Reviewed by Hill (1974c).

Scotoecus pallidus

? *Vespertilio noctulinus* I Geoffroy, 1831: 92, pl. 3; Bengal, India.
S. pallidus Dobson, 1876a: 186, app. D, 208; Mian Mir, Lahore, Punjab.

Range (Map 67). Pakistan (Sind); N India (Punjab, Uttar Pradesh, Bihar). A record from Eritrea, Ethiopia is erroneous (Hill, 1974c).

Remarks. Formerly referred to *Scoteinus* or to *Nycticeius*, this species was transferred to *Scotoecus* by Hill (1974c), who reviewed the history of the possibly prior name *noctulinus*.

NYCTALUS Noctules

Nyctalus Bowditch, 1825: 36, footnote; type species *N. verrucosus* Bowditch, in *N. leisleri*.

Pterygistes Kaup, 1829: 100 (*Vespertilio noctula* Schreiber).
Noctulinia Gray, 1842: 258 (contained two species, one a synonym of *Vespertilio noctula* Schreber).
Panugo Kolenati, 1856: 131 (*Vespertilio noctula* Schreber and *V. leisleri* Kuhl).

Palaearctic and marginally in the north of the Indomalayan Region; c. 6 species of which 3, perhaps 4 in the region (Table 87).

Table 87. The species of *Nyctalus*.

	noctula	*leisleri*	*montanus*
Forearm	48–57	39–46	43–44
Dorsum with hair bases usually darker than tips	0	+	0
Condylobasal length	17.0–19.4	14.8–16.1	15.8–16.3
C-M^3	6.4–7.5	5.5–6.3	6.1–6.5
Anterior upper premolar (P^2) greatly reduced, its crown area much less than that of either upper incisor	+	0	+
P^2 invisible from labial aspect	+	0	+

Nyctalus noctula Noctule

Vespertilio noctula Schreber, 1774: 166, pl. 52; France.
V. labiata Hodgson, 1835a: 700; Nepal.
? *Vesperugo (N.) plancyi* Gerbe, 1880: 71; Pekin, China.
? *N. velutinus* Allen, 1923a: 7; Futsing, Fujian, China.
And other extralimital names.

Range (Map 68). N Pakistan, N India, Burma, S China (Sichuan, Fujian, Zhejiang), ? N Thailand, Vietnam; Taiwan; doubtfully Malaya. Also from W Europe to Japan.

Variation. The subspecies in Pakistan and India is *N. n. labiatus* but the situation in the eastern part of the range remains uncertain. Tate (1942b) treated *plancyi* and *velutinus* as distinct species, but with the reservation that they might be synonymous, and pointed out that *velutinus* might be rather readily distinguished from *noctula* by the presence of small secondary basial pits anterior to the larger primary pits, these secondary pits being absent or only rarely faintly suggested in *N. noctula*. Ellerman & Morrison-Scott (1951) listed *plancyi* and *velutinus* as subspecies of *N. noctula* but Imaizumi & Yoshiyuki (1968) considered *velutinus* (and the Japanese *furvus*) specifically distinct. However, Corbet (1978) pointed out that these authors used mainly average differences and listed *plancyi* as the

Map **68**. *Nyctalus, Hesperoptenus*

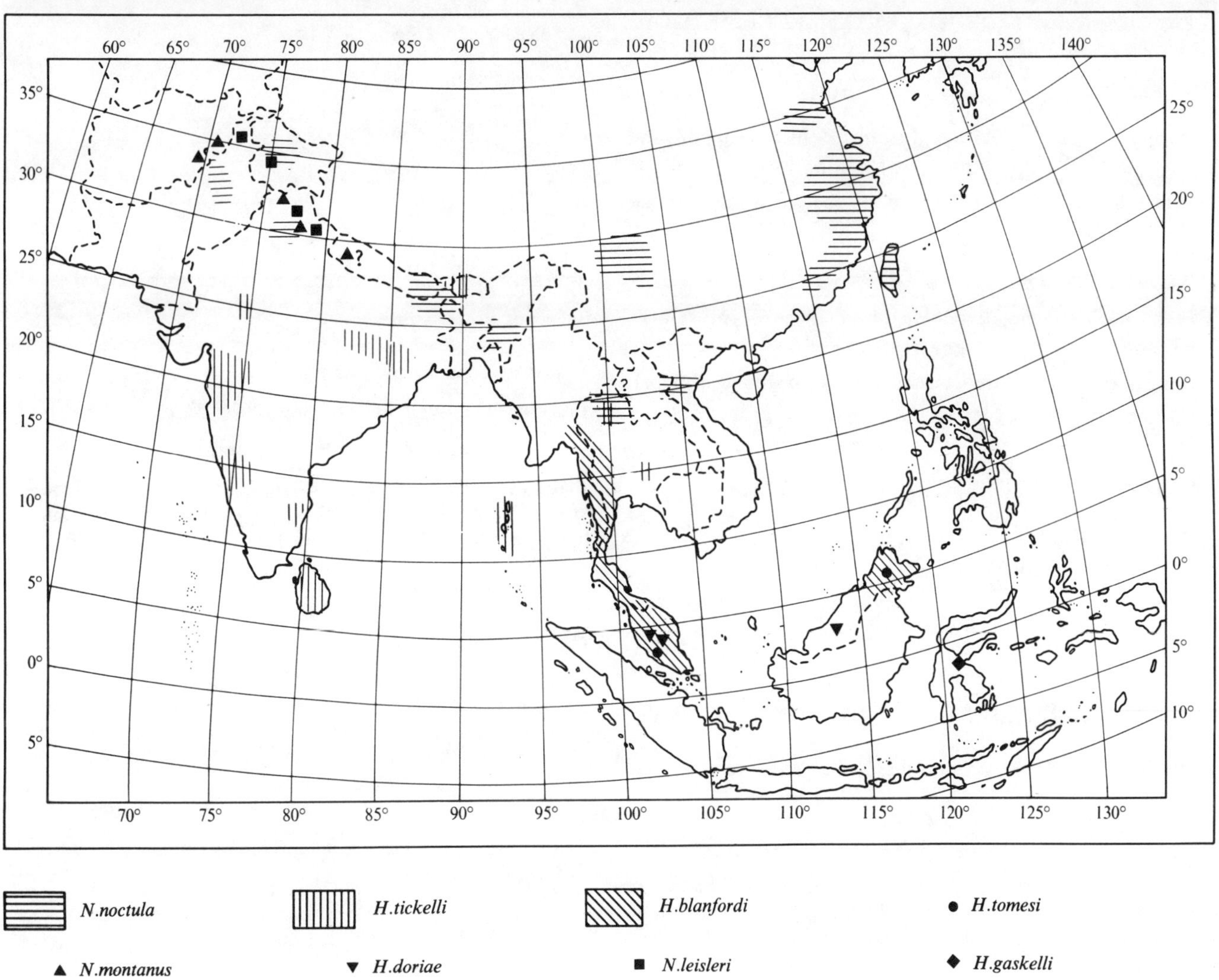

Chinese subspecies of *N. noctula*, with *velutinus* as its probable synonym.

More recently Yu & Chen (1988) have again accorded specific rank to *velutinus*, chiefly on account of its basial pits, smaller size and darker colour when compared with *noctula*, with *plancyi* (an earlier name) as its synonym. Yoshiyuki (1989), who has examined the holotype of *velutinus*, also considered it specifically distinct, but gave little further information.

Specimens from S and E China in the BMNH have clearly defined secondary basial pits. Those from Burma and N India have similar but usually rather shallower pits, while in C Asian examples the pits are shallowly present, faintly indicated, or absent. In the European population the pits are absent or only very faintly indicated. Consequently for the present *plancyi* (with *velutinus*) is considered here as a provisional subspecies of *N. noctula*. The Japanese *furvus* is similar to *velutinus* in size, coloration, large anteorbital foramen and the location of the lachrymal foramen (Yoshiyuki, 1989) but to *noctula* in the presence of shallow secondary basial pits and in its high outer upper incisor. Consequently it too appears in some respects to link *noctula* and *plancyi* (*velutinus*); however there appear to be karyological differences involved (Yu & Chen, 1988; Yoshiyuki, 1989) and clearly further study is needed to establish their exact status.

Nyctalus leisleri Leisler's bat

Vespertilio leisleri Kuhl, 1819: 46; Hanau, Hessen-Nassau, Germany.
= *V. dasykarpos* Kuhl, 1819: 49.
And other extralimital names.

Range (Map 68). N Pakistan, Kashmir, N India (Punjab); and west to S USSR, W Europe, N Africa, Madeira and Canary Is.

Remarks. Gaisler (1970) discussed this species in Kashmir and the Punjab.

Nyctalus montanus

Pterygistes montanus Barrett-Hamilton, 1906: 99; Mussoorie, Kumaon, N India.

Range (Map 68). E Afghanistan; Punjab, N India; ? Nepal.

Remarks. See Gaisler (1970) for a discussion of this species.

PHILETOR

Philetor Thomas, 1902f: 220; type species *P. rohui* Thomas, in *P. brachypterus*.

A monospecific genus distributed from E Nepal to New Guinea. Reviewed by Hill (1966,1971c).

Philetor brachypterus

Vespertilio brachypterus Temminck, 1840: 215, pl. 53; Padang District, Sumatra.
P. rohui Thomas, 1902f: 220; Albert Edward Range, E Papua, 6000ft.
Eptesicus verecundus Chasen, 1940a: 53; Mt Kledang, Perak, Malaya, 2646ft.

Range (Map 67). E Nepal (Koopman, 1983); Malaya, Sumatra; Borneo: Sarawak, Sabah (Hill, 1983; Hill & Francis, 1984); Philippines: Mindanao, Negros; Sulawesi (Hill & Rozendaal, 1989); also to New Guinea and the Bismarck Is. The alleged occurrence of this species in Java is based on mis-identified specimens of *Pipistrellus* (Bauer, in Hill, 1971c; Hill & Rozendaal, 1989). A record from Bangka I does not appear to be authentic.

Variation. Good series in the BMNH suggest that on the whole the forearms of specimens from Malaya are a little longer than those of New Guinea examples; Bornean specimens agree in size with Nepalese and Malayan material (Hill, 1983; Hill & Francis, 1984), but Kock (1981) recorded specimens from New Guinea that are similar in forearm length to those from Malaya.

HESPEROPTENUS

Hesperoptenus Peters, 1869b: 626; type species *Vesperus* (*H.*) *doriae* Peters.
Milithronycteris Hill, 1976: 14 (*Nycticejus tickelli* Blyth).

India to Sulawesi; 5 species in 2 subgenera (Table 88). Reviewed by Hill (1976).

Table 88. The subgenera of *Hesperoptenus*.

	Hesperoptenus	*Milithronycteris*
Species in region	1	4
Ear and tragus fleshy	0	+
2nd phalanx of 3rd digit markedly longer than 1st	+	0
Braincase slightly inflated, rather globose	+	0
Rostrum broadened supraorbitally	0	+
Lower maxillary margin of orbit narrow	+	0
Outer upper incisor intruded	0	+
Maxillary toothrows nearly parallel	+	0

Subgenus *Hesperoptenus*

Hesperoptenus doriae False serotine bat

Vesperus (*H.*) *doriae* Peters, 1869b: 626; Sarawak, Borneo.

Range (Map 68). Malaya, Borneo (Sarawak).

Remarks. Heller & Volleth (1989) recorded two further examples from Malaya, bringing the known representation to four specimens.

Subgenus *Milithronycteris* (Table 89).

Hesperoptenus tickelli Tickell's bat

Nycticejus tickelli Blyth, 1851a: 157; 'central India, Ceylon and the intervening hilly country'; restricted to Chaibassa, Bihar by Anderson (1881) and Wroughton (1918b).
N. isabellinus Kelaart (ex Blyth), 1850: 204 (317 of 1890 reprint); Sri Lanka. *Nomen nudum.*
= *N. isabellinus* Horsfield (ex Blyth), 1851: 38; central India.

Range (Map 68). India, Sri Lanka, Andaman Is, S Burma, Thailand; ? S China.

Variation. Specimens from Sri Lanka and the Andaman Is are a little darker than those from India, and more reddish brown dorsally, especially posteriorly. They also average a little smaller. Thai examples are similar to Indian ones in colour, but in size to those from Sri Lanka and the Andamans.

Hesperoptenus tomesi

H. tomesi Thomas, 1905c: 575; Malacca, Malaya.

Range (Map 68). Malaya, Borneo (Sabah).

Remarks. In Borneo this species was originally recorded as *H. doriae*, from a subadult specimen (Davis, 1962), subsequently recognized as *tomesi* (Hill, 1976). More recently Hill & Francis (1984) have reported adult specimens, while Heller & Volleth (1989) have recorded further Malayan examples.

Hesperoptenus gaskelli

H. gaskelli Hill, 1983: 169; R Ranu, C Sulawesi, 1°51'S, 121°30'E.

Range (Map 68). Central Sulawesi: known only from the type locality.

Hesperoptenus blanfordi Blanford's bat

Vesperugo (H.) blanfordi Dobson, 1877: 312; Tenasserim, S Burma (east of Moulmein, according to Blanford, 1891).

Table 89. The species of *Hesperoptenus* (*Milithronycteris*).

	tickelli	*tomesi*	*gaskelli*	*blanfordi*
Range in region	India - S China	Malaya, Borneo	Sulawesi	Burma - Borneo
Forearm	49–60.5	47.5–51.5	40	24.0–28.4
Dorsum rich, dark brown	0	+	+	0
Well developed pad at base of thumb	0	0	0	+
Greatest length of skull	17.2–19.9	20.1–21.4	15.6	12.1–13.2
$C\text{-}M^3$	7.0–8.1	7.7–8.3	6.0	4.2–4.4
Postorbital region relatively wide	0	0	0	+
Small supraorbital tubercles	0	+	+	+
I^3 with wide cingulum	+	0	0	+
Lower incisors much thickened	0	+	0	0
Lower incisors much imbricated	0	+	+	0

Range (Map 68). S Burma, Thailand, Malaya, Penang I, Borneo (Sabah–Hill & Francis, 1984).

Variation. Specimens from Borneo are smaller in some respects than those from Thailand and Malaya (Hill & Francis, 1984).

Subfamily Miniopterinae

Miniopterinae Miller, 1907: 227; type genus *Miniopterus* Bonaparte.

African, Eurasian and Indo-Australian, with one genus, *Miniopterus*, containing 11 species, of which 7 occur in the Indomalayan Region. It is noteworthy for the elongation of the third digit, its second phalanx nearly three times as long as the first, the lengthened distal part of the digit folding back upon the wing when at rest. The elevation of the subfamily to family status has been suggested (Mein & Tupinier, 1977) but has not gained general acceptance.

MINIOPTERUS Long-fingered bats, Bent-winged bats

Miniopterus Bonaparte, 1837:fasc. 20; type species *Vespertilio ursinii* Bonaparte, in *M. schreibersii*.

A genus of c. 11 species occurring from Europe and Africa to Japan, the Philippines, Solomon Is and Australia; 7 species in the region (Table 90). Classification is confused and uncertain, especially in the Indo-Australian area where there is no agreed consensus.

The genus is arguably one of the most difficult among the Microchiroptera. There is no revision in its entirety although Maeda (1982) attempted a statistical analysis of its Eurasian and Australian component. However his method is one of considerable complexity and has led to results that do not appear to be consistent. Moreover his study takes little account of geographical distribution and its relevance to any classification that may be proposed; thus his phenetic groupings seem at times to have little geographic reality. His work has been heavily and adversely criticized (Hill, 1983; Hamilton-Smith, 1984; Koopman, 1984b; Wilson, 1985)

The classification used here is based chiefly on Hill (1983) and is provisional. It does not appear possible to key *Miniopterus* on obvious morphological features other than size, or certainly not on any that lend themselves readily to key construction. Equally once species are defined, their linear measurements often overlap slightly. However, where two or more species are sympatric it will be found as a general rule that they separate fairly readily into discrete size classes; it is when

Table 90. The species of *Miniopterus*.

	schreibersii	*magnater*	*fuscus*	*medius*	*australis*	*pusillus*	*tristis*
Range in region	Overall	S China-Moluccas	Ryukyu Is	Thailand-Java	Philippines-Java-Moluccas	Thailand-Moluccas	Philippines, Sulawesi
Forearm	42.5–50	47.5–52.5	40.5–47.5	38–44.5	34–40	39–45	52.5–58.5
Condylobasal length	14.6–15.9	15.7–17.3	13.8–14.1	13.8–14.5	12.7–14.0	12.7–13.4	18.4–19.9
Braincase width	7.5–8.5	8.1–9.1	7.4–7.5	7.3–7.8	6.4–7.3	7.0–7.4	8.5–9.2
Mastoid width	8.1–9.1	8.9–9.9	7.9–8.2	7.7–8.4	6.9–7.8	7.3–7.9	9.3–10.1
$M^3\text{-}M^3$	6.3–7.3	7.4–8.0	5.9–6.3	5.8–6.5	5.1–6.0	5.4–5.7	8.3–8.9
$C\text{-}M^3$	5.8–6.7	6.4–7.3	5.4–5.8	5.5–6.0	4.7–5.7	5.0–5.4	7.5–8.3
Well-developed sagittal crest	0	0	0	0	0	0	+

specimens from different, usually widely separated localities are compared that size overlaps obscure identification. Table 90 enables most specimens to be allocated to species but nevertheless some difficulties will occur, notably among the smaller forms.

Miniopterus schreibersii Schreibers' long-fingered bat

Vespertilio schreibersii Kuhl, 1819: 185; 'Columbäzar Höhle', R Danube, Romania.
V. fuliginosa Hodgson, 1835a: 700; Nepal.
V. blepotis Temminck, 1840: 212, pl. 53; Java (restricted by Tate, 1940,1941f).
V. eschscholtzii Waterhouse, 1845: 4; Luzon I, Philippines.
M. ravus Sody, 1930a: 271; Tjilatjap, Babakan, C Java.
M. oceanensis Maeda, 1982: 23; Cable Station, Cape York, Queensland, Australia.
M. haradai Maeda, 1982: 27; Lampang, Chaehom, Thailand.
And many extralimital names.

Range (Map 69). ? Pakistan; India, Sri Lanka, Nepal, N Burma; S China (Zhejiang, Fujian, Hong Kong, Hainan); Taiwan; Thailand, Vietnam, Malaya; Sumatra, Java, Borneo (Sabah, Sarawak); Philippines: Luzon, Marinduque, Mindoro, Tablas, Mactan, Samar, Mindanao; Lombok, ? Timor; C, S, Sulawesi; Moluccas: Ambon, Kai is; Also to W Europe, Africa, N China, Japan, the Solomon Is and Australia.

Variation. The taxonomy of this widely distributed species remains uncertain. Stebbings & Griffith (1984,1986) remarked that 'recent work worldwide suggests that a number of separate species will be identified eventually'. One such example is provided by Maeda (1982) for the Eurasian and Australasian part of its range in which this author recognized six species. However, Maeda's work appears to be unacceptable in many respects (see under *Miniopterus* above); certainly it does not seem possible to base a reasoned interpretation of the many Asian names hitherto included in *M. schreibersii* on his study. The question was discussed in detail by Hill (1983), who gave selected measurements of series of specimens from locations throughout Indo-Australia. The conventional view that *M. schreibersii* is a single, wide-ranging polytypic species is followed here; many subspecies are currently listed, of which the following have been recognized in this region.

M. s. fuliginosus. E Afghanistan to Burma.
M. s. chinensis Thomas, 1908j: 638; 30 miles W of Peking, NE China. C, N China.
M. s. parvipes Allen, G. M., 1923a: 7; Yenping, Fujian, S China. Ranges to Vietnam.
M. s. haradai. Thailand.
M. s. blepotis. Malaya, Sumatra to Java, Lesser Sunda Is and Sulawesi.
M. s. eschscholtzii. Philippines.
M. s. oceanensis. Moluccas; and to Solomon Is and E Australia.

Miniopterus magnater

M. schreibersii magnater Sanborn, 1931: 26; Marienburg, 40 miles up the Sepik R, Papua New Guinea.
M. macrodens Maeda, 1982: 22; Madai Cave, Sabah, Borneo.

Range (Map 69). N Burma; Hong Kong, Hainan; Thailand (including Koh Lak); Malaya; Sumatra, Java, Borneo (Sabah); Nusa Penida I, near Bali; Moluccas: Seram, Ambon; ? Timor; extralimitally to New Guinea and the Bismarck Is.

Variation. Hill (1983) considered *magnater* and *macrodens* conspecific, the latter subspecifically valid, differing in slightly shorter skull, narrower palate and shorter toothrow, and occupying the western part of the range from Burma and S China at least to Java. Maeda (1983) has examined geographic and sexual variation in both. The measurements provided by Goodwin (1979) for specimens from Timor that he referred to *magnater* led Hill (*loc. cit.*) to consider these probably representative instead of *M. schreibersii*. However, Maeda has re-measured these examples, with others from Timor, and has given values which generally exceed those quoted by Goodwin, referring Timorese material to *macrodens*.

Miniopterus fuscus

M. fuscus Bonhote, 1902a: 626; Okinawa I, Ryukyu Is.

Range (Map 69). Ryukyu Is: Amami-oshima, Okinoerabu, Okinawa, Kume-jima, Ishigaki, Iriomote.

Variation. Specimens from the more southerly islands of Ishigaki and Iriomote are slightly larger than those from the more northerly islands and may represent a valid subspecies, *M. f. yayeyamae* Kuroda, 1924: 6 (Ishigaki-Mura, Ishigaki).

Remarks. Evidently allied to *M. medius*, of which it may be the northern representative (Hill, 1983).

Miniopterus medius

M. medius Thomas & Wroughton, 1909d: 382; Kalipoetjang, Tji-Tandoei R, W Java.

Range (Map 69). Thailand (including Tarutau I), Malaya (including Kaban I), Java, Borneo (Sabah, Kalimantan); Anamba Is; also to New Guinea.

Remarks. Discussed by Hill (1983).

Miniopterus australis Little long-fingered bat

M. australis Tomes, 1858d: 125; Loyalty Is, S Pacific; see Hill (1983) for discussion of type material and locality.
M. tibialis Tomes, 1858d: 126; Ambon I, Moluccas.
M. paululus Hollister, 1913a: 311; Guimaras I, Philippines.
M. witkampi Sody, 1930a: 272; Lobang Ipoe, Koetai, E Kalimantan, Borneo.

Range (Map 70). Java, Borneo, Balembangan I; Philippines: Luzon, Polillo, Marinduque, Mactan, Guimaras, Negros, Mindanao; Timor; Sulawesi; Togian Is (Owen et al., 1987; Hill, 1991); Moluccas: Bacan, Seram, Ambon, ? Buru; also to new Caledonia, Loyalty Is and E Australia.

Variation. Specimens from the western part of the range are generally slightly smaller than those from the more easterly localities; a range of measurements was tabulated by Hill (1983). Four subspecies have been recognized in the region, with two further extralimital subspecies to the east.

M. a. paululus. Philippines, Balembangan I.
M. a. witkampi. Borneo.
M. a. shortridgei Laurie & Hill, 1957: 128, renaming of *M. a. minor* Laurie & Hill, 1954: 72; Kalipoetjang, Tji-Tandoei R, S Java, pre-occupied by *M. minor* Peters, 1867. Java; ? Timor.
M. a. tibialis. Sulawesi, Moluccas.

Remarks. Discussed by Hill (1983).

Map **69**. *Miniopterus* (part), *Murina* (part)

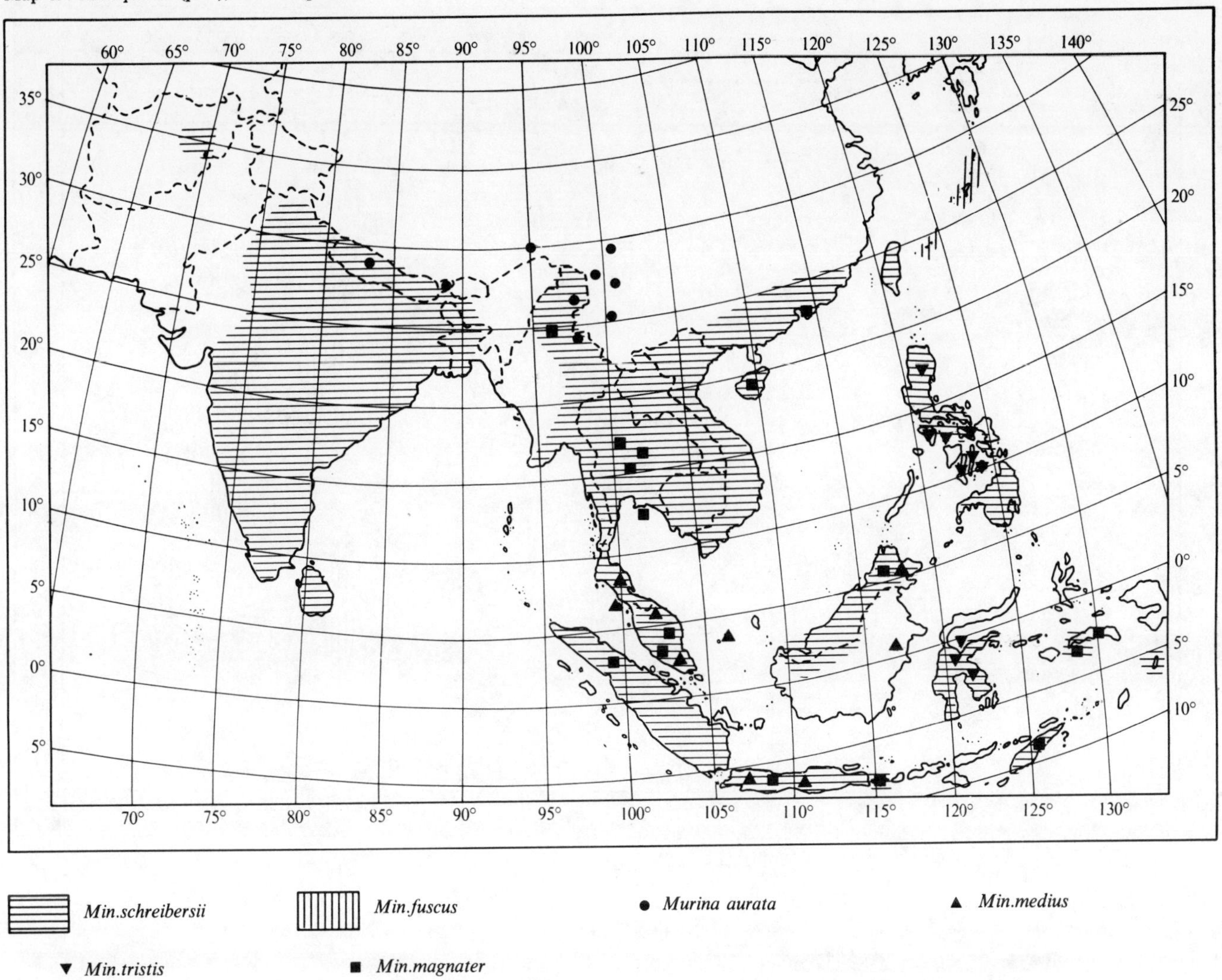

Miniopterus pusillus

M. pusillus Dobson, 1876a: 162; Nicobar Is (Wroughton, 1918b; Hill, 1983).
M. macrocneme Revilliod, 1914: 360; New Caledonia and Loyalty Is.

Range (Map 70). S India; ? Nepal (Maeda, 1982), Hong Kong; Nicobar Is; Thailand (including Koh Lak); Sumatra, Java; Timor; Sulawesi; ? Togian Is (Hill, 1991); Moluccas: Bacan, Seram, Ambon; Kai Is; also to New Caledonia and Loyalty Is.

Variation. Hill (1983) considered *pusillus* and *macrocneme* conspecific, the latter subspecifically valid and differing chiefly in its longer tibiae; *pusillus* extends to Java and Sulawesi, *macrocneme* occupying the eastern part of the range.

Remarks. This species was reviewed by Hill (1983). It is similar in many respects to *M. australis* but in addition to its generally longer forearm may be distinguished by its longer tibiae, which range from 16.5–20mm against 13.5–17 in *M. australis*; longer tail, ranging from 43–52mm against 35–45; a more globose braincase that is higher and more elevated than in *M. australis*, rising more abruptly from the rostrum and higher posteriorly, the occiput nearly equal or equal to the height at the frontal rather than lower; shorter, weaker rostrum; and longer, less massive upper canines.

Miniopterus tristis

Vespertilio tristis Waterhouse, 1845: 3; Philippines.
Including extralimital subspecies and synonyms.

Range (Map 69). Philippines: Luzon, Lubang, Mindoro, Tablas, Samar, Guimaras, Negros, Cebu, Bohol; Sulawesi; Togian Is (Owen *et al.*, 1987; Hill, 1991); also extralimitally in New Guinea, its adjacent islands, the Solomon Is and Vanuatu.

Variation. The nominate subspecies occurs in the Philippines; a second larger subspecies with longer forearm and metacarpals, *M. t. celebensis* Peterson, 1981: 836, 841 (Luwu, Wawondula, Sulawesi, 2°38'S, 121°21'E) is found in Sulawesi and the Togian Is. Three further subspecies occur extralimitally in the eastern part of the range.

Remarks. The classification of this species was discussed by Hill (1983).

Map **70**. *Miniopterus* (part), *Murina* (part)

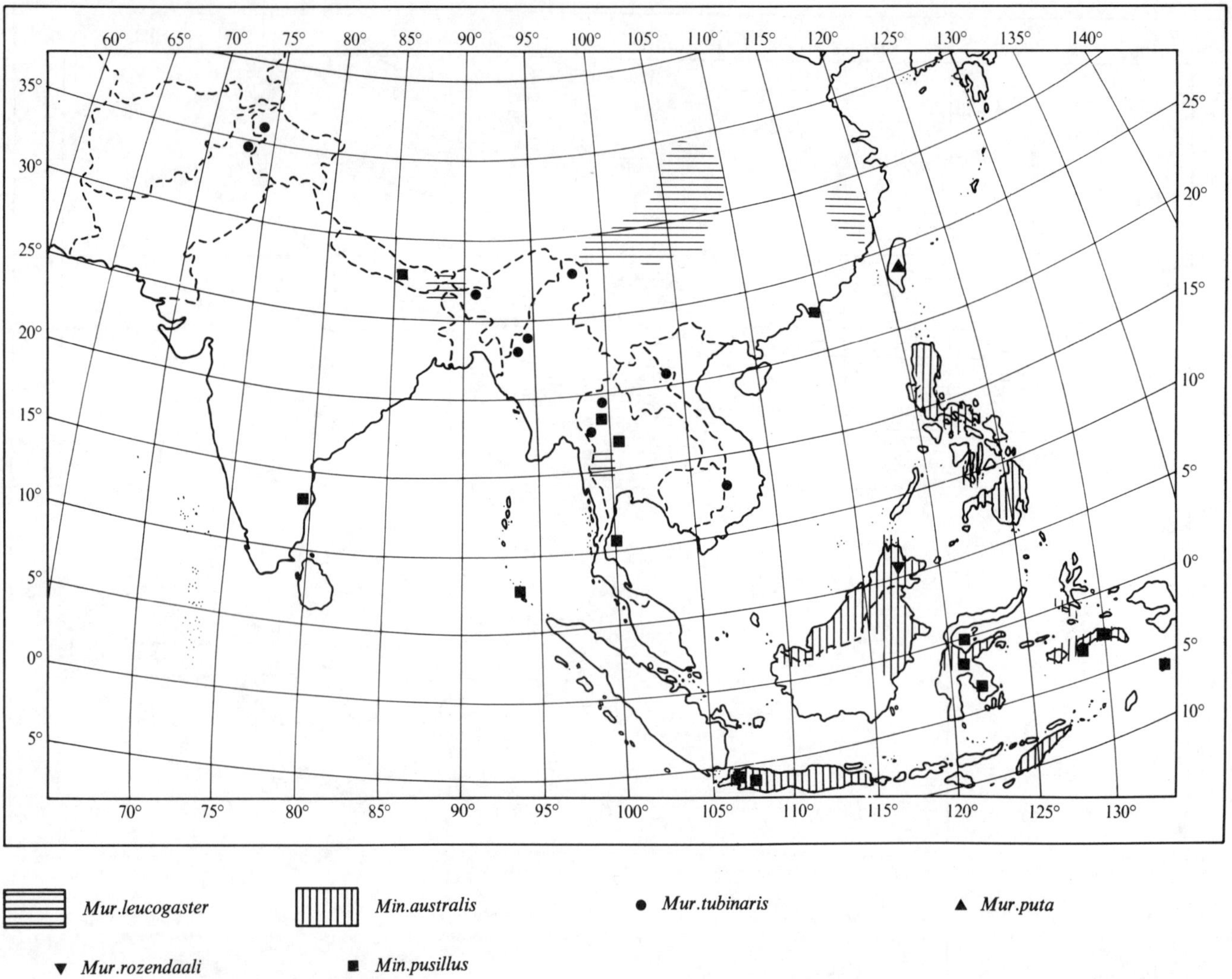

Subfamily Murininae

Tube-nosed bats

Murininae Miller, 1907: 229; type genus *Murina* Gray.

NE Pakistan to E Siberia, Japan, Bismarck Is and NE Australia; 2 genera, both represented in the region (Table 91). They are characterized by distinctly tubular nostrils (only other genus with similarly tubular nostrils is *Nyctimene*, a megachiropteran), and relatively unreduced, unmodified anterior upper premolar.

MURINA

Murina Gray, 1842: 258; type species *Vespertilio suillus* Temminck.
Ocypetes Lesson, 1842: 30 (included *V. suillus*). Pre-occupied by *Ocypetes* Risso, 1826 (Arachnida) and *Ocypetes* Wagler, 1829 (Aves).
Harpiola Thomas, 1915p: 309 (*M. grisea* Peters).

A genus of c. 14 species in 2 subgenera, from NE Pakistan to Japan, the Bismarck Is and Australia (Table 92).

Subgenus *Murina*

The key below does not include *M. puta* from Taiwan which has not been seen, but is apparently close to *M. huttoni*.

1. Maxillary tooth-rows distinctly convergent anteriorly; inner upper incisor (I^2) anterior to the outer (I^3), its postero-external face at most abutting the antero-internal face of the outer tooth, with little or no overlap; crown area of anterior upper premolar (P^2) one half or rather less that of posterior upper premolar (P^4) **2**

Maxillary toothrows not usually markedly convergent, more nearly parallel; I^3 usually more lateral to I^2, its antero-lateral face overlapping the external face of the inner tooth to some extent; crown area of P^2 two-thirds or more that of P^4 ...... **6**

2. Rostrum deep and massive .... ***M. leucogaster***

Rostrum not especially deep or massive **3**

3. Canines much reduced, their height subequal to the height of the posterior premolars ($P^4/_4$), crown area of upper canine about 1/2 that of P^4, of lower canine equal to or less than that of P_4 .... ***M. aurata***

Canines less reduced, their height equal to or more usually exceeding that of $P^4/_4$, crown area of upper canine usually 2/3 or more that of P^4, of lower canine a little greater than that of P_4 .... **4**

4. Larger: forearm 32.5–37, greatest length of skull 15.7–17.3, condylobasal length 14.3–15.8, M^3-M^3 5.4–6.1, C-M^3 5.2–5.8 .... ***M. florium***

Smaller: forearm 28.5–33, greatest length of skull 13.9–15.6, condylobasal length 12.2–14.2, M^3-M^3 5.0–5.3, C-M^3 4.7–5.3 .... **5**

5. Dorsal pelage brownish; (Malaya—New Guinea) ***M. suilla***

Dorsal pelage grey to greyish brown, some hairs so tipped, others totally grey; (NE Pakistan—Vietnam) ... ***M. tubinaris***

6. Mesostyle (central labial cusp) of 1st and 2nd upper molars not greatly reduced; labial face of teeth nearly flat or with shallow concavity, its apex flattened or rounded; hypoconid (posterior labial cusp) and entoconid (posterior lingual cusp) of 1st and 2nd lower molars small but evident, little reduced, separated from protoconid (anterior labial cusp) and metaconid (central lingual cusp) by a wide trough, forming a clear posterior trigonid **7**

Mesostyle of 1st and 2nd upper molars much reduced, labial faces with nearly V-shaped indentation; hypoconid and entoconid of 1st and 2nd lower molars much reduced, sometimes barely evident, not widely separated from protoconid and metaconid, posterior trigonid small or little evident **8**

7. Smaller: forearm 32–33.5, condylobasal length 14.4–14.6, C-M^3 5.2–5.6; dorsal pelage tipped anteriorly with shining golden brown, bronze or ochraceous bronze, otherwise dark brown; upper incisors lying almost on a straight line across longitudinal axis of skull .... ***M. rozendaali***

Larger: forearm 33–37, condylobasal length 15.3–16.7, C-M^3 5.8–6.1; dorsal pelage uniform greyish brown or brown; outer upper incisor (I^3) overlapping inner but slightly postero-lateral to it .... ***M. huttoni***

8. Dorsal pelage dark brown, tinged with bronze, shorter hairs dark brown tipped, longer hairs tipped with bright golden brown or bright golden yellow; venter buffy, posteriorly with ochracous tinge; hypoconid and entoconid of 1st and 2nd lower molars separated from protoconid and metaconid by a moderate trough .... ***M. aenea***

Dorsal pelage predominantly brown or rufous brown, the longer hairs not tipped with brighter contrasting colour; ventral surface greyish white, with brownish tinge; hypoconid and entoconid of 1st and 2nd lower molars separated from protoconid and metaconid by a narrow trough... ***M. cyclotis***

Murina leucogaster Greater tube-nosed bat

M. leucogaster Milne-Edwards, 1872b: 252, pls 37b, 37c; Moupin district, Sichuan, China.
M. rubex Thomas, 1916: 639; Pashok, near Darjeeling, NE India.
And other extralimital names.

Range (Map 70). NE India; W Thailand (McBee *et al.*, 1986); China (Sichuan, Fujian, Shaanxi, SE Tibet); and north to Mongolia, Manchuria, SE Siberia and Japan.

Table 91. The genera of Murininae.

	Murina	*Harpiocephalus*
Species in region	10	2
Forearm	Usually <44	>44
Condylobasal length	Usually <18.5	>18.5
Rostrum relatively short, broad, anteriorly truncate	0	+
Cusp pattern of 1st and 2nd upper molars normal, W-pattern of surface ridges present	+	0
Main cusps of 1st and 2nd upper molars closely approximated, W-pattern largely lost, these teeth premolariform	0	+
3rd upper molar reduced to mere scale, almost lacking any surface structure, or absent	0	+

Table 92. The subgenera of *Murina*.

	Murina	*Harpiola*
Species in region	9	1
Wing attached at base of claw of 1st toe	+	0
Wing attached to base of 1st toe	0	+
Canines much reduced, their height and bulk ≤ those of anterior premolars; upper canines only slightly exceeding upper incisors in height	0	+
Metacone (median posterior cusp) of 1st and 2nd upper molars scarcely higher than paracone (median anterior cusp)	0	+

Variation. Two subspecies in the region:
M. l. leucogaster. S China.
M. l. rubex. NE India. More reddish.

Murina aurata Little tube-nosed bat

M. aurata Milne-Edwards, 1872b: 250, pls 37b, 37c; Moupin, Sichuan, China.
? *Harpiocephalus feae* Thomas, 1891a: 884; Biapo, Karin Hills, Burma.

Range (Map 69). Nepal, NE India, NE Burma, S China (Sichuan, Yunnan, SE Tibet); N Thailand (Hill, 1983; Melville, 1984).

Variation. Maeda (1980) reviewed this species and considered *feae* to be a tentative synonym of *M. aurata*. It differs from *aurata* chiefly in its duller, more brownish colour and in the absence of a distinct fringe of hairs along the edge of the uropatagium.

Murina florium

M. florium Thomas, 1908l: 371; Flores I, Lesser Sunda Is.
M. lanosa Thomas, 1910a: 534; Seram I, Moluccas.
M. toxopei Thomas, 1923a: 254; En-Biloro, Buru I, Moluccas.

Range (Map 71). Lesser Sunda Is: Flores, ? Sumbawa; Sulawesi (Hill & Rozendaal, 1989); Moluccas: Bacan (Hill & Rozendaal, 1989), Buru, Seram, Goram, ? Ambon; also in Bismarck Is and NE Australia.

Map 71. *Murina* (part)

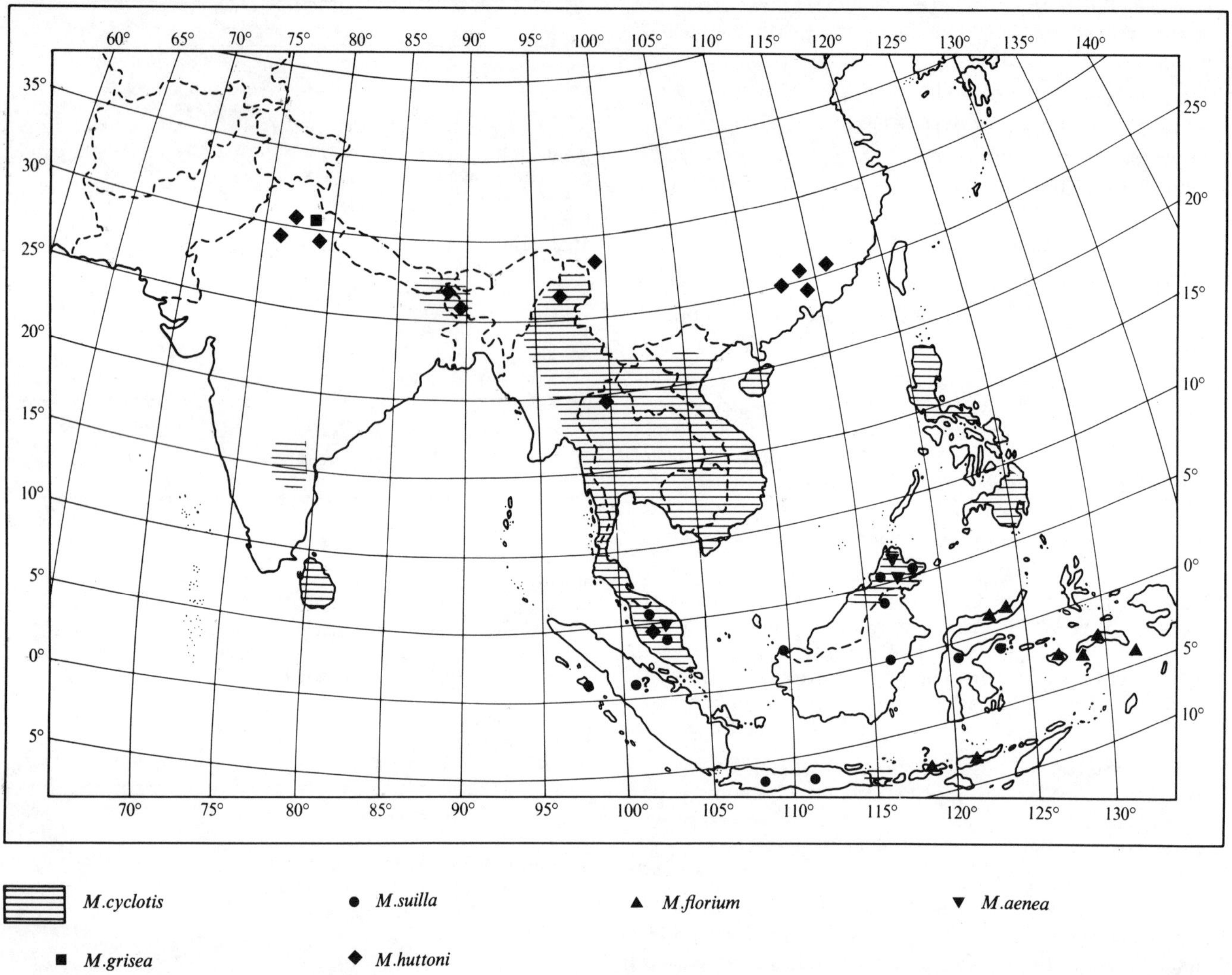

Variation. Two subspecies can be recognized:
M. f. florium. Lesser Sunda Is and Sulawesi.
M. f. lanosa. Moluccas. Tail, hind limbs and uropatagium more densely haired.

Koopman & Danforth (1989) referred small specimens from Sulawesi, Peleng I and New Guinea to *M. florium* but these agree more closely in size with *M. suilla* (*q.v.*) as here understood; moreover Hill & Rozendaal (1989) recorded rather larger specimens with wider palate from Sulawesi that are referable to *M. florium*. The tentative record from Sumbawa is based on Mertens (1936) who recorded *M. suilla* from this island, but this record may refer to *M. florium* (Laurie & Hill, 1954).

Remarks. Koopman & Danforth (1989) considered *florium* possibly conspecific with *suilla*, but in view of uncertainties and inconsistencies in size and geographical variation retained the current specific separation.

Murina suilla Brown tube-nosed bat

Vespertilio suillus Temminck, 1840: 224, pl. 56; Tapos, Java.
? *M. balstoni* Thomas, 1908l: 370; Tasimalaja, Preangar, Java.
M. canescens Thomas, 1923a: 254; Nias I, W Sumatra.

Range (Map 71). Malaya; Sumatra; Nias I off W coast of Sumatra; Java, Borneo; doubtfully in Sulawesi, Peleng I and in New Guinea.

Variation. The characters advanced by Thomas (1908l) to support *balstoni* (generally browner colour, uniformly white venter, flanks more greyish than the centre, sparsely pilose hind limbs and uropatagium, dark cutaneous system, and smaller skull) seem scarcely to justify its separation from *suilla*. In cranial size it stands at the lower part of the range of this species: the other points seem more probably individual variation. Possibly *canescens* from Nias I is a valid subspecies, but it is poorly known.

Remarks. Tentative records from Sulawesi, Peleng I and New Guinea are based on Koopman & Danforth (1989) who discussed small examples of *Murina* from these islands and referred them to *M. florium* but which agree more nearly in size with *M. suilla* as here defined, rather than with this larger species. Possibly, as these authors suggested, *suilla* and *florium* are conspecific, but to accept this view involves a wide variation in size. Relatively few specimens of *Murina* are known from this part of the region and clearly more material is needed to resolve the question.

Murina tubinaris

Harpiocephalus tubinaris Scully, 1881b: 200; Gilgit, Kashmir.

Range (Map 70). NE Pakistan, Kashmir, NE India, Burma; N Thailand (Hill, 1983; Melville, 1984); Laos, Vietnam.

Remarks. Reviewed by Hill (1964). Tate (1941f) indicated that *tubinaris* might be conspecific with *M. huttonii* of which Ellerman & Morrison-Scott (1951) listed it as a tentative subspecies; however Hill (1964) recognised it as distinct. More recently Koopman & Danforth (1989) have suggested that *tubinaris* might be conspecific with *M. suilla* which apart from differences in colour it closely resembles.

Murina rozendaali

M. rozendaali Hill & Francis, 1984: 319; Gomantong, Sabah, Borneo, 5°31'N, 118°4'E.

Range (Map 70). Borneo (Sabah).

Murina huttonii Hutton's tube-nosed bat

Harpyiocephalus huttonii Peters, 1872b: 257; Dehra Dun, Kumaon, NW India.

Range (Map 71). N India, SE Tibet, S China (Fujian), Thailand, Malaya.

Variation. Reviewed by Hill (1964). A second slightly larger, brighter and more brownish subspecies, *M. h. rubella* Thomas, 1914g: 440 (Kuatun, Fujian) can be recognized in S China and may extend to N Thailand.

Murina puta

M. puta Kishida, 1924b: 127,130,132; Taiwan.

Range (Map 70). Taiwan.

Remarks. Tate (1941f) suggested relationship with *M. aurata* and its allies, or with the Japanese *hilgendorfi* (in *M. leucogaster*) but according to Yoshiyuki (1989) *puta* is similar in size and coloration to *M. huttonii rubella*.

Murina aenea

M. aenea Hill, 1964: 57, pls 54,55; Ulu Chemperoh, near Janda Baik, Bentong District, Pahang, Malaya, c. 3°18'N, 101°50'E, 2000ft.

Range (Map 71). Malaya; Borneo (Sabah—Hill & Francis, 1984).

Murina cyclotis Round-eared tube-nosed bat

M. cyclotis Dobson, 1872d: 210; Darjeeling, NE India.
M. eileenae Phillips, 1932b: 329; Mousakande, Gammaduwa, East Matale Hills, Sri Lanka, 3000ft.

Range (Map 71). Sri Lanka; S India (Ghosh, 1989); NE India, W, N Burma, Thailand, Vietnam; Hainan; Malaya; Borneo (Sarawak, Brunei, Sabah); Philippines: Luzon, Mindanao; Lesser Sunda Is: Lombok.

Variation. Reviewed by Hill (1964). Three subspecies can be recognized:
M. c. cyclotis. NE India to Vietnam.
M. c. eileenae. Sri Lanka. Slightly darker and duller.
M. c. peninsularis Hill, 1964: 55, pl. 52,53; Ulu Chemperoh, near Janda Baik, Bentong District, Pahang, Malaya, c. 3°18'N, 101°50'E, 2000ft. Malaya. Relatively wider anterior rostrum, more massive teeth.

Kitchener *et al.* (1990) suggested that the population on Lombok may also be subspecifically distinct.

Remarks. Hill (1964) pointed out that the rostral and dental features of *M. cyclotis* tend toward the extreme condition found in *Harpiocephalus*.

Subgenus *Harpiola*

Briefly reviewed by Tate (1941f) who treated it as a distinct genus.

Murina grisea Peters' tube-nosed bat

M. grisea Peters, 1872b: 258; Jeripanee, Mussoorie, Kumaon, NW India, 5500ft.

Range (Map 71). Apparently known only from the holotype (in BMNH), from Kumaon, India.

Remarks. The skull of the holotype is distorted and in poor condition, and the teeth appear to have been slightly eroded by a strong preserving fluid, perhaps formalin. The features upon which Thomas based *Harpiola* do not appear to be artificial, but to some extent the status of *grisea* must remain uncertain until further material is obtained.

HARPIOCEPHALUS

Harpiocephalus Gray, 1842: 259; type species *H. rufus* Gray = *H. harpia*.

India to S China and the Moluccas; 2 species (Table 93).

Harpiocephalus harpia Hairy-winged bat

Vespertilio harpia Temminck, 1840: 219, pl. 55; SE side of Mt Gede, Java. Lectotype designated by Husson (1955a).
= *H. rufus* Gray, 1842: 259.
Noctulinia lasyura Hodgson, 1847h: 896; 'central hills, sub-Himalayas', restricted to Darjeeling, NE India by Wroughton (1918b).
Lasiurus pearsoni Horsfield, 1851: 36; Darjeeling, NE India.
H. rufulus Allen, G. M.,1913: 214; Lao-Kai, Tonkin, Vietnam.

Harpiocephalus harpia

Range (Map 72). S, NE India; S China (Fujian); Taiwan; Vietnam; Sumatra, Java, Borneo (Sabah); Lombok; Ambon (S Moluccas).

Variation. One subspecific name has been proposed:
H. h. madrassius Thomas, 1923n: 88; Perumal, Palni Hills, S India.

Subspecific variation remains unclear. Hill & Francis (1984) considered *mordax* a distinct species. Das (1986b) reviewed Indian specimens of *Harpiocephalus* and those recorded in the literature, but was unaware of this conclusion. Das considered that pelage colour had little subspecific relevance and compiled measurements that indicated a wide range of size variation and possibly some sexual dimorphism. He synonymized *madrassius* with *lasyurus* and concluded that the genus was likely to prove monospecific and *H. harpia* to be monotypic. However, specimens from NE India listed by Das that have expanded zygomata and wide rostra may be *H. mordax* (*q.v.*).

Harpiocephalus mordax

H. mordax Thomas, 1923n: 88; Mogok, N Burma.

Range (Map 72). Burma; Thailand (McBee *et al.* 1986); Borneo (Sabah–Hill & Francis, 1984).

Remarks. Ellerman & Morrison-Scott (1951) considered *mordax* a possible subspecies of *H. harpia* but Hill & Francis (1984) concluded that it was specifically valid. Three female specimens from NE India (Darjeeling, Sikkim, Assam) listed and measured by Das (1986b) have characteristically expanded zygomata (zygomatic width 14.4–14.5) and wide rostra (C-C 6.9–7.1) and may also represent *H. mordax* rather than *H. harpia* with which he listed them. Examined specimens referred to *mordax* are also female and as Das pointed out, females of *H. harpia* often have larger skulls than males. However, a small series of 2 males and 3 females of *H. harpia* from Java in the BMNH suggests that the degree of dimorphism is less than the observed differences in rostral and tooth size between *harpia* and *mordax*.

Table 93. The species of *Harpiocephalus*.

	harpia	*mordax*
Range	India - Moluccas	?NE India -Borneo
Zygomatic width	12.1–13.9	14.3–14.9
Width across supraorbital tubercles	7.0–7.8	7.9–8.3
Width across anteorbital foramina	5.8–6.5	6.8–7.3
C^1-C^1 (external)	5.8–6.8	6.9–7.5
Incisor and canine teeth massive	0	+

Map 72. *Harpiocephalus*, *Kerivoula* (part)

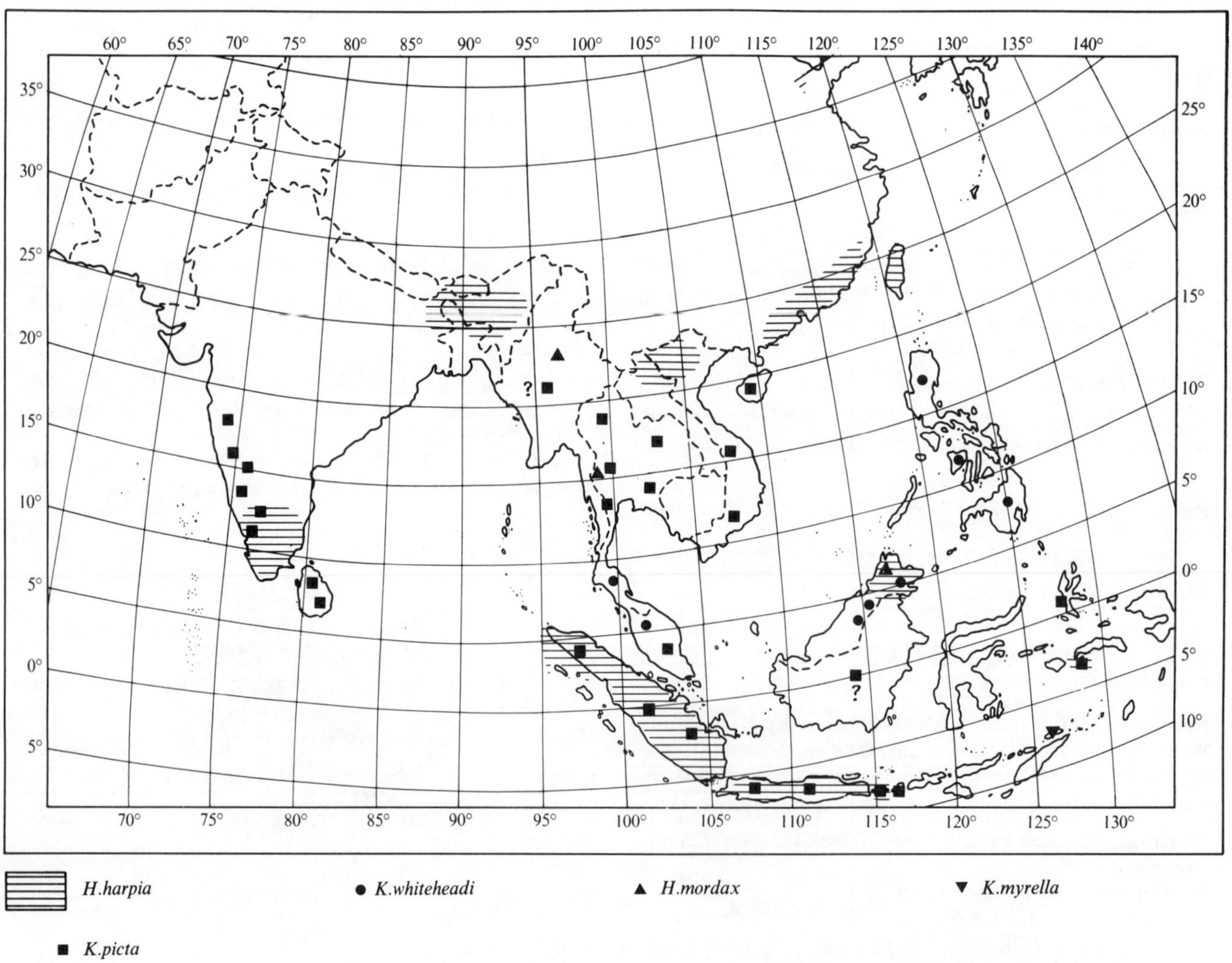

Subfamily Kerivoulinae

Kerivoulinae Miller, 1907: 232; type genus *Kerivoula* Gray.

C, S Africa, India to Philippines and Australia, with 2 genera, both occurring in the region (Table 94), characterized by long, funnel-shaped ears and long, pointed tragus. Familial rank has been suggested (Sigé, 1974; Van Valen, 1979) but has not been generally adopted.

Table 94. The genera of Kerivoulinae.

	Kerivoula	*Phoniscus*
Species in region	9	2
Posterior margin of tragus deeply notched near base	0	+
Postorbital region markedly constricted, not tubular	+	0
Anterior narial emargination usually distinctly longer than wide	+	0
Anterior palatal emargination usually longer than wide	+	0
I^3 not reduced, its greatest diameter subequal to that of I^2	+	0
C^1 with prominent lateral groove	0	+

KERIVOULA

Kerivoula Gray, 1842: 258; type species by subsequent designation (Peters, 1866d) *Vespertilio pictus* Pallas.
Nyctophylax Fitzinger, 1861: 390; apparently based on *Vespertilio tralatitius*, Temminck, 1840 = *V. tralatitius* Horsfield, 1824, an indeterminable name according to Laurie & Hill (1954). According to Palmer (1904) *Nyctophylax* was 'a new name for the barbaric *Kerivoula* Gray, 1842'.
= *Cerivoula* Blanford, 1891: 338.

Africa, India to Philippines, New Guinea etc.; c. 18 species of which 9 occur in the region. Reviewed by Hill (1965).

1. Anterior upper premolar (P^2) narrow, elongate, its length exceeding its width (Fig. 16k, p. 119); breadth of cochleae 3x their distance apart; (forearm 27.5–31, condylobasal length 11.8–11.9, C-M^3 c. 5.0) .... ***K. whiteheadi***

P^2 wide, its length equal to or less than its width (Fig. 16l); breadth of cochleae not exceeding 2x their distance apart .. **2**

2. Bright orange colour extending to ears and flight membranes; interfemoral membrane with prominent fringe on posterior margin; length of 2nd upper premolar (P^3) exceeding its width; (forearm 32.5–39, condylobasal length 12.7–13.8, C-M^3 5.1–6.0) .... ***K. picta***

Pelage, ears and flight membranes not orange; no prominent interfemoral fringe; length of P^3 equal to or less than its width **3**

3. Skull smaller: condylobasal length <11.8, braincase width <6.0, mastoid width <6.7, C-M^3 <5.1 .... **4**

Skull larger: condylobasal length >11.5, braincase width >6.5, mastoid width >6.9, C-M^3 usually >5.1 .... **5**

4. Skull smaller; braincase more globular; palate shorter; condylobasal length 10.0–11.1; palatal length 6.0–6.4; C-M^3 4.1–4.5; (forearm 24.8–29.3)................................ ***K. minuta***

Skull larger; braincase elongate; palate longer; condylobasal length 11.1–11.8; palatal length 6.4–6.7; C-M^3 4.6–5.0; (forearm 26.7–30.7) .... ***K. intermedia***

5. Braincase rising abruptly from rostrum; post-palatal extension narrow, its width 1.4 or less; (forearm 28.5–33.5, condylobasal length 12.5–13.0, C-M^3 5.3–5.6).... ***K. pellucida***

Braincase not rising especially abruptly from rostrum; post-palatal extension wider, its width 1.5–1.8 .... **6**

6. Palate wider, C^{1-1} (external) 4.1–4.9, M^{3-3} (external) 6.0–7.0; (generally larger, forearm >40, condylobasal length >15.0, C-M^3 >6.5 .... ***K. papillosa***

Palate narrower, C^{1-1} 3.1–3.8, M^{3-3} 4.8–5.8; (generally smaller, forearm <40, condylobasal <15.0, C-M^3 <6.5..... **7**

7. Rostrum slightly broadened anteriorly, the width across the canine cingula (C^{1-1}) considerably exceeding width across anterior upper premolars (P^{2-2}); rostrum narrowest across 2nd upper premolars (P^{3-3})....................................... ***K. myrella***

Rostrum not broadened anteriorly, its narrowest width across anterior upper premolars (P^{2-2}), this similar to or only slightly less than width across canine cingula **8**

8. Generally smaller: forearm 30–38.5, condylobasal length 12.2–13.8; zygomatic width 8.0–9.0, braincase width 6.7–7.4, braincase depth 4.9–5.8; C-M^3 4.8–6.0 .... ***K. hardwickii***

Generally larger: forearm 37–39, condylobasal length 14.3–14.9, zygomatic width 9.2–9.8, braincase width 7.4–7.7, braincase depth 6.1–6.4, C-M^3 6.1–6.4 .... ***K. flora***

Kerivoula whiteheadi

K. whiteheadi Thomas, 1894e: 460; Molino, Isabella, NE Luzon I, Philippines.
K. pusilla Thomas, 1894e: 461; Mt Mulu, E Sarawak, Borneo.
K. bicolor Thomas, 1904b: 199; Biserat, Jalor, S Thailand.

Range (Map 72). S Thailand, Malaya; Borneo (Sarawak); Philippines: Luzon, Panay, Mindanao.

Variation. Few specimens are known. A limited sample suggests that *K. w. whiteheadi* from Luzon and Panay is very similar to *K. w. pusilla* from Mindanao and Borneo but possibly slightly larger, while *K. w. bicolor* from the mainland differs in whitish and not greyish underparts, white-tipped wings, and slightly narrower rostrum.

Kerivoula picta Painted bat

Vespertilio pictus Pallas, 1767: 7 (based on Seba, 1734: 91, pl. 56, figs 2,3); probably Ternate I, N Moluccas (Tate, 1941f; Laurie & Hill, 1954).
? *V. p. rubellus* Kerr, 1792: 96; locality unknown.
V. kirivoula Cuvier, F.,1832: 9.

Range (Map 72). India; Sri Lanka; ? Burma; S China (Guangdong, Hainan); Vietnam, Thailand, Malaya; Sumatra, Java, ? Borneo; Bali, Lombok; Moluccas: Ternate, Ambon.

Variation. The nominate subspecies occupies most of the range; a second slightly larger subspecies *K. p. bellissima* Thomas, 1906d: 423 has been described from Pak-hoi, Guangdong, S China and has been recorded from Hainan.

Map 73. *Kerivoula* (part), *Phoniscus* (part)

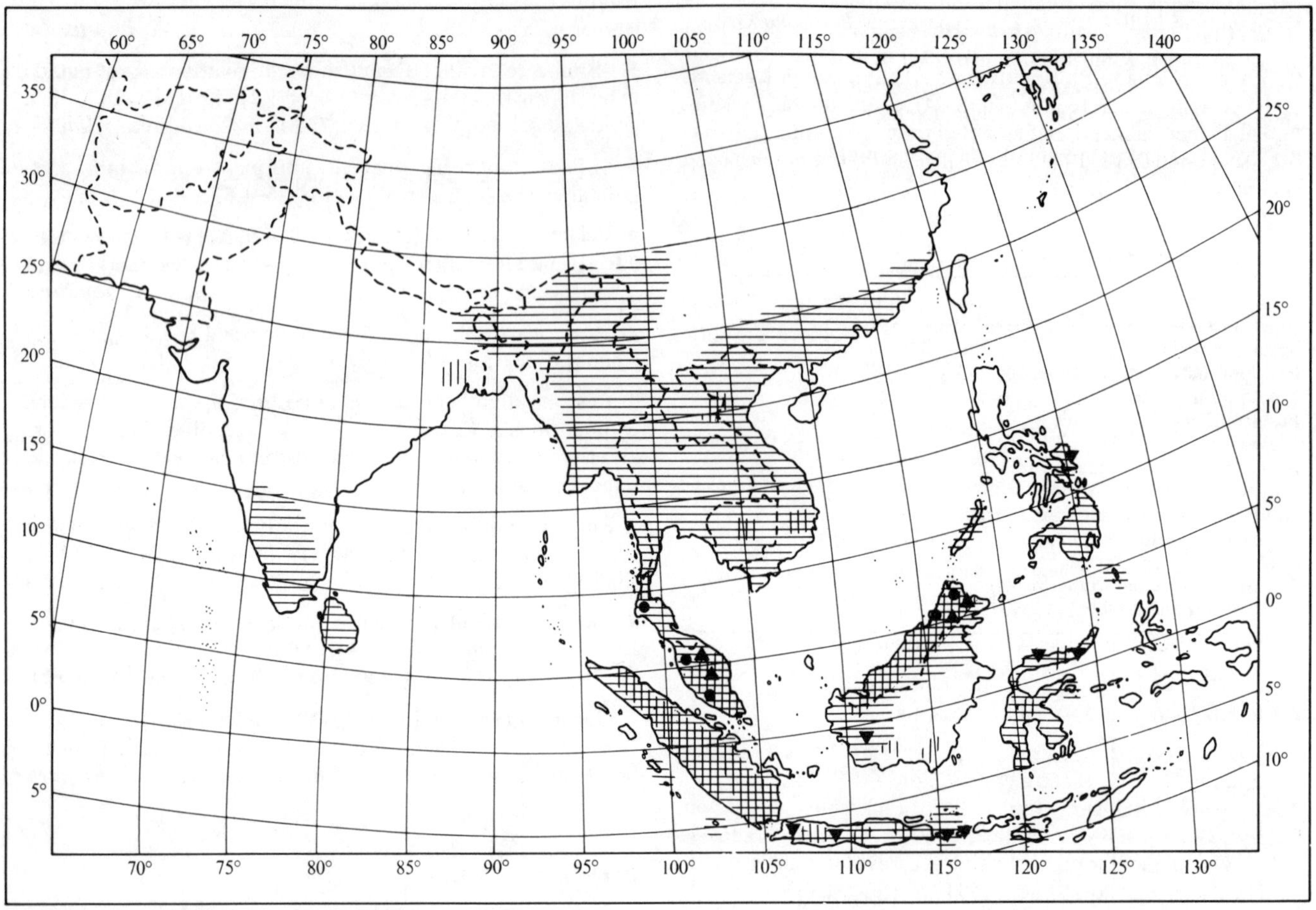

K.hardwickii K.papillosa ● K.minuta ▲ K.intermedia ▼ P.jagorii

Kerivoula minuta Least forest bat

K. minuta Miller, 1898: 321; Lay Song Hong, Trang, S Thailand.

Range (Map 73). S Thailand, Malaya; Borneo (Sabah–Medway, 1977; Hill & Francis, 1984).

Remarks. The smallest examples of this species are little larger than *Craseonycteris thonglongyai*, generally accepted as among the smallest if not the smallest of bats.

Kerivoula intermedia

K. intermedia Hill & Francis, 1984: 323; Lumerau, Sabah, Borneo, 5°12'N, 118°52'E.

Range (Map 73). Malaya; Borneo (Sabah).

Kerivoula pellucida

Vespertilio pellucidus Waterhouse, 1845: 6; Philippines.
K. bombifrons Lyon, 1911a: 134; Along Matam R, W Borneo.

Range (Map 74). Malaya; Sumatra, Java, Borneo; Philippines: Mindoro, Samar, Palawan, Mindanao; Sulu Is: Jolo, Sibutu.

Kerivoula papillosa Papillose bat

K. papillosa Temminck, 1840: 220, pl. 55; Bantam, W Java (restricted by Tate, 1940).
K. lenis Thomas, 1916c: 416; Calcutta, Bengal, India.

Range (Map 73). NE India; Thailand (McBee *et al.*, 1986); Vietnam, Cambodia; Malaya; Sumatra, Java, Borneo; Sulawesi (Hill, 1983).

Variation. One subspecific name has been proposed:
K. p. malayana Chasen, 1940a: 55; Ginting Bedai, Selangor-Pahang boundary, Malaya, 2300ft.

Hill (1965) recognized three subspecies, *K. p. lenis* from NE India, *K. p. malayana* from much of the remainder of the range, and the nominate subspecies from Java. Later, Hill (1983) referred a specimen from Sulawesi to *K. p. malayana*. However, no examples of the nominate subspecies have been seen, although Chasen (1940a) remarked that a few that he had seen were rather greyer than *malayana*. The holotype of *lenis* is a little smaller than more eastern specimens.

Kerivoula hardwickii Hardwicke's forest bat

Vespertilio hardwickii Horsfield, 1824:part 8; Java.
K. fusca Dobson, 1871d: 215; locality unknown.

K. depressa Miller, 1906d: 64; Biapo, Carin Hills, NE of Tounghoo, S Burma.
K. engana Miller, 1906e: 825; Dua I, about 1 mile from Enggano I, SW of Sumatra.
K. crypta Wroughton & Ryley, 1913: 14; Karbidetta forest, Shimoga, S India.
K. malpasi Phillips, 1932b: 331; Kumbalgamuwa, Mulhalkelle dist. (c. 30 miles E of Kandy), Central Prov., Sri Lanka, 3000ft.

Range (Map 73). India; Sri Lanka; Burma, S China (Sichuan, Guangxi, Fujian), Vietnam, Thailand, Malaya; Sumatra; Mentawai Is: Sipura, Siberut; Dua I (off Enggano I); Java, Kangean Is, Borneo; Philippines: Samar, Palawan, Mindanao; Bali, Nusa Penida I, Lombok, ? Sumba; Talaud Is: Karakelang (Hill & Rozendaal, 1989); Sulawesi; Peleng I.

Variation. There appears to be no consistent variation in size over the range although specimens from Lombok I are unusually large (Kitchener *et al.*, 1990); specimens from the western part of the range from N India to Vietnam have relatively flattened braincases.

Remarks. Discussed by Hill (1965,1983) and Hill & Rozendaal (1989). Formerly held to include *flora* (*q. v.*) but this is here considered a distinct species.

Kerivoula flora

K. flora Thomas, 1914g: 441; S Flores, Lesser Sunda Is.

Range (Map 74). Borneo (Sabah–Hill & Rozendaal, 1989); ? Bali; Lesser Sunda Is: Lombok, Flores, ? Sumbawa, ? Sumba.

Remarks. Variously referred to *K. hardwickii* or to *K. papillosa* as a subspecies, *flora* was considered specifically distinct by Hill & Rozendaal (1989), who reported a specimen apparently referable to *flora* from Sabah, in near sympatry with *K. hardwickii*. More recently Kitchener *et al.* (1990) have reported its sympatric occurrence with *K. hardwickii hardwickii* on Lombok I. The species differs from *K. hardwickii* in generally greater size, relatively longer, narrower rostrum and relatively narrower palate, more linear upper toothrows which are not pinched in at the posterior upper premolars, and relatively narrow post-palatal extension; and from *K. papillosa* in smaller size, unwidened rostrum, and unexpanded upper canines. It may occur on Bali, whence Sody (1933b) recorded a similarly large specimen that may represent *K. flora*. Mertens (1936) recorded *flora* (as a subspecies of *K. hardwickii*) from Sumbawa; the listing of *K. hardwickii* from Sumba by Pohle (1950) may also refer to *K. flora*.

Kerivoula myrella

K. myrella Thomas, 1914i: 438; Manus I, Admiralty Is, Bismarck Is.

Range (Map 72). Wetar I (N of Timor) (Hill & Rozendaal, 1989); and in the Bismarck Is.

Remarks. This species is close to *K. hardwickii*, differing chiefly in its slightly widened anterior rostrum and basally more massive canines. Few examples are known: more material may show it to be a subspecies of *K. hardwickii* as Hill & Rozendaal (1989) suggested.

Map **74**. *Kerivoula* (part), *Phoniscus* (part), *Nyctophilus*

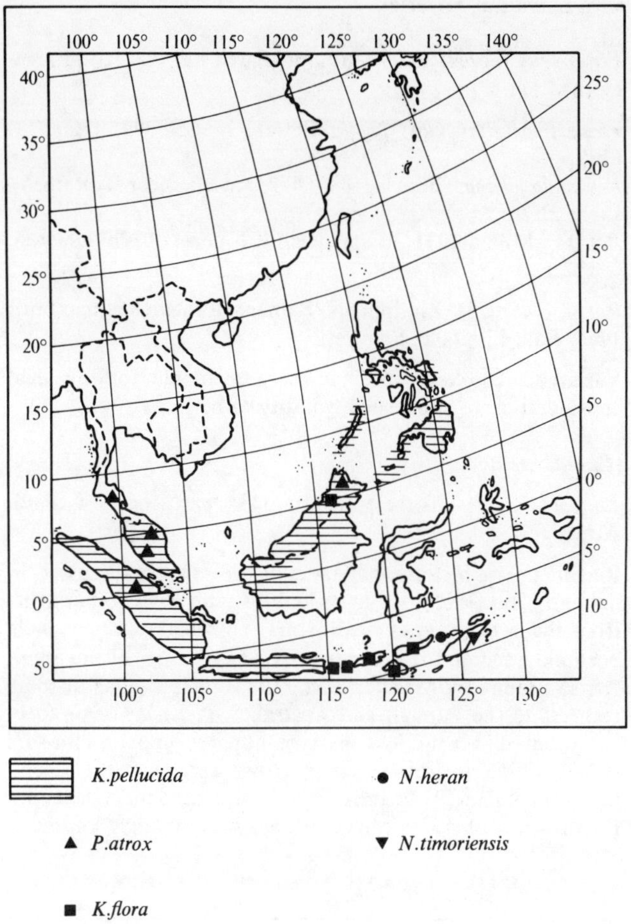

PHONISCUS

Phoniscus Miller, 1905: 229; type species *P. atrox* Miller.

Thailand to Australia; 3 species, with a further named form of uncertain status and provenance, doubtfully attributed to South Africa but possibly Asian in origin; otherwise 2 species in the region (Table 95). Ryan (1965) considered from a limited sample that *Phoniscus* could not be separated either generically or subgenerically from *Kerivoula*, and it was subsumed into *Kerivoula* as a subgenus by Koopman (1982) who considered the Papuan *K. agnella* to be somewhat intermediate. However, *Kerivoula* and *Phoniscus* are separated by a wide range of external, cranial and dental characters, and apart from some shortening of the narial emargination on occasion (Koopman, 1982) *agnella* agrees with *Kerivoula* in the features of the tragus, rostrum, anterior palate and incisors.

Phoniscus atrox Groove-toothed bat

P. atrox Miller, 1905: 230; vicinity of the Kateman R, E Sumatra.

Range (Map 74). S Thailand, Malaya; Sumatra; Borneo (Sabah–Hill & Francis, 1984).

Variation. Specimens from Borneo are slightly smaller than those from the mainland and differ slightly in some cranial and dental features (Hill & Francis, 1984). The Bornean population may be subspecifically separable but none has been compared with Sumatran material.

Phoniscus jagorii Peters' trumpet-eared bat, Gilded groove-toothed bat

Vespertilio (*Kerivoula*) *jagorii* Peters, 1866d: 399; Samar I, Philippines.
Kerivoula javana Thomas, 1880: 472; Kosala, near Bantam, W Java, 2100ft.
P. rapax Miller, 1931: 412; Paleleh I, N coast of NE Sulawesi, c. 122°E.

Range (Map 73). Philippines: Samar; Borneo (Kalimantan); Java, Bali, Lombok; Sulawesi.

Variation. Too few specimens are available for any real consideration of subspecific validity to be made.

[*Phoniscus aerosa*]

Kerivoula aerosa Tomes, 1858b: 333; 'east coast of South Africa'.

Remarks. The history and status of *aerosa* were discussed in detail by Hill (1965). Without doubt it is a *Phoniscus* but apart from the two syntypes (in BMNH) no other examples have been obtained and it is not treated as a member of the South African fauna by Meester *et al.* (1986). It is similar in some respects to the Papuan and Australian *P. papuensis* or to *P. atrox* but differs in less massive anterior upper premolars (P^{2-2}), which are rather less in crown area than the second upper premolars (P^{3-3}) and do not much exceed them in height. Possibly it is of Asian origin and has been wrongly localized.

Table 95. The species of *Phoniscus*.

	atrox	*jagorii*
Range in region	Thailand - Borneo	Philippines - Java, Sulawesi
Forearm	31.5–34.5	37.6–38.5
Condylobasal length	12.9–13.8	15.3–15.9
C-M^3	5.6–6.2	6.7–7.1
Second upper premolar (P^3) elongate	0	+
P^3 with prominent antero-internal prolongation	0	+

Subfamily Nyctophilinae

Nyctophili Peters, 1865c: 524; type genus *Nyctophilus* Leach.

Australia, New Guinea, Lesser Sunda Is and perhaps Timor I; 2 genera and 10 species of which one, perhaps two, may occur in the Indomalayan Region. Readily recognized by large, conjoined ears and a square, truncate muzzle surmounted by a low, fleshy dermal outgrowth or noseleaf.

NYCTOPHILUS

Nyctophilus Leach, 1821: 78; type species *N. geoffroyi* Leach.

A predominantly Australian genus of 9 species, 2 extending to New Guinea, one on Lembata I, Lesser Sunda Is and one possibly on Timor I. *N. heran* is distinguished from *N. timoriensis* by its small size as follows (measurements of *N. timoriensis* in parentheses): forearm <40 (>45), greatest length of skull <17 (>19), C^1-C^1 at alveoli <4.7 (>5.0), C-M^3 <6.1 (>7.0).

Nyctophilus heran

N. heran Kitchener, How & Maharadatunkamsi, 1991a: 98; Kampong Merdeka, Desa Hadakewa, Lembata I, Lesser Sunda Is (Nusa Tenggara), Indonesia, 8°22'S, 123°31'E.

Range (Map 74). Lesser Sunda Is: Lembata I.

Remarks. Known only from the holotype. At least superficially similar to *N. geoffroyi*, the most widely distributed of Australian *Nyctophilus*, to which it may be related.

Nyctophilus timoriensis Greater long-eared bat

Vespertilio timoriensis Geoffroy, 1806: 200; locality unknown, possibly Timor I.
And other, extralimital names.

Range (Map 74). ? Timor I; extralimitally New Guinea and Australia.

Remarks. It is assumed that this species occurs on Timor I, where its presumed holotype may have originated (Rode, 1941; Koopman, 1984b; Bannister *et al.*, 1988), but it has never again been obtained there, and the exact provenance of Geoffroy's specimen remains uncertain. Tomes (1858c) and Goodwin (1979) considered on distributional grounds that it is unlikely that *Nyctophilus* occurs on Timor, but more recently Hill & Pratt (1981) have recorded *N. timoriensis* from New Guinea. Kitchener *et al.* (1991) remarked that the discovery of *N. heran* on Lembata I perhaps supported the similar occurrence of *N. timoriensis* on Timor, and that further collecting is required before it can be excluded from the bat fauna of this island.

Family Molossidae

Free-tailed bats

Molossidae Gill, 1872: 17; type genus *Molossus* Geoffroy.

A family of 13 genera and c. 89 species distributed from SW Canada to Argentina, from S Europe to Japan, SE Asia, the Philippines, N Australia to the Solomon Is and Fiji, chiefly confined to tropical and subtropical regions, with 4 genera and 13 species in the Indomalayan Region (Table 96). Known from the late Eocene in Europe, the late Oligocene or early Miocene in South America, the Miocene of Africa, and from the Pleistocene in Asia, the East Indies and Australia. Molossids are very small to large (forearm 27–85m) and are characterized by a broad, obliquely truncate muzzle, the upper lip often wrinkled, ears that are usually fleshy and often united over the forehead by a band of integument and have a small tragus, by flight membranes that are usually thick and leathery, and by a stout tail that protrudes considerably beyond the edge of a relatively narrow uropatagium.

They are insectivorous and roost in caves, rock crevices, under rocks or shale, in buildings, hollow trees, under bark or in foliage. Some are highly gregarious and form very large colonies.

Table 96. The genera of Molossidae.

	Tadarida	*Mormopterus*	*Otomops*	*Cheiromeles*
Range in region	Throughout	Sumatra, Moluccas	S India, Java	Malaya to Sulawesi, Philippines
Species in region	7	2	2	2
Ears	Rounded	Angular	Rounded	Rounded
Ear integument	Thick	Thin	Thick	Thick
Hairless	0	0	0	+
Distinct pale area over shoulders	0	0	+	0
Postorbital zygomatic projection greatly developed	0	0	+	0
Tympanic bullae very large, elongate	0	0	+	0

TADARIDA

Tadarida Rafinesque, 1814: 55; type species *Cephalotes teniotis* Rafinesque.
Nyctinomus Geoffroy, E.,1818: 114 (*N. aegyptiacus* Geoffroy).
Dinops Savi, 1825: 229 (*D. cestoni* Savi, in *Cephalotes teniotis* Rafinesque).
Dysopes Cretzschmar, 1830–31: 69 (pre-occupied by *Dysopes* Illiger, 1811).
Mops Lesson, 1842;18 (*M. indicus* Lesson, in *T. mops*).
Chaerephon Dobson, 1874a: 144 (*Molossus* (*Nyctinomus*) *johorensis* Dobson).
Philippinopterus Taylor, 1934: 314 (*P. lanei* Taylor, in *Nyctinomus sarasinorum* Meyer).
Also other, extralimital, synonyms.

Much of Americas from British Columbia to Argentina and Antilles; S Europe to Japan; Africa to SE Asia, N Australia, Vanuatu and Fiji; c. 41 species in 5 subgenera; 3 subgenera including 7 species in Region (Table 97). Indo-Australian representatives were reviewed by Hill (1961b).

Subgenus *Tadarida*

Afghanistan to India and Sri Lanka; S China; also to Europe, Africa, NE Asia and the Neotropics (Table 98).

Tadarida teniotis European free-tailed bat

Cephalotes teniotis Rafinesque, 1814: 12; Sicily.
Nyctinomus insignis Blyth, 1862a: 90; Amoy, Fujian, China.
? *T. latouchei* Thomas, 1920f: 283; Chin-wang-tao, NE Chihli, N China.
Also other extralimital names.

Range (Map 75). Afghanistan; probably Pakistan (recorded from extreme E Iran by Roguin, 1988); NE India (Hill, 1963a); S China (Yunnan, Fujian); ? NE Thailand; N Ryukyu Is:Amami-Oshima (Yoshiyuki *et al*., 1989); ? Taiwan; also to S Europe, Madeira, Canary Is, N Africa, Korea, Japan.

Variation. The following subspecies has been described :
T. t. coecata Thomas, 1922a: 392; Mekong Valley, Yunnan, China, c. 28°20'N, 7000ft.

The subspecies in most of the region is probably *T. t. insignis*, which Imaizumi & Yoshiyuki (1965), Yoshiyuki (1989) and Yoshiyuki *et al*. (1989) considered to be a distinct species. An Indian example is slightly paler and more greyish ventrally (Hill, 1963a). Yoshiyuki *et al*. (1989) recorded a small example from Amami-Oshima I in the N Ryukyu Is, referring it to *latouchei* which they considered specifically distinct on grounds of size and colour. Ellerman & Morrison-Scott (1951) considered *latouchei* (known otherwise from the immature holotype) to be a synonym of *T. t. insignis*, but it may apply correctly to this small form on the Ryukyu Is.

Tadarida aegyptiaca Egyptian free-tailed bat

Nyctinomus aegyptiacus Geoffroy, E.,1818: 128, pl. 2; Egypt.
= *Dysopes geoffroyi* Temminck, 1826: 226, pls 19, 23.
N. tragata Dobson, 1874a: 143; Calcutta, India.
T. sindica Wroughton, 1919b: 732; Kashmor, Upper Sind Frontier, Pakistan.

Table 97. The subgenera of *Tadarida*.

Subgenus	*Tadarida*	*Mops*	*Chaerephon*
Species in region	2	2	3
Palatal branch of premaxillae	0	+/0	+
Anterior palatal emargination large, extending posteriorly beyond incisor roots (L) or usually absent or slight (0)	L	0	0
M^3 normal (N) or reduced and flattened antero-posteriorly (R)	N	R	N
M^3: posterior ridge extending lingually to inner angle of anterior ridge (+); or obsolescent or obsolete, extending not less than halfway to inner angle of anterior ridge (0)	+	0	+

Table 98. The species of *Tadarida* (*Tadarida*).

	teniotis	*aegyptiaca*
Range in region	Afghanistan, NW India, S China	Afghanistan to India, Sri Lanka
Forearm	55.5–64	45–53
Width of anterior rostrum cf postorbital width	Usually < or only slightly >	Considerably >
Lower incisors	Usually 3–3	Usually 2–2

T. thomasi Wroughton, 1919b: 732; Bhuj, Cutch, India.
T. gossei Wroughton, 1919b: 733; Sassoon Hospital, Poona, India.
Also other extralimital names.

Range (Map 75). Afghanistan, Pakistan, India, Sri Lanka; also to SW Asia and Africa.

Variation. Three, perhaps four subspecies have been recognized, but their validities are very uncertain. Chaturvedi (1964) demonstrated the identity of *tragata* with *aegyptiaca* but did not examine its validity as a possible subspecies. Sinha (1970) united *thomasi* and *gossei*, while Brosset (1962) considered *sindica, thomasi* and *gossei* probably inseparable.

Subgenus *Mops*

Africa, SW Arabia, Malaya to Philippines and Sulawesi; 9 species, 2 in Region (Table 99).

Tadarida mops Malayan free-tailed bat

Dysopes mops Cuvier, F.,1824: 250 (*nomen nudum*).
? *D. tenuis* Temminck, 1827: 228, pls 19,23,24; Java (not *Nyctinomus tenuis* Horsfield, 1822 according to Tate, 1941b).
? *D. labiatus* Temminck, 1827: 228, footnote; Java.
Molossus mops Blainville, 1840: 101; Sumatra.

Table 99. The species of *Tadarida* (*Mops*).

	mops	*sarasinorum*
Range	Malaya, Sumatra, Borneo	Philippines, Sulawesi
Forearm	43–48	38–45
Skull: greatest length	19.9–21.9	18.4–20.2
C-M^3	7.0–7.7	6.2–6.9
Rostrum	Not especially short	Relatively shorter, narrower
Braincase	Slightly elongate	Narrower, less elongate
Base of canines	Relatively massive	Less massive

Mops indicus Lesson, 1842: 18; 'Inde continentale' (*nomen nudum*).

Range (Map 75). Malaya, Sumatra, Borneo; ? Java.

Remarks. There is no definitive record from Java. However, Tate (1941b) described and measured an alleged "cotype" of *Dysopes labiatus* Temminck, 1827 in the Leiden Museum. This account suggests that the name may refer to a bat of the subgenus *Mops*, possibly *T. mops*. The question was discussed by Hill (1961b); it should be noted that if *labiatus* is conspecific with *T. mops* it antedates this latter name by many years.

Map **75**. *Tadarida*

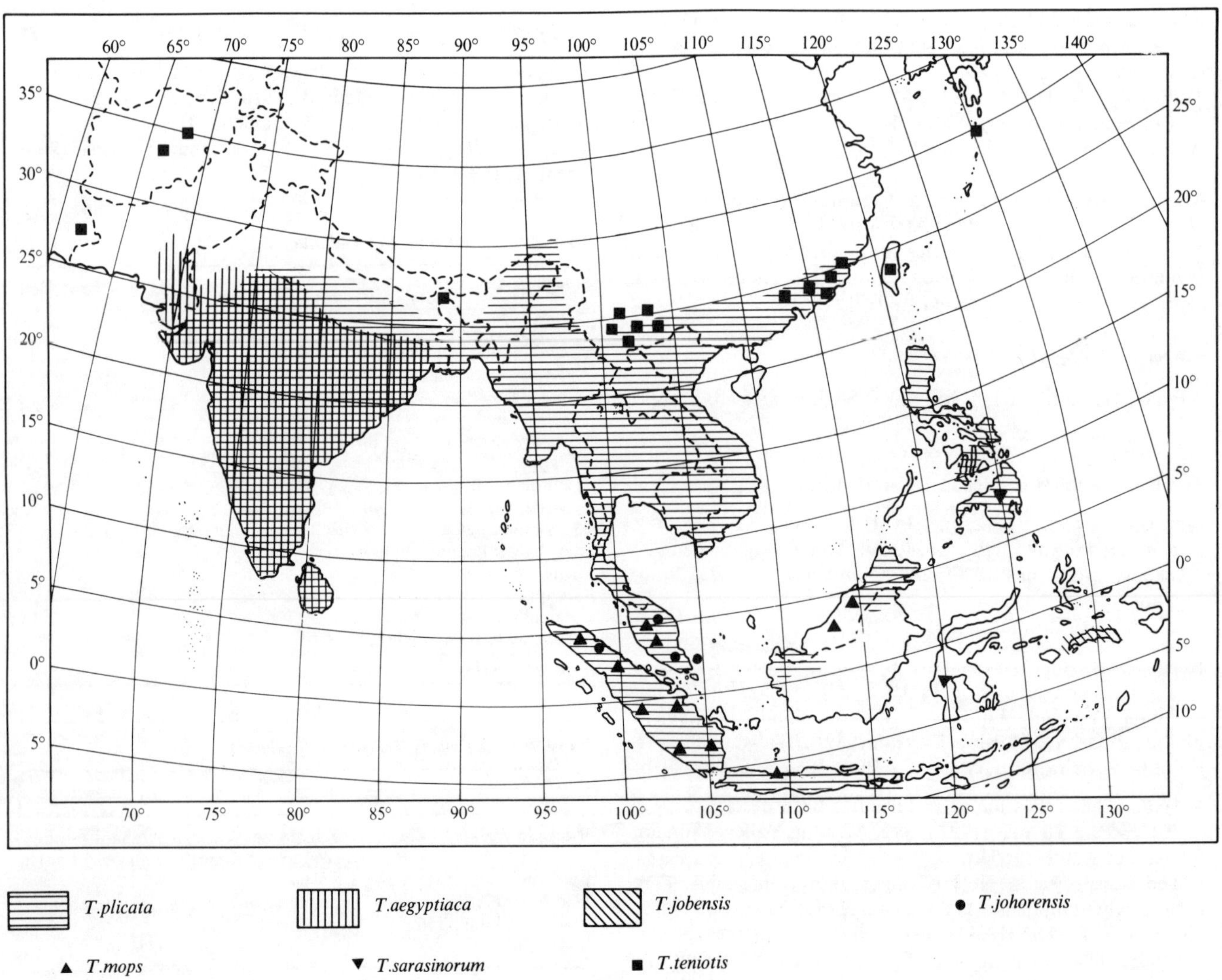

Tadarida sarasinorum Sulawesi mastiff bat

Nyctinomus sarasinorum Meyer, 1899: 16, pl. 4; Batulappa, N of Lake Tempe, central Sulawesi.
Philippinopterus lanei Taylor, 1934: 314; Saub, Cotobato, Mindanao, Philippines.

Range (Map 75). Mindanao (Philippines), Sulawesi and Peleng I (east of Sulawesi).

Variation. Two subspecies may be recognized and are discussed by Hill & Rozendaal (1989):
T. s. lanei. Philippines.
T. s. sarasinorum. Sulawesi and Peleng I. Cranial crests slightly less substantial; rostrum a little lower, wider; narial emargination slightly wider posteriorly; and a little smaller in some respects; but available material is very limited.

Remarks. It is possible that adequate series of specimens might show *sarasinorum* and *mops* to be conspecific.

Subgenus *Chaerephon*

Africa to SW Arabia; India, Sri Lanka to Philippines; Vanuatu, Fiji and N Australia; c. 14 species, 3 in Region (Table 100).

Table 100. The species of *Tadarida* (*Chaerephon*).

	johorensis	*plicata*	*jobensis*
Range	Malaya, Sumatra	India-Lombok, Philippines	Seram
Deep interaural pocket	+	0	0
Interaural band	High, well developed	Narrow	Moderate
Vertical extension of interaural band	+	0	0
Ear integument	Thick	Thick	Thin
Pilosity of internal ear conch	Moderate	Moderate	Sparse
Hair-bases of dorsal pelage	Blackish brown	Blackish brown	Brownish

Tadarida johorensis

Molossus (*Nyctinomus*) *johorensis* Dobson, 1873c: 22; Johore, Malaya.

Range (Map 75). Malaya, Sumatra; Sri Buat I (E coast Johore).

Remarks. This species was reviewed by Hill (1974b) who described its interaural pocket in detail and pointed out that this in females exactly resembles that of males.

Tadarida plicata Wrinkle-lipped free-tailed bat

Vespertilio plicatus Buchannan, 1800: 261, pl. 13; Bengal, India.
Nyctinomus bengalensis Desmarest, 1820: 116.
N. tenuis Horsfield, 1822: pt 5, pl. ; 'Hills of Prowoto' (= Proewoto, east of Semarang, C Java), Java.
N. dilatatus Horsfield, 1822: part 5; W Java (restricted by Chasen, 1940a).
Dysopes murinus Gray, 1830 (in 1830–35): pl. 7.
Chaerephon luzonus Hollister, 1913a: 312; Cagayan, Luzon, Philippines.

Range (Map 75). S Tibet (Lake Palti—BMNH); India, Sri Lanka, Burma; S China: Hong Kong, Hainan; Thailand, Cambodia, Vietnam, Malaya; Sumatra; Borneo; Java; Philippines: Luzon, Leyte, Negros, Cebu, Mindanao; Lesser Sunda Is: Bali, Lombok (Kitchener *et al.*, 1990); Cocos Keeling I.

Variation. The following subspecific names have been proposed:
Chaerephon p. insularis Phillips, 1932b: 334; Kumbalgamuwa, near Mulhalkelle, 30 miles SE of Kandy, Central Prov., Sri Lanka, 3000ft.
C. p. adustus Sody, 1936a: 50; Pangandaran, CS Java.

Five subspecies have been recognized: those in Java were discussed by Sody (1936a), Tate (1941b), Chasen (1940a) and Hill (1961b), who also (1983) reported further Bornean specimens.
T. p. plicata. S Tibet and India to S China, Sumatra and Borneo.
T. p. insularis. Sri Lanka; smaller and darker.
T. p. luzonus. Philippines; smaller and paler.
T. p. dilatata. W Java; darker.
T. p. tenuis. C, E Java, Lesser Sunda Is; smaller, slightly more brightly coloured.

Tadarida jobensis

Nyctinomus jobensis Miller, 1902c: 246; Ansus, Jobi I, Geelvinck Bay, West Irian, New Guinea.
Also other extralimital names.

Range (Map 75). Moluccas: Seram (Durden *et al.*, 1990; BMNH); also to New Guinea, NW, N Australia, Vanuatu and Fiji Is.

Variation. Subspecies in region is *T. j. jobensis*, extending to New Guinea and perhaps the Bismarck Archipelago.

Remarks. Evidently closely related to *T. plicata* with which it may prove conspecific: specimens from Seram, collected by Operation Raleigh in 1987, tend towards *T. plicata* and support this view.

OTOMOPS

Otomops Thomas, 1913e: 90; type species *Nyctinomus wroughtoni* Thomas.

Africa south of the Sahara to Ethiopia; S India; Java; New Guinea; 5 species, 2 in the region (Table 101).

Table 101. The species of *Otomops*.

	wroughtoni	*formosus*
Range	S India	Java
Forearm	62–67.5	59–60
Crown	Brownish, little paler than lower back	Greyish buff, much paler than lower back
Skull: greatest length	23.0–25.1	23.1–24.0
$C\text{-}M^3$	8.4–9.0	8.5–8.7

Otomops wroughtoni Wroughton's free-tailed bat

Nyctinomus wroughtoni Thomas, 1913e: 87; Barapede Cave, near Talewadi, Kanara, India.

Range (Map 76). S India.

Otomops formosus Javan mastiff bat

Otomops formosus Chasen, 1939a: 186; Tjibadak, W Java.

Range (Map 76). Java.

Remarks. No specimen has been seen but it is evident from the description and from Boeadi (1990) that *formosus* is very similar to *wroughtoni*. IUCN: K.

MORMOPTERUS

Mormopterus Peters, 1865a: 258; type species *Nyctinomus jugularis* Peters.

Peru, N Chile; Cuba; Ethiopia, Natal, Madagascar, Mauritius, Réunion, Sumatra; Moluccas to C, SW Australia; 10 species, 2 in region (Table 102).

Table 102. The species of *Mormopterus*.

	doriae	*beccarii*
Range	Sumatra	Moluccas
Lower incisors	3–3	2–2
Anterior lower premolar	Well developed, c.2/3 height of posterior lower premolar	Vestigial or absent

One fossil species has been described from the region: *M.* (*Hydromops*) *nonghenensis* Legendre *et al.*,1988: 284; Phitsanulok Basin, Thailand; Miocene.

Mormopterus doriae

M. doriae Andersen, 1907b: 5, 42; Soekaranda, Deli, NW Sumatra.

Range (Map 76). Sumatra.

Mormopterus beccarii

M. beccarii Peters, 1881: 484; Ambon, Moluccas.
Nyctinomus astrolabiensis Meyer, 1899: 19, pls 10,11; Bongu, Astrolabe Bay, Papua New Guinea.

Range (Map 76). Moluccas: Halmahera (Hill & Rozendaal, 1989), Seram (Dorden *et al.*, 1990; BMNH) and Ambon; also in New Guinea and Queensland.

Variation. Subspecies in region is *M. b. beccarii*, a second slightly larger subspecies with slightly larger teeth, *M. b. astrolabiensis*, occurring in New Guinea and Queensland. Both are discussed by Hill (1983).

CHEIROMELES Hairless bats, Naked bats

Cheiromeles Horsfield, 1824:part 8; type species *C. torquatus* Horsfield.
= *Chiropotes* Gloger, 1841: 49.

Malaya and Sumatra to Philippines and Sulawesi; 2 species, one of doubtful validity (Table 103).

Cheiromeles torquatus Hairless bat, Naked bat

C. torquatus Horsfield, 1824:pt 8; Penang, Malaya.
Dysopes cheiropus Temminck, 1826,1: 218, pls 17,23: Thailand.

Map **76**. *Otomops, Mormopterus, Cheiromeles*

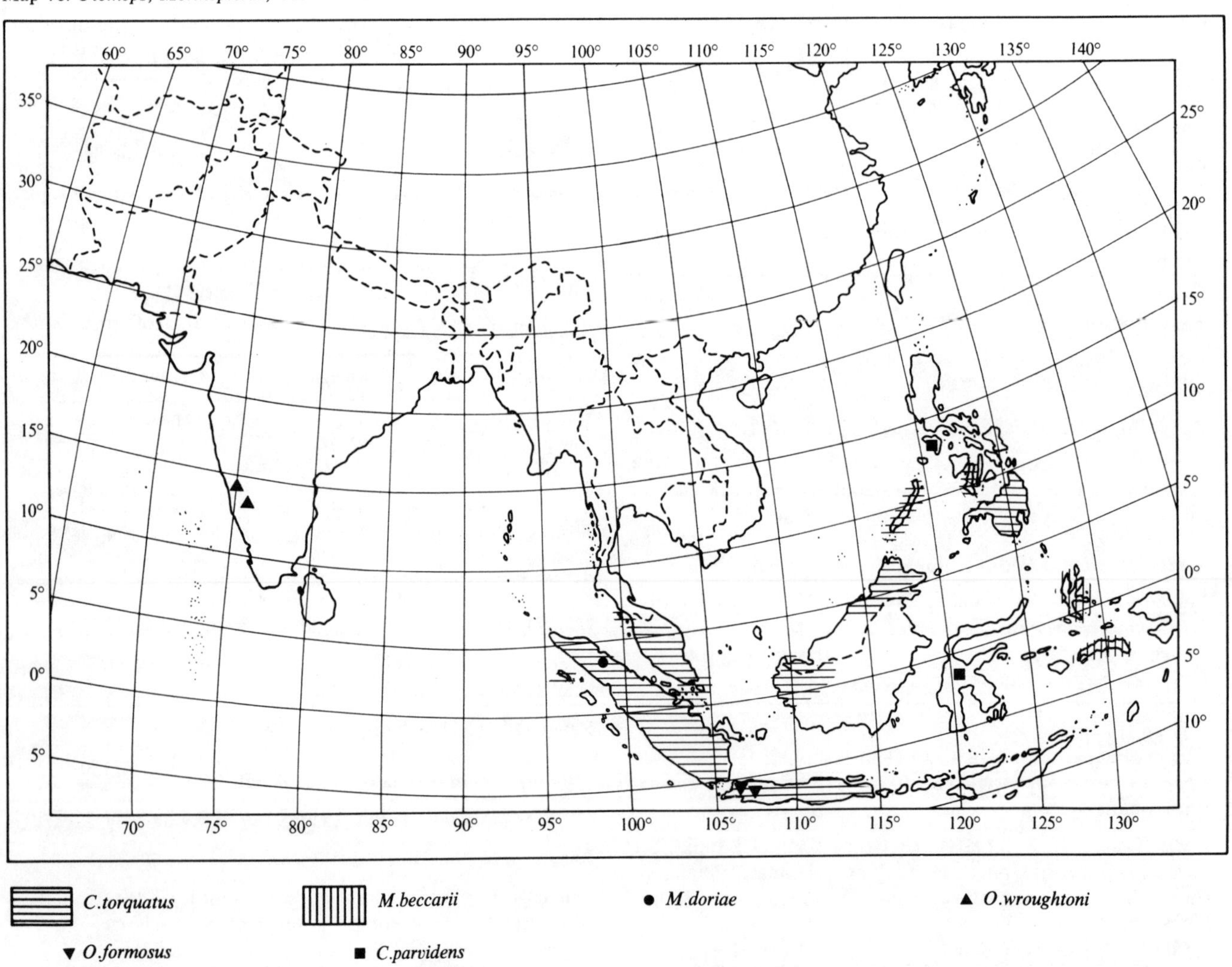

Table 103. The species of *Cheiromeles*.

	torquatus	*parvidens*
Range	Malaya, Sumatra to Philippines	Sulawesi, ?Mindoro
Forearm	73–85	c.72
Condylobasal length	27.7–32.5	c.27.2
C-M^3	10.4–12.9	c.10.0
Teeth	Normal	Disproportionately small

C. caudatus Temminck, 1841a: 348, pl. 66; Bantam, Java (Tate, 1940).
C. jacobsoni Thomas, 1923l: 592; Lugu, Simalur (= Simeulue) I, off W Sumatra.

Range (Map 76). Malaya, Sumatra, Java, Borneo; Philippines: Palawan (Sanborn, 1952a), Negros (Rabor, 1986), Mindanao (Taylor, 1934); the small islands of Tarutau (Peninsular Thailand), Tioman (Malaya), Kundur (Riau Is), Simeuleu (W Sumatra), Bangka. Doubtfully in mainland Thailand and Indochina.

Variation. Three subspecies are recognized in current listings:
C. t. torquatus. Tarutau I, Malaya to Sumatra, Borneo and Philippines.
C. t. jacobsoni. Simeulue I; smaller.
C. t. caudatus. Bangka I, Java; smaller than *torquatus* (Chasen, 1940a).

Remarks. Heaney *et al.* (1987) refer records from Negros and Mindanao in the Philippines to *C. parvidens* but specimens from Mindanao recorded by Taylor (1934) are similar in size to Bornean examples in the collection of the BMNH. No specimens from Bangka I or Java have been examined.

Cheiromeles parvidens

C. parvidens Miller & Hollister, 1921b: 100; Pinedapa, C Sulawesi.

Range (Map 76). Sulawesi; Philippines: Mindoro (Lawrence, 1939).

Remarks. The status of *parvidens* must remain uncertain; no specimens have been examined. It was listed as a distinct species by Honacki *et al.* (1982), but Koopman (in Honacki *et al.*) remarked "Probably conspecific with *torquatus*". Lawrence (1939) recorded a specimen from Mindoro which apart from some small points of coloration seems virtually indistinguishable from the holotype of *parvidens*: on the other hand, examples from the intervening island of Mindanao seem on size grounds to be more appropriately referred to *C. torquatus.*

ORDER PRIMATES

Primates

The primates are well represented in the Neotropical, Afrotropical and Indomalayan regions. Of the 11 families commonly recognized, five are represented in the region, including two, Tarsiidae and Hylobatidae, that are endemic (Table 104). There is general agreement that the family Tupaiidae, formerly treated as primates, should be excluded, as done here (order Scandentia). Primates are characterized by rather primitive, generalized limbs; reduced nasal region; large, forward-facing eyes; complete post-orbital bar; and large brain. Incisors are never more than 2/2. They are mostly arboreal, diurnal and omnivorous, feeding upon fruit, buds and invertebrates, with some, amongst the Colobinae, more folivorous, and the tarsiers (*Tarsius*), insectivorous.

Recent reviews of primates are by Szalay & Delson (1979) and Grine *et al.* (1988)—phylogeny; Wolfheim (1983) and Benirschke (1986)—distribution and conservation; Napier & Napier (1985)— a popular account. Useful reviews of taxonomy are those of Schwarz (1986), Jenkins (1987,1990), Napier (1981,1985), Groves (1989) and, with the emphasis on SE Asia, Weitzel *et al.* (1988). In spite of the vast literature on primates there is no Indomalayan genus for which an adequate taxonomic revision exists, giving diagnoses and ranges of all the taxa now recognized.

All species of primates are on CITES Appendix II except those shown here as on Appendix I.

Family Loridae

Lorises

Loridae Gray, 1821: 298; type genus *Loris* Geoffroy.

Indomalayan and Afrotropical regions; 4 genera (or 8 if the African galagos are included) of which two occur in the Indomalayan Region (Table 105). They differ from all other Indomalayan primates in their slow locomotion, but in this respect are paralleled by the African genera *Perodicticus* (heavily built like *Nycticebus*) and *Arctocebus* (a slender form like *Loris*). Although it has been suggested that *Loris* is more

Table 104. The families of Primates.

	Loridae	Tarsiidae	Cercopithecidae	Hylobatidae	Hominidae
Genera in region	2	1	5	1	2
Head and body (cm)	20–37	9–16	34–83	37–90	125–200
Tail: head & body	0	>	0–160%	0	0
Fore limbs much longer than hind	0	0	0	+	+/0
Hallux and pollex opposable to other digits	+	0	+	+	+/0
Claws on 2nd (C2) and/or 3rd (C3) digits of hind feet, otherwise all digits with nails (N)	C2	C2 & C3	N	N	N
Premolars	3/3	3/3	2/2	2/2	2/2
Lower incisors and canine narrow, forming 6-pronged comb	+	0	0	0	0

Table 105. The species of *Loris* and *Nycticebus* (Loridae).

	L.tardigradus	*N.coucang*	*N.intermedius*	*N.pygmaeus*
Range	S India, Sri Lanka	Assam-Borneo	N Vietnam, C China	Vietnam etc.
Head & body	200–250	265–365	240–260	210–290
Ear (% of skull)	c.40%	29–38%		43–52%
Dorsal colour	Grey-brown	Grey-brown	Dull brown	Bright orange-brown
Legs	Long,slender	Short,stout	Short,stout	Short,stout
Dark dorsal stripe	Faint/absent	Bold	Bold	Faint or absent
Skull: greatest length	45–55	54–67	50–53	48–55
$I^1 : I^2$	= (very small)	>>	>>	>>
Rostrum width:length	=	>	>	>

closely related to the African *Arctocebus* than to the adjacent *Nycticebus* (Schwartz & Tattershall, 1985) there seem to be stronger grounds for considering the two Asian genera to form a closely related pair. This is supported by cranial morphology, immunology and karyotype (Groves, 1989; Rumpler *et al.*, 1987).

LORIS

Tardigradus Boddaert, 1785: 43; type species *T. loris* Boddaert, in *L. tardigradus*. Pre-occupied by *Tardigradus* Brisson, 1762 (Xenarthra) which is however doubtfully valid.

Loris Geoffroy, 1796a: 48 (*L. gracilis* Geoffroy, in *L. tardigradus*).
Stenops Illiger, 1811: 73 (*L. tardigradus* L.).

Contains one species.

Loris tardigradus Slender loris

Lemur tardigradus Linnaeus, 1758: 29; Sri Lanka.
= *Tardigradus loris* Boddaert, 1785: 67.
Loris gracilis Geoffroy, 1796a: 48; Sri Lanka.
Loris ceylonicus Fischer, 1804: 28; Sri Lanka.

Map 77. *Loris, Nycticebus, Tarsius*

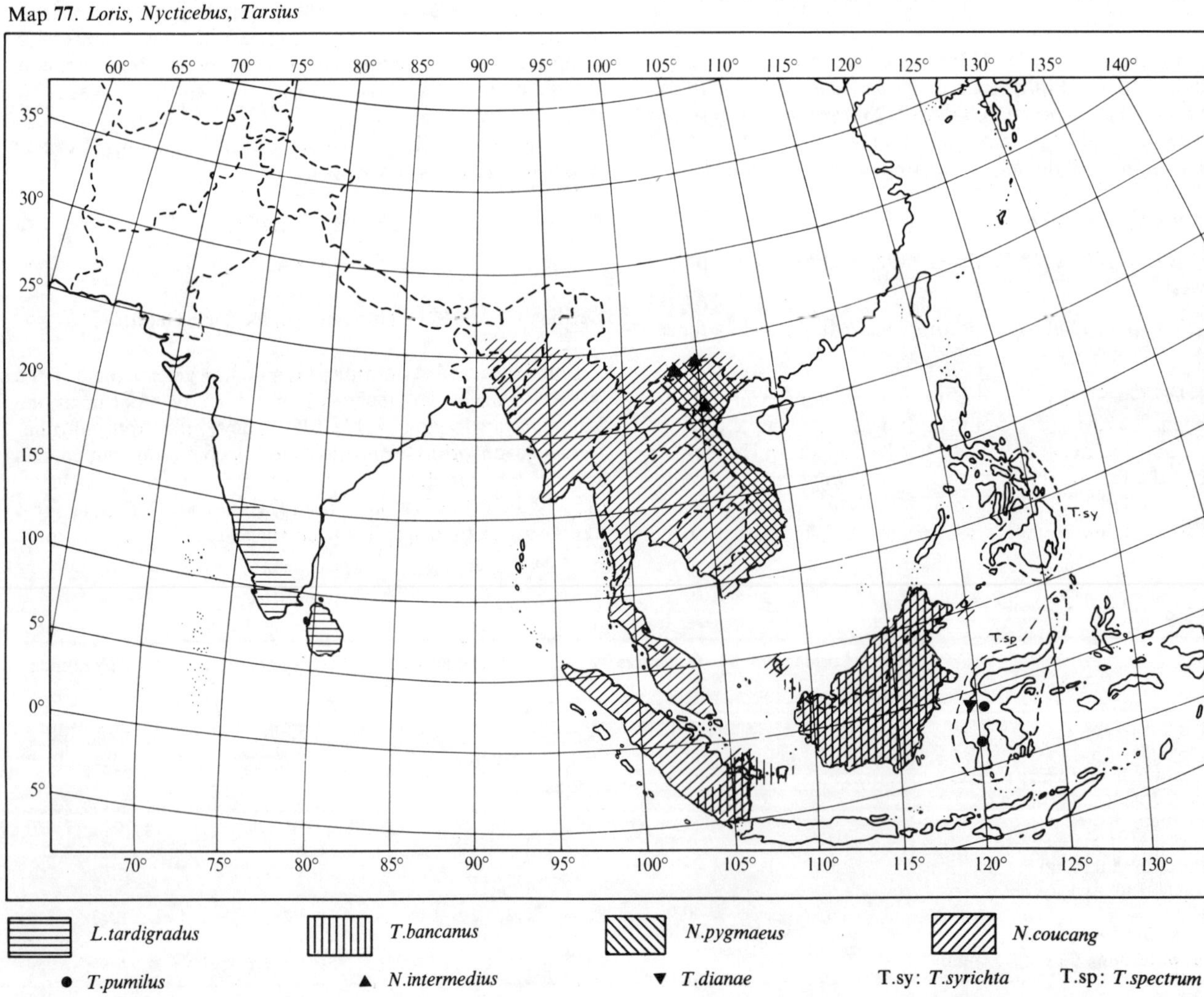

L. lori lydekkerianus Cabrera, 1908: 139; Madras, India.
L. malabaricus Wroughton, 1917: 45; Huvinakadu Estate, Kutta, S Coorg, SW India, 2843ft.

Range (Map 77). Sri Lanka and S India, north to c. 15°N in the Western Ghats; up to 1850m.

Variation. The following subspecific names have been proposed:
L. gracilis zeylanicus Lydekker, 1904c: 346; Peradeniya, Sri Lanka.
L. t. grandis Hill & Phillips, 1932: 111; Mousakande, Gammaduva, Central Prov., Sri Lanka, 2200ft.
L. t. nordicus Hill, 1933: 113,120; Talawa, North Central Prov., Sri Lanka, 50ft.
L. t. nycticeboides Hill, 1942a: 73; Horton Plains, Sri Lanka, 6000ft.

Five subspecies were recognized by Pocock (1939) on the basis of slight variation in the pelage. Those from wetter areas (SW India, S Sri Lanka) tend to have the pelage browner and less grey than those from drier areas. There is no clear distinction between those from Sri Lanka and India.

NYCTICEBUS

Nycticebus Geoffroy, E.,1812: 163; type species *N. bengalensis* Geoffroy, in *N. coucang* Boddaert.

Indochinese and Sunda subregions; 3 species (Table 105). Reviewed by Groves (1971b).

Nycticebus coucang Slow loris

Tardigradus coucang Boddaert, 1785: 67; Bengal.
Lori bengalensis Lacépède, 1800: 68; Bengal.
N. javanicus Geoffroy, E.,1812: 164; Java.
N. cinereus Milne-Edwards, 1867c: 11; Bangkok, Thailand.
N. menagensis Trouessart, 1897: 63; ? Tawitawi I, Sulu Is, Philippines. Based upon the name *menagensis* Nachtrieb, 1892 which was published without a generic name.
N. borneanus Lyon, 1906c: 535; Sakaian R, Sangau Dist, W Borneo.
N. bancanus Lyon, 1906c: 536; Bangka I, E Sumatra.
N. philippinus Cabrera, 1908: 137; Catagan, Mindanao, Philippines (erroneous—see Fooden, 1991b).
N. tenasserimensis Elliot, 1913: 25; Amherst, N Tenasserim, Burma.
N. incanus Thomas, 1921k: 627; Kyeikpadein, Pegu, Burma.
N. ornatus Thomas, 1921k: 627; Batavia, W Java.

Nycticebus coucang

Range (Map 77). E Bangladesh, Assam (Pillai, 1973), Burma, W and S Yunnan and SW Guangxi through Indochina and Thailand to Malaya; Sumatra, Java, Borneo; the smaller islands of Kings (Mergui), Penang, Tioman (Malaya), Singapore (perhaps extinct), Batam, Galang (Riau Is), Banka, Bunguran (N Natuna Is); Tawitawi, Siminul, Sanga Sanga, Bongao (Sulu Is, SW Philippines). A much-repeated record from Mindanao, Philippines is erroneous (Fooden, 1991b).

Variation. The following subspecific names have been proposed:
N. tardigradus var. *malaiana* Anderson, 1881: 95; Malacca, Malaya.
N. c. hilleri Stone & Rehn, 1902: 139; Tanah Datar, Padang Highlands, Sumatra.
N. c. natunae Stone & Rehn, 1902: 140; Bunguran I, N Natuna Is.
N. c. insularis Robinson, 1917b: 101; Tioman I, Malaya.
N. c. buku: Robinson, 1917: 101; Sumatra. Not *Semnopithecus buku* Martin, 1838 which was based upon a monkey (Jenkins, 1987: 144).
N. c. brachycephalus Sody, 1949b: 145; ? Tebing Tinggi I, NE Sumatra (bought as a captive—see Johnson, 1952).

Four subspecies can be recognized (Table 106, adapted from Groves, 1971b).

Nycticebus pygmaeus Lesser slow loris

N. pygmaeus Bonhote, 1907c: 2 (and 1907b: 4); Nahtrang, Annam, Vietnam.

Range (Map 77). Vietnam, Cambodia, Laos and S Yunnan (Quan *et al*., 1987); east of the R Mekong. Sympatric with *N. coucang* in much of its range.

Remarks. IUCN: V.

Nycticebus intermedius Intermediate slow loris

N. intermedius Dào, 1960b: 242; forest of Hoa-bink, N Vietnam, 21°50'N, 105°20'E, 300m.

Range (Map 77). N Vietnam and adjacent part of SE Yunnan, China. Apparently sympatric with both *N. coucang* and *N. pygmaeus* in parts of Yunnan, e. g. where the Yunnan/Vietnam border crosses the Red River.

Remarks. On the basis of Yunnan specimens examined, skulls appear indistinguishable in size and form from those of *N. pygmaeus*. However the pelage characters seem distinct, this species is heavier (500–800g, cf. 250–400g in *N. pygmaeus*) and there are differences in chromosomes (Wang Y-x., pers. comm.).

Table 106. The subspecies of *Nycticebus coucang*.

	bengalensis	*coucang*	*javanicus*	*menagensis*
Range	N of Kra	Malaya, Sumatra	Java	Borneo-Mindanao
Head & body	330–370	265–330	–	265–300
Frosting	Strong, on neck & arms	Slight, flanks	0	Slight, flanks
Fork pattern on head	Unclear	Bold	Thin, dark	Broad, vague
Dorsal stripe	Thin, tawny	Broad, dark	Thin, black	Broad, dark

Family Tarsiidae

Tarsiers

Tarsina Gray, 1825: 338; type genus *Tarsius* Storr.

A distinctive family, with one genus, confined to the Indomalayan Region. They are not closely related to any other living primates and assessment of their affinities has been controversial. Although frequently linked with the prosimians (Loridae etc.), there appear to be somewhat better grounds for treating them as the sister-group of the Anthropoidea, i. e. the monkeys, apes and man (Andrews, 1988; Koop *et al.*, 1989). Tarsiers are mainly insectivorous. They are strictly nocturnal and live in the understorey of forest.

TARSIUS Tarsiers

Tarsius Storr, 1780: 33; type species *Lemur tarsier* Erxleben, in *T. syrichta* L. Conserved by Opinion 122 (1931) and Direction 24 (1955).
Rabienus Gray, 1821: 299 (*Lemur spectrum* Pallas).
Cephalopachus Swainson, 1835: 352 (*T. bancanus* Horsfield).
= *Hypsicebus* Lesson, 1840: 253.

Sunda, Philippine and Sulawesi subregions; 5 species. The recently described *T. dianae* is listed below but not included in the key (Table 107). The classification and many aspects of biology were reviewed in Niemitz (1984).

One fossil species has been described from the region: *T. thailandica* Ginsburg & Mein, 1987: 1213; Li, Lampoon Prov., NW Thailand; Lower Miocene. (Based upon a single lower tooth.)

Tarsius syrichta Philippine tarsier

Simia syrichta Linnaeus, 1758: 29; Philippines ('Luzon', presumably in error).
Lemur tarsier Erxleben, 1777: 71; Philippines.
T. philippensis Meyer, 1894/95: 1; Samar I, Philippines.
T. carbonarius Heude, 1899: 164, pl. 32; Davao Gulf and Pulangui R, Davao, Mindanao, Philippines.
T. fraterculus Miller, 1910: 404; Sevilla, Bohol I, Philippines.

Range (Map 77). Philippine islands of Mindanao, Basilan, Bohol, Dinagat, Leyte, Biliran and Samar.

Remarks. IUCN: E.

Tarsius spectrum Greater Sulawesi tarsier

Lemur spectrum Pallas, 1779a: 275; Macassar, S Sulawesi (Hill, 1933).
= *T. pallasii* Geoffroy, 1796: 49.
T. daubentonii Geoffroy, 1796: 49; Sulawesi.
T. fuscus Fischer, 1804: 37.

Tarsius bancanus

T. fischerii Desmarest, 1804: 32; 'Madagascar'.
T. sangirensis Meyer, 1896/97: 9, pl. 4; Sanghir Is, N of Sulawesi.

Range (Map 77). Throughout Sulawesi and on the adjacent islands of Sangihe, Peleng and Salayar; mainly below 500m but up to 1500m in SE Sulawesi. A record from Savu I (Lesser Sunda Is) has not been substantiated (Musser & Dagosto, 1987).

Variation. The following subspecific names have been proposed:
T. fuscus fuscomanus Fischer, 1804: 37, pl. 3; Macassar, S Sulawesi.
T. f. dentatus Miller & Hollister, 1921b: 103; Laboea Sore, N of Parigi, Sulawesi.
T. f. pelengensis Sody, 1949b: 143; Peleng I, E of Sulawesi.

The insular forms are not distinctive and no subspecies are justified (Niemitz, 1984).

Tarsius pumilus Lesser Sulawesi tarsier

T. pumilus Miller & Hollister, 1921b: 103; Rano Rano, central Sulawesi, 1°30'S, 120°28'E, 1800m.

Range (Map 77). Known only from two localities in the mountains of central Sulawesi, at Latomodjong (3°30'S, 120°5'E, 2200m) and the type locality.

Remarks. Treated as conspecific with *T. spectrum* by Laurie & Hill (1954) and others, but shown to be clearly distinct by Musser & Dagosto (1987). IUCN: I.

Tarsius dianae

T. dianae Niemitz *et al.*,1991: 105; Lore Lindu National Park, Central Sulawesi, 1°10'S, 120°9'E, 700m.

Range (Map 77). Known only from the type locality in Central Sulawesi.

Table 107. The species of *Tarsius* (except *T.dianae*).

	syrichta	*spectrum*	*pumilus*	*bancanus*
Range	Philippines	Sulawesi	Sulawesi	Sumatra, Borneo
Head & body	118–140	112–151	95,98	115–154
Distal 1/4 of tail with hair > diameter of tail	0	+	+	+
Extent of long-haired distal section of tail	–	30–50%		c.25%
Pale spot behind each ear	0	White	Buff/0	0
Skull: greatest length	37–40	35–40	31	37–39
Bulla length	11.4–13.0	10.6–12.0	9.6,9.7	11.9–13.0

Tarsius bancanus Sunda tarsier

T. bancanus Horsfield, 1821:pl. ; Banka I, Sumatra.
T. saltator Elliot, 1910a: 152; Billiton I, E Sumatra.
T. borneanus Elliot, 1910a: 153; Landak R, W Borneo.
T. tarsier natunensis Chasen, 1940a: 86; Sirhassen I, S Natuna Is.

Range (Map 77). S Sumatra, Borneo and the islands of Bangka, Belitung and Serasan (S Natuna Is).

Family Cercopithecidae

Old-world monkeys

Cercopithecidae Gray, 1821: 297; type genus *Cercopithecus* Brünnich, 1772.

Indomalayan and Afrotropical regions, marginally in the Palaearctic; c. 16 genera of which 6 occur in the Indomalayan Region (Table 108). These fall into two clearly defined subfamilies, Cercopithecinae, with one genus, *Macaca*, in the region, and Colobinae, with the remainder. Classification down to subgenera was reviewed by Strasser & Delson (1988) whose conclusions are followed here with regard to genera. All these monkeys are diurnal and most are predominantly arboreal. Although fruit is the most important item in the diet of many species they are mostly omnivorous, taking a wide range of animal and vegetable food. The ecology and systematics of the Sunda species were reviewed by Medway (1970).

The following name should be treated as Cercopithecidae *incertae sedis* (not in *Nycticebus* as recorded by Chasen, 1940a): *Semnopithecus buku* Martin, 1838: 435; 'Sumatra and other Malay islands'.

Subfamily Cercopithecinae

A mainly Afrotropical group with one genus in the Indomalayan Region.

MACACA Macaques

Simia Linnaeus, 1758: 25; type species *S. sylvanus* L. Suppressed by Opinion 114 (1929).
Macaca Lacépède, 1799: 4; type species *Simia inuus* L., in *M. sylvanus* L. (NW Africa).
= *Salmacis* Gloger, 1841: 35.
Inuus Geoffroy, E.,1812: 100 (*I. ecaudatus* Geoffroy, in *M. sylvanus* L.).
Silenus Goldfuss, 1820: 479 (*Cynocephalus silenus* Schreber).
= *Vetulus* Reichenbach, 1862: 125.
Magus Lesson, 1827a: 43 (*M. sylvanus* L. and *M. maurus* Cuvier).

Macaca fascicularis

Cynopithecus Geoffroy, 1836: 16 (*C. niger* Desmarest).
Maimon Wagner, 1839a:iv *bis*, 141 (*Simia nemestrina* L.).
= *Nemestrinus* Reichenbach, 1862: 139. Pre-occupied by *Nemestrinus* Latreille, 1802 (Insecta).
Rhesus Lesson, 1840: 95 (*Cercopithecus mulatta* Zimmermann).
Pithex Hodgson, 1841e: 1212.
Lyssodes Gistel, 1848: 9 (*M. speciosus* Cuvier).
Cynamolgus Reichenbach, 1862: 130 (*M. irus* Cuvier).
Zati Reichenbach, 1862: 130 (*M. radiata* Geoffroy).
Gymnopyga Gray, 1866f: 202 (*M. inornatus* Gray, in *M. maura*).
Pithecus Elliot, 1913: 176. Not of Cuvier & Geoffroy, 1795.
Szechuanopithecus Young & Liu, 1951: 52 (*S. yangtzensis* Young & Liu, in *M. robusta* Young *fide* Koenigswald, 1954).
Cynomacaca Khajuria, 1953: 303 (*Papio ochreatus* Ogilby).

Indomalayan Region, plus one species, *M. sylvanus*, in NW Africa and one, *M. fuscata*, in Japan; c. 14 species in the Indomalayan Region (Tables 109 and 110). The genus is not very clearly defined in relation to African genera such as *Allenopithecus* and *Papio* and it is possible that the Indomalayan (and Japanese) species should not be so closely associated with *M. sylvanus* (NW Africa), the type species of *Macaca*.

The species on the mainland are fairly well delimited but on Sulawesi there is a group of allopatric forms with varying degrees of intergradation or hybridization between them. Classification within the genus has been reviewed by Fooden (1976,1980), Cronin *et al.* (1980), Fooden & Lanyon (1989) and Fa (1989). The forms on Sulawesi were reviewed by Fooden (1969) and by Groves (1980b) whose conclusions, albeit tentative, are followed here. Although the number of species to be recognized on Sulawesi remains uncertain there is now general agreement that together they constitute a monophyletic group within *Macaca*. Many other aspects of biology of the

Table 108. The genera of Cercopithecidae.

	Macaca	*Presbytis*	*Semnopithecus*	*Pygathrix*	*Nasalis*
Species in region	14	8	10	4	2
Head & body (cm)	34–73	42–60	40–78	51–83	50–76
Tail (cm)	2–80	50–84	56–110	51–97	10–75
Cheek-pouches	+	0	0	0	0
Neonatal coat	Brown	White with black cross	Black or golden		
Flap on rim of each nostril	0	0	0	+	0
Stomach complex, sacculated	0	+	+	+	+

Table 109. The species of *Macaca*. Only one species, *M. maura*, for Sulawesi is included; all the Sulawesi species are dealt with separately in Table 110.

	silenus	*nemestrina*	*maura*	*sinica*	*radiata*	*assamensis*	*thibetanus*	*arctoides*	*fascicularis*	*mulatta*	*cyclopis*
Range	SW India	Burma - Borneo	Sulawesi	Sri Lanka	S India	Nepal-Vietnam	S China	S China - Malaya	Vietnam - Gt Sunda Is	Pakistan - China	Taiwan
Head & body: m	51–61	49–60	64–69	42–53	45–60	54–73	54–71	52–65	41–65	48–64	c.56
f	46–48	43–58	50–59	37–45	34–52	44–68	49–63	48–59	39–51	47–53	
Tail (% of h&b)	>50%	25–40%	6–7%	100–140%	95–140%	30–65%	7–15%	3–13%	100–140%	30–75%	70–115%
Cheek ruff & beard prominent	++	0	0	0	0	0	+	0	0	0	0
Forehead almost naked	0	0	0	0	0	0	0	+	0	0	0
Scalp hair direction	Radial	Radial	Normal	Radial, long	Radial	Radial	Radial	Sparse	Normal	Normal	Normal
Rump brighter than shoulders	0	0	0	0	0	0	0	0	0	+	+
Tail with terminal tuft	+	0	0	0	0	0	0	0	0	0	0
Female sexual swelling mainly around anus (a), vulva (v) or slight (s)	v	a	a	s	s	s	s	s	s	s	s
Glans penis: short, blunt	+	+	+	+	+	+	+		+	+	+
short, pointed											
long								+			

genus were reviewed in Lindburg (1980).

The following nominal fossil species have been described from the region:

M. sivalensis Lydekker, 1878a: 70; Siwaliks. = ?*Presbytis sivalensis*:Delson (1980).

M. anderssoni Schlosser, 1924: 3; Mien-Chih-Hsien, Honan, China; Tertiary.

M. robusta Young, 1934: 122; Choukoutien, N China.

Szechuanopithecus yangtzensis Young & Liu, 1951: 52; Kolo-shan, near Chungking, Sichuan; Pleistocene. Allocated to *M. robusta* Young by Koenigswald (1954).

M. youngi Gu, 1980: 324; Zhong Xiang Xian, Hubei, China; Pliocene.

Macaca silenus Lion-tailed macaque

Simia silenus Linnaeus, 1758: 26; 'Ceylon'.
[*Cercopithecus*] *vetulus* Erxleben, 1777: 25.
Simia ferox Shaw, 1793: 69.

Range (Map 78). Hills of SW India, at altitudes of 300–1300m.

Variation. The following subspecific name has been proposed: *Simia (Cercopithecus) s. albibarbatus* Kerr, 1792: 64.

Ecology. Confined to dense evergreen and semi-evergreen forest; mainly arboreal.

Remarks. Viable hybrids have been obtained between this species and both *M. nemestrina* and *M. fascicularis* (Fooden, 1975). Its status and conservation were reviewed by Heltne (1985). CITES: Appendix I; IUCN: E.

Macaca nemestrina Pig-tailed macaque

Simia nemestrina Linnaeus, 1766: 35; Sumatra.
S. platypygos Schreber, 1774: 74; locality unknown.
S. longicrurus Link, 1795: 62.
S. fusca Shaw, 1800: 24, pl. 13.
Simia carpolegus Raffles, 1821: 243; Bencoolen, S Sumatra.
Macacus libidinosus Geoffroy, I.,1826: 589; locality unknown.
Pithecus maimon Blainville, 1839:pl. 10.
M. leoninus Blyth, 1863a: 7; N Arakan, SW Burma.
M. andamanensis Bartlett, 1869: 57; Port Blair, Andaman Is (an individual possibly taken there from Burma, not established).
M. pagensis Miller, 1903a: 61; S Pagi I, Mentawai Is, W Sumatra.
M. broca Miller, 1906i: 558; Sapagaya R, NE Borneo.
M. adusta Miller, 1906i: 559; Champang, Tenasserim, Burma.
M. insulana Miller, 1906i: 560; Chance I, Mergui Is, Burma.

Range (Map 78). From Burma and Vietnam to Malaya; Sumatra, Borneo; and the smaller islands of Chance, Sullivan (Mergui Is); Phuket (Thailand); Penang, Tioman (Malaya); Singapore (introduced); Siberut, Sipura, N Pagai, S Pagai (Mentawai Is); Bangka, Natuna Is (introduced).

Variation. The following subspecific names have been proposed:
M. n. indochinensis Kloss, 1919b: 343; Lat Bua Kao, E Thailand.
Rhesus n. mentaveensis Beaux, 1923: 37; Sipura I, Mentawai Is.
M. n. blythii Pocock, 1931b: 305; locality unknown, probably NE India.
M. n. nucifera Sody, 1936a: 42; Banka I, Sumatra.

Three subspecies can be recognized and were described in detail by Fooden (1975):

Table 110. The species of *Macaca* on Sulawesi.

	maura	*ochreata*	*tonkeana*	*nigra*
Range in Sulawesi	SW	SE	C, NW	NE
Tail (% of h&b)	6–7%	6–7%	2–12%	2–4%
Crest on crown	0	0	0/+	+
Dorsal pelage	Long, silky	Long, fine	Long, silky	Coarse, woolly
Rump paler than back	0	+	+	0
Outer surface of fore legs paler than dorsum	0	+	0	0
Cheek whiskers prominent and pale	0	0	+/0	0

M. n. nemestrina. Borneo, Sumatra, Malaya and north to the Isthmus of Kra. Dorsum black or dark brown, contrasting sharply with the rest of the pelage; tail densely haired, dark above.
M. n. leonina. Mainland north of the Isthmus of Kra, with a small area of intergradation with *M. n. nemestrina*. Dorsum as rest of pelage, black confined to crown and top of tail; cheeks pale; tail densely haired.
M. n. pagensis. Mentawai Is. Dorsum, thighs, shoulders and cheeks dark brown, without agouti banding; fore legs reddish-brown; tail short, pale and nearly naked.

Ecology. A forest species, occurring only very marginally in plantations and other disturbed habitats.

Remarks. Viable hybrids have been obtained with *M. silenus*, *M. fascicularis*, *M. mulatta*, *M. maura* and *M. nigra* (Fooden, 1975). The form on the Mentawai Is (*pagensis*) is distinctive and might warrant recognition as a separate species. CITES: IUCN: E (*M. pagensis*).

Macaca maura Moor macaque

Macacus maurus Schinz, 1825: 257; 'India'.
= *Simia cuvieri* Fischer, 1829: 30.
M. (*Maimon*) *brachyurus* Smith, 1842: 103, pl. 1; 'India'.
M. inornatus Gray, 1866: 202; locality unknown.
Papio (*Inuus*) *hypomelas* Matschie, 1901a: 262; locality unknown.

Map 78. *Macaca* (part)

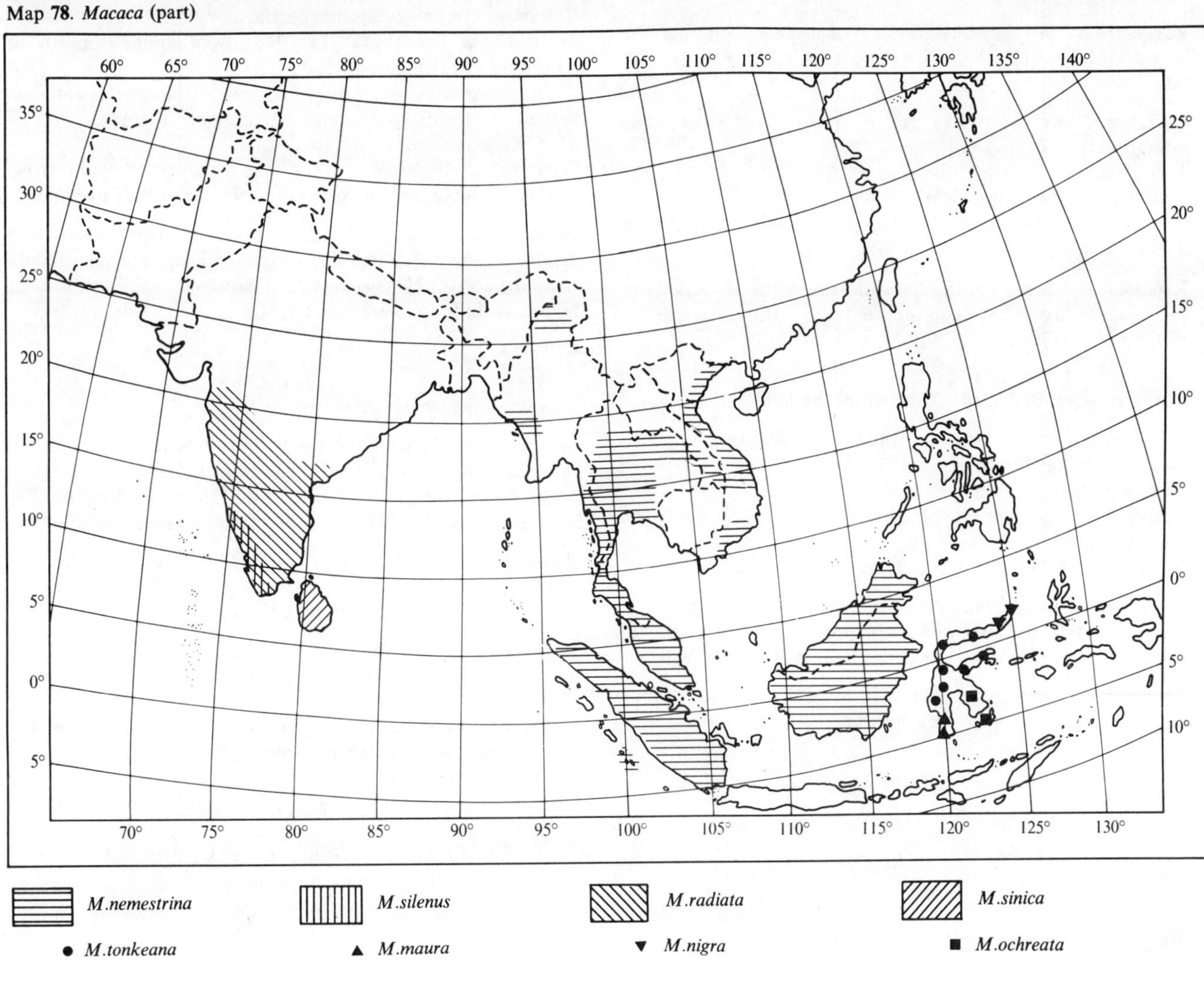

Range (Map 78). SW Sulawesi, north to about 3°50'S.

Variation. The following subspecific name has been proposed: *M. m. majuscula* Hooijer, 1950a: 57; Badjo, 20 km SW of Watampone, central Bone, S Sulawesi; subfossil.

Remarks. Extensive natural hybridization with *M. tonkeana* in the boundary zone has been reported (Supriatna *et al.*, 1990). IUCN: V.

Macaca ochreata Booted macaque

Papio ochreatus Ogilby, 1841: 56; locality unknown.
P. (*Inuus brunnescens*) Matschie, 1901a: 265; Buton I, SE Sulawesi.
M. fusco ater Schinz, 1844: 58; 'Celebes ?'.
Cynocephalus celebensis Elbert, 1911: 148; ? Osso Sohua, Sulawesi.

Range (Map 78). SE peninsula of Sulawesi and the adjacent islands of Buton and Muna.

Macaca tonkeana Tonkean macaque

Macacus tonkeanus Meyer, 1899: 3; Tonkean, Sulawesi.
Papio (*Inuus*) *hecki* Matschie, 1901a: 257; Buol, NW Sulawesi.
P. (*I.*) *tonsus* Matschie, 1901a: 261; locality unknown.
Cynopithecus togeanus Sody, 1949b: 138; Malengi I, Togean Is, NE Sulawesi.

Range (Map 78). Central and NW Sulawesi including the island of Malengi (Togean Is). Parapatric with *M. maura* in SW Sulawesi.

Variation. Two subspecies can be recognized, with some intergradation around 0–1°S:
M. t. tonkeana. Central Sulawesi and the Togian Is. Hind legs dark, as dorsum; tail 5–6cm.
M. t. hecki. NW Sulawesi. Hind legs greyish, much lighter than dorsum; tail 2–3cm.

Remarks. Extensive natural hybridization with *M. maurus* in the boundary zone has been reported (Supriatna *et al.*, 1990). IUCN: V (*M. hecki*).

Macaca nigra Black ape, Crested macaque

Cynocephalus niger Desmarest, 1822: 534; locality unknown.
= *C. malayanus* Desmoulins, 1824: 262; 'Philippines'.
Papio nigrescens Temminck, 1849: 111; Sulawesi.
Cynopithecus aethiops Reichenbach, 1863: 163; 'Celebes, Philippines, Moluccas'.
C. lembicus Miller, 1931: 413; Lembeh I, NE Sulawesi.

Range (Map 78). NE Sulawesi, west to c. 123°E. Any previous contact with *M. tonkeana* to the west probably no longer exists: a gap of c. 30km was reported by Sugardjito *et al.* (1989).

Remarks. Separated as the sole member of the genus *Cynopithecus* by many earlier authors.

Macaca sinica Toque monkey

Simia sinica Linnaeus, 1771: 521; locality unknown.
= *S. sinensis* Cuvier, 1797: 99.
Cynamolgus (*Zati*) *audeberti* Reichenbach, 1862: 132.
M. pileatus: Blyth, 1863a: 9; not of Kerr, 1792.

Range (Map 78). Sri Lanka.

Variation. The following subspecific names have been proposed:
M. s. inaurea Pocock, 1931a: 286; Cheddikulam, N Prov., Sri Lanka.
M. s. aurifrons Pocock, 1931a: 286; Rayagam Korale, W Prov., Sri Lanka.
M. (*Zati*) *s. opisthomelas* Hill, 1942b: 402; Horton Plains, highlands of Sri Lanka.
M. s. longicaudata Deraniyagala, 1965b: 261; Menik Ganga, near Kataragama, Uva Prov., Sri Lanka.

Two subspecies were recognized by Fooden (1979) but with a broad area of intergradation:
M. s. sinica. N Sri Lanka. Crown uniformly coloured.
M. s. aurifrons. S Sri Lanka. Anterior part of crown pale yellow, posterior golden brown.

Ecology. In both deciduous and evergreen broadleaved forest, at all altitudes up to c. 2100m; mainly arboreal.

Macaca radiata Bonnet macaque

Cercocebus radiatus Geoffroy, E.,1812: 98; locality unknown.

Range (Map 78). Peninsular India, north to c. 21°N in the west and c. 17°N in the east. In the north they are generally parapatric with *M. mulatta* but mixed troops have been recorded near the eastern end of the boundary (Fooden *et al.*, 1981).

Variation. The following subspecific name has been proposed: *M. r. diluta* Pocock, 1931a: 278; Boothapundy, north of Aramboly, Travancore, S India.

Two subspecies are usually recognized although it is not clear if there is any sharp discontinuity between them:
M. r. radiata. Most of range.
M. r. diluta. Extreme south of India, south of Palni Hills. Venter unpigmented, dorsal hairs more conspicuously banded.

Ecology. They occupy deciduous forest and are tolerant of disturbed habitats; mainly arboreal. They are marginally sympatric with *M. silenus* which however occurs mostly at higher altitude in evergreen forest.

Remarks. Reviewed in detail by Fooden (1981).

Macaca assamensis Assamese macaque

M. assamensis McClelland, 1839: 148; Assam.
M. pelops Hodgson, 1841e: 1213; N Nepal.
M. problematicus Gray, 1871c: 128; Dhalimkot, Bhutan.
M. rheso-similis Sclater, 1872: 495, pl. 25; 'East Indies'.

Range (Map 79). Himalayas from C Nepal to Assam, W Thailand, Laos, N Vietnam and S China (SE Tibet, S Yunnan, SW Guangxi). At altitudes of 200–2750m.

Variation. The following subspecific name has been proposed: *M. a. coolidgei* Osgood, 1932: 209; Hoi Xuan, Annam, Vietnam.

Two subspecies can be recognized on the basis of tail-length:
M. a. assamensis. Assam and eastwards. Tail of adult males 25–45% of head & body.
M. a. pelops. West of Assam. Tail of adult males 50–63%.

The difference in tail length is less clear in females, with considerable overlap.

Ecology. Mainly confined to broadleaved evergreen forest, arboreal.

Map 79. *Macaca* (part)

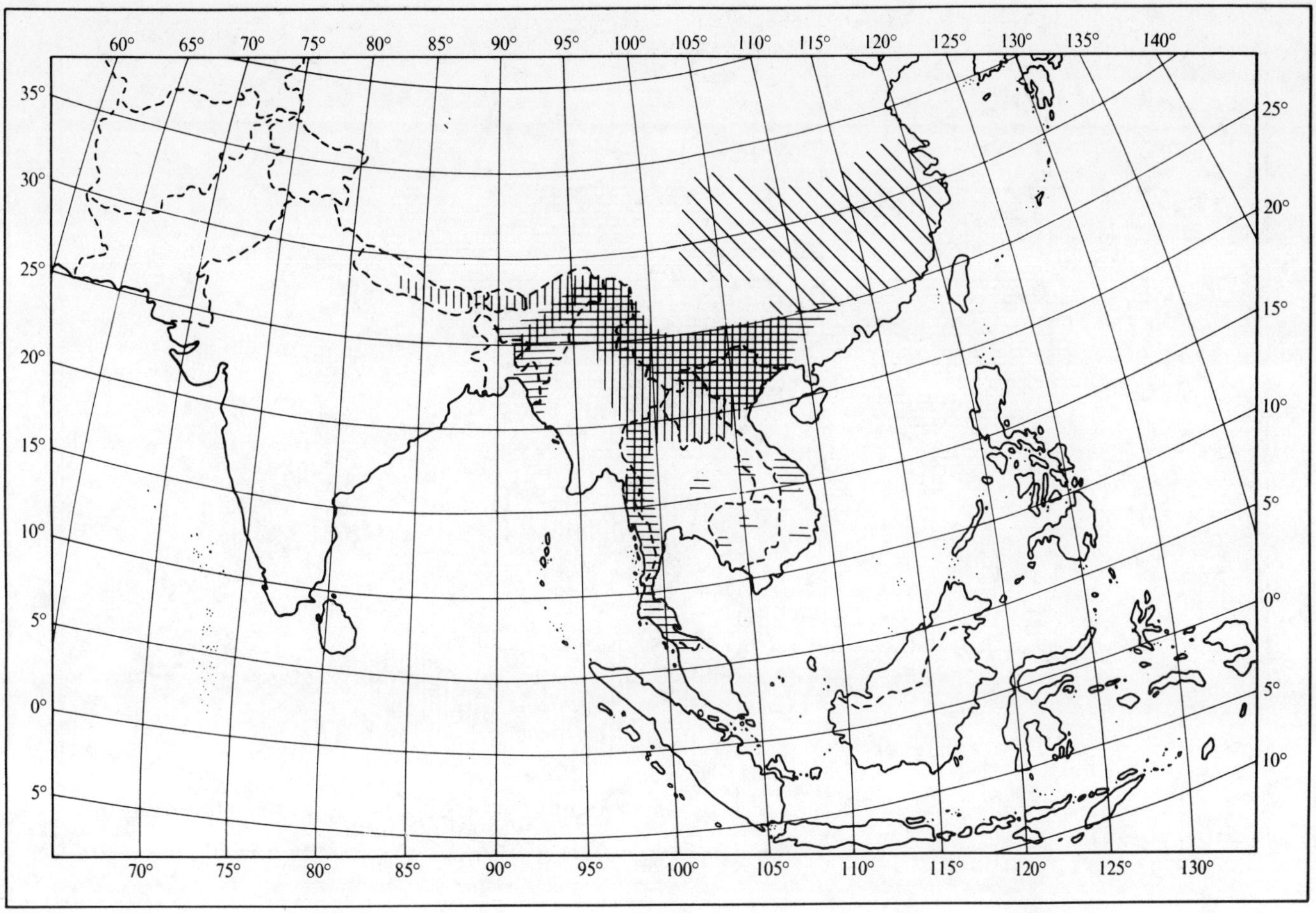

M.arctoides *M.assamensis* *M.thibetana*

Remarks. Viable hybrids have been obtained with *M. arctoides*. This species was reviewed in detail by Fooden (1982,1986,1988).

Macaca thibetana Chinese stump-tailed macaque

Macacus thibetanus Milne-Edwards, 1870a: 341; 20km NNW of Moupin (= Baoxing), Sichuan, China.
M. (Magus) arctoides esau Matschie, 1912b: 309; W of Lochangho, Guangdong, China.
Pithecus pullus Howell, 1928b: 41; near Kuatun, Fujian, China.

Range (Map 79). China from S Gansu, Sichuan and N Yunnan to Fujian, apparently parapatric with *M. assamensis* and *M. arctoides* in the south. At altitudes of 1000–2500m.

Ecology. Mainly in broadleaved evergreen forest; predominantly terrestrial.

Remarks. Closely related to *M. assamensis* and could be treated as a subspecies of it. Reviewed in detail by Fooden (1983,1986,1988). IUCN: K.

Macaca mulatta Rhesus macaque

Cercopithecus mulatta Zimmermann, 1780: 195; 'India'.
Simia (Cercopithecus) fulvus Kerr, 1792: 73; 'India'.
S. rhesus Audebert, 1798: 5; locality unknown.
S. erythraea Shaw, 1800: 33; locality unknown.
M. oinops Hodgson, 1841e: 1212; Nepal Terai.
Inuus sancti-johannis Swinhoe, 1866b: 556; N Lena I, Hong Kong.
M. lasiotus Gray, 1868c: 60, pl. 6; Sichuan, China.
M. tcheliensis Milne-Edwards, 1871a: 227, pls 32,33; mountains east of Tcheli Prov., N China.
M. vestitus Milne-Edwards, 1892: 761; Tengri-nor, Tibet.
Pithecus littoralis Elliot, 1909: 250; Kuatun, Fujian, China.
P. brachyurus Elliot, 1909: 251; Hainan. Pre-occupied by *P. brachyurus* H Smith, 1842 (in *M. maura*).
= *P. brevicaudus* Elliot, 1913: 216.
M. siamica Kloss, 1917g: 247; Meping Rapids, below Chiengmai, Thailand.

Range (Map 80). From E Afghanistan, N Pakistan and N peninsular India to N Thailand and N Vietnam, and throughout most of southern China; an isolated population, perhaps introduced, in NE China near Beijing. Other introduced populations in Florida, USA and on Puerto Rico.

Variation. The following subspecific names have been proposed:
M. rhesus villosus True, 1894a: 2; Lolab, Wular Lake, c. 40 miles NW of Srinagar, Kashmir.

Map **80**. *Macaca* (part)

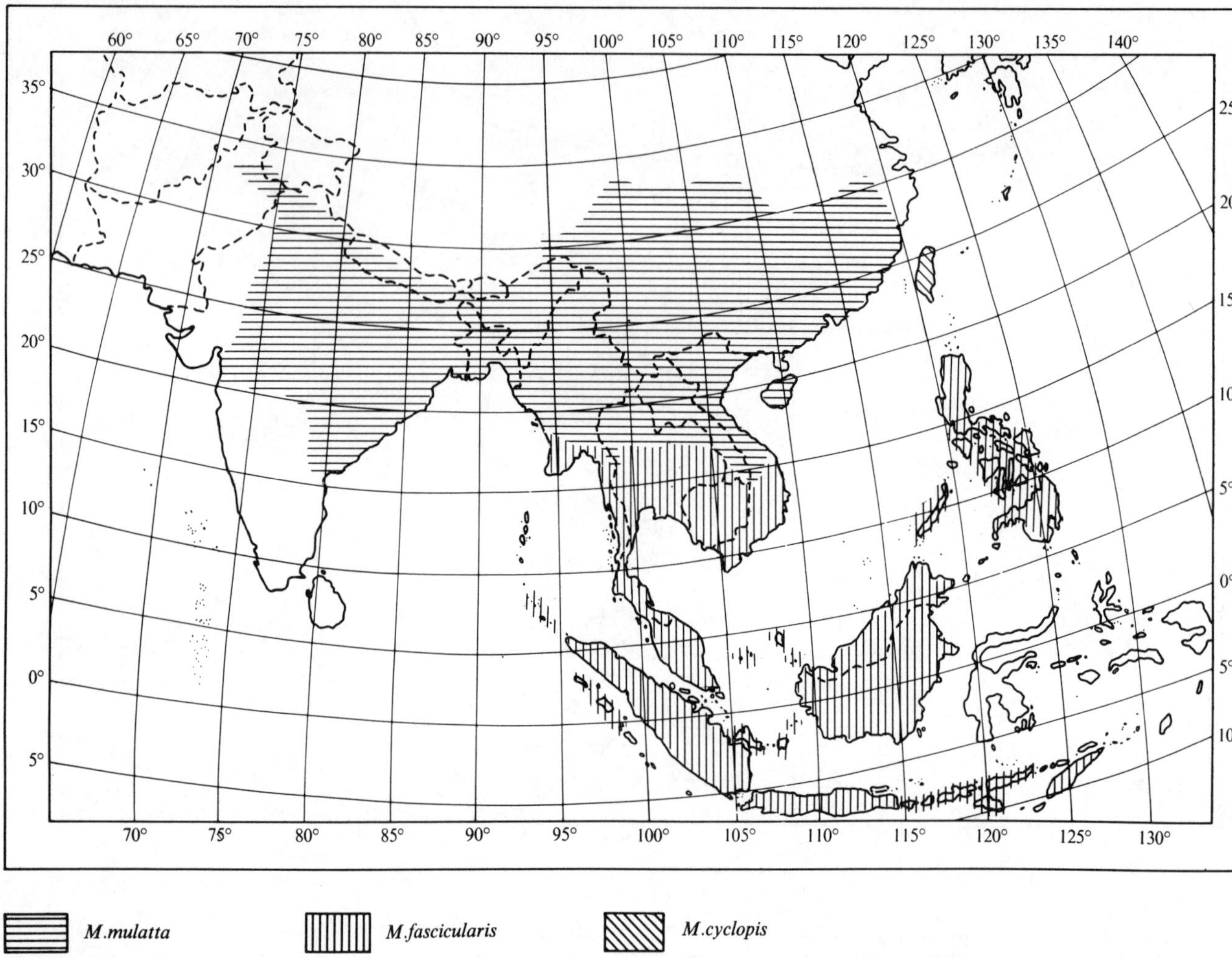

M. m. mcmahoni Pocock, 1932c: 544; Kootai, lower Chitral, Pakistan, 3600ft.

Geographical variation is slight and it is unlikely that discrete subspecies can be recognized. Size is greatest in the northwest (Kashmir, Punjab etc.). Distribution and variation in China were reviewed by Jiang *et al.* (1991).

Remarks. See remarks under *M. fascicularis.*

Macaca fascicularis Crab-eating macaque

Simia aygula Linnaeus, 1758: 27; 'India'. By most authors incorrectly interpreted as a senior synonym of *Presbytis comata* (Desmarest). Suppressed by Opinion 1400 (1986).

[*M.*] *irus* Cuvier, F.,1818: 120; Sumatra. Not proposed in association with a generic name and therefore invalid.

Simia fascicularis Raffles, 1821: 246; Sumatra. Conserved by Opinion 1400 (1986).

M. carbonarius Cuvier, in Geoffroy & Cuvier, 1825:pl. 52; Sumatra.

Semnopithecus kra Lesson, 1830: 20; Sumatra.

M. aureus Geoffroy, 1831: 58,76; Pegu, Burma (Pocock, 1939).

M. philippinensis Geoffroy, 1843a: 568, pl. 33; Manila, Luzon, Philippines.

M. palpebrosus Geoffroy, 1851: 93; Manila, Luzon, Philippines.

M. fur Slack, 1867: 36, pl. 1; Luzon, Philippines.

M. cristatus Gray, 1871c: 30; locality unknown.

M. cynomolgus cumingi Gray, 1871c: 30; Philippines.

M. pumilus Miller, 1900a: 241; Bunoa I, Tambelan Is, W of Borneo.

M. umbrosus Miller, 1902b: 789; Little Nicobar I, Nicobar Is.

M. phaeura Miller, 1903a: 63; Nias I, W Sumatra.

M. fuscus Miller, 1903b: 476; Simalur I, W Sumatra.

Cynomolgus mindanensis Mearns, 1905: 428; Pantar, Davao, Mindanao, Philippines, 1900ft.

C. suluensis Mearns, 1905: 430; foot of Crater Lake Mountain, Jolo I, Sulu Is, Philippines.

C. cagayanus Mearns, 1905: 431; Cagayan Sulu I, Sulu Sea, S Philippines.

M. mordax Thomas & Wroughton, 1909d: 380; Tjilatjap, W Java.

M. resima Thomas & Wroughton, 1909d: 381; Tasikmalaja, W Java.

Pithecus validus Elliot, 1909: 252; Cochin-China, Vietnam.

P. alacer Elliot, 1909: 253; Kundur I, Riau Is.

P. laetus Elliot, 1909: 255; Tinggi I, E Malaya.

P. karimoni Elliot, 1909: 255; Karimon I, Riau Is.

P. dollmani Elliot, 1909: 256; Changi, Singapore I.

P. bintangensis Elliot, 1909: 257; Bintang I, Riau Is.

P. lapsus Elliot, 1910b: 343; Banka I, Sumatra.

P. agnatus Elliot, 1910b: 344; Tuangku I, Banjak Is, W Sumatra.

P. lingungensis Elliot, 1910b: 344; Lingung I, N Natuna Is.
P. lautensis Elliot, 1910b: 345; Laut I, N Natuna Is.
P. sirhassenensis Elliot, 1910b: 345; Sirhassen I, S Natuna Is.
P. vitiis Elliot, 1910b: 346; Domel I, Mergui Is, Burma.
P. carimatae Elliot, 1910b: 346; Karimata I, SW Borneo.
P. mandibularis Elliot, 1910b: 347; Sama R, near Pontianak, Borneo.
P. baweanus Elliot, 1910b: 347; Bawean I, Java Sea.
P. cupidus Elliot, 1910b: 348; Mata Siri I, Java Sea.
P. lingae Elliot, 1910b: 349; Lingga I, Lingga Is, E Sumatra.
P. impudens Elliot, 1910b: 350; Sugi I, Riau Is.
P. capitalis Elliot, 1910b: 350; Trang, S Thailand.
P. mindorus Hollister, 1913: 328; Mindoro I, Philippines.
P. mansalaris Lyon, 1916a: 452; Mansalar I, W Sumatra.

Range (Map 80). From S Burma and Vietnam to Malaya; Sumatra, Java, Borneo; throughout the Philippines; most of the Nicobar and Mergui Islands; most of the islands around Sumatra and Borneo except the Mentawai Is; and the Lesser Sunda Is east to Timor; beyond the region there are introduced populations on Mauritius and on Angaur I (W Pacific).

Variation. The following subspecific names have been proposed:
Cynomolgus mindanensis apoensis Mearns, 1905: 429; Mt Apo, S Mindanao, Philippines, 6000ft.
M. irus argentimembris Kloss, 1911: 116; Penang I, Great Redang Is, E Malaya.
Pithecus f. limitis Schwarz, 1913b: 298; Leloguma, Timor I.
Pithecus fuscus lasiae Lyon, 1916a: 453; Lasia I, near Simalur, W Sumatra.
M. irus atriceps Kloss, 1919b: 347; Kram I, near Cape Liant, SE Thailand.
M. i. sublimitis Sody, 1932b: 338; Sumba I, Lesser Sunda Is.
M. i. tua Kellogg, 1944: 75; Muara Tua I, E coast Borneo.
M. i. karimondjawae Sody, 1949b: 132; Karimon Djawa I, Java Sea.
M. i. submordax Sody, 1949b: 133; Banju Wetan, Bali.

The form on most of the Philippine islands, including Palawan, *M. f. philippinensis*, is darker brown than *M. f. fascicularis* which occurs on the other Greater Sunda Islands, the Sulu Is and SW Mindanao; an intermediate form occurs in the rest of Mindanao and on Negros I (Fooden, 1991b).

Ecology. In spite of the English name they are versatile feeders, taking a wide range of fruit and invertebrates. They are characteristic of forest edge and occur extensively in plantations and other disturbed habitats.

Remarks. Closely related to and parapatric with *M. mulatta*. A limited amount of hybridization in Thailand has led to them being considered conspecific (Fooden, 1964), but any intergradation appears to be sufficiently limited that separate specific status is appropriate, as was subsequently concluded by Fooden (1976,1980).

Macaca cyclopis Taiwan macaque

M. cyclopis Swinhoe, 1863: 350; Taiwan.
M. (radiatus) affinis Blyth, 1863a: 8; Taiwan.

Range (Map 80). Taiwan. Introduced on the Japanese island of Oshima, south of Tokyo.

Ecology. Found in all forest types from lowland forest to montane scrub; feed on the ground and in trees; reviewed by Poirier (1986).

Remarks. This species is closely related to *M. mulatta* but also to the Japanese *M. fuscata*. Superficially it appears more similar to *M. mulatta*. A biochemical study indicated that these three were more closely related to each other than to *M. fascicularis* (Hayasaki *et al.*, 1988).

Macaca arctoides Bear macaque

Macacus speciosus of many authors, not of Geoffroy, 1826 which was based upon a specimen of *M. fuscata* (Japan). Suppressed by Opinion 920 (1970).
Macacus arctoides Geoffroy, 1831: 61; 'Cochinchine', Vietnam.
= *M. ursinus* Gervais, 1854: 93.
Papio melanotis Ogilby, 1839b: 31; locality unknown.
M. rufescens Anderson, 1872a: 204; locality unknown (purchased in Singapore).
M. brunneus Anderson, 1872b: 628; near Bhamo, NE Burma.
M. harmandi Milne-Edwards, 1897a: 10; Chanthaburi Mts, SE Thailand.

Range (Map 79). From Assam and S China (Yunnan, ? Sichuan, Guizhou, Guangxi) to N Malaya (Perlis); at altitudes from below 1500 to 2500m in Yunnan. Mapped in detail by Fooden (1990).

Variation. The following subspecific names have been proposed:
M. (Magus) a. melli Matschie, 1912b: 308; Lianshan Mts, NW Guangdong, China.
M. speciosa subfossilis Jouffroy, 1959: 213; Tung-Lang, Ninh-Binh Prov., N Vietnam; Pleistocene.

In a detailed review Fooden (1990) concluded that no subspecies could be recognized.

Ecology. In evergreen forest; mainly terrestrial.

Remarks. Although superficially rather similar to *M. thibetanus* (they were considered conspecific by Ellerman & Morrison-Scott, 1951) and apparently parapatric with it, this species differs very clearly in some characters, especially of the genitalia, and its relationship to other species of *Macaca* is unclear. It was reviewed in detail by Fooden *et al.* (1985).

[*Theropithecus*]

Theropithecus Geoffroy, 1843a: 576; type species *Macaca gelada* Rüppell.

Ethiopia; one living species, *T. gelada*. One fossil species has been described from India:
T. delsoni Gupta & Sahni, 1981: 69; 1.4km WSW of Mirzapur Rest House, Ropar Dist, Punjab, India; Upper Siwaliks, Pleistocene.

[*Papio*]

Papio Erxleben, 1777: 15; type species *Cynocephalus papio* Desmarest.
Cynocephalus Cuvier & Geoffroy, 1795: 462 (*Simia cynocephalus* L.).

Afrotropical Region; one nominal fossil species in the Indomalayan Region:
Cynocephalus falconeri Lydekker, 1886c: 7; Siwalik Hills, India.

Subfamily Colobinae

Presbytina Gray, 1825: 338. Suppressed by Opinion 1202 (1982).
Semnopithecidae Owen, 1843: 55. Suppressed by Opinion 1202 (1982).
Colobinae Jerdon, 1867: 3; type genus *Colobus* Illiger. Conserved by Opinion 1202 (1982).

The colobines are mostly arboreal monkeys adapted to a diet including a considerable proportion of leaves. They lack cheek pouches and have a complex sacculated stomach and rather high cusps on the cheek-teeth. The taxonomy and many other aspects were reviewed by Napier (1985).

PRESBYTIS Leaf monkeys, Surelis

Presbytis Eschscholtz, 1821: 196; type species *P. mitrata* Eschscholtz. Conserved by Opinion 1202 (1982).
Corypithecus Trouessart, 1879b: 53 (*Semnopithecus frontatus* Müller).
Lophopithecus Trouessart, 1879b: 53 (*Semnopithecus rubicundus* Müller).

Sunda subregion; c. 8 species (Table 111). Formerly included the species here separated in *Semnopithecus* but these are excluded following Groves (1989). A distinctive feature shared by species of *Presbytis* is the pelage of the new-born which are white with a dark mid-dorsal stripe and a dark transverse stripe across the shoulders. The specific distinctiveness of the forms *melalophos, femoralis, thomasi* and *potenziani* is supported by differences in the calls of the adult males (Wilson & Wilson, 1975,1976).

Presbytis melalophos Mitred leaf monkey

Simia melalophos Raffles, 1821: 245; Bencoolen, S Sumatra.
P. mitrata Eschscholtz, 1821: 196; ? S Sumatra.
Semnopithecus flavimanus Geoffroy, in Lesson, 1831: 109, pl. xl; Bencoolen, Sumatra. Restricted by Chasen (1940a).
S. sumatrana Müller & Schlegel, 1840:pl. 10 (and 1841: 61,73); Mt Ophir, N of Padang, W Sumatra.
S. nobilis Gray, 1842: 256; 'India', corrected to Indrapura, Sumatra by Chasen (1940a).
S. ferrugineus Schlegel, 1876: 42; Padang, W Sumatra.
P. batuanus Miller, 1903a: 65; Pinie I, Batu Is, W Sumatra.
P. fusco-murina Elliot, 1906: 49; Telok Betong, Lampongs, S Sumatra.
Pithecus femoralis fluviatilis Chasen, 1940a: 76; Moearadoea, Palembang, Sumatra, 100m.
P. aygula margae Hooijer, 1948a: 234; Serdang, Deli, NE Sumatra.

Range (Map 81). Sumatra, except for the north (where it is replaced by *P. thomasi*) and most of the east coastal plain (where it is replaced by *P. femoralis*).

Variation. Geographical variation on Sumatra is extensive and complex. Ten subspecies were mapped by Wilson & Wilson (1976) and illustrated by Whitten (1987). The characters of some of the more distinctive are summarized in Table 112.

Remarks. Has been held to include also the forms here incorporated in *P. femoralis*.

Presbytis femoralis Banded leaf monkey

Semnopithecus femoralis Martin, 1838 (August): 436; Singapore.
S. chrysomelas Müller, 1838 (October): 138; Pontianak, W Borneo.
S. siamensis Müller & Schlegel, 1841: 60; Malacca, Malaya.
S. nigrimanus Geoffroy, 1843a: 546; 'Java', corrected to Malaya by Chasen (1940a: 74).
S. neglectus Schlegel, 1876: 47; Singapore.
S. cruciger Thomas, 1892e: 475; Miri, Sarawak.
S. natunae Thomas & Hartert, 1894: 652; Bunguran I, N Natuna Is.
P. rhionis Miller, 1903a: 64; Bintang I, Riau Is.
P. cana Miller, 1906b: 275; Kundur I, Riau Is.
P. percura Lyon, 1908b: 671; Kompei, E Sumatra.
P. catemana Lyon, 1908b: 672; Kateman R, E Sumatra.
P. nubigena Elliot, 1909: 268; Malacca, Malaya.
P. dilecta Elliot, 1909: 270; Dusun Tua, Selangor, Malaya.
P. robinsoni Thomas, 1910e: 25; Trang, S Thailand.
P. australis Miller, 1913b: 28; Jambu Luang, E coast of Johore, Malaya.
P. arwasca Miller, 1934a: 126; Miri, Sarawak, Borneo.

Range (Map 81). Peninsular Thailand and Malaya, E Sumatra, W Borneo; and on the smaller islands of Singapore; Bintan, Kundur (Riau Is); Bunguran (N Natuna Is).

Variation. The following subspecific names have been proposed:
P. neglecta keatii Robinson & Kloss, 1911a: 174; Trang, S Thailand.
Pithecus f. paenulatus Chasen, 1940a: 75; Teloek Pandji, Baroemoen R, S of Laboehanbilik, NE Sumatra.

Table 111. The species of *Presbytis*.

	melalophos	*femoralis*	*thomasi*	*potenziani*	*comata*	*hosei*	*frontata*	*rubicunda*
Range	Sumatra	Malaya, Sumatra Borneo	N Sumatra	Mentawai Is	Java	Borneo	Borneo	Borneo
Dorsum	Red-brown - orange or grey	Grey - black	Grey	Black	Grey	Grey	Grey-brown	Reddish
Venter	White/buff	Pale	White	White/ orange-brown	White	White	Light grey	Reddish
Crest on head	+	Small/0	+	Small ridge	+, black	+, low	+	+, low
White patch on forehead	0		+	+	0	+ (female)	+	0
Cheeks	Pale	Dark/pale	Dark	Pale	Dark	White	Black	Reddish
Hands and feet	As dorsum	Black	Black	Black	Blackish	Black	Black	Black/brown
Tail	As dorsum	Blackish	Grey	Black, ± white tip	Dark above, light below	Grey	Grey	As dorsum

Map **81**. *Presbytis*

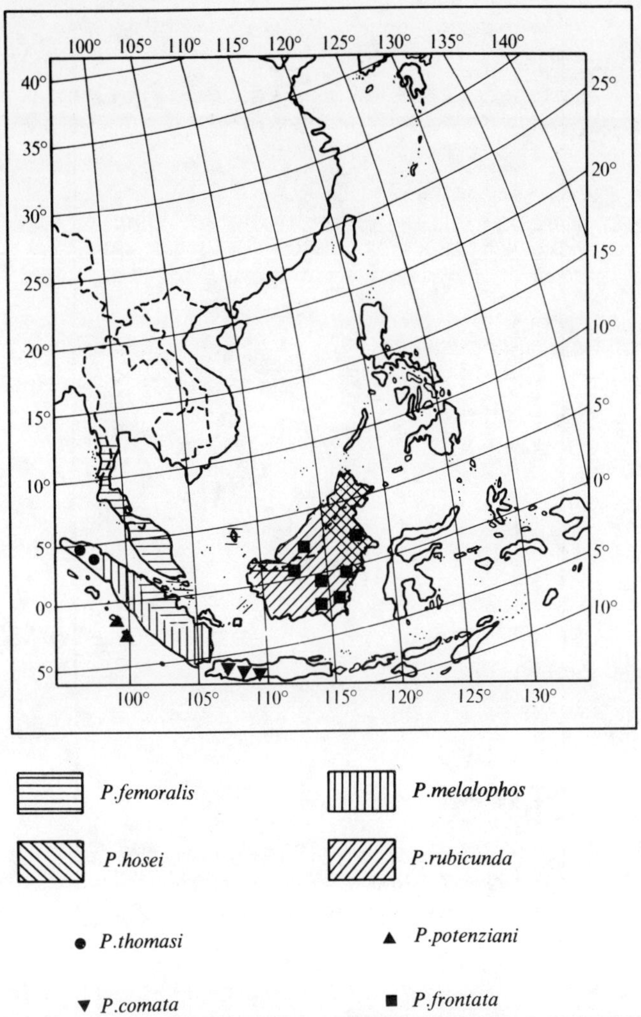

Geographical variation in colour and pattern is extensive. The main elements are tabulated in Table 113. On Borneo a reddish morph ('*cruciger*') occurs within the population of *P. f. chrysomelas*. For more details see Weitzel *et al.* (1988) and Whitten (1987).

Presbytis thomasi North Sumatran leaf monkey

Semnopithecus thomasi Collett, 1893: 613; Langkat, NE Sumatra.

Range (Map 81). N Sumatra, apparently parapatric with *P. melalophos* to the south.

Variation. The following subspecific name has been proposed: *P. t. nubilus* Miller, 1942: 130; Blangnanga, Aceh, N Sumatra, 3600ft (darker and longer-haired).

Remarks. Included in *P. comata* by Brandon-Jones (1984).

Presbytis potenziani Mentawai leaf monkey

Semnopithecus potenziani Bonaparte, 1856: 412; Sipora I, Mentawai Is.
= *P. chrysogaster* Blyth, 1875: 10.

Range (Map 81). The Mentawai islands of Siberut, Sipura, N Pagai and S Pagai.

Variation. The following subspecific name has been proposed: *Pithecus p. siberu* Chasen & Kloss, 1928c: 811, pl. 1; Siberut I, Mentawai Is (venter black or dull brown, not reddish as in *P. p. potenziani*).

Remarks. A distinctive species, probably most closely related to *P. thomasi* which it most closely resembles in voice (Wilson & Wilson, 1975). CITES: Appendix I; IUCN: E.

Presbytis comata Grizzled leaf monkey

P. aygula of many authors, not *Simia aygula* Linnaeus, 1758 which is a synonym of *Macaca fascicularis mordax* (Weitzel & Groves, 1985).

Table 112. Some subspecies of *Presbytis melalophos*.

	mitrata	*melalophos*	*fluviatilis*	*nobilis*	*batuana*
Range in Sumatra	S	SW	SC	C	Batu Is
Dorsum	Grey/brown	Red-brown	Buff or cream	Red	Dark
Legs (external)	Grey/brown	Yellowish red	Buff or cream	Red	Dark
Venter	White, ± reddish	White	White	White ± orange	Narrow white line
Crown	Black/white	Reddish	Cream	Yellowish	

Table 113. The subspecies of *Presbytis femoralis*.

	robinsoni	*siamensis*	*femoralis*	*rhionis*	*catemana*	*percura*	*paenulata*	*chrysomelas*	*natunae*
Range	N Malaya	C Malaya	S Malaya	Riau Is	E Sumatra	E Sumatra	E Sumatra	Borneo	Natuna Is
Dorsum	Black/ dark brown	Black/ dark brown	Blackish	Grey	Grey		Grey	Black	Dark grey/ brown
Abdomen	White	White	White	White-grey	White/grey	White	White	White/pale brown	White
Chest	Dark brown	White	Thin white line	White	White	White median line	White	White	White
Pale area on outside of thigh	0	+	0	+	(+)	0	+	0	0
Proximal, ventral half of tail	Dark	Light brown	Dark	Light brown		White	Grey	White	Dark
Cheeks	Dark	White	Black	White	White		White	Black	White

Semnopithecus comatus Desmarest, 1822: 533; Java.
S. fulvogriseus Desmoulins, 1825: 570; Java.

Range (Map 81). W Java.

Variation. The following subspecific name has been proposed: *Pithecus aygula fredericae* Sody, 1930d: 68; Mt Slamat, W-C Java.

Two subspecies can be distinguished as follows:
P. c. comata. W Java. Venter entirely white.
P. c. fredericae. Mt Slamat. Venter white only on neck and lower abdomen.

Remarks. The Bornean form *hosei* and the Sumatran *thomasi* were included in this species by Brandon-Jones (1984) but were treated as separate species by Payne *et al.* (1985) and Whitten (1987) respectively. CITES: Appendix II; IUCN: E.

Presbytis hosei Grey leaf monkey

Semnopithecus hosei Thomas, 1889b: 159, pl. xvi; Niah, Baram, Sarawak, Borneo.
S. everetti Thomas, 1893f: 582, pl. xli; Mt Kinabalu, N Borneo.
S. sabanus Thomas, 1893d: 230, pl. vii; Paitan, N Borneo.
P. canicrus Miller, 1934b: 15; Dutch NE Borneo.

Range (Map 81). N and E Borneo.

Variation. Three subspecies can be recognized:
P. h. hosei (including *everetti*). Sarawak, S Sabah. White collar; crest black; nose pink.
P. h. sabana. N and E Sabah. No white collar; crest black; nose pink.
P. h. canicra. E Kalimantan. No white collar; crest brown; nose black.

Remarks. Included in *Presbytis comata* by Brandon-Jones (1984).

Presbytis frontata White-fronted leaf monkey

Semnopithecus frontatus Müller, 1838: 136; SE Borneo.
P. nudifrons Elliot, 1909: 266; Bajalong, C Sarawak, Borneo.

Range (Map 81). Borneo.

Variation. Northern animals (*P. f. nudifrons*) are paler and the spot on the forehead is bisected by a black sagittal line.

Presbytis rubicunda Red leaf monkey

Semnopithecus rubicundus Müller, 1838: 137; Mt Sekumbang, SE of Banjermasin, S Borneo.
P. carimatae Miller, 1906a: 65; Karimata I, SW Borneo.
P. ignita Dollman, 1909: 204; Mt Mulu, N Sarawak, Borneo.

Range (Map 81). Borneo, including Karimata I.

Variation. The following subspecific names have been proposed:
Pygathrix r. rubida Lyon, 1911a: 139; mouth of Kendawangan R, SW Borneo.
P. r. chryseus Davis, 1962: 64; E of Kretam, Kechil R, SE end of Dewhurst Bay, Kinabatangan Dist, N Borneo.

Variation on mainland Borneo is slight but the form on Karimata I is more distinctive, characterized by a very robust skull.

SEMNOPITHECUS

Semnopithecus Desmarest, 1822: 532; type species *Simia*

Table 114. The species of *Semnopithecus*.

	entellus	*auratus*	*francoisi*	*vetulus*	*johnii*	*cristatus*	*obscurus*	*phayrei*	*pileatus*	*geei*
Range	India etc.	Java	Indochina	Sri Lanka	S India	Vietnam - Borneo	Malaya etc.	Assam - Vietnam	Assam etc.	Assam
Head & body (cm)	40–80	46–75	48–67	47–70	51–74	41–65	42–68	42–60	49–76	c.50
Tail (cm)	90–100	61–82	82–89	62–92	76–97	56–84	57–86	72–80	83–110	c.70
Dorsum	Brown	Black or red	Black ± white on rump	Black or grey-brown	Black, frosted on rump	Black, ± frosted white	Dark grey brown	Grey, ± paler on rump	Grey, ± brown	Pale straw
Venter	White - pale brown	Black - brown	Black	As dorsum	Black	Dark grey	Brown-grey	Grey	Grey/white/ orange-brown	Straw, tinged orange
Crest on crown	0	Slight	Large	0	0	+	0	0	+	+
White ring round eye	0	0	0	0	0	0	+	0	0	0
Cheek-tufts prominent	0(+)	+	0	+	+	+	+	+	+	+
Head	As dorsum or white	As dorsum	Black or off-white	Paler than dorsum	Brown	Black	As dorsum, paler on nape	As dorsum	As dorsum	As dorsum

Map **82**. *Semnopithecus* (part)

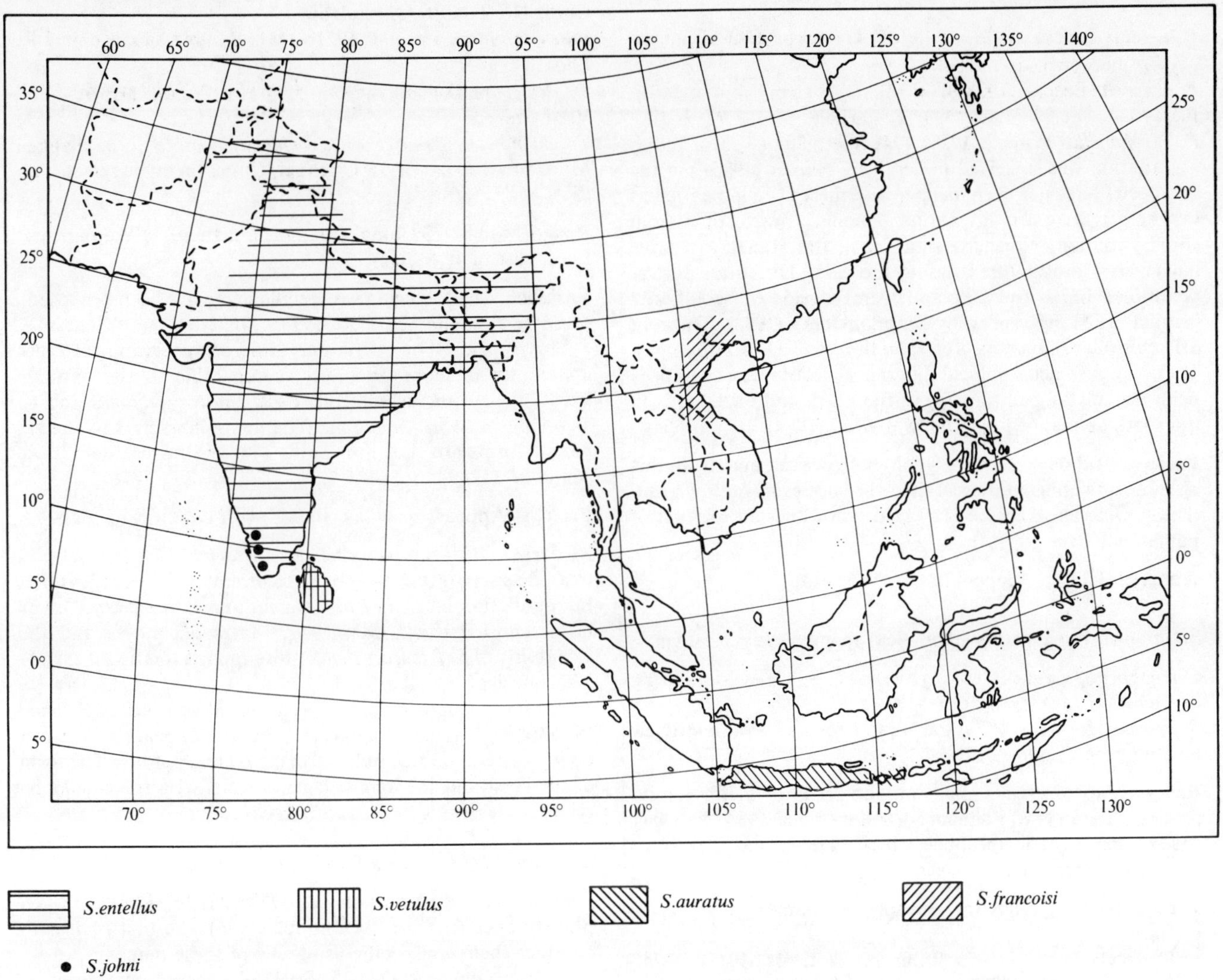

entellus Dufresne. Conserved by Opinion 1202 (1982).
Trachypithecus Reichenbach, 1862: 89 (*S. pyrrhus* Horsfield, in *S. auratus*).
Kasi Reichenbach, 1862: 101 (*Cercopithecus johnii* Fischer).
Presbypithecus Trouessart, 1879b: 56 (*Cercopithecus kephalopterus* Zimmermann, in *S. vetulus*).

Throughout most of the region; 10 species (Table 114). These have been included in *Presbytis* but they appear to form a distinct monophyletic group. Delimitation of some of the species is still very uncertain. All the species except *S. entellus* have been separated in a genus *Trachypithecus*, e. g. by Groves (1989) on the basis of the distinctive skull of *S. entellus* and the black pelage of the neonate. However the close relationship of these groups is shown by the record of a live-born hybrid between *S. entellus* and *S. obscurus* (Hill, 1939) and the apparent hybridization in the wild between *S. entellus* and *S. johnii* (Hohmann, 1988).

The following nominal fossil species have been described from the region:
S. subhimalayanus Meyer, in Bronn, 1848: 1133; Siwaliks.
S. palaeindicus Lydekker, 1884b: 123(=19); India; Siwaliks.
S. asnoti Pilgrim, 1910: 64; Asnot, India; Middle Siwaliks.

Semnopithecus entellus Hanuman langur

Simia entellus Defresne, 1797: 49; Bengal, India. Conserved by Opinion 1202 (1982).
Semnopithecus schistaceus Hodgson, 1841e: 1212; Nepal Terai.
S. hodie Hodgson, 1841e: 1212; Nepal.
= *S. nipalensis* Hodgson, 1841e: 1212.
S. hypoleucos Blyth, 1841c: 839; Travancore, S India.
S. dussumieri Geoffroy, 1843b: 719; Malabar coast, S India.
S. pallipes Blyth, 1844b: 312; S India; *nomen nudum*.
S. priam Blyth, 1844a: 470; Coromandel coast, India.
Presbytis anchises Blyth, 1844a: 470; Deccan, India.
P. thersites Blyth, 1847b: 1271; Trincomalee, Sri Lanka.
P. lania Elliot, 1909: 273; Chumbi Valley, S Tibet.

Range (Map 82). Himalayas from N Pakistan to Assam and south throughout peninsular India; Sri Lanka; up to 3600m in the Himalayas.

Variation. The following subspecific names have been proposed:
Presbytis e. achilles Pocock, 1928: 478; Satthar Hill, Gorkha, 50 miles NW of Katmandu, Nepal, 12000ft.
P. e. ajax Pocock, 1928: 480, pl. 2; Dealah, Chamba, Punjab, 6000ft.

P. e. achates Pocock, 1928: 488; Haunsbhavi, Dharwar, India, 2000ft.
P. e. iulus Pocock, 1928: 490; Jog, Gersoppa Falls, Kanara-Mysore border, India, 1300ft.
P. e. aeneas Pocock, 1928: 492; Makut, S Coorg, India, 250ft.
P. e. elissa Pocock, 1928: 493; Nagarhole, SE Coorg, India.
P. e. priamellus Pocock, 1928: 494; Shernelly, Cochin, India.

Variation was reviewed in detail by Napier (1985) but the subspecies have not been mapped and there is no indication as to whether there are any abrupt boundaries as distinct from clinal variation. Northern animals, in the Himalayas, have longer cheek tufts, often hiding the ears. The forms in SW peninsular India are dark and were considered specifically distinct, as *S. hypoleucos*, by Brandon-Jones (1984). Those in SE peninsular India have a crest on the crown (*S. e. priam* etc.). There are also geographical differences in the way the tail is held—mainly looped forward in the north and backwards in the south as mapped by Roonwal *et al.* (1984).

Ecology. Although feeding upon both leaves and fruit this species is much less arboreal than the species of *Presbytis* and the other *Semnopithecus*. They raid crops and are tolerant of human activity.

Remarks. CITES: Appendix I.

Semnopithecus auratus Ebony leaf monkey, Javan langur

Cercopithecus auratus Geoffroy, E.,1812: 93; 'Les Moluques'. restricted to Java by Elliot (1913).
S. pyrrhus Horsfield, 1823; Pasuruan Prov., E Java (restricted by Robinson & Kloss, 1919b).

Range (Map 82). Java, Bali and Lombok. A claim for its presence also in NW Vietnam (Brandon-Jones, 1984: 407) was based upon a single specimen which is much more likely to represent a form of *T. francoisi.*

Variation. The following subspecific names have been proposed:
Pithecus pyrrhus sondaicus Robinson & Kloss, 1919b: 374; Tjibodas, W Java, 4500ft.
P. p. kohlbruggei Sody, 1931b: 349; Bali.
Trachypithecus p. stresemanni Pocock, 1935: 931; Bali.

The pelage is dichromatic in E Java, with reddish and black morphs; monochromatic black elsewhere. The type of *auratus* was interpreted by Elliot (1913) as an adult female that had, exceptionally, retained the golden, infantile colour.

Semnopithecus francoisi Francois' leaf monkey

S. francoisi Pousargues, 1898: 319; Lungchow, Guangxi, China.
S. (Lophopithecus) poliocephalus Trouessart, 1911: 271; Kai-Chin, NE Tonkin, N Vietnam.
Pithecus laotum Thomas, 1921b: 181; Ban Na Sao, Mekong R, Laos, 17°30'N.
P. delacouri Osgood, 1932: 205; Hoi Xuan, NE Annam, Vietnam.
Trachypithecus (Presbytis) leucocephalus T'an, 1957: 63; Sangen Mt, Chongzuo county, S Guangxi, China (neotype selected by Li & Ma, 1980).

Range (Map 82). S China (Guangxi, Guizhou), N Viẽtnam and Laos, east of R Mekong.

Variation. The following subspecific name has been proposed: *Presbytis f. hatinhensis* Dao, 1970b: 61; Xom-cuc, Vietnam.

This species shows strikingly different geographical races apparently in close proximity, comparable to the extreme situation seen in the squirrels *Callosciurus erythraeus* and *C. finlaysonii*, also in Indochina. The main characters of the six named forms are shown in Table 115. For further details see Ma *et al.* (1989).

Ecology. Appears to be associated with outcrops of rock.

Remarks. The forms *leucocephalus* and *delacouri* are sometimes treated as separate species, e. g. by Brandon-Jones (1984), but intergradation between *leucocephalus* and *P. francoisi s. s.* has been reported (Li & Ma, 1980; Ma *et al.*, 1989), and the variability of *S. f. leucocephalus* also supports treating it as only a subspecies.

The form *poliocephalus* has been treated as a race of *S. johnii* (S India) by Brandon-Jones (1984), but this appears to be based upon a superficial resemblance in colour and pattern. The form of the crest and the long pelage, especially on the shoulders, relate it closely to *P. francoisi.* IUCN: E.

Semnopithecus vetulus Purple-faced leaf monkey

Cercopithecus vetulus Erxleben, 1777: 25; Sri Lanka. Given priority over *C. senex* by Phillips (1926), the first reviser to consider them conspecific using one of these names.
C. senex Erxleben, 1777: 25; Sri Lanka (based on albinos).
C. kephalopterus Zimmermann, 1780: 185; Sri Lanka.
S. nestor Bennett, 1833: 67; Sri Lanka, probably Rayigam.
Presbytis ursinus Blyth, 1851a: 155; Nuwara Eliya, Sri Lanka.
P. albinus Kelaart, 1852: 7; Matale, Central Prov., Sri Lanka.
Pithecus philbricki Phillips, 1927: 57; Kantalai, near Trincomalee, E Prov., Sri Lanka, 200ft.

Range (Map 82). Sri Lanka.

Variation. The following subspecific names have been proposed:
Presbytis cephalopterus var. *monticola* Kelaart, 1850a: 207 (321

Table 115. The subspecies of *Semnopithecus francoisi.*

	francoisi	*leucocephalus*	*poliocephalus*	*hatinhensis*	*delacouri*	*laotum*
Range	N Vietnam, S China	S China	NE Vietnam	C Vietnam	NE Vietnam	Laos
Head	Black	White, ± black on crest	Buffy white	Black	Black	White, crest black
Rump and thighs	Black	Black	White-frosted	Black	White	Black
White stripes extending behind ears	0	–	-	+	+	+
Tail	Black	Distal $\frac{1}{2}$–$\frac{3}{4}$ white	Black	Black	Black	Black
Feet	Black	± white	Black	Black	Black	Black

of 1890 reprint); Nuwara Eliya, Sri Lanka.
Pithecus v. phillipsi Hinton, 1923b: 510; Gonapola, Panadura dist, Sri Lanka.
Presbytis senex harti Deraniyagala, 1955b: 293; Thinakai, N Prov., Sri Lanka.

Geographical variation is considerable. The characters of four subspecies are summarized in Table 116, based upon the more detailed table by Napier (1985).

Semnopithecus johnii Hooded leaf monkey

Cercopithecus johnii Fischer, 1829: 25; Tellicherry, S India (probably based upon a captive specimen from outside the natural range—Jerdon, 1867: 8).
S. cucullatus Geoffroy, 1834: 38, pl. 1; the Ghats, Bombay, India.
S. jubatus Wagner, 1839a: 305; S India.

Range (Map 82). Hills of SW India from Coorg (c. 12°N) south to c. 8°15'N.

Remarks. Closely related to *S. vetulus* of Sri Lanka but probably justifying specific separation. Several animals suspected of being hybrids between this species and *S. entellus* have been recorded in troops of *S. johnii* (Hohmann, 1988). A record from N Vietnam by Brandon-Jones (1984), based upon the form *poliocephalus*, seems much more likely to represent a form of *S. francoisi*. IUCN: E.

Table 116. The subspecies of *Semnopithecus vetulus*.

	vetulus	*nestor*	*monticola*	*philbricki*
Range in Sri Lanka	SW	W	Central	NE
Weight (kg)	– 7	3.5–4.5	c.9	c.10
Pelage	Black	Grey-brown	Grey-brown	Grey-brown
Pale, short-haired patch on rump	+, white	+, beige	0	0
Tail (cm)	Long c.68–75	Long c.70–71	Short, thick 62–67	Long 82–88
Cheek-tufts	Moderate	Moderate	Long	Long

Semnopithecus cristatus Silvered leaf monkey

Simia cristatus Raffles, 1821: 244; Bencoolen, S Sumatra.
S. pruinosus Desmarest, 1822: 533; Sumatra.
S. germani Milne-Edwards, 1877: 8; Cochin-China and Cambodia.
Presbytis margarita Elliot, 1909: 271; Langbian, Annam, Vietnam.
Pygathrix ultima Elliot, 1910b: 351; Mt Dulit, Sarawak, Borneo.
Presbytis vigilans Miller, 1913b: 29; Sirhassen I, S Natuna Is.

Map **83**. *Semnopithecus* (part)

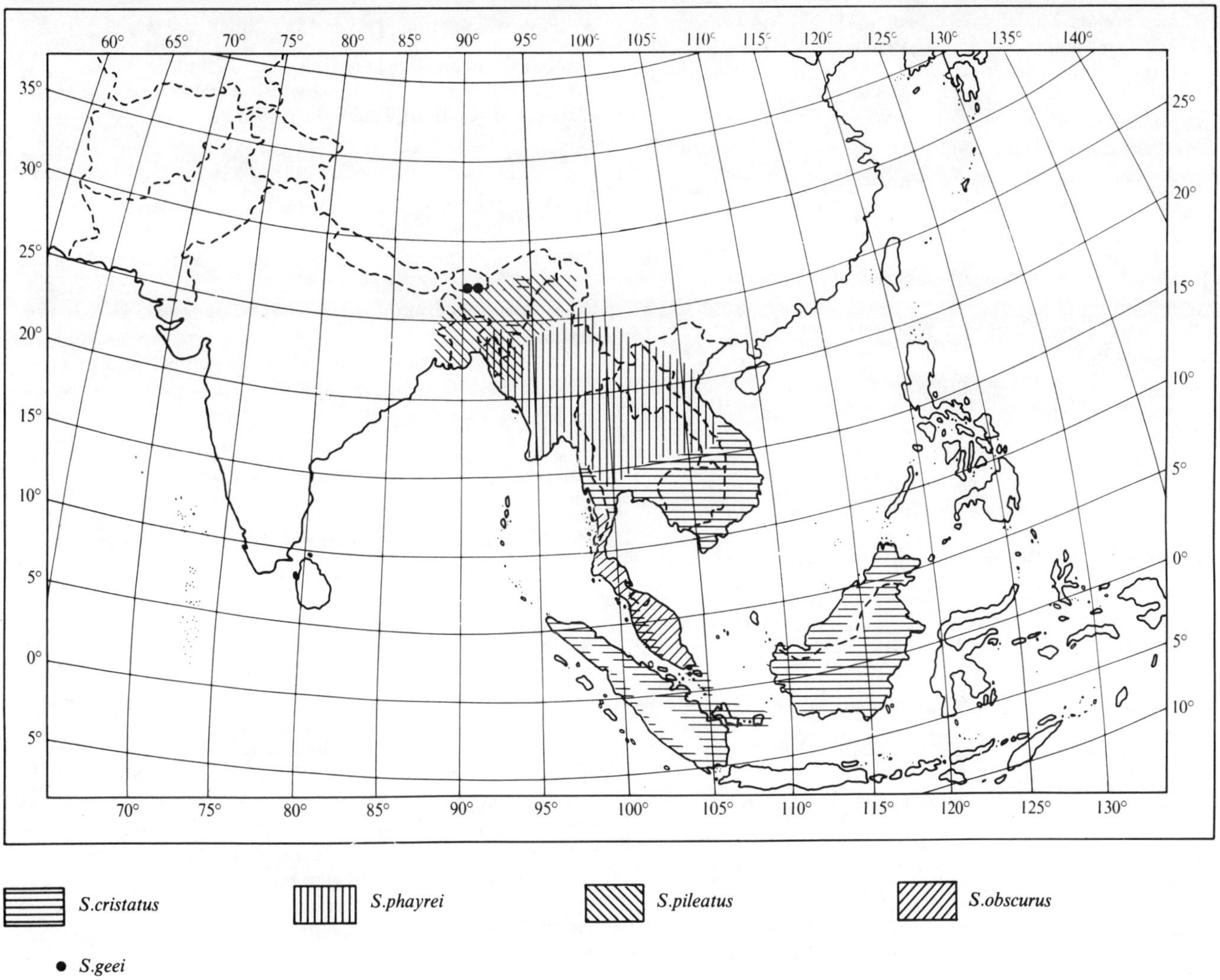

Range (Map 83). From SE Assam and Burma to central Vietnam and Thailand (but apparently absent from the Isthmus of Kra and peninsular Thailand); W coast of Malaya; Sumatra, Borneo; small islands of Batam, Bintan, Bulan, Sugi, Galang, Jombol (Riau Is), Lingga, Bakong, Sebangka (Lingga Is), Bangka, Belitung, Serasen (S Natuna Is).

Variation. The following subspecific names have been proposed:
Presbytis c. pullata Thomas & Wroughton, 1909a: 439; Batam I, Riau Is.
P. germaini mandibularis Kloss, 1916b: 32; Chang I, SE Thailand.
P. c. koratensis Kloss, 1919b: 340; Lat Bua Kao, 30 miles W of Korat, Thailand.
Pithecus pyrrhus atrior Pocock, 1928: 673; Ye Forest, S of Mulmein, Ataran dist, Tenasserim, Burma, 500ft.
Trachypithecus c. robustus Hooijer, 1962: 41; Tegoean, central Java; Pleistocene.
Presbytis c. caudalis Dao, 1977: 325; N Vietnam.

Variation is slight, affecting mainly the amount of grizzling caused by pale hair-tips, which is minimal on Java; on Sumatra it increases northwards (Wilson & Wilson, 1976). A prominent crest is present except in Burma.

Semnopithecus obscurus Dusky leaf monkey

S. obscurus Reid, 1837: 14; Malacca, Malaya.
S. leucomystax Müller & Schlegel, 1841: 59; Malacca.
S. halonifer Cantor, 1845: 235; Penang I, Malaya.
Pygathrix sanctorum Elliot, 1910b: 351; St Matthew I, Mergui Is.
P. flavicauda Elliot, 1910b: 352; Trang, S Thailand.
Presbytis corvus Miller, 1913b: 27; Terutau I, S Thailand.

Range (Map 83). S Thailand, S Burma and Malaya; including some of the Mergui Is (Burma), Teratau I (Thailand), and Penang, Langkawi and E Perhentian Is (Malaya).

Variation. The following subspecific names have been proposed:
Presbytis o. carbo Thomas & Wroughton, 1909b: 534; Terutau I, Thailand.
P. o. styx Kloss, 1911: 116; E Perhentian I, E Malaya.
P. o. smithi Kloss, 1916a: 5; Klong Bang Lai, Patiyu, S Thailand.
Trachypithecus o. corax Pocock, 1935: 944; Tenasserim Town, S Burma.
Pithecus o. seimundi Chasen, 1940a: 80; Pennan I, Thailand.

Semnopithecus phayrei Phayre's leaf monkey

Presbytis phayrei Blyth, 1847c: 733; Arakan, Burma.
? *S. holotephreus* Anderson, 1879: 27; locality unknown.
P. melamera Elliot, 1909: 267; Cadu Ciaung, near Bhamo, NE Burma.
? *P. crepuscula* Elliot, 1909: 271; Mt Mulaiyit, Tenasserim, Burma, 5000ft.
Pithecus shanicus Wroughton, 1917: 47; Se'en, Hsipaw State, Shan States, Burma.
Presbytis argenteus Kloss, 1919b: 338; Lat Bua Kao, west of Korat, Thailand.

Range (Map 83). From Assam and Bangladesh through Burma to S Yunnan, N Vietnam and N Thailand.

Variation. The following subspecific name has been proposed:
Presbytis crepuscula wroughtoni Elliot, 1909: 272; Pachebon, C Thailand.

On the mainland the pelage, especially of the extremities, tends to be paler in the north than in the south. Those on the islands of Penang, Langkawi and Terutau are very dark. The characters of three tentative subspecies are given in Table 117 but these are unlikely to be discrete.

Table 117. The subspecies of *Semnopithecus phayrei*.

	phayrei	*shanicus*	*crepusculus*
Range	Bangladesh - N Burma	C Burma	S Burma - Vietnam
Whorl of hair on crown	0	+	0
Venter	White or buff	Grey	Light grey
White pubic patch	Both sexes	?	Female only

Remarks. Possibly conspecific with *S. obscurus* (and included in it, in part, by Brandon-Jones, 1984), but kept separate by Napier (1985) and by Weitzel *et al.* (1988).

Semnopithecus pileatus Capped leaf monkey

S. pileatus Blyth, 1843b: 174; locality unknown.
S. argentatus Horsfield, 1851: 7; Sylhet, Bangladesh.
Presbytis shortridgei Wroughton, 1915e: 56; Homalin, upper Chindwin, Burma.
P. brahma Wroughton, 1916e: 654; Seajulia, Dafla Hills, N Lakhimpur, upper Assam.
P. durga Wroughton, 1916e: 655; Cahar, Assam.

Range (Map 83). Bangladesh, Assam, N and W Burma; up to 1000m. Sympatric with *S. phayrei* in Cachar and Karimganj districts of Assam (Choudhury, 1990).

Variation. The following subspecific names have been proposed:
Presbytis shortridgei belliger Wroughton, 1915e: 57; Hkamti, upper Chindwin, Burma.
Pithecus p. saturatus Hinton, 1923c: 81; Bara Hapjan, Lakhimpur, upper Assam.
P. p. tenebricus Hinton, 1923c: 81; Matunga R, N Kamrup, Assam.

The characters of five subspecies are summarized in Table 118 (adapted from Napier, 1985).

Remarks. CITES: Appendix I.

Semnopithecus geei Golden leaf monkey

Presbytis geei Gee, 1956 (20 Jan.): 252; Jamduar Forest Rest House, E bank of R Sankosh, Goalpara Dist, Assam, India.

Range (Map 83). S Bhutan and adjacent part of Assam east to Manas River.

Remarks. This species was inadvertently, but validly, described by Gee (1956) before the full description by Khajuria (1956). It appears to be very closely related to *S. pileatus* but it has been reported to coexist with both *S. pileatus* and *S. entellus* in the Manas sanctuary (Ghosh & Biswas, 1976). However in Bhutan it does not occur in areas where *S. pileatus* occurs (Subba, 1989), and the same is probably the case in the Manas sanctuary. It could perhaps be interpreted as a peripheral subspecies of *S. pileatus*, differing only in the retention by the adult of the golden, juvenile pelage found in most species of *Semnopithecus*, a situation foreshadowed by the orange-brown venter in the adults of some races of *S. pileatus*. CITES: Appendix I; IUCN: R.

Table 118. The subspecies of *Semnopithecus pileatus.*

	shortridgei	*brahma*	*pileatus*	*durga*	*tenebricus*
Range	NE Burma	Dafla Hills	Assam-Bangladesh	E Assam, NW Burma	Kamrup
Venter	Grey	Buff	Yellowish white	Orange-brown	Orange/buff
Cheek whiskers	Grey	Light grey	Yellowish white	Orange-brown	Buff
Sharp contrast on flanks	0	0	+	+	+
Nape buff	0	0		0	+

Semnopithecus incertae sedis:

Presbytis barbei Blyth, 1847c: 734; Tipperah Hills, Tripura, India.
It is very doubtful whether this form is an independent species or a form of *S. cristatus*. It was treated as a species by Brandon-Jones (1984) with a range to Burma and S China. For further details see Napier (1985: 53).
Presbytis ruhei Knottnerus-Meyer, 1933: 259; Sangora, S Thailand.

PYGATHRIX

Pygathrix Geoffroy, E.,1812: 90; type species *Simia nemaeus* L.
Rhinopithecus Milne-Edwards, 1872b: 233 (*Semnopithecus roxellana* Milne-Edwards).
Presbytiscus Pocock, 1924b: 17 (*Rhinopithecus avunculus* Dollman).
Megamacaca Hu & Qi, 1978 (*M. lantianensis*, a fossil species, allocated to *Rhinopithecus* by Gu & Jablonski, 1989: *Acta anthropol. sin.* 8:343–6).

S China and Indochina; 5 species (Table 119). *Rhinopithecus* is often recognized as a separate genus but there seems to be no doubt that *Pygathrix* as used here is a monophyletic group, characterized especially by the form of the nose, with flaps of skin on the margins of the nostrils.

Pygathrix nemaeus Douc langur

Simia nemaeus Linnaeus, 1771: 521; Cochin-China.
Semnopithecus nigripes Milne-Edwards, 1871b: 7, pl. 1; Saigon, Vietnam.

Range (Map 84). Vietnam, S Laos and NE Cambodia. A 19th century record from the island of Hainan was based upon an animal that was correctly identified but the origin of which cannot be substantiated (Fooden & Feiler, 1988).

Map **84**. *Pygathrix, Nasalis*

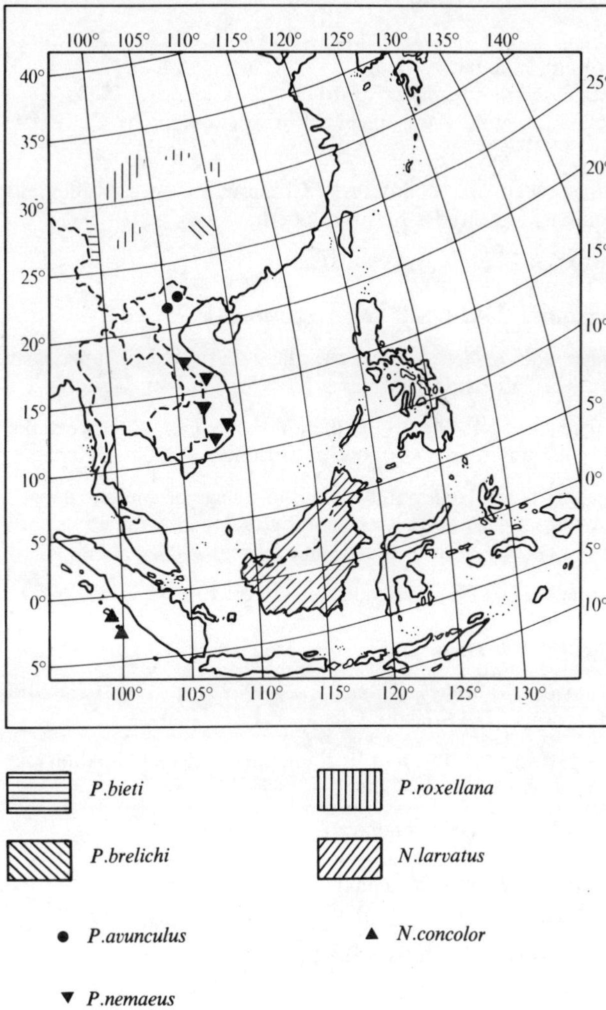

Table 119. The species of *Pygathrix*.

	nemaeus	*roxellana*	*bieti*	*avunculus*	*brelichi*
Range	Vietnam	China	Yunnan	N Vietnam	Guizhou
Head & body (cm)	53–63	66–76	74–83	51–62	c.73
Tail (% of h&b)	c.110%	85–95%	70–90%	140–150%	c.130%
Dorsum	Grey	Dark grey-brown	Dark grey	Blackish	Grey-brown
Venter	Grey	Pale orange	White	Orange-yellow	Yellow-grey
Crest on crown	0	Short	Long	0	0
Cheeks	White	Orange	White	Yellow	Black
Hands and feet	Black	Yellowish	Black	Black	Black
Back of thighs	Black	Light patch		Black	Light
Penis	Pink			Black	

Variation. One subspecific name has been proposed:
Presbytis nemaeus moi Kloss, 1926c: 214; Langbian Peak, S Annam, Vietnam, 5500–6500ft.

Two subspecies can be recognized although the nature of the boundary between them is unknown:
P. n. nemaeus. Central Vietnam and S Laos. Legs reddish; forearms white; face reddish or yellowish brown; scrotum white.
P. n. nigripes. S Vietnam, NE Cambodia. Legs black; forearms grey; face black or bluish; scrotum blue.

Remarks. CITES: Appendix I; IUCN: E.

Pygathrix roxellana Golden snub-nosed monkey

Semnopithecus roxellana Milne-Edwards, 1870a: 341; near Moupin, Sichuan, China.
Rhinopithecus tingianus Matthew & Granger, 1923: 588; Sichuan, China; fossil. Included in *R. roxellana* by Colbert & Hooijer (1953).

Range (Map 84). Mountains of China in S Gansu, Hubei and Sichuan; at altitudes of 1000–3000m.

Remarks. CITES: Appendix I; IUCN: V.

Pygathrix bieti Black snub-nosed monkey

Rhinopithecus bieti Milne-Edwards, 1897b: 157; Kiape, near Atuntze, left bank of Mekong R, NW Yunnan, China.

Range (Map 84). Mountains of N Yunnan, China, between the Mekong and Yangtze rivers, at altitudes of 3300–4000m.

Ecology. They feed mainly upon the foliage of conifers as well as of broad-leaved trees and epiphytic lichens, travelling both in the canopy and on the ground (Yang, 1988).

Remarks. CITES: Appendix I; IUCN: E.

Pygathrix brelichi Grey snub-nosed monkey

Rhinopithecus brelichi Thomas, 1903b: 224, pl. 21; ? Fan Jin Shan range, N Guizhou, China.

Range (Map 84). The Fan Jin Shan mountains, N Guizhou and extreme SE Sichuan, China.

Remarks. CITES: Appendix I; IUCN: E.

Pygathrix avunculus Tonkin snub-nosed monkey

Rhinopithecus avunculus Dollman, 1912: 18,503; Yen Bay, Song-koi R, Tonkin, N Vietnam.

Range (Map 84). Mountains of N Vietnam.

Remarks. CITES: Appendix I; IUCN: E.

NASALIS

Nasalis Geoffroy, E.,1812: 90; type species *N. larvatus* Geoffroy.
Simias Miller, 1903a: 66 (*S. concolor* Miller).

Borneo and the Mentawai Islands; 2 species (Table 120). There are many differences between the two species and they are often placed in separate genera. However Groves (1989) considered them congeneric on the basis of several significant shared characters: the form of the nose (adult *N. concolor* resembling juvenile *N. larvatus*), thin enamel, sagittal crest, narrow rostrum and other cranial features. He also considered *Nasalis* to be the sister-group of all other colobines and suggested a separate subfamily, Nasalinae.

Table 120. The species of *Nasalis.*

	larvatus	*concolor*
Range	Borneo	Mentawai Is
Nose long and pendulous in ad.m	+	0
Pelage	Reddish brown	Brownish black or creamy buff
Face	Brown	Black
Tail (% of h&b)	c.110–120%	c.30%

Nasalis larvatus Proboscis monkey

Cercopithecus larvatus van Wurmb, 1787: 353; Pontianak, W Borneo.
Simia (*Cercop.*) *capistratus* Kerr, 1792: 73; Borneo.
N. recurvus Vigors & Horsfield, 1828: 110; Pontianak, Borneo.

Range (Map 84). Borneo.

Variation. The following subspecific name has been proposed:
N. l. orientalis Chasen, 1940a: 84; Salim Batu, Bulungan, NE Borneo.

Ecology. Mainly in mangrove and riverine forest; arboreal but swimming readily to reach new feeding areas.

Remarks. CITES: Appendix I; IUCN: V.

Nasalis concolor Pig-tailed langur

Simias concolor Miller, 1903a: 67, pl. 14; S Pagi I, Mentawai Is.

Range (Map 84). Mentawai islands of Siberut, Sipura, N Pagai and S Pagai.

Variation. The following subspecific name has been proposed:
Simias c. siberu Chasen & Kloss, 1928c: 813; Siberut I, Mentawai Is (darker and with more conspicuous white cheeks than on the Pagai islands; skull broader and less prognathous).

Pelage is dichromatic, c. 25–35% being creamy buff in contrast to the usual blackish brown.

Remarks. CITES: Appendix I; IUCN: E.

Family Hylobatidae

Gibbons

Hylobatina Gray, 1871c: 4; type genus *Hylobates* Illiger.

Indochinese and Sunda subregions; c. 11 species. One genus, *Hylobates*, is recognized here although *H. syndactylus* has often been given separate generic rank in *Symphalangus*.

Although sometimes included with the great apes in the family Pongidae, the gibbons form a very distinctive group that appears to justify a separate family, characterised especially by the very long arms and hands associated with locomotion by brachiation in the forest canopy. Most species of gibbons feed predominantly upon fruit, augmented by buds, flowers and insects, but *H. syndactylus* is a more specialized folivore, eating rather more leaves than fruit.

Table 121. The subgenera of *Hylobates*.

	Hylobates	*Bunopithecus*	*Nomascus*	*Symphalangus*
No. of species	6	1	3	1
Head & body (cm)	35–65	45–60	45–55	75–90
Sexual dichromatism	0	+	+	0
Juvenile colour	As adult	Whitish	Whitish	As adult
Crown hair: fan-wise (F), erect (E)	F	F	E	F
Laryngeal sac	0	Small	Small, male only	Large, both sexes
Facial profile of skull	Sinuous	Sinuous	Straight	Sinuous
Chromosomes (2n, NF)	44,84	38,	52,94	50,94

HYLOBATES Gibbons

Gibbon Zimmermann, 1777: 405 (*Simia longimana* Schreber, in *Homo lar* L.). Suppressed by Opinion 257 (1954).
= *Cheiron* Burnett, 1829b: 307.
Hylobates Illiger, 1811: 68; type species *Homo lar* Linnaeus. Conserved by Opinion 122 (1931).
= *Laratus* Gray, 1821: 297.
Symphalangus Gloger, 1841: 34 (*S. syndactylus* Gloger).
Siamanga Gray, 1843a: 1 (*H. syndactylus* Raffles).
Bunopithecus Matthew & Granger, 1923: 588 (*B. sericus* Matthew & Granger, a fossil species). Treated as a subgenus of *Hylobates* by Colbert & Hooijer (1953).
Brachitanytes Schultz, 1932: 369 (*Symphalangus klossi* Miller).
Nomascus Miller, 1933: 159 (*Hylobates leucogenys* Ogilby).
Hoolock Hainoff *et al.*,1984: 626 (*Simia hoolock* Harlan).

The taxonomy of gibbons was reviewed by Groves (1972,1984,1989), Marshall & Sugardjito (1986) and by Jenkins (1990). The gibbons of southern China were reviewed by Ma, Wang & Poirier (1988) and their distribution mapped by Bleisch & Nan (1990). Apart from *H. syndactylus* which is broadly sympatric with forms of *H. lar* and *H. agilis*, they form a parapatric series with varying degrees of intergradation or reproductive isolation between adjacent forms. The number of species recognized has therefore been unstable. The treatment here follows that of Groves (1984) except for the inclusion of the Bornean form *albibarbis* in *H. agilis*, giving 11 species in 4 subgenera (Table 121). The territorial calls are distinctive for each species.

The following nominal fossil species have been described from the region:
Bunopithecus sericus Matthew & Granger, 1923: 588; Sichuan, China; Pliocene. Tentatively allocated to *Hylobates hoolock* by Groves (1972).
H. tianganhuensis Lei, 1985: 19; Songlinzhuang, Sihong County, Jiangsu, China; Xiacowan Formation, Middle Miocene.

Subgenus *Hylobates* (Table 122)

The six species of this subgenus have a diploid chromosome number of 44, with nearly identical G-banding pattern (Stanyon *et al.*, 1987).

Hylobates lar White-handed gibbon, Lar gibbon

Hylobates lar

Homo lar Linnaeus, 1771: 521; Bukit Cheraka, Klang, Jeram, Selangor, Malaya. Conserved and neotype designated by Opinion 1219 (1982).
Simia longimana Schreber, 1774: 66, pl. 3; Malacca, Malaya. Suppression has been requested (Jenkins & Groves, 1992).
Pithecus varius Latreille, 1801: 276; Malacca, Malaya.
P. variegatus Geoffroy, E.,1812: 88; Malacca, Malaya.
Simia albimana Vigors & Horsfield, 1828: 107; ? W Sumatra. Suppression has been requested (Jenkins & Groves, 1992).
H. entelloides Geoffroy, 1843b: 717; Malaya, c. 12°N. Conserved by Opinion 1219 (1982).

Range (Map 85). S Yunnan, E Burma and W Thailand to Malaya; N Sumatra; absent from N peninsular Malaya.

Table 122. The species of *Hylobates* (*Hylobates*).

	lar	*pileatus*	*agilis*	*moloch*	*muelleri*	*klossi*
Range	Burma-Malaya, Sumatra	Cambodia etc.	Malaya, Sumatra, S Borneo	Java	Borneo	Mentawai Is
Overall colour: male	Black to buff	Black	Black to buff	Grey	Grey-brown	Black
female	Black to buff	Grey/buff	Black to buff	Grey	Grey-brown	Black
Hands and feet contrasting white	+	+	0	0	0	0
White round muzzle	+	Male only	+	+	Faint/0	0
White eye-brows	+	Male only	Male only	+	+	0

Map **85**. *Hylobates* (part)

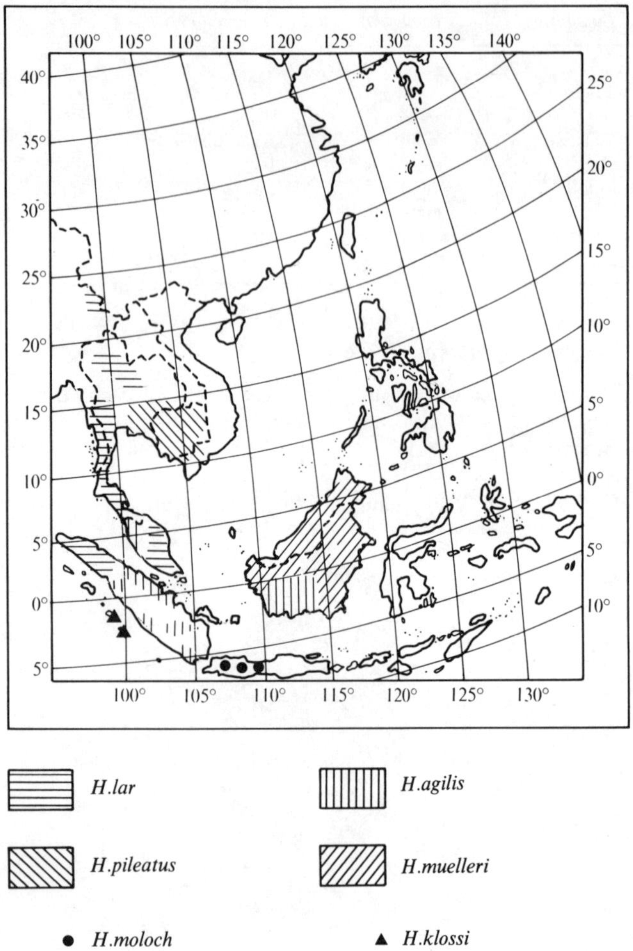

Variation. The following subspecific names have been proposed:
H. albimanus vestitus Miller, 1942: 131; Blangnanga, Aceh, N Sumatra, 3600ft.
H. l. carpenteri Groves, 1968: 625; Mt Angka, N Thailand, 3400ft.
H. l. yunnanensis Ma & Wang, 1986: 402,410; Manglian, SW Yunnan, 2000m.

The overall colour is dichromatic in most populations, with geographical variation but no sexual dichromatism. Four subspecies can be recognized, but boundaries cannot be clearly defined in the mainland part of the range:
H. l. carpenteri. N Thailand, ? Yunnan. Dark phase dark brown; pale phase creamy white; pelage longer than in other subspecies (>8cm on the back).
H. l. entelloides. Thailand. Dark phase black; pale phase honey-coloured.
H. l. lar. S peninsular Thailand and Malaya. Dark phase brown; pale phase creamy. In two segments: pale phase dominant in the north, dark in the south.
H. l. vestitus. N Sumatra. Monochromatic light brown, darker below.

Remarks. A limited amount of interbreeding with *H. pileatus* takes place at the boundary in Thailand, and with *H. agilis* in Malaya. CITES: Appendix I.

Hylobates pileatus Pileated gibbon

H. pileatus Gray, 1861: 136; Phu Quoc I, Cambodia.

Range (Map 85). W Cambodia, SW Laos, and SE Thailand.

Remarks. See Remarks under *H. lar*. CITES: Appendix I; IUCN: E.

Hylobates agilis Agile gibbon

H. agilis Cuvier, in Geoffroy & Cuvier, 1821: 3; W Sumatra.
H. rafflei Geoffroy, E.,1829: 34; W Sumatra.
H. unko Lesson, 1840: 53; Sumatra.
H. albo nigrescens Ludeking, 1862: 36; Agam, W Sumatra.
H. albo griseus Ludeking, 1862: 36; Agam, W Sumatra.
? *H. mülleri albibarbis* Lyon, 1911a: 142; near Sukudana, Matan R, SW Borneo.

Range (Map 85). Central and S Sumatra, NW Malaya; and very marginally in peninsular Thailand; also SW Borneo if *albibarbis* is conspecific.

Variation. There are no significant geographical differences between the Sumatran and Malayan populations. Both are dichromatic, with dark (black to maroon) and light (grey to cream) morphs. In the highlands of Sumatra the two forms are about equally common; elsewhere the black form is dominant (Weitzel *et al.*, 1988). The Bornean form *albibarbis* is characterized by light pelage and darker extremities.

Remarks. Those in SW Borneo (*albibarbis*) have vocalizations different from the adjacent *H. muelleri*. They were allocated to *H. agilis* by Marshall & Sugardjito (1986) but were considered conspecific with the other Bornean gibbons, *H. muelleri*, by Weitzel *et al.* (1988). A limited amount of interbreeding takes place with *H. lar* in Malaya and with *H. muelleri* in Borneo. CITES: Appendix I.

Hylobates moloch Javan gibbon

Pithecus cinereus Cuvier, 1797; Batavia, Java. Pre-occupied by *Simia* (*Papio*) *cinerea* Kerr, 1792.
Simia moloch Audebert, 1799: 1st fam.,sect. 2, pl. 2; 'Moluccas', corrected and restricted to Mt Slamat, Java by Sody (1949b).
S. leucisca Schreber, 1799:pl. 3B; Java.
H. javanicus Matschie, 1893: 62; Java.
H. lar pongoalsoni Sody, 1949b: 123; Kali Kidang, Mt Slamat, C Java, 800m.

Range (Map 85). W Java, east to c. 109°10'E.

Remarks. CITES: Appendix I; IUCN: E.

Hylobates muelleri Bornean gibbon

H. mülleri Martin, 1841: 444; SE Borneo.
H. funereus Geoffroy, 1850: 874; 'l'isle Solo', = Jolo I, Sulu Is, Philippines.
H. cinereus abbotti Kloss, 1929: 119; Landak R, Pontianak, SW Borneo.

Range (Map 85). Borneo, except the southwest.

Variation. Geographical variation was described and mapped by Marshall & Sugardjito (1986). Three subspecies can be provisionally recognized although there is considerable individual variation and intergradation:

H. m. muelleri. SE Borneo. Crown, venter and feet black, contrasting with greyish brown on rest of pelage.
H. m. funereus. N Borneo (Sabah, N Sarawak and NE Kalimantan). Dark, crown darker than body, feet concolorous with body.
H. m. abbotti. SW Borneo (including S Sarawak). Paler, crown and feet usually not contrasting.

Remarks. This form has been included in *H. lar* (e. g. by Groves, 1972) and in *H. moloch* (by Kloss, 1929). There is limited interbreeding between it and *H. agilis albibarbis* in SW Borneo (Marshall & Sugardjito, 1986). CITES: Appendix I.

Hylobates klossi Kloss gibbon, Beeloh

Symphalangus klossi Miller, 1903a: 70, pl. 17; S Pagi I, Mentawai Is.

Range (Map 85). Mentawai islands of Siberut, Sipura, N Pagai and S Pagai.

Remarks. CITES: Appendix I; IUCN: E.

Subgenus *Bunopithecus*

Hylobates hoolock Hoolock gibbon

Simia golok Bechstein, 1799: 70. Suppressed by Opinion 1219 (1982).
Simia hoolock Harlan, 1831: 52, pl. 2; Garo Hills, Assam. Conserved by Opinion 1219 (1982).
H. fuscus Lewis, 1834: 40, pl. 1; 'vicinity of Himalay Mountains'.
H. choromandus Ogilby, 1838a: 69; locality unknown.
H. scyritus Ogilby, in Royle, 1840:lx; Assam.

Range (Map 86). Assam and Bangladesh (east of the Brahmaputra R), S Yunnan and Burma.

Variation. The following subspecific name has been proposed:
H. h. leuconedys Groves, 1967b: 280; Sumprabum, N Burma, 26°35'N, 97°42'E, 4000ft.
Two subspecies can be recognized:
H. h. hoolock. West of R Chindwin. Black phase with black genital tuft; white brow-lines almost continuous.
H. h. leuconedrys. East of R Chindwin. Black phase with white genital tuft; brow-lines well separated.

Map **86**. *Hylobates* (part), *Pongo*

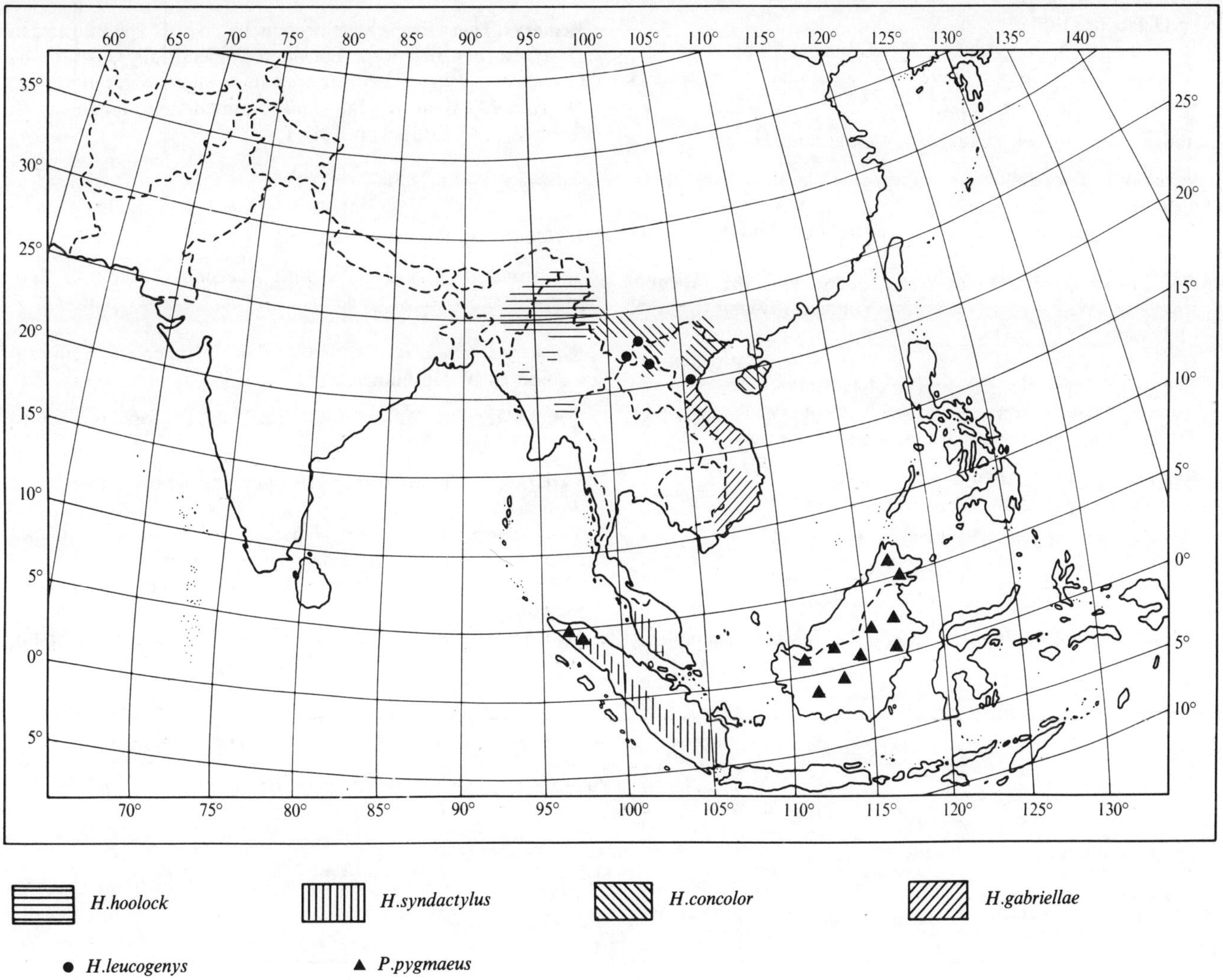

Table 123. The species of *Hylobates* (*Nomascus*).

	concolor	*leucogenys*	*gabriellae*
Range	S China etc.	N Vietnam etc.	Vietnam, Laos
Venter of females	Black or pale	Pale	Pale
Cheeks in males	Dark	Light	Pinkish
Crown patch (female)	Long (10–18cm) pointed behind	Long (>10cm) rounded	Short (<10cm) triangular
Dorsal pelage: length	35–50	40–70	33–45
Pelage: texture	Fine	Coarse	Fine
Baculum	Short (6–9cm) straight	Longer (8–12cm) bent at tip	As *concolor*

Remarks. Included in subgenus *Bunopithecus* by Prouty *et al.* (1983). CITES: Appendix I; IUCN: V.

Subgenus *Nomascus* (Table 123)

Hylobates concolor Black gibbon

Simia concolor Harlan, 1826: 231, pl. 9; 'Borneo'; fixed as Chapa, N Vietnam by Groves (1972: 58).
= *H. harlani* Lesson, 1827b: 111.
= *H. niger* Ogilby, 1840b: 21.
H. nasutus Kunkel d'Herculais, 1884: 86; (incorrectly attributed to A Milne-Edwards); near Along Bay, Tonkin, N Vietnam. Considered to be incorrect and designated as Chapa, Tonkin by Delacour (1951).
H. hainanus Thomas, 1892f: 145; Hainan I, China.
H. henrici Pousargues, 1897: 367; Lai-chau, Tonkin, N Vietnam.

Range (Map 86). N Vietnam, S Yunnan and Hainan I.

Variation. The following subspecific names have been proposed:
H. c. lu Delacour, 1951: 121; Ban Nam-Khueng, Upper Mekong, Laos.
H. c. jingdongensis Ma & Wang, 1986: 399, 409; Wenbu, Jingdong, Wuliang Mts, central Yunnan, 1800m (E of R Mekong).
H. c. furvogaster Ma & Wang, 1986: 402, 410; Menglai, Cangjan, W Yunnan, 2000m (W of R Mekong).

Five subspecies were recognized by Groves & Wang (1990). The principal differences are shown in Table 124.

Remarks. CITES: Appendix I; IUCN: V.

Hylobates leucogenys White-cheeked gibbon

H. leucogenys Ogilby, 1840b: 20; Muang Khi, east of Mekong R, Laos (restricted by Kloss, 1929, corrected by Fooden, 1987).

Range (Map 86). N Vietnam, Laos and S Yunnan (east of the R Mekong). Sympatric with *H. concolor* in parts of N Vietnam and S Yunnan.

Remarks. CITES: Appendix I.

Hylobates gabriellae

H. gabriellae Thomas, 1909c: 112; Langbian, near Nha-trang, inland from Phanrang, S Annam, Vietnam, 1500ft.
H. concolor siki Delacour, 1951: 122; Thua-luu, S of Hue, central Annam, Vietnam.

Range (Map 86). S and C Vietnam, S Laos (E of R Mekong).

Remarks. These forms have been included in *H. concolor* and in *H. leucogenys* but were considered specifically distinct by Groves and Wang (1990). It appears to show closer affinity to *H. concolor* than to the more immediately adjacent *H. leucogenys*. CITES: Appendix I.

Subgenus *Symphalangus*

Hylobates syndactylus Siamang

Simia gibbon Miller, 1779: 170; Bencoolen, Sumatra. Suppressed relative to *S. syndactyla* Raffles, 1821 by Opinion 1224 (1982).
Simia syndactyla Raffles, 1821: 241; Bencoolen, Sumatra. Conserved by Opinion 1224 (1982).

Range (Map 86). Malaya, Sumatra. There are fossil records on Java (Hooijer, 1960).

Variation. The following subspecific names have been proposed:
Symphalangus s. continentis Thomas, 1908e: 301; Semangko Pass, Selangor-Pahang border, Malaya, 3000ft.
[*Siamanga s.*] *volzi* Pohl, 1911: 53; Padang Highlands, Sumatra.
Symphalangus s. subfossilis Hooijer, 1960b: 34; Sibrambang

Table 124. The subspecies of *Hylobates concolor*.

	furvogaster	*jingdongensis*	*concolor*	*nasutus*	*hainanus*	*lu*
Range	Yunnan, W of R Mekong	Yunnan, E of R Mekong	Yunnan - N Vietnam	N Vietnam, N of Red R	Hainan I	W Laos
Body pelage: length	>40	>40	>40		35–40	40–47
Crown tuft of male	c.36	>35	30–35		<30	27
Pelage of female	Dark grey-buff	Gingery buff	Dull buff	Golden brown	Golden brown	
Black on venter of female	+	Slight	+	0	0	+
Light face-ring in female	Narrow, light grey	Broad, light grey	Narrow, white		0	Broad, white

Cave, Padang Highlands, Sumatra; early Holocene. Based upon isolated teeth.

Subspecific separation of the two populations cannot be supported (Groves, 1984). The subfossil teeth from Sumatra are larger, on average, than those of the present population.

Remarks. CITES: Appendix I.

Family Hominidae

Hominidae Gray, 1825: 338; type genus *Homo* Linnaeus.

As used here includes Man along with the four species of great apes (in three genera). Although the precise relationship of *Homo* to the African genera *Gorilla* and *Pan* remains uncertain it seems clear that together these genera form a sister-group of the Indomalayan *Pongo*, reflected here by using two subfamilies following Groves (1989).

Subfamily Ponginae

Simiadae Fleming, 1822: 172; type genus *Simia* L. (suppressed).
Pongidae Elliot, 1913:errata page; type genus *Pongo* Lacépède.

Contains one genus.

PONGO

Simia: Linnaeus, 1760, in part. Suppressed by Opinion 114 (1929).
Ourangus Zimmerman, 1777: 399. Suppressed by Opinion 257 (1954).
Pongo Lacépède, 1799: 4; type species *P. borneo* Lacépède.
= *Lophotus* Fischer, 1813.
Macrobates Billberg, 1828:table A.
Satyrus Lesson, 1840: 40 (*S. rufus*). Pre-occupied by *Satyrus* Meigen, 1803 (Insecta).

Contains one species.

Pongo pygmaeus Orang utan

Simia pygmaeus Linnaeus, 1760: 68; 'Africa' in error, probably Borneo.
Simia satyrus Linnaeus, 1766: 34. Suppressed by Opinion 114 (1929).
Ourangus outangus Zimmerman, 1777: 399; locality unknown.
Pongo borneo Lacépède, 1799: 4; Borneo.
Simias agrias Schreber, 1799:pl. IIC; Borneo.
Pongo wurmbii Tiedemann, 1808: 329; Borneo.
P. abelii Lesson, 1827a: 32; Sumatra.
Simia morio Owen, 1837: 92; Borneo.
Satyrus rufus Lesson, 1840: 40; locality unknown.
Simia gigantica Pearson, 1841: 660; Sumatra.
Pithecus bicolor Geoffroy, 1843a: 526; Sumatra.
Pithecus brookei Blyth, 1853d: 375; Sarawak, Borneo.
P. owenii Blyth, 1853d: 375; Sarawak.
P. curtus Blyth, 1856d: 525; Sarawak.
Satyrus sundaicus Mayer, 1856: 282; locality unknown.
S. knekias Mayer, 1856: 282; Borneo.
Pithecus wallichii Gray, 1871c: 8; Borneo.
= *P. wallacei* Elliot, 1913: 193.
P. tuakensis Selenka, 1896: 385; upper Mrekai R, NW Borneo.
Pithecus sumatranus Selenka, 1896: 386; Sumatra.
Pan vetus Miller, 1915b: 19; locality unknown—the basis of the 'Piltdown Man' forgery (Lowenstein *et al.*, 1982).

Range (Map 86). Borneo and N Sumatra. There are Pleistocene records in Java, N Vietnam and S China.

Variation. The following subspecific names have been proposed:
Pithecus satyrus landakkensis Selenka, 1896: 384; NW Borneo.
P. s. batangtuensis Selenka, 1896: 384; near Batangtu, NW Borneo.
P. s. dadappensis Selenka, 1896: 384; near Dadap Mts, NW Borneo.
P. s. genepaiensis Selenka, 1896: 385; near Genepai Mts, NW Borneo.
P. s. skalauensis Selenka, 1896: 385; W of Dadap Mts, NW Borneo.
P. s. rantaiensis Selenka, 1896: 385; Rantai, NW of Skalau Mts, NW Borneo.
P. sumatranus deliensis Selenka, 1896: 386; NE Sumatra.
= *P. s. langkatensis* Selenka, 1896: 386.
P. sumatranus abongensis Selenka, 1986: 386; near Abongabong Mts, NE Sumatra.
Pongo p. palaeosumatrensis Hooijer, 1948c: 187; Sibrambang Cave, Padang Highlands, C Sumatra; subfossil.
P. p. weidenreichi Hooijer, 1948c: 280; Hoshangtung Cave, Yunnan, China; Pleistocene.

The Sumatran and Bornean populations constitute clearly definable subspecies:
P. p. pygmaeus. Borneo. Smaller; pelage maroon; female without beard; beard of male smaller; muzzle prognathous and 8-shaped in front view; cheek-pads of adult males protruding and almost naked; skull with suborbital fossa.
P. p. abelii. Sumatra. Larger; pelage cinnamon; female with beard; beard of male larger; muzzle less prognathous, O-shaped in front view; cheek-pads of adult males flat and downy; suborbital fossa absent.

Molecular genetic studies have also shown considerable differences between the two subspecies (Janczewski *et al.*, 1990).

Remarks. For a monograph on this species see Schwartz (1989); for a concise account see Groves (1971c). The confused nomenclature was clarified by Groves & Holthius (1985). CITES: Appendix I; IUCN: E.

Subfamily Homininae

HOMO

Homo Linnaeus, 1758: 20; type species *H. sapiens* L.
Pithecanthropus Dubois, 1894 (*Anthropopithecus erectus* Dubois).
Sinanthropus Black, 1927: 21 (*S. pekinensis* Black & Zdansky).
Javanthropus Oppenoorth, 1932: 49 (*H.* (*J.*) *soloensis* Oppenoorth).
Meganthropus Weidenreich, 1944: 480 (*M. palaeojavanicus* Weidenreich).

One Recent species. The following names from the region were allocated to the extinct species *H. erectus* by Groves (1989):
Anthropopithecus erectus Dubois, 1892: 10; Trinil, Java.
H. javanensis primigenius Houze, 1896: 1.
H. pithecanthropus Manouvrier, 1896: 396.
Hylobates giganteus Bumuller, 1899: 124.
Pithecanthropus duboisii Morselli, 1901: 1.
Hylobates gigas Krause, in Bardeleben, 1909.
Homo trinilis Alsberg, 1922: 398.

Sinanthropus pekinensis Black & Zdansky, in Black, 1927: 21; Zhoukoudian, N China.
H. (*Javanthropus*) *soloensis* Oppenoorth, 1932: 49; 10km N of Ngawi, Ngandong, Java.
= *Pithecanthropus ngandongensis* Sartono, 1975: 346.
H. primigenius asiaticus Weidenreich, 1932: 212; Ngandong, Java.
H. modjokertensis Koenigswald, 1936: 1007; Perning, Java.
Meganthropus palaeojavanicus Koenigswald, in Weidenreich, 1944: 460; Sangiran, Java.
Pithecanthropus dubius Koenigswald, in van Bemmelen, 1949: 110; Sangiran, Java.
Sinanthropus officinalis Koenigswald, 1952: 308; source unknown: based upon teeth purchased from a pharmacist in Hong Kong.
Sinanthropus lantianensis Woo, 1965: 1035; Gongwangling, Lantian, Shaanxi, China.
H. (*Sinanthropus*) *erectus yuanmouensis* Hu, 1973: 67; Shangnabang, Yuanmou, Makai Valley, Yunnan.
Pongo brevirostris Krantz, 1975: 369; Sangiran, E Java.
H. erectus trinilensis Sartono, 1976; Java.
H. palaeojavanicus sangiranensis Sartono, 1976; Sangiran, Java.
H. erectus hexianus Huang *et al.*,1982: 251; Hexian, Anhui, China.
= *H. e. hexianensis* Huang, 1982: 109.

The following additional fossil names have been proposed:
H. orientalis Ho, 1988: 362; Zhupeng, Yuanmou, Yunnan, China; Pliocene.
H. habilis zhupengensis Jiang *et al.*,1989: 939; Youanmou, Yunnan, China; Pleistocene.

Homo sapiens Humankind

H. sapiens Linnaeus, 1758: 20.
H. wadjakensis Dubois, 1921: 1019; Wajak, Java; fossil.
H. mincopoeus Pycraft, 1925: 163; Andaman Is.
H. ceylonensis Pycraft, 1925: 163; Sri Lanka.
? *H. sinhaleyus* Deraniyagala, 1957: 7; Radalla, Karangoda, near Kuruvita, Sabaragamura Prov., Sri Lanka; subfossil.
H. erectus mapaensis Kurth, in Heberer, 1965: 383; Maba (= Ma-pa), Guangdong, China; fossil.
H. erectus narmadensis Sonakia, 1984: 159; Hathnora, near Hoshangabad, Narmada Valley, India, 22°52'N, 77°52'E; middle Pleistocene.

Variation. The following subspecific names have been proposed:
H. mincopoeus sumatrensis Pycraft, 1925: 163; Sumatra.
H. s. balangodensis Deraniyagala, 1945b: 140 (footnote) (and 1953b: 125); Ravan-alla cave, Balangoda, Ratnapura, Sabaragamura Prov., S Sri Lanka; subfossil.

Human racial characteristics were reviewed by Baker (1974).

ORDER CARNIVORA

Carnivores

The carnivores constitute a stable and clearly defined order except for uncertainty as to the inclusion or exclusion of the pinnipede families (which are not represented in the Indomalayan Region). Many aspects of carnivore biology were reviewed by contributors to Gittelman (1989), and phylogenies showing the relationships between the families have been published by Wozencraft (1989a) and by Flynn *et al.* (1988). The greatest uncertainty with regard to the delimitation of the families concerns the placement of the pandas, *Ailuropoda* and *Ailurus*. The former is now fairly generally agreed to be best included in the Ursidae; the allocation of *Ailurus* is more uncertain and it is here given separate family rank. Many aspects of classification at lower taxonomic levels were discussed by Wozencraft (1989b). The carnivores of China have recently been reviewed, in Chinese, by Gao *et al.* (1987). Eight families are represented in the Region (Table 125).

Family Canidae

Dogs and foxes

Canini Fischer, 1817: 372; type genus *Canis* L.

A worldwide family with four clearly defined genera represented in the region (Table 126). Most canids are adapted for hunting on relatively open ground and they are therefore poorly represented in the forested areas of SE Asia. Two of the species dealt with here, *Canis lupus* and *Cuon alpinus*, are communal hunters preying mainly upon ungulates; the remainder are solitary hunters preying mainly upon rodents. The family was monographed by Mivart (1890), and classification was reviewed by Clutton-Brock *et al.* (1976) and Van Gelder (1978); many aspects, including classification and chromosome data, were reviewed in Fox (1975).

CANIS Wolves and jackals

Canis Linnaeus, 1758: 38; type species *C. familiaris* L., the domestic dog.

Table 125. The families of Carnivora.

	Canidae	Ursidae	Ailuridae	Mustelidae	Viverridae	Herpestidae	Hyaenidae	Felidae
Genera in region	4	2	1	11	11	1	1	6
Head & body (cm)	50–100	110–200	51–64	12–80	35–95	25–59	c.120	40–200
Tail (% of h&b)	>40%	0	60–70%	>20%	>50%	>50%	c.30%	>30%
Feet digitigrade (D) or plantigrade (P)	D	P	D	P (D)	D/P	P	P	D
Hind digits	4	5	5	5	5	5	4	4
Anal scent glands prominent	0	0	0	+	+ (0)	+	+	0
Post-canine teeth: above	6	6	5	3–5	6	6	5	3–4
below	7(6)	7	6	3–4	5–6	6	4	3
Carnassial teeth (P^4, M_1) trenchant, much larger than adjacent teeth	+	0	0	+	+	+	+	+

Table 126. The genera of Canidae.

	Canis	*Vulpes*	*Nyctereutes*	*Cuon*
Species in region	3	4	1	1
Head & body (cm)	60–100	40–65	50–65	c.90
Black facial mask	0	0	+	0
Pelage reddish	0	+/0	0	+
Muzzle slender	0	+	0	0
Ears pointed (P) or rounded (R)	P	P	R	R
Skull: greatest length	145–240	85–140	110–127	170–200
No. of small lower molars behind large carnassial	2	2	2	1

Vulpicanis Blainville, 1837: 279 (*C. aureus* L.).
= *Sacalius* Smith, 1839: 214.
= *Oxygous* Hodgson, 1841b: 213.

Holarctic and African, with two wild species extending into the northern and western parts of the Indomalayan Region. In addition feral domestic dogs occur through most of the region (Table 127).

The following names have been based upon fossil forms from the region:
Canis curvipalatus Bose, 1880: 134; Punjab; Siwalik beds.
C. cautleyi Bose, 1880: 135; Punjab; Siwalik beds.
C. yuanmoensis You & Qi, 1973: 68; Yuanmou, Yunnan; lower Pleistocene.
C. pinjorensis Gaur, 1987b: 58; NW India; Pliocene/Pleistocene.

Canis lupus Wolf

C. lupus Linnaeus, 1758: 38; Sweden.
C. pallipes Sykes, 1831: 101; Deccan, India.
Many other synonyms from the Palaearctic and Nearctic regions.

Range (Map 87). Pakistan and India, east to Bihar and Orissa; throughout most of China; also throughout most of the Palaearctic and Nearctic regions. It is now very sparsely distributed in most of its range in India and Pakistan, being confined to uncultivated and sparsely inhabited country, e. g. in the desert region of Rajasthan (Shahi, 1982).

Variation. The wolves of the Indian plains and the Deccan, often designated *C. l. pallipes*, are a little smaller and have a shorter winter pelage compared with those at higher altitudes in the Himalayas and China.

The following fossil subspecies has been described:
C. l. variabilis Pei, 1934: 13; Choukoutien, N China. Recorded also at Nanzhao, Henan (Han & Xu, 1985).

Map **87**. *Canis, Nyctereutes*

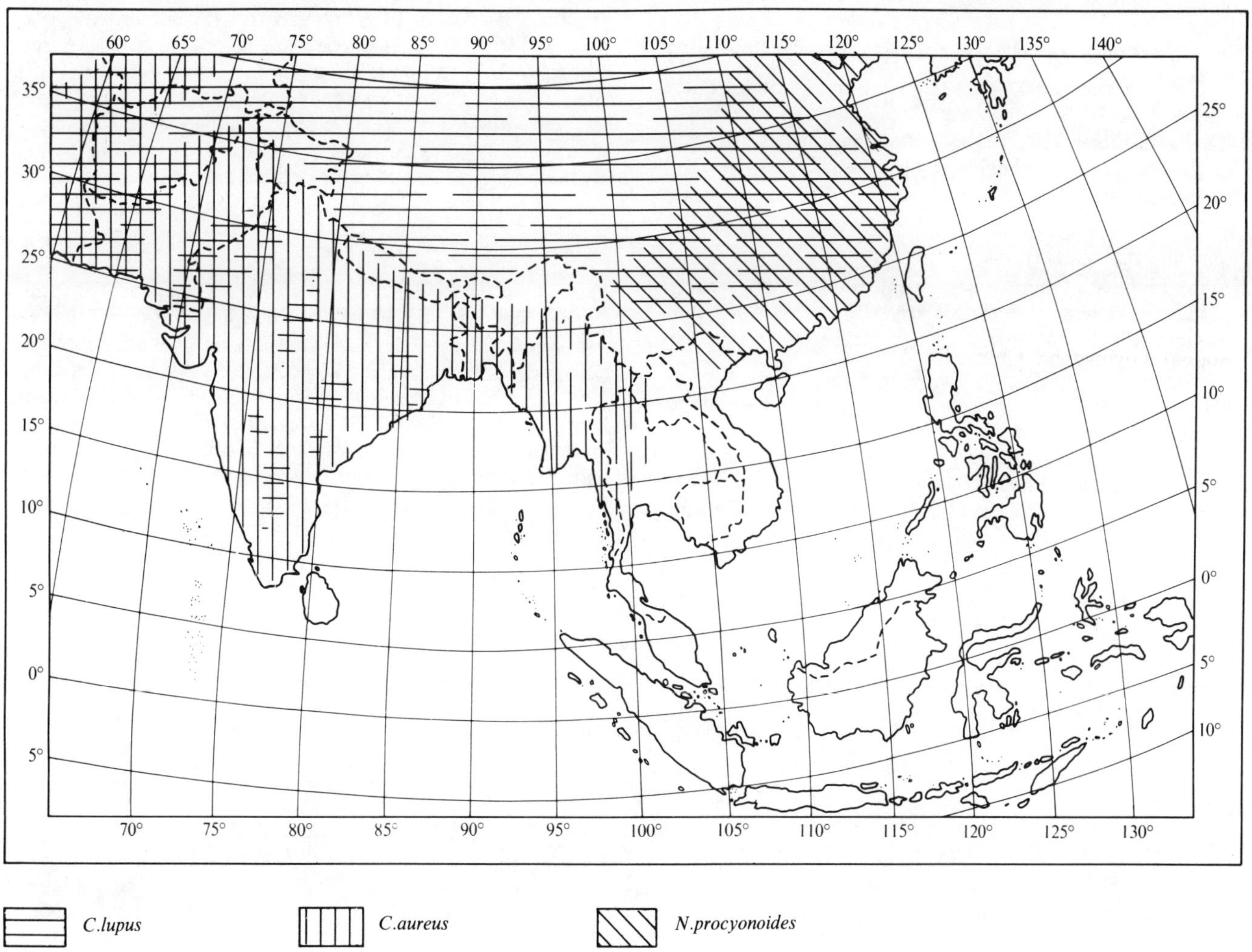

Remarks. Reviewed by Mech (1974). CITES: India, Pakistan, Nepal, Bhutan, Appendix I; China, Appendix II. IUCN: V.

Canis aureus Golden jackal

C. aureus Linnaeus, 1758: 40; Lar Prov., Iran.
? *C. cruesemanni* Matschie, 1900: 145; Menam, Thailand (based on captive specimens).
C. naria Wroughton, 1916c: 651; Virajpet, S Coorg, India, 3000ft.
C. lanka Wroughton, 1916c: 652; Mankeni, East Prov., Sri Lanka.

Range (Map 87). From Thailand and Burma through the whole of India and Pakistan; Sri Lanka; also west through SW Asia to SE Europe, and Africa south to Senegal and Tanzania.

Variation. The following subspecific names have been proposed from the region:
C. a. indicus Hodgson, 1833b: 237; Nepal.
C. indicus kola Wroughton, 1916c: 651; Palanpur, Gujarat, NW India.

Ecology. A very versatile species, occurring mainly in open country and frequently around towns and villages. They prey extensively upon rodents but also feed on carrion.

Remarks. Capable of producing fertile hybrids with the domestic dog (Gray, 1972), but the incidence of such hybridization outside captivity is uncertain.

Canis familiaris Domestic dog, Pariah dog, Pye dog, Dingo

C. familiaris Linnaeus, 1758: 38. Based upon domestic dogs with recurved tails.
C. antarticus Kerr, 1792: 136, pl.; New South Wales, Australia. Suppressed except for homonymy by Opinion 451 (1957).
C. dingo Meyer, 1793: 33; New South Wales.

Range. Feral dogs occur virtually throughout the region, including the Sunda, Philippine and Molucca islands, mainly in the vicinity of human settlement, although they are so frequently ignored in zoological surveys and zoological literature that it is difficult to assess the extent to which they exist independently of Man.

Variation. The following subspecific names have been proposed from the region:
C. f. var. *tenggerana* Kohlbrugge, 1896b: 283; Mt Tengger, E Java. See also Jentink (1897).
C. tenggeranus harappensis Prashad, 1936: 22; Harappa, Punjab, Pakistan; 'prehistoric times'.

Table 127. The species of *Canis*.

	lupus	*aureus*	*familiaris**
Head & body (cm)	75–100	60–75	60–75
Height at shoulder (cm)	40–70	35–43	30–40
Tail (relaxed)	Pendant	Pendant	Usually up-curled
Skull: greatest length	200–240	145–165	c.190
Profile of skull between frontal and nasals	Concave	Straight	Concave
Upper molars with well defined cingulum	0	+	0
Upper carnassial: length	22–28	15–18	18–20

* Characters of those feral dogs most like the wild species.

In spite of the enormous variability amongst domestic breeds of dogs, most feral dogs in the region conform rather closely to a single form, characterized by upright, pointed ears, moderately up-curled tail, size about that of *C. aureus* or rather smaller, and most often a uniformly yellowish brown pelage although many other colour varieties occur. Those in Thailand, and probably elsewhere in the region, are scarcely distinguishable from Australian dingos although they are a little smaller on average and are more synanthropic than in Australia (Corbett, 1985). Those in the interior of Malaya and Borneo are small and uniformly yellow-brown, in contrast to the urban and village pye (or pariah) dogs which tend to be larger and much more variable (Medway, 1977b).

Ecology. There is every intergradation between dogs that are kept as fully domestic pets or working dogs, and those that are truly feral. Most of the latter appear to subsist by scavenging around human habitation, taking human faeces besides other food, but the extent to which they hunt live prey such as rodents is probably underestimated. A road-side survey in Bangladesh found them only near human habitation; in contrast to domestic dogs the feral dogs there came into oestrous only once a year, mostly in September/October (Oppenheimer & Oppenheimer, 1975).

Remarks. The feral dogs of the region are likely to be an anciently domesticated form directly derived from the small Indian form of the wolf, *C. lupus*. Very little information appears to be available on their relationship with the few remaining wolves in the area or with stray domestic dogs of more modern breeds. Although the domestic dog has arisen primarily from the wolf, *C. lupus*, and is interfertile with it, there are good grounds for treating and naming it as a separate, discrete entity. This reflects the actual situation, since hybrids are rare, and avoids the nomenclatural confusion and ambiguity that would result if they were treated as conspecific (Corbet & Clutton-Brock, 1984).

VULPES Foxes

Vulpes Frisch, 1775: 15; type species *Canis vulpes* L. Although Frisch (1775) has been placed on the *Official Index of Rejected and Invalid Works* by the International Commission on Zoological Nomenclature, *Vulpes* Frisch, 1775 was placed on the *Official List of Generic Names* by Opinion 1129 (1979) and thereby made available. (This prevents *Vulpes* from having to be replaced by *Fennecus* if *F. zerda* is considered congeneric with *Vulpes* spp.).

Holarctic and African; one species is Indomalayan and three others extend marginally into the region in the north and northwest (Table 128).

Vulpes vulpes Red fox

Canis vulpes Linnaeus, 1758: 40; Sweden.
C. himalaicus Ogilby, 1837a: 103; Mussooree, Kumaon, NW India.
V. nepalensis Gray, 1837a: 578; Nepal.
V. pusillus Blyth, 1854: 729; Salt Range, Punjab, Pakistan.
V. leucopus Blyth, 1854: 729; Multan, Punjab, Pakistan.
V. griffithi Blyth, 1854: 730; Kandahar, Afghanistan.
V. hoole Swinhoe, 1871: 631; near Amoy, Fujian, S China.
V. lineiventer Swinhoe, 1871: 632; near Amoy, Fujian, S China.
V. aurantioluteus Matschie, 1907: 168; Tatsienlu, Sichuan, China.

Map **88**. *Vulpes, Cuon*

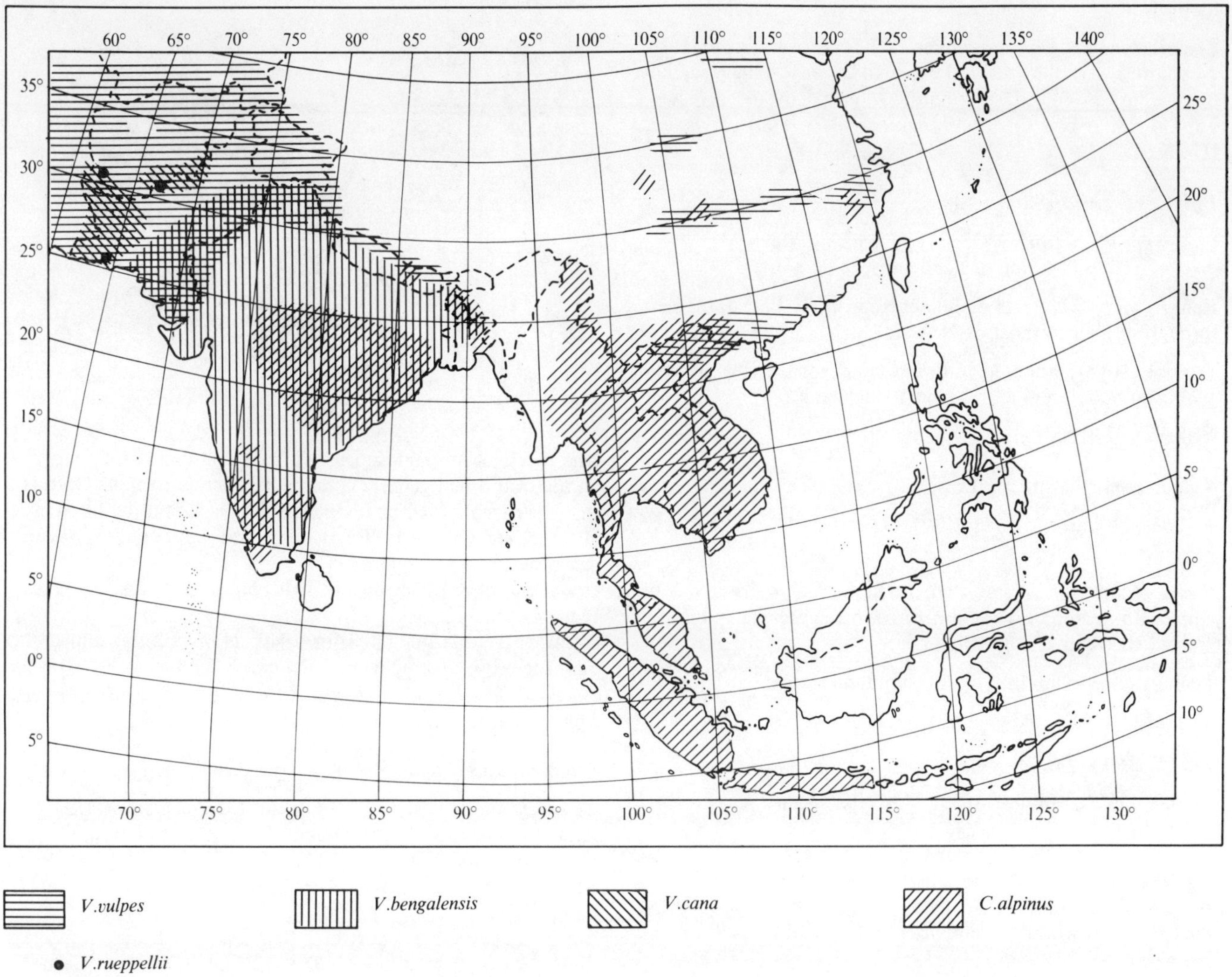

V. ferrilatus eckloni Jacobi, 1923: 6; Bamutang, 3 days SW of Batang, Sichuan, China. Pre-occupied by *Canis eckloni* Przewalski, 1884, a form of *V. corsac*.
Also many synonyms from other parts of the range.

Range (Map 88). N Vietnam north through parts of China; the Himalayas, at altitudes of 1000–4000m, to Pakistan; widespread throughout the less arid parts of the Palaearctic and Nearctic regions.

Table 128. The species of *Vulpes*.

	vulpes	*bengalensis*	*rueppellii*	*cana*
Range in region	N India-S China	India etc	Pakistan	Pakistan
Head & body (cm)	49–65	45–60	40–50	c.42
Hind feet (cm)	11–15	11–12	c.10	8–11
Tail-tip	White	Black	White	Black, ± white
Back of ears black	+	0	0	0
Pads of feet hairy	0	0	+	0
Skull: greatest length	115–145	105–120	95–108	<90

Variation. One subspecific name has been proposed from the region:
Canis v. montana Pearson, 1836: 313; 'lower range of the Himalayan mountains'.

In the Himalayas, as elsewhere in the range, there is considerable individual variation in coat colour, with red, yellow-brown, blackish and hoary forms. The colour becomes lighter towards the more arid western Himalayas, and those in the lowland arid zone of NE India and Pakistan are small, more consistently yellow-brown, and often with the feet and lower legs white.

Vulpes bengalensis Bengal fox

Canis bengalensis Shaw, 1800: 330; Bengal.
C. kokree Sykes, 1831: 101; Deccan, India.
C. (*V.*) *indicus* Hodgson, 1833b: 237; India.
C. (*V.*) *rufescens* Gray, 1834(in 1830–35):pl. 3; India.
C. chrysurus Gray, 1837a: 577; Nepal.
V. hodgsonii Gray, 1837a: 578; Nepal.
V. xanthura Gray, 1838c: 68; Nepal.

Range (Map 88). India northeast to West Bengal, Bangladesh and the Cachar district of Assam (Kock & Posamentier, 1984), and west through Nepal (Johnson *et al.*, 1980) to the Indus valley in Pakistan. Up to at least 1400m in the Himalayan foothills.

Remarks. Although very similar to *V. vulpes* and apparently vicariant with it, the great difference in karyotype suggests that there is no case for considering them conspecific: *V. vulpes* — 2n=34–38, FN=72; *V. bengalensis*—2n=60; FN=72. IUCN: I.

Vulpes rueppellii Sand fox

Canis rüppellii Schinz, 1825: 508; Dongola, Sudan.
Also extralimital synonyms and subspecific names.

Range (Map 88). Parts of Baluchistan in W Pakistan; west through Arabia to most of N Africa.

Ecology. Highly adapted to desert conditions, living in very arid sand-dune areas in lowland Baluchistan.

Remarks. IUCN: K.

Vulpes cana Blanford's fox

V. canus Blanford, 1877a: 321; Gwadar, Baluchistan, W Pakistan.

Range (Map 88). W Pakistan, perhaps also in Kutch, NW India (Ranjitsinh, 1985); also through Afghanistan and NE Iran to the Kopet Dag Mts.

Ecology. Occurs mainly in arid montane steppe.

Remarks. CITES: Appendix II; IUCN: K.

NYCTEREUTES

Nyctereutes Temminck, 1839: 285; type species *N. viverrinus* Temminck, in *N. procyonoides*.

One species. One of the most distinctive genera of canids, with no very close affinity with any other. Unique features are the dark facial mask and the absence of tail-wagging in submissive animals.

Nyctereutes procyonoides Raccoon-dog

Canis procyonoides Gray, 1834(in 1830–35):pl. 1; vicinity of Canton, S China.
N. sinensis Brass, 1904: 22; Yangtze Valley, China.
N. stegmanni Matschie, 1907: 175,180; Hsing-an-fu, Chinkiang, Jiangsu, SE China.
And other synonyms from further north.

Range (Map 87). N Vietnam and throughout most of China; north to the Amur Basin and in Japan. Introduced in much of USSR and E Europe.

Variation. One subspecific name has been proposed from the region:
N. p. orestes Thomas, 1923c: 657; NW flank of Likiang Range, Yunnan, China, c. 10-12000ft.

Remarks. Reviewed by Ward & Wurster-Hill (1990).

CUON

Cuon Hodgson, 1838: 152; type species *Canis primaevus* Hodgson. Conserved by Opinion 384 (1956).
= *Cyon* Agassiz, 1848: 326 (unjustified emendation).
Chrysaeus Smith, 1839: 167 (*Canis dukhunensis* Sykes).
Anurocyon Heude, 1892: 102 (*A. clamitans* Heude).

Cuon alpinus

One species. Although separated from *Canis* and other genera on the traditional generic character of dental formula (in this case absence of M_3), its distinctiveness is supported by many other characters. The following fossil species have been described from the region:
C. sangiranensis Koenigswald, 1940: 58, pl. 3; Sangiran, Java; Pleistocene.
C. dubius Teilhard de Chardin, 1940: 11; N China; Pliocene. Recorded also at Liucheng, Guangxi, S China.
C. simplicidens Young & Liu, 1951: 65; Koloshan, near Chungking, Sichuan, China.

Cuon alpinus Dhole, Red dog, Indian wild dog

Canis alpinus Pallas, 1811: 34; near Udskoi Ostrog, Uda R, Amur region, USSR. Conserved by Opinion 384 (1956).
Canis javanicus Desmarest, 1820: 198; Java.
Canis familiaris var. *sumatrensis* Hardwicke, 1821: 235, pl. 23; Sumatra.
Canis dukhunensis Sykes, 1831: 100; Deccan, India.
Canis primaevus Hodgson, 1833b: 221; Nepal.
Canis rutilans Müller, 1840: 27,51; Java.
Cuon grayiformis Hodgson, in Gray, 1863b: 5; Sikkim.
Cuon lepturus Heude, 1892: 102(footnote); Poyang Lake, south of the Yangtze, Jiangxi, China.
Anurocyon clamitans Heude, 1892: 102(footnote); Taihu, near mouth of the Yangtze, China.
Cyon antiquus Matthew & Granger, 1923: 584; Yen-ching-kao, near Wan-hsien, Sichuan, China; Pleistocene. Made the type of *Crassidens* Kretzoi, 1941, but included in *C. javanicus* (*C. alpinus*) by Colbert & Hooijer (1953).
C. crassidens Koenigswald, 1933: 22; Pandejan, Java; Late Pleistocene. Status *fide* Schütt (1973).
Also other synonyms from the northern part of the range.

Range (Map 88). Throughout most of the forested parts of India and S China south to Malaya, Sumatra and Java; also north to Korea, the Amur Basin and west to the Altai and Pamir Mts. Recorded as fossils in Sri Lanka (Deraniyagala, 1958b), and probably Borneo (Cranbrook, 1988). The recent status and range in India and Burma were reviewed and mapped by Johnsingh (1985).

Variation. The following subspecific names have been proposed from the region:
C. javanicus infuscus Pocock, 1936a: 38; Moulmein, Tenasserim, Burma.

C. a. adustus Pocock, 1941: 156; Upper Burma.
C. javanicus sinhaleyus Deraniyagala, 1958b: 84; Ihala, Minuvandeniya, Gonapitaya, Kuruvita, Sri Lanka; Pleistocene.

Variation on the mainland is likely to be clinal and follows the common pattern of increase in length of the pelage with altitude, and darker and brighter colour in the high rainfall areas of SE Asia.

Ecology. A communal hunter preying mainly upon ungulates in open forested country.

Remarks. Subfossil remains from a midden dated about 10 000 BP have been held to indicate its former presence in Borneo although this was based upon the absence of *Canis* remains of this age throughout SE Asia rather than upon identification by morphology (Cranbrook, 1988). For a detailed taxonomic monograph see Pocock (1936a); and a concise general review, Cohen (1978). CITES: Appendix II; IUCN: V.

Family Ursidae

Bears

Ursini Fischer, 1817: 372; type genus *Ursus* L.

Americas, Palaearctic and Indomalayan regions. The bears constitute a clearly defined family except for the question as to whether the genera *Ailuropoda* and *Ailurus* should or should not be included. There is an increasing consensus that *Ailuropoda* should be included, as is done here; *Ailurus* is probably a more distantly related member of the same group and is here placed in a separate family, Ailuridae.

The true bears form a group of rather closely related species (Table 129) for which the generic classification has been very unstable. A variable number of species have been included in the genus *Ursus* and at one extreme each species has been allocated to a separate genus. It has been suggested (Hoffmann in Wozencraft, 1989b) that *malayanus* is closely related to *thibetanus* and *ursinus* to *arctos*.

It seems best to include all species in *Ursus* except the South American *Tremarctos ornatus* which stands apart from the others in molecular and karyological studies (Nash & O'Brien, 1987; Wayne *et al.*, 1989; Goldman *et al.*, 1989). This is reinforced by the production of viable hybrids in captivity, e. g. between *U. malayanus* and *U. ursinus* (Gray, 1972; Asakura, 1987).

For a popular account of all species, with colour photographs, see Domico (1988).

URSUS Bears

Ursus Linnaeus, 1758: 47; type species *U. arctos* L. Conserved by Opinion 75 (1922).
Melursus Meyer, 1793: 155 (*M. lybius* Meyer, in *Bradypus ursinus* Shaw). Conserved by Opinion 384 (1956).
= *Arceus* Goldfüss, 1809: 301.
= *Prochilus* Illiger, 1811: 109.
= *Chondrorhynchus* Fischer, 1814: 142.
Helarctos Horsfield, 1825b: 221,233 (*H. euryspilus* Horsfield, in *U. malayanus* Raffles). Conserved by Opinion 384 (1956).
Selenarctos Heude, 1901a: 2 (*U. thibetanus* Cuvier).
= *Arcticonus* Pocock, 1917a: 129.
And other synonyms based upon extralimital species.

Holarctic and Indomalayan regions; 6 species. Three species occur extensively in the region; a fourth, *U. arctos*, occurs on the northern boundary, in the Himalayas and SW China, mainly at high altitude. The following nominal fossil species have been described from the region:
U. sivalensis Cautley & Falconer, 1836: 193; Siwalik Hills.
U. namadicus Falconer, 1868: 552, pl. 26; Narbada, India.
U. theobaldi Lydekker, 1884a: 211(=34), pl. 28; Kangra Dist., W India; Pliocene.
U. premalayanus Koenigswald, 1935: 876; ? S China (based on teeth in apothecary shops).
U. (Helarctos) sino-malayanus Thenius, 1948: 12; China; Pliocene.
U. ningzhenensis Liu *et al.*,1982: 4, 25; Chao Xian, Anhui, China.

Ursus thibetanus Asiatic black bear

U. thibetanus Cuvier, G.,1823: 325; Sylhet, Assam, India.
= *U. torquatus* Wagner, 1841a: 144.
U. formosanus Swinhoe, 1864b: 380; Taiwan.
U. gedrosianus Blanford, 1877c: 204; Tump, 70 miles N of Gwadar, Mekran coast, Baluchistan, Pakistan. See also Blanford (1877a).
Selenarctos mupinensis Heude, 1901: 2, pl. 2; Moupin, Sichuan, China.
Selenarctos melli Matschie, in Mell, 1922: 34; Hainan I, China.
U. kokeni Matthew & Granger, 1923: 580; Yen-ching-kao, near Wan-Hsien, Sichuan, China; Pleistocene. (Listed in *U. thibetanus* by Han & Xu, 1985.)

Range (Map 89). From Afghanistan and Pakistan east along the Himalayas (at altitudes of 1000–4000m) to Assam, Burma, N Thailand, Indochina and S China; Hainan and Taiwan; also north to the Amur Basin and Japan.

Table 129. The species of *Ursus* and *Ailuropoda* (Ursidae).

	U.thibetanus	*U.malayanus*	*U.ursinus*	*U.arctos*	*A.melanoleuca*
Range in region	Himalayas-S China	Indochina, Sunda	India etc	N fringe	SW China
Body mainly white	0	0	0	0	+
Ears large and tufted	+	0	+	0	+
White crescent on chest	+	+	+	0	0
Ear pinnae tubular at base	+	0	+	+	0
Length of pelage (mid-dorsal)	40–85	13–35	50–120	100–120	30–50
Fore feet: posterior plantar pads as broad as anterior (B) or very small (S)	B	B	S	S	S
Bullae swollen	0	+	0	0	0
Upper incisors	3:3	3:3	2:2	3:3	3:3

Variation. The following subspecific names have been proposed from the fringes of the region:
U. torquatus var. *arboreus* Gray, 1864b: 688; Darjeeling, India.
U. torquatus macneilli Lydekker, 1909a: 609; Tatsienlu, Sichuan, China.
Selenarctos t. laniger Pocock, 1932a: 115; Aru, upper Lidder Valley, Kashmir.

Those in Baluchistan are apparently smaller than those further east—Roberts (1977) reports an adult male with head & body 167cm compared with about 200cm in Kashmir—and brownish rather than black.

Remarks. CITES: Appendix I; IUCN: V (E).

Ursus malayanus Sun bear

U. malayanus Raffles, 1821: 254; Bencoolen, Sumatra. Conserved by Opinion 384 (1956).
Helarctos euryspilus Horsfield, 1825b: 221; Borneo.
H. annamiticus Heude, 1901: 1, pl. 1; Annam, Vietnam.

Range (Map 89). SW China (W Sichuan, S Yunnan), Assam SE of the Brahmaputra R (Kurup, 1968), Burma and Vietnam to Malaya; Sumatra, Borneo. Recorded up to 2300m in Borneo.

Ursus malayanus

Variation. Bornean animals are small, with head and body up to c. 126cm, compared with up to 140cm in Malaya, but samples are small and it is doubtful if subspecific separation is justified.

Remarks. CITES: Appendix I; IUCN: V.

Ursus ursinus Sloth bear

Bradypus ursinus Shaw & Nodder, 1791:pls 58–59; Patna, Ganges, India.
= *U. labiatus* Blainville, 1817: 74.

Map **89**. *Ursus, Ailuropoda, Ailurus*

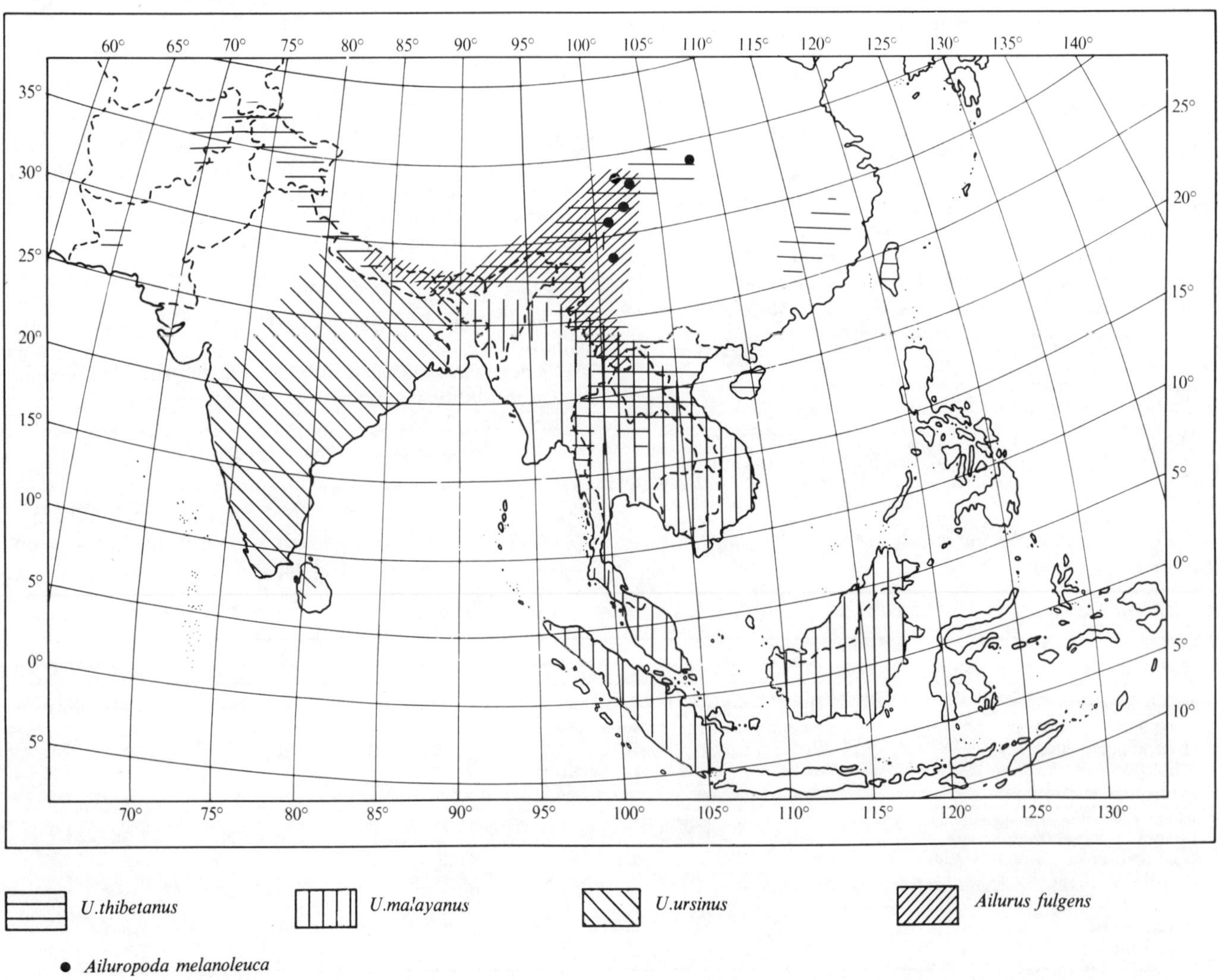

Melursus lybius Meyer, 1793: 156; 'Africa interior'.
Arceus niger Goldfüss, 1809: 301.
U. longirostris Tiedmann, 1820: 11.
M. inornatus Pucheran, 1855: 392; Sri Lanka.

Range (Map 89). India, north to the Himalayan foothills, lowland Nepal, east to Assam (NW of the Brahmaputra R); and Sri Lanka. There has been a massive reduction in range in recent decades but there is little information on present status (Cowan, 1972), except in Sri Lanka where it is now confined to the lowland dry zone (Santiapillai & Santiapillai, 1990).

Variation. One subspecific name has been proposed:
Melursus u. valaha Deraniyagala, 1949: 103; Sri Lanka.
Two subspecies can be recognized (Pocock, 1941):
U. u. ursinus. India. Larger: greatest length of skull (male) 300-345mm, mean 322 (n=6).
U. u. inornatus. Sri Lanka. Smaller: greatest length of skull (male) 277–282mm, mean 280 (n=4).

Ecology. In Nepal they were found to feed about equally on fruit and on insects, especially termites (Laurie & Seidensticker, 1977).

Remarks. Deraniyagala (1949) proposed the name *Melursus u. valaha* for the form in Sri Lanka in the belief that the holotype of *M. inornatus* (the only previous name from Sri Lanka) was an extinct form of *U. malayanus*, having six upper incisors. This was based solely upon the inclusion by Gray (1864b) of *inornatus* as a species of *Ursus* rather than of *Melursus*, but there is no other indication either by Gray or in the original description of *inornatus* that the holotype of *inornatus*, a young animal, had six upper incisors. CITES: Appendix III (India); IUCN: V.

AILUROPODA

Ailuropoda Milne-Edwards, 1870b: 1; type species *Ursus melanoleucus* David.
= *Pandarctos* Gervais, 1870: 161.
= *Ailuropus* Milne-Edwards, in David, 1872: 92.
Aelureidopus Woodward, 1915: 428 (*Ael. baconi* Woodward).

One species. The allocation of *Ailuropoda* to the Ursidae, rather than to the Procyonidae in which it has frequently been placed on the basis of dentition, e. g. by Simpson (1945), was made forcibly in the detailed anatomical monograph of Davis (1964). This has been reinforced by more recent biochemical and karyological studies (e. g. O'Brien *et al.*, 1985; Mayr, 1986).

The following fossil species has been described:
A. microta Pei; *Gigantopithecus* Cave, Liucheng, Guangxi, China; Early Pleistocene.

Ailuropoda melanoleuca Giant panda, Bamboo bear

Ursus melanoleucus David, 1869: 13; Moupin, Sichuan, China.
Aeluropus fovealis Matthew & Granger, 1923: 579; Yen-ching-Kao, near Wan-Hsien, Sichuan; Middle Pleistocene.
Aelureidopus baconi Woodward, 1915: 428; Mogok, N Burma; Pleistocene.

Range (Map 89). Sichuan, extreme S Gansu and S Shaanxi, China. The population is divided in about nine separated segments, each with reserve status, between the Liang Mts in S Sichuan and the Qinling Mts in S Shaanxi. The total population in the early 1980s was estimated at 555–650 animals (Schaller *et al.*, 1985). There are Pleistocene records from throughout most of central and E China from Yunnan to Beijing, and in N Burma (mapped by Schaller *et al.*, 1985: 12).

Variation. The two extinct, nominal species detailed in the synonymy above have been treated as subspecies of *A. melanoleuca* (Colbert & Hooijer, 1953; Wang, 1974).

Ecology. Confined to montane bamboo forest and entirely dependant upon a few species of bamboo for food.

Remarks. For a detailed account see Schaller *et al.* (1985); briefly reviewed by Chorn & Hoffmann (1978); a detailed, but more popular, account by Catton (1990). CITES: Appendix I; IUCN: E.

Family Ailuridae

Red panda

Ailurina Gray, 1843a:xxxi; type genus *Ailurus* Cuvier. Conserved by Direction 53 (1956).

One genus and species. *Ailurus fulgens* was placed in the Procyonidae by Simpson (1945) mainly on the basis of dentition. It is generally agreed to have arisen close to the dichotomy between the Procyonidae and the Ursidae and there is no consensus on its precise position in relation to these families. However most evidence points to it being the sister-group of the Ursidae as understood here. It has been included in the Ursidae as the sole member of a subfamily Ailurinae (Wozencraft, 1989a) but was left '*incertae sedis*' close to the Ursidae by Wozencraft (1989b).

AILURUS

Ailurus Cuvier, in Geoffroy & Cuvier, 1825:liv. 50; type species *A. fulgens* Cuvier. Conserved by Opinion 384 (1956).
= *Arctaelurus* Gloger, 1841: 28.

One species.

Ailurus fulgens Red panda, Lesser panda, Cat-bear

A. fulgens Cuvier, in Geoffroy & Cuvier, 1825:liv. 50; 'East Indies'. Conserved by Opinion 384 (1956).
= *A. refulgens* Milne-Edwards, 1874: 380.
A. ochraceus Hodgson, 1847c: 1118; 'Sub Himalayas', 7000–13000ft.

Range (Map 89). E Himalayas from Nepal (from c. 82°E) through N Burma and Yunnan to Sichuan; at altitudes of 2200–4800m.

Variation. The following subspecific name has been proposed:
A. f. styani Thomas, 1902d: 251; Yangliupa, NW Sichuan, China.
Western specimens are smaller and lighter in colour than those in the east.

Remarks. Reviewed by Roberts & Gittelman (1984), Glatson (1989) and Catton (1990). CITES: Appendix II; IUCN: K.

Family Mustelidae

Weasels etc.

Mustelini Fischer, 1817: 372; type genus *Mustela* L.

The Mustelidae is a discrete family of small to medium sized carnivores. They are very diverse in appearance and most fall clearly into four subfamilies of which three are represented in the Indomalayan Region (Table 130). The most anomalous genus, *Mellivora*, has often been separated in its own subfamily Mellivorinae but is here included in the Mustelinae following most recent authors. Many species are heavily exploited for fur, e. g. in China (Tan, 1989). Accounts, with maps, of the more endangered species were prepared by Schreiber *et al.* (1989).

Table 130. The subfamilies of Mustelidae.

	Mustelinae	Melinae	Lutrinae
Tail thick and muscular	0	0	+
Claws short, partly retractile	+	0	0
P^4 (carnassial) with inner edge reduced	+	0	0
P^4 with broad posterior cingulum	0	+	0
M^1 transversely elongate	+	Slight or 0	0
Bullae inflated	+	+	0

Table 131. The genera of Mustelinae.

	Mustela	*Vormela*	*Martes*	*Mellivora*
Species in region	6	1	2	1
Head & body	120–360	320–360	300–640	640–840
Tail (% of h & b)	20–70%	50–60%	55–80%	20–30%
Hind feet	22–54	c.43	80–90	115–125
Pelage mottled	0	+	0	0
Venter and legs black	0	+	0	+
Skin abnormally thick	0	0	0	+
Condylobasal length	30–70	50–55	75–105	120–140
Premolars	3/3	3/3	4/4	3/3

Subfamily Mustelinae

Weasels and martens

Almost worldwide; 11 genera of which 4 are represented in the region, one very marginally (Table 131). They are long-bodied, short-legged and predominantly terrestrial in habits.

MUSTELA Weasels

Mustela Linnaeus, 1758: 45; type species *M. erminea* L.
Putorius Cuvier, G.,1817: 147 (*M. putorius* L.).
Arctogale Kaup, 1829: 30 (*M. erminea* L.).
Lutreola Wagner, 1841a: 239 (*Viverra lutreola* L.).
Gymnopus Gray, 1865c: 118 (*Mustela leucocephalus* Gray, in *M. nudipes* Desmarest). Pre-occupied by *Gymnopus* Brookes, 1828 (Aves).
= *Plesiogale* Pocock, 1921: 805. Pre-occupied by *Plesiogale* Pomel, 1847, an extinct mustelid.
= *Pocockictis* Kretzoi, 1947: 285.
Kolonokus Satunin, 1911: 264 (*M. sibirica* Pallas).
And other synonyms based upon extralimital species.

Holarctic and Neotropical, extending marginally into the Indomalayan Region; 17 species of which four are wholly Indomalayan and two occur on the northern fringe of the region (Table 132). The following additional Palaearctic species reach the northern border of the region and extend a little below 3000m in places: *M. erminea* (W Himalayas); *M. altaica* (Himalayas, W China); *M. eversmanni* (Kashmir, Tibet, Sichuan, S Shaanxi). Males are substantially larger than females.

The following nominal fossil species has been described from the region:
M. lydekkeri Colbert, 1933b; Chinji, N Punjab, Pakistan; Lower Siwaliks.

Mustela nivalis Weasel

M. nivalis Linnaeus, 1766: 69; Vesterbotten Prov., Sweden.
M. russelliana Thomas, 1911a: 4 (and 1911b: 168); Tatsienlu, Sichuan, China, 12000ft.
M. tonkinensis Björkegren, 1942: 1; Chapa, Tonkin, N Vietnam.
Also many extralimital synonyms.

Table 132. The species of *Mustela*.

	nivalis	*kathiah*	*sibirica*	*strigidorsa*	*nudipes*	*lutreolina*
Range in region	S China, N Vietnam	Himalayas-S China	Himalayas, China	Himalayas, Indochina	Malaya, Sumatra, Borneo	Java
Head & body: male	140–200	230–290	310–380		310–320	300–320
female	120–140	200–230	260–290	c.290		
Tail (% of h & b)	20–45%	50–70%	c.55%	c.60%	c.67%	45–55%
Soles hairy	+	+/0	+	0	0	+
Dorsum	Dark brown	Dark brown	Light brown	Dark brown	Golden brown	Dark brown
Venter	White or yellowish	Light orange-brown	As dorsum	Throat and thin mid-ventral streak white or yellow	As dorsum	As dorsum, some white on throat
Pale dorsal stripe	0	0	0	+	+ (anterior only)	0
Front of head white	0	0	0 (or in front of eyes)	0	+ (to behind ears)	0
Condylobasal length	29–37	42–55	51–68	57–65	50–57	51–60
Bulla with foramen at centre of medial edge	+	+	+	+	0	+
Tip of baculum bifid	0	0	0	+	0	0

Range (Map 90). N Vietnam, and at high altitude in Sichuan, SW China; and throughout most of the Palaearctic and Nearctic regions; approaching this region in Afghanistan but apparently absent from the high Himalayas.

Variation. The apparently isolated form in Sichuan is distinctive and may warrant specific rank as *M. russelliana*. It is characterized by very small size (head and body of adult females 128 and 133mm), pale apricot-coloured venter, and rather long tail (c. 40% of head and body). The form in Vietnam (*tonkinensis*) appears to be known from only a single adult male. It is larger (head and body 200mm) with a long tail (45% of head and body) and the venter only slightly tinged yellow.

Mustela kathiah Yellow-bellied weasel

M. (*Putorices*) *kathiah* Hodgson, 1835b: 702; Kachar, N Nepal.
M. (*P.*) *auriventer* Hodgson, 1837b: 563 (*nomen nudum*).
Putorius dorsalis Trouessart, 1895: 235; Tatsienlu, Sichuan, China.
Arctogale tsaidamensis Hilzheimer, 1910: 309; Tsaidam Mts, W Kokonor, Qinghai, China.
A. melli Matschie, in Mell, 1922: 17; Guangdong, S China.

Range (Map 90). Himalayas from c. 78°E in Uttar Pradesh east to Assam, N Burma, N Vietnam and parts of S and SW China; also on Hainan and Taiwan according to the map in Gao *et al.* (1987) but this species is not included in earlier sources on Taiwan. Mainly between 1000 and 2000m in Nepal.

Remarks. Closely related to *M. altaica* which is parapatric with *M. kathiah*, occurring at higher altitudes. In Nepal reported to have been kept and trained to kill house rats and small game (Hodgson, 1835b).

Mustela sibirica Siberian weasel, Kolinsky

M. sibirica Pallas, 1773: 701; Vorposten Tigerazkoi, near Ust-Kamenogorsk, W Altai Mts, NE Kakakhstan, USSR.
M. hodgsoni Gray, 1843b: 118; Himalayas.
M. itatsi Temminck, 1844: 34, pl. 7; Japan.
Putorius davidianus Milne-Edwards, 1872a: 92 (footnote); Jiangxi, S China.
P. moupinensis Milne-Edwards, 1874: 347, pl. 59; Moupin, Sichuan, China.
Lutreola major Hilzheimer, 1910: 310; near Sungpan, N Sichuan. Pre- occupied by *M. erminea major* Nilsson, 1820 and by *Foetorius pusillus major* Fatio, 1905 (in *M. nivalis*).

Map **90**. *Mustela*

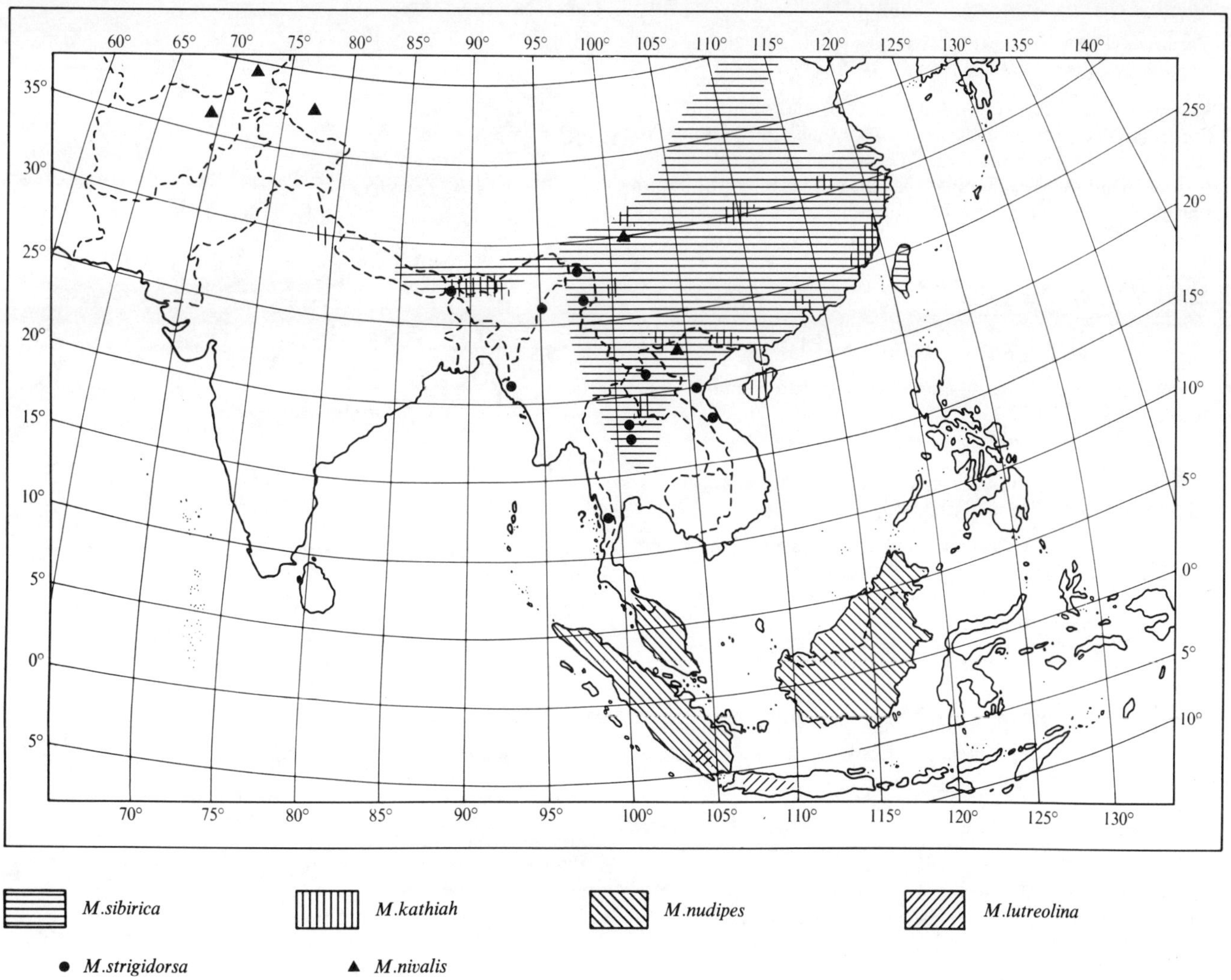

L. tafeli Hilzheimer, 1910: 310; near Sungpan, Sichuan, China.
M. (*Lutreola*) *taivana* Thomas, 1913d: 91; Mt Arizan, Taiwan, 8000ft.
M. hamptoni Thomas, 1921a: 500; Mt Imaw-bum, Kachin Prov., Burma, 9000ft.
Lutreola melli Matschie, in Mell, 1922: 35; Canton region, S China.
And other extralimital synonyms.

Range (Map 90). E Himalayas, mostly above 2400m, east to northern parts of Burma, Thailand, Laos and Vietnam; most of China, Taiwan; also throughout N China, much of Siberia and Japan; introduced on Iriomote I, Ryukyu Is in 1965 to control rats (Obaba, 1967).

Variation. At higher altitudes in the Himalayas and W China the muzzle is white to a variable extent, in some extending back to the eyes but much less in extent than is normal in *M. nudipes*. Those on Taiwan are not clearly distinguishable from mainland animals.

Ecology. Heavily exploited for fur, e.g. in China.

Mustela lutreolina Indonesian mountain weasel

M. lutreolina Robinson & Thomas, 1917: 261; Tjibodas, W Java, 5500ft.

Range (Map 90). Java and S Sumatra, mostly above 1000m.

Variation. The white on the throat is very variable and usually asymmetrical (van Bree & Boeadi, 1978).

Remarks. Localities were listed and skull and baculum were illustrated by Brongersma (1940). Externally this species is scarcely distinguishable from *M. lutreola* or *M. vison*, the European and American minks, but the skull is more similar to that of *M. sibirica* (indeed doubtfully distinguishable from it) than to *M. lutreola* or *M. nudipes*. IUCN: K.

Mustela nudipes Malayan weasel

M. nudipes Desmarest, 1822: 537; 'Java', corrected to W Sumatra by Robinson & Kloss (1919c).
Gymnopus leucocephalus Gray, 1865c: 119; Sumatra and Borneo.
M. hamakeri Dammerman, 1940: 266; Djambi, S Sumatra.

Range (Map 90). Peninsular Thailand, Malaya, Sumatra and Borneo; up to 1700m in Borneo.

Variation. Variation was reviewed by Brongersma & Junge (1942) who could not detect any difference between those from Sumatra and Borneo. Malayan specimens seem equally indistinguishable (Hill, 1960).

Ecology. Mainly terrestrial.

Remarks. Very similar to *M. sibirica* but easily distinguished by the white on the head which extends back behind the ears.

Mustela strigidorsa Back-striped weasel

M. strigidorsa Hodgson in Gray, 1853: 191 (and Horsfield, 1855: 107); Sikkim.

Range (Map 90). Himalayas from E Nepal to Assam, NE Burma, S Yunnan, N Thailand, NW Laos and N Vietnam (Ratajszczak & Cox, 1991). Mainly under 2000m in Nepal; down to 350m in Thailand.

Remarks. This species has the normal three lower premolars as in all *Mustela* spp., not four as reported by Lekagul & McNeely (1977).

VORMELA

Vormela Blasius, 1884: 9; type species *Mustela sarmatica* Pallas, in *Vormela peregusna* Gueldenstaedt. Conserved by Opinion 384 (1956).

One species.

Vormela peregusna Marbled polecat

Mustela peregusna Gueldenstaedt, 1770: 441; R Don, S Russia. Also extralimital synonyms.

Range (Map 91). W Pakistan; and through the subdesert and steppe zones of central Asia from Mongolia to Asia Minor and SE Europe.

Variation. The population in Baluchistan has been referred to *V. p. alpherakii* Birula, 1910 (Transcaspia).

MARTES Martens

Martes Pinel, 1792: 55 (footnote); type species *M. domestica* Pinel, in *M. foina* Erxleben.
Charronia Gray, 1865c: 108 (*M. flavigula* Boddaert).
= *Lamprogale* Ognev, 1928: 26.

Martes flavigula

Holarctic and Indomalayan; 7 species of which one is primarily Indomalayan and another occurs on the northern fringe of the region (Table 133). Martens are less specialized than *Mustela* spp. in body form, with relatively shorter body, longer legs, longer ears and longer muzzle (with four premolars above and below).

Table 133. The species of *Martes*.

	foina	*flavigula*
Range in region	Himalayas, W China	Widespread
Head & body	300–460	400–640
Tail (as % of h & b)	c.60%	c.75%
Dark line above pale throat	0	+
Upper lip with median groove	+	0
Rump and tail very dark	0	+
Skull: greatest length	78–80	85–110
Tip of baculum	Simple	Hooked, with 3–4 prongs

The following fossil species has been described:
M. (*Mustella*) *sinensis* Young & Liu, 1951: 72; Koloshan, near Chungking, Sichuan, China.
Also several others from northern China.

Fossil and Recent species of the entire genus were reviewed by Anderson (1970).

Martes foina Beech marten, Stone marten

Mustela foina Erxleben, 1777: 458; Germany.
Mustela intermedia Severtzov, 1873: 61 (and 1876: 45); basin of the Chu, Tallas and Naryn, E Turkestan, 4000–9000ft.
Also many extralimital synonyms.

Range (Map 91). SW China (W Sichuan, N Yunnan) and Himalayas, at altitudes above 1500m; also from N China west through Turkestan to Israel and much of Europe.

Variation. One subspecific name from the edge of the region is relevant:
M. f. kozlovi Ognev, 1931: 631; Kam, Mekong Valley, E Tibet.

Ecology. Often diurnal; more terrestrial than *M. flavigula*.

Martes flavigula Yellow-throated marten

Mustela flavigula Boddaert, 1785: 88; locality unknown, fixed as Nepal (Pocock, 1941).
Mustela melina Kerr, 1792: 183; locality unknown.
Viverra quadricolor Shaw, 1800: 429; locality unknown.
Mustela leucotis Bechstein, 1800: 375; locality unknown.
Mustela hardwickii Horsfield, 1828: 239, pl. 8; Nepal.
Galidictis chrysogaster Smith, 1842: 167; Mussourie, Kumaon, N India.
Mustela henrici Schinz, 1845: 36; Sumatra.
Martes gwatkinsi Horsfield, 1851: 99; Madras, India.
Martes chrysospila Swinhoe, 1866a: 286; central Taiwan.
Charronia melli Matschie, in Mell, 1922: 17,34; Guangdong, S China.
C. yuenshanensis Shih, 1930b: 3; Yuen Shan, Wuchanghsien, Hunan, China.

Range (Map 91). Much of China (and north to Korea and the Amur Basin), south to Malaya, Sumatra, Bangka, Java and Borneo; west along the Himalayas to Pakistan and E Afghanistan; an isolate in S India (Coorg, Nilgiri Hills and Travancore); Hainan and Taiwan (now very rare). Between about 300 and 3000m in the Himalayas; up to 1700m in Borneo; between 1500 and 2600m in Java.

Variation. The following subspecific names have been proposed:
Martes f. xanthospila Swinhoe, 1871: 623; central Taiwan.
Mustela f. peninsularis Bonhote, 1901g: 346; Bankasun, Tenasserim, Burma.

Map **91**. *Martes, Vormela, Mellivora*

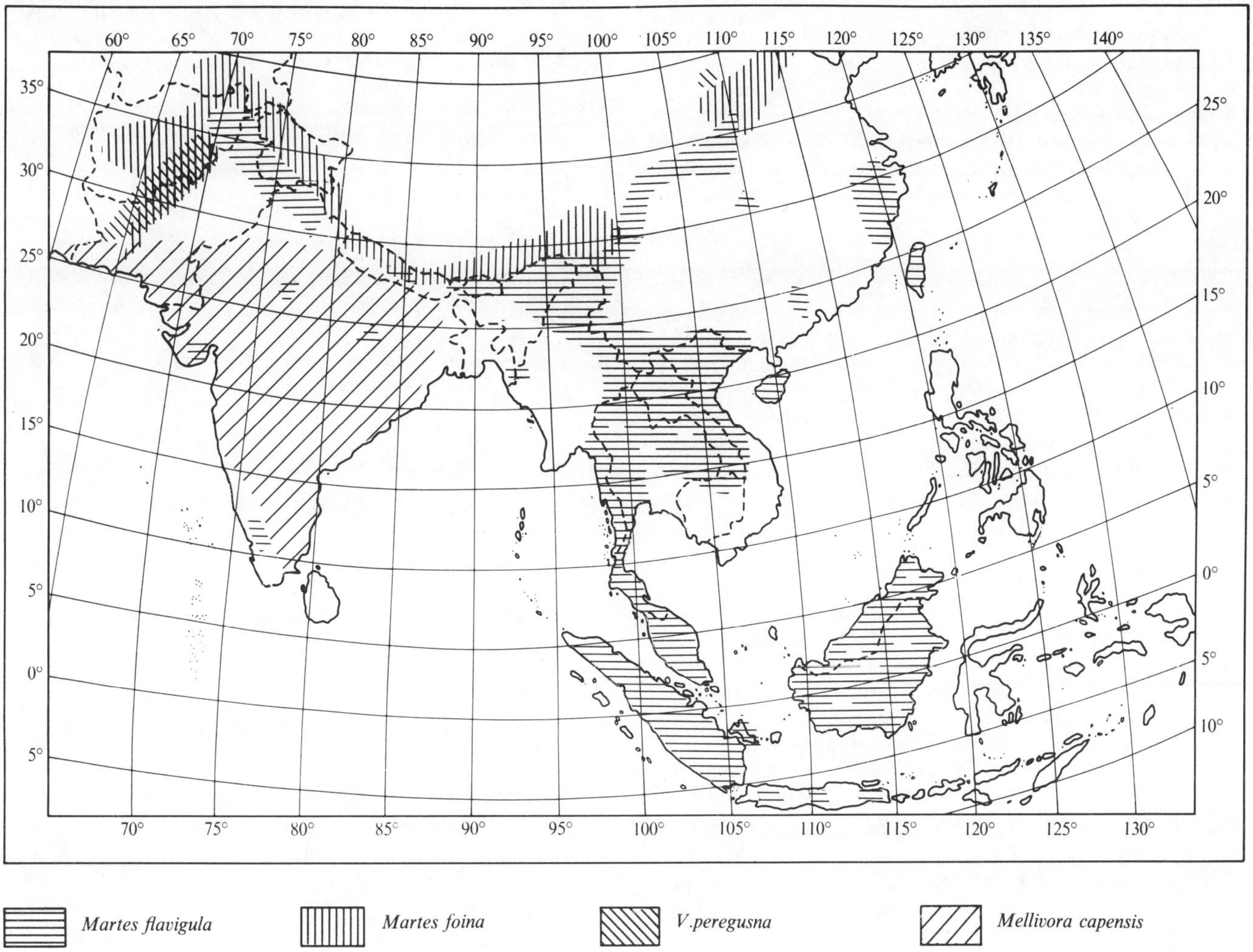

Mustela f. kuatunensis Bonhote, 1901g: 348; Kuatun, NW Fujian, China.
Mustela f. szetchuensis Hilzheimer, 1910: 310; Sungpanting, Sichuan, China.
Martes f. indochinensis Kloss, 1916b: 35; Klong Menao, SE Thailand.
Martes f. saba Chasen & Kloss, 1932: 13; near Sandakan, N Borneo.
Lamprogale f. robinsoni Pocock, 1936b: 403; Tjibodas, W Java, 5000ft.
Charronia f. tyrannus Colbert & Hooijer, 1953: 50; Yenching-kou, Sichuan, China; Pleistocene. Based on large size.
M. f. hainana Hsu & Wu, 1981: 145,148; Xinlong Dist, Hainan I, China. Size small: head & body—male 560mm, female 450mm; greatest length of skull—male 92mm, female 88mm.

Variation was described in detail by Pocock (1936c). Three main groups can be recognized:

China-Thailand-Pakistan (*M. f. flavigula* etc.). Head very dark, contrasting with light nape and shoulders; venter light brown, not contrasting very sharply with pale throat; P^1 and P_1 present (absent in 2 out of c. 80).

Sunda subregion (*M. f. henrici* etc.). Head brown or greyish brown; venter dark brown, contrasting sharply with pale throat. The head is rather greyer in Java. All P^1 and P_1 absent in the only two specimens from Java; present elsewhere.

Taiwan (*M. f. chrysospila*). Intermediate between above two: head brown; venter moderately dark.

S India (*M. f. gwatkinsi*). Pelage as in *M. f. flavigula* except that whole of dorsum is dark; skull, especially the braincase, very flattened above, with a very prominent frontal concavity in profile (not seen elsewhere in range); P^1 and P_1 usually absent (in 4 out of 5).

Ecology. Active from ground to canopy. Mainly diurnal where not disturbed.

Remarks. The isolated southern Indian form was recognized as a separate species *M. gwatkinsi* by Pocock (1941) mainly on the basis of the baculum which has the four terminal prongs less developed than in N. India (but only one available and none from SE Asia), and the rather flattened skull. Separation of this species in the genus *Charronia* is likewise based primarily upon the shape of the baculum. IUCN: I.

MELLIVORA

Mellivora Storr, 1780: 34; type species *Viverra ratel* Sparrmann, in *M. capensis* (Schreber). Conserved by Opinion 384 (1956).
Ratellus Gray, 1827a: 118 (*Viverra capensis* Schreber).
= *Melitoryx* Gloger, 1841: 57.
Ursitaxus Hodgson, 1836e: 61 (*U. inauritus* Hodgson).
Lipotus Sundevall, 1843a: 199 (*Ursus mellivorus* Cuvier).

One Recent species. The following fossil species have been described from the region:
Ursitaxus sivalensis Falconer, 1868: 553; Pinjor zone, Siwaliks.
M. punjabiensis Lydekker, 1884a: 183; Asnot, Punjab, Pakistan; Dhok Pathan zone, Siwaliks.

Mellivora capensis Ratel, Honey badger

Viverra capensis Schreber, 1776:pl. 125 (and 1777: 451); Cape of Good Hope, South Africa.
Ursus indicus Kerr, 1792: 188; India.
Ratelus mellivorus Bennett, 1830: 13; interior of Madras.
Ursitaxus inauritus Hodgson, 1836e: 61; Muckwanpur, S Nepal.
Mellivora ratel Horsfield, 1851: 120; India.
Also other, extralimital, synonyms.

Range (Map 91). The drier parts of peninsular India east to Bengal; Nepal and S Pakistan; west through SW Asia and most of the drier parts of Africa.

Variation. There is little regional variation except for substantially longer pelage at the northern edge of the range, in the Himalayan foothills.

Subfamily Melinae

Badgers

Melina Bonaparte, 1838: 111.

Palaearctic and Indomalayan regions; perhaps also in N America if *Taxidea* is included. Three genera are wholly Indomalayan; one other, *Meles*, extends into China from the Palaearctic (Table 134). Badgers are short-bodied compared with the Mustelinae, with powerful feet and claws adapted for digging. They are omnivorous, feeding upon invertebrates, tubers etc., and the dentition is characterized especially by M^1 which is large and rather flat- crowned (but less so in *Melogale* which in several respects links the Mustelinae and Melinae).

Table 134. The genera of Melinae.

	Meles	*Arctonyx*	*Melogale*	*Mydaus*
No. of species	1	1	4	2
Head & body	c.450	550–700	300–400	330–520
Tail (as % of h & b)	c.25%	20–40%	40–60%	c.10%
Throat pale	0	+	+	0
White ear tufts	+	+/0	0	0
White median stripe snout to:	Nape	Nape	Mid-back	Rump
Venter black	+	+	0	+
Skull: greatest length	110–125	115–170	70–90	80–100
M^1	Rectangular	Rhombic	Broader than long	c.square
Bullae	Prominent	Flat	Prominent	Very small

The classification was reviewed by Long (1978,1981); see also Long & Killingley (1983).

MELES

Meles Brisson, 1762: 13; type species *Ursus meles* L.
= *Taxus* Cuvier & Geoffroy, 1795: 187.
= *Melesium* Rafinesque, 1815: 59.
Meledes Kastschenko, 1925: 21 (*M. taxus arenarius* Satunin).

One species.

Meles meles Eurasian badger

Ursus meles Linnaeus, 1758: 48; Upsala, Sweden.
M. leptorhynchus Milne-Edwards, 1867a: 374; near Pekin, N China.
M. chinensis Gray, 1868b: 207; Amoy, China.
Also many other extralimital synonyms.

Range (Map 92). Most of N, E and S China; SE Tibet; also in Japan and west through S Siberia to Israel and most of Europe.

Variation. Chinese animals differ from European ones mainly in the head stripes which are less clearly defined, the median white one in particular not extending so far back onto the nape.

ARCTONYX

Arctonyx Cuvier, in Geoffroy & Cuvier, 1825:liv. 51; type species *A. collaris* Cuvier.
Trichomanis Hubrecht, 1891: 242 (*T. hoevenii* Hubrecht).

One Recent species. The following fossil species have been described:
A. minor Pei (reference not found).
A. rostratus Matthew & Granger, 1923: 580; Yen-ching-kao, near Wan-hsien, Sichuan, China; Pliocene.

Arctonyx collaris Hog-badger

A. collaris Cuvier, in Geoffroy & Cuvier, 1825:pt 51, pl. ; Bhutan Duars, E Himalayas.
Meles albogularis Blyth, 1853c: 590; 'E Tibet' (probably = Sichuan).
A. taxoides Blyth, 1853c: 591; Assam.
A. isonyx Hodgson in Horsfield, 1856: 398, pl. L; Sikkim Terai.
Meles leucolaemus Milne-Edwards, 1867a: 374; near Pekin, N China.
Meles (*A.*) *obscurus* Milne-Edwards, 1871a: 200; Sichuan, China.
Trichomanis hoevenii Hubrecht, 1891: 242; mountains between Palembang and Bencoolen, S Sumatra.
A. dictator Thomas, 1910b: 424; Lamra, Trang, S Thailand.
A. annaeus Thomas, 1921g: 524; Nha-trang, Annam, Vietnam.

Range (Map 92). Most of China, west to Sikkim, south to peninsular Thailand and Perak in Malaya; Sumatra.

Variation. The following subspecific names have been proposed:
A. c. taraiyensis Gray, 1863b: 7; Sikkim Terai.
A. leucolaemus orestes Thomas, 1911g: 27 (and 1911f: 688); Tsingling Mts, Shaanxi, China, 12000ft.
A. obscurus incultus Thomas, 1922a: 395; Chinteh, Anhui, China.
A. leucolaemus milne-edwardsii Lönnberg, 1923: 322; Minshan, S Gansu, China.
A. c. consul Pocock, 1940a: 465; Thaundaung, near Toungoo, lower Burma, 4500ft.

Map **92**. *Meles, Arctonyx, Melogale, Mydaus*

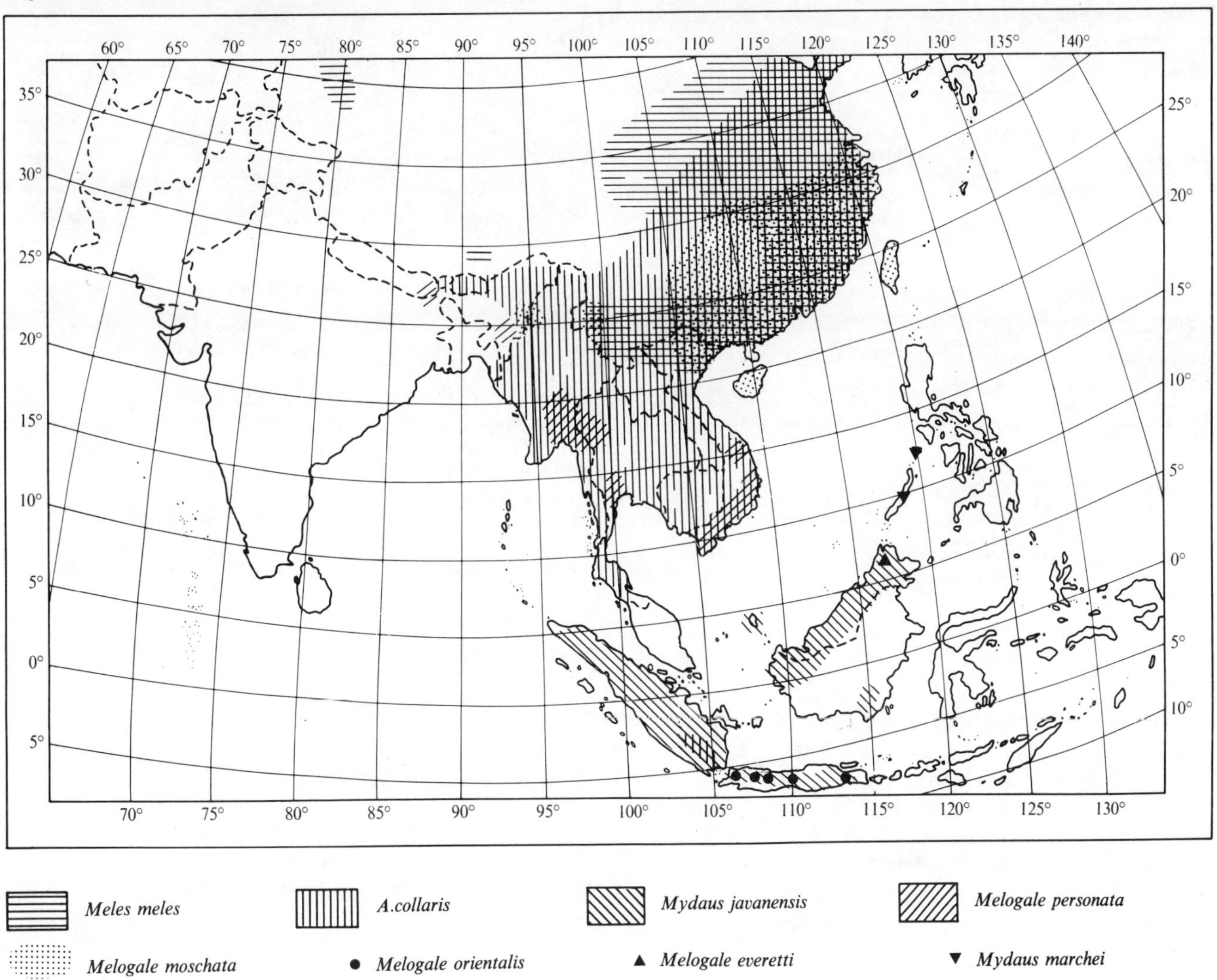

Individual variation in pelage is considerable, with very dark individuals occurring at widely scattered locations in China and in Sumatra. These have the dorsum almost entirely black but retaining pale hair-bases and white face-stripes and tail.

Size appears to vary geographically, being least in the north (China and the E Himalayas) and greatest in the south, in Indochina, Thailand etc.

MELOGALE Ferret-badgers

Melogale Geoffroy, 1831(March): 129; type species *M. personata* Geoffroy.
Helictis Gray, 1831b(August): 94 (*H. moschata* Gray). Conserved by Opinion 384 (1956).
Nesictis Thomas, 1922f: 194 (*Helictis everetti* Thomas).

Indochinese subregion, Java and Borneo; 4 species (Table 135). These are small badgers which in some respects bridge the gap between the Mustelinae and the other Melinae. In particular M^1 is transversely elongate and waisted as in the Mustelinae.

Four species were recognized by Everts (1968) in a detailed metrical analysis but without clear diagnoses. Although the two northern species are very similar externally the teeth are very different: small and slender in *M. moschata*, large and chunky in *M. personatus*. Likewise the forms on Java and Borneo might be considered conspecific on the basis of external appearance but the difference in dentition suggests that specific rank is more appropriate.

Table 135. The species of *Melogale*.

	personata	*orientalis*	*everetti*	*moschata*
Range	Nepal-Thailand	Java	Borneo	Indochina
Tail (as % of h & b)	50–60%	45–50%	28–32%	40–45%
Tail	Pale, ± white distally	Dark, short white tip	Dark	Medium, ± white tip
Venter: pale zone constricted on chest	0	+	+	0
Skull: greatest length	72–90	73–85	66–72	70–83
Teeth	Massive	Slender	Slender	Slender
Rostrum width at P^4	24–28	21–23	17–19	18–22
P^4: outer margin	Convex	Convex	Convex	Straight or concave

Melogale personata Large-toothed ferret-badger

M. personata Geoffroy, 1831: 137, pl. 5; near Rangoon, Burma.
Gulo nipalensis Hodgson, 1836b: 237; Nepal.
Helictis orientalis; Blanford, 1888b: 173. Not of Horsfield, 1821.
H. pierrei Bonhote, 1903d: 592; near Saigon, S Vietnam.
M. tonquinia Thomas, 1922f: 195; Yen-bay, Songkoi R, Tonkin, N Vietnam.

Range (Map 92). Nepal to Assam, Burma, Thailand and Vietnam.

Variation. The following subspecific name has been proposed:
M. p. laotum Thomas, 1922f: 194; Nan, Thailand, 200m.

There is considerable variation in the facial pattern but it is unlikely that any discrete subspecies can be justified.

Remarks. The Javan form *orientalis*, here treated as a separate species, was tentatively included in *M. personata* by Long (1978).

Melogale orientalis Javan ferret-badger

Gulo orentalis Horsfield, 1821:part 1; E Java.
= *Mydaus maccourus* Temminck, 1824,1: 20.

Range (Map 92). Java and Bali (Riffel, 1991).

Variation. The following subspecific name has been proposed:
Helictis o. sundaicus Sody, 1937a: 211; Mt Salak, W Java, 600m.

Remarks. IUCN: K.

Melogale everetti Bornean ferret-badger

Helictis everetti Thomas, 1895a: 331; Mt Kinabalu, N Borneo, 4000ft.

Range (Map 92). N Borneo, known only from Mt Kinabalu, between 1070 and 3000m.

Remarks. Recorded (as *M. orientalis*) as subfossil in the Niah Caves, Sarawak with an estimated age of 22000 BP (Harrisson, 1970). On the basis of dentition *M. everetti* appears to more closely resemble *M. moschata* than the adjacent *M. orientalis* with which is it usually associated. IUCN: K.

Melogale moschata Small-toothed ferret-badger

Melogale moschata

Helictis moschata Gray, 1831b: 94; Canton, Guangdong, S China. Conserved by Opinion 384 (1956).
H. subaurantiaca Swinhoe, 1863: 355; Taiwan.
H. ferreo-griseus Hilzheimer, 1905b: 298; near Hankow, Hubei, China.
H. millsi Thomas, 1922b: 432; Mokokchung, Naga Hills, Assam, India, 5000ft.
H. taxilla Thomas, 1925: 500; Ngai-tio, Tonkin, N Vietnam.

Range (Map 92). S China (Sichuan, Hubei and Zhejiang southwards) to Assam, Burma and Vietnam; Hainan and Taiwan.

Variation. The following subspecific names have been proposed:
Helictis subaurantiaca modesta Thomas, 1922f: 196; Bankoro, Taiwan.
H. taxilla sorella Allen, 1929: 8; Futsing, Fujian, SE China.
M. m. hainanensis Zheng & Xu, 1983: 168; Dali, Hainan I, China.

Small forms (*taxilla, sorella*) occur in SE China and Vietnam. Those on Hainan and Taiwan are not clearly distinguishable from mainland animals. On Taiwan the throat and the dorsal stripe tend to be yellowish but this is not constant.

Ecology. Heavily exploited for fur in China.

MYDAUS Stink badgers

Mydaus Cuvier, in Geoffroy & Cuvier, 1821:liv. 27; type species *M. meliceps*, in *M. javanensis* (Leschenault). Conserved by Opinion 384 (1956).
Suillotaxus Lawrence, 1939: 63 (*M. marchei* Huet).

Sumatra, Java, Borneo and Palawan; 2 species (Table 136). The species on Palawan, *M. marchei*, has been separated generically from the more widespread Sunda species *M. javanensis* on the basis mainly of larger teeth. When variability of *M. javanensis* is taken into account this cannot be upheld and it is even dubious if specific separation is justified.

Table 136. The species of *Mydaus*.

	javanensis	*marchei*
Range	Sumatra, Java, Borneo	Palawan
White dorsal stripe reaching rump	Usually	Rarely
Skull: greatest length	80–100	74–90
M^1-M^1 (at crown)	23–25	22–24

Mydaus javanensis Javan stink badger, Teledu

Mephitis javanensis Leschenault, in Desmarest, 1818: 520; Java.
Mydaus meliceps Cuvier, in Geoffroy & Cuvier, 1821:liv. 27; Java. Conserved by Opinion 384 (1956).
M. lucifer Thomas, 1902e: 442; Borneo, on mainland opposite Labuan I.
M. ollula Thomas, 1902e: 443; Bunguran I, N Natuna Is.
M. luciferoides Lönnberg & Mjöberg, 1925a: 509; Kelabit, NE Sarawak, Borneo.

Range (Map 92). Sumatra, Java and Borneo; and on Bunguran, N Natuna Is.

Variation. The following subspecific name has been proposed: *M. j. montanus* Moulton, 1921: 142; Kelabit, NE Sarawak, Borneo.

The extent of the mid-dorsal white stripe is very variable within each of the island populations; it is always present on the shoulders, nearly always present on the rump, but frequently interrupted in the centre of the back. On Java it is usually interrupted; on Borneo it is most often continuous.

Mydaus marchei Palawan stink badger

M. marchei Huet, 1887: 149; Palawan, Philippines.
M. schadenbergii Jentink, 1895: 46; Calamian Is, Philippines, probably at Malbato, Busuanga I.

Range (Map 92). Palawan and adjacent small islands of Busuanga and Iloc, Philippines.

Subfamily Lutrinae

Otters

Lutrina Bonaparte, 1838: 111; type genus *Lutra* Brisson.

Almost world-wide except for Australasia. Otters share the long slender body and short legs of the Mustelinae but are highly adapted for life in water, with fully webbed feet, thick tapering tail and dense pelage. The delimitation of genera within the Lutrinae has been very unstable. Amongst the four Indomalayan species this only affects the nomenclature of *Lutrogale perspicillata*. Although frequently included in *Lutra* or placed close to it (Van Zyll de Jong, 1987), it has also been considered more closely related to *Aonyx* than to *Lutra*, e. g. by Davis *et al.* (1979) on the basis of chromosomes. *Lutrogale* is therefore used here to allow either of these views without upsetting nomenclature. All four species are compared in Table 137.

Table 137. The species of *Lutra, Lutrogale* and *Aonyx* (Lutrinae).

	Lutra lutra	*Lutra sumatrana*	*Lutrogale perspicillata*	*Aonyx cinerea*
Range in region	Widespread	Indochina, Sumatra, Borneo	Widespread	Widespread
Head & body	540–800	570,620	520–750	360–460
Hind feet	100–110	95–105	100–140	85–95
Rhinarium	Naked	Hairy	Naked	Naked
Feet webbed	Fully	Fully	Fully	Partly
Claws projecting well beyond digits	+	+	+	0
Throat	Off-white	White	Cream	Buff
Pelage: white zone of 'wool'(mm)	5–6	c.5	c.3	5–6
Venter, cf dorsum	Much paler	Scarcely paler	Slightly paler	Slightly paler
Tail with lateral keels	0	0	+	0
Skull: greatest length	100–130	100–110	110–130	85–95
Dorsal profile of braincase	Straight	Straight	Rounded	Rounded
Anterior rim of orbit above:	P^{3-4}	P^3	P^{2-3}	P^{2-3}
Upper premolars	4	4	4	3
Post-orbital: inter-orbital constriction	50–75%	70–75%	80–110%	90–110%

LUTRA

Lutra Brisson, 1762: 13; type species *Mustela lutra* L.
= *Lutra* Brünnich, 1771: 34.
= *Lutris* Duméril, 1806: 12.
= *Lutrix* Rafinesque, 1815: 59.
Barangia Gray, 1865c: 123 (*L. sumatrana* Gray).
Lutronectes Gray, 1867d: 180 (*L. whiteleyi* Gray, in *L. lutra* L.).
And other synonyms based upon extralimital species.

Americas, Palaearctic, Africa and Indomalaya; c. 7 species of which 2 occur in the Indomalayan Region. The American species were separated in the genus *Lontra* by Van Zyll de Jong (1987) who considered them more closely related to *Aonyx* than to *Lutra s. s.* All species feed primarily upon fish, often supplemented by crustaceans.

The following fossil species have been described from the region:
L. sivalensis Lydekker, 1868: 331; Siwalik Hills.
L. palaeindica Falconer, 1868: 552; Upper Siwaliks.
L. bathygnathus Lydekker, 1884a: 193(=16); Punjab; Siwalik Beds; allocated to the genus *Plesiogulo* by Teilhard de Chardin & Leroy (1945b).

Lutra lutra Eurasian otter

Mustela lutra Linnaeus, 1758: 45; Upsala, Sweden.
= *L. vulgaris* Erxleben, 1777: 448.
L. nair Cuvier, F.,1823b: 247; Pondicherry, India.

L. indica Gray, 1837a: 580; Madras, India.
L. chinensis Gray, 1837a: 580; probably near Canton, S China.
L. monticolus Hodgson, 1839b: 320; Nepal.
L. aurobrunneus Hodgson, 1839b: 320; Nepal.
L. kutab Schinz, 1844: 354; Kashmir.
Barangia ?nepalensis Gray, 1865c: 124; Nepal.
L. sinensis Trouessart, 1897: 283.
L. hanensis Matschie, 1907: 150; Hsinganfu, Shaanxi, China.
L. intermedia Pohle, 1920: 37,62; Sumatra.
Also many extralimital synonyms.

Range (Map 93). S and N India (but not central); China to Indochina and Thailand; Sri Lanka, Hainan, Taiwan, Sumatra; doubtfully in Borneo; widespread throughout most of the Palaearctic Region. Records from Java are likely to be erroneous (van Bemmel, 1949b).

Variation. The following subspecific names have been proposed from the region:
L. l. barang Cuvier, F.,1823b: 246; 'Java', more likely Sumatra (Chasen, 1940a: 92).
L. l. ceylonica Pohle, 1920: 72; Nuwara Eliya, Sri Lanka.
L. l. hainana Xu & Liu, in Xu *et al.*,1983: 328; Hainan I, China.

Animals in the southern populations, both in S India/Sri Lanka and in Sumatra, appear to be smaller than those from further north, with greatest skull length <110mm in the south and >110mm in the north.

Ecology. Mainly montane in the region, but also coastal elsewhere.

Remarks. CITES: Appendix I.

Lutra sumatrana Hairy-nosed otter

Barangia sumatrana Gray, 1865c: 123; Sumatra.
Lutra lovii Günther, 1877: 736; Borneo, opposite Labuan I.
L. brunnea Pohle, 1920: 37,44; near Pontianak, Borneo.

Range (Map 93). Vietnam and S Thailand south to Malaya; Sumatra, Borneo; and the islands of Singapore and Bangka (Sumatra).

Remarks. CITES: Appendix II; IUCN: K.

LUTROGALE

Lutrogale Gray, 1865c: 127; type species *Lutra perspicillata* Geoffroy.

One species. The following fossil species were allocated to *Lutrogale* by Willemsen (1986):
Lutra palaeoleptonyx Dubois, 1908: 1267; Java.
Lutra robusta Koenigswald, 1933: 25; Boemiajoe, Java.

Map **93**. *Lutra, Lutrogale, Aonyx*

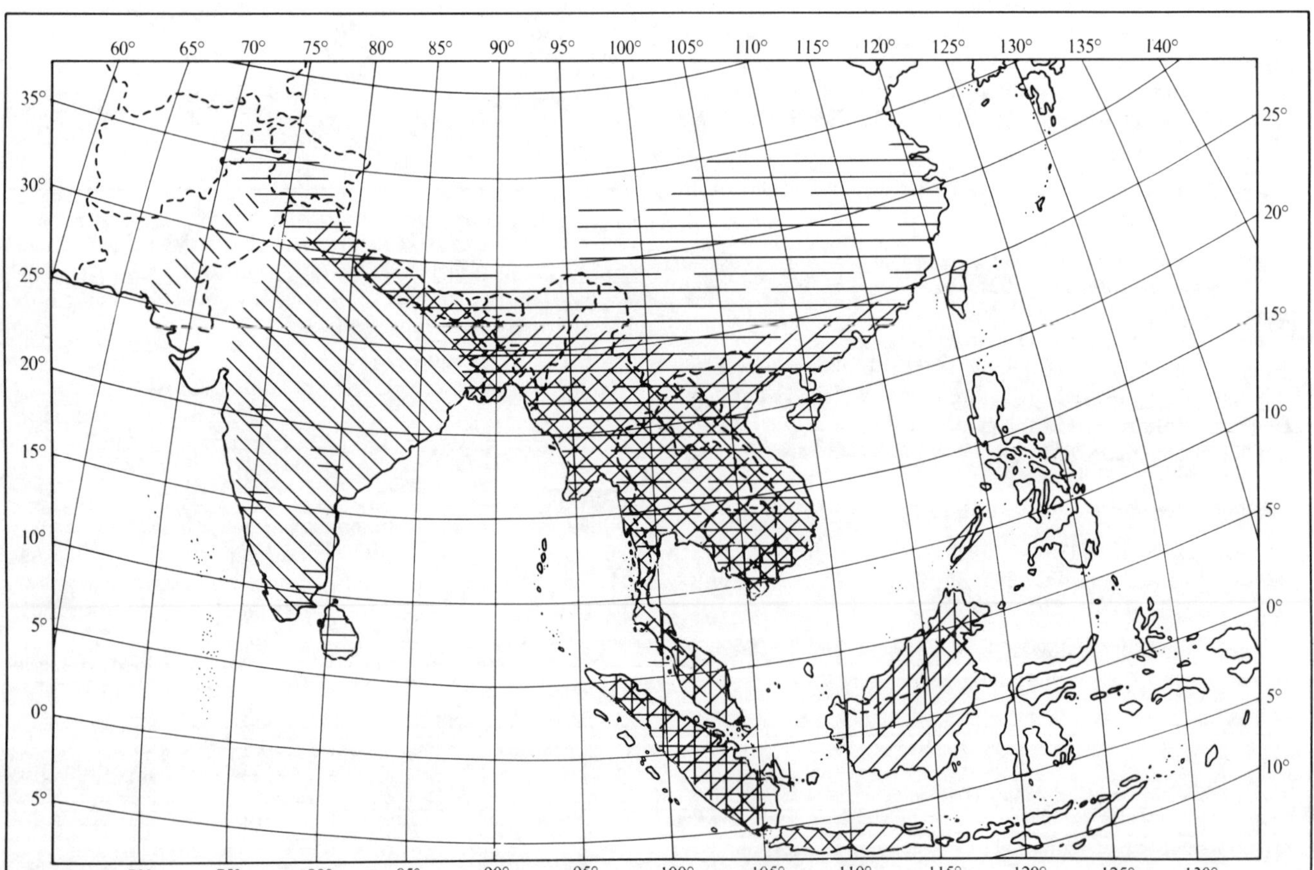

Lutrogale perspicillata Smooth-coated otter

Lutra perspicillata Geoffroy, I.,1826: 519; Sumatra.
Lutra simung Lesson, 1827a: 156; Sumatra.
Lutra tarayensis Hodgson, 1839b: 319; Nepal Terai.
Lutra macrodus Gray, 1865c: 128.
Lutra ellioti Anderson, 1879: 212; Madras, India.

Range (Map 93). Pakistan and most of India, east to SW Yunnan, Indochina and south to Malaya; Sumatra, Java (van Bree, 1968), Borneo; an apparently isolated population in S Iraq.

Variation. The following subspecific names have been proposed:
Lutrogale p. sindica Pocock, 1940b: 517; Chak, Sukkur dist, W Sind, Pakistan.
Lutrogale p. maxwelli Hayman, 1957: 710; Abusakhair, 35 miles SE of Amara, Iraq.

Remarks. The occurrence of this species in the Sunda region was reviewed and mapped by van Bree (1968). In parts of India and Bangladesh otters (probably of this species) are trained to assist with fishing, by herding fish into nets and by catching large fish (Prater, 1971; Biswas, 1973). CITES: Appendix II; IUCN: K.

AONYX Small-clawed otters

Aonyx Lesson, 1827a: 157; type species *A. delalandi* Lesson, in *A. capensis* (Schinz) (Africa).
Amblonyx Rafinesque, 1832: 62 (*A. concolor* Rafinesque). Conserved by Opinion 384 (1956).
Leptonyx Lesson, 1842: 72 (*Lutra leptonyx* Horsfield). Preoccupied by *Leptonyx* Swainson, 1821 (Aves) and *Leptonyx* Gray, 1837 (Mammalia, Pinnipedia).
= *Micraonyx* Allen, J. A.,1920: 24.

Afrotropical and Indomalayan regions; 3 species of which one occurs in the region.

Aonyx cinerea Oriental small-clawed otter

Lutra cinerea Illiger, 1815: 90, 99; Batavia, Java.
L. leptonyx Horsfield, 1823; Java.
Amblonyx concolor Rafinesque, 1832: 62; Garo Hills, Assam.
L. indigitatus Hodgson, 1839b: 320; Nepal.
Aonyx horsfieldi Gray, 1843b: 119; Java.
A. sikimensis Horsfield, 1855: 109; Sikkim.
L. (*Hydrogale*) *swinhoei* Gray, 1867d: 182; Gawkang I, near Amoy, S China.

Range (Map 93). S and N (but not central) India and S China (Yunnan, S Sichuan, Guangdong, Guangxi, Fujian, Hainan) to Malaya; Sumatra, Java, Borneo, Palawan; and on the small islands of Singapore, Riau Is, Lingga Is; Laut (Borneo); Karimun Jawa (Java Sea); up to 1300m in Borneo. Taiwan was included in the range by Honacki *et al.* (1982) but not by previous authors.

Variation. The following subspecific names have been proposed:
Amblonyx c. fulvus Pohle, 1920: 133; Lao Key, Tonkin, N Vietnam.
Amblonyx c. wurmbi Sody, 1933a: 441; near Poeger, Watangan Mts, Besoeki, E Java; 50m.
Amblonyx c. nirnai Pocock, 1940b: 515; Virajpet, S Coorg, India, 3000ft.

In S India the pelage tends to be darker than elsewhere. Considerable individual variation occurs in the presence or absence of the first two premolars (above and below) and in the shape of the posterior margin of the rhinarium (Yoshiyuki, 1971).

Ecology. Diet consists mainly of crustaceans and other invertebrates.

Remarks. CITES: Appendix II; IUCN: K.

Family Viverridae

Civets

Viveridae [*sic*] Gray, 1821: 301; type genus *Viverra* L. Conserved by Direction 53 (1956).

Afrotropical and Indomalayan regions; c. 35 species of which 17, in 11 genera, occur in the Indomalayan Region (Table 139). Viverrids are medium-sized carnivores, many of them arboreal, with rather unspecialized dentition and omnivorous diet, including fruit and invertebrates as well as small vertebrates. Most species have a prominent perineal scent gland situated in front of the anus. Some species are exploited for civet (scent-gland secretion) and for food, especially in China (Tan, 1989).

A checklist for the entire family was produced by Wenzel & Haltenorth (1972). Accounts, with maps, of the rarer species were prepared by Schreiber *et al.* (1989). For a bibliography of literature on the family see Dargel (1990).

VIVERRA

Viverra Linnaeus, 1758: 43; type species *V. zibetha* L. Conserved by Opinion 384 (1956).
Moschothera Pocock, 1933a: 441 (*V. civettina* Blyth).

Indomalayan Region; 4 species (Table 138). Large, terrestrial species with digitigrade feet (the sole hairy behind the plantar pads). The African civet is often included but was separated generically (as *Civettictis civetta*) by Pocock (1939) and Wozencraft (1989b).

Table 138. The species of *Viverra*.

	zibetha	*tangalunga*	*megaspila*	*civettina*
Range	Nepal-Malaya	Sunda	Indochina	SW India
Head & body (cm)	74–86	61–70	75–95	c.76
Tail (as % of h & b)	55–60%	45–55%	30–50%	c.43%
Pattern on tail	Fig.17a	Fig.17b	Fig.17c	Fig.17c
Spots on dorsum (no. of rows on each flank)	Obscure	Small (8)	Large (5)	Large (5)
Front feet with sheaths covering claws 3 and 4	0	?	+	+
Post-orbital processes well developed	+	+	0	0

The following fossil species have been described from the region (or adjacent areas):
V. bakeri Bose, 1880: 131; Siwalik Hills.
V. durandi Lydekker, 1884a: 271(=94); Siwalik Hills.
V. karnuliensis Lydekker, 1886a: 121; Billa Surgam, Karnul Dist, Madras Prov., India.
V. (?) *chinjiensis* Pilgrim, 1932: 99; Kukar Dhok, Chinji, Salt Range, Pakistan; Lower Siwaliks.
V. peii Qiu, 1980: 304; Zhoukoudien, N China.

Table 139. The genera of Viverridae.

	Viverra	*Viverricula*	*Prionodon*	*Paradoxurus*	*Paguma*	*Arctictis*	*Arctogalidia*	*Macrogalidia*	*Hemigalus*	*Cynogale*	*Diplogale*
No. of species	4	1	2	4	1	1	1	1	2	2	1
Head & body	615–950	460–630	350–420	420–690	415–730	610–965	440–575	c.600	500–620	575–820	470–540
Tail (as % of h & b)	30–60%	55–65%	80–95%	70–90%	90–105%	c.80%	100–120%	c.80%	60–70%	c.25%	c.60%
Feet digitigrade (hairy behind plantar pads)	+	+	(+)	0	0	0	0	0	0	0	0
Tail with dark and pale bands	+	+	+	0	0	0	0	(+)	Base only	0	0
Throat with transverse black and white bands	+	+	0	0	0	0	0	0	0/(+)	0	0
Dorsum with multiple longitudinal stripes, at least on rump	0	+	0	0	0	0	(+)	0	0	0	0
Dorsum with transverse dark bands	0	0	+/0	0	0	0	0	0	+	0	0
Mid-dorsal crest	+	0	0	0	0	0	0	0	0	0	0
Black and white pattern on side of face	+	0	0	+	+	0	0	0	+	0	0
Flanks spotted	+	+	+	(+) faintly	0	0	0	0	0	0	0
Nape with hair reversed	0	0	0	+/0	+/0	0	+/0	+	+	+	+
Vibrissae very thick and long	0	0	0	0	0	0	0	0	0	+	0
Pelage very densely woolly	0	0	0	0	0	0	0	0	0	+	0
Perineal scent glands	M,F	M,F	0	M,F	M,F	M,F	F only	?	M,F	M,F	M,F
Skull: greatest length	120–150	90–105	70–80	90–125	102–137	120–145	104–118	135–160	95–115	c.128	90–100
Palate extends behind teeth by x% of toothrow	10%	10%	15–20%	10%	20%	30%	25%	15%	c.12%	7%	6%
Median foramen between paired incisive foramina	0	0	0	0	0	0	0	0	+	+	0
P^{1-2} and P_{1-2} with prominent accessory cusps before and behind the main cusp	0	0	+*	0	0	0	0	0	0	0	+
M^2	+	+	0	+	+	+	+	+	+	+	+
Incisor row strongly curved	+	(+)	0	0	0	0	0	(+)	+/++	0	0
I^{2-3} flattened	0	0	0	0	0	0	0	0	+	0	0

* Except anterior cusp on P^1.

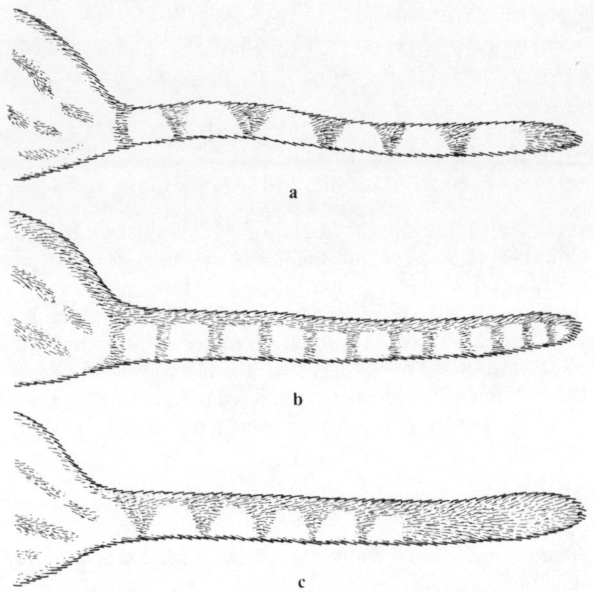

Fig.**17**. *Viverra*. Typical pattern of tail of (a) *V.zibetha*; (b) *V.tangalunga*; (c) *V.megaspila* and *V.civettina*.

Viverra zibetha Large Indian civet

V. zibetha Linnaeus, 1758: 44; Bengal. Conserved by Opinion 384 (1956).
V. undulata Gray, 1830a: 9, pl. 8; Nepal.
V. orientalis or *melanurus* Hodgson, 1841c: 47; Nepal.
V. undulata Hodgson, 1841c: 51; Nepal. Pre-occupied by *V. undulata* Gray, 1830, above.
V. civettoides Hodgson, 1841a: 62.
V. ashtoni Swinhoe, 1864b: 379; Suykaou, Min River, Fukien, China.
V. filchneri Matschie, 1907: 192; Hinganfu, SE Shaanxi, China.

Range (Map 94). Nepal to S China (north to S Shaanxi), south through Indochina to Malaya; Hainan, Singapore. Introduced to the Andaman Is (Lever, 1985). (Not in peninsular India as mapped by Lekagul & McNeely, 1977.)

Variation. The following subspecific names have been proposed:
V. z. picta Wroughton, 1915c: 64; H'Kamti, upper Chindwin R, N Burma, 500ft.
V. z. pruinosus Wroughton, 1915c: 64; Thaget, Little Tenasserim R, S Burma.
V. z. sigillata Robinson & Kloss, 1920: 176; Bang Nara, Patani, peninsular Thailand.

Map **94**. *Viverra*

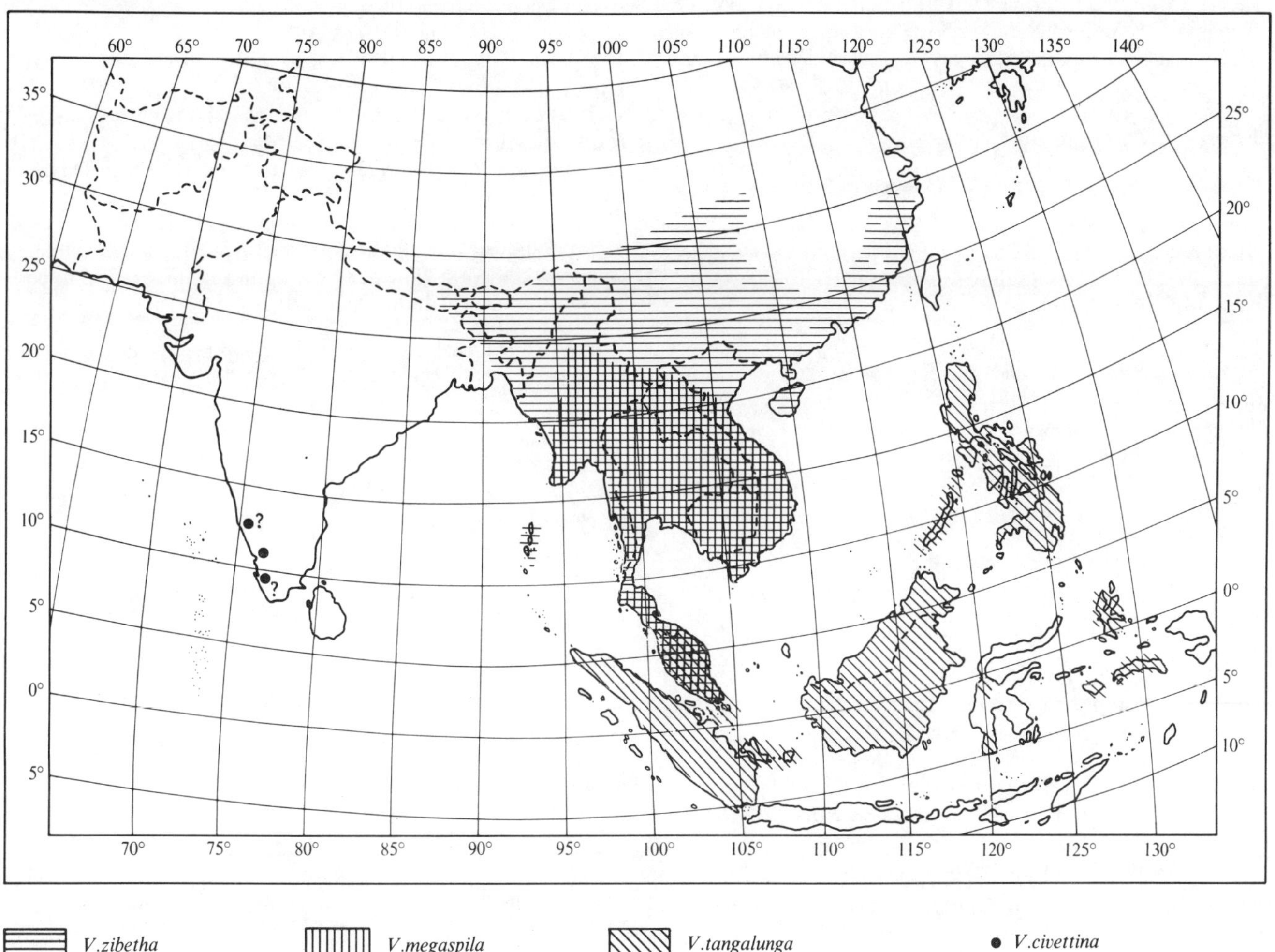

V. z. surdaster Thomas, 1927: 46; Xieng Khouang, Laos.
V. z. expectata Colbert & Hooijer, 1953: 56; Yenchingkou, Sichuan, China; Pleistocene.
V. z. hainana Wang & Xu, in Xu *et al.*,1983: 133; Hainan I, China.

In the north of the range and at high altitude the thick winter pelage tends to obscure the pattern; otherwise there is little regional variation.

Viverra tangalunga Sunda civet, Malay civet

V. tangalunga Gray, 1832: 63; W Sumatra.

Range (Map 94). Malaya, Sumatra, Borneo, Palawan; the smaller islands of Langkawi (Malaya), Singapore, Bintan, Kundur (Riau Is), Lingga, Bangka, Belitung, Karimata, Laut; Calamian Is; also on most of the other Philippine islands, Sulawesi, Sangihe Is and the Moluccan islands of Buru, Seram, Ambon and Halmahera, probably by introduction.

Variation. The following subspecific name has been proposed:
V. t. lancavensis Robinson & Kloss, 1920: 177; Langkawi I, Malaya.

There is very little variation in pelage amongst the island populations.

Viverra megaspila Large-spotted civet

V. megaspila Blyth, 1862b: 331; Prome, lower Burma.

Range (Map 94). Burma, S China (S Yunnan and SW Guangxi— Schreiber *et al.*, 1989), Thailand, Indochina and Malaya; island of Penang; perhaps also Singapore and Sumatra (Chasen, 1940a: 94, footnote).

Viverra civettina Malabar civet

V. civettina Blyth, 1862b: 332; Travancore, S India.

Range (Map 94). Western Ghats of SW India, from c. 14°N in Karnataka south to Cape Comorin. Apparently not recorded since the 19th century until it was rediscovered near Elayur, Kerala in 1987 (Schreiber *et al.*, 1989).

Remarks. Included as a subspecies of *V. megaspila* by Ellerman & Morrison-Scott (1951) and others, but given separate specific rank by Wozencraft (1989b). On the few specimens available the spots are more crowded and irregular than in *V. megaspila* but it is very doubtful if it justifies specific separation. Lindsay (1928) gave several characters that she claimed distinguished these two, but of these only one appears to hold on the material examined, namely the more curved coronoid process of the mandible in *V. civettina*, with the tip extending back to overhang the articular process. IUCN: E.

VIVERRICULA

Viverricula Hodgson, 1838: 152; type species *Civetta indica* Desmarest. Conserved by Opinion 384 (1956).

One species. Distinguished externally from *Viverra* by narrower head and absence of a median crest of long hairs on the back.

Viverricula indica Small Indian civet, Rasse

Viverra malaccensis: of many authors, not Gmelin, 1788: 92 (see Pocock, 1933a).
Viverra indica Desmarest, 1817: 170; India.
Viverra rassae Horsfield, 1823:pl. ; Java.
Viverra bengalensis Gray, 1831 (in 1830–35): 1, pl. 4; Calcutta.
Viverra pallida Gray, 1831a: 17; probably near Canton, Guangdong, China.
Viverra schlegeli Pollen, in Schlegel, 1866: 78; Mayotte and Nossi-Falay Is, NW of Madagascar.
Viverricula hanensis Matschie, 1907: 196; Hankow, S China.

Range (Map 95). From the Indus Valley in Pakistan, India and S China (N to Sichuan and the lower Yangtze), south through Indochina to Malaya; Sumatra, Java, Hainan, Taiwan, Sri Lanka; small islands of Penang (Malaya), Singapore; Kangean, Bawean, Bali; Lombok, Sumbawa (probably introduced); introduced to Madagascar, the Comoro Is and Socotra (NW Indian Ocean). Reported to have been introduced to the Philippines (Lever, 1985) but not listed from there by Heaney *et al.* (1987).

Variation. The following subspecific names have been proposed:
V. malaccensis deserti Bonhote, 1898b: 120; Sambhar, Rajasthan, India.
V. pallida taivana Schwarz, 1911: 637; Taiwan.
V. malaccensis thai Kloss, 1919b: 352; Prapatom, W of Bangkok, Thailand.
V. m. atchinensis Sody, 1931b: 351; Peureula, E Aceh, N Sumatra.
V. m. muriavensis Sody, 1931b: 353; Keling, N of Mt Moeria, N coast of central Java, 100m.
V. m. baliensis Sody, 1931b: 353; Bali.
V. i. mayori Pocock, 1933a: 632; Maha Oya, E Prov., Sri Lanka.
V. i. wellsi Pocock, 1933a: 640; Kangra, Himachal Pradesh, India, 2000ft.
V. i. baptistae Pocock, 1933a: 643; Hasimara, Bhutan Duars, India.
V. i. klossi Pocock, 1933a: 654; Penang I, Malaya.

Geographical variation is slight apart from the longer winter pelage in the north. The dark bands on the throat tend to be more prominent in India than further east.

Ecology. Nocturnal and terrestrial, sleeping in underground burrows.

PRIONODON Linsangs

Prionodon Horsfield, 1822; type species *Felis gracilis* Horsfield (in *P. linsang* Hardwicke).
= *Linsang* Müller, 1840: 28.
= *Priodontes* Lesson, 1842: 60.
Pardictis Thomas, 1925: 498 (*P. pardicolor* Hodgson).

Indochinese and Sunda subregions; 2 species (Table 140). A distinctive genus, characterized by long, slender body, digitigrade feet, absence of scent glands and absence of M^2.

Table 140. The species of *Prionodon*.

	linsang	*pardicolor*
Range	Sunda	Indochina
Pattern on dorsum	Transverse dark bands	Rounded dark spots
Skull: greatest length	70–80	65–75

Prionodon linsang Banded linsang

Viverra linsang Hardwicke, 1821: 236, pl. 24; Malacca.
Felis gracilis Horsfield, 1821:pl. ; E Java.
V. hardwichii Lesson, 1827a: 172; Java.
P. maculosus Blanford, 1878b: 71; Bankachon, S Burma.

Range (Map 95). Peninsular Burma, Thailand and Malaya; Sumatra, Bangka, Belitung, Java, Borneo. (Shown, as *Prionodon s. s.*, on Sulawesi in a map by Michaelis (1969), but other sources on Sulawesi do not mention *Prionodon.*)

Variation. The following subspecific names have been proposed:
P. l. fredericae Sody, 1936a: 43; Bangka I, Sumatra.
P. l. interliniurus Sody, 1949b: 157; Tandjung Pandan, Billiton I, Sumatra.

Remarks. CITES: Appendix II.

Prionodon pardicolor Spotted linsang

P. pardicotor [*sic*] Hodgson, 1841d: 57.

Range (Map 95). E Nepal to N Burma, S China (Yunnan, S Sichuan, Guizhou, SW Guangxi) and northern parts of Thailand, Laos and Vietnam.

Variation. One subspecific name has been proposed:
Pardictis p. presina Thomas, 1925: 499; Ngai-tio, Tonkin, N Vietnam, 4800ft.

Regional differences are very slight and any variation is probably continuous.

Remarks. CITES: Appendix I.

PARADOXURUS Palm civets

Paradoxurus Cuvier, in Geoffroy & Cuvier, 1821:liv. 24; type species *P. typus* Cuvier, in *P. hermaphroditus*. Conserved by Opinion 384 (1956).
Platyschista Otto, 1835: 1089 (*P. pallasii* Otto, in *P. hermaphroditus*).
Bondar Gray, 1865d: 531 (*Viverra bondar* Desmarest, in *P. hermaphroditus*).
Macrodus Gray, 1865d: 536 (*P. macrodus* Gray, in *P. hermaphroditus*).

Indomalayan Region; 4 species (Table 141). Predominantly arboreal.

Paradoxurus hermaphroditus Common palm civet, Toddy cat

Viverra hermaphrodita Pallas, in Schreber, 1777: 426; ?India. Conserved by Opinion 384 (1956).
V. nigra Desmarest, 1820: 208; Pondicherry, India. Preoccupied by *V. nigra* Peale & Beauvois, 1796 (in *Mephitis mephitis* Schreber, the North American striped skunk).
= *P. typus* Cuvier, in Geoffroy & Cuvier, 1821:liv. 24.
= *P. niger* Blanford, 1885b: 792.

Map **95**. *Viverricula, Prionodon*

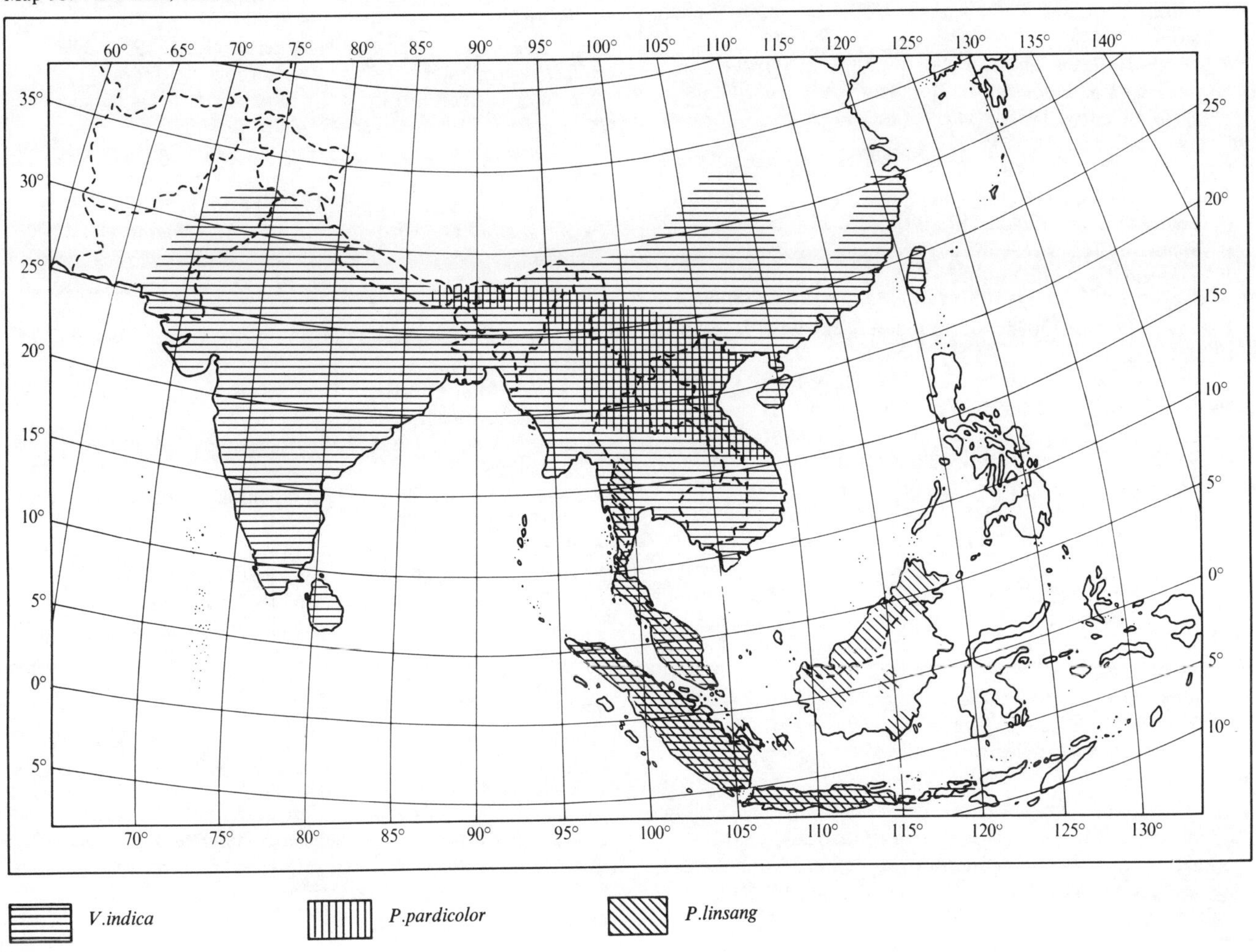

Table 141. The species of *Paradoxurus*.

	hermaphroditus	*zeylonensis*	*jerdoni*	*lignicolor*
Range	Widespread	Sri Lanka	S India	Mentawai Is
Head & body	420–690	460–560	550–610	450–485
Pelage	Mottled	Uniform yellowish brown	Greyish brown ± grizzled on lower back	Uniform yellowish brown
Hair on nape	Normal	Reversed	Reversed	Normal
Skull: greatest length	100–125	90–106	109–115	97–107
Incisive foramina	4.6–5.6	4.5–6.5	7.0–9.6	(3.9)
P^4: crown length of outer margin	7.4–8.6	6.9–8.1	8.2–9.2	(8.5)

V. prehensilis Desmarest, 1820: 208; Bengal. Pre-occupied by *V. prehensilis* Kerr, 1792 (in *Potos flavus* (Schreber), the kinkajou of central America).
V. bondar Desmarest, 1820: 210; Bengal.
V. musanga Raffles, 1821: 252; Sumatra.
P. leucopus Ogilby, 1829: 301; locality unknown.
P. pallasii Gray, 1832: 67; India.
P. pennantii Gray, 1832: 66; Higher Province of Bengal.
P. dubius Gray, 1832: 66; Java.
P. crossi Gray, 1832: 67; India.
P. hirsutus Hodgson, 1836a: 72; Nepal Terai.
P. musangoides Gray, 1837a: 579; Sumatra.
P. philippinensis Jourdan, 1837c: 523; Philippines.
P. felinus Wagner, 1841a: 349; India.
P. setosus Jaquinot & Pucheran, 1853: 25, pl. 6; Seram, Moluccas.
P. strictus Hodgson, in Horsfield, 1855: 105; Nepal Terai.
P. quadriscriptus Hodgson, in Horsfield, 1855: 106; Nepal.
P. nigrifrons Gray, 1865d: 535; India.
P. macrodus Gray, 1865d: 538; ? Java.
P. nictitans Taylor, 1891: 429, pl. ; Kondmals, Orissa, India.
P. minor Bonhote, 1903a: 9; Kampong Jalor, S Thailand.
P. canescens Lyon, 1906a: 597; Billiton I, E Sumatra.
P. brunneipes Miller, 1906b: 269; Kundur I, Riau Is.
P. padangus Lyon, 1908b: 655; Padang I, E Sumatra.
P. minax Thomas, 1909b: 375; Davao, Mindanao, Philippines.
P. torvus Thomas, 1909b: 375; Bungau I, Tawitawi Is, Sulu Is, Philippines.
P. sabanus Thomas, 1909b: 376; Spitang, N Borneo.
P. kangeanus Thomas, 1910c: 385; Kangean I, N of Java.
P. vicinus Schwarz, 1910a: 230; probably Assam.
P. sumbanus Schwarz, 1910b: 422; Sumba I, Lesser Sunda Is.
P. celebensis Schwarz, 1911: 634; Bonthain, SE Sulawesi.
P. cochinensis Schwarz, 1911: 635; Saigon, Vietnam.
P. exitus Schwarz, 1911: 636; Fumai, E of Canton, Gwangdong, China.
P. parvus Miller, 1913b: 1; Simalur I, W Sumatra.
P. ravus Miller, 1913b: 1; Trang, peninsular Thailand.
P. birmanicus Wroughton, 1917: 51; Mingun, near Sagaing, upper Burma.
P. jorandensis Ali *et al.*,1988: 45; Joranda, Simlipal, Mayurbhanj, Orissa, India, 20°45'N, 85°40'E, 440m (a partial albino of *P. hermaphroditus*—holotype examined).

Range (Map 96). N Pakistan (Rana, 1989), India and S China (Yunnan, S Sichuan, Guangdong, Guangxi) to Malaya; Sumatra, Java, Borneo, Palawan, Sri Lanka, Hainan; most associated small islands, including the Mergui and Riau groups and Enggano; also present, probably by introduction, on the Philippines (most islands), Sulawesi, the Lesser Sunda Is to Timor, and the Moluccan islands of Seram, Sula Is and Kai Is (and the Aru Is, New Guinea).

Variation. The following subspecific names have been proposed:
Viverra musanga var. *javanica* Horsfield, 1824:pl. ; Java.
P. typus var. *fuliginosus* Gray, 1832: 65; S India.
P. typus sumatranus Fischer, 1829: 159; Sumatra.
P. h. milleri Kloss, 1908: 143; Tioman I, Malaya.
P. h. hanieli Schwarz, 1912a: 131; Baung, Amarassie, Timor I.
P. h. senex Miller, 1913b: 3; Domel I, Mergui Is, Burma.
P. h. fuscus Miller, 1913b: 3; James I, Mergui Is. Pre-occupied by *P. zeylanicus* var. *fuscus* Kelaart, 1852 (in *P. zeylonensis*).
= *P. h. milleri* Ellerman & Morrison-Scott, 1951: 288. Pre-occupied by *P. h. milleri* Kloss, 1908.
= *P. h. conisbeei* Ellerman & Morrison-Scott, 1952: 399.
P. h. pallens Miller, 1913b: 4; Kisseraing I, Mergui Is.
P. h. sacer Miller, 1913b: 4; St Matthew I, Mergui Is.
P. h. pulcher Miller, 1913b: 5; Clara I, Mergui Is.
P. h. canus Miller, 1913b: 5; Teratau I, SW Thailand.
P. h. simplex Miller, 1913b: 6; Bangka I, Sumatra.
P. h. enganus Lyon, 1916a: 442; Enggano I, W Sumatra.
P. h. laotum Gyldenstolpe, 1917: 26; Chieng Hai, NW Thailand.
P. philippinensis baritensis Lönnberg, 1925: 61; Barito R, central Borneo.
P. h. rindjanicus Mertens, 1929: 333; Rindjani, Lombok, Lesser Sunda Is, 1200m.
P. h. lehmanni Mertens, 1929: 334; Sumbawa-Besar, W Sumbawa, Lesser Sunda Is.
P. h. balicus Sody, 1933a: 439; Boelelang, N Bali.
P. h. scindiae Pocock, 1934a: 176; Guna, Gwalior, India.
P. h. laneus Pocock, 1934a: 178; Gopalpur, Kangra, Himachal Pradesh, India, 5200ft.
P. h. vellerosus Pocock, 1934a: 181; Kashmir.
P. h. cantori Pocock, 1934b: 635; Penang I, Malaya.
P. h. hainanus Wang & Xu, 1981: 446; Dongfang, Hainan I, China. Pre- occupied by *P.* (*Paguma*) *larvatus hainanus* Allen, J. A.,1909 (in *Paguma larvata*).
= ***Paradoxurus hermaphroditus dongfangensis*** Corbet, *nom. nov.* .

In spite of the proliferation of names, geographical variation is not very great. Northern forms tend to have any pattern in the pelage diffuse or absent, especially in long winter coat; in Malaya the pelage is short and the pattern of spots and stripes clearer. Those in the Sunda islands tend to be small, with brownish, agouti dorsum and little dorsal pattern. Some insular forms (e. g. Kangean Is and Philippines) are very dark, e. g. 2/8 examined from Negros, Philippines were a uniform dark brown without pattern or agouti banding, the others being dark but with a pattern visible. Several albino or partial albino individuals have been recorded, especially in Orissa, India (e. g. the types of *nictitans* and *jorandensis*).

The incidence and extent of a white tail-tip is very variable and erratic.

Ecology. Mainly nocturnal. Found in secondary growth, plantations and frequently around human habitation.

Paradoxurus lignicolor Mentawai palm civet

P. lignicolor Miller, 1903a: 44; N Pagi I, Mentawai Is.
P. hermaphroditus siberu Chasen & Kloss, 1928c: 817; Siberut I, Mentawai Is.

Range (Map 96). Mentawai Islands of Siberut, Sipura, N Pagai and S Pagai.

Remarks. This species is a uniform light brown (the hairs without dark bases) and is scarcely distinguishable from *P. zeylonensis*, but it lacks the reversed hair on the nape characteristic of that species in all eight specimens examined. A few individuals of *P. hermaphroditus* from the Philippines (Negros, Basilan) are brownish with minimal pattern, but these have the basal zone of the pelage grey. In spite of the variability of *P. hermaphroditus*, of 556 specimens examined none closely matches *P. lignicolor* in colour of the pelage. Specimens from Enggano I, south of the Mentawai Is (*enganus* Lyon) and from Simeulue I, north of the Mentawai Is (*parvus* Miller) are referable to *P. hermaphroditus*, not to *P. lignicolor*. IUCN: E.

Paradoxurus zeylonensis Golden palm civet

Viverra zeylonensis Pallas, in Schreber, 1778: 451; Sri Lanka.
= *V. zeylanica* Gmelin, 1788:89.
P. aureus Cuvier, F.,1822: 48, pl. 4; locality unknown.

Range (Map 96). Sri Lanka, apart from the dry zones of the north and east.

Variation. The following subspecific names have been proposed:
P. zeylanicus var. *fuscus* or *montanus* Kelaart, 1852: 39–40; Newera Eliya, Sri Lanka.

Remarks. This species coexists with *P. hermaphroditus* on Sri Lanka. In some individuals a very faint trace of darker longitudinal stripes is visible but the otherwise uniform light golden brown pelage and reversal of hair on the nape are constant.

Map **96**. *Paradoxurus, Paguma, Macrogalidia*

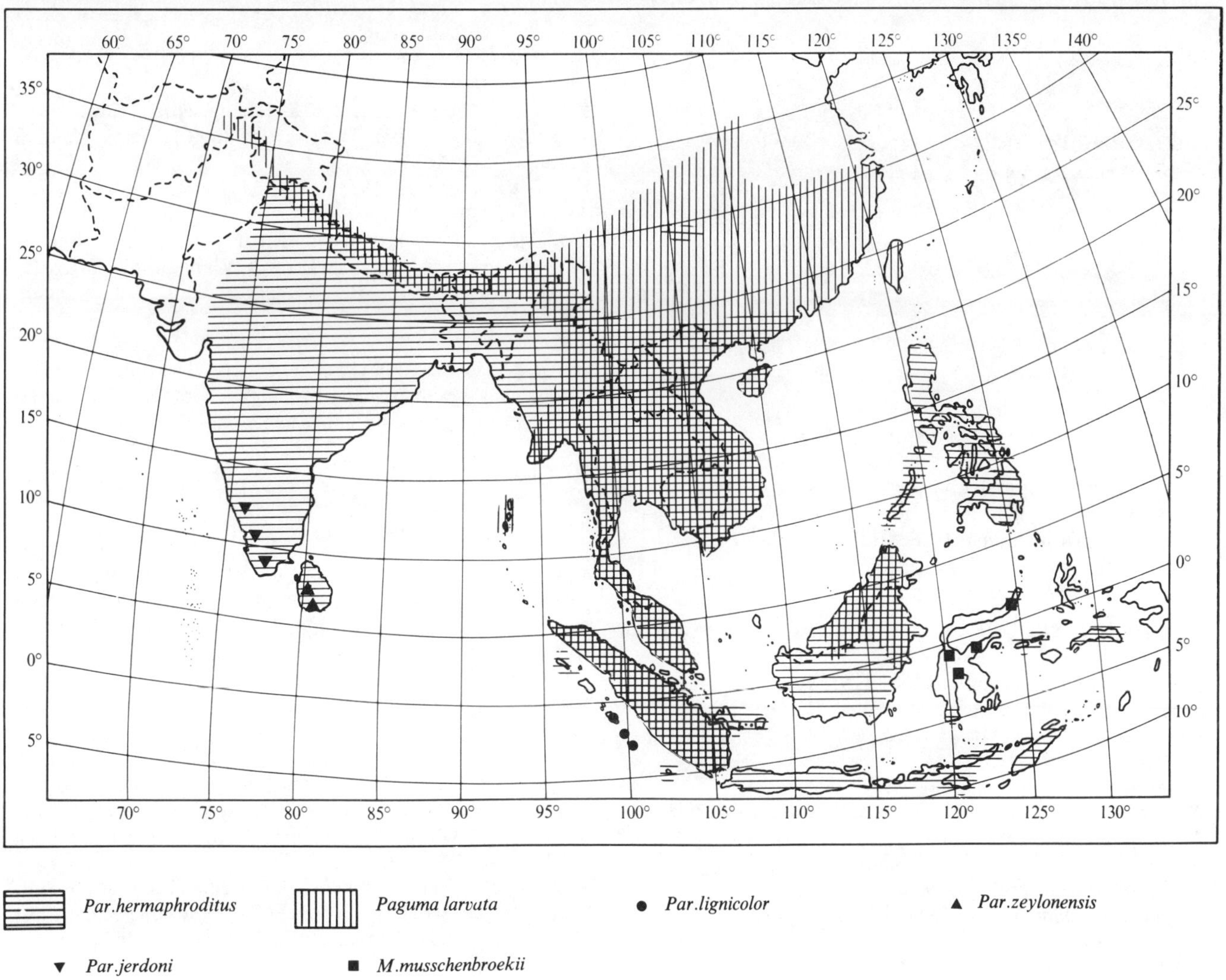

Paradoxurus jerdoni Jerdon's palm civet

P. jerdoni Blanford, 1885a: 613 (and 1885b: 802); Kodaikanal, Palni Hills, S India.

Range (Map 96). SW India from Coorg southwards, apparently co-existing with *P. hermaphroditus*.

Variation. One subspecific name has been proposed:
P. j. caniscus Pocock, 1933c: 865; Virajpet, S Coorg, SW India, 3000ft. This form, at the northern end of the range, has a facial pattern as in *P. hermaphroditus*, although more diffuse, and the posterior half of the dorsum more extensively grizzled and contrasting with the uniform anterior region.

The tail-tip is white in 4 out of 14 examined, but the reversed lie of the hair on the nape is constant throughout the species.

Remarks. This form could with some justification be considered conspecific with *P. zeylonensis* but these both appear to be sympatric with *P. hermaphroditus*. Cranially *P. jerdoni* is distinguishable from both of these by the longer incisive foramina and by the larger size and distinctive shape of P^4 which has the medial lobe broadly confluent with the rest of the tooth, not separated by a posterior constriction. IUCN: I.

PAGUMA

Paguma Gray, 1831b: 95; type species *Gulo larvatus* Smith.
Ambliodon Jourdan, 1837b: 445 (*Paradoxurus jourdanii* Gray).

Paguma larvata

One species. This genus differs from *Paradoxurus* mainly in the absence of spots or stripes in the dorsal pelage (but these are visible in young animals), in the posterior elongation of the palate, and smaller teeth. Cranial and dental characters link it more closely with *Arctictis* than with *Paradoxurus*.

Paguma larvata Masked palm civet

Gulo larvatus Smith, 1827: 281, pl. ; locality unknown, probably S China.
Paradoxurus grayi Bennett, 1835b: 118; India.
Paradoxurus nipalensis Hodgson, 1836a: 76; Nepal.
Paradoxurus lanigerus Hodgson, 1836a: 79; N Nepal, perhaps Tibet.
Paradoxurus leucomystax Gray, 1837b: 88; Sumatra.
P. jourdanii Gray, 1837a: 579; locality unknown, probably Malacca, Malaya.
Viverra (*Paradoxurus*) *aurata* Blainville, 1842:liv. 10, 96; locality unknown, probably Malaya.
Paradoxurus ogilbyi Fraser, 1849: 46, pl. 10; locality unknown, probably Borneo.
? *Paradoxurus rubidus* Blyth, 1858: 275; locality unknown.
Paradoxurus tytlerii Tytler, 1864: 188; Viper I, Port Blair, S Andaman I.
Paguma leucocephala Gray, 1865d: 541; Borneo.
Paradoxurus robustus Miller, 1906g: 26; Trang, peninsular Thailand.
P. reevesi Matschie, 1907: 183; Hing-an-fu, China.

Range (Map 96). Himalayas from N Pakistan east to much of China and south to Malaya; Hainan, Taiwan, Sumatra, Borneo; Andaman Is; introduced to Honshu and Shikoku, Japan. A report of its occurrence on Palawan may be erroneous (Heaney *et al.*, 1987).

Variation. The following subspecific names have been proposed:
P. l. var. *taivana* Swinhoe, 1863: 354; Taiwan.
P. l. hainana Thomas, 1909b(c. 10 Apr.): 377; Five Finger Mts (Wuchih), Hainan I, China.
Paradoxurus (*Paguma*) *l. hainanus* Allen, J. A.,1909a(17 Apr.): 240; Cheteriang, Hainan I, China.
P. l. intrudens Wroughton, 1910: 793; Sima, Myitkyina, upper Burma.
P. grayi wroughtoni Schwarz, 1913a: 289; Gharial, near Murree, N Punjab, Pakistan.
P. l. annectens Robinson & Kloss, 1917: 243; Bukit Gantang, Perak, Malaya.
P. l. vagans Kloss, 1919a: 73; Sikawtur, 40 miles NW of Raheng, Thailand, 1500ft.
P. l. yunalis Thomas, 1921h: 617; Yenyuensien, S Sichuan, China.
P. l. rivalis Thomas, 1921h: 618; Ichang, Hubei, China.
P. leucomystax janetta Thomas, 1928b;101; Bankachon, S Burma.
P. l. neglecta Pocock, 1934a: 334; Mokokchung, Naga Hills, Assam, 4500ft.
P. l. nigriceps Pocock, 1939: 424; Nam Tamai, upper Burma.
P. l. chichingensis Wang, in Peng & Wang, 1981: 167, 175; Chiching, Gaoligong Mts, Gongshan Xian, NW Yunnan, China.

Geographical variation is very extensive. The following main groups can be recognized although there is likely to be extensive intergradation between the mainland forms.

Himalayas and Taiwan (*P. l. grayi* etc.). Head pattern obscure; pale median stripe dull and confined to head; dorsum greyish; hairs on nape usually normal but sometimes reversed.

China, Laos, Vietnam, N Burma (*P. l. larvata* etc.). Head pattern conspicuous: median stripe white and extending from snout to nape or between shoulders, bordered with black; clear white oblique stripe between eye and ear; neck dark; dorsum yellowish brown. The head pattern becomes less bold in N and E China.

Malaya (*P. l. annectens*). Head with diffuse, off-white median stripe, not extending to nape but linking with large off-white cheek stripes; dorsum brown, nape darker; hair on nape reversed; tail with short white tip (2–3cm) in some individuals.

Sumatra (*P. l. leucomystax*). Median stripe absent or obscure; head pale; pale patch in front of ear large and conspicuous; rest of pelage very dark; hair on nape reversed; tail with long white tip (6–12cm; n = 5).

Borneo (*P. l. leucocephala*). As Sumatra but white tail-tip short (2–3cm); dorsum less dark in some.

Andaman Is (*P. l. tytlerii*). As Himalayas but short-haired; hair on nape reversed.

Ecology. Arboreal and terrestrial. Heavily exploited in China as food (Tan, 1989).

MACROGALIDIA

Macrogalidia Schwarz, 1910b: 423; type species *Paradoxurus musschenbroekii* Schlegel.

Sulawesi; one species. Close to *Paradoxurus* and *Paguma*, but with more robust teeth (especially the premolars) and with the surfaces of all the teeth showing a pattern of small, elongate pits, most noticeable on the canines. (This latter feature also occurs in *Paguma* but to a lesser degree.)

Macrogalidia musschenbroekii Brown palm civet

Paradoxurus musschenbroekii Schlegel, 1879: 43; N Sulawesi.

Range (Map 96). Central, E and NE Sulawesi, in lowland and montane forest.

Variation. One subfossil subspecies has been described: *M. m. meridionalis* Hooijer, 1950a: 89; Bola Batoe Cave, near Badjo, Barebo dist, c. 20 km SW of Watampone, central Bone, SW Sulawesi. Distinguished only by larger teeth.

Ecology. Coexists with *Viverra tangalunga* but mainly in primary forest; feeding upon small mammals and fruit (Wemmer & Wathing, 1986).

Remarks. IUCN: R.

ARCTICTIS

Arctictis Temminck, 1824: 21; type species *Viverra binturong* Raffles. Conserved by Opinion 384 (1956).

One species. A specialized relative of *Paradoxurus* and *Paguma*, characterized by large size, very long pelage and prehensile (but wholly furred) tail.

Arctictis binturong Binturong

Viverra? binturong Raffles, 1821: 253; Malacca. Conserved by Opinion 384 (1956).
Paradoxurus albifrons Cuvier, F., 1822: 44, 48; Bhutan, E Himalayas.

Map 97. *Arctictis, Arctogalidia, Hemigalus, Diplogale, Cynogale*

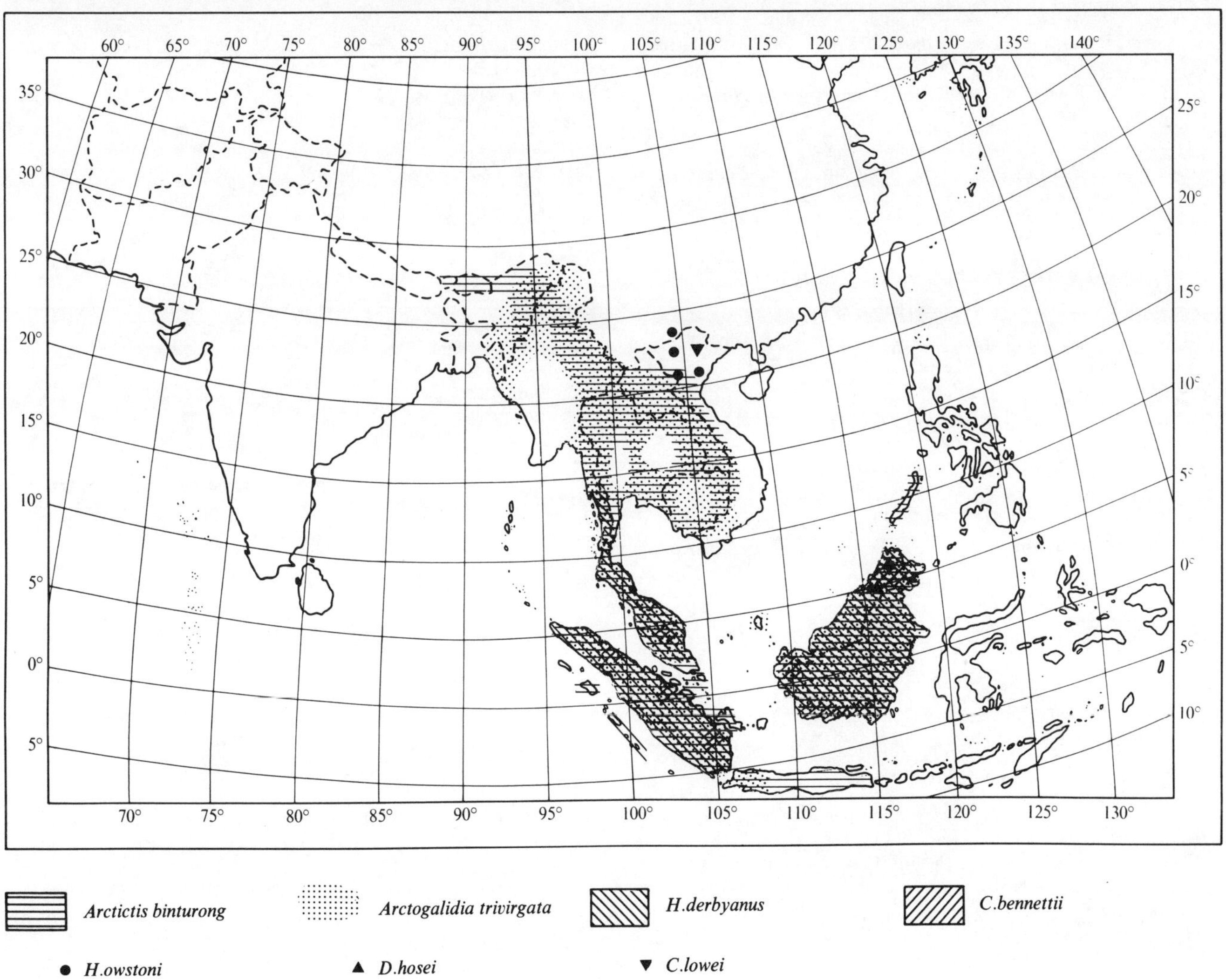

A. penicillatus Temminck, 1841a: 310; Java.
A. pageli Schwarz, 1911: 636; Sandakan, N Borneo.
Arctitis [*sic*] *whitei* Allen, J. A.,1910: 15; Iwahig, Palawan, Philippines.
A. gairdneri Thomas, 1916i: 270; Sai Yoke, SW Thailand.
A. niasensis Lyon, 1916a: 443; Nias I, W Sumatra.

Range (Map 97). From Sikkim to Burma, SW Yunnan and Indochina to Malaya; Sumatra, Java, Borneo, Palawan; and the small islands of Bintan and Kundur (Riau Is), Nias (W Sumatra), Bangka (E Sumatra). Probably once occurred in Nepal (Pocock, 1939: 434) and confirmed in Sikkim in 1982 (Wirth, 1989), but apparently always rare and local west of Burma.

Variation. The following subspecific name has been proposed:
A. b. kerkhoveni Sody, 1936a: 43; Bangka I, E Sumatra.
A. b. menglaensis Wang & Li, in Gao *et al.*,1987: 296; Yunnan, China.

On most of the mainland and Sumatra the pelage is uniformly dark except for a white frosting on the head, throat and fore legs. On Borneo and Java and in Yunnan the frosting extends throughout the whole pelage.

Ecology. Mainly in mature forest; arboreal.

ARCTOGALIDIA

Arctogale Gray, 1865d: 542; type species *Paradoxurus trivirgatus* Gray. Pre-occupied by *Arctogale* Kaup, 1829 (in *Mustela* L.).
= *Arctogalidia* Merriam, 1897: 302. Conserved by Opinion 384 (1956).

One species. An arboreal form, most similar to *Paguma* but differing especially in the absence of a perineal scent gland in the male.

Arctogalidia trivirgata Small-toothed palm civet

Paradoxurus trivirgatus Gray, 1832: 68; locality unknown: 'Moluccas'— W Java designated in error by Jentink (1887: 95)— see van Bemmel (1952: 25). Conserved by Opinion 384 (1956).
P. trilineata Wagner, 1841a: 346; W Java (van Bemmel, 1952: 39–40).
P. leucotis Horsfield, 1851: 66; Tenasserim, Burma. Attributed by van Strien (1986) to Hose (1893), but it is adequately described by Horsfield (1851).
P. stigmaticus Temminck, 1853: 121; Dusun R, S Borneo.
P. prehensilis Desmarest; Sclater, 1877: 681, pl. 71. Not *P. prehensilis* Desmarest, 1820 which is in *P. hermaphroditus*.
A. inornata Miller, 1901a: 131; Bunguran I, N Natuna Is.
A. simplex Miller, 1902a: 156; Lingga I, E Sumatra.
A. major Miller, 1906g: 25; Trang, peninsular Thailand.
A. fusca Miller, 1906b: 269; Kundur I, Riau Is.
A. minor Lyon, 1906a: 599; Billiton I, E Sumatra.
A. tingia Lyon, 1908b: 652; Tebing Tinggi I, E Sumatra.
A. sumatrana Lyon, 1908b: 653; Makapan, E Sumatra.
A. macra Miller, 1913b: 6; Domel I, Mergui Is.
A. mima Miller, 1913b: 7; Batam I, Riau Is.
A. depressa Miller, 1913b: 8; Bintang I, Riau Is.
A. bicolor Miller, 1913b: 7; Klumpang Bay, SE Borneo.
A. bancana Schwarz, 1913b: 297; Simpang, Bangka I, Sumatra.
A. millsi Wroughton, 1921b: 600; Mokokchung, Naga Hills, Assam, 5000ft.

Range (Map 97). From Assam, Yunnan and Indochina to Malaya; Sumatra, W Java (very rare), Borneo; and the small islands of Domel, King's (Mergui Is), Teratau (SW Thailand), Langkawi (Malaya), Singapore, Batam, Bintan, Kundur, Sugi, Bulan (Riau Is), Lingga, Sinkep (Lingga Is), Bangka, Belitung, Tebing Tinggi (E Sumatra); Pinie (Batu Is), Bunguran (N Natuna Is).

Variation. Variation, especially in the Sunda part of the range, was described by van Bemmel (1952) who recognized seven subspecies in the region based on minor features of pelage and size. Three main groups can be recognized:

Mainland north of the Isthmus of Kra (*A. t. leucotis*): pelage paler, more yellowish; dark pelage on head confined to muzzle or forehead; median white streak on muzzle usually present; distal part of ear pale; tail dark.

Malaya, Sumatra, Borneo (*A. t. trivirgata*): darker; whole of head and usually also nape darker than rest of dorsum; white streak on muzzle usually present; ears and tail dark.

Java (*A. t. trilineata*): as *A. t. trivirgata* but pelage long and shaggy; no white streak on muzzle; tail with terminal and/or subterminal light bands.

Ecology. Mainly nocturnal and arboreal, in mature forest.

Remarks. IUCN: I.

HEMIGALUS

Hemigalus Jourdan, 1837b: 442; type species *H. zebra* Gray (in *H. derbianus*).
Chrotogale Thomas, 1912e: 17 (and 1912f: 499) (*C. owstoni* Thomas).

Table 142. The species of *Hemigalus*.

	derbyanus	*owstoni*
Range	Malaya, Sumatra, Borneo	Indochina
Head & body	500–565	560–620
Transverse dark bands extending to flanks	0	+
Fore legs, sides of neck and throat spotted	0	+
Rostrum very elongate, teeth small and widely spaced (Fig.18)	0	+

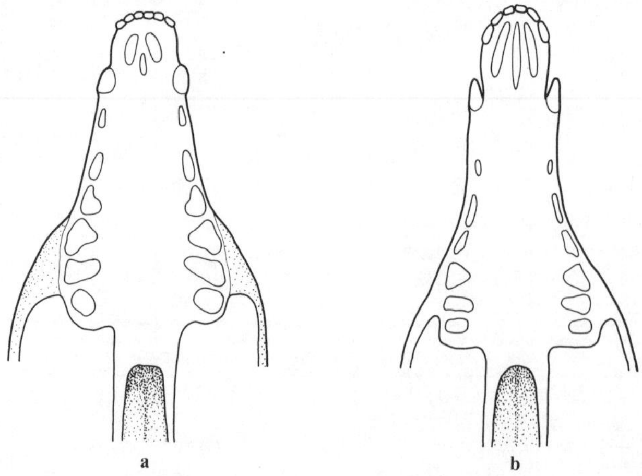

Fig. **18**. *Hemigalus*. Skulls of (a) *H.derbyanus*; (b) *H.owstoni*.

Sunda and Indochinese subregion; 2 species (Table 142). Characterized by a complex pattern of transverse dark bands on the dorsum and proximal part of the tail, and by the curvature of the incisor rows, the posterior incisors being more or less flattened and trenchant. *H. owstoni* is usually given separate generic rank in *Chrotogale* but the near-identity of the complex pattern, the median incisive foramen, and the curvature of the incisor rows all appear to indicate close relationship, although the last character is much more extremely developed in *H. owstoni*, and accompanied by elongation of the rostrum and reduction in size of the teeth (Fig. 18).

Hemigalus derbyanus Banded palm civet

Viverra hardwickii Gray, 1830: 9; Malaya. Pre-occupied by *V. hardwichii* Lesson, 1827, in *Prionodon*.
= *H. derbyanus invisus* Pocock, 1933b: 1006.
Paradoxurus derbyanus Gray, 1837a: 579; Malaya.
= *Viverra derbyi* Temminck, 1841a: 343.
P. zebra Gray, 1837a: 579; locality unknown.
V. boiei Müller, 1838: 144; SE Borneo.
Hemigale minor Miller, 1903a: 43; S Pagi I, W Sumatra.

Range (Map 97). Peninsular Thailand and Malaya; Sumatra, Borneo; islands of Siberut, Sipura and S Pagai (Mentawai Is).

Variation. Two subspecific names have been proposed:
H. d. incursor Thomas, 1915k: 613; Bankachon, Tenasserim, Burma.
Hemigalea d. sipora Chasen & Kloss, 1928c: 817; Sipora I, Mentawai Is.

Those in Malaya, Sumatra and Borneo do not appear to be separable, and the same applies to the form on the Mentawai Is which does not show any tendency towards melanism as in many other taxa on the Mentawai Is.

Ecology. Live mainly on the ground in mature forest, especially by rivers.

Remarks. CITES: Appendix II.

Hemigalus owstoni Owston's banded civet

Chrotogale owstoni Thomas, 1912e: 17 (and 1912f: 500); Yen Bai, Songkoi R, Tonkin, N Vietnam.

Range (Map 97). N Vietnam, N Laos, S Yunnan and SW Guangxi (Schreiber *et al.*, 1989).

Remarks. IUCN: K.

DIPLOGALE

Diplogale Thomas, 1912e: 18; type species *Hemigale hosei* Thomas.

One species. Although frequently included in *Hemigalus* there are many differences. *D. hosei* lacks the presumably derived complex pattern of the pelage, median incisive foramen and the curved incisor rows shared by the species of *Hemigalus*, and has more prominent accesory cusps on the anterior premolars than any other viverrid.

Diplogale hosei Hose's civet

Hemigale hosei Thomas, 1892a: 250; Mt Dulit, Sarawak, Borneo, 4000ft.

Range (Map 97). Mountains of NW Borneo, above 600m.

CYNOGALE

Cynogale Gray, 1837b: 88 (and 1837a: 579); type species *C. bennettii* Gray.
? *Lamictis* Blainville, 1837: 279 (*Viverra carcharias* Blainville).
Potamophilus Müller, 1838: 140 (*P. barbatus* Müller).

Two species (Table 143). A distinctive genus, strongly adapted for aquatic life with dense woolly underfur and very thick, long vibrissae.

Table 143. The species of *Cynogale*.

	bennettii	*lowei*
Range	Malaya, Sumatra, Borneo	Vietnam
Face (below eye) and throat	Dark (lighter only in inter-ramal region)	White
Dorsal guard hairs: underfur (length)	<2x	>2x
Dorsal guard hairs with light bands	+	0
Tail (as % of head & body)	c.25%	c.30%
Vibrissae: colour	White	Mostly brown
: size of longest (mm)	c.90 x 0.32	c.50 x 0.25

Cynogale bennettii Sunda otter-civet

C. bennettii Gray, 1837c: 88; Sumatra.
? *Lamictis carcharias* Blainville, 1837: 279; Java, presumably in error.
Potamophilus barbatus Müller, 1838: 142; Borneo.

Range (Map 97). Malaya; Sumatra, Borneo; doubtfully in peninsular Thailand (based on captive specimen).

Ecology. Aquatic, feeding mainly on fish, invertebrates and fruit.

Remarks. CITES: Appendix II; IUCN: K.

Cynogale lowei Tonkin otter-civet

C. lowei Pocock, 1933b: 1034; Backan, Tonkin, N Vietnam, 500ft.

Range (Map 97). N Vietnam; perhaps also the valley of the Red R in S Yunnan.

Remarks. This species is known by only a single poorly preserved skin in the BMNH. Although it was treated as a subspecies of *C. bennettii* by Ellerman & Morrison-Scott (1951), it differs in so many respects as to make that unlikely. The face, sides of neck, throat and breast are white; the pelage, with short but dense underfur and long, unbanded guard hairs, is not matched by any age-group of *C. bennettii* which all have the protruding parts of the guard hairs much shorter than the underfur and conspicuously banded; the vibrissae are thinner and shorter (but still larger than in other Viverridae); and the tail is relatively longer.

Although not confirmed in China, Professor Wang Yingxiang (pers comm.) saw skins of animals trapped by fishermen at a lake near the Red R, Yunnan in 1978, 1982 and 1985 which he suspected were of this species.

Family Herpestidae

Mongooses

Herpestina Bonaparte, 1845: 1; type genus *Herpestes* Illiger. Conserved by Direction 53 (1956).

Africa and Indomalayan regions; c. 35 species of which 8, all in the genus *Herpestes*, occur in the Indomalayan Region (Table 144). They are long-bodied, short-legged terrestrial carnivores, often diurnal. They have often been included in the family Viverridae but were treated as a separate family by Pocock (1939) and by Wozencraft (1989a). They are characterized especially by the uniquely derived nature of the anal scent gland and the structure of the auditory bullae which have the ectotympanic expanded and a unique circular meatal tube. For a general account, with emphasis on behaviour, see Hinton & Dunn (1967). For a bibliography of literature see Dargel (1990).

The only fossils reported from the region are of Miocene age in the Attock district of Pakistan whence three separate but unnamed species were recorded by Barry (1983).

HERPESTES

Herpertes Illiger, 1811: 135; corrected to *Herpestes* : 302; type species *Viverra ichneumon* L. Conserved by Opinion 384 (1956).
Mangusta Horsfield, 1822 (*Ichneumon javanicus* Geoffroy).
Urva Hodgson, 1837b: 561 (*Gulo urva* Hodgson).
= *Mesobema* Hodgson, 1842: 910.
Calogale Gray, 1865d: 560 (*H. nepalensis* Gray, in *H. javanicus*).
Calictis Gray, 1865d: 564 (*H. smithii* Gray).
Taeniogale Gray, 1865d: 569 (*H. vitticollis* Bennett).
Onychogale Gray, 1865d: 570 (*Cynictis maccarthiae* Gray, in *H. fuscus*).

Herpestes urva

Africa and Indomalayan Region; 12 species of which 8 are Indomalayan (Table 144). All species of *Herpestes* that have been examined have a unique peculiarity of the karyotype (shared only with the African *Atilax paludinosus* which should probably be included in *Herpestes*) in which the functional Y-chromosome has been translocated to an autosome. The males therefore have one fewer chromosome than the females—2n = 35 and 36 in the six Asiatic species examined (Fredga, 1972). On the basis of the karyotype, Fredga considered *edwardsii, auropunctatus* (in *H. javanicus*) and *fuscus* to be

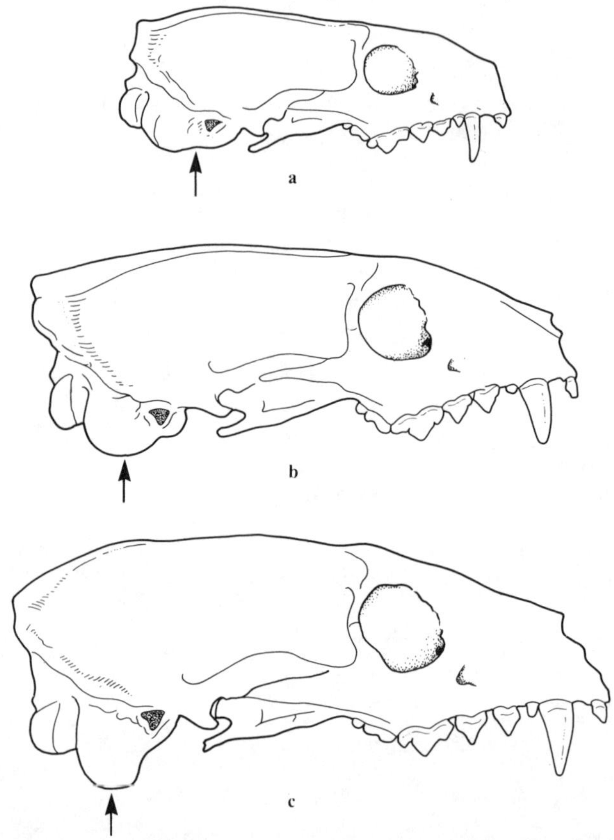

Fig. 19. *Herpestes*. Profile of skulls of (a) *H. javanicus*; (b) *H. fuscus*; and (c) *H. vitticollis*, to show different degrees of inflation of the posterior chamber. (After Pocock, 1941.)

Table 144. The species of *Herpestes*.

	javanicus	*edwardsii*	*smithii*	*fuscus*	*vitticollis*	*urva*	*semitorquatus*	*brachyurus*
Range	Widespread	India etc., Sri Lanka	India, Sri Lanka	S India, Sri Lanka	S India, Sri Lanka	Nepal-Malaya	Sumatra, Borneo	Sunda
Head & body	250–410	360–450	390–470	330–480	430–530	480–590	400–460	365–470
Tail (as % of h & b)	60–80%	90–100%	90–100%	60–70%	50–70%	60–70%	>60%	50–55%
Hind feet	50–70	70–90	80–86	65–87	90–100	90–110	82–93	75–93
Dorsal pelage: length	5–20	20–30	30–40	25–35	50–60	40–50	10–20	15–25
: neat, agouti (A) or shaggy, mottled (M)	A	A	A	A	A	M	A	A
Stripe from ear to shoulder	0	0	0	0	Black, bordered white	White	White	0
Tail darker towards tip	0	0	0	0	+	0	0	0
Legs darker than dorsum	0	0	+	+	+	+	+	+
Skull: greatest length	57–84	70–80	78–92	77–93	103–108	94–102	c.85	84–100
Bulla with posterior chamber greatly inflated (Fig. 19)	0	0	+	+	+	+	+	+

closely related, and likewise *urva* and *brachyurus*. All these appeared more closely related to each other than to any of the African *Herpestes*.

Sulawesi was included in the range of *Herpestes* in a map by Michaelis (1969) but *Herpestes* was not listed for Sulawesi by Laurie & Hill (1954) nor by Musser (1974a).

Herpestes javanicus Small Asian mongoose

Ichneumon javanicus Geoffroy, E.,1818: 138; W Java.
? *I. ruber* Geoffroy, E.,1818: 138.
Mangusta auropunctata Hodgson, 1836b: 235; Nepal.
H. nepalensis Gray, 1837a: 578; N India.
? *H. exilis* Gervais, in Eydoux & Souleyet, 1841: 32, pl. 3; Tourane, Annam, Vietnam.
Mangusta pallipes Blyth, 1845b: 346; Kandahar, Afghanistan.
? *H. rutilus* Gray, 1861: 136; Cambodia.
H. persicus Gray, 1865d: 554; Mohammerah, W Iran.
H. rafflesii Anderson, 1875: 282; Sumatra.
Mungos rubrifrons Allen, J. A.,1909a: 240; Mt Wuchih, Hainan I, China.
Mungos perakensis Kloss, 1917e: 124; Taiping, Perak, Malaya.
Mungos incertus Kloss, 1917e: 125; Ongut, Trang, S Thailand.
Mungos siamensis Kloss, 1917f: 215; Muang Prae, N Thailand.
H. palustris Ghose, 1965: 174; Nalbani, North Salt Lake, 24-Parganas Dist, W Bengal, India.

Range (Map 98). From Pakistan and N India to S China (S Yunnan to Guangdong) and south to Malaya; Hainan; Java; also west to Iran and Iraq. Introduced to Ambon (Moluccas). Also widely introduced beyond the region for rodent control (mainly *H. j. auropunctatus*): in Hawaii and Fiji; on most of the Caribbean islands and N coast of South America; on Mauritius; on Mafia I, E Africa; and on the Adriatic islands of Yugoslavia (Lever, 1985; Nellis, 1989).

Variation. The following subspecific names have been proposed:
H. auropunctatus birmanicus Thomas, 1886a: 84; Pegu, Burma.
Mungos exilis peninsulae Schwarz, 1910a: 231; Bangkok, Thailand.
Mungos auropunctatus helvus Ryley, 1914: 661; Deesa, Palanpur, Gujarat, India.
H. j. orientalis Sody, 1936a: 44; Rambipoedji, Besoeki, E Java, 100m.
H. j. tjerapai Sody, 1949b: 163; Perlak, Aceh, N Sumatra.

Size increases from northwest to southeast, being least in N India and greatest in Java. The following figures for condylobasal length of skull (adult males) are taken from Pocock (1941): Pakistan and NW India mean 64mm (n=5); N India etc. 63 (n=5); Burma 68 (n=4); Thailand 72–76; Vietnam 78–79; Java 80–81.

Map **98**. *Herpestes* (part)

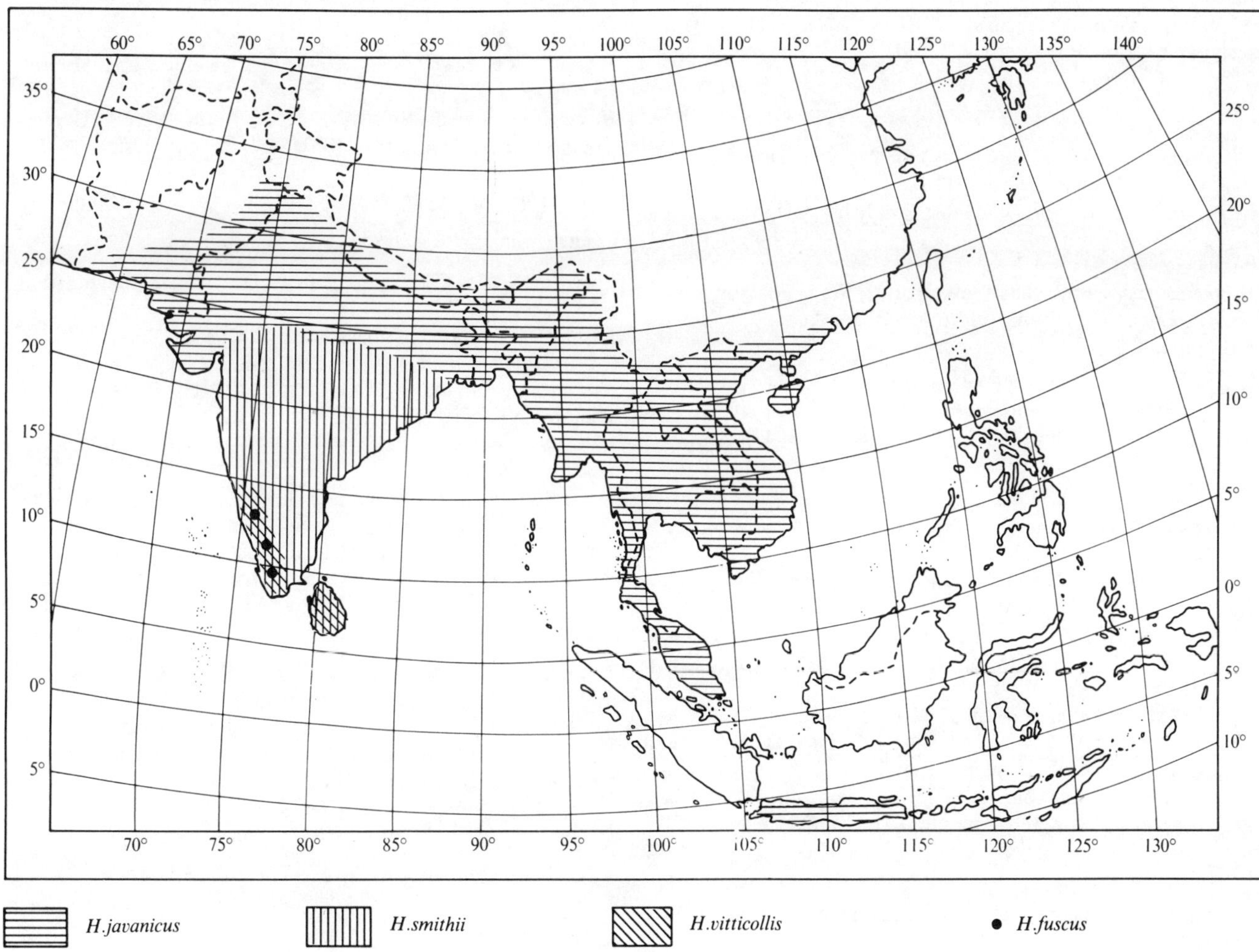

The pelage is palest in the more arid areas of NW India and Pakistan, darker but greyish in Assam and Burma, strongly suffused with reddish in parts of Vietnam and Java.

Ecology. Mainly diurnal, feeding on a wide variety of small vertebrates and large invertebrates. Confined to mesic habitats in the northwest, but frequently around human habitation.

Remarks. Two species, *H. javanicus* and *H. auropunctatus*, distinguished by size, were considered by Chasen (1940a) to be sympatric in Malaya. On this basis Ellerman & Morrison-Scott (1951) also treated the northern *H. auropunctatus* as specifically distinct from *H. javanicus* and have been widely followed. However the situation in Malaya was studied by Wells (1989) who concluded that this was an error due to sexual dimorphism in *H. javanicus*; that there was no evidence for sympatry; and that *H. auropunctatus* had not been validly recorded in Malaya, *perakensis* being a form of *H. javanicus s. s.* rather than of *H. auropunctatus*. Since there appear to be no other grounds for considering these two specifically distinct they are here combined, as was done by Pocock (1941) and by Lekagul & McNeely (1977).

An intensively studied species but mostly on the basis of introduced populations of '*H. auropunctatus*'; reviewed by Nellis (1989).

Herpestes edwardsii Indian grey mongoose

Viverra mungo Gmelin, 1788: 84 of many authors but this name is now reserved for the African banded mongoose, *Mungos mungo*.
Ichneumon edwardsii Geoffroy, E.,1818: 139; 'East Indies', designated as Madras (Pocock, 1933).
I. griseus Geoffroy, E.,1818: 157.
H. frederici Desmarest, 1823: 60; Malacca.
Mangusta malaccensis Fischer, 1829: 164; Malacca.
Mangusta (H.) nyula Hodgson, 1836b: 236; Nepal.
? *H. pallidus* Wagner, 1841a: 311.
H. ponticeriana Gervais, 1841: 32; Pondicherry, India.
H. ferrugineus Blanford, 1874: 661, pl. 18; Larkhana, Sind, Pakistan.
H. andersoni Murray, 1884b: 34; Kotree, Sind, Pakistan.
Mungos lanka Wroughton, 1915d: 53; Cheddikulam, N Prov., Sri Lanka, 177ft.

Range (Map 99). India NE to Assam, west to Pakistan; Sri Lanka; also to Iran and Arabia. Has been introduced in Malaya (but probably extinct there—Wells, 1989), to the Ryukyu Is (in 1910—Takashima, 1954) and to Mauritius. Recorded to at least 2100m in the Himalayas.

Variation. The following subspecific names have been proposed:
Mungos mungo pallens Ryley, 1914: 660; Palanpur, N Gujarat, India.
M. m. moerens Wroughton, 1915d: 52; Ganoor, Nimar, India.
M. m. ellioti Wroughton, 1915d: 52; Dharwar, India. Preoccupied by *H. ellioti* Blyth (in *H. smithii* Gray).
= *H. e. carnaticus* Wroughton, 1921d: 23.
H. griseus montanus Bechthold, 1936: 149; Hazara, N India.

Regional variation is slight; the degree of erythrism is individually variable but is more prevalent in the north, showing especially in the extremities: feet, tail and muzzle.

Ecology. Mainly diurnal. Extends into arid habitats, including sandy desert, in Pakistan. Preys upon a wide variety of vertebrates and invertebrates, including venomous snakes and scorpions.

Remarks. In captivity viable hybrids have been obtained with *H. ichneumon* (Gray, 1972).

Herpestes smithii Ruddy mongoose

H. smithii Gray, 1837a: 578; ? Bombay, India.
H. thysanurus Wagner, 1839b: 439; Kashmir.
H. ellioti Blyth, 1851a: 162; Carnatic region, SE India.
H. jerdonii Gray, 1865d: 550; Madras.
H. monticolus Jerdon, 1867: 135; inland from Nellore, SE India.
H. rubiginosus Kelaart, 1852: 43; not of Wagner, 1841.

Range (Map 98). Peninsular India, north at least to 25°N in Rajasthan and 24°N in Bihar; Sri Lanka.

Variation. The following subspecific names have been proposed:
H. s. zeylanius Thomas, 1921i: 24; Mankeni, E Prov., Sri Lanka.
H. s. rusanus Thomas, 1921i: 25; Sambhar, Rajasthan, India.
H. s. canens Thomas, 1921i: 25; Mt Abu, Rajasthan, India.

Two subspecies can be recognized although the differences are slight:
H. s. smithii. India. Tail 90–110% of head and body; pelage more greyish.
H. s. zeylanius. Sri Lanka. Tail 75–90% of head and body; pelage more reddish.

Herpestes fuscus Indian brown mongoose

H. fusca Waterhouse, 1838b: 55; India.
H. flavidens Kelaart, 1850b: 209 (323 of 1890 reprint); Kandy, Sri Lanka.
H. fulvescens Kelaart, in Blyth, 1851a: 162; Kandy, Sri Lanka.
Cynictis maccarthiae Gray, 1851: 131, pl. 31; Jaffna, N Sri Lanka.
? *H. ceylanicus* Nevill, 1887: 62; Trincomalee, Sri Lanka.

Range (Map 98). SW India, Sri Lanka.

Variation. The following subspecific names have been proposed:
H. flavidens phillipsi Thomas, 1924b: 240; Mousakande Estate, Gammaduwa, Central Prov., Sri Lanka.
H. flavidens siccatus Thomas, 1924b: 240; ? Aripo, near Mannar, N Prov., Sri Lanka.
H. fuscus rubidior Pocock, 1937: 233; Anasigalla, Matugama, W Prov., Sri Lanka.

In India the overall colour is dark greyish brown and the light subterminal bands on the hairs are creamy; in Sri Lanka the overall colour is very variable, from yellowish to reddish brown, rarely as dark as in India and the subterminal bands are always more yellowish or reddish. However too few records are available to establish how much of this is geographical.

Remarks. The inclusion of this form in *H. brachyurus*, as done by Bechthold (1936) followed by Wenzel & Haltenorth (1972), seems quite unjustified on morphological grounds and is not in accord with karyological data which links *H. fuscus* with *H. edwardsii* and *H. javanicus* (Fredga, 1972).

Herpestes vitticollis Stripe-necked mongoose

H. vitticollis Bennett, 1835a: 67; Travancore, S India.
Crossarchus rubiginosus Wagner, 1841a: 329; 'East Indies'.

Range (Map 98). SW India; Sri Lanka.

Variation. The following subspecific name has been proposed:
H. v. inornatus Pocock, 1941: 49; Chipgeri, N Kanara, India.

There is a gradual increase in reddish tones towards the south.

Herpestes urva Crab-eating mongoose

? *Viverra fusca* Gray, 1830:pl. 5.
= *Urva cancrivora* Hodgson, 1837b: 561–4.
Gulo urva Hodgson, 1836b: 238; Nepal.
Urva hanensis Matschie, 1907: 190; Hankow, China.

Range (Map 99). Nepal to Bangladesh (Reza Kahan, 1980; Farid Ahsan, 1989), Assam, S and E China (Yunnan and Sichuan—Jiangsu) and south to peninsular Thailand and Malaya (N Perlis and Selangor—Wells & Francis, 1988); Hainan, Taiwan.

Variation. The following subspecific names have been proposed:
H. u. annamensis Bechthold, 1936: 150; Phu Qui, Annam, Vietnam.
H. u. formosanus Bechthold, 1936: 151; Taiwan.
H. u. sinensis Bechthold, 1936: 152; Kwanting, S China.

Ecology. Nocturnal, found mainly near water where they prey upon fish, frogs, crabs, crayfish etc.

Herpestes semitorquatus Collared mongoose

H. semitorquatus Gray, 1846d: 211; Borneo.

Range (Map 99). Sumatra, Borneo.

Variation. One subspecific name has been proposed:
Mungos s. uniformis Robinson & Kloss, 1919c: 302; Mt Pasaman, Ophir district, W Sumatra.

Remarks. It has been suggested that this form might be conspecific with *H. urva* (Honacki *et al.*, 1982). This seems unlikely: it is much smaller, the pelage is very different, being short, neat and finely agouti-banded compared with very long, shaggy and coarsely mottled in *H. urva*. *H. semitorquatus* has also been included in *H. brachyurus* (Schwarz, 1947) but is distinguishable by its clear neck-stripe, redder pelage, longer tail and, in the Bornean specimens examined, the contrast between the pelage above and below the neck-stripe: dark above, lighter and reddish below.

Map **99**. *Herpestes* (part), *Hyaena*

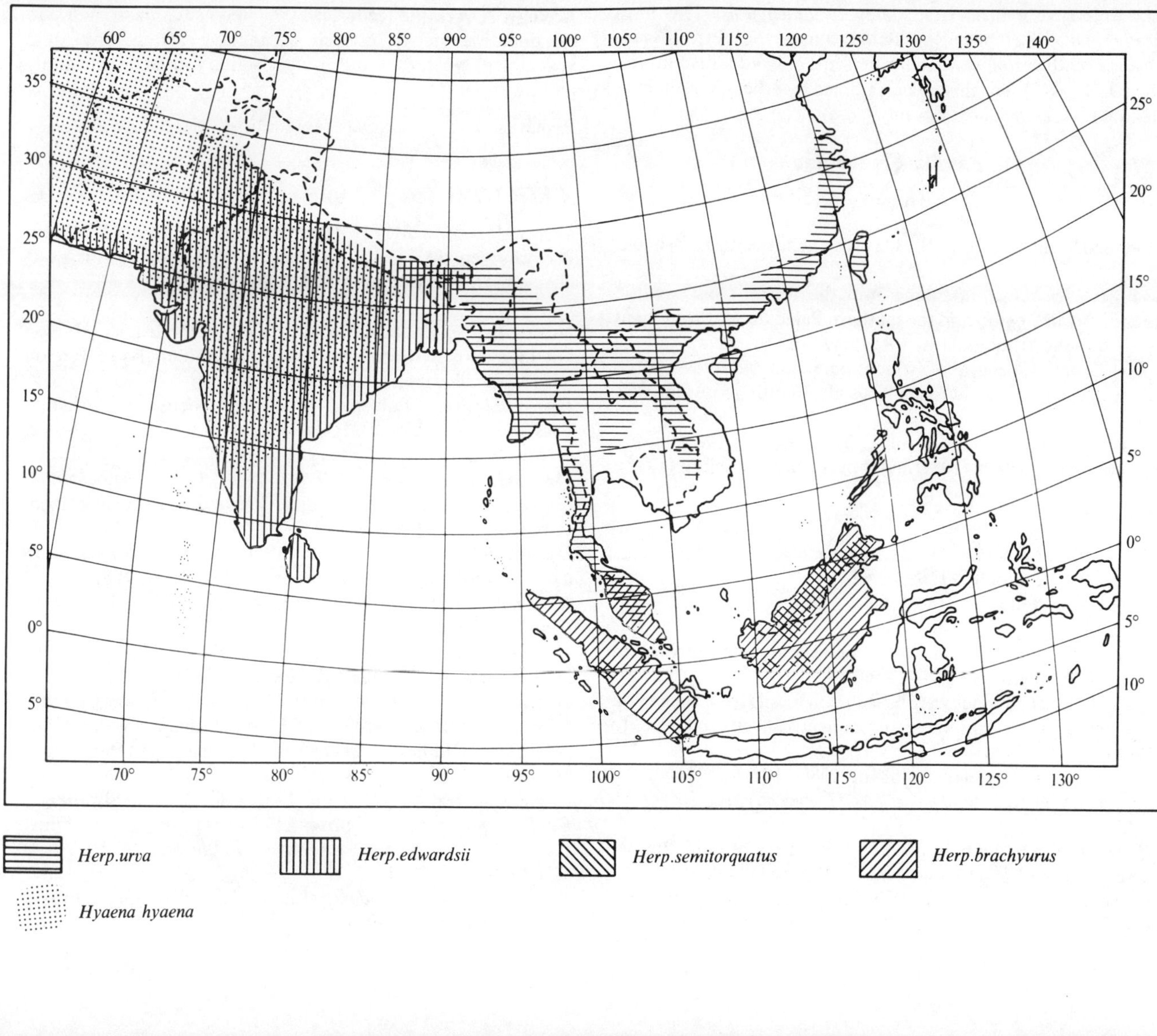

Herpestes brachyurus Short-tailed mongoose

H. brachyurus Gray, 1837c: 88; Malacca.
H. parvus Jentink, 1895: 48; Calamian Is (probably Busuanga), Philippines.
H. hosei Jentink, 1903: 226; Baram R, Sarawak, Borneo.
Mungos palawanus Allen, J. A.,1910: 17; Iwahig, Palawan I, Philippines.

Range (Map 99). Malaya, Sumatra, Borneo, Palawan and the adjacent small island of Busuanga (Calamian Is).

Variation. The following subspecific names have been proposed:
H. b. sumatrius Thomas, 1921j: 134; Deli, NE Sumatra.
H. b. rajah Thomas, 1921j: 135; Sarawak, Borneo.
H. b. dyacorum Thomas, 1921j: 135; Mt Dulit, Sarawak, Borneo.
H. b. javanensis Bechthold, 1936: 150; Java.

Specimens from Malaya, Sumatra, Borneo and Palawan are not clearly separable by pelage. A very faint longitudinal stripe on each side of the neck is detectable in some individuals, caused only by the concentration of pale agouti bands where the forward-lying hair of the throat meets the backward-lying hair of the nape.

Remarks. The Bornean form *hosei*, known only from the holotype, has usually been treated as an independent species but differs very little from other *H. brachyurus*. The form *semitorquatus* has been considered conspecific with *H. brachyurus* on the grounds that it represents only a reddish morph (Schwarz, 1947), but this is contradicted by other characters as detailed under *H. semitorquatus*.

Family Hyaenidae

Hyaenas

Hyaenadae [*sic*] Gray, 1821: 302; type genus *Hyaena* Brisson.

Mainly Africa but extending into the western part of the Indomalayan Region and the southern Palaearctic; throughout most of the Indomalayan Region, SE to Java, in the Pleistocene; 3 genera of which only one survives in the Indomalayan Region but another is abundantly represented as fossils.

Fossil hyaenas of the Siwalik Beds (W Himalayas) were reviewed by Vos *et al.* (1987) and those of Java by Schütt (1972).

HYAENA

Hyaena Brisson, 1762: 13, 168; type species *Canis hyaena* L.
= *Hyaena* Brünnich, 1771: 34, 42.
= *Euhyaena* Falconer, 1868: 464.

Two species in Africa, of which one extends into the western parts of the region.

The following names were based upon fossils from the region (excluding those that have subsequently been allocated to *Crocuta*, listed below):
H. macrostoma Lyddeker, 1884a: 298(=121), pl. 36; Jabi, Punjab. Allocated to *Thalassictis* (*Lycyaena*) by Vos *et al.* (1987).
H. brachygnatha Dubois, 1908: 1265; Java; Middle Pleistocene. Allocated to *H. brevirostris* by Schütt (1972).
H. carnifex Pilgrim, 1913: 312; Chinji, Salt Range, Punjab; Siwaliks. Based upon a mandible previously, but erroneously, associated with the holotype maxilla of *Dissopsalis carnifex* Pilgrim, 1910, a hyaenodontid. Allocated to *Crocuta* by Pilgrim (1932) and to *Percrocuta* by Vos *et al.* (1987).
H. (*Hyaenictis*) *bosei* Matthew, 1929: 493; Siwaliks. Allocated to *Hyaenictis* by Vos *et al.* (1987).
H. brevirostris licenti Pei. Reference not found.
H. brevirostris neglecta Kurten, 1956: 39; Jammu, Punjab.
H. licenti chaoxianensis Liu *et al.*,1982: 5; Chao Xian, Anhui, China; Middle Pleistocene.

Hyaena hyaena Striped hyaena

Canis hyaena Linnaeus, 1758: 40; Benna Mts, Laristan, S Iran.
= *H. striata* Zimmerman, 1780: 256.
= *H. orientalis* Tiedemann, 1808: 350.
= *H. fasciata* Thunberg, 1820: 59.
= *H. antiquorum* Temminck, 1820: 51.
= *H. virgata* Ogilby, in Royle, 1840:lxvi.
H. vulgaris Desmarest, 1820: 215; ? Egypt.

Range (Map 99). India, south to the Nilgiri Hills, north to the lowlands of Nepal (confirmed by Mitchell & Derksen, 1976), west to Pakistan; also through SW Asia and much of Africa south to Kenya.

Variation. One subspecific name is relevant:
H. vulgaris indica Blainville, 1844: 2, Hyènes, 82, pl. 6; India.

Ecology. A versatile feeder, preying upon small mammals and on domestic stock, as well as taking carrion, fruit and vegetables. Nocturnal and usually solitary or in small family groups.

Remarks. The species was reviewed by Rieger (1981).

[*CROCUTA*]

Crocuta Kaup, 1828:1145; type species *Canis crocuta* Erxleben. Pre-occupied by *Crocuta* Meigen, 1800 (Diptera) which has been suppressed (Opinion 678, 1962).

Now with a single species confined to the Afrotropical Region but many fossils from throughout the Indomalayan Region have been allocated to *Crocuta*:
Hyaena sivalensis Falconer, 1868:128; ? between Markanda Pass and Pinjor, Siwalik Hills. Allocated to *Crocuta* by Vos *et al.* (1987).
Hyaena sinensis Owen, 1870:422, pl.28; China. Allocated to *C.crocuta* by Colbert & Hooijer (1953). Recorded also from Cambodia (Bouchud, 1968).
Hyaena felina Bose, 1880:130; Siwalik Hills. Allocated to *Crocuta*, probably in *C. sivalensis*, by Vos *et al.* (1987).
H. colvini Lydekker, 1984a:113/240; Siwaliks; allocated to *Crocuta* by Badam (1979).
H.ultima Matsumoto, 1915;2, pl.1; Sichuan; allocated to *C. crocuta sinensis* by Colbert & Hooijer (1953).
H. gigantea Schlosser, 1903:35, pl.2; Shansi [= Shanxi], China.
C. mordax Pilgrim, 1932:150; Dhok Pathan, Siwalik Hills. Allocated to *Adrocuta exinia* (Roth & Wagner, 1854) by Vos *et al.* (1987).
C. gigantea var. *latro* Pilgrim, 1932:146; Hasnot; Siwaliks.
C. (*Percrocuta*) *grandis* Kurten, 1957:397; Nila, India; Dhok Pathan zone. Allocated to *Allohyaena* (*Dinocrocuta*) by Vos *et al.* (1987).
C. (*Percrocuta*) *macrodonta* Liu *et al.*, 1978:156; Lantian, Shaanxi, China; Pliocene.

Family Felidae

Cats

Felini Fischer, 1817: 372; type genus *Felis* L.

Worldwide except Australasia; c. 36 species of which 11 are wholly or widely Indomalayan and a further 4 occur on the fringes of the region. The family is characterized by short muzzle, with premolars reduced to 3/2 (often to 2/2) and the molars to 1/1; digitigrade feet with usually retractile claws; and large auditory bullae, divided by a septum. They prey mainly upon vertebrates, usually stalking and pouncing from cover and using the front paws and claws to strike.

Although the species are mostly well defined, generic classification has been extremely unstable and there is no current consensus. Amongst the Indomalayan species, the cheetah, *Acinonyx*, is generally accepted as generically distinct. The 'big cats' are also usually separated from *Felis* in the genus *Panthera* although the clouded leopard, here called *Pardofelis nebulosa*, is sometimes also included. The remaining cats have at one extreme all been included in *Felis* and at the other extreme have been distributed amongst many genera. A comprehensive classification recognizing 17 genera (worldwide) was produced by Pocock (1917b). More recent studies are those of Weigel (1961), based upon pelage pattern; Hemmer (1978), a broadly based classification; Kral & Zima (1980), based on karyotypes; Kratochvil (1982), based on karyotypes and bacula; Groves (1982a), based mainly on cranial and dental characters; Collier & O'Brien (1985), using immunological distance but not including all the Indomalayan species; and Herrington (1986), a cladistic study based upon a very wide spectrum of characters.

One of the most unstable aspects has been the relationship of the caracal (here *Felis caracal*) and the lynxes (not represented here). The classification used here is nearest to that of Herrington (1986), differing only in retaining *rubiginosus* in *Prionailurus*, and in recognizing the genus *Catopuma* as distinct from the African *Profelis*, as done by Hemmer (1978) and by Groves (1982a). It differs from that of Groves (1982a) only in including *Mayailurus* in *Prionailurus* and in keeping *Pardofelis* separate from *Panthera* rather than as a subgenus within *Panthera*.

A fairly 'split' classification has been used here, recognizing six genera (Table 145), in order to comply as closely as possible with classifications proposed by those who have undertaken detailed studies of the group. However much of the problem in the family is due to the rather small differences between species and the lack of congruence between different sets of characters. In particular the cranial differences that have played a large part in most classifications are mostly subtle ones of proportion to which it is very difficult to attribute polarity with any confidence and which are very likely to have arisen independently. There is therefore a case for reducing the genera to two or three. This is enhanced by the number of intergeneric hybrids that have been reported, albeit in captivity (Gray, 1972). Following the classification used here, these include viable hybrids between *Prionailurus bengalensis* and *Felis catus* (also recorded by Yosida, 1982); *P. bengalensis* and *F. chaus*; *Panthera pardus* and *Puma concolor*; (as well as intra-generic hybrids between species of *Felis* and *Panthera* as used here). Using the phylogeny of Herrington (1986) for example, this would lead to the inclusion of *Prionailurus* and *Acinonyx* in *Felis*; and *Catopuma* and *Pardofelis* in *Panthera*. The many possible phylogenies could then be explored with less disruption to the nomenclature.

Reviews of ecology and conservation are those of Eaton (1973) and Miller & Everett (1986).

Fossil and Recent Felidae of China, including several fossil species described from N China, were reviewed by Teilhard de Chardin & Leroy (1945b).

Additional Palaearctic species that occur on the northwestern fringe of the region from Pakistan to China, and may in places extend marginally into it, are *Felis manul* (= *Otocolobus manul*), *F. bieti* (China only), *Lynx lynx* and *Panthera uncia*, the last recorded down to 1500m in N Pakistan in winter.

All species are included in CITES Appendix II except those in Appendix I and *F. catus*.

FELIS

Felis Linnaeus, 1758: 41; type species *F. catus* L., the domestic cat. Conserved by Opinion 91 (1926) and Direction 24 (1955).
= *Catus* Fitzinger, 1855: 265.
Chaus Gray, 1843a: 44 (*F. chaus* Güldenstaedt).
= *Catolynx* Severtzov, 1858: 387.

Table 145. The genera of Felidae.

	Felis	*Prionailurus*	*Catopuma*	*Pardofelis*	*Panthera*	*Acinonyx*
Species in region	4	5	2	2	3	1
Head & body (cm)	40–76	40–61	63–84	46–104	107–200	140–150
Tail (as % of h & b)	25–60%	25–60%	50–65%	75–110%	40–60%	c.50%
Pelage predominantly spotted (Sp), striped (St), or uniform (U)	(Sp,St),U	Sp,U	St,U	Sp	Sp,St,U	Sp
Spots in groups forming reticulate pattern	0	0	0	+	0	0
Ears tufted	+/0	0	0	0	0	0
Ears with white patch on back	0	+	0	+/0	+/0	+
Hair on nape	Normal	Normal	Normal/ reversed	Normal	Reversed	Reversed
Hair on throat	Normal	Reversed	Normal/ reversed	Reversed	Reversed	Reversed
Tail white below	0	0	+	0	0	0
Claws retractile	+	+	+	+	+	0
Pupils (contracted)	Vertical	Vertical			Round	Round
Post-orbital bar usually complete	0	+	+/0	+/0	0	0
P^2 usually present	+/0	+/0	+	+/0	+	+/0

Table 146. The species of *Felis*.

	silvestris	*chaus*	*margarita*	*caracal*
Range in region	NW India etc.	India-China	W Pakistan	NW India
Head & body (cm)	46–59	56–72	c.57	64–76
Tail (as % of h & b)	50–60%	40–50%	50%	26–38%
Flanks with spots (Sp) or vertical stripes (St)	Sp	0	St (faint)	0
Double stripes behind and below eye	+	0	(+)	0
Back of ears: as dorsum (D); black (B); orange-brown with black tips (O/B)	D	O/B	O/B	B
Paws white	0	0	+	0
Ears tufted	0/+ (-10mm)	+ (15mm)	Short	+ (c.50mm)
Belly spotted	+	Faint/0	0	Faint/0
Tail banded	+	+	+	0
Legs banded	+	+	+	0
Foot pads covered with hair	0	0	+	0
Skull: greatest length	84–98	96–126	90	110–125
P^2 usually present	+	+	+	0

Caracal Gray, 1843a: 46 (*C. melanotis* Gray, in *F. caracal* Schreber).
= *Urolynchus* Severtzov, 1858: 389.
Eremailurus Ognev, 1926: 356 (*E. thinobius* Ognev, in *F. margarita*).

As used here confined to the Palaearctic and Afrotropical Regions and the northwestern part of the Indomalayan Region; c. 6 species (plus the domestic cat) of which 4 are represented in the Indomalayan Region (Table 146), the others being *F. nigripes* in S Africa and *F. bieti* in the steppes of W China and Mongolia. Of the species included here all but *F. caracal* have been included in *Felis* in most recent classifications. The genus has frequently been extended to include a variable number of the other cats, usually but not always excluding those placed here in *Panthera*. For detailed descriptions of the species and their variation see Pocock (1951).

External characters of the genus are tendencies to transverse (vertical) stripes on the flanks and terminal tufts on the ears.

The following nominal fossil species have been described in the genus:
F. cristata Falconer & Cautley, 1836c: 135, pl. 21; Siwalik Hills, India; Upper Siwaliks. Allocated to *Panthera* by Badam (1979).
F. subhimalayana Bronn, 1848: 492; Upper Siwaliks.
F. grandicristata Bose, 1880: 128 (in *F. cristata*, *fide* Lydekker, 1884a).
F. (*? Cynaelurus*) *brachygnathus* Lydekker, 1884a: 149; Siwaliks.

Felis silvestris Wild cat

F. (*Catus*) *silvestris* Schreber, 1777: 397; Germany. Conserved by Opinion 465 (1957).
F. lybica Forster, in Buffon, 1780: 313; Gafsa, Tunisia.
F. ornata Gray, 1830:pl. 2; India.
F. servalina Jardine, 1834: 232; India.
F. constantina; Pocock, 1939: 286; not *F. constantina* Forster, 1780: 313 (Algeria) which Pocock later (1951: 53) accepted as a synonym of *Leptailurus serval* (Schreber).
Also many extralimital synonyms.

Range (Map 100). Central and NW India, Pakistan; west throughout SW Asia, S and central Europe and most of Africa.

Variation. The Indian form, *F. s. ornata*, differs from those in Africa and Europe especially in having the body spotted, the spots tending to form vertical rows on the flanks (in Africa and SW Asia the body is almost completely un-patterned and in Europe there are transverse stripes). *F. s. ornata* is also smaller than those other forms but appears to intergrade with them (in both pelage and size) in Iran.

Ecology. In India mainly in scrub habitat, feeding predominantly upon rodents, especially *Meriones*, but also *Lepus* and birds (Kumar Sharma, 1980).

Remarks. This species is the ancestor of the domestic cat. It may be that *Felis bieti*, which occurs in the high steppes of Mongolia but reaches south to Sichuan, should be considered conspecific with *F. silvestris*.

Felis catus Domestic cat

Felis catus Linnaeus, 1758: 42; Sweden. Conserved by Opinion 91 (1926) and Direction 24 (1955).
F. torquata Cuvier, in Geoffroy & Cuvier, 1826:pl. 126; Nepal.
F. inconspicua Gray, 1837a: 577; Nepal.
F. megalotis Müller, 1840: 54; Timor, Indonesia.
F. huttoni Blyth, 1846a: 169; Kandahar, Afghanistan.
F. deliensis Hilzheimer, 1906a: 112; Deli, Sumatra.
And other extralimital synonyms.

Range. Domestic cats are kept throughout the region. Although there are feral populations in and around most human settlements the extent to which feral populations occur independently of man is uncertain.

Variation. The following subspecific names have been proposed from the region or are relevant to it:
F. c. domesticus Erxleben, 1777: 520; Sweden.
F. c. sinensis Kerr, 1792: 154; China.
F. c. madagascariensis Kerr, 1792: 155; Madagascar (but Pocock (1939) suggested that this form had been imported from Malaya).
F. libyca siamensis Trouessart, 1904: 273; Thailand.

Remarks. The evolution of the domestic cat was reviewed by Robinson (1982).

Felis chaus Jungle cat

F. chaus Gueldenstaedt, 1776: 483; Terek R, N of Caucasus.
F. affinis Gray, 1830(in 1830–35):pl. 3; Gangootri, Tehri Garhwal, N India.
F. kutas Pearson, 1832: 75; Midnapore, c. 70 miles W of Calcutta, Bengal, India.
Lynchus erythrotus Hodgson, 1836b: 233; Nepal.
F. jacquemontii Geoffroy, 1844: 58, pls 2,3; Kursali, near Dehra Dun, N India, 8500ft.

Range (Map 100). Vietnam, Thailand and SW China (Yunnan, S Sichuan) to India (throughout) and Pakistan; Sri Lanka; west to Egypt, Asia Minor and much of Soviet Turkestan. Fossil finds have been reported in Java (Hemmer, 1971).

Variation. The following subspecific names have been proposed:
F. affinis fulvidina Thomas, 1929b: 834; Kampong Tomb, Annam, Vietnam.
F. c. prateri Pocock, 1939: 298; Jacobabad, Sind, Pakistan.
F. c. kelaarti Pocock, 1939: 300; Cheddikulam, N Prov., Sri Lanka.
F. c. valbalala Deraniyagala, 1955a: 291; Karnool, S India. Based on size average larger than in Sri Lanka but not compared with other Indian forms.

It is unlikely that discrete subspecies can be recognized. Those that have been described have been based mainly upon small differences in size and more substantial ones in the length of the pelage which is longer at higher altitudes and latitudes. Some adults retain a slight trace of vertical stripes on the flanks which are more pronounced in kittens. Also variable is the presence of a mid-dorsal dark, brown stripe, especially on the posterior part of the dorsum. When well developed it tends to be narrowly bordered with black.

Ecology. Less strictly nocturnal than the other cats.

Remarks. Fertile female hybrids have been produced in captivity between *F. chaus* and *F. catus* (Gray, 1972).

Felis margarita Sand cat

F. margarita Loche, 1858: 49, pl. 1; near Negonca, Algeria. Including extralimital synomyms.

Range (Map 100). Extreme west of Pakistan; and west through the deserts of SW Asia and N Africa.

Variation. One subspecific name has been proposed from the region:
F. m. scheffeli Hemmer, 1974b: 32; Nushki Desert, Baluchistan, Pakistan.

Map **100**. *Felis*

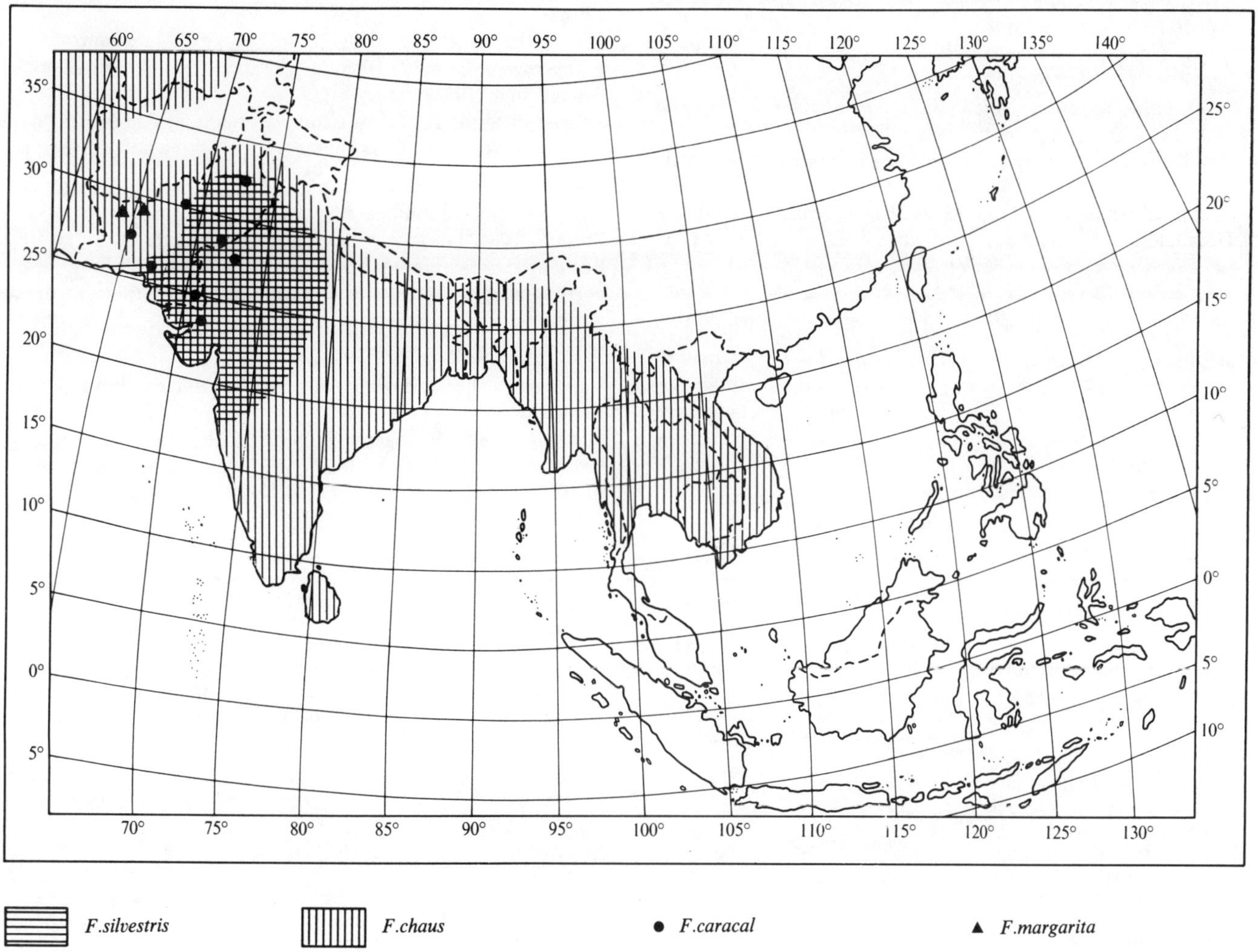

Remarks. Reviewed by Hemmer (1974a). IUCN: (E).

Felis caracal Caracal

F. caracal Schreber, 1776:pl. 110 (and 1777: 413); Table Mt, Cape Town, South Africa.
= *Caracal melanotis* Gray, 1843a: 46.

Range (Map 100). The arid parts of NW India and Pakistan; through SW Asia and most of the non-forested parts of Africa. Formerly east at least to 83°E in India (Mirzapur—Pocock, 1939); now very rare but surviving in parts of Rajasthan at least until 1986 (Stuart, 1984; Divyabhanusinh, 1987); considered very rare in Gujarat in 1987 (Chavan, 1987).

Variation. The following subspecific names have been applied to the Indian population:
F. c. bengalensis Fischer, 1829: 210; Bengal. Pre-occupied by *F. bengalensis* Kerr, 1792, = *Prionailurus bengalensis*.
F. (Caracal) c. schmitzi Matschie, 1912c: 64; Dead Sea region, Israel.

Ecology. A species of open arid country, usually found only remote from human habitation. A predator on small ungulates as well as on birds and small mammals.

Remarks. The taxonomic position of this species has been particularly unstable: on the one hand it has been included in or close to the genus *Felis*, either along with the species of *Lynx* (Groves, 1982a) or without *Lynx* (Kral & Zima, 1980; Herrington, 1986); on the other it has been placed, usually with *Lynx*, close to *Panthera* (Collier & O'Brien, 1985). CITES: Appendix I.

PRIONAILURUS

Prionailurus Severtzov, 1858: 387; type species *Felis pardochrous* Hodgson, in *P. bengalensis*.
Zibethailurus Severtzov, 1858: 387 (*Felis viverrinus* Bennett).
= *Viverriceps* Gray, 1867f: 268.
Ictailurus Severtzov, 1858: 387 (*Felis planiceps* Vigors & Horsfield).
= *Ailurogale* Fitzinger, 1869a: 249.
= *Ailurinus* Gervais, 1870: 159.
Mayailurus Imaizumi, 1967: 75 (*M. iriomotensis* Imaizumi).

Indomalayan Region; 5 species (Table 147). Frequently included in *Felis* but they form a moderately distinctive group characterized externally by a tendency for the spots to form longitudinal stripes, by a white patch on the back of each ear and usually by a reversal of hair on the throat. Cranial and dental characters are less consistent.

P. rubiginosus is usually included but was excluded by Herrington (1986) who considered that it was more closely related to *Felis* than to *Prionailurus*. The other species of *Prionailurus* share a presumed derived position of the foramen rotundum which is situated on the basicranium, not at the rear of the orbit.

Prionailurus bengalensis Leopard cat

Felis bengalensis Kerr, 1792: 151; S Bengal, India.
F. javanensis Desmarest, 1816: 115; Java.
? *F. undata* Desmarest, 1816: 115; Java.
F. sumatrana Horsfield, 1821:pl. ; Bencoolen, Sumatra.
F. minuta Temminck, 1824: 130; Java.
? *F. nipalensis* Horsfield & Vigors, 1829: 382; ? Nepal.
F. chinensis Gray, 1837a: 577; ? Canton, S China.
Leopardus ellioti Gray, 1842: 260; Bombay Presidency, India.
L. horsfieldii Gray, 1842: 260; Bhutan.
L. reevesii Gray, 1843a: 44; China.
F. pardochrous Hodgson, 1844b: 286; Nepal.
F. wagati Gray, 1867e: 400; Tenasserim, Burma.
F. tenasserimensis Gray, 1867e: 400; Tenasserim, Burma.
? *F. herschelii* Gray, 1869: 28; India.
F. scripta Milne-Edwards, 1872a: 92, pls 57,58; Sichuan, China.
F. punctulata Elliot, 1882: 2; Palembang, Sumatra.
F. ricketti Bonhote, 1903e: 374; Foochow, Fujian, China.
F. ingrami Bonhote, 1903f: 474; Van Gin Shan, N Guizhou, China.
F. tingia Lyon, 1908b: 658; Tebing Tinggi I, E Sumatra.
F. microgale Dubois, 1908: 1267; Java; fossil. Status *fide* Brongersma (1935).
F. sinensis Shih, 1930a: 4; Chinsiu, Guangxi, China.
Also some extralimital synonyms from N China and E Siberia.

Range (Map 101). From N Pakistan and throughout most of India and China, south to Malaya; Singapore, Sumatra, Java, Borneo, Bali, Lombok; the Philippine islands of Palawan, Busuanga, Cebu, Negros and Panay; Hainan and Taiwan; also north to Korea and the Amur Basin, but this northern form is sometimes considered specifically distinct, as *P. euptilura* (Elliot, 1871).

Variation. The following subspecific names have been proposed:
P. b. borneoensis Brongersma, 1935: 26; Rantau, SE Borneo.
P. b. trevelyani Pocock, 1939: 273; near Gilgit, Kashmir, 5000ft.

Table 147. The species of *Prionailurus*.

	bengalensis	*iriomotensis*	*viverrinus*	*planiceps*	*rubiginosus*
Range	Widespread	Ryukyu Is	Pakistan-Java	Malaya	India
Head & body (cm)	40–61	55–60	60–75	44–51	40–48
Tail (as % of h&b)	40–60%		<50%	27–34%	47–60%
Ground colour	Yellow-brown or grey-brown	Dark buffy brown	Grey-brown	Dark brown, reddish on head	Grey-brown
Spotted, with some dark lines on back	+	+	+	0	+
Tail banded	+	+	+	0	0
Skull: greatest length	86–102	96–99	123–151	95–100	73–79
Sagittal crest	Short/0	Short	Long	Long	Short
Nasals very narrow behind	0	0	+	+	+

P. b. alleni Sody, 1949b: 181; Nodoa, Hainan I, China. Lectotype selected by Lawrence & Musser (1990).
P. b. koenigswaldi Hemmer, 1971: 370; Sampung, near Ponorogo, central Java; Neolithic.
Felis b. hainana Xu & Liu, in Xu *et al.*,1983: 344; Hainan I, China.

The presence of P^2 is variable throughout the range: it was found to be absent on one or both sides in 40% of a sample of 65 from the northern part of the range (Himalayas and N Vietnam northwards) and in 11% of a sample of 62 from the southern part (Glass & Todd, 1977).

Ecology. A forest species avoiding open country but found also in plantations and suburban habitats.

Remarks. Conservation status was reviewed by Santiapillai & Suprahan (1985). The form *euptilurus* in N China and the Amur Basin has been considered specifically distinct from *P. bengalensis* on the basis of larger size, a pelage pattern more resembling *P. viverrinus* and some cranial characters (Heptner, 1971). However the nature of the boundary between these has not been clarified and Gao *et al.* (1987) continued to treat *euptilurus* as a race of *P. bengalensis*. CITES: Appendix I/II.

Prionailurus iriomotensis Iriomote cat

Mayailurus iriomotensis Imaizumi, 1967: 75; Haimida, Iriomote I, Yaeyama Is, Ryukyu Is, Japan.

Range (Map 101). Known only from the island of Iriomote, Ryukyu Is.

Remarks. The principal characters used by Imaizumi (1967) to justify the genus *Mayailurus* were: the ventral border of the paroccipital process separated from the auditory bulla; an oval disc on each side of the basisphenoid and basioccipital region; P^3 with postero-external corner evenly rounded, without a cingulum cusp; and auditory bullae unusually small. *P. bengalensis* is variable in all these respects and none provides a very clear distinction. It has been considered by most subsequent authors to be a derivative of *P. bengalensis* but its generic status has also been upheld on the grounds that it shows some cranial resemblances to *Catopuma* as well as to *Prionailurus* (Groves, 1982a). The G-banded karyotype is not distinguishable from that of *P. bengalensis* and *P. viverrinus*, nor from the American *Puma concolor* (Wurster-Hill *et al.*, 1987). IUCN: E.

Map **101**. *Prionailurus*

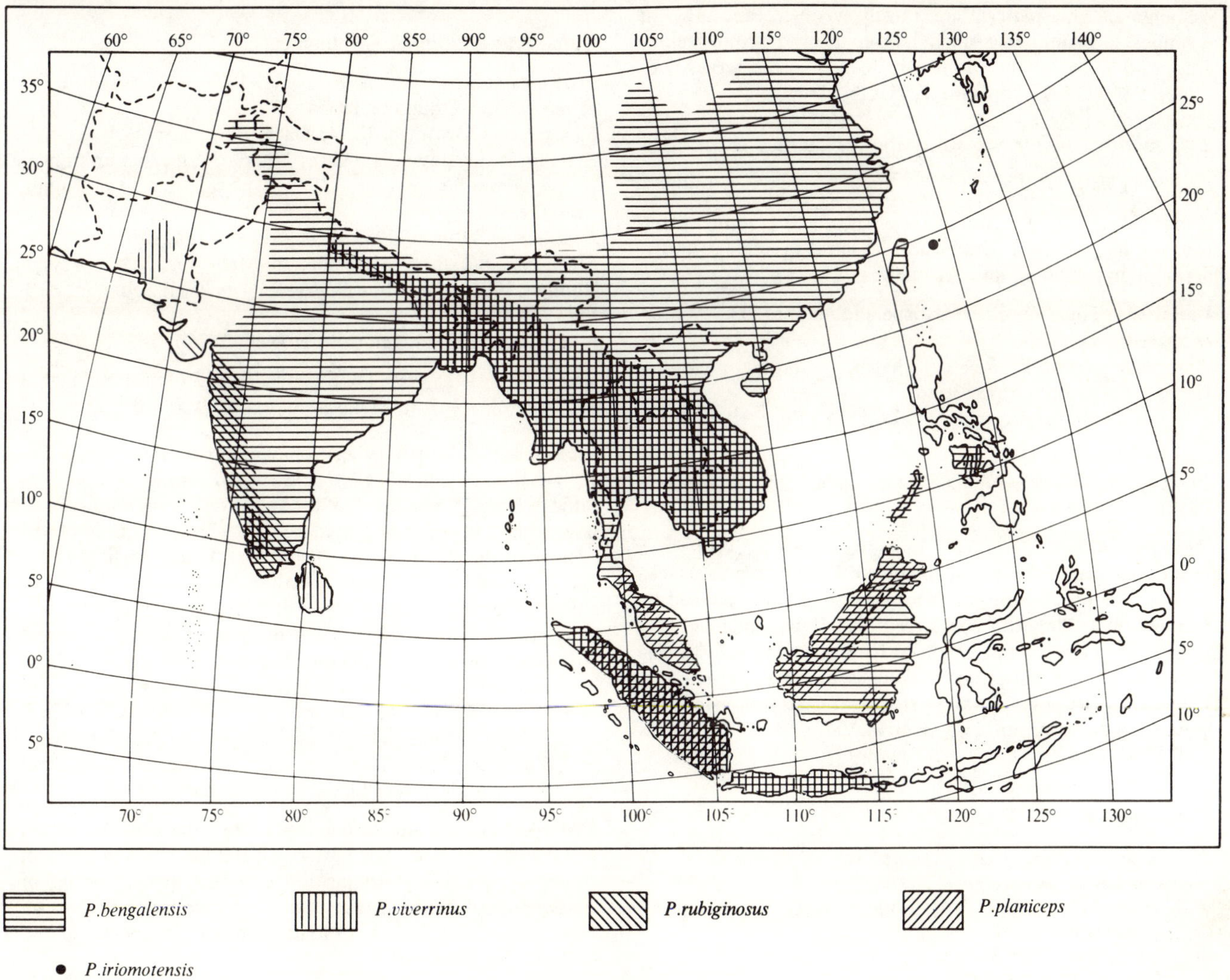

Prionailurus viverrinus Fishing cat

Felis viverrinus Bennett, 1833: 68; India, probably the Malabar coast.
= *Viverriceps bennettii* Gray, 1867f: 268.
F. himalayanus Jardine, 1834: 230, pl. 24; Himalayas.
F. viverriceps Hodgson, 1836b: 232; Nepal.

Range (Map 101). The lower Indus Valley in Pakistan, SW India, and from N India to Vietnam and Thailand; Sumatra, Java, Sri Lanka; records from Taiwan are probably erroneous and based upon traded skins.

Variation. One subspecific name has been proposed:
P. v. rhizophoreus Sody, 1936a: 45; Pamanoekan, Cheribon, N coast of W Java.

Ecology. Mainly in water-side habitats with dense cover, hunting both by swimming (e. g. after wildfowl) and on land (Roberts, 1977).

Prionailurus planiceps Flat-headed cat

Felis planiceps Vigors & Horsfield, 1827: 450; Sumatra.

Range (Map 101). Peninsular Thailand, Malaya, Sumatra and Borneo.

Remarks. Although lacking the bold pattern of other *Prionailurus*, this species shows the white-spotted ears and some elements of the pattern are faintly visible, especially in juveniles, e. g. numerous small dark spots on the hind feet. A unique, but presumably primitive, feature is the large, two-rooted P^2. CITES: Appendix I; IUCN: I.

Prionailurus rubiginosus Rusty-spotted cat

Felis rubiginosa Geoffroy, 1831: 140; Pondicherry, S India.

Range (Map 101). Peninsular India (north to Gujarat in the west— Chavan, 1987), Jammu and Kashmir (a single record—Chakraborty, 1978b), and Sri Lanka.

Variation. The following subspecific names have been proposed:
P. r. phillipsi Pocock, 1939: 278; Mousakanda, Gammaduwa, Central Prov., Sri Lanka, 3000ft.
P. r. koladivius Deraniyagala, 1956: 113; Kathiraveli, E Prov., Sri Lanka.

Ecology. In Gujarat, confined to moist teak forest (Chavan, 1987).

Remarks. Although normally included in *Prionailurus* when that genus is recognized as distinct from *Felis*, this species lacks some of the peculiarities of *Prionailurus* and was excluded by Herrington (1986) who considered that it was more closely related to *Felis ss* but might justify its own genus. The characters of the pelage that it shares with *Prionailurus*—e. g. longitudinal stripes and white patch on the back of each ear—were considered primitive by Herrington. CITES: Appendix I/II; IUCN: K.

CATOPUMA

Catopuma Severtzov, 1858: 387; type species *Felis moormensis* Hodgson, in *C. temminckii*.
Pyrofelis Gray, 1874b: 354 (*Felis temminckii* Vigors & Horsfield).
Badiofelis Pocock, 1932b: 749 (*Felis badia* Gray).

Indomalayan Region; 2 species (Table 148). Most recent classifications agree that these species should be excluded from *Felis*. They have often been included with the African golden cat, *Profelis aurata*, in the genus *Profelis*, and the phylogeny of Herrington (1986) supports that view. However there are considerable differences between these as emphasized by Hemmer (1978) and Groves (1982a) and there may even be a stronger case for combining *Catopuma* with *Prionailurus* and/or with *Pardofelis*.

Table 148. The species of *Catopuma*.

	temminckii	*badia*
Range	Nepal - Sumatra	Borneo
Head & body (cm)	69–84	63–69
Hind feet (cm)	16–18	12–14
Pelage	Uniform reddish or grey, or boldly spotted and striped	Uniform reddish brown or grey
Hair on nape and throat reversed	0	+
Transverse dark lines on throat	+	0
Sagittal crest in male	Long	Short

Catopuma temminckii Golden cat

Felis temminckii Vigors & Horsfield, 1827: 451; Sumatra.
F. moormensis Hodgson, 1831a: 177; Nepal.
F. nigrescens Gray, 1863b: 4; Darjeeling, N India.
F. tristis Milne-Edwards, 1871a: 223, pl. 31; locality unknown.
F. dominicanorum Sclater, 1898: 2, pl. 1; Foochow, Fujian, China.
F. semenovi Satunin, 1904: 524; NE Sichuan, China.
F. (*Catopuma*) *melli* Matschie, in Mell, 1922: 36; Weishi, Yunnan. Pre- occupied by *F.* (*Neofelis*) *melli* Matschie, in Mell, 1922: 35 (in *Pardofelis nebulosa*).
= *F. temminckii badiodorsalis* Howell, 1926b: 143.

Range (Map 102). Nepal to S China (N to S Gansu, S Shaanxi and Anhui), south to Malaya; Sumatra.

Variation. Variation in pelage colour and pattern is extreme but so few precisely localized specimens are available that it is difficult to judge how much of the variation is geographical and how much the result of polymorphism—both certainly occur. Four main forms are frequent, the first three with no pattern except on the head and faint spotting on the venter: (1) overall colour yellowish brown; (2) overall colour light greyish brown; (3) overall colour very dark brown to black; (4) boldly patterned with black spots and stripes on a brown or grey ground-colour. The patterned form (*tristis*) is recorded only from China and Nepal, possibly also N Burma. It is the dominant form in S Sichuan and N Yunnan but is less frequent than the plain forms in S Yunnan (Wang Y-x, pers. comm.). The few seen from Indochina, Malaya and Sumatra are un-patterned.

Remarks. The various colour forms are united by a constant pattern on the head. The pattern on the body, when present, shows some striking similarities with that of the species of *Pardofelis*, especially the presence of narrow dark crescents on each shoulder which are not found in any other species of cat. CITES: Appendix I; IUCN: I.

Catopuma badia Bornean bay cat

Felis badia Gray, 1874a: 322, pl. 49; Sarawak, Borneo.

Range (Map 102). Borneo; known from only a few widely scattered records, mostly in Sarawak.

Remarks. Although usually accepted as a vicariant of *P. temminckii*, this species was separated generically from *C. temminckii* and placed in *Pardofelis* by Leyhausen (1973), on the basis of behavioural characters, as had previously been done by Pocock (1917b) on the basis of cranial characters. However the external resemblance to *C. temminckii* is very close, including the fine agouti speckling in the grey morph and significantly the unique tail pattern with the ventral surface white, a pattern not found in either of the species included here in *Pardofelis* nor in *Prionailurus*. IUCN: R.

PARDOFELIS

Pardofelis Severtzov, 1858: 387; type species *Felis marmorata* Martin.
Neofelis Gray, 1867f: ´265 (*F. macrocelis* Temminck, in *P. nebulosa*).

Indomalayan Region; 2 species (Table 149). Except in very 'lumped' classifications, these two species have usually been placed in separate genera on the basis of differences in size and cranial shape. However the unique and complex pattern of the pelage is unlikely to be independently derived or primitive, and most recent classifications agree in linking them with each other and close to the genus *Panthera* (Hemmer, 1978; Herrington, 1986), or even within *Panthera* (Groves, 1982a). An alternative view was expressed by Leyhausen (1973) who included *badia*, but not *nebulosa*, in *Pardofelis*, on the basis of behavioural characters.

Table 149. The species of *Pardofelis*.

	marmorata	*nebulosa*
Head & body (cm)	46–53	69–104
Tail (as % of h & b)	100–110%	75–95%
Pale grey patch on back of ear	+	0
Skull: greatest length	88–100	140–190
Sagittal & occipital crests	Weak	Strong

Map **102**. *Catopuma, Pardofelis, Acinonyx*

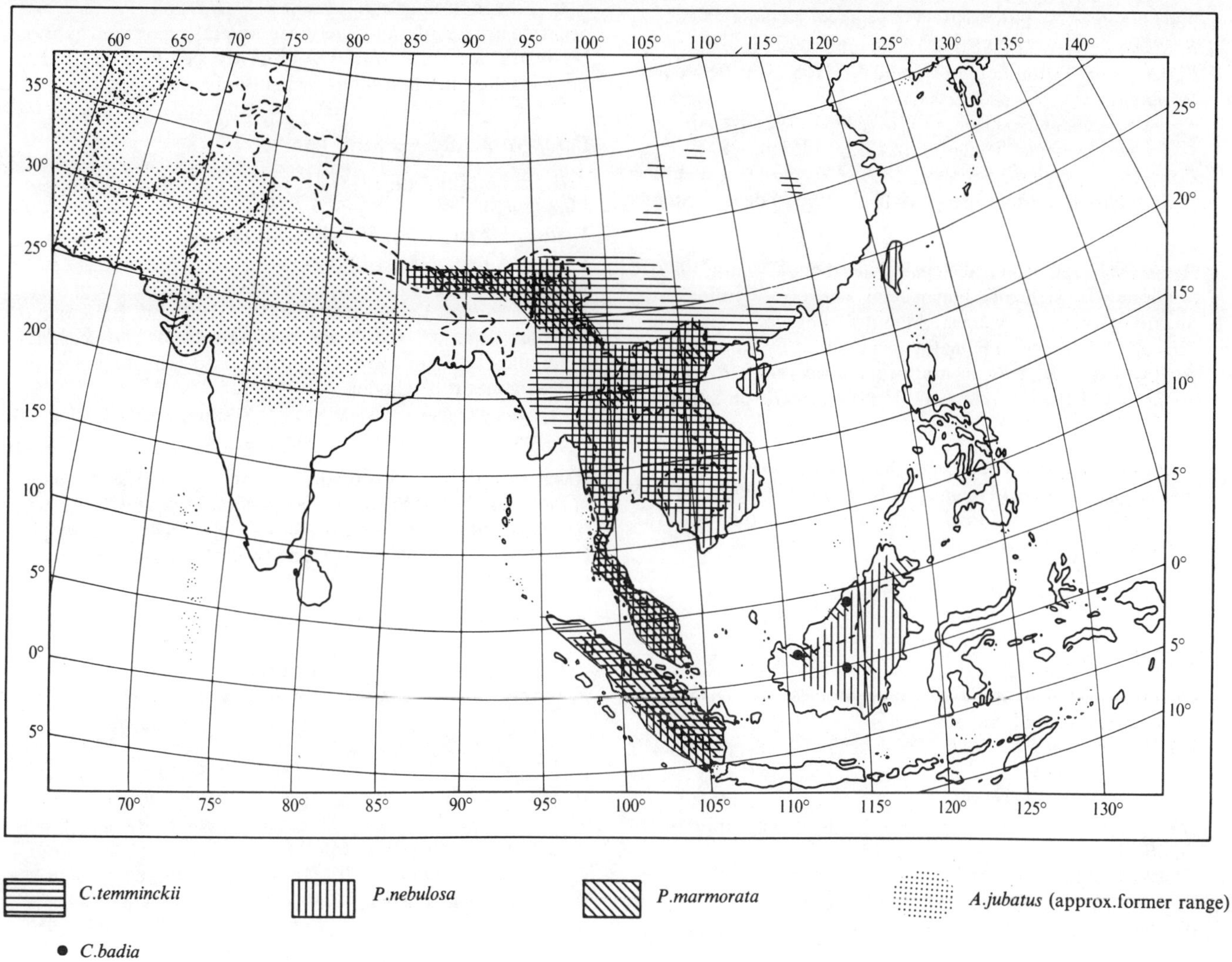

Pardofelis marmorata Marbled cat

Felis marmorata Martin, 1837: 108; Sumatra (Robinson & Kloss, 1919: 261).
? *F. longicaudata* Blainville, 1843: 47.
F. charltonii Gray, 1846d: 211; Darjeeling, N India.
F. ogilbii Hodgson, 1847d: 44; Sikkim.

Range (Map 102). Nepal to Assam, N Burma, NW and C Yunnan and N Vietnam, south to Malaya; Sumatra, Borneo.

Remarks. CITES: Appendix I; IUCN: I.

Pardofelis nebulosa Clouded leopard

Pardofelis nebulosa

Felis nebulosa Griffith, 1821: 37; Canton, S China.
F. diardi Cuvier, G.,1823: 437; Sumatra (Java in error).
F. macrocelis Temminck, in Horsfield, 1825a: 543; Bencoolen, W Sumatra.
F. macroselloides Hodgson, in Gray, 1853: 192; Nepal.
= *F. macrosceloides* Hodgson, in Gray, 1853:pl. 38.
Leopardus brachyurus Swinhoe, 1863: 352, pl. 43; Taiwan.
F. (*Neofelis*) *melli* Matschie, in Mell, 1922: 35; ? near Canton, S China.

Range (Map 102). Central Nepal (Dinerstein & Mehta, 1989) to S China (E Sichuan, Fujian and north to Qinling Mts, Shaanxi), south to Malaya; Sumatra, Borneo; Hainan and Taiwan. The range in Sumatra was mapped by Santiapillai & Ashby (1988). In Taiwan it appears to be very rare but was confirmed in 1983 (Rabinowitz, 1988). Formerly on Singapore I.

Variation. One fossil subspecies has been described:
Neofelis n. primigenia Hemmer & Koenigswald, 1964: 11; ? S China (type purchased in SE China but also recorded in Java); Pleistocene.

Remarks. CITES: Appendix I; IUCN: V.

PANTHERA

Panthera Oken, 1816: 1052; type species *Felis pardus* L. Conserved by Opinion 1368 (1985).
[*Tigris* Oken, 1816: 1066 (*Felis tigris* L.). Suppressed by Opinion 417 (1956).]
= *Tigris* Gray, 1843a: 40.
[*Leo* Oken, 1816: 1070 (*Felis leo* L.). Suppressed by Opinion 417 (1956).]
= *Leo* Brehm, 1829: 637.
Pardus Fitzinger, 1868b: 459 (*Felis pardus* L.).
Feliopsis Stremme, 1911a: 56 (*F. palaeojavanica* Stremme, in *P. tigris*).

As used here, represented in the Neotropical, Afrotropical and Indomalayan Regions, formerly also widespread in the Palaearctic (and in the Pleistocene also in the Nearctic); 5 species of which 3 occur in the Indomalayan Region (Table 150). One other, *P. uncia*, the snow leopard, occurs at high altitude on the northern fringes of the region. The genus has usually been characterized especially by a peculiarity of the hyoid apparatus in which the terminal bones are separated by a long elastic ligament.

Table 150. The species of *Panthera*.

	pardus	*leo*	*tigris*
Range in region	Widespread	NW India	Widespread
Head & body (cm)	−140	−200	−215
Body pelage	Spotted ± obscured by black	Uniform	Striped
Ad.male with mane	0	+	0
Ruff on cheeks	0	+	+
Tuft on tail-tip	0	+	0
Skull: greatest length	c170–200	250–310	270–375
Anterior tip of nasals level with anterior (A) or posterior (P) edge of canine	A	P	A

In zoos hybridization between *P. leo* and *P. tigris* has occurred frequently and there are several reports of hybrids between *P. leo* and *P. pardus*. A cross between *P. pardus* and *P. tigris* resulted in an aborted foetus (Gray, 1972).

Panthera pardus Leopard, Panther

Felis pardus Linnaeus, 1758: 41; Egypt. Conserved by Opinion 1368 (1985).
F. fusca Meyer, 1794: 394; Bengal, India.
F. melas Cuvier, 1809: 152; Java.
F. variegata Wagner, 1841a: 483; Java.
F. longicaudata Valenciennes, 1856: 1036; Sri Lanka and Malabar coast of India. Pre-occupied by *F. longicaudata* Blainville, 1843 (in *Pardofelis marmorata*) but attributed by Valenciennes to F Cuvier.
Leopardus perniger Gray, 1863b: 3; Sikkim, 6000–8000ft.
Also many extralimital synonyms.

Range (Map 103). In historical times throughout the mainland of the region and on Sri Lanka, Singapore, Java, and Kangean I (perhaps introduced on the last—Helvoort *et al.*, 1985). Probably still distributed sparsely throughout the less densely populated parts of its former range but extinct on Singapore.

Variation. The following subspecific names have been proposed from the region:
Felis variegata var. *nigricans* Wagner, 1841a: 483; Java.
F. v. var. *atro-maculata* Wagner, 1841a: 483; Java.
P. variegata var. *nigra* Fitzinger, 1868b: 473; Java.
Felis p. chinensis Brass, 1904: 6; China.
F. p. variegata Allen, G. M.,1912b: 235; Changyanghsien, Hubei, China. Pre-occupied by *Felis variegata* Wagner, 1841.
P. p. sindica Pocock, 1930: 80; Kirthar Range, Sind-Baluchistan border, Pakistan.
P. p. millardi Pocock, 1930: 316, pl. 8; Kashmir.
P. p. delacouri Pocock, 1930: 325, pl. 11; Hue, Annam, Vietnam.
P. p. kotiya Deraniyagala, 1949: 103; Sri Lanka.

P. p. szechuanensis Zukovsky, 1964: 172; mountains of W Sichuan.

Melanic individuals, in which the spots are only faintly visible, occur as rarities in some populations, e. g. in Pakistan, but in others they are frequent, or even in the majority as in Malaya (Medway, 1983).

Ecology. Leopards are versatile hunters, preying upon a wide range of mammals from medium-sized ungulates such as *Axis axis* and calves of domestic cattle to small rodents, as well as birds and reptiles.

Remarks. CITES: Appendix I; IUCN: T.

Panthera tigris Tiger

Felis tigris Linnaeus, 1758: 41; Bengal, India.
= *Tigris striatus* Severtzov, 1858: 386.
= *T. regalis* Gray, 1867f: 263.
F. groeneveldtii Dubois, 1907: 454; Java, fossil.
F. oxygnatha Dubois, 1908: 1266; Java; fossil.
F. trinilensis Dubois, 1908: 1266; Java; fossil.
Feliopsis palaeojavanica Stremme, 1911a: 56; Trinil, Java; fossil.
And other, extralimital, synonyms.

Range (Map 103). Formerly, in historical times, throughout the mainland of the region and on the islands of Sumatra, Java, Bali and probably Borneo (Medway, 1977a); also west to the Caucasus and north to the Altai Mts and the Amur Valley.

Now extinct on Bali (since the 1950s) and perhaps also on Java (last record 1979). In recent years the range has become very fragmented. Within the region tigers survive in a number of isolated populations in India including 15 'Project Tiger' reserves; in Bangladesh and S Nepal; in parts of Burma, Thailand, Vietnam, Laos, Cambodia and S Yunnan; in SE China (widely scattered but very sparse south of the Yangtze in the 1980s—only about 50–80 individuals estimated in 1986); Malaya and Sumatra. Evidence of their presence in Borneo is limited to one subfossil canine and one sight record in 1974 (Medway, 1977a); both need corroboration. Elsewhere probably the only viable population is in NE China and the adjacent Ussuri region of USSR.

There are fossil records from Japan and from Siberia north to the Bering Straits (Herrington, 1987).

Variation. The following subspecific names have been proposed from the region:
Felis t. sumatrana Blainville, 1843:liv. 12, pl. vii; Sumatra. Preoccupied by *F. sumatrana* Horsfield, 1821 (in *Felis bengalensis*).

Map **103**. *Panthera*

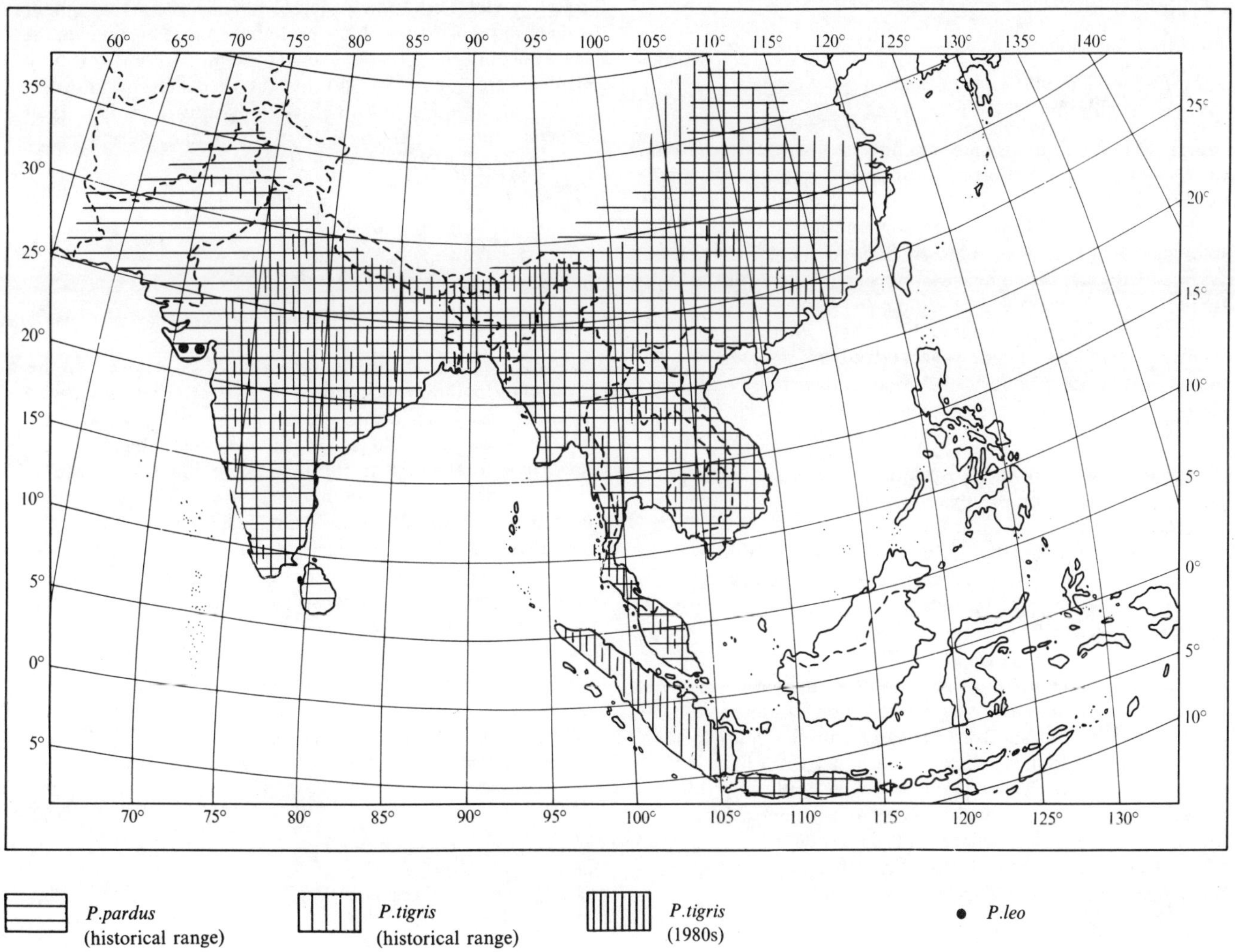

Felis t. sondaicus Temminck, 1844: 43; Java.
? *F. t. ruber* Ludeking, 1862: 41; Agam, W Sumatra.
F. t. var. *amoyensis* Hilzheimer, 1905a: 598; near Hankow, Hubei, China.
F. t. balica Schwarz, 1912b: 325; Bali.
P. t. styani Pocock, 1929: 531; 'Northern China', but more likely from the Yangtze Valley (Allen, 1938: 472).
P. t. sumatrae Pocock, 1929: 535; Deli, NE Sumatra.
Felis t. soloensis Koenigswald, 1933: 14; Ngandong, Java; fossil.
P. t. corbetti Mazak, 1968: 105; Quang-Tri, Annam, Vietnam.

There is geographical variation in size; the largest being in the Amur population and in India, the smallest on the Sunda Islands. There is a tendency for the Indian ones to have a paler ground colour than those in SE Asia. A multivariate analysis of cranial measurements showed the central Chinese form (*amoyensis*) to be the most distinctive but no diagnosis was given (Herrington, 1987).

Remarks. The phylogeny, classification, variation, recent status and many other aspects of the tiger were reviewed by contributors to Tilson & Seal (1987). For a brief review see Mazak (1981). The fossil forms listed in the synonymy were referred to *P. tigris* by Brongersma (1935). On the basis of behaviour Leyhausen (1973) allocated the tiger to the genus *Neofelis*, along with *N. nebulosa*, a view that does not seem to be supported by other character sets. CITES: Appendix I; IUCN: E.

Panthera leo Lion

Felis leo Linnaeus, 1758: 41; Constantine, Algeria.
Leo asiaticus Brehm, 1829: 638; Asia.

Range (Map 103). In Asia now confined to a single population of 200–300 in the Gir Forest, Kathiawar Peninsula, Gujarat, NW India. Also in many parts of the savannah zone of Africa south of the Sahara. Formerly throughout northern India east to Bengal (confirmed by subfossil finds in Bengal at 87°E—Dutta, 1976); and throughout SW Asia. The history of the lion in India was documented by Kinnear (1920). A single fossil tooth from Sri Lanka, probably Late Pleistocene, was identified as *P. leo* by Deraniyagala (1939b) on the basis of size alone. This has been confirmed as *P. leo* rather than *P. tigris* on the basis of other characters although its subspecific distinction cannot be upheld (Hemmer, 1967).

Variation. The following subspecific names have been proposed from the region or adjacent areas:
Felis l. persicus Meyer, 1826: 6; Iran.
F. l. bengalensis Bennett, 1829: 1; Hariana, N India. Preoccupied by *Felis bengalensis* Kerr, 1792 (in *Prionailurus*).
F. l. goojratensis Smee, 1834: 140; Ahmadabad, Gujarat, India.
F. l. indicus Blainville, 1843:pl. 6; India.
P. l. sinhaleyus Deraniyagala, 1939b: 237; Kuruvita, Sri Lanka; Pleistocene; (based upon a single lower carnassial tooth).

The Asian population is a moderately distinct subspecies (*P. l. persicus*) distinguished from African forms especially by the presence of a prominent ridge in the mid-ventral line of the body, by a rather scanty mane in males (although this is matched by many individuals in Africa), a longer tuft on the tail-tip, and more prominent tufts of hair on the elbows. Skulls can only be distinguished by a combination of measurements—for a detailed study see Todd (1966).

Remarks. CITES: Appendix I; IUCN: (E).

ACINONYX

Acinonyx Brookes, 1828: 33; type species *A. venator* Brookes. Conserved by Opinion 384 (1956).
Cynailurus Wagler, 1830: 30 (*Felis jubata* Schreiber).
= *Cynaelurus* Gloger, 1841: 63.

One species. *Acinonyx* has usually been seen as the most aberrant member of the family Felidae, often reflected in placing it as the sole member of a subfamily Acinonychinae (Pocock, 1917b etc.; Leyhausen, 1973). However this has been challenged in most recent classifications, in most extreme degree in the phylogeny of Herrington (1986) which placed *Acinonyx* and *Otocolobus* (the small manul cat of the Tibetan Plateau) as sister groups close to the *Felis* group. Immunological data also placed *Acinonyx* close to many other cats, especially *Leptailurus serval* (Africa) and *Puma concolor* (America) (Collier & O'Brien, 1985).

Acinonyx jubatus Cheetah

Felis jubata Schreber, 1776:pl. 105 (and 1777: 392, 586); Cape of Good Hope, S Africa. Conserved by Opinion 384 (1956).
F. venatica Griffith, 1821: 93; India.
A. venetor Brookes, 1828: 16, 33; India.
And other extralimital synonyms.

Range (Map 102). Until the early 20th century throughout much of central, N and W India and Pakistan; and from there through Soviet Turkestan, much of SW Asia and throughout the drier parts of Africa. Now probably extinct in Asia and in Africa north of the Sahara. The last Indian cheetahs certainly recorded were shot in 1947 but they survived later in Turkestan and Afghanistan. However a possible sighting in India in 1990 has been reported (Anon, 1991). Former and present range were mapped by O'Brien *et al.* (1986).

Remarks. CITES: Appendix I; IUCN: E.

ORDER CETACEA

Whales, dolphins and porpoises

The two very clearly distinct suborders of cetaceans, the Odontoceti (toothed whales) and the Mysticeti (baleen whales) are sometimes given ordinal rank (Table 151). Although all cetaceans are totally aquatic they fall into four categories with respect to habitat and distribution in the Indomalayan Region: (1) the river dolphins, *Platanista gangetica* and *Lipotes vexillifer*, which are confined to fresh or estuarine water (and to the Indomalayan Region); (2) species of shallow coastal waters and estuaries, entering large rivers to some extent: *Sousa chinensis*, *Orcaella brevirostris* and *Neophocaena phocaenoides*, also primarily or entirely Indomalayan in distribution; (3) marine species resident in Indomalayan waters, e. g. *Delphinus delphis*, *Balaenoptera edeni*: these mostly have pantropical or wider

Table 151. The suborders of Cetacea.

	Odontoceti	Mysticeti
Families in region	6	2
Teeth present	+	0
Horny 'baleen' plates on upper jaw forming filter mechanism	0	+
Blowhole single	+	0

Table 152. The families of Odontoceti.

	Platanistidae	Iniidae	Delphinidae	Phocoenidae	Physeteridae	Ziphiidae
Genera in region	1	1	14	1	2	2
Max.length (in region)	2.5m	2.4m	10m	(1.8m)	20m	(7m)
Dorsal fin	Low, obtuse	Low, obtuse	High, usually pointed	0	0/small, pointed	Small, pointed
Mouth behind tip of snout	0	0	0	0	+	0
Throat with 2 grooves	0	0	0	0	0	+
Tail fluke notched in centre	+	+	+	+	+	0
Teeth in each upper jaw	Many	Many	Many[1]	Many	0	0[3]
Teeth in each lower jaw	Many	Many	Many[2]	Many	Many	1[3]
Teeth flattened	0	0	0	+	0	0
Cervical vertebrae fused	None	None	1st 2–6	1st 5	All 7	1st 3 or 4
Maxillae forming high crest on each side, almost meeting in front of blow-hole	+	0	0	0	0	0

1. None in *Grampus*.
2. Only 4 in *Grampus*.
3. Many in one extralimital genus, *Tasmacetus*.

distributions; (4) marine species whose main distribution is in colder waters but that migrate seasonally towards the tropics, entering Indomalayan waters regularly (e. g. *Balaenoptera physalus*) or sporadically or marginally (e. g. *Balaena glacialis*).

The northern Indian Ocean is an important wintering and breeding area for many Antarctic stocks of whales and was declared a sanctuary by the International Whaling Commission in 1979, extending south to 55°S (Holt, 1988).

Useful recent works that give concise data on all species of cetaceans are Watson (1981), Leatherwood & Reeves (1983), Minasion *et al.* (1984), Tinker (1988), Bonner (1989), Evans (1989) and Martin *et al.* (1990). Detailed accounts of the Platanistidae and all the larger whales can be found in Ridgeway & Harrison (1985,1989). Records from around India, Pakistan and Sri Lanka were reviewed by Silva (1988) and Leatherwood & Reeves (1991), and those from Singapore by Sigurdsson & Yang (1990). The nomenclatural data given here are based primarily upon the comprehensive checklist of Hershkovitz (1966), modified by later works.

SUBORDER ODONTOCETI

Toothed whales, dolphins, porpoises

Although this is a well-defined taxon, classification at the family level has been very unstable. A conservative classification is used here, recognizing only six families (Table 152). However the Delphinidae and Physeteridae are frequently split as noted under the heading of these families.

Family Platanistidae

Platanistina Gray, 1846c: 45; type genus *Platanista* Wagler.

Frequently used to include all four genera of river dolphins but here restricted to *Platanista* following Heyning (1989b) who separated the Platanistidae and Iniidae at superfamily level. All the river dolphins have long, slender beaks with numerous small teeth and retain the primitive condition of free cervical vertebrae giving a visible and mobile neck. *Platanista* is unique in the gross development of maxillary crests and the enlargement of the teeth towards the tip of the rostrum.

PLATANISTA

Susu Lesson, 1828:pl. 3 (*S. platanista* Lesson). Suppressed by Opinion 1565 (1989).
Platanista Wagler, 1830: 35; type species *Delphinus gangeticus* Roxburgh. Conserved by Opinion 1565 (1989).

Contains one species (but often treated as two).

Platanista gangetica Ganges-Indus dolphin, Susu

Delphinus gangeticus Roxburgh, 1801: 171; Ganges Delta, near Calcutta. Conserved by Opinion 1565 (1989).
Delphinus gangetica Lebeck, 1801: 280; Hooghly R, near Calcutta, India.
= *Susu platanista* Lesson, 1828:pl. 3.
D. rostratus Shaw, 1801: 514; 'Indian Seas'.
= *D. shawensis* Blainville, in Desmarest, 1817: 153.
P. indi Blyth, 1859b: 493; R Indus, Pakistan. A neotype was designated by Pilleri & Gihr (1977) from Chak on the R Indus.

Range (Map 104). The rivers Ganges, Brahmaputra and Indus, and their major tributaries.

Variation. The following subspecific name has been proposed: *P. g. minor* Owen, 1853: 448; R Indus.

The Ganges and Indus populations were treated as separate species by Pilleri *et al.* (1982): *P. platanista* and *P. indi* (*P. minor* of later authors). The grounds for doing so have been disputed by Kasuya (1972) and by Reeves & Brownell (1989) and it does not seem possible to provide clear differential diagnoses that would justify even subspecific separation.

Remarks. Reviewed by Reeves & Brownell (1989). CITES: Appendix I; IUCN: V.

Family Iniidae

Iniina Gray, 1846c: 25, 45; type genus *Inia* D'Orbigny (S America).
Lipotidae Zhou *et al.*,1979: 72 (*Lipotes* Miller).

As used here, following Heyning (1989b), this family contains three genera of river dolphins, two in South America and one, *Lipotes*, in China.

LIPOTES

Lipotes Miller, 1918: 2; type species *L. vexillifer* Miller.

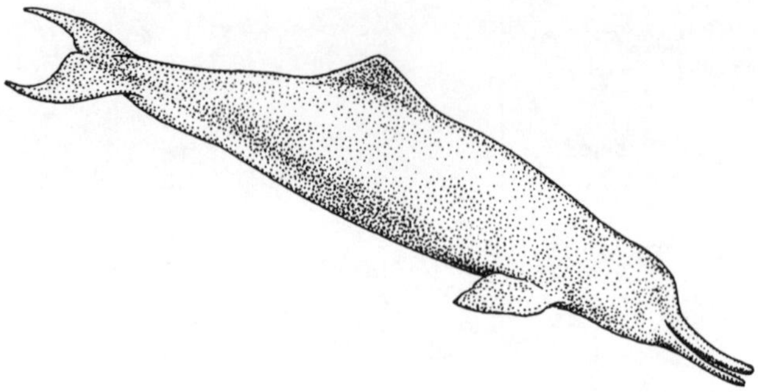

Lipotes vexillifer

Contains one species. This has variously been allocated to a separate family Lipotidae (Zhou *et al.*, 1979), to the Pontoporiidae (Reeves & Brownell, 1989) or included in the Platanistidae.

Lipotes vexillifer Chinese river dolphin, Baiji

L. vexillifer Miller, 1918: 2; Tungting Lake, Hunan, China.

Range (Map 104). The River Yangtze from the mouth upstream as far as Yichang, about 111°20'E; formerly in the adjacent Tungt'ing Lake but no longer present there.

Remarks. CITES: Appendix I; IUCN: E.

Family Delphinidae

Marine dolphins

Delphinidae Gray, 1821: 310; type genus *Delphinus* L.

Worldwide, marine, with only a few species occurring also in fresh water; c. 28 species of which 16 have been recorded in Indomalayan seas. The family has been split into as many as three families: Globicephalidae, including (of the genera listed here) *Globicephala, Orcinus, Pseudorca, Feresa* and *Peponocephala*; Stenidae, including *Steno* and *Sousa*; and Delphinidae, with the remainder. However these are difficult to diagnose, the differences seem slight, and there are records of intergeneric hybrids that cut across these divisions, e. g. between *Tursiops* and *Steno, Pseudorca* and *Globicephalus* (Sylvestre & Tasaka, 1985). On the other hand the Phocoenidae, here treated as a separate family, is sometimes included. Fourteen genera are represented in the region (Tables 153–155). Delphinids are mainly fish-eaters.

Map **104**. *Platanista, Lipotes, Orcaella, Neophocaena*

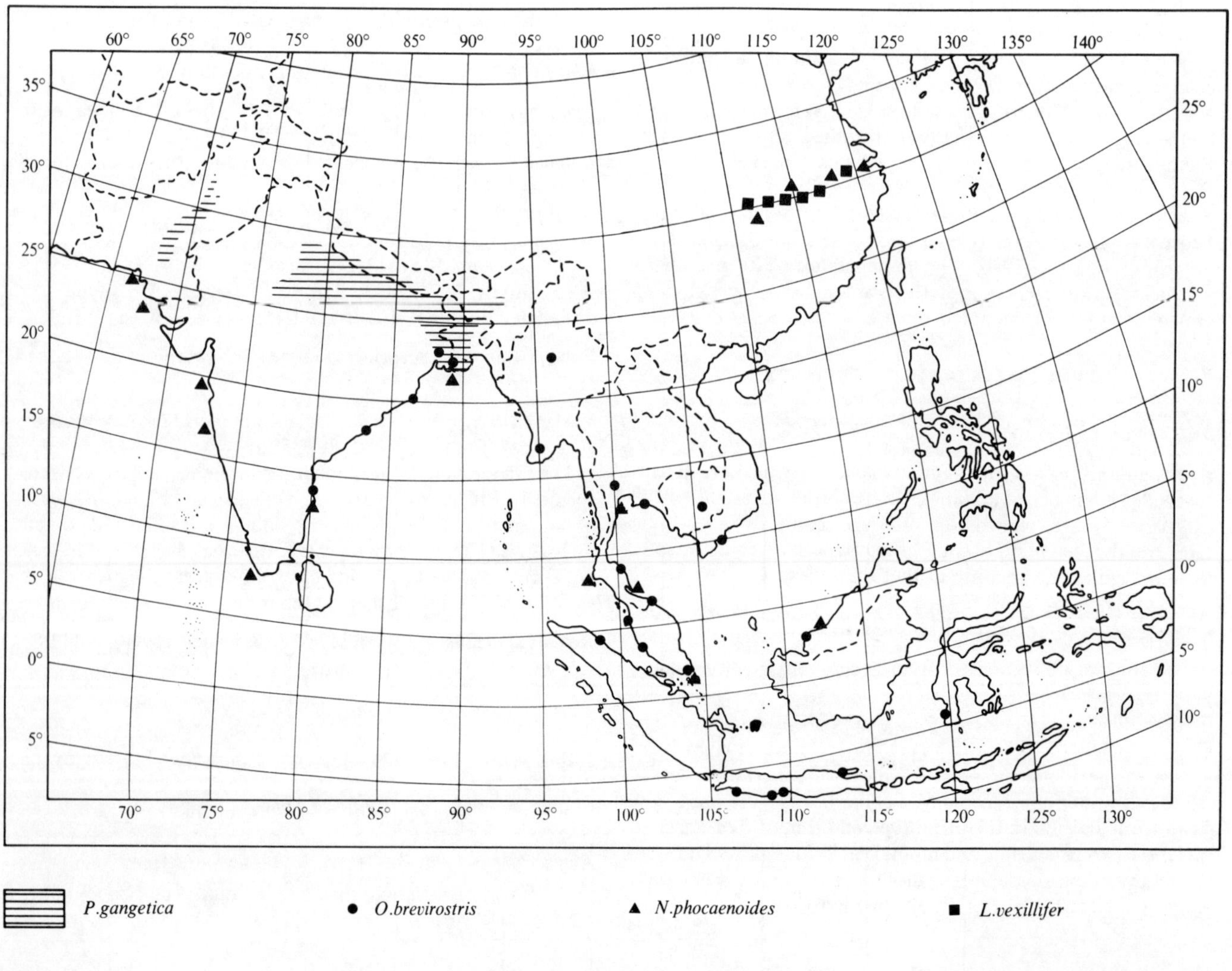

Table 153. The genera and species of small Delphinidae with little or no beak.

	Lagenorhynchus obliquidens	*Lagenodelphis hosei*	*Peponocephala electra*	*Orcaella brevirostris*	*Feresa attenuata*	*Grampus griseus*
Habitat	Oceanic	Oceanic	Oceanic	Coastal	Oceanic	Oceanic
Length: mean (max.) (m)	2.1 (2.4)	2.3 (2.6)	2.2 (2.75)	2.0 (2.3)	2.3 (2.75)	3 (4.25)
Bold pattern on flank	+	0	0	0	0	0
Forehead bulbous	0	0	0	+	0	(+)
Upper teeth, each row	23–33	39–44	21–25	12–19	10–13	0 (upper) 2–7 (lower)

Table 154. The genera of small Delphinidae with slender beak.

	Steno	*Sousa*	*Stenella*	*Delphinus*	*Tursiops*
Species in region	1	1	3	1	1
Length: mean (max.) (m)	2.4 (2.75)	2 (3)	(3.0)	2.1 (2.6)	3 (4.2)
No.of teeth each row	20–27	30–38	35–55	40–57	20–26
Complex pattern on flanks	0	0	+/0	+	0
Beak sharply defined from forehead	0	+	+	+	+
Broad hump below dorsal fin	0	+	0	0	0
Palate with deep groove each side	0	0	0	+	0

Table 155. The genera and species of large Delphinidae.

	Pseudorca crassidens	*Orcinus orca*	*Globicephala macrorhynchus*
Max.length: male	6m	10m	6.8m
female	5.4m	8.5m	5m
White on flanks	0	+	0
Adult male with tall, erect dorsal fin	0	+	0
Forehead bulbous	0	0	+
Notch between forehead and snout	0	+	+
Upper teeth (each side)	7–10	10–13	8–9
Lower teeth (each side)	9–12	10–13	8–9

The generic allocation of the following names is uncertain:
Delphinus malayanus Lesson, 1826: 184; Karimata Strait, between Java and Borneo.
D. (*Steno?*) *maculiventer* Owen, 1866: 21, pl. 6; Madras, India.

STENO

Steno Gray, 1846c: 30; type species *Delphinus rostratus* Cuvier, in *S. bredanensis*.
= *Glyphidelphis* Gervais, 1859: 301.
Stenopontistes Miranda-Ribeiro, 1936: 19, 42 (*S. zambezicus* Miranda-Ribeiro, in *S. bredanensis*).

Contains one species. Sometimes separated from the Delphinidae in a family Stenidae but this seems to have little justification, especially in view of recorded hybridization with *Tursiops* (Sylvestre & Tasaka, 1985).

Steno bredanensis Rough-toothed dolphin

Delphinus rostratus Desmarest, 1817: 160; Paimpol, Brittany, France. Pre-occupied by *D. rostratus* Shaw, 1801, in *Platanista gangetica*.
= *D. bredanensis* Lesson, 1828: 206.
D. reinwardtii Schlegel, 1841: 27, pl. 3; Java.
And several extralimital synonyms.

Range. Worldwide in tropical and subtropical waters, mainly at the edges of the continental shelves. Recorded in Indomalayan seas off Pakistan, Sri Lanka, the Nicobar Is and Java.

Remarks. CITES: Appendix II.

SOUSA Humpback dolphins

Sousa Gray, 1866i: 213; type species *Steno lentiginosa* Gray, in *S. chinensis*.

Tropical oceans and coasts; one species, perhaps two.

Sousa chinensis Indopacific humpback dolphin

Delphinus chinensis Osbeck, 1765: 7; Canton R, China.
= *D. sinensis* Desmarest, 1822: 514.
D. plumbea Cuvier, G.,1829: 288; Malabar, S India.
Steno lentiginosa Gray, 1866i: 213; Waltair, Vizagapatam, Madras, India.
Sotalia borneensis Lydekker, 1901b: 88, pl. 8; Sipang, mouth of Sarawak R, Sarawak, Borneo.
Sotalia fergusoni Lydekker, 1903b: 411, pl. D; Trivandrum, S India.

Range. Coastal waters throughout the region including the lower reaches of the main Chinese rivers, and recorded 250km from the sea in a tributary of the Ganges (Ghosh & Choudhury, 1986); also east to NE Australia, north to N China and west to S Africa; perhaps also in W Africa if *teuszii* is conspecific.

Ecology. Usually in sheltered coastal waters including mangrove swamps and estuaries, rarely out of sight of land.

Remarks. Although the West African form *teuszii* Kukenthal, 1892 is usually considered a separate species, intergradation in S Africa suggests that it should perhaps be included in *S. chinensis* (Ross, 1984). CITES: Appendix I.

STENELLA

Clymene Gray, 1864a: 237 (*Delphinus euphrosyne* Gray, in *S. caeruleoalba*). Pre-occupied by *Clymene* Lamarck, 1818 (Annelida).
= *Clymenia* Gray, 1868d: 6.
= *Prodelphinus* Gervais, in van Beneden & Gervais, 1880: 604.
Stenella Gray, 1866i: 213; type species *Steno attenuatus* Gray.

Worldwide, mainly in warmer waters; 5 species, of which 3 occur in Indomalayan waters (Table 156). These are fast-swimming, social, pelagic dolphins. The taxonomy has been very unstable but was revised, in part, by Perrin *et al.* (1987).

Table 156. The species of *Stenella*.

	coeruleoalba	*attenuata*	*longirostris*
Length (m)	Av.2.4 (–3.0)	M:1.7–2.6 F:1.6–2.4	Av.1.8 (–2.1)
Adults spotted	0	+	0
Stripe from flipper to:	Eye	Gape	Eye
Narrow dark stripe on flank	+	0	0
Beak	Short, thick	Short, thick	Slender
Teeth (each row)	c.44 (43–49)	c.37 (35–40)	c.50 (46–65)

Stenella coeruleoalba Striped dolphin

Delphinus coeruleo-albus Meyen, 1833: 609, pl. 43; E coast of S America.
? *Prodelphinus petersii* Lütken, 1889: 40; Indian Ocean.
And other, extralimital synonyms.

Range. Widespread from tropical to warm temperate waters; large schools have been recorded south of Java (Watson, 1981) and there are records from the Java and Savu Seas, and from Sri Lanka, India and the Maldive Is.

Remarks. CITES: Appendix II.

Stenella attenuata Pantropical spotted dolphin

Delphinus velox Cuvier, G.,1829: 288; between Sri Lanka and the Equator. Considered a *nomen oblitum* by Perrin *et al.* (1987).
Steno attenuatus Gray, 1846c: 44, pl. 28; locality unknown. Conservation has been proposed: *Bull. zool. Nomencl.* 47:case 2717 (1990).
Delphinus brevimanus Wagner, 1846:pl. 361; near Singapore.
= *D. microbrachium* Gray, 1850b: 119.
And other, extralimital synonyms.

Range. Worldwide in tropical and subtropical waters including the Indian Ocean and the Java Sea.

Remarks. The taxonomy of this species was reviewed by Perrin *et al.* (1987). CITES: Appendix II.

Stenella longirostris Spinner dolphin

Delphinus longirostris Gray, 1828: 1; locality unknown.
D. roseiventris Wagner, 1846:pl. 360; Moluccas.
And other, extralimital synonyms.

Range. Worldwide in the tropical oceans, including the Indian Ocean where it has been recorded around Sri Lanka and the Maldive Is, and the W Pacific including the Gulf of Thailand, the Moluccas and the Java Sea.

Variation. Subspecific variation was reviewed by Perrin (1990) who recognized only *S. l. longirostris* in the Indian Ocean and W Pacific, characterized by a distinct tripartite pattern.

Remarks. This is the species of dolphin most frequently caught as a result of gill-netting for fish off SW India (Lal Mohan, 1988). CITES: Appendix II.

DELPHINUS

Delphinus Linnaeus, 1758: 77; type species *D. delphis* L. Conserved by Opinion 75 (1922).
= *Rhinodelphis* Wagner, 1846: 281, 316.
= *Eudelphinus* van Beneden & Gervais, 1880: 600.
= *Delphis* Gray, 1864a: 236.

Delphinus delphis

Contains one species, perhaps two if *D. tropicalis* is distinct from *D. delphis*. Some other genera, e. g. *Tursiops* and *Stenella*, are very similar and doubtfully justify exclusion from *Delphinus*.

Delphinus delphis Common dolphin

D. delphis Linnaeus, 1758: 77; European seas. Conserved by Opinion 75 (1922).
D. longirostris Cuvier, G.,1829: 288; Malabar coast, India. Pre-occupied by *D. longirostris* Gray, 1828, = *Stenella longirostris*.
= *D. dussumieri* Blanford, 1891: 588. Pre-occupied by *D. dussumieri* Fischer, 1829, in *Cephalorhynchus heavisidei* (Gray) (Delphinidae).
= *D. tropicalis* van Bree, 1971: 345.
D. frithii Blyth, 1859b: 492; between India and England.
D. pomeegra Owen, 1866: 23, pl. 6; off coast of Madras, India.

Range. Worldwide from tropical to warm temperate waters; including both coastal and off-shore waters of the Arabian Sea, the Bay of Bengal, Straits of Malacca, around Borneo, the Gulf of Thailand and the Chinese coast.

Variation. Inshore animals around the Indian Ocean tend to have longer beaks and have been considered specifically distinct, as *D. tropicalis* (van Bree & Gallagher, 1978). However most recent authors consider them conspecific with *D. delphis*.

Remarks. A fishery for this species (and *Tursiops truncatus*) has been conducted from Sri Lanka. CITES: Appendix II.

TURSIOPS

Tursio Gray, 1843a:xxiii, 105; type species *T. truncatus* Montagu. Pre-occupied by *Tursio* Fleming, 1822, in *Physeter*.
= *Tursiops* Gervais, 1855: 323.
Gudamu Gray, 1868d: 6 (*Clymenia gadamu* Gray, in *T. truncatus*).

Probably only one species, although others have been recognized.

Tursiops truncatus Bottle-nosed dolphin

Delphinus truncatus Montagu, 1821: 75, pl. 3; R Dart, Devon, England.
D. aduncus Hemprich & Ehrenberg, 1833:sig. k (verso); Belhosse I, Red Sea.
D. perniger Elliot, in Blyth, 1848: 250; Bay of Bengal.
Steno? gadamu Gray, 1866h: 394; India.
T. fergusoni Lydekker, 1903c: 41, pl. 3; Trivundrum, Travancore, India.
? *T. dawsoni* Lydekker, 1909c: 802; off Trivundrum, S India.
And many other, extralimital, names.

Range. Worldwide in warm and temperate seas, mainly but not exclusively in coastal waters; including the northern and eastern coasts of the Indian Ocean and around the Sunda Islands and Taiwan.

Variation. The population in the Indian Ocean has been treated as specifically distinct, *T. aduncus*, characterized by small size and a pattern of spots, especially on the venter. However this was considered no more than of subspecific value in a detailed review by Ross & Cockroft (1990).

Remarks. Fisheries for this species (and *Delphinus delphis*) have been conducted from Sri Lanka and Taiwan. Monographed by Leatherwood & Reeves (1990). CITES: Appendix II.

LAGENODELPHIS

Lagenodelphis Fraser, 1957: 496; type species *L. hosei* Fraser.

Contains one species.

Lagenodelphis hosei Fraser's dolphin

L. hosei Fraser, 1957: 496; mouth of Lutong R, Baram, Sarawak, Borneo.

Range. Recorded from widely scattered localities in the tropical and subtropical oceans, including Sri Lanka, Borneo, Taiwan (Miyakazi & Wada, 1978) and the Philippines, where it is common in the Camotes Sea and Bohol Strait.

Remarks. CITES: Appendix II.

LAGENORHYNCHUS

Lagenorhynchus Gray, 1846a: 84; type species *L. albirostris* Gray (N Atlantic).

Worldwide in temperate waters; 6 species, of which one occurs very marginally in the region.

Lagenorhynchus obliquidens Pacific whitesided dolphin

L. obliquidens Gill, 1865: 177; near San Francisco, California, USA.
And other synonyms.

Range. Temperate N Pacific; extending south, perhaps only sporadically, as far as the Ryukyu Is and Taiwan (Honacki *et al.*, 1982).

Remarks. CITES: Appendix II.

[*CEPHALORHYNCHUS*]

A record from Sarawak, Borneo of *C. hectori* (van Beneden, 1881), otherwise known only from New Zealand, was given by Harrisson (1960), but is likely to be erroneous (van Bree, 1972).

PEPONOCEPHALA

Electra Gray, 1866h: 268; type species *Lagenorhynchus electra* Gray. Pre-occupied by *Electra* Lamouroux, 1816 (Bryozoa).
= *Peponocephala* Nishiwaki & Norris, 1966: 95.

One species. This genus is somewhat intermediate between the Delphinidae *s. s.* and the larger whales of the *Globicephala* group (*Globicephala, Orcinus, Pseudorca, Feresa*) that are sometimes treated as a separate family, Globicephalidae. It shares the head shape of these but has the large number of teeth (20–25 per row) of the Delphinidae *s. s.* .

Peponocephala electra Melon-headed dolphin

Lagenorhynchus electra Gray, 1846c: 35, pl. 13; locality unknown.
Delphinus (*Lagenorhynchus*) *fusiformis* Owen, 1866: 22, pl. 5; Madras, India.
= *Electra obtusa* Gray, 1868d: 7.

Range. A rare species known from a few widely scattered localities mainly in tropical and subtropical waters, including India and Sri Lanka, the Gulf of Thailand and the Savu Sea (Lesser Sunda Is) and especially around the Philippines; it has also been reported to be seen regularly off the Mekran coast of Pakistan (Roberts, 1977).

Remarks. CITES: Appendix II.

ORCAELLA

Orcaella Gray, 1866h: 285; type species *O. brevirostris* Owen.

Contains one species. Although most often included in the Delphinidae it has also been allocated to the family Monodontidae, with the Arctic genera *Monodon* and *Delphinapterus*. However this has not been universally accepted.

Orcaella brevirostris Irrawaddy dolphin

Orca brevirostris Gray, 1866h: 285; Vizagapatam, Madras, India.
O. fluminalis Gray, 1871b: 80; R Irrawaddy, Burma.

Range (Map 104). Coastal waters from E coast of India to Java, Borneo, Sulawesi and S Vietnam; and in the larger rivers including the Ganges, Irrawaddy and Mekong; also around New Guinea and the northern half of Australia.

Remarks. Reviewed by Marsh *et al.* (1989). CITES: Appendix II; IUCN: K.

PSEUDORCA

Pseudorca Reinhardt, 1862: 151; type species *P. crassidens* Owen.
Neorca Gray, 1871b: 81 (*P. meridionalis* Gray, in *P. crassidens*).

Contains one species.

Pseudorca crassidens False killer whale

Phocaena crassidens Owen, 1846: 516; Lincolnshire, England; subfossil.
And extralimital synonyms.

Range. Worldwide in tropical to temperate oceans, including the Indian Ocean: records from Pakistan (Pilleri & Gihr, 1976), Kerala (India), Sri Lanka, the Andaman Is (James, 1984) and from Sumatra to the Banda Sea.

Ecology. Mainly pelagic but frequently stranded, sometimes in large numbers.

Remarks. CITES: Appendix II.

ORCINUS

Orca Gray, 1846c: 33. Pre-occupied by *Orca* Wagler, in *Hyperoodon*.
Orcinus Fitzinger, 1860: 204; type species *Delphinus orca* L.
Ophysia Gray, 1868d: 8 (*Orca capensis* Gray, in *O. orca*).

Contains one species.

Orcinus orca Killer whale

Delphinus orca Linnaeus, 1758: 77; European seas.
And many extralimital synonyms.

Range. Worldwide at all latitudes but more frequent in cooler waters; recorded in Pakistan, Sri Lanka, the Bay of Bengal, around the Lesser Sunda Is and Borneo.

Remarks. Reviewed by Heyning & Dahlheim (1988).

GRAMPUS

Grampus Gray, 1828: 2; type species *Delphinus griseus* Cuvier.

Contains one species.

Grampus griseus Grey dolphin, Risso's dolphin

Delphinus griseus Cuvier, 1812: 13, pl. 1; Brest, France.
= *G. cuvieri* Gray, 1846a: 85.
Globicephalus chinensis Gray, 1866h: 323; near Leuchen, S China.
And many extralimital synonyms.

Range. Worldwide from tropical to warm temperate waters; common in the East China Sea including the NW coast of Borneo, and around Sri Lanka; also recorded near the Maldive Is.

Remarks. Hybridization with *Tursiops truncatus* has been recorded both under natural conditions and in captivity (Sylvestre & Tasaka, 1985), although the differences in skull and dentition are sufficiently great that generic separation seems justifiable on morphological grounds. CITES: Appendix II.

GLOBICEPHALA

Globicephala Lesson, 1828: 276; type species *Delphinus globiceps* Cuvier, in *G. melas*.
= *Cetus* Wagler, 1830: 33. Pre-occupied by *Cetus* Billberg, in *Physeter*.
= *Globiceps* Flower, 1884: 508.

Worldwide; 2 species, only one occurring in the region. (An unconfirmed sighting of the (temperate) long-finned pilot whale, *G. melas*, off Sarawak was recorded by van Strien (1986), probably based on the inclusion of 'Pilot whale' in a list of species in Sarawak by Harrisson (1960)—it can be assumed to be an error for *G. macrorhynchus*).

Globicephala macrorhynchus Short-finned pilot whale

G. macrorhynchus Gray, 1846c: 33; 'South Seas'.
Delphinus carbonarius Wagner, 1846: 305, pl. 352; locality unknown.
G. indicus Blyth, 1852: 358; Hooghly R, near Serampore, W Bengal, India.
And extralimital synonyms.

Range. Worldwide in tropical and subtropical oceans; recorded from the seas around Java, Borneo and Sumatra, the Savu Sea, the Bay of Bengal and around the Maldive Is.

Remarks. CITES: Appendix II.

FERESA

Feresa Gray, 1870: 77; type species *Orca intermedia* Gray, in *F. attenuata*.

Contains one species.

Feresa attenuata Pygmy killer whale

Delphinus intermedius Gray, 1827b: 376; locality unknown. Pre-occupied by *D. intermedius* Harlan, 1827, in *Globicephala melas* (Traill).
= *F. occulta* Jones & Packard, 1956: 167.
F. attenuata Gray, 1875: 52=184; locality unknown.

Range. Worldwide in tropical and subtropical waters, but apparently confined to deep waters and records are few; three were caught off Trincomalee, Sri Lanka in 1983.

Remarks. CITES: Appendix II.

Family Phocoenidae

Porpoises

Phocaenina Gray, 1825: 340; type genus *Phocoena* Cuvier.

A family of three genera, represented mainly in the N Pacific, N Atlantic and S America; one species in the Indomalayan Region (Table 152). Sometimes included in the Delphinidae but distinguished especially by the flattened teeth.

NEOPHOCAENA

Neomeris Gray, 1846c: 30; type species *N. phocaenoides* Cuvier. Pre-occupied by *Neomeris* Costa, 1844 (Annelida).
= *Neophocaena* Palmer, 1899b: 23.

Contains one species.

Neophocaena phocaenoides Finless porpoise

Delphinus phocaenoides Cuvier, G.,1829: 291; ? Cape of Good Hope, South Africa (probably in error).
D. melas Schlegel, 1841: 32; Japan. Pre-occupied by *D. melas* Traill, 1809, = *Globicephala melas*.
= *Neophocaena sunameri* Pilleri & Gihr, 1975: 657.
Delphinapterus molagan Owen, 1866: 24; Madras, India.
Neomeris kurrachiensis Murray, 1884c: 351; Karachi, Pakistan.
N. asiaeorientalis Pilleri & Gihr, 1972: 118; Yangtze R, China.

Range (Map 104). Coastal waters and estuaries throughout most of the region; also west to Iran and very doubtfully S Africa, east to New Guinea and north to S Japan. It has been recorded in several of the main rivers, e. g. to 65km from the mouth of the Indus and in the Yangtze as far as L Tung T'ing, over 1000km from the sea.

Variation. The forms *phocaenoides* (Pakistan to Borneo), *asiaeorientalis* (China) and *sunameri* (Korea and Japan) were treated as three separate species by Pilleri & Gihr (1975), but doubtfully justify even subspecific separation according to Rice (1977).

Remarks. CITES: Appendix I.

Family Physeteridae

Sperm whales

Physeteridae Gray, 1821: 310; type genus *Physeter* L.

Two genera, sometimes placed in separate families (Table 157).

Table 157. The genera and species of Physeteridae.

		Physeter macrocephalus	*Kogia breviceps*	*Kogia simus*
Length: mean (max.) (m)	M	15 (20)	3.0 (3.4)	2.4 (2.7)
	F	11 (17)		
Dorsal fin: height		Obscure	<20cm	>20cm
Upper teeth (each row)		0 *	0	0–3
Lower teeth (each row)		18–29	12–16	8–13

* Very small teeth may be present but rarely erupt.

PHYSETER

Physeter Linnaeus, 1758: 76; type species *P. macrocephalus* L.
= *Catodon* Linnaeus, 1761: 18.
Physalus Lacépède, 1804:xi, 219 (*P. cylindricus* Bonnaterre).
= *Cetus* Billberg, 1828: 39.
Tursio Fleming, 1822: 211 (*T. vulgaris* Fleming).
Meganeuron Gray, 1865f: 439 (*Catodon* (*M.*) *kreffti* Gray).

Contains one species.

Physeter macrocephalus Sperm whale

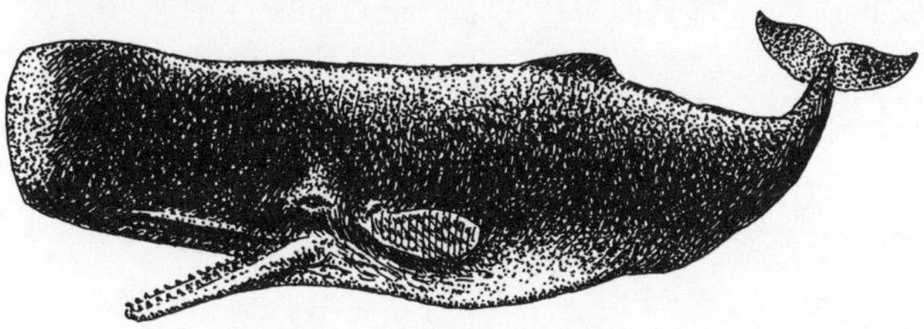

Physeter macrocephalus

P. macrocephalus Linnaeus, 1758: 76; Norwegian seas.
P. catodon Linnaeus, 1758: 76; Orkney Is, Scotland (restricted by Thomas, 1911).
P. australasianus Desmoulins, 1822: 618; Moluccas and New Zealand.
And other extralimital synonyms.

Range. Worldwide, migrating seasonally between breeding grounds in the tropics and cooler waters, only the adult males reaching polar waters. Found throughout the Indian Ocean and from the S China Sea northwards along the Pacific coast.

Ecology. They feed mainly upon squid, including the large deep-water species as well as small surface-living ones.

Remarks. The name of this species has been the subject of much debate. The name *Physeter catodon* of Linnaeus, 1758 seems quite unambiguous, with the description 'fistula in rostro' (blowhole on snout) which excludes all other large whales. In contrast *P. macrocephalus* Linnaeus, 1758 is described as having 'fistula in cervice' (blowhole in neck) which would appear to exclude the sperm whale. It was nevertheless based upon an earlier description of a sperm whale. The case for using *P. macrocephalus* was made by Holthuis (1987) and this has the advantage of having been by far the most frequently used name in spite of objections by Schevill (1987).

For a detailed review of this species see Rice (1989). CITES: Appendix I.

KOGIA

Kogia Gray, 1846c: 22; type species *Physeter breviceps* Blainville.
Euphysetes Wall, 1851: 46 (*E. grayi* Wall, in *K. breviceps*).
Callignathus Gill, 1871: 738 (*Physeter simus* Owen). Preoccupied by *Callignathus* Agassiz, 1846 (Insecta).
= *Callignathula* Strand, 1926: 61.

Worldwide in tropical and subtropical oceans; 2 species (Table 157), as clarified by Handley (1966). They are rarely seen at sea and when they are the species cannot be distinguished; precise locality records are therefore mainly derived from strandings. The genus was reviewed in detail by Caldwell & Caldwell (1989).

Kogia breviceps Pygmy sperm whale

Physeter breviceps Blainville, 1838b: 337, pl. 10; Cape of Good Hope, South Africa.

Range. Apparently throughout the tropical and subtropical oceans. Recorded on the east coast of India and in the S China Sea (Caldwell & Caldwell, 1989) including the coast of Sarawak.

Remarks. CITES: Appendix II.

Kogia simus Dwarf sperm whale

Physeter (*Euphysetes*) *simus* Owen, 1866: 30, pl. 10; Madras, India.

Range. Known from widely scattered records, mostly strandings, throughout the tropical and subtropical regions, including E India, Sri Lanka, the Lesser Sunda Is, Taiwan (Chou, 1989) and Okinawa (Ryukyu Is).

Remarks. Reviewed by Nagorsen (1985). CITES: Appendix II.

Family Ziphiidae

Beaked whales

Hyperoodontina Gray, 1846c: 25.
Ziphiina Gray, 1850b: 56; type genus *Ziphius* Cuvier.

A distinctive family of mainly pelagic whales characterized especially by the absence of teeth in the upper jaw and only one or two pairs in the lower (except in *Tasmacetus*). The teeth are well developed only in adult males in which they often protrude from the closed mouth. Classification within the family was reviewed by Moore (1968). Three species, possibly four, in two genera have been recorded in the region (Table 158).

Table 158. The genera and species of Ziphiidae.

	Mesoplodon ginkgodens	*Mesoplodon densirostris*	*Ziphius cavirostris*
Max.length	5.2m	4.8m	6.4m
Colour	Dark grey	Dark grey	Light brown or grey, head often white
Position of teeth in lower jaw (from tip)	1/4	1/3	At tip
Shape of teeth in section	Flat	Flat	Oval
Jaw bone greatly enlarged below tooth	0	+	0

MESOPLODON

Aodon Lesson, 1828: 155 (*A. dalei* Lesson, in *M. bidens* Sowerby). Pre-occupied by *Aodon* Lacépède, 1789 (Pisces).
= *Nodus* Wagler, 1830: 34.
Dioplodon Gervais, 1850: 16 (*Delphinus densirostris* Blainville).
Mesoplodon Gervais, 1850: 16; type species *Delphinus sowerbensis* Blainville, in *M. bidens* Sowerby. Conserved by Opinion 1289 (1985).
And other names based upon extralimital species.

Worldwide; 12 species of which 2 have been recorded in the region (Table 159). Reviewed by Mead (1989).

Mesoplodon ginkgodens Ginkgo-toothed beaked whale

M. ginkgodens Nishiwaki & Kamiya, 1958: 77; Sagami Bay, near Tokyo, Japan.
M. hotaula Deraniyagala, 1963c: 13 (and 1963b); Ratmalana, near Colombo, Sri Lanka. Allocated to *M. ginkgodens* by Moore & Gilmore (1965).

Range. Indian Ocean, and tropical to temperate N Pacific. Records from Sri Lanka, E Sumatra and Taiwan.

Remarks. CITES: Appendix II.

Mesoplodon densirostris Blainville's beaked whale

Delphinus densirostris Blainville, in Desmarest, 1817: 178; locality unknown.
Ziphius sechellensis Gray, 1846c: 28, pl. 6; Seychelles Is, Indian Ocean.

Range. Worldwide in tropical and subtropical waters. Recorded from the Nicobar Is and from Taiwan (Kasuyu & Nishiwaki, 1971).

Remarks. CITES: Appendix II.

[***Mesoplodon carlhubbsi*** Arch-beaked whale]

An undocumented record of this North Pacific species from the Strait of Malacca by van Strien (1986) cannot be substantiated.

ZIPHIUS

Ziphius Cuvier, G.,1823: 350; type species *Z. cavirostris* Cuvier.
Diodon Lesson, 1828: 123 (*D. desmaresti* Risso, in *Z. cavirostris*).
= *Hypodon* Haldeman, 1841: 127.
= *Aliama* Gray, 1864a: 242.
Petrorhynchus Gray, 1865e: 524 (*Hyperoodon capensis* Gray, in *Z. cavirostris*).
Ziphiorrhynchus Burmeister, 1865: 363 (*Z. cryptodon* Burmeister, in *Z. cavirostris*).

Contains one species.

Ziphius cavirostris Goosebeak whale, Cuvier's beaked whale

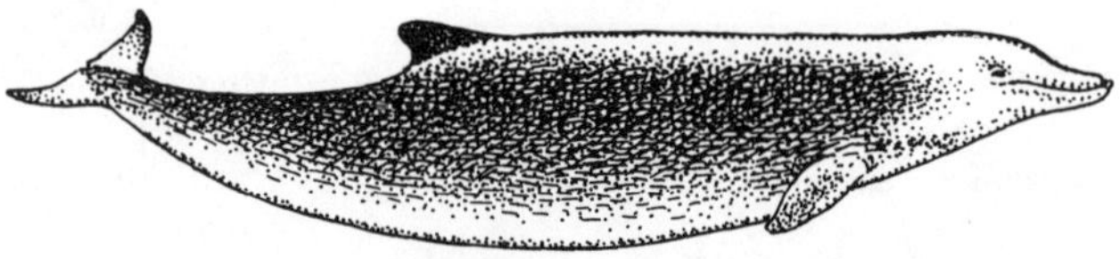

Ziphius cavirostris

Z. cavirostris Cuvier, G.,1823: 350, pl. 27; Dept of Bouches-du-Rhone, S France (fossil).
Including many extralimital synonyms.

Range. Worldwide from tropical to temperate oceans, mainly in deep water. In the region, recorded around India and Sri Lanka and from Javan waters eastwards to the Moluccas.

Remarks. Reviewed by Heyning (1989a). CITES: Appendix II.

[*HYPEROODON*]

A record of *H. planifrons*, the southern bottle-nose whale, in Sri Lanka by Deraniyagala (1960b) has been discredited, probably referring to *Ziphius cavirostris* (Mead, 1989). The species occurs in the temperate parts of the Southern Ocean.

SUBORDER MYSTICETI

Baleen whales

The baleen whales are characterized above all by the absence of teeth and their replacement by multiple flat sheets of keratinous baleen projecting down from the upper jaws to form a filter. Two families are recognized (Table 159).

Table 159. The families and genera of Mysticeti.

	Balaenopteridae		Balaenidae
	Balaenoptera	*Megaptera*	*Balaena*
Species in region	5	1	1
Dorsal fin	+	+	0
Spout	Single, narrow	Single, broad	Double, divergent
Flippers	Narrow, short, mostly dark	Narrow, very long, usually white	Broad, short, dark

Family Balaenopteridae

Rorquals

Balaenopteridae Gray, 1864a: 203; type genus *Balaenoptera* Lacépède.

Worldwide; 2 genera, both represented in the region (Table 159). These are large, oceanic whales, mostly migrating seasonally between cold and warm waters.

Table 160. The species of *Balaenoptera*.

		acutorostrata	*edeni*	*borealis*	*physalus*	*musculus*
Length: mean (max.) (m)	M	8.0 (9.4)	12 (14.3)	15 (18.5)	21 (25)	25 (31)
	F	8.2 (10.2)	13 (14.6)	16 (20)	22 (27)	26 (32)
White on flipper		+	0	0	0	0
Distance of fin from tail relative to total length		1/3	>1/3	>1/3	1/3	1/4
Ridge(s) on top of head		1	3	1	1	1

BALAENOPTERA

Balaenoptera Lacépède, 1804:xxxvi, 114; type species *B. gibbar* Lacépède, in *B. physalus* L.
= *Physalus* Duméril, 1806: 28.
= *Catoptera* Rafinesque, 1815: 61.
Boops Gray, 1821: 310 (*Balaena boops* L., in *B. physalus* L.).
Ptychocetus Gloger, 1841: 174 (*B. musculus* L., designated by Hershkovitz, 1966: 152).
Swinhoia Gray, 1866h: 382 (*B. swinhoii* Gray, in *B. physalus*).
And many others.

Worldwide; 5 species, all recorded in Indomalayan seas (Table 160). The following fossil species has been described from the region:
B. taiwanica Huang, 1966: 184. Shantahu, Chutung-Chen, Hsinchu-Hsien, Taiwan; early Pliocene (based upon a tympanic bone).

Balaenoptera acutorostrata Minke whale

B. acuto-rostrata Lacépède, 1804:xxxvii, 134, pl. 4; Cherbourg, France.
B. bonaerensis Burmeister, 1867:xxiv; near Belgrano, Buenos Aires, Argentina.
And other extralimital names.

Range. Worldwide in temperate and polar seas, migrating towards the equator in winter but rarely entering the tropics. Recorded near Sri Lanka, the Straits of Malacca, Savu Sea and the Philippines.

Variation. The following subspecific name has been proposed from the region:
B. a. thalmaha Deraniyagala, 1963b: 79, pl. 1; Delft, Sri Lanka.

Remarks. Reviewed by Stewart & Leatherwood (1985). CITES: Appendix I.

Balaenoptera edeni Bryde's whale

B. edeni Anderson, 1879: 551; Thaybyoo Choung, Gulf of Martaban, between Sittang and Beeling Rivers, Burma.
B. brydei Olsen, 1912 (and 1913: 1073); Saldanha Bay, W Cape Prov., S Africa.

Range. Worldwide in tropical and subtropical waters, frequently close to shore; there is a population in the Bay of Bengal (from Sri Lanka to Straits of Malacca), another around the Philippines, and fairly frequent records from the Arabian Sea, e. g. off Pakistan.

Remarks. Reviewed by Cummings (1985). CITES: Appendix I.

Balaenoptera borealis Sei whale

B. borealis Lesson, 1828: 342; Schleswig-Holstein, Germany.
Sibbaldius schlegelii Flower, 1865: 408; Java.
B. alba Giglioli, 1870: 74; Java Sea.
And other, extralimital names.

Range. Worldwide in temperate and subtropical waters, showing less clearly defined seasonal migrations than the other large rorquals. Probably not regular anywhere in Indomalayan waters except that the N Pacific population moves south to about 20°N in winter. There are records from Borneo, Java and the Gulf of Thailand, and a recent stranding on the east coast of India (Venkataramanujam *et al.*, 1988).

Remarks. Reviewed by Gambell (1985). CITES: Appendix I.

Balaenoptera physalus Fin whale

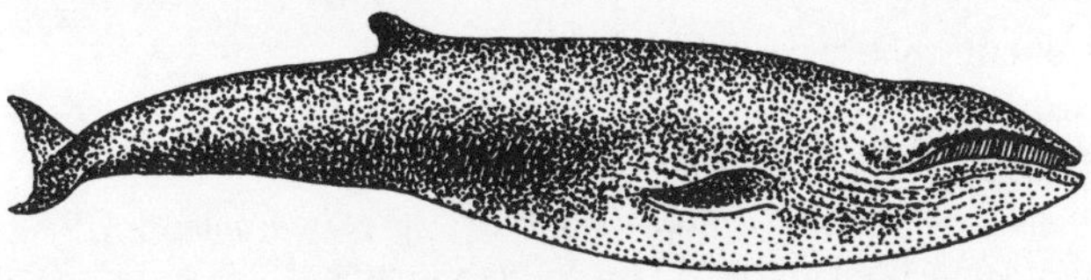

Balaenoptera physalus

Balaena physalus Linnaeus, 1758: 75; European seas.
Balaena boops Linnaeus, 1758: 76; N Atlantic.
Balaenoptera swinhoii Gray, 1866g: 725; Taiwan.
= *Swinhoia chinensis* Gray, 1868d: 3.
B. blythi Anderson, 1879: 564; India.
And many other, extralimital names.

Range. Worldwide, migrating from temperate feeding ground in summer towards the equator in winter. Whales of the N Pacific population extend south to Philippine waters in winter and members of the southern population reach the Moluccas and the south coast of Java in the southern winter. Both northern and southern animals have been reported wintering in the Banda Sea (Moluccas) by Watson (1981).

Remarks. Reviewed by Gambell (1985). CITES: Appendix 1; IUCN: V.

Balaenoptera musculus Blue whale

Balaena musculus Linnaeus, 1758: 76; Firth of Forth, Scotland.
Balaenoptera indica Blyth, 1859b: 488; Sordip, Bay of Bengal, India.
B. intermedia Burmeister, 1872a:vii (and 1872b: 413); near mouth of R Lujan, Buenos Aires, Argentina.
And many other, extralimital names.

Range. Worldwide, feeding in summer in the polar seas and migrating towards the equator in winter. Animals from the N Pacific reach Taiwan in winter. Southern animals enter the

Indian Ocean in winter, but since blue whales have been recorded in the N Indian Ocean at all times of year there may be a population there separate from that in the Southern Ocean. They have been recorded especially around Sri Lanka and the west coast of India, and from Java to the Banda Sea and Moluccas (Watson, 1981).

Variation. One subspecific name has been proposed:
B. m. brevicauda Ichihara, 1966: 109; SW Indian Ocean at 49°59'S, 28°25'E.

Northern and southern populations are marginally separable as subspecies, the southern ones being *B. m. intermedia*. A third form, provisionally accepted as a subspecies *B. m. brevicauda* and characterized by smaller size, occurs in the southern Indian Ocean, possibly also in the northern part. Its relation to *B. m. intermedia* is not clear.

Remarks. Reviewed by Yochem & Leatherwood (1985). CITES: Appendix I; IUCN: E.

MEGAPTERA

Megaptera Gray, 1846a: 83; type species *M. longipinna* Gray, in *M. novaeangliae* Borowski.
Perqualus Gray, 1846c: pl. 32 (*Balaenoptera boops* Fabricius, in *M. novaeangliae* Borowski).
= *Kyphobalaena* Eschricht, 1849: 108.
Poescopia Gray, 1864a: 207 (*Balaena lalandii* Fischer, in *M. novaeangliae* Borowski).

Contains one species.

Megaptera novaeangliae Humpback whale

Balaena novae angliae Borowski, 1781: 21; New England, USA.
B. nodosa Bonnaterre, 1789: 5; New England, USA.
M. indica Gervais, 1883: 1567; Persian Gulf.
And many other, extralimital names.

Range. Worldwide, migrating seasonally between breeding areas in tropical waters and temperate feeding areas. There is a calving area to the west of India and Sri Lanka and another between Taiwan and Japan; records elsewhere in the region are more sporadic.

Remarks. Reviewed by Winn & Reichley (1985). CITES: Appendix I; IUCN: V.

Family Balaenidae

Right whales

Balanadae Gray, 1821: 310; type genus *Balaena* L.
= Balaenidae Gray, 1825: 340.

A family of three species in two genera, characterized by very large head, strongly arched upper jaws with very long baleen, and fused neck vertebrae. Only one species in the region, probably irregularly.

BALAENA Right whales

Balaena Linnaeus, 1758: 75; type species *B. mysticetus* L. Conserved by Opinion 75 (1922).
Eubalaena Gray, 1864a: 201 (*B. australis* Desmoulins).
And other synonyms.

Polar and temperate oceans; 2 species of which one has strayed into Indomalayan waters.

Balaena glacialis Black right whale

B. glacialis Mueller, 1776: 7; North Cape, Norway.
B. australis Desmoulins, 1822: 161; Cape of Good Hope, S Africa.
And many other synonyms.

Range. Coastal regions of north and south temperate oceans; they migrate towards the equator in winter but probably only rarely enter tropical waters. Northern animals reach the Ryukyu Is in winter; southern ones do not normally enter the region but one was reported to have stranded in Gujarat, NW India in 1944 (Moses, 1948).

Remarks. CITES: Appendix I; IUCN: E.

[***Eschrichtius robustus*** Grey whale]

A record of this north Pacific species (as *Rachianectes glaucus*) from Sri Lanka (Deraniyagala, 1948) was based on a scant description of a wounded animal that was caught and released at sea. It was almost certainly wrongly identified and cannot be accepted without further evidence.

ORDER SIRENIA

Sea cows

Tropical America, West Africa and the Indian Ocean; 4 species in 2 families; only one species in the Indomalayan Region. Sea cows are totally aquatic herbivores found in fresh water and coastal seas.

Family Dugongidae

Dugong

Dugongidae Gray, 1821: 309; type genus *Dugong* Lacépède.

Contains one living genus and also the recently extinct *Hydrodamalis* of the NW Pacific.

DUGONG

Dugong Lacépède, 1799: 17; type species *D. indicus* Lacépède, in *D. dugon*.
= *Platystomus* Fischer, 1803: 353.
= *Dugungus* Tiedemann, 1808: 554.
= *Halicore* Illiger, 1811: 140.
= *Dugongidus* Gray, 1821: 309.

Contains one species.

Dugong dugon Dugong

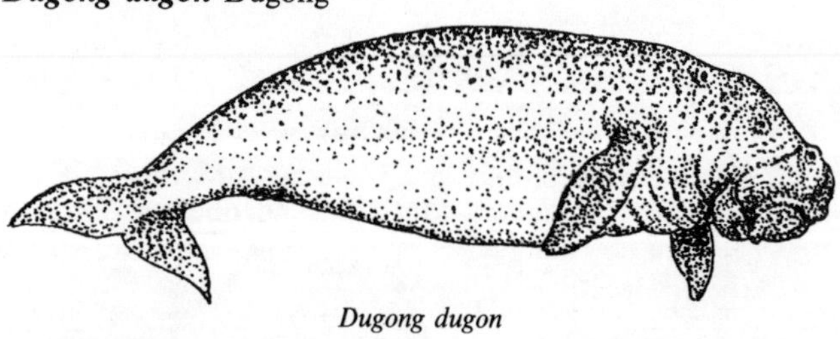

Dugong dugon

Trichecus dugon Mueller, 1776; Cape of Good Hope to Philippines.
= *Trichechus dugung* Erxleben, 1777: 599.
= *D. indicus* Lacépède, 1799: 17.
= *Halicore dugong* Illiger, 1811: 141.
And other synonyms from the Red Sea and Australia.

Range (Map 105). Formerly in coastal waters of much of the region, north to the Ryukyu Is, Taiwan and the adjacent mainland; also west to E Africa, and east to N Australia and the Gilbert and Ellice Is. In recent years limited, in the region, to the Indus Delta and NW India; the waters between S India and Sri Lanka; the Andaman Is; west coast of peninsular Burma and Thailand to Malaya, Singapore and E Sumatra; and the northern Moluccas.

Ecology. Live in sheltered coastal waters where they feed mainly upon sea-grasses of the families Potamogetonaceae and Hydrocharitaceae.

Remarks. Reviewed by Husar (1978) and by Nishiwaki & Marsh (1985). CITES: Appendix I; IUCN: V.

ORDER PROBOSCIDEA

Elephants

A distinctive order with only two extant species, referred to a single family. Formerly much more diverse, with many species becoming extinct during or since the Pleistocene.

Family Elephantidae

Elephants

Elephantidae Gray, 1821: 305; type genus *Elephas* L.

Indomalayan and Afrotropical regions; 2 extant genera, one in the Indomalayan Region.

ELEPHAS

Elephas Linnaeus, 1758: 33; type species *E. maximus* L.
Hypselephas Osborn, 1936: 12 (*E. hysudricus* Falconer & Cautley).

One Recent species. Data from osteology, dentition and immunology all agree in showing that *Elephas* is more closely related to the extinct *Mammuthus* (mammoth) than to the extant African *Loxodonta* (Shoshani *et al.*, 1985).

The following nominal fossil species and subspecies have been described from the region:
E. planifrons Falconer & Cautley, 1846:pl. 2 (and Falconer, 1868: 3); Siwalik Hills, India.

Map **105**. *Dugong, Elephas, Tapirus*

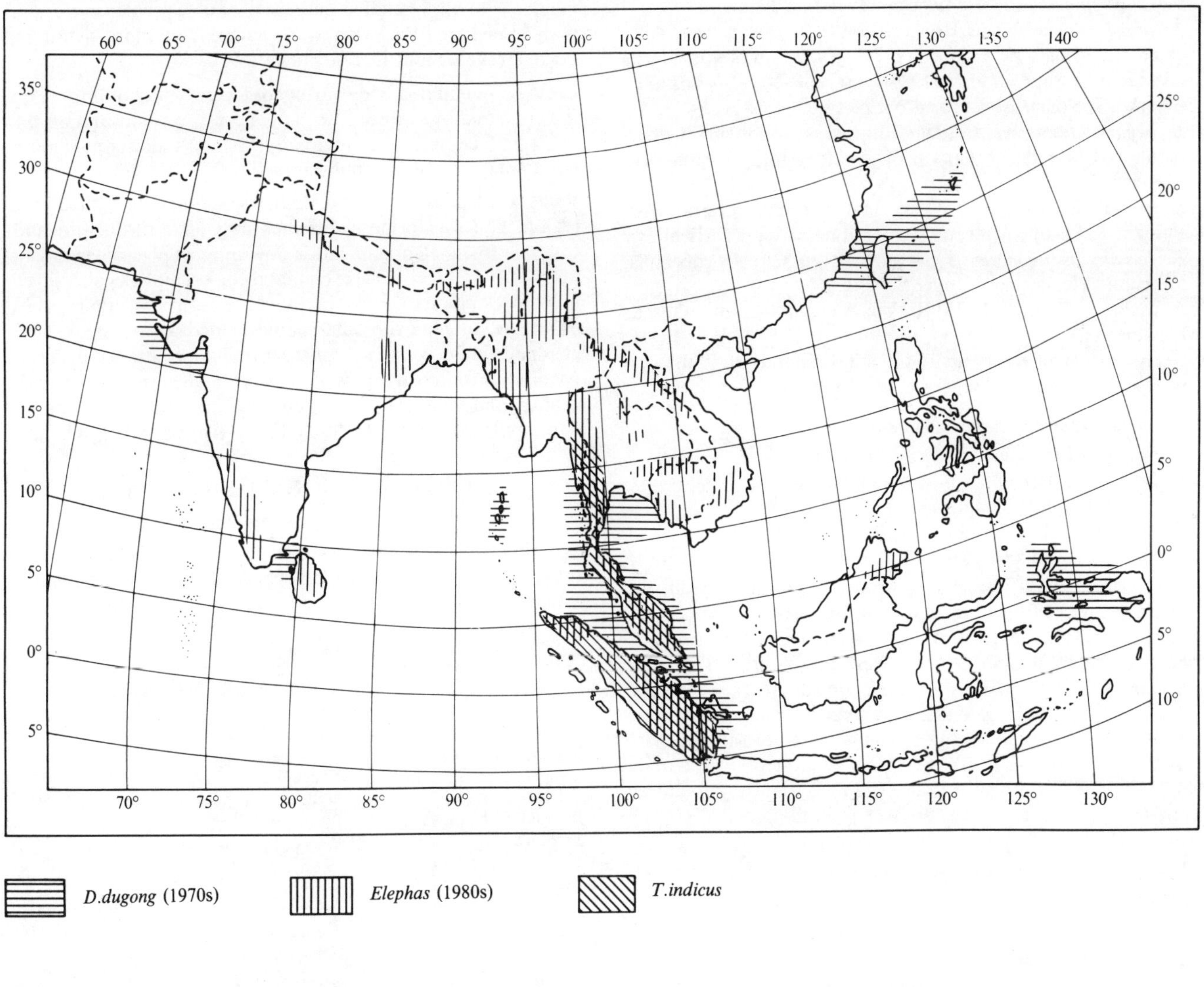

E. insignis Falconer & Cautley, 1846:pl. 2 (and Falconer, 1868: 3, 28); Siwalik Hills, India.
E. ganesa Falconer & Cautley, 1846:pl. 2 (and Falconer, 1868: 4, 33); India.
E. namadicus Falconer & Cautley, 1846:pl. 12 (and Falconer, 1868: 15); Siwalik Hills, India. Usually allocated to *Palaeoloxodon.*
E. hysudricus Falconer & Cautley, 1846: 41, pls 1, 4, 6 (and Falconer, 1868: 2, 6); Siwalik Hills, India. Type of the genus *Hypselephas* Osborn, 1936.
E. cliftii Falconer & Cautley, 1846: 47; near Yenangyaung, Irrawaddy R, Burma. Allocated to *Stegodon elephantoides* by Osborn (1942).
E. bombifrons Falconer & Cautley, 1847:pl. 25 (and Falconer, 1868: 41); Burma.
E. hysudrindicus Dubois, 1908: 1258; Kendeng, Java. Allocated to *Palaeoloxodon* by Osborn (1942).
E. platycephalus Osborn, 1929: 21; near Biswan, Amilee Creek, Simla Hills, India; Upper Pliocene or Lower Pleistocene. Allocated to *Platycephalus* by Osborn (1942).
E. platycephalus angustidens Osborn, 1929: 22; 3 miles W of Chandigarh, Siwalik Hills, India. Allocated to *Hypselephas hysudricus* by Osborn (1942).
Stegodon hypsilophus Hooijer, 1954b: 92; N of Djetis and Perning, E Java; Middle Pleistocene. Allocated, probably erroneously, to *E. celebensis* by Hooijer (1974).
Hypselephas hysudricus sinhaleyus Deraniyagala, 1944: 45; Nagoda, Sri Lanka; Pleistocene. Pre-occupied by *Elephas maximus sinhaleyus* Deraniyagala, 1939 (see below).
Archidiskodon celebensis Hooijer, 1949b: 206; Sulawesi. Allocated to *Elephas* by Hooijer (1974) but this was considered erroneous by Sondaar *et al.* (1989) who believed *celebensis* to be a derivative of *Mastodon* and not close to *Elephas.*
E. beyeri Koenigswald, 1956; Philippines, probably Cabarruyan I. Close to, perhaps a pygmy form of, *E. namadicus* (Groves, 1985a).
E. irshadi Sarwar, 1977: 156; near Sardhok, Gujrat dist, Punjab, Pakistan; Miocene (Pinjor zone of Upper Siwaliks).
E. hysudricus khoharensis Sarwar, 1979: 329; Pabbi Hills, Pakistan; Pleistocene (Upper Siwaliks).
E. youeensis Xue, 1981: 35; R You, Weinan, Shaanxi, China; Pleistocene.

The history of elephants in the Sunda islands and Sulawesi was summarized by Sondaar *et al.* (1989).

Elephas maximus Asian elephant

E. maximus Linnaeus, 1758: 33; Sri Lanka.
E. indicus Cuvier, 1797: 148; Sri Lanka.
E. asiaticus Blumenbach, 1797: 124; Sri Lanka.
E. sumatranus Temminck, 1847: 91; Sumatra.
E. hirsutus Lydekker, 1914: 20; Negri Sembilan, Malaya.

Range (Map 105). Formerly, in historic times, throughout the mainland part of the Region and on the islands of Sri Lanka and Sumatra; in prehistoric times also on Java and Borneo, and beyond the Region west to Iraq and north through most of lowland China. At present scattered populations exist throughout much of the Indochinese subregion from Assam and extreme S Yunnan to Vietnam, Thailand and in Malaya, with isolated populations in SW and NE India, Nepal, Sri Lanka, Sumatra and N Borneo. It is uncertain whether the population on Borneo is relict or the result of re-introduction. Feral on some of the Andaman Is.

Variation. The following subspecific names have been proposed (including those from beyond the region):

E. m. bengalensis Blainville, 1845: 353, pl. iii; Bengal.
E. indicus ceylanicus Blainville, 1845: 353, pl. iii; Sri Lanka.
E. m. sinhaleyus Deraniyagala, 1939a: 130; Sri Lanka; fossil.
E. m. vilaliya Deraniyagala, 1939a: 130; Manampitaya, Mahavili R, E Sri Lanka (extant).
E. m. dakhunensis Deraniyagala, 1950: 8 (of reprint); Travancore, S India (extant).
E. m. borneensis Deraniyagala, 1950: 10 (of reprint); Borneo (based on living, possibly introduced animals).
E. m. sondaicus Deraniyagala, 1950: 10 (of reprint); Java (based upon a 'medieval' stone carving).
E. m. rubridens Deraniyagala, 1950: 10 (of reprint); China (based upon a bronze statuette).
E. m. persicus Deraniyagala, 1950: 10 (of reprint); Iran (based upon an early historical description).
E. m. asurus Deraniyagala, 1950: 11 (of reprint); 'Mesopotamia' (based upon a prehistoric painting in Egypt and other early representations).
E. m. birmanicus Deraniyagala, 1951a: 49; Burma.
E. m. ruber Deraniyagala, 1951a: 51; 'shore of inland sea from Nakama to Sigora', Thailand.
E. m. sichiaoshanensis Wang, J-k.,1978: 123; Kuanshan Commune, Nanhai County, Guangdong, China; Holocene, subfossil ('not later than 2000–2500 years').

The form on Sri Lanka (*E. m. maximus*) is characterized by the absence of tusks. Elsewhere it is unlikely that more than one discrete subspecies can be recognized (*E. m. bengalensis*). Indian animals tend to be larger than those further east; the incidence of unpigmented patches of skin is greater in S India than elsewhere. Variation was reviewed in more detail by Pocock (1943a) and Deraniyagala (1950).

Ecology. Found in a wide variety of forest types but limited on the one hand by the need for daily access to water and on the other by feeding primarily upon grass and therefore avoiding large areas of closed-canopy forest.

Remarks. Widespread as a domestic animal throughout the region. This was formerly based entirely upon the capture and taming of wild elephants but is now in most places dependent upon captive breeding. In Thailand for example legal capture of wild elephants ceased in the 1970s—in 1988 there were estimated to be over 5000 domesticated and 2–3000 wild elephants in the country (Lair, 1988). The species was reviewed concisely by Shoshani & Eisenberg (1982) and its status, ecology and management in detail by Sukumar (1989) and by Santiapillai & Jackson (1990). CITES: Appendix I; IUCN: E.

ORDER PERISSODACTYLA

Odd-toed ungulates

Worldwide except for Australasia and the Nearctic (but present in the latter into the Holocene); Recent species fall into three very distinct families (Table 161). The evolution of

Table 161. The families of Perissodactyla.

	Tapiridae	Rhinocerotidae	Equidae
Genera (species) in region	1 (1)	2 (3)	1 (2)
Shoulder height (max, cm)	120	200	140
Limbs	Short, stout	Short, stout	Long, slender
Dermal horns on snout	0	+	0
Scrotum	0	0	+
Upper canines	+	0	+/0
Incisors, each side	3/3	0–1/0–2	3/3

perissodactyls was reviewed in Prothero & Schoch (1989), in which an earlier view was revived, that the hyraxes, family Procaviidae (not represented in the Indomalayan Region) should also be included.

Family Tapiridae

Tapirs

Taperidae Gray, 1821: 306; type genus *Tapirus* Cuvier.

Contains one genus.

TAPIRUS Tapirs

Tapirus Brisson, 1762: 81; type species *T. terrestris* Brisson (S America); doubtfully available.
= *Tapirus* Brünnich, 1771: 44.
= *Tapir* Blumenbach, 1779.
= *Rhinochoerus* Wagler, 1830: 17.
= *Tapyra* Liais, 1872: 397.
Acrocodia Goldman, 1913: 65 (*T. indicus* Desmarest).
Megatapirus Matthew & Granger, 1923: 588 (*T.* (*M.*)*augustus* Matthew & Granger).

Neotropical and Indomalayan Regions; 4 species, only one in this region.

The following fossil species have been recorded from the region:
T. pentapotamiae Falconer, 1868: 415; Siwalik Hills, India. Allocated to the suid genus *Listriodon* by Pilgrim (1926).
T. sinensis Owen, 1870: 426, pl. 29; Sichuan, China.
T. pandanicus Dubois, 1908: 1265; Kendeng, Java.
T. (*Megatapirus*) *augustus* Matthew & Granger, 1923: 588; Yen-ching-kao, near Wan-hsien, Sichuan, China; ? Upper Pliocene.
T. yunnanensis Shi *et al.*,1981: 7, 13; Zhaoting, NE Yunnan; 27°15'N, 130°38'E; Pliocene.

Tapirus indicus Malayan tapir

T. indicus Desmarest, 1819: 458; Peninsular Malaya.
Hydrochoerus sumatrensis Gray, 1821: 306; Sumatra (*nomen nudum*).
T. malayanus Raffles, 1821: 270; Malacca.
T. bicolor Wagner, 1835: 400; Malacca.

Range (Map 105). Burma and E Thailand from c. 18°N to Malaya; Sumatra. Formerly on Borneo where it survived at least until 8000 BP, probably later (Medway, 1960); in N India: a subfossil record near Allahabad dated between 1165BC and 550AD (Banerjee & Ghosh, 1981); and during the Pleistocene on Java (Sondaar, 1984).

Variation. One subfossil subspecies has been described from the region:
T. i. intermedius Hooijer, 1947: 288; Padang Highlands, C Sumatra; subfossil (based upon large size).

Remarks. The distribution and conservation status in Sumatra were reviewed by Santiapillai & Ramono (1990). CITES: Appendix I; IUCN: E.

Family Rhinocerotidae

Rhinoceroses

Rhynocerotidae Gray, 1821: 306; type genus *Rhinoceros* L. Afrotropical and Indomalayan Regions; 4 genera of which 2, with 3 species, are Indomalayan (Table 162). Classification of the Asian species was reviewed in detail by Groves (1967a) and phylogenies for the family were proposed by Groves (1983) and Prothero *et al.* (1986). The fossil and prehistoric forms were revised by Hooijer (1946). Criteria for the identification of isolated horns were given by Groves (1971a). A comprehesive bibliography with over 3000 references was compiled by Rookmaaker (1983), and a comprehensive popular account of all living species was produced by Penny (1987). For status and conservation see Khan (1989).

Table 162. The species of *Rhinoceros* and *Dicerorhinus* (Rhinocerotidae).

	R.unicornis	*R.sondaicus*	*D.sumatrensis*
Range	Nepal-Assam	Java, Vietnam	Burma-Borneo
Shoulder height (m)	1.6–1.8	1.4–1.7	1.0–1.3
Horns	1	1	2
Base of horn with anterior channel	+	+	0
Females with horns	+	Usually 0	+
Hair on body visible throughout life	0	0	+
Skin folds	Prominent	Prominent	Weak
Skull: greatest length	560–625	440–550	470–570
Occiput	Inclined forward	Inclined forward	Subvertical

RHINOCEROS Asian one-horned rhinoceroses

Rhinoceros sondaicus

Rhinoceros Linnaeus, 1758: 56; type species *R. unicornis* L.
Naricornis Frisch, 1775. Rejected by Opinion 258 (1954).
Eurhinoceros Gray, 1868e: 1009 (*R. javanicus* Cuvier, in *R. sondaicus*).
Monocerorhinus Wüst, 1922: 654 (*R. sondaicus* Desmarest).

Indomalayan Region; 2 species (Table 162). Unique amongst living rhinos in having only one horn but also characterized by prominent skin folds, enlarged lower canines and sloping occiput.

The following nominal fossil species and subspecies have been described from the region:
R. sivalensis Falconer & Cautley, 1847:pl. 73; Upper Siwaliks.
R. platyrhinus Falconer & Cautley, 1847:pls 72, 75; Upper Siwaliks (in *Coelodonta*: Badam, 1979).
R. palaeindicus Falconer & Cautley, 1847:pl. 73; Upper Siwaliks.
R. perimensis Falconer & Cautley, 1847:pls 75, 76; Perim (= Peram) I, Gulf of Cambay, Gujarat, India; Middle Siwaliks (in *Aceratherium*: Hooijer, 1946).
R. namadicus Falconer, 1868: 21; Narbada, India.

R. sinensis Owen, 1870: 424, pl. 29; Sichuan, China.
R. deccanensis Foote, 1874: 1; Gokak, Belgaum dist, India.
R. iravadicus Lydekker, 1876: 18 (= 36); Prome, Burma (in *Aceratherium perimensis*: Hooijer, 1946).
R. planidens Lydekker, 1878a: 95; Padri, Punjab, Pakistan (in *Aceratherium perimensis*: Hooijer, 1946).
R. sivalensis var. *gajensis* Lydekker, 1881: 40; Lower Siwaliks.
R. karnuliensis Lydekker, 1886a: 121; Karnul Caves, S India.
R. kendengindicus Dubois, 1908: 1259; Kendeng, Java.
R. sinhaleyus Deraniyagala, 1939b: 235; Talavitaya, Sri Lanka.
R. philippinensis Koenigswald, 1956; Cagayan, Philippines. Likely to belong to a genus of the subfamily Aceratheriinae, not *Rhinoceros* (Groves, 1985a).
R. kagavena Deraniyagala, 1958: 116,122; Hiriliyädda, Talavitiya, Sabaragamuva Prov., Sri Lanka; Pleistocene.
R. barinagalensis Srivastava & Verma, 1972: 77; Chandigarh, Haryana, India; Pleistocene.
R. (*Gaindatherium*) *vidali* Heissig, 1972: 24; Nagri, Salt Range, Pakistan; Tertiary (Middle Siwaliks).
R. sinensis hayasakai Otsuka & Lin, 1984: 9; Taiwan.

Rhinoceros unicornis Indian rhinoceros, Greater one-horned rhinoceros

R. unicornis Linnaeus, 1758: 56; ? Assam.
R. indicus Cuvier, G.,1817: 239; India.
R. asiaticus Blumenbach, 1830: 107.
R. stenocephalus Gray, 1868e: 1018.
[*R. jamrachii* Sclater, 1876: 650; ? Manipur (Rookmaaker, 1977a). Invalid.]

Range (Map 106). The 'Terai' lowlands of Nepal, Bengal and Assam, with the main concentration now in the Kaziranga National Park, Assam (c. 1250 in 1989) and the Chitwan National Park in Nepal (c. 400 in 1989). Formerly west at least to the Indus Valley as late as the 16th century AD.

Ecology. In Nepal grass was found to constitute 70–90% of the diet although a great variety of other plants were eaten (Laurie, 1982).

Remarks. For a concise review of the species see Laurie *et al.* (1983). CITES: Appendix I; IUCN: E.

Rhinoceros sondaicus Javan rhinoceros, Lesser one-horned rhinoceros

R. sondaicus Desmarest, 1822: 399; Sumatra (Sody, 1946); usually given as Java.
R. javanicus Geoffroy & Cuvier, 1824:liv. 445; Java.
R. javanus Cuvier, G.,1829: 247; Java.
R. inermis Lesson, 1838: 514; 'Les Sundries', i. e. Sunderbunds, Bangladesh/W Bengal.

Map **106**. *Rhinoceros, Dicerorhinus, Equus*

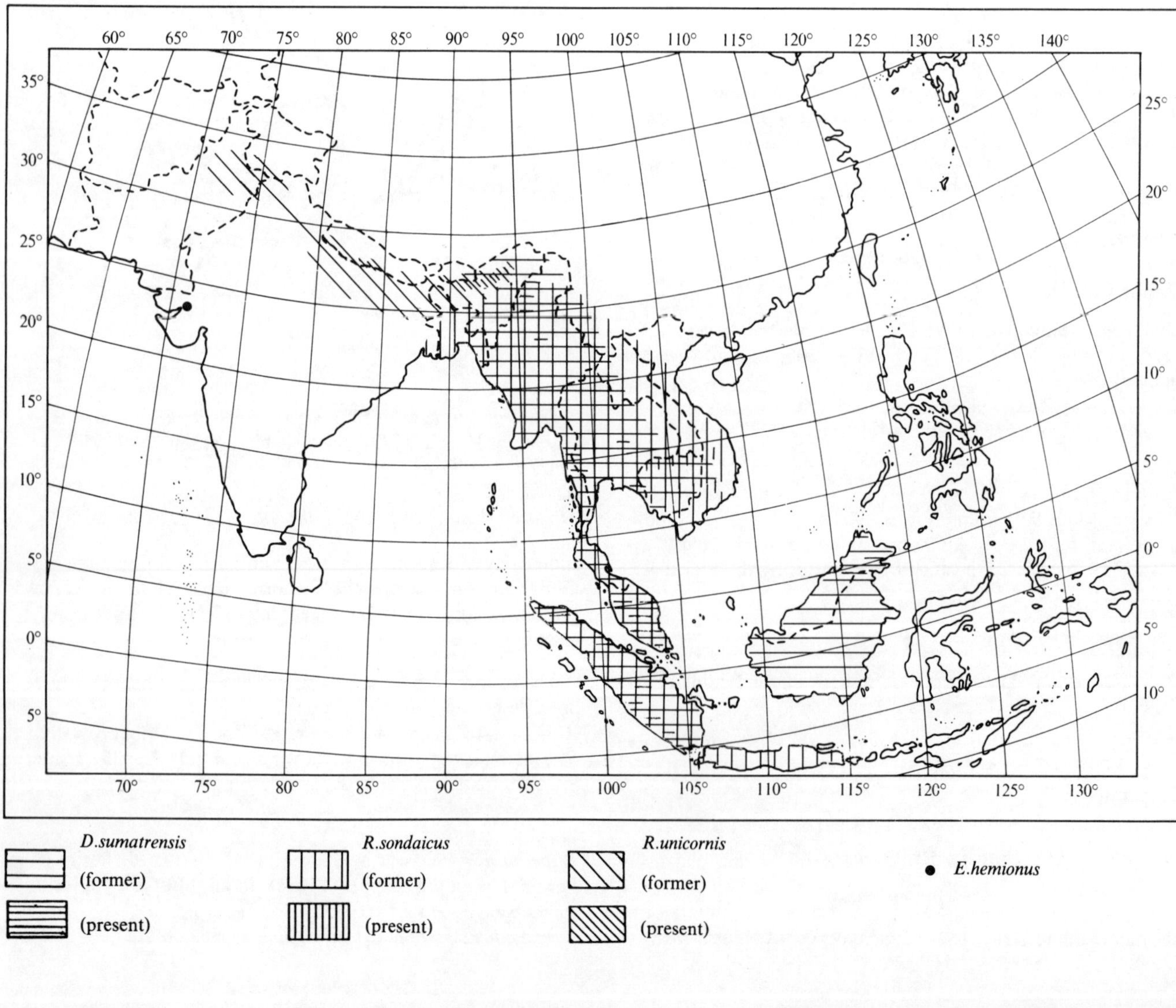

R. nasalis Gray, 1868e: 1015; 'Borneo', probably Java.
R. floweri Gray, 1868e: 1015; Sumatra.
R. sivasondaicus Dubois, 1908: 1245; Kendeng, Java (fossil).
Aceratherium boschi Koenigswald, 1933: 121; Java; Pliocene (*fide* Aimi & Sudijono, 1979).

Range (Map 106). Formerly, in historical times, from Bhutan and West Bengal to Burma, Laos and Vietnam, through Thailand to Malaya; and on Sumatra and Java. Now certainly known to be present only in the Ujung Kulon Reserve in extreme western Java (c. 57 animals in 1989: Santiapillai *et al.*, 1990), and in Vietnam where 10–15 were located c. 130km NE of Saigon in 1989 (Schaller *et al.*, 1990). Perhaps also surviving in Laos and Burma. Recent reports from Cambodia and Laos are likely to be of this species but these require confirmation (Rookmaaker, 1988). Subfossil finds show its former presence in Borneo around the Pleistocene/Holocene boundary and in middens dated c. 8–10000 BP (Cranbrook, 1986). It has been suspected of occurring in New Guinea (Sody, 1959) but this has not been confirmed and is improbable (Heptner, 1960).

Variation. The following subspecific names have been proposed:
R. s. annamiticus Heude, 1892: 75,113, pl. xixA; locality unknown, presumably Annam, Vietnam.
R. s. simplisinus Deraniyagala, 1946: 165; Nivitigala, Sabaragamura Prov., Sri Lanka; fossil.
R. s. guthi Beden & Guérin, 1973: 19; Phnom Loang, Kampot Prov., Cambodia; Pleistocene.

Remarks. For a detailed account see Sody (1959). CITES: Appendix I; IUCN: E.

DICERORHINUS

Dicerorhinus Gloger, 1841: 125; type species *R. sumatrensis* Cuvier = *R. sumatrensis* Fischer. Conserved by Opinion 1080 (1977).
= *Didermocerus* Brookes, 1828: 75. Rejected by Opinion 1080 (1977).
= *Ceratorhinus* Gray, 1868a: 1021.

One species. A primitive genus, more closely related to *Rhinoceros* than to the African genera and with many fossil species recorded in Europe and N Asia.

The following fossil species has been described from the edge of the region:
D. lantianensis Hu & Qi, 1978: 38; Lantian, Shaanxi, China; Pleistocene.

Dicerorhinus sumatrensis Sumatran rhinoceros, Asian two-horned rhinoceros

Rhinoceros sumatrensis Fischer, 1814: 301; Bencoolen dist, S Sumatra. Conserved by Opinion 1080 (1977).
R. crossi Gray, 1854b: 251; locality unknown.
R. lasiotis Buckland, 1872: 89; Chittagong, E Bengal.
Ceratorhinus niger Gray, 1873c: 357, pl. 11; Sunghi-njong dist, Malacca. Pre-occupied by *R. niger* Schintz, 1845, in *Diceros bicornis*, Africa.
C. blythii Gray, 1873c: 360; ? Tenasserim, Burma.

Range (Map 106). Formerly from Assam, Burma and Indochina to Malaya, and on Sumatra and Borneo. Still present in small numbers in Borneo (Sabah, N Sarawak, probably also Kalimantan), Sumatra and Malaya, and perhaps in parts of Thailand, Burma and Indochina. In Malaya there was estimated to be a total of 50–75 individuals in ten isolated areas in 1975–81 (Flynn & Tajuddin Abdullah, 1984); in Sumatra c. 400–750 in 1985; and in Burma c. 10 in 1985.

Variation. One fossil and one Recent subspecies have been described:
D. s. eugenei Sody, 1946: 151; cave at Ngalau Lida Ajer, near Pajakombo, Padang Highlands, C Sumatra; fossil.
Didermocerus s. harrissoni Groves, 1965: 130; Suan-Lambah, Sabah, N Borneo. Based on average smaller size. However subfossils dated c. 8–10000 BC indicate that they were then about 10–20% larger than non-Bornean animals (Cranbrook, 1986).

Ecology. A browser, feeding upon a large variety of plant species but rarely grass other than bamboo (van Strien, 1975).

Remarks. For detailed accounts see van Strien (1975) and Groves & Kurt (1972). The taxonomic history was reviewed by Rookmaaker (1984). CITES: Appendix I; IUCN: E.

Family Equidae

Horses and asses

Equidae Gray, 1821: 307; type genus *Equus* L.

Afrotropical and Palaearctic Regions; c. 7 wild species of which one occurs marginally in the northwest of this region. Only one genus is recognized here, but the Asiatic asses are often separated in the genus *Hemionus*. Classification was reviewed by Groves & Willoughby (1981) and Groves (1986) who recognized only one genus, *Equus*, with six subgenera. For a comprehensive popular account see Groves (1974a).

EQUUS

Equus Linnaeus, 1758: 73; type species *E. caballus* L. Conserved by Opinion 271 (1954).
Asinus Brisson, 1762: 70 (*E. asinus* L.).
= *Onager* Brisson, 1762.
= *Asinus* Gray, 1824: 244.
Hemionus Stehlin & Graziosi, 1935: 3 (*E.* (*Hemionus*) *hemionus* Pallas).
Asinohippus Trumler, 1961: 118 (*E. khur* Lesson).
Praehemionus Trumler, 1961: 119 (*E. sivalensis* Falconer & Cautley— fossil).

Palaearctic and Afrotropical regions; c. 7 species. Although only one species is native (but very marginal) in the region, domestic horses and donkeys, and their hybrid mules, are widely used and are detailed below (Table 163).

Table 163. The species of *Equus*.

	hemionus	*asinus*	*caballus*
Shoulder height (cm)	110–127	80–150	Variable, commonly –170
Length of ears (cm)	19–21	Longer	Shorter
Mid-dorsal stripe	+	+	0
Transverse stripe on shoulders	0	+	0
Horizontal stripes on legs	0	+/0	0
'Chestnut' (horny callous) on each hind leg	0	0	+
Mane	Erect	Erect	Usually lax
Long hair on tail	Distal half	Distal half	From base

The following names have been given to fossil forms in the region:
E. sivalensis Falconer & Cautley, 1849:pl. 81 (and Falconer, 1868: 22); Siwaliks.
E. namadicus Falconer & Cautley, 1849:pl. 81; Siwaliks (included in *E. sivalensis* by Hooijer, 1949a and Badam, 1979).
E. palaeonus Falconer & Cautley, 1849:pl. 82 (and Falconer, 1868: 220); Siwaliks, India (in *E. sivalensis*: Badam, 1979).
E. cautleyi Hopwood, 1936: 907; Upper Siwaliks; Lower Pleistocene (in *E. sivalensis*: Badam, 1979).
E. yunnanensis Colbert, 1940: 1; Makai Valley, Yunnan; Lower Pleistocene.
E. sivalensis minor Gaur & Chopra, 1984: 20; Pinjar Formation, Upper Siwaliks, India.

Equus hemionus Asiatic wild ass, Onager

E. hemionus Pallas, 1775: 394, pl. 7; Tarei-Nor, Dauria, Transbaikalia, USSR.
E. onager Boddaert, 1785: 160; Kasbin, NW Iran.
E. khur Lesson, 1827: 347; Little Rann of Cutch, India.
E. indicus George, 1869: 35.
Also other, extralimital, synonyms.

Range (Map 106). Formerly widespread in the arid zone of NW India (including the present Pakistan), west to Syria and through much of central Asia; now limited in this region to the Little Rann of Kutch Desert in Gujarat where there were about 2000 in 1983 (Smielowski & Praduman, 1988). It probably became extinct in Baluchistan and the extreme south of Pakistan, on the Indian border, during the 1960s.

Variation. One subspecific name has been proposed from the region:
Microhippus h. blanfordi Pocock, 1947: 143; Sham Plains, E Baluchistan, Pakistan, 29°20'N, 69°40'E. Based upon a single specimen shot in 1882, having less white ventrally than *E. h. khur*; now extinct.

Compared with more northern subspecies the Indian form, *E. h. khur*, is characterized by darker summer pelage; white extending further up the flanks; winter pelage short (as summer); dorsal stripe not extending beyond middle of tail; no dark rings round hooves; long ears: 190–210mm (Groves, 1974a, 1986). It has also been included in a more wide-ranging subspecies, *E. h. onager*, extending west originally to Syria, e. g. by Mazak (1963).

Ecology. A desert-adapted grazer, living in herds and feeding on a variety of grasses and sedges.

Remarks. CITES: Appendix I (*E. h. khur* only); IUCN: E.

Equus asinus Domestic donkey, Domestic ass

E. asinus Linnaeus, 1758: 73.

Range. Domesticated animals are widespread in the arid zone of Pakistan, NW India and Nepal, and in parts of China, including the southeast as well as the arid west and north; smaller numbers occur elsewhere in the region. A feral population of dwarf stature exists in NW Sri Lanka (Lever, 1985).

Remarks. Domestication of the ass probably took place around 4000 BC in NE Africa or SW Asia, from the wild *E. africanus* Fitzinger, 1857. Subfossils in Pakistan and NW India indicate its presence there from about 2000 BC. The history and relationships were reviewed by Epstein (1984) and Groves (1986).

Equus caballus Domestic horse

E. caballus Linnaeus, 1758: 73; Europe. Conserved by Opinion 271 (1954).

Range. Horses are kept throughout the region and are still important as draught animals in China and India. Feral populations of horses occur in NW Sri Lanka (Lever, 1985), and on the islands of Flores and Rintja in the Lesser Sunda Is (Pfeffer, 1957).

Remarks. The ancestral wild species, *E. ferus* Boddaert, 1785, occurred in the steppe zones of the Palaearctic.

[ORDER TUBULIDENTATA]

Aardvarks

Now represented by a single species *Orycteropus afer* in the Afrotropical Region, but two nominal fossil species have been described from the Indomalayan Region:

ORYCTEROPUS

Orycteropus Geoffroy, 1796b: 102; type species *Myrmecophaga capensis* Gmelin, in *O. afer* Pallas.

O. browni Colbert, 1933a: 1; Nathot, N Punjab, Pakistan; Nagri zone, Siwaliks.
O. pilgrimi Colbert, 1933a: 6; 7 km W of Hasnot, N Punjab, Pakistan; Nagri zone, Siwaliks.

ORDER ARTIODACTYLA

Even-toed ungulates

The artiodactyls are well represented in the region by the families Suidae, Cervidae and Bovidae, and there are three endemic species of Tragulidae; the Moschidae is represented only marginally in the north, and the Camelidae only by domestic stock (Table 164). The Hippopotamidae and Giraffidae are included on the basis of some extinct forms described in Recent genera. All species were described in the series of British Museum catalogues by Lydekker (1913b, 1915, 1916) and Lydeker & Blaine (1914).

Family Suidae

Pigs

Suidae Gray, 1821: 306; type genus *Sus* L.

Throughout the Old World (but in Australasia only by introduction); 5 genera of which 2 now occur in the Indomalayan Region and one is represented only by fossils (Table 165). The family was reviewed, with a comprehensive revision of the Indomalayan species, by Groves (1981a); and by Sjarmidi & Gerard (1988); the Indonesian forms, fossil and Recent, were reviewed by Hardjasasmita (1987).

SUS

Sus Linnaeus, 1758: 49; type species *S. scrofa* L. Conserved by Opinion 75 (1922).
Porcula Hodgson, 1847e: 423 (*P. salvania* Hodgson).
Scrofa Gray, 1868a: 38 (*Sus domesticus* Brisson).

Table 164. The families of Artiodactyla.

	Suidae	Camelidae	Tragulidae	Moschidae	Cervidae	Bovidae
Genera in region	2	1	2	1	6	11
Permanent horns, at least in males	0	0	0	0	0	+
Deciduous antlers in males	0	0	0	0	+*	0
Pre-orbital glands	0	0	0	0	+	+/0
Interdigital glands		0	0	0	+/0	+/0
Preputial gland enlarged	0	0	0	+	0	0
Metacarpals of 2nd and 5th digits	Complete	Absent	Complete	Rudimentary	Rudimentary	Rudimentary
Cuboid and navicular bones fused	0	0	+	+	+	+
Upper incisors	+	+	0	0	0	0
I_1 angular, much wider than I_2	+	+	0	0	+	+
Teeth hypsodont	0	+	+	+	+	+
C^1	+ (tusks)	+	+ (tusks)	+ (tusks)	+	0
Long diastema between incisors and cheek-teeth	0	+	+	+	+	+
Stomach: no. of chambers	2	3	3	4	4	4

* Except in *Hydropotes*.

Table 165. The species of *Sus* (except *S.domesticus*) and *Babyrousa* (Suidae).

	S.scrofa	*S.barbatus*	*S.verrucosus*	*S.celebensis*	*S.salvanius*	*B.babyrussa*
Range	Widespread	Malaya, Sumatra, Borneo, Philippines	Java etc.	Sulawesi etc.	Nepal etc.	Sulawesi etc.
Head & body (cm)	115–150	120–150	90–160		60–65	85–105
Ears	Large	Small	Large	Very small	Small	Very small
Beard on lower jaw	0	+	0	0	0	0
Disc of snout	Vertical	Oblique	Oblique	Oblique	Vertical	Oblique
Tail with double tuft	0	+	0	0	0	0
Mid-dorsal crest	+/0	+	+	0	0	0
Tuft on crown	0	0	0	+	0	0
Facial warts	0	2 pairs	3 pairs, infra-orbital largest	3 pairs, pre-orbital largest	0	0
All-black as well as agouti hairs	0	+	+	+	0	–
Skull: greatest length (males)	260–415	320–500	350–420	260–330	c.180	
Upper incisors (each side)	3	3	3	3	3	2
Canines angular in section (males)	+	+	+	+	+	0
C^1 emerges:	Laterally	Laterally	Laterally	Laterally	Laterally	Upwards
C_1 of male - tip:	Occludes with C^1	Free	Free	Free	Occludes with C^1	Free

Euhys Gray, 1869: 339 (*S. barbatus* Müller).
Aulacochoerus Gray, 1873d: 435 (*S. vittatus* Müller).
Dasychoerus Gray, 1873d: 435 (*S. verrucosus* Müller & Schlegel).
Sinesus Heude, 1892: 102 (? *Sus scrofa*).

Palaearctic and Indomalayan regions; 5 species, all represented in the Indomalayan Region and all but one confined to it (Table 165). The domestic pig is here treated as an additional species.

The following nominal fossil species and subspecies have been described from the region:
S. giganteus Falconer & Cautley, 1847:pl. 69; Siwaliks.
S. hysudricus Falconer & Cautley, 1847:pl. 70; Siwaliks.
S. pusillus Falconer, 1868: 415; Siwaliks. Allocated to *Sanitherium schlagintweiti* Meyer by Lydekker (1884b).
S. punjabiensis Lydekker, 1878b: 81; Siwaliks.
S. titan Lydekker, 1884b: 59; Siwaliks.
S. falconeri Lydekker, 1884b: 66; Siwaliks.
S. karnuliensis Lydekker, 1886b: 49; Billa Surgam, Karnul dist, Madras State, S India.
S. brachygnathus Dubois, 1908: 1264; Kendeng, Java.
S. macrognathus Dubois, 1908: 1264; Kendeng, Java.
S. advena Pilgrim, 1926: 59, pl. 20; Haritalyangar; Nagri zone, Siwaliks.
S. comes Pilgrim, 1926: 60, pl. 19; Hasnot; Dhok Pathan zone, Siwaliks.
S. adolescens Pilgrim, 1926: 60, pl. 20; Hasnot; Dhok Pathan zone, Siwaliks.
S. praecox Pilgrim, 1926: 60, pl. 19; Hasnot; Dhok Pathan zone, Siwaliks.
S. peregrinus Pilgrim, 1926: 61, pl. 19; no locality; Tatrot zone.
S. bakeri Pilgrim, 1926: 62; no locality.
S. cautleyi Pilgrim, 1926: 63; Siwalik Hills.
S. namadicus Pilgrim, 1926: 64, pl. 20; Narbada Valley, India; Middle Pleistocene.
S. terhaari Koenigswald, 1933: 44; Ngandong, Java.
S. stremmi Koenigswald, 1933: 49; Boemiajoe, Java.
S. vatualangensis Koenigswald, 1933: 52; Watoealang, Java.
S. coerti Koenigswald, 1934: 192; Jetis, E Java; Pleistocene.
S. sinhaleyus Deraniyagala, 1947: 224; Gonapitiya, Kuruvita, Sri Lanka.
S. sangiranensis Koenigswald, 1963: 192; Sangiran, Java; Pleistocene.
S. xiaozhu Han *et al.*,1975: 252; Bijiashan, Liuzhou, Guangxi, China; Quaternary.
S. bijiashanensis Han *et al.*,1975: 253; Bijiashan, Liuzhou, Guangxi, China; Quaternary.
S. jiaoshanensis Zhao, 1980: 301; Jiaoshan, Wuming, Guangxi, China; Pleistocene.

S. subtriquetra Xue, 1981: 40; R You, Weinan, Shaanxi, China; Pleistocene.
S. lydekkeri bantangensis Liu *et al.*, 1982: 13; Chao Xian, Anhui, China; Middle Pleistocene.
S. choprai Gaur, 1987a: 132; 315 km NE of Mirzapur village, NE of Chandigarh, NW India; Pinjor formation, Upper Siwaliks, Pliocene/Pleistocene.
S. gomphotherioides Sarwar, 1988: 402; Jassian, Attock Dist, Punjab, Pakistan; Tatrotian of Upper Siwaliks.

The following names are likely to have been based upon hybrids (Groves, 1981a):
S. ternatensis Rolleston, 1877: 276; Ternate I, Moluccas.
S. ceramensis Rosenberg, 1878: 362; Seram I, Moluccas.
S. papuensis goramensis Beaux, 1924: 431; Goram, SE of Seram, Moluccas.

Sus scrofa Wild boar

S. scrofa Linnaeus, 1758: 49; Germany. Conserved by Opinion 75 (1922).
= *S. setosus* Boddaert, 1785: 157.
S. aper Boddaert, 1785: 157.
S. vittatus Boie, 1828: 240; Sumatra (Jentink, 1905).
S. cristatus Wagner, 1839b: 435 (misprinted as 535); near Calcutta, India (Groves, 1981).
S. leucomystax Temminck, 1842: 6; Japan.
S. indicus Gray, 1843a: 185; Nepal (restricted by Groves, 1981a).
S. timoriensis Müller & Schlegel, 1843:pl. 31 (and 1845: 173); Timor I.
S. affinis Gray, 1847d: 71; Nilgiri Hills, S India (*nomen nudum*).
S. zeylonensis Blyth, 1851a: 173; Sri Lanka.
S. bengalensis Blyth, 1860a: 105; Bengal and Kutak, India.
Porcula taivana Swinhoe, 1863: 360; Taiwan (lectotype selected by Groves, 1981a).
S. moupinensis Milne-Edwards, 1872a: 93; Moupin, Sichuan, China.
S. chirodontus Heude, 1888: 54; Poyang Lake, Jiangxi, China.
S. taininensis Heude, 1888: 54; Cochin China.
S. palustris Heude, 1888: 54; valley of Yangtze, China; preoccupied by *S. palustris* Rütimeyer, 1861 (Europe, subfossil).
S. dicrurus Heude, 1888: 55; divide between Han and Kincha Rivers, Shaanxi, China.
S. curtidens Heude, 1892: 114; locality unknown.
S. laticeps Heude, 1892: 114; locality unknown.
S. collinus Heude, 1892: 114, pl. 19A; Ning-kouo Mts, China.
S. acrocranius Heude, 1892: 114; locality unknown.
S. leucorhinus Heude, 1892: 114; China.
S. paludosus Heude, 1892: 114, pl. 19A; Tchen-kiang, Yangtze, China.
S. melas Heude, 1892: 114; locality unknown.
S. scrofoides Heude, 1892: 114, pl. 19A; Cambodia and Annam.
S. microdontus Heude, 1892: 114, pl. 19A; locality unknown.
S. spatharius Heude, 1892: 114, pl. 19A; locality unknown.
S. planiceps Heude, 1892: 144, pl. 19A; Ho Shan, Anhui, China.
S. flavescens Heude, 1899: 130; Yangtze and Taihu, Jiangsu, China.
? *S. natunensis* Miller, 1901a: 117; Laut I, N Natuna Is. Perhaps based upon feral domestic pigs (Groves, 1981a).
S. milleri Jentink, 1905: 176; Java.
S. jubatus Miller, 1906h: 745; Trang, S Thailand.
S. jubatulus Miller, 1906h: 746; Terutau I, Malaya.
S. peninsularis Miller, 1906h: 749; Mt Pulai, SW Johore, Malaya.
S. rhionis Miller, 1906h: 749; Ungar I, Riau Is.
? *S. niadensis* Miller, 1906h: 751, pl. 62; Nias I, W Sumatra.
? *S. babi* Miller, 1906h: 752; Babi I, W Sumatra. Perhaps based upon feral domestic pigs (Groves, 1981a).
S. andersoni Thomas & Wroughton, 1909a: 441; Batam I, Riau Is.
Also many other extralimital synonyms.

Range (Map 107). Throughout most of the region southeast as far as Sumatra and Java (but not Borneo); and including the islands of Sri Lanka, Taiwan, the Ryukyus (Iriomote, Ishigaki), Andamans, Nicobars and many small inshore islands (but not the Mentawai Is): Teratau, Langkawi, Singapore, most of the Riau Is; probably also on Bali, Lombok, Sumba, Sumbawa and Komodo in the Lesser Sunda Is, and Timor. Determination of the eastern limits is confused by possible introductions, by the presence of feral domestic pigs and by possible hybridization between them and other species, especially *S. barbatus* and *S. celebensis*. Also throughout much of the Palaearctic Region except for the taiga, tundra and extreme desert zones; and by introduction elsewhere, e. g. in South America, New Guinea and much of E and N Australia.

Variation. The following subspecific names have been proposed from the region:
S. aper var. *aipomus* Hodgson, 1842: 911; Nepal. *Nomen nudum.*
S. aper var. *isonotus* Hodgson, 1842: 911; Nepal. *Nomen nudum.*
? *S. babi tuancus* Lyon, 1916a: 453; Tuangku I, Banjak Is, W Sumatra. Perhaps based upon feral domestic pigs (Groves, 1981a).
S. leucomystax riukiuanus Kuroda, 1924: 11; Kabira, Ishigaki-jima, Ryukyu Is.
S. vittatus pallidiloris Mees, 1957: 214; Padang, Sumatra.
S. s. davidi Groves, 1981a: 37; Sind, Pakistan.

Seven subspecies in the region were recognized and described by Groves (1981a), but there is little evidence of discrete boundaries. The most conspicuous difference is in the development of a dorsal crest of long hair: well developed in most parts of the range but short or absent in China, Taiwan and the Ryukyu Is. Overall colour is pale in the arid zone of Pakistan and NW India, much darker elsewhere. Size is greatest in S India and Sri Lanka (greatest length of skull c. 400–420mm, usually <400mm elsewhere). The nasals are relatively shorter in Malaya and the Sunda Islands (*S. s. vittatus*): c. 45–48% of greatest length of skull, cf 48–51% elsewhere.

Ecology. Versatile with respect to habitat, avoiding only very arid areas. Although omnivorous they are mainly vegetarian, concentrating upon roots, rhizomes and tubers, and can be a serious pest of cultivation.

Remarks. The ancestor of the domestic pig. IUCN: (V)— *S. s. riukiuanus*.

Sus domesticus Domestic pig

S. domesticus Brisson, 1762: 74.
S. sinensis Brisson, 1762: 75; China.
S. scrofa domesticus Erxleben, 1777: 179.
S. s. sinensis Erxleben, 1777: 181; China.

The case for treating the domestic pig as a species separate from its ancestor, *S. scrofa*, is perhaps less strong than with many other domesticated animals, but it is nevertheless useful to do so (Corbet & Clutton-Brock, 1984). Throughout most of the range of *S. scrofa* there is probably little interbreeding between wild and domestic pigs but the distinction between wild, feral and domestic is less clear in parts of Indonesia. Domestic pigs are kept throughout the region, but especially in China and in those parts of island SE Asia settled by Chinese. Some feral populations occur although these are only likely to remain identifiable as such in the absence of wild *S. scrofa*, as on some small islands such as the Andamans and Nicobars. The following names are likely to be based on feral domestic pigs (Groves, 1981a):

S. andamanensis Blyth, 1858: 267; Port Blair, Andaman Is.
S. nicobaricus Miller, 1902b: 755; Great Nicobar I, Bay of Bengal.
S. floresianus Jentink, 1905: 178; Flores I, Lesser Sunda Is.
S. babi enganus Lyon, 1916a: 454; Engano I, W Sumatra.

All the main local forms of *S. scrofa*, e. g. in India, China and Indonesia, are likely to have been involved in the production of domestic stock, and Chinese pigs were used in the development of modern breeds in Europe in the late 18th century. The history of domestication was reviewed by Epstein & Bichard (1984) and the confused situation in SE Asia by Groves (1981a).

Sus salvanius **Pygmy hog**

Porcula salvania Hodgson, 1847e: 423, pl. 12, 13; Sikkim Terai, India.

Range (Map 107). Formerly throughout the Terai grassland zone from Nepal to Assam; now probably fragmented and only confirmed in recent years from Assam (Oliver, 1980).

Ecology. Confined to tall-grass savannah.

Remarks. Although superficially very different from *S. scrofa* because of its very small size, its very close relationship to *S. scrofa* is indicated by the karyotype (Bosma *et al.*, 1982) as well as by cranial and dental characters (Groves, 1981a). CITES: Appendix 1; IUCN: E.

Sus barbatus **Bearded pig**

S. barbatus Müller, 1838: 149; Banjermasin, S Borneo.
S. longirostris Nehring, 1885: 347; Kuala Kapuas, W Kalimantan, Borneo.
S. celebensis var. *philippensis* Nehring, 1886: 83; Luzon, Philippines.
S. ahoenobarbus Huet, 1888: 5; Paragua I, Palawan, Philippines.

Map **107**. *Sus, Babyrousa*

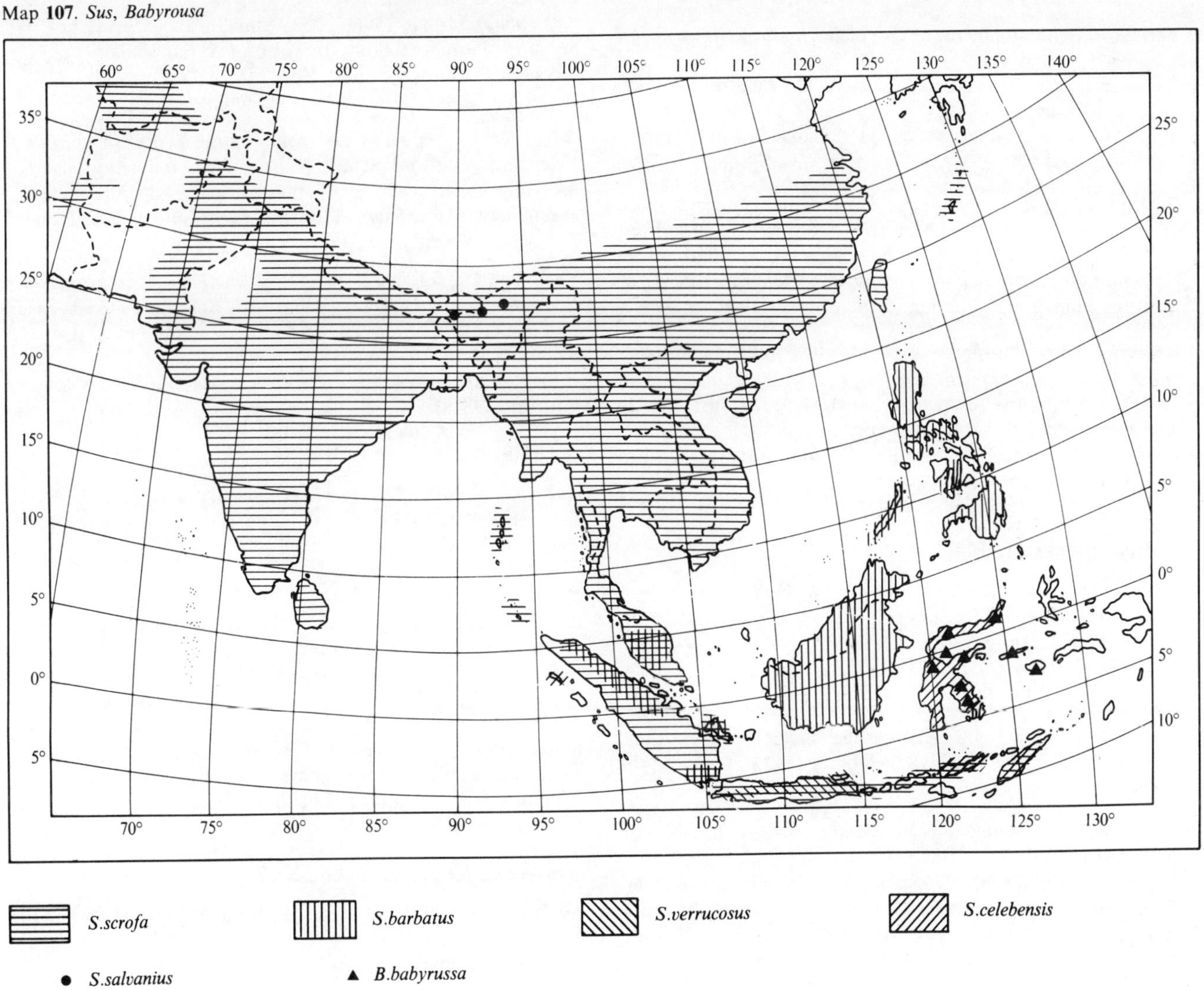

S. marchei Huet, 1888: 6; Jala Jala, La Laguna Prov., Luzon, Philippines.
S. minutus Heude, 1888:pl. 15; 'Mindanao', corrected to Luzon, between La Laguna and Tarlac.
S. effrenus Heude, 1888:pl. 17; Jala-jala, Luzon, Philippines.
S. microtis Heude, 1888:pl. 17; Bataan, Luzon, Philippines.
S. frenatus Heude, 1888:pl. 17; Jala-jala, Luzon, Philippines.
S. joloensis Heude, 1888:pl. 17; Jolo, Sulu Is, Philippines.
S. cebifrons Heude, 1888:pl. 17; Cebu, Philippines.
S. inconstans Heude, 1892:pl. 29C; Mindanao, Philippines.
S. megalodontus Heude, 1892:pl. 29A; Mindanao, Philippines.
S. mainitensis Heude, 1892:pl. 29C; Mainit I, Philippines.
S. crassidens Heude, 1892:pl. 29C; locality unknown.
S. arietinus Heude, 1892: 114; locality unknown.
S. calamianensis Heude, 1892: 114; Calamianes Is, Philippines.
S. verrucosus mindanensis Major, 1897: 524; Ayala, Mindanao, Philippines.
S. oi Miller, 1902: 51; Indragiri R, E Sumatra.
S. gargantua Miller, 1906h: 743; SE Kalimantan, Borneo.
S. verrucosus sumatranus Kelm, 1939: 220; Lampung, S Sumatra.
S. celebensis negrinus Sanborn, 1952a: 145; Pagyabunan, Bais, Negros I, Philippines.

Range (Map 107). Malaya, Sumatra, Borneo and throughout most of the Philippines; also on the Riau Is and Bangka.

Variation. The following subspecific names have been proposed:
S. b. var. *palavensis* Nehring, 1889: 22; Puerto Princesa, Palawan, Philippines.
S. b. balabacensis Major, 1897: 534; Balabac I, Philippines.
S. b. edmondi Sody, 1937a: 212; Lampongs, S Sumatra.

Ecology. Sometimes form large herds which undergo mass movements, related to the erratic but often synchronized fruiting of trees, especially dipterocarps. At these times in particular they are of major economic importance as a source of meat for human consumption (Caldecott & Caldecott, 1985).

Remarks. The Philippine pigs included here have been considered to represent *S. verrucosus*, e. g. by Laurie & Hill (1953), but this was refuted by Groves (1981a). In captivity fertile hybrids, probably of reduced viability, have been produced between *S. barbatus* and *S. scrofa* (Groves, 1981a). IUCN: (V)— *S. b. cebifrons, S. b. oi.*

Sus verrucosus Javan pig

S. verrucosus Boie, 1832: 466; Tjikau, W Java.
= *S. verrucosus* Müller & Schlegel, 1843:pl. 28 (and 1845: 107).
S. mystaceus Gray, 1873b: 62; Java.
? *S. bucculentus* Heude, 1892:pl. 20B; 'Cochin China', i. e. presumably S Vietnam.

Range (Map 107). Java, including the islands of Madura and Bawean; ? Vietnam. On Java they occur below 800m and the range is now very fragmented (mapped by Blouch, 1988).

Variation. The following subspecific names have been proposed:
S. v. var. *ceramica* Gray, 1868a: 24; 'Ceram', perhaps in error.
S. v. borneensis Major, 1897: 534; 'Borneo', probably in error.
S. v. olivieri Sody, 1941b: 393; S Soemenap, E Madura I, Java.
S. v. blouchi Groves, 1981a: 47; Bawean I, Java Sea.

The form on Madura I is not distinguishable from that on the mainland of Java, but that on Bawean is distinguishable by small size and reddish rather than yellowish pelage (Groves, 1981a).

Remarks. On Java some hybridization occurs with *S. scrofa* (Blouch & Groves, 1990). The form *bucculentus* was allied to *S. verrucosus* by Heude, a conclusion supported by Groves (1981a) who considered that there was little doubt that the locality was correctly interpreted as Vietnam. However there has been no subsequent confirmation of the species in Indochina and it may have been an import there, either as an individual or as an ephemeral population. IUCN: V.

Sus celebensis Sulawesi wild boar

S. celebensis Müller & Schlegel, 1843:pl. 28 (and 1845: 172); Manado, Sulawesi.
? *S. verrucosus amboinensis* Major, 1897: 528; 'Amboina'
Microsus maritimus Heude, 1899: 115; Macassar, S Sulawesi.
M. macassaricus Heude, 1899: 115; Macassar, S Sulawesi.
? *M. floresianus* Heude, 1899: 115; Flores, Lesser Sunda Is (perhaps a domestic form of *S. celebensis*).
S. nehringii Jentink, 1905: 186; S Sulawesi.
S. weberi Jentink, 1905: 187; Saleyer I, S Sulawesi.
S. niadensis Miller, 1906h: 753; Nias I, W Sumatra. Perhaps based upon a feral *S. celebensis.*
? *S. mimus* Miller, 1906h: 753; Simalur I (= Simaleue), W Sumatra. Perhaps a domestic form of *S. celebensis.*
? *S. heurni* Hardjasasmita, 1987: 46; near Pota, Manggarai Plain, W Flores, Lesser Sunda Is, 400m.

Range (Map 107). Sulawesi, including the islands of Lembeh, Peleng, Butung and Selayar; also on the following islands, probably as a result of domestication and introduction: Halmahera (Moluccas); Flores (Lesser Sunda Is); Timor; Simaleue (W Sumatra).

Variation. The following subspecific name has been proposed:
S. c. sarasinorum Hooijer, 1950a: 116; Bola Batoe Cave, near Badjo (Berebo dist.), c. 20 km SW of Watampone, C Bone, S Sulawesi; subfossil.

Size is greater in the south than in the north of Sulawesi (Groves, 1981a).

BABYROUSA

Babirussa Frisch, 1775; not available (Opinion 258, 1954).
Babyrousa Perry, 1811:sig. C; type species *B. quadricornua* Perry, in *B. babyrussa* L.

Sulawesi; one species. Characterized especially by the extraordinary form of the upper tusks (C^1) which arise from alveoli that face almost vertically upwards, and are long and slender, not angular in cross-section, and curve backwards in a flat spiral parallel to the sagittal plane of the head, making as much as one complete whorl in some individuals. However there are many other differences and it is generally agreed that *Babyrousa* is only distantly related to all other surviving genera of Suidae (Groves, 1981a; Bosma & Haan, 1981; Sjarmidi & Gerard, 1988).

Babyrousa babyrussa Babirusa

Sus babyrussa Linnaeus, 1758: 50; Buru, Moluccas (Borneo in error).
= *Babirussa alfurus* Lesson, 1827: 338.

Aper orientalis Brisson, 1762: 76; East Indies.
B. quadricornua Perry, 1811:sig. c.
Babirusa celebensis Deniger, 1910: 7; Sulawesi.

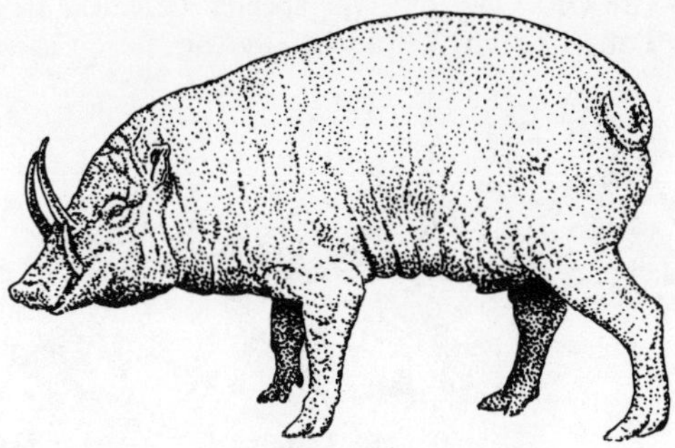

Babyrousa babyrussa

Range (Map 107). Sulawesi, and the adjacent islands of Malengi, Taliabu, Sulabesi and Buru. Likely to have been introduced by human agency to the last three islands (Groves, 1980a).

Variation. The following subspecific names have been proposed:
Babirussa b. frosti Thomas, 1920e: 187; Taliaboe I, Sula Is, E of Sulawesi.
B. b. beruensis Hooijer, 1948b: 1326; Beru, near Cabenge, c. 100km N of Macassar, S Sulawesi; Pleistocene. Allocated to the extinct genus *Celebochoerus* by Hooijer (1954a), but retained as a subspecies of *B. babyrussa* by Groves (1980a).
B. b. togeanensis Sody, 1949b: 187; Malengi I, Togian Is, N Sulawesi.
B. b. bolabatuensis Hooijer, 1950a: 121; Bola Botoe Cave, near Badjo, 20 km SW of Watampone, Barebo dist, Central Bone, S Sulawesi; subfossil.

Remarks. CITES: Appendix I; IUCN: V.

[*POTAMOCHOERUS*]

Potamochoerus Gray, 1854a: 129; type species *Choiropotamus pictus* Gray, in *P. porcus* (L.).

A genus now confined to the Afrotropical Region, with one living species, but represented in the Indomalayan Region by fossil species from the Siwalik beds of NW India:
P. palaeindicus Pilgrim, 1926: 27, pl. 10; Tatrot Plateau; lower Pinjor horizon (Pliocene).
P. theobaldi Pilgrim, 1926: 27, pl. 10; Gagret, Kangra Dist, Himachal Pradesh, India; Upper Siwaliks.
Sus hysudricus, S. giganteus and *S. titan* (listed under *Sus*) have also been allocated to *Potamochoerus* (Matthew, 1929).

[Family Hippopotamidae]

Hippopotamuses

Not now present in the region but the Recent genus *Hippopotamus*, now confined to Africa, was widespread during the Pleistocene. The following nominal species and subspecies were based upon fossils from the region. They were revised by Hooijer (1950b) who recognized only two species, *H. sivalensis* and *H. iravaticus*:

HIPPOPOTAMUS

Hippopotamus Linnaeus, 1758: 74; type species *H. amphibius* L.
Hexaprotodon Falconer & Cautley, 1836a: 51 (*H. sivalensis* Falconer & Cautley).

Hippopotamus (*Hexaprotodon*) *sivalensis* Falconer & Cautley, 1836a: 51; Siwalik Hills, N India; Pleistocene.
H. dissimilis Falconer & Cautley, 1836a: 51; Siwalik Hills, India. Allocated to *Merycopotamus* by Falconer & Cautley (1847) and Colbert (1943).
H. iravaticus Falconer & Cautley, 1847:pl. 57 (and Falconer, 1868: 498); Irawaddy Valley, Burma; Lower Pleistocene.

The following names were all allocated to *H. sivalensis* by Hooijer (1950b):
Hexaprotodon anisiperus McClelland, 1838: 1045, pl. 59; ? Siwalik Hills, India.
Hexaprotodon megagnathus McClelland, 1838: 1045, pl. 59; ? Siwalik Hills, India.
Hexaprotodon platyrhynchus McClelland, 1838: 1045, pl. 59; ? Siwalik Hills, India.
H. (*Hexaprotodon*) *namadicus* Falconer & Cautley, 1847:pl. 57 (and Falconer, 1868: 498); Narbada Valley, C India; Middle/Upper Pleistocene.
H. (*Tetraprotodon*) *palaeindicus* Falconer & Cautley, 1847:pl. 57 (and Falconer, 1868: 497); Narbada Valley, C India; Middle/Upper Pleistocene.
Hexaprotodon sivajavanicus Dubois, 1908: 1265; Java; Lower Pleistocene.
H. (*Hexaprotodon*) *simplex* Koenigswald, 1933: 33; Boemijoe, Java; Lower Pleistocene.
H. (*Hexaprotodon*) *antiquus* Koenigswald, 1934: 192, pl. 4; Java; Lower Pleistocene. Pre-occupied by *H. antiquus* Desmarest, 1822 (from Europe).
= *H. koenigswaldi* Hooijer, 1942: 281.
H. ? sinhaleyus Deraniyagala, 1939b: 232; Ratnapura dist, Sri Lanka; Lower Pleistocene.
H. sivalensis duboisi Hooijer, 1950b: 54; Naliwala, Sirmur State, Punjab; Pleistocene.
H. s. soloensis Hooijer, 1950b: 75; Tinggang, Solo Valley, Java.

Family Camelidae

Camels

Camelidae Gray, 1821: 307; type genus *Camelus* L.

Not represented by wild species in the region, but domesticated camels are widely used in the arid northwest and two nominal fossil species have been described from the region.

CAMELUS Camels

Camelus Linnaeus, 1758: 65; type species *C. bactrianus* L. (Opinion 16, 1910).
Dromedarius Gloger, 1841:xxxiii (*D. africanus* Gloger).

Palaearctic Region; 2 species, both domesticated, only one now found within the region.
The following fossil species have been described from the region:
C. sivalensis Falconer & Cautley, 1836b: 120, pl. 20; Upper Siwaliks.
C. antiquus Lydekker, 1885a: 78; Upper Siwaliks.

Camelus dromedarius One-humped camel, Dromedary

C. dromedarius Linnaeus, 1758: 65.

Range. Widespread as a domestic animal throughout the arid zone of Pakistan and NW India; west throughout SW Asia and circum-Saharan Africa; feral in Australia.

Remarks. Since this species appears to be fully interfertile with the domestic two-humped camel, *C. bactrianus* (Mason, 1984b: 106), it is arguable that they should be considered conspecific. This was in fact done by Bohlken (1961) who gave *dromedarius* as a synonym of *bactrianus*—following him as 'first reviser' the correct name for the combined species would be *C. bactrianus*. However the resulting nomenclature would be so cumbersome or confusing that it seems best to treat both the one-humped and two-humped camels as 'domestic species', *C. dromedarius* and *C. bactrianus*, separate from the surviving wild camel, *C. ferus* Przewalski, 1883 (Corbet & Clutton-Brock, 1984). The one-humped camel has never been known in the wild state and was probably first domesticated in Arabia prior to 1000 BC.

In the Indian region they are used as draught, pack and riding animals. In 1985 the population in India was estimated at 1.1 million, 70% of these in Rajasthan (Khauna *et al.*, 1990). Prior to 1000 AD *C. bactrianus* was used as a draught animal in the same region. The history and distribution of camels were reviewed by Mason (1984b) and all aspects by Köhler-Rollefson (1991).

Family Tragulidae

Mouse-deer, Chevrotains

Tragulidae Milne-Edwards, 1864: 157; type genus *Tragulus* Brisson.

Indomalayan and Afrotropical regions; 3 genera of which 2 are Indomalayan (Table 166). The smallest of the ungulates, sharing with the Moschidae enlarged upper canines in the males, absence of antlers or horns, and absence of preorbital glands (all probably primitive characters); differing in the triangular cross-section of the upper canines, the less complex stomach, with three chambers, lobulate kidneys and complete lateral metacarpals. The Indian species, *meminna*, is often included in *Tragulus* but shows such a mixture of characters linking it to both *Tragulus s. s.* and to the African *Hyemoschus* (e. g. the patterned pelage) that it is best placed in a separate genus, *Moschiola*, as done by Groves & Grubb (1987). The characters of the four species were reviewed by Smit-van Dort (1989) and the relationships of the family by Janis (1984).

Table 166. The species of *Tragulus* and *Moschiola* (Tragulidae.)

	T.javanicus	*T.napu*	*M.meminna*
Range	Indochina-Borneo	Indochina-Borneo	India, Sri Lanka
Head & body	195–410	500–600	460–600
Tail	65–80	80–110	20–60
Hind foot	110–125	135–150	125–155
Dorsal colour	Red-brown, ± uniform	Red-brown, coarsely mottled	Greyish, finely speckled
Flanks and rump with white spots and stripes	0	0	+
No.of white lines on throat	3	5	5/7
Nape with black stripe	+	+	0
Throat with inter-ramal gland	+	+	0
Scrotum	Naked	Naked	Hairy
Skull: condylobasal length	80–100	90–115	85–110

TRAGULUS Malay mouse-deer

Tragulus Brisson, 1762: 65; type species *T. indicus* Brisson.
Tragulus Pallas, 1779b: 27 (doubtfully valid)

Tragulus javanicus

Indochinese and Sunda subregions; 2 species (Table 166). These are very similar—where they are sympatric they differ clearly in size, but the larger species, *T. napu*, shows a reduction in size on some small islands where the other is absent (Dort, 1988). *T. napu* can usually be distinguished also by the more mottled appearance of the dorsal pelage and by the presence on the throat of a projection of dorsal colour on each side interrupting the smooth upper edge of the white line from nape to lower lip (Fig. 20).

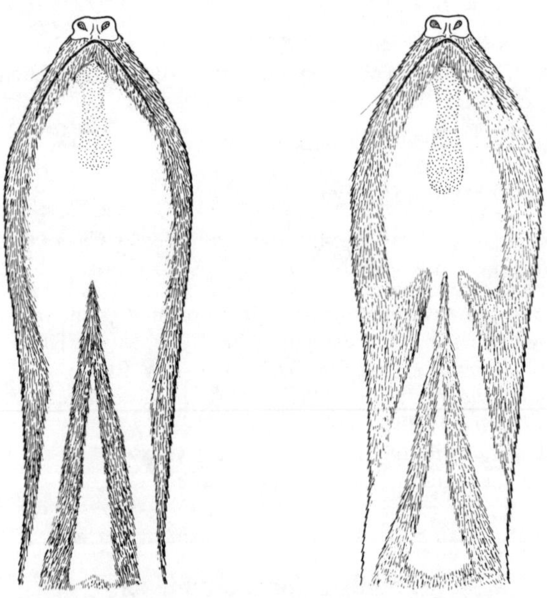

Fig.**20**. *Tragulus*. Typical pattern on throat of (a) *T.javanicus*; (b) *T.napu*. The naked glandular area on the chin is stippled.

The following fossil species has been described from the Region:
T. sivalensis Lydekker, 1882: 30; near Hasnot, Pakistan; Siwaliks.

Tragulus javanicus Lesser Malay mouse-deer

? *T. indicus* Brisson, 1762: 65.
Cervus javanicus Osbeck, 1765: 357; Udjon Kulon Peninsula, W Java.
Moschus kanchil Raffles, 1821: 262; Sumatra.
M. pelandoc Smith, 1827: 302; locality unstated, restricted to Cheribon, N coast of Java by Dobroruka (1967).
M. fulviventer Gray, 1836: 15; Singapore I (Kloss, 1918).
T. fuscatus Blyth, 1858: 278; Java.
T. affinis Gray, 1861: 138; Cambodia.
T. pallidus Miller, 1901a: 116; Laut I, N Natuna Is.
? *T. ravus* Miller, 1902e: 173; Trang, S Thailand.
T. russulus Miller, 1903a: 3; Tana Balu I, Batu Is, W Sumatra.
T. focalinus Miller, 1903e: 35; Buitenzorg, W Java.
T. virgicollis Miller, 1903e: 37; Mt Dulit, Sarawak, Borneo, 3000ft.
T. natunae Miller, 1903e: 38; Bunguran I, Natuna Is.
T. subrufus Miller, 1903e: 39; Sinkep I, Lingga Is.
T. rubeus Miller, 1903e: 40; Bintang I, Riau Is.
T. ravulus Miller, 1903e: 41; Adang I, Butang Is, Thailand.
T. lancavensis Miller, 1903e: 41; Langkawi I, W Malaya.
T. lampensis Miller, 1903e: 42; Lampi I (= Sullivans I), Mergui Is, Burma.
T. brevipes Miller, 1903b: 443; Bangkaru I, Banjak Is, W Sumatra.
T. russeus Miller, 1903b: 444; Tuangku I, Banjak Is, W Sumatra.
T. luteicollis Lyon, 1906a: 579; Banka I, Sumatra.
T. carimatae Miller, 1906a: 55; Karimata I, SW Borneo.
T. fulvicollis Lyon, 1908b: 630; Bengkalis I, E Sumatra.
T. pinius Lyon, 1916a: 455; Pinie Is, Batu Is, W Sumatra.

Range (Map 108). From S Yunnan (Shi & Chen, 1989) and Indochina to Malaya; Sumatra, Java, Borneo; and on the following smaller islands: Lampi, King's (Mergui Is); Pipidon, Phuket (Thailand); Langkawi, Penang, Redang (Malaya); Singapore; Riau Is, Lingga Is; Tana Balu, Tana Masa, Pinie, Bangkuru, Tuangku, Bengkalis (Sumatra); Bangka; Karimata; Laut, Bunguran (N Natuna). Up to 600m in Borneo.

Variation. The following subspecific names have been proposed:
T. kanchil pierrei Bonhote, 1903g: 293; Bien Hoa, S Vietnam.
T. k. hosei Bonhote, 1903g: 293; Baram, Sarawak, Borneo.
T. k. everetti Bonhote, 1903g: 295; Bunguran I, N Natuna Is.
T. k. longipes Lyon, 1908b: 628; Little Siak R, E Sumatra.
T. russulus masae Lyon, 1916a: 456; Tana Masa I, Batu Is, W Sumatra.
T. kanchil williamsoni Kloss, 1916g: 88; Me Song Forst, Pre, N Thailand.
T. k. penangensis Kloss, 1918b: 253; Penang I, W Malaya.

Map **108**. *Tragulus, Moschiola, Moschus*

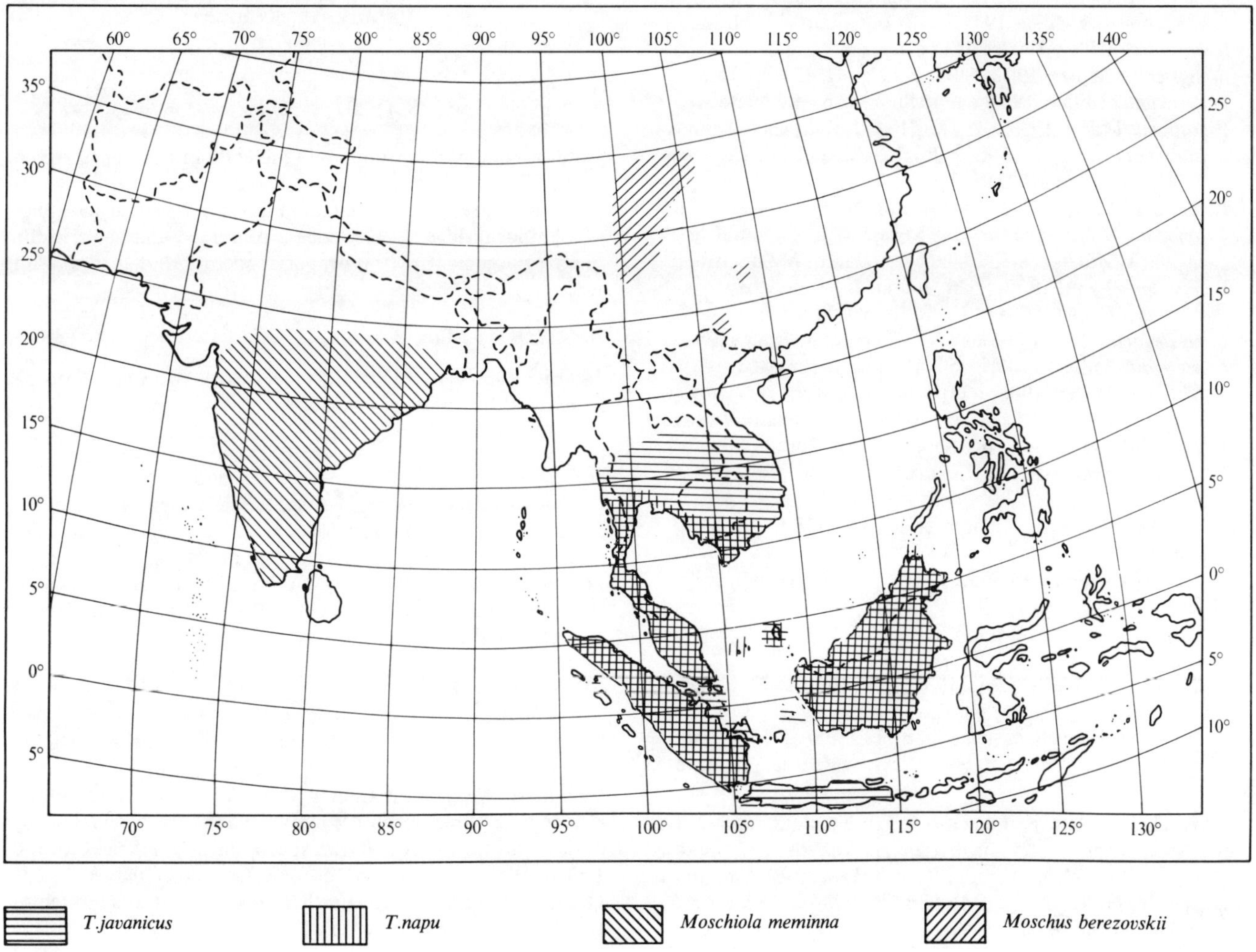

T. k. angustiae Kloss, 1918b: 254; Bankachon, Victoria Point, Tenasserim, S Burma.
T. k. mergatus Thomas, 1923m: 85; King's I, Mergui Is, Burma.
T. k. klossi Chasen, 1935a: 98; near Sandakan, N Borneo.
T. k. insularis Chasen, 1940a: 198; Junk Seylon I (Phuket), SW Thailand.
T. k. pidonis Chasen, 1940a: 198; Pipidon I, SW Thailand.
T. k. pumilus Chasen, 1940a: 199; Great Redang I, E Malaya.

Differences between the major segments of the range in Malaya, Sumatra, Java and Borneo are slight: all have the dorsum suffused with black and a clear dark stripe on the nape, but in Java there is no median dark streak on the chest. North of the Isthmus of Kra the pelage is more uniform, the dorsum paler, greyish brown, the dark stripe on the nape indistinct, and no median stripe on the chest (*T. j. affinis*). The forms on the small islands are not highly differentiated but that on Penang (Malaya) is rather dark.

Tragulus napu Greater Malay mouse-deer

T. javanicus: Chasen (1940a) *et al.*, not of Osbeck, 1765 (see van Bemmel, 1949).
Moschus napu Geoffroy & Cuvier, 1822:liv. 37; S Sumatra (Sody, 1931b).
M. stanleyanus Gray, 1836: 65; locality unknown, selected as Batam I, Riau Is by Chasen (1940a).
T. nigricans Thomas, 1892a: 250; Balabac I, Philippines.
T. canescens Miller, 1900f: 185; Trang, S Thailand.
T. umbrinus Miller, 1900e: 191; Langkawi I, Malaya.
T. rufulus Miller, 1900a: 227; Tioman I, Malaya.
T. bunguranensis Miller, 1901a: 113; Bunguran I, N Natuna Is.
T. pretiosus Miller, 1902a: 144; Lingga I, Lingga Is.
T. nigricollis Miller, 1902a: 145; Singkep I, Lingga Is.
T. borneanus Miller, 1902e: 174; Suanlamba R, N Borneo.
T. batuanus Miller, 1903a: 2; Tana Bala I, Batu Is, W Sumatra.
T. lutescens Miller, 1903e: 32; Sugi Bawa I, Riau Is.
T. flavicollis Miller, 1903e: 33; Sugi I, Riau Is.
T. formosus Miller, 1903e: 34; Bintang I, Riau Is.
T. amoenus Miller, 1903b: 439; Mansalar I, W Sumatra.
T. jugularis Miller, 1903b: 440; Mansalar I, W Sumatra.
T. bancanus Lyon, 1906a: 576; Banka I, Sumatra.
T. billitonus Lyon, 1906a: 578; Billiton I, Sumatra.
T. nigrocinctus Miller, 1906b: 250; Kundur I, Riau Is.
T. pretiellus Miller, 1906b: 253; Bakong I, Lingga Is.
T. perflavus Miller, 1906b: 251; Batam I, Riau Is.
T. versicolor Thomas, 1910a: 535; Nhatrang, Annam, Vietnam.
T. sebucus Lyon, 1911a: 64; Sebuko I, SE Borneo.
T. kanchil siantanicus Chasen & Kloss, 1928b: 31; Siantan I, Anamba Is.
T. kanchil anambensis Chasen & Kloss, 1928b: 32; Mata I, Anamba Is.
T. javanicus neubronneri Sody, 1931b: 355; Peureulak, E Aceh, N Sumatra.
T. javanicus bangui Chasen & Kloss, 1932: 16; Banguey I, N Borneo.
T. javanicus abjectus Chasen, 1935b: 11; Sirhassen I, S Natuna Is.
T. kanchil abruptus Chasen, 1935b: 12; Subi I, S Natuna Is.
T. javanicus hendersoni Chasen, 1940a: 196; Jimaja I, Anamba Is.

Range (Map 108). S Vietnam and S Thailand to Malaya; Sumatra, Borneo; and on the smaller islands of Langkawi, Tioman (Malaya); Singapore; Riau Is, Lingga Is, Anamba Is, Bunguran (N Natuna Is), S Natuna Is; Balabac (Palawan). Up to 1000m in Borneo.

Variation. The following subspecific names have been proposed:
T. canescens terutus Thomas & Wroughton, 1909b: 536; Terutau I, SW Thailand.
T. pretiellus parallelus Miller, 1911c: 165; Sebang I, Lingga Is.
T. napu niasis Lyon, 1916a: 455; Nias I, W Sumatra.

Where this species is sympatric with *T. javanicus* the condylobasal length of the skull is over 95mm, but on some small islands in the absence of the smaller species it is smaller, with condylobasal length down to 90mm (Dort, 1988).

MOSCHIOLA

Moschiola Hodgson, 1844b: 292; type species *T. mimenoides* Hodgson, in *T. meminna* Erxleben.

One species.

Moschiola meminna Indian chevrotain

Moschus meminna Erxleben, 1777: 322; Sri Lanka.
= *Meminna indica* Gray, 1843a: 172. Not of Brisson, 1762.
Tragulus mimenoides Hodgson, 1842: 914; Nepal Terai.
Meminna malaccensis Gray, 1843a: 172; 'Singapore'.

Range (Map 108). Peninsular India, NW to Rajasthan (Tehsin, 1981); Sri Lanka. Early records from Nepal and the Himalayan foothills of NW India have not been confirmed (Champion, 1929).

Family Moschidae

Musk deer

Moschidae Gray, 1821: 307; type genus *Moschus* L.

Contains one genus. Commonly included in the Cervidae but treated as a separate family in many recent studies, e. g. Groves & Grubb (1987) who considered it likely to be the sister-group of (Antilocapridae + Cervidae). A unique character is the great enlargement of the preputial scent gland in the adult male.

MOSCHUS Musk deer

Moschus Linnaeus, 1758: 66; type species *M. moschiferus* L. Conserved by Opinion 75 (1922).
= *Odontodorcus* Gistel, 1848: 82.

E Asia, both in the Palaearctic and in the Indochinese part of the Indomalayan Region; 3–6 species of which only one, *M. berezovskii*, occurs in this region. Of the others one, *M. moschiferus*, is wholly Palaearctic and the others occur at high altitude on the Palaearctic fringe of the Himalayas to W China, mainly above 3000m. These comprise *M. chrysogaster* Hodgson, 1839 in the E Himalayas and W China (including *sifanicus* Büchner, 1891); *M. fuscus* Li, 1981 in W Yunnan, N Burma and SE Tibet; with *M. leucogaster* Hodgson, 1839 perhaps a separate species in the Himalayas.

All forms have at times been considered conspecific, e. g. by Ellerman & Morrison-Scott (1951), but there is now a consensus that there are at least three species, probably more. While recognizing three species, Flerov (1952) used the name *M. moschiferus* for the Himalayan species and *M. sibiricus* for the Palaearctic one. However the type locality of *moschiferus* is clearly in the range of the northern species and was limited to the Altai Mts by Heptner *et al.* (1961).

Table 167. The genera of Cervidae.

	Cervus	*Elaphurus*	*Axis*	*Hydropotes*	*Muntiacus*	*Elaphodus*
Species in region	7	1	4	1	7	1
Shoulder height (cm)	65–160	120–140	65–100	45–55	45–75	50–70
Antlers	+	+	+	0	+	+
Frontal glands	0	0	0	0	+	0
Glandular cavity in fore leg	0	0	0	+	0	0
Pedal glands on hind feet	0	0	+	+	+	0
Metatarsal glands	+/0	+/0	+	0	0	0
Inguinal glands	0	0	0	+	0	0
Width of $I_1:I_2\text{-}C_1$	=	=	>	=	>	>
C^1	+/0	+	0	+	+	+
C^1 long, tusk-like in male	0	0	–	+	+	+

There has been no comprehensive revision and the delimitation of some species is still very uncertain. Himalayan forms were reviewed by Grubb (1982) and Chinese ones by Gao (1985). Musk deer are browsers and are mainly montane in the region. All have been heavily exploited for musk and many populations are probably endangered.

One fossil form has been named from the region:
M. moschiferus plicodon Colbert & Hooijer, 1953: 108; Yenchingkou, Sichuan, China; Pleistocene.

Moschus berezovskii Chinese forest musk deer

M. berezovskii Flerov, 1929: 519; Ho-tsi-how Pass, near Lungan, Sichuan, China.
M. moschiferus caobangis Dao, 1969; Trung Khanh, Cao Bang, N Vietnam. (See also Dao, 1977.)
M. moschiferus anhuiensis Wang, S., Hu & Yan, 1982: 133, 137; Changling region, Jinzhai County, Anhui, China, 31°11'N, 115°54'E, 500m. (Allocated to *M. berezovskii* by Groves & Feng, 1986.)

Range (Map 108). S and central China from Anhui, S Shaanxi and S Gansu to Yunnan and Guangxi, and adjacent parts of N Vietnam.

Recognition. This species is distinguished from the other, high altitude, species by dark, finely speckled, olive-brown pelage, nearly black on the rump, with orange-brown on the legs and venter; three pale, longitudinal stripes under the throat and neck; ears black at the tip, orange-brown at the base. Juveniles are spotted, as in all species of *Moschus*.

Remarks. In China this species is extensively farmed for the production of musk.

Family Cervidae

Deer

Cervina Goldfüss, 1820:xx, 374; type genus *Cervus* L.
= Cervidae Gray, 1821: 307.

The Cervidae constitute a clearly defined family apart from the question as to whether the genus *Moschus* should be included or separated in a distinct family as done here. Classification was reviewed by Groves & Grubb (1987) and that of the Philippine forms by Grubb & Groves (1983). Karyological data were reviewed by Neitzel (1982) and (for Chinese species) by Wang & Du (1988). The distribution of deer in China was reviewed, with maps, by Ohtaisha & Gao (1990). Six genera in the Indomalayan Region are recognized here (Table 167).

CERVUS

Cervus Linnaeus, 1758: 66; type species *C. elaphus* L.
= *Elaphus* Smith, 1827: 307.
Rusa Smith, 1827: 105 (*C. unicolor* Kerr).
= *Hippelaphus* Bonaparte, 1836: 4.
Harana Hodgson, 1838: 154 (*C. wallichii* Cuvier, in *C. elaphus*).
= *Pseudocervus* Hodgson, 1841e: 914.
Rucervus Hodgson, 1838: 154 (*C. elaphoides* Hodgson, in *C. duvaucelii*).
Panolia Gray, 1843a: 180 (*P. acuticornis* Gray, in *C. eldi*).
Hippelaphus Sundevall, 1846: 177 (*C. hippelaphus* Smith, in *C. unicolor*). Pre-occupied by *Hippelaphus* Reichenbach, 1835 and by *Hippelaphus* Bonaparte, 1836 (see above).
Sika Sclater, 1870a: 115 (*C. sika* Temminck, in *C. nippon*).
= *Elaphoceros* Fitzinger, 1874: 347, 352.
= *Sikaillus* Heude, 1898: 98.
Pseudaxis Gray, 1872: 70 (*C. taiouanus* Blyth, in *C. nippon*).
Sambur Heude, 1888: 8 (*C. aristotelis* Cuvier, in *C. unicolor*).
Thaocervus Pocock, 1943b: 554 (*Rucervus schomburgki* Blyth).

Nearctic, Palaearctic and Indomalayan regions; c. 10 species of which 7 occur in the Indomalayan Region (Table 168). Two Palaearctic species, *C. elaphus* and *C. albirostris*, occur on the northwestern fringes of the region, at high altitudes in the Himalayas and W China. *Rusa* is often recognized as a distinct genus, with *C. unicolor, C. timorensis, C. alfredi* and *C. mariannus*. However the differences are slight and there is no suite of concordant differences to justify such a division.

The following nominal fossil species have been described from the region:
C. triplidens Lydekker, 1876: 67(=49), pl. 8; ? Potwar dist., Siwaliks.
C. simplicidens Lydekker, 1876: 51, pl. 8; Siwaliks.
C. latidens Lydekker, 1876: 65(=47), pl. 8; near Hasnot, Punjab; Middle Siwaliks.
C. sivalensis Lydekker, 1880:xvii; Siwaliks.
C. lydekkeri Martin, 1886: 63, pl. 7; Java.
C. liriocerus Dubois, 1907: 454; Kendeng, Java.
C. kendengensis Dubois, 1908: 1259; Kendeng, Java. Allocated to *C.* (*Rusa*) *hippelaphus* Cuv. (in *C. timorensis*) by Koenigswald (1933).
C. palaeomendjangan Dubois, 1908: 1260; Java.
C. astylodon Matsumoto, 1926: 21; Shimajiri Dist, Okinawa I, Ryukyu Is; Pleistocene. In *Pseudaxis* according to Teilhard & Leroy (1942).
C. (*Rucervus*) *riukiuensis* Matsumoto, 1926: 22; Shimajiri Dist., Okinawa I, Ryukyu Is; Pleistocene.

Table 168. The species of *Cervus*.

	nippon	*duvaucelii*	*eldii*	*unicolor*	*timorensis*	*mariannus*	*alfredi*
Range	China, Taiwan	N India	Assam-Indochina	India-Sumatra	Sulawesi, Java etc.	Philippines	Philippines
Shoulder height (cm)	65–110	120–135	115–120	120–160	100–110	c.70	c.70
Adults spotted	(+)	(+)	(+)	0	0	0	+
Young spotted	+	+	+	(+)	0	+	+
Rump pale	+	+	0	0	0	0	0
Venter: darker (D) or lighter (L) than dorsum	L	L	L	D	L		L
Tail	Short	Short	Short	Long, bushy, black	Less bushy, as dorsum	Short, short-haired	Short
Ears	Small	Small	Small	Large, broad	Large, broad	Small	Small
Metatarsal glands	+	0	+	+			
Antlers: max.no.of points on each	5	6	6	3	3(4)	3	3
Angle between brow tine and main beam of antler	c.80°	c.90–120°	c.180°	c.30–60°	c.50°	c.50°	
Lateral incisiform teeth	Narrow	Broad		Broad	Broad	Narrow	Narrow
Bullae	Very large	Large	Large	Small	Small	Small	Small

C. punjabensis Brown, 1926: 1; 2 miles W of Chandigarh, Punjab; Upper Siwaliks.
C. (*Rusa*) *oppenoorthi* Koenigswald, 1933: 76; Semboengan, Java.
C. (*Rusa*) *stehlini* Koenigswald, 1933: 78; Boemiajoe, Java.
C. problematicus Koenigswald, 1933: 80; Boemiajoe, Java.
C. zwaani Koenigswald, 1933: 81; Boemiajoe, Java.
C. (*Sika*) *sintikuensis* Shikama, 1937: 78; Tikutô, Sintiku-syû, Taiwan.
C. (*Deperetia ?*) *kokubuni* Shikama, 1937: 83; Syatin, near Tainan, SW Taiwan. Allocated to *Elaphurus* by Otsuka & Shikama (1978).
C. (*Deperetia ?*) *syatinensis* Shikama, 1937: 83; Syatin, near Tainan, SW Taiwan. Allocated to *Elaphurus* by Otsuka & Shikama (1978).
C. colberti Azzaroli, 1954: 81; 9 miles W of Kulka, Punjab; Pleistocene.
Rusa yunnanensis Lin, Pan & Lu. Reference not found.

Cervus nippon Sika deer

C. nippon Temminck, 1836: 22; Japan.
C. sika Temminck, 1844: 54, pl. 17; Japan.
? *C. pseudaxis* Gervais, in Eydoux & Souleyet, 1841: 64, pl. 12; locality unknown, probably Vietnam (Glover, 1956).
C. taiouanus Blyth, 1860a: 90; Taiwan.
C. kopschi Swinhoe, 1873a: 574; Kien-chang, Jiangxi, China.
C. frinianus, C. gracilis, C. lachrymosus, C. ignotus Heude, 1882: 185; right bank of Yangtze, below L Poyang.
C. andreanus Heude, 1882: 186; *C. joretianus* Heude, 1882: 187; Ningkwofu, 75 miles S of Nanking, China.
C. devilleanus Heude, 1882: 187; Taiwan.
Sika brachyrhinus, S. cycloceros Heude, 1885: 2; *S. grilloanus, S. pouvrelianus* Heude, 1885: 3; *S. oxycephalus* Heude, 1885: 4; *S. yuanus* Heude, 1885: 5; right bank of Yangtze, below L Poyang.
S. dominicanus, S. novioninus, S. schulzianus, S. morrisianus Heude, 1885: 6; Taiwan.
S. schlegeli, S. fuscus, S. hollandianus Heude, 1885: 7; 'small islands south of Japan'.
S. granulosus, S. surdescens Heude, 1888: 2, pl. 0; locality not recorded.
Sikelaphus soloensis Heude, 1894: 147; Jolo I, Sulu Is, Philippines.
S. riverianus Heude, 1894: 153; L Poyang, China.
S. dugenneanus Heude, 1894: 156; locality not recorded.
S. arietinus Heude, 1894: 162; locality not recorded.
And other, extralimital, synonyms.

Range (Map 109). Probably at one time throughout China and south to N Vietnam (Dao, 1977); Taiwan and the Ryukyu Is; introduced on Jolo (Sulu Is, Philippines). In S China now survives as small isolated populations in Guangxi, Jiangxi, near Shanghai and in Sichuan. On Taiwan survives only in captivity and as a recent reintroduction. Also in NE China, adjacent parts of USSR and in Japan, and widely introduced elsewhere, e. g. in Europe, USSR, USA and New Zealand.

Variation. The following subspecific names have been proposed:
Sika n. keramae Kuroda, 1924: 12; Zamamishima, Kerama group, middle Ryukyu Is.
C. n. swinhoei Glover, 1956: 103; Newchwang, Amoy, China. Pre-occupied by *C. swinhoii* Sclater, 1862 (in *C. unicolor*).
C. n. sichuanicus Guo *et al.*,1978: 187, 192; Ruoergai dist, Sichuan, China, 2900m.

Ten subspecies were recognized and diagnosed by Groves & Grubb (1987) of which three occur in this region:
C. n. taiouanus. Taiwan. Brightly coloured, spotted throughout the year; rump patch small.
C. n. pseudaxis. Brightly coloured; spotted in summer, faintly in winter; dark dorsal stripe; ears very white inside.
C. n. soloensis. Sulu Is. Spotted throughout the year; red-brown dorsal stripe.

Remarks. Throughout China and on Taiwan they are farmed in large numbers to produce antlers in velvet for medicinal use. Viable, fertile hybrids between this species and *C. elaphus* have occurred on a number of occasions, e. g. in Britain and Ireland and in the Soviet Far East. The species was reviewed by Feldhammer (1980). IUCN: (E)—*C. n. keramae, C. n. pseudaxis, C. n. taiouanus*; I—*C. n. sichuanus*.

Cervus duvaucelii Swamp deer, Barasingha

C. duvaucelii Cuvier, G.,1823: 505; locality not recorded, presumably N India.
C. bahrainja Hodgson, 1834: 99; Nepal (*nomen nudum*).
= *C. elaphoides* Hodgson, 1835c: 648.

C. smithii Gray, 1837d: 45; N India.
C. dimorphé Hodgson, 1843b: 897; Saul forests of Morung, Nepal.
C. euceros Gray, 1850a:pl. 40; India (= *C. euryceros* in text: 76).
Cervus or *Rucervus schomburgki* Blyth, 1863c: 155; Thailand.
C. eucladoceros Falconer, 1868: 587; W bank of Ganges, S of Hardwar, United Provinces, India.

Range (Map 109). N India from Kumaon to Assam, and S Nepal; and in central India, in E Madhya Pradesh (now confined to Kanha National Park). Formerly also in C Thailand (*C. d. schomburgki*) where it was last recorded in 1938, and in the 19th century also in the Indus Valley in Pakistan. Introduced in Texas, USA.

Variation. The following subspecific names have been proposed:
Rucervus d. branderi Pocock, 1943b: 558; Mandla, Central Provinces, India.
C. d. ranjitsinhi Groves, 1982c: 626; Gauhati, Assam.

Geographical variation was reviewed by Groves (1982c) who recognized three surviving and one extinct subspecies:

C. d. duvaucelii. NW India and Nepal. Antlers long and slender, not compressed nor palmated; nasals short in relation to snout.
C. d. ranjitsinhi. Assam. Antlers short and thick, branching low, with very short anterior branch, somewhat compressed and with a tendency to palmation; nasals longer.
C. d. branderi. Madhya Pradesh. Antlers long, slender and much-branched, with long anterior branch; nasals long; feet more densely haired above hooves.
C. d. schomburgki. Thailand (extinct). Antlers with short beam and long multiple branches.

Ecology. The populations north of the Ganges occupy swampy habitats, but those south of the Ganges are found on drier grasslands.

Remarks. CITES: Appendix I; IUCN: E.

Cervus eldii Thamin, Brow-antlered deer

C. eldii M'Clelland, 1842a: 417, pl. xii; Manipur, E India.
= *C.* (*Rusa*) *frontalis* M'Clelland, 1843: 401.
= *Panolia acuticauda* Blyth, 1864: 370.
= *C. e. typicus* Lydekker, 1898a: 200.

Map **109**. *Cervus*

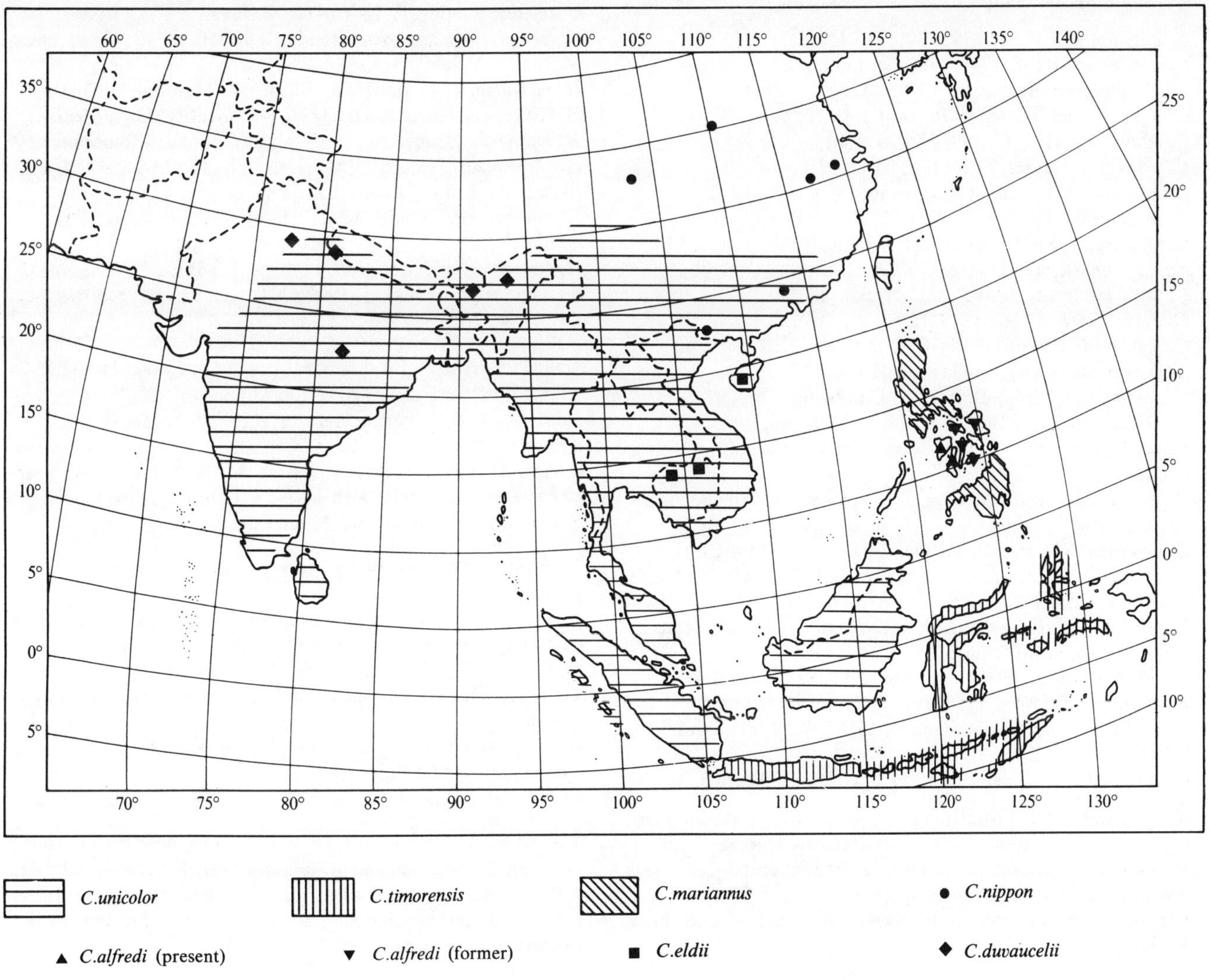

Panolia acuticornis Gray, 1843a: 180; Manipur.
Panolia platyceros Gray, 1843a: 181; Thailand. Pre-occupied by *C. platyceros* Cuvier, 1798 = *C. dama* L.
= *C. e. siamensis* Lydekker, 1915: 104.
C. lyratus Schinz, 1845: 395.
Rucervus thamin Thomas, 1918c: 364: Pegu, Burma.

Range (Map 109). Formerly from Manipur through Burma and Thailand to Vietnam and SE China; Hainan I. Early records on Taiwan were errors for Hainan (Thomas, 1918c: 365). Now limited to three isolated segments: Manipur; in Cambodia and S Laos; and on Hainan I. Subfossil record, estimated c. 1000 BC, on Java (Dammerman, 1934).

Variation. The following subspecific names have been proposed:
C. e. cornipes Lydekker, 1901a: 257; Manipur, India.
Rucervus platyceros hainanus Thomas, 1918c: 364; Hainan I, China.
R. thamin brucei Thomas, 1918c: 366; Thimbaung-Gwin Plain, Ruby Mines district, upper Burma.

Ecology. At one time found in large herds. The Manipur population lives in swampy lowlands but the others occupy drier habitats although all are predominantly grazers.

Remarks. CITES: Appendix I; IUCN: V/E.

Cervus unicolor Sambar

C. axis unicolor Kerr, 1792: 300; Sri Lanka.
C. axis major Kerr, 1792: 300; Sri Lanka.
= *C. albicornis* Bechstein, 1799: 112.
C. niger Blainville, 1816: 76; probably N India.
C. equinus Cuvier, G.,1823: 45; Sumatra.
C. aristotelis Cuvier, G.,1823: 503; Nepal.
C. leschenaulti Cuvier, G.,1823: 506; Coromandel, India.
C. malaccensis Cuvier, F.,1824:pl. 10; Malacca.
C. hippelaphus Smith, 1827: 105; ? Bengal. Pre-occupied by *C. elaphus hippelephas* Erxleben, 1777.
C. jarai Hodgson, 1831b: 321; Nepal.
C. heterocerus Hodgson, 1841f:pl. opp. p. 722.
C. nepalensis Hodgson, 1841f:pl. opp. p. 722.
Axis pennantii Gray, 1843a: 180; India.
C. cambojensis Gray, 1861: 138; Cambodia.
C. swinhoii Sclater, 1862: 152, pl. 17; Taiwan.
Sambar curvicornis, S. longicornis Heude, 1888: 42; Cochin-China.
S. planidens, S. colombertinus, S. combalbertinus Heude, 1888: 43; Cochin-China.
S. lignarius, S. lemeanus Heude, 1888: 44; Cochin-China.
S. errardianus, S. joubertianus, S. latidens, S. planiceps Heude, 1888: 45; Cochin-China.
S. officialis, S. simoninus, S. brachyrhinus, S. verutus Heude, 1888: 46; Cochin-China.
C. brookei Hose, 1893b: 206; Mt Dulit, Sarawak, Borneo.
Rusa dejeani Pousargues, 1896: 12; Sichuan, China.
Hippelaphus hamiltonianus Heude, 1896: 49, pl. 11; Sandakan, Borneo.
C. spatulatus Thomas, 1901a: 54; ? Sarawak, Borneo.

Range (Map 109). From India to S China (Yunnan, Sichuan, Guangdong) and south to Malaya; Sumatra, Borneo; Hainan, Taiwan; and many small islands including Siberut, Pagai, Nias (W Sumatra); formerly on Singapore. Up to c. 3000m in the Himalayas in summer. Introduced in Australia and New Zealand.

Variation. The following subspecific names have been proposed:
Rusa equina pennantii Fitzinger, 1875: 296; Borneo. Pre-occupied by *Axis pennantii* Gray, 1861 (above).
C. u. oceanus Chasen & Kloss, 1928c: 818; Siberut I, Mentawai Is.
C. u. hainana Xu, in Xu *et al.*,1983: 395; Hainan I, China.
Size is greatest in India and Sri Lanka, least in the Sunda subregion and in Taiwan.

Ecology. Live in small groups, usually in woodland, and feed mainly by browsing.

Remarks. Viable hybrids have been produced in captivity between this species and *C. timorensis* (van Mourik & Schurig, 1985).

Cervus timorensis Timor deer

C. timorensis Blainville, 1822: 267; Timor I.
C. hippelaphus Cuvier, G.,1825: 40, pl. 5; Java. Pre-occupied by *C. elaphus hippelaphus* Erxleben, 1777.
C. peronii Cuvier, G.,1825: 46; Timor.
? *C. tunjuc* Horsfield, 1830: 645; *nomen nudum.*
C. moluccensis Quoy & Gaimard, 1830: 133, pl. 24; Buru I, Moluccas.
C. russa Müller & Schlegel, 1845b: 217, pl. xliii; Java.
C. lepidus Sundevall, 1846: 180; E Java.
Hippelaphus macassaricus Heude, 1896: 50, pl. 13; Gowa, near Macassar, S Sulawesi.
H. menadensis Heude, 1896: 50, pl. 12; Menado, N Sulawesi.
C. (*Rusa*) *celebensis* Rörig, 1896: 49; locality not recorded.
H. buruensis Heude, 1897: 92, pl. 16; Buru I, Moluccas.
H. floresiensis Heude, 1897: 92, pl. 15; Adonara I and Ilimandiri, Flores, Lesser Sunda Is.
H. sumbavanus Heude, 1897: 92, pl. 15; Sumbawa I, Lesser Sunda Is.
H. hoevellianus Heude, 1897: 94, pl. 16; Buru I, Moluccas.
? *C. tavistocki* Lydekker, 1900: 204; locality not recorded (see Grubb & Groves, 1983: 122).

Range (Map 109). From Java through the Lesser Sunda Is to Timor; Sulawesi and most of the Moluccan islands (including Buru, Ambon, Saparua, Seram, Halmahera, Banda, Bacan, Ternate). Recorded in Borneo in the 19th century but not confirmed since. Probably native in Java and introduced elsewhere. Introduced also in New Guinea, Australia, New Zealand, Madagascar and Mauritius.

Variation. The following subspecific names have been proposed:
C. russa javanicus Müller & Schlegel, 1845b: 217; Java. Pre-occupied by *C. javanicus* Osbeck, 1762 (= *Tragulus javanicus*).
C. t. tunjuc Lydekker, 1915: 66; Sumatra.
C. hippelaphus renschi Sody, 1932b: 335; Sendang, W Bali I, Lesser Sunda Is.
Rusa t. laronesiotes van Bemmel, 1949a: 235; Meeuwen I, between Java and Sumatra.
Rusa t. djonga van Bemmel, 1949a: 255; Muna I, SE Sulawesi.

Remarks. Viable hybrids have been produced in captivity between this species and *C. unicolor* (van Mourik & Schurig, 1985). This species is often called 'Rusa deer' but this is misleading and ambiguous since *C. unicolor* is the type of the generic name *Rusa*.

Cervus mariannus Philippine rusa

C. mariannus Desmarest, 1822: 436; Guam, Mariana Is, W Pacific Ocean.
C. philippinus Smith, 1827: 147; Philippines.
C. nigricans Brooke, 1876: 57, pl. 9; Philippines.
Ussa crassicornis, U. gonzalinus Heude, 1888: 32, 35; locality unrecorded, probably Philippines.
Ussa gorrichianus, U. tuasoninus, U. guidoteanus, U. roxasianus, U. longicuspis, U. microdontus, U. telesforianus, U. brachyceros Heude, 1888: 21–36; Batangas Prov., Luzon, Philippines.
U. hippolitianus, U. chrysotrichos Heude, 1888: 39; Batangas and La Laguna provinces, Luzon.
U. nublanus, U. spatharius Heude, 1888: 24,25; La Laguna Prov., Luzon.
U. ramosianus, U. ambrosianus, U. macarianus, U. garcianus, U. rosarianus, U. verzosianus Heude, 1888: 27–37; Nueva Ecija Prov., Luzon.
U. rubiginosa Heude, 1888: 37; Bataan and Nueva Ecija provinces, Luzon.
U. elorzanus Heude, 1888: 28; Bataan Prov., Luzon.
U. maraisianus, U. dialliardianus Heude, 1888: 31,32; Jala Jala forest, Luzon.
U. corteanus Heude, 1888: 37; Mariveles, north of Manila Bay, Luzon.
U. guevaranus Heude, 1888: 40; Mariquina, Luzon.
U. marzaninus Heude, 1888: 33; Infanta, Luzon.
U. barandanus Heude, 1888: 22; Mindoro I, Philippines.
U. francianus Heude, 1888: 24; Mati, Mindanao I, Philippines.
Melanaxis ? elegans Heude, 1888: 49; Philippines.
Melanaxis basilanensis Heude, 1888: 49; Basilan I, Philippines.
C. steerii Elliot, 1896: 72, pl. 8; Basilan I, Philippines.
U. baryceros Heude, 1899: 139; Batangas and La Laguna provinces, Luzon.
U. atheneensis Heude, 1899: 138; La Laguna Prov., Luzon.
U. michaelinus, U. villemerianus, U. vidalinus Heude, 1899:pl. 32A; San Miguel de Murcia, Tarlac, Luzon.
? *C.* (*Rusa*) *unicolor boninensis* Lydekker, 1905b: 391; Bonin Is, W Pacific.
Rusa nigellus Hollister, 1913a: 332; Mt Malindang, Mindanao, Philippines.

Range (Map 109). On most Philippine islands except Cebu, Masbate, Negros, Panay and Palawan; also on the Pacific islands of Guam (Mariana Is), Ponape (Caroline Is); and Bonin (SE of Japan) if *boninensis* is conspecific.

Variation. The following subspecific name has been proposed: *Rusa nigellus apoensis* Sanborn, 1952a: 155; Mt Apo, Mindanao, Philippines.

Grubb & Groves (1983) recognized four subspecies as follows but on slight metrical differences and necessarily small samples, without clear diagnoses:
C. m. mariannus. Luzon (and Guam and Ponape).
C. m. barandanus. Mindoro.
C. m. nigricans. Mindanao and Basilan.
C. m. nigellus. Highlands of Mindanao. Small.

Remarks. Reviewed by Grubb & Groves (1983).

Cervus alfredi Philippine spotted deer

C. alfredi Sclater, 1870b: 381, pl. 28; Philippines.
Melanaxis masbatensis Heude, 1888: 47, pl. 3; Masbate I, Philippines.
M. breviceps Heude, 1888: 47; Masbate I, Philippines.
M. ? elegans Heude, 1888: 49; locality not recorded.

Range (Map 109). Philippine islands of Negros and Panay; formerly also on the adjacent islands of Cebu, Guimaras, Masbate, Bohol and Siquijor (Oliver *et al.*, 1991).

Recognition. Distinguished from other Philippine deer by its spotted pelage and from all other Indomalayan deer by the very dense and soft texture of the pelage.

Remarks. A very poorly known species. Reviewed by Grubb & Groves (1983) who considered it a member of the subgenus *Rusa*, not *Axis* as might be suggested by its spotted pelage. IUCN: E.

ELAPHURUS

Elaphurus Milne-Edwards, 1866: 382; type species *E. davidianus* Milne-Edwards.

China; 1 species. Close to *Cervus*, but with a number of peculiar characters such as the dominant anterior branch of the antlers, the long tail and horse-like incisors. Viable hybrids with *C. elaphus* have been produced in captivity by artificial insemination (Asher *et al.*, 1988), raising doubts as to the justification of separating it generically from *Cervus*. The geographical and fossil history were reviewed by Ji (1985).

Elaphurus davidianus Mi-lu, Père David's deer

E. davidianus Milne-Edwards, 1866: 382; Imperial Hunting Park, Beijing, China (based upon captive animals).
Cervus tarandoides David, 1867: 28 (*nomen nudum*).
C. (*Rucervus*) *menziesianus* Sowerby, 1933: 141; near Anyang, Henan, China; subfossil.

Range (Map 109). Formerly, as shown by fossil and subfossil finds, much of the lowlands of E China, south to the Yangtze Basin, and on Taiwan. They were domesticated in the Zhou dynasty (prior to 200 BC) and appear to have become extinct in the wild after the Han dynasty, i. e. after c. 200 AD (Cao, 1987). As living animals, known only from captive stock that survived in an enclosed hunting park south of Beijing until 1900. The present stock derives entirely from animals taken to England in 1900 and subsequently dispersed to zoos elsewhere. Reintroduction to China began in 1985 and by 1989 there were populations of c. 100 in Nanhaizi Park near Beijing, and c. 70 in the Da Feng Reserve north of Shanghai (Ohtaishi & Gao, 1990).

Recorded from Hainan I in the 19th century but it is unclear whether these were captive or wild (Groves & Grubb, 1987).

Remarks. Reviewed by Beck & Wemmer (1983). IUCN: E.

AXIS

Axis Smith, 1827: 312; type species *Cervus axis* Erxleben.
Hyelaphus Sundevall, 1846: 180 (*C. porcinus* Zimmermann).

Indomalayan Region; 4 species of which 2 are doubtfully distinct (Table 169). The genus is often included in *Cervus* but differs in a number of ways, especially the presence of pedal glands in the hind feet (but not in the fore feet also as in *Dama*). Fertile hybrids have been reported in captivity between *A. axis* and *A. porcinus*, with back-crosses to *A. axis* (Gray, 1972).

The following nominal fossil species have been described from the region:
Axis shanseius Teilhard & Trassaert, 1937: 44; SE Shanxi, China; recorded in Yunnan.

Table 169. The species of *Axis*.

	axis	*porcinus*	*kuhlii*	*calamianensis*
Range	India, Sri Lanka	India-Indochina	Bawean I (Java)	Calamian Is (Philippines)
Shoulder height (cm)	c.90	60–75	60–70	60–70
Adults spotted	+	0	0	0
Dorsum darker then flanks	0	+	0	0
Muzzle	Black	Black	White	White
Legs darker than body	0	0	0	+
Tail bushy	0	0	+	+
Pre-orbital gland with hair tuft (when everted)	+	0	0	0
Antlers: max. length (cm)	90	45	50	30

Axis yunnanensis Lin & Zhang, 1978: 82; Lijiang Basin, Yunnan.
Axis weiningensis Wu & Zhang, 1983: 322; near Wangjiayuanzi, Caohai region, SW Guizhou, China; ? Upper Pleistocene.

Axis axis Chital, Spotted deer

Cervus axis Erxleben, 1777: 312; banks of the R Ganges, India.
C. nudipalpebra Ogilby, 1831: 136; banks of R Ganges, India.
A. minor Hodgson, 1842: 914.

Range (Map 110). Throughout most of India; Sri Lanka; introduced to the Andaman Is. Also introduced in many other parts of the world, e. g. in Europe, Texas, S America, Hawaii, Australia and New Zealand.

Variation. The following subspecific names have been proposed:
Cervus a. maculatus Kerr, 1792: 300; banks of R Ganges, India.
C. a. var. *indicus* Fischer, 1829: 619.
C. a. var. *ceylonensis* Fischer, 1829: 619; Sri Lanka.
C. (*Rusa*) *a. zeylanicus* Lydekker, 1905a: 947; Sri Lanka.

Map **110**. *Axis, Elaphurus, Elaphodus, Hydropotes*

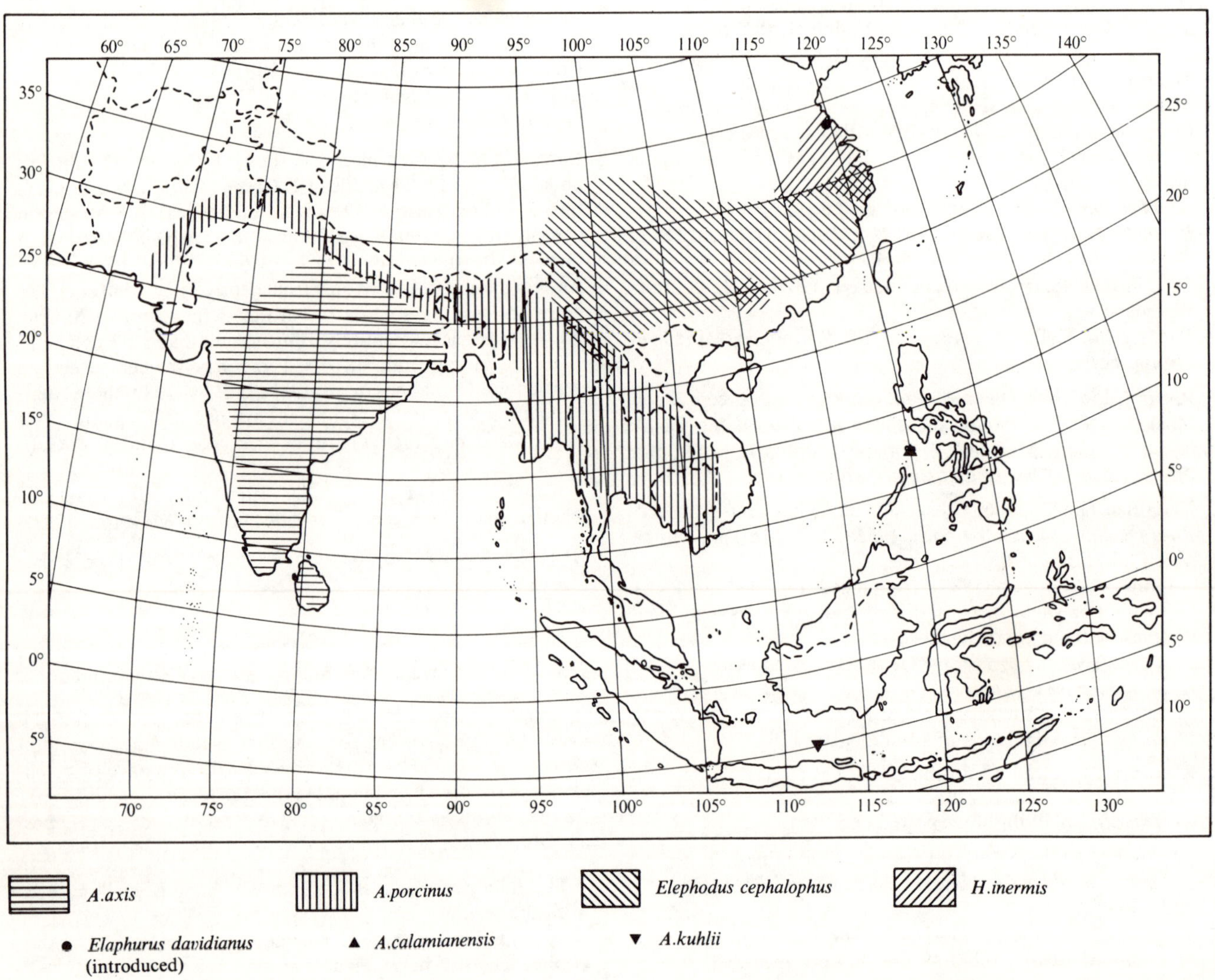

C. (Axis) a. javanicus Koenigswald, 1933: 65; Ngandong, Java; fossil. Pre-occupied by *C. javanicus* Osbeck, 1765 (= *Tragulus javanicus*).
= *Axis sunda* Kretzoi, 1947: 286.

Axis porcinus Hog deer

Cervus porcinus Zimmermann, 1777: 532; Bengal.
C. porcinus Zimmerman, 1780: 131; Bengal.
C. porcinus Schreber, 1784: pl. 251; Bengal.
? *C. pumilio* Smith, 1827: 120; locality not recorded.
? *C. dodur* Royle, 1834: pl. 5 (and Ogilby 1840c: lxxiii); ? Kashmir.
A. oryzus Kelaart, 1852: 83; Sri Lanka.
C. minor Sclater, 1883: 169; India. Pre-occupied by *C. minor* Hodgson, 1842 (in *A. axis*).
Hyelaphus annamiticus Heude, 1888: 50; Baria, ? Indochina.

Range (Map 110). From Pakistan and N India to Burma, SW Yunnan, Thailand, Laos and S Vietnam; introduced in Sri Lanka. Introduced also in Victoria, Australia, and in USA.

Variation. The following subspecific name has been proposed: *Cervus p. hecki* Lydekker, 1908a: 583; Thailand.

Ecology. Mainly associated with habitats of tall grass and marshland, grazing and browsing.

Remarks. CITES: Appendix I (*A. p. annamiticus*).

Axis kuhlii Bawean deer

Cervus kuhlii Temminck, 1836: 11; Bawean I, Java Sea.

Range (Map 110). Known only on Bawean I, north of Java.

Remarks. Clearly a derivative of *A. porcinus*, but as with *A. calamianensis* it is uncertain whether it is a natural relict (represented by the extinct *A. oppenoorthi* in Java), or as seems more likely, represents an early introduction of *A. porcinus* by human agency (Groves & Grubb, 1987). CITES: Appendix I; IUCN: R.

Axis calamianensis Calamian deer

Hyelaphus calamianensis Heude, 1888: 49, pl. 4; Calamian I, Philippines.
Cervus culionensis Elliot, 1897: [159]; Culion I, Calamian Is, Philippines.

Range (Map 110). Islands of Busuanga and Culion, Calamian Is, Philippines.

Remarks. Perhaps an introduced form of *A. porcinus*—see Remarks under *A. kuhlii* above. CITES: Appendix I; IUCN: V.

MUNTIACUS Muntjacs

Muntiacus Rafinesque, 1815: 56; type species *Cervus muntjak* Zimmermann. Conserved by Opinion 460 (1957).
= *Cervulus* Blainville, 1816: 74.
= *Stylocerus* Smith, 1827: 319.
Prox Ogilby, 1837b: 135 (*Cervus moschatus* Blainville, in *M. muntjak*).
Procops Pocock, 1923a: 207 (*Cervulus feae* Thomas & Doria).

Indomalayan Region; c. 7 species, but the delimitation of species is still provisional (Table 170). The classification was reviewed by Ma *et al.* (1986). Some species have an abnormally small number of chromosomes (see Table 170). Although often separated, with *Elaphodus*, from other deer in a subfamily Muntiacinae, this was refuted by Groves and Grubb (1987) who considered them close to most of the other Old-world deer. Unique characters are the elongate pedicles of the antlers, raised ridges on either side of the frontal bones, and large frontal glands.

The following nominal fossil species have been described:
Cervulus kendengensis Stremme, 1911a: 58; Trinil, C Java.
M. bumiajuensis Koenigswald, 1933: 61; Boemiajoe, Java.
M. nanus Teilhard & Trassaert, 1937: 24; SE Shanxi, China; Pliocene. Recorded in Yunnan.
M. lacustris Teilhard & Trassaert, 1937: 21; SE Shanxi, China; Pliocene. Recorded at Yuanmou, Yunnan.
Cervulus szechuanensis Young & Liu, 1951: 78; Koloshan, near Chungkiang, Sichuan.
M. gaoshanensis Xu & Wei, 1980: 176; Long Dong, Fuyang Gaoshan, Zhejiang, China.

Muntiacus muntjak Indian muntjac, Barking deer

Cervus muntjak Zimmermann, 1780: 131; Java. Conserved by Opinion 460 (1957).
C. vaginalis Boddaert, 1785: 136; Bengal.
= *Stylocerus muntjacus* Kelaart, 1852: 85.
? *C. subcornutus* Blainville, 1816: 77.

Table 170. The species of *Muntiacus*.

	muntjak	*reevesi*	*crinifrons*	*feae*	*rooseveltorum*	*gongshanensis*	*atherodes*
Range	India-Borneo	S China, Taiwan	SE China	Burma etc.	Indochina	Yunnan	Borneo
Head & body (cm)	98–120	64–87	113–132	88–100	99–102	95–105	86–92
Dorsum	Chestnut	Reddish	Blackish brown	Dark brown	Dark red-brown	Dark brown	Yellowish brown
Dorsal stripe on neck	0	+	0	0	0	0	+
Tuft on forehead	0	0	Long	Short, f only	0	0	0
Frontal glands well developed	+	+	(+)	(+)	+		
Tail: upper surface	As dorsum	As dorsum	As dorsum	Black	As dorsum	Black	Black
Antlers: length	−130	−85					−42
Pedicels				Long, thin		Short	Short, thin
Skull: greatest length	197–220	146–164	203–230	188–216	c.175	195–205	172–201
Chromosomes: 2n - m, f	7, 6 or 8	46, 46	9, 8	15, 13		9, 8	

C. moschatus Blainville, 1816: 77; Bencoolen, Sumatra.
C. aureus Smith, 1826:pl. opp. p. 148; S India.
C. ratwa Hodgson, 1833a: 139; Nepal.
C. melas Ogilby, in Royle, 1840:lxxiii; India.
= *C. styloceros* Schinz, 1845: 549.
C. albipes Wagner, 1844a: 394; Bombay and Poona, India.
Cervulus tamulicus Gray, 1872: 94; Deccan, India.
C. curvostylis Gray, 1872: 94; Pachebon, Thailand.
C. pleiharicus Kohlbrugge, 1896a: 192; Pleihari, SE Borneo.
M. bancanus Lyon, 1906a: 582; Banka I, Sumatra.
M. rubidus Lyon, 1911a: 73; Pamuking Bay, SE Borneo.

Range (Map 111). Throughout India, marginally in NE Pakistan, and east to S China (S Tibet and Yunnan—Fujian), south to Malaya; Sumatra, Java, Borneo; Sri Lanka; Hainan; and on the smaller islands of Middle Andaman (introduced), Pangkor (Malaya), Singapore (extinct), Bintan (Riau Is), Bangka, Belitung; Nias (W Sumatra), Bawal, Matasiri (Borneo); Bali, Lombok (probably introduced). Introduced to Texas, USA.

Variation. The following subspecific names have been proposed:
Cervulus m. grandicornis Lydekker, 1904a: 780; Thouagyen Forest, Amherst Dist, Burma.
M. m. peninsulae Lydekker, 1915: 18; Pangkor I, Perak, Malaya.
M. m. robinsoni Lydekker, 1915: 18; Bintang I, Riau Is.
M. m. malabaricus Lydekker, 1915: 24; Nagarhol, Coorg, S India.
M. m. montanus Robinson & Kloss, 1918a: 69; Korinchi Peak, Sumatra, 7300ft.
M. m. annamensis Kloss, 1928: 399; Langbian Peak, S Annam, Vietnam.
M. m. nigripes Allen, 1930b: 11; Nodoa, Hainan I, China.
M. m. nainggolani Sody, 1932b: 337; Sendang, W Bali I.
M. m. margae Hooijer, 1951: 4; Yenchingkou, Wanhsien, Sichuan, China; Pleistocene.
M. m. menglalis Wang & Groves, in Ma, Wang & Groves, 1988: 96; Pujiao, Mengla county, S Yunnan.
M. m. yunnanensis Ma & Wang, in Ma, Wang & Groves, 1988: 101; Wokang Dashan, Menglai, Cangyuan Vazu Zizhixian, W Yunnan, 23°20'N, 99°13'E, 2200m.

Subspecific variation was reviewed by Ma, Wang & Groves (1988) who diagnosed ten subspecies but without demonstrating the existence of any discrete boundaries between the mainland forms. The Sunda animals (*M. m. muntjak*) are redder than those further north (*M. m. vaginalis*).

Map **111**. *Muntiacus*

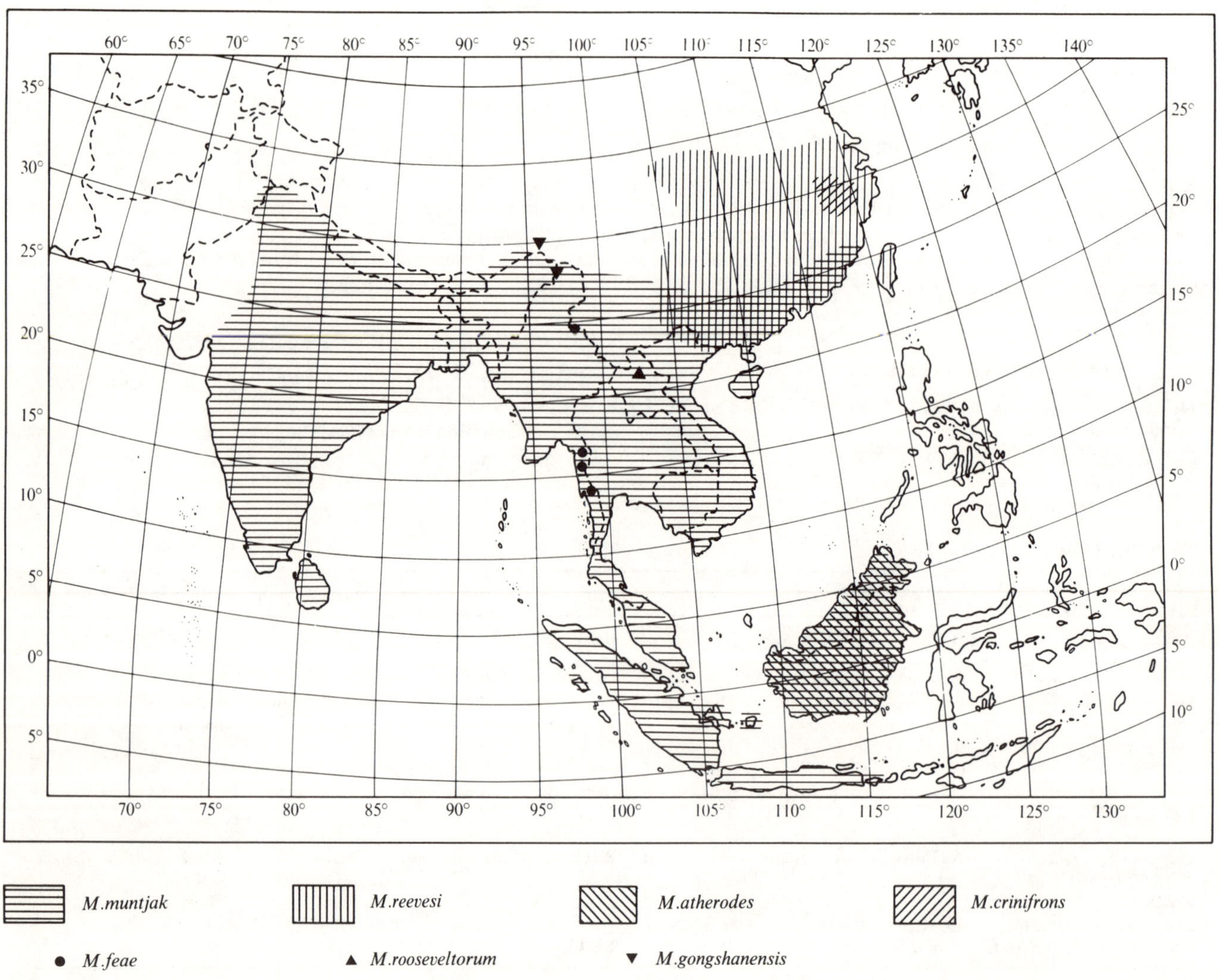

Remarks. There are karyological differences between northern and Malayan forms that might support their recognition as separate species (Wurster-Hill & Atkins, 1972) but this needs confirmation on the basis of larger samples and additional localities.

Muntiacus reevesi Chinese muntjac, Reeve's muntjac

Muntiacus reevesi

Cervus reevesi Ogilby, 1839a: 105; near Canton, Gwangdong, China.
Cervulus lachrymans Milne-Edwards, 1872a: 93; Moupin, Sichuan, China.
C. sclateri Swinhoe, 1873b: 814; near Ningpo, Zhejiang, China.
C. micrurus Sclater, 1875: 421, pl. 51; ? Taiwan.
C. sinensis Hilzheimer, 1905b: 297; ? Hwei Shan, Anhui, China.
C. bridgemani Lydekker, 1910a: 38 (and 1911: 989); Hwei Shan, Anhui, China.

Range (Map 111). Throughout SE China, northwest to S Gansu; Taiwan. Introduced in England and France.

Variation. The following subspecific names have been proposed:
Cervulus r. pingshiangicus Hilzheimer, 1906b: 169; Pingshiang, Anhui.
M. lachrymans teesdalei Lydekker, 1915: 27; Tatung, Yangtze Valley. Holotype confirmed as *M. reevesi* (P Grubb, pers. comm.).

The form on Taiwan (*M. r. micrurus*) is distinguishable by small size and reddish on the back of the ears; there is little variation on the mainland (Groves & Grubb, 1987).

Remarks. It has been suggested that the form *teesdalei* from the lower Yangtze might represent *M. rooseveltorum* rather than *M. reevesi* (Ma *et al.*, 1986) but the holotype of *teesdalei* has been confirmed as representing *M. reevesi* (P Grubb, pers. comm.).

Muntiacus crinifrons Black muntjac

Cervulus crinifrons Sclater, 1885: 1, pl. 1; near Ningpo, Zhejiang, China.

Range (Map 111). Zhejiang and adjacent parts of Anhui and perhaps of Fujian and Jiangsu. According to Ohtaishi & Gao (1990) formerly west along the coastal provinces to Guangxi and E Yunnan, but the evidence for this is not clear. The population has been estimated to be c. 5000–6000 (Ohtaishi & Gao, 1990).

Remarks. Distribution, status and ecology were reviewed by Lu & Sheng (1984). It occurs mainly at higher altitude (around 1000m) than *M. reevesi*. CITES: Appendix I; IUCN: V.

Muntiacus feae Fea's muntjac

Cervulus feae Thomas & Doria, 1889: 92; Thagata Juva, SE of Mt Mulaiyit, Tenasserim, Burma.

Range (Map 111). The border region of W Thailand and E Burma between about 13°30' and 17°N; and SW Yunnan.

Remarks. Recorded also from E Tibet (Chan *et al.*, 1984), but these are more likely to represent *M. gongshanensis* than *M. feae* (Wirth & Groves, 1988). A little known species, clearly distinct from the adjacent or perhaps sympatric *M. muntjak* and possibly closer to *M. crinifrons* (Grubb, 1977). IUCN: E.

Muntiacus rooseveltorum Roosevelts' muntjac

M. rooseveltorum Osgood, 1932: 332; Muong Yo, Laos.

Range (Map 111). Laos; perhaps also in Vietnam and adjacent parts of SE Yunnan.

Remarks. On the basis of some cranial characters it has been suggested that the form *teesdalei* from the lower Yangtze, here listed under *M. reevesi*, might be an outlier of *M. rooseveltorum* (Ma *et al.*, 1986), but P Grubb (pers. comm.) believes the holotype of *teesdalei* to be *M. reevesi*. A report on the karyotype of *M. rooseveltorum* (Wurster-Hill & Seidel, 1985) concluded that it was indistinguishable from that of *M. muntjak vaginalis* but provided no evidence for identifying the animal concerned as *M. rooseveltorum*. It is however possible that *rooseveltorum* is conspecific with *M. feae*.

Muntiacus gongshanensis Gongshan muntjac

M. gongshanensis Ma, in Ma *et al.*,1990: 47; Mijiao, Puladi, Gongshan County, NW Yunnan, 27°35'N, 98°47'E.

Range (Map 111). Known only from the type locality in W Yunnan, China, and from Bome County in E Tibet (Wirth & Groves, 1988).

Muntiacus atherodes Bornean yellow muntjac

M. atherodes Groves & Grubb, 1982: 210; Cocoa Research Station, Tawau, Sabah, Borneo, 800ft.

Range (Map 111). Throughout most of Borneo.

Remarks. This form was for long known as *M. pleiharicus* Kohlbrugge or as a subspecies of *M. muntjak*. Groves & Grubb (1982) showed that the type of *pleiharicus* is a form of *M. muntjak*, but that the yellowish form is a distinct species which is sympatric with *M. muntjak* in several parts of Borneo.

ELAPHODUS

Elaphodus Milne-Edwards, 1872a: 93; type species *E. cephalophus* Milne-Edwards.
Lophotragus Swinhoe, 1874: 453 (*L. michianus* Swinhoe).

One species. This resembles *Muntiacus* in the small antlers, loss of most elements of the lateral digits, and greatly reduced chromosome number (which is not due to Robertsonian fusion). It differs in having the muzzle broad and short, the lacrymal pits very large (larger than the orbits) and short pedicles with rudimentary antlers.

Elaphodus cephalophus Tufted deer

E. cephalophus Milne-Edwards, 1872a: 93; Moupin, Sichuan, China.
Lophotragus michianus Swinhoe, 1874: 453, pl. 59; near Ningpo, Zhejiang, China.
E. ichangensis Lydekker, 1904b: 169; Ichang, Hubei, China.

Range (Map 110). S China from E Tibet to Fujian and Zhejiang, and adjacent parts of Burma.

Variation. The following subspecific names have been proposed:
E. michianus fociensis Lydekker, 1904b: 169; Fing-ling, Fujian.
E. c. megalodon Hooijer, 1951: 11; Yenchingkou, Wanhsien, Sichuan, China; Pleistocene.

[*CAPREOLUS*]

Capreolus Frisch, 1775: 3 (not available).
Capreolus Gray, 1821: 307; type species *Cervus capreolus* L.

A Palaearctic genus, with probably two species, one of which, *C. pygargus* (Pallas, 1771), may occur marginally in the region, in S Shaanxi and NW Sichuan, China. One nominal fossil species has been described from the region but has subsequently been allocated to *Elaphurus* (Otsuka & Shikama, 1978):
C. (*?*) *formosanus* Shikama, 1937: 84; Syatin, near Tainan, Taiwan; Pleistocene.

HYDROPOTES

Hydropotes Swinhoe 1870d: 264 (and 1870c: 90); type species *H. inermis* Swinhoe.
= *Hydrelaphus* Lydekker, 1898a: 219.

China; 1 species. A distinctive genus characterized by absence of antlers and presence of long tusk-like upper canines in the males (both presumably primitive features), but distinguished from *Moschus* by the presence of pre-orbital glands as in most Cervidae. It has been considered the sister-group of all other cervids by Groves & Grubb (1987).

Hydropotes inermis Chinese water deer

H. inermis Swinhoe, 1870d: 264 (and 1870c: 90); Deer I, Yangtze R, a few miles upstream from Chinkiang, Jiangsu, China.
H. affinis Brooke, 1872: 524; Yangtze R, c. 40 miles from Shanghai.
H. kreyenbergi Hilzheimer, 1905b: 298; Chinkiang, Jiangsu.
H. argyropus Heude, 1884: 1017; Korea.

Range (Map 110). Lower Yangtze Basin; and in S China, in Guangxi, Guangdong and S Hunan; also in Korea. Introduced population in England.

Remarks. IUCN: R.

[Family Giraffidae]

Giraffes

Not now present in the region but the Recent genus *Giraffa*, now confined to Africa, is represented by fossils.

GIRAFFA

Giraffa Brisson, 1762: 37; type species *G. giraffa* Brisson, in *G. camelopardalis* (L.).
Giraffa Brünnich, 1771: 36, 46 (*Cervus camelopardalis* L.).
= *Camelopardalis* Schreber, 1784:pl. 255.
= *Trachelotherium* Gistel, 1848.

Camelopardalis sivalensis Falconer & Cautley, 1843: 246; Siwaliks.
C. affinis Falconer & Cautley, 1843: 246; Siwaliks.
G. punjabiensis Pilgrim, 1910: 69; Punjab; Middle Siwaliks; *nomen nudum.*
G. priscilla Matthew, 1929: 537; Chinji, Pakistan.
G. sivalensis sinhaleyus Deraniyagala, 1945a: 80; Kuruvita, Sri Lanka. = *Bibos gaurus sinhaleyus*: (Deraniyagala, 1951).

Family Bovidae

Cattle, antelopes etc.

Bovidae Gray, 1821: 308; type genus *Bos* L.
Antilopidae Gray, 1821: 307 (*Antilope* Pallas).
Capridae Gray, 1821: 307 (*Capra* L.).

A clearly defined family, characterized by the presence of a pair of non-deciduous horns on bony frontal cores. These are always present in males, but species vary as to whether they are equally developed, less developed or absent in females. The family is most abundant in the Afrotropical Region, but is well represented in the Indomalayan Region, with 11 genera of which 5 are confined to the region (Table 171).

One additional genus, *Pseudois*, is represented by a single species, *P. nayaur*, on the Tibetan Plateau and this extends marginally below 3000m in places, e. g. in the upper Yangtze valley where a small form, *P. n. schaeferi* appears to be isolated below the forest at altitudes of c. 2600-3500m.

The fossil Bovidae of India were reviewed by Pilgrim (1939).

BOS

Bos Linnaeus, 1758: 71; type species *B. taurus* L. Conserved by Opinion 75 (1922).
= *Taurus* Rafinesque, 1814: 30.
Bibos Hodgson, 1837a: 499 (*Bibos subhemachalus* Hodgson, in *Bos gaurus*).
Gaveus Hodgson, 1847g: 705 (*B. frontalis* Lambert).
Zebus Blyth, 1860b: 283 (*Z. gibbosus* Blyth, in *B. taurus*).
Gauribos Heude, 1901: 3 (*G. laosiensis* Heude, probably in *B. gaurus*).
Uribos Heude, 1901: 5 (*U. platyceros* Heude, probably in *B. gaurus*).
Bubalibos Heude, 1901: 6 (*Bubalibos annamiticus* Heude, probably in *B. gaurus*).
Novibos Coolidge, 1940: 425 (*B.* (*Bibos*) *sauveli* Urbain).

Table 171. The genera of Bovidae.

	Bos	*Bubalus*	*Boselaphus*	*Tetracerus*	*Antilope*	*Gazella*	*Budorcas*	*Naemorhedus*	*Hemitragus*	*Capra*	*Ovis*
Species in region	5	5	1	1	1	2	1	5	2	3	1
Shoulder height (cm, male)	120–200	70–190	120–150	c.60	c.80	c.65	100–130		70–100	85–105	70–90
Females with horns	+	+	0	0	0	+/0	+	+	+,large	+,small	+,small
Two pairs of horns	0	0	0	+	0	0	0	0	0	0	0
Horn bases wide apart	+	+	0	0	0	0	0	0	0	0	0
Horns: x-section	Circular or oval	Triangular			Circular	Flat				Flattened	Triangular
Beard on chin/throat	0	0	+	0	0	0	+	0	+	+	0
Dorsal crest on neck	0	0	+	0	0	0	0	0	0	0	0
Tail with terminal tuft	+	+	0	0	0	0	0	0	0	0	0
Underside of tail naked	0	0	0		0	+	0	0		+	0
Pre-orbital glands	0	0	+, shallow	+	+ (large)	+	0	+	0	0	+
Inguinal glands	0	0	0	0	+	+	0		0	0	+
Pedal glands	0	0		+	0	0	0	+	0	0(+)	+
Teats (pairs)	2	2	2	2	1	1	2		1/2	1	1

Nearctic, Palaearctic and Indomalayan regions; 5 surviving wild species, of which 3 are Indomalayan (Table 172). Two additional domesticated forms are treated as species.

The following nominal species have been based upon fossils from the region:

B. palaeindicus Falconer, 1859: 130; Narbada, India. = *Bubalus* [*Alcelaphus*] *palaeindicus*: (Lydekker, 1884) *et al.*
B. namadicus Falconer, 1859: 232; Narbada, India.
B. acuticornis Falconer, 1868: 23; Siwalik Hills; Upper Siwaliks.
B. occipitalis Falconer, 1868: 280; Siwaliks. = *Hemibos occipitalis*: Lydekker, 1884.
B. sivalensis Falconer, 1868: 555; Siwaliks. = *Bison sivalensis*: (Lydekker, 1884).
B. acutifrons Lydekker, 1877: 30; Siwaliks.
B. planifrons Lydekker, 1877: 30; Siwaliks.
B. (*Bibos*) *palaeogaurus* Rütimeyer, 1878: 154.
Leptobos falconeri Rütimeyer, 1878: 157; Siwaliks. Allocated to *Bos* by Lydekker (1898c).
Leptobos (*Bibos ?*) *frazeri* Rütimeyer, 1878: 165; Narbada Valley, India; Pleistocene. Allocated to *B. falconeri* by Lydekker (1898c).
B. platyrhinus Lydekker, 1878a: 119(=32); Kangra Dist, Siwaliks.
Leptobos dependicornis Dubois, 1908: 1261; Java. Allocated to *Bibos palaeosondaicus* by Hooijer (1958).
Bibos palaeosondaicus Dubois, 1908: 1262; Java.
Bibos protocavifrons Dubois, 1908: 1262; Java. Allocated to *Bibos palaeosondaicus* by Hooijer (1958).
Bibos geron Matsumoto, 1915: 21, pl. 9; Sichuan; Lower Pleistocene.
B. sondaicus fossilis van der Maarel, 1932: 42, pl. 2; Java; *nomen nudum.* Allocated to *Bibos palaeosondaicus* by Hooijer (1958).
Bibos cosijni Koenigswald, 1934: 198; Djeti, E Java. Allocated to *Epileptobos groeneveldtii* (Dubois) by Hooijer (1958).
Bibos frontalis sinhaleyus Deraniyagala, 1939b: 235; Sri Lanka. A *nomen nudum.*
= *Bibos sinhalayus* Deraniyagala, 1963a: 16, pl. 4; Thotipitiyage vatta, near Harakada ovita, Pahalagama, Ellavala, Sri Lanka.

Bos taurus Domestic cattle

B. taurus Linnaeus, 1758: 71; Sweden.
B. indicus Linnaeus, 1758: 72; China.
B. pusio Swainson, 1835: 283; ? China.
Zebus gibbosus Blyth, 1860b: 284; India.
Also synonyms from other parts of the world.

Range. Cattle of this form are kept throughout the region, but in parts of Indonesia are subordinate to cattle derived from *B. javanicus*.

Remarks. Except for some recent introductions the *taurus* cattle of the region are mainly of the humped form or zebu, distinguished from the European form by a hump over the shoulders, larger dewlap, sloping rump and more upright horns. It has been proposed that this was derived from a wild ancestor, *B. namadicus*, that was specifically distinct from the Palaearctic *B. primigenius*, the ancestor of *B. taurus*. However this has been disputed, e. g. by Epstein & Mason (1984) who considered *namadicus* and *primigenius* conspecific, although it is nevertheless likely that the zebu arose by independent domestication in India (Naik, 1978). There is archaeological and subfossil

Table 172. The species of *Bos*.

	taurus	*gaurus*	*frontalis*	*javanensis*	*sauveli*
Range	Widespread, domestic	India-Malaya	Assam etc., domestic	Burma-Borneo	Indochina
Shoulder height (cm, male)	Variable	170–200	? -170	120–170	170–190
Colour of ad.male	Variable	Black	Black	Blackish or brown	Black
Colour of ad.female	Variable	Blackish brown	Blackish brown	Reddish brown	Grey
Dewlap in male	+/0	+/0	Very large	Small	Very large
Lower legs white	0	+	+	+	+
White patch on rump	0	0	0	+	0
White band round muzzle	0	0	0	+	0
Dorsal ridge from nape to withers	0/+	+	+	+	+
Horns frayed near tips	0	0	0	0	+
Horns connected by horny skin		0		+	0

evidence of both humped and non-humped cattle in the Indus Valley during the 2nd millenium BC. In most of the region these cattle are used primarily as draught animals and for milk.

When it is considered desirable to use scientific terminology for domestic cattle, the recommended form is '*Bos taurus* (Jersey)', '*Bos taurus* (zebu)' etc.

Bos sauveli Kouprey

B. (*Bibos*) *sauveli* Urbain, 1937: 307; near Tchep, N Cambodia.

Range (Map 112). Confined to a small area comprising N Cambodia, S Laos and immediately adjacent parts of Thailand and Vietnam. It has been recorded since 1975 throughout most of its historically known range.

Ecology. Confined to open forest; mainly a grazer but taking some browse.

Remarks. The distribution, status and ecology were reviewed by MacKinnon & Stuart (1989); and its relationship to other bovids by Groves (1981b) who concluded that it was more closely related to *B. primigenius* (the ancestor of *B. taurus*) than to either *B. gaurus* or *B. javanicus*. This is an enigmatic form. On the one hand it might be construed as a vicariant of *B. javanicus* but it has also been suggested that it is a hybrid of domestic origin or even a remnant of the otherwise extinct *B. namadicus* (Grigson, 1989). CITES: Appendix I; IUCN: E.

Bos gaurus Gaur

B. gaurus Smith, 1827: 399; Mainpat, Sarguja Tributary States, India, [c. 23°N, 83°E] (Harper, 1940). Conserved by Opinion 1348 (1985).
B. gour Hardwicke, 1827: 231; 'district of Ramgurh and tableland of Sirgoojas', India.
Bibos subhemachalus Hodgson, 1837a: 499; Saul Forest, Nepal. = *Bos cavifrons* Hodgson, 1837c: 747.
B. asseel Horsfield, 1851: 181.
Gauribos laosiensis Heude, 1901: 3; Laos.
G. brachyrhinus Heude, 1901: 4; Pursat, Cambodia.
G. sylvanus Heude, 1901: 4; Mois, Tonkin, N Vietnam.
G. mekongensis Heude, 1901: 5; Kratie, Mekong Valley.
Uribos platyceros Heude, 1901: 5; between Tourane Bay and the rivers of Hué, C Vietnam.
Bubalibos annamaticus Heude, 1901: 6; Annam, Vietnam.
B. (?) *leptoceros* Heude, 1901: 7; Kampot, Cambodia.
Bibos (?) *fusicornis* Heude, 1901: 9; Tonkin, N Vietnam.

Range (Map 112). Formerly throughout the forested parts of India and S Nepal, and east to Vietnam and south to Malaya. Range much reduced, but present in parts of India, Nepal, S Yunnan and in scattered populations from there to Malaya.

Variation. The following subspecific names have been proposed:
B. g. readei Lydekker, 1903a: 266; Myitkyina, Upper Burma.
B. g. hubbacki Lydekker, 1907a: 64; Pahang, Malaya.
Bibos g. grangeri Colbert & Hooijer, 1953: 122; Yenchingkou, Sichuan; Pleistocene.

Remarks. The synonyms of Heude (1901) were tentatively allocated to *B. gaurus* by Bohlken (1961). This species has been called *B. frontalis*, i. e. including the domesticated gayal. However the latter is probably sufficiently distinctive to justify treatment as a separate domestic species which avoids a great deal of nomenclatural ambiguity. CITES: Appendix I; IUCN: V.

Bos frontalis Gayal, Mithan (domestic)

B. frontalis Lambert, 1804: 57; ? NE Chittagong, Bangladesh.
B. gavaeus Colebrooke, 1805: 488; India. Based upon animals in a herd belonging to the Governor General.
B. sylhetanus Geoffroy & Cuvier, 1824:liv. 42.

Range (Map 112). Confined to the highlands of Assam and adjacent parts of Bangladesh and Burma.

Remarks. Although much disputed in the past, the evidence seems overwhelming that this form is a domesticated derivative of *B. gaurus* with which it is fully interfertile, and this is supported by data from biochemical and karyological studies (Simoons, 1984). It is however clearly distinguishable from wild *B. gaurus* and for purposes of nomenclature is best treated as a separate domestic species rather than part of *B. gaurus* (Corbet & Clutton-Brock, 1984). Besides the features shown in Table 172 it differs from *B. gaurus* also in having shorter legs, a less prominent dorsal ridge, shorter head and less curved horns. It is kept primarily for religious and social purposes.

Bos javanicus Banteng

B. javanicus d'Alton, 1823:[7], pl. viii; Java (Hooijer, 1956).
B. leucoprymnus Quoy & Gaimard, 1830: 140; Batavia, Java.
B. sondaicus Blyth, 1842d: 445; based upon *B. sondaicus* Müller, 1840, a *nomen nudum*.

Map **112**. *Bos*

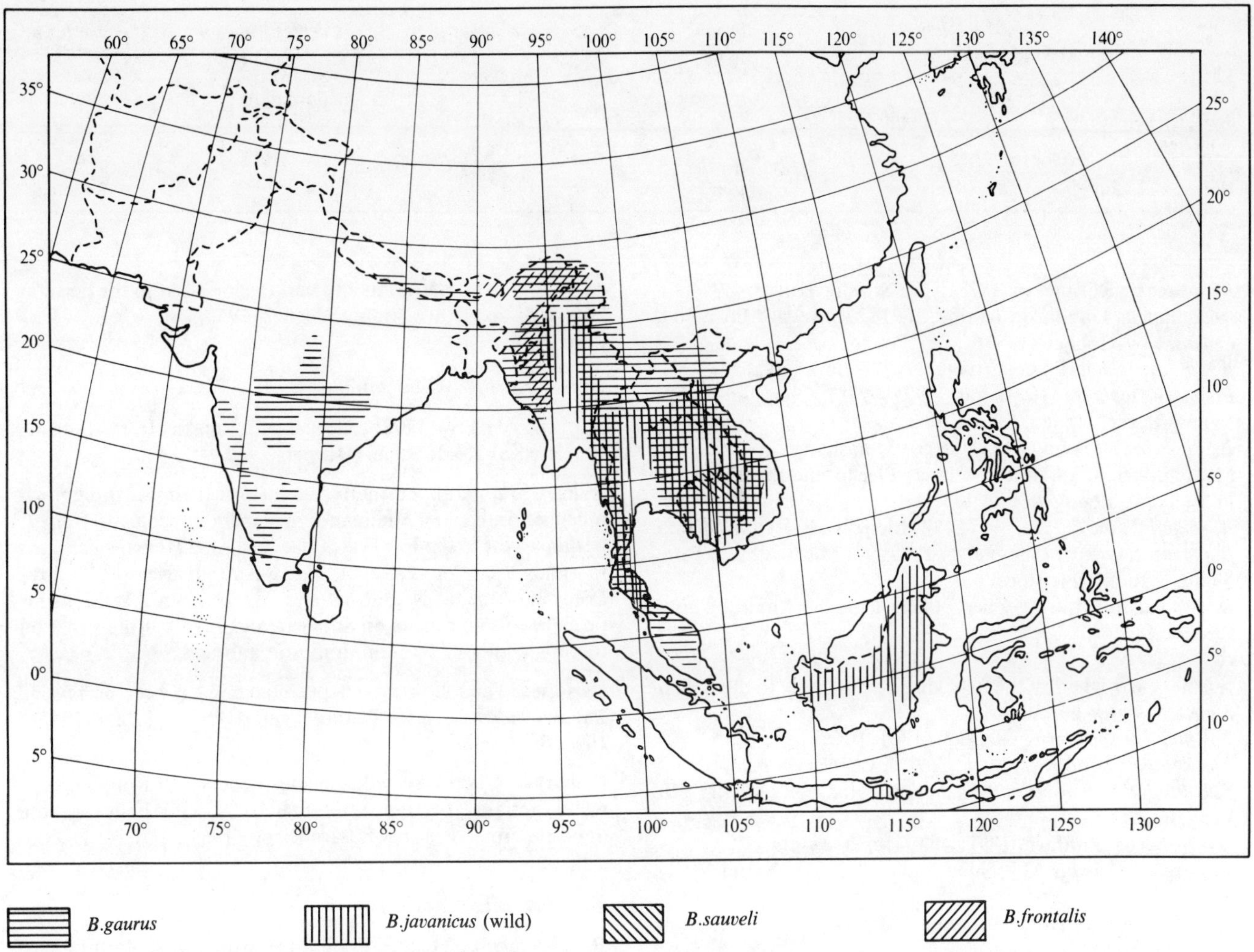

B. banteng Wagner, 1844a: 517; based upon *B. banteng* Temminck, 1836, a *nomen nudum*.
B. bantiger Schlegel & Müller, 1845: 198; Java.
Bibos discolor Heude, 1901: 8; Indochina; Bohlken (1961).
Bibos longicornis Heude, 1901: 9; Indochina; Bohlken (1961).

Range (Map 112). From Burma, Thailand and Vietnam to Malaya; Java, Borneo. Records on Sumatra may refer to feral Bali cattle (see Remarks). Introduced (probably feral domestic) on the islands of Enggano (Sumatra) and Sangihe (Sulawesi).

Variation. The following subspecific names have been proposed:
B. sondaicus birmanicus Lydekker, 1898b: 277; Burma.
? *B. s. butleri* Lydekker, 1905d: 151; Perak, Malaya. Perhaps a domesticated form.
B. s. porteri Lydekker, 1909b: 669; Thailand.
B. s. lowi Lydekker, 1912: 906; Sarawak, Borneo.

Mainland animals have the white rump patch less extensive than on the islands, and the adult males are brown rather than black.

Remarks. A domesticated form, Bali cattle, is the dominant domestic form of cattle on the islands of Bali, Lombok and Timor, and also occurs in parts of Java, Sumatra, Malaya, Borneo, Sulawesi, New Guinea and N Australia. It is smaller than the wild form, with shorter, thicker horns. They are used for traction, riding and meat. Reviewed by Rollinson (1984). IUCN: V.

BUBALUS Asian buffalos

Bubalus Frisch, 1775: 1. Not available (Opinion 581, 1960).
Bubalus Smith, 1827: 371; type species *Bos bubalus* Gmelin = *B. bubalus* L.
Anoa Smith, 1827: 355; type species *Antilope depressicornis* Smith.
Buffelus Rütimeyer, 1865: 334 (*Buffelus indicus* Rütimeyer, in *B. bubalis*).
Probubalus Rütimeyer, 1865: 334 (*B. sivalensis* Rütimeyer).

Indomalayan Region; 4 wild species plus the domestic water buffalo (Table 173). More closely related to the African *Syncerus* than to the other Indomalayan cattle, a conclusion that has been supported by biochemical data (Hartl *et al.*, 1988). The species on Sulawesi were reviewed by Groves (1969a).

The following nominal species have been based upon fossils from the region:
B. (*Hemibos*) *triquetricornis* Rütimeyer, 1867: 23; Siwaliks.
B. platyceros Lydekker, 1877: 31; Siwaliks.

Table 173. The species of *Bubalus*.

	arnee	*bubalis*	*mindorensis*	*depressicornis*	*quarlesi*
Range	Assam etc.	Throughout (domestic)	Mindoro, Philippines	Sulawesi (lowland)	Sulawesi (highland)
Shoulder height (cm)	160–190	160 190	100–105	c.80–100	c.70
White gorget	+/0	+/0	+/0	0	
Legs white	+	+		+	0
Horns: x-section	Flat	Flat	Flat	Flat	Rounded

B. sivalensis Rütimeyer, 1878: 138; Siwalik Hills.
Anoa santeng Dubois, in Jentink, 1891: 221; Java. Allocated to *Duboisia* by Hooijer (1958).
B. palaeokerabau Dubois, 1908: 1263; Trinil, Java.
B. mephistopheles Hopwood, 1925: 238; Chang-te-ho, on Yellow River, Henan, China.
B. brevicornis Young, 1937: 505; Miaoshang, 15 li NE of Mienchihhsien, W Honan, China; Pleistocene (recorded also in Sichuan: Young & Liu, 1951).
B. youngi Chow & Hsu, 1957: 457; Menghsien, Henan, China.
B. maruvattoorensis Ghose *et al.*,1972: 97; Maruvattoor, Tamil Nadu, S India; Pleistocene.
B. triangulatus Liu & Zhen, 1981: 219; N Anhui, China; Pleistocene.

It is not clear whether the following names apply to the wild or domestic water buffalo:
Bos buffelus spirocerus Hodgson, 1841: 912.
Bos kerabau Sundevall, 1846: 202; Java. *Nomen nudum.*
Bubalus buffelus (or *arna*) *macrocerus* Hodgson, 1847g: 710; Bengal.
Bos b. fulvus Blanford, 1891: 492; Upper Assam.
B. b. septentrionalis Matschie, 1912a: 70 (103); N India.

Bubalus bubalis Domestic water buffalo

Bos bubalis Linnaeus, 1758: 72; Rome, Italy.
B. buffelus Blumenbach, 1788: 123; Asia.
Buffelus indicus Rütimeyer, 1865: 334; India.
B. mainitensis Heude, 1894: 205; Mindanao, Philippines.
B. moellendorffi Nehring, 1894: 185; Busuanga I, Calamian Is, Philippines.

Range. Used as a domestic animal throughout the region.

Variation. The following subspecific names have been proposed:
Bos b. sondaicus Schlegel & Müller, 1845: 195, pl. 40; E Indian Islands.
Bos b. hosei Lydekker, 1898c: 126; Sarawak, Borneo.
B. kerabau ferus Matschie, 1898: 43; Nueva Ecija and Cagayan, Philippines.

Two main types occur: swamp buffalo, with widely spreading horns as in the wild species, usually two pale chevrons on the throat, otherwise grey—mainly in S China and Indochina; and river buffalo, with shorter, more tightly curved horns, usually all-black—mainly in India.

Remarks. An important domestic animal throughout the region. Probably domesticated prior to 2000 BC in SW Asia. They are used as draught animals, for all kinds of agricultural traction, and, especially in India, for milk production. Reviewed by Cockrill (1984). On the island of Sumba (Lesser Sunda Is) exceptionally long, straight horns are induced by a combination of castration and making incisions in the horns at an age of about two years (Wegner, 1950).

Bubalus arnee Water buffalo (wild)

Bos arnee Kerr, 1792: 336; 'India, north from Bengal', restricted to Kuch Bihar (Harper, 1940).

Range (Map 113). Formerly, in historical times, throughout much of India and mainland SE Asia; and on Sri Lanka; perhaps west to the Euphrates (Cockrill, 1984) . Now confined to small areas of S Nepal, the Bastar district of Madhya Pradesh, Assam, N Burma and W Thailand. Wild forms probably also occurred on Sumatra and Java but the situation there is confused by feral domestic animals.

Variation. The following subspecific name has been proposed: *Bubalus bubalis migona* Deraniyagala, 1952: 198 (and 1953a: 103); Sri Lanka.

Remarks. Where the wild species occurs cows of domestic water buffalo are frequently mated by wild bulls but the opposite cross is probably rare (Scott, 1969). IUCN: E.

Bubalus mindorensis Tamarau

B. mindorensis Heude, 1888: 4; Mindoro I, Philippines. Published before August (Hooper, 1941).
Anoa mindorensis Steere, in Sclater, 1888 (16 Aug): 364 (and Steere, 1889: 415); Mindoro I, Philippines.

Range (Map 113). Confined to the island of Mindoro, Philippines.

Remarks. Has been treated as a subspecies of *B. arnee*, e. g. by Bohlken (1958), but best treated as a separate species which is however more closely related to *B. arnee* than to the species on Sulawesi (Groves, 1969a). The population fell from an estimated 10000 in 1900 to 120 in 1975, rising to 370 in 1987 (Petocz, 1989). CITES: Appendix I; IUCN: E.

Bubalus depressicornis Lowland anoa

Antilope depressicornis Smith, 1827: 293; Sulawesi. Conserved by Opinion 1349 (1985) over *Bos bubalus anoa* Kerr, 1792: 239.
Probubulus celebensis Rütimeyer, 1865: 331, 334; Sulawesi.

Range (Map 113). The lowlands of Sulawesi. Recorded throughout most of the island except the SW peninsula where there are however subfossil records.

Variation. The following subspecific name has been proposed: *Bos (Bubalus) d. fergusoni* Lydekker, 1905e: 378; based upon animals in zoological gardens at Trivandrum, India.

Remarks. IUCN: E.

Map **113**. *Bubalus, Boselaphus, Tetracerus*

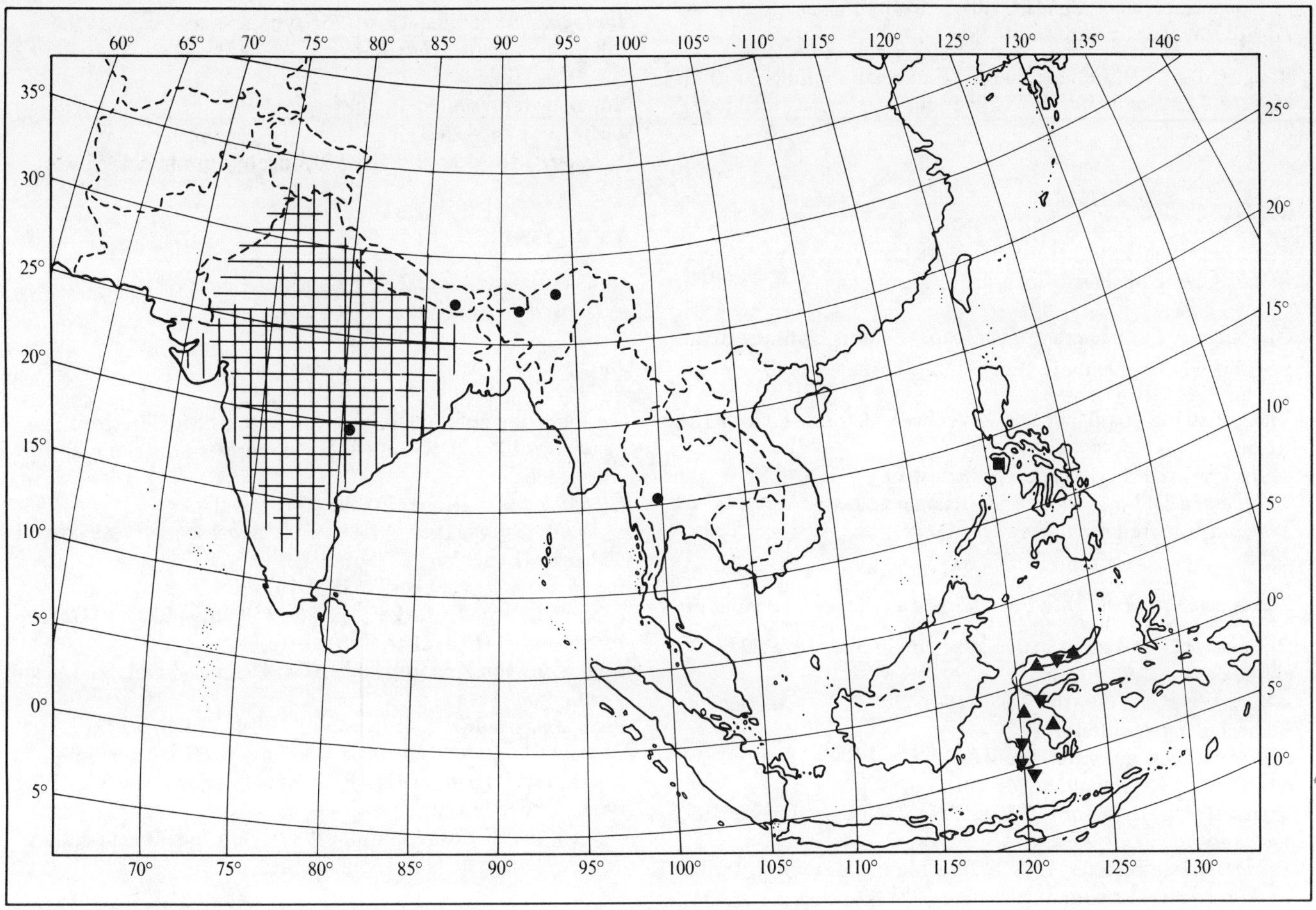

Boselaphus tragocamelus — *T.quadricornis* — ● *Bubalus arnee* — ▲ *Bubalus depressicornis* — ▼ *Bubalus quarlesi* — ■ *Bubalus mindorensis*

Bubalus quarlesi Mountain anoa

Anoa quarlesi Ouwens, 1910: 7; Toradja Mts, Sulawesi. Conserved by Opinion 1349 (1985) over *Bos bubalus anoa* Kerr, 1792.

Range (Map 113). Mountains of Sulawesi and on the adjacent island of Buton.

Remarks. The specific status of this form was demonstrated by Groves (1969a). CITES: Appendix I; IUCN: E.

BOSELAPHUS

Boselaphus Blainville, 1816: 75; type species *Antilope picta* Pallas, in *B. tragocamelus*.
Portax Smith, 1827: 366 (*Damalis risia* Smith, in *B. tragocamelus*).

One species.

The following fossil species have been described from the region:
Portax namadicus Rütimeyer, 1878: 89; Narbada, India.
B. lydekkeri Pilgrim, 1910: 70; Punjab; Middle Siwaliks. = *Selenoportax lydekkeri*: (Pilgrim, 1937).

Boselaphus tragocamelus Nilgai

Antilope tragocamelus Pallas, 1766: 5; plains of peninsular India.
A. albipes Erxleben, 1777: 280; India.
A. picta Pallas, 1777: 14; India.

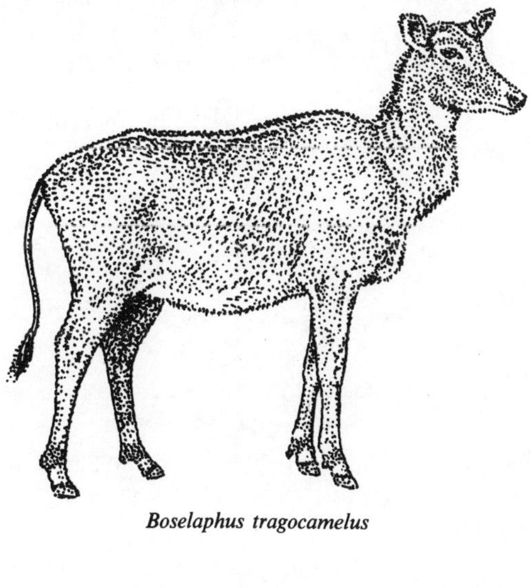

Boselaphus tragocamelus

= *Damalis risia* Smith, 1827: 363.
= *Tragelaphus hippelaphus* Ogilby, 1837b: 138.
Portax tragelaphus Sundevall, 1846: 198.

Range (Map 113). India from the Himalayan foothills south to Mysore. Marginally into Pakistan in the northwest. Introduced in Texas, USA.

TETRACERUS

Tetracerus Leach, in Hardwicke, 1825a: 524; type species *Antelope chickara* Hardwicke, in *T. quadricornis*.

One species. The presence of two pairs of horns is unique in the Bovidae.

The following fossil species have been described from the Region:
T. daviesi Lydekker, 1886c: 19; Siwaliks.
T. kroesenii Dubois, 1907: 454; Kendeng, Java. Allocated to *Duboisia* by Stremme (1911a).

Tetracerus quadricornis Four-horned antilope, Chousingha

Cerophorus (*Cervicapra*) *quadricornis* Blainville, 1816: 75, 78; plains of peninsular India.
Antilope chickara Hardwicke, 1825a: 520, pl. 15; W Bengal, Bihar and Orissa, India.
Tetraceros striatocornis Brookes, 1828: 64; locality not recorded.
A. tetracornis Hodgson, 1836c: 525; locality not recorded; *nomen nudum*.
A. sub-4-cornutus Elliot, 1839: 225; S Mahratta country, India.
= *T. subquadricornutus* Hodgson, 1847f: 89.
T. subquadricornis Gray, 1843a: 159.
T. iodes Hodgson, 1847f: 90; 'Saul forests beneath the Sub-Himalayas'.
T. paccerois Hodgson, 1847f: 90; as above.

Range (Map 113). Most of peninsular India except for the southwestern region.

Remarks. CITES: Appendix III (Nepal).

[*TRAGELAPHUS*]

Tragelaphus Blainville, 1816: 75; type species *Antilope sylvatica* Sparrman, in *T. scriptus* Pallas.
Strepsiceros Smith, 1827: 305 (*A. strepsiceros* Pallas).

Not now represented in the region but the following nominal fossil species has been described:
Stripsiceros (*?*) *falconeri* Lydekker, 1885b: 170; Perim I, NW India; Siwaliks.

[*KOBUS*]

Kobus Smith, 1840:pl. 28; type species *Antilope ellipsiprymnus* Ogilby (Africa).
= *Cobus* Buckley, 1876: 284.

Not now represented in the region but the following nominal fossil species has been described:
Cobus (*?*) *palaeindicus* Lydekker, 1886d: 53; Siwaliks.

[*ALCELAPHUS*]

Alcelaphus Blainville, 1816: 75; type species *Antilope bubalis* Pallas, in *A. buselaphus* Pallas.

Not now represented in the region, but the following fossil species has been described from the region:
A. bakeri Lydekker, 1885b: 170; locality unknown; Siwaliks.

ANTILOPE

Antilope Pallas, 1766: 1; type species *Capra cervicapra* L.
= *Cervicapra* Sparrman, 1780: 281.

One species, closely related to *Gazella*.

The following nominal fossil species have been described from the region although few are likely to represent the genus as now understood:
A. palaeindica Falconer, 1859: 154; Siwaliks.
A. sivalensis Lydekker, 1878b: 154; Siwaliks. = *Hippotragus sivalensis*: (Lydekker, 1885b).
A. patulicornis Lydekker, 1878b: 157; Siwaliks.
A. porrecticornis Lydekker, 1878b: 158; Siwaliks. = *Gazella porrecticornis*: (Lydekker, 1884b).
A. saatensis von Koenigswald, 1934: 192, pl. 3; Kali Saät, near Boemiadjoe, Java.
A. madjokertensis Koenigswald, 1934: 192, pl. 3; Java. Allocated to *Duboisia santeng* (Dubois) by Hooijer (1958).
A. subtorta Pilgrim, 1937: 812; near Siswan, Siwalik Hills, India; Upper Siwaliks.
(*cf. A.*) *planicornis* Pilgrim, 1939: 40; locality uncertain, probably Perim I, NW India.

Antilope cervicapra Blackbuck

Capra cervicapra Linnaeus, 1758: 69; inland of Trivandrum, Travancore, India (Zukovsky, 1927).
A. rupicapra Mueller, 1776: 56; Bengal.
A. bilineata Gray, 1830(in 1830–35):pl. 12; India.
Cervicapra bezoartica Gray, 1843a: 159; India.
A. hagenbecki Zukovsky, 1927: 125; Bengal.
A. rajputanae Zukovsky, 1927: 125; near Bahawalpur, Rajputana/Punjab.
A. centralis Zukovsky, 1928: 60; Gwalior, India.

Range (Map 114). Throughout the drier parts of peninsular India and north to SE Pakistan and W Bengal. Marginally in Nepal. They became extinct in Pakistan in the 1970s, but have been reintroduced to Suhandra and Kalabagh National Parks (Anon, 1987). The present and former ranges were mapped by Ranjitsinh (1989). Introduced in Texas, Argentina and W Australia.

Variation. Geographical variation was reviewed by Groves (1982b) who recognized two discrete subspecies, although the nature and position of the boundary have not been defined:
A. c. cervicapra. South and east India. Small (skull length of adult male under 230mm); no grey sheen; pelage short and fine; leg-stripes well marked; eye-ring narrowed above eye.
A. c. rajputanae. NW India. Larger (skull of adult males over 230mm); adult males with grey sheen in breeding season; pelage long and rough; leg-stripes obscure on shanks; eye-ring broad all round eye.

Map **114**. *Antilope, Gazella, Budorcas*

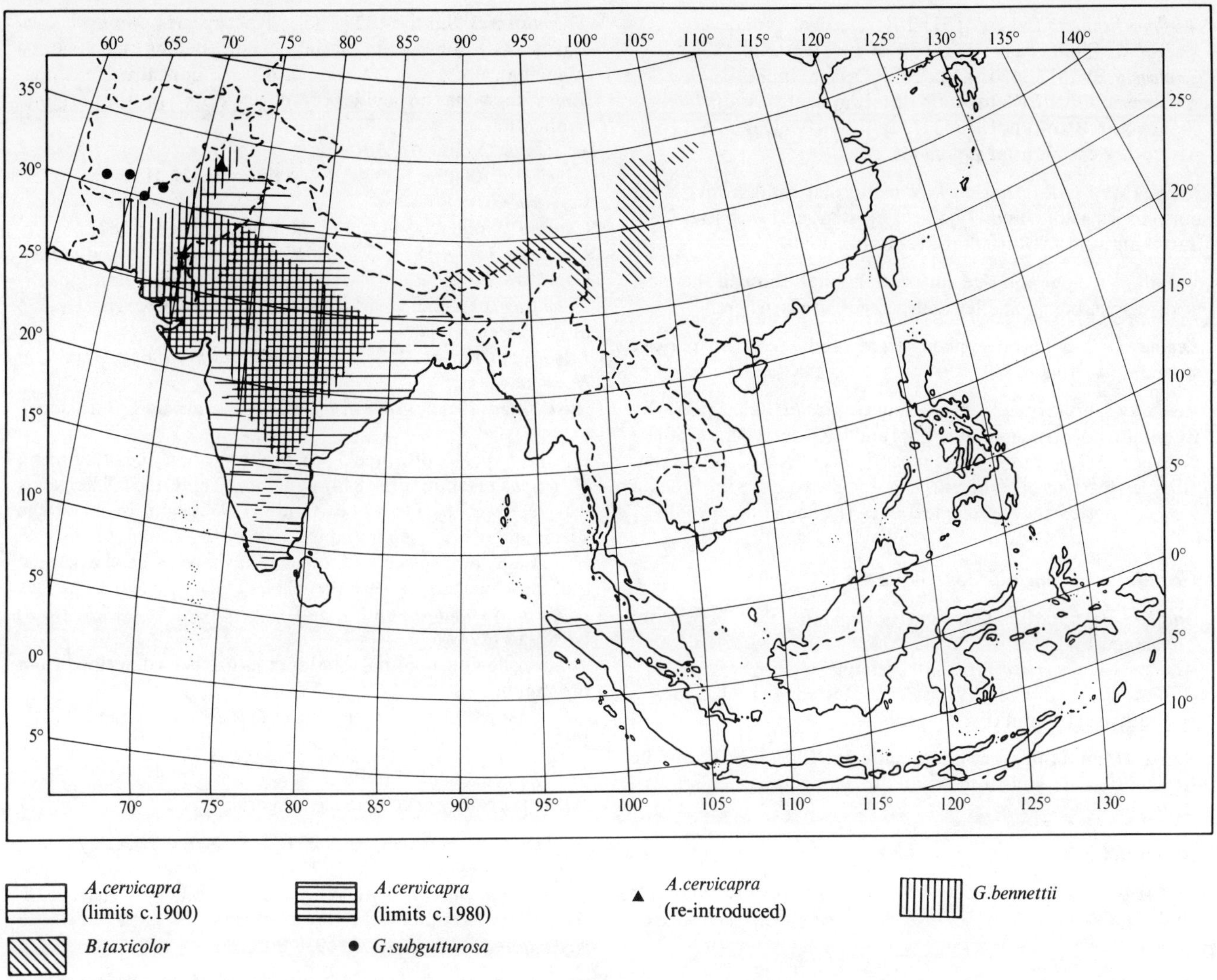

Ecology. Blackbuck are grazers, forming large herds on grassland and in open thorn forest, and capable of surviving in semidesert habitats.

Remarks. CITES: Appendix III (Nepal).

GAZELLA Gazelles

Gazella Pallas, 1769: 468; Lichtenstein, 1814: 152; Rafinesque, 1815: 56. All suppressed by Direction 23 (1955).
Gazella Blainville, 1816: 75; type species *Capra dorcas* L. Conserved by Opinion 108 (1929).
= *Dorcas* Gray, 1821: 307.
Tragops Hodgson, 1847g: 695 (*A. bennettii* Sykes). Pre-occupied by *Tragops* Wagler, 1830 (Reptilia).
= *Tragopsis* Fitzinger, 1869b: 157.
Trachelocele Ellerman & Morrison-Scott, 1951: 389 (*Antilope subgutturosa* Güldenstaedt).

Africa to central Asia and marginally to the arid northwestern part of the Indomalayan Region; c. 15 species of which 2 are represented in the region, one very marginally (Table 174). Gazelles are small bovids adapted to fast running in open habitats. Males (and sometimes females) have tranversely ringed horns forming an open spiral. There are antorbital, inguinal and pedal glands; a short tail without a terminal tuft; and usually a horizontal black and white stripe on each side of the muzzle.

Table 174. The species of *Gazella*.

	bennettii	*subgutturosa*
Range in region	India, Pakistan	W Pakistan
Shoulder height (cm)	c.65	52–65
Horns (male)	Almost straight	Lyrate
〃 (female)	Well developed	Usually absent *
Nose and face	Brown	Mostly white
Male with cartilaginous swelling on throat during rut	0	+

* Present in Arabian form.

The following nominal fossil species have been described from the region:
G. lydekkeri Pilgrim, 1937: 800; Dhok Pathan, Punjab; Middle Siwaliks.
? *G. superba* Pilgrim, 1939: 36, pl. 1; Hasnot, India; Dhok Pathan stage, Middle Siwaliks.

Gazella bennettii Indian gazelle, Chinkara

Antilope bennettii Sykes, 1831: 104; Deccan, India.
A. arabica Elliot, 1839: 223. Not of Lichtenstein, 1827.
G. christii Blyth, 1842d: 452; Thar Desert, India.
A. hazenna Geoffroy, 1844: 74; Malwa, central India.
G. fuscifrons Blanford, 1873: 317; Jalk, Seistan Desert, SE Iran.
Also other extralimital synonyms.

Range (Map 114). Drier parts of peninsular India from c. 16°S north to Rajasthan and Gujarat; Pakistan; and west to E Iran. Surviving well in Rajasthan (Rahmini, 1990).

Variation. Colour and size vary with habitat, those in the most arid regions being smaller and paler than elsewhere.

Ecology. Occur on open plains from sandy desert to plateau steppe up to 1500m.

Remarks. This form was included in *G. gazella* (Pallas, 1766) by Ellerman & Morrison-Scott (1951) and in *G. dorcas* (L.,1758) by Groves (1969b) but has more recently been treated as distinct, albeit a vicariant of *G. dorcas*, e. g. by Groves (1985b, 1988), a view supported by karyological data (Furley *et al.*, 1988).

Gazella subgutturosa Goitred gazelle

Antilope subgutturosa Gueldenstaedt, 1780: 251; NW Iran. Conserved by Direction 23 (1955).
A. dorcas var. *persica* Gray, 1843a: 160.
G. seistanica Lydekker, 1910a: 202; Seistan, E Iran.
Also other extralimital synonyms.

Range (Map 114). Very marginally in W Pakistan, on the Afghan border; and from Arabia to Mongolia.

BUDORCAS

Budorcas Hodgson, 1850: 65; type species *B. taxicolor* Hodgson.

One species. Reviewed by Neas & Hoffmann (1987). The following fossil species has been described from just beyond the northern border of the region:
B. teilhardi Young, 1948: 157; Yueshe, Shanxi, China; Pleistocene.

Budorcas taxicolor Takin

B. taxicolor Hodgson, 1850: 65, pl. 1; Mishmi Hills, N Assam.
B. bedfordi Thomas, 1911g: 27 (and 1911f: 693, pl. 29); Taipai Shan, Shaanxi, China, 10000ft.

Range (Map 114). From Bhutan through N Assam, N Burma and Yunnan to Sichuan, Shaanxi, S Gansu and adjacent parts of Tibet and Qinghai, China; at altitudes of 1500–4500m.

Variation. The following subspecific names have been proposed:
B. t. var. *tibetana* Milne-Edwards, 1874: 367, pl. 74; Moupin, Sichuan.
= *B. t. sinensis* Lydekker, 1907a: 350.
B. t. whitei Lydekker, 1907b: 887; Bhutan.
B. t. mitchelli Lydekker, 1908b: 790; Tatsienlu, Sichuan.
B. t. lichii Young, 1948: 157; Anyang, N Henan, China; subfossil.

Remarks. CITES: Appendix II; IUCN: (R/I).

NAEMORHEDUS Gorals and serows

Naemorhedus Smith, 1827: 352; type species *Antilope goral* Hardwicke. Subsequent variations of spelling such as the etymologically correct *Nemorhaedus* are unjustified emendations; there is no evidence of an error in the original publication.
= *Kemas* Ogilby, 1837b: 138.
Capricornis Ogilby, 1837b: 139 (*Antilope thar* Hodgson, in *N. sumatraensis*).
Caprina Wagner, 1844a: 457. Not of d'Orbigny, 1822.
Urotragus Gray, 1871: 372 (*Antilope caudata* Milne-Edwards).
Capricornulus Heude, 1898: 13 (*Antilope crispa* Temminck).
Nemotragus Heude, 1898: 13 (*Capricornis erythropygius* Heude, in *N. sumatraensis*).
Lithotragus Heude, 1898: 13 (*Capricornis maritimus* Heude, in *N. sumatraensis*).
Austritragus Heude, 1898: 14 (*Antilope sumatraensis* Bechstein).

E Asia; 6 species of which 5 occur in the region, the other being *N. crispus* (Temminck, 1844) in Japan (Table 175). The genus was reviewed by Groves & Grubb (1985) who included the species more often separated in *Capricornis* (*N. sumatraensis, N. crispus* and *N. swinhoei*) although it has always been accepted that these two groups were very close. The forms here treated as *N. goral, N. caudatus* and *N. baileyi* were reviewed (as *N. goral*) by Mead (1989).

The following nominal fossil species has been described from the region:
(*aff. Capricornis*) *roylei* Pilgrim, 1939: 57; ? Hundes, Tibet; Pleistocene.

Naemorhedus goral Himalayan goral

Antilope goral Hardwicke, 1825a: 518; Nepal Himalaya.
? *A. duvaucelii* Smith, 1827: 279; locality not recorded.
Urotragus bedfordi Lydekker, 1905c: 83; ? Dharmsala, Himalayas.
N. hodgsoni Pocock, 1908: 195; Sikkim.

Range (Map 115). The Himalayas from N Pakistan and Kashmir east to Assam; at altitudes of 820–2000m in Pakistan.

Ecology. Primarily a grazer.

Remarks. The following two species have often been included in *N. goral* but are excluded here following Groves & Grubb (1985). CITES: Appendix I.

Naemorhedus caudatus Chinese goral

Antilope caudata Milne-Edwards, 1867a: 377; Bureja Mts, Amurland, USSR.
N. griseus Milne-Edwards, 1872a: 93; Moupin, Sichuan, China.
Antilope (*N.*) *cinerea* Milne-Edwards, 1874: 362, pl. 70; Moupin, Sichuan, China.
Kemas arnouxianus Heude, 1888: 3; Kihsien, Chekiang, SE China.
K. henryanus Heude, in Henry, 1890: 93; near Ichang, Hubei, China; *nomen nudum*.
K. xanthodeiros Heude, 1894: 243; W Sichuan.
K. iodinus Heude, 1894: 243; E Sichuan.
K. galeanus, K. vidianus Heude, 1894: 243; Yu Ho Mts, S Shaanxi, China.
K. niger, K. fargesianus Heude, 1894: 241; *K. initialis,*

K. curvicornis, K. versicolor Heude, 1894: 244; Chenkouting, Sichuan.
K. aldridgeanus Heude, 1894: 244; Hubei, China.
K. fantozatianus Heude, 1894: 245; mountains of Kium-tcheou, right bank of middle Han, Hubei, China.
Urotragus evansi Lydekker, 1905c: 83; Mt Victoria, Pokokku dist, Arakan, Burma.
And other, extralimital, synonyms.

Range (Map 115). Most of China and south to Burma and Thailand; also north through Korea and Manchuria to the Ussuri region of the Soviet Far East.

Naemorhedus baileyi Red goral

N. baileyi Pocock, 1914: 32; Dre, Yigrong Tso (Lake), Po Me, Tibet, 9000ft.
N. cranbrooki Hayman, 1961: 319; Adung Valley, N Burma, 28°10'N, 97°40'E, 8000ft.

Range (Map 115). N Burma and adjacent parts of Assam, W Yunnan and SE Tibet. It appears to occur in very close proximity to *N. caudatus griseus* in SE Tibet (Feng *et al.*, 1986: 239).

Remarks. IUCN: V.

Table 175. The species of *Naemorhedus*.

	goral	*caudatus*	*baileyi*	*sumatraensis*	*swinhoei*
Range	Himalayas	China-Thailand	N Burma etc.	S China-Sumatra	Taiwan
Shoulder height (cm)	65–70			85–94	
Tail (mm)	c.75	c.125		80–140	
Dorsal colour	Greyish brown	Greyish or yellowish brown	Reddish brown	Grey	Brown
Dorsal stripe	Short or 0	+	+	+	Faint
Pre-orbital gland	Small, no cavity	Small, no cavity	Small, no cavity	Large, deep cavity	Large

Map **115**. *Naemorhedus*

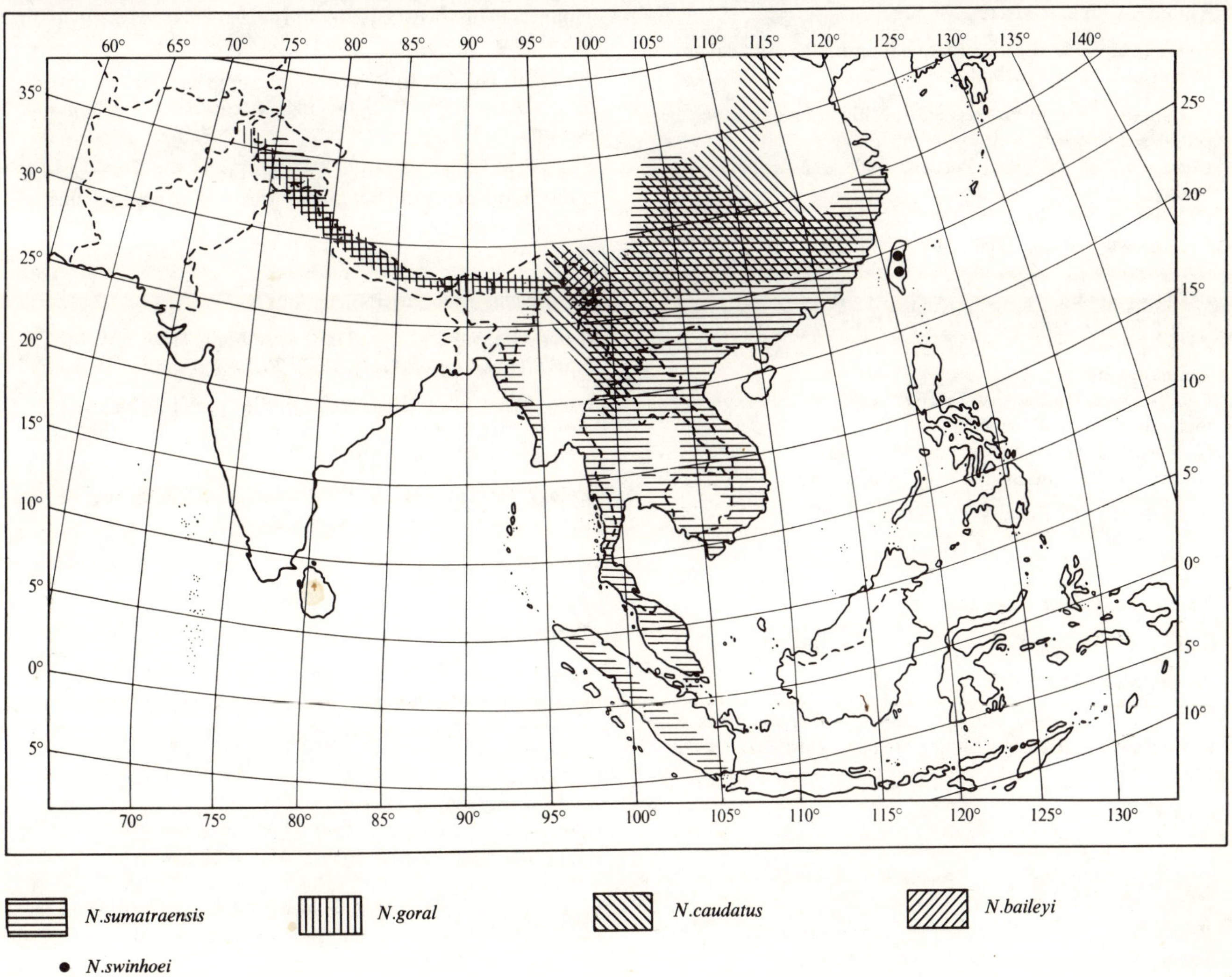

Naemorhedus sumatraensis Southern serow

Antilope sumatraensis Bechstein, 1799: 98; Sumatra.
A. interscapularis Lichtenstein, 1814: 165; Sumatra.
A. thar Hodgson, 1831b: 324; Nepal Himalaya.
A. bubalina Hodgson, 1832: 12; Nepal.
Capricornis rubida Blyth, 1863a: 174; Arakan Hills, Burma.
C. (*Antilope*) *milne-edwardsii* David, 1869: 10; Moupin, Sichuan, China.
N. edwardsii David, 1872: 90.
C. argyrochaetes Heude, 1888: 4; subprefecture of Tchou-ki, Zhejiang, China.
C. maritimus Heude, 1888: 4; Baie d'Along, Tonkin, N Vietnam.
C. rocherianus Heude, 1894: 225; *C. benetianus* Heude, 1894: 227; Baie d'Along, Tonkin, Vietnam.
C. maxillaris Heude, 1894: 229; Chaohing, Zhejiang, China.
C. platyrhinus, C. cornutus, C. erythropygius, C. microdontus, C. ungulosus, C. nasutus, C. vidianus, C. fargesianus, C. brachyrhinus, C. pugnax, C. longicornis, C. chrysochaetes Heude, 1894: 232, 233; Sichuan etc., China.
C. marcolinus Heude, 1897: 151; Tonkin, N Vietnam.
C. berthetianus Heude, 1898: 8; Tonkin, N Vietnam.
C. gendrelianus Heude, 1899: 210; Tonkin, N Vietnam.
C. collasinus Heude, 1899: 211; Guangdong, China.
N. swettenhami Butler, 1900: 675; Larut Hills, Perak, Malaya.
C. osborni Andrews, 1921: 1; Hui-Yao, 20 miles from Tengueh, Yunnan, China.

Range (Map 115). S China (N to S Gansu), the Himalayas west to Kashmir, south to Malaya; Sumatra.

Variation. The following subspecific names have been proposed:
Capricornis s. robinsoni Pocock, 1908: 12, 185; Selangor, Malaya.
C. s. humei Pocock, 1908: 178; Kashmir.
C. s. rodoni Pocock, 1908: 180; Chamba, Punjab.
C. s. jamrachi Pocock, 1908: 183; Kalimpong, near Darjeeling, N Bengal, India.
C. s. annectens Kloss, 1919b: 391; Koh Lak, SW Thailand.
C. s. montinus Allen, 1930a: 5; Likiang Range, Snow Mts, Yunnan, China.
C. s. kanjerus Colbert & Hooijer, 1953: 127; Yenghingkou, Sichuan, China; Pleistocene.
C. s. guilinensis Huang *et al.*,1983: 152; Chuanshan Cave, Guilin County, Guangxi, China; Pleistocene.

Remarks. CITES: Appendix I; IUCN: (E).

Naemorhedus swinhoei Taiwan serow

Capricornis swinhoei Gray, 1862b: 320; Taiwan.

Range (Map 115). Taiwan.

Ecology. Occurs at all altitudes from 200–3870m but especially in montane forest and grassland between 1000 and 2000m (Lue, 1987).

Remarks. IUCN: V.

HEMITRAGUS Tahrs

Hemitragus Hodgson, 1841b: 218; type species *Capra quadrimammis* vel *jharal* Hodgson, in *H. jemlahicus*.
Kemas Gray, 1852: 146 (*Capra* (*Ibex*) *warryato* Gray, in *H. hylocrius*). Not of Ogilby, 1837.

Arabia and India; 3 species of which 2 occur in the region (Table 176).

Table 176. The species of *Hemitragus*.

	jemlahicus	*hylocrius*
Range	Himalayas	SW India
Mane on shoulders, throat and chest	Long, shaggy	Short, bristly
Males with contrasting pale 'saddle'	0	+
Anterior face of horn	Narrow, with prominent keel	Flat, keel at inner edge

Hemitragus jemlahicus Himalayan tahr

Capra jemlahica Smith, 1826:plate opp. p. 308; Jemla Hills, Nepal.
= *C. tubericornis* Schinz, 1845b: 467.
C. jharal Hodgson, 1833c: 129; Nepal.
C. quadrimammis Hodgson, 1836d: 710; Nepal.

Range (Map 116). Himalayas from Kashmir to SW Bhutan (Holmes, 1970); between 2500 and 4400m in Nepal. Introduced in S Africa and New Zealand.

Variation. The following subspecific name has been proposed: *H. j. schaeferi* Pohle, 1944: 184; 10km NW of Chuntang, Sikkim.

Ecology. In Nepal they range from the broad-leaved forest zone to the alpine meadows but favour grassy cliffs with patches of trees (Schaller, 1973).

Hemitragus hylocrius Nilgiri tahr

Kemas hylocrius Ogilby, 1838b: 81; Nilgiri Hills, SW India.
Capra (*Ibex*) *warryato* Gray, 1842: 267; Nilgiri Hills, SW India.

Range (Map 116). Hills of SW India from the Nilgiri Hills south to Periyar. Greatly reduced in numbers and estimated at c. 2200 in 1987 (Rice, 1988).

Ecology. Live mainly on cliff ledges and on the fringes of the open grassy plateaus, mainly c. 2000m.

Remarks. IUCN: V.

CAPRA Goats

Capra Linnaeus, 1758: 68; type species *C. hircus* L. Conserved by Opinion 91 (1926) and Direction 24 (1955).
= *Hircus* Brisson, 1762: 12.
= *Aries* Link, 1795: 96.
= *Tragus* Schrank, 1798: 78.
= *Aegoceros* Pallas, 1811: 224.
Orthaegoceros Trouessart, 1905: 738 (*C. falconeri* Wagner).
Also other names based upon extralimital species.

Palaearctic Region; c. 6 wild species of which 2 occur on the northwestern fringe of the Indomalayan Region; and the domestic goat (Table 177).

The following nominal fossil species have been described from the region:
C. sivalensis Lydekker, 1878b: 169; Siwaliks.
C. perimensis Lydekker, 1878b: 170; Perim I, NW India. = *Tragocerus perimensis*: (Pilgrim, 1910).

Capra aegagrus Wild goat, Bezoar

C. aegagrus Erxleben, 1777: 160; Daghestan, Caucasus.
C. blythi Hume, 1875: 240; Sind, Pakistan. *Nomen nudum.* = *C. hircus blythi* Lydekker, 1898: 264.
C. falconeri chialtanensis Lydekker, 1913b: 171; Chialtan Range, near Quetta, Baluchistan, Pakistan.
Also other, extralimital, synonyms.

Range (Map 116). SW Pakistan; and west to Asia Minor.

Variation. Pakistan animals tend to have the pelage shorter and paler, and the horns more slender, than those further west. Those in the Chialtan Range have slightly twisted horns which have led them to be included in *C. falconeri*, but this was convincingly refuted by Schaller (1977: 29).

Ecology. Confined to rocky habitats at all altitudes up to 3300m. A versatile feeder, both grazing and browsing.

Table 177. The wild species of *Capra* and *Ovis*.

	C.aegagrus	*C.falconeri*	*O.vignei*
Range in region	SW Pakistan	Pakistan	N,W Pakistan
Shoulder height (male)	85–95	90–105	70–90
Dorsal crest	+	(+)	0
Beard	Male only	Both sexes	0
Horns curve in one plane (1) or two (2)	1	2	2
Keel on anterior face of horn	+	+	0

Capra hircus Domestic goat

C. hircus Linnaeus, 1758: 68; Sweden. Conserved by Opinion 91 (1926) and Direction 24 (1955).

Range. A domestic animal of major importance throughout the region.

Variation. Throughout most of the region the goats are small and short-eared with small, simply curved horns. In N India and Pakistan a larger form with twisted horns and long, pendant ears is common (as in most of SW Asia).

Map **116**. *Hemitragus, Capra, Ovis*

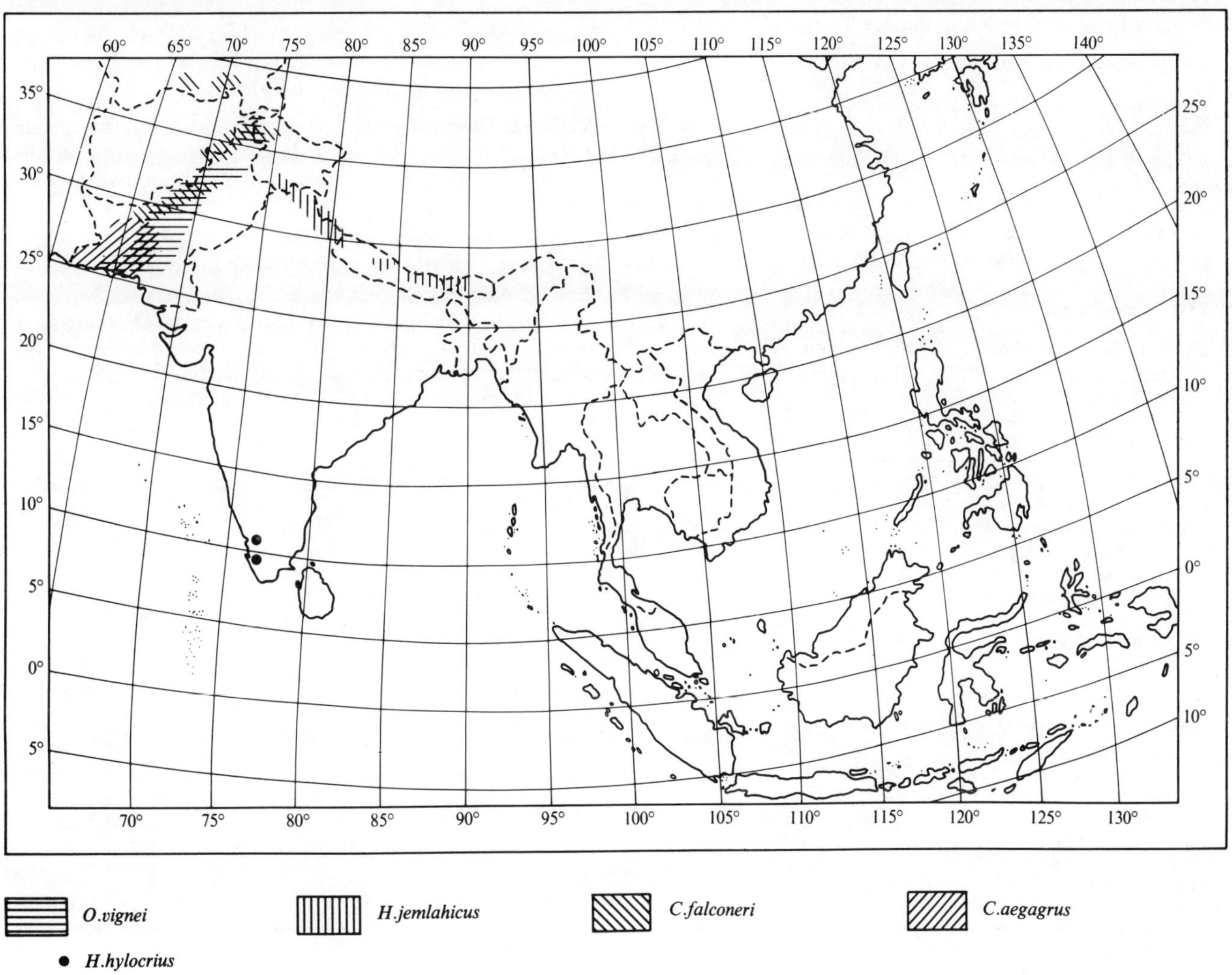

Remarks. Derived from *C. aegagrus*, probably around 8000–7000 BC in the region of Iran/Iraq. Domestic goats were known in India by 100 BC but probably did not spread in most of SE Asia until the Arab colonizations in the 14th century AD. Goats are important producers of milk, meat, hair and leather. Reviewed by Mason (1984c).

Capra falconeri Markhor

Aegoceros (*Capra*) *falconeri* Wagner, 1839b: 430; Kashmir (restricted to Astor by Lydekker, 1913b: 163).
C. megaceros Hutton, 1842: 535, pl. xx; Kandahar, Afghanistan.
C. jerdoni Hume, 1875: 240; Suleman Range, Punjab.

Range (Map 116). Kashmir and Pakistan; and northwest to Afghanistan and SE Tadjikistan; at altitudes of 600–3600m in Pakistan. In parts of SW Pakistan they are marginally sympatric with *C. aegagrus* without interbreeding (Schaller, 1977: 28).

Variation. The following subspecific name has been proposed from the region:
C. f. cashmiriensis Lydekker, 1898c: 290; Pir-Panjal Range, Kashmir.

In spite of the amount of variation in the shape of the horns it is doubtful if discrete subspecies can be recognized. There is a cline linking small animals with short ruffs and straight tightly twisted horns in the south and larger ones with longer ruffs and horns with an open spiral in the north.

Remarks. CITES: Appendix I/II; IUCN: V/E.

OVIS Sheep

Ovis Linnaeus, 1758: 70; type species *O. aries* L.

Nearctic and Palaearctic regions, marginally in the arid northwest of the Indomalayan Region; 6 species, of which one, plus the domestic sheep, occurs in the region.

Ovis vignei Urial

O. vignei Blyth, 1841b: 70; Astor, Kashmir.
O. cycloceros Hutton, 1842: 514, pl. xix; Hazara Hills, Afghanistan.
O. blanfordi Hume, 1877: 327, pl. 4; hills above Bolan Pass, near Kelat, Baluchistan, Pakistan.

Range (Map 116). North and west Pakistan, west of the R Jhelum; west to Iran and S Soviet Turkestan.

Variation. The following subspecific name has been proposed from the region:

O. v. punjabiensis Lydekker, 1913a: 10; Salt Range, Punjab, Pakistan.

Those in the Salt Range are smaller and more reddish than elsewhere.

Ecology. A gregarious grazer, mainly on the steppes of the lower hills but up to 3000m in places.

Remarks. This form could be treated as a subspecies of *O. orientalis* (Gmelin, 1774) which ranges from Asia Minor to Iran, with some hybrid populations between *orientalis s. s.* and *vignei* in Iran. *O. vignei* differs from *O. orientalis* in having a longer ruff on the throat and in chromosome number: 2n=58, compared with 2n=54 in *O. orientalis* (and in the domestic sheep). CITES: Appendix I.

Ovis aries Domestic sheep.

O. aries Linnaeus, 1758: 70; Sweden. Conserved by Opinion 75 (1922).

Range. Sheep are kept in large numbers in Pakistan, Nepal, most of India and Sri Lanka; and in Java and Sulawesi, with smaller numbers in Sumatra and Malaya.

Variation. Fat-tailed breeds are dominant in W Pakistan and in Java and Sulawesi. Primitive hair sheep predominate in S India and Sri Lanka; elsewhere there are mainly thin-tailed wool breeds.

Remarks. The principal ancestor of the domestic sheep was almost certainly the mouflon, *O. orientalis*, which ranges from Iran to Asia Minor, but there is a considerable degree of interfertility amongst the various species of *Ovis* and it is possible that *O. vignei* might have contributed to the development of some breeds. The history and diversification of domestic sheep were reviewed by Ryder (1984).

Table 178. The families of Rodentia.

	Sciuridae	Pteromyidae	Muridae	Gliridae	Zapodidae	Dipodidae	Hystricidae
No.of genera in region	17	11	78	2	2	3	3
Gliding membrane on each flank, from wrist to ankle	0	+	0	0	0	0	0
No.of functional digits (fore/hind)	4/5	4/5	4/5	4/5	4/5	4/3–5	5/5
Hind limbs elongate	0	0	0	0	+	++	0
Tail usually:	Long, bushy	Long, bushy	Short-haired	Long, slightly bushy	Long, nearly naked	Long, with terminal tuft	With flat or hollow bristles
Dorsal pelage with thick, hollow quills	0	0	0	0	0	0	+
No.of cheek-teeth (upper/lower)	4–5/4	4–5/4	3/3*	4/4	3–4/3	3–4/3	4/4
Cheek-teeth hyposodont	0	0	0/+	0	0	0	0/+
Infra-orbital foramen large, transmitting muscle	0	0	+	+	+	+	+
Fibula reduced and fused with tibia	0	0	+	+	+	+	0

* Rarely fewer.

ORDER RODENTIA

Rodents

This is by far the largest order of mammals both worldwide and in the Indomalayan Region. Seven families are dealt with here, including three that are primarily Palaearctic or Holarctic and occur only marginally in this region (Table 178).

Rodents are characterized especially by the possession of a single pair of incisors above and below that are chisel-edged and ever-growing, separated by a long diastema (without canines) from a compact row of three to five (rarely one or two) cheek-teeth. Although the incisors are probably adapted primarily to feeding on seeds (including their extraction from hard-shelled nuts) which form the principal component of the diet of many species, they enable many other kinds of tough vegetable food to be utilized, e. g. bark and rhizomes, and in some species are used as forceps to grasp insects.

Rodent classification is unstable at a number of levels: the inter-relationships of the families; the rank to be accorded to some taxa at family or subfamily level; the delimitation of genera (especially where species are numerous as in *Rattus* and related genera); and the limits of individual species, given the frequency of sibling species differing in subtle ways that are masked by intraspecific variation. Many aspects of classification within the Rodentia have been reviewed by contributors to the volume edited by Luckett & Hartenberger (1985), and the phylogeny of the major subgroups was reviewed by Jaeger (1988). The economically important rodents of India were reviewed, with an extensive bibliography, by Roonwal (1987).

The fossil rodents from the productive Siwalik beds of NW India and Pakistan were reviewed by Black (1972), and those of the whole of southern Asia by Jacobs *et al.* (1989).

Family Sciuridae

Non-flying squirrels

Sciuridae Gray, 1821: 304; type genus *Sciurus* Linnaeus.

Throughout Eurasia, Africa and the Americas; c. 220 species of which about 60 occur in the Indomalayan Region, in 17 genera (Tables 179 and 180).

Reviews of classification of the Sciuridae (and Pteromyidae) were made by Gray (1867a), Thomas (1915b) and Pocock (1923b). For a review of the squirrels of Nepal see Mitchell (1979); the ecology of squirrels in Malaya was described by MacKinnon (1978).

A revised classification, based heavily upon cranial characters, was produced by Moore (1959a) who recognized 17 Indomalayan genera. These are used here except for the inclusion of *Glyphotes* in *Callosciurus*. The sequence of Moore's classification is also used except to keep *Exilisciurus* beside *Nannosciurus*, i.e. ignoring Moore's subtribe Hyosciurina, following Heaney (1985); and to associate *Sciurotamias* with *Ratufa*. Moore's tribes, raised to subfamily rank, are used here although not all can be satisfactorily defined.

An earlier revision by Zahn (1942), covering the genera *Ratufa, Funambulus, Tamiops, Lariscus, Menetes, Rhinosciurus, Dremomys* and *Rheithrosciurus* as used here, introduced many radical changes in classification which have been convincingly refuted by subsequent authors. However several of the genera used here are very poorly defined and it is likely that a

Table 179. The genera of unstriped Sciuridae.

	Ratufa	*Sciurotamias* (part)	*Callosciurus* (part)	*Sundasciurus*	*Dremomys*	*Rhinosciurus*	*Nannosciurus*	*Exilisciurus*	*Hyosciurus*	*Prosciurillus*	*Rubrisciurus*
No.of species	4	1	5	9	5	1	1	3	2	4	1
Range	India-Sunda	China	Indochina, Sunda	Sunda	Indochina, Borneo	Sunda	Sunda	Sunda, S Philippines	Sulawesi	Sulawesi	Sulawesi
Head & body (max.)	470	250	270	250	230	240	90	95	240	140	260
Venter rufous	0	0	+/0	+/0	+/0	0	0	0	0	+/0	+
Ear tufts absent (0), <ears (S), >ears (L)	0/S	0	0	0	0	0	0	0/L	0	S/0	S
Hair on nape reversed	0	0	0	0	0	+	0	0	+	0	0
Pairs of teats	3	3/4	2/3	3	3	2	2	2	2	3	3
Muzzle long and pointed	0	0	0	0	+/0	++ (Fig.23)	0	0	++	0	0
Septae in auditory bullae	0	3	1	1	1	1	1	0	0	0/1	0
Bullar septae Y-shaped, forming antero-mesial lobe	–	0	0	+	0	0	0	–	–	0	–
Baculum with accessory blade (Fig.21)	0	0	+	+	+	+	+	+	+	+	+
Blade of baculum serrated	–	–	0	0	0	0	0	+	0	0	0
P^3 absent (0), small and lateral to P^4 (S), large and anterior to P^4 (L)	0	S	L	L	S	L	L	L	L	L	L
Sagittal crest	0	0	0	+/0	0	0	0	0	+	0	+
Upper incisors pro-odont (P), orthodont (Or) or opisthodont (Op)	Or/Op	Op	Or	Or	Or	P (Fig.23)	P	P	Or	P	Or

Table 180. The genera of Sciuridae with dorsal or lateral stripes.

	Sciurotamias (part)	*Funambulus*	*Callosciurus* (part)	*Tamiops*	*Lariscus*	*Menetes*	*Rheithrosciurus*	*Tamias*
No.of species	1	5	10	4	3	1	1	1
Range	China	India, Sri Lanka	Indochina, Sunda	Indochina	Sunda	Indochina	Borneo	China
Stripes on flanks only	+	0	+	0	0	0	+	0
Mid-dorsal stripe light (L) or dark (D)	–	L	–	D*	D/L	D		D
Head & body (max.)	230	200	270	160	230	230	360	150
Ear tufts	0	0	0	0	0	0	++	0
Pairs of teats	?	2	2/3	3	3	3	4	4
Incisors grooved	0	0	0	0	0	0	+	0
Upper incisors pro-odont (P), orthodont (Or) or opisthodont (Op)	Op	Or	Or/P	Or	Or	Or	Op (Fig.26)	Or
Septae in bullae	2	1/2	1	1	1	1	2	2
P^3 absent (0), small (S), or large (L)	0	S	L	S	L	L	0	S
Baculum with accessory blade (Fig.21)	0	0	+	+	+	+	0	0
Sagittal crest	0	0	0	0	+	+	0	0

* In *T.rodolphi* the mid-dorsal dark stripe is bisected by a very faint, thin light stripe.

comprehensive revision, taking into account all the available characters, would make significant changes.

In addition to the 16 Indomalayan genera, one Holarctic genus, *Tamias*, extends marginally into the region, in China, and is included here. Other Palaearctic species that come close to the borders of the Indomalayan Region are *Marmota caudata* which occurs in Pakistan and NW India above the tree-line (above 2900m); *Marmota himalayana* (possibly a race of *M. bobak*) in the Himalayas from Pakistan to Bhutan and in Yunnan and Sichuan, also above the tree-line and mostly above 4000m; and *Spermophilus dauricus* which occurs in the steppes of N China south to about 36°N in Gansu.

All species north of the Isthmus of Kra were reviewed in detail by Moore & Tate (1965). The ecology of diurnal squirrels in Malaya was reviewed by Payne (1979b) and of those on the island of Siberut (Mentawai Is) by Whitten (1981).

Subfamily Ratufinae

Ratufini Moore, 1959a: 167; type genus *Ratufa* Gray.

RATUFA Giant squirrels

Ratufa Gray, 1867a: 273; type species *Sciurus indicus* Erxleben.
Rukaia Gray, 1867a: 275 (*S. macrourus* Pennant).
Eosciurus Trouessart, 1880b;291 (*S. bicolor* Sparrman).

A distinctive genus of four Indomalayan species, placed by Moore (1959a) as the sole member of a tribe, Ratufini (Table 181). The zoogeography of the Indian species was discussed by Moore (1960) and the ecology of the Malayan species by Payne (1979a). All species generally forage and nest high in the canopy of forest trees.

One fossil species has been described:
R. maelongensis Mein *et al.*, 1990; Thailand; Lower Miocene.

Ratufa macroura Sri Lankan giant squirrel

Sciurus macrourus Pennant, 1769: 1; Highlands of Sri Lanka.
S. ceylonicus Erxleben, 1777: 416; Sri Lanka.
S. ceilonensis Boddaert, 1785: 117; Sri Lanka.
S. tennentii Blyth, 1849: 600.

Range (Map 117). S Peninsular India (S of 12°N) and Sri Lanka. In India known to be surviving at only two localities in Tamil Nadu: Sriviliputtur Forest and Chinnar Wildlife Sanctuary (Ramachandran, 1989).

Variation. The following subspecific names have been proposed:
Sciurus m. montanus Kelaart, 1852: 50; Sri Lanka.
R. m. melanochra Thomas & Wroughton, 1915a: 36; Kottawa, S Prov., Sri Lanka.
R. m. dandolena Thomas & Wroughton, 1915a: 36; Wellawaya, Uva, Sri Lanka.
R. m. sinhala Phillips, 1931: 215; Nikawewa, Veddichchai Game Reserve, 10 miles S of Kantalai, E Prov., Sri Lanka.

Three subspecies can be recognized (Moore & Tate, 1965):
R. m. dandolena (including *sinhala*). India and most of Sri Lanka. Dorsal pelage mostly brownish grey; tail with white bands; venter off-white.
R. m. macroura. Highlands of Sri Lanka. Dorsum black; tail black with some white-tipped hairs; venter yellowish; grey line on each flank.
R. m. melanochra. Lowlands of SW Sri Lanka. As *R. m. macroura* but tail all black and no grey line on flank.

Recognition. The white on the lower parts of the hind legs and on the hind feet, contrasting with black on at least the distal parts of the hind digits, is unique to this species.

Remarks. CITES: Appendix II.

Ratufa indica Indian giant squirrel, Malabar squirrel

Sciurus indicus Erxleben, 1777: 420; Bombay Presidency, India.
S. purpureus Zimmermann, 1777: 518; Bombay.
S. maximus Schreber, 1784:pl. 217B (and 1785: 785); Malabar, India.
S. bombayus Boddaert, 1785: 117.
S. malabaricus Scopoli, 1786: 85.
S. elphinstoni Sykes, 1831: 103; Deccan, India.

Table 181. The species of *Ratufa*.

	macroura	*indica*	*bicolor*	*affinis*
Range	S India, Sri Lanka	Pen.India	Himalayas-Java	Malaya, Sumatra, Borneo
Head & body (max.)	410	420	470	380
Dorsum	Black or grey-brown	Reddish brown	Black or brown	Brown*
Ears	Brown or black	Red	Black	Black or brown
Ear tufts	0	+	+/0	0
Hind feet	Pale, digits black	As dorsum	Black, ± white patch	Pale or dark
Nape with transverse pale band	+	+	0(+)	0
Thigh pale	0	0	0	+
Black marks on chin	0	0	+	0
Maxillary tooth-row	12.5–14.4	13.8–16.8	13.7–15.2	12.1–12.9

* May be black with pale feet in Borneo.

Range (Map 117). Most of the forested parts of peninsular India south of about 22°N, mainly in the Western Ghats, with isolates northeast to S Bihar.

Variation. The following subspecific names have been proposed:
Sciurus i. dealbatus Blanford, 1897: 299; Mahal Dangs, Surat.
S. i. bengalensis Blanford, 1897: 303; India.
R. i. superans Ryley, 1913: 436; Wotekolli, S Coorg, 2800ft.
R. i. centralis Ryley, 1913: 436; Bori, Hoshangabad, Central Provincc, 1600ft.

There is considerable variation in colour and pattern, as illustrated in colour by Abdulali & Daniel (1952) who showed eight colour forms. There is a stepped cline in western India from a uniformly pale buff or white form (*dealbatus*) in the north, through forms with increasing amounts of reddish brown to forms with a considerable amount of black in the extreme southwest. Moore & Tate (1965) grouped these in four subspecies, three of which meet with intergradation in the region of the Nilgiri Hills in SW India:
R. i. dealbata. Surat district, NW India (probably isolated from rest of species). Uniform pale creamy buff to almost white.
R. i. indica. W Ghats from about 18°30' to 10°N. Dorsum reddish brown; tail brown proximally, white distally in north but all-brown in south.
R. i. maxima. South of 10°N. Dorsum with black on nape, shoulders and rump; tail black (brown distally in north).
R. i. centralis. Central and eastern India. Dorsum brown with black on fore legs; tail black with pale tip.

Remarks. CITES: Appendix II; IUCN: (E)—*R. i. dealbata, R. i. elphinstonei.*

Ratufa bicolor Black giant squirrel

Sciurus bicolor Sparrman, 1778: 70; Anjer, W Java.
S. javensis Zimmermann, 1780: 342; Java.
S. albiceps Desmarest, 1817: 105; E Java.
S. leschenaultii Desmarest, 1822: 335; Java.
S. humeralis Coulon, 1836: 122; Java.
S. giganteus McClelland, 1839: 150; Assam.
S. macruroides Hodgson, 1849: 775; Bengal—probably a *nomen nudum.*
R. melanopepla Miller, 1900c: 71; Telibon I, S Thailand.
R. anambae Miller, 1900a: 215; Jimaja I, Anamba Is.
R. tiomanensis Miller, 1900a: 216; Tioman I, Malaya.
R. angusticeps Miller, 1901a: 130; Lingung I, N Natuna Is.
R. palliata Miller, 1902a: 147; Indragiri R, E Sumatra.
R. laenata Miller, 1903b: 449, pl. 19; Tuangku I, Banjak Is, W Sumatra.
R. phaeopepla Miller, 1913b: 25; R Balik, S Tenasserim.
R. celaenopepla Miller, 1913b: 26; Domel I, Mergui Is, Burma.
R. felli Thomas & Wroughton, 1916b: 226; Yin, E bank lower Chindwin R, Burma.

Range (Map 117). From E Nepal through Assam, Burma, S & W Yunnan, Thailand and Indochina to Malaya; also on Hainan, Sumatra, Java and Bali, and the following smaller islands: Domel (Mergui); Kut, Samui, Talibon (Thailand); Tioman, Langkawi, Penang (Malaya); Tana Bala, Tuangku (E of Sumatra); Jamaja (Anamba Is); Lingung (N Natuna Is). From 500 to 2000m in Nepal.

Variation. The following subspecific names have been proposed:
Sciurus b. sondaica Müller & Schlegel, 1844: 85; Java.
S. b. indica Müller & Schlegel, 1844: 85; Malacca (not of Erxleben, 1777).
R. gigantea hainana Allen, 1906: 472; Cheteriang, Hainan I.
R. melanopepla fretensis Thomas & Wroughton, 1909b: 535; Langkawi I, Malaya.
R. b. major Miller, 1911b: 28; Mt Gede, Tjibodas, W Java, 5000ft.
R. melanopepla penangensis Robinson & Kloss, 1911b: 245; Penang I, Malaya.
S. m. peninsulae Miller, 1913b: 25; Lay Song Hong, Trang, S Thailand.
R. b. baliensis Thomas, 1913b: 506; Bali.
R. melanopepla dicolorata Robinson & Kloss, 1914a: 227; Samui I, S Thailand.
R. palliata batuana Lyon, 1916a: 445; Tana Bala I, Batu Is, W Sumatra.
R. gigantea lutrina Thomas & Wroughton, 1916b: 226; Tatkon, W bank Chindwin R, Burma.
R. phaeopepla marana Thomas & Wroughton, 1916b: 227; Mt Popa, Burma.

Map **117**. *Ratufa*

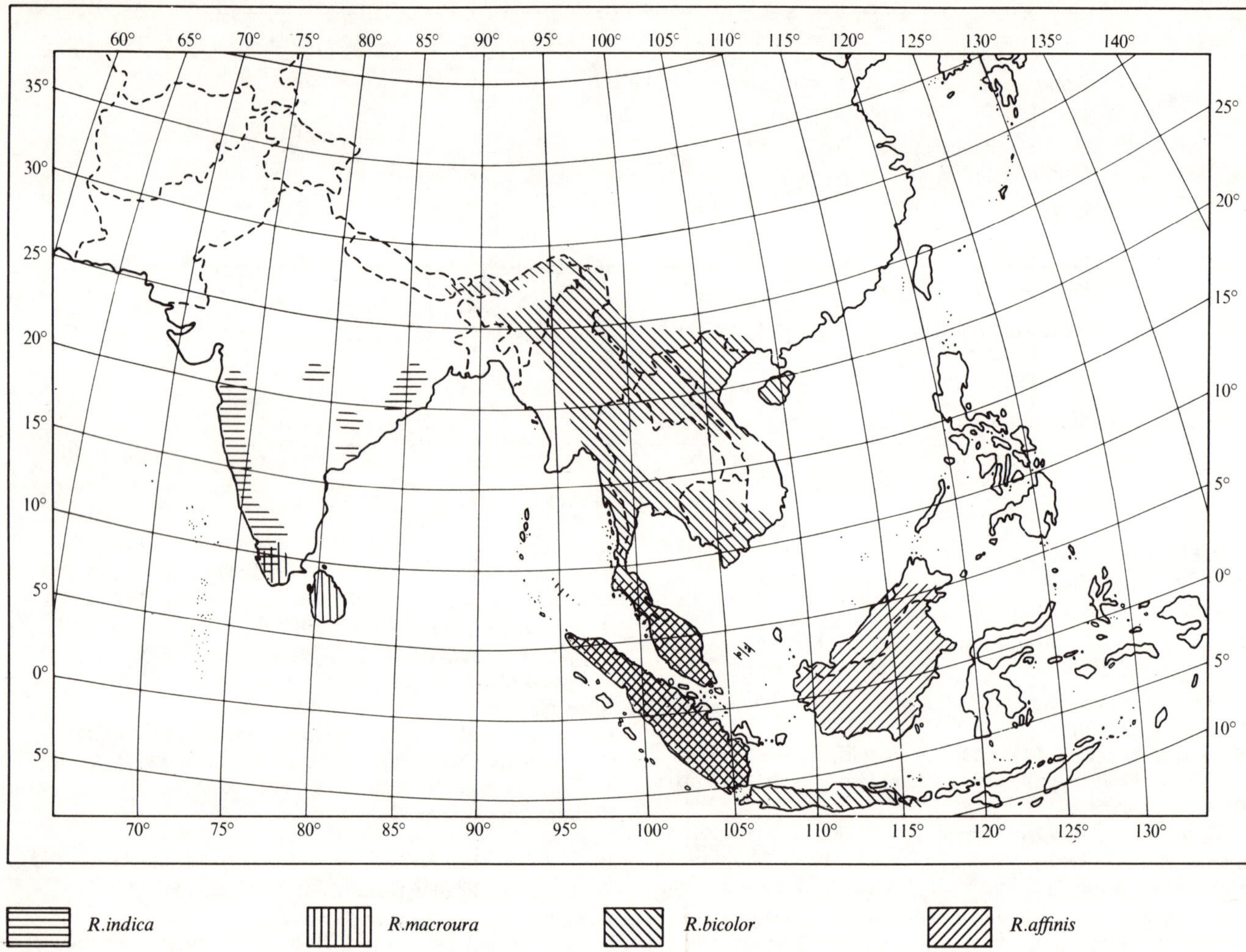

R. melanopepla leucogenys Kloss, 1916b: 43; Lem Ngop, SE Thailand.
R. m. sinus Kloss, 1916b: 44; Kut I, Thailand.
R. melanopepla condorensis Kloss, 1920c: 71; Main I, Condore Is, Vietnam.
R. b. smithi Robinson & Kloss, 1922: 89; Langbian Peaks, S Annam, Vietnam.
R. gigantea stigmosa Thomas, 1923i: 86; Doi Sritepe, Chiengmai, Thailand.

North of the Isthmus of Kra, Moore & Tate (1965) recognized and mapped the following mainland subspecies:
R. b. gigantea. Nepal to Yunnan. Dorsum and tail black; ears tufted; fore legs black on anterior surface; venter pale buff.
R. b. felli. Central Burma (W of R Irrawaddy). Dorsum brown to pale straw or a mixture of these; no ear tufts; buff patch on anterior surface of fore legs.
R. b. hainana. NE Burma, Yunnan, N Thailand, Laos, N Vietnam, Hainan. Dorsum and tail black; ears tufted; venter rusty yellow.
R. b. phaeopepla. C & S Burma, W Thailand. Dorsum and tail uniformly black; ear tufts absent or obscure; fore legs with buff on anterior surface.
R. b. leucogenys. S Thailand, Cambodia. Dorsum uniformly black; no ear tufts; fore *and* hind feet with large pale patches.
R. b. smithi. S Vietnam. Dorsum with black obscured by long pale hair-tips; extensive white on fore legs; slight white patch on hind feet.
R. b. peninsulae. Peninsular Thailand and Malaya. Dorsum and tail black; venter cream to dull orange with dark hair-bases showing through to a variable extent.

To these can be added:
R. b. palliata. Sumatra. Dorsum light brown, darker on head and tail; fore legs with pale patch; venter creamy with dark bases showing.
R. b. bicolor. Java. Dorsum dark brown with white frosting, especially on head; often a buffy central 'saddle'; fore legs with pale patch; tail white to straw with dark hair-bases showing especially at base and tip.

The forms on small islands are not highly differentiated from those on the adjacent mainland, e. g. Penang, Langkawa and Tioman as Malayan Peninsula; Batu Is as Sumatra; Bali similar to Java.

Ecology. The ecology in Malaya was studied in detail by Payne (1979a).

Recognition. Amongst the great variation, the black or very dark hind feet, with at most a small white patch, distinguish this species from all but a few island forms of *R. affinis*, in which the feet are pale brown to white.

Remarks. CITES: Appendix II.

Ratufa affinis Pale giant squirrel

Sciurus affinis Raffles, 1821: 259; Singapore I.
S. hypoleucos Horsfield, 1823; Bencoolen, W Sumatra.
S. auriventer Geoffroy, 1831: 150; Malacca; Java in error.
S. ephippium Müller, 1838: 147; SE Borneo.
S. bicolor bunguranensis Thomas & Hartert, 1894: 658; Bunguran I, N Natuna Is.
S. bicolor nanogigas Thomas & Hartert, 1895: 491; Laut I, N Natuna Is.
R. pyrsonota Miller, 1900c: 75; Trang, S Thailand.
R. notabilis Miller, 1902a: 150; Lingga I.
R. femoralis Miller, 1903b: 447; Tuangku I, Banjak Is, W Sumatra.
R. nigrescens Miller, 1903b: 448; Mansalar I, W Sumatra.
R. insignis Miller, 1903a: 4; Sugi I, Riau Is.
R. conspicua Miller, 1903a: 5; Bintang I, Riau Is.
R. balae Miller, 1903a: 6; Tana Bala I, Batu Is, W Sumatra.
R. piniensis Miller, 1903a: 7; Pinie I, Batu Is, W Sumatra.
R. masae Miller, 1903a: 7; Tana Masa I, Batu Is, W Sumatra.
R. carimonensis Miller, 1906b: 257; Great Karimon I, Riau Is.
R. condurensis Miller, 1906b: 258; Kundur I, Riau Is.
R. confinis Miller, 1906b: 259; Sinkep I, Lingga Is.
R. polia Lyon, 1906a: 585; Billiton I.
R. arusinus Lyon, 1907g: 442; Aru Bay, NE Sumatra.
R. catemana Lyon, 1907g: 443; Kateman R, SE Sumatra.
R. bulana Lyon, 1909b: 482; Bulan I, Riau Is.
R. cothurnata Lyon, 1911a: 93; Mt Palung, near Sukadana, W Borneo.
R. griseicollis Lyon, 1911a: 94; Panebangan I, W Borneo.
R. vittata Lyon, 1911a: 94; Laut I, SE Borneo.
R. vittatula Lyon, 1911a: 95; Sebuku I, SE Borneo.

Range (Map 117). Peninsular Thailand, Malaya, Sumatra, Borneo; small islands of Singapore; Bintan, Sugi, Karimun, Bulan (Riau Is); Lingga, Singkep (Lingga Is); Mansalar, Tanahbala, Tanahmasa, Tuangku, Pini (W Sumatra); Belitung; Panebangan, Laut, Sebuku (Borneo). A record from S Vietnam (Cao, 1984) seems more likely to be based upon misidentified *R. bicolor smithi.*

Variation. The following subspecific names have been proposed:
R. ephippium baramensis Bonhote, 1900c: 496; Baram dist, Sarawak.
R. e. sandakanensis Bonhote, 1900c: 497; Sandakan, N Borneo.
R. e. sirhassenensis Bonhote, 1900c: 498; Sirhassen I, S Natuna Is.
R. polia bancana Lyon, 1906a: 587; Bangka I.
R. a. johorensis Robinson & Kloss, 1911b: 244; Padang Tuan, Segamat, N Johore, Malaya.
R. ephippium lumholtzi Lönnberg, in Lönnberg & Mjöberg, 1925b: 514; Pipoh Bulungan, NE Borneo.
R. e. dulitensis Lönnberg & Mjöberg, 1925b: 514; foot of Mt Dulit, Sarawak.
R. a. banguei Chasen & Kloss, 1932: 22; Banguey I, N Borneo.
R. a. interposita Kloss, 1932: 2; Ulu Gombak, Selangor, Malaya.
R. a. frontalis Kloss, 1932: 2; Taiping, Perak, Malaya.
R. a. klossi Hill, 1960: 47; Lam-ra, Trang, peninsular Thailand.

In peninsular Malaya the dorsal pelage is uniformly brown, becoming paler in the south, and the forms on the islands of Bunguran, Serasan and Laut (Natuna Is) are very similar. On Borneo most animals have the dorsal pelage and tail very dark brown or black but the feet remain entirely pale, a feature not found in any *R. bicolor*. A few are uniformly brown.

In *R. a. banguei* (Banggi I, N Borneo) the pelage is similar to the dark form on mainland Borneo except that the tail has more pronounced pale rings.

Ecology. The ecology in Malaya was studied in detail by Payne (1979).

Remarks. CITES: Appendix II.

SCIUROTAMIAS Chinese rock squirrels

Sciurotamias Miller, 1901d: 23; type species *Sciurus davidianus* Milne-Edwards.
Rupestes Thomas, 1922a: 398 (*R. forresti* Thomas).

China; 2 species (Table 182). This genus was tentatively placed by Moore (1959a) in the tribe Tamiasciurini, associated only with the Nearctic red squirrels, *Tamiasciurus*, based upon the possession of three septae in each auditory bulla. However study of the reproductive tract of *S. davidianus* suggests a closer affinity with *Ratufa* (Callahan & Davis, 1982), although the presence of cheek pouches is a significant difference from *Ratufa*.

Both species are mainly terrestrial, being found on rocky hillsides, often amongst scrub rather than forest.

Table 182. The species of *Sciurotamias*.

	davidianus	*forresti*
Range	Sichuan-NE China	Yunnan
Pale stripe on flank	0	+
Muzzle and cheeks orange brown	0	+
Pale (grey) patch behind ear	+	0
P^3 (small tooth anterior to 4 large cheek-teeth)	+	0

Sciurotamias davidianus Père David's rock squirrel

Sciurus davidianus Milne-Edwards, 1867b: 196; mountains near Peking, China.
Sciurus consobrinus Milne-Edwards, 1874: 305; Moupin, Sichuan.
Dremomys latro Heude, 1898: 55, pl. 12; ? Shantung.
D. collaris Heude, 1898: 55, pl. 12.
D. saltitans Heude, 1898: 55, pl. 12; Hupeh.
Sciurotamias owstoni Allen, J. A.,1909b: 428; Taipai Shan, Shaanxi.

Range (Map 118). Low mountains of China from S Gansu and Hebei in the north to Sichuan, NW Yunnan and extreme eastern Tibet; also an apparently isolated population in SW Guizhou and Guangxi; earlier records from Shandong are probably erroneous.

Variation. One subspecific name has been proposed:
Sciurotamias d. thayeri Allen, G. M.,1912b: 231; Washan, W Sichuan.

The following two contiguous subspecies were recognized and mapped by Moore & Tate (1965) but it is unlikely that they are discrete:
S. d. davidianus. Shanxi and Hebei. Dorsum greyish; venter greyish white, faintly washed with buff.

Map **118**. *Funambulus, Sciurotamias*

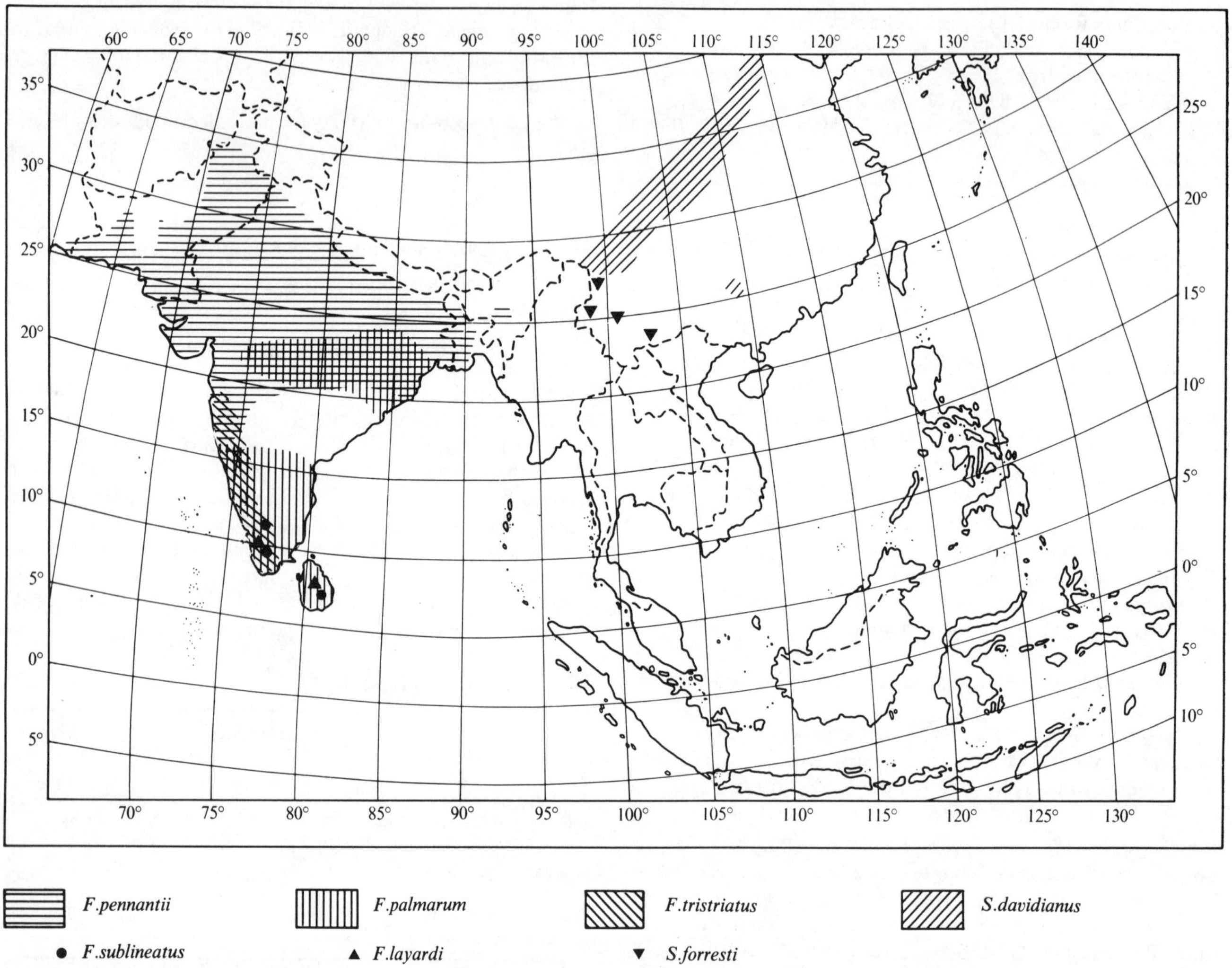

S. d. consobrinus. Shaanxi to Sichuan. Dorsum darker; venter more strongly washed with buff.

There is an apparent cline in the size and shape of the female genital bone, smallest in Sichuan, largest in the northeast (Callahan & Davis, 1982).

Sciurotamias forresti Forrest's rock squirrel

Rupestes forresti Thomas, 1922a: 399; Mekong-Yangtze divide, Yunnan, China, 27°N.

Range (Map 118). Yunnan, China (in W, NW, central and SE— specimens in Kunming Institute of Zoology).

Subfamily Funambulinae

Funambulinae Pocock, 1923b: 237; type genus *Funambulus* Lesson.

FUNAMBULUS Indian striped squirrels

Funambulus Lesson, 1835:pl. 43; type species *F. indicus* Lesson, in *F. palmarum*.
= *Palmista* Gray, 1867a: 279.
Tamiodes Pocock, 1923b: 215 (*Sciurus tristriatus* Waterhouse).
Prasadsciurus Moore & Tate, 1965: 71 (*F. pennanti* Wroughton).

Indian subregion including Sri Lanka; 5 species (Table 183). A distinctive genus that has been linked with the African genera *Funisciurus, Paraxerus* and *Myosciurus* in a tribe Funambulini by Moore (1959a). This contradicted an earlier classification by Zahn (1942) who included *Tamiops* as a subgenus of *Funambulus* on the basis of the striped pelage, ignoring major differences in the baculum and skull. Chromosome data for three species were reviewed by Aswathanarayana (1987).

Funambulus palmarum Indian palm squirrel

Sciurus palmarum Linnaeus, 1766: 86; Madras, India.
S. penicillatus Leach, 1814: 6; Madras.
S. indicus Lesson, 1835:pl. 43. Pre-occupied by *S. indicus* Erxleben, 1777 = *Ratufa indica*.
S. brodiei Blyth, 1849: 602; Point Pedro, N Sri Lanka.
S. kelaarti Layard, in Blyth, 1851a: 166; Hambanlotte, Sri Lanka.
F. robertsoni Wroughton, 1916b: 647; Pachmarhi, Hoshangabad, C Provinces, India.
F. bengalensis Wroughton, 1916a: 648; Haziribagh, Bihar.
F. gossei Wroughton & Davidson, 1919: 730; Kotagiri, Nilgiri Hills, India, 4100–4500ft.

Table 183. The species of *Funambulus*.

	palmarum	*striatus*	*layardi*	*sublineatus*	*pennanti*
Range	Pen.India	W India	SW India, Sri Lanka	SW India, Sri Lanka	N India, Pakistan
Pale stripes	3	3	3	3	5
Stripes bold (B) or obscure (O)	B	B	B	O	B
Median stripe bright (orange)	0	0	+/0	0	0
Venter bright rufous or orange (B); drab and dull (D); pale (P)	P	P	B	D	P
Tail with red mid-ventral line	+	+	+	0	0
Skull: greatest length	<40	>40	<40	<37	
Baculum simple (S) or bifurcate (B)	B	B	(B)	B	S
Chromosomes: 2n	46/48	46			54
FNa	66/68	70			70

Range (Map 118). Peninsular India in two segments: Madhya Pradesh, Bihar and Parulia district of West Bengal (Agrawal & Bhattacharyya, 1987) in the north; Andra Pradesh and Karnataka southwards; and in Sri Lanka.

Variation. The following subspecific names have been proposed:
F. p. comorinus Wroughton, 1905: 411; Trivandrum, Travancore, India.
F. p. favonicus Thomas & Wroughton, 1915b: 39; Udugama, S Prov., Sri Lanka.
F. p. olympius Thomas & Wroughton, 1915b: 41; Urugalla, C Sri Lanka, 1600ft.
F. p. bellaricus Wroughton, 1916b: 647; Vijayanagar, Bellary, India.
F. p. matugamensis Lindsay, 1926 :239; Anasigalla, Matugama, W Prov., Sri Lanka.

Geographical variation is slight. In Sri Lanka the lateral pale stripes, and sometimes the central one also, tend to be a richer orange than in India although the southernmost mainland form, *comorinus*, tends in that direction.

Funambulus tristriatus Jungle striped squirrel

Sciurus tristriatus Waterhouse, 1837: 496; Madras, India.
S. (*Tamias*) *dussumieri* Milne-Edwards, 1867b: 226; Malabar.
F. wroughtoni Ryley, 1913: 437; Srimangala, Coorg, India, 2782ft.
F. thomasi Wroughton & Davidson, 1919: 729; Khandalla, Bombay Presidency, India, 2000ft.

Range (Map 118). Western Ghats of India north to Bombay.

Variation. The following subspecific names have been proposed:
F. t. numarius Wroughton, 1916b: 646; Helwak, Satara Dist, India.
F. t. annandalei Robinson, 1917a: 41; Sasthancotta, W side W Ghats, Travancore, India.

Geographical variation is slight but the pale stripes tend to be wider and lighter in the north.

Funambulus layardi Layard's striped squirrel

Sciurus layardi Blyth, 1849: 602; Ambegamoa Hills, Sri Lanka.

Range (Map 118). Sri Lanka (central and SW); Kerala, SW India.

Variation. Three subspecies can be recognized although it is likely that the two on Sri Lanka intergrade.
F. l. layardi. Central highlands of Sri Lanka. Median stripes narrow (c. 3mm) and pale buff; head as dorsum; venter dull chestnut.
F. l. signatus Thomas, 1924b: 241 (Rakwana, Ratnapura dist, Sri Lanka). SW Sri Lanka. Median stripe wider (c. 5mm) and brighter orange; head and venter as in *F. l. layardi.*
F. l. dravidianus Robinson, 1917a: 42 (Western Ghats, Travancore, India). SW India. Median stripe narrow (c. 2mm), light orange buff; head and cheeks rich rufous orange; venter yellowish.

Funambulus sublineatus Dusky striped squirrel

Sciurus sublineatus Waterhouse, 1838d: 19; Nilgiri Hills, S India.
S. delesserti Gervais, 1841: 171; Nilgiri Hills.
S. trilineatus Kelaart, 1852: 54; Sri Lanka.
= *F. kathleenae* Thomas & Wroughton, 1915b: 38; Kottawa, S Prov., Sri Lanka.
S. palmarum var. *obscura* Pelzeln & Kohl, 1886: 525; uplands of Sri Lanka.

Range (Map 118). SW India and S Sri Lanka.

Variation. Two subspecies can be recognized as follows:
F. s. sublineatus. India. Inner dark stripes narrow (3–5mm).
F. s. obscurus. Sri Lanka. Inner dark stripes wider (6–8mm).

Funambulus pennantii Northern palm squirrel

Funambulus pennantii Wroughton, 1905: 411; Mandvi Taluka, Surat dist, Bombay Presidency, India.

Range (Map 118). Iranian Baluchistan, Pakistan, Nepal (up to 600m) and N India, including most of Rajasthan (Ghose, 1976a), S to c. 16°N in the Western Ghats, east to the Khasi-Junta Hills, Assam (Kurup, 1968). The fact that this conspicuous species had not previously been recorded east of the Brahmaputra suggests that this last might represent an introduction. Introduced to the Andaman Is (Saha, 1980); and to Perth, W Australia (Seebeck, (1989).

Variation. The following subspecific names have been proposed;
F. p. argentescens Wroughton, 1905: 413; Rawalpindi, N Punjab.

F. p. lutescens Wroughton, 1916a: 430; Deesa, Palanpur, Gujarat.

It is doubtful if discrete subspecies can be recognized. In the drier parts of the range, in the northwest (*argentescens*), the pelage is paler and there is a single black band on the hairs of the tail (two elsewhere) (Moore & Tate, 1965).

Recognition. The five pale stripes distinguish this species from all other *Funambulus* although the outer pale stripes are often rather obscure and are separated from the equally pale ventral pelage by a band of only slightly darker pelage. The simple baculum, without a bifurcated tip, is also unique in the genus and led Moore & Tate (1965) to separate *F. pennantii* in a subgenus *Prasadsciurus*.

Subfamily Callosciurinae

Contains most of the Indomalayan genera of Sciuridae (and confined to the Indomalayan Region), characterized by the unique form of the baculum which has a sharp-edged accessory blade attached to its dorsal surface (Fig. 21).

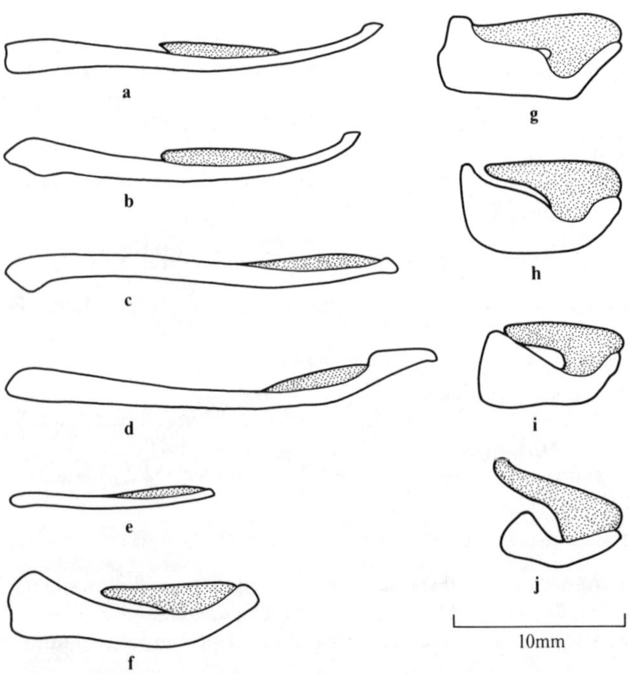

Fig. **21**. *Callosciurus*. Lateral aspect of baculum of (a) *C. erythraeus*; (b) *C.finlaysonii*; (c) *C.prevostii*; (d) *C.notatus*; (e) *C.simus*; (f) *C.melanogaster*; (g) *C.caniceps*; (h) *C.phayrei*; (i) *C.pygerythrus*; (j) *C.inornatus*. The distal end is on the right and the accessory blade is shaded. (e) from Hill (1959a), others original.

CALLOSCIURUS

Callosciurus Gray, 1867a: 277; type species *Sciurus rafflesii* Vigors & Horsfield (in *C. prevostii*).
Baginia Gray, 1867a: 279 (*S. notatus* Boddaert).
Erythrosciurus Gray, 1867a: 285 (*S. ferrugineus* Cuvier, in *C. finlaysonii*).
Heterosciurus Trouessart, 1880b: 292 (*S. erythraeus* Pallas).
Glyphotes Thomas, 1898b: 250 (*G. simus* Thomas).
Tomeutes Thomas, 1915b: 385 (*S. lokroides* Hodgson, in *C. pygerythrus*).
Hessonoglyphotes Moore, 1959b: 5 (*Glyphotes* (*H.*) *canalvus* Moore, in *C. orestes*).

E Himalayas and S China to Greater Sunda Islands and marginally in Philippines and Sulawesi; 15 species (Tables 184 and 185).

The species here included in *Tamiops*, *Sundasciurus* and *Rubrisciurus* have been included in *Callosciurus*. The karyotypes of *C. erythraeus* (as *C. flavimanus*), *C. finlaysonii*, *C. caniceps*, *C. prevostii*, *C. notatus* and *C. nigrovittatus* from the Malay Peninsula are very similar, with 2n=40 and a characteristic pair of 'marker' acrocentric autosomes with satellites in the short arms (Yong *et al.*, 1975; Nadler *et al.*, 1975).

The two cotypes of the name *Macroxus chinensis* Gray, 1867a: 282, originally believed to have been from China, were considered to belong to *Sciurus* (*Guerlinguetus*) from central or South America by Moore & Tate (1959).

Callosciurus erythraeus Pallas's squirrel

Sciurus erythraeus Pallas, 1779a: 377; Garo Hills, Assam (restricted by Moore & Tate, 1965).
S. flavimanus Geoffroy, 1831: 148; Tourane, Annam, Vietnam.
S. atrodorsalis Gray, 1842: 263; Moulmein, Burma (not Bhutan as erroneously given by Gray).
S. castaneoventris Gray, 1842: 263; probably near Canton, China, but given by Ellerman & Morrison-Scott (1951) as Hainan.
S. erythrogaster Blyth, 1843d: 970; Manipur.
S. griseopectus Blyth, 1847a: 873; locality unknown, probably Burma (Moore & Tate, 1965), possibly SE China.

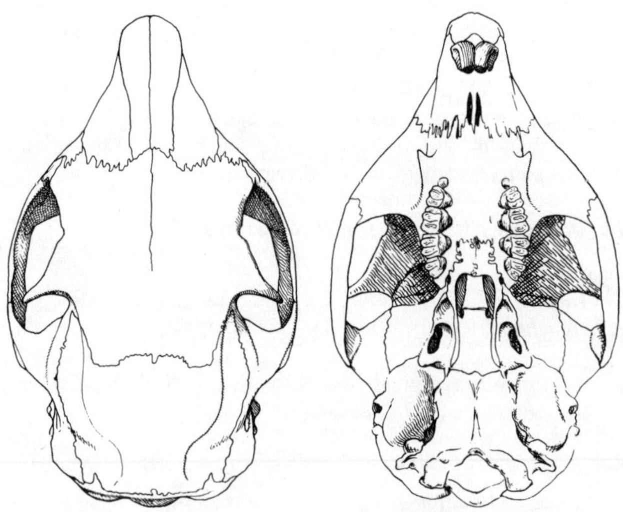

Fig.**22**. *Callosciurus erythraeus*.

S. hyperythrus Blyth, 1856b: 474; ? Sittang Valley, Burma.
S. siamensis Gray, 1860c: 500; Thailand.
S. griseimanus Milne-Edwards, 1867b: 195; near Saigon, Vietnam.
Macroxus leucopus Gray, 1867a: 282; Cambodia.
Macroxus punctatissimus Gray, 1867a: 283; Cachar, Assam.
S. sladeni Anderson, 1871: 139; Thizyain, upper Burma.
S. gordoni Anderson, 1871: 140; Bhamo, upper Burma.
S. styani Thomas, 1894b: 363; between Shanghai and Hangchow, probably Kahing, SE China.
S. thaiwanensis Bonhote, 1901a: 165; Baksa, S Taiwan.

S. rubeculus Miller, 1903a: 22; Trang, S Thailand.
S. haringtoni Thomas, 1905a: 314; Moungkan, upper Burma.
S. tsingtanensis Hilzheimer, 1905b: 298; 'Tsingtao, China', actually Nimrod Sound, near Ningpo, *fide* Allen, 1940: 632.
Herpestes leucurus Hilzheimer, 1905b: 299; China.
= *H. albifer* Hilzheimer, 1906b: 177.
S. vassali Bonhote, 1907b: 9, footnote; Ninh Hoa, Annam, Vietnam.
S. kemmisi Wroughton, 1908a: 491; Katha, upper Irawaddy, Burma.
C. crumpi Wroughton, 1916a: 425; Sedonchen, Sikkim.
C. ferrugineus phanrangis Robinson & Kloss, 1922: 91; Tour Cham, near Phanrang, Annam, Vietnam.
C. caniceps canigenus Howell, 1927a: 81; Hayanhsien, Hangchow Bay, Zhejiang, China.
C. ferrugineus primus Allen & Coolidge, 1940: 157; Mae Wan R, Mt Sonket, N Thailand.

Range (Map 119). From Sikkim, Bhutan and Assam east through Burma to S China, including Sichuan north of the Yangtze, and south to Indochina, Thailand and Malaya (but absent from much of central Thailand occupied by *C. finlaysonii*). Also on Taiwan and Hainan. Introduced from Taiwan to the islands of Izuo, Oshima and Tomogashima, Japan; introduced to S France (Jouanin, 1986). Up to 3600m in NW Yunnan.

Variation. The following subspecies have been described:
Sciurus gordoni var. *intermedia* Anderson, 1879: 241; Dikrang Valley, Assam.
S. e. bhutanensis Bonhote, 1901a: 161; Bhutan.
S. castaneoventris ningpoensis Bonhote, 1901a: 163; Ningpo, Zhejiang, SE China.
S. thaiwanensis roberti Bonhote, 1901a: 166; NW Taiwan.
S. t. centralis Bonhote, 1901a: 166; Lak-ku-li, C Taiwan.
S. e. insularis Allen, 1906: 473; Lei-Mui-Mon, Hainan, China.
S. leucopus fumigatus Bonhote, 1907c: 2; Ninh Hoa, Veitnam.
S. castaneiventris bonhotei Robinson & Wroughton, 1911b: 234; Chinchiensan, Sichuan, China.
S. c. michianus Robinson & Wroughton, 1911b: 234; Meechee, Yunnan.
S. c. haemobaphes Allen, G. M.,1912a: 177; Chiping, SE Yunnan, China.
S. atrodorsalis shanicus Ryley, 1914: 663; Gokteik, N Shan States, Burma, 2133ft.

Table 184. The species of *Callosciurus* north of the Isthmus of Kra.

	erythraeus	*quinquestriatus*	*finlaysonii*	*caniceps*	*inornatus*	*pygerythrus*	*phayrei*
Range	Indochinese	NE Burma, W Yunnan	Thailand etc.	Thailand-Malaya	N Laos, N Vietnam	Nepal-Burma	Burma
Head & body	200–260	c.210	190–230	180–240	220–230	180–230	190–230
Dorsum	Agouti	Agouti	Usually red ± agouti[1]	Agouti	Agouti	Agouti	Agouti
Venter	Red or orange-brown[2]	3 black and 2 white stripes	Dark red-white	Grey-agouti, at least on sides of thorax	Uniform grey	Grey, cream or orange-buff	Buff/orange-brown
Pale patch on hips	0[3]	0	0	0	0	+	0
Lateral dark stripe	0[4]	+	0	0	0	0	+ (diffuse)
Tail with pale ring near base	0	0	+/0	0	0	0	0
Skull: greatest length	48–57	50–52	45–57	44–57	c.52	45–50	46–54
Maxillary tooth-row	8.2–10.2	9.7–10.1	8.0–10.2	8.2–10.5	9.6–9.7	7.9–9.0	7.7–10.3
Baculum long, slender (L), or short and stout (S) (Fig.21)	L		L	S	S	S	S

1. Black, white and piebald morphs also occur.
2. Locally cream in Chindwin Valley, Burma.
3. Present in parts of Vietnam and Laos.
4. Occasionally a very thin one.

Table 185. The species of *Callosciurus* south of the Isthmus of Kra.

	prevostii	*baluensis*	*notatus*	*nigrovittatus*	*adamsi*	*melanogaster*	*orestes*	*simus*
Range	Widespread	Borneo	Widespread	Malaya, Java, Sumatra	Borneo	Mentawai Is	Borneo	Borneo
Head & body	200–270	210–255	150–225	170–240	150–170	195–230	130–170	95–145
Hind foot	45–61	48–55	38–47	40–50	33–40	43, 51	32–40	27–33
Lateral pale stripe	+(0)	+	+	+	+	Faint	+	+
Lateral black stripe	+/0	+	+	+	+	Obscure	+	+
Venter	Reddish	Reddish	Orange-brown	Grey	Orange brown	Black or dull brown	Grey	Orange-brown
Tail: head & body	=	=	<	=	<	<	<	<
Tail banded	+/0	0	(+)	+	+	0	+	+
White or yellow tuft behind ears	0	0	0	0	+	0	+	+
Baculum (Fig. 21)	Slender		Slender			Medium		Slender

NB. See Table 184 for two additional northern species, *C.erythraeus* and *C.caniceps*, that occur in Malaya.

S. sladeni midas Thomas, 1914e: 198; Myitkyina, upper Burma.
S. s. rubex Thomas, 1914e: 198; Yin, lower Chindwin, Burma.
S. s. bartoni Thomas, 1914e: 199; Uyu R, 50 miles E of Homalin, upper Chindwin, Burma.
S. harringtoni solutus Thomas, 1914e: 199; Homalin, upper Chindwin, Burma.
S. e. youngi Robinson & Kloss, 1914a: 224; Mt Tahan, Pahang, Malaya.
C. e. nagarum Thomas & Wroughton, 1916b: 228; Sadiya, Assam.
C. e. crotalius Thomas & Wroughton, 1916b: 229; Hkamti, W bank Chindwin R, Burma.
C. e. kinneari Thomas & Wroughton, 1916b: 229; Tatkon, W bank Chindwin R, Burma.
C. sladeni shortridgei Thomas & Wroughton, 1916b: 232; Hkamti, upper Chindwin, Burma.
C. s. fryanus Thomas & Wroughton, 1916b: 232; Minsin, upper Chindwin, Burma.
C. s. careyi Thomas & Wroughton, 1916b: 233; Tamanthe, upper Chindwin, Burma.
C. s. millardi Thomas & Wroughton, 1916b: 234; Pyaungbyin, 40 miles N of Kindat, Chindwin, Burma.
C. atrodorsalis zimmeensis Robinson & Wroughton, 1916: 91; Chiengmai, N Thailand.
S. e. pranis Kloss, 1916d: 178; Koh Lok, Pran, S Thailand.
S. a. tachin Kloss, 1916d: 178; Tachin, C Thailand.
S. a. thai Kloss, 1917a: 285; Raheng, C Thailand.
C. e. gloveri Thomas, 1921a: 502; Nagchuka, W Sichuan, China.
C. castaneoventris aquilo Wroughton, 1921b: 601; Dibong R, Adiya, Assam, 600ft.
C. e. wellsi Wroughton, 1921c: 775; Shangpung, Jaintia Hills, Assam.
C. flavimanus quantulus Thomas, 1927: 51; Xieng Khouang, Laos.
C. f. contumax Thomas, 1927: 52; Kontoum, S of Dakto, Vietnam.
C. f. dactylinus Thomas, 1927: 52; Dakto, Vietnam.
C. f. pirata Thomas, 1929b: 836; Napi, Laos.
C. e. woodi Harris, 1931: 1; Lungtau, 25 miles E of Nangking, Jiangsu, China.
C. e. hendeei Osgood, 1932: 270; Chapa, Tonkin, Vietnam.
C. f. bolovensis Osgood, 1932: 276; Paksong, Boloven Plateau, Laos.
C. e. nigridorsalis Kuroda, 1935: 281; Riran, Taito, SE Taiwan.
C. sladeni vernayi Carter, 1942: 1; Tapa Hka, N Burma, 26°9'N, 96°16'E.
C. e. cucphuongis Dao, 1965: 1243; Kuk-phuong, near Hanoi, Vietnam.

Map **119**. *Callosciurus* (part)

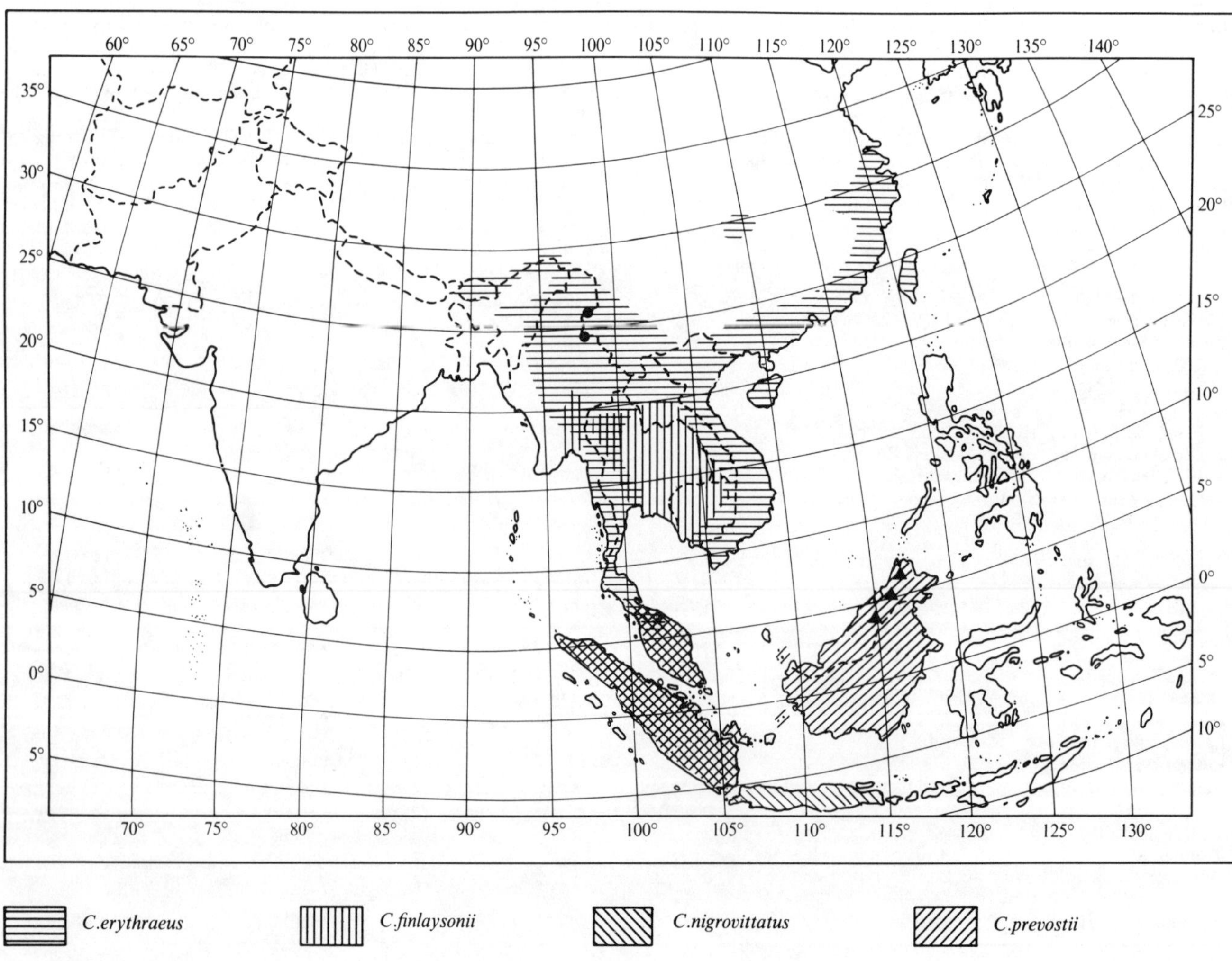

C. e. wuliangshanensis Li & Wang, 1981: 71; Huangcaolin, Mt Wuliangshan, Yunnan, China, 1800m.
C. e. gongshanensis Wang, in Peng & Wang, 1981: 170; Jiulida, E flank of Gaoligong Mts, Gongshan Xian, NW Yunnan.
C. e. qinlingensis Xu & Chen, 1989: 289, 301; Shanyang County, S Qinling Mts, Shaanxi, China, 620m.
C. e. dabshanensis Xu & Chen, 1989: 291, 301; Wanyuan County, S of Dabashan Mts, Sichuan, China, 650m.

Geographical variation is very considerable, with, in some areas, very different colour forms in close proximity, sometimes, but not always, separated by obvious barriers such as the major rivers. Detailed descriptions were given by Moore & Tate (1965) who demonstrated many areas of intergradation and mapped the main divisions. However a full understanding of the complex variation will require mapping on a much more detailed scale.

The major aspects of the variation are as follows:

C. e. erythraeus group. From Malaya through W Thailand, Burma west of the R Chindwin, and Assam to Sikkim, China and N Vietnam, Hainan and Taiwan. Also south in Burma between the Irrawaddy and Salween rivers (*gordoni, shanicus*). Dorsal surface of feet with agouti speckling as on back, with or without darkening towards the digits. Ventral pelage usually reddish brown; in some parts of Assam (*punctatissimus*), in Bhutan (*bhutanensis*) and parts of Yunnan (*gongshanensis*) dark maroon. Ventral pelage sometimes wholly agouti, as in Sikkim (*crumpi*), S Burma (*shanicus* etc.) and in some specimens in Taiwan (*thaiwanensis*), more often with a wedge of agouti dividing the red of the thorax, sometimes continuing as a narrow mid-ventral line on to the abdomen (normal in Malaya—*rubeculus* etc.). (This last feature often varies considerably in one area, e. g. on Taiwan.) Tail as dorsum but sometimes with the tip reddish (*erythraeus*), black (*bhutanensis, erythrogaster, thaiwanensis*), or with numerous white hairs, usually associated with faint transverse banding (*castaneiventris, insularis*). A wide black stripe in the centre of the lower back is present in S Burma (*atrodorsalis, shanicus*). In Tripura and parts of Assam the pelage is almost completely melanic (Agrawal & Bhattacharyya, 1977b).

C. e. flavimanus group. S Vietnam, Cambodia, Laos. Feet paler than dorsum, and usually also paler than venter (e. g. light orange brown when venter is rich red-brown) and sometimes agouti; venter varying from cream to rich red-brown, sometimes with a slight trace of a mid-ventral agouti or paler line; muzzle often slightly paler than back. Often with a pale patch on the side of each thigh, as in *C. pygerythrus*. Intergrades in N Vietnam with a form that is virtually indistinguishable from *C. e. erythraeus* of Assam. For example in a series of six from Phouc Mon, central Vietnam (Field Museum, Chicago) the hind feet range from light orange-brown to almost entirely grey.

C. e. sladeni group. Between the rivers Chindwin and Irrawaddy in Burma. Feet coloured as venter, not agouti nor darkened distally. Muzzle and often forehead also non-agouti, coloured as venter. An extremely variable group: in the north of the area (e. g. *vernayi*) and in the south (e. g. *kemmisi*) the dorsal pelage is agouti and the venter dark red-brown as in most *C. erythraeus*, but between these, on the east bank of the R Chindwin, there is a very variable group of forms that are predominantly buff, cream or straw-coloured, usually with a narrow dark line on the flank (*harringtoni, fryanus, careyi, solutus*). Some otherwise dark individuals have pale straw-coloured tails (*millardi*). However in all these the muzzle and feet are detectably closer to the ventral rather than the dorsal colour.

This group has been treated as a separate species, *C. sladeni*, but Moore & Tate (1965) have shown that in the north they intergrade with *C. e. intermedius* which is a typical member of the *erythraeus* group of races. The extraordinary variation in this form was interpreted by McCann (1978) as cryptic adaptation to different forest types, on the basis of his field observations. Coloured illustrations of seven forms from the Chindwin Valley were published by Thomas & Wroughton (1916b), and their variation and distribution along the valley were further discussed by Mackenzie (1951).

C. e. styani group. E China. Feet agouti but not darkened. Venter pale cream with greyish bases, or dull orange-brown. Dorsum rather pale and greyish. Small. Locally variable, with southern form (*ningpoensis*) much darker below than *styani* from near Shanghai.

The insular forms, on Taiwan and Hainan, are not highly differentiated. On Taiwan they are very variable, with the venter ranging from all-agouti (as in the type of *thaiwanensis*) to maroon with (type of *centralis*) or without (type of *roberti*) a central agouti stripe, the last two forms being the most usual. A melanic individual has been recorded in Taiwan (as *C. flavimanus*) by Jones & Tseng (1968). The form on Hainan (*insularis*) is scarcely separable from that on the mainland of N Vietnam.

Variation in the western forms has also been reviewed by Chakraborty (1985), and the characters of the Chinese forms were tabulated by Xu & Chen (1989).

Recognition. In spite of the enormous variation most forms can be recognized as *C. erythraeus* by the combination of reddish venter (albeit varying from dark maroon to pale creamy buff) and olive- brown agouti dorsum. Although some forms have the entire ventral pelage agouti this is always tinged with red or yellow and is never pure grey on the sides of the thorax as in *C. caniceps*. The dorsum is never entirely reddish brown without agouti speckling as in some *C. finlaysonii*.

Remarks. The forms included here have usually been divided between two species, *C. erythraeus* and *C. flavimanus*, but in very different ways. Ellerman & Morrison-Scott (1951) limited *C. flavimanus* to the forms with pale feet in Indochina, and between the Chindwin and Irrawaddy (*sladeni* etc.), but they included in *C. flavimanus* the form *phayrei* here given specific rank. Moore & Tate (1965) limited *C. erythraeus* to forms west of the Irrawaddy (including *sladeni* etc.) and allocated all others, including those from China, Indochina, Thailand and Malaya, to *C. flavimanus*. They were unable to diagnose these two species, but based the distinction entirely upon the juxtaposition, in NE Burma and Yunnan, of two very dissimilar forms, *quinquestriatus* to the east of the upper Irrawaddy/Nmai, and *C. e. intermedius* to the west of these rivers, and the supposition that the river could not act as an effective barrier. This led to the anomaly of separating, in *C. erythraeus* and *C. flavimanus*, forms that were not only virtually indistinguishable, e. g. from Assam (*C. e. erythraeus s. s.*) and from N Vietnam, but were connected by a chain of very similar forms, e. g. *intermedius, michianus*.

In view of the evidence of intergradation presented by Moore & Tate (1965), e. g. between *flavimanus* and *hendeei* in Vietnam, and between *sladeni* and *intermedius* in Burma, there seems no alternative but to include *flavimanus, sladeni* and related forms in one species, *C. erythraeus.*

Chromosomes of specimens in Yunnan (recorded as *C. erythraeus castaneoventris* and *C. e. michianus*) were reported as 2n=40, FN=72 (Wang *et al.*, 1980).

Callosciurus quinquestriatus

Sciurus quinquestriatus Anderson, 1871: 142, pl. x; Ponsee, Kakhyen Hills, Yunnan/Burma border.
S. beebei Allen, J. A.,1911: 338; Sansi Gorge, Burma/Yunnan border (Sarawak in error).

Range (Map 119). Between the upper Irrawaddy/Nmai (NE Burma) and the upper Salween (Yunnan).

Variation. The following subspecific names have been proposed:
C. q. imarius Thomas, 1926b: 640; W flank of Mt Imaw Bum, Kachin, N Burma, 7000ft.
C. q. sylvester Thomas, 1926b: 641; Schweli-Salween divide, W Yunnan, 9000ft.

While most specimens have the pale ventral bands pure white, a substantial minority have these faintly tinged with pinkish buff and a small minority more strongly so but still very different from the deep orange-brown of the sympatric *C. erythraeus gordoni.*

Remarks. This form was treated as a subspecies of *C. erythraeus* (with which it appeared to be parapatric) by Moore & Tate (1965) on the ground of intergradation with *C. e. gordoni* to the south. However the specimens that led to that conclusion have been examined and are very much closer to *C. e. gordoni* than to *C. quinquestriatus.* The latter is so distinctive, with three bold longitudinal ventral black stripes separated by white stripes, at most lightly tinged with orange-brown, that it seems best to retain it as a species until intergradation can be demonstrated. However the slight indication of a '*quinquestriatus*' pattern in some *C. e. gordoni* does suggest that *C. quinquestriatus* is a local derivative of *C. erythraeus.* Specimens in the Kunming Institute also show a wide gulf between those *C. quinquestriatus* with most colour on the venter and the few *C. e. gordoni* that show a slight trace of dark bands (without dilution of the orange-brown colour). The two forms are sympatric over a considerable area of W Yunnan west of the Salween R between 24 and 26°N, but with *C. quinquestriatus* forming a small minority (Wang, Y-x, pers. comm.).

Callosciurus finlaysonii Finlayson's squirrel, Variable squirrel

Sciurus finlaysonii Horsfield, 1823:pt 7; Koh Chang I, Thailand.
S. ferrugineus Cuvier, F.,1829:pl. 238; Pegu, Burma.
S. keraudrenii Reynaud, in Lesson, 1830b:pl. 1; Burma.
S. cinnamomeus Temminck, 1853: 250; Cambodia.
S. splendens Gray, 1861: 137; S Cambodia.
S. bocourtii Milne-Edwards, 1867b: 193; Ayutha, Thailand.
S. leucogaster Milne-Edwards, 1867b: 193 (not of Cuvier, 1831).
S. germanii Milne-Edwards, 1867b: 193; Condor I, Vietnam.
S. harmandi Milne-Edwards, 1877: 8; Phu Quoc I, Cambodia.
S. leucocephalus Bonhote, 1901c: 54; Cheimat, R Menam, Thailand.
S. floweri Bonhote, 1901d: 455; Klong Morn, near Bangkok, Thailand.
S. nox Wroughton, 1908b: 397; coast SE of Bangkok, Thailand.
S. albivexilli Kloss, 1916: 47; Koh Kut I, SE Thailand.
C. cockerelli Thomas, 1928a: 100; Nan, N Thailand.

Range (Map 119). Thailand, Cambodia and Laos; an isolated population in Burma, east of the lower Irrawaddy, if *ferrugineus* is conspecific.

Variation. The following nominal subspecies have been described:
Sciurus bocourti sinistralis Wroughton, 1908b: 399; Pichit, Menam R, Thailand.
S. b. dextralis Wroughton, 1908b: 400; Kampeng, lower Meping Valley, Thailand.
S. b. lylei Wroughton, 1908b: 401; Chengmai, Thailand.
S. f. portus Kloss, 1915a: 158; Koh Chang (I), Thailand.
S. f. folletti Kloss, 1915a: 159; Koh Phai (I), Thailand.
C. f. tachardi Robinson, 1916a: 36; R Mee Nan, 30 miles NE of Utaradit, N Thailand.
S. ferrugineus frandseni Kloss, 1916b: 46; Koh Chang (I), Thailand.
S. f. trotteri Kloss, 1916d: 178; Koh Lan (I), Thailand.
S. bocourti grutei Gyldenstolpe, 1917: 37; Bang Hue Pong, N Thailand.
C. f. prachin Kloss, 1920b: 103; Krabin, C Thailand.
C. f. rajasima Kloss, 1920b: 103; Lat Bua Kao, E Thailand.
C. ferr. williamsoni Robinson & Kloss, 1922: 90; Xieng Khouang, Mekong R, Laos.
C. ferr. herberti Robinson & Kloss, 1922: 90; Hup Bon, near Sriracha, SE Thailand.
C. ferr. pierrei Robinson & Kloss, 1922: 91; Phu Quoc I, Cambodia.
C. ferr. annellatus Thomas, 1929b: 839; Angkor, Cambodia.
C. ferr. menamicus Thomas, 1929b: 839; Nan, N Thailand.
C. f. boonsongi Moore & Tate, 1965: 184; Phu Phan, Sakon Nakhon Prov., E Thailand.

Regional variation is so extreme, from all-white to all-red to all-black and many other variants, that it is very doubtful whether all these forms constitute a discrete species. Although much of the variation has a regional basis, there is also polymorphism within some populations. The following subspecies were recognized (and diagnosed and mapped) by Moore & Tate (1965), but only a brief indication of their main characteristics is given here. The Burmese *ferrugineus* was given specific rank by Moore & Tate (1965) but is included here as a subspecies.

Mainland.
C. f. nox. A small area of coastal Thailand, near Cape Liant, SE of Bangkok. Entirely black.
C. f. cinnamomeus (including *splendens* and *herberti*). SE Thailand and SW Cambodia. All red or with a small amount of olive agouti on head and limbs.
C. f. annellatus. C & E Cambodia and S Laos. Mainly reddish brown with a cream band near the base of the tail.
C. f. williamsoni. Laos. Red to orange above, sharply demarcated from chestnut below (as in typical *C. erythraeus*); tail maroon.
C. f. boonsongi. NE Thailand. Mostly black or very dark, with white rim on ear and grey venter, but a small proportion of creamy white and red individuals occur.

C. f. bocourtii (including *leucogaster, leucocephalus, floweri, tachardi, prachin, rajasima* and *cockerelli*). C Thailand. Dorsum black, rest, including head, white or cream; entirely cream in some peripheral populations. Many intermediates occur, with a mixture of white and black hairs on the dorsum and tail.
C. f. sinistralis (including *dextralis, lylei* and *grulei*). C to NW Thailand. Mostly red but dorsum reddish agouti and a cream band on the tail as in *C. f. annellatus*.
C. f. menamicus. NC Thailand. All red with buffy white tip to tail.
C. f. ferrugineus. Burma. Doubtfully distinguishable from *C. f. menamicus* but geographically isolated.

Small islands.
C. f. finlaysonii. Si Chang I, Thailand. All white, tinged yellow especially on back.
C. f. folletti. Phai I, Thailand. White with grey hair-bases showing.
C. f. trotteri. Lan I, Thailand. Grey with white tail and blackish feet.
C. f. frandseni. Chang I, Thailand. Red with grey flanks.
C. f. albivexilli. Tut I, Thailand. Black with tip of tail white.
C. f. harmandi (including *pierrei*). Phu Quoc, Vietnam. Dark reddish brown above, light orange-red below, tail greyish white.
C. f. germaini. Condor I, Vietnam. All-black.

Recognition. This species is so variable that no simple diagnosis applies. Those with agouti dorsum are usually very grey with white or cream ventral pelage (*bocourti*), but some forms, e. g. of *sinistralis*, can only be distinguished from the adjacent *C. erythraeus atrodorsalis* by the pale base of the tail. Reddish forms never have the mid-dorsal zone agouti although the flanks may be agouti (*cockerelli*). White or near-white forms always have the venter the same or paler than the dorsum, not darker cream or orange as in the pale forms of *C. erythraeus*. Black forms are usually a uniform shiny black without any hint of the lateral bands found in other dark forms such as *C. melanogaster*.

Remarks. The concept of this species used here is that of Moore & Tate (1965) except for the inclusion of the Burmese form *ferrugineus* (including *keraudrenii*). This form was treated as a separate species by Moore & Tate solely on the grounds of its geographical separation and to accord with their hypothesis of the mode of speciation. However it is virtually indistinguishable from *C. f. menamicus*, from many *C. f. cinnamomeus* and from the type of *pierrei*, which are widely scattered within the range of *C. finlaysonii*.

This species is predominantly parapatric with *C. erythraeus*. It appears to be marginally sympatric with *C. erythraeus* in some areas, e. g. *C. f. sinistralis* with *C. e. zimmeensis* in N Thailand, but is suspected of interbreeding with it elsewhere, e. g. *C. f. sinistralis* with *C. e. thai* in W Thailand (Moore & Tate, 1965). There is also at least one locality in SW Laos (Ban Nong Bua) where individuals with non-agouti red dorsum (cf. *C. f. williamsoni*) appear to coexist with others with agouti dorsum and paler feet and muzzle referable to *C. e. flavimanus*, with intermediate individuals from nearby localities (specimens in Singapore collection). More detailed study in the field in areas of suspected sympatry or intergradation will be necessary in order to establish whether the forms here included in *C. finlaysonii* are better treated as such or as further subspecies of *C. erythraeus*.

Callosciurus caniceps Grey-bellied squirrel

Sciurus caniceps Gray, 1842: 263; N Tenasserim, Burma (originally Bhutan in error).
S. chrysonotis Blyth, 1847a: 873; Amherst, Tenasserim, Burma.
S. bimaculatus Temminck, 1853: 251; Malayan peninsula.
S. concolor Blyth, 1856b: 474; Malacca, Malaya.
S. epomophorus Bonhote, 1901b: 272; Junk Seylon (= Phuket I), Thailand.
S. lancavensis Miller, 1903a: 16; Langkawi I, Straits of Malacca.
S. adangensis Miller, 1903a: 17; Adang I, Butung Is.
S. sullivanus Miller, 1903a: 17; Sullivan I, Mergui Is, Burma.
S. domelicus Miller, 1903a: 18; Domel I, Mergui Is.
S. bentincanus Miller, 1903a: 19; Bentinck I, Mergui Is.
S. matthaeus Miller, 1903a: 19; St Matthews I, Mergui Is.
S. lucas Miller, 1903a: 20; St Luke I, Mergui Is.
S. casensis Miller, 1903a: 20; Chance I, Mergui Is.
S. altinsularis Miller, 1903a: 21; High I, Mergui Is.
S. helgei Gyldenstolpe, 1917: 34; S of Koh Lak, SW Thailand.
C. erubescens Cabrera, 1917: 518; Selangor, Malaya.
C. moheius Thomas & Robinson, 1921: 122; Pulau Mohea N, SW Thailand.

Range (Map 120). Throughout most of Thailand, peninsular Burma and Malaya and the adjacent small islands, from the Mergui Is to Langkawi on the west.

Variation. The following nominal subspecies have been described:
Sciurus epomophorus davisoni Bonhote, 1901b: 273; Bankason, Tenasserim, Burma.
S. concolor terutavensis Thomas & Wroughton, 1909b: 535; Terutau I, S Thailand.
S. epomophorus fluminalis Robinson & Wroughton, 1911b: 233; Meping Rapids, N Thailand.
S. e. milleri Robinson & Wroughton, 1911b: 233; Trang, S Thailand.
S. concolor fallax Robinson & Kloss, 1914a: 225; Koh Pennan (I), S Thailand.
S. c. samuiensis Robinson & Kloss, 1914a: 226; Koh Samui, S Thailand.
S. epomophorus inexpectatus Kloss, 1916d: 178; Koh Lak, Pran, SW Thailand.
C. e. nakanus Thomas & Robinson, 1921: 119; Koh Naka, S Thailand.
C. e. panjius Thomas & Robinson, 1921: 119; Panjang I, S Thailand.
C. e. mapravis Thomas & Robinson, 1921: 120; Koh Maprau, S Thailand.
C. e. panjioli Thomas & Robinson, 1921: 120; Panjong Anak I, S Thailand.
C. e. tacopius Thomas & Robinson, 1921: 121;Koh Rah, SW Thailand.
C. e. pipidonis Thomas & Robinson, 1921: 121; Koh Pipidon, S Thailand.
C. concolor telibius Thomas & Robinson, 1921: 121; Pulau Telibun, S Thailand.
C. moheius mohillius Thomas & Robinson, 1921: 122; Pulau Mohea S, SW Thailand.
C. epomophorus tabaudius Thomas, 1922c: 1067; Tavoy I, Mergui Is.
C. e. hastilis Thomas, 1923e: 377; Hastings I, Mergui Is.
C. c. helvus Shamel, 1930: 72; Koh Tao, S Thailand.

Map **120**. *Callosciurus* (part)

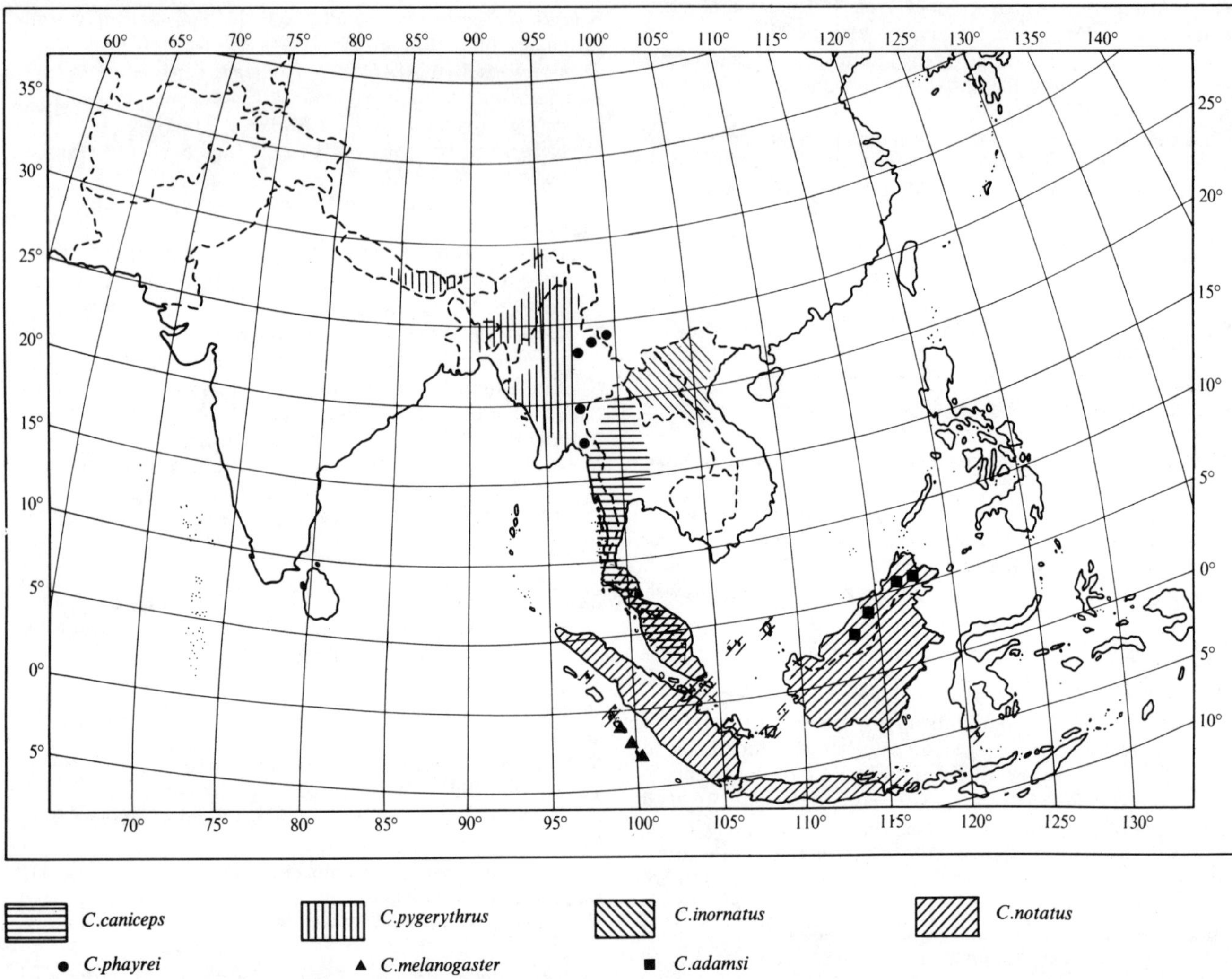

Variation is slight compared with most species of *Callosciurus*. Three subspecies can be recognized on the mainland although they intergrade (Moore & Tate, 1965). From north to south these are:

C. c. caniceps. Main part of Thailand and adjacent part of Burma. Back and flanks yellowish or orange-brown during drier season (Nov.–May), grey in wet season (June—Oct.); tail-tip black.

C. c. bimaculatus (including *davisoni, milleri*). Peninsular Burma and Thailand, south to c. 7°N. Back grey, flanks reddish, tail-tip black.

C. c. concolor (including *erubescens*). Malaya south of c. 7°N. Back grey with reddish tinge, flanks grey, no black tail-tip.

The island forms are generally similar to those on the adjacent mainland but the following are more distinctive (north to south):

Domel, Bentinck and Kisseriang Is, Mergui Archipelago. *C. c. domelicus*. Dark all over—deep red on flanks and dark grey below.

Chance I, W coast Thailand. *C. c. casensis*. Pale grey below, orange-brown on flanks strongly marked.

Butang, Teratau and Mohea Is. *C. c. adangensis*. Mid-ventral line of tail bright buff; otherwise intermediate between mainland *C. c. concolor* and *C. c. bimaculatus*, with pale venter of former and black tail-tip of latter.

Recognition. Although very similar to some forms of *C. erythraeus*, this species can be recognized by the pure grey venter, at least on either side of the mid-line on the thorax. The mid-ventral line is often marked by an inconspicuous stripe of darker grey or brown with agouti speckling, and the abdomen is often suffused with reddish brown. The feet, and usually the legs, are grey, even when the dorsum is yellowish brown. The baculum is totally different from that of *C. erythraeus* and *C. finlaysonii*.

Remarks. The forms *crumpi* (Sikkim), *thaiwanensis* (Taiwan) and *shanicus* (E Burma) were included in *C. caniceps* by Ellerman & Morrison-Scott (1951). Although these all have the venter agouti it is tinged with red or yellow, and these forms intergrade with more normally coloured, red-bellied forms of *C. erythraeus* to which species they are here allocated.

Callosciurus inornatus

Macroxus inornatus Gray, 1867a: 282; Laos.

C. imitator Thomas, 1925: 502; Thai-nien, Tonkin, Vietnam.

Range (Map 120). N Laos, N Vietnam and S Yunnan (Stroganova, 1957), northwest to the R Mekong; sympatric with the larger *C. erythraeus hendeei.*

Remarks. This form was included as an isolated eastern subspecies of *C. pygerythrus* by Ellerman & Morrison-Scott (1951) on the basis of its overall similarity to the grey-bellied race *C. p. stevensi.* However *C. inornatus* completely lacks the light patches on the hips characteristic of *C. pygerythrus* (200 skins of *C. inornatus* examined). It is probably most closely related to *C. caniceps* with which it is allopatric, separated by the Mekong River (Moore & Tate, 1965).

Callosciurus pygerythrus Irrawaddy squirrel

Sciurus pygerythrus Geoffroy, 1831: 145, pl. 7; Pegu, Burma.
S. lokroides Hodgson, 1836: 232; Nepal.
S. blythii Tytler, 1854: 172; Dacca, Bangladesh.
Macroxus similis Gray, 1867a: 281; Sikkim.
S. stevensi Thomas, 1908f: 246; Beni-Chang, Abor-Miri Hills, N Assam, 4000ft.

Range (Map 120). Central Nepal (west at least to 85°E) to Burma north and west of the R Sittang; Yunnan.

Variation. The following nominal subspecies have been described:
Sciurus lokroides mearsi Bonhote, 1906c: 338; Chinbyit, W of Chindwin R, Burma.
S. p. janetta Thomas, 1914e: 203; Mandalay, Burma, 200ft.
Tomeutes similis owensi Thomas & Wroughton, 1916b: 236; Minsiu, E bank of upper Chindwin R, Burma.
T. mearsi virgo Thomas & Wroughton, 1916c: 419; Tatkon, upper Chindwin, Burma.
T. mearsi bellona Thomas & Wroughton, 1916c: 419; Kin, W bank Chindwin, Burma.

Seven subspecies were recognised by Moore & Tate (1965) but these are much less differentiated than in other species of *Callosciurus* in the region. In most areas there is a seasonal alternation between a brighter summer (wet-season) pelage, with a light patch on the side of each hip, and a duller winter pelage without such patches.

C. p. pygerythrus. Between the Sittang and lower Irrawaddy. Venter cinnamon-rufous, buff anteriorly.
C. p. janetta. Between Irrawaddy and lower Chindwin. Dorsum greyer, venter light buff; patch on hip white and confluent with light venter, present all year.
C. p. owensi. Between Irrawaddy and upper Chindwin. More rufous in summer; patch on hip buff, separated from ventral pelage, seasonal.
C. p. mearsi (including *bellona* and *virgo*). Burma W of Chindwin, Manipur and Nagaland. Venter buff; buff eye-ring; patch on hip white, seasonal.
C. p. stevensi. W of upper Chindwin, NE Assam. Venter grey, patch on hip buff, seasonal.
C. p. blythi. Assam and Bangladesh S of Brahmaputra. Venter pale greyish buff, patch on hip white or cream, seasonal.
C. p. lokroides (including *assamensis* and *similis*). Bhutan to Nepal. Venter greyish with a little buff on abdomen and axillae; patch on hip buffy orange, seasonal.

Recognition. Smaller than the sympatric *C. erythraeus* and distinguished also by the characteristic light patch on the side of each hip. This is sometimes separated from the pale ventral pelage and sometimes confluent with it but forming a noticeable blip in the otherwise straight line of demarcation between dorsal and ventral pelage. It is only present seasonally in most subspecies.

Remarks. The form *inornatus*, allocated to this species by Ellerman & Morrison-Scott (1951), is here given specific rank.

Callosciurus phayrei

Sciurus phayrei Blyth, 1856b: 472,476; Martaban, Burma.
S. blanfordii Blyth, 1862b: 333; Ava, upper Burma.
C. griseimanus heinrichi Tate, 1954: 1; Maymyo, Mandalay, Burma, 800m.

Range (Map 120). Burma, between the lower Sittang and Salween rivers and in SW Yunnan (specimens in Kunming Institute). Marginally sympatric with *C. finlaysonii ferrugineus* in the Sittang Valley.

Recognition. Unique amongst *Callosciurus* from north of Malaya in having a diffuse, blackish stripe, often up to 1cm wide, between the agouti dorsum and the pale buff to orange-brown venter. In other respects it is very similar to some of the pale-bellied, pale-footed forms of *C. erythraeus*, e. g. *shortridgei* and *contumax*. The black tail-tip is constant (as in northern *C. caniceps*).

Remarks. This form was included in *C. flavimanus* by Ellerman & Morrison-Scott (1951), but I agree with Moore & Tate (1965) in treating it as a distinct species, most closely related to *C. pygerythrus*. The baculum agrees with those of *C. pygerythrus* and *C. caniceps* and is quite different from that of *C. erythraeus*.

Callosciurus prevostii Prevost's squirrel.

Sciurus prevostii Desmarest, 1822: 335; Malacca, Malaya.
S. rafflesii Vigors & Horsfield, 1828: 113, pl. iv; Sumatra, probably Bencoolen.
? *S. redimitus* Boon Mesch, 1829: 243; Sumatra.
S. rufogularis Gray, 1842: 263; 'China', corrected to Sumatra (Bonhote, 1901e).
S. rufonigra Gray, 1842: 263; 'India', restricted to Labuan I, Borneo by Chasen & Kloss (1925).
S. atricapillus Schlegel, 1863: 27; Kapuas R, W Borneo.
S. piceus Peters, 1866a: 429; N Sumatra—Tenasserim in error.
Macroxus sarawakensis Gray, 1867a: 277; Sarawak.
M. pluto Gray, 1867a: 283; Sarawak, more likely from British N Borneo (Chasen, 1940a).
S. mimellus Miller, 1900a: 218; Wai I, Tambelan Is, W of Borneo.
S. mimiculus Miller, 1900a: 219; St Barbe I, Tambelan Is, S China Sea.
S. caroli Bonhote, 1901e: 173; Marudi R, Baram, Sarawak.
S. melanops Miller, 1902a: 151; Indragiri R, E Sumatra.
S. erebus Miller, 1903b: 456; Tapanuli Bay, NW Sumatra.
S. carimatae Miller, 1906a: 57; Karimata I, SW Borneo.
S. condurensis Miller, 1906b: 260; Kundur I, Riau Is.
S. carimonensis Miller, 1906b: 261; Great Karimon I, Riau Is.
S. mendanauus Lyon, 1906a: 589; Mendanau I, W of Billiton, Sumatra.
S. sanggaus Lyon, 1907d: 554; Sanggau, W Borneo.
S. prosperinae Lyon, 1907c: 275; Temaju I, W Borneo.
S. nyx Lyon 1908b: 658; Rupat I, E Sumatra.
S. armalis Lyon, 1911a: 82; Panebangan I, W Borneo.
S. pelapius Lyon, 1911a: 82; Pelapis I, W Borneo.

Range (Map 119). Peninsular Thailand, Malaya, Sumatra, Borneo; N Sulawesi (probably introduced); the small islands of Kundur, Gt Karimon (Riau Is); Rupat, Penjalei, Bangka, Mendanau (Sumatra); Wai, St Barbe (Tambelan Is); Serasan (S Natuna Is); Karimata, Sanggau, Temaju, Panebangan, Pelapis, Labuan, Bangi, Balembangan (Borneo).

Variation. The following nominal subspecies have been described:
Sciurus rafflesii var. *indica* Müller & Schlegel, 1844: 86; Malacca, Malaya.
S. r. var. *borneoensis* Müller & Schlegel, 1844: 86; Pontianak, W Borneo.
S. p. var. *sumatranus* Schlegel, 1863: 25; Sumatra.
S. p. var. *bangkanus* Schlegel, 1863: 26; Bangka I.
S. p. humei Bonhote, 1901e: 170; Klang, Selangor, Malaya.
S. p. kuchingensis Bonhote, 1901e: 170; Kuching, Sarawak.
S. p. navigator Bonhote, 1901e: 171; Sirhassen I, S Natuna Is.
S. caroli griseicauda Bonhote, 1901e: 174; Mt Kalulong, Baram, Sarawak, 2000ft.
S. p. harrisoni Stone & Rehn, 1902: 132; Mt Sugi, Lampongs, S Sumatra.
S. borneoensis palustris Lyon, 1907d: 553; N bank Kupuas R, W Borneo.
S. melanops penialius Lyon, 1908b: 637; Penjalei I, E Sumatra.
S. p. wrayi Kloss, 1910: 148; Kuala Lipis, Pahang, Malaya.
S. atricapillus atrox Miller, 1913b: 23; Mt Talisaian, SE Borneo.
S. p. caedis Chasen & Kloss, 1932: 25; Balambangan I, N Borneo.
S. p. banksi Chasen, 1933b: 195; Baram dist, Sarawak, Borneo.
C. p. waringensis Sody, 1949a: 103; Riam, Kotawaringin, SW Borneo, c. 100m.
C. p. coomansi Sody, 1949a: 103; Pematang Tudjuh, N bank of Pungur Besar R, Pontianak dist, W Borneo.

Geographical variation is very considerable, both between islands and within the major islands, sometimes with intergradation but in some places apparently discrete forms are separated by rivers.

Malaya (mainland). *C. p. prevostii.* Dorsum and tail black; lateral stripe white and broad, extending from thighs forward to neck and cheek (but often lightly suffused with grey on the face). The white stripe is unbroken across the shoulder in the south (*prevostii* s. s.) but is interrupted by reddish brown on the shoulder in the west and north (*humei*, including *wrayi*).

Sumatra (mainland). In southern two-thirds (*C. p. rafflesii*) a white lateral stripe is present, the shoulders are rufous, the sides of the neck and face dark grey and the sides of the muzzle white. In the north (*C. p. piceus*) there is no white stripe, the black dorsal pelage abutting directly on the dark reddish brown venter. An intermediate is represented by *rufoniger*.

Borneo (mainland). Conspicuously variable. Most are illustrated in colour by Payne *et al.* (1985).
C. p. pluto. North. Dorsum and tail black, pale lateral stripe absent or very thin. It is doubtful if this form is distinguishable from *C. p. piceus* of N Sumatra which is the earlier name.
C. p. caroli. N Sarawak. Dorsum grey, tail banded light and dark grey. Lateral stripe including thighs pale grey, shoulders grey.
C. p. atricapillus. S Sarawak, C & E Kalimantan. Dorsum and legs brown agouti, tail banded, lateral stipe white, restricted to flank, with black stripe below it.
C. p. borneoensis. S Sarawak, extreme W Kalimantan. Dorsum blackish, tail banded, lateral stripe white, confined to flank, shoulder reddish, thighs grey.
C. p. palustris. SW (near Pontianak). As *borneoensis* but shoulders grey.
C. p. sanggaus. SW Kalimantan. Dorsum and tail blackish, white stripe on flank; shoulders and thighs grey tinged red.
C. p. waringensis. S Kalimantan. As *C. p. sanggaus* but less red on shoulder, and thighs paler.
C. p. atrox. E Kalimantan. Dorsum agouti olive brown, tail, feet and muzzle black, lateral stripe narrow and short, thighs grey.

Small islands.

Riau Archipelago (*condurensis, carimonensis*). Doubtfully distinct from *C. p. rafflesii* of S Sumatra.

Rupat (E of Sumatra) (*nyx*). Intermediate between N and S forms on Sumatran mainland.

Banggi and Balembangan, N Borneo (*caedis*). As *C. p. pluto* on adjacent mainland.

Serasan (W of Borneo) (*C. p. navigator*). As *C. p. humei* of Malaya but white stripe suffused with orange on both face and thighs.

Size varies considerably amongst the islands, being smallest (down to c. 200mm head and body) on the smallest islands, e. g. St Barbe I in the Tambelan Is, reaching a maximum of around 260-275mm on larger islands, e. g. Kundur, Penjalei, Rupat and Banka, and variable, but generally large to medium, on Borneo, Sumatra and the Malayan peninsula (Heaney, 1978).

Remarks. In spite of the extreme variability the limits of this species are fairly clear. The forms *baluensis* and *baramensis* in N Borneo were included in *C. prevostii* by Chasen (1940a) but are here treated as a separate species, *C. baluensis*, following Medway (1977a).

Callosciurus baluensis Kinabalu squirrel

Sciurus baluensis Bonhote, 1901e: 174; Mt Kinabalu, N Borneo, 1000ft.
S. prevostii baramensis Chasen, 1940a: 131; Mt Dulit, Sarawak, Borneo, 5000ft.
? *S. erythromelas* Temminck, 1853: 248; Gorantalo and Kema, Sulawesi.
= *S. schlegeli* Gray, 1867a: 278.

Range (Map 119). Mountains of NW Borneo, in Sabah and Sarawak, at altitudes of 300–2000m. Also on Sulawesi (probably introduced) if *erythromelas* is conspecific.

Variation. The following subspecies has been described:
C. b. medialis Allen & Coolidge, 1940: 156; Mt Tibang, central Borneo.

In *C. b. baramensis* the tail is speckled like the body, in *C. b. baluensis* it is uniformly dark.

Recognition. Distinguished from the adjacent form of *C. prevostii* on Borneo (*C. p. pluto*) by the agouti dorsum and the presence of a distinct light stripe on the flank; and from the more distant white-striped races of *C. prevostii* by the absence of any white or grey on the thighs.

Remarks. This form was included in *C. prevostii* by Chasen (1940a), but has subsequently been treated as a separate species, e. g. by Medway (1977a), on the grounds that its range overlaps the upper range of *C. prevostii.* The type of *erythromelas* (Leiden Museum) matches Bornean *C. baluensis* rather closely in pelage but has more orange-brown on the cheeks.

Callosciurus notatus Plantain squirrel

Sciurus notatus Boddaert, 1785: 119; W Java.
? *S. dschinschicus* Gmelin, 1788: 151; 'Pondicherry, India', corrected to W Java by Kloss (1921).
= *S. badjing* Kerr, 1792: 262.
= *S. ginginianus* Shaw, 1801: 147.
S. plantani Ljungh, 1801: 99; W Java.
S. bilineatus Geoffroy, in Desmarest, 1817: 106; W Java (Kloss, 1921).
S. vittatus Raffles, 1821: 259; Bencoolen, W Sumatra.
S. bivittatus Desmarest, 1822: 543; Sumatra.
Macroxus toupai Lesson, 1827a: 238; Sumatra.
? *S. diardii* Jentink, 1879b: 38; Nusa Kambangan I, S coast of Java.
S. (*Rheithrosciurus*) *microtis* Jentink, 1879b: 40; Salayar I (S of Sulawesi).
S. andrewsii Bonhote, 1901d: 456; Tjigombong, W Java.
S. baluensis suffusus Bonhote, 1901e: 175; Tutong R, NW Borneo.
S. tenuirostris Miller, 1900a: 221; Tioman I, Malaya.
S. anambensis Miller, 1900a: 223; Siantan I, Anamba Is.
S. abbottii Miller, 1900a: 224; Big Tambelan I, W of Borneo.
S. lutescens Miller, 1901a: 124; Sirhassen I, S Natuna Is.
S. seraiae Miller, 1901a: 125; Seraia I, Natuna Is.
S. rutiliventris Miller, 1901a: 126; Midae I, Natuna Is.
S. rubidiventris Miller, 1901a: 127; Bunguran I, N Natuna Is.
S. lautensis Miller, 1901a: 128; Laut I, N Natuna Is.
S. pemangilensis Miller, 1903a: 9; Pemanggil I, Malaya.
S. peninsularis Miller, 1903a: 10; N bank Endau R, SE Pahang, Malaya.
S. aoris Miller, 1903a: 10; Aor I, Malaya.
S. pannovianus Miller, 1903a: 11; Panau I, Atas Is, S China Sea.
S. ictericus Miller, 1903a: 12; Tana Bala I, Batu Is, W Sumatra.
S. saturatus Miller, 1903b: 453; Mansalar I, W Sumatra.
S. pretiosus Miller, 1903b: 454; Bungkuru I, Banjak Is, W Sumatra.
S. ubericolor Miller, 1903b: 455; Tuangku I, Banjak Is, W Sumatra.
S. tedongus Lyon, 1906a: 591; Tandjong Tedong, Banka I, Sumatra.
S. billitonus Lyon, 1906a: 593; Billiton I, Sumatra.
S. serutus Miller, 1906a: 58; Serutu I, Karimata Is, Borneo.
S. director Lyon, 1909c: 509; Direction I, S China Sea.
S. lamucotanus Lyon, 1911a: 85; Lamukotan I, W Borneo.
S. datus Lyon, 1911a: 86; Datu I, W Borneo.
S. arendsis Lyon, 1911a: 87; Arends I, Java Sea.
S. siriensis Lyon, 1911a: 87; Mata Siri I, Java Sea.
S. poliopus Lyon, 1911a: 88; Pamukuang Bay, S Borneo; preoccupied by *S. poliopus* Fitzinger, 1867 (in *Sciurus aureogaster* Cuvier, C America).
= *S. conipus* Lyon, 1911b: 98.
S. marinsularis Lyon, 1911a: 89; Laut I, SE Borneo.
? *S. atristriatus* Miller, 1913b: 22; Lo Bon Bon, SE Borneo.
C. albiculus Miller, 1942: 139; Blangnanga base camp, Atjeh, N Sumatra, 3600ft.

Range (Map 120). Peninsular Thailand, Malaya, Sumatra, Java, Bali, Borneo; also on most of the adjacent small islands north to Penang and the Perhentian Is in Malaya, east to Lombok (Lesser Sunda Is) and Salayar (S of Sulawesi).

Variation. The following nominal subspecies have been described:
Sciurus n. miniatus Miller, 1900b: 79; Trang, S Thailand.
S. n. albescens Bonhote, 1901f: 446; Acheen, N Sumatra.
S. vittatus dulitensis Bonhote, 1901f: 451; Mt Dulit, Sarawak, 1000ft.
S. v. tarussanus Lyon, 1907f: 279; Tarussan Bay, W Sumatra.
S. v. tapanulius Lyon, 1907f: 280; Tapanuli, W Sumatra.
S. n. rupatius Lyon, 1908b: 640; Rupat I, E Sumatra.
S. v. nesiotes Thomas & Wroughton, 1909a: 439; Batam I, Riau Is.
S. v. subluteus Thomas & Wroughton, 1909a: 440; Si Karang, SE Johore, Malaya.
S. n. madurae Thomas, 1910c: 386; Madura I, Java.
S. n. balstoni Robinson & Wroughton, 1911b: 234; Tjilatjap, SC Java.
S. v. plasticus Kloss, 1911: 117; Great Redang I, Malaya.
S. v. scottii Kloss, 1911: 117; Little Redang I, Malaya.
S. v. perhentiani Kloss, 1911: 118; W Perhentian I, Malaya.
S. v. proteus Kloss, 1911: 118; E Perhentian I, Malaya.
S. v. watsoni Kloss, 1911: 118; Lantinga I, Malaya.
S. v. famulus Robinson, 1912: 592; Dayang I, Malaya.
S. n. stresemanni Thomas, 1913b: 505; Bali.
S. dulitensis dilutus Miller, 1913b: 23; Tanjong Batu, Borneo, 2°15'N.
S. v. maporensis Robinson, 1916b: 64; Mapor I, Riau Is.
S. v. singapurensis Robinson, 1916c: 73; Singapore I.
S. n. tamansari Kloss, 1921a: 230; Tamansari, Idjen Masif, E Java, 1600ft.
S. v. lighti Chasen & Kloss, 1924: 58; Penang I, Malaya.
S. v. stellaris Chasen & Kloss, 1924: 58; Bintang I, Riau Is.
S. v. lunaris Chasen & Kloss, 1924: 58; Bulan I, Riau Is.
S. n. guillemardi Kloss, 1926a: 260; Tenggol I, Malaya.
C. n. vanheurni Sody, 1929a: 327; Tjipanas, near Garoet, W Java.
C. n. verbeeki Sody, 1929a: 330; Bandar, Padangan dist, Rembang, Java.
S. n. malawali Chasen & Kloss, 1932: 26; Mallewalle I, N Borneo.
C. n. nicotianae Sody, 1936c: 217; Silalas, near Medan, Deli, N Sumatra, 50m.
S. n. percommodus Chasen, 1940b: 489; Lesten, Atjeh, N Sumatra, 700m.
C. n. kalianda Sody, 1949a: 88; Kalianda, SE Lampongs, S Sumatra, 100m.
C. n. prinsulae Sody, 1949a: 89; Prinsen I, (= Panaitan I), W Bantam, W Java.
C. n. magnificus Sody, 1949a: 92; Nusa Barung I, S of Besuki, E Java.
C. n. vinocastaneus Sody, 1949a: 93; Kali Tjempaga, Sampit, SW Borneo, <100m.
C. n. raptor Hill, 1960: 59; Little Karimon I, Riau Is.
C. n. tinggius Hill, 1960: 61; Tinggi I, E of Johore, Malaya.

In spite of the enormous number of named forms, geographical variation is slight compared with that in most other widespread species of *Callosciurus*. The dorsal pelage and tail are very uniform with only a little local darkening, and most variation concerns the colour of the ventral pelage (pale grey to dark chestnut), the clarity of the lateral dark and light stripes, and the extent, or absence, of reddish brown hairs in the tail-tip.

Malaya (*C. n. miniatus* etc.). Venter grey with slight brown tinge to dull orange-brown, pale flank stripe narrow. Tail-tip reddish brown, especially in the west.

Sumatra (*C. n. vittatus*). Venter usually orange-brown, tail-tip not usually reddish, but many not distinguishable from Malayan individuals.

Java (see Hill, 1960). Venter grey in the west (*C. n. notatus*), with only a slight suffusion of buff on the abdomen; the amount of buff in the ventral pelage and around the eye increases eastwards (*balstoni, tamansari*).

Borneo. Most closely resemble the Malayan forms but the ventral colour is dark in the northwest, dark chestnut or maroon in the extreme form (*suffusus*).

Small islands. Most have the ventral pelage a light orange-brown and the lateral stripes fairly prominent. Exceptions are:
Tana Bala I (*C. n. ictericus*). Dorsum dark, venter dark orange-brown, lateral pale stripe obscure.
Bali (*C. n. stresemanni*). Venter very dull, pale orange-brown.
Madura (*C. n. madurae*). Venter yellow-buff. Stripes dull.
Mallewalle I, N Borneo (*C. n. malawali*). Black lateral stripes very wide, encroaching on venter which is dull yellow-brown with a dark mid-line.
Salayar, S of Sulawesi (*C. n. microtis*). Lateral stripes dull; venter dull buff (closely resembles some Javan specimens).

A reddish brown tail-tip is present in many island forms.

Recognition. Grey-bellied forms can be distinguished from *C. nigrovittatus* by at least a little suffusion of buff or brown on the abdomen. Those with darkest venters, in NW Borneo, are very similar to *C. baluensis*, but have the lateral pale stripe less developed and the tail faintly banded.

Remarks. The very dark-bellied form *suffusus* from NW Borneo was included in *C. prevostii* by Chasen (1940a) and closely resembles *C. baluensis*, but examination of the type suggest that it is best included in *C. notatus* which it resembles in the banded tail, rather dull lateral stripes and small size. The name *atristriatus* from SE Borneo was based upon a young individual (M^3 half erupted) and has been allocated to *C. orestes* (Payne *et al.*, 1985, following Chasen, 1940a, who included it in *C. nigrovittatus orestes*), but it lacks the characteristic lower incisors of *C. orestes*.

Callosciurus nigrovittatus Black-striped squirrel

Sciurus nigrovittatus Horsfield, 1823: 149; W Java (restricted by Kloss, 1921a).
S. grisei-venter Geoffroy, 1831: 147; W Java.
S. klossii Miller, 1900a: 225; Saddle I, Tambelan Is, W of Borneo.
S. bilimitatus Miller, 1903a: 8; Trengganu, Malaya.

Range (Map 119). Peninsular Thailand, Malaya, Sumatra, Java; and on the small islands of Tioman and Saddle I (Tambelan Is). A record from Cochin China, Vietnam (Cao, 1984) seems improbable and requires verification.

Variation. The following nominal subspecies have been described:
Sciurus (*bilimitatus*) *microrhynchus* Kloss, 1908: 144; Tioman I, Malaya.
S. n. johorensis Robinson & Wroughton, 1911a: 166; Pelepak, Johore, Malaya.
S. n. bocki Robinson & Wroughton, 1911a: 167; Pajo, Padang Highlands, Sumatra.
S. n. besuki Kloss, 1921a: 231; Tamansari, Idjen Massif, E Java, 1600ft.
C. n. madsoedi Sody, 1929b: 163; Rahtowu, Gingsir, Mt Moeria, C Java, 480m.
C. n. acraeus Miller, 1942: 138; Blangnanga base camp, Atjeh, N Sumatra, 3600ft.
C. n. salakensis Sody, 1949a: 98; Tjianten, Mt Salak, W Java, 1000m.
C. n. bantamensis Sody, 1949a: 99; Tjikudjang, Bantam, W Java, <100m.
C. n. phoenicurus Sody, 1949a: 99; Mt Tjerimai, Cheribon, W Java, 1000m.
C. n. tenggerensis Sody, 1949a: 99; Ranu Pani, Tengger Mts, E Java, 2100m.

Variation is slight. All have some brown on the throat and some also under the neck, but in parts of Malaya this is a richer orange-brown and in some individuals extends back as a narrow line in the centre of the thorax (*bilineatus*). This is found also in some Javan specimens, e.g. of *salakensis* and *bantamensis*.

Callosciurus adamsi Ear-spot squirrel

Sciurus adamsi Kloss, 1921b: 151; Long Mujam, 150 miles up Baram R, Baram, Sarawak, 700–900ft.

Range (Map 120). N & W Borneo, up to 900m.

Remarks. Included by Chasen (1940a) in '*C. albescens*' (Sumatra) which is here included in *C. notatus*, *C. adamsi* was treated as a separate species by Medway (1977a) which seems fully justified. It differs from the adjacent *C. orestes* in the dull orange-brown venter, longer rostrum and orthodox lower incisors with convex anterior surfaces, and appears to replace *C. orestes* at lower altitudes.

Callosciurus melanogaster Mentawai squirrel

Sciurus melanogaster Thomas, 1895d: 668; Sipora I, Mentawai Is.
S. atratus Miller, 1903a: 13; N Pagi I, Mentawai Is.

Range (Map 120). Sipura, Siberut and N Pagai Islands, Mentawai Is, Sumatra.

Variation. Three subspecies seem to be clearly separable:
C. m. melanogaster. Sipura I. Dorsum dark; venter almost uniformly black, slightly mottled by lighter hair-bases.
C. m. mentawi Chasen & Kloss, 1928c: 822. Siberut I. Dorsum slightly lighter and more reddish; venter chestnut brown, brighter at sides forming short lateral bands in the position of the black bands in *C. nigrovittatus*. Some individuals have a little blackish especially in the mid-line.
C. m. atratus. N Pagai I. Dorsum as in *C. m. mentawi* but usually with scattered light hair-tips giving a coarsely grizzled appearance; venter mostly suffused with black but with a varying amount of light hairs giving a grizzled appearance.

Ecology. On Siberut this species occupies mature forest and secondary growth, is semi-crepuscular and has a mixed diet of fruit and arthropods (Whitten, 1981).

Remarks. The baculum is intermediate between those of *C. caniceps* and *C. prevostii* (Fig. 21f).

Callosciurus orestes Bornean black-banded squirrel

Sciurus notatus orestes Thomas, 1895e: 529; Mt Dulit, Sarawak, Borneo, 4000ft.

S. nigrovittatus venetus Chasen, 1940a: 139; Kenokok, Mt Kinabalu, Borneo, 3300ft.
Glyphotes (*Hessonoglyphotes*) *canalvus* Moore, 1959b: 5; Mt Dulit, Borneo, 3°3'N, 114°11'E.

Range (Map 121). Borneo; mainly montane, above 1000m.

Recognition. Distinguished from all other Bornean species by the all-grey underparts, short rostrum, and the form of the lower incisors which have concave anterior faces; and from the very similar but allopatric *C. nigrovittatus* also by smaller size and presence of a pale spot behind each ear.

Remarks. This form was originally described as a subspecies of *C. notatus* but was listed as a subspecies of *C. nigrovittatus* by Chasen (1940a) and by Medway (1977a). It was given specific rank by Payne *et al.* (1985), influenced in particular by the skull and teeth which tend towards the condition found in *C. simus*. In fact this species is so intermediate between the *notatus* group of *Callosciurus* and '*Glyphotes simus*' as to justify including the latter in *Callosciurus*. In erecting a new subgenus *Hessonoglyphotes* and species *canalvus* in the genus *Glyphotes*, Moore (1959b) made comparison with *Callosciurus adamsi* and *Glyphotes simus* but not with *C. orestes* with which all the characters and measurements of *canalvus* agree.

Callosciurus simus Sculptor squirrel

Glyphotes simus Thomas, 1898b: 251; Mt Kinabalu, Borneo.

Range (Map 121). NW Borneo, montane, between 1300 and 1700m.

Recognition. Externally closely resembles *C. adamsi* and *C. orestes*, but much smaller and with a very short rostrum and peculiarly flattened, slightly divergent lower incisors.

Remarks. This little known species was reviewed by Hill (1959a) and Moore (1959b). It was described in a new genus *Glyphotes* on the basis of the very short rostrum, peculiar lower incisors and small size, but these features are shown to a lesser degree in *C. orestes* which is so intermediate between *Callosciurus* of the *notatus* group and *C. simus* as to indicate strongly that the latter is a specialized offshoot of the *notatus* group. It shares with *C. notatus* not only the complex pattern of lateral stripes but also the nearly identical baculum (Fig. 31d, e). The auditory bullae also clearly relate it to *Callosciurus* rather than to *Sundasciurus*.

Map **121**. *Callosciurus* (part), *Tamiops*

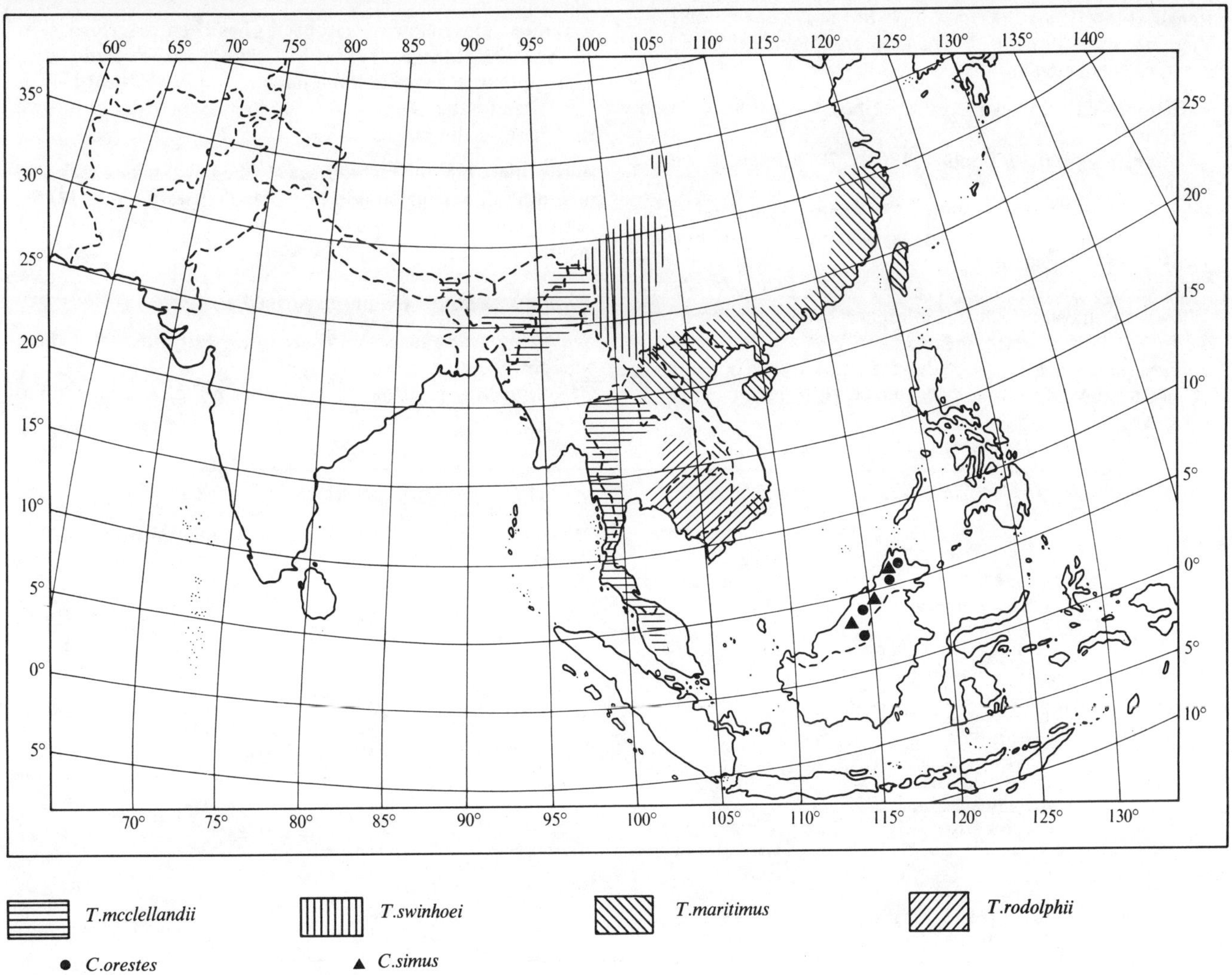

TAMIOPS

Tamiops Allen, 1906: 475; type species *T. macclellandi hainanus* Allen.

A genus of four species in the northeastern part of the Indomalayan Region (Table 186). It has been included in *Callosciurus*, e. g. by Ellerman (1940) and Moore (1959a) but was excluded by Moore & Tate (1965) who gave a detailed account with maps and whose classification is followed here. The four species recognized by Moore & Tate (1965) are difficult to define and this arrangement must be regarded as tentative. The species are parapatric with perhaps very marginal sympatry in places. The forms in Vietnam and parts of Laos were also mapped in detail by Dao (1970a) who however treated them as races of a single species *C. mcclellandii*.

Tamiops mcclellandii Himalayan striped squirrel

Sciurus mcclellandii Horsfield, 1840: 152; Assam.
S. pembertoni Blyth, 1843a: 887; Bhutan.
S. barbei Blyth, 1847a: 875; Ye, Tenasserim, Burma.
Tamias leucotis Temminck, 1853: 252; Malacca Peninsula.
S. novemlineatus Miller, 1903c: 147; Trang, S Thailand, 1500ft.
T. inconstans Thomas, 1920b: 306; Yunnan, probably near Mongtse = Mengtz.

Range (Map 121). Sikkim to N Burma, Yunnan and N Vietnam, south through W Thailand and Malaya (where it is montane, above 600m).

Variation. The following subspecific names have been proposed:
Sciurus m. manipurensis Bonhote, 1900b: 51; Aimole, Manipur, India.
S. m. kongensis Bonhote, 1901c: 55; Raheng, Thailand.
T. m. collinus Moore, 1958a: 1; Maymyo, upper Burma, 22°05'N, 96°30'E, 800m.

Six subspecies were recognized by Moore & Tate (1965) as follows, the most significant difference being between *T. m. mcclellandii* and the remainder:
T. m. mcclellandii. Sikkim—N Burma and extreme W Yunnan. Stripes narrow and obscure; venter greyish ± creamy tips.
T. m. inconstans. Yunnan and N Vietnam. As above but venter brighter orange buff and mid-dorsal stripe brown (black in all other subspecies).
T. m. collinus. C Burma—N Thailand. Intermediate between *T. m. mcclellandii* and *T. m. kongensis* but inner dark stripes darker than in either of these.
T. m. kongensis. W Thailand. Background pelage pale, greyish; lateral pale stripes broad and pale cream; venter tinged orange.
T. m. barbei. S Burma. Stripes bold, lateral pale stripes broad and orange-brown; venter tinged orange.
T. m. leucotis. S Thailand—Malaya. Rather pale and small but doubtfully distinct from *T. m. barbei.*

Tamiops rodolphii

Sciurus (Tamias) rodolphii Milne-Edwards, 1867b: 227; Cochin-China.
T. macclellandi liantis Kloss, 1919b: 370; Satahip, near Cape Liant, SE Thailand.
T. lylei Thomas, 1920b: 307; coast 50 miles S of Bangkok, Thailand. Pre-occupied by *Sciurus bocourti lylei* Wroughton, 1908, a form of *Callosciurus* in which *Tamiops* has been included.
= *Callosciurus holti* Ellerman, 1940: 355.
T. macclellandi dolphoides Kloss, 1921c: 101; Kampong Som Bon, Cambodia.

Range (Map 121). E Thailand, Laos, C & S Vietnam.

Variation. The following subspecies has been described:
T. r. elbeli Moore, 1958a: 4; Ban Lad, Pookeio dist, Chaiyaphum Prov., central Thailand, c. 16°40'N, 101°45'E. Crown and nape bright yellowish compared with dull brown elsewhere in the range.

Recognition. Unique features are (1) the light stripes are equal in width (all narrow) and in brightness (all very light); (2) the central dark stripe is usually bisected by a very faint lighter median stripe.

Tamiops swinhoei Swinhoe's striped squirrel

Sciurus mcclellandii var. *swinhoei* Milne-Edwards, 1874: 308; Moupin, Sichuan, China.
T. vestitus Miller, 1915a: 115; 65 miles NE of Peking, China.

Table 186. The species of *Tamiops*.

	mcclellandii	*rodolphii*	*swinhoei*	*maritimus*
Range	E Nepal-Malaya	Thailand-S Vietnam	China	SE China, Taiwan
Size	Smaller	Smaller	Larger	Larger
head & body	112–122	105–120	120–155	100–145
hind feet	28–32	28–30	30–38	30–35
Mid-dorsal dark line with thin, pale central stripe	0	+	0	0
Inner pair of light stripes less conspicuous than outer	+	0	+	+
Outer pale stripes continuous with cheek stripes	+	+(0)	0(+)	0
Ventral pelage	Grey, tipped cream or dull orange	Dull orange	Grey, ± cream tips	Grey, ± cream tips
Skull: greatest length	31–34	c.33	35–37	35–37
Maxillary tooth-row	4.8–5.6	5.4–5.7	5.7–6.4	5.7–6.3

T. clarkei Thomas, 1920b: 304; upper Yangtze Valley at 27°20'N, 8000ft.
T. maritimus forresti Thomas, 1920b: 305; Lichiang Range, Yunnan, 27°20'N, 11000ft.
T. spencei Thomas, 1921a: 503; N Kachin Prov., N Burma, 28°22'N, 97°40'E, 10000ft.
T. macclellandi russeolus Jacobi, 1923: 11; Tsalila Pass, between upper Yangtze and Mekong rivers, Sichuan/Yunnan, 3500m.
T. monticolus olivaceus Osgood, 1932: 292; Lo Gui Ho, Tonkin, Vietnam, 5000ft.

Range (Map 121). Yunnan and Sichuan, and adjacent parts of Gansu, Tibet, Burma and N Vietnam; also a very isolated population in N Hebei, NE China. Mainly montane, between 1500 and 3500m.

Variation. The following subspecies has been described:
T. s. chingpingensis Lu & Qyan, in Lu *et al.*,1965: 288; Chingping, Suang-Kiang District, W Yunnan, 2600m.

The isolated population in NE China (*T. s. vestitus*) is rather pallid, especially in winter pelage. Moore & Tate (1965) did not distinguish subspecies within the main part of the range although the variation is considerable both in background colour—very dark in Vietnam (*olivaceus*)—and in the prominence of the stripes—the lateral stripes are very broad and white in parts of Yunnan (*clarkei*) and very obscure in N Burma (*spencei*).

Tamiops maritimus

Sciurus mcclellandii maritimus Bonhote, 1900b: 51; Foochow, Fujian, China.
S. mc. monticolus Bonhote, 1900b: 52; Ching-feng-ling, Fujian.
S. mc. formosanus Bonhote, 1900b: 52; N Formosa.
? *T. mc. hainanus* Allen, 1906: 476; Lei Mui Mon, mountains of Hainan I, China.
T. mc. riudoni Allen, 1906: 477; Riudon, lowland Hainan I.
T. sauteri Allen, J. A.,1911: 339; Chip-Chip, N Taiwan.
T. mc. laotum Robinson & Kloss, 1922: 92; Pak Hin Bun, Mekong R, Laos.
T. mc. moi Robinson & Kloss, 1922: 92; Langbian Peaks, S Annam, Vietnam, 5500–6500ft.

Range (Map 121). E China from near Shanghai through Fujian, Guangdong and Guangxi to Vietnam and Laos; an isolated record from Hubei; and on Taiwan and Hainan.

Variation. The form on Taiwan is not distinguishable from the nominate form in mainland China; the form on Hainan is smaller (hind feet 28–30mm, cf 32–35 on the mainland) but is not distinguishable by pelage (Moore & Tate, 1965). Specimens in Laos (*laotum*) are rather pale.

Remarks. On the basis mainly of its small size Wang Yingxiang (pers. comm.) considers *T. hainanus* specifically distinct from both *T. maritimus* and *T. swinhoei* and represented in S Yunnan as well as Hainan I. However it seems very doubtful if it or *T. maritimus* can justify separation as species from *T. swinhoei*.

SUNDASCIURUS Sunda squirrels

Sundasciurus Moore, 1958b: 2; type species *Sciurus lowii* Thomas.
Aletesciurus Moore, 1958b: 3 (*S. hippurus* Lyon).

Malaya, Sunda Islands and Philippines; 13 species (Table 187). These were included in *Callosciurus* by Ellerman (1940) and in *Sciurus* by Chasen (1940a). The genus was based primarily upon the form of the auditory bulla within which a single Y-shaped septum isolates an anteromesial compartment which may or not be inflated, forming a lobe visible on the external surface of the bulla; in *S. hippurus* and some of the Philippine species (subgenus *Aletesciurus* of Moore, 1958b) the Y-shaped septum can be seen only through the meatus or by dissection. The baculum, where known, is short with a triangular accessory blade as in the *caniceps* group of *Callosciurus*.

Table 187. The species of *Sundasciurus*.

	lowei	*tenuis*	*jentinki*	*brookei*	*hippurus*	*philippinensis*	*steerei*	*moellendorffi*	*rabori*
Range	Malaya, Sumatra, Borneo	Malaya, Sumatra, Borneo	Borneo	Borneo	Malaya, Sumatra, Borneo	Mindanao etc.	Palawan etc.	Calamian Is	Palawan
Head & body	132–157	130–155	120–144	140–205	213–238	190–215	190–220	190–220	163–186
Tail (% of h & b)	60–70%	85–95%	95–100%	70–105%	c.100%	85–100%	85–100%	90–100%	c.85%
Hind feet	32–37	32–35	30–34	32–38	52–61	40–53	45–50	49–52	43–44
Tail banded	0	(+)	+	+	0/+	+	(+)		+
Tail thick, bushy (B); average (A); or thin (T)	B	T	T	A	B	A	A	A	
Dorsum	Brown, agouti	Brown, agouti	Reddish brown	Brown, agouti	Reddish brown	Brown	Chestnut	Light brown	Dark brown
Venter	White/buff	Grey, white or buff tips	Grey, white or buff tips	Grey, white tips	Red or white	Grey	Chestnut	Dull brown	Grey, cream tips
Orange between hind legs (male)	0	0	0	+	0	+/0	(+)	0	
Eye-ring reddish or buff (R); white (W)	R	R	W	R	R	R	R		
Ear margin white	0	0	+	0	0	0	0		
Bulla: antero-mesial compartment visible externally	+	+	+	+	0	0	+	+	+
Sagittal crest	0	0	0	0	+	+	+	+ (short)	+
Maxillary tooth-row	c.6.0–7.0	6.9–7.7	6.5	7.8	c.10.0–10.6	9.2–9.7	10.2–10.7	10.5–11.3	9.3–9.8

Sundasciurus lowii Low's squirrel

Sciurus lowii Thomas, 1892a: 253; Lumbidan, Sarawak.
Sc. fraterculus Thomas, 1895d: 669; Sipora I, Mentawai Is.
Sc. lingungensis Miller, 1901a: 123; Lingung I, near Bunguran I, Natuna Is.
Sc. robinsoni Bonhote, 1903a: 24; Bukit Besar, Patani, peninsular Thailand.
Sc. balae Miller, 1903a: 14; Tana Bala I, Batu Is, W Sumatra.
Sc. piniensis Miller, 1903a: 14; Pinie I, Batu Is, W Sumatra.
Sc. pumilus Miller, 1903a: 15; S Pagi I, Mentawai Is.
Sc. seimundi Thomas & Wroughton, 1909a: 440; Kundur I, Riau Is.
Sc. humilis Miller, 1913b: 24; Kateman R, E Sumatra.
Sc. vanakeni Robinson & Kloss, 1916: 270; Barong Bharu, Korinchi, Sumatra, 4000ft.

Range (Map 122). Peninsular Thailand and Malaya, Sumatra, Borneo and adjacent small islands of Kundur (Riau Is); Pini, Tana Bala (Batu Is), Siberut, Sipura, S Pagai (Mentawai Is), Laut, Bunguran, Lingung, Serasan (Natuna Is), Banggi (N Borneo).

Variation. The following subspecific names have been proposed:
Sciurus l. natunensis Thomas, 1895b: 26; Sirhassen I, Natuna Is.
Sc. robinsoni alacris Thomas, 1908e: 306; Selangor-Pahang border, Malaya, 3000ft.
Sc. l. bangueyae Thomas, 1910c: 387; Banggi I, N Borneo.
Sc. l. siberu Chasen & Kloss, 1928c: 824; Siberut I, Mentawai Is.

Geographical variation is slight. Specimens from Borneo, lowland Sumatra and Malaya appear indistinguishable; those from Mt Korinchi, Sumatra (c. 1200m) have the ventral pelage greyer and occupying a narrower zone. Populations on some small islands are essentially as on the mainland (e. g. Kundur, Banggi), others are slightly differentiated in the colour of the ventral pelage: Sipura and Siberut (*S. l. fraterculus*)—venter grey, with orange-brown tinge anteriorly; Serasan, Natuna Is (*S. l. natunensis*)— venter with a slight pinkish brown flush, especially on chest and groin.

Ecology. On Siberut this species forages on the ground and lower levels of the forest, feeding on bark and arthropods (Whitten, 1981).

Sundasciurus tenuis Slender squirrel

Sciurus tenuis Horsfield, 1823; Singapore.
Sc. modestus Müller, 1840: 55; Mt Singgalang, Padang Highlands, Sumatra.
Sc. procerus Miller, 1901a: 122; Bunguran I, N Natuna Is.
Sc. parvus Miller, 1901c: 33; Nulu, Sarawak, 1000ft.
Sc. bancarus Miller, 1903b: 451; Bangkaru I, Banjak Is, W Sumatra.
Sc. mansalaris Miller, 1903b: 451; Mansalar I, Tapanuli Bay, W Sumatra.
Sc. tahan Bonhote, 1908: 6; Mt Tahan, Pahang, Malaya, 5300ft.

Range (Map 122). Peninsular Thailand and Malaya; Sumatra, Borneo; and small islands of Great Redang, Tioman (Malaya); Singapore; Mansalar, Tana Bala, Tana Masa, Bangkaru (W Sumatra); Jimaja, Mobur, Siantan (Anamba Is); Bunguran (N Natuna Is). Up to at least 1650m in Borneo.

Variation. The following subspecific names have been proposed:
Sc. t. surdus Miller, 1900b: 80; Trang, peninsular Thailand.
Sc. t. sordidus Kloss, 1911: 119; Great Redang I, Malaya.
Sc. t. gunong Robinson & Kloss, 1914b: 119; Kao Nawng, Bandon, peninsular Thailand, 3500ft.
Sc. t. altitudinis Robinson & Kloss, 1916: 269; Korinchi Peak, Sumatra, 7300ft.
Sc. mansalaris batus Lyon, 1916a: 443; Tana Bala I, Batu Is.
Tomeutes t. tiomanicus Robinson, 1917b: 103; Tioman I, Malaya.
Sc. t. siantanicus Chasen & Kloss, 1928b: 33; Siantan I, Anamba Is.

Geographical variation is slight, with little difference between Malayan and Bornean forms and those on the islands of Singapore, Tioman, Great Redang and Siantan.

Remarks. The form *altitudinis* (Mt Kerinci, Sumatra) was treated as a separate species by Miller (1942) who considered the two marginally sympatric, with *S. altitudinis* distinguished by longer rostrum and smaller bullae. Comparison of the type series with others from Sumatra does not support this view.

Map **122**. *Sundasciurus*

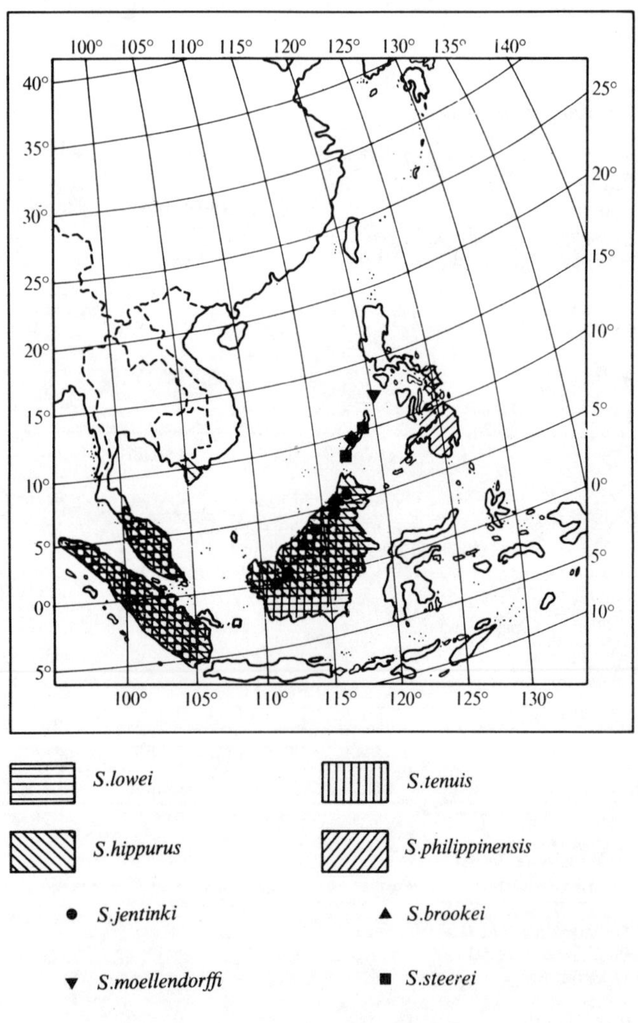

Sundasciurus jentinki Jentink's squirrel

Sciurus jentinki Thomas, 1887a: 128; Mt Kinabalu, N Borneo.

Range (Map 122). Borneo, montane, above 900m (to 3140m on Mt Kinabalu).

Variation. One subspecies has been described:
Sciurus j. subsignanus Chasen, 1937a: 80; Telen R, E Borneo, above 1172m. Distinguished by buff rather then white tips to the hairs on the tail.

Remarks. This species appears to be the ecological equivalent of *S. tenuis* at higher altitudes in NW Borneo (Payne *et al.*, 1985).

Sundasciurus brookei Brooke's squirrel

Sciurus brookei Thomas, 1892a: 253; Mt Dulit, Sarawak.

Range (Map 122). Mountains of N, C and W Borneo, from 600 to 1500m.

Remarks. This form was treated as a subspecies of *Callosciurus nigrovittatus* by Sody (1949a), but it has the auditory bullae characteristic of *Sundasciurus* and completely lacks the lateral stripes characteristic of *C. nigrovittatus* and its allies.

Sundasciurus hippurus Horse-tailed squirrel

Sciurus hippurus Geoffroy, 1831: 149; Malacca (Java in error).
Sc. rufogaster Gray, 1842: 263; Malacca.
Sc. pryeri Thomas, 1892b: 214; Sapugaia R, near Sandakan, N Borneo.
Sc. hippurosus Lyon, 1907h: 26; Tarussan Bay, W Sumatra.
Sc. hippurellus Lyon, 1907h: 27; Batu Ampar, Landak R, W Borneo.

Range (Map 122). Malaya, Sumatra and Borneo; doubtfully in S Vietnam (Dao & Cao, 1990) and S Thailand; up to at least 1000m in Malaya and 1500m in Borneo.

Variation. The following subspecific names have been proposed:
Macroxus rufogaster var. *borneensis* Gray, 1867a: 283; Sarawak.
= *Sciurus h. grayi* Bonhote, 1901e: 171 (footnote). The latter name was proposed on the mistaken assumption that *borneensis* Gray, 1867 was pre-occupied by *Sciurus rafflesii* var. *borneoensis* Müller & Schlegel, 1844, a form of *Callosciurus prevostii.*
Sc. pryeri inquinatus Thomas, 1908f: 247; Lawas R, Sarawak.
S. h. ornatus Dao & Cao, 1990: 234; Nguyen Phich, Thaibink, Minh Hai Prov., S Vietnam.

There is considerable geographical variation in Borneo as shown in Table 188.

Recognition. Constant features are the grey head, shoulders and fore legs. Some forms very closely resemble some forms of *Callosciurus erythraeus* but in Malaya, where both species occur, *C. erythraeus* always has some agouti brown on the venter and the tail is lighter and less bushy.

Sundasciurus philippinensis

Sciurus philippinensis Waterhouse, 1839a: 117; Mindanao, Philippines.
Sc. mindanensis Steere, 1890: 29; Mindanao, Philippines.
Sc. samarensis Steere, 1890: 30; Samar and Leyte, Philippines.
Sc. cagsi Meyer, 1890: 599; Davao, S Mindanao.
Callosciurus davensis Sanborn, 1952a: 117; Davao, Mindanao.

Range (Map 122). Mindanao, Basilan, Samar, Leyte, Bohol and adjacent small islands, Philippines.

Variation. Those on Leyte differ from those on Mindanao in lacking rufous hairs under the base of the tail (although the males have a slight rusty tinge around the genitalia as on Mindanao but not on Samar). The form *davensis* on Mindanao has the venter dull orange-brown rather than grey, thus approaching the condition seen in *S. steerii* on Palawan although *davensis* lacks the orange-brown tinge on the face and behind the ears seen in *S. steerii.*

Remarks. The following two species may prove to be better interpreted as subspecies of *S. philippinensis.*

Sundasciurus steerii

Sciurus steerii Günther, 1877: 735; Balabac I and Palawan, Philippines.
? *S. juvencus* Thomas, 1908g: 498; Palawan, Philippines.

Range (Map 122). Palawan and Balabac I, Philippines, in lowland forest and plantations.

Sundasciurus moellendorffi

Sciurus möllendorffi Matschie, 1898: 41; Culion I, Calamian Is, Philippines.
? *Sc. albicauda* Matschie, 1898: 42; Calamian Is.
Callosciurus hoogstraali Sanborn, 1952a: 115; Dimaniang, Busuanga I, Calamian Is, Philippines.

Range (Map 122). Calamian Is, Philippines (on the islands of Culion, Busuanga, Linapacan, Iloc and Tampel).

Sundasciurus rabori Palawan montane squirrel

S. rabori Heaney, 1979: 281; Magtaguimbong, Mt Mantalingajan, Palawan, Philippines, 8°48'N, 117°40'E, 3680–4350ft.

Table 188. The subspecies of *Sundasciurus hippurus.*

	Malaya, Sumatra (*hippurus*)	Borneo C,SE (*borneensis*)	SW (*hippurellus*)	NC (*inquinitus*)	Sabah (*pryeri*)
Dorsal pelage	Brown agouti	Chestnut	Brown agouti	Brown agouti	Brown agouti
Venter	Chestnut	Chestnut	Chestnut	White	Light orange-brown
Hind legs	Grey	Grey	Grey	Brown	Brown
Tail	Black	Black	Black	Banded grey and black	Banded grey and black

Range (Map 122). Known only from the type locality on Palawan, Philippines.

Remarks. Heaney (1979) considered this species a member of the subgenus *Aletesciurus*, closely related to *S. steerii*.

DREMOMYS Plain long-nosed squirrels

Dremomys Heude, 1898: 54; type species *Sciurus pernyi* Milne-Edwards.
Zetis Thomas, 1908f: 245 (*S. rufigenis* Blanford).

Himalayas to S China and Malaya, and on Borneo, always montane; 5 species (Table 189). The species included in *Dremomys* are sufficiently similar that there is no reasonable doubt that they form a monophyletic group but it is very doubtful if there is any justification for separating this genus from *Callosciurus*. Moore & Tate (1965) distinguish it only on the basis of longer rostrum—'nasal bone exceeds the least interorbital width'—but in one species, *D. lokriah*, these measurements are equal.

A detailed revision by Moore & Tate (1965) recognized five species. They considered *D. pyrrhomerus* (China) specifically distinct from *D. rufigenis* (Burma to Malaya) without identifying any single character that allowed clear separation; they are here tentatively united, but *D. gularis*, treated as a subspecies of *D. pyrrhomerus* by Moore & Tate, is here considered a separate species. The species are basically allopatric or parapatric, but perhaps with some marginal sympatry. Three species have been recorded, apparently from the same altitude, in the Mokokchung distict of Nagaland, India.

They forage on the ground as well as in trees; besides fruit and seeds invertebrates comprise a significant part of the diet.

Dremomys lokriah Orange-bellied Himalayan squirrel

Sciurus lokriah Hodgson, 1836b: 232; Nepal.
D. macmillani Thomas & Wroughton, 1916b: 238; Tatkon, W Burma.

Range (Map 123). Central Nepal (west to c. 84°30'E) to Assam, Burma and W Yunnan, at altitudes of 1000 to 2700m in Nepal.

Variation. The following subspecific names have been proposed:
D. l. bhotia Thomas, 1916c: 418 (and Wroughton, 1916a: 426); Sedonchen, Sikkim.
D. l. subflaviventris Thomas, 1922b: 429; Assam (validating a *nomen nudum* of Gray, 1843a: 144).
D. l. garonum Thomas, 1922b: 430; Tura, Garo Hills, Assam.
D. l. pagus Moore, 1956: 1; Mt Victoria, Pakokku, Chin Hills, W Burma, 1400m.
D. l. motuoensis Cai & Zhang, 1980a: 444; Hanni, Motuo region, Xizang (Tibet).

Although Moore & Tate (1965) recognized four subspecies, including *D. l. garonum, D. l. pagus* and *D. l. macmillani*, variation is likely to be continuous and affects mainly the ventral pelage in which the yellowish hair-tips are brighter and occupy a wider band in the east and south than in Nepal and Sikkim. *D. l. macmillani* (Naga Hills, Manipur etc.) is unique in usually having a narrow mid-dorsal dark stripe from the nape to the centre of the back although this is variable in marginal populations. It also has the venter pale creamy yellow, a rusty patch around the anus and a more conspicuous orange-buff patch behind each ear. These last three characters all give *macmillani* a closer resemblance to *D. pernyi* than to *D. lokriah*, especially to *D. p. howelli* which occurs in the same area (and also shows a dark stripe). This would appear to indicate some intergradation. However the skulls appear to show a more clear-cut difference, *macmillani* having the short rostrum of *D. lokriah*, seen especially in the shorter nasal bones, contrasting with the more elongate rostrum of *D. pernyi*.

Dremomys pernyi Perny's long-nosed squirrel

Sciurus pernyi Milne-Edwards, 1867b: 230, pl. 19; Sichuan, China.
Zetis owstoni Thomas, 1908f: 248; Mt Arizan, C Taiwan.
D. senex Allen, G. M.,1912b: 229; Nantu, Ichang, Hubei, China.
D. rufigenis lentus Howell, 1927a: 80; Wenchuanhsien, Sichuan.

Range (Map 123). S China (S Gansu, Sichuan, Yunnan, Hubei, Fujian) to N Vietnam (Cao, 1984), Burma, Assam and Manipur; Taiwan. To 3900m in NW Yunnan.

Variation. The following subspecific names have been proposed:
D. p. flavior Allen, G. M.,1912a: 178; Mengtsz (= Mongtse), S Yunnan.
D. p. griselda Thomas, 1916b: 392; Nagchuka, W Sichuan.
D. p. modestus Thomas, 1916b: 393; Suiyang, Guizhou.
D. p. chintalis Thomas, 1916b: 394; Chinteh, Anhui.
D. p. calidior Thomas, 1916b: 394; Kuatun, Fujian.
D. p. howelli Thomas, 1922a: 401; Machangkai, 25 miles SW of Tengyueh, SW Yunnan.
D. p. mentosus Thomas, 1922a: 401; 6 miles W of Kindat, W Burma, 6000ft.

Table 189. The species of *Dremomys*.

	lokriah	*pernyi*	*rufigenis*	*gularis*	*everetti*
Range	Nepal-Burma	Assam-S China	Malaya-Yunnan	S China, N Vietnam	Borneo
Venter	Yellow-orange tips	White-yellow tips	Grey, ± white tips	Dark grey	Grey, short white tips
Rusty patch around anus	0(+)	+	+	+	+/0
Tail red below	0	0	+	+	0
Red on head	0	0	+/0	+	0
Conspicuous red patch on outside of thigh	0	0	+/0	+/0	0
Throat orange-brown back to a line between fore legs	0	0	0	+	0
Length of nasals	15–16	15–18	17–20	18–20	14–17

Map **123**. *Dremomys*

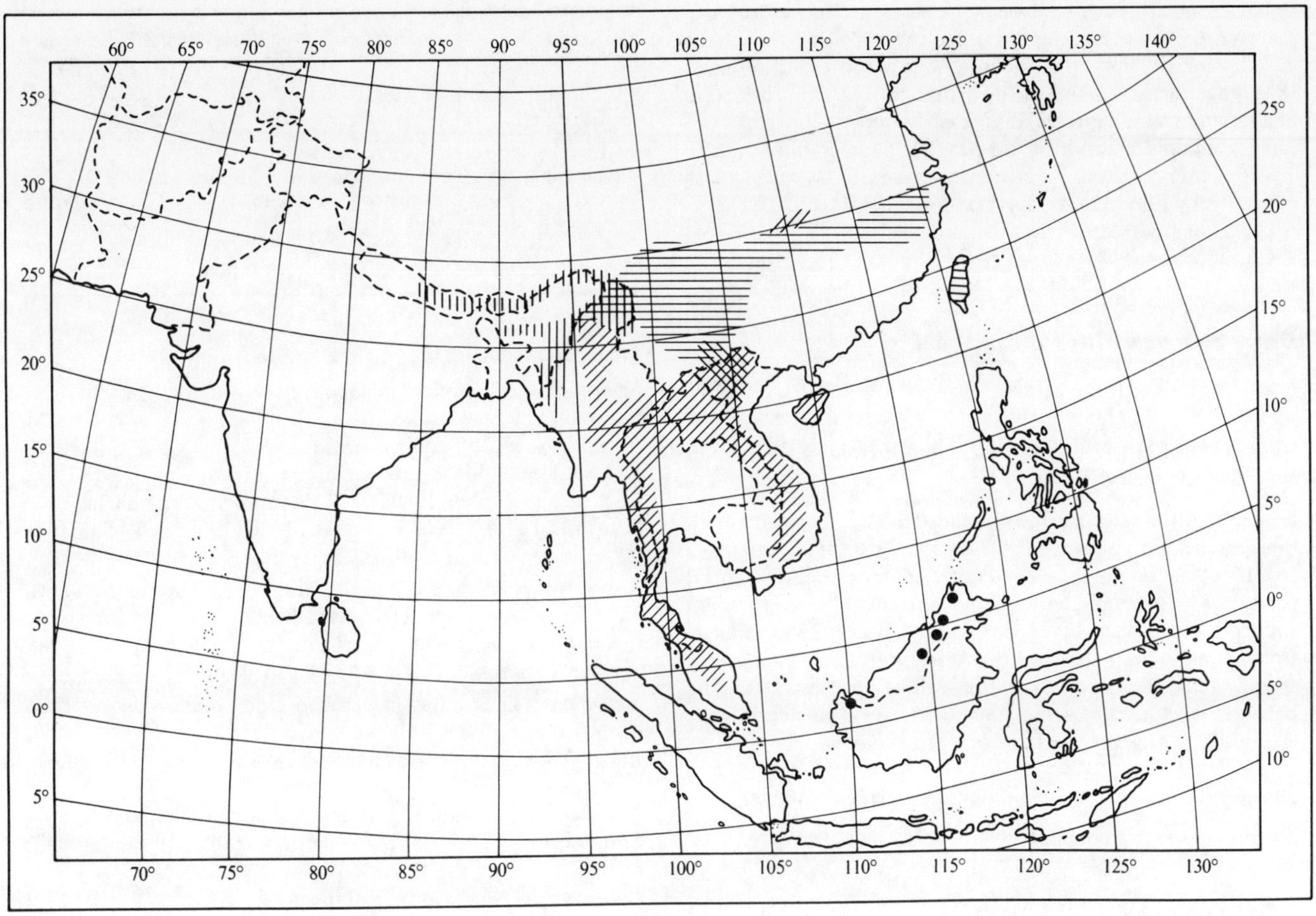

D.pernyi — *D.lokriah* — *D.gularis* — *D.rufogenis*

● *D.everetti*

D. p. imus Thomas, 1922a: 402; W side of Mt Imaw Bum, N Burma, 7000ft.

D. p. lichiensis Thomas, 1922a: 403; Likiang Range, Yunnan.

Geographical variation is slight, involving mainly the depth of colour of the dorsal pelage (pale and greyish in Sichuan, more yellowish in E China) but with more extreme forms in the southwest and in Taiwan:

D. p. howelli. SW Yunnan, Burma and Assam. A faint mid-dorsal stripe as in *D. lokriah macmillani* in the same area.

D. p. owstoni. Taiwan. Dorsal pelage dark; ventral pelage with creamy yellow hair-tips; post-auricular patch less prominent. All these characters give this form a closer external resemblance to *D. lokriah* than to *D. pernyi* but the muzzle is long as in typical *D. pernyi*.

Dremomys rufigenis Red-cheeked squirrel

Sciurus rufigenis Blanford, 1878a: 156, pl. viii; Mt Mulaiyit, Tenasserim, S Burma.

S. pyrrhomerus Thomas, 1895c: 242; Ichang, Hubei, China.

Funambulus riudonensis Allen, 1906: 472; Riudon, Hainan I, China.

D. melli Matchie, in Mell, 1922: 23; mountains E of Siudsau, Guangdong, China.

Range (Map 123). Malay peninsula north to Assam, Yunnan and Vietnam; isolates in central and SE China and on Hainan. Mainly montane (above 900m in Malaya).

Variation. The following subspecific names have been proposed:

Funambulus r. fuscus Bonhote, 1907c: 2 (and 1907b: 10); Nhatrang, Bali, Annam, Vietnam.

F. r. belfieldi Bonhote, 1908: 9; Mt Ulu Kali, Selangor, Malaya.

D. r. adamsoni Thomas, 1914d: 25; Maymyo, Burma.

D. r. ornatus Thomas, 1914d: 26; near Mengtsz (= Mongtse), S Yunnan.

D. r. opimus Thomas & Wroughton, 1916b: 237; Hkamti, upper Chindwin, Burma.

D. r. laomache Thomas, 1921b: 182; Ban Hoi Mak, near Pak Hin Bun, Mekong R, Laos.

In the main part of the range variation is slight and probably continuous, involving especially the extent of reddish colour on the head and thighs. In parts of Burma and Assam there is a very slight hint of a mid-dorsal stripe as in *D. lokriah macmillani*. The eastern Chinese form can be treated as a discrete subspecies:

D. r. pyrrhomerus. China including Hainan. Prominent reddish patch on the outside of each thigh; rostrum on average longer than in *D. r. rufogenis*.

Remarks. The isolated SE Chinese form has been treated as a separate species, *D. pyrrhomerus*, including *riudonensis, melli* and *gularis* (Moore & Tate, 1965). However this appeared to be based primarily upon the discontinuous range and differences in average measurements of the orbit and rostrum. They were unable to give diagnoses that clearly separated these two species, mainly because of the peculiar characters of *gularis*. The Chinese *pyrrhomerus* is here included in *D. rufigenis* as done also by Ellerman & Morrison-Scott (1951), but *D. gularis* is provisionally treated as a distinct species.

Chromosomes: 2n=38; FN=62 (Nadler & Hoffmann, 1970).

Dremomys gularis Red-throated squirrel

D. pyrrhomerus gularis Osgood, 1932: 284; Mt Fan Si Pan, near Chapa, Tonkin, Vietnam.

Range (Map 123). The valley of the Red R from N Vietnam to central Yunnan.

Remarks. Although previously included in *D. pyrrhomerus* or *D. rufigenis* this form is rather distinctive by virtue of the extensive reddish colour on the throat and neck and the dark grey ventral pelage. These characters seem constant (40 Vietnamese specimens examined by Moore & Tate, 1965; 22 specimens from Yunnan in Kunming Institute) and appear to justify specific rank unless intergradation can be demonstrated. The baculum is as in *D. rufigenis* in Malaya, but comparison with that of *D. r. pyrrhomerus* has not been possible.

Dremomys everetti Bornean mountain ground squirrel

Sciurus everetti Thomas, 1890b: 171; Mt Penrissen, W Sarawak.

Range (Map 123). N and W Borneo, montane, from 980 to 3400m.

Variation. Some individuals, perhaps only males, have a patch of orange-brown in the anal region, as in *D. pernyi* but more discretely defined.

Remarks. This species does not appear to be very closely related to any one of the mainland species of *Dremomys*.

LARISCUS Striped ground squirrels

Laria Gray, 1867a: 276; type species *Sciurus insignis* Cuvier. Pre-occupied by *Laria* Scopoli, 1763, a lepidopteran insect.
= *Lariscus* Thomas & Wroughton, 1909d: 389.
Paralariscus Ellerman, 1947b: 259 (*Sciurus hosei* Thomas).

Sunda subregion; 3 species (Table 190). Mainly terrestrial.

Lariscus insignis

Table 190. The species of *Lariscus*.

	insignis	*obscurus*	*hosei*
Range	Sunda	Mentawai Is	Borneo
Longitudinal dark stripes	3	3 (faint)	4
Two pale stripes on each flank	0	0	+
Venter	White or pale buff, ± grey hair-bases	Grey	Orange-brown (to roots)
Length of nasals	15.0–16.6	16.8–18.0	11.7, 12.0

Lariscus insignis Three-striped ground squirrel

Sciurus insignis Cuvier, in Geoffroy & Cuvier, 1821:liv. 34; Lampongs, S Sumatra.
Funambulus niobe Thomas, 1898b: 249; Pajo, ? = Pajokombo, Padang Highlands, W Sumatra.
F. castaneus Miller, 1900a: 217; Siantan I, Anamba Is.
F. rostratus Miller, 1903a: 24; Tana Bala I, Batu Is, W Sumatra.
F. peninsulae Miller, 1903a: 25; Trang, S Thailand.

Range (Map 124). Peninsular Thailand and Malaya, Sumatra, Java, Borneo; and small islands of Penang, Tioman, Singapore (probably extinct), Bintang (Riau Is), Tana Bala (Batu Is), Siantan (Anamba Is). Up to 370m in Borneo.

Variation. The following subspecific names have been proposed:
Funambulus i. diversus Thomas, 1898b: 248; Baram Dist, Sarawak.
F. i. jalorensis Bonhote, 1903a: 24; Jalor, S Thailand.
Laria i. javana Thomas & Wroughton, 1909c: 19 (and 1909d: 389); Buitenzorg (= Bogor), W Java.
Lariscus i. meridionalis Robinson & Kloss, 1911a: 172; Singapore.
L. i. fornicatus Robinson, 1917b: 102; Tioman I, Malaya.
L. niobe vulcanus Kloss, 1921a: 233; Ongop Ongop, Idjen Massif, Besoeki, E Java, 5700ft.
L. i. saturatus Chasen, 1935a: 99; Bintang I, Riau Is.
L. i. murianus Sody, 1937a: 219; Mt Moeriah, Semliro, C Java, 800m.
L. i. diversoides Sody, 1949a: 114; Sanggul, Benkulen, W Sumatra, c. 500m.
L. i. atchinensis Sody, 1949a: 115; Atang Putar, Atjeh, N Sumatra, 1000m.

The form on Borneo is most distinctive; those on the other main islands are only marginally separable from the mainland Malayan form:
L. i. jalorensis. Malaya. Dorsum rather greyish; flanks slightly tinged brown; venter white (hair all-white), tinged orange especially in axillae and groin. Tail short: 85–108mm.
L. i. insignis. Sumatra and Java. Variable but tending to be darker above; ventral hair grey, tipped buff, occupying a narrower zone than in *L. i. jalorensis* (but bright orange-brown, without grey bases, in *diversoides* in S Sumatra). Some on Java are not clearly separable from Malayan individuals. Montane forms on Sumatra (including *niobe*) are darker but much less so than *L. obscurus* and with short nasals typical of *L. insignis*.
L. i. diversus. Borneo. Flanks orange-brown; otherwise as Malaya but tail longer (120–140mm).

Those on the smaller islands of Penang, Tioman, Singapore, Bintang etc. are scarcely differentiated from the Malayan form. There is a tendency for the flanks to be browner but much less so than in Borneo.

Map **124**. *Lariscus, Menetes, Rhinosciurus*

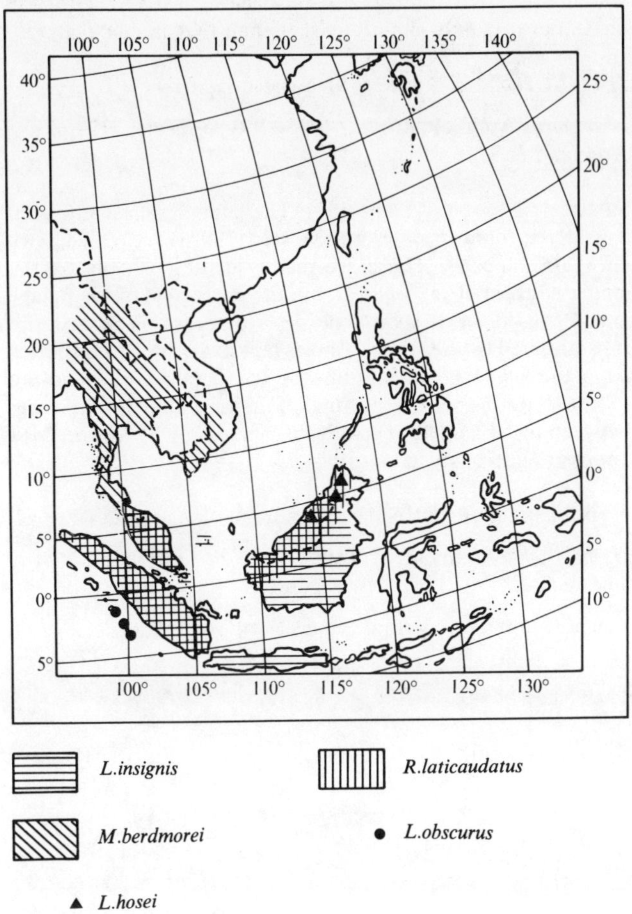

Lariscus obscurus Mentawai three-striped squirrel

Funambulus obscurus Miller, 1903a: 23; S Pagi I, Mentawai Is.
L. niobe siberu Chasen & Kloss, 1928c: 827; Siberut I, Mentawai Is.
L. insignis auroreus Sody, 1949a: 114; N Pagi I, Mentawai Is.

Range (Map 124). Islands of Siberut, Sipura, N Pagai and S Pagai, Mentawai Islands.

Recognition. The Mentawai forms are all separable from all forms of *L. insignis* by dark dorsum, making the stripes obscure, dark grey venter with no buff or just a little brownish suffusion posteriorly and sometimes a paler silvery central area, and long muzzle (nasals 17–18mm). The baculum is not distinguishable from that of *L. insignis*.

Ecology. Terrestrial, feeding mainly upon fallen fruit (Whitten, 1981).

Lariscus hosei Four-striped ground squirrel

Sciurus hosei Thomas, 1892b: 215; Mt Batu Song, Baram dist, Sarawak, Borneo, 5000ft.

Range (Map 124). NW Borneo, up to 1530m on Mt Kinabalu.

Recognition. This species is readily distinguished by the brown mid-dorsal stripe and the orange-brown venter (without grey hair-bases).

Remarks. CITES: Appendix II.

MENETES

Menetes Thomas, 1908f: 244; type species *Sciurus berdmorei* Blyth.

Indochinese subregion; 1 species. It is very doubtful if this justifies separation from *Lariscus* with which it is parapatric. It is characterized especially by the unusually high relief on the cheek-teeth. The upper incisors are less robust than in *Dremomys* and *Lariscus*. On the basis of the karyotype Petit & Dutrillaux (1985) proposed to separate *Menetes* from *Callosciurus* in a new tribe Menetini, but this was based upon only one species of *Callosciurus* and did not take into account any other Indomalayan species of the tribe Callosciurini.

Menetes berdmorei Indochinese ground squirrel

Sciurus berdmorei Blyth, 1849: 603; Thoungyeen dist, Tenasserim, S Burma.
S. mouhotei Gray, 1861: 137; Cambodia.
S. pyrrocephalus Milne-Edwards, 1867b: 225; Cochin-China.

Range (Map 124). Central Burma, SW Yunnan and C Vietnam south to peninsular Thailand.

Variation. The following subspecific names have been proposed:
Lariscus b. amotus Miller, 1913b: 24; Domel I, Mergui Is, Burma.
M. b. consularis Thomas, 1914d: 24; Nan, N Thailand, 200m.
M. b. decoratus Thomas, 1914d: 24; Mt Popa, Burma.
M. b. moerescens Thomas, 1914d: 25; Bali, near Nhatrang, Annam, Vietnam.
M. b. umbrosus Kloss, 1916b: 49; Koh Chang (I), Thailand.
M. b. rufescens Kloss, 1916b: 50; Koh Kut (I), Thailand.
M. b. koratensis Gyldenstolpe, 1917: 39; Sakerat, near Korat, E Thailand.
M. b. peninsularis Kloss, 1919b: 375; Ban Kok Klap, Nakon Sritamarat, near Bandon, peninsular Thailand.

Although there is considerable geographical variation affecting the presence and prominence of the dark stripes, it is unlikely that discrete subspecies can be recognized. The dark stripes, especially the median one, are faint or absent in the drier areas of Thailand and Cambodia but prominent in Burma and peninsular Thailand. There is some indication that there may also be a seasonal difference, the dark stripes being less prominent in the dry season (Moore & Tate, 1965; Askins, 1977). In peninsular Thailand the pale stripes on either side of the median line are rufous, in Burma they are agouti-brown.

Ecology. A ground-living squirrel found in savanna as well as in forest.

Remarks. Chromosomes: 2n=62; FN=76 (Nadler & Hoffmann, 1970).

RHINOSCIURUS

Rhinosciurus Blyth, 1856b: 477; type species *R. tupaioides* Blyth (derived from Gray, 1843a: 195 in which it is a *nomen nudum*).

Sunda subregion; 1 species. A distinctive genus with long, slender rostrum and very small, pro-odont upper incisors (Fig. 23). *Hyosciurus* should perhaps be included.

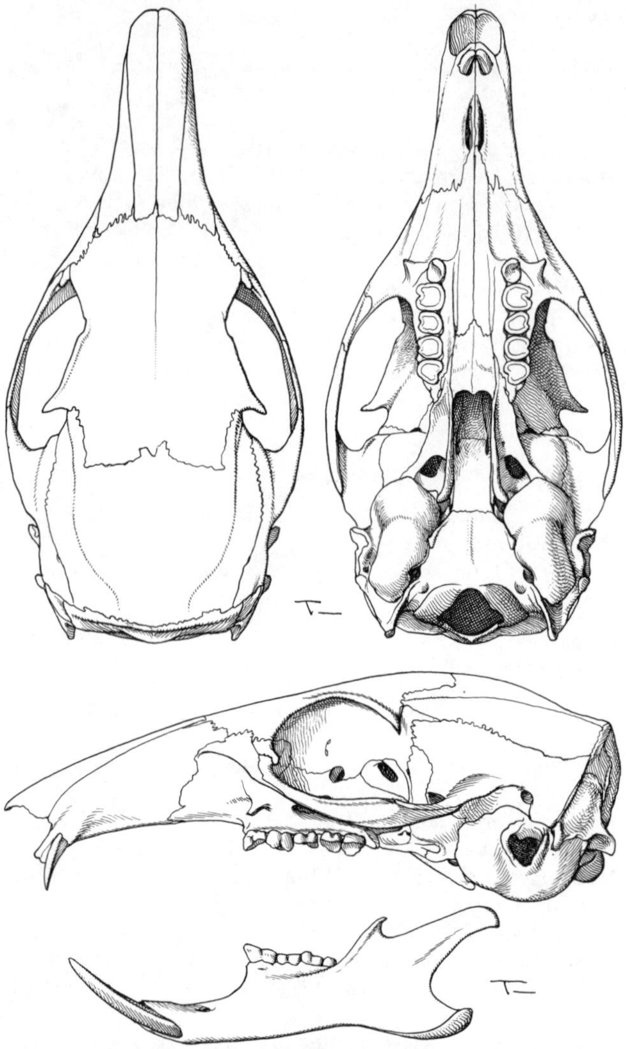

Fig.23. *Rhinosciurus laticaudatus.*

Rhinosciurus laticaudatus Shrew-faced squirrel

Sciurus laticaudatus Müller, 1840: 34; Müller & Schlegel, 1841:pl. 15 (and 1844: 100); Pontianak, W Borneo.
= *R. tupaioides* Gray, 1843a: 195.
R. robinsoni Thomas, 1908k: 104; Tioman I, Malaya.
R. peracer Thomas & Wroughton, 1909a: 440; Maxwell's Hill, Perak, Malaya.
R. leo Thomas & Wroughton, 1909a: 440; Singapore.
R. incultus Lyon, 1916a: 444; Tuangku I, Banjak Is, W Sumatra.

Range (Map 124). ? S Thailand; Malaya, Sumatra, Borneo; and small islands of Tioman, Singapore, Bintang, Batan, Kundur, Karimon (Riau Is), perhaps Lingga (Chasen, 1940a), Serasan (S Natuna Is), Bunguran (N Natuna Is). Up to 750m in Malaya.

Variation. The following subspecific names have been proposed:
Sciurus robinsoni alacris Thomas, 1908e: 306; Semangko Pass, Selangor-Pahang boundary, Malaya; 3000ft.
R. leo rhionis Thomas & Wroughton, 1909a: 441; Karimon I, Riau Is.
R. l. saturatus Robinson & Kloss, 1919a: 274; Rimbo Pengadang, Barisan Range, W Sumatra, 3200ft.

Geographical variation is very slight and it is doubtful whether the recognition of any subspecies is justified. The form on Sumatra is only slightly darker than that in Malaya.

HYOSCIURUS Sulawesi long-nosed squirrels

Hyosciurus Archbold & Tate, 1935: 2; type species *H. heinrichi* Archbold & Tate.

Sulawesi; 2 species (Table 191). The genus has usually been considered monospecific, but Musser (1987a) recognized two parapatric species. These are readily distinguishable by the longer rostrum of *H. heinrichi*, but the characters of the pelage and P^3 used in the original description of *ileile* are not substantiated by the larger samples now available. Both species have the hair in the mid-line of the nape reversed, often followed by a median parting of hair, about 3–4cm long, between the shoulders, a condition shared only by *Rhinosciurus* amongst Indomalayan squirrels.

Hyosciurus heinrichi Sulawesi montane long-nosed squirrel

H. heinrichi Archbold & Tate, 1935: 2; Latimodjong Mts, Sulawesi, 2300m.

Range (Map 125). Central Sulawesi, above 1500m.

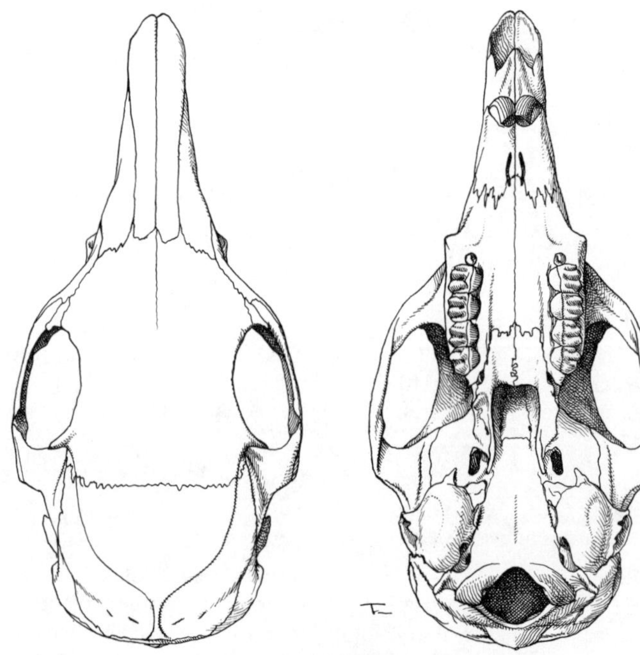

Fig.24. *Hyosciurus heinrichi.*

Map **125**. *Nannosciurus, Exilisciurus, Hyosciurus, Rubrisciurus*

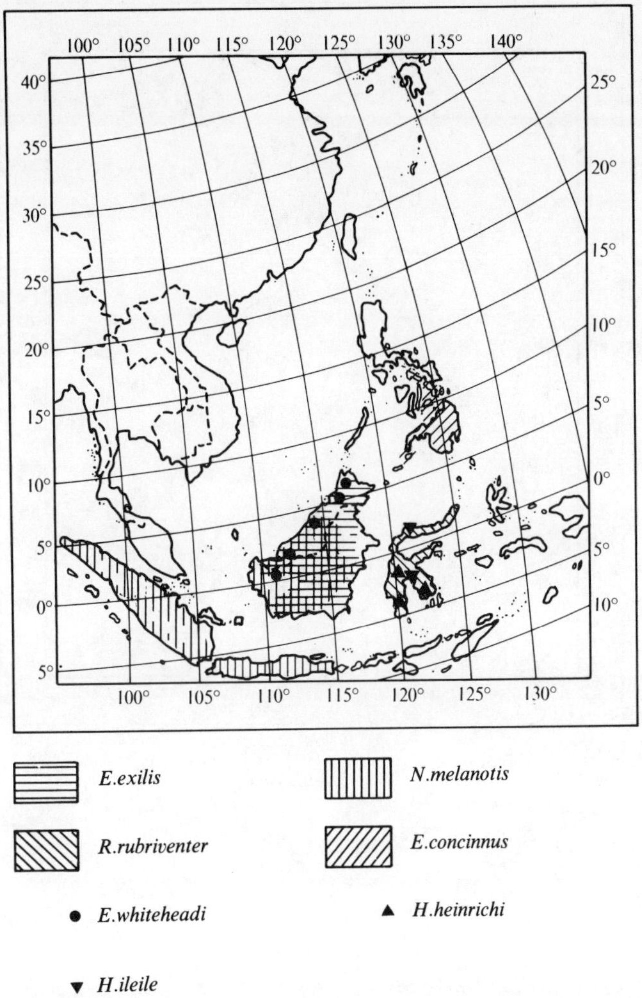

Table 191. The species of *Hyosciurus*.

	heinrichi	*ileile*
Range in Sulawesi	Above 1500m	Below 1700m
Head & body	195–240	195–250
Tail	90–108	110–135
Snout	Longer	Shorter
Skull: greatest length	63–67	57–65
Nasals: length	25–27	19–23

Hyosciurus ileile Sulawesi lowland long-nosed squirrel

H. heinrichi ileile Tate & Archbold, 1936: 1; Ile Ile, N Sulawesi, 1700m.

Range (Map 125). Sulawesi, in the lowlands and up to 1700m.

NANNOSCIURUS Pygmy squirrels

Nannosciurus Trouessart, 1880b: 292; type species *Sciurus melanotis* Müller & Schlegel.

Greater Sunda Islands; 1 species. Other species formerly included in *Nannosciurus* were separated in the genus *Exilisciurus* by Moore (1958b).

Nannosciurus melanotis Black-eared pygmy squirrel

Sciurus melanotis Müller & Schlegel, 1841 :pl. 14 (and 1844: 87); Java (restricted by Lyon (1906a); lectotype selected by Heaney, 1985: 25).
S. soricinus Waterhouse, 1838c: 46 (*nomen nudum*); Anderson, 1878b: 265; Java.
N. pulcher Miller, 1902a: 153; Singkep I, Lingga Is.
N. sumatranus Lyon, 1906b: 53; Tarussan Bay, W Sumatra.
N. borneanus Lyon, 1906b: 54; Sanggau, W Borneo.
N. bancanus Lyon, 1906b: 55; Bangka I, Sumatra.

Range (Map 125). Java, Sumatra, Borneo and small islands of Sinkep and Bangka (Sumatra), below 1000m.

Variation. One subspecies has been described:
N. m. pallidus Chasen & Kloss, 1928a: 43; Long Poehoes, E Borneo.

On Java (*N. m. melanotis*) the venter is dark grey [n=10]; on Borneo (*N. m. borneanus*), Sumatra [n=6] and Bangka [n=4] the venter is tinged with orange-brown to a variable extent.

EXILISCIURUS

Exilisciurus Moore, 1958b: 4; type species *Sciurus exilis* Müller.

Borneo and Philippines; 3 species (Table 192). Formerly included in *Nannosciurus* but separated by Moore (1958b), supported by Heaney (1985) who gave a detailed review. However Heaney demonstrated that together these two genera are likely to form a monophyletic group; their recognition as separate genera is therefore somewhat arbitrary.

Table 192. The species of *Exilisciurus*.

	exilis	*whiteheadi*	*concinnus*
Range	Borneo	Borneo	S Philippines
Head & body	62–82	83–93	82–90
Hind feet	18–23	17–22	24–27
Ear tufts	0	Long	0
Skull: greatest length	22–23	25–26	26–27
Maxillary tooth-row	3.2–3.6	3.6–3.9	3.4–4.2
Blade of baculum serrated along entire length (E); or distal 2/3 only (D)	D	D	E

Exilisciurus exilis Least pygmy squirrel

Sciurus exilis Müller, 1838: 148; 'Batang Singalang', Sumatra, corrected to Kapuas basin, W Borneo by Chasen & Kloss (1928a); lectotype from Tanah Laut (=Pleihari) dist., S Kalimantan, Borneo designated by Heaney (1985).

Range (Map 125). Borneo, up to at least 1700m; also on Banggi I, N of Borneo. The lectotype (Leiden Museum) is labelled 'Mt Singalang, Sumatra' but no further specimens have been recorded from Sumatra (nor from Malaya) and this was presumed to be erroneous by Chasen & Kloss (1928a) and Heaney (1985).

Variation. Two subspecies have been described but are unlikely to be justified:
Nannosciurus e. retectus Thomas, 1910c: 387; Banguey I, N Borneo.
N. e. sordidus Chasen & Kloss, 1928a: 44; Long Temelan, Telen R, E Borneo.

Exilisciurus whiteheadi Tufted pygmy squirrel

Sciurus whiteheadi Thomas, 1887a: 127; Mt Kinabalu, N Borneo, 3000ft.

Range (Map 125). N and W Borneo, from 600–3000m.

Exilisciurus concinnus Philippine pygmy squirrel

Sciurus concinnus Thomas, 1888b: 407; Isabella, Basilan I, Philippines.
Nannosciurus samaricus Thomas, 1897a: 1 (and 1898c: 389, pl. 30); Samar I, Philippines.
N. surrutilus Hollister, 1913a: 313; Mt Bliss, Mindanao, Philippines.
N. lunceforđi Taylor, 1934: 373; Saub, Cotabato Prov., Mindanao, Philippines.

Range (Map 125). Philippine islands of Mindanao, Basilan, Siargao, Dinagat, Bohol, Leyte and Biliran; up to 2000m on Mindanao.

Variation. Differences between the island populations are not sufficient to justify the recognition of subspecies (Heaney, 1985).

PROSCIURILLUS Sulawesi dwarf squirrels

Prosciurillus Ellerman, 1947b: 259; type species *Sciurus murinus* Forsten.

Sulawesi; 4 species (Table 193). The relationship of *Prosciurillus* with the genera *Nannosciurus, Exilisciurus* and *Sundasciurus* was reviewed by Heaney (1985).

Table 193. The species of *Prosciurillus*.

	murinus	*abstrusus*	*leucomus*	*weberi*
Range in Sulawesi	Throughout	SE	Throughout	Central
Head & body	c.130	c.134	140–180	c.160
Hind foot	27–30	33	38–41	c.38
Mid-dorsal black stripe	0	0	0	+
Back of ears white (W); black (B) or as dorsum (D)	D	W	B/D	B
Venter (overall colour)	Grey	Grey	Rufous or buff	Dark chestnut
Skull: greatest length	31–37	38–39	43–47	45
Maxillary tooth-row	5.5–6.1	7.7–8.6	6.8–7.1	7.0

Prosciurillus murinus

Sciurus murinus Müller & Schlegel, 1844: 87; 'Celebes', restricted to NE Sulawesi by Sody (1949b).
? *S. evidens* Miller & Hollister, 1921b: 99; Lembeh I, off NE Sulawesi.

Range (Map 126). Throughout most of Sulawesi, including the small island of Lembeh.

Variation. Two subspecies have been described:
Sciurus m. necopinus Miller & Hollister, 1921b: 98; Goenoeng Lehio (SW of Lake Lindoe), central Sulawesi.
Sciurillus m. griseus Sody, 1949a: 77; Bumbulan, Menado Dist, N Sulawesi (S Coast). Ventral pelage light.

Ecology. Forages on the ground and in the lower canopy (G G Musser, pers. comm.).

Map **126**. *Prosciurillus, Rheithrosciurus, Tamias*

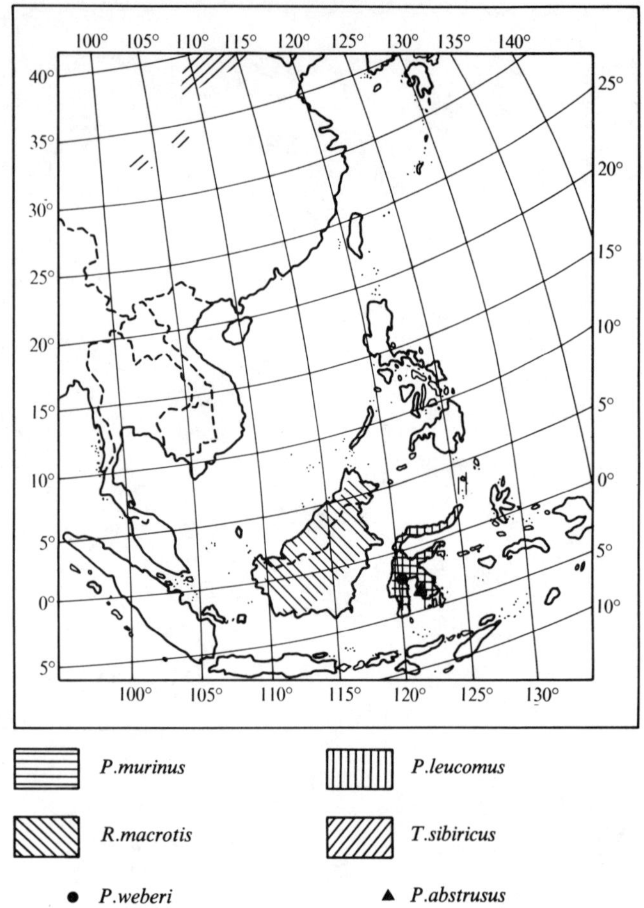

Prosciurillus leucomus

Sciurus leucomus Müller & Schlegel, 1844: 87; 'Celebes', restricted to Minahassa, N Sulawesi by Meyer (1898).
S. rosenbergii Jentink, 1879b: 37; Siao, Sanghir Is.
? *S. alstoni* Anderson, 1879: 252, pl. xxi; locality unknown, 'probably Borneo'.
S. tonkeanus Meyer, 1896/97: 25, pl. 10; Tonkean, SE central Sulawesi.
S. tingahi Meyer, 1896/97: 27, pl. 10; Tagalunda and Ruang, Sanghir Is.
S. sarasinorum Meyer, 1898: 1 (and 1899: 21, pl. 5); Ussu, N end of Gulf of Boni, Sulawesi.
S. topapuensis Roux, 1910: 516; Mt Topapu, C Sulawesi, 1550m.
S. mowewensis Roux, 1910: 516; Mowewe, SE Sulawesi.
S. elbertae Schwarz, 1911: 639; Eempuhu, E Kabaena I, SE Sulawesi.

Range (Map 126). Sulawesi, at all altitudes, and on the islands of Siau, Tagalund and Ruang (Sanghir Is), Buton, E Kabaena (SE Sulawesi).

Variation. The following subspecific names have been proposed:
Sciurus l. occidentalis Meyer, 1898: 2; between Kottabangon and Bolang Mongondo, NW Sulawesi.
Callosciurus l. hirsutus Hayman, 1946: 576; Tamalanti, W Sulawesi, 3300ft.

The form on the Sangihe Is, *P. l. rosenbergi*, is distinctive, lacking the grey patches behind the ears, with the ear-fringes concolorous with the dorsum rather than black.

There is considerable geographical variation in colour but the limits of the forms on the mainland are unknown. The following are distinctive:

P. l. leucomus. N Sulawesi. A prominent grey patch behind each ear; ear fringes black; venter orange-brown, the hairs with pale grey bases; tail as dorsum.

P. l. occidentalis. NW Sulawesi. As above but a conspicuous black patch behind each ear and the grey patches less prominent.

P. l. rosenbergi. Sangihe Is. No grey patches; ear fringes brown as dorsum; venter buff; tail dark.

P. l. hirsutus. W Sulawesi. As *P. l. rosenbergi* but pelage very long and dense.

P. l. sarasinorum. SE Sulawesi. No grey patches; ear tufts white; venter orange-brown; tail dark.

Ecology. Occupy the canopy of primary forest (Musser, 1987a).

Prosciurillus weberi

Sciurus weberi Jentink, 1890a: 115; Luwu, near Palopo, C Sulawesi.

Range (Map 126). Central Sulawesi.

Remarks. Included in *P. leucomus* by Laurie & Hill (1954) but it is more distinctive than the other forms of *P. leucomus* especially by virtue of a bold mid-dorsal black stripe and would appear to justify specific rank.

Prosciurillus abstrusus

P. abstrusus Moore, 1958c: 3; Tanke Salokko Mt, Mengkoka Mts, SE Sulawesi, 3°40'S, 121°13'E, 1500m.

Range (Map 126). Known only from the type locality where it coexists with both *P. murinus* and *P. leucomus*.

Recognition. More similar to *P. murinus* than to *P. leucomus*, but differs in larger size and white-backed ears. Distinguished from the adjacent *P. leucomus sarasinorum* by smaller size, darker dorsum, shorter and thinner tail and untufted ears.

RUBRISCIURUS

Rubrisciurus Ellerman, in Laurie & Hill, 1954: 94; type species *Sciurus rubriventer* Müller & Schlegel.

Sulawesi; 1 species. This was described as a subgenus of *Callosciurus* mainly on the basis of its large size. Its relationship to *Callosciurus, Sundasciurus* and *Prosciurillus* is unclear.

Rubrisciurus rubriventer Sulawesi giant squirrel

Sciurus rubriventer Müller & Schlegel, 1844: 86. Minahassa, N Sulawesi.

Range (Map 125). Sulawesi.

Ecology. Forages in the canopy but nests in the low understorey (Musser, 1987a and pers. comm.).

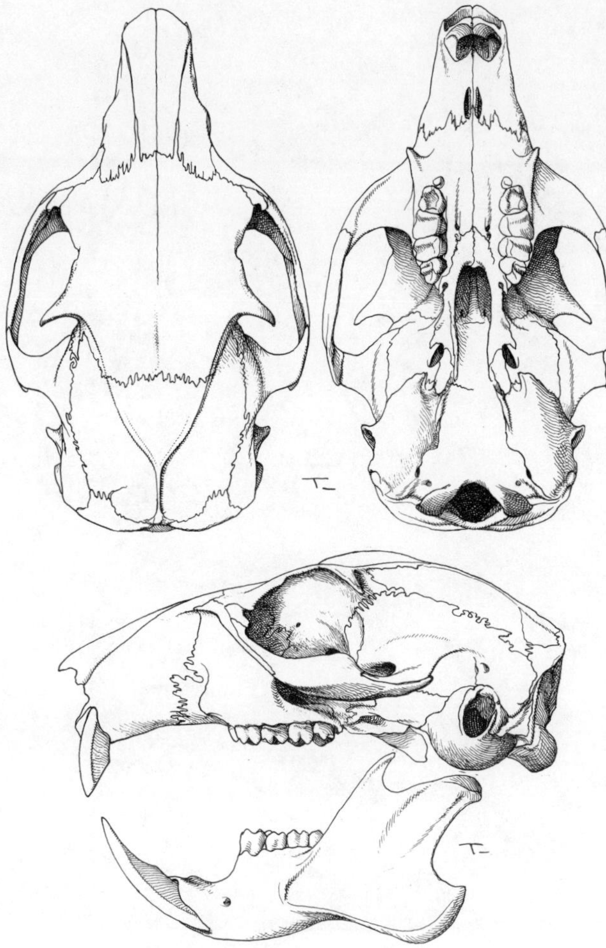

Fig.**25**. *Rubrisciurus rubriventer*.

? Subfamily Sciurinae

RHEITHROSCIURUS

Rheithrosciurus Gray, 1867a: 271; type species *Sciurus macrotis* Gray.

Borneo; 1 species. A very distinctive genus, characterized by extremely opisthodont upper incisors which are very deep and narrow (c. 6x2.5mm) with fine parallel ridges and grooves (c. 10) on the anterior surfaces (Fig. 26). The lower incisors each have c. 7 ridges. The rostrum is extremely deep, narrow and parallel-sided. The baculum is unique amongst Indomalayan squirrels but is hardly distinguishable from that of the Holarctic *Sciurus*, e. g. *S. vulgaris*, with a complex, asymmetrical spatulate tip (two specimens in BMNH). This has led to the inclusion of *Rheithrosciurus* in the tribe Sciurini in spite of the geographical isolation (Pocock, 1923b; Moore, 1958a). It resembles *Sciurus* also in having four pairs of teats, tufted ears and two septae in each auditory bulla.

Rheithrosciurus macrotis Tufted ground squirrel

Sciurus macrotis Gray, 1856: 1437 (and 1857: 341, pl. 46); Sarawak.

Range (Map 126). Borneo, up to c. 1000m.

Ecology. Found mainly on the ground or lower tree trunks (Payne *et al*., 1985).

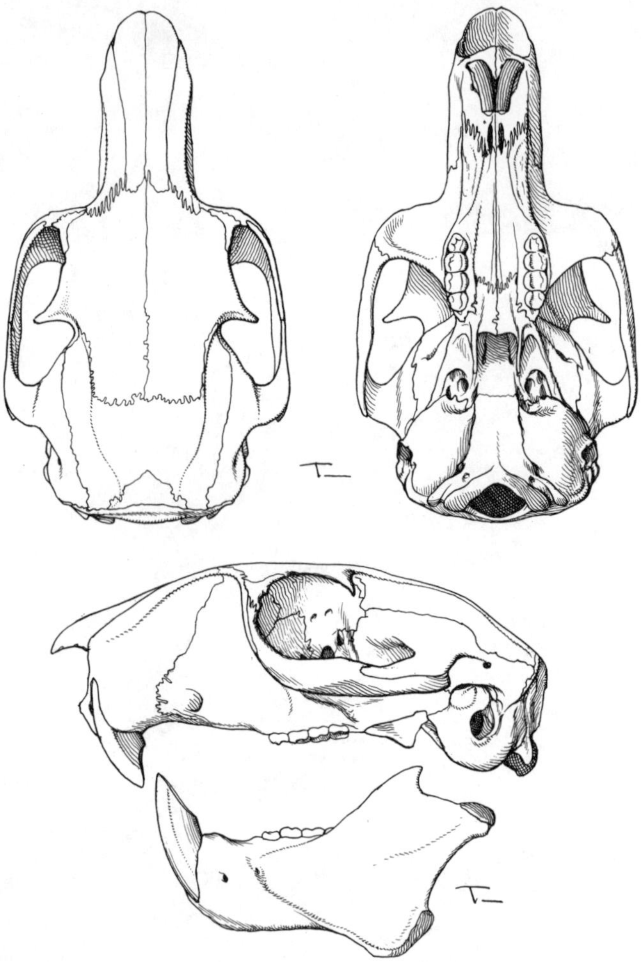

Fig.**26**. *Rheithrosciurus macrotis*.

Subfamily Marmotinae

Marmotinae Pocock, 1923b: 240; type genus *Marmota* Frisch.

TAMIAS Chipmunks

Tamias Illiger, 1811: 83; type species *Sciurus striatus* Linnaeus.
Eutamias Trouessart, 1880c: 86 (*S. striatus asiaticus* Gmelin).

N America (20 species) and N Palaearctic (1 species). The Palaearctic species extends south to Sichuan and is therefore included here. Additional fossil species have been described from China and Pakistan:
Eutamias sihongensis Qiu & Liu, 1986: 200; Xiacaowan, Jiangsu; Miocene.
Eutamias urialis Munthe, 1980: 4; Sumbalenwala, Mianwali Dist., Pakistan, 32°56'N, 71°39'E; Miocene.

Tamias sibiricus Siberian chipmunk

Sciurus sibiricus Laxmann, 1769: 69; Barnaul, W Siberia.
S. striatus asiaticus Gmelin, 1788: 150; E Siberia.
Eutamias albogularis Allen, J. A.,1909b: 429; Taipai Shan, S Shaanxi, China.
And other names from the Palaearctic Region.

Range (Map 126). Coniferous forest zone of Siberia and N China, south to N Sichuan; Hokkaido.

Variation. The form in Sichuan, S Gansu and S Shaanxi, *T. s. albogularis*, is darker than those from further north.

Family Pteromyidae

Flying squirrels

Pteromyini Brandt, 1855: 157; type genus *Pteromys* Cuvier, 1800.
Petauristidae Miller, 1912: 940.

A predominantly Indomalayan group with 11 genera in the region, plus one in the Palaearctic and one in N America (Table 194). The flying squirrels have by some been believed to be polyphyletic in relation to the non-flying squirrels, but there is little foundation for this and the anatomy of the wrist, and the way it supports the gliding membrane, points strongly to a monophyletic origin, being quite different from that of the other, unrelated groups of gliding mammals (Thorington, 1984). Family rank has been supported on the grounds that the differences in dentition are as clear-cut in Oligocene fossils as they are now (Bruijn & Uney, 1989). Some of the genera are very weakly defined.

The complexity of the teeth and the size of the caecum are correlated with the importance of leaves in the diet, as reviewed by Muul & Lim (1978) for the Malaysian species.

The following fossil species has been described:
Pteromys lopingensis Young, 1947: 166; Loping, Jiangxi, China; Pleistocene. Probably referable to *Petaurista*, perhaps *Trogopterus*.

TROGOPTERUS

Trogopterus Heude, 1898: 46; type species *Pteromys xanthipes* Milne-Edwards.
Belomys Thomas, 1908c: 2 (*Sciuropterus pearsonii* Gray).

Himalayas, Indochina, S China; 2 species (Table 195), characterized especially by complex, semi-hypsodont teeth (Fig. 27). *Belomys* has usually been treated as a separate genus but there seems little doubt that these two species are very closely related.

Trogopterus xanthipes Complex-toothed flying squirrel

Pteromys xanthipes Milne-Edwards, 1867a: 376; NE Hebei, China.
T. mordax Thomas, 1914c: 230; Ichang, Hubei.
T. himalaicus Thomas, 1914c: 231; Gyantse, Chumbi Valley, S Tibet.
T. edithae Thomas, 1923c: 658; NW flank of Likiang Range, Yunnan.
T. minax Thomas, 1923c: 660; near Wan Chuen, upper Min R, Sichuan.

Range (Map 127). China: in Hebei, S Gansu, Shaanxi, Hubei, Sichuan, Yunnan and SE Tibet.

Variation. There is variability in the colour of the tail (dark in Yunnan) and the amount of orange tinge in the ventral pelage (maximum in specimens from Beijing and Ichang) but specimens are too few to define subspecies clearly.

Ecology. A folivore, eating leaves including those of conifers such as *Thuja* and *Pinus* (Wang, F., 1985).

Table 194. The genera of Pteromyidae.

	Trogopterus	*Pteromyscus*	*Petaurista*	*Eupetaurus*	*Aeretes*	*Biswamoyopterus*	*Iomys*	*Hylopetes*	*Petinomys*	*Petaurillus*	*Aeromys*
No. of species	2	1	9	1	1	1	2	10	7	2	2
Range	China, Himalayas	Sunda	Widespread	Himalayas	China	Assam	Sunda	Widespread	Widespread	Malaya, Borneo	Malaya, Borneo, Sumatra
Head & body (max.)	300	290	580	610	310	405	230	300	300	90	425
Hind feet	30–60	35–44	60–82	85–87	60–65	78	30–40	22–65	21–52	16–20	60–78
Tail distichous	+	+	0	+	+	0	++	++	++	+	0
Interfemoral membrane	0	0	+	+	0	+	0	0	0	0	+
Pelage very dense, woolly	0	0	0/+	+	0	0	0	0	0	0	0
Upper incisors grooved	0	0	0	0	+	0	0	0	0	0	0
Incisors brownish orange (O) or pale yellow (Y)	O	O	O	Y	O	Y	O	O	O	O	Y
P^3 present	+	(+)	+	+	+	+	0	+/0	+	+	+
P^4: M^1	>>	=	=/>	>>	=	>	=	=	=	<	=
M_3: M_2	=/>	>	=	=			>	=	>	=	>
Cheekteeth complex	++ (Fig.27)	++	+	0	+	0	0	0	0	0	0
" hypsodont	+	0	0/+	++	0	0	0	0	0	0	0
Septae in auditory bullae (number); H = honeycomb pattern	8–10/H	c.5	1–4	2/3	3	H	0/2	2/3	5–7/H	2	3

Table 195. The species of *Trogopterus*.

	xanthipes	*pearsoni*
Range	China	Himalayas-S China
Head & body (max.)	300	200
Hind feet	c.60	31–39
P^4: M^1	> >	>
M_3: M_2	=	>
Septae in bullae	8–10	Honeycomb

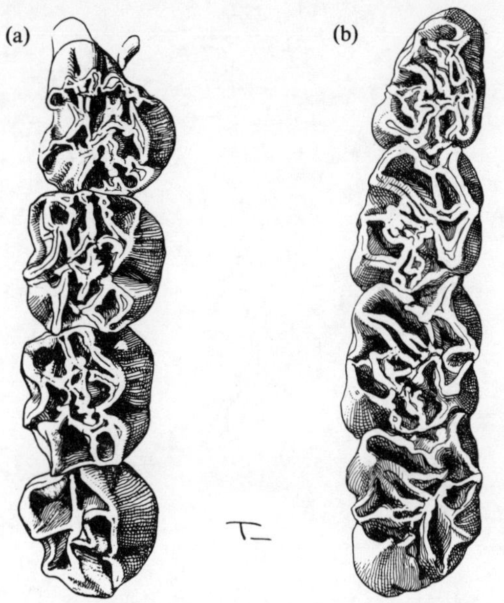

Fig.**27**. *Trogopterus xanthipes*. Cheekteeth: (a) upper right; (b) lower left.

Trogopterus pearsonii Hairy-footed flying squirrel

Sciuropterus pearsonii Gray, 1842: 263; Darjeeling, India.
S. villosus Blyth, 1847a: 866; Upper Assam.
S. kaleensis Swinhoe, 1863: 359; N Taiwan.
Belomys trichotis Thomas, 1908c: 7; Machi, Manipur, India.

Range (Map 127). Himalayas from E Nepal east to N Burma, Thailand, Vietnam, Yunnan and Guangdong; and on Hainan and Taiwan; between 1500 and 2400m in Nepal, down to 900m in Sikkim.

Variation. There is little variation in dorsal colour but some specimens from Manipur and Burma (*trichotis*) have the underside pure white, without grey bases. The following subspecies has been described:
Belomys p. blandus Osgood, 1932: 269; Muong Moun, S of Lai Chau, Tonkin, Vietnam.

PTEROMYSCUS

Pteromyscus Thomas, 1908c: 3; type species *Sciuropterus pulverulentus* Günther.

Malaya, Sumatra, Borneo. One species, closely related to *Trogopterus*.

Pteromyscus pulverulentus Smoky flying squirrel

Sciuropterus pulverulentus Günther, 1873: 413, pl. 38; Penang I, Malaya.
P. borneanus Thomas, 1908c: 7; Baram dist, Sarawak.

Map **127**. *Trogopterus*, *Pteromyscus*, *Petaurista* (part)

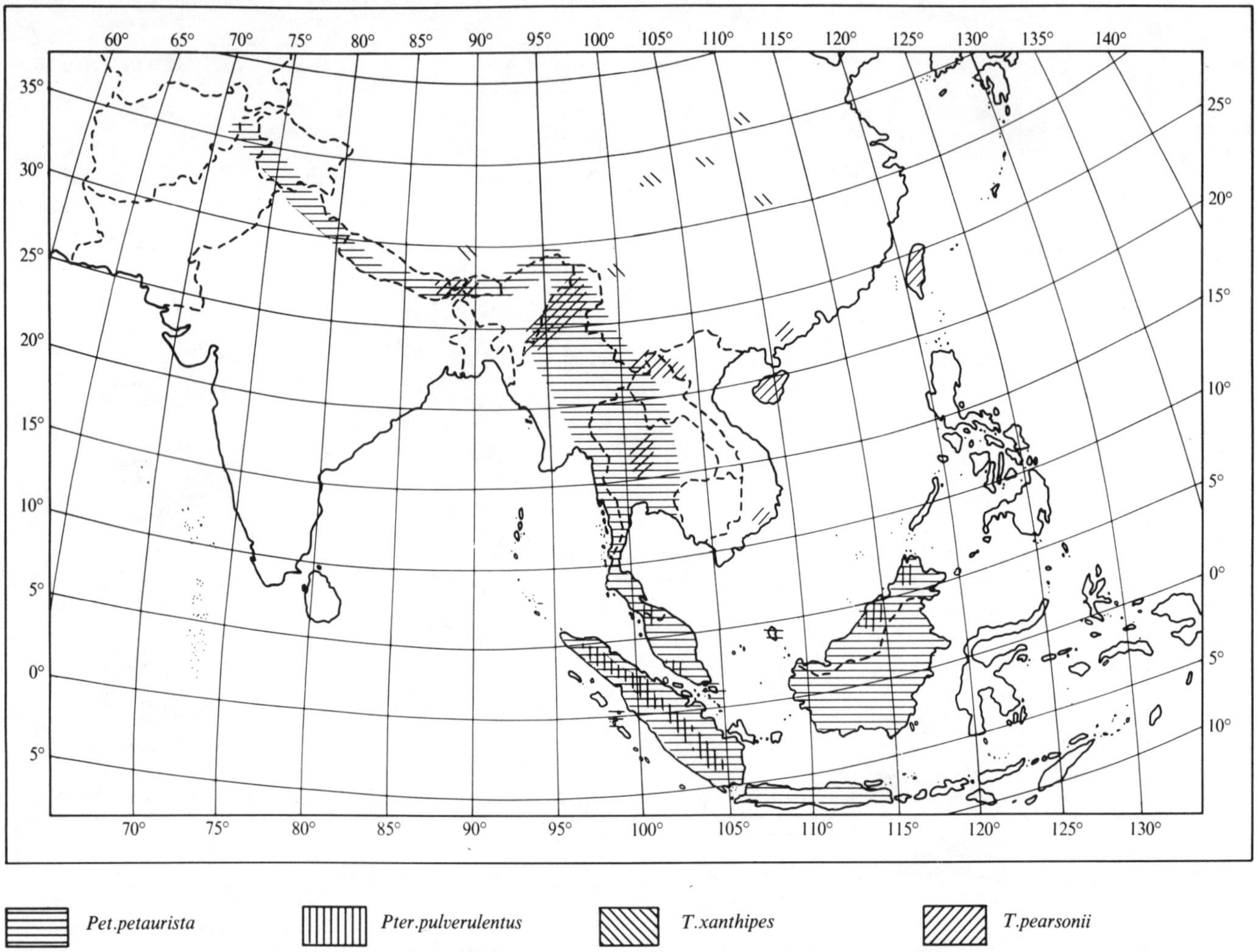

Range (Map 127). Peninsular Thailand and Malaya, Sumatra and Borneo (N & W); and on the island of Penang (Malaya).

Variation. The Bornean form was distinguished by shorter white flecks in the dorsal pelage, whiter underside and larger size, but is doubtfully distinct.

Ecology. Confined to undisturbed primary forest and therefore very vulnerable in many parts of its range (Muul, 1989).

PETAURISTA Giant flying squirrels

Petaurista Link, 1795: 78; type species *Sciurus petaurista* Pallas.

Throughout the mainland and the Greater Sunda Islands, and on Taiwan and Japan. Ten species, of which one, *P. leucogenys* Temminck, 1827, is outside the Region, in Japan (Table 196). Delimitation of the species has been unstable; the arrangement adopted here differs substantially from that of Ellerman (1949) in recognizing *P. xanthotis* as distinct from *P. leucogenys*; *P. nobilis* as distinct from *P. magnificus* (Ghose & Saha, 1981); *P. caniceps* and *P. sybilla* as distinct from *P. elegans*; and *P. philippensis* as distinct from *P. petaurista*, although that group badly needs a comprehensive revision to clarify the relationship between these forms, especially in southern China, Burma and Thailand. Although some chromosome data are available they are too scanty to throw any light on the problem.

Middle Pleistocene fossils from Koloshan, Sichuan and elsewhere in S China have been identified as *P.* cf. *brachyodus* Young, 1934, described from Zhoukoudian, N China (Young & Liu, 1951).

Petaurista philippensis

Table 196. The species of *Petaurista*.

	petaurista	*philippensis*	*elegans*	*sybilla*	*caniceps*	*alborufus*	*magnificus*	*nobilis*	*xanthotis*
Range	Himalayas-Sunda	S China-Indochina-Sri Lanka	S China-Borneo	Yunnan, Burma	Himalayas, SW China	S China, Taiwan	Nepal	Sikkim	SW China
Head & body (max.)	520	490	370	c.400	370	c.580	420	490	430
Tail (max.)	630	550	375	c.370	400	c.430	480	460	345
Hind feet	75–100	65–90	59–67	c.60	55–65	c.78	72–74	70–81	65–68
Head pure white	0	0	0	0	0	+	0	0	0
Head grey	0	0	0	0	+	0	0	0	0
Yellowish shoulder patches	0	0	0	0	0	0	+	+	0
Dark dorsal 'saddle', ± thin mid-dorsal pale stripe	0	0	0	0	0	0	0	+	0
Dorsal pelage with bold white spots	0	0	+	0	0	0	0	0	0
Dorsal pelage hoary with white hair-tips	+/0	+/0	+/0	0	0	0	+	0	+
Tail with discrete black tip	+	0	+/0	0	+	+/0	+	+/0	+
Skull: max.length	66–73	65–81	<65	c.60	62–65	78–83	65–70	68–75	65–67
Maxillary tooth-row	14.6–17.4	15.0–17.8	13.5–14.6	14.0–14.4	13.4–15.0	15.2–16.4	14.9–15.8	15.0–16.8	15.0–18.0

Petaurista petaurista Red flying squirrel

Sciurus petaurista Pallas, 1766: 54; W Java, restricted to Preanger Regencies by Robinson & Kloss (1918c: 221).
P. taguan Link, 1795; ? Java.
Pteromys nitidus Desmarest, 1818: 403; Java.
Pt. albiventer Gray, 1834(in 1830–35):pl. 18; Nepal.
Pt. melanotus Gray, 1837a: 584; Selangor; originally Nepal in error.
Pt. inornatus Geoffroy, 1844: 62; N India.
P. nitidula Thomas, 1900a: 592; Bunguran I, N Natuna Is.
P. batuana Miller, 1903a: 27; Tana Bala, Batu Is, W Sumatra.
P. terutaus Lyon, 1907b: 17; Terutau I, Straits of Malacca.
P. birrelli Wroughton, 1911a: 1014; Muree, Hazara, Punjab.
P. fulvinus Wroughton, 1911a: 1014; Simla, Punjab.
P. candidula Wroughton, 1911a: 1014; Kindat, W Burma.
P. mimicus Miller, 1913b: 27; Rupat I, E Sumatra.
P. taylori Thomas, 1914e: 205; Bankasun, S Tenasserim, Burma.
P. annamensis barroni Kloss, 1916c: 33; Hup Bon, Sriracha, SE Thailand.

Range (Map 127). Sumatra, Java, Borneo and Malaya, through Thailand and Burma to the Himalayan foothills, west to Pakistan and NE Afghanistan; including adjacent small islands of Tioman, Tarutau, Penang, Singapore, Rupat (E Sumatra), Bintan (Riau Is), Tana Bala (Batu Is), Bunguran. In Nepal it occurs at altitudes between 150 and 1500m; in Pakistan from 1350m up to the tree-line.

Variation. The following nominal subspecies have been described:
P. nitida rajah Thomas, 1908d: 251; Mt Dulit, Baram Dist, Sarawak.
P. nitida marchio Thomas, 1908d: 251; Si Rambi, Sumatra.
P. nitida cicur Robinson & Kloss, 1914a: 223; Ban Kok Klap, Bandon, NE Malaya.
P. p. nigricaudatus Robinson & Kloss, 1918b: 223; Idjen Massif, E Java, 5700ft.
P. p. penangensis Robinson & Kloss, 1918b: 224; Penang I, Malaya.
P. p. lumholtzi Gyldenstolpe, 1920: 28; Poeroek Tjahoe, Barito R, central Borneo.
P. p. stellaris Chasen, 1940a: 113; Bintang I, Riau Is.
P. p. rufipes Sody, 1949a: 68; Kluang, Palembang, SE Sumatra. Pre-occupied by *P. p. rufipes* Allen, 1925 from China (in *P. philippensis*).
= *P. p. sodyi* Harris, 1951: 234.
P. p. interceptio Sody, 1949a: 69; Mt Tjerimai, W Java, 700m.
P. p. nigrescens Medway, 1965b: 116; Sandakan, Sabah, Borneo.

Geographical variation in colour is considerable but there is no comprehensive description or revision. The major aspects can be described as follows:

Java (*P. p. petaurista* etc.). Very dark rufous or greyish brown to black above, more rufous in the east; tail dark grey or black, sometimes with rufous in the centre; venter greyish buff to orange-tawny. (See Sody, 1949a, for further details.)

Sumatra (*P. p. marchio*). Dark rufous above, dull buffy brown below, tail rufous with black tip and some black at base.

Borneo and Malayan Peninsula, including Bunguran, Riau and Penang islands. (*P. p. melanotus* etc.). Bright orange brown above, similar but much paler below; feet and tail-tip black. Darker forms occur locally in Borneo but are unlikely to constitute discrete subspecies. The form on Bunguran I (*P. p. nitidula*) is very similar but has some white on the cheeks and above the eyes —it is clearly a form of *P. petaurista* and not *P. elegans* as proposed by Ellerman (1949) on the basis of size alone.

Batu Is (W of Sumatra) (*P. p. batuana*). Black above, very dark grey-brown below.

Tarutao I (S Thailand) (*P. p. terutaus*). As Malaya but head speckled agouti and trace of agouti speckling on back.

S Burma (Tenasserim), W Thailand etc. (*P. p. taylori*). Very similar to *P. p. terataus* but light speckling on head and back more pronounced. This form, along with that on Tarutao I, therefore shows a transition between the uniformly rufous form in peninsular Malaya and the heavily speckled forms in N Burma and the Himalayas.

Burma and Thailand (*P. a. candidula* etc.). Pale rufous with strong white speckling, especially on head; tail pale brown or grey with short black tip; throat white; a pale line over each shoulder.

W Yunnan—Himalayan foothills (*P. a. albiventer* etc.). Brown above, with white speckling; membrane and tail dark brown, not rufous; venter off-white to pale buff, throat usually white.

Recognition. South of the Isthmus of Kra this species is clearly distinguished from sympatric species by the combination of unspotted rufous pelage, black feet and black tail-tip; north of the Isthmus the dorsal pelage, especially the head, is increasingly speckled with white but the discrete black tail-tip, white anterior margins of the ears and usually a pale line on each shoulder distinguish it from sympatric forms of *P. philippensis*.

Ecology. Occurs in plantations as well as in forest.

Remarks. The forms here included in *P. philippensis* were included in *P. petaurista* by Ellerman & Morrison-Scott (1951) and by Ellerman (1961). The differences between the forms of *P. petaurista* in Malaya and peninsular Thailand (e. g. *P. p. taylori*) and those in Burma and Thailand allocated here to *P. philippensis* seem extreme. However much remains to be done to clarify the situation, especially to identify the position and nature of the boundary on the mainland and to map the variation in detail.

The forms *candidula, barroni* and *taylori* seem much more similar to the Himalayan forms typified by *P. p. albiventer* than to the white-headed *P. alborufus* of China in which they were included by Ellerman (1949). They also lack the enlarged P^4 characteristic of *P. alborufus*. Hill (1960) has already excluded *taylori* from *P. alborufus*. These forms appear to form a remarkably smooth cline linking the Himalayan forms with the uniformly rufous Sunda forms. However this cline may be sharply broken at the Isthmus of Kra, with only limited hybridization between two dissimilar forms, justifying recognition of the northern form as a separate species, *P. albiventer* (I Muul, pers. comm.).

Chromosomes (Selangor, Malaya): 2n=38, FNa=72 (Yong & Dhaliwal, 1976).

Petaurista philippensis

Pteromys philippensis Elliot, 1839: 217; near Madras, India.
Pt. oral Tickell, 1842: 401; Singhbum dist, Orissa, S Bihar, India.
Pt. griseiventer Gray, 1843a: 133.
Pt. petaurista var. *cineraceus* Blyth, 1847a: 865; Arakan, Burma.
Pt. grandis Swinhoe, 1863: 358; Taiwan.
Pt. yunanensis Anderson, 1875: 282; Momein, Yunnan.
P. lylei Bonhote, 1900a: 192; Doi Sritepe, Chiengmai, Thailand.
P. lanka Wroughton, 1911a: 1014; Sri Lanka.
P. cinderella Wroughton, 1911a: 1014; The Dangs, Surat Dist, Bombay Presidency, India.
P. annamensis Thomas, 1914e: 204; Bali, Nhatrang, S Annam, Vietnam.
P. mergulus Thomas, 1922c: 1067; Ross I, Mergui Is, Burma.
? *P. rufipes* Allen, 1925a: 13; Yungau, Fujian, China.
P. hainana Allen, 1925a: 14; Nam Fong, Hainan.
P. rubicundus Howell, 1927a: 82; Mapientung, c. 60 miles NW of Suifu, Sichuan, China.

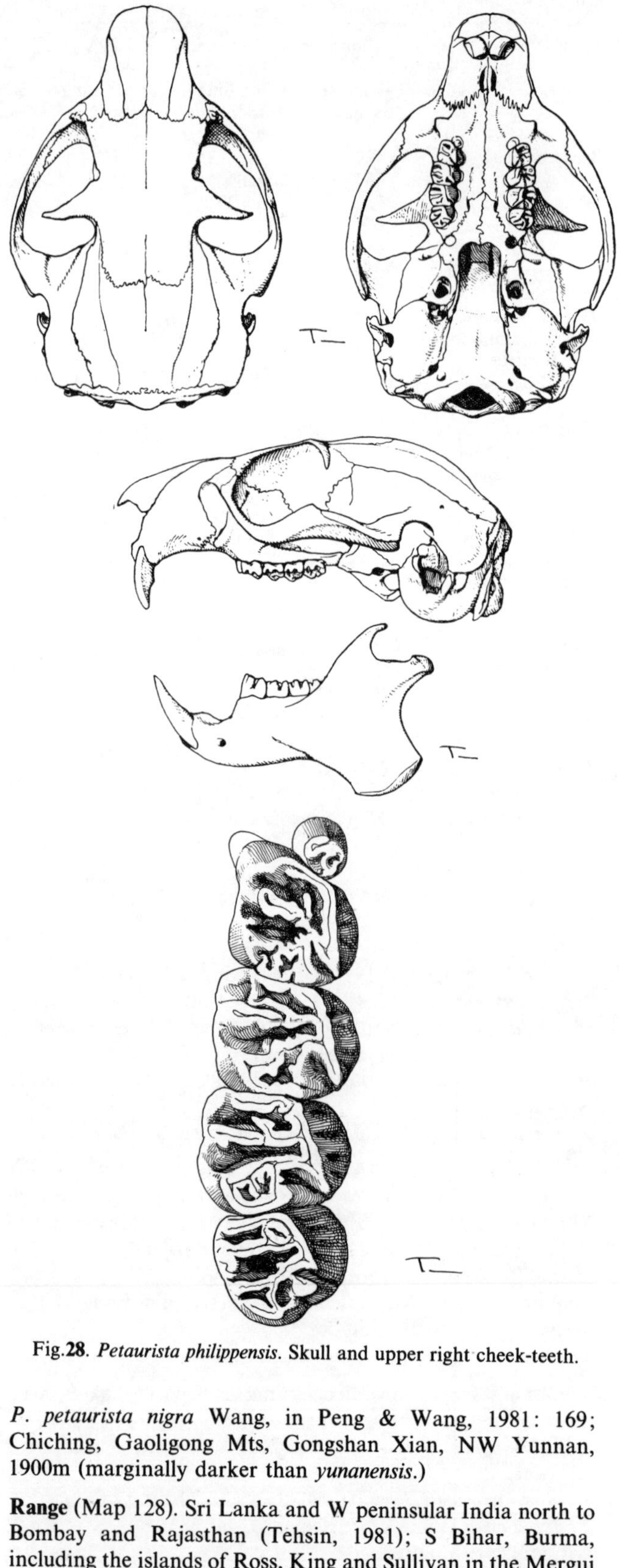

Fig.**28**. *Petaurista philippensis*. Skull and upper right cheek-teeth.

P. petaurista nigra Wang, in Peng & Wang, 1981: 169; Chiching, Gaoligong Mts, Gongshan Xian, NW Yunnan, 1900m (marginally darker than *yunanensis*.)

Range (Map 128). Sri Lanka and W peninsular India north to Bombay and Rajasthan (Tehsin, 1981); S Bihar, Burma, including the islands of Ross, King and Sullivan in the Mergui Is, and Thailand through Indochina to S China (Yunnan and S Gansu to Fujian and S Zhejiang); also on Hainan and Taiwan. In Thailand this species appears to coexist with *P. petaurista* at some localities, e. g. Huai Ko Mong (Mae Salieng dist, Mae

Hongson prov.) from where both species are represented in the collection of the TISTR, Bangkok.

Variation. The following nominal subspecies have been described:

P. lylei venningi Thomas, 1914d: 27; Kalaw, S Shan States, Burma.

P. l. badiatus Thomas, 1925: 501; Ngai-tio, Tonkin, Vietnam.

P. mergulus reguli Thomas, 1926a: 22; King I, Mergui Is, Burma.

P. m. primrosei Thomas, 1926a: 22; Sullivan I, Mergui Is, Burma.

P. cineraceus stockleyi Carter, 1933: 1; Melamoong, NW Thailand, 2000ft.

P. lylei miloni Bourret, 1942b: 28; Diem-her, Lang-son Prov., Indochina.

Geographical variation is very extensive and has not been adequately described nor mapped; the main trends are as follows:

Peninsular India and Sri Lanka (*P. p. philippensis*, etc.). Grey above, strongly speckled; pale grey below, without brown or orange tinge.

W Burma (*cineraceus*). Very similar to *philippensis*. Dorsal pelage with speckling rather fine; tail and membranes grey.

Mergui Is, Burma (*P. a. mergulus* etc.). Grey above, with yellow or brown tinge and fine agouti speckling, quite different from long-tipped, hoary effect in most mainland forms; venter off-white; tail grey.

E Burma, Thailand (*lylei*, including *venningi*). Dorsal pelage very dark grey, heavily frosted with white; dorsal membrane, feet and tail black; venter buff; ears red-fronted.

Vietnam, S China (*annamensis*, including *badiatus* and *miloni*). Dorsal pelage brown, heavily frosted with white; venter pale orange-brown.

SW China, N Assam (*yunanensis* etc.). Very dark rufous above with little speckling; tail black or dark rufous.

Taiwan (*grandis*). Dark brown above, dull orange-brown below; tail grey-brown darkening gradually towards tip; less sleek than similar form of *P. petaurista* in Sumatra and slight hint of agouti speckling.

Recognition. Where this species coexists with *P. petaurista*, e. g. in parts of Thailand, it is distinguished especially by the tail which is either uniformly coloured (usually grey or black) or darkens gradually towards the tip without a discrete black tip;

Map **128**. *Petaurista* (part)

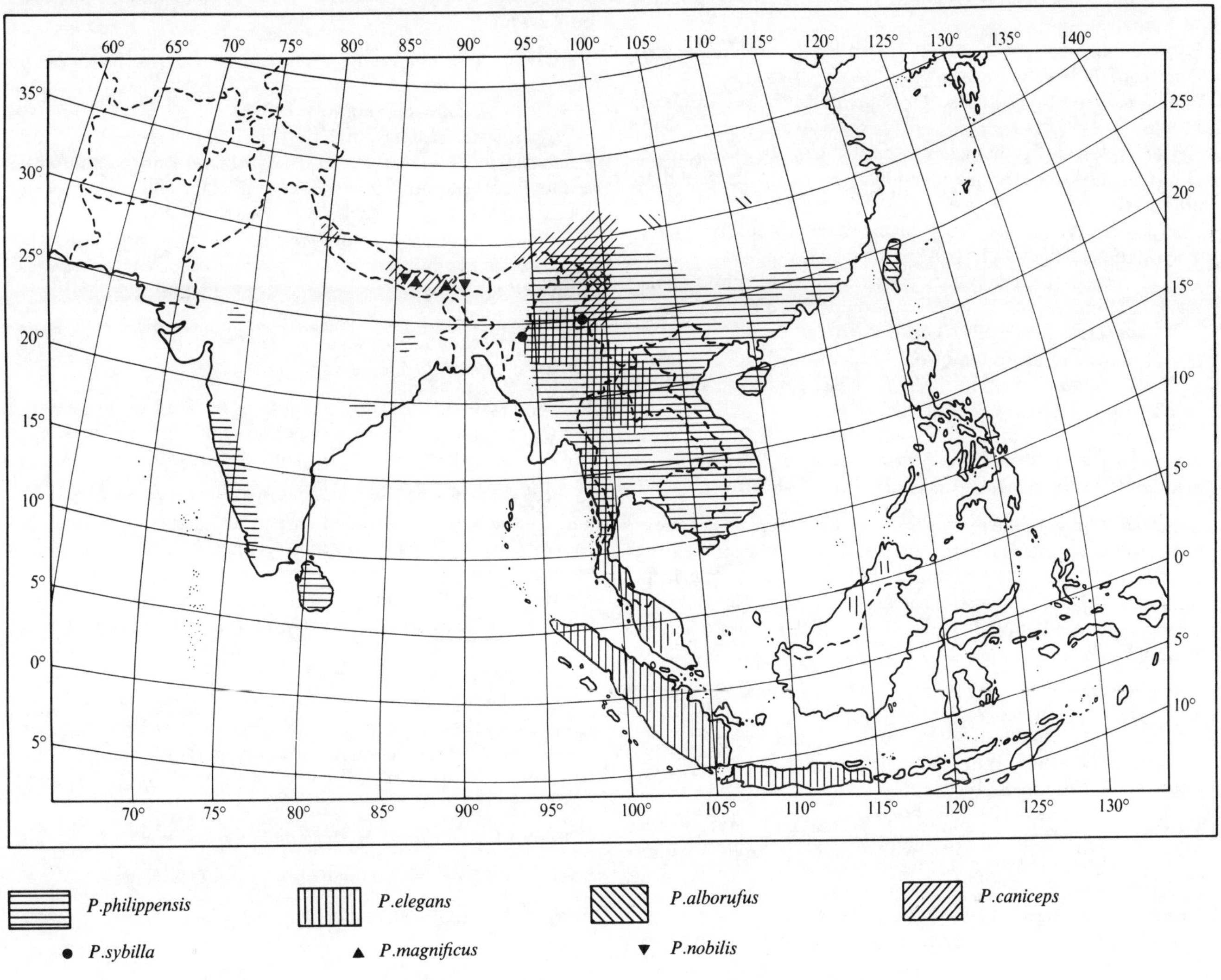

also by having the ears uniformly coloured or with the anterior margins reddish brown. The head is not lighter than the rest of the dorsum.

Ecology. Sympatric with *P. alborufus* on Taiwan, but mainly at lower altitude and using lower layer of the canopy; feed on bark of the conifer *Cryptomeria japonica* (Lee *et al.*, 1986).

Remarks. This species seems clearly separable from *P. petaurista* in Burma, Thailand and Indochina, but much remains to be done to clarify their relationship and some forms included here may warrant separate specific rank.

Petaurista elegans Lesser giant flying squirrel

Pteromys elegans Temminck, 1836:xii; Nusa Kumbangan I, S Java.
Pt. punctatus Gray, 1846d: 211; Malacca, Malaya.
P. marica Thomas, 1912d: 687; ? near Mongtsze, Yunnan, China.

Range (Map 128). S China (Yunnan) south through Burma to Malaya, Sumatra, Java and NW Borneo; island of Nusa Kombangan (S Java). Mainly montane in the north; down to c. 200m in Malaya; 1000–1700m in Borneo.

Variation. The following nominal subspecies have been described:
P. punctata sumatrana Kloss, 1921a: 230; Padang Highlands, W Sumatra.
P. punctata banksi Chasen, 1933a: 194; Lumu Lumu, Mt Kinabalu, N Borneo, 5500ft.
P. e. slamatensis Sody, 1949a: 70; Kaligua, Mt Slamat, C Java, 1400m.

The following principal subspecies can be recognized, although those on the mainland will probably be found to intergrade:

P. e. elegans (Java). Back black, dark rufous on rump, unspotted (but fine white frosting on black area).

P. e. banksi (Borneo). Back black with bold white spots, black extending back in centre of otherwise rufous rump.

P. e. punctata (Malaya). As on Borneo, but ground colour of back lighter and more uniform. Feet and tail black.

P. e. sumatrana (Sumatra). As Malaya, but white spotting much less, confined to central area of back.

P. e. marica (S China, Thailand, Indochina). Brown, with white spots much fewer than in Malaya, normally confined to head and centre of back; feet and tail reddish brown.

Remarks. Although the extremes are very different they appear to be linked by intermediates supporting the case for treating them as conspecific. Ellerman (1949) included *nitidula* from Bunguran I (Natuna Is) in this species on the basis of small size. Externally it is a typical *P. petaurista* to which species it is here allocated.

Petaurista sybilla

P. sybilla Thomas & Wroughton, 1916a: 424; 50 miles W of Kindat, Chin Hills, W Burma.

Range (Map 128). Mountains of Burma and adjacent W Yunnan, apparently sympatric with *P. caniceps* in W Yunnan (Wang, Y-x., pers. comm.).

Remarks. Although it lacks white spots, this form is in other respects much more similar to *P. elegans marica* than to *P. caniceps*. It appears to coexist with a spotted form in W Yunnan but it is not clear whether these are truly distinct and sympatric.

Petaurista caniceps Grey-headed flying squirrel

Sciuropterus caniceps Gray, 1842: 262; Nepal.
S. senex Hodgson, 1844a: 68; Nepal.
P. clarkei Thomas, 1922a: 396; Mekong Valley at 28°N, Yunnan, China.
S. gorkhali Lindsay, 1929b: 566; Apoon Sottidanda, Gorkha, Nepal, 12000ft.

Range (Map 128). Himalayas from Nepal east to E Yunnan and E Tibet.

Remarks. Included in *P. elegans* by Ellerman & Morrison-Scott (1951), but there appear to be two forms sympatric in W Yunnan, an unspotted one, *clarkei*, referable to *C. caniceps* and the boldly spotted form *marica* referable to *P. elegans* (specimens in Kunming Institute).

Petaurista alborufus Red-and-white flying squirrel

Pteromys alborufus Milne-Edwards, 1870a: 342; Moupin, Sichuan, China.
? *Pt. pectoralis* Swinhoe, 1871: 634; SW Taiwan.
P. lena Thomas, 1907b: 522; Tapposha, central Taiwan.

Range (Map 128). S China in Sichuan, Yunnan and Hubei; also on Taiwan.

Variation. The following nominal subspecies have been described:
P. a. leucocephalus Hilzheimer, 1905b: 298 (? Tibet).
P. a. castaneus Thomas, 1923d: 172 (Ichang, Hubei).
P. a. ochraspis Thomas, 1923d: 172 (Likiang Range, Yunnan).

The form *lena* on Taiwan is very distinct and might merit specific rank:

Mainland: Anterior half of back rufous, posterior half pale straw-coloured; tail dark rufous, paler at base; venter orange-brown or white; throat and neck usually white; feet and legs rufous or black.

Taiwan: Back uniformly rufous; tail dull brown; whole of venter and parts of feet white.

Ecology. Sympatric with *P. philippensis* on Taiwan but mainly at higher altitude and using upper layer of canopy (Lee *et al.*, 1986); and in parts of W Yunnan.

Remarks. The forms *candidula, barroni* and *taylori* from Burma and Thailand were included in this species by Ellerman & Morrison-Scott (1951), presumably on the basis of the pale speckling on the muzzle. However they show no other resemblance and in both external and dental characters they seem much closer to *P. petaurista* in which they are included here.

The name *pectoralis* was applied to a specimen from Taiwan that was a rich rufous colour with white breast and a white streak along the centre of the abdomen, but without mention of a white head. Its status remains uncertain. Nothing like it has been seen during the examination of 64 specimens of *P. a. lena*.

Petaurista magnificus Hodgson's flying squirrel

Sciuropterus magnificus Hodgson, 1836b: 231; Nepal.

Range (Map 128). Nepal to Darjeeling Dist, W Bengal, ? S Tibet.

Variation. The following subspecies was based upon a single specimen with darker colour and paler shoulder patch:
P. m. hodgsoni Ghose & Saha, 1981: 99; Ghoombhanjang, Darjeeling Dist, W Bengal, India (c. 2117m).

Remarks. This species was held to include also the forms *nobilis* and *chrysotrix* by Blyth (1863a) and most subsequent authors, but these are excluded following Ghose & Saha (1981). Chromosomes 2n = 38, FNa = 72 (Chatterjee & Majhi, 1975). It is not clear whether the record from S Tibet, near the Nepalese border (Feng *et al.*, 1986: 297) refers to *P. magnificus s. s.* or to *P. nobilis*.

Petaurista nobilis

Sciuropterus nobilis Gray, 1842: 263; Darjeeling.
S. chrysotrix Hodgson, 1844a: 67.

Range (Map 128). C Nepal to Bhutan (c. 1670–3000m).

Variation. The following subspecies was described on the basis of larger size, richer colour and usual absence of a pale mid-dorsal stripe:
P. n. singhei Saha, 1977: 27; Gomchu, Gomchu Valley, Bhutan (c. 2286m).

Remarks. Included in *P. magnificus* by most authors but seems clearly distinct, as discussed in detail by Ghose & Saha (1981).

Petaurista xanthotis

Pteromys xanthotis Milne-Edwards, 1872b: 301; ? Moupin, Sichuan, China.
Pt. filchnerinae Matschie, 1907: 208; Siningfu, upper Hwangho, Gansu, China.
Pt. buechneri Matschie, 1907: 210; Ssi-ku, between Gansu and Sichuan.

Range (Map 129). W China from Yunnan and E Tibet to Gansu.

Remarks. This form was included in the Japanese *P. leucogenys* by Ellerman & Morrison-Scott (1951), but is in fact very distinct; the cheek-teeth are very much more complex than in *P. leucogenys* (or indeed in any other *Petaurista*) and the distinctive white streak in front of and below the ear in *P. leucogenys* is lacking. The teeth were illustrated in detail by McKenna (1962). Externally it is very similar to some Chinese forms of *P. caniceps* but has black feet and an orange-brown patch behind each ear.

Map **129**. *Petaurista* (part), *Iomys*, *Eupetaurus*, *Biswamoyopterus*, *Aeretes*

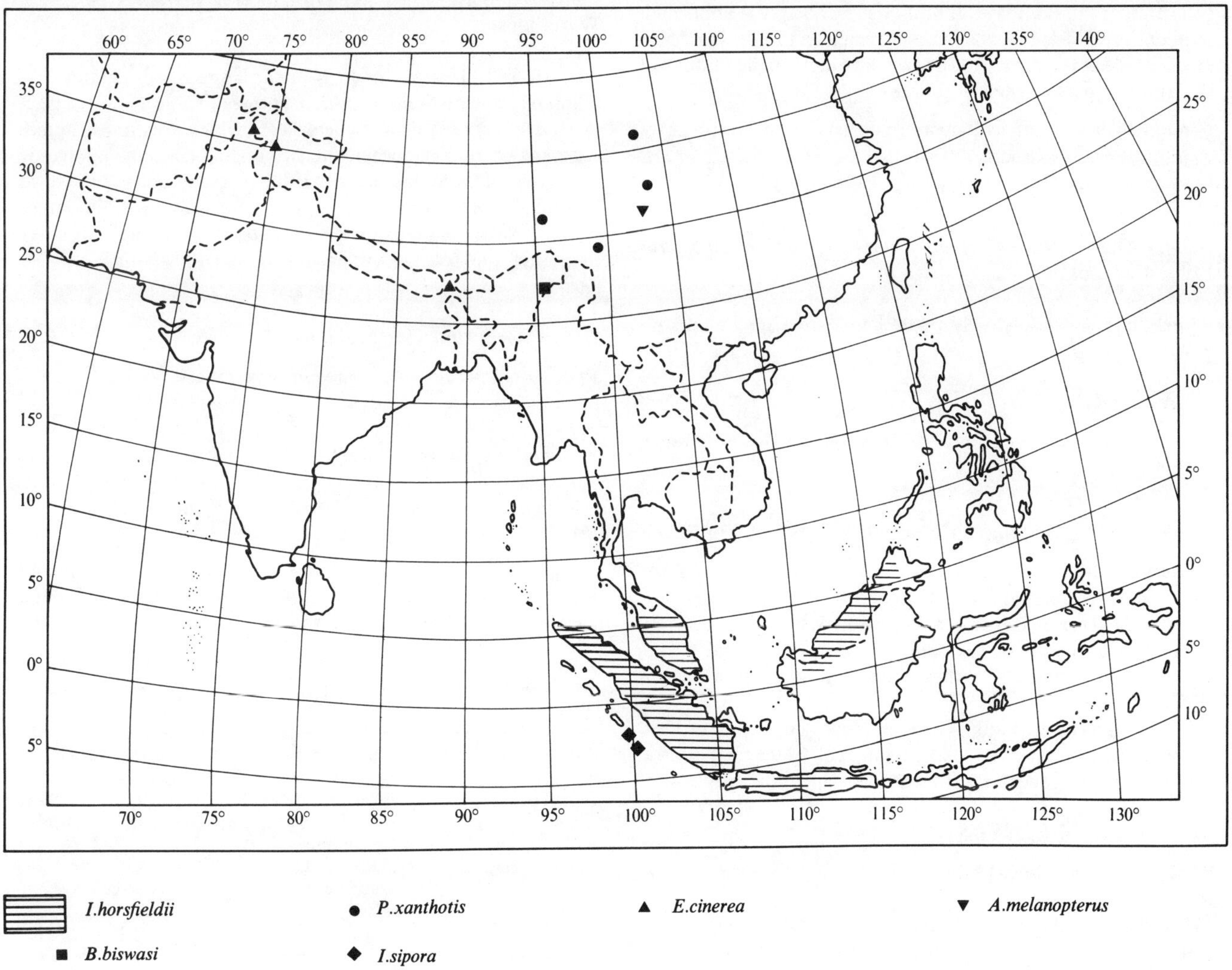

EUPETAURUS

Eupetaurus Thomas, 1888c: 256; type species *E. cinereus* Thomas.

Himalayas; one species. This has the most hypsodont teeth of any flying squirrel and on the basis of dentition alone was included in a separate family Eupetauridae by Schaub (1953) who considered it related to the hystricoid rodents and far removed from the Sciuridae. However this is quite unjustified and *Eupetaurus* appears to be closely related to *Petaurista*, especially *P. xanthotis*, as argued in detail by McKenna (1962) who illustrated the teeth.

Eupetaurus cinereus Woolly flying squirrel

E. cinereus Thomas, 1888c: 258; Gilgit Valley, Kashmir.

Range (Map 129). Known for certain only from a few localities at high altitude in Chitral (N Pakistan) and Gilgit (Kashmir), and in N Sikkim at 3000m (Agrawal & Chakraborty, 1970). It has been included in a list of the mammals of Yunnan (Wang & Yang, 1986: 135), based upon two skins, without skulls, in the Kunming Institute that appear to be this species. There is a specimen labelled 'Tibet' in the Leiden Museum (with skull, suggesting that it is not a trade skin), but the species was not recorded from Tibet by Feng *et al.* (1986).

Variation. Of nine skins examined, one from Chitral in the collection of the Bombay Natural History Society is abnormal, with uniform dark chocolate-brown pelage, a tuft of white hair on each side of the throat and a few white hairs in the midline of the anterior part of the the thorax.

Ecology. The habitat in Pakistan consists of sparse forest of *Pinus* and *Picea* (Roberts, 1977).

AERETES

Aeretes Allen, 1940: 745; type species *Pteromys melanopterus* Milne-Edwards.

China; one species, closely related to *Petaurista* but distinguished by broad grooved upper incisors and smaller M^3.

Aeretes melanopterus

Pteromys melanopterus Milne-Edwards, 1867a: 375; NE Hebei, China.
P. sulcatus Howell, 1927a: 82; NE of Beijing, 3000ft.

Range (Map 129). China: Sichuan and S Gansu; and in Hebei.

Variation. Although based on small samples (three of each) the two disjunct populations seem to be recognisable as subspecies:
A. m. melanopterus. Hebei. Smaller (head & body 275–328mm, condylobasal length of the sole skull 53mm); dorsal pelage lighter (dull sandy-buff), ventral pelage off-white with a slight buff wash.
A. m. szechuanensis Wang, Y-c. *et al.*, 1966: 86; Hei-shui Dist, Sichuan. Larger (h & b 334–355mm, cbl 64–66mm); dorsal pelage darker (greyish drab), ventral pelage dull buffy-white.

BISWAMOYOPTERUS

Biswamoyopterus Saha, 1981: 331; type species *B. biswasi* Saha.

NE India. One species. Although closely resembling species of *Petaurista* in most features, the sole specimen of *B. biswasi* was placed in a new genus mainly on the basis of simpler cheek-teeth, as in most of the genera of small flying squirrels, and unpigmented incisors. In fact the cheek-teeth, which are little worn, are not greatly different from those of *Petaurista* spp. with similar wear, while the incisors have a pale yellow pigment as in *Aeromys*. *Biswamoyopterus* does however differ strikingly from *Petaurista* in having multiple septae in the auditory bullae, giving a honeycomb effect with about 10–12 cells, a condition found in *Trogopterus* and *Petinomys* but not in any species of *Petaurista*.

Biswamoyopterus biswasi Namdapha flying squirrel

B. biswasi Saha, 1981: 333; Deban, 26km E of Miao, Namdapha, Tirap Dist, Arunachal Pradesh, India (350m).

Range (Map 129). Known only from the type locality in NE India.

Recognition. It differs from all other large, rufous flying squirrels, except *Petaurista alborufus lena* on Taiwan and some *P. philippensis yunanensis*, in having the entire underside white, with only a faint orange-rufous tint on the patagium. The white speckling in the dorsal pelage distinguishes it from *P. a. lena*.

Remarks. Known only from the holotype.

IOMYS

Iomys Thomas, 1908c: 1; type species *Pteromys horsfieldii* Waterhouse.

Malaya and Greater Sunda Islands; 2 species, including one allopatric but distinctive insular species (Table 197). Characterized by very distinctive cheek-teeth which are almost square with a cusp at each corner, the two lingual cusps being deeply separated. The bullae are small but very bulbous. They also lack small upper premolars (P^3) but these are very rudimentary in many other flying squirrels and absent in a small minority including the sole specimens of *Hylopetes winstoni* and *H. bartelsi*.

Iomys horsfieldii Horsfield's flying squirrel

Pteromys (*Sciuropterus*) *horsfieldii* Waterhouse, 1838a: 87; Java or Sumatra.
Sciuropterus davisoni Thomas, 1886a: 84; Malacca, Malaya.
S. thomsoni Thomas, 1900b: 275; Bakong R, Baram dist, Sarawak.
I. lepidus Lyon, 1911a: 78; Batu Jurong, Kendawangan R, SW Borneo.

Range (Map 129). Malaya, Sumatra, Borneo (N & W); and islands of Penang (Malaya) and Singapore (perhaps extinct on last).

Table 197. The species of *Iomys*.

	horsfieldii	*sipora*
Range	Malaya, Sumatra, Java, Borneo	Mentawai Is
Dorsal pelage	Brown or brownish grey, with light tips to hairs	Very dark, few or no light tips
Ventral pelage	White, tinged orange-brown	Grey, some ochraceous tips
Septae in bullae	1 or 2	0 or 1

Variation. The following three subspecies can be tentatively recognized but too few specimens have been seen from Sumatra and Java:
I. h. davisoni. Malaya, Singapore. Dorsal pelage dark brownish grey, tail dark brown above, mottled with reddish brown at edges. Tooth-row short (8.7–9.5mm, n=16).
I. h. penangensis Chasen, 1940a: 115; Penang I, W Malaya. Dorsal pelage including tail lighter and brighter than on mainland, tooth-row longer (9.4–10.0mm, n=7).
I. h. thomsoni. Borneo. Doubtfully separable from *I. h. davisoni* but tail more uniformly red-brown.

Ecology. Occurs in plantations as well as forest.

Iomys sipora

I. horsfieldi sipora Chasen & Kloss, 1928c: 819; Sipora I, Mentawai Is, W Sumatra.

Range (Map 129). Only known from Sipura I and N Pagai I in the Mentawai Is, W of Sumatra.

Remarks. While retaining this form in *I. horsfieldii* Chasen (1940a) described it as 'a remarkably distinct form' and commented that 'only those who take the broadest view of the trinominal function will link it to *horsfieldii*'. Its melanism in relation to *Iomys* from mainland Sumatra is paralleled in many other species on the Mentawai Islands.

HYLOPETES

Hylopetes Thomas, 1908c: 6; type species *Sciuropterus everetti* Thomas (a race of *H. spadiceus* Blyth).
Eoglaucomys Howell, 1915: 109 (*S. fimbriatus* Gray).

Himalayas to Greater Sunda Islands; 10 species (Table 198). Some of those in SE Asia were revised by Hill (1960,1962b) involving substantial changes to the list of Chasen (1940a). The genus is very similar to the Palaearctic *Pteromys* Cuvier, 1800 and the North American *Glaucomys* Thomas, 1908c.

Fossil species have been described, e. g. from Europe, but these cannot be distinguished from *Petinomys* on the basis of the teeth (Bouwens & Bruijn, 1986).

Hylopetes spadiceus Red-cheeked flying squirrel

Sciuropterus spadiceus Blyth, 1847a: 867; Arakan, Burma.
S. everetti Thomas, 1895b: 27; Bunguran I, N Natuna Is.
S. harrisoni Stone, 1900: 462; Menbuang R, Sarawak, Borneo.
S. amoenus Miller, 1906b: 264; Kundur I, Riau Is.
S. (*H.*) *belone* Thomas, 1908e: 305; Teratau I, Straits of Malacca.
H. sagitta sumatrae Sody, 1949a: 71; Redelong, E Atjeh, N Sumatra, 1300m.

Range (Map 130). Burma and Indochina to Malaya, Sumatra and Borneo; small islands of Con Son (Vietnam), Tarutau (Straits of Malacca), Kundur (Riau Is), Bunguran.

Variation. Geographical variation is very slight. The following subspecies has been described:
H. harrisoni caroli Gyldenstolpe, 1920: 29; Boelongean Dist, NE Borneo.

Remarks. Called *H. sagitta* by Chasen (1940a) but the type of *sagitta* has been claimed to be a *Petinomys* (see Hill, 1960,1962b).

Table 198. The species of *Hylopetes*.

	spadiceus	*lepidus*	*sipora*	*alboniger*	*nigripes*	*phayrei*	*fimbriatus*	*baberi*	*bartelsi*	*winstoni*
Range	Burma-Sunda	Sunda	Mentawai Is	Himalayas-Vietnam	Palawan	Indochina, Taiwan	Kashmir etc.	Kashmir etc.	Java	N Sumatra
Size	Small	Small	Small	Medium	Medium	Medium	Large	Large	Small	Small
head & body	133–184	117–150	140	175–225	250–290	144–197	240–290	c.300	133–145	142
hind feet	24–35	22–30		36–45	50–58	30–36	50–65	60–62	c.28	30
Venter	Grey, ± buff tinge	Grey	Grey, buff tips	Grey	Grey	White or grey	White, ± grey bases	White, ± grey bases	White, light grey bases	Dark grey, white or buff tips
Contrasting edge at base of tail	Orange-brown	Grey or white	0	0	0	0	0	0	Pale	Pale
White 1/2 collar	0	0	0	+	+	+	0	0	0	0
Cheeks (below eye)	Orange	Grey	Orange	Grey	Grey	White	Grey	Grey	Buff	Buff
Throat	White	White	As venter	White	White	White	As venter	As venter	As venter	As venter
Skull: greatest length	33–39	29–38	35–37	42–51	45–54	37–43	55–61	c.62	c.34	36
Maxillary tooth-row	6.0–7.7	5.3–7.2	7.3	8.5–10.5	10.7–12.2	7.8–8.7	10.5–12.3	c.12.2	6.5	6.8
Septae in bullae	2	2	3	2	2	2	2	2	3	2

Map **130**. *Hylopetes*

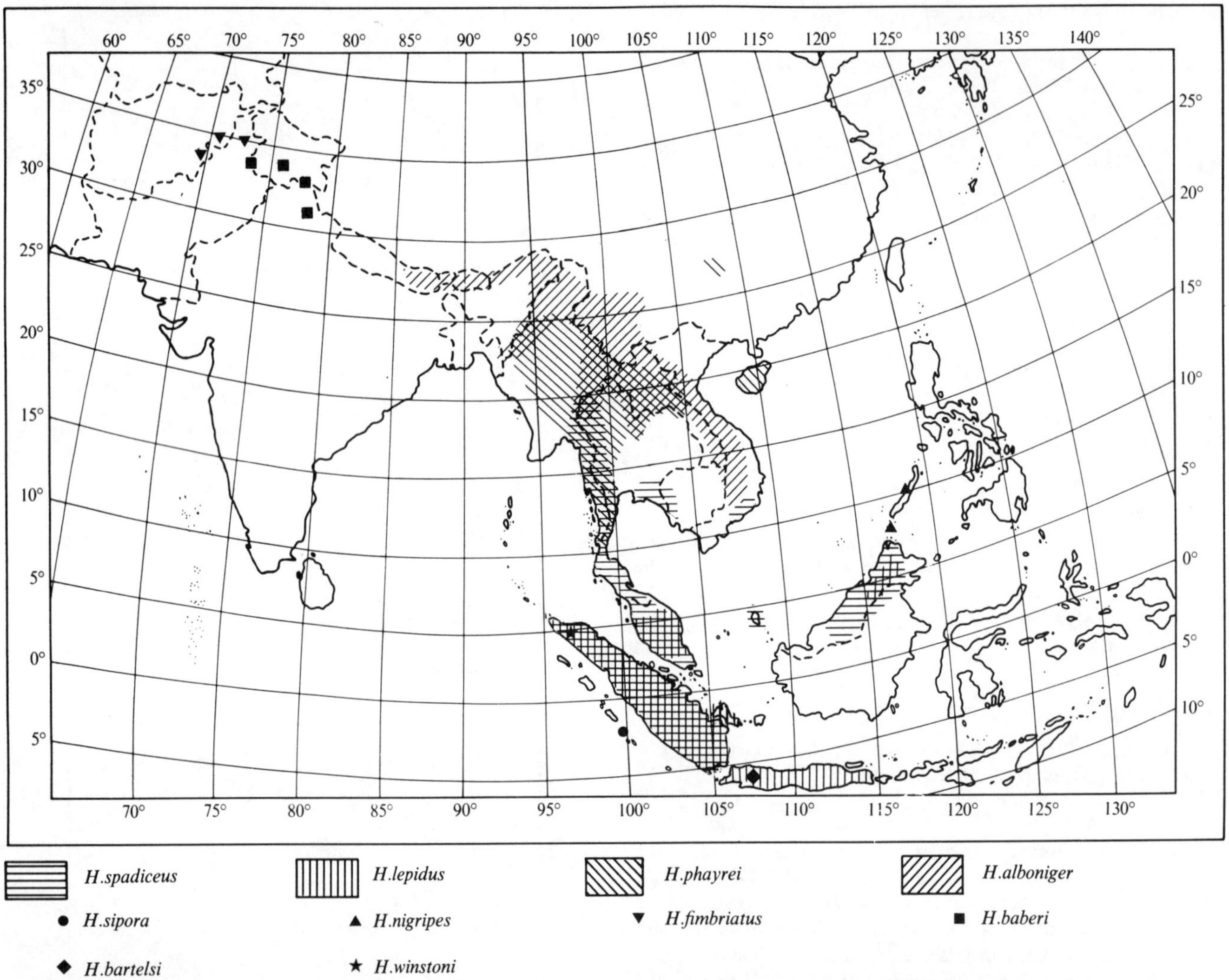

Hylopetes lepidus Grey-cheeked flying squirrel

Pteromys lepidus Horsfield, 1822: Java.
P. aurantiacus Wagner, 1841a: 135; Bangka I, Sumatra.
Sciuropterus platyurus Jentink, 1890b: 145; Deli, NE Sumatra.

Range (Map 130). ? S Thailand; Malaya, Sumatra (including Bangka I), Java, NW Borneo (Mt Kinabalu and Kelabit uplands only).

Variation. Malayan and Bornean specimens appear to be rather brighter brown above, those on Java and Sumatra being greyer.

Remarks. Reviewed by Hill (1962b).

Hylopetes sipora Sipura flying squirrel

H. sagitta sipora Chasen, 1940a: 117; Sipora I, Mentawai Is, Sumatra.

Range (Map 130). Sipura I, Mentawai Is, W of Sumatra.

Remarks. Given specific rank by Hill (1962b) who considered that it was not clearly referable to either *H. spadiceus* or *H. lepidus*. The tail is much broader and bushier than in these species. The bullae also differ, having three septae and being rather less inflated than in other *Hylopetes* (but more so than in most *Petinomys*).

Hylopetes alboniger Particoloured flying squirrel

Sciuropterus alboniger Hodgson, 1836b: 231; Nepal.
S. turnbulli Gray, 1838c: 68; 'India'.
? *Pteromys leachii* Gray, 1837a: 584.
P. (Hylopetes) leonardi Thomas, 1921a: 501; Kachin Prov., N Burma, 8000ft, 28°5'N, 97°23'E.

Range (Map 130). Himalayas from Nepal through Assam to S China (Yunnan, Sichuan, S Zhejiang), Burma, N Thailand and Vietnam; and on Hainan. At altitudes of 1500–2800m in Nepal, up to 3300m in Yunnan.

Variation. The following nominal subspecies have been described:
Pteromys (Hylopetes) a. orinus Allen, 1940: 723; Likiang Range, Yunnan, c. 7800ft (larger).
H. a. chianfengensis Wang & Lu, in Shaw *et al*, 1966: 270; Chianfeng Mt, Hainan I, China.

Remarks. The chromosome number is 2n = 38, all bi-armed, based upon three specimens from near Darjeeling, India (Garg & Sharma, 1972).

Hylopetes nigripes Palawan flying squirrel

Sciuropterus nigripes Thomas, 1893c: 30; Puerta Princesa, Palawan, Philippines.

Range (Map 130). Palawan and adjacent small island of Bancalan, Philippines.

Variation. The following insular subspecies was described on the basis of smaller size:
Sciuropterus n. elassodontus Osgood, 1918: 1; Bancalan Is, (S of Palawan), Philippines.

A considerable proportion of specimens on Palawan show varying degrees of albinism from small irregular spots to almost completely white.

Remarks. In spite of its very isolated geographical position this species seems very similar to *H. alboniger* and it is doubtful if its separation can be justified.

Hylopetes phayrei Phayre's flying squirrel

Sciuropterus phayrei Blyth, 1859a: 278; Rangoon, Burma.
Pteromys (*Petinomys*) *electilis* Allen, 1925a: 16; Namfong, Hainan I, China.

Range (Map 130). Burma, Thailand, Laos and NW Vietnam (Dao & Cao, 1990); Hainan Island and Guizhou, China (specimen from Shiquian County, Guizhou, c. 27°30'N, 108°30'E in Kunming Institute); perhaps also Fujian, China (Honacki *et al.*, 1982).

Variation. The following subspecific names have been proposed, but recognition of discrete subspecies is not justified apart from *H. p. electilis* on Hainan:
Sciuropterus p. probus Thomas, 1914d: 28; Mt Poppa, Burma (white on feet).
S. p. laotum Thomas, 1914d: 28; Laos Mountains (larger).
Pteromys p. anchises Allen & Coolidge, 1940: 153; Mt Angka, N Thailand, 4300ft (bullae small, darker).

Remarks. Although described in *Petinomys*, the form *electilis* is clearly a *Hylopetes*, having only two septae in each bulla. Overall, it is very similar to mainland *H. phayrei*, differing in having the venter grey with creamy hair-tips and a pure white patch in each axilla and on the throat, rather than predominantly white as in most *H. phayrei*.

Hylopetes fimbriatus Small Kashmir flying squirrel

Sciuropterus fimbriatus Gray, 1837a: 584; Simla, Punjab.

Range (Map 130). N Punjab (Pakistan) and Kashmir east to Simla (Himachal Pradesh).

Hylopetes baberi

Sciuropterus baberi Blyth, 1847a: 866; Nijrow, Kohistan, Afghanistan.

Range (Map 130). NE Afghanistan, Hazara (N Pakistan) and Kashmir, from 2000m up to the tree-line.

Remarks. Included in *H. fimbriatus* by Ellerman (1940) and subsequent authors, but given specific rank by Chakraborty (1981) on the grounds that it is sympatric with the smaller *H. fimbriatus* in Kashmir.

Hylopetes bartelsi

Petinomys bartelsi Chasen, 1939a: 185; Mt Pangrango, W Java.

Range (Map 130). Known only from the type locality in W Java.

Remarks. Although superficially resembling some species of *Petinomys* this species differs from all *Petinomys* in lacking P^3 and in having only three septae in each auditory bulla.

Hylopetes winstoni

Iomys winstoni Sody, 1949a: 75: Baleg, E Atjeh, N Sumatra, 1200m.

Range (Map 130). Known only from the type locality in Aceh, N Sumatra.

Remarks. This is known only from the holotype. It could perhaps be considered conspecific with *H. bartelsi* but there are sufficient differences to justify keeping them separate until larger samples of each can be assessed. The upper incisors are orthodont (pro-odont in *H. bartelsi*) and the bullae are smaller, separated by about 4.5mm (3.3mm in *H. bartelsi*). Although it agrees with *Iomys* in lacking P^3, the form of the molars is normal, without the very characteristic separation of two lingual cusps found in *Iomys*.

PETINOMYS

Petinomys Thomas, 1908c: 6; type species *Sciuropterus lugens* Thomas.
Olisthomys Carter, 1942: 2 (*Pteromys* (*O.*) *morrisi* Carter, in *P. setosus*).

Throughout most of the Region. Seven species (Table 199), closely related to *Hylopetes* but distinguished by having multiple septae in the auditory bullae, sometimes with transverse as well as vertical septae giving the appearance of a honeycomb. In most species the bullae are also rather flat.

Petinomys hageni Hagen's flying squirrel

Sciuropterus hageni Jentink, 1888c: 26; Tandjong-Morawa, Serdang, Deli, NE Sumatra.

Range (Map 131). Sumatra; W Borneo (known only from a single specimen from Pontianak).

Variation. The following was based upon a single specimen with faded skin and damaged skull:
P. h. ouwensi Sody, 1949a: 74; Kubu region, Pontianak, W Borneo. Rostrum shorter and broader than on Sumatra (but length of upper tooth-row, 10.0mm, equals the smallest of ten *P. h. hageni*, 10.0-11.1mm).

Petinomys lugens

Sciuropterus lugens Thomas, 1895d: 666; Sipora I, Mentawai Is.
S. maerens Miller, 1903a: 26; N Pagi I, Mentawai Is.

Range (Map 131). The islands of Sipura and N Pagai (Mentawai Is).

Variation. The form on N Pagai I, *maerens*, was described on the basis of a single specimen with lighter colour and shorter ears compared with *P. l. lugens*. Four of the nine specimens examined from Sipura and two of four from N Pagai I have a light, creamy streak in the mid-ventral line.

Remarks. This form was treated as a subspecies of *P. hageni* by Chasen (1940a), but the much smaller size, much darker pelage and bushier tail suggest that specific rank is justified. *P. lugens* is the type species of *Petinomys*.

Table 199. The species of *Petinomys*.

	hageni	*lugens*	*fuscocapillus*	*crinitus*	*genibarbis*	*setosus*	*vordermanni*
Range	Sumatra, Borneo	Mentawai Is	S India, Sri Lanka	Philippines	Sunda	Burma-Sunda	Malaya, Borneo
Head & body	230,250	230–270	c.300	310–400	c.180–195	106–128	120–123
Tail	235,250	210–235	c.250	300–400	150–180	95–120	110–120
Hind feet	c.47	41–45	50–55	c.60	27–31	21–23	c.21
Dorsal pelage	Brown, slightly grizzled	Very dark brown	Medium brown	Greyish	Brown, grey on head	Grey	Brown
Venter	Buff	Dark grey brown, ± white streak	Grey or white	Buff, grey bases	Light orange-buff	Creamy white	White, tinged buff
Cheeks: colour	Brown	Dark brown	Brown	Grey	Orange-buff	Grey	Orange-buff
″ : tuft of long hair very prominent	0	0	0	0	+	0	0
Black eye-ring	+	0	0	(+)	0	(+)	+
Tail with contrasting pale sides at base	0	0	0	0	0	White	Orange-buff
Pale 1/2 collar behind ears	0	0	0	(+)	0	+	0
Hair fringe at end of carpal spur mainly white (W) or dark (D)	W	D	W	W	D	D	D
Skull: greatest length	50–54	49	c.57	53–68	c.38	27–31	28–31
Maxillary toothrow	9.8–11.1	9.4–10.0	11.1	11.6–14.7	c.7.6	4.5–5.7	5.3–5.6
Septae in bullae (H = honeycomb pattern)	H	H	6–10/H	7–10, some branched	c.7	4–6	H

Petinomys fuscocapillus Travancore flying squirrel

Sciuropterus fuscocapillus Jerdon, in Blyth, 1847a: 867; S India.
S. layardi Kelaart, 1850b: 215; Sri Lanka.

Range (Map 131). S India and Sri Lanka. In India known only from Travancore and a recent specimen from Pelhanamthitta district, Kerala (9°23'N, 76°37'E), reported by Kurup (1989).

Variation. Very few specimens are known but the ventral pelage appears to be white, tinged rufous in India and grey in Sri Lanka.

Remarks. This species has the auditory bullae more inflated than in other species of *Petinomys*, including the otherwise very similar *P. hageni* in Sumatra.

Petinomys crinitus

Sciuropterus crinitus Hollister, 1911b: 185; Basilan I, Philippines.
? *S. mindanensis* Rabor, 1939: 389; Badiangan, Gingood, Misamis Oriental, Mindanao, Philippines.

Range (Map 131). Islands of Basilan, Mindanao, Dinagat and Siargao, Philippines.

Variation. One subspecies has been described:
P. c. nigricaudus Sanborn, 1954: 285; Tuod, c. 15 miles inland from Mantikaw, Misamis Oriental, Mindanao, Philippines. Larger and darker than typical *crinitus*, but described without reference to *mindanensis*.

Specimens from Dinagat and Siargao are nearly indistinguishable from those from Mindanao (Heaney & Rabor, 1982).

Remarks. The form *mindanensis* is undoubtedly a *Petinomys*, with a honeycomb pattern of septae in the bullae (paratype in US Nat. Mus.). It appears to differ from other *P. crinitus* only in having a black nape.

Petinomys genibarbis Whiskered flying squirrel

Pteromys genibarbis Horsfield, 1822; Java.
? *Sciurus sagitta* Linnaeus, 1766: 88; Java.

Range (Map 131). Malaya, Sumatra, Java, Borneo.

Variation. The following nominal subspecies have been described:
Sciuropterus g. malaccanus Thomas, 1908e: 304; Malacca, Malaya.
S. g. borneoensis Thomas, 1908e: 304; Bakong R, Baram, Sarawak.

Remarks. It has been suggested, e. g. by Ellerman & Morrison-Scott (1955) on the basis of examination of the type skull by Dr G H H Tate, that *Sciurus sagitta* L. is a *Petinomys*, not a *Hylopetes* as previously thought. However until the identity of this specimen can be confirmed it is better not to allow it to usurp the well-known name of this species.

Petinomys setosus Temminck's flying squirrel

Pteromys (Sciuropterus) setosus Temminck, 1844: 49; Padang, W Sumatra.
Pteromys (Olisthomys) morrisi Carter, 1942: 2; Taro, N Burma, 26°21'N, 96°11'E.

Range (Map 131). N Burma and N Thailand; Malaya, Sumatra and Borneo.

Variation. *P. s. morrisi* is slightly larger and has a white tip to the tail.

Map **131**. *Petinomys*

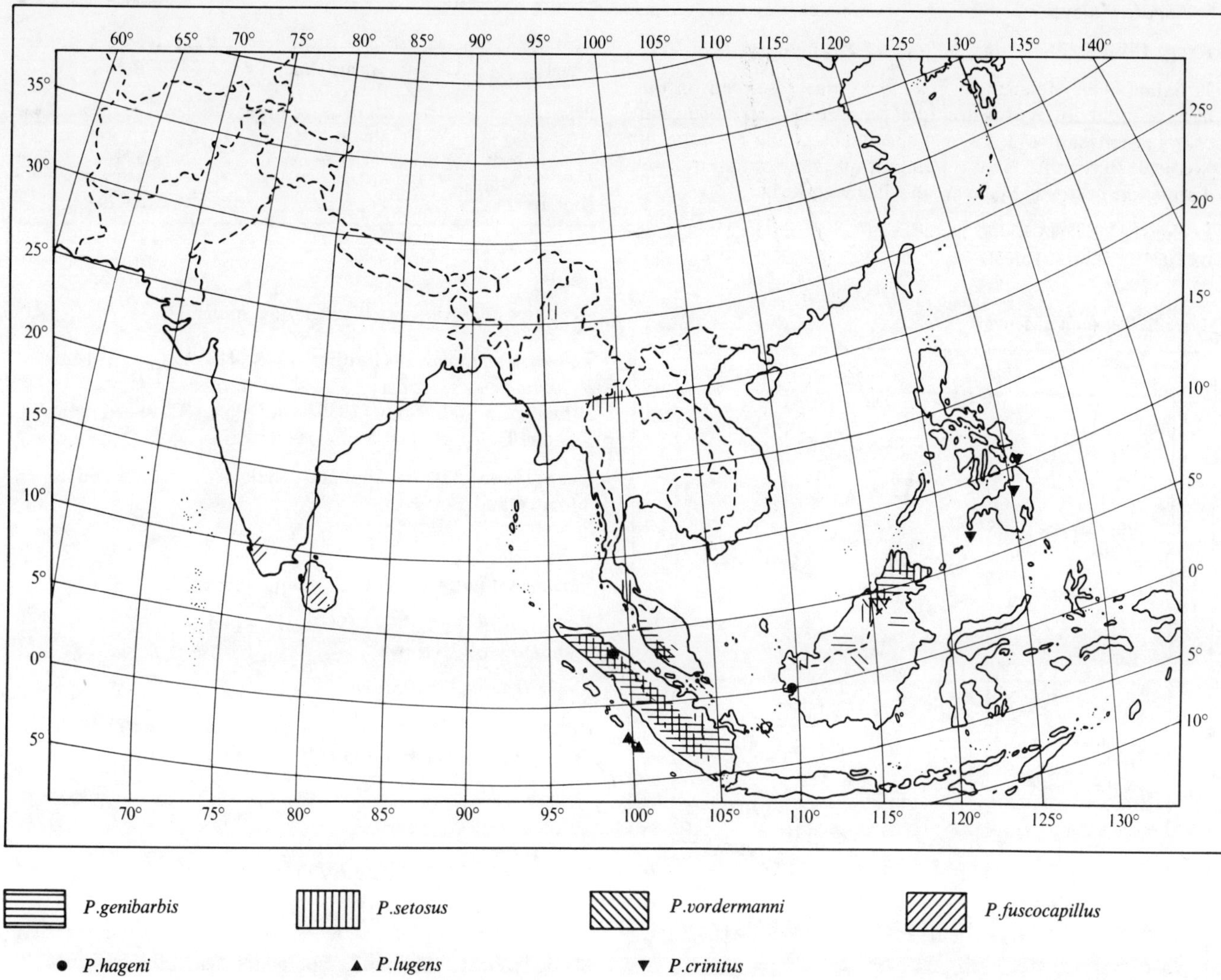

Remarks. Ellerman & Morrison-Scott (1951) included *phipsoni* Thomas from Tenasserim but this form was allocated to *P. vordermanni* by Hill (1962b). The form *morrisi* was included in *P. setosus* by Muul & Thonglongya (1971). This species and the next two are very similar to *Hylopetes spadiceus*, but the auditory bullae are of the *Petinomys* type.

Petinomys vordermanni Vordermann's flying squirrel

Sciuropterus vordermanni Jentink, 1890c: 150; Billiton I, E of Sumatra.
Pteromys (*Petinomys*) *phipsoni* Thomas, 1916a: 422; Tenasserim village, S Burma.

Range (Map 131). S Burma, S Thailand (Muul, 1980), Malaya, Borneo, Belitung I and Galang I (Riau Is).

Remarks. The form *phipsoni* was placed in *P. setosus* by Ellerman & Morrison-Scott (1951) but was allocated to this species by Hill (1962b). McKenna (1962) considered this species sufficiently distinctive to justify a separate genus although he did not name one. Characters included extreme crenulation of the cheek-teeth and complex 'honeycomb' septae in the auditory bullae, as in *Trogopterus*.

PETAURILLUS Pygmy flying squirrels

Petaurillus Thomas, 1908c: 3; type species *Sciuropterus hosei* Thomas.

Malaya and Borneo. Two species, but known from very few specimens (Table 200).

Table 200. The species of *Petaurillus*.

	hosei	*emiliae*
Range	Borneo, Malaya	Borneo
Head & body	75–90	68, 72
Tail	79–98	67
Hind foot	18–20	16, 17
Skull length (condylobasal)	23.1–25.2	21.2
Maxillary tooth-row	4.0–4.4	3.8
Cheeks	White or greyish	Buffy white

Petaurillus hosei Hose's pygmy flying squirrel

Sciuropterus hosei Thomas, 1900b: 275; Toyut R, Baram dist, Sarawak.

S. (Petaurillus) kinlochii Robinson & Kloss, 1911a: 171; Kapar, Selangor, Malaya.

Range (Map 132). W and N Borneo; Selangor, Malaya.

Variation. The Malayan form *kinlochii* was described on the basis of slight differences in cranial proportions and colour, but examination of five additional specimens at the Institute of Medical Research, Kuala Lumpur, does not support any distinction between Malayan and Bornean forms.

Ecology. Has been found in rubber plantations in Malaya as well as in natural forest.

Map **132**. *Petaurillus, Aeromys*

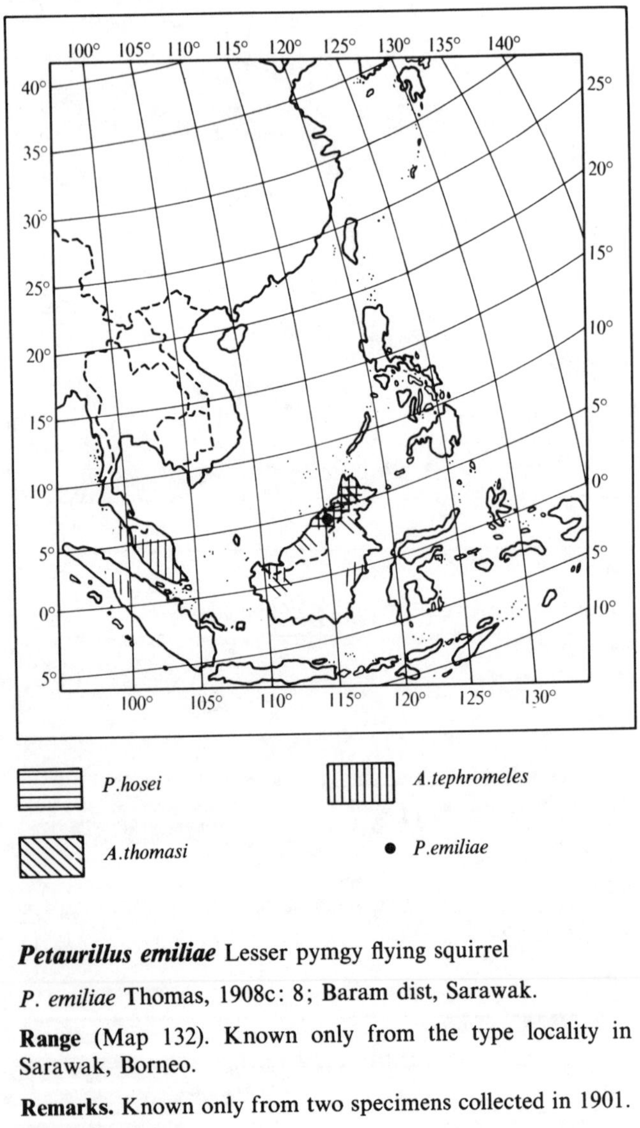

Petaurillus emiliae Lesser pymgy flying squirrel

P. emiliae Thomas, 1908c: 8; Baram dist, Sarawak.

Range (Map 132). Known only from the type locality in Sarawak, Borneo.

Remarks. Known only from two specimens collected in 1901.

AEROMYS

Aeromys Robinson & Kloss, 1915: 23; type species *Pteromys tephromelas* Günther.

Malaya, Sumatra, Borneo. Two species (Table 201). Formerly included in *Hylopetes*, but distinguished by narrow, terete tail and short, broad baculum (as in *Pteromyscus*).

Table 201. The species of *Aeromys*.

	tephromelas	*thomasi*
Range	Malaya, Sumatra, Borneo	Borneo
Pelage	Dark grey-brown, almost black	Reddish brown
Head & body	355–426	350–403
Tail	410–505	340–430
Hind feet	67–78	60–70

Aeromys tephromelas Black flying squirrel

Pteromys tephromelas Günther, 1873: 413; Penang I, Malaya.
P. phaeomelas Günther, 1873: 413; Borneo.
Petaurista bartelsi Sody, 1936: 146; Pagar Djawa, Pematang Siantar, Deli dist, NE Sumatra.

Range (Map 132). ? S Thailand; Malaya (including Penang I), Sumatra and Borneo.

Aeromys thomasi Thomas's flying squirrel

Petaurista thomasi Hose, 1900: 215; Silat R, c. 70 miles S of Claudetown, E Sarawak.

Range (Map 132). Borneo.

Remarks. Included in *Hylopetes* by Chasen (1940a) but placed in *Aeromys* by Ellerman (1940).

Family Muridae

Mice, rats, etc.

Muridae Gray, 1821: 303; type genus *Mus* Linnaeus.

As used here, in its widest sense, a world-wide family of over 1100 species. The dental formula is 1.0.0.3/1.0.0.3 (but the molars reduced to 2 or even 1 in a very few species by loss of the posterior teeth). The infra-orbital foramen is enlarged above to transmit muscle, and narrowed below. The anterior root of the zygomatic arch is flattened, forming a 'zygomatic plate' for muscle attachment.

There are many discrete groups within this complex that have variously been treated as subfamilies of Muridae or as separate families, e. g. the Arvicolinae (voles etc.) and Gerbillinae (gerbils). Chaline *et al.* (1977) for example recognized, but did not diagnose, eight families. However the contents of some of these are unstable and controversial and they are here treated as 14 subfamilies of which seven are represented in the region, five of these rather marginally (Table 202). This is to some extent a conservative and interim arrangement pending a greater degree of consensus on the limits of the included taxa, but it also reflects the very close overall similarity between some members of the different groups, e. g. amongst the Murinae, Gerbillinae and the American Sigmodontinae, differing mainly by small details of the molar cusp pattern.

The characters and relationships of the subfamilies were reviewed by Carleton & Musser (1984).

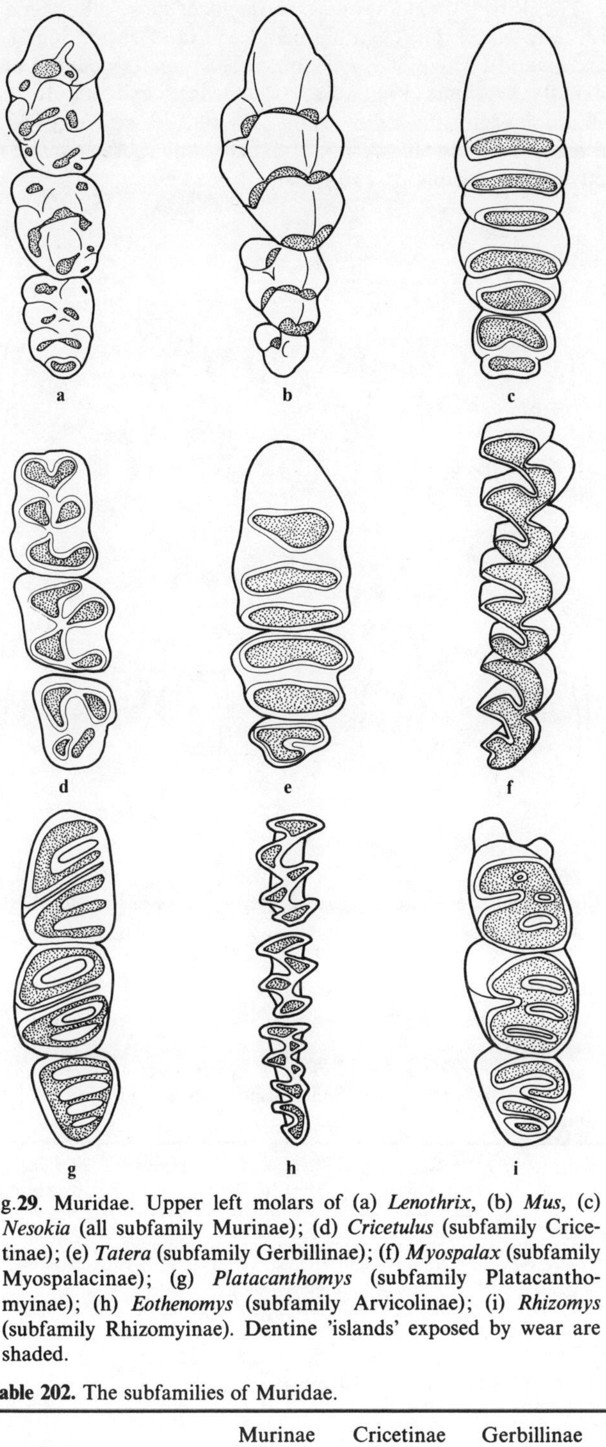

Fig.**29**. Muridae. Upper left molars of (a) *Lenothrix*, (b) *Mus*, (c) *Nesokia* (all subfamily Murinae); (d) *Cricetulus* (subfamily Cricetinae); (e) *Tatera* (subfamily Gerbillinae); (f) *Myospalax* (subfamily Myospalacinae); (g) *Platacanthomys* (subfamily Platacanthomyinae); (h) *Eothenomys* (subfamily Arvicolinae); (i) *Rhizomys* (subfamily Rhizomyinae). Dentine 'islands' exposed by wear are shaded.

Subfamily Murinae

Mice and rats

Eurasia, Africa, Australasia; c. 450 species of which 213 occur in the Indomalayan Region. These are mainly forest-adapted mice and rats. As a group they are fairly clearly distinguished from all other subfamilies of the Muridae except for a few genera in the Philippines that have variously been included in the Murinae, the Hydromyinae or, in the case of *Rhynchomys*, a third subfamily Rhynchomyinae. Neither of these groups is easy to define and both are here included in the Murinae.

The subfamily is characterized especially by the pattern of the upper molars in which the cusps form three longitudinal rows. In most species these remain detectable even after moderate wear when the cusps in each transverse row coalesce to form a transverse lamina, usually curved, with the outer cusps behind the central one. In those with the most hypsodont teeth, e. g. *Nesokia*, the pattern of cusps is lost very early to form discrete transverse laminae.

Generic classification has been very unstable, especially with regard to the genus *Rattus* and its relatives. In recent years Misonne (1969) and Musser (1969–1991) have redefined many genera in a more convincing way than hitherto but many problems remain. Some Indian genera, e. g. *Millardia* and *Cremnomys*, are difficult to define and need to be reviewed in relation to African genera as well as to other Indomalayan genera.

Surprisingly little attention has been paid to the complex characters of the male reproductive tract in Indomalayan Murinae although study of those in New Guinea (Lidicker, 1968) has shown their potential for the resolution of phylogenetic problems and 13 species in China were described and figured by Yang & Fang (1988). Sperm morphology of 19 species has been described with some interesting results, e. g. the three species of *Maxomys* were quite different from *Rattus* spp. (Breed & Yong, 1986). Another character that would repay detailed study is the nature and incidence of flattened bristles or spines in the pelage. Although variable, probably seasonally, in some species, they could nevertheless be of value in classification above the species.

Attempts to resolve phylogenies by electrophoresis of proteins have so far given very inconclusive results (Bonhomme *et al.*, 1985). Data on karyotypes covering a variety of genera have been published by Yong (1969), Duncan & Van Peenen (1971), Duncan *et al.* (1974), Duncan (1976) and Gadi & Sharman (1983). These are summarized in some of the identification tables but they need to be used with caution and should not be used for critical analysis without reference to the original sources.

Table 202. The subfamilies of Muridae.

	Murinae	Cricetinae	Gerbillinae	Myospalacinae	Platacanthomyinae	Arvicolinae	Rhizomyinae
Genera in region	61	3	4	1	2	5	2
Head & body	50–400	70–130	65–190	150–245	80–140	85–140 (400)	150–480
Tail (% of h & b)	60–200%	25–120%	80–160%	20–30%	70–145%	8–70%	15–50%
Tail hairs: all short	+	+		+		+	+
: slightly longer at tip	(+)		+				
: long and distichous					+		
: bushy throughout	(+)						
Cheek pouches	0	+	0	0	0	0	0
Skull: greatest length	16–85	23–41	27–51	35–60	23–26; 32–34	20–65	40–85
I[1] grooved	0 (+)	0	+ (0)	0	0	0 (+)	0
Upper molars (Fig.29)	a,b,c	d	e	f	g	h	i
Molars rootless	0	0	0	+	0	+/0	0

Data on the origin and fossil history of the Murinae were reviewed by Jacobs (1977).

There are too many uncertainties preventing the arrangement of genera in a sequence reflecting relationships. Instead a geographical sequence has been adopted as follows:

* Genera represented in Pakistan, peninsular India and Sri Lanka.
* Additional genera represented in the Indochinese subregion (S China and Assam to Thailand).
* Additional genera represented in the Sunda subregion (peninsular Malaya and Greater Sunda Is, including Palawan).
* Additional genera endemic to the Philippines (except Palawan).
* Additional genera endemic to Sulawesi.
* Additional genera represented on other island groups, namely the Lesser Sunda Is, the Moluccas and the Ryukyu Is.

The large number of genera and the subtlety of their distinguishing characters make it impracticable to tabulate the characters in manageable subgroups. Some genera can be easily recognized by more conspicuous characters as follows:

Dorsal pelage consisting mostly of spines: *Acomys*.

First (i. e. medial) digits of hind feet with short flat nails instead of the usual elongate, sharply pointed claws (arboreal forms): see Table 205.

Tail bushy: *Phloeomys, Crateromys* (Philippines).

Each upper incisor with a longitudinal groove on anterior face: *Golunda, Lenothrix* (faintly seen in *Diomys, Uromys, Echiothrix, Carpomys* and *Tryphomys*).

Third molars absent: *Rhynchomys, Chrotomys silaceus, Hydromys*, some *Mus*.

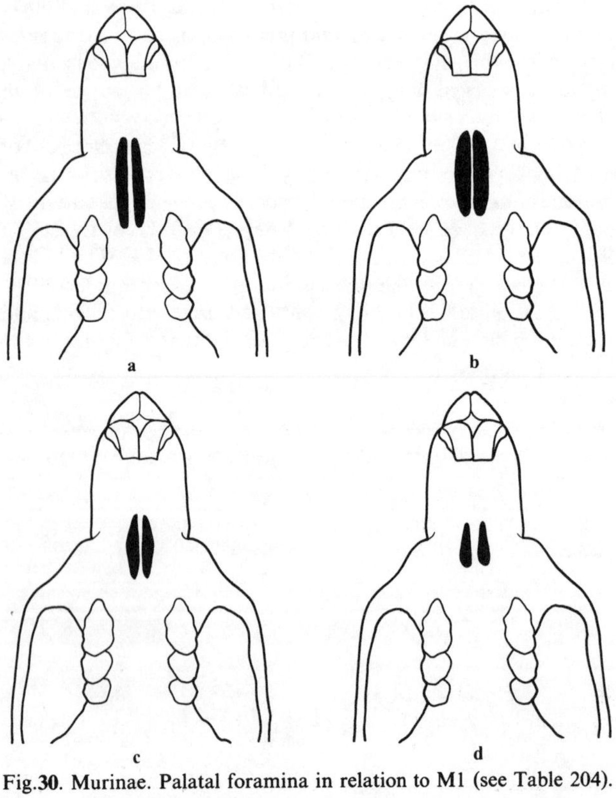

Fig.30. Murinae. Palatal foramina in relation to M1 (see Table 204).

The principal external measurements of those other genera represented on the Asian mainland and the Greater Sunda Is are given in Table 203, and some corresponding cranial and dental characters in Table 204. Genera endemic to the Philippines are treated in Table 206 and the key on p. 325; those endemic to Sulawesi in Table 207; and those endemic to other small islands in Table 208.

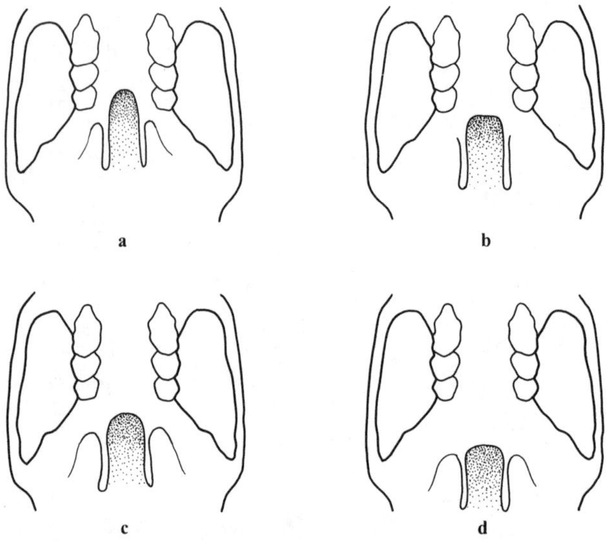

Fig.31. Murinae. Posterior edge of bony palate in relation to M3 (see Table 204).

Table 203. Some external characters of the genera of Murinae on the Asian mainland and Greater Sunda Islands (excluding those in Table 205).

	Head & body	Tail as % of h&b	Hind feet	Flat spines in pelage
Mus	65–120	80–130%	14–25	+/0
Acomys	95–110	c.95%	18–19	++
Rattus	140–250	80–150%	30–46	+/0
Millardia	80–200	80–100%	18–34	0
Cremnomys	100–200	105–150%	23–38	0
Golunda	100–170	80–100%	21–27	0
Nesokia	130–210	65–80%	27–36	0
Bandicota	160–345	80–110%	30–70	0
Vernaya	58–75	170–200%	16–17	0
Micromys	50–70	100–120%	13–16	0
Apodemus	90–110	90–140%	20–27	0
Hadromys	100–120	c.115%	23–28	0
Dacnomys	230–290	115–140%	50–56	0
Diomys	100–135	c.100%	24–28	0
Niviventer	100–200	100–180%	20–40	+/0
Leopoldamys	180–290	110–180%	42–55	0
Maxomys	100–225	85–130%	24–45	+
Berylmys	135–300	70–120%	33–60	+
Haeromys	55–75	140–200%	15–19	0
Lenothrix	170–235	100–150%	30–36	0
Sundamys	185–300	100–140%	37–61	0
Palawanomys	125–160	80–105%	32–34	0

Key to the genera of Murinae on the Asian mainland, the Greater Sunda Islands, Sri Lanka and Taiwan.

1. First (medial) digit of hind foot with a flat nail; (tail > head and body) **2**

First digit of hind foot with an elongate, sharply pointed claw **7**

2. Smaller: upper molar row <5.5mm (head & body usually <125m, hind feet <25mm) **3**

Larger: upper molar row >5.5mm (head & body usually >125mm, hind feet usually >25mm) **4**

3. 5th (outermost) digits of fore and hind feet without claws .... ***Vandeleuria***

5th digits with claws .... ***Chiropodomys***

4. Flat spines in dorsal pelage **5**

No flat spines in dorsal pelage **6**

5. Dorsum yellowish brown; smaller: head & body <160mm, hind feet <30mm, greatest length of skull <40mm, upper molar row <7.5mm (Indochina) .... ***Chiromyscus***

Dorsum dark brown, agouti-banded; larger than above (Java) .... ***Kadarsanomys***

6. Tail noticeably hairy towards tip; terminal pads of digits bifid; dorsal pelage greyish brown; teats: 2 pairs pectoral, 2 pairs abdominal; M^{1-2} symmetrical, with 3 outer cusps well developed and similar to inner cusps .... ***Hapalomys***

Tail naked throughout; terminal pads of digits not bifid; dorsal pelage reddish brown; no pectoral teats, 2 pairs abdominal; M^{1-2} asymmetrical, outer cusps much smaller than inner .... ***Pithecheir***

7. Pelage of posterior half of dorsum consisting mainly of spines; bony palate extending far behind M^3, about 3mm or half way from M^3 to bullae, with a median ridge; (Pakistan only) .... ***Acomys***

Dorsal pelage with spines absent or fewer than soft hairs; palate not extending so far behind teeth, if at all **8**

8. Very small: head & body <80mm, hind feet <22mm, upper molar row <4.0mm **9**

Larger than above **12**

9. Tail <135% of head & body **10**

Tail >135% of head & body **11**

10. Smaller: head & body <70mm, hind feet <17mm; ears very short (<12mm); M^1 with 5 roots, anterior row of cusps not greatly distorted .... ***Micromys***

Table 204. Some cranial and dental characters of the genera of Murinae on the Asian mainland and Greater Sunda Islands (excluding those in Table 205)

	Greatest length of skull	Upper molar row	Incisive foramina (Fig.30)	Posterior palate (Fig.31)
Mus	19–30	3.0–4.6 (Fig.29b)	a-c	c
Acomys	c.31	4.4–4.6	a	d
Rattus	29–56	4.3–9.5	a/b	c
Millardia	24–45	4.7–8.0	a	b/c
Cremnomys	31–47	8.2–11.7	a	b/c
Golunda	29–35	5.6–7.0 (Fig.33a)	c	a
Nesokia	36–47	6.6–8.7 (Fig.29c)	c/d	b
Bandicota	35–69	5.9–11.4	a/b	b/c
Vernaya	c.22	3.3–3.4	b	b
Micromys	18–22	2.8–3.5 (Fig.33b)	b	c
Apodemus	24–30	3.6–5.0	b/c	b/c
Hadromys	28–32	5.4–6.0	b	b
Dacnomys	53–61	11.1–12.4	c	a
Diomys	29–33	5.1–5.6	a	c
Niviventer	29–46	5.0–8.3	b	b
Leopoldamys	52–59	8.3–10.6	c/d	a/b
Maxomys	29–50	5.0–8.1	d	a
Berylmys	36–63	6.2–10.6	b/c	a/b
Haeromys	22–25	3.0–3.5	b/c	c
Lenothrix	37–42	7.9–8.6 (Fig.29a)	d	a
Sundamys (Fig.34)	44–63	8.2–11.7	b	c
Palawanomys	36–40	6.7–7.0	b	c

Table 205. The genera of arboreal Murinae with flat nails on the first digits of the hind feet.

	Vandeleuria	*Chiropodomys*	*Hapalomys*	*Chiromyscus*	*Pithecheir*	*Kadarsanomys*	*Abditomys*
No.of species	3	5	2	1	2	1	1
Range	India-China	S China-Java	Burma-Malaya	Indochina	Malaya, Java	Java	Philippines
Head & body	60–125	65–125	120–165	145–160	150–180	160–210	215–230
Tail (as % of h & b)	100–160%	120–160%	115–125%	135–140%	c.115%	140–150%	105–125%
Hind feet	17–21	16–29	22–32	27–29	26–32	37–44	45–47
Flat spines in dorsal pelage	0	0	0	+/0	0	Few, soft	0
Dorsum	Yellowish brown	Variable	Mid-brown	Yellowish brown	Yellowish brown	Dark brown, agouti	Mid-brown, course agouti
Venter	White or grey	White or grey	Off-white, grey bases	Cream to roots	White, ± apricot tinge	White to roots	Buff, pale grey bases
5th digits of fore and hind feet with claw	0	+	+	+	+	+	+
Skull: greatest length	20–25	20–30	32–41	38,39	39–46	40–47	49–51
Upper molar row	3.0–3.7	2.9–5.1	5.9–8.0	c.6.8	7.4–9.4	7.8–8.4	9.7–9.9
Auditory bullae	Inflated	Small,globular	Inflated	Small	Inflated	Large	Large (9.2)

Table 206. The genera of Murinae endemic to the Philippines (*Mus* and *Rattus* also present).

	No.of species	Head & body	Tail (%)	Hind feet	Greatest length of skull	Upper molar row	Molars laminate
Bullimus	2	230–330	<h&b	46–53	50–72	9.0–11.0	+
Limnomys	1	100–125	120–150%	27–30	28–33	5.0–5.6	(+)
Tarsomys	1	c.135	c.100%	30	c.37	6.0,6.2	0
Tryphomys	1	130–175	c.100%	35–42	36–44	7.6–9.2	0
Abditomys	1	215–230	105–125%	45–47	49–51	9.7–9.9	+
Apomys	8	75–170	80–140%	20–40	24–41	4.4–7.5	(+)
Carpomys	2	190–200	90–95%	30–34	40,44	6.1–8.8	+
Batomys	3	175–200	55–95%	35–37	43–46	7.7–9.4	(+)
Anonymomys	1	125	120–165%	28–31	32–34	5.7–6.0	(+)
Phloeomys	2	c.480	90–100%	72–78	83–89	7.0–8.5	+
Crateromys	3	240–390	90–100%	50–82	53–71	11.8–17.3	0
Crunomys	3*	95–130	65–75%	23–27	28–32	3.7–4.2	0
Archboldomys	1	125	<h&b	28	28	4.8	0
Chrotomys	3	180–195	50–70%	32–40	36–40	3.7–6.5	0
Rhynchomys	2	190–195	55–75%	37–42	43–49	2.1–2.3	0

* Also one species on Sulawesi.

Table 207. The genera of Murinae endemic to Sulawesi.

	Bunomys	*Lenomys*	*Eropeplus*	*Margaretamys*	*Taeromys*	*Paruromys*	*Echiothrix*	*Melasmothrix*
No.of species	4	1	1	3	6	1	1	3
Head & body	145–200	180–270	195–230	95–200	180–250	220–260	200–220	110–155
Tail (% of head & body)	80–100%	80–105%	c.130%	120–170%	80–120%	c.120%	c.90%	75–150%
Hind feet	34–42	41–44	40–45	23–39	40–54	50–55	45–50	26–38
Dorsal pelage with spines	0	0	0	+/0	0	0	+	0
" " : texture	Long, soft	Long	Long, soft	Soft	Long, soft	Soft, brush-like	Long, coarse	Short, soft
" " : colour	Red- or grey-brown	Grey-brown	Grey-brown	Grey-brown	Brown	Medium brown	Medium brown	Dark brown
Venter	Grey, buff tips	Buff, ± grey bases	Grey, cream tips	Grey or white	Grey or white	White to roots	White to roots	Grey or brown
Tail: extent of white distally	0–30%	c.50%	50–70%	0	50–100%	60–70%	30–60%	0–10%
Tail well-haired distally, with terminal tuft	0	0	0	+	0	0	0	(+)
Skull: greatest length	36–46	53–56	47,48	28–43	43–53	56–63	51–55	31–40
Upper molar row	5.7–8.2	10.6–13.1	9.1–10.2	4.5–7.5	6.6–10.7	9.6–11.0	6.3–6.8	4.8–6.0
Incisive foramina (Fig.30)	c	c	b/c	b	a/b/c	c	c	c
Posterior edge of palate (Fig.31)	a/b	a	a/b	b	b	c	c	b/c
Incisors	Orange	Orange	Orange	Orange	Orange	Orange	White or pale yellow	White or pale yellow

Table 208. The genera of Murinae endemic (within the Region) to the Lesser Sunda Is, the Moluccas, the Ryukyu Is and Sri Lanka.

	Papagomys	*Komodomys*	*Melomys*	*Uromys*	*Hydromys*	*Diplothrix*	*Tokudaia*	*Srilankamys*
Species in region	1	1	5	1	1	1	1	1
Range in region	Flores I	Komodo Is	Moluccas	Kei Is	Kei Is	Ryukyu Is	Ryukyu Is	Sri Lanka
Head & body	260–435	130–200	115–170	c.280	c.240	c.200	125–175	125–160
Tail (% of head & body)	80–100%	80–95%	75–140%	c.85%	c.105%	c.120%	70–80%	110–125%
Hind feet	71–88	31–39	26–34	c.50	c.56	c.49	32–34	28–33
Dorsal pelage with flat spines	+	+	0	0	0	+	++	0
Skull: greatest length	68–82	34–41	33–39	c.63	c.53	c.51	39–42	38–39
Upper molar row	14.7–15.3	7.0–7.7	6.1–7.8	c.11.4	c.8.0	c.10.2	5.9–6.3	4.8–5.2

Larger: head & body usually >70mm; ears usually >11mm; M^1 with 3 roots, medial cusp of anterior row far behind central cusp........ ***Mus*** (part)

11. M^1 with postero-internal cusp (3 cusps in inner row); incisive foramina long, c. 4.4mm; frontal bones usually with two perforations on dorsal surface; (China)........ ***Vernaya***

M^1 without postero-internal cusp (2 cusps in inner row); incisive foramina shorter, 3.0–3.7mm; frontal bones intact; (Borneo, Palawan)........ ***Haeromys***

12. Upper incisors each with a longitudinal groove in anterior face........ **13**

Upper incisors without grooves (or very faintly grooved in *Diomys*)........ **14**

13. Larger: head & body >170mm, hind feet >28mm, skull >40mm, upper molar row >8.0mm; pelage with harsh guard hairs over soft, woolly contour hairs; distal 30–75% of tail white; M^1 with postero-internal cusp; (Malaya, Borneo)........ ***Lenothrix***

Smaller than above; pelage entirely soft; tail without white tip; M^1 without postero-internal cusp; (India etc.)... ***Golunda***

14. Dorsal pelage including some flat spines........ **15**

No flat spines........ **19**

15. Bony palate extending behind M^3........ ***Rattus*** (part)

Bony palate not extending behind M^3........ **16**

16. Incisive foramina broad and short, ending far in front of M^1........ ***Maxomys***

Incisive foramina narrow and longer, reaching or almost reaching M^1........ **17**

17. M^1 over half length of tooth-row, M^3 very small; (head & body usually <110mm; hind feet usually <23mm) ***Mus*** (part)

M^1 under half length of tooth-row, M^3 less reduced; (head & body usually >110mm; hind feet usually >23mm)........ **18**

18. Pelage soft (apart from a small number of flat spines)........ ***Niviventer***

Pelage coarse........ ***Berylmys***

19. Molars with transverse laminae; upper incisors broad **20**

Molars remaining cuspidate........ **21**

20. Tail <80% of head & body; teats: 2 pairs pectoral, 2 pairs abdominal; incisive foramina short, 12–14% of skull, not reaching M^1........ ***Nesokia***

Tail >80% of head & body; teats: 3 pairs pectoral, 3 pairs abdominal, or up to 9 pairs not divided into sets; incisive foramina longer, over 18% of skull, reaching M^1.. ***Bandicota***

21. Dorsal pelage, when parted, showing long black or grey basal zone sharply demarcated from brown tips, pelage usually soft; (smaller: head & body usually <200mm, hind feet usually <40mm; tail usually sharply bicoloured, dark above, pale below)........ **22**

Dorsal pelage with dark basal zone, if present, not sharply demarcated, pelage often harsh; (many large, head & body up to 300mm, hind feet up to 60mm; tail usually not paler below) **28**

22. 5th (outer) digit of hind foot longer, tip of claw almost reaching base of claw of 4th digit; (tail longer than head & body; incisors narrow)........ **23**

5th digit much shorter........ **24**

23. Incisive foramina very long and narrow, extending between M^{1-1}; (peninsular India)........ ***Cremnomys***

Incisive foramina not extending between M^{1-1}........ ***Niviventer*** (part)

24. Upper incisors white or pale yellow, pro-odont; (dorsal pelage long and dark; upper surfaces of feet white; incisive foramina extending between M^{1-1})........ ***Diomys***

Upper incisors orange, orthodont or opisthodont........ **25**

25. Tail usually < head & body; incisive foramina very long, extending between M^{1-1}; (feet pale above)........ ***Millardia***

Tail usually ⩾ head & body; incisive foramina shorter, not extending between M^{1-1}........ **26**

26. Venter and upper surfaces of hind feet brown or greyish brown; upper incisors wider, combined width >2.5mm... **27**

Venter pale grey; feet pale; upper incisors narrow, combined width c. 1.5mm........ ***Apodemus***

27. Hind feet <30mm; dorsum brown with coarse agouti speckling........ ***Hadromys***

Hind feet >30mm; dorsum dark chocolate brown, not speckled........ ***Palawanomys***

28. Incisors yellow; (venter white; tail white below; incisive foramina very short and wide, combined width >$^1/_2$ length)........ ***Srilankamys***

Incisors orange........ **29**

29. Molars very large, tooth-row 20–22% of skull length; (very large, greyish brown above and below, tail uniformly brown)........ ***Dacnomys***

Molars of normal size, <20% of skull length........ **30**

30. Incisive foramina short and broad, not reaching M^{1-1}; (venter white)........ ***Leopoldamys***

Incisive foramina long and narrow, reaching M^{1-1}........ **31**

31. Rostrum wide and parallel-sided; palate not extending behind M^3; (larger: head & body >185mm, hind feet >37mm)........ ***Sundamys***

Rostrum narrower and tapering; palate usually extending behind M^3........ ***Rattus*** (part)

Key to the endemic genera of Murinae on the Philippine islands other than Palawan (*Mus* and *Rattus* also occur).

1. Tail densely haired, hairs equal to or exceeding diameter of tail; large: head & body >250mm, hind feet >50mm........ **2**

Tail sparsely haired; smaller........ **3**

2. Molar cusps discrete........ ***Crateromys***

Molar cusps coalescing to form transverse laminae........ ***Phloeomys***

3. Mid-dorsal buff stripe, bordered by black, from head to rump .... ***Chrotomys*** (part)

No dorsal stripe **4**

4. Muzzle and rostrum long and slender; incisors small and pro-odont; 3rd molars absent or very small above and below ... **5**

Muzzle, rostrum and incisors normal; 3rd molars well developed **6**

5. Muzzle very long and slender; rostrum in front of zygomatic plate c. 50% of length of skull; 3rd molars absent; M^{1-2} 2.1-2.3mm, without trace of cusps .... ***Rhynchomys***

Muzzle less elongate; rostrum c. 30% of length of skull; 3rd molars present or absent; M^{1-2} c. 3.7–4.2mm ...... ***Chrotomys***

6. 1st (medial) digit of hind foot with a flat nail or short, blunt claw; incisors very wide (combined width of upper incisors c. 6mm) .... ***Abditomys***

1st digit of hind foot with a narrow, pointed claw; incisors narrower **7**

7. Dorsal pelage with spines **8**

Dorsal pelage without spines **9**

8. Tail much longer than head & body .... ***Anonymomys***

Tail shorter than head & body .... ***Crunomys***

9. Larger: head & body >175mm; (tail shorter than head & body) **10**

Smaller than above **12**

10. Very large: head & body >220mm, greatest length of skull >48mm; hind feet long and narrow, >45mm; prominent mid-ventral glandular area .... ***Bullimus***

Smaller; hind feet <40mm; no mid-ventral glandular area **11**

11. Hind feet <35mm; molars complex, with narrow laminae .... ***Carpomys***

Hind feet >35mm; molars simpler, not laminate ***Batomys***

12. Pelage coarse; posterior edge of palate with broad median ridge and lateral vacuities; (5 pairs of teats: 3 pectoral, 2 abdominal) .... ***Tryphomys***

Pelage soft; palate normal **13**

13. Tail densely haired, obscuring scales, >120% of head & body; venter off-white, sharply demarcated from dark flanks .... ***Limnomys***

Tail not densely haired **14**

14. Claws long, c. 2.5mm; molars high-cusped ***Archboldomys***

Claws normal; molars slightly laminate **15**

15. Very dark all over, including venter, feet and tail; pelage coarse .... ***Tarsomys***

Venter and/or feet lighter than dorsum; pelage soft; (teats: 2 pairs, inguinal only) .... ***Apomys***

Murinae *incertae sedis*

Mus kakhyensis Anderson, 1879: 307; Ponsee, Kakhyen Hills, W Yunnan, China (unidentifiable *fide* Marshall, 1977b).

MUS

Mus Linnaeus, 1758: 59; type species *M. musculus* L.
= *Musculus* Rafinesque, 1814: 13.
Leggada Gray, 1837a: 586 (*M. booduga* Gray).
Pyromys Thomas, 1911d: 996 (*P. priestleyi* Thomas, a form of *M. saxicola*).
Leggadilla Thomas, 1914f: 682 (*M. platythrix* Bennett).
Coelomys Thomas, 1915c: 414 (*C. mayori* Thomas).
Oromys Robinson & Kloss, 1916: 270 (*O. crociduroides* Robinson & Kloss; pre-occupied by *Oromys* Leidy, 1853, a fossil caviomorph rodent).
= *Mycteromys* Robinson & Kloss, 1918a: 57.
Tautatus Kloss, 1917c: 279 (*T. thai* Kloss, in *M. cookii*).
Gatamyia Deraniyagala, 1965a: 214 (*G. weragami* Deraniyagala, probably in *M. booduga*).

Eurasia and Africa; about 40 species, of which 17 occur in the Indomalayan Region. Classification has been very unstable but the Asian forms were revised by Marshall (1977b, 1979,1986) making use of karyological as well as morphological data and the examination of all available types. These papers provided a very valuable clarification of a very confused situation but much remains to be done before the phenetic and geographical limits of the species are adequately known. The maps given here are therefore very approximate and incomplete. Cytogenetics of species in Thailand were studied by Markvong (1977) and the karyology of the entire genus was reviewed by Capanna (1985).

Mus musculus

Asian species have been divided amongst three subgenera, *Mus, Pyromys* and *Coelomys*, and the Afrotropical species placed in a fourth, *Nannomys* Peters, 1876 (type species *M. setulosus* Peters from Cameroun). It is very doubtful if the Asian subgenera are monophyletic but they are used here for convenience (Table 209). It has been argued, mainly on the basis of biochemical studies, that *Pyromys, Coelomys* and *Nannomys* justify generic rank (Bonhomme, 1986) but this is premature, without taking into account a larger sample of the relevant species, including the type species.

The following fossil species have been described from the Indomalayan Region:

M. auctor Jacobs, 1978: 38; Dhok Pahan Formation, Punjab, Pakistan; Late Miocene.
M. elegans Sen, 1983: 37; Pul-e Charkhi, Kabul Basin, E Afghanistan; Pliocene.

Table 209. The subgenera of *Mus*.

	(*Mus*)	(*Pyromys*)	(*Coelomys*)
No. of species in region	6	5	5
Pelage spiny	0	+*	+/0
Eye	Larger	Larger	Very small
Supra-orbital ridges prominent and protruding (as in *Rattus*)	0	+	0
Interorbital width	<4mm	3.5–4.5mm	>4mm
Palatal foramina (as % of occipito-nasal length)	Long, >20%	Long, >20%	Short, <20%

* Except in *M.saxicola gurkha*

The following names are unlikely to represent independent species but cannot be allocated to a species:
M. lepidus Elliot, 1839: 216; S Mahratta country, India.
M. albidiventris Blyth, 1852: 351; near Calcutta.
M. spinulosus Blyth, 1854: 734; Punjab.

Subgenus *Mus* (Table 210).

Mus musculus House mouse

M. musculus Linnaeus, 1758: 62; Upsala, Sweden.
M. domesticus Rutty, 1772: 281; Dublin, Ireland; a *nomen nudum*.
M. praetextus Brants, 1827: 125; Syria.
M. castaneus Waterhouse, 1843a: 134; Philippines.
M. mollosinus Temminck, 1844: 51, pl. 15; Japan.
M. homourus Hodgson, 1845: 268; Nepal.
M. dubius Hodgson, 1845: 268; Nepal; pre-occupied by *M. dubius* Fischer, 1829.
M. urbanus Hodgson, 1845: 269; Katmandu, Nepal.
M. bactrianus Blyth, 1846a: 140; Kandahar, Afghanistan.
M. manei Kelaart, 1852: 64; Sri Lanka (based upon *M. manei* Gray, 1843 which was a *nomen nudum*).
M. gerbillinus Blyth, 1853a: 410; Punjab.
? *M. theobaldi* Blyth, 1853b: 583; Punjab.
M. tytleri Blyth, 1859a: 296; Dehra Dun, N India.
M. rama Cantor, in Blyth, 1865: 194; Penang, Malaya.
M. viculorum Anderson, 1879: 308; Ponsee, W Yunnan, China.
M. commisarius Mearns, 1905: 449; Davao, S Mindanao, Philippines (in building).
Also other extralimital synonyms.

Range (Map 133). Within the Indomalayan Region out-door forms are limited to E Afghanistan, Pakistan, the Himalayan foothills and parts of S China, Taiwan and the Philippines, i. e. mostly in areas not occupied by other species of the subgenus *Mus*. Indoor forms are found throughout the region but in the southeast are mainly in ports and larger towns. Beyond the Indomalayan Region out-door forms are also found west to N Africa, and indoor forms are almost world-wide.

Variation. The following subspecific names have been proposed:
M. m. sinicus Cabrera, 1922: 166; Ningpo, Zhejiang, China.
M. molossinus yonakuni Kuroda, 1924: 8; Yonakuntjima, Ryukyu Is.
M. bactrianus tantillus Allen, 1927a: 9; Wanhsien, Sichuan, China.
M. musculus taiwanus Horikawa, 1929: 80; N Taiwan.
M. m. fredericae Sody, 1933a: 438; Poso, central Sulawesi, sea-level.
M. m. domesticus Schwarz & Schwarz, 1943: 65; Dublin, Ireland. Name attributed to Rutty (1772) where it was a *nomen nudum*. On Official List of Specific Names, to be given precedence over all names, with the exception of *musculus* L.,1758, that are considered to be synonyms of it (Opinion 1607, 1990).
M. m. pygmaeus Biswas & Khajuria, 1955: 28; Pangboche, Imje Valley, Khumbu, E Nepal, c. 14000ft. Pre-occupied by *M. citellus* var. *pygmaeus* Pallas, 1779 (in *Spermophilus*), *M. jaculus* var. *pygmaea* Pallas, 1792 (in *Alactagulus pumilio*) and *M. pygmaeus* Milne-Edwards, 1874 (in *Micromys minutus*).
= *M. m. khumbuensis* Biswas & Khajuria, 1968: 218.

Table 210. The species of *Mus* (*Mus*)

	musculus	*booduga*	*dunni*	*cookii*	*cervicolor*	*caroli*
Range	Widespread	India etc.	India, Sumatra	India-Thailand	Himalayas, Indochina	S China-Java
Head & body	70–85	65–75	65–75	70–100	80–100	c.75
Hind feet	16–19	14–16	14–16	16–21	16–18	c.18
Tail: head & body	Variable	<	=	=	<	=
Venter	Brown or white	White or grey	White, grey bases	Light grey, slate bases	White/grey	Whitish
Feet	Brown or white		White	Grey or brown	White	Dark or white
Tail bicoloured	+/0	+	+	+	+, pale above	+, dark above
Skull: greatest length	20–22	19–20	19–20	23–25	21–23	19–21
Upper molar row	3.0–3.5	3.0–3.3	3.0–3.3	3.5–4.2	3.4–3.8	3.3–3.5
I^1 pro-odont (P), orthodont (Or), or opisthodont (Op)	Op	Op	Or	Op	P	P
I^1 with notch	+	+	0	+	0	0/+
Nasals overhanging incisors in dorsal view	+	+	+	+	+	0
Incisive foramina reach 1st or 2nd cusp of M^1	2nd	2nd	2nd	1st	2nd	1st
Rostrum short and deep (<25% of skull length)	+	0	0	0	0	0

Map **133**. *Mus* (part)

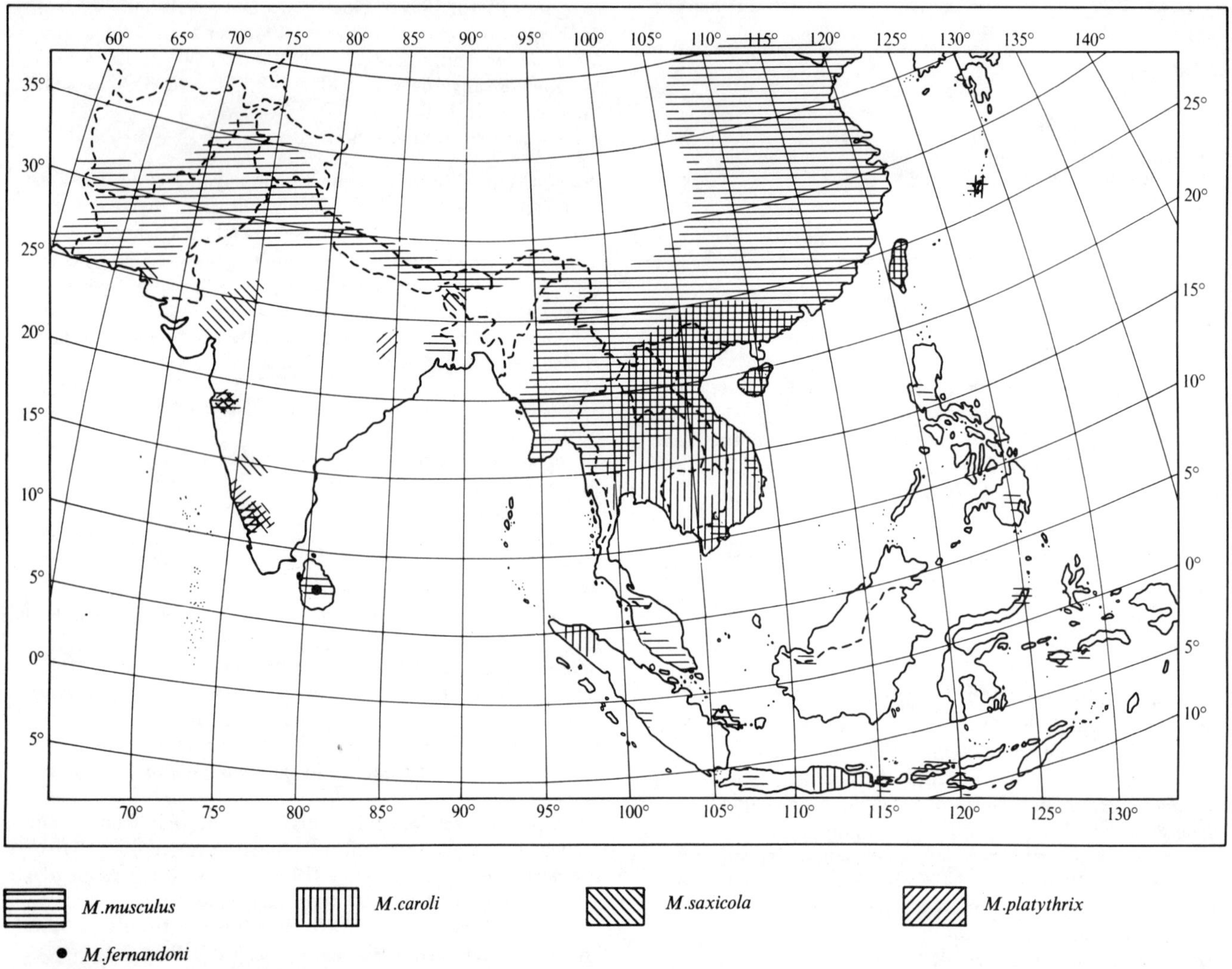

Geographical variation has been intensively studied, involving morphology, karyology, isozymes and DNA (e. g. Marshall, 1986), but no comprehensive integration and mapping of the results have been produced. The following account of subspecies is based upon the summary by Marshall (1986) and includes the two principal European subspecies and the Japanese one in view of their importance in relation to the origin of the laboratory forms and the continuing uncertainty about many of the relationships involved. The boundary between *M. m. musculus* and *M. m. domesticus* in Europe has been intensively studied and shown to be rather sharp, with very limited interbreeding. No such sharp boundaries have been detected elsewhere and intergradation is likely to be considerable. Some aspects of genetic variation throughout the species were reviewed by Moriwaki *et al.* (1986).

The characters of the principal subspecies are shown in Table 211 and their ranges summarized below.

M. m. musculus. Europe north and east of a line from Denmark to Greece, through W Siberia to N China and Korea, mainly indoors.

M. m. domesticus. Europe west and south of the above line, including the British Isles, mainly indoors. Also world-wide by introduction, e. g. in N and S America, Australia, New Zealand, S Africa. The principal ancestor of domestic and laboratory mice.

M. m. praetextus (including *bactrianus*). NW Africa to Pakistan.

M. m. homourus. Himalayas, out-doors, in cultivation.

M. m. castaneus. India to Yunnan and SE Asia; indoors, except where other species of the subgenus *Mus* are absent, as in the Philippines. Linked to *M. m. praetextus* by a pale (but indoors) form, *tytleri*, in N India. The chromosomes of *M. m. castaneus* show a closer affinity with European *M. m. domesticus* than with Japanese *M. m. molossinus* (Seth & Saxena, 1977).

M. m. molossinus (including *manchu* and *yamashimai*). Japan, N China, Taiwan; an out-door form in the south (e. g. on Taiwan), indoors elsewhere. Intergrades with *M. m. castaneus* in Sichuan. Perhaps originated by hybridization between *M. m. musculus* and *M. m. castaneus* (Yonekawa *et al.*, 1986).

M^3 and M_3, which are always small in the genus *Mus*, are absent (sometimes unilaterally or only above or below) in a

Table 211. The subspecies of *Mus musculus*.

	musculus	*domesticus*	*praetextus*	*homourus*	*castaneus*	*molossinus*
Tail: head & body	$<$	$\geq$	$=$	$=$	$>$	$<$
Dorsum	Medium brown	Medium brown	Pale brown	Dark brown	Greyish brown	Greyish brown
Venter	Brownish grey	Brown	Pure white	White, grey bases	Brown	White grey bases
Hind feet	Brown	Brown	White	White	Brown	White
Tail bicoloured (paler below)	+	0	+	+	0	+
Zygomatic plate - anterior edge	Straight	Straight	Slightly convex	Straight	Strongly convex	Strongly convex

proportion of indiviudals. In most populations the proportion lacking one or more of these teeth is very small, e. g. less than 2% in Britain, but in a sample of 48 from Singapore 9% lacked M^3 (Berry, 1963), while 5 out of 13 from Kuala Lumpur, Malaya lacked M_3 at least on one side (Zoological Reference Collection, Singapore). In a sample of 53 from various localities in Yunnan, 5 lacked M3 above and below, 2 lacked both M^3 but not M_3, and 4 lacked a single one of these teeth. Limiting the analysis to clearly brown-bellied *M. m. domesticus* the frequency was 5/20 lacking all four teeth and one lacking both M^3.

Remarks. The domestic and laboratory mice were derived mainly from *M. m. domesticus* as described in detail by Berry (1984). In view of the rather sharp boundary between it and *M. m. musculus* in Europe, with only limited interbreeding, the western European form has by some been treated as a distinct species *M. domesticus* (e. g. by Berry, 1984). Several forms in the western Palaearctic that have been included in *M. musculus* (e. g. *spicilegus* and *spretus*) have been convincingly shown to be distinct species.

Mus booduga Little Indian field mouse

Leggada booduga Gray, 1837a: 586; S Mahratta country, India.
M. terricolor Blyth, 1851a: 172; S India.
M. fulvidiventris Blyth, 1852: 351; Trincomali, Sri Lanka.
M. beavanii Peters, 1866b: 559; Manbhoum, India.
Leggada lepidoides Fry, 1931: 921; Mt Popa, Burma.
? *Gatamyia weragami* Deraniyagala, 1965a: 214; Dasgiriya, near Laggala, N Central Prov., Sri Lanka (based upon a single, very young individual—see Hill, 1980a: 205).

Range (Map 134). Pakistan, India (perhaps including Kashmir: Lal, 1976), Burma and Sri Lanka. Recorded up to 3695m in Uttar Pradesh (Mandal, 1984). Probably widespread in India but range poorly known because of confusion with *M. dunni*.

Remarks. Chromosomes: 2n=40, very similar in G-band pattern to those of *M. musculus* and *M. dunni* but differing in distribution of constitutive heterochromatin (Sen & Sharma, 1983).

Mus dunni

Leggada dunni Wroughton, 1912a: 339; Ambala, Punjab, Haryana, India, 900ft.

Range (Map 134). India and Sumatra (probably introduced in the latter). Probably widespread in India but range poorly known because of confusion with *M. booduga*.

Variation. Although the chromosome number is constant at 2n=40, there is considerable variation in the bi-armed autosomes giving FN=42–48 (Aswathanarayana & Manjunatha, 1981a).

Remarks. This form has been included in *M. booduga* but is sympatric with it and has a distinctive karyotype.

Mus cookii

M. cookii Ryley, 1914: 664; Gokteik, Shan States, Burma, 2133ft.
Tautatus thai Kloss, 1917c: 280; Raheng, Thailand.
Leggada rahengis Kloss, 1920a: 61; Metaw, 40 miles NW of Raheng, Thailand, 1500ft.
L. nagarum Thomas, 1921c: 597; Gologhat, Naga Hills, Assam, 300ft.
L. palnica Thomas, 1923h: 87; Shambagama, Palni Hills, S India, 6000ft.

Range (Map 134). India and Nepal to Burma, Thailand and Vietnam, mostly in mountains.

Ecology. Occurs in grassy areas in conifer and broadleaf forest as well as in adjacent cultivated land.

Mus cervicolor Fawn-coloured mouse

M. cervicolor Hodgson, 1845: 268; Nepal.
M. strophiatus Hodgson, 1845: 268; Nepal.
M. cunicularius Blyth, 1856c: 721; Cherrapunji, Khasi Hills, Assam.
M. nitidulus Blyth, 1859a: 294; Schwegyin, Burma.
Leggada nagarum imphalensis Roonwal, 1948: 386; 16 miles N of Imphal, Manipur, India, 3500ft.

Range (Map 134). Kashmir (Chakraborty, 1983) and Nepal to Burma, Thailand, Laos and Vietnam; S Andaman I, Sumatra and Java; at altidudes of 1150–3200m in Nepal (Abe, 1977).

Variation. The following subspecific names have been proposed:
Leggada nitidula poppaea Thomas, 1919a: 420; Mt Popa, Burma.
Tautatus thai annamensis Robinson & Kloss, 1922: 99; Dalat, Langbian Plateau, Annam, Vietnam, 4500ft.

The forest form *poppaea* in Burma is larger (head & body c. 90mm, hind feet c. 17.5mm) than the widespread synanthropic form (head & body c. 80, hind feet c. 16).

Map **134**. *Mus* (part)

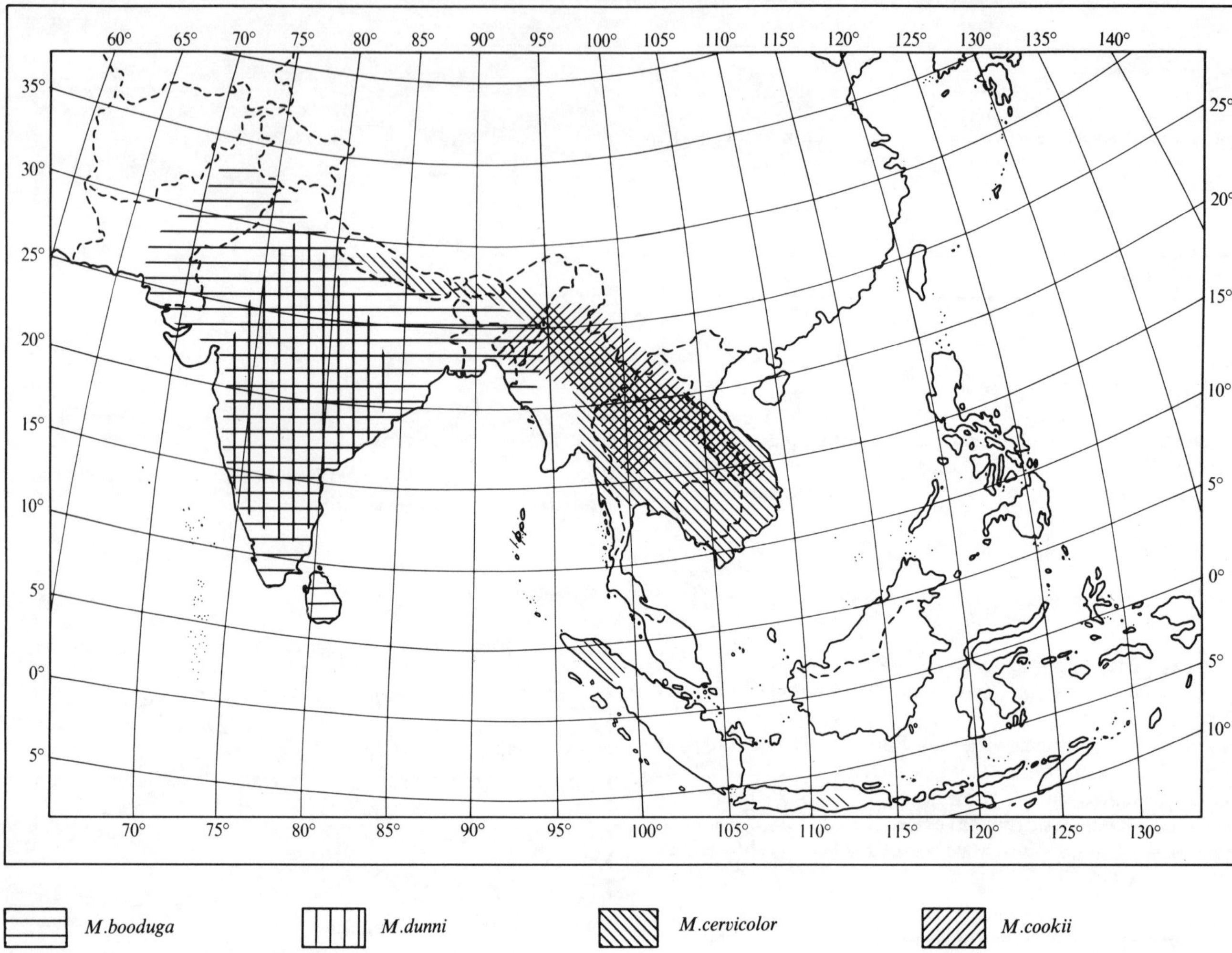

Mus caroli

M. caroli Bonhote, 1902a: 627; Okinawa, Ryukyu Islands.
? *M.* (*musculus ?*) *ouwensi* Kloss, 1921d: 120; Probolingo, E Java.
M. formosanus Kuroda, 1925: 16; Taihoku, Taiwan.

Range (Map 133). Ryukyu Islands, Taiwan, Hainan, S China (Fujian, Yunnan) to Malaya (Langham & Lam, 1977); perhaps also on Sumatra, Java (including Madura) and Flores if *ouwensi* is conspecific.

Variation. In S Thailand the venter is pure white (as in some *ouwensi* in Java) whereas further north the ventral pelage is grey-based.

Ecology. Occurs in rice and hay fields, sometimes sympatrically with *M. cervicolor*, but also in grassy areas in deciduous forest in Burma.

Remarks. Female hybrids with *M. musculus* have been obtained but appeared to be sterile (West *et al.*, 1978). The inclusion of Javan *ouwensi* in *M. caroli* is doubtful. It differs in having the upper incisors orthodont, longer incisive foramina and pale grey or white venter.

Subgenus *Pyromys* (Table 212).

Mus platythrix Indian brown spiny mouse

M. platythrix Bennett, 1832: 121; Dukhun (= Deccan), peninsular India.
Leggada bahadur Wroughton & Ryley, 1913: 18; Karwar, Kanara dist, India.
L. grahami Ryley, 1913: 434; Wotekolli, S Coorg, India, 2000ft.
L. hannyngtoni Ryley, 1913: 435; Makut, S Coorg, India, 250ft.

Range (Map 133). Peninsular India and north to West Bengal (Mandal, 1986).

Variation. Karyological data suggests that this is really a complex of sibling species, those in the northwest (Pune) having 2n=30 and those in southern India having 2n=26 (Marshall, 1977b; Capanna, 1985) but the limits of these forms and any correlation with morphological characters have not been defined.

Ecology. Found mainly in uncultivated hilly country.

Table 212. The species of *Mus* (*Pyromys*).

	platythrix	*saxicola*	*phillipsi*	*shortridgei*	*fernandoni*
Range	Pen.India	India	Pen.India	Burma-Indochina	Sri Lanka
Head & body	110–120	80–90	80–90	c.122	c.100
Ventral pelage with grey bases	0	0	0	+	0
Teats (pairs)	5	6	5	5	5
Hind feet with dark stripe	+	0	0	0	0
Skull: greatest length	26–30	c.25	21–23	c.28	c.24
I^1 orthodont (Or) or opisthodont (Op)	Op	Op	Op	Or	Op
I^1 notched	0	0	0 (+)	+	0 (+)

Mus saxicola

M. saxicola Elliot, 1839: 215; Madras, India.
M. (*Leggada*) *ramnadensis* Bentham, 1908: 385; Ramnad, Madura, Madras, India.
Pyromys priestleyi Thomas, 1911d: 996; Varawah, Sind—name based on a skin of *Mus* accidentally associated with a skull of *Millardia*; the skin was designated as lectotype by Bishop (1974).
Leggada platythrix sadhu Wroughton, 1911b: 1001; Virawah, Sind, India.
L. cinderella Wroughton, 1912b: 770; Bhuj, Cutch, India.
? *Leggadilla gurkha* Thomas, 1914e: 200; Jerna, Ramnagar, Kumaon, N India, 1500ft.

Range (Map 133). S India to Pakistan, the Himalayan foothills and West Bengal (Agrawal & Bhattacharya, 1987).

Variation. The chromosomes show Robertsonian polymorphism, with 2n=22–26 (Rishi & Puri, 1978). Although the pelage is usually spiny, in the northwestern form *gurkha* and in individuals elsewhere it not only lacks spines but is particularly soft. Although Marshall (1977b) considered these conspecific with *M. saxicola* this needs confirmation.

Mus phillipsi

M. phillipsi Wroughton, 1912b: 772; Asigarh, Nimar, Central Prov., India, 1500ft.
Leggadda (*sic*) *surkha* Wroughton & Ryley, 1913: 17; Vijayanagar, Bellary, India.
Leggada siva Thomas & Ryley, 1913b: 242; Sivasamudram, S Mysore, India, 2500ft.

Range (Map 135). Peninsular India N to Rajasthan; perhaps also in the foothills of Nepal, to 800m (Abe, 1977, as *M. cervicolor phillipsi*).

Mus shortridgei

Leggadilla shortridgei Thomas, 1914d: 30; Mt Popa, Burma.
? *M. platythrix nghialoensis* Dào, 1966: 439; Phiêng ban, Nghia lo, N Vietnam.

Range (Map 135). Burma, Thailand, Cambodia and Vietnam.

Ecology. In Thailand they live in long grass and pygmy bamboo in deciduous forest.

Remarks. In Thailand the karyotype shows a complex polymorphism: 2n=46–49; NF=48–52 (Gropp *et al.*, 1973).

Mus fernandoni

Leggadilla fernandoni Phillips, 1932b: 325; Kumbalgamuwa, C Prov., Sri Lanka, 3000ft.

Range (Map 133). Sri Lanka.

Subgenus *Coelomys* (Table 213).

Mus mayori

Coelomys mayori Thomas, 1915c: 416; Pattipola, C Sri Lanka, 6120ft.
C. bicolor Thomas, 1915d: 49; Kottawa, S Prov., Sri Lanka, 250ft. Pre-occupied by *M. bicolor* Tichomirow & Kortchagin, 1889, a form of *M. musculus*.
= *M. m. pococki* Ellerman, 1947a: 382.

Range (Map 135). Sri Lanka.

Variation. In SW Sri Lanka, at lower altitudes (under 1000m), the spines are wider and more prominent, and the venter, feet and underside of the tail are white; in the central highlands, around 1500–2000m, the spines are thinner and less prominent, and the venter is variably marked with brown. It is unlikely that there is any sharp division between these forms.

Mus pahari Sikkim mouse

M. pahari Thomas, 1916c: 415; Batasia, Sikkim, 6000ft.
Leggada jacksoniae Thomas, 1921c: 596; Laitkynsao, Khasi Hills, Assam.
L. cookii meator Allen, 1927a: 6; Taipingpu, Shweli R, Yunnan, China.

Range (Map 135). Darjeeling and Sikkim to Burma, Yunnan, S Sichuan, Thailand, Laos and Vietnam.

Variation. The following subspecies have been described:
Leggada p. gairdneri Kloss, 1920a: 60; Me-taw, 40 miles NW of Raheng, Thailand, 1500ft.
M. p. mocchauensis Dào, 1978: 390; Xôm lôm, Môc chau, Sòn la, N Vietnam.

In Thailand and Indochina the dorsal pelage is greyer and more strongly spiny than further west.

Remarks. Chromosomes in Thailand: 2n=48, NF=48 (Gropp *et al.*, 1973).

Mus famulus

M. famulus Bonhote, 1898a: 99; Coonoor, Nilgiri Hills, S India, 5000ft.

Range (Map 135). Nilgiri Hills, S India, around 1500m.

Mus crociduroides Sumatran shrew-mouse

Oromys crociduroides Robinson & Kloss, 1916: 271; Mt Korinchi, W Sumatra, 10000ft.

Range (Map 135). Sumatra, in high montane forest (2200–3000m).

Mus vulcani Javan shrew-mouse

Mycteromys crociduroides vulcani Robinson & Kloss, 1919b: 378; Kandang Badak, Mt Gede, W Java, 7900ft.

Range (Map 135). Java, in montane forest.

Table 213. The species of *Mus* (*Coelomys*).

	mayori	*pahari*	*famulus*	*crociduroides*	*vulcani*
Range	Sri Lanka	Himalayas-Vietnam	S India	Sumatra	Java
Head & body	c.98	80–100	c.90	90–100	82,95
Tail: h & b	=	=	<	>>	=
Hind feet	c.25	17–23	c.21	21–23	22,23
colour	White	Light brown	Dark brown	Light brown	Dark brown
Pelage spiny	+	+/0	0	0	0
Venter	White, ±brown	Silvery	Medium brown	Grey, pale silvery tips	Grey, buff tips
Teats (pairs)	5	5	3	3	3
Skull: greatest length	c.30	23–26	c.24	24–26	24,26
Upper molar row	c.4.6	3.5–3.9	c.3.7	3.9–4.1	c.3.5

Map **135**. *Mus* (part)

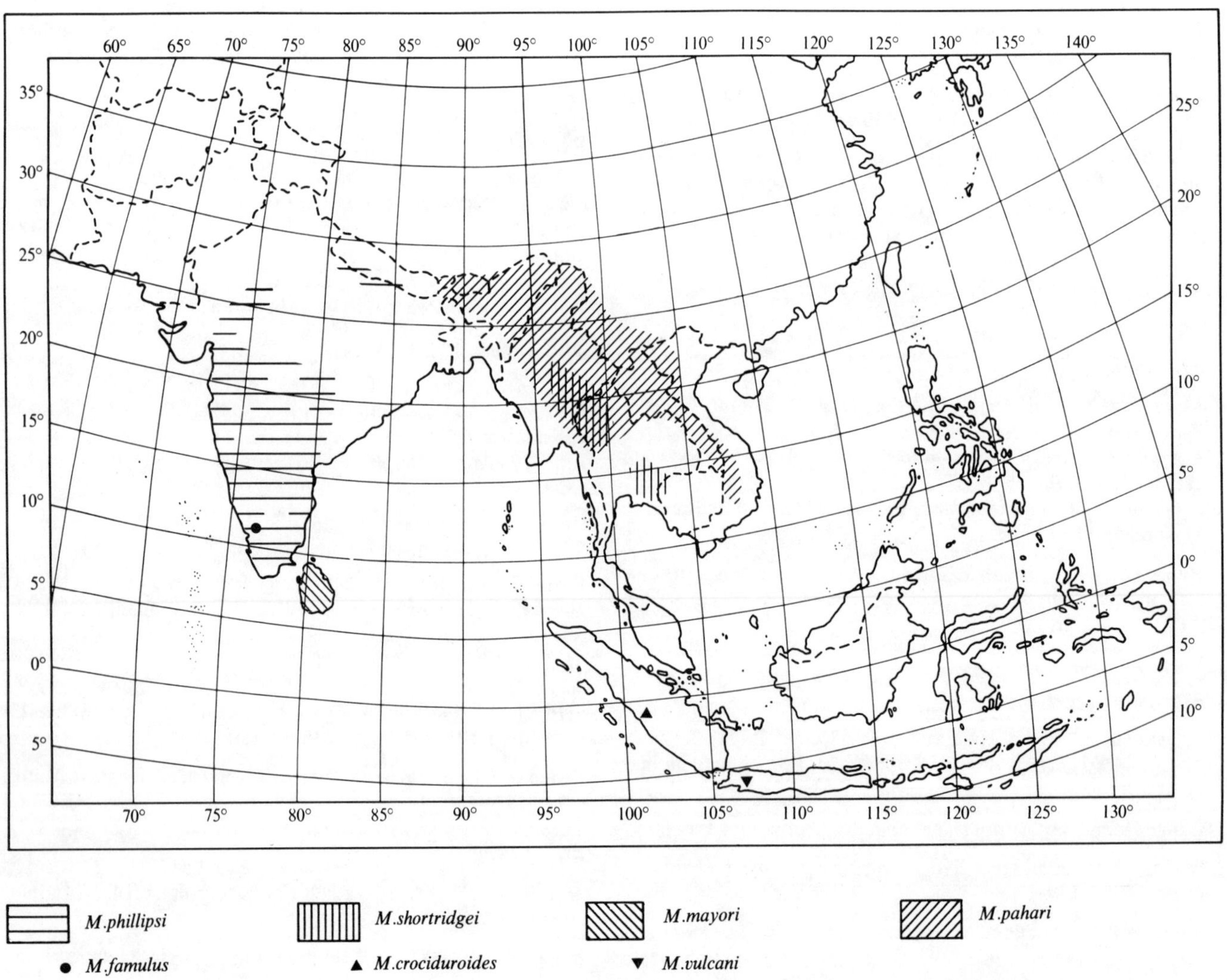

VANDELEURIA Long-tailed tree mice

Vandeleuria Gray, 1842: 265; type species *Mus oleraceus* Bennett.

Indian and Indochinese subregions; 2 or perhaps 3 species (Table 214). All forms have been treated as conspecific (Ellerman, 1961) but the highland form in Sri Lanka, *nolthenii*, is rather distinctive and probably warrants specific rank (Musser, 1979). It was described as a race of *V. nilagirica* from the Nilgiri Hills, S India. The latter is very poorly known but is here treated tentatively as a separate species.

Fossil records (not allocated to species) have been reported from SW China (Zheng, 1988).

Vandeleuria oleracea Indian long-tailed tree mouse

Mus oleraceus Bennett, 1832: 121; Deccan, India.
M. dumeticola Hodgson, 1845: 268; Nepal.
M. povensis Hodgson, 1845: 269; Nepal.
? *M. badius* Blyth, 1859a: 295; Tenasserim Prov., Burma.
V. rubida Thomas, 1914e: 202; Bageswar, Kumaon, N India, 3200ft.
V. sibylla Thomas, 1914e: 202; Chantabun, Thailand.

Range (Map 136). Forested parts of peninsular India and in the Himalayan foothills from Kumaon to Yunnan and most of Indochina; west to C Rajasthan (Advani & Mathur, 1982); also in the lowlands of Sri Lanka; up to c. 2500m in Nepal.

Variation. The following subspecific names have been proposed:
V. o. spadicea Ryley, 1914: 659; Lunwa, Palanpur, Gujarat, India.
V. o. modesta Thomas, 1914e: 202; Ramnagar, Kumaon, N India, 1100ft.
V. o. marica Thomas, 1915e: 54; Koira, Chaibassa, Orissa, India.
V. dumeticola scandens Osgood, 1932: 320; Muong Boum, Tonkin, Vietnam.

Table 214. The species of *Vandeleuria.*

	oleracea	*nolthenii*	*nilagirica*
Range	China-Sri Lanka (lowland)	Sri Lanka (highlands)	SW India
Head & body	60–100	120–125	? 105
Tail	95–130	120–125	? 142
Hind feet	17–20	20–21	21
Ventral pelage	White, ± fulvous wash	Grey	White with grey bases
Skull: greatest length	20–24	24–25	?

Map **136**. *Vandeleuria, Acomys*

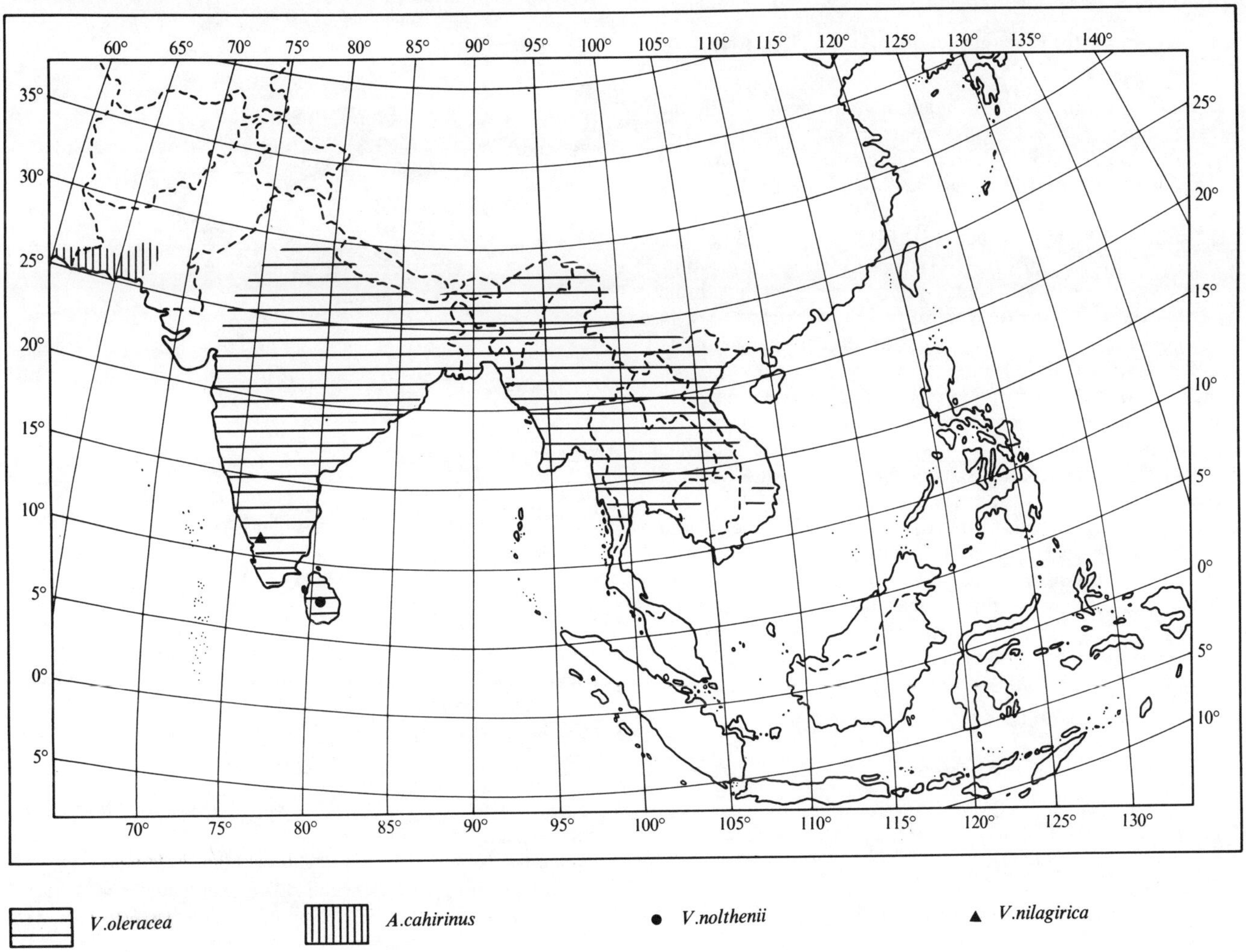

In India the pelage is brighter in the north, pale in Gujarat and darker and duller in the south; there is also some geographical variation in size (Agrawal & Chakraborty, 1980).

There is considerable variation in chromosomes. Samples from N India (Varanasi, Uttar Pradesh and Sirmoni dist, Himachal Pradesh) have been claimed to have an additional sex chromosome giving 2n = 29 (Sharma & Rajiva Raman, 1972; Gropp *et al.*, 1972). This has not been found in samples from N and NE Thailand—2n = 26 and 28 (Gropp *et al.*, 1972; Winking *et al.* 1979) nor in Mysore, S India—2n = 28 (Satya Prakash & Aswathanarayana, 1973,1976). Although these Mysore specimens were named *V. o. nilagirica* this was on purely geographical grounds and they cannot be assumed to be *V. nilagirica* as used here.

Vandeleuria nolthenii

V. nilagirica nolthenii Phillips, 1929b: 165; West Haputale, Ohiya, Sri Lanka, 6000ft.

Range (Map 136). Highlands of Sri Lanka.

Remarks. I concur with the view of Musser (1979) that this form should be given specific rank distinct from *V. oleracea*. It seems better not to include it in *V. nilagirica* of S India until the status of that form is clarified.

Vandeleuria nilagirica Nilgiri long-tailed tree mouse

Mus nilagiricus Jerdon, 1867: 203; Ootacamund, S India.

Range (Map 136). Probably confined to the Nilgiri Hills, SW India.

Remarks. This form was included in *V. oleracea* by Ellerman (1961) and by Agrawal & Chakraborty (1980). There is no type. The original description is poor but distinguished it from *V. oleracea* by having the underside 'bright fawn-yellow' compared with white in *V. oleracea*, and a longer tail: 5 inches (127mm) against 4–4$^1/_2$ inches (101–114mm). A single specimen in the British Museum (NH) from Kutta, S Coorg, altitude 2843ft (no. 13.8.22.70) stands apart from all other Indian *Vandeleuria* in having the ventral pelage with grey bases (but less grey than in *V. nolthenii*) and an exceptionally long tail (142mm). Another specimen from the same locality and altitude, and also one from Coonor, Nilgiri Hills, are normal *V. oleracea* with the ventral pelage entirely white and shorter tails. It therefore seems possible that the grey-bellied individual could represent a separate species coexisting with *V. oleracea*, and that the correct name is *V. nilagirica*.

Further specimens will be required to determine whether Phillips (1929b) was correct in allocating the highland form in Sri Lanka, *nolthenii*, to *V. nilagirica*. However if *V. nilagirica* is a distinct species, it is unlikely that it includes most of the specimens allocated to *V. o. nilagirica* by Ellerman (1961) and by Agrawal & Chakraborty (1980).

ACOMYS Spiny mice

Acomys Geoffroy, 1838: 126; type species *Mus cahirinus* Desmarest.
Acosminthus Gloger, 1841: 95 (*Mus dimidiatus* Cretzschmar).
Acanthomys Lesson, 1842: 135 (*Mus hispidus* Brants, in *M. dimidiatus* Cretzschmar).

Africa and Arabian region, east to Pakistan; c. 9 species of which one reaches the Indomalayan Region. This is a clearly defined genus, characterized especially by strongly spiny dorsal pelage and a unique posterior extension of the bony palate. The species are poorly defined.

Acomys cahirinus Cairo spiny mouse

Mus cahirinus Desmarest, 1819: 70; Cairo, Egypt.
M. dimidiatus Cretzschmar, 1827: 37, taf. 13; Sinai, Egypt.
A. flavidus Thomas, 1917a: 205; Laki Hills, Sehwan, Sind, Pakistan.
Also many extralimital synonyms.

Range (Map 136). SW Pakistan, west of the R Indus; west through Arabia to Egypt and parts of Africa south of the Sahara.

Variation. *A. flavidus* was described solely on the basis of pale colour and lack of a sharp line of demarcation between dorsal and ventral pelage. However, three further specimens from Pakistan are no different in colour from many in the Arabian region and Israel, and like them have a sharp line of demarcation.

Remarks. This species is little known in Pakistan. Both the taxonomy and nomenclature are somewhat uncertain since the name *cahirinus* refers to a grey, synanthropic form and *dimidiatus* to a brown non-synanthropic form, the relationship between these being obscure. There is considerable karyological variation (Nevo, 1985) but no data appear to be available from the population in Pakistan.

Acomys cahirinus

RATTUS

Ruttus Fischer, 1803: 128; type species *Mus decumanus* Pallas (in *R. norvegicus*). There is no evidence within this publication that the spelling *Ruttus* was an error. However it has, very sensibly, been universally accepted as an error for *Rattus* and it would serve no useful purpose to revert to *Ruttus*. Thomas (1916k) argued that the type species of this name is *Mus rattus* on the basis that the French vernacular 'Rat' implied this species, but *Mus decumanus* is the only species given with a latinized name and must stand as the type species according to the Code (Article 67(g)(i)).
Rattus Gray, 1821: 303 (*Mus rattus* L.).
= *Epimys* Trouessart, 1881: 117.
Stenomys Thomas, 1910d: 507 (*Mus verecundus* Thomas, a species in New Guinea).
Christomys Sody, 1941a: 260 (*Mus macleari* Thomas).
Mollicomys Sody, 1941a: 260 (*Mus hoffmanni* Matschie). Doubtfully valid since it was based only upon a mammary formula shared with *Cironomys* and *Pullomys*.

Cironomys Sody, 1941a: 260 (*Rattus hoogerwerfi* Chasen).
Pullomys Sody, 1941a: 260 (*Mus pulliventer* Miller, in *R. rattus*).
Octomys Sody, 1941a: 261 (*Mus concolor* Blyth, in *R. exulans*). Doubtfully valid—based solely upon a mammary formula shared with *Octopodomys* (= *Vernaya*). Pre-occupied by *Octomys* Thomas, 1920 (Rodentia, Octodontidae).
Togomys Dieterlen, 1986: 12 (*T. melanoderma* Dieterlen, in *R. exulans*).

Throughout the region, but probably only by introduction in some parts; c. 33 species (Tables 215–221). The genus *Rattus* was listed with 84 Indomalayan species in 7 subgenera by Ellerman (1949), along with many African and Australasian species. There appears to be considerable agreement that the African species should be separated, in the genera *Praomys, Mastomys, Hylomyscus, Aethomys* etc. Most of the Asiatic subgenera recognised by Ellerman, on rather dubious grounds—especially relative size of the auditory bullae, palate and incisive foramina, have been substantiated by study of a wider range of characters but with considerable changes in the included species. These are here treated as genera along with other groups that Ellerman included in *Rattus* (*Maxomys, Leopoldamys, Berylmys, Cremnomys, Apomys, Niviventer, Bunomys, Batomys, Taeromys, Lenothrix, Bullimus* and *Diplothrix*).

The remaining species, corresponding to the subgenus *Rattus* and part of the subgenus *Stenomys* of Ellerman, are retained

Table 215. The species of *Rattus* on the Asian mainland with some broad, flat spines in the dorsal pelage (all have the tail uniformly dark).

	rattus	*tiomanicus*	*argentiventer*	*losea*	*hoxaensis*	*exulans*	*ranjiniae*
Head and body	140–200	125–190	150–210	120–190	c.175	110–125	160–165
Tail (as % of h & b)	90–150%	95–120%	90–125%	c.100%	98%	c.110%	109, 115%
Venter	White, or grey with buff tips	White	Light grey	Dull grey, ± buff or white tips	White, thin medial dark line	Grey, ± brown tips	Dirty white
Hind feet	Usually dark	Dark	Dark	Dark	White	Dark	Light
Teats: pairs (pectoral/abdominal)	2 or 3*/3	2/3	3/3	2/3		2/2	
Pads of hind feet smooth (S) or finely lamellated (L)	L	L	S	S		L	
Skull: greatest length	37–47	35–47	36–46	33–43	41	29–32	42–45
Upper molar row	6.0–8.0	6.0–8.0	6.7–8.2	6.0–7.5	7.3	4.5–6.2	8.2–8.3

* If three, post-axial two are very close together, <5mm apart.

Table 216. The species of *Rattus* on the Asian mainland without broad flat spines in the dorsal pelage.

	norvegicus	*nitidus*	*remotus*	*turkestanicus*	*osgoodi*	*annandalei*
Head & body	200–250	150–200	170–225	160–180	125–170	145–220
Tail (as % of h & b)	80–90%	90–110%	115–125%	110–130%	c.85%	100–150%
Hind feet	c.44	33–36	33–40	26–31	25–36	35–40
Venter	Grey	Grey	White or cream	White	Dark grey-brown	White
Tail lighter below than above	0	0	0	+	0	0
Teats: pairs (pectoral/abdominal)	3/3	3/3	3/3	3/3	2/3	2 or 3/2 or 3
Skull: greatest length	36–56	40–45	44–49	32–45	31–36	43,51
Upper molar row	6.8–8.8	6.5–7.3	c.7.4	6.0–7.3	5.3–6.0	7.0–8.0
Incisive foramina	8.6–9.2	7.3–8.8	c.7.6	6.6–8.5	5.9–7.2	6.4,8.0

Table 217. The species of *Rattus* endemic to Sri Lanka and the Greater Sunda Islands (all have soft, dense dorsal pelage).

	montanus	*baluensis*	*korinchi*	*hoogerwerfi*
Range	Sri Lanka	Borneo	Sumatra	N Sumatra
Head & body	160–180	158–188	166,169	170–196
Tail (% of h & b)	110–140%	c.110%	125–135%	c.120%
Hind feet	31–34	28–33	32–33	34–37
Venter	Grey, buff tips	Grey, buff tips	Pale grey, white tips	Grey, buff tips
Tail white distally	0	0	0	+ (50–60%)
Teats: pairs (pectoral/abdominal)		2/3	1/2	1/3
Skull: greatest length	42,43	39–44	41,42	42,44
Upper molar row	7.6,7.8	6.3–6.9	7.6–7.7	7.0–7.7

Table 218. The species of *Rattus* endemic to Indian Ocean islands.

	palmarum	*stoicus*	*lugens*	*adustus*	*enganus*	*nativitatis*	*macleari*
Islands	Nicobars	Andamans	Mentawai	Enganno	Enggano	Christmas (extinct)	Christmas (extinct)
Head & body	275	220–250	190–220	c.180	c.230	250–275	c.255
Tail (% of h & b)	78%	80–90%	95–105%	82%	c.115%	c.65%	c.90%
Hind feet	48	c.41	36–40	c.42	c.40	51–53	45
Dorsal pelage with flat spines	+	+	+ (few, narrow)	+	0	0	0
Venter	Pale grey	Pale grey/buff	Dark grey	Dark grey	Light grey	Dark brown	Grey/buff
Tail lighter below	0	+	0	0	0	0	0
Tail white distally	0	0	0	0	0	0	+ (30–60%)
Skull: greatest length	49–54	46–55	43–51	46	45	51	51
Upper molar row	8.9,9.0	7.6–8.6	7.3,8.5	8.3,8.7	c.8.0	7.0,7.2	9.0

here in *Rattus*. It has been suggested (Musser, 1981a) that the rats of New Guinea and Australia currently included in *Rattus* might not be sufficiently related to the Asiatic species to warrent inclusion in *Rattus*. This is supported by a study combining electrophoretic analysis of proteins with morphological data (Gemmeke & Niethammer, 1984) which indicated that *R. villosissimus* of Australia is more distantly related to *Rattus s. s.* than are the species of *Berylmys* and *Bandicota*.

The Indomalayan species have never been comprehensively revised, with diagnoses, and much remains to be done to test the species limits adopted here, which must be treated as very provisional. Many of the differences between species are subtle and are confused by subspecific variation, especially where synanthropic species have been introduced and show character-displacement depending upon the presence or absence of competing indigenous species. Most taxonomic work has been on too local a scale to adequately resolve the problems, and the potentially valuable data on chromosomes and genetic relationships have not yet been integrated with the morphological data, often because of failure to preserve or record the morphological evidence. A detailed but very unorthodox review of the '*Rattus rattus*' group by Schwarz & Schwarz (1967) included most of the species recognized here in *R. rattus* but gave detailed descriptions of many of them.

Table 219. The species of *Rattus* endemic to Sulawesi (excluding *R.mollicomulus* and *R.koopmani* which are similar to *R.hoffmanni*).

	hoffmanni	*xanthurus*	*bontanus*	*marmosurus*
Head & body	140–210	194–265	185–235	175–230
Tail (% of h & b)	95–105%	115–125%	120–130%	110–150%
Hind feet	34–41	40–46	40–46	34–40
Dorsal pelage	Soft	Coarse	Soft	Soft
Tail white distally	0	+/0	+	+
Tail with terminal brush	0	+	+	+
Skull: greatest length	38–48	45–55	44–50	42–47
Upper molar row	6.4–8.2	7.4–8.4	7.7–8.8	6.5–7.3

The following fossil species has been described:
R. trinilensis Musser, 1982b: 65, Trinil, central Java; middle Pleistocene.

The following recently described species has not been critically assessed nor included in the keys:
Rattus hainaldi Kitchener, How & Maharadatunkamsi, 1991c: 557; Mt Ranakah, above Kampong Robo, Desa Longko, 8km SSE of Ruteng, W Flores, Indonesia, 8°35′S, 120°33′E, 1300m.

Rattus rattus Roof rat, House rat, Ship rat

Mus rattus Linnaeus, 1758: 61; Sweden.

Synonyms from mainland range:
Musculus frugivorus Rafinesque, 1814: 13; Sicily, Italy (based upon a white-bellied morph).
Mus alexandrinus Desmarest, 1819: 47; Alexandria, Egypt (based upon a morph with brown dorsum and grey venter).
M. rufescens Gray, 1837a: 585; Dharwar, India.
? *M. asiaticus* Gray, 1837a: 585; India.
M. flavescens Elliott, 1839: 214; Dharwar, India.
M. brunneus Hodgson, 1845: 266; Nepal.
M. brunneusculus Hodgson, 1845: 267; Nepal.
M. arboreus Horsfield, 1851: 141; Bengal.
? *M. crassipes* Blyth, 1859a: 295; India.
M. flavipectus Milne-Edwards, 1872a: 93; Moupin, Sichuan, China.
M. ouang-thomae Milne-Edwards, 1872a: 93; Kiangsi, China.

Table 220. The species of *Rattus* endemic to the Philippines.

	everetti	*tyrannus*	*mindorensis*	*tawitawiensis*
Range	Central	Central	Mindoro	Sulu Is
Head & body	200–250	270–310	c.150–190	170,208
Tail (% of h & b)	90–110%	c.90%	c.90%	c.90%
Hind feet	40–50	c.60	30–33	39–40
Dorsal pelage	Long, coarse	Long, coarse	Short	Short
Venter	Dense, soft	Sparse, coarse	Dense, soft	Dense, soft
Skull: greatest length	45–53	66	35–40	42,47
Upper molar row	8.0–10.0	10.3	6.1–7.1	7.9,8.0
Incisive foramina	9.1–10.1		6.3,6.5	8.1–8.6

Table 221. The species of *Rattus* endemic to the Moluccas.

	feliceus	*ceramicus*	*morotaiensis*	*elaphinus*
Islands	Seram	Seram	Morotai	Taliabu
Head & body	210	120–135	160–200	208
Tail (% of h & b)	82%	c.100%	110–120%	90–110%
Hind feet	43	28–30	32–38	34
Dorsal pelage with spines	+	0	++	0
Venter	White	Dark grey, brown tips	Creamy buff, rust on throat	Grey-buff
Tail	Pallid	Dark	Dark	Dark
Tail: scales/cm.	6–7	c.12	8–9	10–11
Skull: greatest length	51,55	37,38	37–45	49
Upper molar row	8.5	5.8–6.2	6.2–6.5	7.1–8.0
Incisive foramina reach M^1	+	0	+	+

? *M. germani* Milne-Edwards, 1872b: 289; Condor I, S Vietnam.
M. sladeni Anderson, 1879: 305; Ponsee, Kakhyen Hills, W Yunnan, 3500ft.
M. yunnanensis Anderson, 1879: 306; Ponsee, Yunnan, China.
M. griseiventer Bonhote, 1903a: 30; Perak, Malaya.
R. molliculus Robinson & Kloss, 1922: 97; Daban, Phanrang Prov., S Annam, Vietnam.

Synonyms from major islands:
M. kandianus Kelaart, 1850b: 212; Nuwara Eliya, Sri Lanka.
M. ceylonus Kelaart, 1850b: 213; Sri Lanka.
M. tetragonurus Kelaart, 1850c: 217; Hendala, near Colombo, Sri Lanka.
M. nemoralis Blyth, 1851a: 168; Sri Lanka. Pre-occupied by *M. nemoralis* Sélys Longchamps, 1841, a form of *R. rattus* in Switzerland.
? *M. canna* Swinhoe, 1871: 636; near Tamsuy, Taiwan.
M. diardii Jentink, 1880b: 13; W Java.
M. neglectus Jentink, 1880b: 14; Banjermasin, S Borneo.
M. mindanensis Mearns, 1905: 442; Todaya, Mt Apo, Davao, Mindanao, Philippines, c. 1232m.
M. zamboangae Mearns, 1905: 443; Zamboanga, W Mindanao, Philippines.
M. kelleri Mearns, 1905: 444; Davao, S Mindanao, Philippines.
Epimys benguetensis Hollister, 1913a: 323; Camp John Hay, Baguio, Benguet, N Luzon, Philippines. (Based upon a composite specimen: a skin of *R. rattus* and a skull of *R. nitidus*; the skin was designated the holotype by Musser, 1977a: 6.)
E. kelaarti Wroughton, 1915b: 48; Pattipola, highlands of Sri Lanka.
R. dammermani Thomas, 1921d: 247; Wadjo, Sulawesi (given by Thomas as NW corner of Gulf of Tomini, N Celebes, but this is probably erroneous and the type locality was restricted to Wadjo, SW Sulawesi, 4°3'S, 120°9'E by Musser, 1971a).
R. lalolis Tate & Archbold, 1935a: 2; Lalolis, 30 miles SE of Mengkoka Mts, SE Sulawesi, 300m.
R. toxi Sody, 1941a: 309; Wadjo, SW Sulawesi, 4°3'S, 120°9'E (restricted by Musser, 1971a).

Synonyms from small islands:
M. (*Leggada*) *andamanensis* Blyth, 1860a: 103; Andaman Is.
M. pannosus Miller, 1900e: 190; Adang I, Butang Is, E Sumatra.
? *M. flebilis* Miller, 1902b: 762; Henry Lawrence I, Andaman Is.
? *M. pulliventer* Miller, 1902b: 765; Great Nicobar I.
M. atratus Miller, 1902b: 767; Barren I, Andaman Is. Pre-occupied by *Mus atratus* Philippi, 1900.
= *M. atridorsum* Miller, 1903d: 50.
Epimys coloratus Hollister, 1913a: 317; Basilan I, Philippines.
E. robiginosus Hollister, 1913a: 318; Cagayan I, Philippines.
E. panellus Miller, 1913b: 8; Rawi I, Butang Is, E Sumatra.
R. palelae Miller & Hollister, 1921a: 69; Palelah I, N Sulawesi.
R. masaretes Sody, 1937a: 217; Buru I, Moluccas (a synonym of *R. r. moluccarius* according to Musser, 1970c).
R. sapoensis Sody, 1941a: 306; Malengi, Togian I, E of Sulawesi.

Range (Map 137). Ubiquitous throughout the Region in and around human habitation, including most islands, large and small. In the Indian and Indochinese subregions it is also common in cultivated crops and often in secondary forest, scrub and mangrove swamps. Also introduced throughout the tropical and subtropical parts of the world and more locally in temperate regions where it is more restricted to urban habitats.

Variation. The following subspecific names have been proposed:

Mainland:
R. r. thai Kloss, 1917a: 286; Raheng, Thailand.
R. r. tistae Hinton, 1918: 68; Pashok, Sikkim.
R. r. bhotia Hinton, 1918: 72; Hasimara, Bhutan Duars, NE India.
R. r. narbadae Hinton, 1918: 77; Sakot, Hoshangabad, Central Provinces, India, 1200ft.
R. r. girensis Hinton, 1918: 83; Sasan, Junagadh, Kathiawar, India, 400ft.
R. r. satarae Hinton, 1918: 87; Ghatmatha, Satara dist, India, c. 2000ft.
R. r. wroughtoni Hinton, 1919: 384; Coonoor, Nilgiri Hills, India, 6000ft.
R. r. gangutrianus Hinton, 1919: 389; Ranibagh, Naini Tal, Kumaon, India, 2500ft.
R. r. khyensis Hinton, 1919: 398; 25 miles W of Kindat, Chin Hills, W Burma, 600ft.
R. r. tikos Hinton, 1919: 400; Tenasserim Town, S Burma.
R. r. tatkonensis Hinton, 1919: 402; Tatkon, W bank of Chindwin R, Burma.
R. r. bullocki Roonwal, 1948: 385; c. 16 miles N of Imphal, Manipur, India, c. 3500ft.
? *R. r. yeni* Dao, 1960a: 234; Vinh-Kim, Vinh-Linh, Quang-tiu Prov., central Vietnam.
R. r. povolnyi Niethammer & Martens, 1975: 344; Dar-i-Nur, Djalalabad, E Afghanistan, 1200m.

Major islands:
Epimys r. turbidus Miller, 1913b: 12; Tanggarong, S bank of Mahakam R, SE Borneo.
R. r. palembang Tate & Archbold, 1935a: 1; Morcarah Duwa, Palembang, Sumatra.
R. r. makassarius Sody, 1941a: 266; Macassar, S Sulawesi.
R. r. argyraceus Sody, 1941a: 275; Mapangat, Manado, N Sulawesi.
R. r. barussanoides Sody, 1941a: 276; Macassar, S Sulawesi.

Small islands:
Epimys r. dentatus Miller, 1913b: 14; Hastings I, Mergui Is, Burma.

E. r. insulans Miller, 1913b: 14; Helfer I, Mergui Is, Burma.
E. r. exsul Miller, 1913b: 15; James I, Mergui Is, Burma.
E. r. fortunatus Miller, 1913b: 15; Chance I, Mergui Is, Burma.
E. r. portus Kloss, 1915b: 221; Chang I, Thailand.
E. r. poenitentiarii Kloss, 1915b: 222; Phai I, Thailand.
E. r. rangensis Kloss, 1916b: 56; Rang I, Thailand.
E. r. klumensis Kloss, 1916b: 56; Klum I, Thailand.
E. r. makensis Kloss, 1916b: 56; Mak I, Thailand.
E. r. kraensis Kloss 1916b: 57; Kra I, Thailand.
R. r. lanensis Kloss, 1919b: 378; Lan I, Inner Gulf of Siam, Thailand.
R. r. kramensis Kloss, 1919b: 379; Kram I, Thailand.
R. r. mesanis Kloss, 1919b: 379; Mesan I, near Cape Liant, SE Thailand.
R. r. sumbae Sody, 1930c: 98; Mao-Marroe, Sumba I, Lesser Sunda Is.
R. r. santalum Sody, 1932a: 159; Waingapoe, Sumba I, Lesser Sunda Is.
R. r. samati Sody, 1932a: 159; Bali.
R. r. moluccarius Sody, 1933a: 437; Buru I, Moluccas.
R. r. septicus Sody, 1933a: 437; Banda I, Moluccas.
R. mindanensis tablasi Taylor, 1934: 439; Odiongan, Tablas I, Philippines.
R. r. panjius Chasen, 1937b: 85; N Panjang I, W peninsular Thailand.
R. r. alangensis Chasen, 1937b: 87; Alang Yai I, W peninsular Thailand.
R. r. lontaris Chasen, 1937b: 88; Lontar I, W peninsular Thailand.
R. r. kadanus Chasen, 1937b: 89; Kadan I, W peninsular Thailand.
R. r. moheius Chasen, 1937b: 91; Mohea I, W peninsular Thailand.
R. r. pipidonis Chasen, 1937b: 92; Pipidon I, W peninsular Thailand.
R. r. robinsoni Chasen, 1940a: 154; Samui I, E peninsular Thailand.
R. r. auroreus Sody, 1941a: 266; N Pagi I, Mentawai Is, Sumatra.
R. r. pelengensis Sody, 1941a: 267; Peleng I, E of Sulawesi.
R. r. obiensis Sody, 1941a: 268; Obi I, Moluccas.
R. r. talaudensis Sody, 1941a: 275; Liroeng, Talaud Is, NE of Sulawesi.
R. r. amboinensis Laurie & Hill, 1954: 102; Pokka, Ambon I, Moluccas.
? *R. r. holchu* Chaturvedi, 1966: 141; Car Nicobar I, Nicobar Is.

Map **137**. *Rattus* (part)

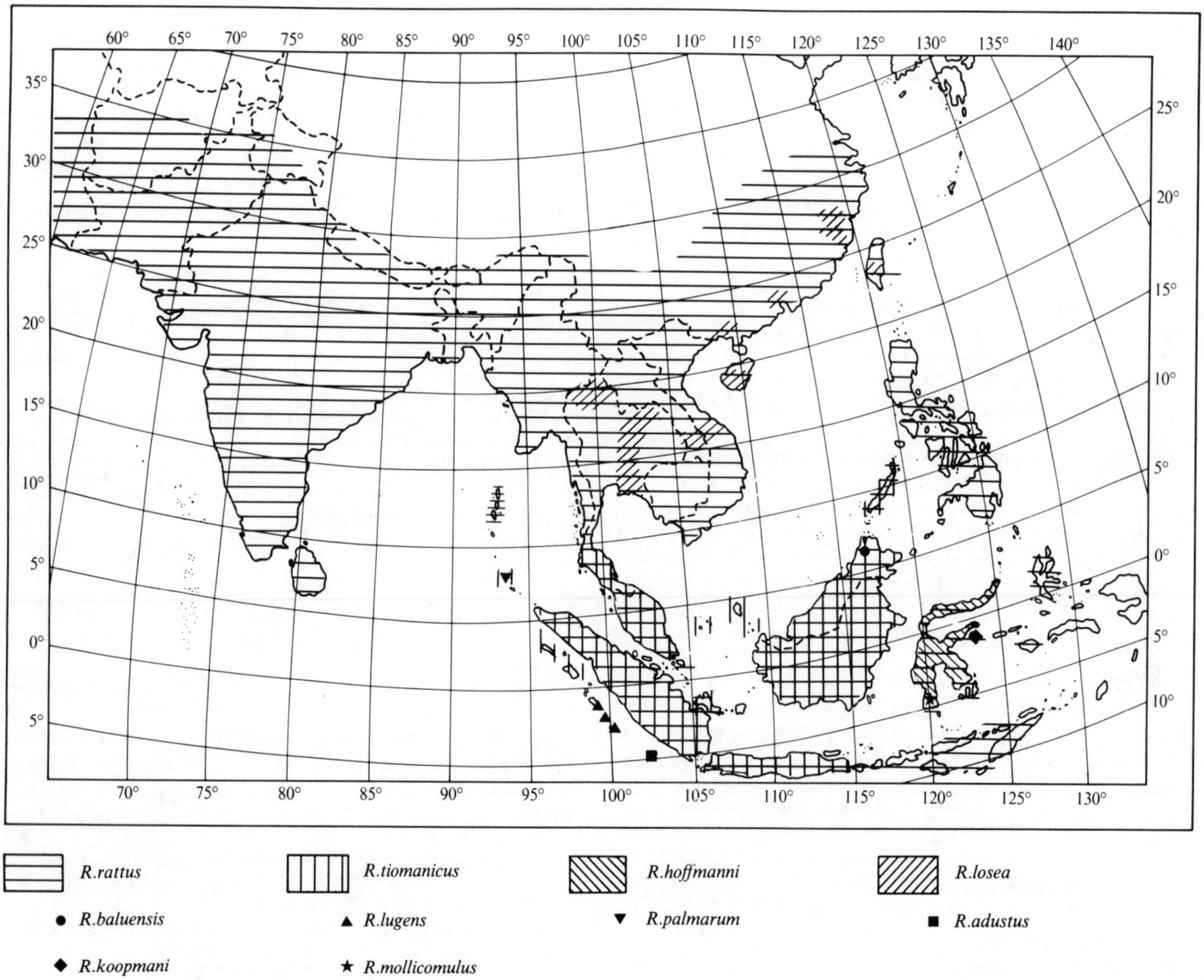

Throughout much of India, S China and adjacent areas there appears to be a mosaic of two main forms, both with brown upperparts: one with the venter greyish, more or less tipped with buff, associated especially with buildings, often called *R. r. rufescens* in India and *R. r. flavipectus* in China; and one with the venter white, sometimes with a medial streak of grey or with irregular greyish patches, associated especially with agricultural habitats, secondary forest etc., e. g. *R. r. brunneusculus* (Nepal) and *R. r. wroughtoni* (S India). Montane forms tend to be similar but with longer, softer pelage with few or no spines, at least in winter. In addition melanic individuals occur, especially in urban or industrial areas. These are usually referred to as *R. r. rattus* (the normal dark form of Europe) which is widely carried by shipping, but some may represent melanic forms of indigenous Asiatic subspecies.

The relationship between dark and white-bellied forms is complex. In Europe, and other parts of the world, they appear to behave as morphs in the same population and this appears to be the case also in Egypt, the type locality of the form *alexandrinus* (brown above, grey-brown below) which is similar to many in the Indomalayan Region. The same dimorphism appears to occur in some parts of India (Niethammer, 1975) but on the whole the two main colour forms are distinct and have been treated by some as species, *R. rufescens* for the grey-bellied and *R. rattus* for the white-bellied form, e. g. by Tiwari *et al.* (1971). They mapped the distribution of white-bellied subspecies in India and claimed that they lived sympatrically with *rufescens* without hybridization. However, the basis for accepting the white-bellied rather than the dark-bellied form as being conspecific with *R. rattus s. s.* is unclear.

The common form in Malaya and the Greater Sunda islands, usually known as *R. r. diardii*, is normally grey-bellied and is doubtfully distinguishable from similar forms in N India and China, although the tail tends to be shorter, and they are generally large. Even larger forms occur in the Philippines (*mindanensis*). The incidence of a white-bellied morph in these populations is difficult to assess because of confusion with other species, especially *R. tiomanicus*. The house rats on the Lesser Sunda islands (*R. r. sumbae*) are larger than *R. r. diardii* of the Greater Sunda islands and also tend to be paler above and below (Musser, 1972b).

Karyological and genetic variation have been extensively studied (Yosida, 1980) but failure to record the pelage or other morphological features of the samples used makes it difficult to integrate the results with knowledge of morphological variation. The karyotype throughout most of the region is constant at 2n=42, but in Sri Lanka a highland form (referred with little foundation to *R. r. kandianus*) has 2n=40 and a lowland form has 2n=38. The last is also the normal form in much of peninsular India and in most other parts of the world where *R. rattus* has been introduced. The differences are explicable in terms of centric fusion, a single fusion producing the form with 2n=40 and the same plus an additional one accounting for the form with 2n=38. Estimates of genetic distance by electrophoresis of izozymes have shown that in spite of these differences the forms of *R. rattus* are more closely related to each other than any is to *R. norvegicus, R. losea* or *R. exulans* (Baverstock *et al.*, 1983).

In India the distinction between the chromosome forms with 2n=38 and 2n=42 is not entirely congruent with the colour forms, both being found in dark-bellied populations referred to *R. r. rufescens* as well as in white-bellied populations referred to *R. r. wroughtoni* in S India (Gadi & Sharma, 1983). An earlier report giving 2n=38 for dark-bellied *R. r. rufescens* and 2n=42 for white-bellied *R. r. wroughtoni* in adjacent sites at Sagar, Mysore State, was based upon only three of each form (Lakhotia *et al.*, 1973). The identity of the form with 2n=40 on Sri Lanka also needs confirmation since other distinctive forms, e. g. *R. r. kelaarti* and *R. montanus* also occur at high altitudes there. Supernumary chromosomes have been found sporadically in the form with 2n=42, giving 2n up to 48 in some individuals in Thailand and up to 46 in Malaya (Yosida, 1980).

No subspecific classification is likely to be stable and useful until the following conditions are met: it is based on the entire range; the limits of diagnosable subspecies are mapped; the nature of the relationship between syntopic or adjacent colour forms is clarified; and data from karyological and genetic studies are integrated with the morphological and geographical data. The main trends of variation can be summarized as follows but this is no more than a general impression gained by scanning the literature and a superficial sampling of the main collections:

1. Near-melanic forms, *cf R. r. rattus*. Occur locally and sporadically, mainly in ports.
2. Forms with brown dorsum and grey-brown venter (caused by grey bases and brown tips to the hairs), with rather short, harsh pelage and many soft, flattened spines. *R. r. rufescens*, along with *flavipectus* (China), *diardii* (Java etc.), *kandianus* (Sri Lanka), *brunneus* (Nepal). Mainly in and around buildings. Such populations often include a small proportion of white-bellied but otherwise similar animals.
3. Forms with brown dorsum, white (or creamy) venter and usually soft, flat spines in the pelage: *R. r. brunneusculus*, including *wroughtoni* and many other names. Widespread, mainly lowland, in out-door but mainly synanthropic habitats, also in secondary forest.
4. Forms with brown dorsum, white venter and long soft pelage without spines, at least in winter: *R. r. kelaarti* (Sri Lanka), *arboreus* (Bengal), *tistae* (Sikkim), *gangutrianus* (Kumaon). Mainly in montane habitats, apparently including some natural forest. Some of these have light grey bases to the ventral hair but the longer, softer pelage distinguishes them from the *rufescens* group.
5. Some local peculiarities occur, e. g. *R. r. satarae* in the northern parts of the Western Ghats in India, with an unusually long tail, about 150% of head and body length, but otherwise similar to the white-bellied lowland form.

Recognition. Although the number of teats varies from 5 to 6 pairs, the variation affects only the post-axial ones which may be represented on each side by a single teat or by a pair very close together, generally within 5mm. This helps to distinguish *R. rattus* from other species, for example *R. remotus*, with 6 pairs since in these there are two widely separated post-axial teats on each side, at least 10mm apart. Distinction from *R. tiomanicus* is dealt with under that species.

Ecology. A serious pest through the consumption and contamination of growing crops and stored food, and also as a reservoir of plague (*Yersinia*) and scrub typhus (*Rickettsia*). Primarily an arboreal species, reflected in its depredations on tree crops like coconut and oil-palm as well as its habit of living in roofs and upper storeys of buildings.

Remarks. Many forms that were described as subspecies of *R. rattus*, or have been included in it, have been convincingly shown to represent distinct species, especially *R. argentiventer*, which is clearly distinct, and *R. tiomanicus*. The latter has been shown to be distinct from, and sympatric with *R. rattus* in

Malaya and the Greater Sunda islands, but the allocation of some small-island forms to these species is uncertain.

It has been suggested, e. g. by Medway & Yong (1976), that the Sunda form *diardii* is specifically distinct from *R. rattus* but the limits of these and their relationship to the Indian forms (*rufescens* etc.) have not been clarified. Likewise it has been proposed that in India the form *rufescens* (with dark venter) is specificaly distinct from the white-bellied races of *R. rattus* (Tiwari *et al.*, 1971) but this also needs clarification in relation to the dark European form on which *R. rattus* was based and in relation to the available karyological data. In S China the form *flavipectus* has often been given specific rank e. g. by Allen (1940). However again this was in contrast to European *R. rattus* and no means of distinguishing it from similar Indian forms like *rufescens* were suggested.

An extreme view, which has been universally rejected, was taken by Schwarz & Schwarz (1967) who included most of the species here included in the genus *Rattus* as subspecies of *R. rattus*.

Rattus tiomanicus Malaysian wood rat

Mus tiomanicus Miller, 1900a: 209; Tioman I, Malaya.
M. siantanicus Miller, 1900a: 210; Siantan I, Anamba Is.
M. tambelanicus Miller, 1900a: 212; Great Tambelan I, Tambelan Is.
? *M. burrus* Miller, 1902b: 768; Trinkut I, Nicobar Is.
? *M. burrulus* Miller, 1902b: 770; Car Nicobar I, Nicobar Is.
? *M. burrescens* Miller, 1902b: 771; Great Nicobar I, Nicobar Is.
M. jalorensis Bonhote, 1903a: 29; Nong Chik, Pattani, peninsular Thailand.
M. julianus Miller, 1903a: 34; St Julian I, near Tambelan Is.
? *M. simalurensis* Miller, 1903b: 458; Simalur I, NW Sumatra.
M. jarak Bonhote, 1905b: 69; Jarak I, Straits of Malacca.
M. rattus rhionis Thomas & Wroughton, 1909a: 441; Bintan I, Riau Is.
M. r. rumpia Robinson & Kloss, 1911a: 169; Rembia I, Sembilan Is, W of Malaya.
Epimys neglectus ducis Lyon, 1911a: 99; Datu I, W Borneo.
E. n. lamucotanus Lyon, 1911a: 100; Lamukotan I, SW Borneo.
E. maerens Miller, 1911b: 26; Nias I, W Sumatra.
E. tingius Miller, 1913b: 9; Tinggi I, E Malaya.
E. roa Miller, 1913b: 10; Aur I, E Malaya.
E. mara Miller 1913b: 10; Maratua I, E Borneo.
E. tua Miller, 1913b: 12; Maratua I, E Borneo.
E. rattus pauper Miller, 1913b: 13; Sirhassen I, S Natuna Is.
E. r. viclana Miller, 1913b: 13; Langkawi I, W Malaya.
R. r. batin Robinson, 1916b: 66; Mentigi, Mapor I, Riau Is.
R. r. roquei Sody, 1929b: 163; Semarang, Gedangan, central Java, 65m.
R. r. kundurensis Chasen & Kloss, 1931: 77; Kundur I, Riau Is.
R. r. jemuris Chasen & Kloss, 1931: 78; Jemur I, Aroa Is, Straits of Malacca.
R. r. payanus Chasen & Kloss, 1931: 79; Paya I, Straits of Malacca.
R. r. mangalumis Kloss, 1931a: 88; Mangalum I, NW Borneo.
R. r. banguei Chasen & Kloss, 1932: 35; Banggi I, N Borneo.
R. r. luxuriosus Chasen, 1935b: 20; Bunguran I, N Natuna Is.
R. r. perhentianus Chasen, 1940a: 155; E Perhentian I, E Malaya.
R. r. pemanggis Chasen, 1940a: 156; Pemanggil I, E Malaya.
R. r. vernalus Sody, 1940a: 395; Enggano I, W Sumatra.
R. r. sebesianus Sody, 1941a: 273; Sebesi I, between Sumatra and Java.
R. r. larusius Sody, 1941a: 274; Meeuwen I, W Java.
R. r. delirius Sody, 1941a: 274; Deli I, SW Java.
R. r. generatius Sody, 1941a: 274; Nusa Kambangan, S Java.
R. blangorum Miller, 1942: 145; Blangnanga base camp, Atjeh, N Sumatra, 3600ft.
R. r. pharus Hill, 1960: 75; Pisang I, Straits of Malacca.
R. r. sribuatensis Hill, 1960: 76; Sribuat I, Pahang, Malaya.
R. r. kabanicus Hill, 1960: 77; Kaban I, Johore Is, Malaya.
R. r. terutavensis Hill, 1960: 79; Telok Udang, Terutau I, SW Thailand.
R. r. ambersoni Schwarz & Schwarz, 1967: 130; Sangka Laki I, E Borneo.
R. r. piperis Schwarz & Schwarz, 1967: 132; Mata Siri I, S of Borneo.

Range (Map 137). Peninsular Thailand, Malaya, Sumatra, Java, Borneo and Palawan; and on most of the adjacent small islands. Perhaps also on the Nicobar Is (if *burrus, burrulus* and *burrescens* are conspecific).

Variation. The following subspecific names have been proposed:
R. simalurensis lasiae Lyon, 1916a: 446; Lasia I, Banyak Is, W Sumatra.
R. s. babi Lyon, 1916a: 447; Babi I, Banyak Is, W Sumatra.
R. t. sabae Medway, 1965b: 119; near Tenom, Sabah, Borneo.
R. t. tenggolensis Yong, 1971: 89; Tenggol I, E Malaya.

The ventral pelage varies from white to pale grey but only approaches the dark greyish buff commonly found in *R. rattus* on some islands, e. g. the Nias and Maratua Is (Musser & Califia, 1982).

Recognition. This species can be difficult to distinguish from *R. rattus* because of the variability in both, but the following characters are useful if comparative specimens are available (Musser & Califia, 1982): dorsal pelage sleeker, guard hairs scarcely protruding beyond contour hairs; incisive foramina and rostrum shorter; supraorbital ridges weaker; sides of braincase sloping outwards from the temporal ridges rather than vertically downwards; palatal bridge wider. The ventral pelage is usually entirely white (but greyish buff in *R. t. sabae* in Borneo); the hind feet smaller, usually under 34mm *s. u.*

Ecology. Mainly in scrub, gardens, plantations (e. g. of oil-palm) and secondary forest. Frequently enter houses, especially in the absence of *R. rattus*, and also found in undisturbed forest on islands where indigenous forest rats are absent. An agile climber but less arboreal than *R. rattus*. Molluscs have been shown to form an important part of the diet in Malaya (Lim, 1966).

Remarks. Although earlier treated as a form of *R. rattus*, this has been accepted as a distinct species by recent authors following Harrison (1961), Dhaliwal (1962), Medway & Lim (1966) and others. The synonymy above follows Musser & Califia (1982).

R. tiomanicus has been recognized and diagnosed mainly in relation to the sympatric forms of *R. rattus* in the Sunda Subregion. It has been suggested that some forms from the Indian and Indochinese subregions that are currently included in *R. rattus* may also represent *R. tiomanicus* and may predate that name, e. g. *brunneusculus* Hodgson, 1845 in Nepal (Medway, 1977a). However there is no evidence for this (G G Musser, pers. comm.).

The inclusion of the forms on the Andaman Is is tentative, although the three forms described from there, *burrus, burrulus* and *burrescens*, are very similar to each other and probably conspecific.

Rattus baluensis

Mus baluensis Thomas, 1894c: 454; Mt Kinabalu, N Borneo, 8000ft.

Range (Map 137). Known only in the montane forest of Mt Kinabalu, N Borneo, at altitudes of 2100–3360m.

Remarks. Chasen (1940a) included the Sumatran montane form *korinchi* in this species but this was refuted by Musser (1986) who gave a detailed account of this species and considered it most closely related to *R. tiomanicus*.

Rattus palmarum

Mus palmarum Zelebor, 1869: 26, pl. 3; Nicobar Is.

Range (Map 137). Nicobar Islands (particular islands unknown).

Remarks. A further description was given by Musser & Heaney (1985). It is probably most closely related to *R. tiomanicus*.

Rattus lugens

Mus lugens Miller, 1903a: 33; N Pagi I, Mentawai Is, W Sumatra.
R. rattus mentawi Chasen & Kloss, 1928c: 831; Sipora I, Mentawai Is.

Range (Map 137). Mentawai Islands of Siberut, Sipura and N Pagai.

Remarks. Closely allied to *R. tiomanicus* (Musser, 1986).

Rattus adustus

R. adustus Sody, 1940a: 397; Kiojoh, Enggano I, SW Sumatra, sea-level.

Range (Map 137). Known only from the island of Enggano, SW of Sumatra.

Remarks. Closely allied to *R. tiomanicus* (Musser, 1986).

Rattus hoffmanni

Mus rattus var. *celebensis* Hoffmann, 1887: 17, 18. Preoccupied by *M.* (*Gymnomys*) *celebensis* Gray, 1867 (= *Taeromys celebensis*).
= *M. hoffmanni* Matschie, 1901a: 284.
R. mollicomus Miller & Hollister, 1921a: 71; Mt Klabat, NE Sulawesi, 6500ft.
R. tatei Ellerman, 1941: 215; Tamalanti, central Sulawesi, 3300ft.
R. biformatus Sody, 1941a: 306; Malengi, Togian Is, E of Sulawesi.

Range (Map 137). Sulawesi, in montane and lowland forest.

Variation. The following subspecific names have been proposed:
R. h. linduensis Miller & Hollister, 1921a: 70; Tomado, Lake Lindoe, central Sulawesi.
R. h. mengkoka Tate & Archbold, 1935a: 3; Wawo, Mengkoka Mts, SE Sulawesi, 50m.

Ecology. Occurs in primary forest as well as in plantations and secondary growth, foraging mainly on or near ground level and feeding primarily upon fruit (Musser, 1984).

Remarks. The above synonymy follows Musser & Holden (1991) who gave measurements and detailed descriptions.

Rattus mollicomulus

R. mollicomulus Tate & Archbold, 1935a: 4; Wawa Karaing, Mt Lompobatang, SW Sulawesi, 1500m.

Range (Map 137). Mt Lompobatang, SW Sulawesi, at c. 1000–2000m.

Remarks. Close to *R. hoffmanni* but treated as a distinct species by Musser & Holden (1991). It is paler, with whitish feet, and smaller: hind feet 33–35mm (including claws) cf. 37–42 in *R. hoffmanni*.

Rattus koopmani

R. koopmani Musser & Holden, 1991: 389; Peleng I, Banggai Is, Sulawesi.

Range (Map 137). Peleng I, Sulawesi.

Rattus tawitawiensis

R. tawitawiensis Musser & Heaney, 1985: 5; Batu Batu, Tawitawi I, Sulu Is, Philippines.

Range (Map 138). Tawitawi I, Sulu Is, Philippines.

Remarks. Musser & Heaney (1985) considered this species to be most closely related to *R. hoffmanni* (Sulawesi) and the '*R. palmarum*' group on the Nicobar and Mentawai Islands.

Rattus losea

Mus losea Swinhoe, 1871: 637; Taiwan.
R. sakeratensis Gyldenstolpe, 1917: 46; Sakerat, E Thailand. Based upon a skin of *R. losea* mismatched with a skull of *Maxomys whiteheadi*: the skin has been designated as lectotype (Marshall, 1977a).
R. rattus exiguus Howell, 1927b: 44; 70 miles SW of Yenpingfu, Fujian, China, 500ft.

Range (Map 137). Thailand, Laos, central Vietnam and coastal China north to Fujian; Hainan, Taiwan and the adjacent Pescadores Is.

Variation. Two subspecies can be recognised:
R. l. losea. Taiwan. Larger and darker.
P. l. sakeratensis. Hainan and mainland. Smaller and lighter.

Ecology. Mainly a synanthropic species in rice-fields and other cultivated land, burrowing into the soil; but has also been found in grassy areas in natural pine forest at 850m in Thailand and in coastal mangrove forest (Marshall, 1977a).

Remarks. Reviewed in detail by Musser & Newcomb (1985). A detailed biometric study was made by Aoki & Tanaka (1938). In attempts at interbreeding with *R. norvegicus* no implantation occurred, suggesting a greater genetic distance than between *R. annandalei* and *R. norvegicus* or between *R. rattus* and *R. norvegicus* (Yosida, 1980).

Rattus osgoodi

R. osgoodi Musser & Newcomb, 1985: 18; Langbian Peak, Tuyen Duc Prov., S Vietnam, 12°3'N, 108°26'E, 5000ft.

Range (Map 138). Mountains of S Vietnam between 900 and 2000m.

Remarks. Most closely related to *R. losea* (Musser & Newcomb, 1985).

Rattus argentiventer Ricefield rat

Epimys rattus argentiventer Robinson & Kloss, 1916: 274; Pasir Ganting, W Sumatra, 2°7'S.
R. r. brevicaudatus Horst & Raadt, 1918: 69; Java.
R. pesticulus Thomas, 1921d: 248; Menado, N Sulawesi.
R. rattus bali Kloss, 1921d: 123; Lambolan Amok and Kloengkoeng, Bali, Indonesia.
R. r. saturnus Sody, 1941a: 269; Melolo, Sumba I, Lesser Sunda Is. (A synomym of *R. a. bali* according to Musser, 1972b.)
R. r. chaseni Sody, 1941a: 269; Krian, Perak, Malaya.
R. r. umbriventer Kellogg, 1945b: 121; Mindoro, Philippines.

Range (Map 138). From S Vietnam and Thailand to Malaya; Sumatra, Java, Bali, Borneo, most of the Lesser Sunda Islands (Lombok, Sumbawa, Komodo, Rintja, Flores, Sumba, Timor), Sulawesi, Philippines (Mindoro, Mindanao), Talaud Is (Feiler, 1990), Seram; New Guinea. Also on the small islands of Samui (Thailand) and Kangean (Java). It is likely that it has been extensively introduced by human agency and all the populations east of the Sunda Shelf are likely to have arisen in this way (Musser, 1973c).

Variation. There is relatively little geographical variation throughout the range.

Recognition. Externally distinguishable from *R. rattus* and *R. tiomanicus* by shorter tail and ears, more boldly speckled and usually lighter dorsal pelage, and clear pale grey venter. The pads on the hind feet are smaller and have only fine striations, not prominent lamellae. The skull is distinguished by the shorter and wider rostrum, angled more sharply downwards in relation to the roof of the cranium. The incisive foramina penetrate a little way between the anterior roots of M[1] and the bullae are larger.

Ecology. Known only as a synanthropic species, found mainly in ricefields, grassland and scrub, but also in plantations e. g. of oil palm. They are mainly terrestrial and make burrows. They feed on both leaves and grain of rice, and also extensively on insects and other invertebrates, especially molluscs (Lim, 1966).

Map **138**. *Rattus* (part)

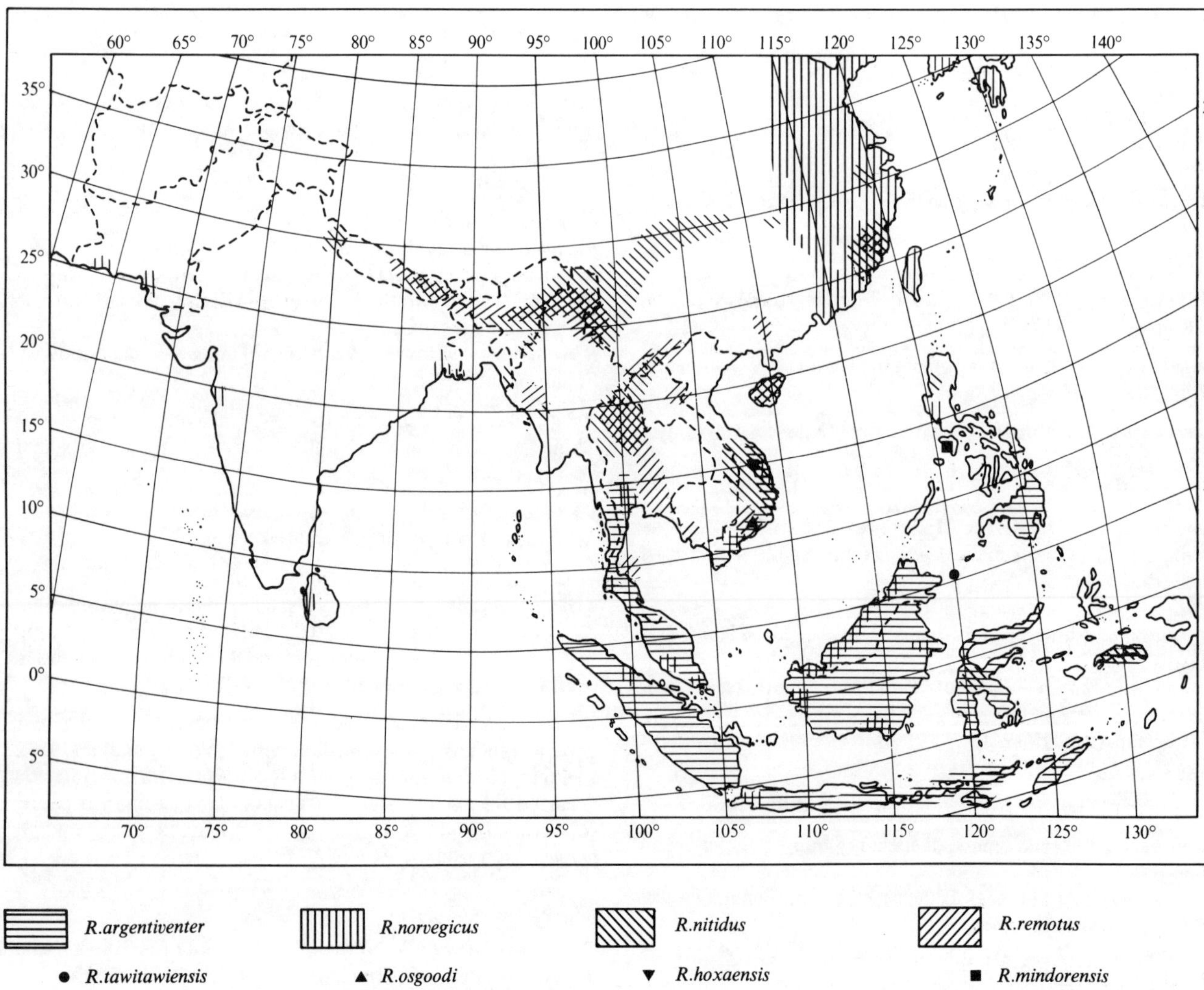

Remarks. This has been treated as an 'ecological form' of *R. rattus* but is quite distinct from that species and does not interbreed where they coexist. It has been reviewed by Musser (1973c). It appears to be most closely related to *R. losea* (Musser & Newcomb, 1985) with which it is parapatric, not being found together although occupying similar habitats (Marshall, 1977a).

Rattus hoxaensis

R. hoxaensis Dào, 1960a: 236; Ho-xa, Vinh-linh, central Vietnam.

Range (Map 138). Central Vietnam.

Remarks. Possibly conspecific with *R. argentiventer* (Cao & Tran, 1985).

Rattus norvegicus Norway rat

Rattus norvegicus

Mus norvegicus Berkenhout, 1769: 5; Great Britain.
M. decumanus Pallas, 1779a: 91; Europe.
M. javanus Hermann, 1804: 63; Batavia, Java; status *fide* Sody (1941a: 285).
M. griseipectus Milne-Edwards, 1872a: 93; Sichuan, China.
M. magnirostris Mearns, 1905: 441; Zamboanga, Mindanao, Philippines.
Also other synonyms from N China, Japan, Europe etc.

Range (Map 138). Widespread as a synanthropic species throughout all the temperate regions of the world, including N and central China and Taiwan where it occurs in cities and in agricultural habitats. In the Indomalayan Region it is mainly confined to ports and large cities, e. g. Bombay, Calcutta, Bangkok, Singapore, Jakarta and Manila, but more widespread in China; the original range is likely to have been in NE China.

Variation. The following subspecific names have been proposed:
Mus decumanus var. *major* Hoffmann, 1887: 17,18; Gorantalo, N Sulawesi and Macassar, S Sulawesi. Pre-occupied by *M. citellus* var. *major* Pallas, 1779 (= *Spermophilus major*) and *M. major* Severtzov, 1873 (in *M. musculus*).
= *Epimys n. hoffmanni* Trouessart, 1904: 365. Pre-occupied by *Mus hoffmanni* Matschie, 1901 (= *Rattus hoffmanni*).
= *Mus n. praestans* Trouessart, 1904: 546.
R. n. sulfureo-ventris Kuroda, 1952: 3; Platas (= Pratas) I, SE of Hong Kong, 20°42'N, 116°43'E. A large form with yellowish white venter.

Subspecific variation in China was described by Wu (1982) (in Chinese). Melanic individuals are frequent in most Asiatic populations and the dominant form in some, e. g. Prachumburi and Chumphon in Thailand (Marshall, 1977a).

Ecology. *Rattus norvigicus* is less of a climber than *R. rattus* and is usually found at ground level, e. g. in grain warehouses and rubbish dumps. It also frequents wet ground and swims readily.

Rattus nitidus Himalayan rat

Mus nitidus Hodgson, 1845: 267; Nepal.
M. horeites Hodgson, 1845: 268; Nepal.
M. aequicaudalis Hodgson, in Horsfield, 1849: 203; Nepal.
M. robustulus Blyth, 1859a: 294; Schwegyin, Tenasserim, Burma (status *fide* Marshall, 1977a).
? *M. rubricosa* Anderson, 1879: 306; Ponsee and Hotha, Kakhyen Hills, W Yunnan, China.
M. ruber Jentink, 1880b: 18; Doreh, Vogelkop, NW New Guinea.
R. manuselae Thomas, 1920a: 424; Mt Manusela, Seram, Moluccas, 4000ft.
R. hoffmanni subditivus Miller & Hollister, 1921a: 70; Toware, Bada, central Sulawesi (status *fide* Musser, 1984).
R. griseiventer rahengis Kloss, 1919a: 74; Raheng, Thailand.
R. vanheurni Sody, 1933a: 435; near Manokwari, Vogelkop, NW New Guinea.
Mus guhai Nath, 1952: 85; Nongpoh, Khasi Hills, Assam, India. Based upon a litter of young *R. nitidus* (Marshall, 1977b).

Range (Map 138). From Kumaon to S China (Yunnan, Sichuan, S Gansu, Fujian), Burma, Thailand and Vietnam; and on Hainan. Also present, probably introduced, in the Philippines (Luzon), central Sulawesi, Seram, the Vogelkop Peninsula of W New Guinea and Palau I (E of the Philippines). Up to 2500m in Nepal and 2700m in Yunnan, with one exceptional capture in Yunnan at 3900m (Wang, Y- x. *et al.*, 1985). Although the identity of specimens from Kumaon listed by Ellerman (1961) was considered doubtful by Niethammer & Martens (1975) these seem to be true *R. nitidus* as understood here.

Variation. The following subspecific name has been proposed:
R. n. obsoletus Hinton, 1919: 415; 50 miles W of Kindat, Chin Hills, W Burma, 5000ft.

Recognition. This species has been frequently confused with *R. rattus* and *R. turkestanicus*. It is distinguishable externally by the soft, spineless pelage, with long, uniform dark-grey hair-bases, and especially the short guard hairs which scarcely protrude beyond the contour hairs even on the rump; the very pale, almost white hind feet; and from *R. rattus* by the presence of six pairs of teats, including two well-separated post-axillary ones on each side. The snout is long, reflected also in long nasal bones (16–18mm).

Ecology. A synanthropic species in most of its range, found mainly in village houses rather than urban habitats.

Remarks. The New Guinea form *ruber*, but not the other New Guinea forms usually associated with *ruber*, was included in *R. nitidus* by Taylor *et al.* (1982). The synonymy above includes the only two extralimital names.

Rattus remotus

Epimys remotus Robinson & Kloss, 1914a: 231; Samui I, E peninsular Thailand.
R. rattus sikkimensis Hinton, 1919: 394; Pashok, Sikkim, India, 3500ft.

? *R. macmillani* Hinton, 1919: 409; Hkamti, W bank of upper Chindwin R, Burma.
R. rattus koratensis Kloss, 1919b: 379; Lat Bua Kao, E Thailand.
R. r. hainanensis Allen, 1926: 3; Namfong, Hainan I, China.
R. sladeni of many authors, but not of Anderson, 1879 which is a synonym of *R. rattus* as used here.

Range (Map 138). Central Nepal to S China, Burma, Thailand and Indochina; also on the islands of Hong Kong, Hainan; Klum (SE Thailand); and Tau, Pangan, Samui and Kra (E of pensular Thailand). The entire range was mapped by Musser & Heaney (1985) under the name *R. sikkimensis*.

Recognition. Similar to *R. rattus* but longer black hair in the mid-dorsal line; venter always completely white; 3 normally-spaced pairs of anterior teats (2, or one plus a closely opposed pair in *R. rattus*); molars black-rimmed; rostrum wide; sides of brain-case not vertical.

Ecology. A species of both primary and secondary forest. Reported (as *R. koratensis*) to be the dominant rat in forest in Vietnam (Cao & Nguyen, 1989).

Remarks. This species has been the source of much confusion. Chasen (1940a) listed the insular *remotus* as a subspecies of *R. annandalei* and Hill (1960) as a subspecies of *R. muelleri* (= *Sundamys muelleri*). Marshall (1977a) treated *R. remotus* as an insular species separate from *R. koratensis* on mainland Thailand. Musser & Heaney (1985) argued that *remotus* is 'really an insular population of *R. sikkimensis*', apparently overlooking the fact that the prior name is *R. remotus*.

Rattus turkestanicus

Mus rattoides Hodgson, 1845: 267; Nepal. Pre-occupied by *Mus rattoides* Pictet & Pictet, 1844, a form of *Rattus rattus* in Brazil (Schlitter & Thonglongya, 1971).
? *M. pyctoris* Hodgson, 1845: 267; Nepal.
M. turkestanicus Satunin, 1903: 588; Assam-bob, Ferghana, Uzbekistan, USSR (published April 1903).
M. vicerex Bonhote, 1903c: 473; Simla, N India (published May 1903).
Epimys rattus shigarus Miller, 1913e: 198; Shigar, Baltistan, Kashmir, 9000ft.
R. rattus khumbuensis Biswas & Khajuria, 1955: 27; Phorcha, Dudh Kosi Valley, Khumbu, Nepal, c. 4300m.

Range (Map 139). From Yunnan through the Himalayas to Kashmir and Pakistan; and west to E Iran, Kirghizia and S Kazakhstan. Montane, at altitudes of 1200–4250m in Nepal.

Variation. The following subspecies has been described:
R. rattoides gilgitianus Akhtar, 1959: 41; Gilgit, N Pakistan.

Differences between samples from Afghanistan and Nepal were documented by Niethammer & Martens (1975), involving ventral pelage (white in Afghanistan, grey in Nepal), some cranial details and chromosomes.

Ecology. Occurs mainly, but not entirely, in and around houses, extending considerably higher than *R. rattus* (Niethammer & Martens, 1975).

Rattus exulans Polynesian rat

Mus exulans Peale, 1848: 47; Tahiti, Pacific Ocean.
M. concolor Blyth, 1859a: 295; Schwegyin, Burma.

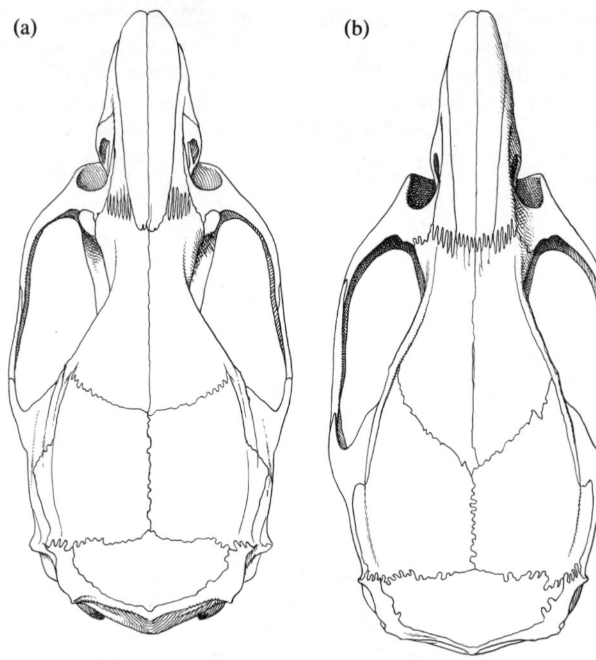

Fig.32. (a) *Rattus turkestanicus*; (b) *Cremnomys blanfordi*.

M. bocourti Milne-Edwards, 1872b: 291 (footnote); Thailand (status *fide* Musser, 1970b).
M. ephippium Jentink, 1880b: 15; Sumatra.
M. wichmanni Jentink, 1890a: 120, pl. 10; Sikka, Flores, Lesser Sunda Is.
M. aemuli Thomas, 1896a: 249; Aemuli, Jampui I, Saleyer Is, S Sulawesi.
M. obscurus Miller, 1900a: 213; Tioman I, Malaya. Pre-occupied by *M. obscurus* Waterhouse, 1837 (= *Akodon obscurus*).
= *M. pullus* Miller, 1901e: 178.
M. surdus Miller, 1903b: 460; Simalur I, W Sumatra.
M. todayensis Mearns, 1905: 445; Todaya, Mt Apo, SE Mindanao, Philippines, 4000ft.
M. vulcani Mearns, 1905: 446; Mt Apo, Davao, Mindanao, Philippines, 7600ft.
M. pantarensis Mearns, 1905: 448; Pantar, Mindanao, Philippines, 1907ft.
M. clabatus Lyon, 1906a: 596; Klabat Bay, Banka I, E Sumatra.
M. luteiventris Allen, J. A.,1910: 14; Iwahig, Palawan, Philippines.
M. buruensis Allen, J. A.,1911: 336; Buru I, Moluccas.
Epimys calcis Hollister, 1911a: 89; Baguio, Benguet, Luzon, Philippines, 1537m.
E. querceti Hollister, 1911a: 90; Haight's Place, Benguet, Luzon, Philippines, c. 2153m.
Epimys mayonicus Hollister, 1913a: 319; Mt Mayon, Albay Prov., Luzon, Philippines, c. 1230m.
E. leucophaetus Hollister, 1913a: 320; Bagamanoc, Catanduanes I, NE Philippines.
E. vigoratus Hollister, 1913a: 321; Mt Halcon, Mindoro, Philippines.
E. basilanus Hollister, 1913a: 322; Basilan I, Philippines.
E. ornatulus Hollister, 1913a: 322; Cagayancillo I, Cagayan Is, Philippines.
E. stragulum Robinson & Kloss, 1916: 274; Korinchi Peak, W Sumatra, 7300ft.

R. raveni Miller & Hollister, 1921a: 68; Toli Toli, N Sulawesi.
R. schuitemakeri Sody, 1933a: 431; Pontianak, W Borneo.
Togomys melanoderma Dieterlen, 1986: 12; 'Togo, W Africa', corrected to presumably Sumatra (Dieterlen, 1989).

Range (Map 139). From Bangladesh through lowland Burma, Thailand, Indochina, Malaya; Sumatra, Java and Borneo; and on most of the adjacent small islands; widespread but probably introduced in the Philippines and Sulawesi; on some of the Lesser Sunda islands (e. g. Flores and Lombok) and the Moluccas (e. g. Buru); also in New Guinea, New Zealand and on the majority of tropical and subtropical Pacific islands.

Variation. The following subspecific names have been proposed:
Mus ephippium negrinus Thomas, 1898c: 403; Canlaon Vulcano, Negros, Philippines, 2011m.
M. vulcani apicis Mearns, 1905: 447; summit of Mt Apo, S Mindanao, Philippines, 9700ft.
R. raveni eurous Miller & Hollister, 1921a: 69; Molengkapoti, Kwandang, N Sulawesi.
R. concolor equile Robinson & Kloss, 1927: 209; Ongop Ongop, Idjen Massif, E Java, 5700ft.
R. c. otteni Kopstein, 1931: 3; Cheribon, N Java.
R. c. malengiensis Sody, 1941a: 281; Malengi I, Togian Is, E of Sulawesi.
R. c. meringgit Sody, 1941a: 283; Batoe Meringgit, Bali.
R. c. solatus Kellogg, 1945a: 65; Morotai I, Halmahera, Moluccas.

Geographical variation is slight, affecting mainly the ventral colour which varies from very light to dark grey with a variable amount of brown in the hair-tips. Montane forms tend to be slightly darker, with longer, softer pelage with flattened spines either absent (e. g. *R. e. stragulum*, Kerinci, Sumatra; *R. e. aquile*, E Java) or sparse (e. g. *R. e. querceti*, Luzon).

Ecology. A common pest species occurring in houses, granaries, cultivation, scrub and forest edge, usually on the ground.

Rattus annandalei Annandale's rat

Mus annandalei Bonhote, 1903a: 30; Sungkai, S Perak, Malaya.
M. bullatus Lyon, 1908b: 646; Rupat I, E Sumatra.
M. villosus Kloss, 1908: 146; Singapore I.

Range (Map 139). Malaya (north to Perak but rather localized); E Sumatra; Singapore; islands of Rupat and Padang (E Sumatra). Recorded from Trang, peninsular Thailand, by

Map **139**. *Rattus* (part)

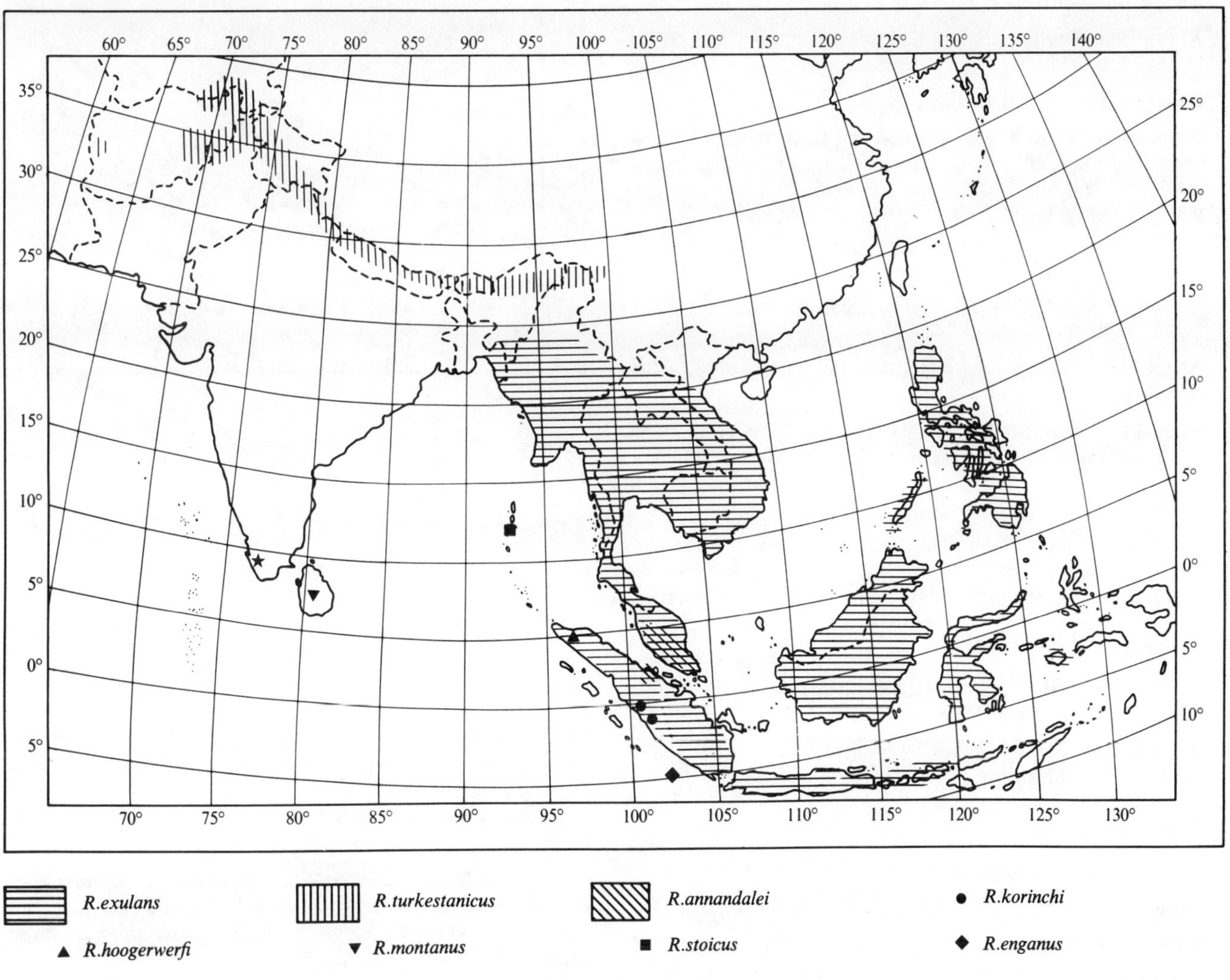

Chasen (1940a), but this was thought to be based upon a misidentified *R. tiomanicus* by Marshall (1977a) who considered that there were no authenticated records of *R. annandalei* from Thailand.

Ecology. Occurs in secondary forest and scrub in Malaya and Singapore; not a pest of crops.

Remarks. Chasen (1940a) included the form *remotus* from Samui I, Thailand in this species, but it is here considered an independent species following Marshall (1977a). Attempts at interbreeding with *R. norvegicus* produced low viability of embryos and no births, suggesting a greater genetic distance than between *R. rattus* and *R. norvegicus* (Yosida, 1980).

Rattus korinchi

Epimys korinchi Robinson & Kloss, 1916: 275; Korinchi Peak, Sumatra, 7300ft.

Range (Map 139). Mt Kerinci and Mt Talakmau, Sumatra.

Remarks. This form was treated as a subspecies of the Bornean *R. baluensis* by Chasen (1940a) but was considered a distinctive species by Musser (1986) who detailed the considerable differences between these two species and between *R. korinchi* and *R. hoogerwerfi*.

Rattus montanus

R. montanus Phillips, 1932b: 323; West Haputale, Ohiya, Sri Lanka, 5200–6000ft.

Range (Map 139). Sri Lanka.

Remarks. It has been suggested that this name might be based upon 'a sample of extremely large *R. rattus kelaarti* males' (McKay, 1984) but the teeth are very distinctive and preclude that possibility, a conclusion supported by Musser (1986).

Rattus hoogerwerfi

R. hoogerwerfi Chasen, 1939b: 207; Blang Kedjeren, Atjeh, N Sumatra, 800m.

Range (Map 139). Known only from the type locality in the mountains of N Sumatra.

Remarks. This species is quite distinct from the other montane Sumatran species *R. korinchi* as detailed by Musser (1986).

Rattus stoicus

Mus stoicus Miller, 1902b: 759; Henry Lawrence I, Andaman Is.
M. taciturnus Miller, 1902b: 762; S Andaman I, Andaman Is.
M. rogersi Thomas, 1907d: 206; north of Ike Bay, W coast of S Andaman I, Andaman Is.

Range (Map 139). Andaman Is, recorded from S Andaman I, Little Andaman I, and Henry Lawrence I.

Remarks. This species was redescribed, with the above synonymy, by Musser & Heaney (1985).

Rattus nativitatis

Mus nativitatis Thomas, 1889a: 533; Christmas I, S of Java.

Range. Known only from Christmas I, S of Java; extinct.

Remarks. Although recorded as abundant in 1897 this species could not be found in 1908 and has not been recorded since (Harper, 1945).

Rattus xanthurus

Mus xanthurus Gray, 1867b: 598; Tondano, N Sulawesi, 3600ft.
M. faberi Jentink, 1883: 176; NW Sulawesi (for status see Musser, 1971b).
? *Taeromys paraxanthus* Sody, 1941a: 313; Toelap West, Tonsealama, Tonsea dist, N Sulawesi. Based upon a composite specimen: skin of *R. xanthurus* and skull of *Taeromys callitrichus*; the skin was designated as lectotype by Musser (1971e).

Range (Map 140). N Sulawesi, in montane and lowland forest; perhaps also in SE Sulawesi and the island of Peleng.

Map **140**. *Rattus* (part)

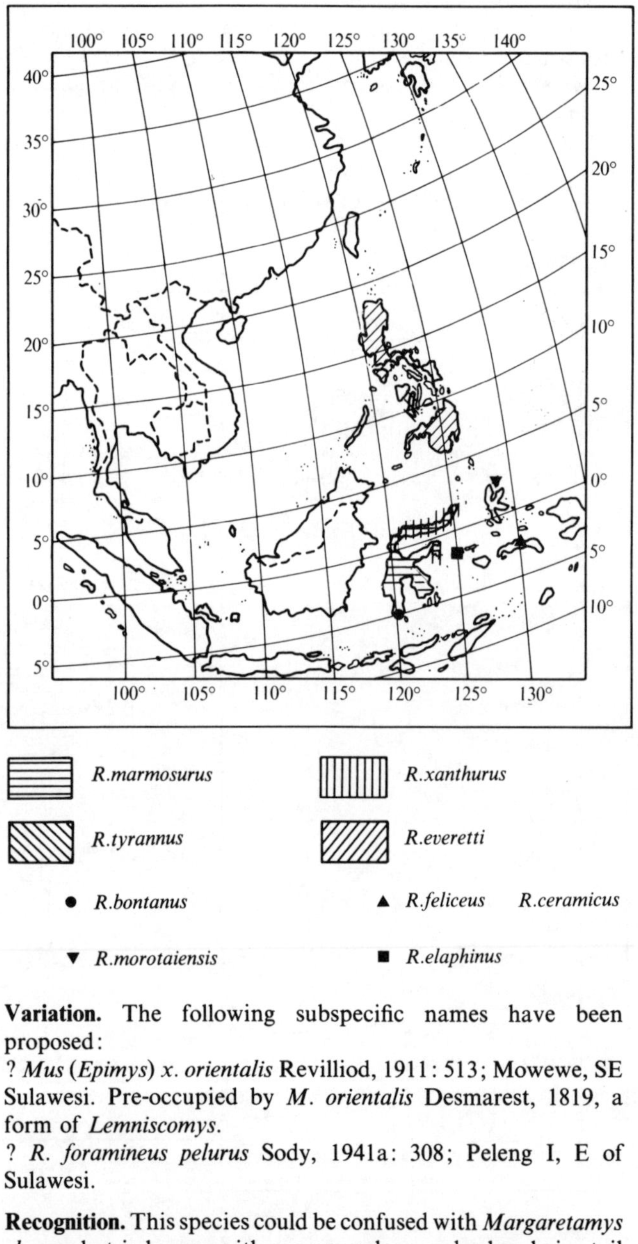

Variation. The following subspecific names have been proposed:
? *Mus* (*Epimys*) *x. orientalis* Revilliod, 1911: 513; Mowewe, SE Sulawesi. Pre-occupied by *M. orientalis* Desmarest, 1819, a form of *Lemniscomys*.
? *R. foramineus pelurus* Sody, 1941a: 308; Peleng I, E of Sulawesi.

Recognition. This species could be confused with *Margaretamys elegans* but is larger, with coarser pelage and a less hairy tail (although slightly tufted at the tip) and a narrower, more discrete dark eye-ring.

Remarks. The taxonomy was discussed by Musser (1984).

Rattus bontanus

R. bontanus Thomas, 1921d: 245; Indrulaman, Mt Bonthain, S Sulawesi, 2000ft.
R. foramineus Sody, 1941a: 308; Boeloekoemba, S Sulawesi, sea-level.

Range (Map 140). S Sulawesi.

Remarks. Possibly conspecific with *R. xanthurus* but differing in longer and softer pelage.

Rattus marmosurus

R. marmosurus Thomas, 1921d: 246; Mt Masarang, Minahassa, N Sulawesi, 2000ft.
R. facetus Miller & Hollister, 1921b: 96; Mt Lehio, SW of Lake Lindoe, central Sulawesi, above 6000ft.
R. tondanus Sody, 1932a: 158; Tondano, NE Sulawesi (for status see Musser, 1971d).
R. salocco Tate & Archbold, 1935a: 7; Tanka Salocco, Mengkoka Range, SE Sulawesi, 1500m.

Range (Map 140). Sulawesi, in montane forest.

Rattus macleari

Mus macleari Thomas, 1887c: 513; Christmas I, S of Java.

Range. Christmas Island, S of Java; extinct.

Remarks. Although recorded as abundant in 1897, this species could not be found in 1908 and has not been recorded since (Harper, 1945).

Rattus enganus

Mus enganus Miller, 1906e: 821; Enggano I, SW of Sumatra.

Range (Map 139). Known only from Enggano I, SW of Sumatra.

Rattus feliceus

R. feliceus Thomas, 1920a: 423; Mt Manusela, Seram, Moluccas, 6000ft.

Range (Map 140). Known only from the island of Seram, Moluccas.

Recognition. Easily distinguishable by the coarse, spiny pelage and short, pallid tail with large, conspicuous scales (6–7 per cm).

Remarks. Treated as a subspecies of the New Guinea species *R. ruber* by Laurie & Hill (1954), but considered a separate species related to *R. jobiensis* and *R. praetor* of New Guinea by Musser & Newcomb (1983).

Rattus morotaiensis

R. morotaiensis Kellogg, 1945a: 66; Morotai I, Gilolo group, Moluccas.

Range (Map 140). Known only from the island of Morotai, north of Halmahera, Moluccas.

Remarks. This form was considered by Kellogg (1945a) to be most similar to *R. ringens* of New Guinea (placed in *R. ruber* by Laurie & Hill, 1954 and in *R. leucopus* by Taylor *et al.*, 1982).

Rattus elaphinus

R. elaphinus Sody, 1941a: 307; Taliabu I, Sula Is, Moluccas.

Range (Map 140). Known only on Taliabu I, Moluccas.

Rattus ceramicus

Stenomys ceramicus Thomas, 1920a: 425; Mt Manusela, Seram I, Moluccas, 6000ft.

Range (Map 140). Known only from the island of Seram, Moluccas.

Remarks. The type species of the generic name *Nesoromys* Thomas which was retained as a genus by Laurie & Hill (1954). It is here included in *Rattus* following Musser (1981a: 168) who considered that it was most closely related to *R. niobe*, a widespread montane species in New Guinea. The palate is unusually long, extending behind M^3 by about twice the length of M^3.

Rattus everetti

Mus everetti Günther, 1879a: 75; 'Philippines', probably Dinagat or N Mindanao.
M. tagulayensis Mearns, 1905: 439; Tagulaya, Davao Gulf, foot of Mt Apo, Mindanao, Philippines, sea level.
M. albigularis Mearns, 1905: 440; Mt Apo, Davao, Mindanao, Philippines, 2075–2430m.
Epimys gala Miller, 1910: 398; Alag R, Mindoro, Philippines.

Range (Map 140). Philippines, on all the main islands except Palawan.

Ecology. Confined to forest, mainly primary (Heaney *et al.*, 1987).

Rattus tyrannus

Epimys tyrannus Miller, 1910: 397; Ticao I, central Philippines.

Range (Map 140). Philippine islands of Ticao and Negros, perhaps also Samar, Dinagat and others.

Remarks. Tentatively included in *R. everetti* by Musser (1982d) but the larger size and very different ventral pelage—coarse and sparse—appear to indicate that it is a distinct species.

Rattus mindorensis

Mus mindorensis Thomas, 1898c: 402; Mindoro, Philippines.
Limnomys picinus Hollister, 1913a: 325; Mt Halcon, Mindoro, 4500ft (based on a composite specimen: the skin was selected as the holotype by Musser (1977b) who considered it a synonym of *R. mindorensis*).

Range (Map 138). Mindoro I, Philippines, especially at high altitude.

Recognition. Almost as dark as *R. rattus rattus* above and below but the pelage is softer, with no yellowish spines and fewer emergent guard hairs. The tail is also shorter than the head and body.

Rattus ranjiniae

R. ranjiniae Agrawal & Ghosal, 1969: 41; Trivandrum, Kerala, S India.

Range (Map 140). Known only from the type locality in S India.

Remarks. The large hind feet and short palate (not extending behind M^3) appear to distinguish this species from any other Indian *Rattus*, but specimens have not been examined and its relationship to species elsewhere is not clear. Although it was described as a member of the subgenus *Rattus*, its inclusion in the genus *Rattus* as restricted here must be considered open to question. It was described as a field rat, coexisting with the more arboreal *R. rattus wroughtoni*.

Rattus timorensis

R. timorensis Kitchener, Aplin & Boeadi, 1991: 446; Gunung Mutis, c. 7km E of Desa Nenas, SW Timor, c. 9°31'S, 126°16'E, c. 1900m.

Range. Known only from the type locality in SW Timor, Indonesia.

Remarks. Not included in the keys. A small species, with one post-axial and two inguinal pairs of teats.

MILLARDIA Soft-furred rats

Millardia Thomas, 1911d: 998; type species *Golunda meltada* Gray.
Grypomys Thomas, 1911d: 999 (*Mus gleadowi* Murray).
Guyia Thomas, 1917a: 201 (*Millardia kathleenae* Thomas).
= *Millardomys* Sody, 1941a: 261.

Indian subregion; 4 species (Table 222). A genus of very doubtful validity and affinities, included in *Rattus* as a subgenus by Ellerman (1961). The genus was established primarily on the basis of short hind feet with fewer than six plantar pads, i. e. only one or no posterior pad in addition to the four interdigital pads. It was reviewed by Agrawal (1970). However the addition of the more recently described *M. kondana*, which in other respects seems very close to *M. meltada*, nullifies this as a diagnostic character. On the basis of the karyotype Gadi & Sharma (1983) considered *Millardia* clearly distinct from *Rattus* and other related genera. It is probably most closely related to *Cremnomys* and the African *Praomys*.

A fossil species, *M. coppensi* Sabatier, 1982, was described from Ethiopia, but has more recently been allocated to *Acomys* (Denys, 1990).

Millardia meltada Metad

Golunda meltada Gray, 1837a: 586; Dharwar, India.
= *Mus mettada* Wroughton, 1907: 997 (unjustified emendation).
Mus lanuginosus Elliot, 1839: 212.
M. listoni Wroughton, 1907: 998; Konkan, Kolaba Dist, W India.
M. comberi Wroughton, 1907: 999; Nasik, Bombay.

Range (Map 141). India NE to Bihar and W Bengal; Pakistan, lowland Nepal, Sri Lanka.

Variation. Three subspecific names have been proposed:
M. m. pallidior Ryley, 1914: 659; Lunwa, Palanpur, Gujarat, India, 150ft.
M. m. dunni Thomas, 1917a: 202; Handiserah, Ambala, Punjab.
M. m. singuri Mandal & Ghosh, 1981: 235; Singur, Hugli Dist, W Bengal (based upon small differences in colour).

Ecology. It has been described as the 'most abundant rodent in the Indian subcontinent', being found in both crop fields and scrub grassland (Rana, 1986).

Millardia kondana

M. kondana Mishra & Dhanda, 1975: 77; Singharh, Poona Dist, Maharashtra, India, 18°23'N, 73°42'E.

Range (Map 141). Known only from the type locality, SE of Bombay, India.

Remarks. In describing *M. kondana* the authors did not mention the presence or absence of *M. meltada* in the same or adjacent localities, although the latter has been recorded from Pune and from Dharwar where they are of normal size, much smaller than *M. kondana*.

Millardia kathleenae

M. kathleenae Thomas, 1914d: 29; Pagan, Burma.

Range (Map 141). Central Burma.

Millardia gleadowi Sand-coloured rat

Mus gleadowi Murray, 1885: 809; Clifton Plain, Karachi, Pakistan.

Range (Map 141). Pakistan and NW India, south to Gujarat and Kathiawar.

Table 222. The species of *Millardia*.

	meltada	*kondana*	*kathleenae*	*gleadowi*
Range	India, Sri Lanka	NW India	Burma	Pakistan, NW India
Head & body	110–155	150–200	130–165	80–115
Tail (% of h & b)	80–90%	85–95%	90–100%	<80%
Hind feet	20–29	27–34	25–30	18–20
Plantar pads	4/5	6	4	4
Teats: pairs (pectoral/abdominal)	2/2	2/2	0/2	1/2
Skull: greatest length	28–36	37–45	34–36	24–27
Upper molar row	5.2–5.9	7.0–8.0	5.3–6.0	4.7–5.8
Incisive foramina	6.9–8.9	8.5–10.0	7.8–8.5	5.6–6.8
Bullae: length	5.4–6.7	6.3–7.0	6.6–7.2	4.7–5.8
: % of skull	c.17–18%	c.13–14%	c.20%	>20%

Map **141**. *Millardia*

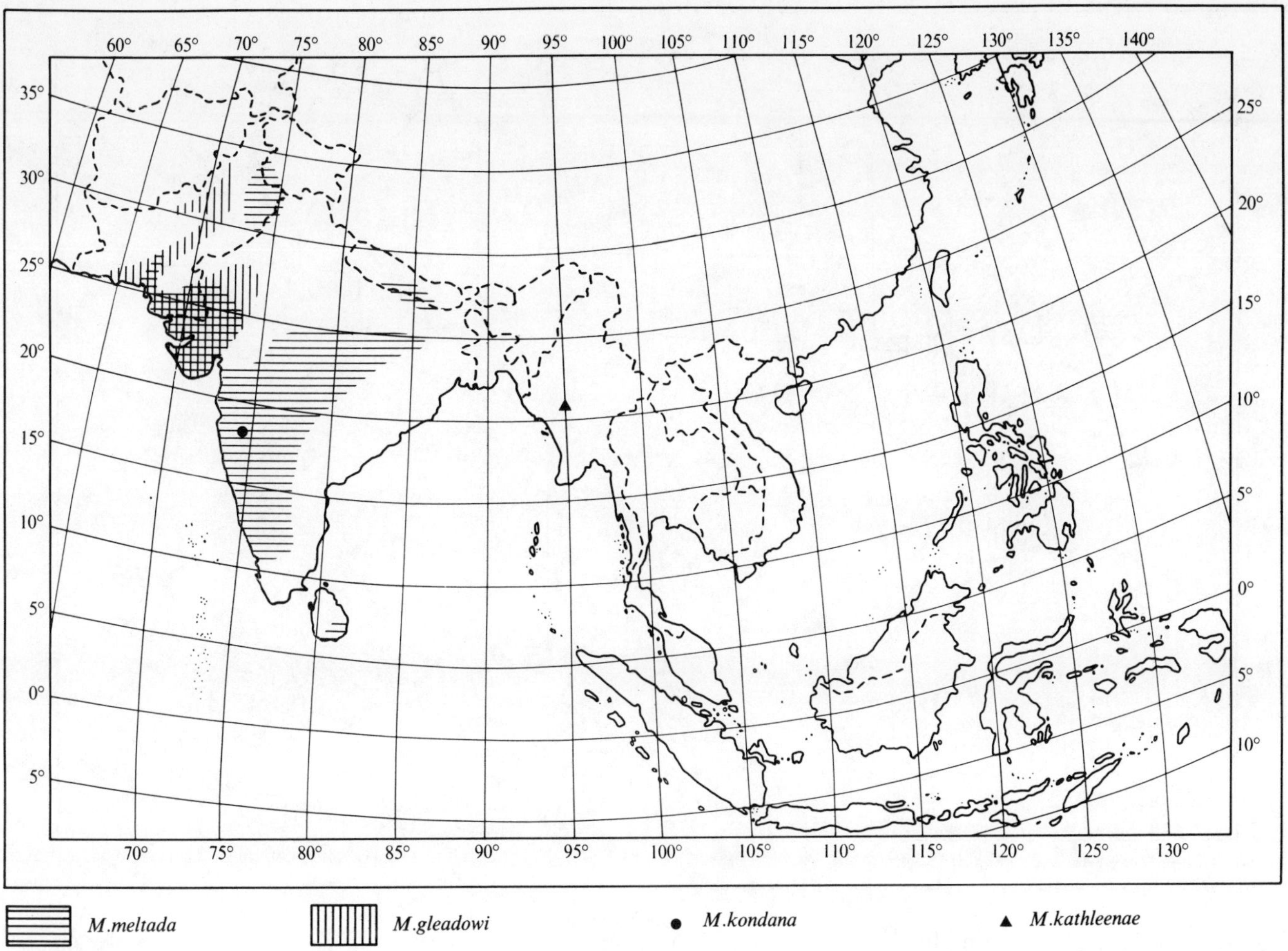

CREMNOMYS

Cremnomys Wroughton, 1912a: 340; type species *C. cutchicus* Wroughton.

Madromys Sody, 1941a: 260 (*Mus blanfordi*). Doubtfully valid since it was based solely upon a mammary formula shared with *Taeromys*.

India and Sri Lanka; 3 species (Table 223). Characterized especially by very long incisive foramina. *C. cutchicus* and *C. elvira* were included in *Rattus* (*Cremnomys*) by Ellerman (1961 etc.) and *C. blanfordi* in *Rattus* (*Rattus*). The inclusion of *blanfordi* in *Cremnomys* is supported by data on chromosomes (Rishi & Puri, 1984) but is nevertheless very dubious. *Cremnomys* appears to be more similar to some African genera, e. g. *Praomys*, than to *Rattus* and other Indomalayan genera, with the possible exception of *Millardia*. *Cremnomys* differs from *Millardia* in having a long 5th toe, six large plantar pads and the squamosal root of the zygomatic arch arising high on the side of the brain-case.

All species appear to be found mainly on rocky hills with trees or scrub.

Cremnomys cutchicus Cutch rat

C. cutchicus Wroughton, 1912a: 340; Dhonsa, Cutch, Gujarat, India.

Table 223. The species of *Cremnomys*.

	cutchicus	*elvira*	*blanfordi*
Range	India	S India	Pen.India Sri Lanka
Head & body	100–150	125–150	150–195
Hind feet	23–29	30–32	33–38
Tail: paler below	Slightly	Slightly	0
distal 1/3 white	0	0	+
Skull: greatest length	31–36	38–41	38–47
Upper molar row	4.8–5.7	5.9–6.1	5.9–6.8
Bullae: length	4.3–5.5	5.5–6.0	7.1–8.4

C. medius Thomas, 1916f: 240; Kudia, Junagadh, Kathiawar, India, 2500ft.

C. australis Thomas, 1916f: 242; Vijayanagar, Bellary, India, 1500ft.

Range (Map 142). India, in three segments: in the northwest in Gujarat and Rajasthan; in Bihar; and in the south between Bellary and Bangalore.

Variation. The following subspecific names have been proposed:

C. medius caenosus Thomas, 1916f: 241; Singar, Gaya, Bihar and Orissa, India, 1400ft.

Map **142**. *Cremnomys, Golunda*

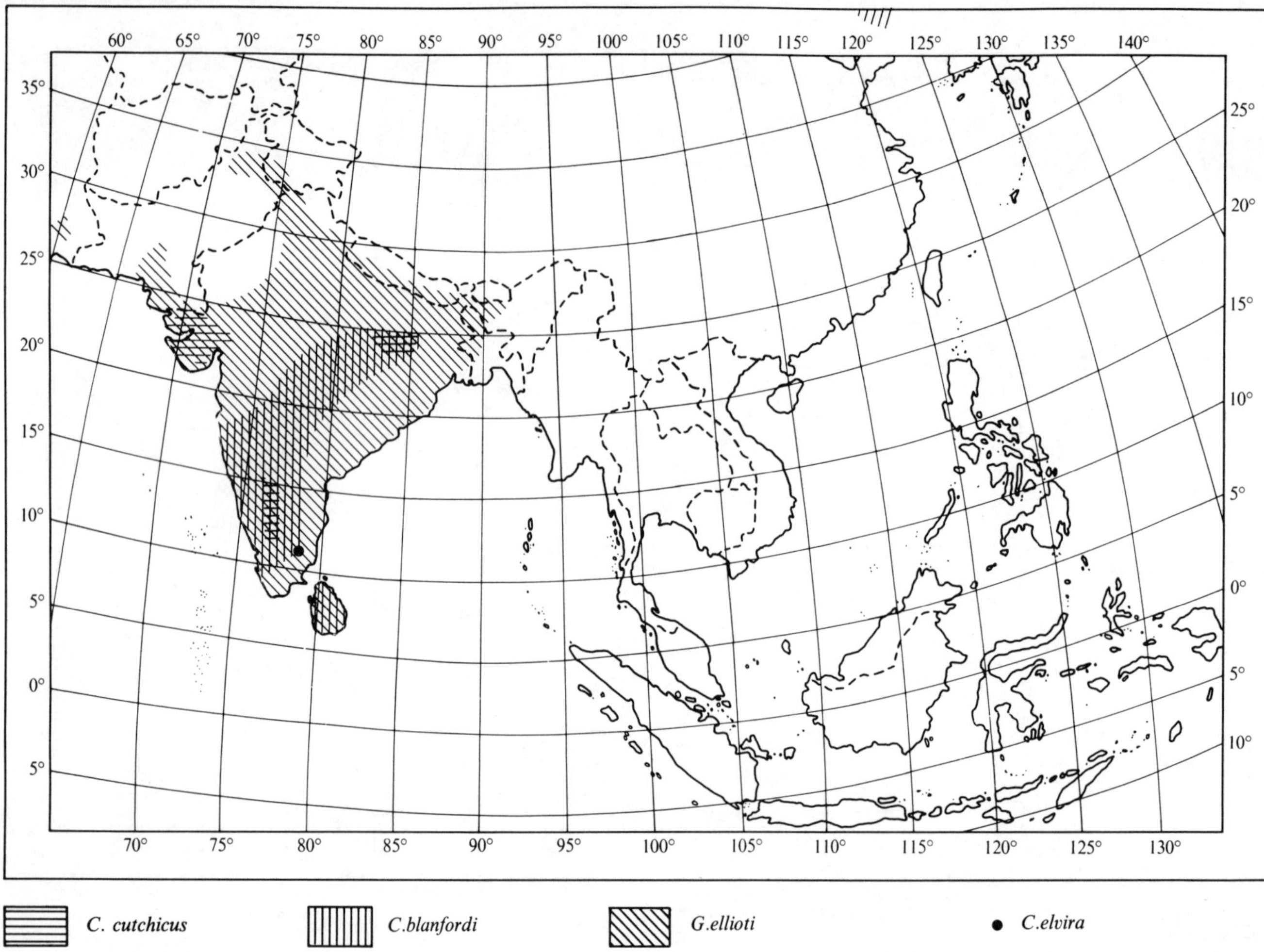

C. m. rajput Thomas, 1916f: 241; Mt Abu, Rajputana, India, 4300ft.
C. australis siva Thomas, 1916f: 242; Sivasumadram, S Mysore, India, 2500ft.
C. c. leechi Harrison, 1974: 681; Barkachha, Windham R, Uttar Pradesh, India, 25°5'N, 82°37'E. (Based mainly on small size.)

Animals in the south tend to be larger (head & body >130mm) and paler below than those in the north. A chromosome complement of 2n=36 has been reported for *C. c. cutchicus* from Rajasthan (Rishi & Puri, 1984). (A report of 2n=42 by Sobti & Gill (1984) was probably erroneous since the locality was given as Rajauri, Jammu & Kashmir, and the species described as a 'house rat'.)

Cremnomys elvira

Rattus (*Cremnomys*) *elvira* Ellerman, 1947c: 207; Kurumbapatti, Salem district, E Ghats, India.

Range (Map 142). S India, known only from the type locality in Tamil Nadu.

Cremnomys blanfordi Blanford's rat

Mus blanfordi Thomas, 1881: 24; Kadapa, Madras, India.

Range (Map 142). Peninsular India north to near Bombay in the west and to Bihar in the east; Sri Lanka.

Ecology. Habits, including burrowing, have been described by Khajuria (1981).

[*MASTOMYS* Multimammate rats]

Mastomys Thomas, 1915f: 477; type species *Mus coucha* Smith.

An African genus, also represented in S Arabia, included in *Rattus* by Ellerman (1941) and more recently in *Praomys*. One fossil species has been described from the Indomalayan Region:
M. colberti Lewis, 1939: 341; 1/2 mile S of Taraum village, NW Bilaspur Kehloor State, Punjab, India.

This was allocated to the extinct genus *Parapodemus* by Misonne (1969) but was considered indeterminate, but possibly referable to the extinct genus *Karnimata*, by Jacobs (1978). It seems certain that it is not a *Mastomys*.

GOLUNDA

Golunda Gray, 1837a: 586; type species *Golunda ellioti* Gray.

Indian subregion including Sri Lanka; one Recent species. It is characterized especially by grooved upper incisors; tall, discrete cusps of the upper molars, the cusps remaining discrete, albeit in the form of concave basins, even after considerable wear (Fig. 33a); and relatively short fifth digits on the hind feet.

A fossil species has been described from Ethiopia and it has been suggested that the members of the African genus *Pelomys* should be included in *Golunda* (Simpson, 1945). One fossil species has been described from the Indomalayan Region: *G. kelleri* Jacobs, 1978: 67; Pabbi Hills, Punjab, Pakistan; Neogene, Upper Siwalik Beds.

Golunda ellioti Indian bush rat

Golunda ellioti Gray, 1837a: 586; Dharwar, India.
Mus hirsutus Elliot, 1839: 213; southern Mahratta country, India.
M. myothrix Hodgson, 1845: 267; Kahulia Powa, Nepal.
M. newara Kelaart, 1850b: 213; Newara Eliya, Sri Lanka.
G. coffaeus Kelaart, 1850b: 213; Sri Lanka.
Pelomys watsoni Blanford, 1876: 181; Kirthar Range, Sind.

Range (Map 142). From Pakistan, Nepal and Assam (west of the Brahmaputra) through peninsular India; Sri Lanka. Also recorded in SE Iran (Misonne, 1990).

Variation. The following subspecific names have been proposed:
G. e. limitaris Thomas, 1923j: 373; 11 miles W of Kohat, NW Frontier, Pakistan, 2200ft.
G. e. paupera Thomas, 1923j: 374; Handisera, near Ambala, Punjab, India, 500ft.
G. e. gujerati Thomas, 1923j: 374; Lunwa, Palanpur, Gujarat, India, 150ft.
G. e. bombax Thomas, 1923j: 375; Andheri, Salsette I, Bombay, India.
G. e. coraginis Thomas, 1923j: 375; Wotekolli, Coorg, India, 2000ft.
G. e. coenosa Thomas, 1923j: 376; Hasimara, Bhutan Duars, India, 300ft.

Geographical variation was reviewed by Agrawal & Chakraborty (1982) who recognized only three subspecies:
G. e. ellioti. Most of range. Dorsum greyish brown (agouti); venter grey; upper molar row 5.8–6.9mm (n=84).
G. e. gujerati. NW India and Pakistan. Dorsum as above; venter white; upper molar row 5.8–6.7mm (n=35).
G. e. newara. Sri Lanka. Dorsum blackish; venter grey; larger: upper molar row 6.8–7.4mm (n=9, BMNH).

Ecology. Occurs in grassland and scrub; can be a serious pest, e. g. by feeding on the buds and fruit of coffee.

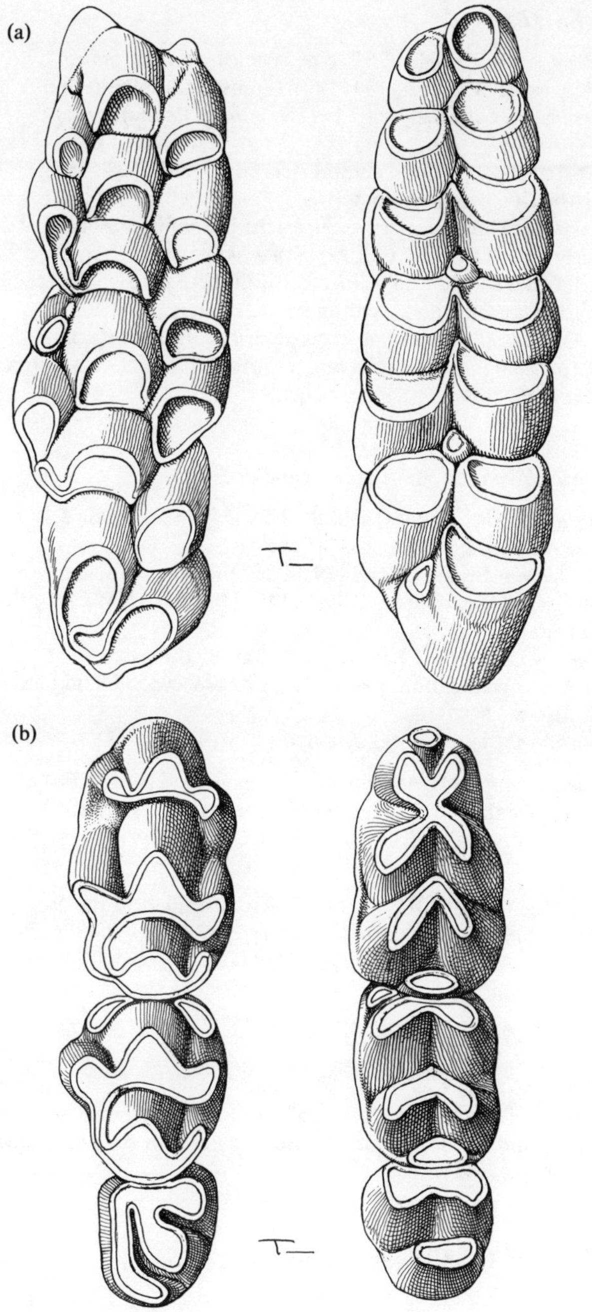

Fig.33. Upper right and lower left molars of (a) *Golunda ellioti*; (b) *Micromys minutus* (not to scale).

[*ARVICANTHIS*]

Arvicanthis Lesson, 1842: 147; type species *Lemmus niloticus* Geoffroy.

Africa and S Arabia; c. 6 species. No living representatives are found in the Indomalayan Region (although *Hadromys humei* is superficially similar). One fossil species has been described from the Indus Basin in Afghanistan:

Arvicanthis magnus Sen, Brunet & Heintz, 1979: 66; Hadja Rona, 5km N of Sarobi (60km E of Kabul), E Afghanistan; Pliocene.

[*PELOMYS*]

Pelomys Peters, 1852: 275; type species *Mus* (*Pelomys*) *fallax* Peters.

Africa south of the Sahara; c. 8 species. It is doubtful if these justify separation from *Arvicanthis.* No living species in the Indomalayan Region, but one fossil species has been described from the Indus Basin in Afghanistan:

Pelomys orientalis Sen, Brunet & Heintz, 1979: 69; Hadji Rona, 5km N of Sarobi (60km E of Kabul), E Afghanistan; Pliocene.

NESOKIA

Nesokia Gray, 1842: 264; type species *Arvicola indica* Gray.
= *Nesocia* Blanford, 1891: 421 (unjustified emendation).
Spalacomys Peters, 1860: 139 (*S. indicus* Peters).

Arid parts of SW and central Asia; one, perhaps 2, species. A distinctive genus characterized by very hypsodont teeth with transverse laminae, the cusp pattern becoming obscured at an early age. It is likely that *Erythronesokia bunnii* Khajuria, 1981 from Iraq should also be included, and perhaps also the species of *Bandicota* as discussed under that genus.

One fossil species has been described from the region:
N. panchkulaensis Raghavan, 1990: 247; Panchkula, Pinjor, Haryana, India.

Nesokia indica Short-tailed bandicoot-rat

Arvicola indica Gray, 1830(in 1830–35):pl. xi; 'India'.
Mus hardwickei Gray, 1837a: 585.
M. huttoni Blyth, 1846a: 139; Kandahar, Afghanistan.
N. griffithi Horsfield, 1851: 145; Pushut, NW Frontier, Pakistan.
Spalacomys indicus Peters, 1860: 143; E India.
N. beâba Wroughton, 1908d: 741; Pithoro, central Sind Desert, Pakistan.
Also other extralimital synonyms.

Range (Map 144). Afghanistan, Pakistan and NW India east to Bihar (Sinha, 1981) and central Nepal (Ghosal, 1973); also from Xinjiang and southern Soviet Turkestan to Israel and Egypt.

Variation. One subspecies has been described in the Region:
N. i. chitralensis Schlitter & Setzer, 1973: 171; 4 miles N of Chitral, Chitral State, Pakistan, 4800ft.

It is unlikely that discrete subspecies can be recognized but the pelage is paler in the west (*huttoni*) and darker in N Pakistan (*chitralensis*).

Ecology. A burrowing species found in both natural grassland and cultivated fields, foraging mainly underground on roots, bulbs and grass. Frequently used as an item of human food (Gheyara, 1956).

BANDICOTA Bandicoot-rats

Bandicota Gray, 1873a: 418; type species *B. gigantea* Hardwicke, in *B. indica* (Bechstein).
Gunomys Thomas, 1907d: 203 (*Arvicola bengalensis* Gray).

Indian and Indochinese subregions (elsewhere probably introduced); 3 species (Table 224). This genus is closely related to *Nesokia* (the prior name) but the lamination of the molars is much less extreme. An electrophoretic study by Radtke & Niethammer (1985) suggested that *Bandicota indica* is more closely related to *Nesokia indica* than to the other species of *Bandicota*, indicating that inclusion of *Bandicota* in *Nesokia* would be appropriate. However this was based upon only eight loci and needs to be reassessed on a larger sample. The cranial and dental differences are as great as between many genera of Murinae. Such a combination of genera would cause considerable nomenclatural confusion since the following species would become *Nesokia indica* (Bechstein) and the present *Nesokia indica* (Gray) would become *N. hardwickei* (Gray).

A further species in Thailand has not been assessed:
B. bangchakensis Boonsong & Felten, 1989: 202; Bang Chak, Amphoe Muang, Prov. Nakhon Si Thammarat, Thailand, 8°30′N, 100°00′E.

Table 224. The species of *Bandicota*.

	bengalensis	*savilei*	*indica*
Range	India, Java, Sumatra, Sri Lanka	Burma-Vietnam	India-Vietnam, Sri Lanka, Java, Taiwan
Head & body	160–250	180–240	210–345
Tail (% of h & b)	80–100%	75–100%	80–110%
Hind feet	30–44	37–43	50–70
Skull: greatest length	36–44	46–50	54–69
Upper molar row	6.0–8.0	7.8–8.7	8.2–11.4

Bandicota indica Large bandicoot-rat

Mus indicus Bechstein, 1800: 497; Pondicherry, India.
? *M. bandicota* Bechstein, 1800: 498; east coast of India.
M. malabarica Shaw, 1801: 54; Malabar, India.
M. perchal Shaw, 1801: 55; Pondicherry, India.
M. giganteus Hardwicke, 1804: 306; Hardwar, United Provinces, India.
M. setifer Horsfield, 1824; Java.
M. (*Rattus*) *nemorivagus* Hodgson, 1836b: 234; Nepal.
M. macropus Hodgson, 1845: 268; Nepal.
M. (*Nesokia*) *elliotanus* Anderson, 1878b: 231; Calcutta.
B. mordax Thomas, 1916d: 642; Chiengmai, Thailand.
B. siamensis Kloss, 1919b: 382; Tachin, central Thailand.
Rattus eloquens Kishida, 1926: 144; Taiwan.
B. jabouillei Thomas, 1927: 54; Tourane, Annam, Vietnam.

Range (Map 143). Most of Pakistan, India, Bangladesh and lowland Nepal, east to Yunnan, Guangxi and Fujian; Burma, Thailand and Vietnam; isolates in Sri Lanka, Taiwan, Hong Kong, N Malayan Peninsula and Java. Probably introduced in the last three areas. Records from Sumatra are probably erroneous (Musser & Newcomb, 1983).

Variation. Two subspecific names have been proposed:
Nesokia nemorivaga taiwanus Tokuda, 1941: 74; Taihoka, Taiwan.
B. i. sonlaensis Dao, 1972 or 1975; N Vietnam.

Ellerman (1961) retained only *B. i. nemorivaga* (Nepal and Assam) as a subspecies, based upon shorter nasals and medium size, but Tiwari *et al.* (1971) maintained that the population in the Western Ghats (SW India) is a distinctive subspecies characterized by larger size: head and body 250–335mm, hind feet 54–55mm, occipitonasal length of skull 56–63mm.

Ecology. In Thailand, Malaya and Java this species is confined to cultivated habitats. In India it is an important pest in rice fields. In S China (Guangxi) it is the dominant rodent in hilly country. They make extensive burrows.

Remarks. On the basis of larger size and longer nasals *B. gigantea* has been recognized as a species separate from *B. indica* but widespread in India and Burma (Pradhan *et al.*, 1989). However these authors did not demonstrate that these characters showed a bimodal distribution generally or at any one locality, and the distinction remains dubious. If there is a specific difference it is more likely to be between two allopatric

Map **143**. *Bandicota, Srilankamys*

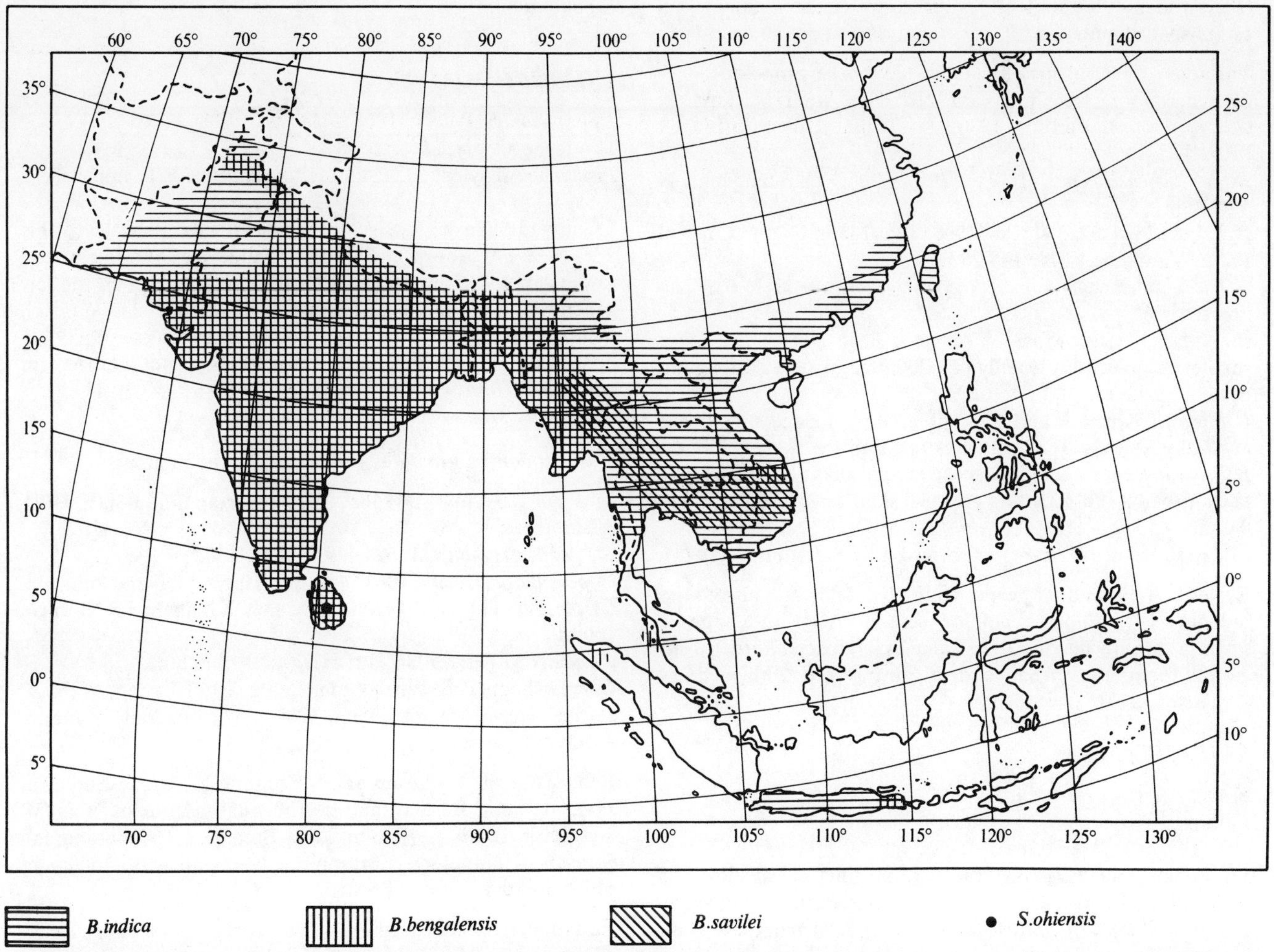

species: *B. indica s. s.* in peninsular India and Sri Lanka (skull >58mm, nasals >21mm) and *B. nemorivaga* in Bengal, Nepal, Assam and eastwards (skull <58mm, nasals <21mm). Differences between occipitonasal length and condylobasal length as used by Pradhan *et al.* are of no significance.

Bandicota savilei Burmese bandicoot-rat

B. savilei Thomas, 1916d: 641; Mt Popa, Burma, c. 2500ft.

Range (Map 143). Central Burma and central Thailand to S Vietnam.

Variation. One subspecies has been described:
B. s. curtata Thomas, 1929c: 205; Raheng, Thailand. Tail shorter than in Burma (c. 75% of head and body).

Ecology. Occurs in grass in open teak forest as well as in cultivated fields (Marshall, 1976).

Remarks. This form was treated as a subspecies of *B. indica* by Ellerman (1961) and others, but it has been shown to be distinct from *B. indica* in Thailand where the two species can be found in the same field (Marshall, 1976).

Bandicota bengalensis Lesser bandicoot-rat

Arvicola bengalensis Gray, 1835(in 1830–35):pl. 21; Bengal, India.

Mus kok Gray, 1837a: 585; Dharwar, SW India.
M. (Neotoma) providens Elliot, 1839: 209; S Mahratta country, India.
M. dubius Kelaart, 1850c: 217 (330 in 1890 reprint); Sri Lanka. (Not of Hodgson, 1845.)
M. daccaensis Tytler, 1854: 173; Deccan, India.
M. tarayensis Hodgson, in Horsfield, 1855: 112; Nepal.
M. plurimammis Hodgson, in Horsfield, 1855: 112; Nepal.
M. morungensis Hodgson, in Horsfield, 1855: 112; Nepal.
M. (Nesokia) blythianus Anderson, 1878b: 227; Bengal, India.
M. (Nesokia) barclayanus Anderson, 1878b: 229; Guna, central India.
Nesokia gracilis Nehring, 1902: 116; Sri Lanka.
Gunomys varius Thomas, 1907d: 204; Georgetown, Penang I, Malaya.
G. varillus Thomas, 1907d: 205; Georgetown, Penang I, Malaya.
G. wardi Wroughton, 1908d: 745; Pandritton, Kashmir, 5500ft.
G. lordi Wroughton, 1908d: 746; Kolaba dist, Konkan, Bombay, India.
G. sindicus Wroughton, 1908d: 746; Pithoro, central Sind Desert, Pakistan.

Range (Map 143). Most of India, S and NE Pakistan, lowland Nepal and Bangladesh, east to Burma; Vietnam; Sri Lanka; also present, probably introduced, on Penang I (Malaya), N

Sumatra and E Java; reported from Patta I, Kenya in 1957 (Fleetwood, 1963) but it is uncertain whether this is an established population.

Variation. Four subspecific names have been proposed:
Gunomys b. sundavensis Kloss, 1921d: 116; Atjeh, N Sumatra.
G. kok insularis Phillips, 1936: 95; Thinney, near Jaffna, N Prov., Sri Lanka.
B. b. hichensis Dao, 1961b: 305; Hich, Thai-nguyœn, N Vietnam.
B. b. giaraiensis Dao & Cao, 1990: 235; Mang Giang, Gia Lai Prov., Vietnam, 14°N, 105°25'E.

Three subspecies can be recognized (Agrawal & Chakraborty, 1976):
B. b. bengalensis. Most of range. Pelage short, not silky; smaller: head & body usually 140–200mm, greatest skull length usually 36-42mm.
B. b. wardi. Kashmir and adjacent territory. Pelage long, dense and silky, especially in winter; size as above.
B. b. varia. Penang I. Pelage short and not silky; larger: head & body usually 190–235mm, greatest skull length usually 44–48mm.

Variation in Pakistan was reviewed by Smiet *et al.* (1978).

Ecology. An abundant pest of cultivated fields and also found indoors in warehouses holding cereals. Also in mangrove swamps. Mainly nocturnal. In Calcutta it replaced *Rattus rattus* as the dominant urban rodent during the 20th century (Prakash, 1977).

SRILANKAMYS

Srilankamys Musser, 1981b: 268; type species *Rattus ohiensis* Phillips.

Sri Lanka; one species. This very distinctive species was included, with considerable reservations, in *Rattus* (*Apomys*) by Ellerman (1949), in *Rattus* (*Lenothrix*) by Ellerman (1961) and in *Rattus* (*Leopoldamys*) by Misonne (1969).

Srilankamys ohiensis

Rattus ohiensis Phillips, 1929b: 167; West Haputale, Ohiya, Sri Lanka, 6000ft.

Range (Map 143). Highlands of Sri Lanka, in Uva Province, at around 1800m.

CHIROPODOMYS

Chiropodomys Peters, 1868: 448; type species *C. penicillatus* Peters (in *C. gliroides*).
Insulaemus Taylor, 1934: 469 (*I. calamianensis* Taylor).

S China to Malaya and Greater Sunda Islands; 5 species (Table 225). These are forest mice found at all levels from the ground to the canopy, often associated with bamboo. One of the few genera having a blunt nail rather than a sharp claw on each hallux. This genus was reviewed in detail by Musser (1979). A new fossil species has been reported, but not named nor described, from SW China (Zheng, 1988).

Chiropodomys gliroides Pencil-tailed tree mouse

Mus gliroides Blyth, 1856c: 721; Cherrapunji, Khasi Hills, Assam.
M. peguensis Blyth, 1859a: 295; Schwegyin, S Burma.
C. penicillatus Peters, 1868: 448; probably Malayan peninsula.
C. pusillus Thomas, 1893a: 345; Mt Kinabalu, N Borneo, 1000ft.
C. niadis Miller, 1903a: 40; Nias I, W Sumatra.
C. anna Thomas & Wroughton, 1909c: 19; Tjilatjap, W Java.
C. jingdongensis Wu & Deng, 1984: 207; Jingdong, Yunnan, 2430m.

Range (Map 144). Assam and S China to Malaya, S Sumatra, Java, Bali and Borneo; and on the small islands of Nias (W Sumatra), Tujuh Is (E Sumatra), Bunguran (N Natuna Is), Serasan (S Natuna Is). Mainly lowland but up to 1600m in Malaya and 1220m in Borneo.

Variation. Five subspecies were recognized by Musser (1979). Those north of the Isthmus of Kra (*C. g. gliroides*) tend to be duller, with relatively longer incisive foramina. The various Sunda populations only differ in average measurements.

Ecology. Found in primary and secondary forest, especially, but not exclusively, where there is bamboo.

Table 225. The species of *Chiropodomys*.

	gliroides	*major*	*calamianensis*	*muroides*	*karlkoopmani*
Range	C China-Sunda	Borneo	Palawan etc.	Borneo	Mentawai Is
Head & body	69–102	94–114	109–122	66–88	84, 107
Tail (% of h & b)	c.135%	c.120%	125–140%	c.125%	146, 160%
Hind feet	16–21	21–28	16–20	15–17	24, 29
Dorsum	Pale chestnut	Grey-brown	Bright reddish brown	Buff-brown	Grey-brown
Venter	White	White or cream, ± buff wash	Orange-brown ± white	Dark grey, slight buff tips	Pale grey or white
Flank stripe	Ochraceous	0	Buff-orange	0	0
Cheeks (below eyes)	Buff	White	Buff	Grey	Grey-brown
Tail: distal 1/3 white	0	0	0	0	+
distichous distally	+/0	+	0	+	0
Skull: greatest length	20–27	27–30	28–29	20–22	27, 29
Upper molar row	3.3–4.2	4.2–5.1	4.0–4.6	2.9–3.1	4.6
Incisors orange (O) or pale yellow (Y)	O	O	O	Y	O
M^{1-2} with postero-internal cusp rounded (as other internal cusps) - (R), or long and thin (L)	L	R	R	L	R

Map **144**. *Chiropodomys, Nesokia*

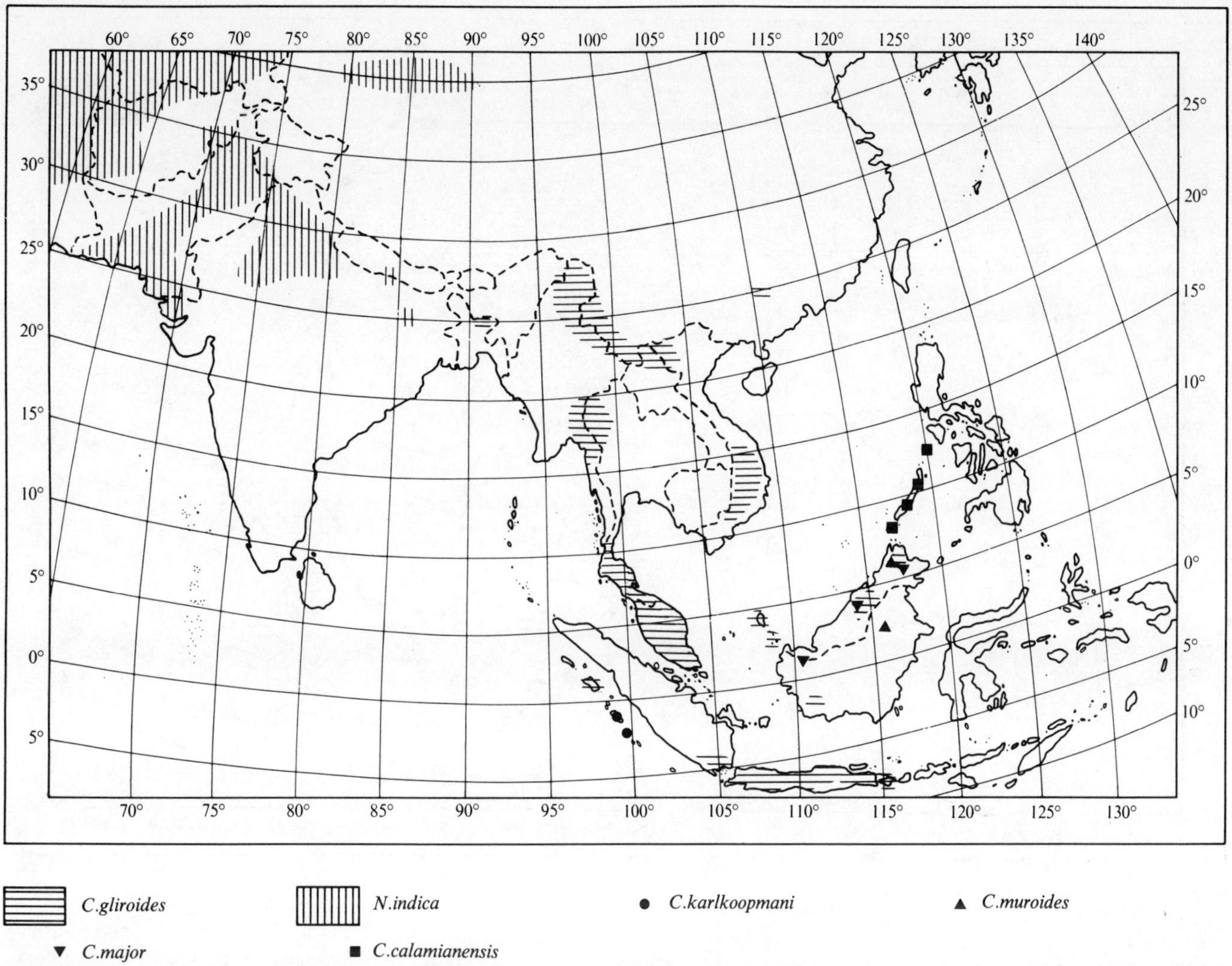

Remarks. In Malaya the karyotype has been reported as 2n = 42, NF = 48 without very close similarity to that of related genera (Yong, 1973). The measurements and pelage characters of *jingdongensis* appear to fall within the range of *C. gliroides* except for the slightly longer, more distichous hairs on the distal part of the tail.

Chiropodomys major

C. major Thomas, 1893a: 344; Sadong, Sarawak, Borneo.
C. legatus Thomas, 1911e: 206; Mt Kinabalu, N Borneo.
C. pictor Thomas, 1911e: 207; Mt Kinabalu, N Borneo.

Range (Map 144). Sabah and Sarawak, Borneo; up to 1500m on Mt Kinabalu.

Chiropodomys calamianensis

Insulaemus calamianensis Taylor, 1934: 470; Minuit, Busuanga I, Calamian Is, Philippines.

Range (Map 144). Palawan and adjacent islands of Busuanga and Balabac, Philippines.

Chiropodomys muroides

C. muroides Medway, 1965b: 133; Bundu Tuhan, Mt Kinabalu, N Borneo, 4000ft.

Range (Map 144). Known only from Mt Kinabalu (at 1200m) and one other locality in central Borneo, at Long Petak.

Chiropodomys karlkoopmani

C. karlkoopmani Musser, 1979: 389; N Pagi I, Mentawai Is, Sumatra.

Range (Map 144). N Pagai and Siberut Islands, Mentawai Islands (Jenkins & Hill, 1982)

VERNAYA

Vernaya Anthony, 1941: 110; type species *Chiropodomys fulvus* Allen.
= *Octopodomys* Sody, 1941a: 261.

Sichuan, Yunnan and Burma; 1 species, known from very few individuals. Four additional fossil species have been reported from S China but not yet described (Zheng, 1988).

Map **145**. *Hapalomys, Micromys, Vernaya*

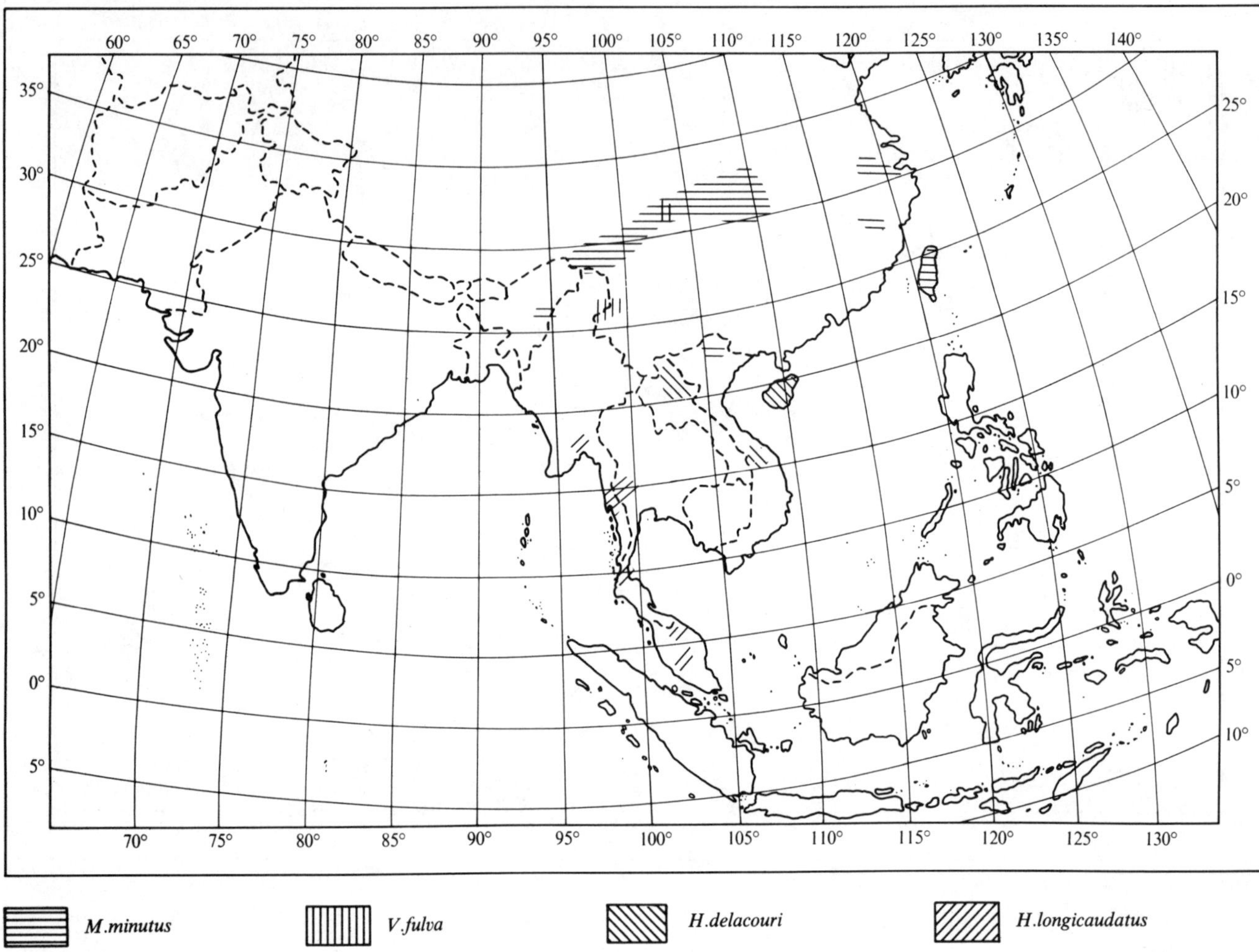

Vernaya fulva Vernay's climbing mouse

Chiropodomys fulvus Allen, 1927a: 11; Yinpankai, Mekong R, Yunnan, China.
V. foramena Wang, Y. *et al.*,1980: 393; Pingwu, Wang-lang Dist, N Sichuan, China, 2490m.

Range (Map 145). N Sichuan, Yunnan and N Burma, at altitudes of 2100–2700m.

Recognition. This species differs from *Vandeleuria* spp. in retaining a claw on the hallux, and from *Micromys* in being more yellowish above, especially on the flanks and cheeks, the grey on the venter more completely concealed by creamy buff hair tips, and the hind feet longer with enlarged digital pads. M^3 is unusually short and wide, with two transverse laminae.

Remarks. Allen (1940) considered *fulva* a synonym of *Vandeleuria dumeticola* (here in *V. oleracea*) but this was refuted by Ellerman (1949) and is clearly wrong. *V. foramena* was distinguished from *V. fulva* primarily by the presence of a pair of vacuities in the frontal bones of the skull. However these are present also in most *V. fulva*, including the holotype.

HAPALOMYS Marmoset-mice

Hapalomys Blyth, 1859a: 296; type species *H. longicaudatus* Blyth.

Indochinese subregion and Malaya; 2 species (Table 226). These are specialized arboreal rats, with a nail instead of a claw on each hallux, short blunt snout and unique molars with the cusps prominent and forming an evenly spaced rectangular grid. Three fossil species have been reported from S China but not yet described (Zheng, 1988). The genus was reviewed in detail by Musser (1972a).

Hapalomys longicaudatus Greater marmoset-mouse

H. longicaudatus Blyth, 1859: 296; Sitang R, Tenasserim, Burma.

Table 226. The species of *Hapalomys*.

	longicaudatus	*delacouri*
Range	Assam - Malaya	Laos, Vietnam, Hainan
Size	Larger	Smaller
Head & body	162–165	121–136
Tail	193–202	140–171
Hind foot	28–32	22–24
Dorsal pelage	Greyish brown	Ochraceous
Cheeks	As dorsum	White
Maxillary tooth-row	7.9–8.0	5.9–6.7

Range (Map 145). S Burma, SE Thailand and Malaya.

Ecology. In Malaya occurs above 360m, always associated with bamboo which provides both food, mainly the flowering shoots and seeds, and nest-sites, in the internodes.

Remarks. The karyotype in Malaya has been reported as 2n=50 with very large metacentric X-chromosome and subacrocentric Y as in *Pithicheir* (Yong *et al.*, 1982).

Hapalomys delacouri Lesser marmoset-mouse

H. delacouri Thomas, 1927: 55; Dakto, Annam, Vietnam.
H. pasquieri Thomas, 1927: 57; Xieng Khouang, Laos.
H. marmosa Allen, 1927a: 12; near Nodoa, Hainan I, China.

Range (Map 145). Central Vietnam, N Laos, Hainan.

Variation. Two subspecies can be recognized on the mainland:
H. d. delacouri. C Vietnam. Tail more scantily haired; nasals longer (11.7–12.0); maxillary tooth-row longer (c. 6.3).
H. d. pasquieri. Laos. Tail with longer hair distally, forming terminal brush 5–7mm long; nasals shorter (c. 10.5); maxillary tooth-row shorter (c. 5.9).

The sole specimen known from Hainan, by a skin only, is not distinguishable from *H. d. delacouri* (Musser, 1972a).

MICROMYS

Micromys Dehne, 1841: 1; type species *M. agilis* Dehne, in *M. minutus*.

Palaearctic and south to N Burma and Assam; one species is generally recognized but others are probably valid.

Micromys minutus Harvest mouse

Mus minutus Pallas, 1771: 454; Simbirsk, R Volga, USSR.
M. erythrotis Blyth, 1856c: 721; Cherrapunji, Khasi Hills, Assam.
M. pygmaeus Milne-Edwards, 1872b: 291; Moupin, Sichuan, China.
Plus other, Palaearctic, synonyms.

Range (Map 145). From Assam, N Burma and N Vietnam to China, and much of the Palaearctic Region from Japan to Britain; also on Taiwan.

Variation. The following subspecific names have been proposed from the Indomalayan Region:
M. m. berezowskii Argyropulo, 1929: 253; Hotszihow, near Lunfanfu, N Sichuan, China.
M. m. takasagoensis Tokuda, 1941: 78; Sikayau, Taiwan.
M. m. pianmaensis Peng, in Peng & Wang, 1981: 172; Pianma, Lushui Xian, W Yunnan, 1930m.

All mainland Indomalayan specimens differ from European ones in being greyish brown above and grey below, lacking the orange-brown tones and white venter of most adult European specimens. They also have the tail longer than the head and body, seen in extreme form in the single specimen available from Assam (head and body 60mm, tail 90mm). Although the body length of this specimen is probably underestimated, the absolute tail-length exceeds that in all available specimens from China and Vietnam (53–82mm, n=24).

The form on Taiwan has the venter white, without grey hair-bases, and the tail longer than head and body.

APODEMUS Palaearctic wood mice

Apodemus Kaup, 1829: 154; type species *Mus agrarius* Pallas.
Sylvaemus Ognev, 1924: 143 (*Mus sylvaticus* L.).
= *Nemomys* Thomas, 1924c: 889.
Alsomys Dukelski, 1928: 42 (*Mus sylvaticus major* Radde, ? in *A. peninsulae* Thomas).

Palaearctic Region and marginally in the Indomalayan Region in S China and the Himalayas; c. 11 species of which 7 are represented in the Indomalayan Region (Table 227). Two groups are commonly accepted as subgenera: *Apodemus* with *A. agrarius* and *A. chevrieri*; and *Sylvaemus* with the remainder. Electrophoretic analysis of proteins has not only supported this division but has suggested that these are as distantly related to each other as each is to *Mus, Rattus* and other genera (Bonhomme *et al.*, 1985). However formal recognition of *Sylvaemus* as a genus should wait until this has been confirmed using a wider variety of species. Study of the complex morphology of the penis of four Chinese species suggests that these two groups are much more similar than either is to *Mus, Micromys* or *Rattus* (Yang & Fang, 1988).

In addition to the species detailed here, *A. peninsulae* (Thomas, 1907c) occurs in N China and at high altitude in

Table 227. The species of *Apodemus*.

	agrarius	*chevrieri*	*sylvaticus*	*draco*	*orestes*	*latronum*	*gurkha*	*peninsulae*
Range in region	S China, Taiwan	SW China	W Himalayas	S China-Assam, Taiwan	S China-Assam	SW China	C Nepal	N & W China
Head & body	75–95	90–110	90–105	90–105	80–100	95–110	100–115	90–110
Hind feet	19–21	21–25	22–23	20–23	23–25	24–27	23–25	<25
Dorsum yellowish-brown (Y) or greyish (G)	Y	Y	Y	Y	Y	Y	G	Y
Narrow mid-dorsal stripe	+	0	0	0	0	0	0	0
Yellowish spot on chest	0	0	+/0	+/0	0/+	0	0	0
Ears: dorsum	Same	Same	Same	Darker	Darker	Darker	Same	Same
Teats: pairs	4	4	3	4	4	3	4	4
M^3: medial lobes	2	2	3	3	3	3	3	3
M^1: medial roots	2	1	2	1(2)	1	(2)	1	1
Skull: greatest length	24–28	26–30	25–27	24–28	25–28	28–30	27–29	25–27
Upper molar row	3.8–4.2	4.0–4.6	3.6–4.1	3.7–4.2	3.8–4.3	4.6–5.0	4.4–4.7	3.8–4.3
Supraorbital ridges prominent	+	+	0	0	0	0	0	0

Qinghai, Tibet, SW Sichuan and W Yunnan and may therefore occur marginally in the region covered here. It is included in the table for comparison.

Subgenus *Apodemus*

Apodemus agrarius Striped field mouse

Mus agrarius Pallas, 1771: 454; Simbirsk, R Volga, Russia.
M. ningpoensis Swinhoe, 1871: 637; Ningpo, Zhejiang, S China.
M. harti Thomas, 1899: 774; Kuatun, Fujian, S China.
Plus many Palaearctic synonyms.

Range (Map 146). E China from Fujian, Guizhou and extreme NE Yunnan northwards, and on Taiwan; also from Korea through S Siberia to C Europe.

Variation. Two subspecies have been described:
A. a. pallidior Thomas, 1908h: 8; near Chefoo, Shandong, China.
A. a. var. *insulaemus* Tokuda, 1941: 84; lowlands of Taiwan.

The black mid-dorsal stripe is faint but still clearly visible in south China (*ningpoensis*) and clearer in the north (*pallidior*), with a clinal transition; these two forms are not separable by measurements (Liu *et al.*, 1988).

Remarks. The montane Chinese form *chevrieri* was included in *A. agrarius* by Allen (1940) and some subsequent authors but is here treated as a separate species following Xia (1984).

Apodemus chevrieri

Mus chevrieri Milne-Edwards, 1872b: 288; Moupin, Sichuan, China.
A. fergussoni Thomas, 1911a: 4 (and 1911b: 172); Wenhsien, S Gansu.

Range (Map 146). Sichuan, Yunnan, W Hubei and S Gansu, China, from c. 1800 to 3000m. Possibly parapatric or marginally sympatric with *A. agrarius* at Wenhsien, S Gansu.

Remarks. Included in *A. agrarius* by Allen (1940) and others but treated as specifically distinct following Xia (1984). This seems fully justified—it is much larger, darker and has no trace of a dorsal stripe, and there is no indication of intergradation where the two forms meet in S Gansu.

Subgenus *Sylvaemus*

Apodemus sylvaticus Wood mouse

Mus sylvaticus Linnaeus, 1758: 62; Upsala, Sweden.
M. arianus griseus True, 1894a: 8; C Kashmir. Pre-occupied by

Map **146**. *Apodemus* (part), *Diomys*

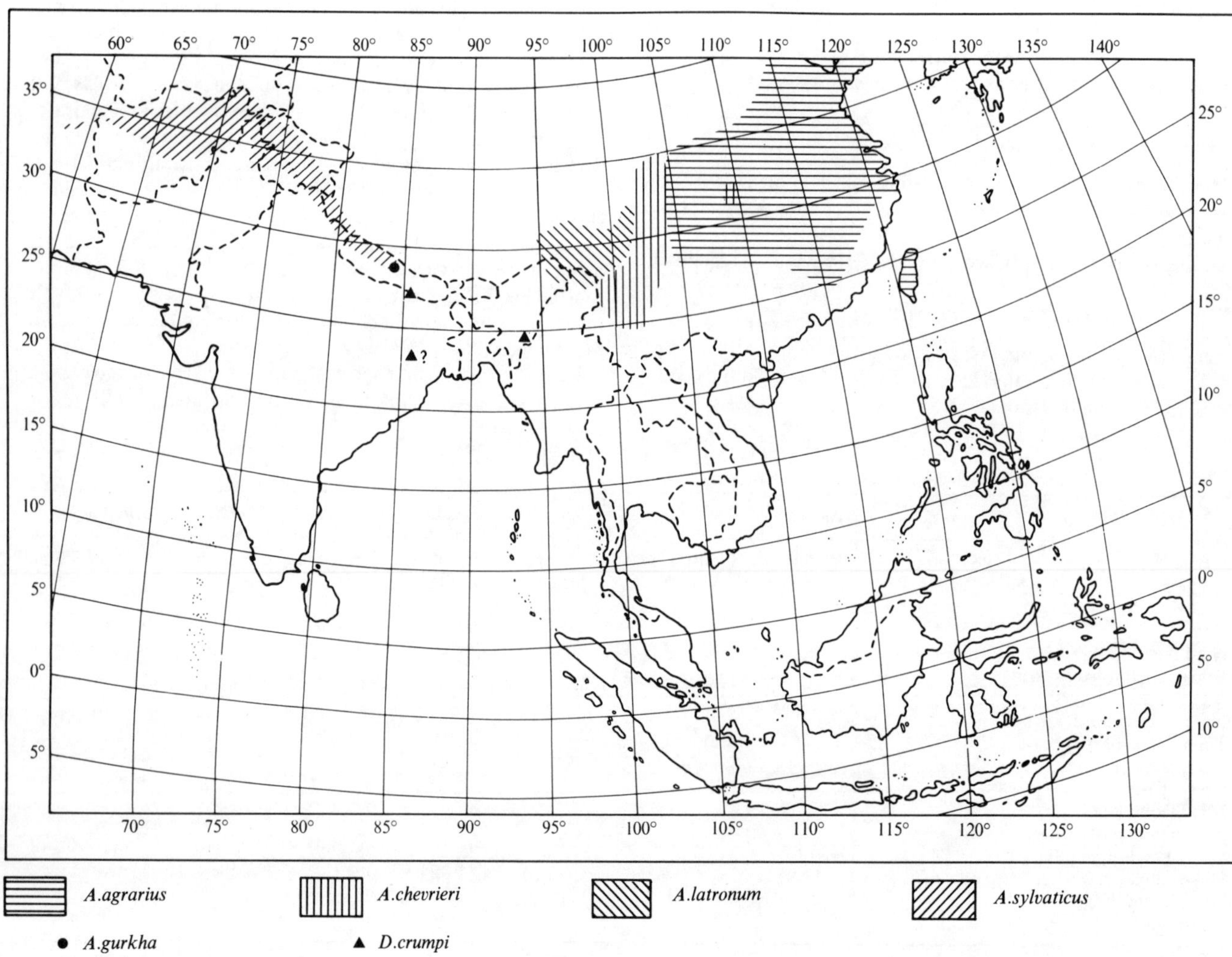

Mus griseus Mina Palumbo, 1868 (a synonym of *A. sylvaticus* from Sicily).
= *A. flavicollis rusiges* Miller, 1913d: 81.
Plus many Palaearctic synonyms.

Range (Map 146). Himalayas from C Nepal (c. 84°E) west to Pakistan and Afghanistan; also widespread through the western Palaearctic except for the desert, taiga and tundra zones. In Pakistan it occurs from 1850 to 3600m; in Nepal it appears to be confined to the drier alpine zone above 3000m, above *A. gurkha* (Martens & Niethammer, 1972).

Variation. The following subspecific names are relevant:
Micromys s. wardi Wroughton, 1908c: 282; Saspul, Ladak, Kashmir.
M. s. pentax Wroughton, 1908c: 283; Thandiani, Punjab.

Study of protein polymorphism in a sample from Nepal showed close similarity with samples from Iran but differences from European forms, casting some doubt on the validity of including the Nepalese form in *A. sylvaticus*, but separating all these clearly from *A. gurkha* (Gemmeke & Niethammer, 1982).

Apodemus draco

Mus sylvaticus draco Barrett-Hamilton, 1900: 417; Kuatun, Fujian, SE China.
? *A. semotus* Thomas, 1908i: 447; Mt Arizan, Taiwan.
? *A. ilex* Thomas, 1922a: 404; Mekong-Salween divide, Yunnan, China.

Range (Map 147). Assam and N Burma to S China (Yunnan, Hubei and Fujian) and north to Sichuan, Gansu, Shanxi and Hebei; perhaps also on Taiwan.

Variation. The form on Taiwan is rather dark (above and below and on the feet). Most individuals in the lowlands of China (*draco s. s.*) have a prominent yellowish chest-spot but this is generally absent or very faint in the montane forms in Sichuan and Yunnan.

Remarks. It seems quite clear that this is not conspecific with the western Palaearctic *A. sylvaticus* as was assumed by Allen (1940) and Ellerman & Morrison-Scott (1951). However its limits are rather unclear. Ellerman (1961) interpreted short-tailed individuals in Assam and Burma as '*A. sylvaticus orestes*' with broken tails. Examination of these suggests that they are more likely to be intact *A. draco* coexisting with the very long-tailed *A. orestes*.

Apodemus orestes

A. speciosus orestes Thomas, 1911c: 49 (and 1912b: 136); Mt Omei Shan, Sichuan, China.

Map **147**. *Apodemus* (part), *Hadromys*, *Dacnomys*

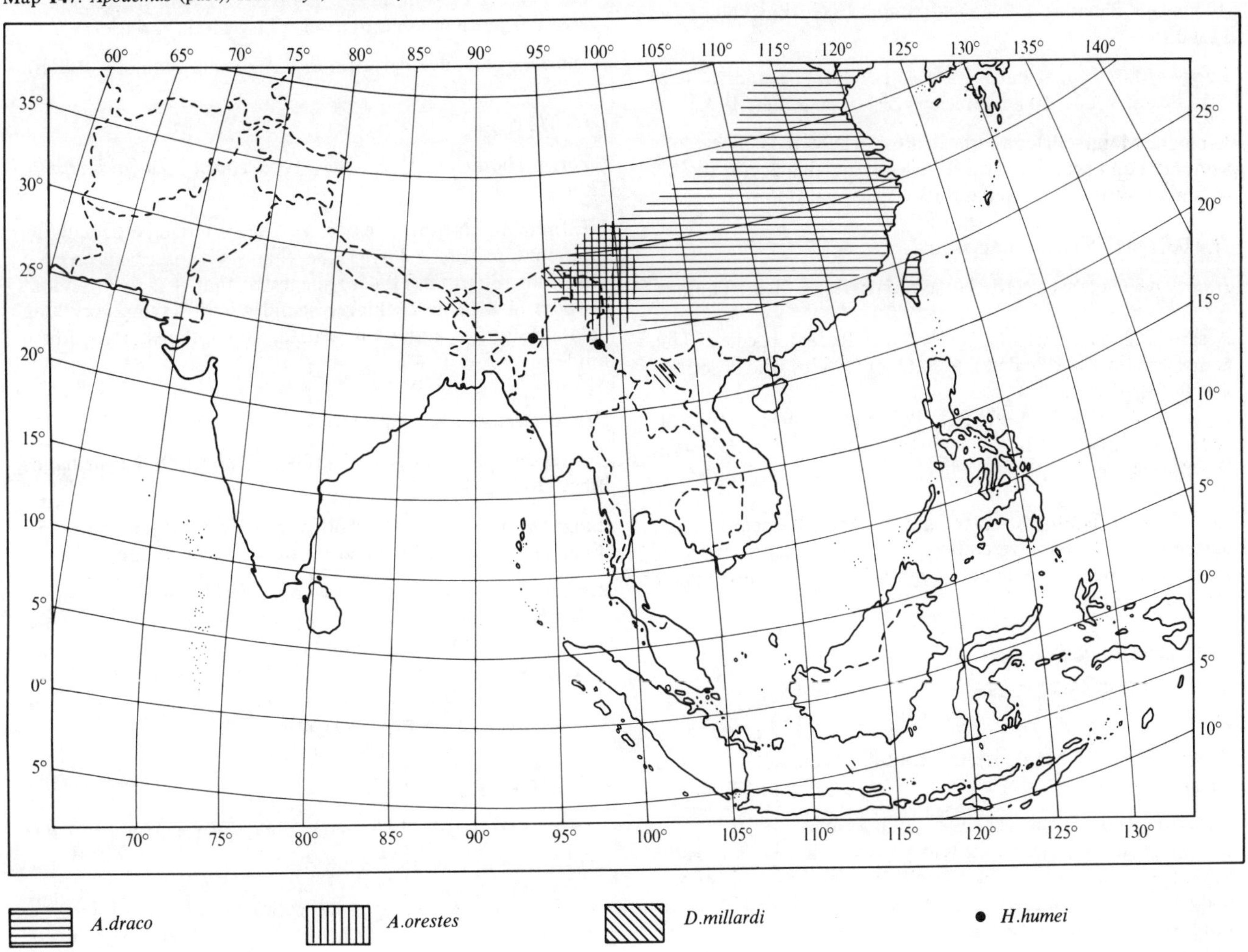

Range (Map 147). Sichuan and Yunnan to N Burma and Assam, at altitudes of up to 3400m in Yunnan.

Variation. Most individuals have no yellow chest-spot or only a very slight mark; however one very long-tailed individual (tail 135% of head and body) from the Mishmi Hills, Assam has the chest-spot large and forming a collar as in European *A. flavicollis*.

Remarks. This species is sympatric (or possibly parapatric) with forms referable to *A. draco* in Sichuan, mainly at higher altitudes. It is clearly distinguishable by darker colour and very long tail (usually over 120% of head and body).

Apodemus latronum

A. speciosus latronum Thomas, 1911c: 49 (and 1912b: 137); Tatsienlu, W Sichuan, China.

Range (Map 146). Sichuan, Yunnan, E Tibet and N Burma; in Yunnan mainly between 2700 and 3600m but up to 4000m.

Remarks. This form was included in the European *A. flavicollis* by Ellerman & Morrison-Scott (1951) but all subsequent authors are agreed that *A. flavicollis* does not occur in E Asia. It was considered conspecific with *A. draco* by Feng *et al.* (1986), but is substantially larger. It is sympatric with *A. orestes* in much of its range.

Apodemus gurkha

A. gurkha Thomas, 1924c: 888; Laprak, Gorkha, Nepal, c. 11500ft.

Range (Map 146). Central Nepal, between 83 and 84°E at altitudes of 2200–3600m (Martens & Niethammer, 1972).

Ecology. Mainly in coniferous forest where it is marginally sympatric and syntopic with *A. sylvaticus* which occurs mainly at higher altitudes in Nepal (GBC, personal observation).

HADROMYS

Hadromys Thomas, 1911d: 999; type species *H. humei* Thomas.

Assam and Yunnan; one, perhaps 2, Recent species. This genus was reviewed in detail by Musser (1987b) who described a new fossil species:
H. loujacobsi Musser, 1987b: 12; Pabbi (or Kharian) Hills, SE edge of Potwar Plateau, Punjab region, N Pakistan; Pleistocene.

In the broad, laminate molars and in some characters of the pelage there are some resemblances to the African genera of grass rats, *Arvicanthis* etc., although this may be convergence (Musser, 1987b).

Hadromys humei Hume's rat

Mus humei Thomas, 1886a: 84 (and 1886b: 63); Moirang, Manipur, India.

Range (Map 147). Manipur, Assam and W Yunnan, at altitudes up to 1300m.

Variation. One subspecies has been described:
H. h. yunnanensis Yang & Wang, 1987: 46; Hulan, Ruili County, W Yunnan, 24°9'N, 97°51'E, 1300m. Large (head and body 123–140, cf 98–120 in *H. h. humei*; occipitonasal length of skull 31.5–33.1 cf 28.4–32.2); venter pure white (cf fawn or reddish); ochraceous patch in front of ear. These differences seem considerable and suggest that *yunnanensis* might be specifically distinct from *H. humei*. There is no tendency towards a white venter nor an ochraceous patch on the cheek in any of the 30 specimens of *H. humei* examined from Manipur and Assam.

DACNOMYS

Dacnomys Thomas, 1916c: 404; type species *Dacnomys millardi* Thomas.

NE India and Laos; one species. It is very similar to *Rattus* and *Niviventer* but has unusually long molar tooth-rows.

Dacnomys millardi Millard's rat, Large-toothed rat

D. millardi Thomas, 1916c: 405; Gopaldhara, near Darjeeling, India, 3440ft.
D. wroughtoni Thomas, 1922b: 430; Dreyi, Mishmi, N Assam, 6000ft.

Range (Map 147). Darjeeling district of West Bengal, NE Assam and Laos.

Variation. One subspecies has been described:
D. m. ingens Osgood, 1932: 315; Phong Saly, Laos.

Specimens from Assam are larger than those from Darjeeling (upper molar row at alveolus 13.1mm in the holotype of *wroughtoni*, cf 11.1mm in the holotype of *millardi*) but too few specimens are available to assess the overall variation.

Remarks. A detailed description was given by Musser (1981b).

DIOMYS

Diomys Thomas, 1917a: 203; type species *Diomys crumpi* Thomas.

NE India and Nepal; one species. Characterized especially by pro-odont incisors and five rather than six plantar pads on each hind foot. Ellerman (1947c) suggested that it is 'a pro-odont offshoot of *Millardia* which resembles it in having very long palatal foramina and a long palate, also in the shortened fifth hind-toe'.

Diomys crumpi Crump's mouse

Diomys crumpi Thomas, 1917a: 204; ? Paresnath, Hazaribagh, Bihar, India.

Range (Map 146). Known from only a few localities in Assam, Bihar and S Nepal (mapped by Ingles *et al.*, 1980).

Remarks. The type skull was mismatched with a skin of *Millardia meltada*. The species has not subsequently been found at the type locality in Bihar and it remains doubtful if the type locality was correctly recorded. Musser & Newcomb (1983) compared *Diomys* with *Berylmys* in some detail and concluded that *Diomys* is a distinctive genus.

NIVIVENTER

Niviventer Marshall, 1976: 402; type species *Mus niviventer* Hodgson.

Himalayas and China to Greater Sunda Islands; 15 species, probably somewhat fewer (Tables 228 and 229). The genus is

Table 228. The species of *Niviventer* north of the Isthmus of Kra.

	niviventer	*confucianus*	*tenaster*	*fulvescens*	*coninga*	*brahma*	*eha*	*langbianis*	*hinpoon*	*andersoni*	*excelsior*
Range	Himalayas	China-Indochina	Assam-Indochina	Himalayas-Indochina	Taiwan, ? Burma	Assam, Burma	E Himalayas	Assam-Indochina	Thailand	SW China	Sichuan
Head & body	120–140	125–170	146,160	120–160	160–180	145–155	100–140	125–140	140	150–200	130–175
Tail (% of h & b)	100–115%	125–135%	130%	115–135%	105–125%	150–160%	140–170%	c.140%	100%	140–150%	125–140%
Hind feet	25–29	28–32	33,34	26–32	31–36	28–32	25–28	25–28	26.5	31–40	31–33
Ears	18–21	20–23	23,24	19–25	23–25	24–25	19–22	20–23	19	22–28	22–27
Dorsum medium brown (B); or yellowish (Y), or greyish (G) brown	G	G/Y	Y	Y	B	B	B	G	Y	G	Y
Dorsum with flat spines	+/0	+ (summer only ?)	+	+/0	+	0	0	+/0	+	0	0
Ventral pelage with grey bases	0	0	0	0	0	+	+	0	+	0	0
Sides of muzzle	Buff	Buff		Buff	Orange-brown	Buff	Buff/grey		As flanks	White	White
Tail: discrete white tip (extent)	+ (10–20%)	+/0		0	0(+)	0	0	0	0	+ (40%)	+ (30%)
Tail lighter below than above	+	+	0	+	(+)	(+)	+	0	+	+	+
Tail: hairs increasing towards tip (length of terminal pencil)	+ (2–3)	+ (c.5)	+(2–3)	0	+	+	(+)	0	0	(+)	+
Teats: pairs (pectoral/abdominal)	2/2	2/2	2/2	2/2	2/2	1/2	1/2	2/2	2/2	2/2	2/2
Skull: greatest length	32–35	34–41	41–42	33–39	40–44	36–38	29–33	33–36	35–36	39–46	38–40
Upper molar row	4.7–5.2	5.6–6.9	6.4,6.7	5.2–5.8	6.2–7.0	6.3–6.8	4.9–5.2	6.1–6.5	5.7	7.2–8.3	6.7–7.2
Temporal ridges prominent	+	+	+	+	+	+	(0)	+	+	0	0

Table 229. The species of *Niviventer* in the Sunda subregion (south of the Isthmus of Kra).

	rapit	*bukit*	*lepturus*	*cremoriventer*
Range	Malaya, Sumatra, Borneo	Malaya, Sumatra, Java	Java	Throughout
Head & body	130–170	130–150	130–155	130–165
Tail (% of h & b)	130–180%	115–130%		110–150%
Hind feet	30–38	25–27	28–32	20–30
Ears	18–25	18–21	22–25	15–21
Dorsum orange or reddish brown	0	0	+	+
Dorsum with flat spines	+	+	0	+
Tail with discrete white tip	0	0(+)	+ (30–50%)	0
Tail: hairs longer towards tip	(+)	0	+	0
Skull: greatest length	40–43	34–37	35–39	31–40
Upper molar row	6.3–7.2	5.4–5.8	6.5–7.0	5.3–6.8 (7.5)

included here in its entirety, including those forms (of *N. confucianus*) from Palaearctic China.

These species were included in *Rattus* by Ellerman (1941) and Chasen (1940a), in *Rattus* (*Maxomys*) by Ellerman (1949) and in *Maxomys* by Misonne (1969). However the type species of *Maxomys, Rattus bartelsi* Jentink, is very different from the others included here and *Niviventer* is the valid name for this group.

The species of *Niviventer* are characterized by soft, dense pelage without long projecting guard hairs but often with flattened spines (although these appear to be seasonal in some species). The venter is uniformly white or pale cream in most species, grey-based in some. In some species the white turns to a bright sulphur yellow in preserved skins. The tail is moderately well haired towards the tip and often white-tipped. The incisive foramina are long and narrow, almost or quite reaching the alveoli of the first molars, the bullae are small, and the molars rather narrow with M^3 just over half the length of M^2. Using characters of the dentition, palatal ridges and karyotype, Niethammer (1977) concluded that *N. niviventer* and *N. eha* (placed by him in *Maxomys*) were more distantly related to *Rattus* than are *Bandicota* and *Nesokia*.

The genus was reviewed briefly by Musser (1981b) who has been followed with respect to the allocation of names, but there has been no detailed revision and many of the species limits remain uncertain. Several species cannot yet be diagnosed in a way that allows all marginal populations to be satisfactorily allocated, as is apparent from the table.

Niviventer niviventer White-bellied rat

Mus (*Rattus*) *niviventer* Hodgson, 1836b: 234; Katmandu, Nepal.
Epimys lepcha Wroughton, 1916a: 429; Chuntang, Sikkim, 5350ft.

Range (Map 148). Himalayas from Pakistan to Sikkim, at altitudes of 2100–3200m in Nepal (Abe, 1977).

Variation. The following subspecific name has been proposed:
Rattus n. monticola Ghose, 1964: 193; Palmajua, Darjeeling Dist, W Bengal, India, c. 2250m.

In Sikkim the available specimens (n=6) all have a prominent dark grey patch on the anterior thorax, extending as a median stripe to the abdomen.

Remarks. This species was extended by Ellerman & Morrison-Scott (1951) to include also the forms listed here in *N. confucianus*, and by Niethammer & Martens (1975) to include *N. fulvescens*, as discussed under these species. Its status and nomenclature remain very uncertain; the type locality, Kathmandu, is considerably lower than the altitudinal range reported by Abe (1977).

Niviventer confucianus

Mus confucianus Milne-Edwards, 1872a: 93; Moupin, Sichuan, China.
Epimys zappeyi Allen, G. M.,1912b: 225; Washan, Sichuan, 9000ft.
Rattus mentosus Thomas, 1916d: 643; Hkamti, upper Chindwin, Burma, 500ft.
? *R. culturatus* Thomas, 1917b: 198; Mt Arizan, Taiwan, 8000ft.
? *R. bukit champa* Robinson & Kloss, 1922: 96; Langbian Peaks, S Annam, Vietnam.
R. elegans Shih, 1931: 7; Yao Shan, Guangdong, China.
R. niviventer yushuensis Wang, S. & Zheng, 1981: 3 (and 8); Jiegu, Yushu County, Qinghai Prov., China, 3500–4000m.
R. n. naoniuensis Zhang & Zhao, 1984; Naoniu Mt, Taoin County, Jilin, NE China.

Range (Map 148). N China to N Burma, N Thailand and Vietnam; Hainan, and perhaps Taiwan if *culturatus* is conspecific.

Variation. The following subspecific names have been proposed (but most are extralimital to the Indomalayan Region):
Mus c. sacer Thomas, 1908h: 6; Chefoo, Shandong, China.
M. c. luticolor Thomas, 1908a: 45 (and 1909a: 972); Yenanfu, Shaanxi, China.
Epimys c. canorus Thomas, 1911f: 690; Wenhsien country, S Gansu, China.
Rattus c. chihliensis Thomas, 1917b: 199; Imperial Tombs, 65 miles E of Peking, China.
R. c. littoreus Cabrera, 1922: 167; Foochow, Fujian, China.
R. c. lotipes Allen, 1926: 11; Nodoa, Hainan.
R. c. yaoshanensis Shih, 1930a: 6; Loshiang and Kutchen, Guangxi, China.

Map **148**. *Niviventer* (part)

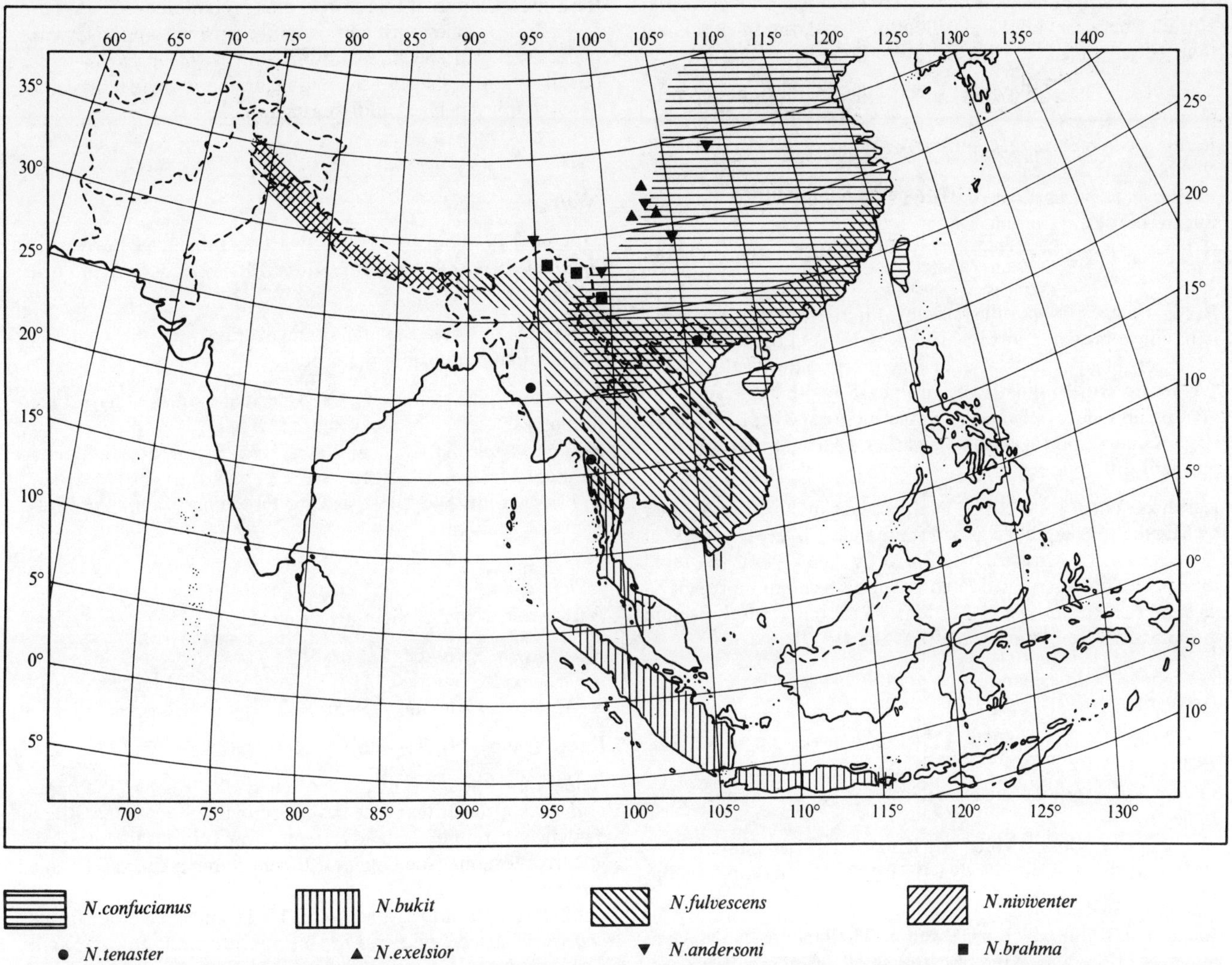

R. c. sinianus Shih, 1931: 3; Yao Shan, Guangdong, China.

The pelage is lighter, more yellowish brown in north China, darker, greyish brown in the southern forms. The tail is usually dark above to the tip but is white distally in a minority of specimens. In China spines are present in the pelage only in summer. The form on Taiwan, *culturatus*, is substantially larger and darker than mainland animals and its inclusion in *N. confucianus* is tentative.

Ecology. This is an abundant and adaptable species in most of China with a wide habitat range from forest to cultivation.

Remarks. The southern boundary of this species and its discrimination from *N. niviventer, N. tenaster* and *N. fulvescens* remain very uncertain.

Niviventer tenaster

Epimys tenaster Thomas, 1916e: 425; Mt Muleyit, Tenasserim, Burma, 5000–6000ft.

Range (Map 148). S Burma, Assam and Vietnam, montane.

Remarks. This form was included as a 'distinct, large race' of *Rattus cremoriventer* by Ellerman (1949). It was treated as a form of *R. niviventer* by Musser (1973b), but later (1981b) considered to be either a southern, montane outlier of *N. confucianus* or a distinct species. Its relationship with all neighbouring forms still requires clarification.

Niviventer fulvescens Chestnut rat

Mus fulvescens Gray, 1847b: 18; Nepal.

M. caudatior Hodgson, in Horsfield, 1849: 203; Nepal.

M. cinnamomeus Blyth, 1859: 294; pre-occupied by *M. cinnamomeus* Pictet, 1844.

M. octomammis Gray, 1863b: 10.

? *Leggada jerdoni* Blyth, 1863b: 350; Sikkim.

M. huang Bonhote, 1905a: 19 (and 1906a: 387); Kuatun, Fujian, China.

= *Rattus flavipilis* Shih, 1930c: 2.

Mus ling Bonhote, 1905a: 19 (and 1906a: 388); Chungfengling, Fujian.

Epimys lepidus Miller, 1913b: 20; Bok Pyin, S Tenasserim, Burma.

E. gracilis Miller, 1913b: 21; Mt Mulaiyit, Tenasserim, Burma.

E. orbus Robinson & Kloss, 1914a: 228; Khao Nawng, Bandon, S Thailand, 3500ft.

Rattus blythi Kloss, 1917d: 8; Schwegyin, Tenasserim, Burma.

R. wongi Shih, 1931: 6; Yao Shan, Guangdong, China.

Range (Map 148). Himalayan foothills from N Pakistan east to Assam, S China (east to Fujian), Thailand and Indochina. At altitudes between 1000 and 2500m in Nepal, but mostly lower than *N. niviventer*.

Variation. The following subspecific names have been proposed:
Rattus blythi mekongis Robinson & Kloss, 1922: 96; Pak Mat, Mekong R, Laos, 18°53'N.
R. huang vulpicolor Allen, 1926: 14; Namting R, Yunnan-Burma border.
R. flavipilis minor Shih, 1930c: 7; Kutchen, Loshiang, Guangxi, China.

Recognition. In Nepal this species differs from *N. niviventer* in having a browner dorsum at all ages, longer tail (usually over 120% of head and body), a narrower skull (reflected in zygomatic width and in the braincase width relative to its height), and more opisthodont upper incisors (Abe, 1977). Both species have spines absent or rather sparse, probably with seasonal differences.

Remarks. The form *orbus* in Thailand was included in *N. rapit* by Marshall (1976), but was considered more likely to represent *N. fulvescens* by Musser (1981b) and by Abe (1983). In Nepal this species has been considered to be only a colour morph of *N. niviventer* (Neithammer & Martens, 1975) but several discrete differences were demonstrated by Abe (1977).

Niviventer bukit

Mus bukit Bonhote, 1903b: 125; Bukit Besar, Jalor, Malaya, 2500ft.
Epimys jerdoni pan Robinson & Kloss, 1914a: 229; Samui I, NE Malaya.
E. j. marinus Kloss, 1916b: 50; Koh Chang (I), Thailand.
Rattus lepturus besuki Sody, 1931a: 214; Ongkop-ongkop, Idjen, E Java, 1850m.

Range (Map 148). S Burma and S Thailand to N Malaya, Sumatra, Java and Bali; also the small islands of Con Son (Vietnam), Chang (Thailand) and Samui (Malaya).

Variation. The following subspecific names have been proposed:
Rattus b. treubii Robinson & Kloss, 1919b: 376; Tjibodas, Mt Gede, W Java, 5000ft.
R. b. temmincki Kloss, 1921a: 233; Banjoewangi, Besoeki, E Java.
R. b. condorensis Kloss, 1926b: 358; Condor I, S Vietnam.
R. b. baturus Sody, 1932b: 334; Mt Agung, Bali, 1500m.
R. b. lepturoides Sody, 1934: 174; Mt Lawu, central Java, 2000m.
R. b. jacobsoni Bartels, 1937b: 121; Mt Tanggamoes, Lampoengs, S Sumatra, c. 700m.
R. b. lieftincki Chasen, 1939b: 208; Atang Poetar, Atjeh, N Sumatra, 1000m.

In Malaya the dorsal pelage is dull greyish brown with more yellowish brown on the flanks; on Java the pelage of the dorsum, and especially of the flanks, is slightly brighter orange-brown, indistinguishable from that of some *N. fulvescens* (e. g. from Darjeeling, W Bengal). On Sumatra the pelage is intermediate in colour between that in Malaya and Java.

Remarks. This species is very similar to *N. fulvescens* and was considered conspecific with it by Abe (1983). However, this was based only upon specimens from Thailand without consideration of topotypical *bukit* from Malaya nor any from the Sunda Islands. It therefore seems preferable to retain these as separate species until more comprehensive comparisons are made. Malayan specimens differ from *N. fulvescens* in duller dorsal pelage, lacking any strong orange-brown tone, more densely spiny pelage, and shorter tail.

Niviventer rapit

Mus rapit Bonhote, 1903b: 123; Mt Kinabalu, N Borneo.
Epimys fraternus Robinson & Kloss, 1916: 273; Korinchi Peak, W Sumatra, 4700ft.

Range (Map 149). Malaya (known only from the Cameron Highlands), Sumatra and Borneo.

Variation. The following subspecific names have been proposed:
Rattus r. cameroni Chasen, 1940a: 176; Cameron Highlands, Pahang, Malaya, c. 5000ft.
R. r. atchinensis Sody, 1941a: 295; Pajatoong Kalan, Atjeh, N Sumatra, 2000m.

Niviventer coninga Swinhoe's rat

Mus coninga Swinhoe, 1864a: 185; Taiwan.
= *M. coxinga* Swinhoe, 1871: 636 (unjustified emendation).
= *M. coxingi* Thomas, 1892c: 939 (unjustified emendation).

Range (Map 149). Taiwan and perhaps N Burma.

Remarks. Although now more often spelled *coxingi*, the name *coninga* is spelled thus twice in the original publication, with no evidence therein of an inadvertent error. It is not therefore an 'incorrect original spelling' and should stand as the valid name (article 32d of Code).

Ellerman & Morrison-Scott (1951) included the forms *andersoni, zappeyi* and *moi* in this species (as *Rattus coxingi*). These forms are all very different from *N. coninga* and are here allocated to *Niviventer andersoni, N. confucianus* and *Maxomys moi* respectively.

Niviventer lepturus

Mus lepturus Jentink, 1880b: 17; Mt Gede, W Java.

Range (Map 149). W & C Java.

Variation. The following subspecific names have been proposed:
Rattus l. fredericae Sody, 1931a: 212; Mt Salak, W Java, 2000m.
R. l. maculipectus Sody, 1934: 173; Mt Tjereme, W Java, 2600m.

Niviventer brahma

Epimys brahma Thomas, 1914c: 232; Anzong Valley, Mishmi Hills, Assam, India, 6000ft.

Range (Map 148). N Assam, Burma and W Yunnan (Gong *et al.*, 1989).

Remarks. A detailed account of this species was given by Musser (1970d, 1973a).

Map **149**. *Niviventer* (part)

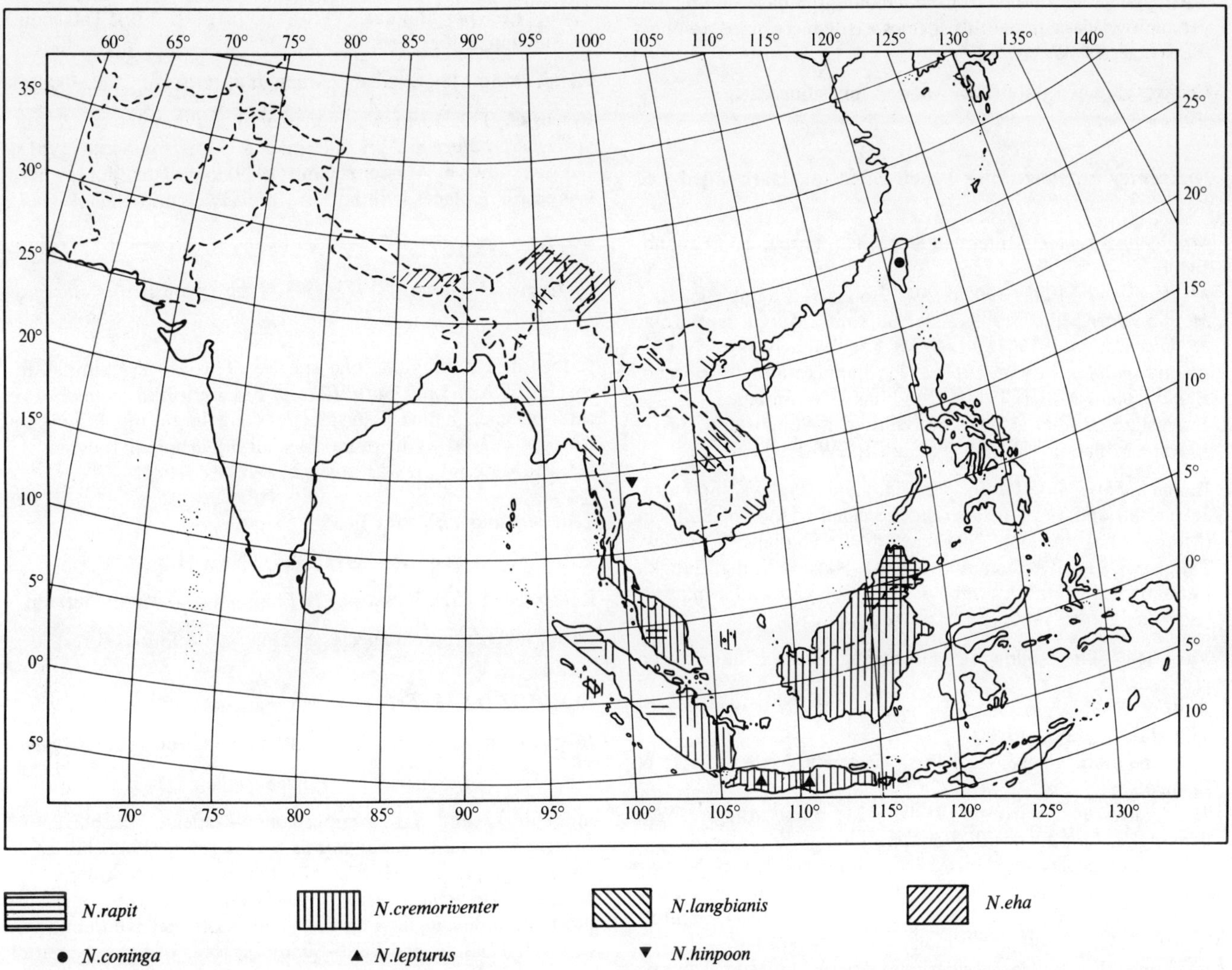

Niviventer eha Smoke-bellied rat

Epimys eha Wroughton, 1916a: 428; Lachen, Sikkim, 8800ft.

Range (Map 149). Himalayas from Nepal (west to about 84°E) to N Burma and Yunnan; at altitudes of 2600–3700m in Nepal.

Variation. The following subspecific name has been proposed: *Rattus e. ninus* Thomas, 1922a: 404; Kiuchiang-Salween divide, Yunnan, China, 28°N, 11000ft.

Some specimens have a patch of brown in the centre of the abdomen, perhaps associated with glandular staining but also associated with absence of grey hair bases.

Ecology. In E Nepal they occur in wet forest dominated by *Abies spectabilis* at the upper edge of the altitudinal range (Gregori & Petrov, 1976).

Remarks. Although clearly distinct and much smaller, this species is most closely related to *N. brahma.*

Niviventer langbianis

Rattus cremoriventer langbianis Robinson & Kloss, 1922: 90; Langbian Peaks, Annam, Vietnam.
R. indosinicus Osgood, 1932: 307; Chapa, Tonkin, Vietnam.

Range (Map 149). E Assam and parts of Burma, Thailand, Laos and Vietnam; up to 2800m in Burma.

Variation. The following subspecific name has been proposed: *Rattus indosinicus vientianensis* Bourret, 1942b: 29; Vientianne region, Laos.

Recognition. From sympatric *N. fulvescens* and from *N. cremoriventer* to the south by duller dorsal pelage, showing bright orange-brown only on the flanks. From *N. cremoriventer* also by longer ears (19-25mm), large bullae, long incisive foramina, extending between the first molars, and four, not five, roots on M^1.

Remarks. Included as a subspecies of *R. cremoriventer* by Marshall (1976) and by Dao & Cao (1990) but considered a distinct species by Musser (1981b).

Niviventer hinpoon Limestone rat

Rattus hinpoon Marshall, 1976: 459; Saraburi Prov., Kaengkhoi Dist, Thailand, 14°35'N, 101°8'E, 200m.

Range (Map 149). Central Thailand.

Recognition. Most similar to *N. bukit* but the small size, short tail (equal to head and body), dull colour, grey-based venter and very spiny pelage make this species easily recognized amongst the species of *Niviventer*.

Ecology. Known only from wooded limestone cliffs.

Niviventer cremoriventer Pencil-tailed rat, Dark-tailed tree rat

Mus cremoriventer Miller, 1900d: 144; Trang, S Thailand, 3000ft.
M. flaviventer Miller, 1900a: 204; Jimaja I, Anamba Is.
M. gilbiventer Miller, 1903a: 35; Sullivan I, Mergui Is, Burma.
M. kina Bonhote, 1903b: 124; Mt Kinabalu, N Borneo.
Epimys spatulatus Lyon, 1911a: 111; Lamukotan I, W Borneo.
E. barussanus Miller, 1911b: 26; Nias I, W Sumatra.
E. mengurus Miller, 1911b: 27; Billiton I, Sumatra.
E. solus Miller, 1913b: 22; Teratau I, SW Thailand.

Range (Map 149). Peninsular Thailand, Malaya, Sumatra, Java, Bali and Borneo; also on the small islands of Sullivan (Mergui Is); Adang, Tarutau (SW Thailand); Buraw, Langkawi, Penang, Tioman (Malaya); Nias, Belitung, Bangka (Sumatra); Jemaja (Anamba Is); Lamukotau and Malawali (Borneo). Up to 1900m in Malaya.

Variation. The following subspecific names have been proposed:
Rattus c. cretaceiventer Robinson & Kloss, 1919b: 377; Tjibodas, W Java, 4500ft.
R. c. malawali Chasen & Kloss, 1932: 32; Mallewalle I, N Borneo.
R. c. sumatrae Bartels, 1937b: 123; Bandjarnegri, Wai Semangka, Lampoengs, S Sumatra.

Five subspecies were recognized and described by Musser (1973b). Only that on Nias I (*N. c. barussanus*) is clearly distinct, by virtue of its large size: hind feet 30–34, greatest length of skull 38–40, upper molar row (at alveoli) 6.7–7.5. The form on Bali was also considered distinctive by Musser, being larger, paler and more spiny than those available from Java, but specimens for comparison were not available from eastern Java.

Ecology. Nocturnal and more arboreal than most species of *Niviventer*, mainly in secondary forest, but also occasionally at ground level.

Remarks. The Indochinese form *langbianis* was treated as a subspecies of *N. cremoriventer* by Marshall (1976) but it can be distinguished by several features justifying specific rank (Musser, 1973b, 1981b).

Niviventer andersoni

Epimys andersoni Thomas, 1911a: 4 (and 1911b: 171); Omei-San, Sichuan, China.

Range (Map 148). Shaanxi, S Gansu, Sichuan, Yunnan and E Tibet, at altitudes of 1800–3000m.

Remarks. This form was considered conspecific with '*Rattus coxingi*' (= *N. coninga*) from Taiwan by Ellerman & Morrison-Scott (1951), but was shown by Musser & Chiu (1979) to be a distinctive species, sympatric in places with both *N. confucianus* and *N. excelsior*.

Niviventer excelsior

Epimys excelsior Thomas, 1911a: 4 (and 1911b: 170); Tatsienlu, W Sichuan, China.

Range (Map 148). Sichuan, China, at altitudes of 2300–3000m.

Remarks. This form was considered a synonym of '*Rattus niviventer confucianus*' by Ellerman & Morrison-Scott (1951), but was shown by Musser & Chiu (1979) to be a distinct species, sympatric in places with both *N. confucianus* and *N. andersoni*.

CHIROMYSCUS

Chiromyscus Thomas, 1925: 503; type species *Mus chiropus* Thomas.

Indochinese subregion; one species. This is very similar to species of *Niviventer* but differs in having a nail instead of a claw on each hallux. The toe pads are large and the claws strongly curved as in other specialized arboreal rodents. A detailed description was given by Musser (1981b).

Chiromyscus chiropus Fea's tree rat

Mus chiropus Thomas, 1891a: 884; Karin Hills, E Burma.

Range (Map 151). E Burma, N Thailand, Laos and Vietnam.

Ecology. An arboreal species, mainly in deciduous forest.

LEOPOLDAMYS Long-tailed giant rats

Leopoldamys Ellerman, 1947b: 261; type species *Mus sabanus* Thomas.

Indochinese and Sunda subregions; 4 species (Table 230). Treated as a subgenus of *Rattus* by Ellerman (1949) but as a distinct genus by Musser (1981b) who gave a detailed diagnosis. It is characterized by very large size, very long tail, short, sleek pelage, long narrow skull, short incisive foramina, very small bullae, and high squamosal roots of the zygomatic arches. In all species there are a few flattened spines in the pelage.

A detailed comparison of *L. sabanus* and *L. edwardsi* in Malaya was made by Yong (1970), including observations of morphology, breeding, ecology, karyology and serology, and supporting their recognition as distinct species.

Leopoldamys sabanus

? *Echimys macrourus* Jentink, 1879d: 97; ? Sumatra. Originally described as an *Echimys* in the belief that it came from Surinam (Husson, 1963).
Mus sabanus Thomas, 1887b: 269; Mt Kinabalu, Borneo.
Mus vociferans Miller, 1900d: 138; Trang, peninsular Thailand, 1000ft.
M. strepitans Miller, 1900a: 207; Siantan I, Anamba Is.
M. fremens Miller, 1902a: 154; Singkep I, Lingga Is.
M. stridens Miller, 1903a: 28; Tioman I, E Malaya.
M. stridulus Miller, 1903a: 29; Bentinck I, Mergui Is, Burma.
M. matthaeus Miller, 1903a: 29; St Matthew I, Mergui Is, Burma.
M. lucas Miller, 1903a: 30; St Luke I, Mergui Is, Burma.
M. masae Miller, 1903a: 32; Tana Masa I, Batu Is, W Sumatra.
M. balae Miller, 1903a: 33; Tana Bala I, Batu Is, W Sumatra.
Epimys nasutus Lyon, 1911a: 104; Panebangan I, W Borneo.
E. luta Miller, 1913b: 18; Laut I, SE Borneo.

Table 230. The species of *Leopoldamys*.

	sabanus	*edwardsi*	*neilli*	*siporanus*
Range	Indochina, Sunda	S China-Malaya, Sumatra	Thailand	Mentawai Islands
Head & body	180–275	215–255	217	292
Tail (% of h & b)	120–180%	120–165%	124%	113%
Hind feet	42–52	45–55	42	52
Dorsal pelage with spines	+	+	0	+
Flanks	Chestnut	Dull brown	Dull brown	As dorsum
Tail with discrete white tip	0	0	0	+ (30–70%)
Tail slightly paler below and distally	(+)	0	+	–
Skull: greatest length	c.54	c.58	c.52	58,59
Upper molar row	9.1–10.0	c.10.3	8.3–8.8	10.2,10.6
Bullae	c.5.0	c.6.5	c.4.5	c.5.5
Chromosomes (2n/FN)	42/54,56	42/56	44/52	

E. stentor Miller, 1913b: 19; James I, Mergui Is, Burma.
E. ululans Robinson & Kloss, 1916: 272; Siolak Daras, Korinchi Valley, SW Sumatra.

Range (Map 150). Indochina and Thailand to Malaya; Sumatra, Java and Borneo; also on most of the Mergui Islands (Burma) and on the islands of Terutau (Thailand), Langkawi, Penang, Tioman (Malaya); Tanahmasa, Tanahbala, Mansalar, Tuangku, Bangkaru (Sumatra); Singkep (Lingga Is), Siantan, Jemaja (Anamba Is); Bunguran (N Natuna Is); Laut, Panebangan (Borneo).

In Malaya it occurs generally below 1000m, but up to 1300m in the absence of *L. edwardsi*.

Variation. The following subspecific names have been proposed:
Mus vociferans lancavensis Miller, 1900e: 188; Langkawi I, Malaya.
M. v. tersus Thomas & Wroughton, 1909b: 535; Teratau I, SW Thailand.
Epimys v. insularum Miller, 1913b: 19; Domel I, Mergui Is, Burma.
E. v. clarae Miller, 1913b: 20; Clara I, Mergui Is, Burma.
E. v. herberti Kloss, 1916a: 25; Pak Jong, E Thailand.
Rattus v. tapanulius Lyon, 1916b: 209; Tapanuli Bay, NW Sumatra.
R. fremens mansalaris Lyon, 1916a: 450; Mansalar I, W Sumatra.
R. f. tuancus Lyon, 1916a: 451; Tuangku I, Banjak Is, W Sumatra.
R. s. mayapahit Robinson & Kloss, 1919b: 375; Tjibodas, W Java, 5000ft.
R. s. revertens Robinson & Kloss, 1922: 95; Daban, Phanrang Prov., S Annam, Vietnam.
R. s. bunguranensis Chasen, 1935b: 17; Bunguran I, N Natuna Is.
R. s. dictatorius Chasen, 1940a: 165; Penang I, Malaya.
R. s. salanga Chasen, 1940a: 166; Junk Seylon I, SW Thailand.
R. s. heptneri Dao, 1961a: 295; forest of Cam-Ha, Hon-Gay, N Vietnam.

Ecology. In all types of forest; arboreal but mainly in the lower strata and also active on the ground.

Leopoldamys edwardsi

Mus edwardsi Thomas, 1882: 587; mountains of W Fujian (probably Kuatun), China.
M. ciliata Bonhote, 1900d: 879; Mt Inas, Perak, Malaya.
M. gigas Satunin, 1903: 562; near Lunganfu, Chodsigou Valley, Sichuan, China.
Epimys listeri Thomas, 1916c: 406; Pashok, Darjeeling, India, 3500ft.
E. setiger Robinson & Kloss, 1916: 271; Barong Bharu, Barisan Range, Korinchi, W Sumatra, 4000ft.
Mus melli Matschie, in Mell, 1922: 26; Mahutze Shan, Guangdong, China.

Range (Map 150). E Himalayan foothills (west to Darjeeling district) to S China (S Gansu, Sichuan, Yunnan, Guangdong, Fujian), N Thailand and N Vietnam; isolated montane populations in Malaya (above 750m) and Sumatra; Hainan I.

Ecology. In Malaya this species occurs in montane forest above the range of *L. sabanus*.

Variation. The following subspecific names have been proposed:
Rattus listeri garonum Thomas, 1921f: 27; Tura, Garo Hills, Assam, 1400ft.
R. e. milleti Robinson & Kloss, 1922: 94; Dalat, Langbian Plateau, Annam, Vietnam.
R. e. hainanensis Xu & Yu, 1985: 130; Bawangling, Hainan I, China. (A small form.)

Leopoldamys neilli

Rattus neilli Marshall, 1976: 485; Kaengkhoi Dist, Saraburi Prov., Thailand, 14°35'N, 101°8'E, 200m.

Range (Map 150). Central Thailand.

Ecology. Known from two localities, both on wooded limestone cliffs.

Leopoldamys siporanus

Mus siporanus Thomas, 1895d: 670; Sipora I, Mentawai Is, Sumatra.
M. soccatus Miller, 1903a: 30; N Pagi I, Mentawai Is.

Map **150**. *Leopoldamys*

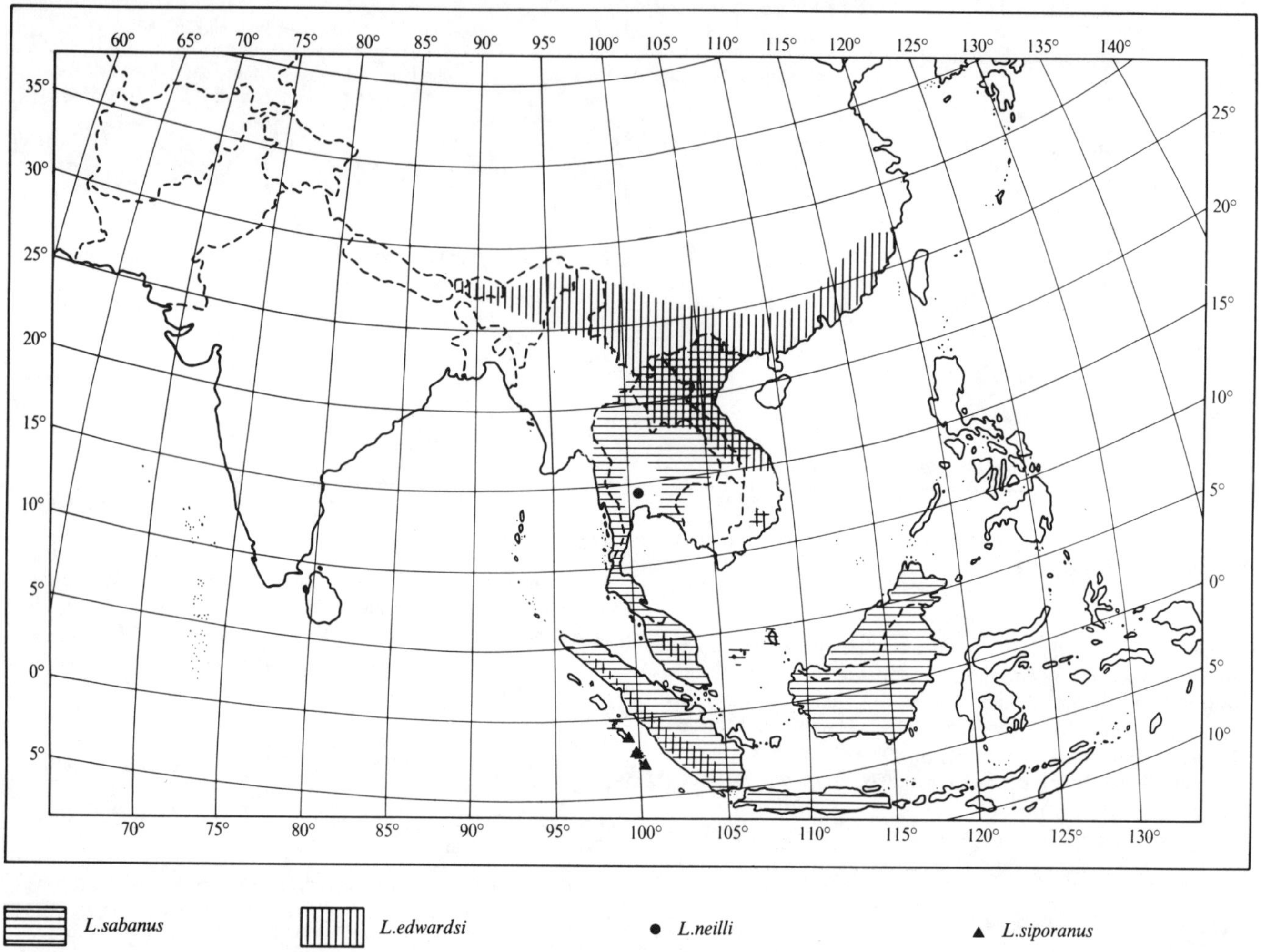

Range (Map 150). Siberut, Sipura, N and S Pagai Islands, Mentawai Is.

Remarks. Included in *Rattus sabanus* by Chasen (1940a) but considered a distinct species by Musser (1981b). Like most Mentawai endemics it is melanic in comparison with other members of its genus.

MAXOMYS

Maxomys Sody, 1936a: 55; type species *Mus bartelsii* Jentink.

Indochinese and Sunda subregions, Sulawesi; 17 species (Tables 231–233). The arrangement used here follows the review by Musser *et al.* (1979). Earlier, e. g. by Misonne (1969), the genus *Maxomys* was held to include the species here included in *Niviventer*. Two additional, undescribed species occur in Sulawesi (Musser, 1987a).

Maxomys bartelsii

Mus bartelsii Jentink, 1910b: 69; Mt Pangrango-Gede, W Java.

Range (Map 151). West and central Java, east to Mt Slamat at c. 109°E.

Variation. The following subspecific names have been proposed:

Rattus b. tjibuniensis Sody, 1933a: 430; Mt Patuha, near Bandung, Tjiboeni, W Java, 1350m.

R. b. obscuratus Bartels, 1938: 323; Mt Slamat, central Java.

These cannot be retained as subspecies on the basis of larger samples. About a third of individuals have an indistinct, cinnamon-coloured gorget between the fore legs. Only a small minority have spines in the pelage (Van Peenen *et al.*, 1974).

Remarks. This species was reviewed by Van Peenen *et al.* (1974).

Maxomys rajah

Mus rajah Thomas, 1894c: 451; Mt Batu Song, Sarawak, Borneo.

M. pellax Miller, 1900d: 147; Trang, peninsular Thailand.

M. lingensis Miller, 1900a: 206; Lingga I, Sumatra.

Epimys similis Robinson & Kloss, 1916: 272; Siolak Daras, Korinchi Valley, W Sumatra, 3100ft.

Range (Map 151). Malaya, Sumatra, Borneo; and on the smaller islands of Durian (Riau Is) and Sirasan (S Natuna Is).

Table 231. The species of *Maxomys* on the Asian mainland (± the Greater Sunda Islands).

	rajah	*surifer*	*whiteheadi*	*inas*	*moi*
Range	Malaya, Sumatra, Borneo	Indochina-Gt Sunda Is	Malaya, Sumatra, Borneo	Malaya	Vietnam, Laos
Head & body	165–225	155–205	100–125	125–160	140–210
Tail (% of h & b)	c.100%	c.100%	85–90%	c.100%	c.105%
Hind feet	35–43	32–47	24–28	31–33	37–44
Dorsal pelage with flat spines	+	+	+	+	0
Flanks with sharp transition between dorsal and ventral colour	+	+	0	(+)	+
Ventral pelage with grey bases	0	0	Light	+	0
Dark mid-ventral streak	+	+/0	0	0	0
Skull: greatest length	41–49	39–48	29–38	36–40	41–46
Upper molar row	6.9–8.1	5.8–7.5	5.1–6.2	6.0–6.7	5.8–6.6
Incisive foramina: length	5.8–7.5	5.5–7.3	4.0–5.4	5.2–6.4	6.5–7.9
Posterior ends of nasals extend behind premaxilla/frontal suture	+	0	0	0	0

Table 232. The species of *Maxomys* confined to the Greater Sunda Islands.

	bartelsii	*alticola*	*ochraceiventer*	*baeodon*	*hylomyoides*	*inflatus*	*panglima*	*pagensis*
Range	Java	Borneo	Borneo	Borneo	Sumatra	Sumatra	Palawan	Mentawai Is
Head & body	125–180	140–175	140–170	125–140	115–130	160–195	165–220	170–220
Tail (% of h & b)	c.100%	110–120%	c.100%	90–95%	c.95%	c.100%	c.100%	c.90%
Hind feet	31–36	34–37	30–35	25–28	28–31	37–43	40–43	41–45
Dorsal pelage with spines	Few, soft	+	+	+	0	+	+	+
Sharp transition between dorsal and ventral colour	+	0	0	0	0	0	+	0
Ventral pelage with grey bases	+	0	+	+	+	0	0	0
Dark mid-ventral streak	0	0	0	0	0	+	0	0
Skull: greatest length	35–40	37–42	35–40	33–35	31–34	41–47	45–49	45–49
Upper molar row	5.0–5.8	5.4–6.1	5.4–6.2	4.1–5.0	5.9–6.5	6.3–7.2	7.1–7.8	6.8–7.9
Incisive foramina	5.8–7.0	5.4–6.4	4.7–5.7	4.2–4.9	4.6–5.5	6.9–8.2	7.0–8.0	6.1–7.9
Posterior ends of nasals extend behind premaxilla/frontal sutures	+	+	0	0	0	+	+	+

Variation. The following subspecific name has been proposed: *Rattus r. hidongis* Kloss, 1921d: 122; Sirhassen I, S Natuna Is.

Table 233. The species of *Maxomys* on Sulawesi.

	musschenbroekii	*hellwaldii*	*dollmani*	*wattsi*
Head & body	130–155	180–215	150–175	165–185
Tail (% of h & b)	c.90%	80–100%	c.120%	c.80%
Hind feet	29–34	39–46	36–37	32–37
Dorsal pelage with flat spines	+	0	0	0
Buff spot on chest	+	0	0	0
Tail with white tip	0	0	+ (c.40%)	0
Skull: greatest length	34-39	45–50	41–44	40–44
Upper molar row	6.1–6.9	6.5–7.2	6.0–6.3	6.4–6.8
Incisive foramina	4.9–6.2	6.7–7.4	6.0–6.4	5.7–6.4
Posterior ends of nasals extend behind premaxilla/frontal sutures	0	+	0	0

Maxomys surifer

Mus surifer Miller, 1900d: 148; Trang, peninsular Thailand.
M. anambae Miller, 1900a: 205; Jimaja I, Anamba Is.
M. luteolus Miller, 1903a: 36; St Matthew I, Mergui Is, Burma.
M. umbridorsum Miller, 1903a: 37; Loughborough I, Mergui Is, Burma.
M. casensis Miller, 1903a: 38; Chance I, Mergui Is, Burma.
M. bentincanus Miller, 1903a: 38; Bentinck I, Mergui Is, Burma.
M. domelicus Miller, 1903a: 39; Domel I, Mergui Is, Burma.
M. catellifer Miller, 1903b: 464; Mansalar I, W Sumatra.
M. carimatae Miller, 1906a: 59; Karimata I, Borneo.
M. serutus Miller, 1906a: 59; Serutu I, Karimata Is, Borneo.
Epimys perflavus Lyon, 1911a: 108; Laut I, SE Borneo.
E. saturatus Lyon, 1911a: 109; Panebangan I, W Borneo.
E. ubecus Lyon, 1911a: 109; Sebuko I, SE Borneo.
E. ravus Robinson & Kloss, 1916: 272; R Kumbang, Korinchi, Sumatra, 4700ft.
Rattus lingensis banacus Lyon, 1916a: 449; Bangkuru I, Banjak Is, W Sumatra.
R. l. mabalus Lyon, 1916a: 449; Tana Masa I, Batu Is, W Sumatra.
R. l. antucus Lyon, 1916a: 449; Tuangku I, Banjak Is, W Sumatra.
R. l. pinacus Lyon, 1916a: 450; Pinie I, Batu Is, W Sumatra.
R. rajah siarma Kloss, 1919a: 75; Sikawtur, 40 miles NW of Raheng, Thailand.
R. r. koratis Kloss, 1919b: 376; Lat Bua Kao, E Thailand.

Map **151**. *Maxomys* (part), *Chiromyscus*

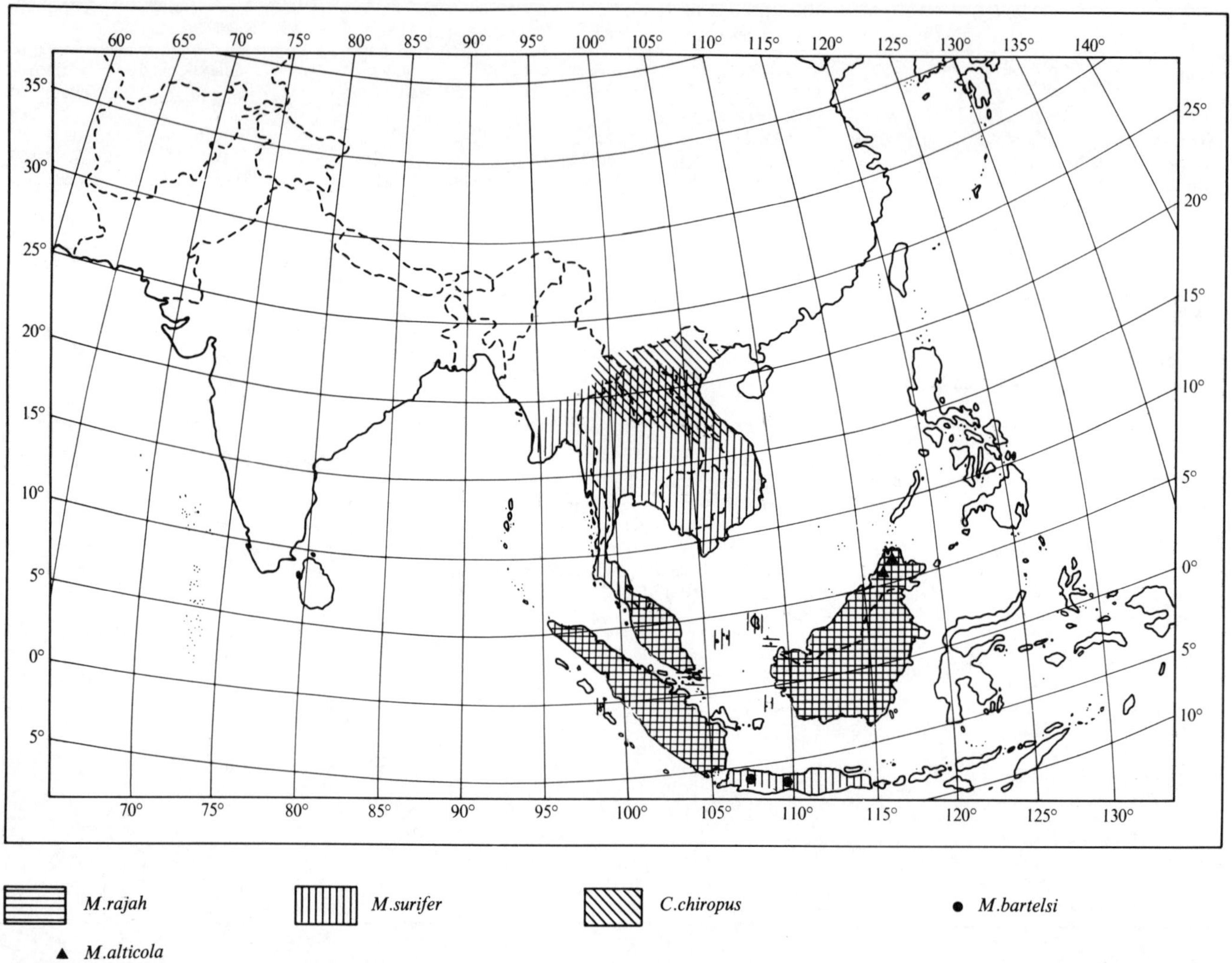

R. r. kramis Kloss, 1919b: 377; Koh Kram, inner Gulf of Thailand.
Rattus bandahara Robinson, 1921: 235; Mt Kinabalu, N Borneo.
R. rajah verbeeki Sody, 1930b: 130; Gedangan, Res. Semarang, central Java, 65m.

Range (Map 151). S Indochina (S Burma, Thailand, Laos, Cambodia, Vietnam), to Malaya; Sumatra, Java, Borneo; and on most adjacent islands, north to the Mergui Is.

Variation. The following subspecific names have been proposed:
Mus s. flavidulus Miller, 1900e: 189; Langkawi I, Malaya.
M. s. butangensis Miller, 1900e: 190; Adang I, Butang Is, W Thailand.
M. s. microdon Kloss, 1908: 145; Tioman I, Malaya (preoccupied by *M. microdon* Peters, 1852).
= *Epimys s. binominatus* Kloss, 1915b: 223.
M. s. flavigrandis Kloss, 1911: 119; E Perhentian I, E Malaya.
M. s. grandis Kloss, 1911: 119; Great Redang I, E Malaya.
M. s. leonis Robinson & Kloss, 1911a: 170; Singapore I.
Epimys s. pemangilis Robinson, 1912: 593; Pemanggil I, E Malaya.
E. s. aoris Robinson, 1912: 594; Aor I, E Malaya.
E. s. manicalis Robinson & Kloss, 1914a: 230; Koh Pennan (I), NE Malaya.
E. s. spurcus Robinson & Kloss, 1914a: 230; Koh Samui (I), NE Malaya.
E. s. finis Kloss, 1916b: 51; Klong Menao, SE Thailand.
E. s. changensis Kloss, 1916b: 52; Koh Chang (I), Thailand.
E. s. kutensis Kloss, 1916b: 52; Koh Kut (I), Thailand.
E. s. pelagius Kloss, 1916b: 53; Koh Rang (I), Thailand.
E. s. eclipsis Kloss, 1916b: 53; Koh Kra (I), Thailand.
E. s. connectens Kloss, 1916b: 53; Koh Mak (I), Thailand.
E. s. tenebrosus Kloss, 1916b: 54; Koh Klum (I), Thailand.
Rattus s. solaris Sody, 1934: 170; Mt Gedeh, W Java, 1350m.
R. s. puket Chasen, 1940a: 169; Junk Seylon I, W Thailand.
R. s. telibon Chasen, 1940a: 170; Telibon I, W Thailand.
R. s. muntia Chasen, 1940a: 170; Muk (=Muntia) I, Trang, W Thailand.
R. s. pidonis Chasen, 1940a: 171; Pipidon I, W Thailand.
R. s. natunae Chasen, 1940a: 173; Bunguran I, N Natuna Is.

It is doubtful if the forms in the main segments of the range, i. e. the mainland and the main Sunda islands, are clearly distinguishable, although those on Java and Borneo appear smaller than those on Sumatra and the mainland.

Most of the small-island forms differ only slightly from the mainland (or large-island) counterparts. The amount of buff suffusion on the venter is variable; some forms have patches of grey on the venter and in one, *eclipsis*, the venter is entirely grey, with a dull brown gorget. *M. s. aoris* from Aor I, Malaya has the naso-lachrymal capsules on the rostrum more developed than in others, but much smaller than in *M. inflatus*. On Muk I, W Thailand (*M. s. muntia*) a high proportion of individuals are melanic (Hill, 1960: 97).

Recognition. From *M. rajah* distinguishable only by colour and pattern: the dorsum is a brighter orange- or reddish brown, always speckled with an admixture of black-tipped and brown-tipped hairs, whereas in *M. rajah* the colour, although varying from buff to a darker greyish brown, is more uniform in any one individual. In *M. rajah* there is never an orange-brown gorget on the throat and there is usually a large mid-ventral greyish brown streak, most prominent in adult males (although a smaller streak is present in some *M. surifer*).

Ecology. Mainly terrestrial, living in burrows in forests, but also found in rice-fields and gardens, especially where these are adjacent to forest.

Remarks. The name *surifer* and its synonyms were included in '*Rattus rajah*' by Ellerman & Morrison-Scott (1951) but these were shown to represent distinct species by Hill (1960), Yong (1972) and subsequent authors, supported by karyological and serological studies.

Maxomys whiteheadi

Mus whiteheadi Thomas, 1894c: 452; Mt Kinabalu, N Borneo.
M. asper Miller, 1900d: 145; Trang, peninsular Thailand.
Rattus klossi Bonhote, 1906b: 9; Johore, Malaya.
Mus batamanus Lyon, 1907a: 654; Batam I, Riau Is.
M. mandus Lyon, 1908b: 644; R Mandau, tributary of Siak R, E Sumatra.
Epimys batus Miller, 1911b: 27; Pinie I, Batu Is, W Sumatra.
Rattus melinogaster Cabrera, 1920: 211; Mt Kinabalu, Borneo.
R. melanurus Shamel, 1940a: 76; Miang Besar I, Sangkulirang Bay, E Borneo.

Range (Map 152). Peninsular Thailand and Malaya, Sumatra, Borneo; and the small islands of Batam (Riau Is), Pinie (W Sumatra), Serasan (S Natuna Is), Malawali, Balembangan, Banggi (N Borneo).

Variation. The following subspecific names have been proposed:
Epimys w. perlutus Thomas, 1911e: 205; Balingean, Sarawak, Borneo.
R. w. piratae Chasen, 1940a: 181; Mallewalle I, N Borneo.
R. w. subitus Chasen, 1940a: 182; Sirhassen I, S Natuna Is.
R. w. coritzae Sody, 1941a: 299; Riam, Kotawaringin, SW Borneo, 470m.

There is no constant difference between the main segments of the range in Malaya, Sumatra and Borneo (Robinson & Kloss, 1918a: 49–50). Those on the island of Banggi (N Borneo) are dull, lacking the warm orange-brown tones seen elsewhere. The form *piratae* from the adjacent island of Malawali is dark above and has the venter much darker grey than elsewhere; the incisive foramina are very wide and extend back level with the anterior edges of the molar alveoli.

Ecology. A terrestrial rat of primary and secondary forest and scrub, occasionally in adjacent fields.

Map **152**. *Maxomys* (part)

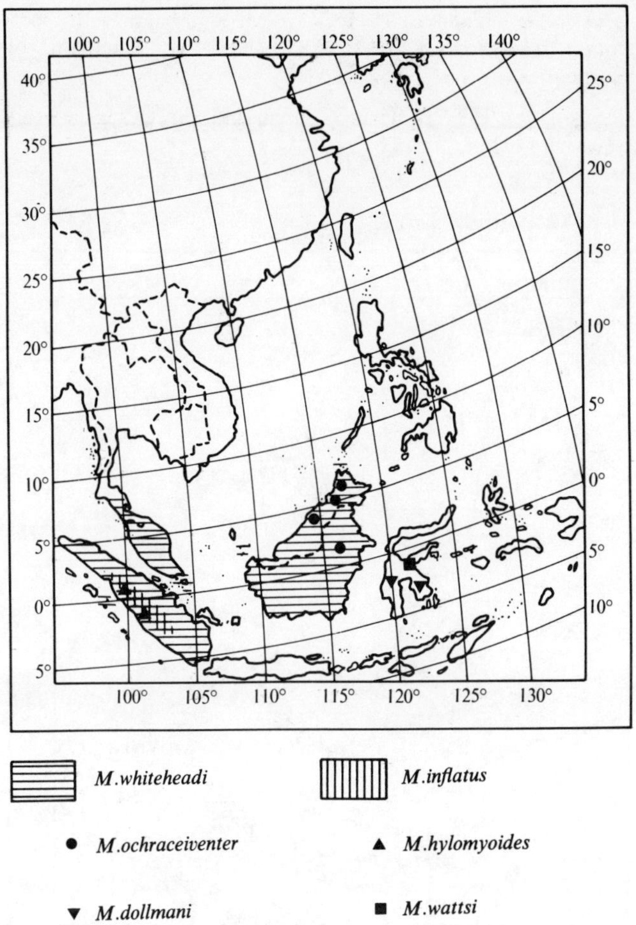

Remarks. The name *Rattus sakeratensis* Gyldenstolpe, 1917 has been associated with this species because of a mismatched skin and skull. The skin, which is *Rattus losea*, has been designated as the lectotype (Marshall, 1977a).

Maxomys alticola Mountain spiny rat

Mus alticola Thomas, 1888b: 408; Mt Kinabalu, N Borneo, 8600ft.

Range (Map 151). N Borneo, known only from Mt Kinabalu and Mt Trus Madi, between 1000 and 3400m.

Variation. The following subspecific name has been proposed: *Rattus a. kinabaluensis* Chasen, 1937a: 81; Mt Kinabalu, N Borneo, 5500ft.

Maxomys ochraceiventer Chestnut-bellied spiny rat

Mus ochraceiventer Thomas, 1894c: 451; Mt Kinabalu, N Borneo, below 3000ft.
Rattus perasper Shamel, 1940a: 76; Mt Batu, E Borneo.

Range (Map 152). N Borneo, up to 1700m on Mt Kinabalu; apparently sympatric with *M. alticola* around 1000m without any intergradation.

Recognition. Distinguishable from *M. alticola* by uniform buffy venter; hind feet, with claw, under 35mm; ear under 20mm; and narrow auditory bullae (width under 3.7mm) (Medway, 1977a); and from *M. surifer* by short tail and buffy venter.

Maxomys baeodon

Mus baeodon Thomas, 1894c: 452; Mt Kinabalu, N Borneo.
Rattus trachynotus Cabrera, 1920: 212; Mt Kinabalu, N Borneo.

Range (Map 153). N Borneo (Sabah and Sarawak), between 900 and 1400m.

Map **153**. *Maxomys* (part)

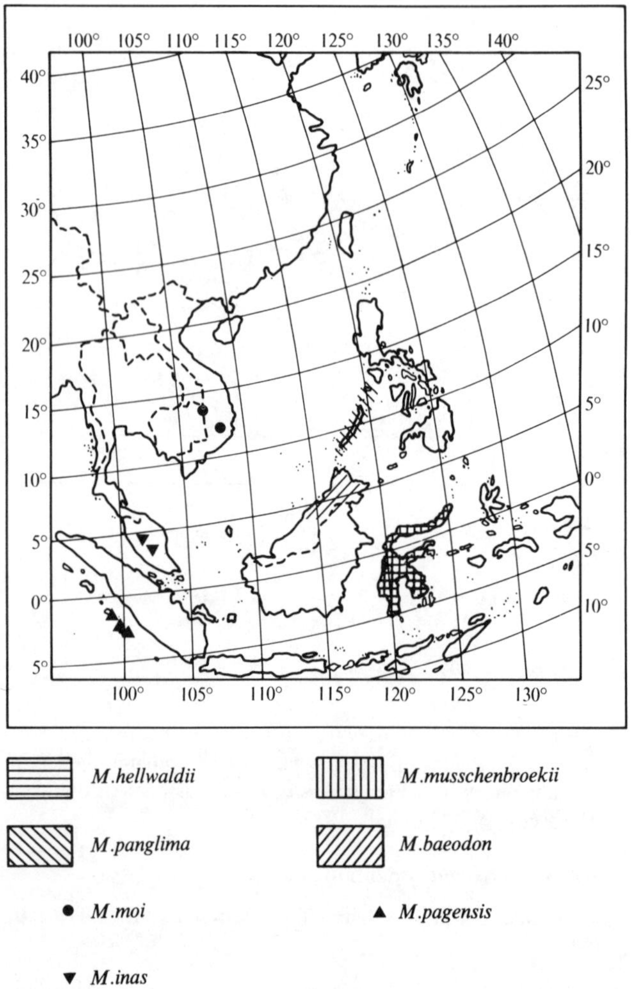

Maxomys inas

Mus inas Bonhote, 1906b: 9; Mt Inas, Perak, Malaya.

Range (Map 153). Peninsular Malaya, montane, mostly above 1000m.

Ecology. Mainly terrestrial.

Remarks. This form was treated as a subspecies of the Bornean '*Rattus alticola*' by Chasen (1940a) but was shown to be a distinctive species by Medway (1965a).

Maxomys hylomyoides

Epimys hylomyoides Robinson & Kloss, 1916: 273; Mt Korinchi, W Sumatra, 7300ft.

Range (Map 152). Sumatra.

Maxomys inflatus

Epimys inflatus Robinson & Kloss, 1916: 273; Mt Korinchi, Sumatra, 4700ft.

Range (Map 152). Sumatra.

Recognition. Very similar to *M. surifer* and *M. rajah* with which it appears to be sympatric, but distinguishable externally by diffuse orange-brown on chest extending forwards to the mouth, not forming a discrete gorget as in many *M. surifer*. The skull is easily recognized by the unique extent of the inflation of the naso-lachrymal capsule on each side of the rostrum, forming a thin-walled hemispherical bubble of bone behind the premaxilla/maxilla suture.

Maxomys moi

Rattus moi Robinson & Kloss, 1922: 95; Arbre Broye, Langbian Mts, Annam, Vietnam.

Range (Map 153). Vietnam and Laos.

Remarks. Although this form was included in '*Rattus coxingi*' (*Niviventer coninga* as used here) by Ellerman & Morrison-Scott (1951), these are very different both externally and cranially and there is no doubt that *moi* should be included in *Maxomys* as used here.

Maxomys panglima

Rattus panglima Robinson, 1921: 235; Palawan I, Philippines.
Rattus palawanensis Taylor, 1934: 416; Sir J Brooke Point, Palawan.

Range (Map 153). Palawan (Philippines) and adjacent smaller islands of Balabac, Busuanga and Culion.

Ecology. In primary and secondary forest at all altitudes.

Maxomys pagensis

Mus pagensis Miller, 1903a: 39; S Pagi I, Mentawai Is.

Range (Map 153). Siberut, Sipura, N and S Pagai Is, Mentawai Is.

Remarks. Externally this species is scarcely distinguishable from *M. surifer* except by the presence of a median dark line on each hind foot.

Maxomys musschenbroekii

Mus musschenbroekii Jentink, 1879a: 10; Menado, N Sulawesi.
Rattus aspinatus Tate & Archbold, 1935a: 9; Ile Ile, N Sulawesi.

Range (Map 153). Sulawesi, throughout, in lowland and montane forest.

Variation. The following subspecific names have been proposed:
Rattus m. tetricus Miller & Hollister, 1921a: 68; Gimpoe, SW of L Lindoe, central Sulawesi.
R. m. lalawora Sody, 1941a: 305; Tangka Salocco, Mengkoka Mts, SE Sulawesi, 2000m.

Ecology. Live on the ground in primary forest (Musser, 1984).

Remarks. The form *aspinatus* was shown to be based upon juvenile *M. musschenbroekii* by Musser (1969a).

Maxomys hellwaldii

Mus hellwaldii Jentink, 1879a: 11; Menado, N Sulawesi.

Range (Map 153). Sulawesi, throughout, in lowland forest.

Variation. The following subspecific names have been proposed:
Rattus h. localis Miller & Hollister, 1921a: 74; Laboea Sore (N of Parigi), Sulawesi.
R. h. cereus Miller & Hollister, 1921a: 74; Toli Toli, NW Sulawesi.
R. h. griseogenus Sody, 1941a: 305; Masembo, Mengkoka Mts, SE Sulawesi, 550m. (The spelling *griseogenys* is an unjustified emendation.)

Maxomys dollmani

Rattus hellwaldi dollmani Ellerman, 1941: 218; Rantekaroa, Quarles Mts, central Sulawesi.

Range (Map 152). Central and SE Sulawesi, in montane forest.

Remarks. This form was shown to be specifically distinct from *M. hellwaldii* by Musser (1969c).

Maxomys wattsi

M. wattsi Musser, 1991: 20; Tambusisi Damar, on slopes of Mt Tambusisi, C Sulawesi, 1°38'S, 121°23'E, 4700ft.

Range (Map 152). Known only from the vicinity of the type locality in central Sulawesi, between 1425 and 1820m.

BERYLMYS

Berylmys Ellerman, 1947b: 261; type species *Epimys manipulus* Thomas.

Indochinese subregion, Malaya and NE Sumatra; 4 species (Table 234). Ellerman (1949 etc.) included only *B. manipulus* and *B. berdmorei* in *Rattus* (*Berylmys*) and allocated *bowersii* (including *mackenziei*) to *Rattus* (*Stenomys*). The genus *Berylmys* as used here was diagnosed and the species revised by Musser & Newcomb (1983). They are large, terrestrial, burrowing rats with distinctive dense, grey dorsal pelage, the hairs rather uniform in length giving a brush-like texture. In *B. manipulus* and *B. berdmorei* the incisors are unusual amongst rodents in being mainly white with just a slight trace of yellow pigment. They occur in forest, mainly in hilly country, and are not usually damaging to agriculture.

Berylmys manipulus

Epimys manipulus Thomas, 1916c: 413; Kampat, Kabaw Valley, 20 miles W of Kindat, W Burma, 600ft.

Range (Map 154). Assam, N and central Burma, and Yunnan (Yang & Wu, 1979).

Variation. One subspecific name has been proposed:
Rattus m. kekrimus Roonwal, 1948: 386; Kekrima (= Chekrima), c. 10 miles SE of Kohima, Naga Hills District, Assam, India, 5390ft.

Berylmys berdmorei

Mus berdmorei Blyth, 1851a: 173; Mergui, Burma.

Range (Map 154). S Burma, Yunnan (Yang & Wu, 1979), N and SE Thailand, Cambodia, N Laos, S Vietnam; also Con Son I (S Vietnam).

Variation. Two subspecific names have been proposed:
Epimys b. magnus Kloss, 1916b: 57; Klong Manao, SE Thailand.
E. b. mullulus Thomas, 1916c: 413; Thagata, Mulaiyit Range, Tenasserim, Burma.

Berylmys mackenziei

Epimys mackenziei Thomas, 1916c: 410; Haingyan, 50 miles W of Kindat, Chin Hills, Burma.

Range (Map 154). Assam, Burma and S Vietnam.

Variation. One subspecific name has been proposed:
Rattus m. feae Thomas, 1916c: 412; Thagata, Mulaiyit Range, Tenasserim, Burma.

Berylmys bowersii

Mus bowersii Anderson, 1879: 304; Hotha, Kakhyen Hills, W Yunnan, China, 4500ft.
M. latouchei Thomas, 1897b: 113; Kuatun, NW Fujian, China.

Table 234. The species of *Berylmys*.

	manipulus	*berdmorei*	*mackenziei*	*bowersii*
Range	Burma etc.	Indochina	N Indochina	Indochina, Malaya, Sumatra
Head & body	135–185	190–255	155–245	230–300
Tail (% of h & b)	c.105%	c.80%	c.110%	105–110%
Hind feet	32–39	36–45	42–51	50–59
Dorsal pelage with flat spines	Few	Few	Abundant	Abundant
White tail-tip	Short	0	50–60%	c.20%
Teats: pairs (pectoral/abdominal)	3/2	3/2	3/2	2/2
Skull: greatest length	36–41	38–50	40–50	51–63
Upper molar row	6.2–6.7	6.6–8.0	8.1–9.8	8.8–10.6
Upper incisors	Pale yellow	Pale yellow	Mid orange	Pale orange
Bullae (% of skull length)	14–15%	17%	13–14%	13–14%

Map **154**. *Berylmys, Haeromys, Lenothrix*

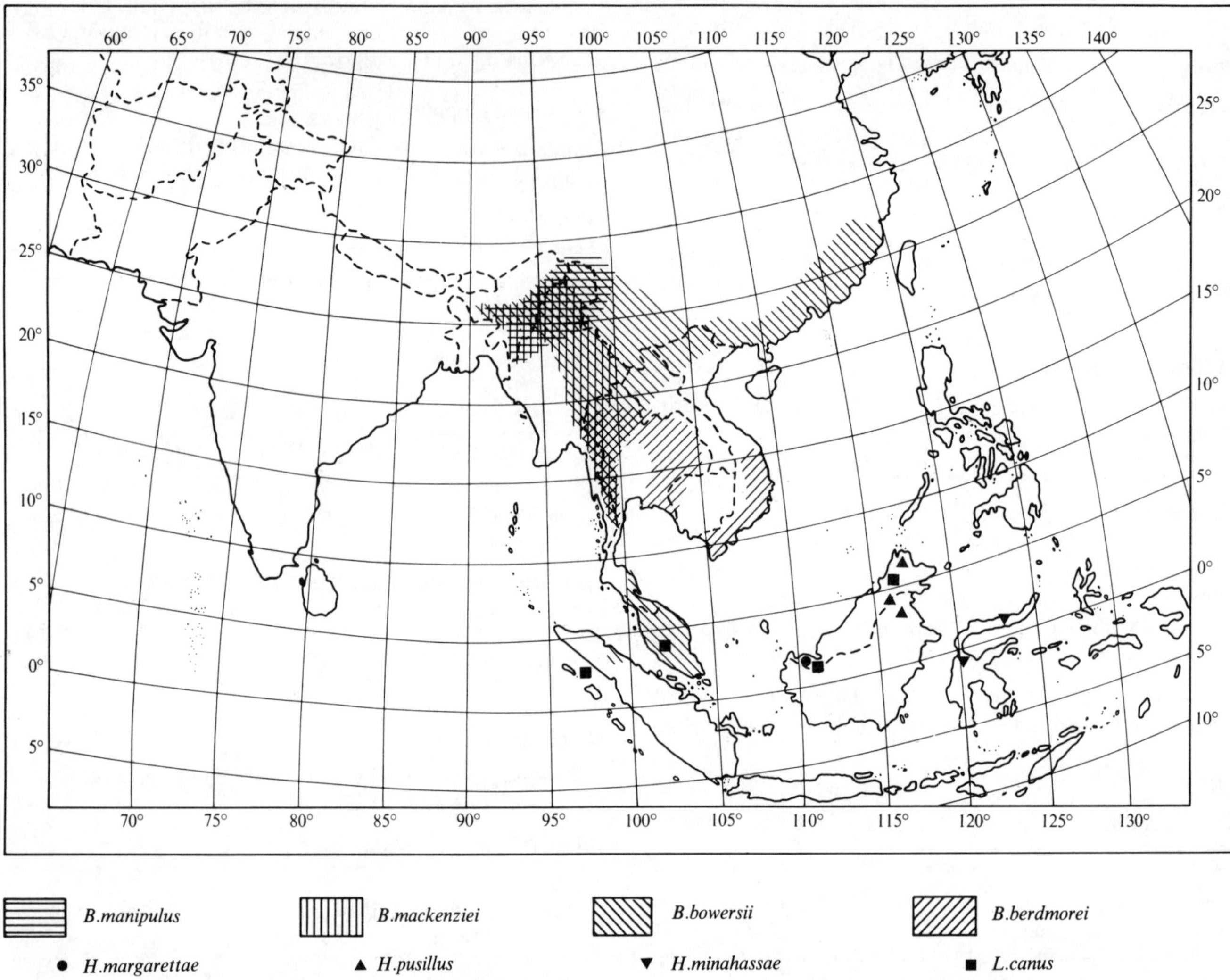

M. ferreocanus Miller, 1900d: 140; Trang, peninsular Thailand, 3000ft.
Rattus kennethi Kloss, 1919c: 81; Sikawtur, 40 miles NW of Raheng, Thailand.
R. wellsi Thomas, 1921f: 26; Mawphlang, Khasi Hills, Assam, 5500ft.

Range (Map 154). Assam to S China and much of Indochina; Malaya and N Sumatra; mainly montane, between 1000 and 2500m but down almost to sea-level in N Vietnam.

Variation. The following subspecific names have been proposed:
Rattus b. lactiventer Kloss, 1919c: 80; Sikawtur, 40 miles NW of Raheng, Thailand.
R. b. totipes Dao, 1966: 438; Ha Tinh, N Vietnam, 18°19'N, 105°40'E, 80m.

The extent of white on the tail varies considerably, a short white tip (c. 20%) being frequent in Assam and Burma and a longer white tip (up to 50%) frequent in Malaya; in most individuals from eastern Indochina and S China the tail is uniformly brown.

Table 235. The species of *Haeromys*.

	margarettae	*pusillus*	*minahassae*
Range	Borneo	Borneo	Sulawesi
Head and body	c.70	c.55–75	c.70–78
Tail	c.143	c.120	c.125–140
Hind feet	18	15–17	19
Ventral pelage with grey hair-bases	0	0	+
Feet: dorsal surface entirely pale (P) or brown with white digits (B)	P	P	B
Skull: greatest length	25.3	21.7–22.8	23–24
Upper molar row	3.5	3.0–3.3	3.2–3.3
Incisive foramina: length	3.7	3.0–3.3	2.9–3.5

HAEROMYS

Haeromys Thomas, 1911e: 207; type species *Mus margarettae* Thomas.

Borneo and Sulawesi; 3 species (Table 235). The characters of the genus were described and compared with other genera by Musser & Newcomb (1983: 557–560). There are further species, as yet undescribed, on Palawan, Philippines (Musser & Newcomb, 1983) and in the highlands of Sulawesi (Musser, pers. comm.).

Haeromys margarettae Greater ranee mouse

Mus margarettae Thomas, 1893a: 346; Penrissen Hills, West Sarawak.

Range (Map 154). Borneo, known only from the type locality in W Sarawak.

Haeromys pusillus Lesser ranee mouse

Mus margarettae pusillus Thomas, 1893d: 232; Mt Kinabalu, N Borneo.

Range (Map 154). Borneo, in Sabah, Sarawak and E Kalimantan; up to 900m on Mt Kinabalu.

Remarks. This form was included in *H. margarettae* by Payne *et al.* (1985) on the grounds that some individuals bridged the size gap between the two forms. However the figures given by them still leave the type of *H. margerettae* substantially larger than the remaining specimens and it seems preferable to retain *H. pusillus* as a species (as done by Chasen, 1940a and Medway, 1965b) until further specimens are available.

Haeromys minahassae Sulawesi pygmy tree rat

Mus minahassae Thomas, 1896a: 247; Rurukan, Minahassa, N Sulawesi.

Range (Map 154). North and central Sulawesi, in lowland forest.

Ecology. Observed to feed on small fig seeds (Musser, 1979: 438).

PITHECHEIR

Pithechir [*sic*] Müller, 1839: 12 (*nomen nudum*).
Pithecheir Lesson, 1840: 264; type species *P. melanurus* Lesson.

Malaya and Java; 2 species (Table 236). A distinctive genus with no very close relatives. They have short, broad hind feet with very large pads as in other arboreal murids and a short nail on each hallux, but the tail is unique in being prehensile although this is not apparent from its structure. The large inflated auditory bullae are also unusual. The characters of the genus were described and its relationships discussed by Musser & Newcomb (1983: 543–547). The recognition of two species was advocated by Muul & Lim (1971) and supported by Musser & Newcomb (1983).

Table 236. The species of *Pithecheir*.

	parvus	*melanurus*
Range	Malaya	Java
Head & body	148–180	159–176
Tail	172–215	190–213
Hind feet	26–30	28–32
Skull: greatest length	39–44	43–46
Upper molar row (alveoli)	7.4–8.2	8.3–9.4
Length of bullae as % of greatest length of skull	c.25%	c.20%

Pithecheir melanurus Red tree rat

Pithechir [*sic*] *melanurus* Müller, 1840: un-numbered table (*nomen nudum*); Java.
Pithecheir melanurus Lesson, 1840: 265; ? Java. Presumed by Lesson to be from Bengal or Sumatra; Ellerman & Morrison-Scott (1955) suggested restriction to W Sumatra but since there are no certain records of occurrence in Sumatra it will better serve stability to accept Java, as recorded by Chasen (1940a). It was listed under Java, and not under Sumatra, by Müller (1840).

Range (Map 155). Java. Records from Sumatra are probably erroneous (Musser, 1982b).

Pithecheir parvus Monkey-footed rat

P. melanurus parvus Kloss, 1916e: 250; Bukit Kutu, Selangor, Malaya, 3400ft.

Range (Map 155). Malaya, up to 1000m.

Remarks. Originally described as a subspecies of *P. melanurus* on the basis of only one immature individual, but shown to justify specific rank on the basis of larger samples (Muul & Lim, 1971; Musser & Newcomb, 1983). Further details were provided by Lim & Muul (1975).

Map **155**. *Sundamys, Pithecheir, Kadarsanomys*

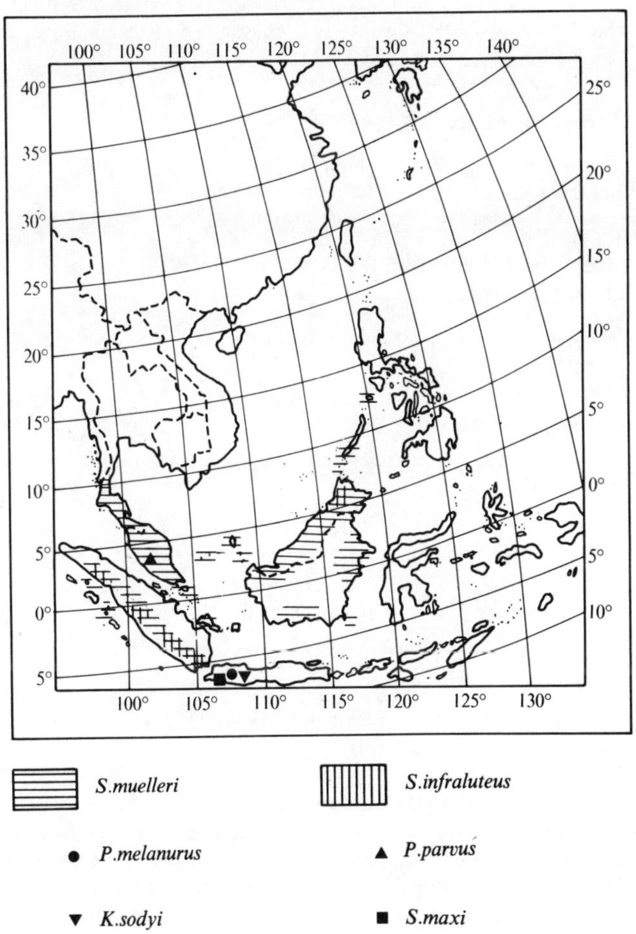

KADARSANOMYS

Kadarsanomys Musser, 1981c: 5; type species *Rattus canus sodyi* Bartels.

Java; one species. An arboreal rat with a nail on each hallux. This genus was considered by Musser & Newcomb (1983) to be distantly related to *Rattus*, rather than to the other arboreal *Rattus*-like genera.

Kadarsanomys sodyi

Rattus canus sodyi Bartels, 1937a: 45; Pasir Datar, Tjiparoj, Mt Pangrango-Gede, W Java, 1000m.

Range (Map 155). Known alive only from the type locality in W Java; subfossil specimens have been found in central and eastern Java (Musser & Newcomb, 1983: 547).

Ecology. This species has arboreal adaptations and appears to be associated especially with bamboo.

LENOTHRIX

Lenothrix Miller, 1903b: 466; type species *L. canus* Miller.

Malaya, parts of Borneo and Tuangku I (Sumatra); one species. Although treated as a subgenus of *Rattus* by Ellerman (1949 etc.) this arboreal rat has been recognized as a distinctive genus by subsequent authors (Misonne, 1969; Medway & Yong, 1976; Musser, 1981b). The dentition shows several primitive features, e. g. presence of a postero-internal cusp (t7) on M^1 and M^2; erect, well separated molar cusps, not overlapping those of the following teeth. Twenty-two characters distinguishing it from *Rattus* were detailed by Musser (1981b).

Lenothrix canus Grey tree rat

L. canus Miller, 1903b: 466, pl. 18; Tuangku I, Banjak Is, Sumatra.

Range (Map 154). Malaya, Borneo (known only from two localities in Sabah and S Sarawak), and the island of Tuangku off NW Sumatra.

Variation. The Malayan form has been described as a subspecies:
Rattus c. malaisia Kloss, 1931b: 105; near Kuala Lumpur, Selangor, Malaya.

Ecology. Occurs in primary and secondary forest, climbing in the lower strata.

Remarks. The form *sodyi* from Java, described as a race of '*Rattus canus*' by Bartels (1937), was shown to be a distinct species by Medway & Yong (1976) and was made the sole member of the genus *Kadarsanomys* by Musser (1981c).

PALAWANOMYS

Palawanomys Musser & Newcomb, 1983: 335; type species *P. furvus* Musser & Newcomb.

Palawan, Philippines; one species. This genus was based mainly upon a number of primitive features of the dentition: M^1 with only one lingual root; molars erect with little overlap; M^1 with cusp t4 (median lingual cusp) well developed and isolated; M^1 with posterior cingulum; M_1 with antero-central cusp; M_2 with posterior lingual cusplet. The pelage is long and soft, c. 18mm on the dorsum, with no trace of spines.

It appears to be distantly related to *Rattus* (Musser & Newcomb, 1983).

Palawanomys furvus

P. furvus Musser & Newcomb, 1983: 335; Mt Mantalingajan, Brooke's Point Municipality, Palawan I, Philippines, 4500ft.

Range (Map 156). Known only from the type locality near the southern end of Palawan I, Philippines.

SUNDAMYS

Sundamys Musser & Newcomb, 1983: 401; type species *Mus muelleri* Jentink.

Sunda subregion except central and E Java; 3 species (Table 237). Large, terrestrial, forest rats with coarse pelage, long dark tails, small auditory bullae and relatively large M^3. Revised in detail by Musser & Newcomb (1983).

Sundamys muelleri

Mus mülleri Jentink, 1880b: 16; Batang Singgalang, Padang Highlands, W Sumatra.
M. validus Miller, 1900d: 141; Khow Sai Dow, Trang, peninsular Thailand.
M. integer Miller, 1901a: 119; Serasan I, Natuna Is.

Map **156**. *Bullimus, Palawanomys, Tryphomys, Limnomys, Abditomys*

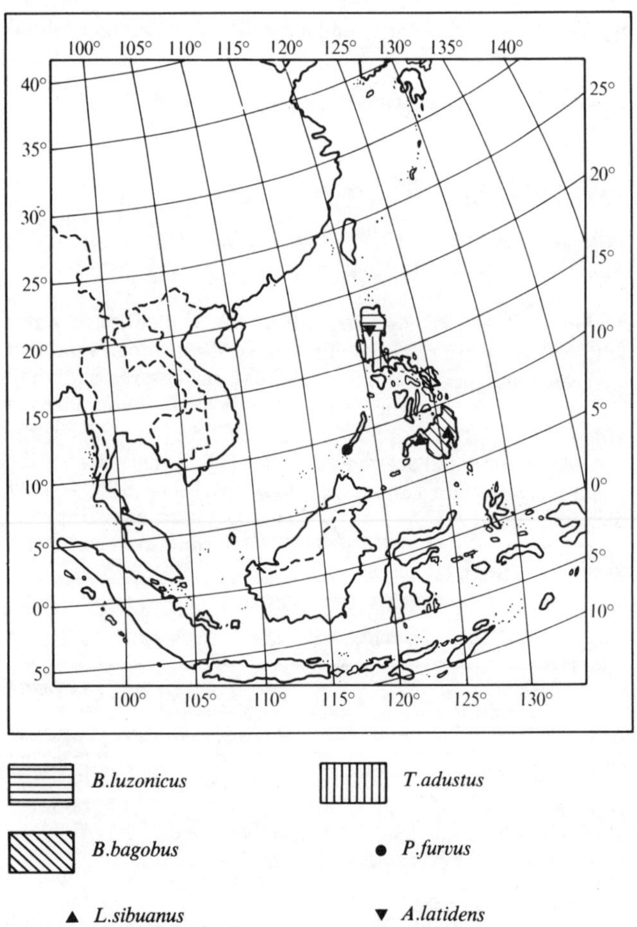

Table 237. The species of *Sundamys*.

	muelleri	*infraluteus*	*maxi*
Range	Malaya, Sumatra, Borneo, Palawan	Borneo, Sumatra	W Java
Head & body	185–300	230–295	220–270
Tail (% of h & b)	100–140%	110–130%	c.120%
Hind feet	35–53	53–59	50–53
Dorsum	Sleek	Shaggy	Normal
Venter	White, light grey or buff	Dark grey, tinged buff	Grey, ± buff wash
Teats: pairs (pectoral/abdominal)	2/2	1/2	2/2
Skull: greatest length	44–62	56–63	56–62
Upper molar row	8.2–11.6	10.5–11.7	10.2–11.7

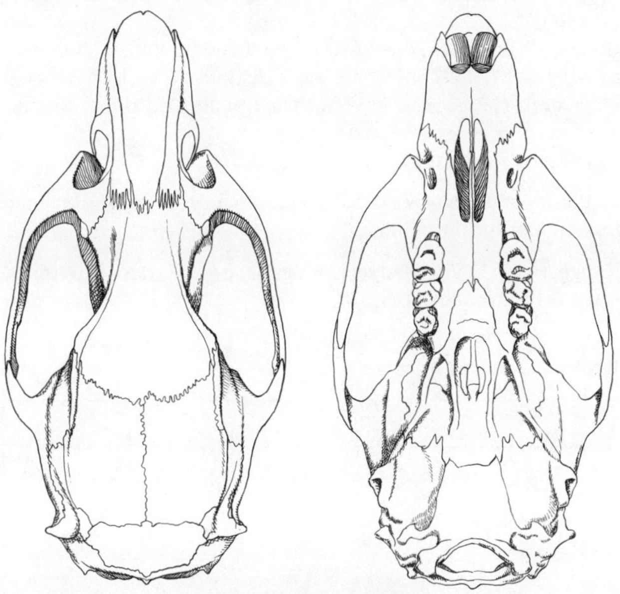

Fig.**34**. *Sundamys muelleri.*

M. firmus Miller, 1902a: 155; Lingga I, SE Sumatra.
M. domitor Miller, 1903b: 461; Mansalar I, W Sumatra.
M. chomboiis Lyon, 1909b: 484; Chombol I, Riau Is, Sumatra.
Epimys sebucus Lyon, 1911a: 102; Sebuku I, SE Borneo.
E. crassus Lyon, 1911a: 103; Lamukotan I, W Borneo.
E. borneanus Miller, 1913b: 15; Telok Karang Tigau, SE Borneo.
Rattus victor Miller, 1913b: 16; near mouth of Rompin R, Pahang, Malaya.
Epimys potens Miller, 1913b: 17; Tuangku I, Banjak Is, W Sumatra.
E. pollens Miller, 1913b: 17; Banka I, SE Sumatra.
E. valens Miller, 1913b: 18; Bankaru I, Banjak Is, W Sumatra.
Rattus virtus Lyon, 1916b: 210; Siak R, near mouth of Gasip R, E Sumatra.
R. balmasus Lyon, 1916a: 447; Tana Bala I, Batu Is, W Sumatra.
R. pinatus Lyon, 1916a: 448; Pini I, Batu Is, W Sumatra.
R. culionensis Sanborn, 1952a: 131; Culion I, Calamian Is, Philippines.

Range (Map 155). Peninsular Burma and Thailand, Malaya, Sumatra, Borneo and Palawan; and on the small islands of Siantan (Anamba Is), most of the Riau Is, Tuangku and Bangkuru (Banjak Is), Mansalar (W Sumatra), Pinie, Tanahmasa and Tanahbala (Batu Is), Banka, Bunguan and Serasan (Natuna Is), Sebuku (SE Borneo), Balembangan and Banggi (N Borneo), Balabac, Culion, Busuanga (Palawan). Up to c. 1000m in Malaya and 1650m in Borneo. Mapped in detail by Musser & Newcomb (1983).

Variation. The following subspecific names have been proposed:
Rattus m. credulus Chasen, 1940a: 162; Bunguran I, Natuna Is.
Mus m. foederis Robinson & Kloss, 1911b: 245; Ulu Temengor, Perak, Malaya.
Epimys m. campus Robinson & Kloss, 1916: 275; Pasir Ganting, W Sumatra.
Rattus validus terempa Chasen & Kloss, 1928b: 36; Siantan I, Anamba Is.
R. m. otiosus Chasen, 1935a: 98; Balembangan I, N Borneo.
R. m. waringensis Sody, 1941a: 289; Riam, Kotawaringin dist, SW Borneo.
R. m. balabagensis Sanborn, 1952a: 131; Mt Balabag, Palawan, Philippines.

The forms on the small islands are not strongly differentiated and only two subspecies are worthy of recognition (Musser & Newcomb, 1983):
S. m. muelleri. Sumatra, Borneo, Palawan and associated islands. Smaller: skull usually 46–54mm, upper molar row usually 8.6-10.3mm.
S. m. validus. Malaya. Larger: greatest length of skull usually 54–59mm, upper molar row usually 9.9–10.7mm.

Sundamys infraluteus

Mus infraluteus Thomas, 1888b: 409; Mt Kinabalu, N Borneo.
Rattus atchinus Miller, 1942: 152; Blangbeke, Aceh Prov., N Sumatra, 6800ft.

Range (Map 155). Sabah (Borneo) and Sumatra; at altitudes of 900–2930m in Borneo and 700–2400m in Sumatra. Marginally sympatric with *S. muelleri* in both Borneo and Sumatra.

Recognition. In the areas of sympatry with *S. muelleri* this species can be recognized by its larger size, especially of the hind feet (over 55mm, with claw; under 50mm in *S. muelleri*), and long, thick, dark, shaggy pelage.

Sundamys maxi

Rattus maxi Sody, 1932a: 157; Patua Mt, Tjiboeni, Bandung, W Java, 1350m.

Range (Map 155). Mountains of W Java between 900 and 1350m.

Remarks. Listed as a subspecies of *Rattus infraluteus* by Chasen (1940a), but treated as a distinct species by Musser & Newcomb (1983).

BULLIMUS

Bullimus Mearns, 1905: 450; type species *B. bagobus* Mearns.

Philippines; 2 species (Table 238). These species were included in the '*xanthurus*' group of *Rattus* (*Rattus*) by Ellerman (1941). Misonne (1969) recognized *Bullimus* as a subgenus of *Rattus* and included species from the Sunda Islands and the Asian mainland. Musser (1982d: 90) treated *Bullimus* as a genus confined to the Philippines.

Table 238. The species of *Bullimus*.

	luzonicus	*bagobus*
Range	Luzon	Mindanao
Head & body	240–290	230–330
Tail	190–210	150–230
Hind feet	46–50	49–53
Dorsal pelage	Soft	Coarse
Venter	Dark grey, buff tips	White, ± pale grey bases
Skull: greatest length	50–54	56–72
Upper molar row	9.0–10.3	9.3–11.0

Bullimus luzonicus

Mus luzonicus Thomas, 1895f: 163; Mt Data, NW Luzon, Philippines, 2460m.

Range (Map 156). N Luzon, Philippines.

Bullimus bagobus

Bullimus bagobus Mearns, 1905: 450: Todaya, Mt Apo, Mindanao, Philippines, 1203m.
Rattus rabori Sanborn, 1952a: 130; Katipunan, Zamboanga, Mindanao, Philippines, 2500ft.

Range (Map 156). Mindanao and Samar Is, Philippines.

Variation. One subspecies has been described:
Rattus b. barkeri Johnson, 1946: 318; Mercedes, 5 miles N of Guinan, SE Samar I, Philippines, 200ft. (Smaller, sparser pelage, grey venter.)

Remarks. This form was treated as a subspecies of '*Rattus luzonicus*' by Sanborn (1954).

LIMNOMYS

Limnomys Mearns, 1905: 451; type species *L. sibuanus* Mearns.

Mindanao, Philippines; one species. *Limnomys* has been treated as a subgenus of *Rattus* (Ellerman, 1941) and as a synonym of *Rattus* (*Bullimus*) by Misonne (1969). Musser (1977b) considered that it should probably be excluded from *Rattus*.

Limnomys sibuanus

L. sibuanus Mearns, 1905: 452; Mt Apo, Mindanao, Philippines, 6600ft.
L. mearnsi Hollister, 1913a: 324; Grand Malindang Peak, W Mindanao, Philippines, 9000ft.

Range (Map 156). Mindanao, Philippines, in montane forest at altitudes of 2000–2700m.

Variation. The diagnostic features of *L. mearnsi* were based upon young individuals of *L. sibuanus* (Musser, 1977b).

Remarks. *Limnomys picinus* Hollister, 1913 from Mindanao, Philippines, was based upon a skin of *Rattus mindorensis* mismatched with a skull of *Limnomys mearnsi*; the skin was designated the holotype of the name *L. picinus* which therefore became a synonym of *Rattus mindorensis* (Musser, 1977b).

TARSOMYS

Tarsomys Mearns, 1905: 453; type species *T. apoensis* Mearns.

Mindanao, Philippines; one species. Included in *Rattus* (*Bullimus*) by Misonne (1969), but listed as a distinct genus by Musser (1981a).

Tarsomys apoensis

T. apoensis Mearns, 1905: 453; Mt Apo, Mindanao, Philippines, 2075m.

Range (Map 157). Highlands of S Mindanao, Philippines, around 2000m.

TRYPHOMYS

Tryphomys Miller, 1910: 399; type species *T. adustus* Miller.

Luzon, Philippines; one species. Included in *Rattus* (*Bullimus*) by Misonne (1969) but treated as a distinct genus by Musser & Newcomb (1983) who gave measurements and illustrations.

Tryphomys adustus

T. adustus Miller, 1910: 399; Haights-in-the-Oaks, Benguet, Luzon, Philippines, 2460m.

Range (Map 156). Benguet, Laguna and Tarlac Provinces, Luzon, Philippines.

Map **157**. *Apomys, Tarsomys*

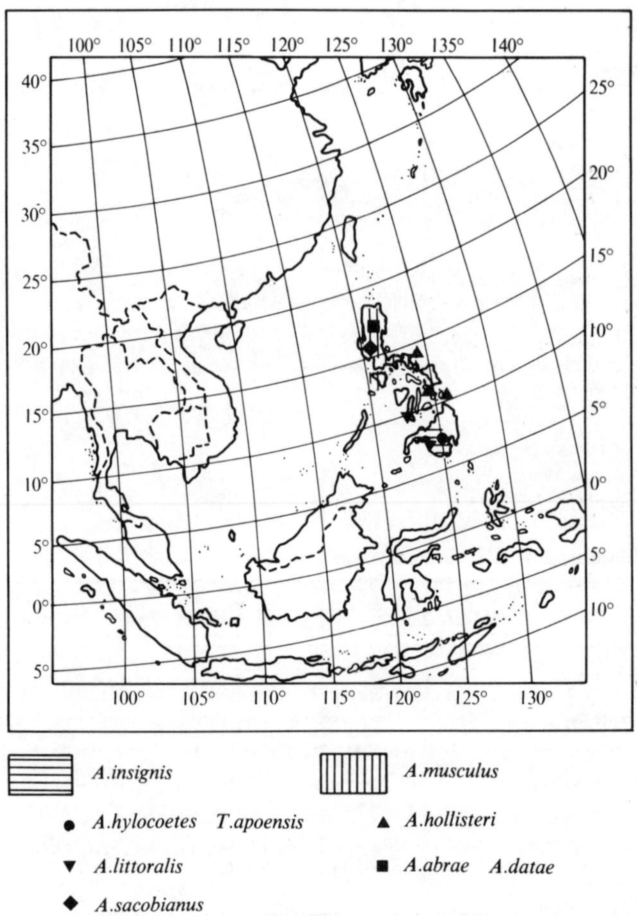

ABDITOMYS

Abditomys Musser, 1982a: 3; type species *Rattus latidens* Sanborn.

Luzon, Philippines; one species. Previously included in *Rattus* but distinguished especially by very wide incisors, rather simple, laminate, but high-crowned, molars and a nail on each hallux. Musser (1982a) concluded that the nail on the hallux is likely to have arisen independently from that in the other genera with this feature, e. g. *Chiropodomys*, all of which have a much larger number of primitive characters than does *Abditomys*. The nail could perhaps be better described as a short, blunt claw since it exceeds the digit and is longer than broad.

Abditomys latidens

Rattus latidens Sanborn, 1952a: 125; Mt Data, Luzon, Philippines, 7500ft.

Range (Map 156). Luzon, known from only a few montane localities.

APOMYS

Apomys Mearns, 1905: 455; type species *A. hylocoetes* Mearns.

Philippines; 8, perhaps fewer, species (Table 239). Treated as a subgenus of *Rattus* by Ellerman (1949) but shown to have a number of distinctive features that exclude it from *Rattus* by Musser (1982c), who reviewed the genus in detail. Some of the species are known from very few localities and specimens, and the delimination of species used here must be considered provisional.

Apomys hylocoetes

A. hylocoetes Mearns, 1905: 456; Mt Apo, Mindanao, Philippines, 6000–6750ft.
A. petraeus Mearns, 1905: 458; Mt Apo, Mindanao, Philippines, 7600ft.

Range (Map 157). Known only from Mt Apo, Mindanao, Philippines, at altitudes of 1800–2300m.

Remarks. This species is sympatric with *A. insignis* but is clearly distinguishable from it by its relatively shorter tail and shorter, broader hind feet.

Apomys insignis

A. insignis Mearns, 1905: 459; Mt Apo, Davao Prov., SE Mindanao, Philippines, 4000–6000ft.
A. bardus Miller, 1910: 402; Mt Bliss, Grand Malindang Mts, E Mindanao, Philippines, 5750ft.

Range (Map 157). Highlands of Mindanao, Philippines, at altitudes of 900–2100m.

Remarks. It is possible that *A. hollisteri* might prove to justify inclusion in *A. insignis*.

Apomys hollisteri

A. microdon Hollister, 1913a: 327; Biga, Catanduanes I, off SE Luzon, Philippines.
= *Rattus hollisteri* Ellerman, 1949: 76 (replacement for *microdon* on the grounds that it is pre-occupied by *Mus microdon* Peters, 1852, a form of *Mastomys* which, like *Apomys*, was included by Ellerman (1949) in *Rattus*. The name *Apomys microdon* is permanently invalid following Article 59(b) of the Code (3rd edition, 1985).

Range (Map 157). The islands of Catanduanes, Leyte and Dinagat, Philippines.

Remarks. This form seems closely related to both *A. musculus* and *A. insignis* and could conceivably be considered conspecific with *A. insignis*.

Apomys musculus

A. musculus Miller, 1910: 403; Baguio, Benguet, Luzon, Philippines, 5000ft.

Range (Map 157). Luzon and Mindoro, Philippines, at altitudes between 1400 and 2000m.

Apomys littoralis

Rattus littoralis Sanborn, 1952a: 134; Bugasan, Cotabato Prov., Mindanao, Philippines, on coastal plain.

Range (Map 157). SW Mindanao and Negros I, Philippines, from sea level to about 1500m.

Variation. Although known from very few specimens, those from Negros are more brightly coloured than those from Mindanao (Musser, 1982c).

Apomys abrae

Rattus abrae Sanborn, 1952a: 133; Abra, Abra Prov., NW Luzon, 3500ft.

Range (Map 157). N Luzon, Philippines, known at altitudes of 900-2500m.

Table 239. The species of *Apomys*.

	hylocoetes	*insignis*	*hollisteri*	*musculus*	*littoralis*	*abrae*	*sacobianus*	*datae*
Range	Mindanao	Mindanao	Catanduanes, Leyte, Dinagat	Luzon, Mindoro	Mindanao, Negros	Luzon	Luzon	Luzon
Head & body	110–115	70–123	90–107	75–125	95–103	120–125	141	140–170
Tail (% of h & b)	105–115%	130–140%	120–140%	80–125%	115–120%	105–110%	106%	c.90%
Hind feet	28–30	30–35	26–31	20–25	23–27	30–34	40	32–36
Upper surface of feet	Brown	Brown	Brown	Buffy white	Brown	Brown	White	Brown
Tail conspicuously paler below	0	0	0	0	0	+	+	(+)
Skull: greatest length	30–32	30–33	27–31	24–27	28–30	33–37	38	37–41
Upper molar row (alveoler)	5.6–5.7	5.5–6.6	5.2–5.7	4.4–4.9	4.9–5.2	6.1–7.0	7.1	6.4–7.5

Apomys sacobianus

A. sacobianus Johnson, 1962: 319; Sacobia R, Clark Air Base, Pampanga Prov., central Luzon, Philippines.

Range (Map 157). Central Luzon.

Remarks. Possibly a large-bodied lowland subspecies of *A. abrae* (Musser, 1982c), distinguished also by paler venter and feet.

Apomys datae

Mus datae Meyer, 1899: 25; Lepanto, N Luzon, Philippines.
A. major Miller, 1910: 402; Haights-in-the-Oaks, Luzon.

Range (Map 157). N Luzon, Philippines, known at altitudes of 1600–2500m.

Remarks. This species is clearly distinguishable from all other *Apomys* by the form of the stapedial and internal maxillary arteries which are as in *Rattus* and most murids, the internal maxillary artery being a continuation of the stapedial, whereas in all other *Apomys* the internal maxillary artery arises from the internal carotid quite independently of the stapedial artery and far anterior to it. This is reflected in the skull by the very small size of the stapedial foramina on the bullae in these other species of *Apomys* (Musser, 1982c).

CARPOMYS

Carpomys Thomas, 1895f: 161; type species *C. melanurus* Thomas.

Luzon, Philippines; 2 species (Table 240). These are average sized rats with slightly hairy tails a little shorter than the head and body; short, wide rostrum; and broad, laminate molars with narrow, closely spaced laminae.

Carpomys melanurus

C. melanurus Thomas, 1895f: 162; Mt Data, NW Luzon, Philippines, 7000–8000ft.

Range (Map 158). NW Luzon, around 2100–1500m.

Carpomys phaeurus

C. phaeurus Thomas, 1895f: 162; Mt Data, NW Luzon, Philippines.

Range (Map 158). NW Luzon.

Table 240. The species of *Carpomys*.

	melanurus	*phaeurus*
Dorsal pelage	Very soft	Normal
Venter	Grey with cream-buff tips	White or buff to roots
Tail: extent of body pelage on base	c.30mm	c.10mm
Tail: hairs obscuring scales for whole length (H); or only towards tip (T)	H	T
Ears	Black	Medium brown
Upper incisors: width	1.5mm (each)	1.1mm (each)
Upper incisors grooved	+ (faintly)	0
Upper molar row	8.8	6.1

BATOMYS

Batomys Thomas, 1895f: 162; type species *B. granti* Thomas.
Mindanaomys Sanborn, 1954: 287 (*M. salomonseni* Sanborn).

Philippines; 3 species (Table 241). A further, undescribed species has been reported on Dinagat I, Philippines (Heaney *et al.*, 1987). Externally these are very similar to species of *Carpomys* but the rostrum is long and slender, and the molars are more rounded, with broader laminae.

Batomys granti

B. granti Thomas, 1895f: 162; Mt Data, Luzon, Philippines.

Range (Map 158). Known only from Mt Data, NW Luzon, Philippines.

Remarks. IUCN: I.

Map **158**. *Carpomys, Batomys, Anonymomys*

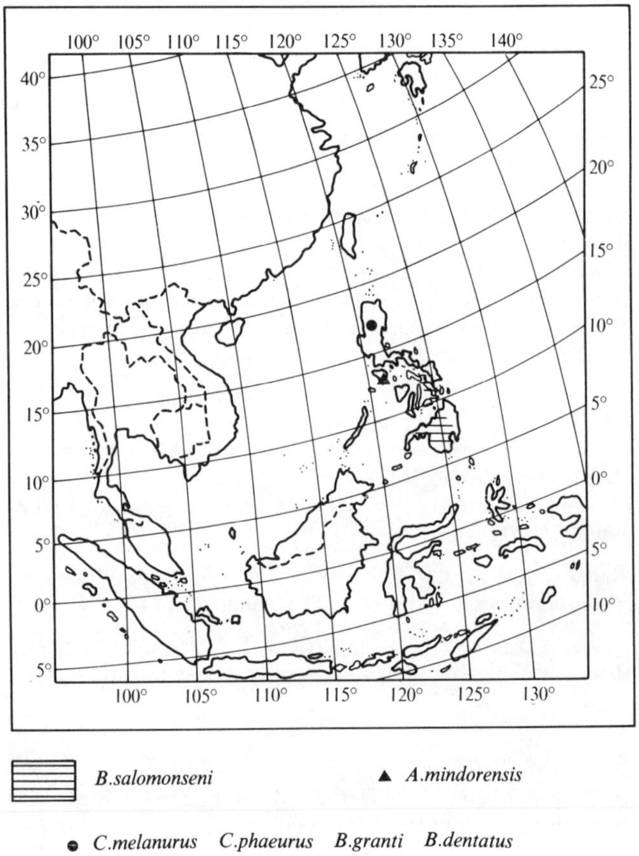

Table 241. The species of *Batomys*.

	granti	*dentatus*	*salomonseni*
Range	Luzon	Luzon	Mindanao, Leyte
Head & body	c.200	195	165–190
Tail (% of h & b)	55–75%	95%	70–85%
Hind feet	35–36	36	34–40
Tail pelage	Long (10–15)	Sparse, short	Sparse, short (3)
Skull: greatest length	46	45	40–45
Upper molar row	7.7	9.4	7.1–8.1

Batomys dentatus

B. dentatus Miller, 1910: 400; Haights-in-the-Oaks, Benguet, Luzon, Philippines, 7000ft.

Range (Map 158). Benguet Prov., NW Luzon, Philippines.

Batomys salomonseni

Mindanaomys salomonseni Sanborn, 1954: 287; Mt Katanglad, Mindanao, Philippines.

Range (Map 158). Philippine islands of Mindanao, Leyte and Biliran.

Remarks. *Mindanaomys* was included in *Batomys* by Musser (1981a: 136) and the differences between the respective type species, *B. granti* and *B. salomonseni*, do indeed seem slight.

ANONYMOMYS

Anonymomys Musser, 1981b: 300; type species *A. mindorensis* Musser.

Mindoro, Philippines; one species. Considered by Musser (1981b) to be an isolated genus on the basis of a unique combination of cranial and dental traits found in other murines rather than any unique characters. It is a light brown, spiny rat with tail about equal to head and body and white venter.

Anonymomys mindorensis Mindoro rat

Anonymomys mindorensis Musser, 1981b: 300; Ilong Peak, Halcon Range, Mindoro, Philippines, 4500ft.

Range (Map 158). Known only from the type locality on Mindoro, Philippines.

PHLOEOMYS

Phloeomys Waterhouse, 1839c: 108; type species *Mus* (*P.*) *cumingi* Waterhouse.

Luzon, Philippines; 2 species (Table 242).

Phloeomys cumingi

Mus (*Phloeomys*) *cumingi* Waterhouse, 1839c;108; Luzon, Philippines.
Capromys [*sic*] *elegans* Cabrera, 1901: 372; ? Philippines.

Range (Map 159). Southern Luzon, Philippines.

Remarks. The cuticular pattern of the hairs consists of broad scales quite different from those of *Crateromys schadenbergi* (Keller, 1988). Chromosomes: 2n=40, FN=60 (Jotterand-Bellomo & Schauenberg, 1988).

Table 242. The species of *Phloeomys*.

	cumingi	*pallidus*
Range	S Luzon	N Luzon
Pelage pattern: all brown (B); or white-tipped, with dark brown on neck, shoulders and muzzle (W)	B	W
Pelage texture: coarse and short (C); or soft and long (S)	C	S
Incisive foramina (lateral margins)	Rounded	Angular
Posterior edge of palate with shallow indentation (S); or long, narrow slit (L)	S	L

Phloeomys pallidus

P. pallidus Nehring, 1890: 105; Luzon, Philippines.

Range (Map 159). Northern Luzon, Philippines.

CRATEROMYS

Crateromys Thomas, 1895f: 163; type species *Phloeomys schadenbergi* Meyer.

Philippines; 3 species (Table 243). Resembling *Phloeomys* in some respects but the molars are strongly cusped, without developing transverse laminae as in *Phloeomys*.

Map **159**. *Phloeomys, Crateromys*

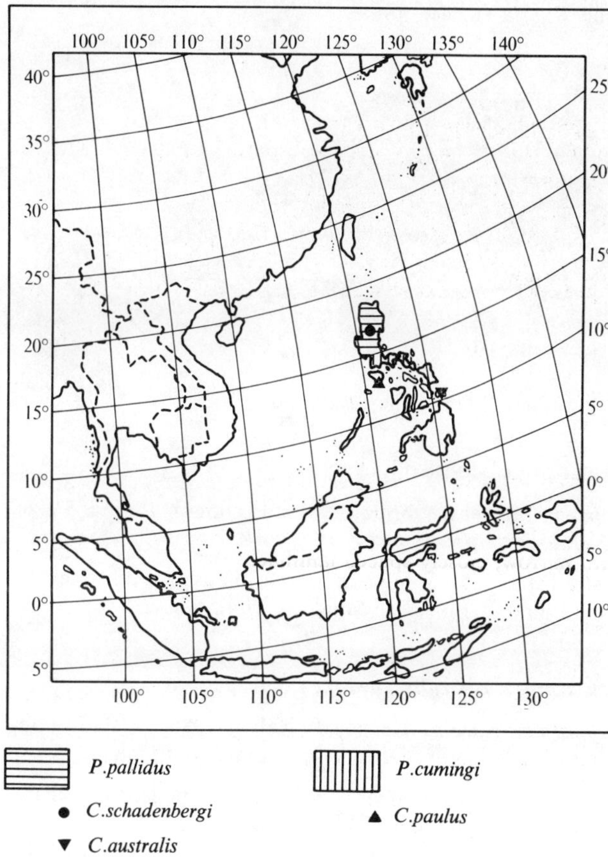

Table 243. The species of *Crateromys*.

	schadenbergi	*australis*	*paulus*
Range	Luzon	Dinagat I	Ilin I
Head & body	343–389	265	255
Tail	363–474	281	215
Hind feet	73–82	54	50
Crest on head	+	0	0
Dorsal pelage	Dark brown or piebald Long, shaggy	Tawny,spotted black Short	Dark brown
Ventral pelage	Dark brown or white	Orange-brown to roots	Cream, grey bases on sides
Tail thick and bushy	+	0	0
Tail: extent of distal white zone	0	$^1/_2$	$^1/_8$
Skull: greatest length	67–71	57	53
Upper molar row	14.6–17.3	12.1	11.8

Crateromys schadenbergi Bushy-tailed cloud rat

Phloeomys schadenbergi Meyer, 1896: 1; Mt Data, NW Luzon, Philippines.

Range (Map 159). Mountains of NW Luzon, Philippines.

Variation. The pelage is highly polymorphic: dark brown all over; dark brown with a broad white 'collar' over the shoulders meeting a white venter (but with brown throat and inguinal region) with or without a white tail-tip; grey all over with dark brown on the head and neck.

Remarks. The cuticular pattern of the hairs consists of narrow lozenge-shaped scales quite different from those of *Phloeomys cumingi* but remarkably similar to those of some squirrels of the genus *Sciurus* (Keller, 1988).

Crateromys schadenbergi

Crateromys paulus Ilin cloud rat

C. paulus Musser & Gordon, 1981: 515; Ilin (= Iling) I, SW of Mindoro, Philippines.

Range (Map 159). Known only from the type locality on the small island of Ilin, near Mindoro, Philippines.

Crateromys australis Dinagat cloud rat

C. australis Musser, Heaney & Rabor, 1985: 3; Balitbiton, Loreto Municipality, Dinagat I, central Philippines.

Range (Map 159). Known only from the type locality on the island of Dinagat, central Philippines.

CRUNOMYS Small shrew-rats

Crunomys Thomas, 1898c: 393; type species *C. fallax* Thomas.

Philippines and Sulawesi; 4 species (Table 244). Revised by Musser (1982d).

Crunomys fallax

C. fallax Thomas, 1898c: 394; Isabella Prov., N Luzon, Philippines, 1000ft.

Range (Map 160). Known only from the type locality in northern Luzon, Philippines.

Remarks. This species has been referred to as the 'Philippine swamp rat' but the only known specimen was collected in forest near a stream and there is nothing about it to suggest aquatic habits (Musser, 1982d).

Table 244. The species of *Crunomys.*

	fallax	*melanius*	*rabori*	*celebensis*
Range	Luzon	Mindanao	Leyte	Sulawesi
Head & body	105	98, 122		115–127
Tail	79	68, 79	90	80–84
Hind feet	23	25,26	27	25–27
Dorsal pelage spiny	+	+	+	0
Dorsum	Greyish brown	Dark chestnut	Very dark brown	Dark brown
Venter	Pale grey	Dark grey	Pale brownish grey	Slightly paler than dorsum
Skull: greatest length	–	28,30	32	29–30
Upper molar row	3.7	4.1, 4.2	4.0	3.7–3.9
Interorbital width	4.6	5.5, 5.7	6.0	5.6–5.8
Upper incisor: width at tip	1.4	1.5, 1.7	1.8	2.2, 2.3

Map **160**. *Crunomys, Archboldomys, Chrotomys, Rhynchomys*

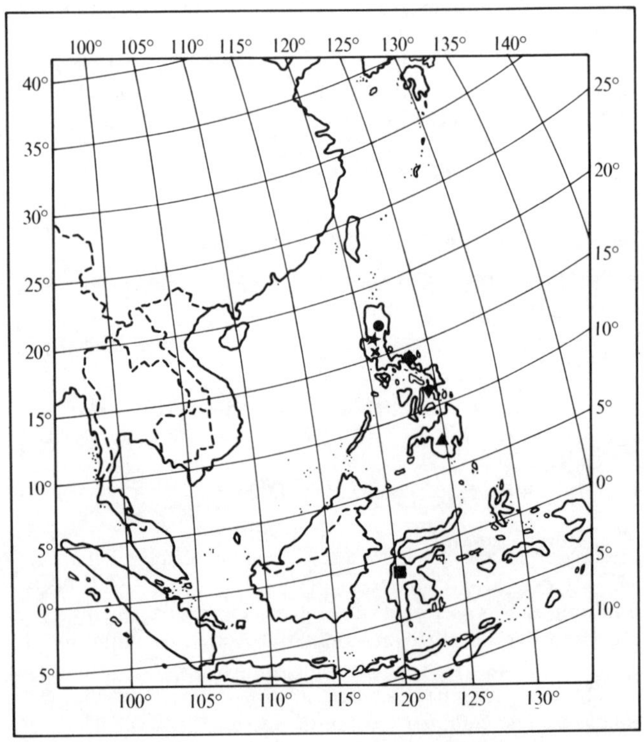

● *Cr.fallax* ▲ *Cr.melanius*

▼ *Cr.rabori* ■ *Cr.celebensis*

★ *Ch.whiteheadi Ch.silaceus R.soricoides*

◆ *A.luzonensis R.isarogensis* ✕ *Ch.mindorensis*

Crunomys melanius

C. melanius Thomas, 1907e: 5 (and 1907f: 141); ? Mt Apo, Mindanao, Philippines.

Range (Map 160). Known only from two specimens from Mindanao, Philippines, one at sea-level in Cotabato Prov., and one at 900m in 'North Central Mindanao', probably on Mt Apo.

Crunomys rabori

C. rabori Musser, 1982d: 14; Barrio Buri, Municipality of Burauen, Mt Lobi Range, Leyte I, Philippines.

Range (Map 160). Known only from the type locality on the island of Leyte, central Philippines.

Crunomys celebensis

C. celebensis Musser, 1982d: 16; near village of Tomado, valley of Danau Lindu, Sulawesi Tengah Prov., central Sulawesi, 1°19'S, 120°5'E, 3500ft.

Range (Map 160). Known only at and near the type locality in central Sulawesi, at altitudes of 800–1050m.

Remarks. Of the Philippine species of *Crunomys*, *C. celebensis* most closely resembles *C. melanius* although there are many differences.

ARCHBOLDOMYS

Archboldomys Musser, 1982d: 30; type species *A. luzonensis* Musser.

Luzon, Philippines; one species, most closely related to *Crunomys*.

Archboldomys luzonensis

A. luzonensis Musser, 1982d: 30; Mt Isarog, SE Luzon, Philippines, 6560ft.

Range (Map 160). Known only from the type locality in SE Luzon, Philippines at c. 2000m.

CHROTOMYS

Chrotomys Thomas, 1895f: 161; type species *C. whiteheadi* Thomas.
Celaenomys Thomas, 1898c: 390 (*Xeromys silaceus* Thomas).

Philippines; 3 species (Table 245). Distinguished by long slender rostrum, with pro-odont incisors, and small, simple molars with M3 very small, sometimes absent. *C. whiteheadi* and *C. mindorensis* have a boldly striped dorsal pelage unique in Indomalayan Muridae. A fourth species has recently been described but has not been assessed nor included in the table: *Chrotomys gonzalesi* Rickart & Heaney, 1991: 387; Mt Isarog, SE Luzon, Philippines.

Table 245. The species of *Chrotomys*.

	whiteheadi	*mindorensis*	*silaceus*
Range	Luzon, >2000m	Luzon, Mindoro, < 500m	Luzon
Dorsum with dark and light stripes	+	+	0
Venter	Dark grey, ± cream tips	Pale grey, white tips	Dark grey, cream tips
Upper molar row	4.5–5.6	5.5–6.5	3.7–4.2
Lower molar row	4.9–5.6	5.9–6.9	3.5,3.7
Third molars usually	Present	Present	Absent

Chrotomys whiteheadi

C. whiteheadi Thomas, 1895f: 161; Mt Data, NW Luzon, Philippines, 8000ft.

Range (Map 160). NW Luzon, at around 2100–1500m.

Chrotomys mindorensis

C. whiteheadi mindorensis Kellogg, 1945b: 123; 3 miles SSE of San Jose, Mindoro, Philippines, 200ft.

Range (Map 160). Mindoro and central Luzon, Philippines, below about 500m.

Chrotomys silaceus

Xeromys silaceus Thomas, 1895f: 161; Lepanto, Mt Data, NW Luzon, Philippines, 8000ft.

Range (Map 160). NW Luzon, Philippines, at altitudes around 2400m.

Remarks. Usually included in a monospecific genus *Celaenomys* because of the absence of M3. However these are very small in *Chrotomys s. s.* and present in one of the four *C. silaceus* examined; generic separation is therefore not justifiable. Although lacking bold dorsal stripes as seen in the other *Chrotomys*, one specimen shows a faint indication of a pale stripe on the nape.

RHYNCHOMYS

Rhynchomys Thomas, 1895f: 160; type species *R. soricoides* Thomas.

Luzon, Philippines; 2 species (Table 246). Revised by Musser & Freeman (1981). The species can be accepted as marginally distinct but they are based upon very few specimens. The molar teeth show an extreme reduction amongst murines of the

Table 246. The species of *Rhynchomys*.

	soricoides	*isarogensis*
Range	NW Luzon	SE Luzon
Head & body	188,195	187
Tail	143,132	105
(% of head and body)	76,68%	56%
Hind feet	41,42	37
Feet: colour of dorsal surface	Brown	White
Tail	All dark	Pale below with dark mid-ventral stripe
Skull: greatest length	47,49	43
Upper molar row (alveolar)	2.2,2.3	2.1

region, with only two small, simple teeth in each row. The genus has been separated from all other murids in a subfamily Rhynchomyinae.

Rhynchomys soricoides

R. soricoides Thomas, 1895f: 160; Mt Data, NW Luzon, Philippines.

Range (Map 160). Known only from Mt Data, NW Luzon, at altitudes of 2400–2600m.

Rhynchomys isarogensis

R. isarogensis Musser & Freeman, 1981: 154; Mt Isarog, SE Luzon, Philippines, 5500ft.

Range (Map 160). Known only from the type locality in SE Luzon.

BUNOMYS

Bunomys Thomas, 1910d: 508; type species *Mus coelestis* Thomas.
Frateromys Sody, 1941a: 260 (*Mus fratrorum* Thomas). Doubtfully valid since it was based solely upon a mammary formula shared by *Arcuomys* (in *Taeromys*).
Floresomys Musser, 1981a: 111 (*F. naso* Musser). Pre-occupied by *Floresomys* Fries *et al.*,1955, a fossil sciuravid rodent.
= *Paulomys* Musser, in Musser *et al.*,1986: 2.

Sulawesi and Flores; 8 species (Table 247). Included in *Rattus* (*Rattus*) by Ellerman (1941,1949) and in *Rattus* (*Bullimus*) by Misonne (1969). Treated as a '*Rattus*-like' genus endemic to Sulawesi by Musser (1981a: 115–117). The species are rather similar with long soft pelage, grey venter lightly tipped with buff and a moderately long rostrum. *Paulamys* is here included in *Bunomys* since in a detailed study of the first living specimen Kitchener, How & Maharadatunkamsi (1991b) were unable to detect any differences that would support separate generic rank.

Bunomys chrysocomus

Mus chrysocomus Hoffmann, 1887: 17; Minahassa, N Sulawesi.
Rattus nigellus Miller & Hollister, 1921a: 72; Bumbaroed-jaba, near Toboli, NC Sulawesi.
R. rallus Miller & Hollister, 1921a: 73; Gimpoe, C Sulawesi.
R. brevimolaris Tate & Archbold, 1935a: 7; Lalolis, 30 miles SE of Mengkoka Mts, SE Sulawesi, 300m.

Range (Map 161). Sulawesi.

Remarks. The forms *nigellus, rallus* and *brevimolaris* were listed in *Rattus adspersus* by Laurie & Hill (1954) but were listed in *B. chrysocomus* by Musser (1981a: 115–117). Chromosomes: 2n=42, NF=58 (Duncan, 1976).

Bunomys coelestis

Mus coelestis Thomas, 1896a: 248; Bonthain Peak, S Sulawesi.

Range (Map 161). SE and SW Sulawesi.

Variation. The following subspecies has been described:
B. coelestis koka Tate & Archbold, 1935b: 1; Tanka Salocco, Mengkoka Mts, SE Sulawesi, 1500m.

Remarks. Included in *B. chrysocomus* by Musser (1981a) but treated as a separate species by Musser (1991).

Table 247. The species of *Bunomys*.

	chrysocomus	*fratrorum*	*andrewsi*	*penitus*	*prolatus*	*coelestis*	*heinrichi*	*naso*
Range (in Sulawesi except *naso*)	Throughout	N	Throughout except N	C,SE, montane	Central, montane	SW	NW	Flores I
Head & body	147–173	157–182	155–175	170–200	155–180	145–180	140–160	164
Tail (% of head & body)	c.85%	c.85%	c.85%	c.100%	c.80%	c.95%	c.92%	71%
Hind feet	34–35	35–40	34–39	39–44	31–34	34–37	31–35	39
Tail white-tipped	0	+	0	+	0(+)	0	0	0
Skull: greatest length	36–40	41–46	39–43	41–46	40–43	39–41	38–40	42
Upper molar row	5.7–6.7	6.8–8.1	6.9–7.4	6.5–7.4	6.3–6.8	5.7–6.7	6.5–7.3	7.1
Incisive foramina	5.2–6.5	6.3–8.0	8.0–8.6	7.3–9.1	6.3–7.1	6.6–8.0	7.2–7.9	7.3

Map **161**. *Bunomys*

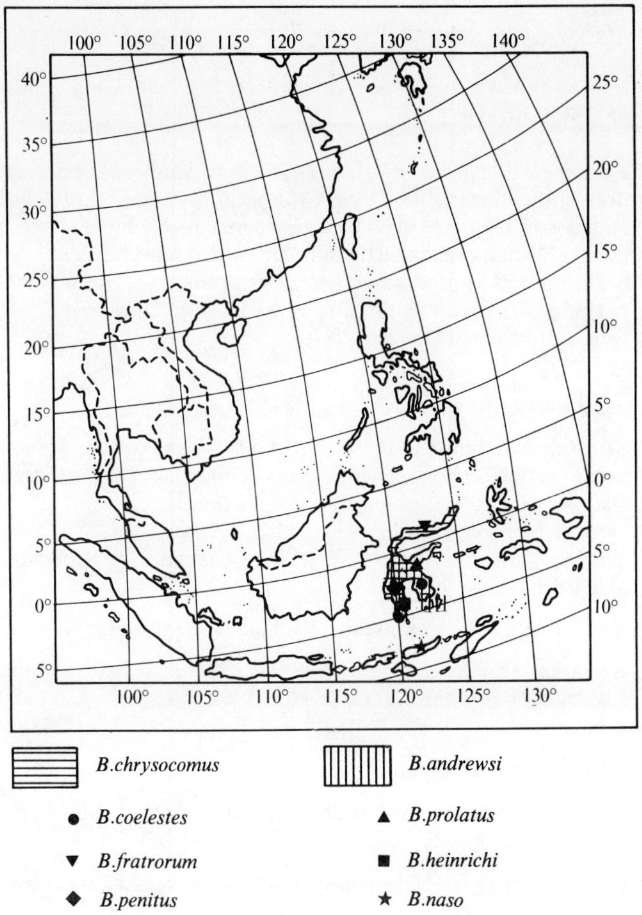

Bunomys prolatus

B. prolatus Musser, 1991: 4; Mt Tambusisi, central Sulawesi, 1°38'S, 121°23'E, 6000ft.

Range (Map 161). Known only from the type locality in central Sulawesi, at 1820m.

Bunomys fratrorum

Mus fratrorum Thomas, 1896a: 246; Rurukan, N Sulawesi.

Range (Map 161). N Sulawesi.

Remarks. Treated as a synonym of *Rattus chrysocomus* by Laurie & Hill (1954) but as a separate species by Musser (1981a: 117).

Bunomys andrewsi

Mus andrewsi Allen, J. A.,1911: 336; Butung I, SE Sulawesi.
Rattus adspersus Miller & Hollister, 1921a: 72; Mt Lehio, SW of Lake Lindoe, C Sulawesi.
R. penitus inferior Tate & Archbold, 1935a: 6; Wawo, Mengkoka Mts, SE Sulawesi, 50m.

Range (Map 161). Sulawesi, except NE, but including the island of Buton; in lowland forest.

Remarks. The above synonymy follows Musser (1981a: 117) with the exclusion of *heinrichi*. This species is also known from subfossils in SW Sulawesi.

Bunomys heinrichi

Rattus penitus heinrichi Tate & Archbold, 1935a: 6: Lambasang, Mt Lampobatang, S Sulawesi, 1100m.

Range (Map 161). SW Sulawesi.

Remarks. Included in *B. andrewsi* by Musser (1981c) but treated as a separate species by Musser (1991).

Bunomys penitus

Rattus penitus Miller & Hollister, 1921a: 72; Goenoeng Lehio, SW of Lake Lindoe, C Sulawesi, 6000ft.
R. sericatus Miller & Hollister, 1921a: 73; Rano Rano, E of Lake Lindoe and N of L Paso, C Sulawesi, above 6000ft.

Range (Map 161). Central and SW Sulawesi, in montane forest.

Remarks. The above synonymy follows Musser (1981a: 117).

Bunomys naso

Floresomys naso Musser, 1981a: 112; Liang Toge cave, near Warukia, 1km S of Lepa, Menggarai Prov., W Flores, Lesser Sunda Is; subfossil (indirectly dated to c. 3550 ± 525 BP).
= *Paulomys naso*: Musser, in Musser *et al.*, 1986:2.

Range (Map 161). Known alive from only a single specimen trapped at Kelimutu, central south Flores; and from subfossil material in W Flores.

Remarks. This species was originally based upon subfossil dentaries and lower teeth. It was redescribed (as '*Paulamys* sp., cf *P. naso*') by Kitchener, How & Maharadatunkamsi (1991b) on the basis of a single trapped animal. In a phylogenetic analysis these authors could determine no reliable characters to distinguish *Paulamys* from *Bunomys*. Although they refrained from uniting the genera there appear to be no grounds for keeping them apart, while accepting that the generic classification of *Bunomys* in relation to '*Rattus*' in this and the Australian Region is still very provisional.

LENOMYS

Lenomys Thomas, 1898c: 409; type species *Mus meyeri* Jentink.

Sulawesi; one species, characterized especially by very high-crowned cuspidate molars. An additional species *L. callitrichus*, was tentatively recognized by Tate (1936) followed by Laurie & Hill (1954), but this was allocated to the genus *Rattus* by Musser (1970a) and subsequently to *Taeromys*.

Lenomys meyeri Trefoil-toothed giant rat

Mus meyeri Jentink, 1879a: 12; Menado, N Sulawesi.
L. longicaudus Miller & Hollister, 1921b: 95; Gimpoe, central Sulawesi.

Range (Map 162). Sulawesi (recorded from N, central and SW), in lowland and montane forest.

Variation. One subspecies has been described:
L. m. lampo Tate & Archbold, 1935b: 5; Wawa Karaing, Mt Lampobatang, S Sulawesi, 2200m. This appears to be distinguishable from the northern and central samples only by longer incisive foramina: 8.6–8.8mm, cf. 6.8–8.1mm (Hooijer, 1952).

Recognition. Externally this species is very similar to *Eropeplus canus* but differs in having the dorsal pelage dull greyish brown, with the bases of the hairs medium grey and the tips cream; and the sides of the face as the dorsum.

Remarks. Subfossil remains, indistinguishable from Recent samples, were reported from SW Sulawesi by Hooijer (1952) and Musser (1984).

Map **162**. *Lenomys, Eropeplus, Margaretamys, Papagomys*

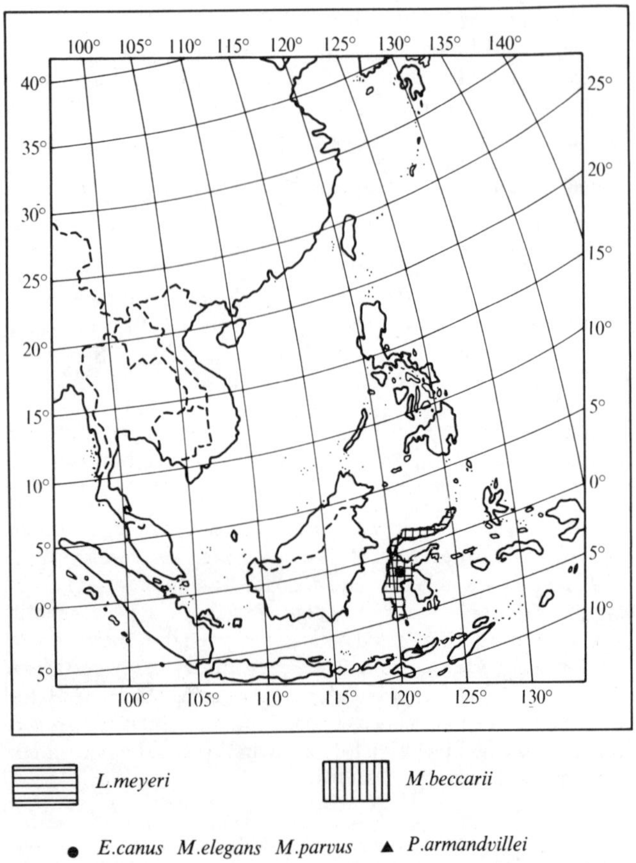

EROPEPLUS

Eropeplus Miller & Hollister, 1921b: 94; type species *E. canus* Miller & Hollister.
Sulawesi; one species. Probably closely related to *Lenomys*, but differing in having the molars low-crowned, the cusps quickly wearing to form broad laminae. The name *Mus callitrichus* Jentink was treated as representing a second species of *Eropeplus* by Tate (1940) and Musser (1969a) but was later shown to be a species of *Taeromys* (Musser, 1970a).

Eropeplus canus Sulawesi grey rat

E. canus Miller & Hollister, 1921b: 94; Mt Lehio (SW of Lake Lindoe), central Sulawesi, above 6000ft.

Range (Map 162). Known only from central Sulawesi, in montane forest between 1800 and 2200m, apparently replacing *Lenomys meyeri* at these altitudes (Musser, 1984).

Recognition. Very similar externally to *Lenomys meyeri* but dorsal pelage softer and browner with very dark grey hair-bases and buff tips, and the sides of the face grey.

MARGARETAMYS

Margaretamys Musser, 1981b: 275; type species *Mus beccarii* Jentink.
(Not = *Margaritamys* Mein & Adrover, 1982 which is a genus of fossil Gliridae.)

Sulawesi; 3 species (Table 248). The type species was previously allocated to *Rattus* (*Maxomys*) by Ellerman (1949) and Laurie & Hill (1954). They are arboreal rats, with long tails, noticeably hairy in the distal third and with a terminal pencil of 6–7mm; and a dark smudge around the eye, sometimes extending forwards on the side of the muzzle. They feed upon fruit and insects (Musser, 1981b).

Margaretamys beccarii

Mus leucopus Jentink, 1879a: 8; Menado-Langkowah, Minahassa region, NE Sulawesi. Pre-occupied by *Acanthomys leucopus* Gray, 1867, later allocated to *Rattus*.
= *Mus beccarii* Jentink, 1880a: 11.
Rattus thysanurus Sody, 1932a: 157; Toemaratas, Sapoetan Mts, Minahassa, N Sulawesi.

Range (Map 162). Central and NE Sulawesi, in lowland forest.

Remarks. This species was allocated to *Rattus* (*Maxomys*) by Ellerman (1949) and Laurie & Hill (1954).

Margaretamys elegans

M. elegans Musser, 1981b: 286; Mt Nokilalaki, C Sulawesi, 6500ft.

Range (Map 162). Known only from the type locality in central Sulawesi between 1600 and 2270m.

Recognition. Could be confused with *Rattus xanthurus* but it is smaller, with softer pelage and the tail appears hairy throughout its length.

Margaretamys parvus

M. parvus Musser, 1981b: 294; Mt Nokilalaki, C Sulawesi, 7400ft.

Range (Map 162). Known only from the type locality in central Sulawesi, in montane forest between 1800 and 2270m.

Table 248. The species of *Margaretamys*.

	beccarii	*elegans*	*parvus*
Range in Sulawesi	C & NE, lowland	C, montane	C, montane
Head & body	120–150	140–200	95–115
Tail	150–200	220–285	155–185
as % of head & body	c.120–150%	c.130–150%	c.150–170%
Hind feet	25–31	34–39	23–25
Dorsal pelage	Semi-spinous, short (10–15)	Soft, long (20–30)	Soft, short (10–13)
Dorsal colour	Greyish brown	Greyish brown	Red-brown
hair bases	Light fawn		Dark grey
Ventral colour	Cream to roots	Grey, ± cream tips	Grey
Distal 30–50% of tail white	0	+/0	0
Skull: greatest length	32–36	37–43	28–30
Upper molar row	4.6–5.5	6.2–7.5	4.5–5.1

TAEROMYS

Taeromys Sody, 1941a: 260; type species *Mus* (*Gymnomys*) *celebensis* Gray.
Arcuomys Sody, 1941a: 260 (*Rattus arcuatus* Tate & Archbold). Although this name was proposed above *Taeromys* on the same page, *Taeromys* can be accepted as having priority following Musser (1981a: 137) as 'first reviser'. However both these names are of doubtful validity since they were based solely upon mammary formulae shared with other genera proposed on the same page: *Taeromys* with *Madromys* (here in *Cremnomys*) and *Arcuomys* with *Frateromys* (here in *Bunomys*).

Sulawesi; 6 species (Table 249). Included in *Rattus* (*Rattus*) by Ellerman (1949) and Laurie & Hill (1954) but treated as a distinct genus confined to Sulawesi by Musser & Newcomb (1983) although without a diagnosis. The genus was originally based solely upon the mammary formula (1 pair anterior, 2 pairs posterior) which is also found in some *Millardia*. They are moderately large rats with soft, agouti dorsal pelage (bases of hairs medium grey) and the distal half or more of the tail white. They are terrestrial, forest rats except for *T. celebensis* which also forages in the crowns of trees in the understorey (Musser, 1984).

Taeromys celebensis

Mus (*Gymnomys*) *celebensis* Gray, 1867b: 598; Menado, N Sulawesi.

Range (Map 163). Central, N and SE Sulawesi, in lowland forest; also recorded from SW Sulawesi as subfossils (Musser, 1987a).

Recognition. Distinguished from *T. callitrichus* by shorter, lighter pelage and from all other species by very long tail.

Taeromys callitrichus

Mus callitrichus Jentink, 1879a: 12; Menado, Sulawesi (lectotype designated by Tate, 1940: 7).
Rattus microbullatus Tate & Archbold, 1935a: 8; Tanka Salocco, Mengkoka Mts, SE Sulawesi, 1500m.
Rattus maculipilis Laurie & Hill, 1954: 115; Rurukan, Minahassa, N Sulawesi, 3500ft.

Range (Map 163). Central, N and SE Sulawesi, in lowland and montane forest.

Variation. One subspecies has been described:
Rattus maculipilis jentinki Laurie & Hill, 1954: 115; Koelawi, central Sulawesi, 2500ft. Based upon longer tail (longer than head and body) and slight differences in colour.

Remarks. The name *callitrichus* has been the source of much confusion because of the labelling of twelve specimens in Leiden Museum, representing five different species, as cotypes (Musser, 1970a).

Taeromys hamatus

Rattus hamatus Miller & Hollister, 1921b: 97; Mt Lehio, central Sulawesi.

Range (Map 163). Central Sulawesi, in montane forest.

Map **163**. *Taeromys*

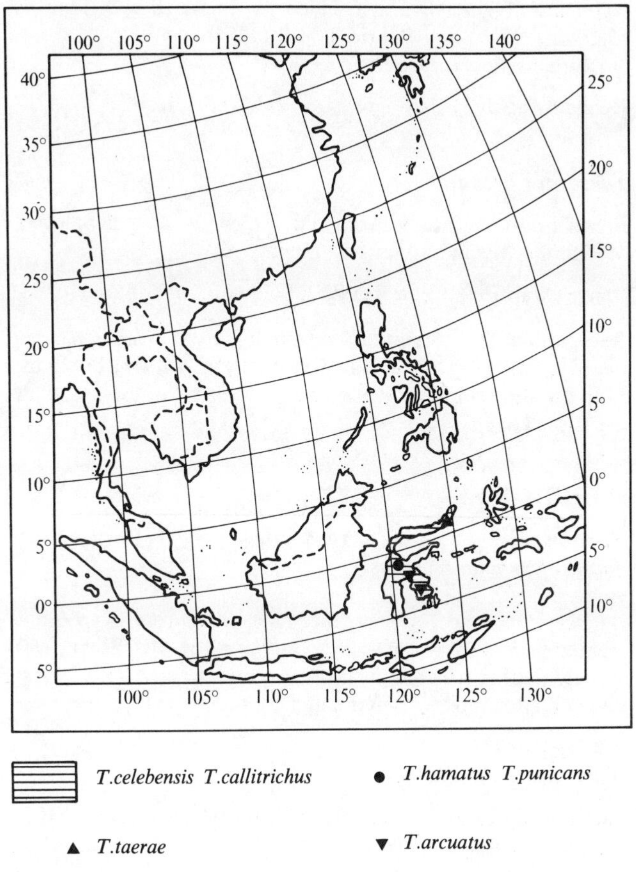

Table 249. The species of *Taeromys*.

	celebensis	*callitrichus*	*hamatus*	*punicans*	*taerae*	*arcuatus*
Range in Sulawesi	C,N,SE	C,N,SE	Central	Central	N	C,SE
Head & body	220–250	200–240	180–210	185,207	200–225	200–212
Tail	255–305	220–260	180–205	156,186	205–220	200–235
Hind feet	43–53	44–54	40–45	45,46	44–46	40–44
Dorsum	Light brown	Dark brown	Dark grey-brown	Dark brown	Dark brown	Dark brown
Venter	White or cream	Buffy grey	Grey, cream tips	Dark buff	Grey, cream tips	Grey, cream tips
Tail: extent of white	60–100%	50–70%	c.50%	<10%	c.50%	40–60%
Skull: greatest length	48–53	49–52	43–48	46	50–52	50–52
Upper molar row (alveoli)	6.6–8.4	9.5–10.7	8.0–8.9	8.4–9.7	8.8–9.9	8.7–9.6
Incisive foramina	6.4–7.6	6.7–8.2	7.6–8.8	5.8	8.4–9.1	6.7–8.0

Taeromys punicans

Rattus punicans Miller & Hollister, 1921b: 98; Pinedapa, central Sulawesi.

Range (Map 163). Central Sulawesi, in lowland forest; also recorded in SW Sulawesi as subfossils (Musser, 1984,1987a).

Taeromys taerae

Rattus taerae Sody, 1932a: 158; Lembean, E of Tondano, N Sulawesi, 600m.
T. tatei Sody, 1941a: 313; Toelap West, Tonsealama, Tonsea Dist, N Sulawesi. Based on a composite specimen: skin of *T. taerae* and skull of *Bunomys fratrorum*; the skin was designated as lectotype of *T. tatei* by Musser (1971e). Pre-occupied by *Rattus tatei* Ellerman, 1941, here treated as a synonym of *R. hoffmanni*, if *Taeromys* is included in *Rattus* as was done by Ellerman (1949) and Laurie & Hill (1954).
= *Rattus simpsoni* Ellerman, 1949: 191.

Range (Map 163). N Sulawesi, in lowland forest.

Taeromys arcuatus

Rattus arcuatus Tate & Archbold, 1935a: 9; Tanka Salocco, Mengkoka Mts, SE Sulawesi, 1500m.

Range (Map 163). Central and SE Sulawesi, in montane forest.

Recognition. Very similar externally to *T. hamatus* but distinguishable by longer ears. Externally as *T. taerae* but in the skull the shorter incisive foramina and sloping zygomatic plate are distinctive.

PARUROMYS

Paruromys Ellerman, in Laurie & Hill, 1954: 117; type species *Rattus dominator* Thomas.

Sulawesi; one species. Described as a subgenus of *Rattus* and treated as a synonym of *Rattus* (*Bullimus*) by Misonne (1969), but treated as a genus by Musser & Newcomb (1983) who considered it most closely related to *Taeromys*.

Paruromys dominator

Rattus dominator Thomas, 1921d: 244; Mt Masarang, Minahassa, N Sulawesi, 4000ft.
R. frosti Ellerman, 1941: 216; Tamalanti, C Sulawesi.

Range (Map 164). Sulawesi, in both lowland and montane forest.

Variation. The following subspecies have been described:
Rattus d. camurus Miller & Hollister, 1921b: 96; Pinedapa, C Sulawesi.
Taeromys d. ursinus Sody, 1941a: 312; Wawa Karaeng, Lampobatang, S Sulawesi, 2200m.

Remarks. *Rattus frosti* was shown to be a synonym of *P. dominator* by Musser (1971c). *Rattus microbullatus* Tate & Archbold, which was also included in the subgenus *Paruromys* by Laurie & Hill (1954), was considered a subspecies of *Taeromys callitrichus* (Jentink) by Musser (1970a) and Musser & Newcomb (1983). The skull and teeth of *P. dominator* were illustrated in detail by Musser & Newcomb (1983: 504–509).

Map **164**. *Paruromys, Echiothrix, Melasmothrix, Komodomys*

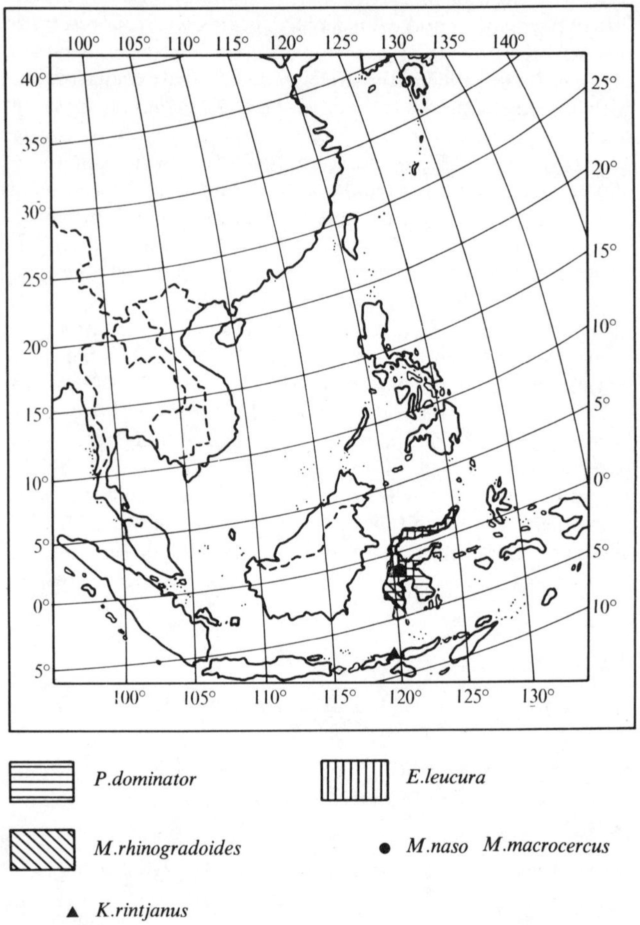

ECHIOTHRIX

Echiothrix Gray, 1867b: 599; type species *E. leucura* Gray.
= *Craurothrix* Thomas, 1896a: 246.

Sulawesi; one species. Distinguished by very long, slender rostrum with small but orthodont incisors, and by spiny dorsal pelage, the spines being flattened, concave above and convex below.

Echiothrix leucura Greater Sulawesi shrew-rat

E. leucura Gray, 1867b: 600; 'Australia', probably N Sulawesi.
E. centrosa Miller & Hollister, 1921a: 67; Winatoe (between Koelawi and Gimpoe), central Sulawesi.
E. brevicula Miller & Hollister, 1921a: 67; Pinedapa (about 5 miles inland from the Gulf of Tomini, near Mapane), central Sulawesi.

Range (Map 164). North and central Sulawesi, in lowland forest.

Ecology. A terrestrial species, feeding on earthworms.

MELASMOTHRIX Lesser Sulawesi shrew-rats

Melasmothrix Miller & Hollister, 1921b: 93; type species *M. naso* Miller & Hollister.
Tateomys Musser, 1969b: 2 (*T. rhinogradoides* Musser).

Sulawesi; 3 species (Table 250). They are terrestrial and nocturnal, feeding on earthworms in montane forest. Although *M. rhinogradoides* and *M. macrocercus* are clearly more similar to each other than to *M. naso*, the differences between them and *M. naso* seem sufficiently slight to justify their inclusion in a single genus.

Table 250. The species of *Melasmothrix*.

	naso	*rhinogradoides*	*macrocercus*
Range in Sulawesi	Central	SW,Central	Central
Head & body	110–125	137–156	110–120
Tail length	85–95	152–170	160–175
as % of h & b	c.75%	c.110%	c.140–150%
Hind feet	26–29	36–38	30–31
Skull: greatest length	31–33	38–40	31–33
Incisive foramina	4.4–5.2	5.8–6.1	4.0–4.5
Upper molar row	4.8–5.7	4.9–6.0	5.0–5.6

Melasmothrix naso

M. naso Miller & Hollister, 1921b: 93; Rano Rano, central Sulawesi.

Range (Map 164). Central Sulawesi, in montane forest.

Melasmothrix rhinogradoides

Tateomys rhinogradoides Musser, 1969b: 3; Latimodjong Mts, SW peninsula, Sulawesi, 3°30'S, 120°5'E, 2200m.

Range (Map 164). SW and central Sulawesi, at altitudes of 2200–2300m.

Melasmothrix macrocercus

Tateomys macrocercus Musser, 1982d: 64; SW slope of Mt Nokilalaki, NE of Danau Lindu, Sulawesi Tengah Prov., C Sulawesi, 7500ft.

Range (Map 164). Known only from the type locality in central Sulawesi, at altitudes of 2000–2250m.

PAPAGOMYS

Papagomys Sody, 1941a: 322; type species *Mus armandvillei* Jentink.

Flores; one Recent species. Reviewed by Musser (1981a) who described an additional subfossil species:
P. theodorverhoeveni Musser, 1981a: 95; Liang Toge cave, near Wakuria, 1 km S of Lepa, Menggarai Prov., W Flores, Indonesia; associated with deposits with a radiocarbon date of 3550 ± 525 BP.

Although often associated with the genus *Mallomys* from New Guinea, recent authors are agreed, as reviewed by Musser (1981a), that there is no close relationship to *Mallomys* and that it is more closely related to *Rattus*.

Papagomys armandvillei Flores giant rat

Mus armandvillei Jentink, 1892: 79, pl. 5; Flores.
P. verhoeveni Hooijer, 1957: 302; Liang Toge cave, near Wakuria, 1 km S of Lepa, Menggarai Prov., Flores (subfossil).

Range (Map 162). Flores, Lesser Sunda Islands.

Variation. One subspecific name has been proposed on the basis of subfossil remains:
P. a. besar Hooijer, 1957: 301; Liang Toge cave, Flores. This was based on slightly larger average size of the teeth and was reviewed in detail by Musser (1981a).

KOMODOMYS

Komodomys Musser & Boeadi, 1980: 397; type species *Rattus rintjanus* Sody.

Komodo Is, Lesser Sunda Islands; one species, probably most closely related to *Papagomys* of Flores.

Komodomys rintjanus Komodo rat

Rattus rintjanus Sody, 1941a: 310; Lohoboeaja, Rintja I, Komodo Is, Lesser Sunda Is (Nusa Tenggara), Indonesia.

Range (Map 164). Islands of Rinca and Padar, Komodo Is, Lesser Sunda Is. Also subfossil on Flores (Musser, pers. comm.).

Remarks. The relationships of this species were further discussed by Musser (1981a).

MELOMYS Naked-tailed rats

Melomys Thomas, 1922d: 261; type species *Uromys rufescens* Alston.

New Guinea, NE Australia and some of the Molucca islands; c. 22 species of which 5 are endemic on the Moluccas (Table 251). These were formerly included in *Uromys* Peters but are distinguished by much smaller size and relatively larger auditory bullae. The classification used here is that of Tate (1951b) except to include *M. talaudium* in *M. fulgens*. It is possible that some of these Moluccan forms are conspecific with species in New Guinea, but that, and their relationship to New Guinea and Australian species, will require a comprehensive revision of the genus. However the three species on the island of Seram are very clearly distinct from each other.

Melomys fulgens

Uromys fulgens Thomas, 1920a: 426; Teleoti Bay, S Seram, Moluccas.
U. talaudium Thomas, 1921d: 248; Liroeng, Talaud Is.

Range (Map 165). Seram (lowlands) and the Talaud island of Salebabu.

Remarks. The two forms included here were treated as separate subspecies of *M. leucogaster* (Jentink, 1908) of New Guinea by Rümmler (1938). Tate (1951b) treated them as two separate species while accepting their close relationship to *M. leucogaster*. The very bright yellow-brown pelage does not appear to occur in any forms of *M. leucogaster*.

Melomys caurinus

Uromys caurinus Thomas, 1921e: 112; Talaud Is.

Range (Map 165). Known only from the holotype from one of the Talaud Islands, NW of Halmahera in the Moluccas.

Remarks. Rümmler (1938) presumed that the difference in colour between the type of *caurinus* and examples of *M. fulgens* was due to it having been in alcohol. There is nothing to

Table 251. The species of *Melomys*.

	fulgens	*caurinus*	*aerosus*	*obiensis*	*fraterculus*
Range	Seram, Talaud Is	Talaud Is	Seram	Obi I	Seram
Head & body	c.150	170	c.150	c.130	c.115
Tail (% of head & body)	120–140%	c.75%	c.92%	c.115%	c.135%
Hind feet	30,34	30	31	27	26
Dorsum	Bright yellow-brown	Dull yellow-brown	Very dark brown	Dull brown	Dull brown
Venter	White, wide	White, narrow	Dark grey, brown tips	Dirty white	Mid grey, cream tips
Skull: greatest length	39	39	37		33
Upper molar row	7.0–7.6	7.8	7.5	6.1	6.7
Width of M^1	2.0–2.2	2.4	2.4	1.8	1.8

Map **165**. *Melomys, Uromys, Hydromys, Diplothrix, Tokudaia*

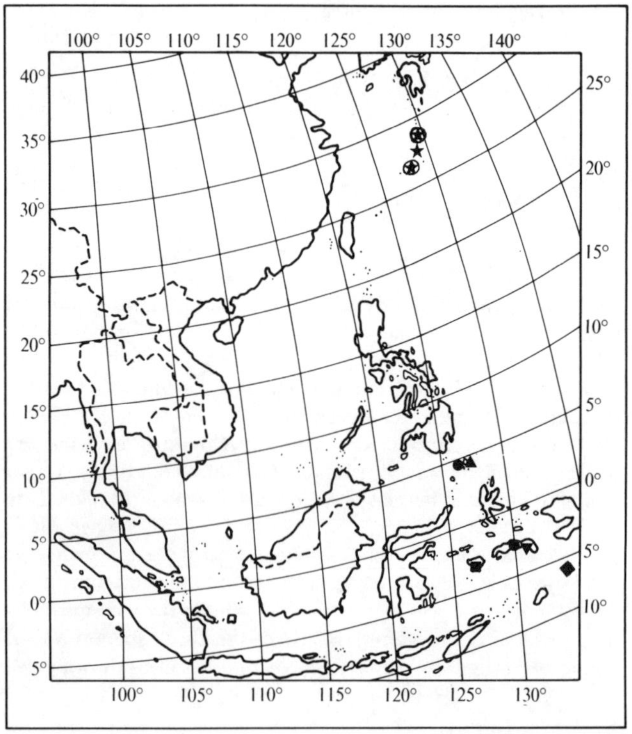

● *M.fulgens* ▲ *M.caurinus*

▼ *M.aerosus M.fraterculus* ■ *M.obiensis*

◆ *U.caudimaculatus H.chrysogaster* ★ *D.legata*

⊛ *T.osimensis D.legata*

indicate that this specimen has been treated differently from the type of *talaudium* which has the same history and shows the bright yellow-brown pelage of *M. fulgens*. Also this would not account for the other peculiarities of *M. caurinus*, i. e. the short tail, narrow white zone on the venter and broad molars.

Melomys aerosus

Uromys aerosus Thomas, 1920a: 428; Mt Manusela, central Seram, Moluccas, 6000ft.

Range (Map 165). Seram Island, Moluccas, at 800–1600m.

Remarks. This species was allocated to the subgenus *Paramelomys* by Rümmler (1938).

Melomys obiensis Obi naked-tailed rat

Uromys obiensis Thomas, 1911e: 208; Obi I, Moluccas.

Range (Map 165). Known only from the island of Obi, Moluccas.

Remarks. Considered by Tate (1951b) to be most closely related to *M. cervinipes* (Gould, 1852) of Australia which differs however in having the ventral pelage grey.

Melomys fraterculus

Uromys fraterculus Thomas, 1920a: 428; Mt Manusela, central Seram, Moluccas, 6000ft.

Range (Map 165). Island of Seram, Moluccas, at about 1500–1800m.

Remarks. This species was allocated to the subgenus *Pogonomelomys* by Rümmler (1938) but was retained in *Melomys*, not *Pogonomelomys*, by Tate (1951b) followed by Laurie & Hill (1954). It is very similar to *M. obiensis* but with a sufficient number of differences to justify retaining it as a separate species.

Uromys caudimaculatus

UROMYS Giant naked-tailed rats

Uromys Peters, 1867c: 343; type species *Mus macropus* Gray, in *U. caudimaculatus* Krefft.

Kai Is, New Guinea, Solomon Is, Queensland; 5 species of which one occurs on the Kai Islands in the Moluccas.

Uromys caudimaculatus

Mus macropus Gray, 1866b: 221; Port Albany, N Queensland. Pre-occupied by *M. macropus* Hodgson, 1845, a synonym of *Bandicota indica*.

Hapalotis caudimaculatus Krefft, 1867: 316; Cape York, N Queensland.

U. aruensis Gray, 1873a: 418; Aru Is, New Guinea.
U. siebersi Thomas, 1923k: 442; Daab Mt, Great Kei I, Moluccas.

Range (Map 165). Kai Islands (Moluccas); also on the Aru Is, throughout New Guinea and in NE Queensland.

Variation. Rümmler (1938) recognized *U. c. siebersi* as an endemic subspecies on the Kai Islands, distinguished from the adjacent *U. c. aruensis* on the Aru Is by greyer flanks and lighter feet. However Tate (1951b) included it in *U. c. aruensis* with a distribution throughout southern New Guinea.

HYDROMYS Australasian water rats

Hydromys Geoffroy, 1804: 253; type species *H. chrysogaster* Geoffroy.

Australia and New Guinea etc. ; 4 species of which one extends west to the Kai Islands (Moluccas).

Hydromys chrysogaster Australian water rat

H. chrysogaster Geoffroy, 1804: 254; Bruni I, Tasmania.
H. beccarii Peters, 1874: 303; Weri, Kei Is, Moluccas.
Also many extralimital synonyms.

Range (Map 165). Kai Islands; and on Aru Is, most of lowland New Guinea and much of Australia.

Variation. All forms in New Guinea and the Aru and Kai Islands were allocated to *H. c. beccarii* by Tate (1951b).

DIPLOTHRIX

Diplothrix Thomas, 1916c: 404 (footnote); type species *Lenothrix legata* Thomas.

Ryukyu Is; one species. This was treated as a species of *Rattus* by Ellerman (1949 etc.) but was given generic rank by Musser & Boeadi (1980) and by Kawamura (1989) who nevertheless considered it closer to *Rattus s. s.* than to *Lenothrix* or *Lenomys*.

Diplothrix legata

Lenothrix legata Thomas, 1906a: 88; Amamioshima, Ryukyu Is.
Mus bowersi okinavensis Namiya, 1909: 455; Okinawa I, Ryukyu Is.

Range (Map 165). The Ryukyu islands of Amamioshima, Tokunoshima and Okinawa. Late Pleistocene fossils also recorded from Miyako I as well as Okinawa (Kawamura, 1979).

TOKUDAIA

Acanthomys Tokuda, 1941: 93; type species *Rattus jerdoni osimensis* Abe. Pre-occupied by *Acanthomys* Lesson, 1842, a synonym of *Acomys*.
= *Tokudaia* Kuroda, 1943: 61.
= *Tokudamys* Johnson, 1946a: 169.

Ryukyu Islands; one species. It was considered to be most closely related to *Apodemus* amongst Recent genera by Kawamura (1989) who made detailed comparisons with many other genera.

Tokudaia osimensis

Rattus jerdoni osimensis Abe, 1934: 107; Sumiyo-mura, Amami-Oshima, Ryukyu Is.

Range (Map 165). Amami and Okinawa Islands, Ryukyu Islands. Fossils (Late Pleistocene or Holocene) have been recorded from Ie I as well as Okinawa (Kawamura, 1989).

Variation. The form on Okinawa was described as a subspecies mainly on the basis of larger size (head & body 150mm, greatest length of skull 40mm) and shorter tail (73% of head & body): *T. o. muenninki* Johnson, 1946a: 170; Hentona, N Okinawa I, Ryukyu Is.

On karyological grounds the two populations were considered unlikely to be conspecific (Honda *et al.*, 1977).

Remarks. IUCN: (E)—*T. o. muenninki.*

Subfamily Cricetinae

Hamsters

Cricetinorum Fischer, 1817: 410; type genus *Cricetus* Leske.

Palaearctic; c. 25 species of which 5, in 3 genera, extend marginally into the Indomalayan Region as defined here, in Afghanistan/Pakistan and in central China (Table 252). They form a clearly defined group of small, terrestrial, seed-eating rodents adapted to steppe habitats, mostly with short tail and internal cheek pouches. The genus *Calomyscus* is included here tentatively although it differs in many ways from typical cricetines.

Table 252. The genera of Cricetinae.

	Cricetulus	*Cansumys*	*Calomyscus*
Species in region	3	1	1
Head and body	75–160	120–130	70–90
Tail (% of h & b)	25–45%	c.85%	90–120%
Tail well-haired (longer than diameter of tail)	0	+	+
Skull: greatest length	23–41	c.33	23–27
Upper molar row	3.3–5.7	6.4,6.6	3.1–3.6
Molars hypsodont	0 (Fig.35)	+	0
M^3	Complex	Complex	Simple, ring-shaped

CRICETULUS Dwarf hamsters

Cricetulus Milne-Edwards, 1871a: 136; type species *C. griseus* Milne-Edwards.
Tscherskia Ognev, 1914: 102 (*C. triton nestor* Thomas).

S Palaearctic; c. 12 species of which 3 extend marginally into this region in Afghanistan and Pakistan, and in central China (Table 253). A further species, *C. alticola* Thomas, occurs in N Kashmir but only at high altitudes, above 3000m. *C. triton* is often separated in the genus *Tscherskia*.

Cricetulus migratorius Grey hamster

Mus migratorius Pallas, 1773: 703; lower R Ural, W Siberia. Including many extralimital synonyms.

Range (Map 166). Kashmir and N Pakistan (but only at high altitudes, over 3000m); W Pakistan (W of the Indus) and E

Afghanistan, down to c. 1350m in Baluchistan; also through Iran and Soviet Turkestan to S Russia, SE Europe and Israel.

Ecology. Mainly in arid montane steppe but extending into cultivated valleys in Baluchistan.

Table 253. The species of *Cricetulus.*

	migratorius	*longicaudatus*	*triton*
Range in region	Afghanistan, Pakistan	C China	C China
Head & body	80–100	75–95	120–150
Tail	23–30	40–48	70–95
Hind feet	15–17	16–17	19–24
Venter	White, ± obscure grey bases	White, grey bases	Grey

Cricetulus longicaudatus Lesser long-tailed hamster

Cricetus (Cricetulus) longicaudatus Milne-Edwards, 1871a: 136, pl. 12; ? near Saratsi, N Shanxi, China.
C. andersoni Thomas, 1908j: 642; 100 miles NW of Tayuenfu, Shanxi, China.
Also other extralimital synonyms.

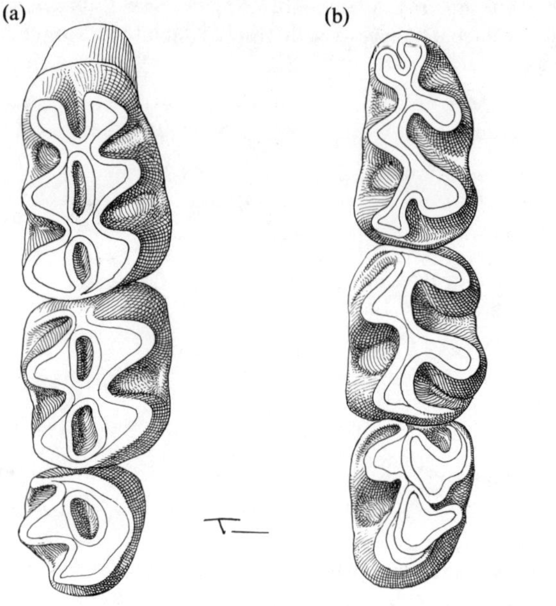

Fig. 35. *Cricetulus migratorius.* Molars: (a) upper right; (b) lower left.

Map **166**. *Cricetulus, Calomyscus, Cansumys*

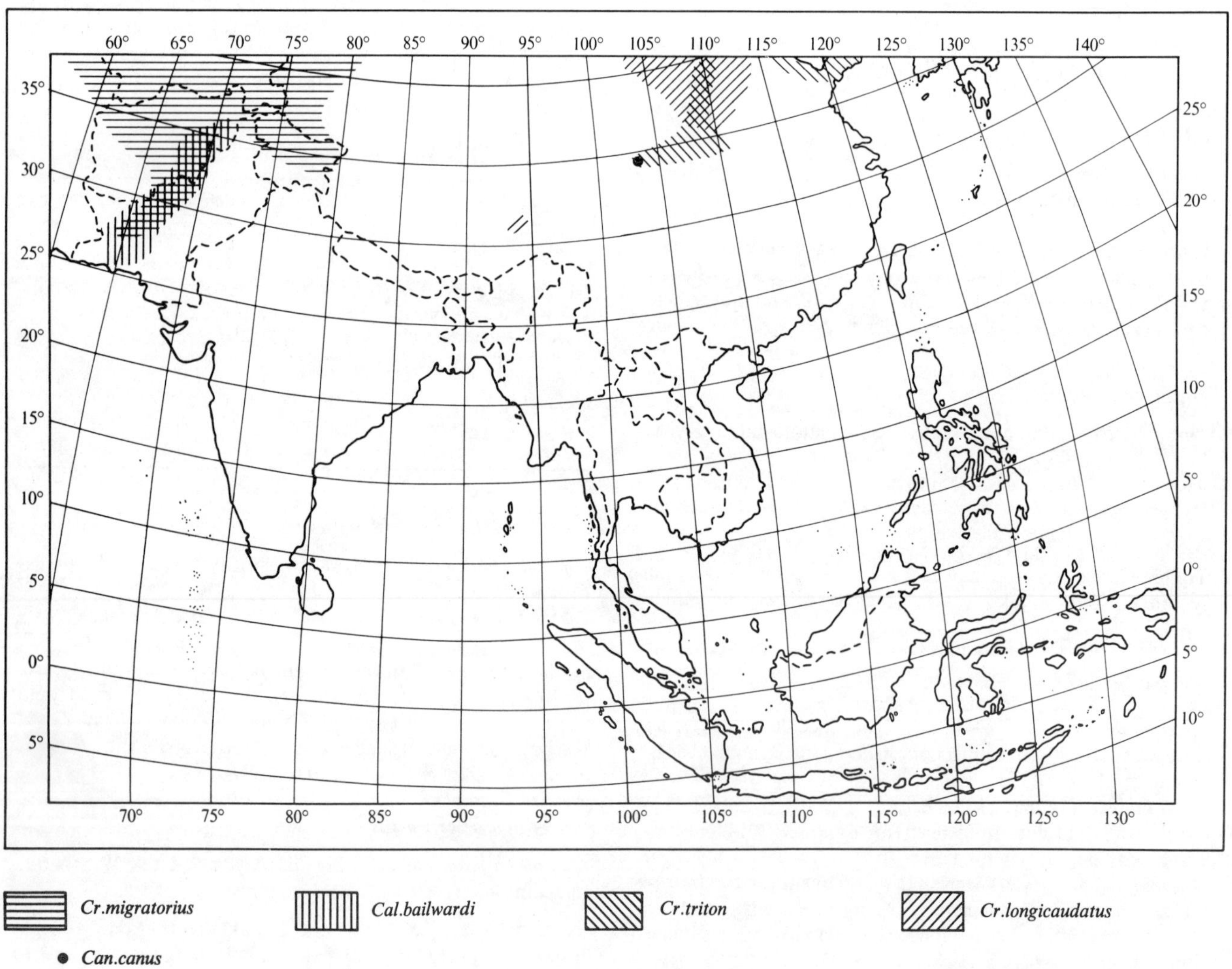

Range (Map 166). N Sichuan, S Shaanxi and S Gansu, China; also from Shanxi westwards through much of Mongolia and some adjacent parts of USSR.

Cricetulus triton Greater long-tailed hamster

Cricetus (*Cricetulus*) *triton* de Winton, 1899: 575; N Shantung, China.
Including several extralimital synonyms.

Range (Map 166). From S Shaanxi (both north and south of the Qinling Mts) and Jiangsu, through NE China to Heilongjiang; at altitudes of 1050–1380m on the south side of the Qinling Mts.

Variation. The southern population that extends into the region is:
C. t. collinus Allen, 1925b: 5; base of Taipai Shan, Qinling Mts, Shaanxi. Characterized by darker pelage than the more northern forms.

A further subspecies has been described from the same region:
C. t. ningshaanensis Song, 1985: 137; Ningshan dist, Shaanxi, 1050m. Characterized by longer tail (66% of head and body) and more extensive white distal zone of tail (34–52mm).

CANSUMYS

Cansumys Allen, 1928b: 244; type species *C. canus* Allen.

C China; one species. A distinctive and atypical hamster characterized especially by having a long, hairy tail and very hypsodont but rooted teeth. Included in *Cricetulus triton* by Argyropulo (1933), followed by Ellerman (1941) and others. Externally it somewhat resembles a young, greyish *C. triton*, but the extreme hypsodonty of the cheek-teeth is a significant difference. Although described as having a short nail-like claw on the hallux, it is only marginally shorter than in *C. triton* and is still claw-like, exceeding the tip of the digit.

Cansumys canus

C. canus Allen, 1928b: 245; Choni, S Gansu, China [34°33'N, 103°20'E].

Range (Map 166). Known only from the type locality in S Gansu, China, probably around 2500m.

Recognition. The two specimens examined (holotype and paratype) both have white patches under each ear and on the side of the neck, and a white line in the centre of the throat, not seen in any *Cricetulus triton*.

CALOMYSCUS Mouse-like hamsters

Calomyscus Thomas, 1905b: 23; type species *C. bailwardi* Thomas.

Turkmenistan to Pakistan; a complex of allopatric forms that have been considered conspecific, e. g. by Ellerman & Morrison- Scott (1951), but have more recently been treated as five separate species, influenced by differences in chromosomes between forms in Turkmenistan and Transcaucasia (Vorontsov *et al.*, 1979). These authors recognized two species in Pakistan and E Afghanistan, differing only in size, the larger *C. baluchi* (including *mustersi*) in the north and the smaller *C. hotsoni* in S Baluchistan. However additional material from Pakistan suggests that these intergrade and they are here treated as forms of *C. bailwardi.*

Calomyscus is an aberrant genus, very doubtfully included in the Cricetinae. On the basis of the auditory ossicles it has been linked with the American Reithrodontomyinae rather than with the Cricetinae (Pavlinov, 1980a), but the genitalia are very different from both (Vorontsov & Potapova, 1979).

They occur in arid country at all altitudes up to c. 3400m, feeding on seeds, which they store, as well as buds and other green plant material.

Calomyscus bailwardi Mouse-like hamster

C. bailwardi Thomas, 1905c: 23 (and 1906b: 525); Malamir, 70 miles NE of Ahwaz, Iran.
C. baluchi Thomas, 1920d: 939; Kelat, Baluchistan, Pakistan.
C. hotsoni Thomas, 1920d: 939; Panjgur dist, Baluchistan, Pakistan.

Range (Map 166). Pakistan, east of the Indus, and SE Afghanistan; also in Iran.

Variation. Two subspecies can be recognised:
C. b. baluchi (including *hotsoni*). Baluchistan. Dorsal pelage sandy brown; smaller on average (skull 23–26mm); tail usually longer than head and body.
C. b. mustersi Ellerman, 1948: 804; Paghman, 17 miles W of Kabul, Afghanistan. E Afghanistan. Dorsal pelage dark greyish brown; larger (skull 25–27mm); tail about equal to head and body.

Subfamily Gerbillinae

Gerbils, Jirds

Gerbillina Gray, 1825: 342; type genus *Gerbillus* Desmarest.

Africa and S Palaearctic, and marginally in the Indian Subregion; c. 95 species in 14 genera of which 4 are represented in the Indomalayan Region (Table 254). A clearly defined group of terrestrial rodents adapted to arid habitats. The tail is usually long and well-haired, forming a more extensive terminal brush than in most Murinae. The auditory bullae are inflated, often extremely so, and the incisors are grooved. For a comprehensive review and classification see Pavlinov *et al.* (1990).

Table 254. The genera of Gerbillinae.

	Gerbillus	*Tatera*	*Meriones*	*Rhombomys*
Species in region	3	1	4	1
Head & body	65–105	140–190	100–170	170–185
Tail	95–145	150–200	80–180	150–175
as % of h & b	125–160%	105–130%	80–110%	85–95%
Hind feet	20–32	34–40	24–41	39
Ear	11–17	21–31	10–26	14–15
Sole	Hairy or naked	Naked	Hairy or naked	Hairy
Tail: sides paler than dorsal and ventral	0	+	0	0
Skull: greatest length	27–30	39–51	31–43	42–46
Upper molar row	3.2–3.9	5.0–6.8	3.9–5.7	6.8–7.3
I^1: no. of grooves	1	1	1	2
Molars rooted	+	+	+	0

GERBILLUS Gerbils

Gerbillus Desmarest, 1804: 22; type species *G. aegyptius* Desmarest, in *G. gerbillus* (Olivier).
Dipodillus Lataste, 1881: 506 (*G. simoni* Lataste).
Endecapleura Lataste, 1882: 127 (*G. garamantis* Lataste).
= *Hendecapleura* Lataste, 1884: 258 (unjustified emendation).
Petteromys Pavlinov, 1982: 30 (*Meriones dasyurus* Wagner).

SW Asia, N and E Africa; c. 50 species, of which 3 reach Pakistan and Afghanistan (Table 255). These are small gerbils with tail much longer than the head and body, and long, slender hind feet. *G. nanus* has been placed in a separate genus *Dipodillus* based upon naked soles, but this has been shown to be an unsatisfactory basis for dividing the genus *Gerbillus* when all species and other characters are taken into account (Lay, 1983). The genus was reviewed by Lay & Nadler (1975) and Lay (1983).

Gerbillus nanus Baluchistan gerbil

G. nanus Blanford, 1875b: 312; Gedrosia, W of Gwadar, Mekran Dist, Baluchistan, Pakistan.
Dipodillus indus Thomas, 1920d: 935; Gambat, Khairpur, Sind, Pakistan.

Range (Map 167). From NE India (Rajasthan and Gujarat) through most of lowland Pakistan and S Afghanistan; west through Iran and the Arabian peninsula to Israel and N Africa, as far as Morocco.

Ecology. Found in arid country, usually on rocky or stony ground with thorn scrub, avoiding sand-dunes.

Remarks. The form *indus* was included in *G. dasyurus* by Ellerman (1961) but has subsequently been shown to be a form of *G. nanus* (Harrison, 1972).

Table 255. The species of *Gerbillus*.

	nanus	*gleadowi*	*cheesmani*
Head & body	65–90	75–100	90–105
Tail	95–125	120–145	110–150
as % of h & b	140–160%	c.150%	c.130%
Hind feet	20–26	26–32	26–31
Black hairs in dorsal pelage	+	0	+
Soles of hind feet	Naked	Hairy	Hairy
Skull: greatest length	22–28	27–30	26–29
greatest height *	8.5–11.1	11.0–11.9	12.1–13.5
greatest width across both bullae	12.0–13.8	14.0–15.3	15.3–17.1
Chromosomes: 2n	52	51 (male) 50 (female)	38
no.of acrocentrics	42	28	2

* With skull resting on plane surface.

Map **167**. *Tatera, Gerbillus*

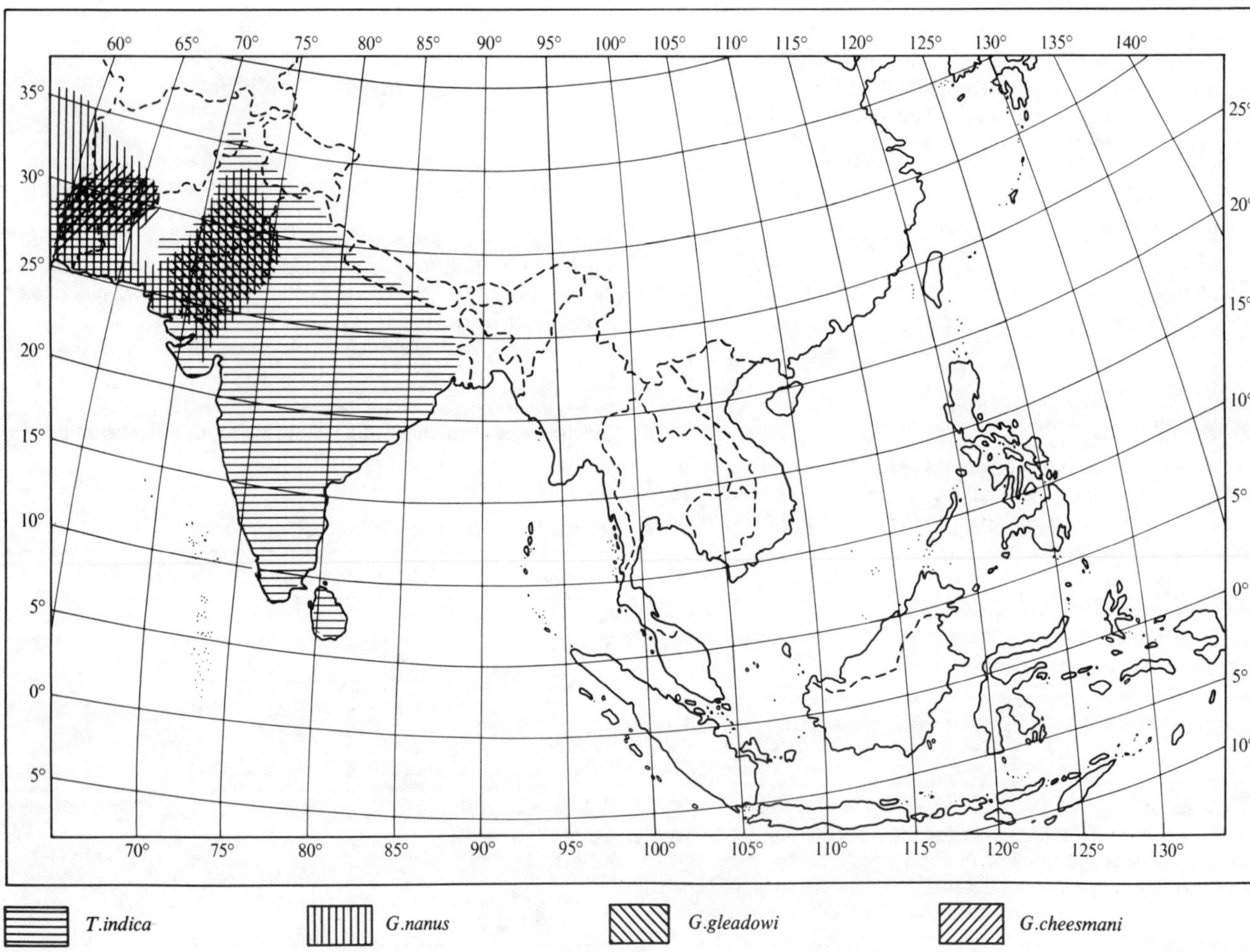

Gerbillus gleadowi Indian hairy-footed gerbil

G. gleadowi Murray, 1886: 246; Beruto, 15 miles SW of Rehti, Mirpur-Drahraki Taluka of Rohri dist, Upper Sind, Pakistan.

Range (Map 167). E Pakistan (mostly E of the R Indus) and adjacent parts of NW India, in W Rajasthan and Gujarat.

Ecology. Found in sand-dunes and arid plains, avoiding cultivation.

Gerbillus cheesmani Cheesman's gerbil

G. cheesmani Thomas, 1919c: 748; near Basra, lower Euphrates, Iraq.

Range (Map 167). NE Baluchistan, in Pakistan and S Afghanistan; also west through parts of Iran and Iraq to the Arabian peninsula.

Variation. Two subspecific names have been proposed that are relevant to the region:
G. c. aquilus Schlitter & Setzer, 1973: 167; 60km W of Kerman, Iran.
G. c. subsolanus Schlitter & Setzer, 1973: 168; 36km E of Nok Kundi, Pakistan, c. 28°48'N, 62°46'E.

Ecology. Occurs on sand-dunes and clay flats in semidesert conditions.

Remarks. The population in Afghanistan/Pakistan/E Iran was treated as a species, *G. aquilus*, separate from *G. cheesmani s. s.* of Iraq and Arabia by Lay & Nadler (1975) and Lay (1982). However this was based upon a small difference in karyotype and average differences in some cranial dimensions, and subspecific rank seems more appropriate. The justification for retaining *G. cheesmani* as a species separate from the allopatric *G. gleadowi* is marginal in terms of morphology but substantial when the great difference in karyotype is considered (see Table 255).

TATERA

Tatera Lataste, 1882: 126; type species *Dipus indicus* Hardwicke.
Also synonyms based upon African species.

Africa to India and Sri Lanka; c. 11 species of which only one reaches the Indomalayan Region. The genus was limited to *T. indica* by Pavlinov *et al.* (1990) who placed the African species in a separate genus, *Gerbilliscus* Thomas, 1897. They are large, rat-like rodents, but distinguished from all Indian *Rattus* and other Murinae by the length of hair on the distal part of the tail.

One fossil species has been described from the region:
Tatera sinhaleya Deraniyagala, 1960a: 13, pl. 1; Marlakkara vila, Muvagama, near Ratnapura, Sri Lanka; Pleistocene. (Based only upon a lower incisor which was purported to be thicker and more curved than Recent specimens in Sri Lanka.)

Tatera indica Indian gerbil, Antelope rat

Dipus indicus Hardwicke, 1807: 279; between Benares and Hardwar, United Provinces, N India.
Gerbillus otarius Cuvier, 1838: 144, pl. 26; peninsular India.
G. hardwickei Gray, 1843a: 132; Dharwar, India (based upon Elliot, 1839: 211).
T. ceylonica Wroughton, 1906: 477; Sri Lanka.
T. sherrini Wroughton, 1917: 43; Jacobabad, Sind, Pakistan.
T. dunni Wroughton, 1917: 43; Ambala, Punjab, India.
And other, extralimital synonyms.

Range (Map 167). Throughout lowland India and Nepal, east to the Ganges delta; west through Pakistan and S Afghanistan; Sri Lanka; also through S Iran and Iraq to Syria.

Variation. Those in peninsular India and Sri Lanka are larger (hind feet usually >40mm) than those from N India and Pakistan, and the tympanic bullae are on average a little smaller in the south. These have been recognized as subspecies, *T. i. indica* in the north and *T. i. cuvieri* in the south (Agrawal & Chakraborty, 1981; Bates, 1988), but there is no evidence of a discrete boundary between the two forms and it is likely that variation is clinal.

Chromosome polymorphism has been reported in samples from Mysore with 13–16 submetacentric chromosomes giving NF=82–86 but 2n constant at 68 (Aswathanarayana & Manjunatha, 1981b).

Ecology. An adaptable species occurring in both arid, uncultivated habitats and in cultivated ground where it can be a serious pest, e. g. of sweet potato and grain. Also occurs in grain stores, especially in villages but also in some cities, e. g. Karachi. They feed upon green plant material and insects as well as grain and roots (Prasad, 1954).

MERIONES Jirds, Sand rats

Meriones Illiger, 1811: 82; type species *Mus tamariscinus* Pallas.
Cheliones Thomas, 1919d: 265 (*Gerbillus hurrianae* Jerdon).
Parameriones Heptner, 1937: 190 (*Gerbillus persicus* Blanford).
And other synonyms based upon extralimital species.

Central Asia to N Africa; c. 16 species of which 4 occur in Pakistan (and one of these in NW India) (Table 256). Rat-like rodents with tail about equal to head and body. The molar teeth are more hypsodont than in *Gerbillus* and *Tatera* (but develop roots), the cusps wearing quickly to form a single flat wearing surface on each tooth. The genus was revised by Chaworth-Musters & Ellerman (1947).

Meriones persicus Persian jird

Gerbillus persicus Blanford, 1875b: 312; Kohrud, 150 miles N of Isfahan, Iran.
And other extralimital synonyms.

Range (Map 168). W Pakistan and Afghanistan; and west through Iran and S Soviet Turkestan to the S Caucasus.

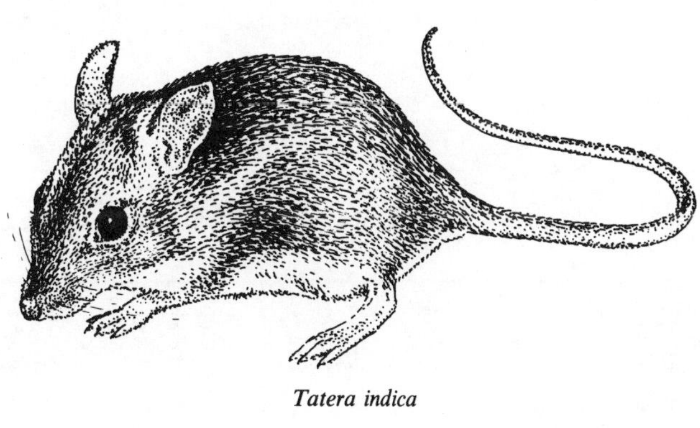

Tatera indica

Variation. One subspecific name has been proposed:
M. p. baptistae Thomas, 1920d: 934; Pasht Kuh, Baluchistan, Pakistan, 27°2'N, 65°12'E.

Ecology. Mainly in rocky montane country, but including cultivated ground in such areas; nocturnal.

Table 256. The species of *Meriones*.

	persicus	*hurrianae*	*libycus*	*crassus*
Head & body	130–170	115–145	115–170	100–125
Tail	145–180	100–150	125–160	80–115
as % of h & b	c.110%	c.100%	c.100%	80–90%
Hind feet	35–41	27–36	30–38	24–29
Ears	21–26	10–14	17–22	13–21
Soles of hind feet	Naked	Naked	Hairy	Hairy
Claws	Pale	Dark	Dark	Dark
Venter	White	Light yellowish grey	White	White
Dorsal pelage: short (5–8mm): S long (>15mm): L	L	S	L	L
Skull: greatest length	40–43	31–37	39–40	32–35
Upper molar row	5.5–5.7	4.5–5.5	c.5.1	3.9–4.5
Bullae (% of skull)	30–31%	27–32%	34–35%	38–40%

Meriones hurrianae Indian desert jird

Gerbillus hurrianae Jerdon, 1867: 186; Hissar, Hariana, India.

Range (Map 168). NW India (Gujarat, Rajasthan), through the Indus plains and along the Baluchistan coast in Pakistan; also marginally in SE Afghanistan.

Variation. One subspecific name has been proposed:
Cheliones h. collinus Thomas, 1919b: 726; Kohat, NW Frontier Province, Pakistan, 1000–1700ft.

A study of variation in the Indian Desert has shown that with increasing aridity body size tends to be smaller and the extremities (tail, feet, ears) relatively longer (Prakash & Kumari, 1978).

Ecology. An abundant species of the Indus plains, found mainly in desert and thorn scrub and on the edge of cultivation, but not usually a serious agricultural pest. They are colonial and diurnal.

Meriones libycus Libyan jird

M. libycus Lichtenstein, 1823a: 5; near Alexandria, Egypt.
Gerbillus erythrourus Gray, 1842: 266; Sahlabad, c. 12 miles SW of Kandahar, Afghanistan.
Also many extralimital synonyms.

Map **168**. *Meriones, Rhombomys*

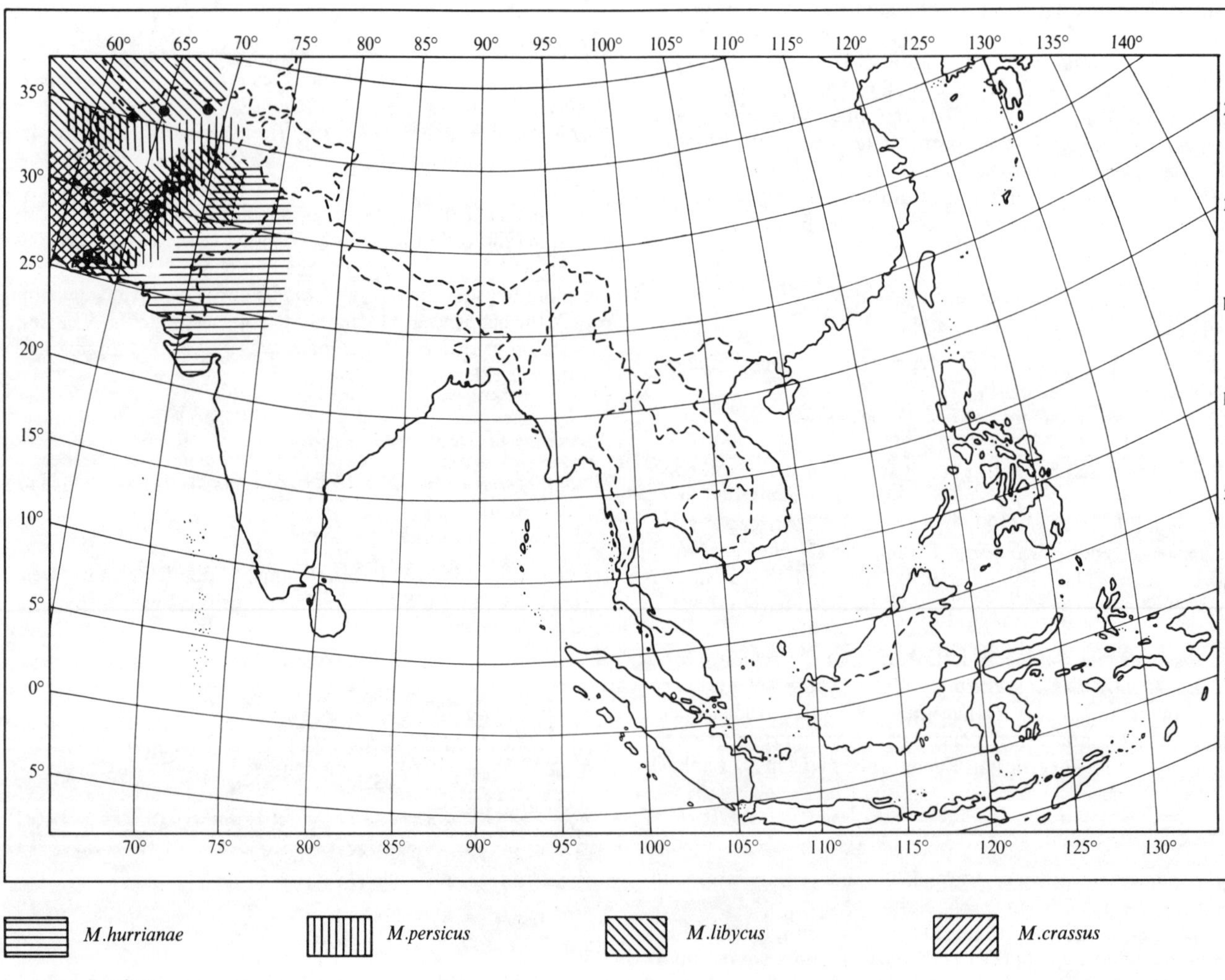

Range (Map 168). Baluchistan, in W Pakistan and E Afghanistan; also from W China to the Sahara. From sea level to 1680m in Baluchistan.

Ecology. Mainly on sandy plains and upland valley bottoms, including the edge of cultivation; a pest of green crops in places.

Meriones crassus Sundevall's jird

M. crassus Sundevall, 1842: 233; Ain Musa, Sinai, Egypt.
Gerbillus swinhoei Scully, 1881a: 228; Gatai, between Kandahar and Kojak Pass, c. 10 miles N of Chaman, Afghanistan.
Also many extralimital synonyms.

Range (Map 168). E Pakistan and S Afghanistan; also west to Arabia and the Sahara.

Ecology. Found on arid stony plains away from cultivation.

Remarks. Reviewed in detail by Koffler (1972).

RHOMBOMYS

Rhombomys Wagner, 1841c: 129; type species *R. opimus* Wagner.

Central Asia; one species, which reaches the western border of Pakistan. Characterized by large size, very hypsodont, rootless molars, and *two* grooves on each upper incisor.

Rhombomys opimus Great gerbil

Meriones opimus Lichtenstein, 1823b: 122; between Orenburg and Bokhara, Uzbekistan, USSR.
Also extralimital synonyms.

Range (Map 168). N Baluchistan, in a very small part of Pakistan and extreme S Afghanistan; also widespread in central Asia from the Caspian Sea east to Mongolia and N Gansu, China. In Pakistan in sand-dune habitat around 1000m.

Subfamily Myospalacinae

Zokors

Myospalacini Lilljeborg, 1866: 25; type genus *Myospalax* Laxmann.

Central Asia to N and W China; one genus. These are burrowing rodents with rootless, hypsodont teeth resembling those of the Arvicolinae (Fig. 29f) but differing from that group in the structure of the skull, especially the infra-orbital foramen and zygomatic plate which resemble those in the Cricetinae.

MYOSPALAX

Myospalax Laxmann, 1769: 75; type species *Mus myospalax* Laxmann.
Myotalpa Kerr, 1792: no. 516 in unpaginated catalogue preceding text (*Mus aspalax* Pallas).
= *Siphneus* Brants, 1827: 19.
Eospalax Allan, 1940: 921 (*Siphneus fontanierii* Milne-Edwards).
= *Zokor* Ellerman, 1941: 541.
And other synonyms of the subgenus *Myospalax*.

The sole genus of Myospalacinae; c. 6 species of which 3 occur on the margins of this region (Table 257). As in other groups of fossorial mammals, e. g. the Spalacidae, variation is complex and there is a tendency for allopatric forms to show karyological differences suggesting specific barriers where morphological differences would indicate only subspecific rank. Two subgenera are usually recognized of which only *Eospalax* is represented in this region. It was revised by Fan & Shi (1982) who recognized and mapped five allopatric species. On the basis of additional material these were reduced to three by Li & Chen (1989) who are followed here.

One fossil species has been described from the edge of the region:
Syphneus tingi Young, 1927: 45; Honan, China; Lower Pleistocene.

Table 257. The species of *Myospalax*.

	fontanierii	*smithii*	*rothschildi*
Head & body	160–245	160–225	150–165
Tail	32–46	34–55	30–45
Hind feet	25–40	25–33	23–27
Claws on fore feet (length & width of longest; dorsal)	Robust c.14x2mm	Robust c.12x2mm	Delicate. c.9x1.5mm
Skull: greatest length	41–59	42–51	36–41
Upper molar row	9.0–13.0	9.1–10.7	7.7–9.7
Parietal ridges converge in mid-line	0	+	0

Myospalax fontanierii Common Chinese zokor

Siphneus fontanierii Milne-Edwards, 1867a: 376; near Peking, N China.
Myotalpa cansus Lyon, 1907i: 134; Taocheo, S Gansu.
Myotalpa rufescens Allen, J. A., 1909b: 428; foot of Taipashing, Shaanxi.
Myospalax baileyi Thomas, 1911h: 727; Rama Song, between Nagchuka and Tatsienlu, W Sichuan.
And other synonyms from N and W China.

Range (Map 169). Central and NE China, from central Sichuan through Shaanxi and Gansu, and NE to SE Heilongjiang.

Remarks. The western forms *cansus* and *baileyi* were treated as independent species by Fan & Shi (1982) but as subspecies of *M. fontanierii* by Li & Chen (1989).

Myospalax rothschildi Rothschild's zokor

M. rothschildi Thomas, 1911i: 722; 40 miles SE of Taochow, Gansu.
M. minor Lönnberg, 1926: 6; near Ashuan, Murshan, Gansu.

Range (Map 169). W Hubei, S Shaanxi and S Gansu, China.

Variation. One subspecific name has been proposed:
M. r. hubeinensis Li & Chen, 1989: 95; 'E region of Daba Mt, Shaanxi Prov., and Badueng and Shenuengjia region, Hubei Prov. '

Myospalax smithii Smith's zokor

Myospalax smithii Thomas, 1911i: 720; 30 miles SE of Taochow, Gansu, China.

Range (Map 169). S Gansu and W Shaanxi, China.

Map **169**. *Myospalax, Typhlomys, Platacanthomys, Ellobius, Hyperacrius*

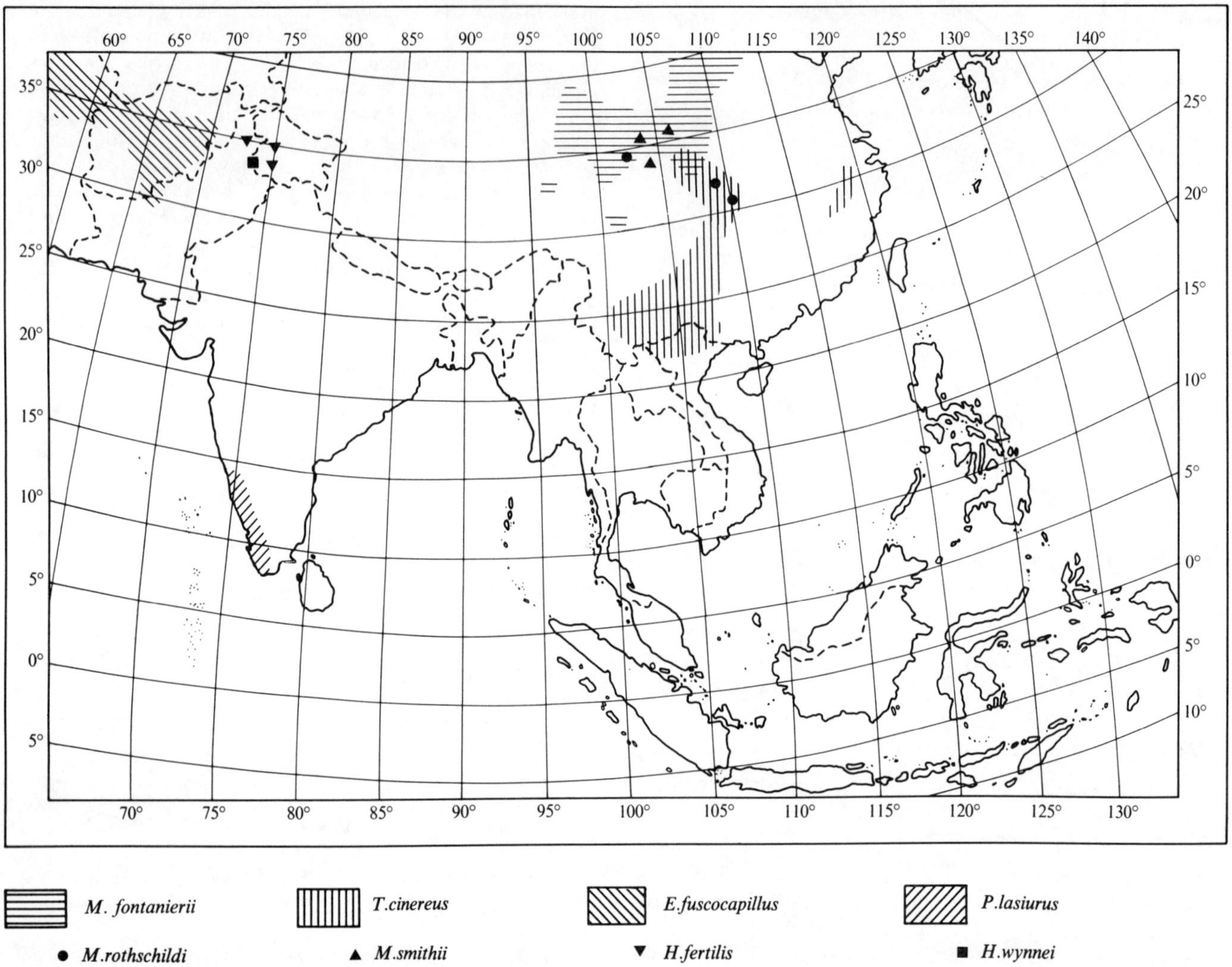

Subfamily Platacanthomyinae

Platacanthomyinae Alston, 1876: 81; type genus *Platacanthomys* Blyth.

S India and S China; 2 monospecific and very distinctive genera (Table 258). These were formerly associated with the Gliridae (dormice) as a subfamily or separate family, but were treated as a subfamily of the Cricetidae by Chaline *et al.* (1977). The two genera are very different externally but have similar teeth, rather hypsodont (but rooted) with diagonal laminae (Figs 36–38).

PLATACANTHOMYS

Platacanthomys Blyth, 1859a: 288; type species *P. lasiurus* Blyth.

One Recent species. The following fossil species has been described from the region:
P. dianensis Qiu, 1989: 270; Shihiuba, Lufeng, Yunnan, China; Upper Miocene.

Table 258. The genera and species of Platacanthomyinae.

	Platacanthomys lasiurus (Fig.37)	*Typhlomys cinereus* (Fig.38)
Range	India	China, Vietnam
Head & body	130–140	80–90
Hind feet	24–26	20–22
Tail	Bushy, brown, < head & body	Short-haired at base, tufted in distal half, > head & body
Dorsal pelage	Spiny, light brown	Soft, grey
Venter	Creamy	Grey
Supra-orbital ridges	Prominent	Weak or absent
Post-palatal foramina (between molars)	Single, large	Multiple, small

Platacanthomys lasiurus Malabar spiny mouse

P. lasiurus Blyth, 1859a: 289; Alipi, Malabar, S India.

Range (Map 169). SW peninsular India, north at least to 14° N in Shimoga District, Karnataka (Rajagopalan, 1968).

Platacanthomys lasiurus

(a)

(b)

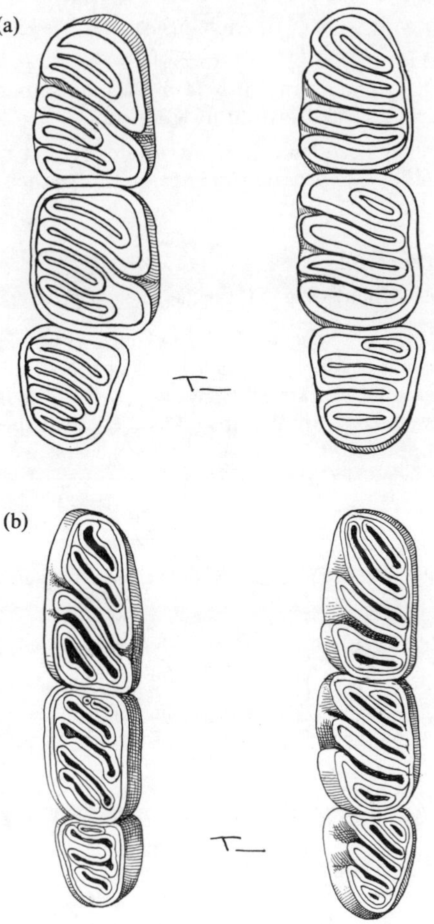

Fig.**36**. Upper right and lower left molars of (a) *Platacanthomys lasiurus*; (b) *Typhlomys cinereus*.

TYPHLOMYS

Typhlomys Milne-Edwards, 1877: 9; type species *T. cinereus* Milne-Edwards.

One Recent species. The following fossil species have been described from the region:
T. primitivus Qiu, 1989: 273; Shihiuba, Lufeng, Yunnan, China; Upper Miocene.
T. hipparionum Qiu, 1989: 276; as above.

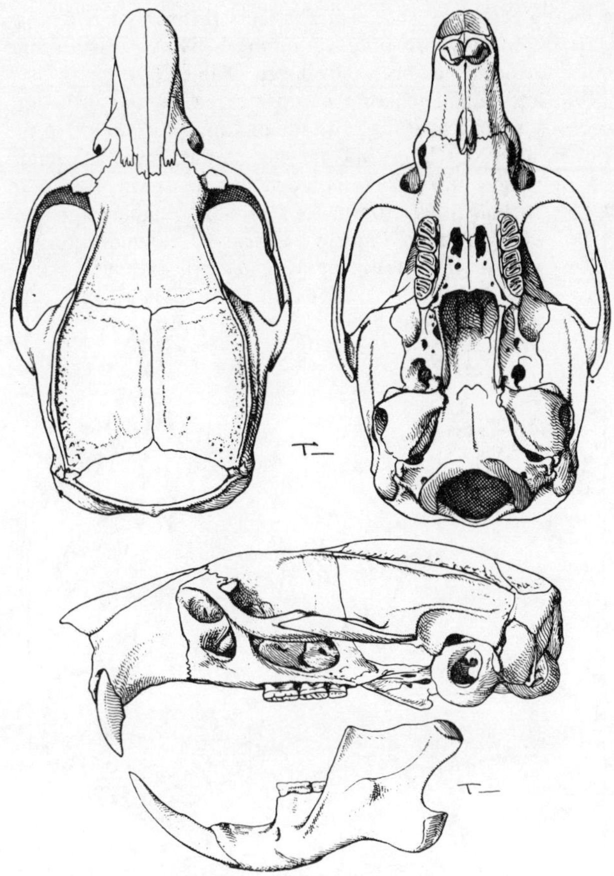

Fig.**37**. *Platacanthomys lasiurus*.

Typhlomys cinereus

T. cinereus Milne-Edwards, 1877: 9; W Fujian, China.

Range (Map 169). S China from S Gansu and S Shaanxi south to Yunnan, Guangxi and Fujian; and in N Vietnam.

Variation. Four, possibly five subspecies have been recognized (Wang & Li, in press; Table 259):
T. c. cinereus. N Fujian, N Zhejiang, S Anhui.
T. c. subsp. Guizhou to Hubei, S Shaanxi and S Gansu.
T. c. jindongensis Wu & Wang, 1984: 213; Jindong Dist, Yunnan, 2430m. Central Yunnan.
T. c. chapensis Osgood, 1932: 298; Chapa, Tonkin, N Vietnam.
Perhaps also an additonal subspecies in central Guangxi.

Subfamily Arvicolinae

Voles

Arvicolidae Gray, 1821: 303; type genus *Arvicola* Lacépède, 1799.
Microtidae Cope, 1891: 90 (*Microtus* Schrank, 1798).

Holarctic; c. 120 species in c. 20 genera. The dominant group of small, herbivorous rodents in the Palaearctic Region, characterized by hypsodont, often rootless, molars. The wearing surfaces form a characteristic pattern of alternating triangles (Fig. 39). One genus, *Hyperacrius*, occurs entirely within the borders of the Indomalayan Region (albeit close to the boundary); another, *Proedromys*, is endemic to central China, including Sichuan; *Eothenomys*, occurs throughout S China; while *Ellobius* is represented very marginally in Pakistan; one genus, *Ondatra*, has been introduced (Table 260).

Two additional genera are represented by the following species which occur at high altitude (mainly over 3000m) in the Himalayas and China and are not treated here in detail, although some have been recorded a little below 3000m in places: *Alticola roylei* (down to 2000m in Afghanistan and 2440m in Pakistan); *Pitymys sikkimensis* (including *irene*) (down to 2700m in parts of Nepal); *P. leucurus* (Himalayas); *P. juldaschi* (Pakistan); *P. afghanus* (down to 2400m in E Afghanistan).

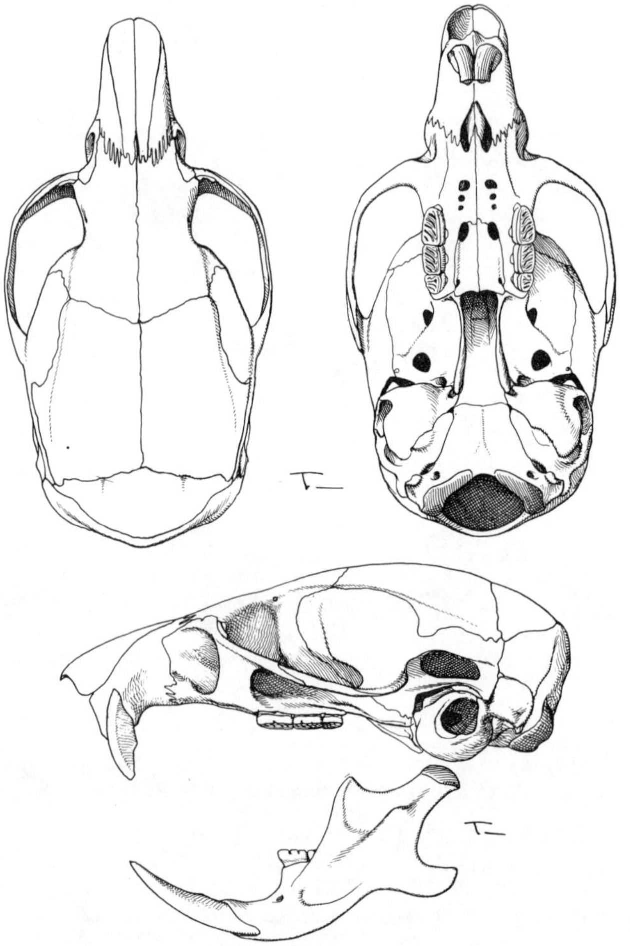

Fig.**38**. *Typhlomys cinereus.*

ELLOBIUS Mole-voles

Ellobius Fischer, 1814: 72; type species *Mus talpinus* Pallas.

Steppe zone of central Asia; c. 5 species, but number somewhat uncertain due to the recognition of allopatric forms with distinctive karyotypes. One species extends into E Afghanistan and Pakistan. They are the most subterranean of the Arvicolinae, with very pro-odont, white incisors, rudimentary ear pinnae, very small eyes, and very short tail.

Ellobius fuscocapillus Afghan mole-vole

Georhychus fuscocapillus Blyth, 1842b: 887; Quetta, Baluchistan, Pakistan.
Also synonyms from NW Afghanistan and Turkmenistan.

Range (Map 169). NE Pakistani Baluchistan and adjacent parts of E Afghanistan. Also from Afghanistan through S Soviet Turkestan and N Iran to Kurdistan.

Ecology. They occur in arid montane steppe, but need fairly deep soil. They forage mainly in underground tunnels for bulbs, rhizomes etc.

HYPERACRIUS

Hyperacrius Miller, 1896: 54; type species *Arvicola fertilis* True.

Kashmir etc. ; 2 species. Probably most closely related to *Alticola* which occurs at high altitudes in the Himalayas and central Asia, but characterized by simpler dental pattern. The genus was reviewed by Phillips (1969). The two species seem quite distinct (Table 261). A Pleistocene species has been described from N China.

Hyperacrius wynnei

Arvicola wynnei Blanford, 1881: 244; Murree, Rawalpindi, Pakistan.

Range (Map 169). Murree region of NE Pakistan and adjacent part of Kashmir, at altitudes of 1850–3050m.

Variation. One subspecific name has been proposed:
H. w. traubi Phillips, 1969: 463; Yakh Tangai, Swat, Pakistan, 6600ft. Smaller.

Ecology. Mainly in coniferous forest and associated grassland.

Table 259. The subspecies of *Typhlomys cinereus*; mean values; modified from Wang & Li (in press).

	cinereus	S Sichuan etc.	*jindongensis*	*chapensis*
Head & body	77	82	75	86
Tail (% of h & b)	131%	143%	144%	140%
Hind feet	19.5	22.4	22.0	22.4
Fore feet	Whitish	Brownish in centre	Brownish in centre	Whitish
Hind feet	Whitish	Dull brown	Dull brown	Dusky
Venter	Greyish white	Greyish white	Greyish white	Dull buff
Posterior palatal vacuities	1–2	3–4	2–3	3–4
Posterior edge of nasals cf posterior edge of premaxillae: level (L); anterior (A)	L	A	L	L
Skull: greatest length	22.7	25.0	23.6	25.6
Length of nasals	6.5	7.4	7.5	7.5
Upper molar row	3.6	3.9	3.7	4.0

Table 260. The genera of Arvicolinae.

	Ellobius	*Hyperacrius*	*Eothenomys*	*Microtus*	*Proedromys*	*Ondatra*
Species in region	1	2	9	3	1	1
Head & body	c.140	95–130	85–125	90–135	103	c.300–400
Tail (% of head & body)	c.9%	24–40%	25–70%	45–75%	40%	65%
Hind feet	21–23	16–22	14–23	17–25	18	c.60–80
Ears	Rudimentary	8–13	10–15		13	20–25
Head darker than body	+	0	0	0	0	0
Skull: greatest length	35–40	23–28	21–29	24–33	27	52–65
Upper molar row	7.5–8.0	5.5–7.2	4.7–7.0	5.5–7.8	6.8	13–15
Upper incisors: pro-odont	++	0	0	0	0	0
grooved	0	0	0	0	+	0
pigmented	0	+	+	+	+	+
Molars remaining rootless	0	+	+	+	+	+
M^3: no.of medial ridges	2	3	3–5	4	2	3
no.of outer ridges	3	2/3	3–5	3	3	3/4

Table 261. The species of *Hyperacrius*.

	wynnei	*fertilis*
Altitude range	1850–3050m	2450–3600m
Head & body	110–130	95–110
Ear	8–11	10–13
Pelage: texture	Long, lax (10–12mm)	Shorter, stiffer, denser (8–9mm)
Skull: greatest length	25–28	23–25
Upper molar row	6.2–7.2	5.5–6.2
Baculum: width of base as % of length	c.90%	c.40%

Hyperacrius fertilis True's vole

Arvicola fertilis True, 1894a: 10; Pir Panjal Mts, Kashmir, 8500ft.
M. (*H.*) *aitchisoni* Miller, 1897: 141; Gulmerg, Kashmir, 9000ft.
Microtus (*H.*) *brachelix* Miller, 1899: 290; Nagmarg, Kashmir, 9000ft.

Range (Map 169). N Pakistan, S Kashmir and adjacent parts of Indian Punjab, at altitudes of 2450–3600m.

Variation. One subspecific name has been proposed:
H. f. zygomaticus Phillips, 1969: 469; 6 miles SW of Utror, Swat Pakistan, 8900ft.

Ecology. Normally above the tree-line, in subalpine scrub and meadows.

EOTHENOMYS

Eothenomys Miller, 1896: 45; type species *Arvicola melanogaster* Milne-Edwards.
Anteliomys Miller, 1896: 47 (*Microtus chinensis* Thomas).

China, Korea and Japan; c. 11 species of which c. 9 occur in S China (Table 262). Very similar to *Alticola* and *Clethrionomys* but with rootless molars. The southern forms are badly in need of revision, especially to establish the limits of *E. melanogaster* in which Ellerman & Morrison-Scott (1951) included a great diversity of forms that seem unlikely to be conspecific. The following arrangement is very tentative.

Eothenomys melanogaster

Arvicola melanogaster Milne-Edwards, 1872a: 93 (footnote); Moupin, Sichuan, China.
? *Microtus* (*E.*) *aurora* Allen, G. M., 1912b: 211; Changyanghsien, Hubei, China.
M. (*E.*) *mucronatus* Allen, G. M., 1912b: 214; Tachiao, W Sichuan.
M. (*E.*) *bonzo* Cabrera, 1922: 168; Foochow, Fujian, China.
E. kanoi Tokuda, 1937: 1118; Taiwan.

Range (Map 170). S China from Sichuan and Fujian to Yunnan, and N Assam; Taiwan.

Variation. The following subspecific names have been proposed:
Microtus (*E.*) *m. colurnus* Thomas, 1911e: 209; Kuatun, Fujian, China.
? *M.* (*E.*) *m. eleusis* Thomas, 1911c: 50 (and 1912b: 139); E of Chaotungfu, Yunnan, China. A long-tailed form very doubtfully referable to *E. melanogaster*.
E. m. confinii Hinton, 1923a: 151; Kiuchiang-Salween divide, Yunnan, China, 28°N, 11000ft.
E. m. libonotus Hinton, 1923a: 151; Dreyi, Mishmi Hills, N Assam, India, 5140ft.

Eothenomys miletus

Microtus (*E.*) *melanogaster miletus* Thomas, 1914h: 474; 10 miles W of Yangpi, W Yunnan, China.
? *M.* (*E.*) *cachinus* Thomas, 1921a: 504; Mt Imaw Bum, Kachin Prov., Burma, 9000ft.
E. fidelis Hinton, 1923a: 150; W flank Likiang Range, Yunnan, China, 27°30'N, 13000–14000ft.

Range (Map 170). W Yunnan and adjacent parts of Burma; perhaps also extreme N Thailand.

Remarks. Chromosomes: 2n=56, all telocentric (Wu *et al.*, 1989).

Eothenomys olitor

Microtus (*E.*) *olitor* Thomas, 1911c: 50 (and 1912b: 139); Chaotungfu, Yunnan, China.

Range (Map 170). W Yunnan, at around 2000m.

Eothenomys proditor

E. proditor Hinton, 1923a: 152; Likiang Range, Yunnan, China, 27°30'N, 13000ft.

Range (Map 170). Sichuan and NW Yunnan, China, at altitudes of 2800–4200m.

Table 262. The species of *Eothenomys*.

	melanogaster	*miletus*	*olitor*	*proditor*	*chinensis*	*wardi*	*custos*	*inez*	*eva*
Range	S China	Yunnan	Yunnan	Yunnan, Sichuan	Sichuan	Yunnan	Yunnan, Sichuan	Shaanxi	Sichuan etc
Head & body	90–96	110–125	80–92	105–115	110–125	105–110	90–105	90–95	85–100
Tail (% of h & b)	25–35%	35–50%	35–43%	25–30%	60–70%	55–60%	40–60%	40–45%	55–65%
Hind feet	15–17	18–21	14–16	17–19	20–23	19–20	16–18	15–17	16–18
Dorsal pelage	Blackish brown, slightly frosted	Medium brown	Dark greyish brown, frosted	Medium yellowish brown	Medium brown slightly frosted	Medium brown slightly frosted	Medium brown	Yellowish brown	Medium brown
Ventral pelage	Dark grey, slightly frosted	Medium grey	Medium grey frosted	Grey, tinged buff	Medium grey, slightly frosted	Medium grey, slightly frosted	Medium grey, slightly frosted	Grey, strongly suffused buff	Medium grey slightly tinged buff
Skull: greatest length	23–26	27–28	21–24	25–28	26–29	23–27	23–25	22–24	22–24
Upper molar row	5.5–6.6	6.2–7.0	5.1–5.8	5.9–6.5	6.2–6.7	5.5–6.0	5.4–6.3	5.2–5.9	4.7–5.8
M^1: no.of ridges on inner side	4	4	3	3	3	3	3	4	3
M_1: adjacent triangles confluent (Con) or closed (Cl)	Con	Con	Con	Con	Con	Con	Con	Cl	Cl
M^3: no.of ridges on inner side	3	4	4	3	4 or 5	4 or 5	4 or 5	3	3

Map **170**. *Eothenomys* (part), *Proedromys*

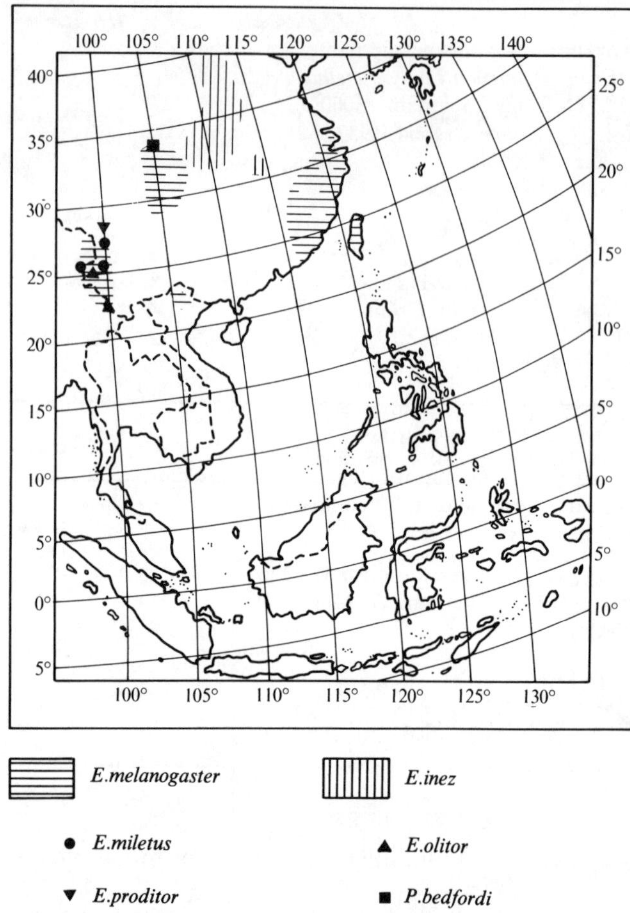

Eothenomys chinensis

Microtus chinensis Thomas, 1891c: 117; Kiatingfu, Sichuan, China.

Range (Map 171). Sichuan, at altitude of 1800–3000m.

Variation. One subspecific name has been proposed: *Microtus (Anteliomys) c. tarquinius* Thomas, 1912c: 517; 23 miles SE of Tatsienlu, Sichuan, China.

Eothenomys wardi

Microtus (Anteliomys) wardi Thomas, 1912c: 516; Chamutong, W of Atunsi, NW Yunnan, China, 13000ft.

Range (Map 171). NW Yunnan, in the Mekong and Salween valleys, at altitudes of 2400–4250m.

Remarks. This form was included in *E. chinensis* by Allen (1940), followed by Ellerman & Morrison-Scott (1951), but is readily distinguished by the shorter tail and especially by the very much smaller auditory bullae which are about 6.7mm long and c. 1.7mm apart at the basio-occipital suture, compared with 9.1 and 1.0 in *E. chinensis*.

Eothenomys custos

Microtus (Anteliomys) custos Thomas, 1912c: 517; Atunsi, NW Yunnan, China, 11500–12500ft.

Range (Map 171). W Yunnan and Sichuan, at altitudes of 2700–3900m.

Variation. Two subspecific names have been proposed:
Microtus (Anteliomys) c. rubellus Allen, 1924: 5; Ssu-shan, Likiang Range, Yunnan, 13000ft.
E. (A.) c. hintoni Osgood, 1932: 321; Wushi, SW of Tatsienlu, Sichuan, 12000ft.

Eothenomys inez

Microtus (E.) inez Thomas, 1908a: 45; 12 miles NW of Kolanchow, Shanxi, China, 7000ft.
M. (E.) nux Thomas, 1910e: 26; Shangchow, SE Shaanxi, China, 3300ft.

Range (Map 170). From N Sichuan and SE Shaanxi through Shanxi and perhaps further east.

Recognition. The dorsal pelage is very uniform in colour and appears sleek and tidy compared with other species of *Eothenomys*.

Remarks. This form was erroneously included in *Clethrionomys rufocanus* by Ellerman & Morrison-Scott (1951), following Hinton (1926).

Map **171**. *Eothenomys* (part), *Microtus*

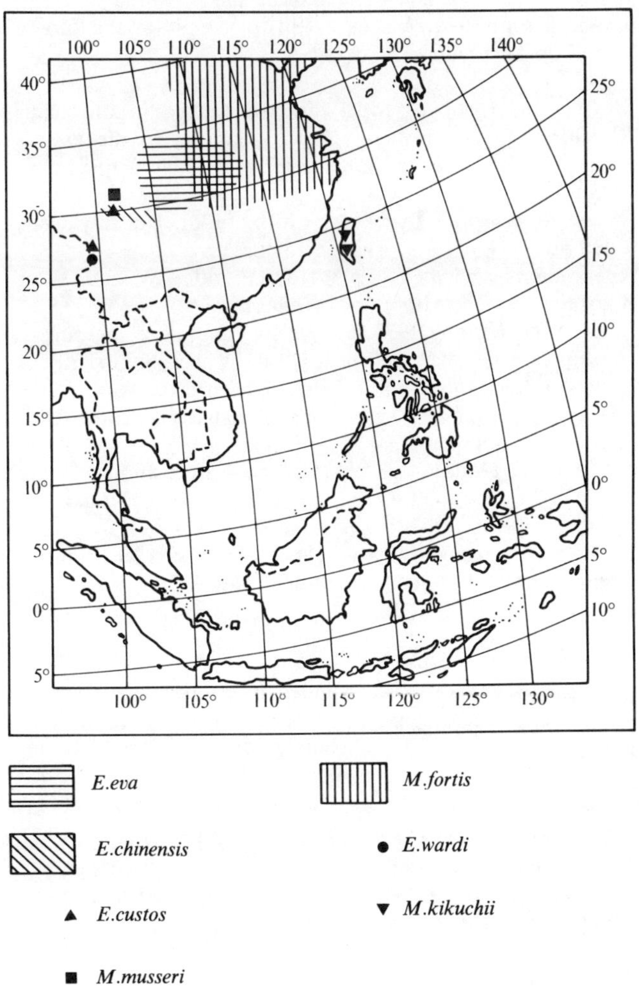

Eothenomys eva

Microtus (Caryomys) eva Thomas, 1911a: 4; SE of Taochow, Gansu, China, 10000ft.
? *M. (C.) alcinous* Thomas, 1911c: 50; Weichoe, Siho R, W Sichuan, China, 8000–10000ft.
Craseomys aquilus Allen, G. M.,1912b: 216; Showlungtan, Hubei, China, 7000–9000ft.

Range (Map 171). Sichuan, Hubei and S Gansu, China, at altitudes of 2400–3600m.

Variation. The southern form *alcinous* is very much darker than typical *eva*, both above and below, with a slight frosting, closely resembling *E. olitor* in pelage (but distinguished by longer tail).

MICROTUS Field voles

Microtus Schrank, 1798: 72; type species *M. terrestris* Schrank (in *M. arvalis* Pallas).
Including many synonyms based upon Palaearctic species.

Holarctic; c. 44 species (or c. 67 if *Pitymys* is included). They have rootless molars with sharply angled triangles. Three species occur in S China and Taiwan (Table 263); others occur at high altitude or just to the north of the boundary in China, namely *M. transcaspicus* (2500–3000m in E Afghanistan); *M. clarkei* (Yunnan, N Burma); *M. millicens* (Sichuan); *M. mandarinus* (S Shaanxi); *M. oeconomus* (S Shaanxi). The following additional fossil species have been recorded south of 35° N in China:
M. brandtioides Young, 1934: 95; Choukoutien, N China; 'Upper Polycene'. Recorded at Hexian (Anhui), Middle Pleistocene (Han & Xu, 1985).
M. epiratticeps Young, 1934: 101; Choukoutien, N China; 'Upper Polycene'. Recorded at Hsintsai, Honan, Middle/Late Pleistocene (Pei, 1956).

Table 263. The species of *Microtus*.

	fortis	*kikuchii*	*musseri*
Range	S China	Taiwan	C Sichuan
Head & body	120–135	120–130	90–130
Tail	50–65	70–95	50–70
as % of h & b	45–50%	65–75%	c.60%
Hind feet	22–25	20–23	17–22
Skull: greatest length	27–33	26–31	24–27
Upper molar row	6.7–7.8	6.3–7.0	5.5–6.7
M_1: no. of closed triangles in front of posterior loop	5	5	4
M^3 (right): posterior lobe (typical form)	Fig.39a	Fig.39b	Fig.39c
M_3 (left): antero-external ridge	Fig.39d	Fig.39e	Fig.39f

Microtus fortis Reed vole

M. fortis Büchner, 1889: 99; N loop of Hwangho R, Ordos Desert, Inner Mongolia, China.
M. calamorum Thomas, 1902c: 167; N bank of lower Yangtse Kiang, near Nanking, Jiangsu, China.
And other extralimital synonyms.

Range (Map 171). Lowland China from Fujian and NE Hunan north to Korea, Manchuria and the Amur Basin, and west to L Baikal.

Variation. Two subspecific names have been proposed:
M. calamorum superus Thomas, 1911g: 27 (and 1911f: 691); 30 miles S of Fenghsiangfu, S Shaanxi, China.
M. f. fujianensis Hong, 1981: 444; Jianyang County, Fujian, China.

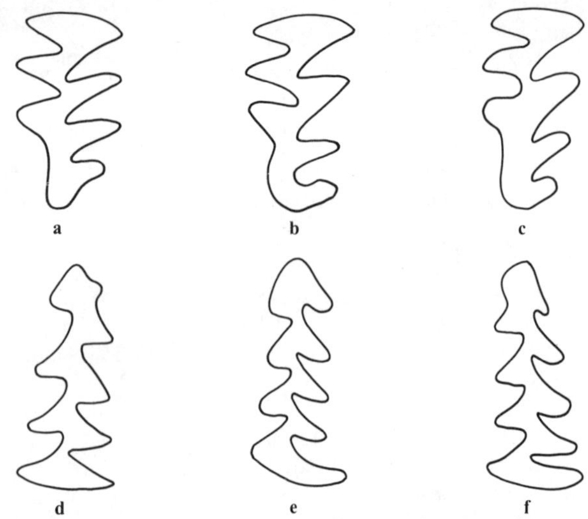

Fig. 39. *Microtus.* Typical occlusal pattern of right M^3 (above) and left M_1 (below) of *M.fortis* (a,d); *M.kikuchii* (b,e); *M.musseri* (c,f).

Microtus kikuchii

M. kikuchii Kuroda, 1920: 36; Mt Morrison, Taiwan, 10000ft.

Range (Map 171). Taiwan, at altitudes of 1900–3000m.

Remarks. Probably most closely related to *M. fortis* from which it differs mainly in having a longer tail and simpler M^3 which usually lacks a posterior extension behind the fourth inner ridge. Detailed description and measurements were given by Kaneko (1987).

Microtus musseri Musser's vole

M. musseri Lawrence, 1982: 6; Qionglai Shan, 48 km W of Wenquan, central Sichuan, China, 2750m.

Range (Map 171). Known only from the vicinity of the type locality in the mountains NW of Chengdu, Sichuan, between 2135 and 3660m.

Remarks. Most closely related to *M. millicens* from higher altitude in Sichuan, but placed in a separate, new genus *Volemys* by Zagorodnyuk (1990).

PROEDROMYS

Proedromys Thomas, 1911b: 177; type species *P. bedfordi* Thomas.

One species in C China, sometimes included in *Microtus* but distinguished by grooved upper incisors.

Proedromys bedfordi

P. bedfordi Thomas, 1911a: 4 (and 1911b: 177); 60 miles SE of Minchow, Gansu, China.

Range (Map 170). S Gansu and Sichuan, China (recorded in Sichuan by Hu & Wang, 1984).

ONDATRA

Ondatra Link, 1795: 76; type species *Castor zibethicus* L. Conserved by Opinion 55 (1913).
= *Fiber* Cuvier, 1800:tab. 1.

Contains one species.

Ondatra zibethicus Muskrat

Castor zibethicus Linnaeus, 1766: 79; E Canada.
And many extralimital synonyms.

Range. Indigenous only in N America; introduced, mainly by escape from fur farms, in several parts of S China, and throughout much of the Palaearctic Region.

Subfamily Rhizomyinae

Bamboo rats

Rhizomyini Winge, 1887: 109; type genus *Rhizomys* Gray.

Africa and SE Asia; 3 genera of which two are Indomalayan (Table 264). These are large, burrowing rodents. However the Indomalayan species forage above ground by digging for roots and rhizomes, especially of bamboo, and are less completely subterranean than the African *Tachyoryctes*. Although monophyly of the Asiatic and African forms has been questioned, it has been supported by recent finds of fossils in the Indian Siwalik beds (Flynn *et al.*, 1990).

Most of the following fossil forms described in the genus *Rhizomys* have subsequently been separated in the extinct genera *Rhizomyides* and *Brachyrhizomys* (Flynn, 1982a, b, 1983):
Rhizomys sivalensis Lydekker, 1878a: 100; Haritalyangar, Simla Hills, N India; ? Pliocene. (In *Rhizomyides*).
R. punjabiensis Colbert, 1933c: 1; 23 miles W and N of Bilaspun, Punjab; Miocene. (In *Rhizomyides.*)
R. lydekkeri Hinton, 1933: 621; synonym of *R. sivalensis.*
R. nagrii Hinton, 1933: 621; near Haritalyangar, Simla Hills, N India; Middle Siwaliks. (In *Brachyrhizomys.*)
R. pilgrimi Hinton, 1933: 621; near Haritalyangar, Simla Hills, N India; Middle Siwaliks. (In *Brachyrhizomys.*)
R. pinjoricus Hinton, 1933: 621; Simila area, India; Upper Siwaliks, Lower Pinjor Zone, early Pleistocene. (In *Brachyrhizomys* or *Anepsirhyzomys*—Flynn *et al.*, 1990).
R. mirzadi Lang & Lavocat, 1968: 80; Bassin de Bamian, Afghanistan; ? Pliocene. (In *Rhizomyides.*)

RHIZOMYS

Rhizomys Gray, 1831b: 95; type species *R. sinensis* Gray.
Nyctocleptes Temminck, 1832: 5, pl. 1 (*Mus sumatrensis* Raffles).

Indochinese Region to Malaya and Sumatra; 3 species (Table 264). Three fossil species have been described from China:
R. schlosseri Young, 1927: 49; Honan; Lower Pleistocene.
R. provestitus Young & Liu, 1951: 55; Koloshan, near Chungking, Sichuan; Pleistocene.
R. szechuanensis Young & Liu, 1951: 57; Koloshan, Sichuan; Pleistocene.

Table 264. The species of *Rhizomys* and *Cannomys* (subfamily Rhizomyinae).

	R.sinensis	*R.pruinosus*	*R.sumatrensis*	*C.badius*
Range	S China etc.	China-N Malaya	Burma-Malaya, Sumatra	Nepal-Indochina
Head & body	230–450	270–350	260–480	150–260
Tail	50–90	100–125	100–200	60–75
as % of h & b	14–24%	34–40%	35–50%	
Hind feet	38–60	45–60	46–67	30–35
Dorsal pelage white-tipped	0	+	0	0
Pelage texture	Dense, soft	Dense, soft	Sparse, harsh	Dense, soft
Throat lighter than rest of venter	+	0	+/0	0
Sides of face rufous	0	0	+	+
Teats (pairs, ant./post.)	1/3	1 or 2/3	2/3	2/2
Foot pads	Granular	Granular	Granular	Smooth
Hind foot: no. of post. pads	2	2	1	2
Skull: greatest length	65–85	56–66	71–78	40–45
Upper molar row	14.0–18.8	13.5–15.2	14.0–15.2	8.8–11.7
Upper incisors slightly pro-odont	0	+	+	+
M^1:M^2	$<$	$<$	$=$	$>$

Rhizomys sinensis Chinese bamboo rat

Rhizomys sinensis Gray, 1831b: 95; near Canton, Guangdong, China.
R. chinensis Swinhoe, 1871: 637; China.
R. vestitus Milne-Edwards, 1872a: 92; W of Moupin, Sichuan, China.
R. davidi Thomas, 1911a: 5 (and 1911b: 179); Kuatun, Fujian, China.
R. wardi Thomas, 1921a: 504; Mt Imaw Bum, Kachin Prov., N Burma, 9000ft.
R. troglodytes Matthew & Granger, 1923: 574; Yen-ching-kao, near Wan-hsien, Sichuan, China; Upper Pliocene? Included in *R. sinensis* by Cao (1985).

Range (Map 172). S China north to S Shaanxi (Qinling Mts), S Gansu and Hubei; south to N Burma and N Vietnam.

Variation. One subspecies has been described:
R. s. reductus Dao & Cao, 1990: 234; Son dong, Ha Bac Prov., N Vietnam, 21°17'N, 106°48'E.

Rhizomys pruinosus Hoary bamboo rat

Rhizomys pruinosus Blyth, 1851: 519; Cherrapunji, Khasi Hills, Assam, India.
R. latouchei Thomas, 1915i: 59; Swatow, Guangdong, S China.
R. pannosus Thomas, 1915i: 60; Chantabun, S Thailand.
R. senex Thomas, 1915j: 313; near Mongtse ?, S Yunnan, China.
R. umbriceps Thomas, 1916g: 445; Bukit Gantang, Larut, Perak, Malaya.

Range (Map 172). From S China (Yunnan, S Sichuan, Guangxi, Guangdong), through E Burma and Indochina to Thailand and Malaya (S to Perak); mainly montane, at altitudes from 1000 to 4000m.

Rhizomys sumatrensis Large bamboo rat

Mus sumatrensis Raffles, 1821: 258; Malacca, Malaya.
Spalax javanus Cuvier, G.,1829: 211; 'Iles de la Sonde'.
Nyctocleptes dekan Temminck, 1832: 7; Malacca.
R. cinereus McClelland, 1841b: 456; Mulmein, Tenasserim, Burma.
R. erythrogenys Anderson, 1877b: 150; Salween Hills, Burma.
Nyctocleptes insularis Thomas, 1915i: 58; Padang Brahrang, Deli, Sumatra.

Range (Map 172). Burma and S Yunnan (Kao *et al.*, 1962) south through Thailand and S Vietnam to Malaya; Sumatra.

Variation. One subspecific name has been proposed:
R. s. padangensis Brongersma, 1936b: 154; Koto Gadang, (Singgalung), SW Sumatra.

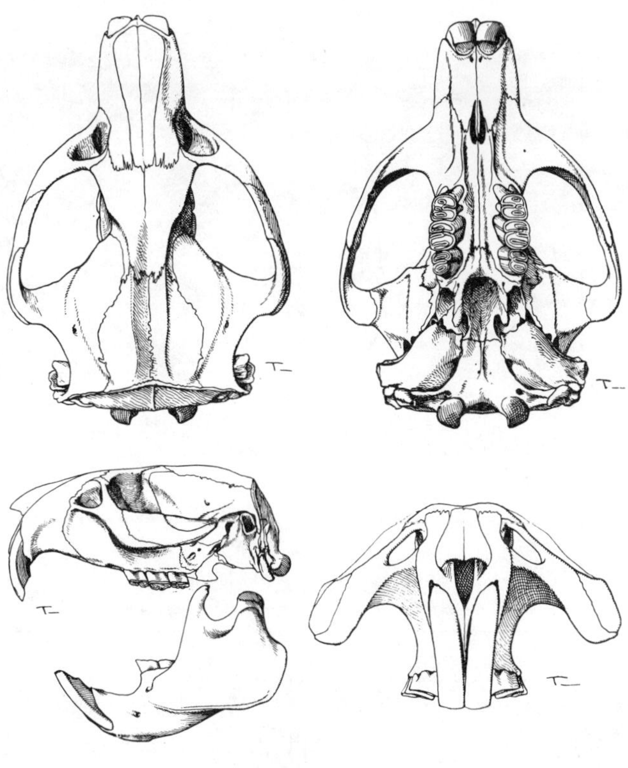

Fig.**40**. *Rhizomys pruinosus.*

Ecology. In Malaya occur in secondary forest, feeding on roots of bamboo but also of cultivated tapioca and sugar cane.

Remarks. Chromosomes: 2n = 50, mostly subtelocentrics and small metacentrics (Hsu & Johnson, 1963).

CANNOMYS

Cannomys Thomas, 1915i: 57; type species *Rhizomys badius* Hodgson.

One species.

Cannomys badius Lesser bamboo rat

Rhizomys badius Hodgson, 1841a: 60; Nepal.
R. minor Gray, 1842: 266; S Thailand, probably Pachaburi, SW of Bangkok.
R. castaneus Blyth, 1843a: 1007; probably Arakan, Burma.
C. pater Thomas, 1915j: 315; Mt Popa, Burma.

Range (Map 172). From Nepal (Terai) and Bangladesh (Kock & Posamentier, 1983) through Assam to Yunnan, Burma, NW Vietnam and Thailand.

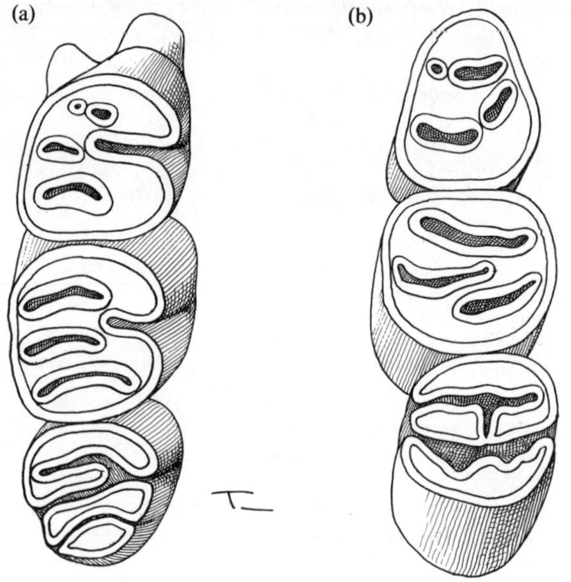

Fig.**41**. *Rhizomys pruinosus*. Molars: (a) upper right; (b) lower left.

Map **172**. *Rhizomys, Cannomys, Dryomys, Chaetocauda*

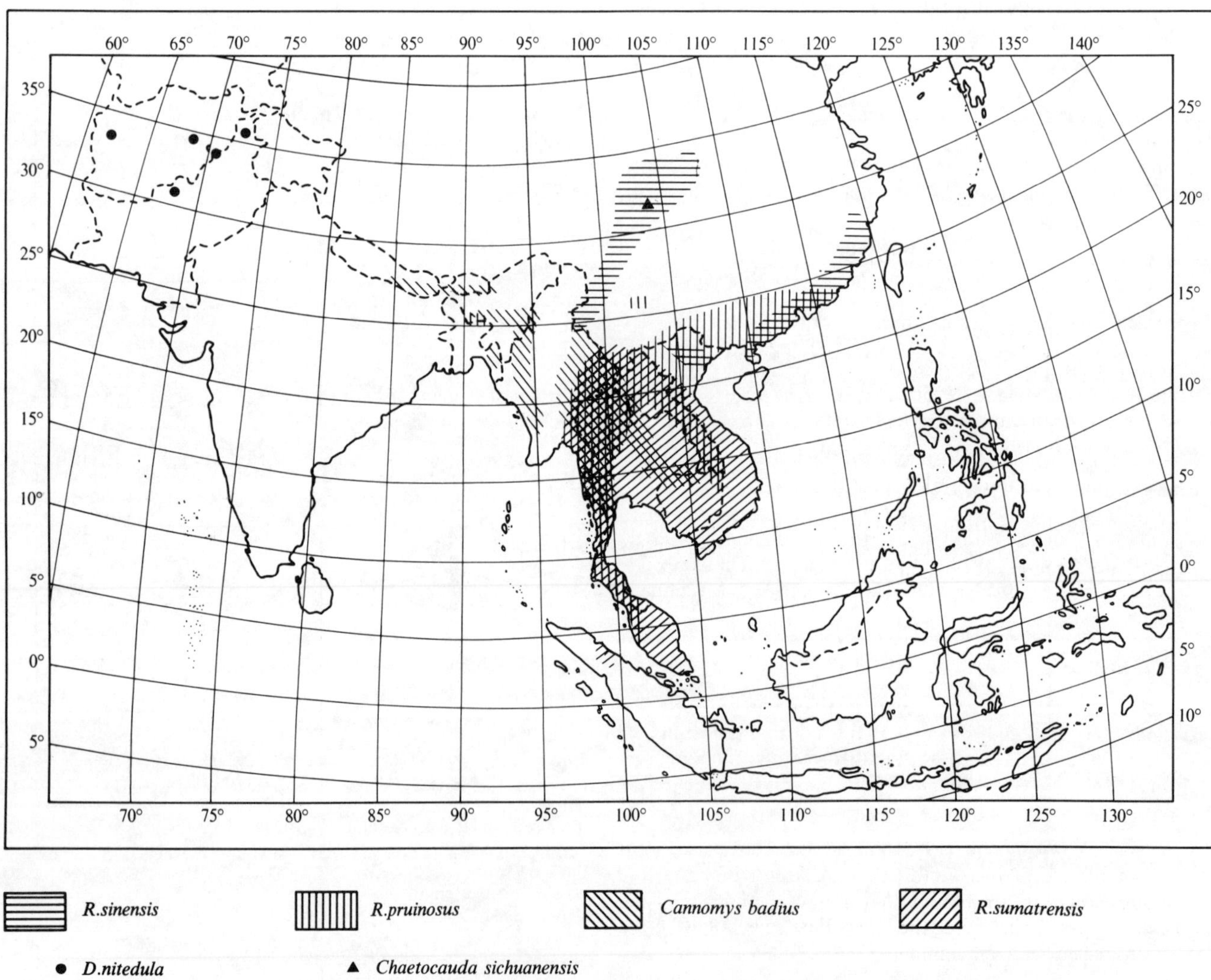

Variation. Two subspecific names have been proposed:
C. b. plumbescens Thomas, 1915j: 315; Gokteik, N Shan States, Burma.
C. minor lonnbergi Gyldenstolpe, 1917: 47; Sakerat, E Thailand.

Family Gliridae

Dormice

Gliridae Thomas, 1897c: 1016; type genus *Glis* Brisson (of doubtful validity).
= Myosidae Gray, 1821: 303 (*Myoxus* Gmelin).
Muscardinidae Palmer, 1899a: 413 (*Muscardinus* Kaup).
The most frequently used name is used here pending resolution of a complex nomenclatural problem.

Palaearctic and Africa; 8 genera of which only 2 now occur in the region: *Dryomys* marginally in E Afghanistan and Pakistan; *Chaetocauda* in Sichuan (Table 265). Dormice are small rodents with usually bushy trails. The dental formula is 1.0.1. 3/1.0.1. 3 and the molars normally have transverse ridges. One additional Recent genus is represented by a fossil species in Pakistan:

Myomimus sumbalenwalicus Munthe, 1980: 12; Mianwali Dist, Pakistan; Upper Miocene.

DRYOMYS

Dryomys Thomas, 1906c: 348; type species *Mus nitedula* Pallas.
= *Dyromys* Thomas, 1907g: 406.

Two species of which one occurs marginally in the region.

Dryomys nitedula Forest dormouse

Mus nitedula Pallas, 1779a: 88; Lower Volga, Russia.
Myoxus pictus Blanford, 1875b: 311; Kohrud, N Iran.
And other, extralimital synonyms.

Range (Map 172). NW Pakistan and E Afghanistan; also from Chinese Turkestan to central Europe.

Variation. Specimens from Pakistan and Afghanistan have been allocated to *D. n. picta*, characterized by larger ears than European forms. Those in Baluchistan are very pale greyish brown on the dorsum.

Ecology. Found in scrub in montane steppe.

CHAETOCAUDA

Chaetocauda Wang, Y.,1985: 67; type species *C. sichuanensis* Wang.

One species. Wang, Y. (1985) considered this genus most closely similar to *Myomimus* (SE Europe, Asia Minor etc.) and combined them in a subfamily Myomiminae. It differs from *Myomimus* mainly in having the tail well-haired (as in all other dormice) and the molars simpler.

Chaetocauda sichuanensis Sichuan dormouse

C. sichuanensis Wang, Y.,1985: 67; Pinwu county, N Sichuan, China.

Range (Map 172). Apparently known only from the type locality in N Sichuan, China, where it occurs in mixed subalpine forest.

Remarks. Not seen.

Family Zapodidae

Jumping mice

Sminthinae Murray, 1866: 360 (*Sminthus* Nordmann, in *Sicista*).
Zapodidae Coues, 1876: 253; type genus *Zapus* Coues.
Sicistinae Allen, 1901: 185 (*Sicista* Gray).

Palaearctic; 2 genera occur very marginally in the region (Table 266). The members of this family have frequently been included in the Dipodidae, e. g. by Klingener (1984). They are small, long-tailed mice, with long hind feet but lacking the bipedal gait of the Dipodidae.

SICISTA Birch mice

Sicista Gray, 1827a: 228; type species *Mus subtilis* Pallas.
Sminthus Nordmann, 1840: 49 (*Sminthus loriger* Nathusius, in *Sicista subtilis*).

Palaearctic; c. 11 species of which one occurs very marginally in this region, in central China and in Kashmir/N Pakistan. The species limits are still somewhat uncertain.

Table 265. The genera of Gliridae.

	Dryomys	*Chaetocauda*
Range in region	Pakistan, Kashmir	Sichuan
Black mask on face	+	0
Tail distichous	+	0
Incisive foramina	Short, stop far in front of M^1	Long, extend behind middle of M^1
Bullae (% of greatest length of skull)	c.28%	c.33%

Table 266. The genera and species of Zapodidae.

	Sicista concolor	*Zapus setchuanus*
Range	C China W Himalayas	Sichuan
Head & body	60–85	68–80
Tail	95–125 (c.140%)	115–145 (145–170%)
Hind feet	18–22	25–31
Flanks contrast with dorsum and venter	0	+
Pelage texture	Soft	Harsher
Skull: greatest length	18–23	21–24
Upper molar row	2.9–3.2	3.5–4.0
I^1 grooved	0	+

Sicista concolor Chinese birch mouse

Sminthus concolor Büchner, 1892a: 267; Guiduisha, N slope of Sining Mts, Gansu, China.
Sminthus leathami Thomas, 1893e: 184; Krishnye Valley, Wardwan, Kashmir.
Sminthus flavus True, 1894b: 341; central Kashmir, 11000ft.
Sicista weigoldi Jacobi, 1923: 15; Hsueshan, near Sungpan, Sichuan, China.

Range (Map 173). Gansu, Sichuan and Yunnan (around 3000m); also in Kashmir and N Pakistan from 2100–4000m if *leathami* and *flavus* are conspecific; and in parts of W China if *tianshanicus* is conspecific.

ZAPUS

Zapus Coues, 1876: 253; type species *Dipus hudsonius* Zimmermann (*Z. hudsonius*, N America).
Eozapus Preble, 1899: 37 (*Zapus setchuanus* Pousargues).

N America (2–4 species) and China (one species). *Eozapus* is frequently given generic rank but the difference from N American *Zapus*, especially *Z. trinotatus*, is very slight. *Eozapus* was based upon the structure of M[1] relative to *Z. hudsonicus*, but *Z. trinotatus* (W coast of USA) tends to bridge the gap.

Zapus setchuanus Sichuan jumping mouse

Z. setchuanus Pousargues, 1896: 13; Tatsienlu, W Sichuan, China.

Range (Map 173). Mountains of central China, in Yunnan (Wang, Y-x. *et al.*, 1985), W Sichuan, Qinghai and S Gansu, at altitudes of 3000–4000m.

Variation. One subspecies has been described:
Zapus s. vicinus Thomas, 1912a: 402; 46 miles SE of Taochow, Gansu, China. Venter entirely white (buff in centre in *Z. s. setchuanus*).

Recognition. Easily distinguished from all other Palaearctic or Indomalayan mice by the characteristic sharply three-coloured body pelage: black-brown on the dorsum, sharply separated from the yellow-brown flanks which in turn are sharply separated from the white venter.

Map **173**. *Allactaga, Jaculus, Salpingotus, Sicista, Zapus*

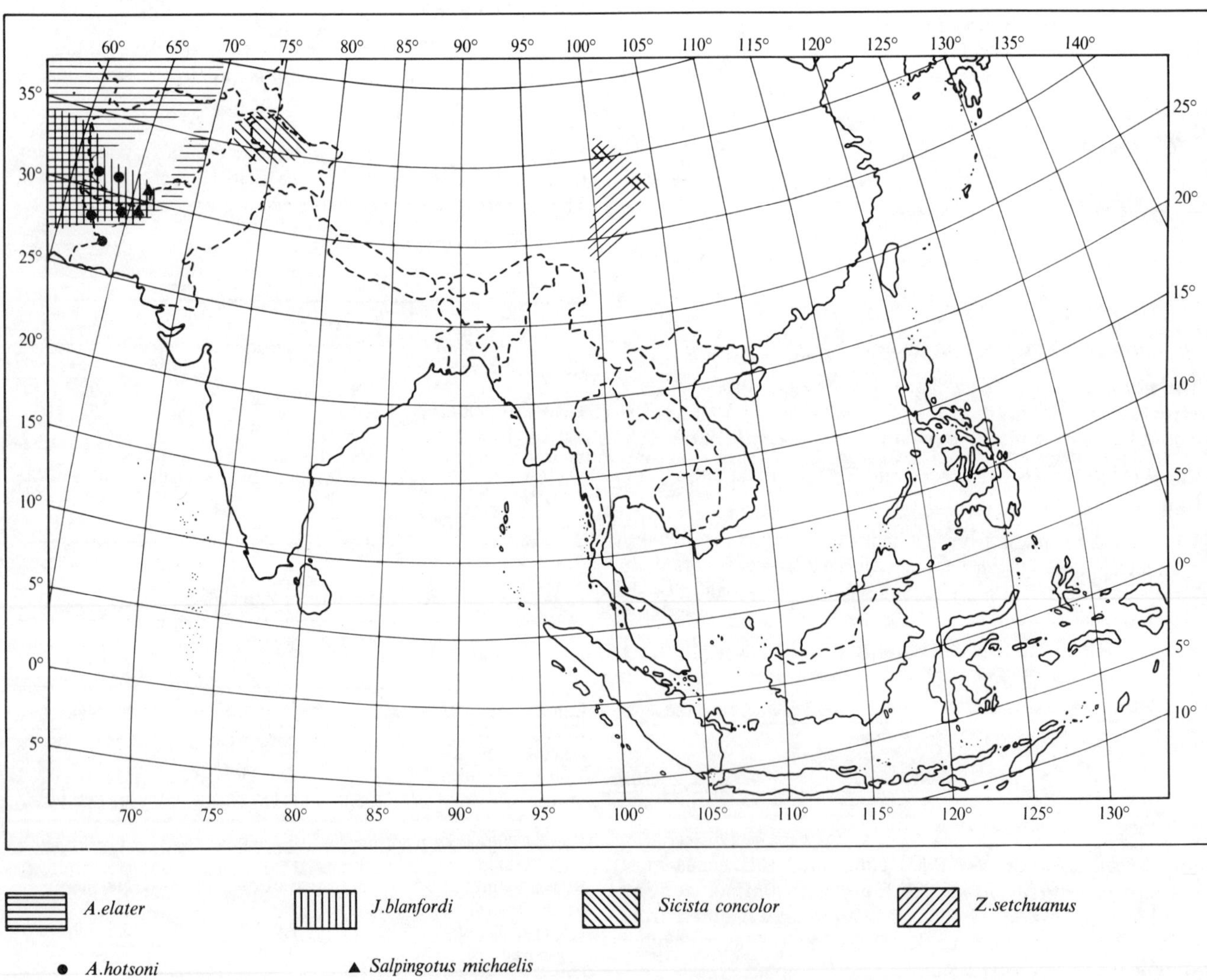

Family Dipodidae

Jerboas

Dipsidae Gray, 1821: 303; type genus '*Dipsus* Gmelin', presumably a *lapsus* for *Dipus* Gmelin.
Dipodidae Waterhouse, 1842: 203 (*Dipus* Zimmermann).

Palaearctic; 10 genera of which 3 occur marginally in this region, in the deserts of S Afghanistan and W Pakistan (Table 267). These are bipedal rodents with elongate hind legs and feet, long tail, usually long ears and often extremely hypertrophied auditory bullae.

SALPINGOTUS Pygmy jerboas

Salpingotus Vinogradov, 1922: 540; type species *S. kozlovi* Vinogradov.
Salpingotulus Pavlinov, 1980b: 47 (*Salpingotus michaelis* Fitzgibbon).

Central Asia; 5 species of which one occurs marginally in W Pakistan. Very small jerboas, with the three central metatarsal bones separate. The genus was reviewed by Vorontsov & Shenbrot (1984).

Salpingotus michaelis Baluchistan pygmy jerboa

Salpingotus michaelis Fitzgibbon, 1966: 431; Nushki Desert, NW Baluchistan, W Pakistan.

Range (Map 173). NW Pakistani Baluchistan.

Ecology. Occupy sand and gravel plains with sparse *Haloxylon* at altitudes of 1000–1500m.

ALLACTAGA

Allactaga Cuvier, 1836: 141; type species *Mus jaculus* Pallas, in *A. major* Kerr.
= *Scirtetes* Wagner, 1841b: 413.
And other synonyms based upon extralimital species.

NE Africa to central Asia; 11 species of which two reach W Pakistan (Table 267). Large jerboas, the hind feet with the three central metatarsals fused but with one or two small lateral digits present. A third species, *A. euphratica* Thomas, 1897, reaches NE Afghanistan.

Table 267. The genera and species of Dipodidae.

	Salpingotus michaelis	*Allactaga elater*	*Allactaga hotsoni*	*Jaculus blanfordi*
Head & body	41–54	100–125	110–160	115–135
Tail	72–110	145–175	135–190	185–200
Hind feet	18–19	50–56	57–59	64–72
Ear	8–11	30–36	39–46	22–26
Hind digits	3	5	5	3
Metatarsals: central 3 fused	0	+	+	+
Tail flag				
black & white	0	+	+	+
shape	Slight	Flat	Terete	Terete
Skull: greatest length	18–20	c.27	c.30	32–39
Bullae grossly inflated	+	0	0	+
I¹ pro-odont	0	+	+	0
grooved	0	0	0	+

Allactaga elater Small five-toed jerboa

Dipus elater Lichtenstein, 1825: 155; E Kakakhstan, USSR.
Alactaga [*sic*] *indica* Gray, 1842: 262; Simkoh Hills, Afghanistan.
= *Allactaga bactriana* Blyth, 1863a: 110.

Range (Map 173). N Baluchistan (W Pakistan) and SE Afghanistan; west through Iran and Soviet Turkestan to Asia Minor.

Ecology. Found on upland gravel plains and arid mountain slopes.

Allactaga hotsoni Hotson's five-toed jerboa

A. hotsoni Thomas, 1920d: 936; Kant, 20 miles SW of Sib, Iranian Baluchistan, 3950ft.

Range (Map 173). Pakistani Baluchistan, S Afghanistan and adjacent parts of SE Iran.

Ecology. Found on barren stony plains where hardly any other rodents occur.

Remarks. Perhaps conspecific with the larger *A. euphratica* Thomas, 1881 (including *williamsi*) which reaches NE Afghanistan near Kabul and extends west to Iraq and Turkey (Hassinger, 1973).

JACULUS

Jaculus Erxleben, 1777: 404; type species *Mus jaculus* L. Conserved by Opinion 730 (1965).
Including synonyms based upon extralimital species.

N Africa to W Pakistan; 5 species of which one reaches W Pakistan.

Jaculus blanfordi Blanford's jerboa

Dipus blanfordi Murray, 1884a: 98; Bushire, Iran.
And extralimital synonyms.

Range (Map 173). Pakistani Baluchistan and S Afghanistan; and in most of Iran.

Ecology. Found on sand and gravel plains, up to 1120m in Pakistan.

Remarks. Although intermediate in many respects between *J. jaculus* and *J. orientalis* the structure of the penis is distinctive and justifies recognition of *J. blanfordi* as a separate species (Didier & Petter, 1960). The form *turcmenicus* Vinogradov & Bondar from Turkmenia was included in *J. blanfordi* by Heptner (1975).

Family Hystricidae

Old-world porcupines

Hystricini Fischer, 1817: 372; type genus *Hystrix* Linnaeus.

Africa, S Palaearctic and most of the Indomalayan Region; 3 genera of which one, *Atherurus*, is often separated in a subfamily Atherurinae (Table 268). The heavy armour of thick spines is unique amongst Indomalayan mammals. For a key, and brief diagnoses, for all species see van Weers (1983). The overall classification within the family was reviewed by van Weers (1979).

Table 268. The genera of Hystricidae.

	Hystrix	*Trichys*	*Atherurus*
Range in region	Widespread	Sunda	Assam-Malaya
Species in region	6	1	1
Head & body	400–900	350–480	380–570
Tail (% of h & b)	<20%	c.50%	25–50%
Terminal quills (Fig.42)	Goblet-shaped	Slender, flat	Beaded
Crest	+/0	0	0

HYSTRIX

Hystrix Linnaeus, 1758: 56; type species *H. cristatus* L. Conserved by Opinion 75 (1922).
Acanthion Cuvier, F.,1823a: 425, 431 (*A. javanicum* Cuvier).
Odocephalus Gray, 1866e: 308 (*Acanthion cuvieri* Gray, in *H. cristatus* L.).
Thecurus Lyon, 1907e: 582 (*T. sumatrae* Lyon).

Africa to SE Asia; 8 species of which 6 occur in the Indomalayan Region (Table 269). Characterized especially by the unique modification of the terminal 'rattle-quills' of the tail which are expanded and hollow distally, the tip breaking off to produce the shape of an open, elongate, stemmed goblet (Fig. 42a). The forms in SE Asia (subgenus *Acanthion*) were revised in detail by van Weers (1979).

The following names were based on fossils from the region:
H. sivalensis Lydekker, 1878a: 100; Asnot, Punjab, Pakistan; Siwaliks.
H. crassidens Lydekker, 1886b: 37, pl. 8; Billa Surgam, Karnul Dist, Madras, India; Pleistocene.
H. (*Acanthion*) *lagrelii* Lönnberg, 1924: 1; 10 li N of Mien-Chih-Hsien, c. 65km W of Honan-Fu, Honan, China; Pleistocene.
H. kiangsenensis Wang, 1931; Zhejiang, China; post-Tertiary.
Acanthion sivalensis sinhaleyus Deraniyagala, 1951b: 122; Gorak goha deniya, Gonapitiya, Kuruvita, Sri Lanka.
H. gigantea van Weers, 1985: 115; Sangiran, Java; Pleistocene?
H. magna Pei, 1987: 89; Liucheng, Guangxi, China; Pleistocene.

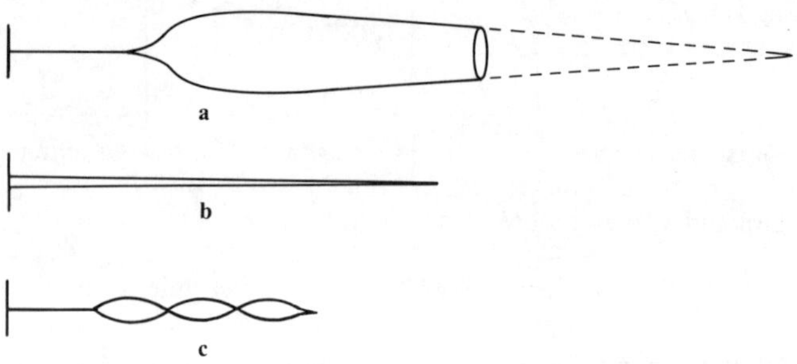

Fig.**42**. Hystricidae. Typical tail-quills of (a) *Hystrix*; (b) *Trichys* and (c) *Atherurus*. In *Hystrix* the tip breaks off to leave a hollow cylinder.

Hystrix indica Indian crested porcupine

H. cristatus var. *indica* Kerr, 1792: 213; India.
H. leucurus Sykes, 1831: 103; Deccan, India.
H. hirsutirostris Brandt, 1835: 39; Talysh, Transcaucasia.
H. zeylonensis Blyth, 1851a: 171; Sri Lanka.
H. malabarica Day, 1863: 446; near Trichoor, Western Ghauts, S India.
H. cuneiceps Wroughton, 1912b: 771; Nokania, Cutch, India.
Also other, extralimital, synonyms.

Range (Map 174). Throughout peninsular India, NE to Bihar, and in S Nepal and Pakistan; Sri Lanka; also west to Israel and north to southern Soviet Turkestan. Recorded in Tibet by Honacki *et al.* (1982) but not in the detailed review of the mammals of Tibet by Feng *et al.* (1986) who record only *H. hodgsoni* (here included in *H. brachyura*).

Variation. The following subspecific name has been proposed (in addition to others from elsewhere in the range):
H. hirsutirostris blanfordi Müller, 1911: 121; Jalk, Baluchistan, Pakistan, 3000m.

Ecology. A versatile species, occurring in many habitats from arid rocky hillsides in Baluchistan to monsoon forest. They feed on rhizomes, tubers, bark etc., causing damage to crops, and to forestry by barking trees at ground level.

Hystrix brachyura

H. brachyura Linnaeus, 1758: 57; Malacca, Malaya.
H. longicauda Marsden, 1811: 118; Sumatra.
Acanthion hodgsoni Gray, 1847c(June): 101; Nepal.
H. alophus Hodgson, 1847b(August): 771; Himalayas.
H. bengalensis Blyth, 1851a: 170; Bengal.
Acanthochoerus grotei Gray, 1866e: 310; Malacca, Malaya.
H. subcristata Swinhoe, 1871: 638; Foochow, Fujian, China.

Table 269. The species of *Hystrix*.

	indica	*brachyura*	*javanica*	*crassispinis*	*sumatrae*	*pumila*
Range in region	India etc.	Nepal-Malaya, Sumatra, Borneo	Java, Lesser Sunda Is	Borneo	Sumatra	Palawan
Head & body	600–900	450–850	550–660	550–660	450–560	400–550
Rattle quills on tail: length of hollow part	35–60	15–34	15–20	12–16	12–16	10–11
Crest on head & nape	Very long	Slight	0	0	0	0
Thickest quills on back: max. diameter	5–7	5–7	3.5–6	6–8	5–7	4–5
Long quills on back: no. of dark bands	2+	1	1	1	1	1
length of white tip	-260	-200	-100	0–120	-70	0

Map **174**. *Hystrix*

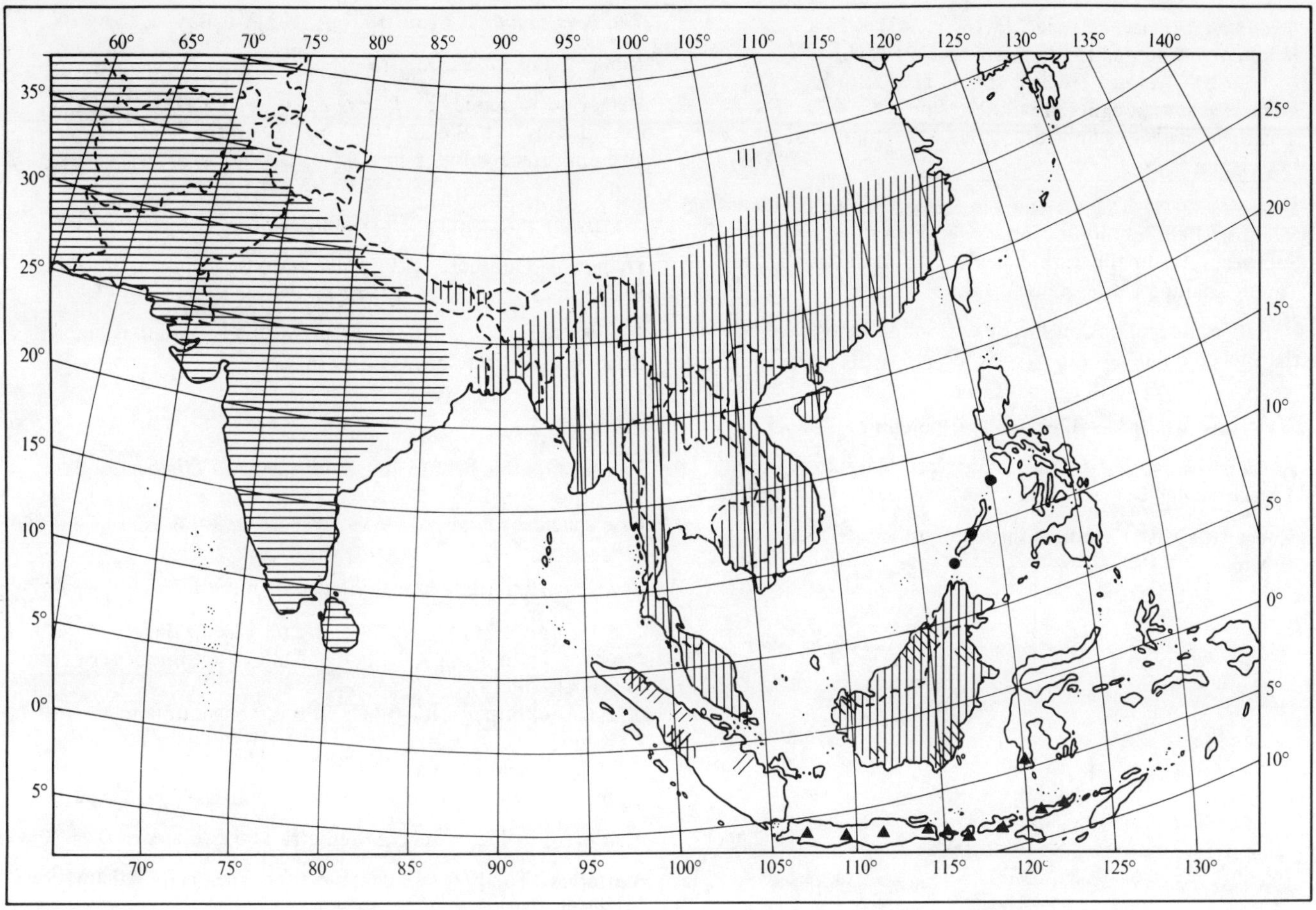

H.indica — *H.brachyura* — *H.crassispinis* — *H.sumatrae*

● *H.pumila* ▲ *H.javanica*

H. mülleri Marshall, in Sclater, 1871: 235; Sumatra.
H. yunnanensis Anderson, 1879: 332; mountains E of Kakhyen Hills, W Yunnan, China.
Acanthion klossi Thomas, 1916h: 139; Tenasserim Town, S Burma.
A. millsi Thomas, 1922b: 431; Sangrachu, Naga Hills, Assam, India.

Range (Map 174). From E Nepal, S Tibet and Bangladesh throughout S China and south through Indochina to Malaya; Sumatra and Borneo; islands of Hainan and Singapore. The northern limit is in S Gansu and S Shaanxi from where Pleistocene fossils have also been recorded, as *H. subcristata* (Chow *et al.*, 1965).

Variation. One subspecific name has been proposed:
Acanthion subcristata papae Allen, 1927: 3; Nodoa, Hainan I, China.

Three subspecies were recognized and diagnosed by van Weers (1979) who however considered them very provisional and likely to intergrade. Any boundaries assigned to them are likely to be arbitrary.

H. b. brachyura. Malaya, Sumatra and Borneo. Crest rudimentary; nasals small (40–50% of skull).
H. b. subcristata. Thailand to S China and Assam. Crest moderately developed; nasals longer, c. 50–60% of skull.
H. b. hodgsoni. Sikkim to Nepal. Small; quills thin; crest rudimentary; nasals c. 52–55% of skull.

Hystrix brachyura

Hystrix javanica Javan porcupine

Acanthion javanicum Cuvier, F.,1823a: 431; Java.
H. torquata van der Hoeven, 1836: 110; Java.
H. ecaudata van der Hoeven, 1836: 110; Java.
H. brevispinosa Wagner, 1843: 20; Java.
Acanthion sumbawae Schwarz, 1911: 639; Dompu, Sumbawa I, Lesser Sunda Is.

Range (Map 174). Java and the Lesser Sunda Is, including Madura, Bali, Lombok, Sumbawa, Flores and Rintja; SW Sulawesi (presumably introduced). Also on the small island of Tanah Djampa (N of Flores).

Variation. Specimens on the Lesser Sunda Is average smaller than those on Java (van Weers, 1979).

Hystrix crassispinis Thick-spined porcupine

H. crassispinis Günther, 1877: 736; Borneo, opposite Labuan I.
Thecurus major Schwarz, 1939: 246; near Samarinda, E Borneo.

Range (Map 174). Borneo, throughout the lowlands and up to 1200m.

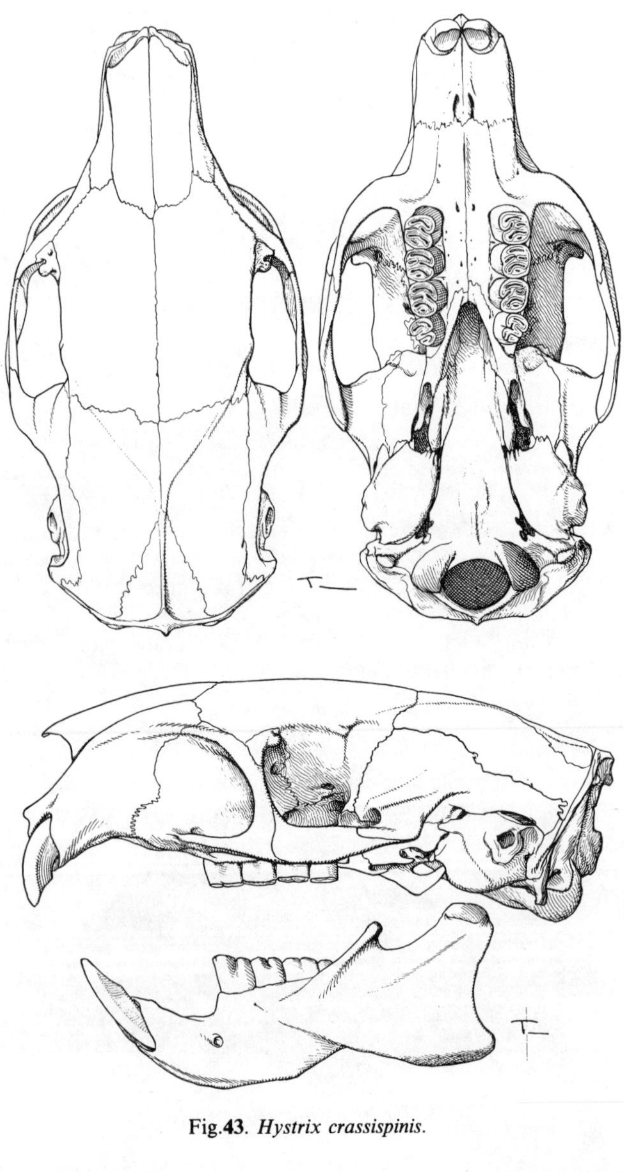

Fig.**43**. *Hystrix crassispinis.*

Hystrix sumatrae

Thecurus sumatrae Lyon, 1907e: 583; Aru Bay, E Sumatra.

Range (Map 174). Sumatra.

Remarks. Included in *H. crassispinis* by Chasen (1940a) but excluded by van Weers (1978) mainly on the basis of smaller size and much thinner spines.

Hystrix pumila Palawan porcupine

H. pumila Günther, 1879b: 106; Puerto Princesa, Palawan, Philippines.

Range (Map 174). Palawan and adjacent small islands of Busuanga and Balabac, Philippines.

TRICHYS

Trichys Günther, 1877: 739; type species *T. lipura* Günther.

One species. Reviewed by van Weers (1976).

Trichys fasciculata Long-tailed porcupine

Hystrix fasciculata Shaw, 1801: 11, pl. 124; Malacca, Malaya. Neotype from Runuk Tanjong, Malaya designated by van Weers (1976: 19).
T. lipura Günther, 1877: 739; Borneo, on mainland opposite Labuan I.
= *T. guentheri* Thomas, 1889c: 235.
T. macrotis Miller, 1903b: 469; Tapanuli Bay, W Sumatra.

Range (Map 175). Malaya, Sumatra and Borneo.

Variation. There is no evidence for subspecific distinction between the three populations.

ATHERURUS Brushtailed porcupines

Atherurus Cuvier, F.,1829b: 483; type species *Hystrix macrourus* L.

Africa and SE Asia; 2 species of which one occurs in the region. Characterized by the unique beaded structure of the tail quills (Fig. 42c).

One fossil species has been described from the region: *Atherurus karnuliensis* Lydekker, 1886b: 38, pl. 8; Billa Surgam, Karnul Dist, Madras, India; Pleistocene.

Atherurus macrourus Asiatic brush-tailed porcupine

Hystrix macrourus Linnaeus, 1758: 57; Malacca, Malaya.
A. zygomatica Miller, 1903a: 42; Aor I, E of Malaya.
A. hainanus Allen, 1906: 470; Hainan I, China.
A. terataus Lyon, 1907e: 587; Teratau I, SW Thailand.
A. tionis Thomas, 1908k: 105; Juara Bay, Tioman I, E of Malaya.
A. assamensis Thomas, 1921c: 598; Cherrapunji, Assam, India.
A. stevensi Thomas, 1925: 505; Ngai-Tio, Tonkin, Vietnam.
A. retardatus Mohr, 1964: 105; origin unknown, from Rotterdam Zoo.
A. angustiramus Mohr, 1964: 108; Doi Sutep, Chengmai, Thailand.

Range (Map 175). From Assam, S China (Yunnan, S Sichuan, Guangxi) and N Burma through Indochina to Malaya; and on the islands of Hainan (China); Aor, Pemanggil and Tioman (E Malaya); Tarutau (SW Thailand).

Remarks. This species was revised and mapped in detail by van Weers (1977).

[Family Ctenodactylidae]

Gundis

A family of five Recent species in four genera with a circum-Saharan distribution. One fossil species from the Indomalayan Region has been described in the Recent genus *Pectinator*, now with a single species *P. spekei* in Ethiopia and Somalia. However the sole Indomalayan species has subsequently been allocated to the extinct genus *Sayimys* by Black (1972):

Sayimys sivalensis

Pectinator sivalensis Hinton, 1933: 622; near Chinji, Attock Dist, Salt Range, Pakistan; Lower Siwalik Beds, Late Miocene.

ORDER LAGOMORPHA

Hares, rabbits and pikas

A clearly defined order whose relationships to other orders are debatable. Two families, both of which are represented in the region although one of them, the Ochotonidae, only very marginally (Table 270). Distribution and current status were reviewed by Chapman & Flux (1990).

Table 270. The families of Lagomorpha.

	Leporidae	Ochotonidae
No.of genera in region	5	1
Hind legs much longer than fore legs	+	0
Hind feet	>60	<40
Ears	Elongate, >30	Rounded, <30
Tail bushy	+	0

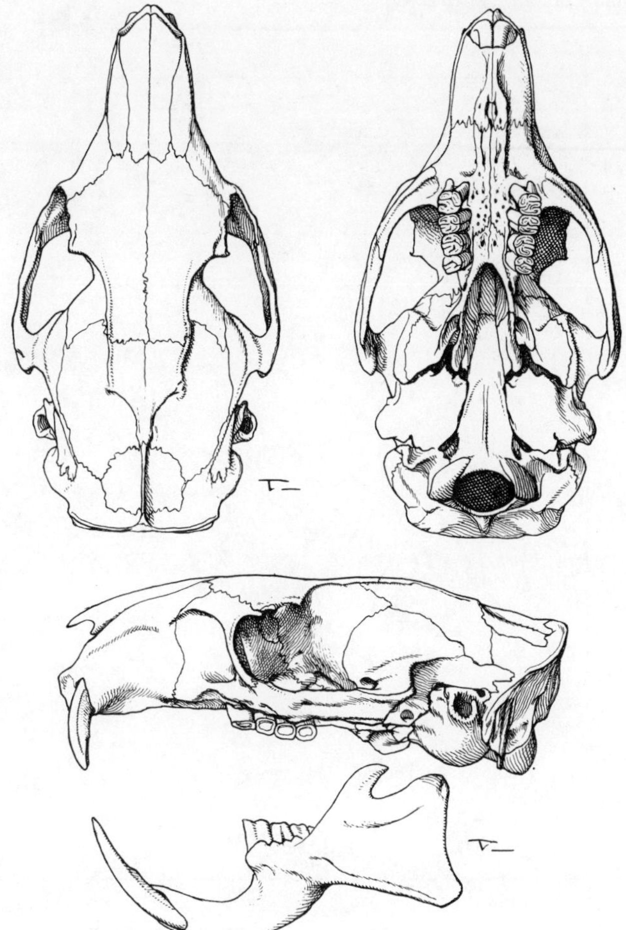

Fig.**44**. *Trichys fasciculata*.

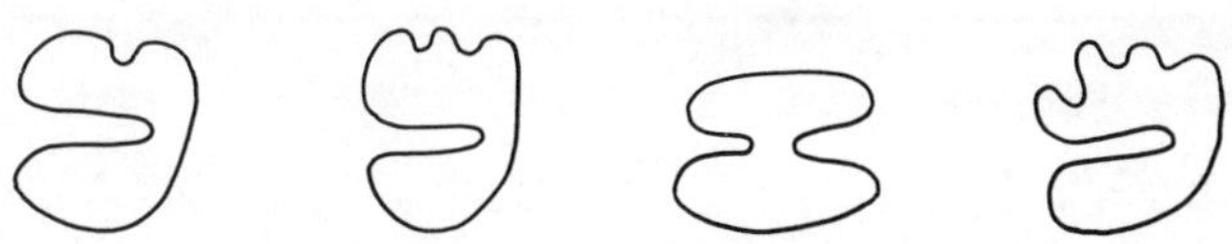

Fig.**45**. Leporidae. Variation of the anterior lower right cheek- tooth (P_3) (see Table 271).

Family Leporidae

Hares and rabbits

Leporinorum Fischer, 1817: 409; type genus *Lepus* L.

Worldwide except Australasia and Madagascar; 11 genera of which 4 are represented in the region (Table 271). The generic classification was reviewed by Corbet (1983). The European rabbit, *Oryctolagus cuniculus* (L.) is included since it is widely kept in domestication and frequently escapes.

LEPUS Hares

Lepus Linnaeus, 1758: 57; type species *L. timidus* L. Conserved by Opinion 91 (1926) and Direction 24 (1955).
Eulagos Gray, 1867c: 222 (*L. mediterraneus* Wagner, in *L. capensis* L.
And other synonyms based upon extralimital species.

N America, Africa and Eurasia; c. 23 species but classification rather unstable. Four species occur wholly within the region and another extends marginally into it (Table 272). The genus is characterized especially by long hind legs and wide nasal passage, both adaptations to running in open habitats. The Chinese species were reviewed (in Chinese) and mapped in detail by Luo (1988).

Lepus nigricollis Indian hare

L. nigricollis Cuvier, F.,1823b: 307; Madras, India.
L. ruficaudatus Geoffroy, I.,1826: 381; Bengal.
L. melanauchen Temminck, 1836: 13; Java.
L. macrotus Hodgson, 1840: 1183; Gangetic Plain, India.
L. aryabertensis Hodgson, 1844b: 293; Madhyades, Nepal (*nomen nudum*).
L. tytleri Tytler, 1854: 176; Dacca, Bangladesh.
L. dayanus Blanford, 1874: 663; Sukkur, Sind, Pakistan.

Map 175. *Trichys, Atherurus*

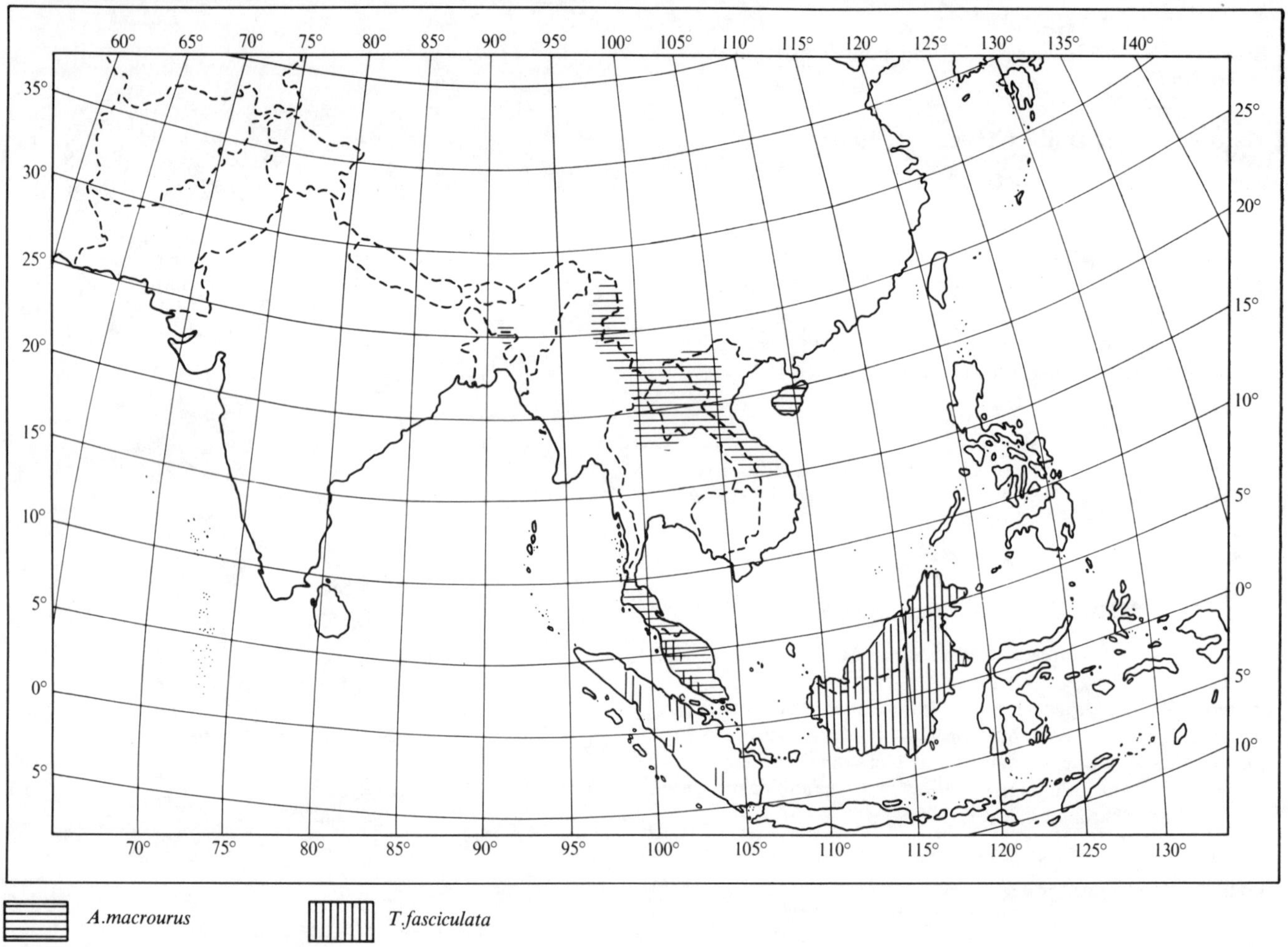

Table 271. The genera of Leporidae.

	Lepus	*Caprolagus*	*Pentalagus*	*Nesolagus*	*Oryctolagus*
Species in region	4	1	1	1	1
Range in region	Mainland, Taiwan, Hainan, Java	Nepal-Assam	Ryukyu Is	Sumatra	Domestic
Head & body	330–530	c.475	430–500	360–400	350–450
Hind feet	85–130	c.80	c.80	c.65	c.90
Ear (to crown)	70–130	c.55	c.44	c.35	c.75
Dorsal pelage striped	0	0	0	+	0
Venter dark	0	0	+	0	0
Pelage texture	Soft or normal	Harsh	Normal	Normal	Soft
Nape contrasting with dorsum	Rufous	0	0	0	Rufous
Skull: greatest length	65–100	c.85	c.80	c.65	c.70
Bony palate (length as % of skull)	7–9%	12%	15%	14%	9%
P_3 (Fig.45)	a	b	d	a or c	a

L. joongshaiensis Murray, 1884b: 51; Joongshai, Sind, Pakistan.

L. simcoxi Wroughton, 1912a: 338; Edalabad, Khandesh, India.

L. mahadeva Wroughton & Ryley, 1913: 15; Dhaim, Mahadeo Hills, Central Provinces, India, 2300ft.

L. rajput Wroughton, 1918a: 361; Sambhar Lake, Rajputana, India.

L. cutchensis Kloss, 1918a: 91; Bhuj, Cutch, Gujarat, India.

L. sadiya Kloss, 1918a: 95; Kobo, c. 15 miles W of Sadiya, NE Assam, India.

Range (Map 176). E Pakistan and peninsular India, east through S Nepal to W Bengal, Bangladesh and Tripura (Agrawal & Bhattacharyya, 1977a); Sri Lanka; also in Java where it seems likely to have been introduced although this has been queried on the basis of subfossil finds dated to the Neolithic period (Dawson, 1971). Introduced in Mauritius and Seychelles (Indian Ocean).

Variation. One subspecific name has been proposed:

L. n. singhala Wroughton, 1915a: 42; Kumbukkan, Sri Lanka.

Geographical variation in pelage is considerable but

Table 272. The species of *Lepus*.

	nigricollis	*sinensis*	*comus*	*capensis*	*peguensis*
Range in region	India, Sri Lanka, Java	S China	SW China	China, Pakistan	Indochina, Hainan
Head & body	330–530	c.400	350–440	400–500	360–500
Hind feet	85–115	c.105	c.130	110–130	90–105
Ear (to notch)	80–120	c.80	100–120	85–130	65–85
Ears black-tipped	0 (+)	(+)	+	+	+
Nape	Rufous or black	Rufous	Grey brown	Rufous	Rufous
Tail: dorsal	Brown	Brown	Black, grizzled white	Black	Black, narrow
sides	White	Grey/brown	Grey	White	White
ventral	White	Off-white	Grey	White	White
Pelage texture	Normal	Normal	Normal	Very soft	Normal
Skull: greatest length	80–100	65–85	83–95	80–90	76–92
I^1: shape of groove	Y	V	V (Y)	V	Y
± cement	+	0	+	0	+
P_3: anterior wall of talonid crenulated	+	+	0 or slightly	0	+

probably clinal. The main groups can be identified as follows:
L. n. nigricollis. S peninsular India. Nape black or dark grey.
L. n. dayanus. Pakistan and NW India. Nape pale buff; dorsum rather pale and greyish.
L. n. ruficaudatus. NE India, Nepal. Nape pale buff; dorsum with stronger yellow/brown tones.

Intermediate forms with grey or greyish brown napes occur in northern peninsular India, e. g. *simcoxi* and *mahadevi*. Variation in this area was described by Ghose (1967). Some specimens in Nepal have the dorsal surface of the tail very dark, approaching the condition seen in *L. peguensis* but less sharply delimited from the white on the rest of the tail.

Ecology. Found in all kinds of open ground, including cultivated land and open scrub; in the northwest also extending into desert. In Nepal it does not extend into the long-grass habitat occupied by *Caprolagus hispidus*, being confined to short grass, open forest and scrub (Bell, 1986). Up to 2400m in Kumaon.

Remarks. Most closely related to the widespread African species *L. victoriae* (including *L. whytei* and often, but incorrectly, called *L. crawshayi*). This was included in *L. nigricollis* by Petter (1961) but excluded by Angermann (1983) who nevertheless accepted a very close relationship between these. Petter (1961) also included in *L. nigricollis* the form here separated as *L. peguensis*.

Lepus peguensis Burmese hare

L. peguensis Blyth, 1856b: 471; Upper Pegu, Burma.
? *L. hainanus* Swinhoe, 1870b: 233, pl. 18; Hainan I, China.
L. siamensis Bonhote, 1902b: 40; Chiengmai, Thailand.
L. vassali Thomas, 1906d: 425; Nhatrang, Annam, Vietnam.

Range (Map 176). SE Burma, Thailand and Indochina; Hainan, if *hainanus* is conspecific.

Variation. The form on Hainan differs in having short ears (as in *L. sinensis*), but in other respects, e. g. pelage, tail and dentition, resembles mainland *L. peguensis*.

Remarks. This species has been included in *L. nigricollis* (e. g. by Petter, 1961) and is certainly very closely related to it. It is probably isolated from it by the forested parts of Assam and W Burma and is readily distinguished by the black-and-white tail. The form on Hainan has been given specific rank (Allen, 1938b) but seems clearly referable to *L. peguensis* on the basis of tail colour and especially the pattern of the upper incisors.

Lepus sinensis East Chinese hare

L. sinensis Gray, 1832 (in 1830–35): 2, pl. 20; near Canton, S China.
L. formosus Thomas, 1908i: 449; Baksa, Taiwan.
L. yuenshanensis Shih, 1930b: 3; Yuen Shan, Wukanghsien, Hunan, China.

Range (Map 176). SE China from Guangxi to S Anhui; Taiwan; and in Korea if *coreanus* is conspecific.

Variation. The following subspecific name has been proposed:
L. s. coreanus Thomas, 1892d: 146; Seoul, Korea. (Doubtfully referable to *L. sinensis*.)
L. s. flaviventris Allen, 1927c: 5; Chunganhsien, Fujian, China.

Lepus comus Yunnan hare

L. comus Allen, 1927c: 9; Tengueh, Yunnan, China, 5500ft.

Range (Map 176). Yunnan and W Guizhou, China.

Variation. Two subspecies have been described:
L. c. pygmaeus Wang & Feng, in Wang, Y-x. *et al.*,1985: 106; Mt Yulong, Lijiang, Yunnan, 27°N, 100°15'E, 2400m.
L. c. peni Wang & Luo, in Wang, Y-x. *et al.*,1985: 106; Kunming, Yunnan.

Differences between the three named forms are slight.

Remarks. This form was included in *L. oiostolus* Hodgson, 1840 (Tibet etc.) by Allen (1938b) and Ellerman & Morrison-Scott (1951) but was given specific rank by Wang, Y-x. *et al.* (1985) and Luo (1988).

Lepus capensis Cape hare

L. capensis Linnaeus, 1758: 58; Cape of Good Hope, South Africa.
L. tolai Pallas, 1778: 17; Adinscholo Mt, near Tchinden, Borsja R, Onon R, E Siberia, USSR.
L. tibetanus Waterhouse, 1841: 7; upper Indus Valley, Baltistan, Kashmir.
? *L. biddulphi* Blanford, 1877b: 324; Yassin, Gilgit, Kashmir.
L. swinhoei Thomas, 1894b: 364; Chefoo, Shandong, China.
L. filchneri Matschie, 1907: 217; Hinganfu, S Shaanxi, China.
L. stegmanni Matschie, 1907: 221; Kiauchow, near Tsingtao, Shandong.
L. aurigineus Hollister, 1912: 181; Kiukiang, N Jiangxi, China.

Map **176**. *Lepus* (part)

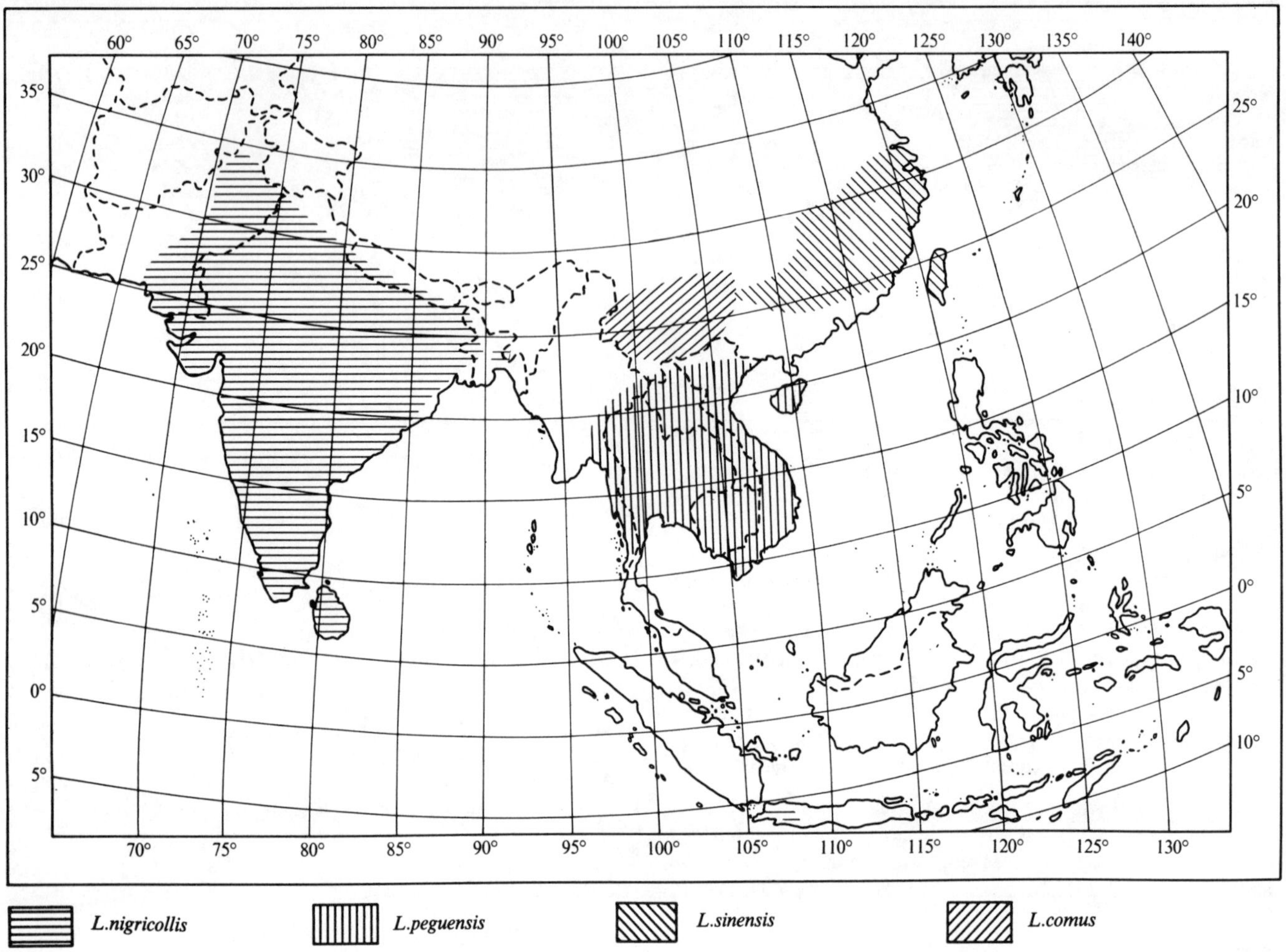

L. europaeus cinnamomeus Shamel, 1940a: 77; Suifu, Sichuan, China.

And many other extralimital synonyms.

Range (Map 177). From central China (Yangtze, Guizhou and N Yunnan) west through central Asia and from the Indus Basin west through SW Asia and most of the arid and savanna zones of Africa, but limits uncertain because of the taxonomic problems in delimiting the species.

Ecology. In Baluchistan found in semi-arid *Artemisia* steppe, especially at higher altitudes but down to 600m.

Remarks. The European *L. europaeus* has been included in this species but there is now fairly general agreement that it should be excluded. The inclusion of *tolai* (and therefore of all the Chinese forms listed above) in *L. capensis* is more controversial but it is difficult to diagnose these as separate species when taking into acount the great regional variation in each.

CAPROLAGUS

Caprolagus Blyth, 1845a: 247; type species *Lepus hispidus* Pearson.

One species. This genus has been extended by the inclusion of *sinensis* and other species, but this does not appear to be justified (Corbet, 1983). It is characterized by harsh pelage, small M^3, and heavy skull with small orbits, and has more in common with some other genera of rabbits than with the hares of the genus *Lepus*.

The following fossil species were considered to be *Caprolagus s. s.* by Dawson (1971) who recorded them from Java.

C. sivalensis Major, 1899: 514, pl. 37; Upper Siwaliks.

Lepus lapis Hooijer, 1964: 76; Sangiran, c. 12km N of Surakarta, central Java; Middle Pleistocene.

Caprolagus hispidus Assam rabbit

Lepus hispidus Pearson, in McClelland, 1839: 152; Himalayan foothills in N Assam.

Range (Map 177). From SW Nepal east to Assam (Bell *et al.*, 1990).

Ecology. Found in tall riverine grassland along the edge of the foothills; sympatric with *Lepus nigricollis* but separated by habitat (Bell, 1986).

Remarks. CITES: Appendix I; IUCN: E.

Map **177**. *Lepus* (part), *Pentalagus, Caprolagus, Nesolagus, Ochotona*

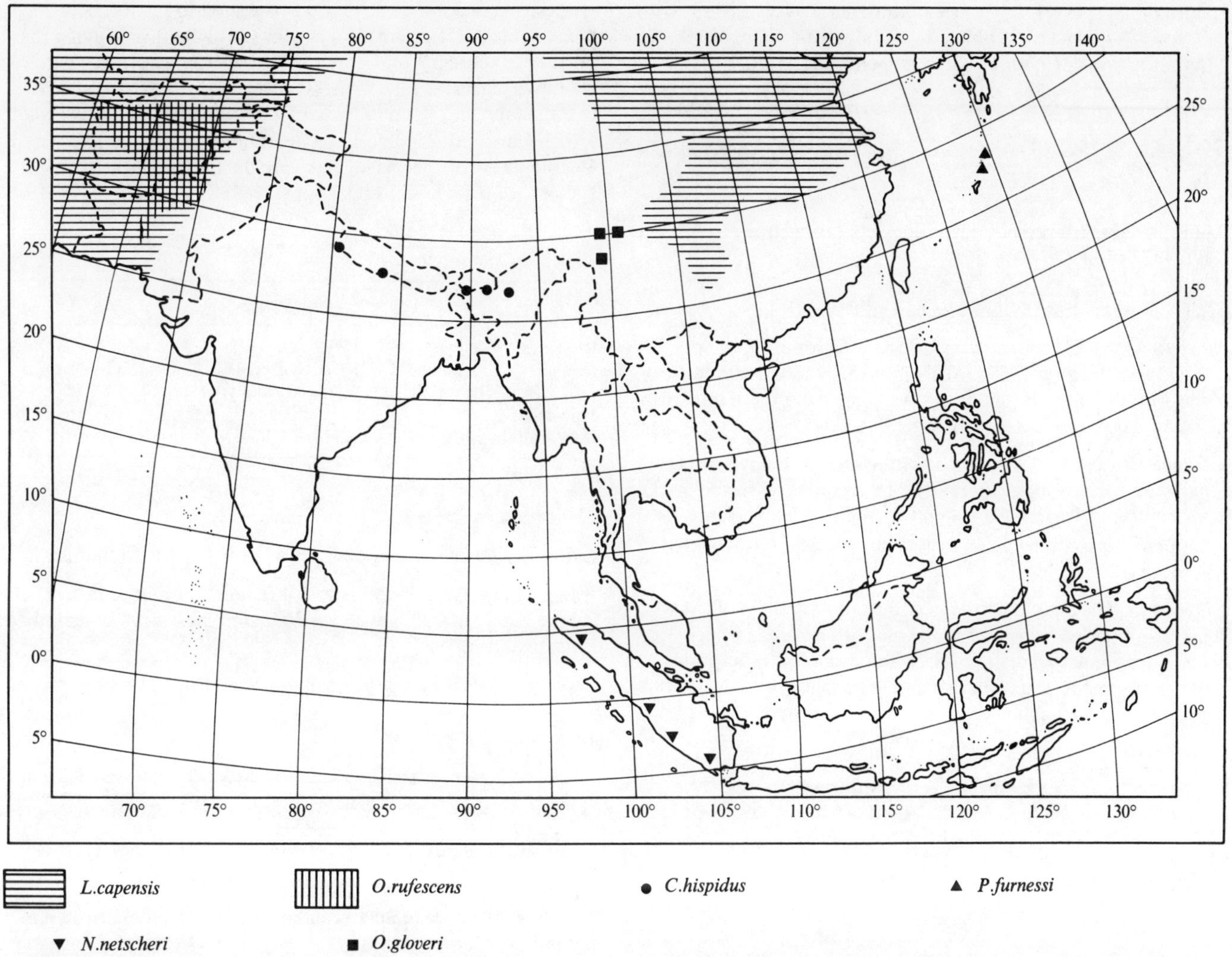

PENTALAGUS

Pentalagus Lyon, 1904: 428; type species *Caprolagus furnessi* Stone.

One species. One of the more distinctive of the many monospecific genera of rabbits, characterized by a suite of apparently primitive characters, e. g. short ears and hind legs, symmetrical P_3, but with some peculiarities, e. g. the extreme crenulation of the cheek-teeth.

Pentalagus furnessi Ryukyu rabbit

Caprolagus furnessi Stone, 1900: 460; Amami-Oshima, Ryukyu Is.

Range (Map 177). Amami and Tokuno Is, Ryukyu Is.

Remarks. IUCN: E.

NESOLAGUS

Nesolagus Major, 1899: 514; type species *Lepus netscheri* Schlegel.

One species. Another distinctive monospecific genus, not obviously closely related to the other genera of rabbits.

Nesolagus netscheri Sumatran rabbit

Lepus netscheri Schlegel, 1880: 59, 62; near Padang-Panjang, Padang Highlands, Sumatra, 2000ft.

Range (Map 177). Sumatra; montane. Known from only six localities, at altitudes of 600–1600m, and confirmed in recent years only from one, in Gunung Leuser Park, N Sumatra (Flux, 1990).

Nesolagus netscheri

Remarks. Efforts to determine the range and status of this very vulnerable species have been hindered by confusion with *Oryctolagus cuniculus* which is widely kept in captivity in Sumatra (Flux, 1990). CITES: Appendix II; IUCN: I.

ORYCTOLAGUS

Oryctolagus Lilljeborg, 1874: 417; type species *Lepus cuniculus* L.

One species. Marginally justifying separation from *Sylvilagus* Gray, 1867 of the Americas.

Oryctolagus cuniculus European rabbit

Lepus cuniculus Linnaeus, 1758: 58; S Europe.
L. nigripes Bartlett, 1857a: 610 (and 1857b: 160); a domesticated form, thought originally to have come from the Himalayas, but actual origin unknown.

Range. Native to SW Europe; introduced in many temperate parts of the world. Apparently not recorded as feral in the Indomalayan Region except on the Maldive Is, but other feral populations are likely to exist since the species is widely kept in captivity.

Remarks. The characters given in the table are those of the wild form, to which most feral populations revert (except that melanics tend to persist). Most domesticated rabbits are of domestic breeds, usually larger and of a variety of colours, often white, black or piebald.

Family Ochotonidae

Pikas

Ochotonidae Thomas, 1897c: 1026.

One genus.

OCHOTONA Pikas

Ochotona Link, 1795: 74; type species *Lepus ogotona* Pallas, in *O. dauurica* Pallas.
Lagomys Cuvier, 1800: 1. Pre-occupied by *Lagomys* Storr, 1780 (= *Marmota* Blumenbach).
And other synonyms based upon extralimital species.

Central and NE Asia, western N America; c. 17 species. Two of these extend sufficiently into this region to justify inclusion here (Table 273); several others occur at high altitude along the northern boundary and some of these extend a little below 3000m in places, e. g. *O. roylei* (Ogilby, 1839), down to c. 2600m in Pakistan and 2700m in Nepal; *O. thibetana* (Milne-Edwards, 1872), down to 2500m in Sichuan and 2400m in Qinghai. Classification at species level is very unstable, but the Chinese species were reviewed and keyed by Feng & Zheng (1985).

Table 273. The species of *Ochotona*.

	rufescens	*gloveri*
Range in region	E Pakistan	SW China
Dorsum	Light grey, tinged rufous	Grey-brown
Pale collar	+	0
Interorbital width	4.2–4.7	5.0–6.2

Ochotona rufescens Afghan pika

Lagomys rufescens Gray, 1842: 266; near Kabul, Afghanistan.

Range (Map 177). E Pakistan; Afghanistan and Iran to the Kopet Dag Mts. Down to 900m in parts of Pakistani Baluchistan.

Ecology. Occurs in dry juniper forest and in open steppe.

Ochotona gloveri

O. gloveri Thomas, 1922e: 190; Nagchuka, W Sichuan, China, 10000ft.
O. erythrotis brookei Allen, 1937: 341; a few miles NW of Jyekundo, Kham, E Tibet.

Range (Map 177). NW Yunnan, W Sichuan and E Tibet; mainly at altitudes of 2000–2400m in Yunnan but up to 3000m (Wang, Y-x. *et al.*, 1985).

Variation. The following subspecific name has been proposed: *O. brookei calloceps* Pen & Feng, in Pen *et al.*,1962: 119,131; Atung, NW of Tehching (Atungtze), Yunnan, 2000m.

Remarks. The form *brookei* was given specific rank by Wang, Y-x. *et al.* (1985) but was included in *O. gloveri* by Feng & Zheng (1985).

Appendix 1

List of 'Recent' species in the Indomalayan Region

Species that occur in the region but are not predominantly or extensively Indomalayan in distribution or are not extant, native, wild species are indicated as follows:

Af—mainly Afrotropical
D —domestic species
E —extinct in Region but surviving elsewhere
I —introduced species
NG—mainly in New Guinea, ± Australia
NP—North Pacific (Cetacea)
P —mainly Palaearctic
SO—southern Ocean (Cetacea)

The number of species listed is shown in parentheses after the name of each higher taxon.

MARSUPIALIA (10)

PERAMELINA (2)

Peramelidae (2)
Echymipera rufescens NG
Rhynchomeles prattorum

DIPROTODONTIA (8)

Phalangeridae (7)
Phalanger orientalis NG
P. ornatus
P. pelengensis
P. rothschildi
Stigocuscus celebensis
Spilocuscus maculatus NG
Ailurops ursinus

Petauridae (1)
Petaurus breviceps NG

EUTHERIA (1031)

PHOLIDOTA (3)

Manidae (3)
Manis pentadactyla
M. crassicaudata
M. javanica

INSECTIVORA (77)

Erinaceidae (12)
Erinaceinae (6)
Erinaceus amurensis P
Hemiechinus auritus P
H. collaris
H. dauuricus P
Paraechinus micropus
P. hypomelas P
Galericinae (6)
Echinosorex gymnurus
Hylomys suillus
H. sinensis
H. hainanensis
Podogymnura truei
P. aureospinula

Talpidae (10)
Uropsilinae (3)
Uropsilus soricipes
U. gracilis
U. andersoni
Talpinae (7)
Talpa micrura
T. grandis
T. longirostris
T. leucura
T. insularis
Scaptonyx fusicauda
Scapanulus oweni

Soricidae (55)
Soricinae (18)
Sorex cylindricauda
S. bedfordiae
Blarinella quadraticauda
Soriculus nigrescens
S. caudatus
S. fumidus
S. leucops
S. macrurus
S. hypsibius
S. lamula
S. salenskii
S. smithii
S. parca
Anourosorex squamipes
Chimarrogale himalayica
C. styani
C. phaeura
Nectogale elegans
Crocidurinae (37)
Suncus murinus
S. montanus
S. zeylanicus
S. stoliczkanus
S. dayi
S. ater
S. mertensi
S. etruscus
Feroculus feroculus
Crocidura fuliginosa
C. orientalis
C. baluensis
C. tenuis
C. elongata
C. mindorus
C. nigripes
C. rhoditis
C. lasiura P
C. hispida
C. nicobarica
C. andamanensis
C. jenkinsi
C. attenuata
C. orii
C. neglecta
C. grayi
C. miya
C. pergrisea P
C. zarudnyi
C. gueldenstaedtii P
C. suaveolens P
C. horsfieldi
C. monticola
C. lea
C. levicula
C. paradoxura
Solisorex pearsoni

SCANDENTIA (16)

Tupaiidae (16)
Tupaiinae (15)
Tupaia glis
T. belangeri
T. nicobarica
T. splendidula
T. montana
T. javanica
T. minor
T. gracilis
T. picta
T. tana
T. dorsalis
Anathana ellioti
Dendrogale murina
D. melanura
Urogale everetti
Ptilocercinae (1)
Ptilocercus lowii

DERMOPTERA (2)

Cynocephalidae (2)
Cynocephalus variegatus
C. volans

CHIROPTERA (307)
MEGACHIROPTERA (74)

Pteropodidae (74)
Pteropodinae (62)
Pteropus hypomelanus
P. faunulus
P. griseus
P. pumilus
P. dasymallus
P. mariannus
P. caniceps
P. melanotus
P. melanopogon
P. chrysoproctus
P. lombocensis
P. leucopterus
P. personatus
P. temmincki
P. giganteus
P. intermedius
P. vampyrus
P. lylei
P. alecto
P. conspicillatus NG
P. ocularis
Acerodon celebensis
A. macklotii
A. leucotis
A. humilis
A. jubatus
A. lucifer
Rousettus egyptiacus Af
R. leschenaulti
R. amplexicaudatus
R. spinalatus
R. celebensis
R. bidens
Neopteryx frosti
Styloctenium wallacei
Dobsonia minor
D. exoleta
D. moluccense NG
D. peronii
D. viridis
Cynopterus brachyotis
C. sphinx
C. titthaecheilus
C. horsfieldii
Megaerops ecaudatus
M. kusnotoi
M. niphanae
M. wetmorei
Ptenochirus jagorii
P. minor
Dyacopterus spadiceus
D. brooksi
Chironax melanocephalus
Latidens salimalii
Penthetor lucasi
Thoopterus nigrescens
Sphaerias blanfordi
Balionycteris maculata
Aethalops alecto
Haplonycteris fischeri
Alionycteris paucidentata
Otopteropus cartilagonodus
Harpyionycterinae (2)
Harpyionycteris whiteheadi
H. celebensis
Nyctimeninae (4)
Nyctimene minutus
N. rabori
N. cephalotes

N. albiventer NG
Macroglossinae (6)
Eonycteris spelaea
E. major
Macroglossus minimus
M. sobrinus
Syconycteris australis NG
S. carolinae

MICROCHIROPTERA (233)

Rhinopomatidae (3)
Rhinopoma microphyllum P, Af
R. hardwickii P, Af
R. muscatellum P

Emballonuridae (11)
Emballonura beccarii NG
E. monticola
E. alecto
E. raffrayana NG
E. nigrescens NG
Taphozous perforatus Af
T. melanopogon
T. longimanus
T. theobaldi
T. nudiventris P, Af
T. saccolaimus

Craseonycteridae (1)
Craseonycteris thonglongyai

Nycteridae (2)
Nycteris javanica
N. tragata

Megadermatidae (2)
Megaderma spasma
M. lyra

Rhinolophidae (42)
Rhinolophus philippinensis
R. sedulus
R. trifoliatus
R. luctus
R. rex
R. paradoxolophus
R. marshalli
R. macrotis
R. euryotis
R. creaghi
R. canuti
R. arcuatus
R. coelophyllus
R. shameli
R. inops
R. rufus
R. subrufus
R. pearsonii
R. yunanensis
R. acuminatus
R. lepidus
R. pusillus
R. cornutus P
R. subbadius
R. monoceros
R. cognatus
R. imaizumii
R. blasii P
R. mehelyi P
R. megaphyllus
R. celebensis
R. virgo
R. borneensis
R. nereis
R. anderseni
R. malayanus
R. stheno
R. thomasi
R. rouxii
R. affinis
R. ferrumequinum P
R. hipposideros P

Hipposideridae (40)
Hipposideros bicolor
H. pomona
H. macrobullatus
H. ater
H. fulvus
H. cineraceus
H. durgadasi
H. halophyllus
H. nequam
H. coronatus
H. ridleyi
H. dyacorum
H. sabanus
H. doriae
H. obscurus
H. pygmaeus
H. galeritus
H. cervinus
H. breviceps
H. coxi
H. papua NG
H. pratti
H. lylei
H. armiger
H. turpis
H. larvatus
H. speoris
H. lankadiva
H. schistaceus
H. diadema
H. lekaguli
H. dinops
H. inexpectatus
Triaenops persicus Af, P
Asellicus stoliczkanus
A. tricuspidatus
Asellia tridens Af, P
Coelops frithii
C. robinsoni
Paracoelops megalotis

Vespertilionidae (119)
Vespertilioninae (86)
Myotis blythii P
M. chinensis P
M. altarium
M. sicarius
M. pequinius P
M. formosus
M. hermani
M. frater P
M. siligorensis
M. rosseti
M. ridleyi
M. ater
M. mystacinus P
M. muricola
M. patriciae
M. montivagus
M. annectans
M. oreias
M. macrotarsus
M. stalkeri
M. fimbriatus
M. longipes
M. daubentonii P
M. horsfieldii
M. hasseltii
M. adversus
M. ricketti
Plecotus auritus P
P. austriacus P
Barbastella leucomelas P
Otonycteris hemprichii P
Scotomanes ornatus
Scotophilus kuhlii
S. heathii
Eptesicus serotinus P
E. bottae P
E. pachyotis
E. demissus
E. gobiensis P
E. nasutus P
Vespertilio murinus P
V. superans P
V. orientalis P
Ia io
Tylonycteris pachypus
T. robustula
Eudiscopus denticulus
Pipistrellus pipistrellus P
P. ceylonicus
P. minahassae
P. paterculus
P. abramus P
P. javanicus
P. coromandra
P. tenuis
P. kuhlii Af, P
P. pulveratus
P. savii P
P. imbricatus
P. macrotis
P. vordermanni
P. cadornae
P. lophurus
P. kitcheneri
P. stenopterus
P. joffrei
P. anthonyi
P. affinis
P. petersi
P. mordax
P. circumdatus
P. societatis
P. cuprosus
Glischropus tylopus
G. javanus
Scotozous dormeri
Scotoecus pallidus
Nyctalus noctula P
N. leisleri P
N. montanus
Philetor brachypterus
Hesperoptenus doriae
H. tickelli
H. tomesi
H. gaskelli
H. blanfordi
Miniopterinae (7)
Miniopterus schreibersii
M. magnater
M. fuscus
M. medius
M. australis NG
M. pusillus
M. tristis NG
Murininae (13)
Murina leucogaster P
M. aurata
M. florium
M. suilla
M. tubinaris
M. rozendaali
M. huttonii
M. puta
M. aenea
M. cyclotis
M. grisea
Harpiocephalus harpia
H. mordax
Kerivoulinae (11)
Kerivoula whiteheadi
K. picta
K. minuta
K. intermedia
K. pellucida
K. papillosa
K. hardwickii
K. flora
K. myrella
Phoniscus atrox
P. jagorii
Nyctophilinae (2)
Nyctophilus timoriensis NG
N. heran

Molossidae (13)
Tadarida teniotis P
T. aegyptiaca P, Af
T. mops
T. sarasinorum
T. johorensis
T. plicata
T. jobensis NG
Otomops wroughtoni
O. formosus
Mormopterus doriae
M. beccarii NG
Cheiromeles torquatus
C. parvidens

PRIMATES (61)

Loridae (4)
Loris tardigradus
Nycticebus coucang
N. pygmaeus
N. intermedius

Tarsiidae (5)
Tarsius syrichta
T. spectrum
T. pumilus
T. dianae
T. bancanus

Cercopithecidae (39)
Cercopithecinae (14)
Macaca silenus
M. nemestrina
M. maura
M. ochreata
M. tonkeana
M. nigra

M. sinica
M. radiata
M. assamensis
M. thibetana
M. mulatta
M. fascicularis
M. cyclopis
M. arctoides
Colobinae (25)
Presbytis melalophos
P. femoralis
P. thomasi
P. potenziani
P. comata
P. hosei
P. frontata
P. rubicunda
Semnopithecus entellus
S. auratus
S. francoisi
S. vetulus
S. johnii
S. cristatus
S. obscurus
S. phayrei
S. pileatus
S. geei
Pygathrix nemaeus
P. roxellana
P. bieti
P. brelichi
P. avunculus
Nasalis larvatus
N. concolor

Hylobatidae (11)
Hylobates lar
H. pileatus
H. agilis
H. moloch
H. muelleri
H. klossi
H. hoolock
H. concolor
H. leucogenys
H. gabriellae
H. syndactylus

Hominidae (2)
Pongo pygmaeus
Homo sapiens

CARNIVORA (83)

Canidae (9)
Canis lupus
C. aureus
C. familiaris D
Vulpes vulpes P
V. bengalensis
V. rueppellii P
V. cana P
Nyctereutes procyonoides
Cuon alpinus

Ursidae (4)
Ursus thibetanus P
U. malayanus
U. ursinus
Ailuropoda melanoleuca

Ailuridae (1)
Ailurus fulgens

Mustelidae (22)
Mustelinae (10)
Mustela nivalis P
M. kathiah
M. sibirica
M. lutreolina
M. nudipes
M. strigidorsa
Vormela peregusna P
Martes foina
M. flavigula
Mellivora capensis P, Af
Melinae (8)
Meles meles P
Arctonyx collaris
Melogale personata
M. orientalis
M. everetti
M. moschata
Mydaus javanensis
M. marchei
Lutrinae (4)
Lutra lutra
L. sumatrana
Lutrogale perspicillata
Aonyx cinerea

Viverridae (20)
Viverra zibetha
V. tangalunga
V. megaspila
V. civettina
Viverricula indica
Prionodon linsang
P. pardicolor
Paradoxurus hermaphroditus
P. lignicolor
P. zeylonensis
P. jerdoni
Paguma larvata
Macrogalidia musschenbroekii
Arctictis binturong
Arctogalidia trivirgata
Hemigalus derbyanus
H. owstoni
Diplogale hosei
Cynogale bennettii
C. lowei

Herpestidae (8)
Herpestes javanicus
H. edwardsii
H. smithii
H. fuscus
H. vitticollis
H. urva
H. semitorquatus
H. brachyurus

Hyaenidae (1)
Hyaena hyaena

Felidae (18)
Felis silvestris P, Af
F. catus D
F. chaus
F. margarita P
F. caracal P, Af
Prionailurus bengalensis
P. iriomotensis
P. viverrinus
P. planiceps
P. rubiginosus
Catopuma temminckii
C. badia
Pardofelis marmorata
P. nebulosa
Panthera pardus
P. tigris
P. leo Af
Acinonyx jubatus E

CETACEA (32)
ODONTOCETI (25)

Platanistidae (1)
Platanista gangetica

Iniidae (1)
Lipotes vexillifer

Delphinidae (16)
Steno bredanensis
Sousa chinensis
Stenella coeruleoalba
S. attenuata
S. longirostris
Delphinus delphis
Tursiops truncatus
Lagenodelphis hosei
Lagenorhynchus obliquidens NP
Peponocephala electra
Orcaella brevirostris
Pseudorca crassidens
Orcinus orca
Grampus griseus
Globicephala macrorhynchus
Feresa attenuata

Phocoenidae (1)
Neophocaena phocaenoides

Physeteridae (3)
Physeter macrocephalus
Kogia breviceps
K. simus

Ziphiidae (3)
Mesoplodon ginkgodens
M. densirostris
Ziphius cavirostris

MYSTICETI (7)

Balaenopteridae (6)
Balaenoptera acutorostrata
B. edeni
B. borealis
B. physalus
B. musculus
Megaptera novaeangliae

Balaenidae (1)
Balaena glacialis NP, SO

SIRENIA (1)

Dugongidae (1)
Dugong dugon

PROBOSCIDEA (1)

Elephantidae (1)
Elephas maximus

PERISSODACTYLA (7)

Tapiridae (1)
Tapirus indicus

Rhinocerotidae (3)
Rhinoceros unicornis
R. sondaicus
Dicerorhinus sumatrensis

Equidae (3)
Equus hemionus P
E. asinus D
E. caballus D

ARTIODACTYLA (61)

Suidae (7)
Sus scrofa
S. domesticus D
S. salvanius
S. barbatus
S. verrucosus
S. celebensis
Babyrousa babyrussa

Camelidae (1)
Camelus dromedarius D

Tragulidae (3)
Tragulus javanicus
T. napu
Moschiola meminna

Moschidae (1)
Moschus berezovskii

Cervidae (21)
Cervus nippon
C. duvaucelii
C. eldii
C. unicolor
C. timorensis
C. mariannus
C. alfredi
Elaphurus davidianus E, I
Axis axis
A. porcinus
A. kuhlii
A. calamianensis
Muntiacus muntjak
M. reevesi
M. crinifrons
M. feae
M. rooseveltorum
M. gongshanensis
M. atherodes
Elaphodus cephalophus
Hydropotes inermis

Bovidae (28)
Bos taurus D
B. sauveli
B. gaurus
B. frontalis D
B. javanicus
Bubalus bubalis D
B. arnee
B. mindorensis
B. depressicornis
B. quarlesi
Boselaphus tragocamelus
Tetracerus quadricornis
Antilope cervicapra
Gazella bennettii
G. subgutturosa P
Budorcas taxicolor
Naemorhedus goral

N. caudatus
N. baileyi
N. sumatraensis
N. swinhoei
Hemitragus jemlahicus
H. hylocrius
Capra aegagrus P
C. hircus D
C. falconeri
Ovis vignei P
O. aries D

RODENTIA (369)

Sciuridae (62)
Ratufinae (6)
Ratufa macroura
R. indica
R. bicolor
R. affinis
Sciurotamias davidianus
S. forresti
Funambulinae (5)
Funambulus palmarum
F. tristriatus
F. layardi
F. sublineatus
F. pennantii
Callosciurinae (49)
Callosciurus erythraeus
C. quinquestriatus
C. finlaysonii
C. caniceps
C. inornatus
C. pygerythrus
C. phayrei
C. prevostii
C. baluensis
C. notatus
C. nigrovittatus
C. adamsi
C. melanogaster
C. orestes
C. simus
Tamiops mcclellandii
T. rodolphi
T. swinhoei
T. maritimus
Sundasciurus lowii
S. tenuis
S. jentinki
S. brookei
S. hippurus
S. philippinensis
S. steerii
S. moellendorffi
S. rabori
Dremomys lokriah
D. pernyi
D. rufigenis
D. gularis
D. everetti
Lariscus insignis
L. obscurus
L. hosei
Menetes berdmorei
Rhinosciurus laticaudatus
Hyosciurus heinrichi
H. ileile
Nannosciurus melanotis
Exilisciurus exilis
E. whiteheadi
E. concinnus
Prosciurillus murinus
P. leucomus
P. weberi
P. abstrusus
Rubrisciurus rubriventer
Sciurinae (1)
Rheithrosciurus macrotis
Marmotinae (1)
Tamias sibiricus P

Pteromyidae (38)
Trogopterus xanthipes
T. pearsonii
Pteromyscus pulverulentus
Petaurista petaurista
P. philippensis
P. elegans
P. sybilla
P. caniceps
P. alborufus
P. magnificus
P. nobilis
P. xanthotis
Eupetaurus cinereus
Aeretes melanopterus
Biswamoyopterus biswasi
Iomys horsfieldii
I. sipora
Hylopetes spadiceus
H. lepidus
H. sipora
H. alboniger
H. nigripes
H. phayrei
H. fimbriatus
H. baberi
H. bartlesi
H. winstoni
Petinomys hageni
P. lugens
P. fuscocapillus
P. crinitus
P. genibarbis
P. setosus
P. vordermanni
Petaurillus hosei
P. emiliae
Aeromys tephromelas
A. thomasi

Muridae (253)
Murinae (213)
Mus musculus I
M. booduga
M. dunni
M. cookii
M. cervicolor
M. caroli
M. platythrix
M. saxicola
M. phillipsi
M. shortridgei
M. fernandoni
M. mayori
M. pahari
M. famulus
M. crociduroides
M. vulcani
Vandeleuria oleracea
V. nolthenii
V. nilagirica
Acomys cahirinus P, Af
Rattus rattus
R. tiomanicus
R. baluensis
R. palmarum
R. lugens
R. adustus
R. hoffmanni
R. mollicomulus
R. koopmani
R. tawitawiensis
R. losea
R. osgoodi
R. argentiventer
R. hoxaensis
R. norvegicus I
R. nitidus
R. remotus
R. turkestanicus
R. exulans
R. annandalei
R. korinchi
R. montanus
R. hoogerwerfi
R. stoicus
R. nativitatis
R. xanthurus
R. bontanus
R. marmosurus
R. macleari
R. enganus
R. feliceus
R. morotaiensis
R. elaphinus
R. ceramicus
R. everetti
R. tyrannus
R. mindorensis
R. ranjiniae
R. timorensis
Millardia meltada
M. kondana
M. kathleenae
M. gleadowi
Cremnomys cutchicus
C. elvira
C. blanfordi
Golunda ellioti
Nesokia indica
Bandicota indica
B. savilei
B. bengalensis
Srilankamys ohiensis
Chiropodomys gliroides
C. major
C. calamianensis
C. muroides
C. karlkoopmani
Vernaya fulva
Hapalomys longicaudatus
H. delacouri
Micromys minutus P
Apodemus agrarius P
A. chevrieri
A. sylvaticus P
A. draco
A. orestes
A. latronum
A. gurkha
Hadromys humei
Dacnomys millardi
Diomys crumpi
Niviventer niviventer
N. confucianus
N. tenaster
N. fulvescens
N. bukit
N. rapit
N. coninga
N. lepturus
N. brahma
N. eha
N. langbianis
N. hinpoon
N. cremoriventer
N. andersoni
N. excelsior
Chiromyscus chiropus
Leopoldamys sabanus
L. edwardsi
L. neilli
L. siporanus
Maxomys bartelsii
M. rajah
M. surifer
M. whiteheadi
M. alticola
M. ochraceiventer
M. baeodon
M. inas
M. hylomyoides
M. inflatus
M. moi
M. panglima
M. pagensis
M. musschenbroekii
M. hellwaldii
M. dollmani
M. wattsi
Berylmys manipulus
B. berdmorei
B. mackenziei
B. bowersii
Haeromys margarettae
H. pusillus
H. minahassae
Pithecheir melanurus
P. parvus
Kadarsanomys sodyi
Lenothrix canus
Palawanomys furvus
Sundamys muelleri
S. infraluteus
S. maxi
Bullimus luzonicus
B. bagobus
Limnomys sibuanus
Tarsomys apoensis
Tryphomys adustus
Abditomys latidens
Apomys hylocoetes
A. insignis
A. hollisteri
A. musculus
A. littoralis
A. abrae
A. sacobianus
A. datae
Carpomys melanurus
C. phaeurus

Batomys granti
B. dentatus
B. salomonseni
Anonymomys mindorensis
Phloeomys cumingi
P. pallidus
Crateromys schadenbergi
C. paulus
C. australis
Crunomys fallax
C. melanius
C. rabori
C. celebensis
Archboldomys luzonensis
Chrotomys whiteheadi
C. mindorensis
C. silaceus
Rhynchomys soricoides
R. isarogensis
Bunomys chrysocomus
B. coelestis
B. prolatus
B. fratrorum
B. andrewsi
B. heinrichi
B. penitus
B. naso
Lenomys meyeri
Eropeplus canus
Margaretamys beccarii
M. elegans
M. parvus
Taeromys celebensis
T. callitrichus
T. hamatus
T. punicans
T. taerae
T. arcuatus
Paruromys dominator
Echiothrix leucura
Melasmothrix naso
M. rhinogradoides
M. macrocercus
Papagomys armandvillei
Komodomys rintjanus
Melomys fulgens
M. caurinus
M. aerosus
M. obiensis
M. fraterculus
Uromys caudimaculatus NG
Hydromys chrysogaster NG
Diplothrix legata
Tokudaia osimensis

Cricetinae (5)
Cricetulus migratorius P
C. longicaudatus P
C. triton P
Cansumys canus
Calomyscus bailwardi P

Gerbillinae (9)
Gerbillus nanus P
G. gleadowi
G. cheesmani P
Tatera indica
Meriones persicus P
M. hurrianae
M. libycus P
M. crassus P
Rhombomys opimus P

Myospalacinae (3)
Myospalax fontanierii P
M. rothschildi P
M. smithii P

Platacanthomyinae (2)
Platacanthomys lasiurus
Typhlomys cinereus

Arvicolinae (17)
Ellobius fuscocapillus P
Hyperacrius wynnei
H. fertilis
Eothenomys melanogaster
E. miletus P
E. olitor
E. proditor P
E. chinensis
E. wardi P
E. custos P
E. inez P
E. eva
Microtus fortis
M. kikuchii
M. musseri
Proedromys bedfordi P
Ondatra zibethicus I

Rhizomyinae (4)
Rhizomys sinensis
R. pruinosus
R. sumatrensis
Cannomys badius

Gliridae (2)
Dryomys nitedula P
Chaetocauda sichuanensis

Zapodidae (2)
Sicista concolor P
Zapus setchuanus P

Dipodidae (4)
Salpingotus michaelis
Allactaga elater P
A. hotsoni P
Jaculus blanfordi P

Hystricidae (8)
Hystrix indica
H. brachyura
H. javanica
H. crassispinis
H. sumatrae
H. pumila
Trichys fasciculata
Atherurus macrourus

LAGOMORPHA (11)

Leporidae (9)
Lepus nigricollis
L. peguensis
L. sinensis
L. comus
L. capensis P, Af
Caprolagus hispidus
Pentalagus furnessi
Nesolagus netscheri
Oryctolagus cuniculus D, I

Ochotonidae (2)
Ochotona rufescens P
O. gloveri P

Total number of species: 1041.

Appendix 2

New names proposed in this work.

No new taxa are described here, but it has been necessary to propose the following new names to replace those found to be pre-occupied.

Rhinolophus megaphyllus thaianus Hill, *nom. nov.* for *R. robinsoni siamensis* McFarlane & Blood, 1986 (p. 101).

Paradoxurus hermaphroditus dongfangensis Corbet, *nom. nov.* for *P. h. hainana* Wang & Xu, 1981 (p. 208).

Appendix 3

Gazetteer

This gazetteer includes only names that appear frequently in the text but are not to be found in most standard modern atlases. They fall into two main categories: names of type localities from which a substantial number of taxa were described; and obsolete names that appear frequently in the literature on mammals.

Amboina. = Island of Ambon, Moluccas.
Amoy. = Hsia-men = Xiamen, Fujian, China, 24°26'N, 118°5'E.
Anhwei. = Anhui Prov., China.
Batavia. = Jakarta, West Java, Indonesia.
Bencoolen. = Bengkulu, SW Sumatra.
Bliss, Mt. Grand Malindang Mts, NW Mindanao, Philippines.
Buitenzorg. = Bogor, West Java, Indonesia.
Canton. = Kuang-chou = Kwangchow = Guangzhou, Guangdong, China, 23°8'N, 113°20'E.
Ceram. = Seram I, Moluccas.
Chapa. = Cha Pa, Tonkin, N Vietnam, 22°20'N, 103°50'E.
Chekiang. = Zhejiang Prov., China.
Condor I. = Con Son I, S Vietnam, 8°41'N, 106°36'E.
Data, Mt. Benguet Prov., Luzon, Philippines, c. 16°51'N, 120°52'E.
Formosa. = Taiwan.
Fukien. = Fujian Prov., China.
Gede, Mt. Peak adjacent to Mt Pangrango, near Cibodas, W Java, 6°48'S, 106°56'E.
Honan. = Henan Prov., China.
Hupeh. = Hubei Prov., China.
Junk Ceylon I. = Puket = Phuket, SW Thailand, 7°52'N, 98°22'E.
Kansu. = Gansu Prov., China.
Kelabit Highlands. NE Sarawak, Borneo.
Kiangsi. = Jiangxi Prov., China.
Kiangsu. = Jiangsu Prov., China.
Korinchi Peak. = Kerinci, Sumatra.
Kwangsi. = Guangxi Prov., China.
Kwangtung. = Guangdong Prov., China.
Kweichow. = Guizhou Prov., China.
Langbian Peak. = Mt Lang Bian, Tuyen Duc Prov., Vietnam, 12°3'N, 108°26'E.
Liu Kiu Is. = Ryukyu Is (= Nansei-shoto), Japan.
Macassar. = Makasar = Ujong Pandang, SW Sulawesi, 5°9'S, 119°28'E.
Moupin. c. 100km W of Chengdu, Sichuan, China, 30°30'N, 102°40'E.
Omi San = Omei Shan, Sichuan, China, 29°35'N, 103°11'E.
Pinedapa. Near Mapane, C Sulawesi.
Poppa, Mt. C Burma, 20°55'N, 95°15'E.
Poona. = Pune, Maharashtra, India, 18°34'N, 73°58'E.
Raheng. = Rahaeng = Tak, W Thailand, 16°51'N, 99°8'E.
Salsette I. Bombay, India.
Shansi. = Shanxi Prov., China.
Shensi. = Shaanxi Prov., China.

Singalang, Mt. Sumatra, 0°24'S, 100°21'E.
Sirhassen I. = Serasan I, Bunguran Selatan Is (= S Natuna Is), S China Sea.
Szechuan. = Sichuan Prov., China.
Taochao. = Taochow = Lin-t'an, Gansu, China, 34°39'N, 103°40'E.
Tatsienlu. = K'ang-ting = Kangding, Sichuan, China, 30°5'N, 102°4'E.
Tj----. Names in Indonesia that were spelled using 'Tj' by the Dutch administration are now spelled using 'C', e. g. *Tjibodas* becomes *Cibodas*.
Tonkin region. In N Vietnam.
Tsinling Mts. = Qinling Mts, S Shaanxi, China.

Appendix 4

Biographical notes

These biographical notes refer to the persons who have contributed most significantly to the study of mammals in the Indomalayan Region, such as collectors or describers, and to those whose names have been commemorated in the names of species or genera that are recognized here. The list is far from comprehensive but aims to include the names that appear most frequently in the text. Their published output can be found in the Bibliography and the scientific names by which they are commemorated can be traced through the Index.

Abbott, Dr William Louis, 1860–1936. American physician and traveller; collected mammals in the Himalayas and Thailand in the 1890s and in Sumatra, Borneo and the adjacent small islands between 1902 and 1908, for the US National Museum.

Adams, C. D. District Officer, Baram, Sarawak.

Agrawal, V. C. Zoologist of the Zoological Survey of India, Calcutta.

Aitken, Edward Hamilton, 1851–1909. Indian civil servant; co-founder of the Bombay Natural History Society; wrote using the pen-name Eha (*Niviventer eha*).

Alfred (Ernest Albert), Duke of Edinburgh, 1844–1900. Second son of Queen Victoria and naval officer; sent living animals from the Philippines to London Zoo, c. 1870.

Ali, Dr Salim, 1896–1987. Indian ornithologist.

Allen, Glover Morrill, 1879–1942. Mammalogist at the Museum of Comparative Zoology, Cambridge, USA from 1907 (formally appointed curator 1925).

Allen, Joel Asaph, 1838–1921. Curator of birds and mammals, American Museum of Natural History, New York, 1885–1921.

Andersen, Knud. Danish zoologist, employed by the British Museum (Natural History) from 1904 to prepare a catalogue of bats; disappeared mysteriously in 1918 (*Rhinolophus canuti*).

Anderson, Dr John, 1833–1900. Superintendant of the Indian Museum and Professor of Comparative Anatomy, Calcutta 1865–86; naturalist and medical officer on the British expeditions from Calcutta to Yunnan, 1867–8 and 1875.

Anderson, Malcolm Playfair, 1879–1919. American naturalist; leader of the Duke of Bedford's Exploration in Eastern Asia to collect for the British Museum and London Zoo; collected in the Philippines 1906 and in China 1907–11.

Andrews, Roy Chapman, 1884–1960. Assistant (1909) then Curator of Mammalogy (1918), later (1935–42) Director of American Museum of Natural History, New York; took part in the cruise of the 'Albatross' in the Malayan Archipelago, 1909; led a series of expeditions to central Asia in 1922–30.

Annandale, Thomas Nelson, 1876–1924. Travelled in Malaya 1899–1902; superintendant of the Indian Museum, Calcutta 1906–16.

Anthony, Harold Elmer, 1890– . Assistant, then Curator of Mammalogy, American Museum of Natural History, New York, 1915 (1921)–1958.

Archbold, Richard, 1907–1976. American traveller who sponsored and participated in expeditions from the American Museum of Natural History to New Guinea etc. 1933–39.

Armandville, Le Cocq d'. Dutch priest on island of Flores, c. 1890.

Baber, = Zahir ud-Din Mohammed, 1483–1530. Founder of the Mogul Empire in India.

Bailey, Frederick Marshman, 1882–1967. British Army officer and diplomat; travelled and collected in Tibet etc. 1903–13.

Bailward, Arthur Churchill, –1923. British Army officer (to Brigadier-General); travelled extensively in Persia 1905 and 1907.

Bartels, Max Eduard Gottlieb, 1871–1936. German; went to Java 1895; plantation manager (*Maxomys, Hylopetes bartelsi*).

Bartels, Max (Junior), 1902–1943. Son of above; collected in Java 1932–42 (*Crocidura maxi, Hylomys suillus maxi*).

Beccari, Dr Odoardo, 1843–1920. Italian botanist, director of Botanic Garden and Herbarium, Florence; collected in Sarawak 1865–68, in the Moluccas and New Guinea 1871–75.

Becker, Margareta. Worked with G. G. Musser in Sulawesi in 1960s.

Bedford, 11th Duke of, 1858–1940. President of the Zoological Society of London 1899–1936; sponsored the collection of mammals in eastern Asia 1904–11; pioneered captive breeding of endangered ungulates including Père David's deer.

Bedford, Duchess of, 1865–1937. Wife of 11th Duke; ornithologist and pioneer aviator.

Belanger, Charles, 1805–1881. French traveller in India (1825), Burma and Java.

Bennett, Edward Turner, 1797–1836. Secretary of the Zoological Society of London 1831–36.

Berdmore, Major, –1859. Collected at Moulmain, Burma c. 1850.

Berezovsky, M. M., –1911. Russian traveller in China.

Biet, Mgr. French apostolic bishop of Tibet.

Biswas, Dr Biswamoy. Joint Director, Zoological Survey of India c. 1960–80 (*Biswamoyopterus*).

Biswas, Dr Shyamrup. Zoologist, Zoological Survey of India, leader of Namdapha Expedition, 1981.

Blainville, Henri Marie Ducrotay de, 1777–1850. Professor of Comparative Anatomy, Paris, succeeding G. Cuvier in 1831.

Blanford, William Thomas, 1832–1905. Member of Geological Survey of India who helped found the Indian Museum, Calcutta and the *Fauna of British India* publications.

Blasius, Johann Heinrich, 1809–1870. Professor of Natural Sciences and Director of Museum, Brunswick, Germany.

Blumenbach, Johann Friedrich, 1752–1840. Professor of Medicine at Göttingen, Germany from 1776.

Blyth, Edward, 1810–1873. Curator of the Museum of the Asiatic Society in Calcutta, 1842–64.

Bonaparte, Charles Lucien, Prince of Musignano, 1803–1857. Nephew of Napoleon Bonaparte; studied mammals at all the major museums in Europe.

Bonhote, John James Lewis, 1875–1922. Ornithologist and aviculturalist; unofficial worker in the Mammal Section of the British Museum (Natural History) with Oldfield Thomas; Zoological Gardens, Giza, Egypt 1913–19.

Bourret, René, c. 1900–1960. French zoologist; director of Ecole Superieur des Sciences, University of Hanoi, Vietnam.

Brisson, Mathurian Jacques, 1723–1806. Curator of the private museum of Réaumur, later Professor of Physics at Navarre, France; teacher of Etienne Geoffroy St Hilaire.

Brongersma, Leo Daniel, 1907–. Dutch zoologist; Leiden Museum 1934– (Director 1958–).

Brooke, Sir James, ('Rajah Brooke'), 1803–1868. Officer of British India Company, Governor of Sarawak from 1841; Governor of Labuan and Consul-General to the Sultan of Borneo from 1846.

Buffon, Georges Louis Leclerc, Compte de, 1707–1788. Keeper of the Jardin du Roi, Paris from 1739; one of the first encyclopaedists of natural history.

Cabrera, Angel, 1879–1960. Spanish zoologist; later in Argentina.

Cadorna, Count Luigi, 1850–1928. Chief-of-Staff of Italian army in 1914–18 war.

Canning, Lt John. British diplomat, Resident in Burma 1803–11.

Caroline. See Rozendaal.

Carolus [*Mus*]. See Rothschild, N. C.

Cautley, Colonel Sir Proby Thomas, 1802–1871. Engineer (Indian canals) and officer of the Bengal Artillery; member of the Council of India; amateur palaeontologist.

Chasen, Frederick Nutter, –1942. Curator (1921–32) then Director (1932–42) of the Raffles Museum, Singapore.

Cheesman, Col. Robert Ernest, 1878–1962. British Army officer and diplomat serving in India, Iraq (1920–23), Ethiopia and Sudan.

Chévrier, Brother. French missionary; accompanied Père David on first journey to Inner Mongolia.

Colbert, Edwin H., 1905—. American vertebrate palaeontologist, at American Museum of Natural History 1930–.

Cook, J. Pemberton. Burma Teak Corporation; contributed collection from Moulmein, Burma to the Mammal Survey of the Bombay Natural History Society, 1913.

Coolidge, Harold Jefferson, 1904–1985. Mammalogist at the Museum of Comparative Zoology, Harvard, USA; organizer of the Asiatic Primate Expedition to Siam, French Indochina, Borneo and Sumatra, 1937.

Cox, E. A. W. Civil servant in Sarawak.

Coxinga = Kuo Hsing Yeh. Chinese Ming refugee, settled and ruled Taiwan, 1661– (*Niviventer coninga*).

Cranbrook, 4th Earl of, 1900–1978. Collected in upper Burma 1931.

Cranbrook, 5th Earl of (formerly Lord Medway), 1933–. Assistant Sarawak Museum 1956–58; Senior lecturer in Zoology at University of Malaya 1961–70.

Creagh, Charles Vandeleur, 1842–1917. Governor of North Borneo 1888–91; of Labuan 1891–96.

Cretzschmar, P. J., 1786–1845. German surgeon and naturalist; founded the Senckenberg Natural History Society in 1817, the museum of which survives as the Senckenberg Museum in Frankfurt, Germany.

Crump, C. A. Collector for the Bombay Natural History Society's mammal survey.

Cuming, Hugh, 1791–1865. English naturalist and private traveller, especially in the Pacific and the Philippines.

Cuvier, Frédéric, 1773–1838. French zoologist (younger brother of Georges); keeper of the menagerie at the Museum d'Histoire Naturelle, Paris from 1804, later Professor of Comparative Physiology.

Cuvier, Georges Leopold Dagobert, Baron, 1769–1832. French zoologist and statesman.

Dammermann, Dr Karel Willem, 1885–1951. Director, Zoological Museum, Buitenzorg (= Bogor), Java, 1919–1940.

Daubenton, Louis Jean Marie, 1716–1800. French naturalist-physician, assistant to Buffon at the Jardin de Roi, Paris, later Professor of Natural History at the College de France.

David, Père Armand. 1826–1900. French Lazarist missionary, travelled in China 1864–1870, collecting especially in Sichuan.

Day, Francis, 1829–1889. Surgeon-General, Indian Army and medical officer in Madras and Cochin; Inspector-General of Fisheries for India.

Delacour, Jean Theodore, 1890–1985. French ornithologist; travelled in Indochina 1923–40; then went to USA, as associate of the American Museum of Natural History, later as Director of the Los Angeles County Museum.

Deraniyagala, Paulus Edward Peiris, 1900–1974. Director, National Museums of Ceylon.

Derby, Edward Smith Stanley, 13th Earl of, 1775–1851. English naturalist and collector; established the 'Knowlesley Menagerie' in Lancashire and bequeathed his collection to the city of Liverpool.

Desmarest, Anselme-Gaëtan, 1784–1838. Professor of Veterinary Science at Alfort, France.

Diard, Pierre Médard, 1795–1863. French; travelled with Duvaucel in India and Sumatra; joined the Natural History Commission of the Dutch Indies in 1829 and became head of the Commission in 1832.

Dobson, George Edward, 1848–1895. Surgeon in Indian Army, later Superintendant of the Military Hospital at Netley, Hampshire, England.

Dollman, John Guy, 1886–1942. Mammalogist at the British Museum (Natural History) 1907–1942, concerned mainly with exhibitions and large mammals.

Doria, Marchese Giacomo, 1840–1913. Founder of the Genoa Museum.

Dunn, Henry Nason, 1864–1952. An officer in the British Royal Army Medical Corps, posted to the North West Frontier, India in 1908.

Duvaucel, Alfred. 1796–1824. Stepson of Georges Cuvier; went to India as 'Naturaliste du Roi', travelling with Diard and collecting also for Raffles in Sumatra.

Eden, Sir Ashley, 1831–1887. British civil servant and diplomat in India; Chief British Commissioner of Burma 1871–74; Lieutenant-governor of Bengal 1877–82.

Edwards (*Herpestes, Leopoldamys*). See Milne-Edwards.

Eha. See Aitken.

Eld, Lieut Percy. Assistant to the Commissioner of Assam, 1841.

Ellerman, Sir John Reeves, 1909–1973. English ship-owner and amateur mammalogist, working at the British Museum (Natural History) c. 1932–1950.

Elliot, Daniel Giraud, 1835–1915. Curator of Mammals, Field Columbian Museum, Chicago, 1894–1908.

Elliot, Sir Walter, 1803–1887. Indian civil servant, and amateur archaeologist and naturalist.

Emily. See Hose.

Erxleben, Johann Christian Polycarp, 1744–1777. German veterinarian.

Everett, Alfred Hart, 1848–1898. English naturalist; collected in Sarawak from 1869, subsequently serving as an officer of the Sarawak Government and the British North Borneo Company from 1872–1890; collected in the Philippines 1877–8 for the Marquis of Tweeddale.

Falconer, Hugh, 1808–1865. Professor of Botany, Medical College, Calcutta and amateur palaeontologist; pioneered the introduction of tea into India.

Fea, Leonardo, 1852–1903. Italian zoologist, at Genoa Museum; collected in Burma 1885–87.

Felix. See Pratt.

Fernando, E. C. Taxidermist at the Colombo Museum, Sri Lanka (1932).

Finlayson, George, 1790–1823. Surgeon, British East India Co. ; naturalist on a mission (led by John Crawford) from the Governor-General of India to the courts of Siam and Cochin-China, 1822.

Fischer (von Waldheim), Johann Gotthelf, 1771–1853. Professor of Natural History and Director of the Natural History Museum at Moscow Academy from 1804; he acquired the title 'von Waldheim' in 1835.

Fitzinger, Leopold Joseph, 1802–1884. Curator of the Vienna Museum.

Fontanier, M. French Consul at Peiping (Beijing) at the time of Père David's travels in China.

Fooden, Jack. Professor of Zoology, Chicago State University and Associate at the Field Museum, Chicago.

Forrest, George, 1873–1932. Scottish plant collector for the Royal Botanic Garden, Edinburgh; made eight expeditions, especially to Yunnan, beginning 1904.

Frith, R. W. G. Collected in Bengal 1848.

Frost, Wilfred J. C. Collected in Dutch East Indies c. 1919–1930s.

Furness, Dr W. H. Collected in the Ryukyu Is, 1896.

Gabrielle. See Vassal.

Gee, Edward Pritchard, 1904–1968. Tea planter and pioneer of conservation in India.

Geoffroy-Saint-Hilaire, Etienne, 1772–1844. Professor of Zoology, Museum d'Histoire Naturelle, Paris.

Geoffroy-Saint-Hilaire, Isidore, 1805–1861. Son of Etienne; assistant, later (1841) Director, Paris Museum.

Germain, Rodolphe. Collected in Indochina.

Gervais, Paul François-Louis, 1816–1879. Professor of Zoology, Paris.

Gleadow, F. Collected in Pakistan in 1880s.

Glover. See Allen, G. A.

Gmelin, Johann Friedrich, 1748–1804. Professor of Chemistry, Göttingen, Germany.

Granger, Dr Walter, 1872–1941. Curator of fossil mammals and palaeontologist of the Central Asiatic Expeditions of the American Museum of Natural History; in Sichuan 1921–26.

Grant, William Robert Ogilvie-, 1863–1924. Ornithologist at the British Museum (Natural History) 1885–1918; organized many expeditions, including that of John Whitehead (*q.v.*) to the Philippines.

Gray, John Edward, 1800–1875. Assistant (1824) later (from 1840) Keeper of Zoology at British Museum; principal founder of the Museum's mammal collection.

Groves, Colin Peter. Professor of Anthropology, Australian National University, Canberra.

Gueldenstaedt, A. I., 1745–1781. Travelled in the southern parts of the Russsian Empire for St Petersburg Academy of Science, 1768–75.

Hagen, B. A., 1853–1919. Assistant curator at Botanic Gardens, Buitenzorg, Java; collected in Sumatra.

Hardwicke, Major General Thomas, 1756–1835. Commander of the Bengal Artillery, Indian Army.

Harrisson, Thomas Harnett, 1911–1976. Anthropologist; curator (1947) then Director (1956) of the Sarawak Museum, Kuching.

Hasselt, Dr Johan Coenraad van, 1797–1823. Went to Java 1n 1820 under the auspices of the Natural History Commission of the Dutch Indies, died Buitenzorg 1823.

Heinrich, Gerd, 1896–. German zoologist; travelled in Sulawesi 1930–32.

Heude, Pierre Marie, 1836–1902. French missionary in Shanghai.

Hinton, Martin Alister Campbell, 1883–1961. Amateur palaeontologist specializing in rodents; joined staff of the British Museum (Natural History) 1921, becoming Keeper of Zoology 1936–45.

Hodgson, Brian Houghton, 1800–1894. British diplomat and official Resident in Kathmandu, Nepal 1833–43.

Hollister, Ned, 1876–1924. Field collector with Biological Survey of USA and curator at US National Museum 1901–1916; then Superintendant of the National Zoological Park, Washington DC.

Hoogerwerf, Andries, 1907–1977. Taxidermist at the Zoological Museum, Buitenzorg; later Officer for Hunting and Nature Preservation at the Forestry Department, Java.

Hooijer, Dirk Albert. Palaeontologist at Leiden Museum, Netherlands.

Horsfield, Dr Thomas, 1773–1859. American-born geologist and physician; travelled in Java 1802–19; co-founder and first Vice-Secretary of the Zoological Society of London 1826–28; curator of the museum of the East India Company.

Hose, Charles, 1863–1929. English physician; government official in Sarawak 1884–1907.

Hose, Mrs Emily. Wife of Charles Hose.

Hotson, Sir (John) Ernest (Buttery), 1877–1944. Indian Civil Service; Acting Governor of Bombay 1931.

Howell, Alfred Brazier, 1886–1961. American naturalist and self-taught anatomist; studied Chinese mammals at the US National Museum, Washington, 1925–28.

Hume, Allan Octavian, 1829–1912. Indian Civil Service 1867–82; ornithologist and big game hunter; bequeathed large collection of ungulates to the British Museum (Natural History).

Hutton, Thomas, 1807–1874. British Army officer in the Afghan war 1839–40.

Illiger, Johann Carl Wilhelm, 1775–1813. Professor of Zoology and a founder of the Zoological Museum in Berlin.

Imaizumi, Yoshinori, 1914–. Chairman of Dept of Zoology, National Science Museum, Tokyo.

Jagor, A. Feodor, 1816–1900. German ethnologist, travelled in Malayan region 1857–93; collected for Berlin Museum.

Jenkins, Paulina Deirdre. Mammalogist at British Museum (Natural History) 1966–.

Jentink, Fredericus Anna, 1844–1913. Curator at the Leiden Museum 1875–1913.

Jerdon, Thomas Coverhill, 1811–1875. Surgeon-Major in Indian Army.

Joffre, Joseph Jacques Césaire, 1852–1931. Commander in Chief of the French Army 1914–16.

Kathleen. See Ryley, K. V.

Kelaart, E. F., 1818–1860. Army Medical Service, Ceylon.

Kikuchi, Y., 1869–1921. Curator of the Taipei Museum, Taiwan.

Kinnear, Sir Norman Boyd, 1882–1957. Curator of the Bombay Natural History Society 1907–; joined staff of British Museum (Natural History) in 1920, becoming Keeper of Zoology in 1945 and Director 1947–1950; primarily an ornithologist.

Kloss, Cecil Boden, 1877–1949. Director, Raffles Museum, Singapore, 1923–32.

Koenigswald, Gustav Heinrich Ralph von, 1902–1982. German palaeontologist and anthropologist; Geological Survey of Java 1931–41; Professor of Palaeontology at Utrecht, Netherlands 1948–68.

Koopman, Dr Karl F., 1920–. Mammalogist at the Field Museum, Chicago 1959–61, then at the American Museum of Natural History, New York, specializing in Chiroptera.

Kuhl, Dr Heinrich, 1797–1821. Assistant of Temminck, went to Java 1820 with van Hasselt and died in Buitenzorg.

Kuroda, Nagamichi, 1889–1978. Japanese mammalogist and ornithologist; President of the Mammalogical Society of Japan 1949–65.

Kusnoto, Setyadiwirja. Director of Botanic Gardens and founder of the Academy of Biology, Bogor, Indonesia.

Lacépède, Bernard (Germain-Etienne), Compte de, 1756–1825. French naturalist, assistant to Buffon; Keeper at the Cabinet de Roi. Sometimes spelled 'La Cepède'.

La Touche, John David Digues, 1861–1935. British customs officer in S China 1882–1921.

Layard, Edgar Leopold, 1824–1900. Civil servant in Ceylon, later in S Africa; founded and curated the South African Museum, Cape Town.

Lekagul, Boonsong, ?1906–1992. Thai physician and conservationist; founded the Association for the Conservation of Wildlife in Thailand in 1950.

Leschenault de la Tour, Jean Baptiste, 1773–1826. Member of French voyage to Australia, collected in Java 1803–06; travelled in India and Ceylon 1817–22, collecting for the Paris Museum.

Lesson, René Primavère, 1794–1849. Naval apothecary, naturalist on the Duperrey expedition to the Pacific on La Coquille, 1821–1825, collecting in the Moluccas.

Lichtenstein, Martin Heinrich Carl, 1780–1857. First Director, Zoological Museum, Berlin, 1815.

Lieftinck, M. A. Director of Zoological Museum, Buitenzorg (Bogor), Java, 1940–.

Linnaeus, Carl, 1701–1778. Swedish naturalist, founder of the current system of biological nomenclature.

Low, Sir Hugh, 1824–1905. Collected in Sarawak and Labuan 1845–77; British Resident of Perak 1877–89.

Lucas, Frederic A. ? Curator of Ward's Museum, Rochester, New York, USA.

Lydekker, Richard, 1849–1915. English zoologist and palaeontologist; served in the Geological Survey of India 1874–82; subsequently unofficial worker at the British Museum (Natural History), specializing in ungulates and exhibitions.

Lyle, Sir Harold, 1873–1927. British consul in Bangkok 1918–24.

Lyon, Marcus Ward, 1875–1942. Assistant Curator of Mammals, US National Museum, Washington 1898–1912; subsequently held university and hospital posts in bacteriology and pathology.

Mackenzie, J. M. D. Burmese Forest Service, contributor to the Mammal Survey of the Bombay Natural History Society.

Macklot, Dr Heinrich Christian, 1799–1832. Joined Leiden Museum 1822 (Osteology Dept); went to Java 1825; studied geology of Timor 1829; died in fight with Chinese rebels in Java.

Maclear, John Pearse, 1838–1907. British naval officer; commander of the scientific survey vessels 'Challenger' and later the 'Flying Fish' which visited Christmas I in 1886.

Marche, Alfred A., 1844–1898. French anthropologist and zoologist; in Philippines 1879–85.

Margaret, the Ranee of Sarawak, –1936. Wife of Rajah Charles Brooke; (*Haeromys*).

Margareta (*Margaretamys*). See Becker, M.

Marshall, Joe T. American zoologist and associate of the National Museum of Natural History, Washington, specializing in the taxonomy of medically important rodents; worked with the SEATO Medical Research Laboratory, Thailand.

Matschie, Paul, 1862–1926. Curator and mammalogist at the Berlin Museum.

Max. See Bartels.

Mayor, Major E. W. Collected in Ceylon, 1913–14, for the Bombay Natural History Society's mammal survey.

McClelland, Dr John, 1805–1875. Geologist and amateur naturalist; visited Assam in 1839 to study tea for the East India Company; founded the Calcutta Magazine of Natural History 1840.

Mearns, Dr Edgar Alexander, 1856–1916. US Army surgeon; collected in the Philippines 1903–07; subsequently led expeditions in Africa.

Mertens, Dr Robert, 1894–1975. Zoologist at Senckenberg Museum, Frankfurt, Germany.

Meyer, Adolph Bernhard, 1840–1911. German naturalist; travelled in the Malayan Archipelago 1870–73; Director, Dresden Zoological Museum 1874–1905.

Millard, Walter Samuel, 1867–1952. Hon. Secretary, Bombay Natural History Society, 1906–20; prime mover, with Oldfield Thomas, in initiating the Society's Mammal Survey.

Millard, Mrs Sybil. Wife of W. S. Millard.

Miller, Gerrit Smith, 1869–1956. Mammalogist at the US National Museum, Washington.

Milne-Edwards, Alphonse, 1835–1900. Professor in School of Pharmacy, Paris, later (1891–1900) Director of the Museum d'Histoire Naturelle.

Moellendorf, Otto Franz von, 1848–1903. German consul in China and Philippines.

Moore, Joseph Curtis. Curator of Mammals, Field Museum, Chicago, 1960s.

Morrison-Scott, Sir Terence Charles Stuart, 1908–1991. Head of the Mammal Section, British Museum (Natural History) 1936–56; Director 1960–68.

Müller, Salomon, 1804–64. Joined Leiden Museum 1823; went to Java 1825, collecting in Java, Sumatra and Borneo and returning to Leiden 1837 (one of the few members of the Natural History Commission for the Dutch Indies to do so).

Musschenbroek. See von Musschenbroek.

Musser, Guy G. Archbold Curator, Department of Mammalogy, American Museum of Natural History, specializing in murid rodents.

Nehring, Alfred, –1904. Professor of Zoology, Berlin.

Neill, William A. Public health researcher working in Thailand in 1970s.

Netscher, E. A member of the Council of the Dutch East Indies, resident at Padang-Pandjang, Sumatra around 1880.

Niphan. See Ratanaworabhan.

Ogilby, W., 1808–1873. Secretary of the Zoological Society of London, 1839–46.

Osgood, Dr Wilfred Hudson, 1875–1947. Field Museum, Chicago 1909–41.

Ouwens, P. A. Supervisor of the Zoological Museum, Buitenzorg (Bogor), Java 1905–1920.

Owen, Sir Richard, 1804–1892. English anatomist and palaeontologist; Hunterian Professor at the Royal College of Surgeons, London; Superintendent of Natural History, British Museum.

Owston, Alan, 1853–1915.

Pallas, Peter Simon, 1741–1811. German naturalist; lived in Russia 1768–1774; founded the museum at St Petersburg; travelled in Siberia as far as the Amur 1793–4.

Panglima Polem. Moslem resistance leader in Aceh, Sumatra; surrendered to Dutch in 1903.

Pearson, John Thomas. Fellow medical student of J. E. Gray, who subsequently collected in India (*Trogopterus pearsonii, Rhinolophus pearsonii*).

Pearson, Dr Joseph. Director of the Colombo Museum, Ceylon in 1920s (*Solisorex pearsoni*).

Pennant, Thomas, 1726–1798. British naturalist, author of one of the first books on Indian zoology.

Perny, Paul-Hubert, 1818–1907. French missionary in China; in Guizhou 1850–60.

Peron, F., 1775–1810. French naturalist on voyage to Australia in 'La Geographie' and 'La Naturaliste', 1800–1804.

Peters, W. C. H., 1815–1884. Professor of Zoology and associate of the Berlin Museum.

Phayre, Sir Arthur Purves, 1812–1885. British commissioner at Pegu, Burma 1852–; Chief Commissioner of Burma 1862–67.

Phillips, R. M. District Superintendant of Police, Dharwar, India; assisted C. A. Crump in collecting for the Mammal Survey of India. (*Mus*).

Phillips, Major William Watt Addison, 1892–1981. Tea planter in Sri Lanka and amateur mammalogist.

Pocock, Reginald Innes, 1863–1947. Superintendent of London Zoo 1904–23; then unofficial worker at the British Museum (Natural History).

Pratt, Charles, Felix and Joseph. Collected in the Moluccas 1919–20 (*Rattus feliceus, Melomys fraterculus*).

Prévost, Florent, –1870. Assistant naturalist and artist, Paris Museum, 1832–.

Quoy, J. P. C., 1790–1869. Naturalist, (with J. P. Gaimard) on the French voyage of the Astrolabe (formerly La Coquille) 1826–29, commanded by J. Dumont d'Urville.

Rabor, Dioscoro S., 1911–. Professor of Wildlife Biology, University of the Philippines.

Raffles, Sir (Thomas) Stamford, 1781–1826. British administrator and zoologist; Lieutenant-Governor of Java (on behalf of the East India Company) 1811–16; Lieutenant-Governor of Bencoolen (Sumatra) 1818–24; founded Singapore 1819; founded Zoological Society of London (with Sir Humphry Davy) 1825.

Rafinesque, Constantin Samuel, 1783–1840. Born Turkey, lived in Sicily 1805–15 then emigrated to USA; a widely ranging naturalist, with a special interest in medicinal plants.

Ratanaworabhan, Dr Niphan. Officer-in-charge of Thai National Reference Collection, Bangkok c. 1973–83.

Reeves, J. 1774–1856. Inspector of tea for the East India Company at Canton, China, succeeded by his son, J. R. (1804–1877).

Rickett, C. B., 1851–1943. Collected at Kuatun, Fujian, China.

Ridley, H. N. 1855–1956. Director of Gardens and Forests, Straits Settlements 1888–; pioneered the rubber industry in Malaya.

Robinson, Herbert Christopher, 1874–1928. Director, Federated Malay States Museums (including Raffles Museum, Singapore) 1908–1926.

Rodolphe. See Germain, R.

Roosevelt, Kermit, 1889–1943. Brother of Theodore (below); accompanied him on the Asiatic Expedition of the Field Museum, 1928–29.

Roosevelt, Theodore, 1887–1944. Son of US President Roosevelt; leader of Kelley-Roosevelts Asiatic Expedition of the Field Museum, Chicago 1928–29; Governor-general of the Philippines 1932–33.

Rosset, C. W. Collected in Cambodia (in 1880s ?).

Rothschild, Nathanial Charles, 1877–1923. Younger brother of Lord Rothschild (below); specialist on fleas; (*Mus caroli*).

Rothschild, Walter, Lord, 1868–1937. Founder of Zoological Museum, Tring, Hertfordshire, England, bequeathed to the British Museum (Natural History).

Roux. French ornithologist.

Roxellana. A Russian who became a wife of the Turkish Sultan Suileman the Magnificent in 1523.

Rozendaal, Caroline. Wife of F. G. Rozendaal (below).

Rozendaal, F. G. Collected bats in Borneo, Sulawesi and Moluccas on behalf of Leiden Museum.

Rudolphi, Carl Asmund, 1771–1832. A founder of the Zoological Museum, Berlin.

Rüppell, Wilhelm Peter Edward Simon, 1794–1884. German naturalist; travelled in NE Africa supported by the Senckenberg Museum, Frankfurt, 1821–34.

Ryley, Kathleen V. Unofficial worker at the British Museum (Natural History) working for Thomas on the Bombay Natural History Society collections.

Salomonsen, Dr Finn, 1909–1983. Leader of the Danish Philippine Expedition for the Zoologisk Museum, Copenhagen, 1951–52.

Sanborn, Colin C. –1962. Curator of mammals, Field Museum of Natural History, Chicago.

Sarasin, Paul B. (1856–1929) & K. Fritz (1859–1942). Swiss researchers (cousins) in Celebes 1893–1903 under the auspices of the Royal Netherlands Geographical Society.

Sauvel. French veterinarian and big game hunter.

Savile, Sir Leopold Halliday, 1870–1953. Chief Engineer of the Bombay Port Trust 1904–19; Honorary Treasurer of the Bombay Natural History Society.

Schadenberg, Dr A. German; collected in Luzon and Mindanao, Philippines in 1880s.

Schlegel, Herman, 1804–84. Curator of vertebrates, later (succeeding Temminck in 1858) second director of the Rijksmuseum van Natuurlijke Historie, Leiden, Netherlands.

Schomburgk, Sir Robert H., 1804–1865. British minister in Bangkok, 1857–64.

Schreibers, Karl F. A. von, 1775–1852. Director of the Vienna Museum 1806–51.

Schwarz, Ernst, –1961. Zoologist in US Armed Forces Institute of Pathology.

Sclater, Philip Lutley, 1829–1913. English lawyer, later Secretary of the Zoological Society of London 1859–1902.

Shamel, H. Harold. Mammalogist at the US National Museum, Washington.

Shortridge, Guy C., 1880–1949. Collected for the British Museum (Natural History) in Java and later for the Indian Mammal Survey; later director of the Kaffrarian Museum, King William's Town, South Africa, 1921–49.

Siebold, Philipp Franz von, 1796–1866. German medical doctor who went to Java and Japan in the service of the Netherlands government, collecting plants especially.

Sladen, Lt Col. Sir Edward Bosc, 1827–1890. British officer and diplomat, serving mostly in Burma, 1852–87; leader of the First Yunnan Expedition, 1867–8.

Smith, Dr J. A. C. Missionary in Shaanxi, China; accompanied Malcolm Anderson (*q. v.*) in collecting in Shaanxi 1909–11.

Sody, Henry Jacob Victor, 1892–1959. Emigrated from the Netherlands to Java in 1918 as a tea planter; Lecturer at the Agricultural College, Bogor 1920–41; researched on rodents based at the Bogor Museum 1941–49.

Sody, Taera. Daughter of H. J. V. Sody (*Taeromys*).

Sowerby, Arthur de Carle, 1885–1954. Founder of the Natural History Museum, Shanghai; accompanied Malcolm Anderson (*q. v.*) in China in 1907.

Stalker, W., 1879–1910. Member of British Ornithological Union's expedition to New Guinea, 1909, who drowned shortly after arriving in New Guinea.

Steere, Joseph Beal, 1842–1940. American zoologist and traveller; professor at Ann Arbor, Michigan, USA; collected in Taiwan (1873) and the Philippines (1887–8).

Stevens, Herbert, 1877–1964. Resident at Darjeeling; collected in Tonkin, Vietnam in 1923 and 1924 for the British Museum (Natural History) funded by the Percy Sladen and Godman Exploration Funds.

Stoliczka, Dr Ferdinand, 1838–1874. Palaeontologist in the Geological Survey of India; died on return journey from the 2nd Mission to Yarkand.

Styan, Frederick W., 1838–1934. Tea merchant and British official Resident at Shanghai 1877–1904.

Swinhoe, Robert, 1836–1877. British consular official in China 1865–75, based successively at Amoy, Formosa, Amoy, Ningpo and Chefoo; collected mammals for the London Zoo and the British Museum.

Sybil. See Millard, Mrs S.

Taera. See Sody.

Tate, George H. H. 1894–1953. Assistant then Curator of Mammalogy, American Museum of Natural History, New York, 1921(1945)–1953.

Taylor, Edward Harrison, 1889–1978. American zoologist, mainly working in herpetology; resident in Philippines 1912–24, subsequently at University of Kansas, Lawrence.

Temminck, Coenraad J. 1778–1858. Founder (1820) and first director of the Rijksmuseum van Natuurlijke Historie, Leiden, Netherlands.

Theobold, W. Collected bats in Burma, 1855.

Thomas, Michael Rogers Oldfield, 1858–1929. Curator of mammals at the British Museum (Natural History) from 1878; primarily responsible for building the mammal collection, and for instigating the Mammal Survey of the Bombay Natural History Society.

Thonglongya, Kitti, 1928–1974. Curator of mammals at the Thai National Reference Collection, Bangkok.

Tokuda, Dr Mitoshi, 1906–1975. Japanese mammalogist; Associate Professor at the Zoological Institute of Kyoto University.

Tomes, Robert Fisher, 1823–1904. English naturalist.

Trouessart, Edouard-Louis, 1842–1927. Director of the Natural History Museum, Anger, France 1870–85; Professor of Mammals and Birds at the Museum National d'Histoire Naturelle, Paris 1905–26.

True, Frederick William, 1858–1914. Curator, Dept of Mammals, US National Museum, Washington 1883–1911, then Assistant Secretary of the Smithsonian Institution, Washington.

van Musschenbroek, Samual Cornelis Jan Willem, 1827–1883. Dutch civil servant in Java; official Resident in N Celebes 1875–76.

Vassal, Mrs Gabrielle. Travelled and collected in Annam, Vietnam with her husband Dr Vassal.

Vernay, Arthur S. Trustee of the American Museum of Natural History who sponsored and participated in an expedition to NE Burma in 1938–39.

Vordermann, Dr A. G., 1844–1902. Naval and civil medical officer in Dutch East Indies.

Wagler, Johann, 1800–1832. Professor of Zoology, University of Munich.

Wagner, Johann Andreas, 1797–1861. German zoologist.

Wallace, Alfred Russel, 1823–1913. English naturalist; travelled in the Malay Archipelago 1854–62; co-author with Charles Darwin of the first theory of evolution by natural selection.

Wang, Ying-xiang, 1938–. Professor and head of Vertebrate Division, Institute of Zoology, Kunming, Yunnan, China.

Ward, F. Kingdon-, 1885–1958. British botanist and explorer; travelled in W China, Tibet, upper Burma etc. 1909–50.

Waterhouse, George Robert, 1810–1888. Curator of the museum of the Zoological Society of London; later (1843–80) at the British Museum, latterly as Keeper of Mineralogy.

Watts, Dr C. H. S. Chief Scientist, South Australian Museum, Adelaide, Australia.

Weber, Max Wilhelm Carl, 1852–1937. Professor of Zoology and Director of the Zoological Museum, Amsterdam 1883–1922; travelled in the Dutch East Indies 1888–9.

Wetmore, Alexander, 1886–1978. American ornithologist; Assistant Secretary (1925–) then Secretary (1945–52) of the Smithsonian Institution, Washington.

Whitehead, John, 1860–1899. English traveller and collector; in Borneo from 1888, Philippines from 1893, died of fever in Hainan.

Witkamp, H., 1880–1953. Mining engineer in Royal Batavian Oil Co. ; collected bats in Borneo 1928.

Wroughton, Robert Charles, 1849–1921. Indian Forest Service 1871–1904; later an unofficial worker at the British Museum (Natural History), working on the Bombay Natural History Society collections.

Wynn, A. B. Indian Geological Survey 1880.

Zarudny, Dr N. A., 1859–1919. Curator of Tashkent Museum (Uzbekistan); travelled in Persia 1885–1903.

Appendix 5

Glossary and abbreviations

Agouti. Of a hair, showing a pattern of dark and light bands on the distal part.
Allopatric. Occupying separate (and separated) ranges (cf. parapatric and sympatric).

Baculum. The bone found in the penis of some mammals (*os penis*).
BMNH. British Museum (Natural History), also known as the Natural History Museum, London.
BP. Before the present.

Cline. A pattern of geographical variation within a species involving a gradual and progressive change in one or more characters over a large area.
Commensal. Living in close association with another species (as used here, man) without disadvantage to the 'host' species. (Therefore correctly applied to the shrew *Suncus murinus* but not to the synanthropic rodents such as *Rattus rattus*.)
Condylobasal length (of skull). From the foremost points of the premaxillae to the posterior surfaces of the occipital condyles (Fig. 9).
Dao. Island (China).

Dorsum. As used here, the dorsal surface of the body, excluding head and tail.

Feral. Denoting an animal or population that has reverted to the wild state after domestication.
FN. Fundamental number (of chromosomes). The total number of arms in the diploid chromosome complement.
FNa. Fundamental number of autosomes. The Fundamental Number excluding the sex chromosomes.

Gnathion. The anterior point of the premaxillae in the sagittal line.
Gunung (= Goenung). Mountain (Malaysia and Indonesia).

Hypsodont. High-crowned (teeth).

Karyotype. The complement of chromosomes characteristic of an individual or species.
Ko, Koh. Island (Thailand).

Maxillary tooth-row. The upper teeth behind and including the canine, born on the maxillary bone.
Mesic. (Habitats) with moderate supply of moisture.
Nomen nudum. A name published without a description (or equivalent) and therefore not available in nomenclature unless subsequently validated.

n. v. Non vidimus (original publication not seen).

Opisthodont. Of the upper incisors (especially of rodents), curving backwards, the tips posterior to the points of emergence from the bone. Cf orthodont and pro-odont.
Orthodont. Of the upper incisors (especially of rodents), pointing downwards, with the tips directly below the points of emergence from the bone.

Palation. The posterior point of the bony palate in the sagittal line.
Parapatric. Occupying separate but contiguous ranges.
Phenetic. Relating to overall similarity irrespective of phylogenetic history.
Polymorphic. (A population) showing a number of discrete variants of a character in considerable numbers (rather than one normal condition and rare variants).
Pro-odont. Of the upper incisors (especially of rodents), projecting forwards, the tips anterior to the points of emergence from the bone.
Pulau, Pulu. Island (Malaysia and Indonesia).

Rostrum. The anterior part of the skull, in front of the orbits.

Sungai. River (Malaysia and Indonesia).
Sympatric. With overlapping ranges.
Synanthropic. Living in close association with human settlements.

Taxon (plural taxa). A taxonomic group irrespective of rank (species, genus etc.).
Tragus. In some bats, a lobe developed from the ventral rim of the ear, extending across the conch.

Venter. As used here, the ventral surface of the body excluding head and tail.

Appendix 6

Additions to maps

The following elements of distribution are mentioned in the text but were omitted from the maps.

Map 4. *Spilocuscus maculatus* — add Kai Is.

Map 9. *Talpa leucura* — add N Laos.

Map 35. *Chironax melanocephalus* — add C Kalimantan.

Map 39. *Emballonura alecto* — add Anamba, Bunguran, Karimata and Kai Is; *Emballonura monticola* — add Karimata Is and S and SE Sulawesi.

Map 40. *Taphozous melanopogon* — add Talaud Is.

Map 41. *Taphozous saccolaimus* — add Talaud Is.

Map 48. *Rhinolophus borneensis* — add S Natuna Is; *R. megaphyllus* — add Wetar I; *R. celebensis* — add Madura I and Kangean Is.

Map 52. *Hipposideros cervinus* — add Mindanao.

Map 53. *Hipposideros lylei* — add S Yunnan.

Map 59. *Myotis hasseltii* — add Sumatra and ?N Sulawesi.

Map 61. *Ia io* — add Assam and N Laos.

Map 63. *Pipistrellus abramus* — add Taiwan.

Map 64. *Pipistrellus pulveratus* — add Hainan I.

Map 66. *Pipistrellus petersi* — add Sabah and N Sulawesi.

Map 67. *Philetor brachypterus* — add Negros and Mindanao Is (Philippines).

Map 68. *Hesperoptenus tickelli* — add S Burma/Thai border.

Map 69. *Murina aurata* — add N Thailand.

Map 70. *Miniopterus pusillus* — add N Sumatra and Timor I.

Map 72. *Kerivoula picta* — add Guangdong, S China.

Map 73. *Kerivoula papillosa* — add W Thailand.

Map 75. *Tadarida plicata* — add Hainan I; *T. sarasinorum* — add Peleng I.

Map 76. *Cheiromeles torquatus* — add Bangka I.

Map 132. *Petaurillus hosei* — add central peninsular Malaya.

BIBLIOGRAPHY

Abdulali, H. & Daniel, J. C. 1952. Races of the giant squirrel (*Ratufa indica*). *Journal Bombay nat. Hist. Soc.* 50: 467–474.

Abe, H. 1977. Variation and taxonomy of some small mammals from central Nepal. *Journal mammal. Soc. Japan* 7: 63–74.

----- 1982. Ecological distribution ... of small mammals in central Nepal. *Mammalia* 46: 477–503.

----- 1983. Variation and taxonomy of *Niviventer fulvescens* and notes on *Niviventer* group of rats in Thailand. *Journal mammal. Soc. Japan* 9: 151–161.

Abe, H. *et al.* 1991. A new mole from Uotsuri-jima, the Ryukyu Islands. *Journal mammal. Soc. Japan* 15: 47–60.

Abe, Y. 1934. On the Amami spinous rat (*Rattus jerdoni osimensis* subsp. n.) ... *Journal Sci. Hiroshima Univ. Zool.* 3: 107–114, pls 1,2.

Advani, R. & Mathur, R. P. 1982. On the occurrence of *Vandeleuria oleracea* in a rural residential habitat in ... Rajasthan. *Journal Bombay nat. Hist. Soc.* 79: 408.

Aellen, V. 1959. Contribution à l'étude de la faune d'Afghanistan. 9. Chiroptères. *Revue suisse Zool.* 66: 353–385.

Agassiz, L. 1848. *Nomenclatoris zoologici index universalis.* Soloduri, 1135pp.

Agranoff, D. *et al.* 1984. Cambridge Sumatra Expedition 1984. *Cambridge Exped. J.* 1984: 69–78.

Agrawal, V. C. 1970. Taxonomic status of *Millardia* Thomas ... *Mammalia* 34: 496–504.

----- 1973. A note on the taxonomic status of *Erinaceus blanfordi* ... *Journal zool. Soc. India* 25: 158–159.

----- 1975. Taxonomic study of the Indo-Burmese subspecies of the common tree-shrew, *Tupaia glis* Diard. Pp. 385–394 in Tiwari, K. K. & Srivastava, C. B. (eds), *Dr B. S. Chauhan commemoration volume 1975.* Orissa: Zool. Soc. India.

Agrawal, V. C. & Bhattacharyya, T. P. 1977a. Report on a collection of mammals from Tripura. *Record zool. Surv. India* 73: 135–157.

----- 1977b. Taxonomic status of *Macroxus punctatissimus* Gray ... *Proceedings zool. Soc. Calcutta* 30: 57–59.

----- 1987. New mammal records from West Bengal. *Bulletin zool. Surv. India* 8: 221–222.

Agrawal, V. C. & Chakraborty, S. 1970. Occurrence of the woolly flying squirrel, *Eupetaurus cinereus* ... in North Sikkim. *Journal Bombay nat. Hist. Soc.* 66: 615–616.

----- 1976. Revision of the subspecies of the lesser bandicoot rat *Bandicota bengalensis* ... *Record zool. Surv. India* 69: 267–274.

----- 1980. Intraspecific variation in the Indian long-tailed tree-mouse, *Vandeleuria oleracea* ... *Bulletin zool. Surv. India* 3: 77–85.

----- 1981. Notes on intraspecific geographical variation in the Indian antelope-rat, *Tatera indica* ... *Record zool. Surv. India* 79: 83–92.

----- 1982. Intraspecific geographical variation in the Indian bush rat, *Golunda ellioti* ... *Record zool. Surv. India* 79: 521–530.

Agrawal, V. C. & Ghosal, D. K. 1969. A new field-rat from Kerala, India. *Proceedings zool. Soc. Calcutta* 22: 41–45.

Ahmad, M. F. & Ghalib, S. A. 1979. A checklist of mammals of Pakistan. *Record zool. Surv. Pakistan* 7: 1–34.

Aimi, M. & Sudijono 1979. On the problematical species *Aceratherium boschi* von Koenigswald, 1933. *Bulletin geol. Res. Dev. Cent.* 1: 37–45.

Akhtar, S. A. 1959. A new rat from Gilgit (Kashmir). *Pakistan J. sci. Res.* 11: 41–53, 4pls.

Alcala, A. C. & Alviola, P. 1970. Notes on the birds and mammals of Borocay ... islands. *Silliman J.* 17: 444–454.

Alexander, P. S. *et al.* 1987. Ecological notes on two sympatric mountain shrews ... in Taiwan. *Journal Taiwan Mus.* 40,2: 1–7.

Ali, S. M. *et al.* 1988. A new species of palm civet from Orissa, India. *Record zool. Surv. India* 85: 45–48.

Allen, G. M. 1908. Notes on Chiroptera. *Bulletin Mus. comp. Zool.* 52: 25–61, 1pl.

----- 1912a. Mammals from Yunnan and Tonkin. *Proceedings biol. Soc. Wash.* 25: 177–180.

----- 1912b. Some Chinese vertebrates: Mammalia. *Memoirs Mus. comp. Zool. Harvard* 40: 201–247.

----- 1913. A new bat from Tonkin. *Proceedings biol. Soc. Wash.* 26: 213–214.

----- 1922. Bats from Palawan, Philippine Is. *Occasional Pap. Mus. Zool. Univ. Mich.* 110: 5pp.

----- 1923a. New Chinese bats. *American Mus. Novit.* 85: 8pp.

----- 1923b. New Chinese insectivores. *American Mus. Novit.* 100: 11pp.

----- 1924. Microtines collected by the Asiatic expeditions. *American Mus. Novit.* 133: 13pp.

----- 1925a. Squirrels collected by the American Museum Asiatic Expeditions. *American Mus. Novit.* 163: 16pp.

----- 1925b. Hamsters collected by the American Museum Asiatic Expeditions. *American Mus. Novit.* 179: 7pp.

----- 1926. Rats (genus *Rattus*) from the Asiatic expeditions. *American Mus. Novit.* 217: 16pp.

----- 1927a. Murid rodents from the Asiatic expeditions. *American Mus. Novit.* 270: 12pp.

----- 1927b. Porcupines from China. *American Mus. Novit.* 290: 4pp.

----- 1927c. Lagomorphs collected by the Asiatic Expeditions. *American Mus. Novit.* 284: 11pp.

----- 1928a. New Asiatic mammals. *American Mus. Novit.* 317: 5pp.

----- 1928b. A new cricetine genus from China. *Journal Mammal.* 9: 244–246.

----- 1929. Mustelids from the Asiatic Expeditions. *American Mus. Novit.* 358: 12pp.

----- 1930a. Bovidae from the Asiatic Expeditions. *American Mus. Novit.* 410: 11pp.

----- 1930b. Pigs and deer from the Asiatic Expeditions. *American Mus. Novit.* 430: 19pp.

----- 1936a. Two new races of Indian bats. *Records Indian Mus.* 38: 343–346.

----- 1936b. The status of *Vespertilio pilosus* Peters. *Journal Mammal.* 17: 168–169.

----- 1937. Second preliminary report on the results of the Second Dolan Expedition to west China and Tibet ... *Proceedings Acad. nat. Sci. Philad.* 89: 341–342.

----- 1938a. A new pygmy fruit bat from Borneo. *Journal Mammal.* 19: 496–498.

----- 1938b, 1940. *The mammals of China and Mongolia.* New York: American Museum of Natural History, 620,620pp.

Allen, G. M. & Coolidge, H. J. 1940. Mammal and bird collections of the Asiatic Primate Expedition: mammals. *Bulletin Mus. comp. Zool.* 87: 131–166.

Allen, H. 1864. Monograph of the bats of North America. *Smithsonian misc. Collns* 7,165:xxi, 85pp.

Allen, J. A. 1901. Notes on the names of a few South American mammals. *Proceedings biol. Soc. Wash.* 14: 183–185.

----- 1906. Mammals from the island of Hainan, China. *Bulletin Am. Mus. nat. Hist.* 22: 463–490.

----- 1909a. Further notes on mammals from the island of Hainan, China. *Bulletin Am. Mus. nat. Hist.* 26: 239–242.

----- 1909b. Mammals from Shen-si Province, China. *Bulletin Am. Mus. nat. Hist.* 26: 425–430.

----- 1910. Mammals from Palawan Island, Philippine Islands. *Bulletin Am. Mus. nat. Hist.* 28: 13–17.

----- 1911. Mammals collected in the Dutch East Indies ... *Bulletin Am. Mus. nat. Hist.* 30: 335–339.

----- 1920. Preliminary note on African Carnivora. *Journal Mammal.* 1: 23–31.

Alsberg, P. 1922. *Das Menschheitsratsel.* Dresden, 515pp.

Alston, E. R. 1876. On the classification of the order Glires. *Proceedings zool. Soc. Lond.* : 88–97, pl. 21.

Alton, E. d' 1823. Die Skelete der Wiederkäuer ... Lief. iv: 10pp, 8pls in Pander, C. H. & Alton, E. d' *Vergleichende Osteologie.* Bonn: Weber.

Andersen, K. 1905a. On the bats of the *Rhinolophus philippinensis* group ... *Annals Mag. nat. Hist.* 16: 243–256.

----- 1905b. On the bats of the *Rhinolophus macrotis* group ... *Annals Mag. nat. Hist.* 16: 289–292.

----- 1905c. On *Hipposideros diadema* and its closest allies. *Annals Mag. nat. Hist.* 16: 497–507.

----- 1905d. A list of the species and subspecies of the genus *Rhinolophus* ... *Annals Mag. nat. Hist.* 16: 648–662.

----- 1905e. On the bats of the *Rhinolophus arcuatus* group ... *Annals Mag. nat. Hist.* 16: 281–288.

----- 1905f. On some bats of the genus *Rhinolophus* ... *Proceedings zool. Soc. Lond.* 2: 75–145, 2pls.

----- 1906a. On the bats of the *Hipposideros armiger* and *commersoni* types. *Annals Mag. nat. Hist.* 17: 35–48.

----- 1906b. On some bats of the genus *Rhinolophus* ... *Proceedings US natn. Mus.* 29: 657–659.

----- 1906c. On some new or little-known bats of the genus *Rhinolophus* ... *Annali Mus. civ. Stor. nat. Giocomo Doria* 2: 173–193, 1pl.

----- 1907a. On *Pterocyon, Rousettus* and *Myonycteris. Annals Mag. nat. Hist.* 19: 501–515.

----- 1907b. Chiropteran notes. *Annali Mus. civ. Stor. nat. Giacomo Doria* 3: 5–45.

----- 1907c. Chiropteran notes. *Annali Mus. civ. Stor. nat. Giacomo Doria* 3: 473–478.

----- 1908. Twenty new forms of *Pteropus. Annals Mag. nat. Hist.* 2: 361–370.

----- 1909a. Notes on the genus *Acerodon* ... *Annals Mag. nat. Hist.* 3: 20–29.

----- 1909b. On the fruit bats of the genus *Dobsonia. Annals Mag. nat. Hist.* 4: 528–533.

----- 1909c. In Thomas, O. & Wroughton, R. C., Diagnoses of new mammals collected by Mr H. C. Robinson in the islands of the Straits of Malacca. *Annals Mag. nat. Hist.* 4: 534–536.

----- 1910. Ten new fruit-bats of the genera *Nyctimene, Cynopterus* and *Eonycteris. Annals Mag. nat. Hist.* 6: 621–625.

----- 1911. Six new fruit-bats of the genera *Macroglossus* and *Syconycteris. Annals Mag. nat. Hist.* 7: 641–643.

----- 1912a. *Catalogue of the Chiroptera in the collection of the British Museum, 2nd. ed. 1: Megachiroptera*. London: British Museum (Natural History).
----- 1912b. A new *Cynopterus* from Borneo. *Annals Mag. nat. Hist.* 10: 640.
----- 1912c. Brief diagnoses of eight new *Petalia* ... *Annals Mag. nat. Hist.* 10: 546–550.
----- 1918. Diagnoses of new bats of the families Rhinolophidae and Megadermatidae. *Annals Mag. nat. Hist.* 2: 374–384.
Andersen, K. & Wroughton, R. C. 1907. On bats of the family Megadermatidae. *Annals Mag. nat. Hist.* 19: 129–145.
Anderson, E. 1970. Quaternary evolution of the genus *Martes. Acta zool. fenn.* 130: 1–132.
Anderson, J. 1871. On three new species of squirrels from Upper Burmah ... *Proceedings zool. Soc. Lond.* : 139–142, pl. x.
----- 1872a. Further remarks on the external characters and anatomy of *Macacus brunneus. Proceedings zool. Soc. Lond.* : 203–212.
----- 1872b. [Letter] *Proceedings zool. Soc. Lond.* (1871): 628–629.
----- 1873. On the species and dentition of the southern Asiatic shrews ... *Proceedings zool. Soc. Lond.* : 227–235.
----- 1875. Description of some new Asiatic mammals ... *Annals Mag. nat. Hist.* 16: 282–285.
----- 1877a. Description of some new and little known Asiatic shrews ... *Journal Asiat. Soc. Bengal* 46,2: 261–283.
----- 1877b. [Exhibition of living bamboo rat] *Proceedings Asiat. Soc. Bengal* : 148–150.
----- 1878a. On the Indian species of the genus *Erinaceus. Journal Asiat. Soc. Bengal* 47,2: 195–213.
----- 1878b. On *Arvicola indica*, Gray, and its relations to the subgenus *Nesokia* ... *Journal Asiat. Soc. Bengal* 47,4: 214–234.
----- 1879. *Anatomical and zoological researches: comprising an account of the two expeditions to western Yunnan*. London: Quaritch, 2 vols.
----- 1881. *Catalogue of Mammalia in the Indian Museum, Calcutta*. Calcutta.
Anderson, J. & de Winton, W. E. 1902. *Zoology of Egypt. Mammalia*. London.
Anderson, S. & Jones, J. K. (eds) 1984. *Orders and families of Recent mammals of the world*. New York etc. : Wiley, 700pp.
Andrews, C. W. 1900. *A monograph of Christmas Island (Indian Ocean)* ... London.
Andrews, P. 1988. A phylogenetic analysis of the primates. Pp. 143–175 in Benton, M. J. (*q. v.*).
Andrews, R. C. 1921. Description of a new species of serow from Yunnan Province, China. *American Mus. Novit.* 6: 3pp.
Angermann, R. 1983. The taxonomy of Old World *Lepus. Acta zool. fenn.* 174: 17–21.
Angst, R. & Mann, P. 1971. Zur Variabilität von *Urogale everetti. Folia primatol.* 15: 148–158.
Anon. 1987. Pakistan restores blackbuck. *Oryx* 21: 251.
----- 1991. Cheetah sighting in India. *Oryx* 25: 69.
Ansell, W. F. H. & Topál, G. 1976. The type locality of *Miniopterus schreibersi* ... *Vertebrata hung.* 17: 15–17.
Anthony, H. E. 1941. Mammals collected by the Vernay-Cutting Burma expedition. *Publications Field Mus. nat. Hist. Zool.* 27: 37–123.
Aoki, B. & Tanaka, R. 1938. Biostatistical research on *Rattus losea* ... *Memoirs Fac. Sci. Agric. Taihoku imp. Univ.* 23: 1–74.
Aplin, K. P. & Archer, M. 1987. Recent advances in marsupial systematics ... Pp. xv-lxxii in Archer, M. (*q. v.*).
Archbold, R. & Tate, G. H. H. 1935. A new genus and species of squirrel from Celebes. *American Mus Novit.* 801: 6pp.
Archer, M. (ed). 1987. *Possums and opossums: studies in evolution*. Chipping Norton, New South Wales: Beatty, 2 vols, 788pp.
Argyropulo, A. I. 1929. Sur une sous-espèce nouvelle de *Micromys minutus* Pall. ... *Doklady Akad. Nauk SSSR* 1929A: 253–255.
----- 1933. Die Gattungen und Arten der Hamster ... *Zeitschrift Säugetierk.* 8: 129–149.
Arnbäck-Christie-Linde, A. 1908. A collection of bats from Formosa. *Annals Mag. nat. Hist.* 2: 235–238.
Arrighi, F. E. *et al.* 1969. Chromosomes of the tree shrews (Tupaiidae). *Cytogenetics* 8: 199–208.
Asakura, S. 1987. A note on a bear hybrid ... *International Zoo Yb.* 9: 88, 1pl.
Asher, G. W. *et al.* 1988. Hybridization of Père David's deer (*Elaphurus davidianus*) and red deer (*Cervus elaphus*) ... *Journal Zool. Lond.* 215: 197–203.
Askins, R. A. 1977. Family Sciuridae. Pp. 337–387 in Lekagul & McNeely (*q. v.*).
Aswathanarayana, N. V. 1987. Evolutionary trends in the genus *Funambulus* ... *Current Science* 56: 1298–1301.
Aswathanarayana, N. V. & Manjunatha, K. R. 1981a. Studies on the chromosomes of the genus *Mus*: karyotype variation in *Mus dunni. Mammalian Chromosome Newsl.* 22: 52–53.
----- 1981b. Studies on the chromosomes of the genus *Tatera* ... *Mammalian Chromosome Newsl.* 22: 55–56.
Aswathanarayana, N. V. *et al.* 1987. Karyology of the pygmy shrew, *Suncus etruscus perrotteti* ... *Current Science* 56: 911–913.
Audebert, J. B. 1798–9. *Histoire naturelle des singes*. Paris, 63 pls.
Azzaroli, A. 1954. Critical observations upon Siwalik deer. *Proceedings Linn. Soc. Lond.* 165: 75–83.

Badam, G. L. 1979. *Pleistocene fauna of India*. Pune: Deccan College, 250pp, 57pls.
Baker, J. R. 1974. *Race*. Oxford: Oxford Univ. Press.
Baker, R. J. *et al.* 1991. On the monophyly of bats. *Systematic Zool.* 40: 216–231. (See also pp. 239–243.)
Banerjee, S. & Ghosh, M. 1981. Prehistoric fauna of Kausambi, near Allahabad, U. P., India. *Record zool. Surv. India* 78: 113–119.
Bangs, O. 1901. Notes on a small collection of mammals from the Liu Kiu Islands. *American Nat.* 35: 561–562.
Banks, E. 1931. A popular account of the mammals of Borneo. *Journal Malay Br. r. Asiat. Soc.* 9,2: 1–139, pls 11–19.
Bannister, J. L. *et al.* 1988. *Zoological catalogue of Australia. 5. Mammalia*. Canberra: Australia Government Publishing Service.
Barbehenn, K. R. 1974. Recent invasions of Micronesia by small mammals. *Micronesica* 10: 41–50.
Bardeleben, K. von 1909. *Handbuch Anat. Menschen* 1,3 (*n. v.*).
Barghoorn, S. F. 1977. New material of *Vespertiliavus* Schlosser and suggested relationships of emballonurid bats ... *American Mus. Novit.* 2618: 29pp.
Barnes, L. G. 1985. Fossil pontoporid dolphins ... *Contributions Sci. Los Angeles Co. Mus.* 363: 34pp.
Barrett-Hamilton, G. E. H. 1900. On geographical and individual variation in *Mus sylvaticus* and its allies. *Proceedings zool. Soc. Lond.* : 387–428, pl. xxv.
----- 1906. Decription of two new species of *Pterygistes. Annals Mag. nat. Hist.* 17: 98–100.
----- 1907. Description of two new species of *Plecotus. Annals Mag. nat. Hist.* 20: 520–521.
Barry, J. C. 1983. *Herpestes* ... from the Miocene of Pakistan. *Journal Paleont.* 57: 150–156.
Bartels, M. 1937a. A new rat from Java. *Treubia* 16: 45–46.
----- 1937b. On two new Muridae from Sumatra ... *Natuurkundig Tijdschr. Ned.-Indië* 97: 121–124.
----- 1938. A new subspecies of *Rattus bartelsii* (Jentink) from central Java. *Treubia* 16: 323–324.
Bartlett, A. D. 1857a (27 June). *Literary Gazette* 2110: 610 (*n. v.*).
----- 1857b (21 Oct.). Description of a rabbit said to be found on the Himalayan Mountains. *Proceedings zool. Soc. Lond.* : 159–160.
----- 1869. Andaman monkey (*Macacus andamanensis*). *Land and Water* 8: 57.
Bates, P. J. J. 1988. Systematics and zoogeography of *Tatera* ... *Bonner zool. Beitr.* 39: 265–303.
Baverstock, P. R. *et al.* 1983. Genetic differentiation among karyotypic forms of the black rat, *Rattus rattus. Genetics* 105: 969–983.
Beaux, O. de 1923. Contributo allo studio dei macachi. *Atti Soc. Lig. Sci. Lett.* 1: 22–39.
----- 1924. Beitrag zur Kenntnis der Gattung *Potamochoerus* Gray. *Zoologische Jb. Abt. Syst.* 47: 379–504, pls 4–6.
Bechstein, J. M. 1799,1800. *Thomas Pennant's allgemeine Uebersicht der vierfüssiger Thiere*. Weimar.
Bechthold, G. 1936. Einige neue Unterarten asiatischer Herpestiden. *Zeitschrift Säugetierk.* 11: 149–153.
Beck, B. B. & Wemmer, C. (eds) 1983. *Biology and management of an extinct species/Père David's deer*. Park Ridge, NJ: Noyes, 193pp.
Becking, J. H. 1989. *Henri Jacob Victor Sody (1892–1959). His life and work* ... Leiden: Brill.
Beden, M. & Guérin, C. 1973. Le gisement de vertébrés du Phnom Loang ... Cambodge ... *Travaux Docum. ORSTOM* 27: 1–93,6 pls.
Bell, D. J. 1986. A study of the hispid hare *Caprolagus hispidus. Dodo* 23: 24–31.
Bell, D. J. *et al.* 1990. The hispid hare *Caprolagus hispidus*. Pp. 128–136 in Chapman, J. A. & Flux, J. E. C. (*q. v.*).
Bell, T. 1836. Cheiroptera. In Todd, R. B. (ed.), *The cyclopaedia of anatomy and physiology*. Vol. 1, pt 2. London.
----- 1839. Insectivora. Pp. 994–1006 in Todd, R. B. (ed.), *The cyclopaedia of anatomy and physiology*, vol. 2. London: Longman.
Benirschke, K. (ed.) 1986. *Primates: the road to self-sustaining populations*. New York etc. : Springer, 1044pp.
Bennett, E. T. 1829. *The Tower Menagerie* ... London, xviii, 241pp.
----- **(ed.)** 1830. *The gardens and menagerie of the Zoological Society delineated ... Quadrupeds, vol. 1*. London: Tegg, 308pp.
----- 1832. In: Meeting of June 26,1832. *Proceedings zool. Soc. Lond.* : 121–127.
----- 1833. [Meeting of June 11, 1833.] *Proceedings zool. Soc. Lond.* 68–69.
----- 1835a. [A collection of skins of Mammalia ... presented by Philip Poole, Esq.] *Proceedings zool. Soc Lond.* : 66–67.
----- 1835b. [A paradoxure now living in the Garden.] *Proceedings zool. Soc. Lond.* : 118.
Bentham, T. 1908. Description of a new species of mouse from the Madura district of Madras. *Records Ind. Mus.* 2: 385–386.

Benton, M. J. 1988a. The relationships of the major groups of mammals . . . *Trends Evol. Ecol.* 3: 40–45.
----- 1988b. *The phylogeny and classification of tetrapods, vol. 2: mammals.* Oxford: Clarendon Press, 344pp.
Bergmans, W. 1978. On *Dobsonia* Palmer 1898 from the Lesser Sunda Islands. *Senckenbergiana biol.* 59: 1–18.
Bergmans, W. & Hill, J. E. 1980. On a new species of *Rousettus* . . . from Sumatra and Borneo . . . *Bulletin Br. Mus. nat. Hist.* (*Zool.*) 38: 95–104.
Bergmans, W. & Rozendaal, F. G. 1982. Notes on *Rhinolophus* Lacépède, 1799 from Sulawesi . . . *Bijdragen dierk.* 52: 169–174.
----- 1988. Notes on collections of fruit bats from Sulawesi . . . *Zoologische Verhand.* 248: 1–74.
Bergmans, W. & van Bree, P. J. H. 1986. On a collection of bats and rats from the Kangean Islands . . . *Zeitschrift Säugetierk.* 51: 329–344.
Berkenhout, J. 1769. *Outlines of the natural history of Great Britain and Ireland.* Vol. 1.
Berry, R. J. 1963. Epigenetic polymorphism in wild populations of *Mus musculus*. *Genetical Res.* 4: 193–220.
----- 1984. House mouse. Pp. 273–284 in Mason, I. L. 1984a (*q. v.*).
Bhattacharyya, T. P. 1976. Occurrence of Indian pipistrelle, *Pipistrellus coromandra* (Gray) . . . in Car Nicobar . . . *Journal Bombay nat. Hist. Soc.* 73: 516.
Bianchi, V. 1917. Notes préliminaires sur les chauve-souris ou chiroptères de la Russie. *Ezhegodnik zool. Muz.* 21 (1916): lxxiii-lxxxii.
Billberg, G. J. 1828. *Synopsis faunae Scandinaviae . . . 1 (1) Mammalia.* Holmia, xii, 56, x pp., tables A-E.
Birula, A. 1910. Contributions à la classification et à la distribution géographiques des mammifères. *Ezhegodnik zool. Muz.* 15: 318–333.
Bishop, I. R. 1974. The status of *Pyromys priestlyi* Thomas, 1911. *Mammalia* 38: 139–141.
Biswas, B. & Ghose, R. K. 1970. Taxonomic notes on the Indian pale hedgehogs of the genus *Paraechinus* . . . *Mammalia* 34: 467–477.
Biswas, B. & Khajuria, H. 1955. Zoological results of the Daily Mail Himalayan Expedition 1954. Four new mammals from Khumbu, eastern Nepal. *Proceedings zool. Soc. Calcutta* 8: 25–30.
----- 1968. A new name for the Khumbu house mouse. *Science Culture* 34: 218.
Biswas, R. N. 1973. On the domestication of the otter by fishermen in Bangladesh. *Journal Bombay nat. Hist. Soc.* 70: 374.
Björkegren, B. 1942. On a new weasel from northern Tonkin. *Arkiv Zool.* 33B, 15: 4pp.
Black, C. C. 1972. Review of fossil rodents from the Neogene Siwalik beds of India and Pakistan. *Palaeontology* 15: 238–266.
Black, D. 1927. On a lower molar hominoid tooth from the Chou Kou Tien deposit. *Palaeontologia sin.* D, 7,1: 24pp, 2pls.
Blainville, H. M. D. de 1816. Sur plusieurs espèces d'animaux mammifères . . . *Bulletin Sci. Soc. philom. Paris* : 73–82.
----- 1817. Sur le paresseaux à cinq doigts (*Bradypus ursinus* de Shaw). *Bulletin Sci. Soc. philom. Paris* : 74–76.
----- 1822. Sur les caractères distinctifs des espèces de cerfs. *Journal Phys. Chim. Hist. nat.* 94: 254–283.
----- 1837. Rapport sur un memoire de M. Jourdan concernant deux nouvelles espèces de mammifères de l'Inde. *Annales Sci. nat. Zool.* 8,2: 270–281.
----- 1838a. Recherches sur l'ancienneté des mammifères insectivores . . . *Compte r. hebd. Seanc. Acad. Sci. Paris* 6: 738–744.
----- 1838b. Sur les cachelots. *Annales fr. Anat.* 2: 335–337.
----- 1839–1864. *Ostéographie, ou description iconographique des mammifères* . . . Paris (in 26 separately paginated parts).
Blanford, W. T. 1873. Note on the gazelles of India and Persia . . . *Proceedings zool. Soc. Lond.* : 313–319.
----- 1874. On two species of *Herpestes* and a hare . . . in Sind. *Proceedings zool. Soc. Lond.* : 661–664, pl. lxxxi.
----- 1875a. List of Mammalia collected . . . in Kashmir. *Journal Asiat. Soc. Bengal* 44,2: 105–112.
----- 1875b. Description of new Mammalia from Persia and Baluchistan. *Annals Mag. nat. Hist.* 16: 309–313.
----- 1876. Description of *Pelomys Watsoni*, a new species of mouse from Sind. *Proceedings Asiat. Soc. Bengal* : 181–182.
----- 1877a. Note on two species of Asiatic bears . . . *Journal Asiat. Soc. Bengal* 46,2: 315–323.
----- 1877b. On an apparently new hare . . . from Gilgit. *Journal Asiat. Soc. Bengal* 46,2: 323–327.
----- 1877c. Note on two species of Asiatic bears . . . (abstract). *Proceedings Asiat. Soc. Bengal* : 204–205.
----- 1878a. On some mammals from Tenasserim. *Journal Asiat. Soc. Bengal* 47,2: 150–167.
----- 1878b. Description of two apparently new mammals from Tenasserim. *Proceedings Asiat. Soc. Bengal* : 71–72.
----- 1881. Description of an *Arvicola* from the Punjab Himalayas. *Journal Asiat. Soc. Bengal* 49 (1880): 244–245.
----- 1885a. [Meeting of June 16, 1855.] *Proceedings zool. Soc. Lond.* : 612–613.
----- 1885b. A monograph of the genus *Paradoxurus*, F. Cuv. *Proceedings zool. Soc. Lond.* : 780–808, pls 49,50.
----- 1888a. Critical notes on the nomenclature of Indian mammals. *Proceedings zool. Soc. Lond.* (1887): 620–638.
----- 1888b, 1891. *Fauna of British India . . . Mammalia*, part 1 (1888): 1–250; part 2 (1891): 251–617. London: Taylor & Francis.
----- 1897. The large Indian squirrel (*Sciurus indicus* Erx.) . . . *Journal Bombay nat. Hist. Soc.* 11: 298–305, pl. 4.
Blasius, J. H. 1857. *Fauna der Wirbelthiere Deutschlands . . . 1. Säugethiere.* Brunswick.
Blasius, W. 1884. Der japanische Nörz, *Foetorius Itatsi* . . . *Berichte naturf. Ges. Bamberg* 13: 1–34.
Bleisch, W. & Nan, C. 1990. Conservation of the black-crested gibbon in China. *Oryx* 24: 147–156.
Blood, B. R. & McFarlane, D. A. 1988. Notes on some bats from northern Thailand . . . *Zeitschrift Säugetierk.* 53: 276–280.
Blouch, R. A. 1988. Ecology and conservation of the Javan warty pig *Sus verrucosus* Müller, 1840. *Biological Conserv.* 43: 295–307.
Blouch, R. A. & Groves, C. P. 1990. Naturally occurring suid hybrids in Java. *Zeitschrift Säugetierk.* 55: 270–275.
Blumenbach, J. F. 1779. *Handbuch der Naturgeschichte*, 1st ed. Gottingen, xi, 559pp, 2pls.
----- 1788. *Ibid*, 3rd ed.
----- 1797. *Ibid*, 5th ed.
----- 1830. *Ibid*, 12th ed.
Blyth, E. 1840. Mammalia, birds, reptiles. In *Cuvier's animal kingdom* . . . London, 670pp.
----- 1841a. Descriptions of three Indian species of bat, of the genus *Taphozous*. *Journal Asiat. Soc. Bengal* 10: 971–977.
----- 1841b. An emended list of the species of the genus *Ovis*. *Proceedings zool. Soc. Lond.* (1840): 62–79.
----- 1841c. Report for the month of September. *Journal Asiat. Soc. Bengal* 10: 836–842.
----- 1842a. Descriptive notice of the bat described as *Taphozous longimanus* . . . *Journal Asiat. Soc. Bengal* 11: 784–786.
----- 1842b. [Report of Curator for April 1842] *Journal Asiat. Soc. Bengal* 11: 444–470.
----- 1843a. [Report from the Curator] *Journal Asiat. Soc. Bengal* 11: 880–891.
----- 1843b. Memorandum from the Zoological Curator. *Journal Asiat. Soc. Bengal* 12: 166–182.
----- 1843c. *Rhizomys* from Arracan. *Journal Asiat. Soc. Bengal* 12: 1006–1007.
----- 1843d. Report from the Curator. *Journal Asiat. Soc. Bengal* 11: 969–970.
----- 1844a. Notices of various Mammalia . . . *Journal Asiat. Soc. Bengal* 13: 463–494.
----- 1844b. [Letter to Richard Taylor Esq.] *Annals Mag. nat. Hist.* 13: 312–313.
----- 1845a. Description of *Caprolagus* . . . *Journal Asiat. Soc. Bengal* 14: 247–249.
----- 1845b. Footnotes, in Hutton, T, Rough notes on the zoology of Candahar. *Journal Asiat. Soc. Bengal* 14: 340–354.
----- 1846a. Footnotes in Hutton, T. Rough notes on the zoology of Candahar. *Journal Asiat. Soc. Bengal* 15: 135–169.
----- 1846b. Additions and corrections to former notes. *Journal Asiat. Soc. Bengal* 15: 169–170.
----- 1846c. Notes on the fauna of the Nicobar Islands. *Journal Asiat. Soc. Bengal* 15: 367–379.
----- 1847a. Supplementary report by the Curator . . . *Journal Asiat. Soc. Bengal* 16: 861–880, pl. 37.
----- 1847b. Report of Curator, Zoological Department. *Journal Asiat. Soc. Bengal* 16: 1271–1276.
----- 1847c. Supplementary report of the Curator. *Journal Asiat. Soc. Bengal* 16: 728–737, 1pl.
----- 1848. Report of curator . . . *Journal Asiat. Soc. Bengal.* 17: 247–255.
----- 1849. Note on the Sciuri inhabiting Ceylon. . . *Journal Asiat. Soc. Bengal* 18: 600–603.
----- 1850. Description of a new species of mole (*Talpa leucura*, Blyth). *Journal Asiat. Soc. Bengal* 19: 215–217.
----- 1851a. Report on the Mammalia . . . inhabiting Ceylon. *Journal Asiat. Soc. Bengal* 20: 153–185.
----- 1851b. Notice of a collection of Mammalia . . . [from] Cherra Punji in the Khasia Hills . . . *Journal Asiat. Soc. Bengal* 20: 517–524.
----- 1852. Report of the Curator, Zoological Department. *Journal Asiat. Soc. Bengal* 21: 341–361.
----- 1853a. Report of the Curator, Zoological Department. *Journal Asiat. Soc. Bengal* 22: 408–417.
----- 1853b. Report of the Curator, Zoological Department for August meeting. *Journal Asiat. Soc. Bengal* 22: 580–584.
----- 1853c. Report of Zoological Curator for September meeting. *Journal Asiat. Soc. Bengal* 22: 589–594.

----- 1853d. Remarks on the different species of orang-utan. *Journal Asiat. Soc. Bengal* 22: 369–383, 10pls.
----- 1854. Report of Curator, Zoological Department, for September, 1854. *Journal Asiat. Soc. Bengal* 23: 729–740.
----- 1855. A memoir on the Indian species of shrews. *Journal Asiat. Soc. Bengal* 24: 24–38.
----- 1856a. Report for May meeting, 1855. *Journal Asiat. Soc. Bengal* 24: 359–363.
----- 1856b. Report of the Curator, Zoological Department, for July 1855. *Journal Asiat. Soc. Bengal* 24: 469–481.
----- 1856c. [Report for October meeting, 1855] *Journal Asiat. Soc. Bengal* 24: 711–723.
----- 1856d. Further remarks on the different species of orang- utan. *Journal Asiat. Soc. Bengal* 24: 518–528.
----- 1858. Report of Curator, Zoological Department for May 1858. *Journal Asiat. Soc. Bengal* 27: 267–290.
----- 1859a. Report of the Curator. *Journal Asiat. Soc. Bengal* 28: 271–298.
----- 1859b. On the great rorqual of the Indian Ocean . . . *Journal Asiat. Soc. Bengal* 28: 481–498.
----- 1860a. Report of the Curator, Zoological Department. *Journal Asiat. Soc. Bengal* 29: 87–115.
----- 1860b. On the flat-horned taurine cattle of S. E. Asia . . . *Journal Asiat. Soc. Bengal* 29: 282–306.
----- 1862a. Report of Curator . . . *Journal Asiat. Soc. Bengal* 30 (1861): 90–98.
----- 1862b. Report of the Curator, Zoological Department, February 1862. *Journal Asiat. Soc. Bengal* 31: 331–345.
----- 1863a. *Catalogue of the Mammalia in the Museum Asiatic Society*. Calcutta, 187, xiii pp.
----- 1863b. A memoir on the rats and mice of India. *Journal Asiat. Soc. Bengal* 32: 327–353.
----- 1863c. [Some horns and other specimens . . .] *Proceedings zool. Soc. Lond.* : 155–156.
----- 1864. [Letter] *Proceedings zool. Soc. Lond.* (1863): 370–371.
----- 1865. Indian rats and mice. *Journal Asiat. Soc. Bengal* 34,2: 192–194.
----- 1875. Catalogue of the mammals and birds of Burma. *Journal Asiat. Soc. Bengal* 44,2 (extra no.): 167pp.
Bobrinskii, N. A. 1918. *Fauna and flora of Russia* 15: 12–16 (*n.v.*).
----- 1926. Note preliminaire sur les chiroptères de l'Asie centrale. *Doklady Akad. Nauk SSSR* A: 95–98.
Boddaert, P. 1767–1770. *Dierkundig Mengelwerk* . . . 2.
----- 1785. *Elenchus animalium* vol. 1. Rotterdam: Hake, 173pp.
Boeadi, B. 1990. On two new specimens of the insectivorous bat species *Otomops formosus* Chasen, 1939. *Mammalia* 54: 489.
Boeadi [B] & Bergmans, W. 1987. First record of *Dobsonia minor* from Sulawesi . . . *Bulletin zool. Mus. Univ. Amsterdam* 11,8: 69–74.
Boeadi [B] & Hill, J. E. 1986. A new subspecies of *Aethalops alecto* from Java. *Mammalia* 50: 263–266.
Bohlken, H. 1958. Vergleichende Untersuchungen an Wildrindern . . . *Zoologische Jb.* (*Allg.*) 68: 113–202.
----- 1961. Haustiere und zoologische Systematik. *Zeitschrift Tierzücht. Züchtungsbiol.* 76: 107–113.
Boie, F. 1830. *Leuconoe* Boie, eine neue Fledermausgattung. *Isis, Jena* 23: 256–257.
Boie, H. 1828. Uittreksels uit brieven van Heinrich Boie van Java. *Bijdragen Natuurk. Wet. Amst.* 3: 231–252.
----- 1832. Briefe von Heinrich Boie geschrieben aus Ostindien. *Neues Staatsb. Mag. Schleswig* 1: 440–500.
Bonaparte, C. L. (J. L.) 1836–41. *Iconografia della fauna italica* Fasc. i, xv. Rome.
----- 1838. Synopsis vertebratorum systematis. *Nuovi Ann. Sci. nat. Bologna* 2,1: 105–133.
----- 1845. *Catalogo metodica dei mammiferi europei*. Milan: Pirola, 36pp.
----- 1856. Excursions dans les divers musées d'Allemagne . . . *Compte r. Acad. Sci.* 43: 410–421.
Bonhomme, F. 1986. Evolutionary relationships in the genus *Mus*. Pp. 19–34 in Potter *et al.* 1986 (*q. v.*).
Bonhomme, F. *et al.* 1985. Electromorphs and phylogeny in muroid rodents. Pp. 671–684 in Luckett, W. P. & Hartenberger, J. L. (eds), *Evolutionary relationships amongst rodents: a multidisciplinary analysis.* New York: Plenum Press.
Bonhote, J. L. 1898a. Description of a new species of *Mus* from south India. *Journal Bombay nat. Hist. Soc.* 12: 99–100, plate.
----- 1898b. On the species of the genus *Viverricula*. *Annals Mag. nat. Hist.* 1: 119–122.
----- 1900a. On a collection of mammals from Siam . . . *Proceedings zool. Soc. Lond.* : 191–195, pl. 18.
----- 1900b. On squirrels of the *Sciurus MacClellandi* group. *Annals Mag. nat. Hist.* 5: 50–54.
----- 1900c. On the squirrels of the *Ratufa* (*Sciurus*) *bicolor* group. *Annals Mag. nat. Hist.* 5: 490–499.
----- 1900d. On the mammals collected during the Skeat Expedition to the Malay Peninsula, 1899–1900. *Proceedings zool. Soc. Lond.* : 869–883., pl. lvi.
----- 1901a. On the squirrels of the *Sciurus erythraeus* group. *Annals Mag. nat. Hist.* 7: 160–167.
----- 1901b. On *Sciurus caniceps* and allied species. *Annals Mag. nat. Hist.* 7: 270–275.
----- 1901c. On a second collection of mammals . . . in Siam. *Proceedings zool. Soc. Lond.* : 52–56.
----- 1901d. On two new species of squirrels from the East Indies. *Annals Mag. nat. Hist.* 7: 455–457.
----- 1901e. On the squirrels of the *Sciurus Prevosti* group. *Annals Mag. nat. Hist.* 7: 167–177.
----- 1901f. On *Sciurus notatus* and allied species. *Annals Mag. nat. Hist.* 7: 444–455.
----- 1901g. On the martens of the *Mustela flavigula* group. *Annals Mag. nat. Hist.* 7: 342–349.
----- 1902a. On some mammals . . . from Okinawa. *Novitates zool.* 9: 626–628.
----- 1902b. On a further collection of mammals made by Mr Th. H. Lyle in Siam . *Proceedings zool. Soc. Lond.* : 38–41.
----- 1903a. Report on the mammals. *Fasciculi malayenses Zool.* 1: 1–45, pls 1–3.
----- 1903b. On new species of *Mus* from Borneo and the Malay Peninsula. *Annals Mag. nat. Hist.* 11: 123–125.
----- 1903c. On a new rat of the *Mus rufescens* group from Simla. *Annals Mag. nat. Hist.* 11: 473–474.
----- 1903d. On a new species of *Helictis*. *Annals Mag. nat. Hist.* 12: 592–594.
----- 1903e. On a new species of cat from China. *Annals Mag. nat. Hist.* 11: 374–376.
----- 1903f. On two new species of cat from China. *Annals Mag. nat. Hist.* 11: 474–476.
----- 1903g. On three new races of *Tragulus kanchil* . . . *Annals Mag. nat. Hist.* 11: 291–296.
----- 1905a. [The mammalian fauna of China, Murinae] *Abstracts Proc. zool. Soc. Lond.* 19: 19.
----- 1905b. On a new species of *Mus* from Pulan Jarak. *Journal Fed. Malay St. Mus.* 1: 69.
----- 1906a. The mammalian fauna of China—part 1, Murinae. *Proceedings zool. Soc. Lond.* (1905): 384–397.
----- 1906b. On mammals from south Johore and Singapore . . . *Proceedings zool. Soc. Lond.* : 4–11, pl. 1.
----- 1906c. On a new race of *Sciurus lokroides* from Burma. *Annals Mag. nat. Hist.* 18: 337–338.
----- 1907a. *Tupaia concolor* sp. n. *Abstracts Proc. zool. Soc. Lond.* : 2.
----- 1907b. On a collection of mammals made . . . in Annam. *Proceedings zool. Soc. Lond.* 1907: 3–11, pl. 2.
----- 1907c. [On a collection of mammals from Annam.] *Abstracts Proc. zool. Soc. Lond.* 1907: 2.
----- 1908. Report on the mammals [Gunong Tahan Expedition]. *Journal Fed. Malay St. Mus.* 3: 1–11, pl. 1.
Bonnaterre, Abbé 1789. *Tableau encyclopédique et méthodique . . . Cétologie*. Paris, xli, 28pp, 12pls.
Bonner, N. 1989. *Whales of the world*. London: Blanford, 191pp.
Boon Mesch, H. C. van der 1829. Beschrijving en afbeelding van eenen nieuwen eekhoren . . . *Verhandelingen K. ned. Inst. Amst.* 2: 241–244, 2 pls.
Boonsong, P. & Felten, H. 1989. Remarks on the genus *Bandicota* in Thailand. *Thai J. agric. Sci.* 22: 197–211.
Borkhausen, M. B. 1797. *Deutsche Fauna . . . Vol. 1 Säugethiere und Vogel*. Frankfurt am Main.
Borowski, G. H. 1781. *Gemeinnüzzige Naturgeschichte der Thierreiches*, vol. 2. Berlin, 196pp.
Bose, P. N. 1880. Undescribed fossil Carnivora from the Siwalik Hills. *Quarterly J. geol. Soc. Lond.* 36: 119–136.
Bosma, A. A. & Haan, N. A. de 1981. The karyotype of *Babyrousa babyrussa* . . . *Acta zool. path. antverp.* 76: 17–27.
Bosma, A. A. *et al.* 1982. The karyotype . . . of the pygmy hog *Sus* (*Porcula*) *salvanius* . . . *Genetica* 61: 99–106.
Bouchud, J. 1968. Sur la présence de *Crocuta crocuta sinensis* Owen dans le Pleistocène inférieure de Phnom-Loang (Cambodge). *Compte r. hebd. Séanc. Acad. Sci. Paris* 267D: 2291–2293.
Bourret, R. 1942a. *Compte r. Conseil Rech. sci. Indochine* 2: 20–27.
----- 1942b. Sur quelques petits mammiferes du Tonkin et du Laos. *Compte r. Conseil Rech. sci. Indochine* : 27–30.
----- 1942c. Les mammifères de la collection du laboratoire de zoologie . . . *Notes Trav. Ecol. sup. Sci. Hanoi* 1: 1–44.
----- 1944. Mammifères récemment entrés dans les collection du laboratoire de zoologie . . . *Notes Trav. Ecol. sup. Sci. Hanoi* 3: 1–17.
----- 1951. Une nouvelle chauve-souris du Tonkin, *Rhinomegalophus paradoxolophus*. *Bulletin Mus. natn. Hist. nat. Paris* 33: 607–609.
Bouwens, P. & de Bruijn, H. 1986. The flying squirrels *Hylopetes* and *Petinomys*

and their fossil record. *Proceedings k. ned. Akad. Wet.* (*Palaeont etc.*) 89: 113–123.

Bowdich, T. 1825. *Excursions in Madeira and Porto Santo* . . . London.

Brandon-Jones, D. 1984. Colobus and leaf monkeys. Pp. 398–408 in Macdonald, D. (ed.) *The encyclopaedia of mammals* vol. 1. London: Allen & Unwin.

Brandt, J. F. 1835. Mammalium exoticorum novorum. *Memoires Acad. imp. Sci. St Petersb.* 6,2, 3: 106pp.

----- 1836. In: Melanges. *Bulletin scient. Acad. imp. Sci. St Petersb.* 1: 32.

----- 1855. Beitrage zur nahern Kenntniss der Säugethiere Russland's. *Memoires Acad. Sci. St Petersb.* 9,1: 1–365.

Brants, A. 1827. *Het geslacht der Muizen*. . . Berlyn, xii, 190pp.

Brass, E. 1904. *Nutzbare Tiere Ostasiens.* Neudamm: Neumann, 130pp.

Breed, W. G. & Yong, H. S. 1986. Sperm morphology of murid rodents from Malaysia . . . *American Mus. Novit.* 2856: 9pp.

Brehem, von. 1829. Der Löwe keine Katze. *Isis, Jena* : 636–639.

Brisson, M. J. 1762. *Regnum Animale in classes IX distributum, etc.* . . . Paris, vi, 382pp.

Brongersma, L. D. 1935. Notes on some Recent and fossil cats . . . *Zoologische Meded. Leiden* 18: 1–89.

----- 1936a. Some comments upon H. C. Raven's paper "Wallace's Line". *Archives néerl. Zool.* 2: 240–256.

----- 1936b. On the subspecies of *Rhizomys sumatrensis* (Raffles). *Zoologische Meded. Leiden* 19: 138–164.

----- 1940. Note on *Mustela lutreolina* . . . *Temminckia, Leiden* 5: 257–263.

----- 1958. *The animal world of Netherlands New Guinea.* Groningen: Wolters, 70pp.

Brongersma, L. D. & Junge, G. C. A. 1942. On the variation of *Mustela* (*Lutreola*) *nudipes* Desm. *Zoologische Meded.* 23: 149–170, pls v-viii.

Bronn, H. G. 1848. *Index palaeontologicus* . . . Stuttgart.

Brooke, V. 1872. On *Hydropotes inermis* and its cranial characters. *Proceedings zool. Soc. Lond.* : 522–525.

----- 1877. On the deer of the Philippine Islands. *Proceedings zool. Soc. Lond.* : 51–60, pls 8–10.

Brookes, J. 1828. *A catalogue of the anatomical and zoological museum of J. Brookes . . . Pt. 1.* London, 76pp.

Brosset, A. 1962. The bats of central and western India. *Journal Bombay nat. Hist. Soc.* 59: 1–57, 1pl.

Brown, B. 1926. A new deer from the Siwaliks. *American Mus. Novit.* 242: 6pp.

Bruijn, H. de & Uney, E. 1989. Petauristinae . . . from the Oligocene of Spain, Belgium and Turkish Thrace. *Science Ser. nat. Hist. Mus. Los Angeles Cty* 33: 139–145.

Brünnich, M. T. 1771. *Zoologiae fundamenta* . . . Hafnia & Lipsia.

----- 1782. *Dyrenes Historie og Dyre-Samlingen udi Universetets Natur-Theater. Band 1.* Kobenhavn.

Buchannan, F. 1800. Description of the *Vespertilio plicatus*. *Transactions Linn. Soc. Lond.* 5: 261–263, 1pl.

Büchner, E. 1889. *Säugethiere von N. M. Przewalskis Reisen nach Central Asien.* St Petersburg.

----- 1891. Die Säugethiere der Gansu-Expedition (1884–87). *Mélanges biol. Bull. Acad. Sci. St Petersb.* 13: 143–164.

----- 1892a. Ueber eine neue *Sminthus*-art aus China. *Mélanges biol. Bull. Acad. Sci. St Petersb.* 13: 267–271.

----- 1892b. Uber eine neue *Sminthus*-Art aus China. *Bulletin Acad. imp. Sci. St Petersb.* 35: 107–111.

Buckland, F. 1872. A new rhinoceros at the Zoological Gardens. *Land & Water* 1872,2: 89.

Buckley, T. E. 1876. On the past and present geographical distribution of the large mammals of South Africa. *Proceedings zool. Soc. Lond.* : 277–293.

Buffon, G. L. L. de 1780. *Naturgeschichte der vierfüssiger Thiere.* Band 6 (by G Forster). Berlin: Pauli, 397pp.

Bumuller, J. 1899. *Das menschliche Femur nebst Beitragen zur Kenntnis der Affenfemora.* Augsburg [pp. 124–138].

Burmeister, H. 1865. *Revista formaceutica* 4: 363.

----- 1867. Sesion del 8 de Mayo. *Actas Soc. palaeont.* :xxiv-xxv.

----- 1872a. *Boletin Mus. Buenos Aires* :xi-xiv (*n. v.*).

----- 1872b. On *Balaenoptera patachonica* and *B. intermedia*. *Annals Mag. nat. Hist.* 10: 413–418.

Burnett, G. T. 1829a. Illustrations of the Alipeda . . . *Quarterly Jl Lit. Sci. Arts Lond.* : 262–269.

----- 1829b. Illustrations of the manupeda, or apes . . . *Quarterly Jl Lit. Sci. Arts Lond.* (1828): 301–307.

Butler, A. L. 1900. On a new serow from the Malay Peninsula. *Proceedings zool. Soc. Lond.* : 675–676.

Butler, P. M. 1980. The tupaiid dentition. Pp. 171–204 in Luckett, W. P. (*q.v.*).

----- 1988. Phylogeny of the insectivores. Pp. 117–141 in Benton, M. J. (*q.v.*).

Cabrera, A. 1901. Descripcion de tres nuevos mamiferos americanos. *Boletin R. Soc. esp. Hist. nat.* 1: 367–373.

----- 1908. Sobre los loris, y en especial sobre la forma filipina. *Boletin R. Soc. esp. Hist. nat.* 8: 135–139.

----- 1909. Un nuevo "*Rhinolophus*" filipino. *Boletin R. Soc. esp. Hist. nat.* 9: 304–306.

----- 1917. Dos ardillas exoticos nuevas. *Boletin R. Soc. esp. Hist. nat.* 17: 516–519.

----- 1920. Dos nuevas ratas de la isla de Borneo. *Boletin R. Soc. esp. Hist. nat.* 20: 211–213.

----- 1922. Sobre algunos mammiferos de la China oriental. *Boletin R. Soc. esp. Hist. nat.* 22: 162–170.

----- 1924. Una nueva forma de caguan de la isla de Borneo. *Boletin R. Soc. esp. Hist. nat.* 24: 128–130.

----- 1925. *Genera mammalium. Insectivora, Galeopithecia.* Madrid.

Cai, G-q. & Zhang, N-z. 1980a. On new subspecies of the Blanford's fruit bat and the orange-bellied Himalayan squirrel from Xizang. *Acta zootaxon. sin.* 5: 443–446.

----- 1980b. On mammalian fauna in Medoq region, Xizang . . . *Proceedings Symp. Qinhai-Xizang (Tibet) Plateau* 2: 1021–1026.

Caldecott, J. & Caldecott, S. 1985. A horde of pork. *New Scientist* 15 Aug. 1985: 32–35.

Caldwell, D. K. & Caldwell, M. C. 1989. Pygmy sperm whale—*Kogia breviceps* . . . dwarf sperm whale *Kogia simus* . . . Pp. 235–260 in Ridgeway, S. H. & Harrison, R. (*q. v.*).

Callahan, J. R. & Davis, R. 1982. Reproductive tract and evolutionary relationships of the Chinese rock squirrel, *Sciurotamias davidianus*. *Journal Mammal.* 63: 42–47.

Campbell, C. B. G. 1974. On the phyletic relationships of the tree shrews. *Mammal Review* 4: 125–143.

Cantor, T. 1845. Observations on two Malayan species of *Semnopithecus*. *Proceedings Linn. Soc. Lond.* 1: 235–236.

----- 1846. Catalogue of Mammalia inhabiting the Malayan Peninsula and islands . . . *Journal Asiat. Soc. Bengal* 15: 171–203.

Cao, K. 1987. Gradual increase and decline of mi-deer *Elaphurus davidianus* in China. *Investigatio Stud. Nat.* 7: 1–6.

Cao, V. S. 1984. Inventaire des rongeurs du Vietnam. *Mammalia* 48: 391–395.

----- 1985. Sur les rats du bambou (Rhizomyidae) au Vietnam. *Mammalia* 48: 606–607.

Cao V. S. & Nguyen M. T. 1989. Number and dynamic population of rodents in tropical forest ecosystem. *Abstracts int. theriol. Congr. 5, Rome* : 1018.

Cao, V. S. & Tran, V. M. 1985. Karyotypes et systematique des rats (genre *Rattus*) du Vietnam. *Mammalia* 48 (1984): 557–564.

Capanna, E. 1985. Karyotype variability and chromosome resilience in rodents: the case of the genus *Mus*. Pp. 643–669 in Luckett, W. P. & Hartenberger, J. L. (*q. v.*).

Carey, D. E. *et al.* 1971. Thottapalayan virus: a presumptive arbovirus isolated from a shrew in India. *Indian J. med. Res.* 59: 1758–1760.

Carleton, M. D. & Musser, G. G. 1984. Muroid rodents. Pp. 289–379 in Anderson, S. & Jones, J. K. (*q. v.*).

Carter, D. C. & Dolan, P. G. 1978. Catalogue of type specimens of Neotropical bats in selected European museums. *Special Publs Mus. Texas Tech. Univ.* 15: 136pp.

Carter, T. D. 1933. A new flying squirrel of the genus *Petaurista* from northwestern Siam. *American Mus. Novit.* 674: 1p.

----- 1942. Three new mammals of the genera *Crocidura*, *Callosciurus* and *Pteromys* from northern Burma. *American Mus. Novit.* 1208: 2pp.

Carter, T. D. *et al.* 1944. *Mammals of the Pacific World.* New York.

Caslick, J. W. 1956. Colour phases of the roof rat. *Journal Mammal.* 37: 255–257.

Catton, C. 1990. *Pandas.* London: Christopher Helm, 152pp.

Catzeflis, F. *et al.* 1985. Unexpected findings on the taxonomic status of East Mediterranean *Crocidura russula* auct. . . . *Zeitschrift Säugetierk.* 50: 185–201.

Cautley, P. T. & Falconer, H. 1836. Note on the *Ursus sivalensis*, a new fossil species from the Siwalik Hills. *Asiatic Res.* 19: 193–200.

Chakraborty, S. 1978a. A new species of the genus *Crocidura* Wagler from Wright Myo, South Andaman Island, India. *Bulletin zool. Surv. India* 1: 303–304.

----- 1978b. The rusty-spotted cat, *Felis rubiginosa* . . . in Jammu and Kashmir. *Journal Bombay nat. Hist. Soc.* 75: 478–479.

----- 1981. Studies on *Sciuropterus baberi* Blyth . . . *Proceedings zool. Soc. Calcutta* 32: 57–63.

----- 1983. Contribution to the knowledge of the mammalian fauna of Jammu and Kashmir. *Record zool. Surv. India, occ. Pap.* 38: 129pp.

----- 1985. Studies on the genus *Callosciurus* . . . *Record zool. Surv. India, occ. Pap.* 63: 93pp.

Chaline, J. *et al.* 1977. Les grandes lignes d'une classification evolutive des Muroidea. *Mammalia* 41: 245–252.

Champion, F. W. 1929. The distribution of the mouse-deer (*Moschiola memmina*). *Journal Bombay nat. Hist. Soc.* 33: 985–986.

Chan, Z-z. *et al.* 1984. On the Fea's muntjac from Xizang (Tibet), China. *Acta theriol. sin.* 4: 88–106.

Chapman, J. A. & Flux, J. E. C. (eds) 1990. *Rabbits, hares and pikas/ Status survey and conservation action plan.* Gland, Switzerland: IUCN, 168pp.

Chasen, F. N. 1933a. A new flying-squirrel from Borneo. *Bulletin Raffles Mus.* 8: 194.

----- 1933b. Notes on Bornean mammals. *Bulletin Raffles Mus.* 8: 195–198.

----- 1935a. Three new Malaysian mammals. *Bulletin Raffles Mus.* 9 (1934): 98–100.

----- 1935b. On a collection of mammals from the Natuna Islands. *Bulletin Raffles Mus.* 10: 5–42.

----- 1936. New and noteworthy bats from the Malaya Peninsula. *Bulletin Raffles Mus.* 12: 133–134.

----- 1937a. Two new mammals from Borneo. *Bulletin Raffles Mus.* 13: 80–81.

----- 1937b. On insular *Rattus rattus* from the North-west coast of the Malay Peninsula. *Bulletin Raffles Mus.* 13: 81–93.

----- 1939a. Four new mammals from Java. *Treubia* 17: 185–188.

----- 1939b. Two new mammals from north Sumatra. *Treubia* 17: 207–208.

----- 1940a. A handlist of Malaysian mammals. *Bulletin Raffles Mus.* 15: 209pp. (For supplement see Ellerman & Morrison-Scott, 1955.)

----- 1940b. The mammals of the Netherlands Indian Mt Leuser Expedition 1937 to north Sumatra. *Treubia* 17: 479–502.

Chasen, F. N. & Kloss, C. B. 1924. Three new races of *Sciurus vittatus*. *Journal Malay Br. r. Asiat. Soc.* 2: 57–58.

----- 1925. Remarks on the black and red squirrels allied to *Sciurus prevostii*. *Journal Malay Br. r. Asiat. Soc.* 3,1: 97–99.

----- 1928a. On some Carnivora, Rodentia and Insectivora ... from Borneo. *Journal Malay Br. r. Asiat. Soc.* 6,1: 38–49.

----- 1928b. On a collection of mammmals from the Anamba Islands. *Journal Malay Br. r. Asiat. Soc.* 6,3: 28–42.

----- 1928c. Spolia Mentawiensia: mammals. *Proceedings zool. Soc. Lond.* 97 (for 1927): 797–840.

----- 1929a. Two new races of *Galeopithecus*. *Bulletin Raffles Mus.* 2: 11–12.

----- 1929b. Notes on flying lemurs (*Galeopterus*). *Bulletin Raffles Mus.* 2: 12–22.

----- 1931. On "*rattus*" rats of the coasts and islands of Malacca Straits. *Bulletin Raffles Mus.* 5: 76–79.

----- 1932. On a collection of mammals from ... North Borneo. *Bulletin Raffles Mus.* 6 (for 1931): 1–82.

Chatterjee, K. & Majhi, A. 1975. Chromosomes of the Himalayan flying squirrel *Petaurista magnificus*. *Mammalia* 39: 447–450.

Chaturvedi, Y. 1964. Taxonomic status of *Tadarida tragata* ... *Journal Bombay nat. Hist. Soc.* 61: 432–437.

----- 1966. A new house rat ... from the Andaman and Nicobar islands. *Proceedings zool. Soc. Calcutta* 19: 141–144.

Chavan, S. A. 1987. Status of wild cats in Gujarat. *Tigerpaper* 14,4: 21–24.

Chaworth-Musters, J. L. & Ellerman, J. R. 1947. A revision of the genus *Meriones*. *Proceedings zool. Soc. Lond.* 117: 478–504.

Cheesman, R. E. 1921. Report on a collection of mammals made ... in Persia. *Journal Bombay nat. Hist. Soc.* 27: 573–581.

Chimimba, C. T. & Kitchener, D. J. 1991. A systematic revision of Australian Emballonuridae ... *Record W. Aust. Mus.* 15: 203–265.

Chorazyna, H. & Kurup, G. U. 1975. Observations on the ecology and behaviour of *Anathana ellioti* in the wild. Pp. 342–344 in Kondo, S. *et al.* (eds) *Contemporary primatology*. Basel etc. : Karger.

Chorn, J. & Hoffmann, R. S. 1978. *Ailuropoda melanoleuca*. *Mammalian Species* 110: 6pp.

Chou, W-l. 1989. First record of dwarf sperm whale (*Kogia simus*) from Taiwan. *Bulletin natn. Mus. nat. Sci.* 1: 23–37.

Choudhury, A. 1990. Overlapping distribution of capped langur *Trachypithecus pileatus* and Phayre's leaf monkey *T. phayrei*. *Journal Bombay nat. Hist. Soc.* 87: 133.

Chow, M. M. & Hsu, Y. 1957. A new species of fossil *Bubalis* from Menghsien, Honan. *Acta palaeont. sin.* 5: 457–466.

Chow, M. M. *et al.* 1965. Mammalian fossils associated with the hominid skull cap of Lantian, Shensi. *Scientia sin.* 14: 1037–1048.

Clark, G. A. *et al.* 1988. Indomalayan Region: a substitute name for Wallace's Oriental Region. *Ibis* 130: 447–448.

Clutton-Brock, J. 1981,1987. *Domesticated animals from early times*. London: Heinemann/British Museum (Natural History), 208pp. (Paperback ed. 1987 with title:*The natural history of domesticated mammals*. Cambridge Univ. Press/BM(NH), 192pp.)

Clutton-Brock, J. *et al.* 1976. A review of the family Canidae ... *Bulletin Br. Mus. nat. Hist.* (*Zool.*) 29: 117–199.

Cockrill, W. R. 1984. Water buffalo. Pp. 52–63 in Mason, I. L. (*q. v.*).

Cohen, J. A. 1978. *Cuon alpinus*. *Mammalian Species* 100: 3pp.

Cohn, L. 1910. Die papuasischen *Perameles*-Arten. *Zoologischer Anz.* 35: 718–728.

Colbert, E. H. 1933a. The presence of tubulidentates in the Middle Siwalik beds of northern India. *American Mus. Novit.* 604: 10pp.

----- 1933b. A new mustelid from the Lower Siwalik beds of northern India. *American Mus. Novit.* 605: 3pp.

----- 1933c. Two new rodents from the Lower Siwalik Beds, India. *American Mus. Novit.* 633: 6pp.

----- 1935. Siwalik mammals in the American Museum of Natural History. *Transactions Am. phil. Soc.* 26: 1–401.

----- 1940. Pleistocene mammals from the Ma Kai Valley of northern Yunnan, China. *American Mus. Novit.* 1099: 4pp.

----- 1943. Pleistocene vertebrates collected in Burma ... *Transactions Am. phil. Soc.* 32: 395–420, 14pls.

Colbert, E. H. & Hooijer, D. A. 1953. Pleistocene mammals from ... Szechuan, China. *Bulletin Am. Mus. nat. Hist.* 102: 1–134, 40pls.

Colebrooke, H. T. 1805. Description of a species of ox named Gayal. *Asiatic Res.* 8: 487–501.

Collett, R. 1893. On a new monkey from north-east Sumatra. *Proceedings zool. Soc. Lond.* (1892): 613–617, 1pl.

Collier, G. E. & O'Brien, S. J. 1985. A molecular phylogeny of the Felidae ... *Evolution* 39: 473–487.

Conisbee, L. R. 1953. *A list of the names proposed for genera and subgenera of Recent mammals...* London: British Museum (Natural History).

Coolidge, H. J. 1938. A new tree shrew of the genus *Tana* ... *Proceedings New England zool. Club* 17: 45–47.

----- 1940. The Indo-chinese forest ox or kouprey. *Memoirs Mus. comp. Zool. Harvard* 54: 421–531, 11pls.

Cope, E. D. 1891. *Syllabus of lectures on geology and paleontology. Part 3* ... Philadelphia: Ferris, 90pp.

Corbet, G. B. 1978. *The mammals of the Palaearctic Region*. London: British Museum (Natural History), 314pp.

----- 1983. A review of classification in the family Leporidae. *Acta zool. fenn.* 174: 11–15.

----- 1984. *The mammals of the Palaearctic Region ... Supplement*. London: British Museum (Natural History), 45pp.

----- 1988. The family Erinaceidae ... *Mammal Review* 18: 117–172.

Corbet, G. B. & Clutton-Brock, J. 1984. Appendix: taxonomy and nomenclature. Pp. 434–438 in Mason, I. L. (*q. v.*).

Corbet, G. B. & Hill, J. E. 1980. *A world list of mammalian species*. London: British Museum (Natural History). (2nd ed. 1986, 3rd ed. 1991).

Corbett, L. K. 1985. Morphological comparisons of Australian and Thai dingoes ... *Proceedings ecol. Soc. Aust.* 13: 277–291.

Coues, E. 1876. Some account ... of *Zapus hudsonicus*. *Bulletin US geol. geogr. Surv. Terr.* 1: 253–262.

Coulon, L. 1836. Description de quelques animaux nouveaux ou peu connus qui se trouvent au musée de Neuchatel. *Mémoires Soc. Sci. nat. Neuchatel* 1 (for 1835): 122–125, pls 8–13.

Cowan, I. M. 1972. The status and conservation of bears of the world. Pp. 347–367 in Herrero, S. (ed.) *Bears—their biology and management*. Morgues: IUCN Publications new Series 23.

Cox, R. 1987. The Philippine spotted deer and the Visayan warty pig. *Oryx* 21: 37–42.

Cranbrook, [4th] Earl of, 1962. The identity of the Malayan mole. *Journal Bombay nat. Hist. Soc.* 59: 942–945.

Cranbrook, [5th] Earl of, 1981. The vertebrate faunas. Pp. 57–69 in Wetmore, T. C. (ed.) *Wallace's line and plate tectonics*. Oxford: Clarendon Press.

----- 1986. A review of fossil and prehistoric remains of rhinoceroses of Borneo. *Sabah Mus. Arch. J.* 1: 50–110.

----- 1988. The contribution of archaeology to the zoogeography of Borneo, with the first record of a wild canid of early Holocene age. *Fieldiana Zool.* 42: 7pp.

Cretzschmar, J. 1826–1831. Säugethiere, pp. 1–78 in *Atlas zu der Reise im nördlichen Afrika von Edouard Rüppell*. Frankfurt.

Cronin, J. E. *et al.* 1980. Molecular evolution and systematics of the genus *Macaca*. Pp. 31–50 in Linburg, D. G. (*q. v.*).

Cummings, W. C. 1985. Bryde's whale—*Balaenoptera edeni*. Pp. 137–154 in Ridgeway, S. H. & Harrison, R. (*q. v.*).

Cuvier, F. 1818. Du macaque de Buffon. *Mémoires Mus. Hist. nat. Paris* 4: 109–120.

----- 1822. Du genre Paradoxure et de deux espèces nouvelles ... *Mémoires Mus. Hist. nat. Paris* 9: 41–48, pl. 4.

----- 1823a. Examen des espèces du genre porc-épic ... *Mémoires Mus. nat. Hist. Paris* 9 (1822): 413–484, pls 1–22.

----- 1823b. Lièvre, *Lepus*. 26: 299–316; Loutre, *Lutra*. 27: 237–250. *Dictionnaire des sciences naturelle*. Strasbourg & Paris.

----- 1824. *Des dents des mammifères* ... Strasbourg & Paris: Lerrault.

----- 1829. *Dictionnaire des sciences naturelles* 59. Strasbourg & Paris: Lerroult, 520pp.

----- 1832. Essai de classification naturelle des vespertilions. *Nouvelle Arch. Mus. Hist. nat.* 1: 1–21, 2 pls.

----- 1836. [On the jerboas and gerbillas] *Proceedings zool. Soc. Lond.* : 141–142.

----- 1838. Mémoire sur les gerboises et les gerbilles. *Transactions zool. Soc. Lond.* 2: 131–148, pls 22–26.

Cuvier, G. [1797]. *Tableau élémentaire de l'histoire naturelle des animaux*. Paris:xvi, 710pp, 14pls. Published prior to Dec. 1797.

----- 1800. *Leçons d'anatomie comparée, vol. 1*. Paris.

----- 1809. Recherches sur les espèces vivantes de grands chats. *Annales Mus. Hist. nat. Paris* 14: 136–164, pls 15,16.

Cuvier, G. 1812. Raport . . . sur divers cétacés . . . *Annales Mus. Hist. nat. Paris* 19: 1–16, pl. 1.

----- 1817. *Le règne animale* . . . . Vol. 1. Paris: xxxvii, 540pp.

----- 1823–24. *Recherches sur les ossemens fossile* . . . . Paris.

----- 1829. *Le règne animale.* 2nd ed.,1. Paris, xxxviii, 584pp.

Cuvier, G. F. & Geoffroy-St Hilaire, E. 1795. Method mammalogique, in *Magasin encyclopédique* 2. (*n.v.*).

Cuy, L. S. 1980. Nycteribiidae . . . of the Philippines. *Kalikasan* 9: 137–168.

----- 1981. Geographical distribution of Philippine Nycteribiidae and their hosts. *Kalikasan* 10: 197–213.

Dammerman, K. W. 1934. On prehistoric mammals from the Sampoeng Cave, Central Java. *Treubia* 14: 477–486.

----- 1940. On the Indo-Australian weasels . . . *Temminckia* 5: 266–272, pl. xv.

Dào Van Tiên, 1960a. Recherches zoologiques dans la region de Vinh-Linh . . . Vietnam. *Zoologischer Anz.* 164: 221–239.

----- 1960b. Sur une nouvelle espèce de *Nycticebus* au Vietnam. *Zoologischer Anz.* 164: 240–243.

----- 1961a. Notes sur une collection de micromammifères de la région de Hon-Gay. *Zoologischer Anz.* 166: 290–298.

----- 1961b. Recherches zoologiques dans la région de Thai-Nguêyn (Nord-Vietnam). *Zoologischer Anz.* 166: 298–308.

----- 1962. Materiaux sur la faune des vertebrés de Vietnam. *Zoologicheskii Zh.* 41: 724–735.

----- 1965. On the forms of the squirrel *Callosciurus erythraeus* and their distribution in Vietnam. *Zoologicheski Zh.* 44: 1238–1244.

----- 1966. Sur deux rongeurs nouveaux . . . au Nord-Vietnam. *Zoologischer Anz.* 176: 438–440.

----- 1969. *Thông báo khoa hoc. Sinh vat hoc, Dai hoc tông hop Ha Noi* 4: 49–53.

----- 1970a. Sur les formes de l'écureuil de McClelland, *Callosciurus mcclellandi* . . . au Vietnam. *Mitteilungen zool. Mus. Berl.* 46: 53–60.

----- 1970b. Sur les formes de semnopithèque noir, *Presbytis francoisi* . . . au Vietnam . . . *Mitteilungen zool. Mus. Berlin* 46: 61–65.

----- 1972 (or 1975). Ve các dang chuot dúi lón (*Bandicota indica*) . . . o Viet Nam . . . *Thong bao khoa hoc, Sinh vat hoc, Dai hoc tong hop Ha Noi* 6: 43–51 (*n. v.*, from Dao, 1985).

----- 1977. Sur quelques rares mammifères au nord du Vietnam. *Mitteilungen zool. Mus. Berlin* 53: 325–330.

----- 1978. Sur une collection de mammifères du plateau de Môc Chau . . . Nord-Vietnam. *Mitteilungen zool. Mus. Berl.* 54: 377–391.

----- 1985. *Scientific results of some mammal surveys in North Vietnam.* Ha Noi: Scientific and Technical Publishing House, 330pp. (Checklist pp. 258–277.)

Dào Van Tiên & Cao Van Sung 1990. Six new Vietnamese rodents. *Mammalia* 54: 233–238.

Dargel, B. 1990. A bibliography on viverrids. *Mitteilungen Hamburg zool. Mus. Inst.* 87, suppl. : 184pp.

Das, P. K. 1986a. Studies on the taxonomy . . . of bats obtained by the Silent Valley (Kerala, India) Expedition, 1980. *Records zool. Surv. India* 8: 259–276.

----- 1986b. Taxonomic status of *Harpiocephalus harpia madrassius* Thomas . . . *Journal Bombay nat. Hist. Soc.* 83: 311–316.

----- 1987. On the number of specimens, repository, and the heterogeneity of the type-series of *Vespertilio montivagus* . . . *Bulletin zool. Surv. India* 8: 39–45.

----- 1990. Occurrence of *Pipistrellus camortae* . . . in the Andaman Islands . . . *Journal Bombay nat. Hist. Soc.* 87: 135–137.

David, A. 1867. Journal d'un voyage en Mongolie . . . *Nouvelles Arch. Mus. Hist. nat. Paris, 3 Bull.*: 18–96.

----- 1869. Voyage en Chine. *Nouvelles Arch. Mus. Hist. nat. Paris 5 Bull.*: 1–13.

----- 1872. Rapport addressé à Mm les Professeurs-Administrateurs de Museum d'Histoire Naturelle. *Nouvelles Arch. Mus. Hist. nat. Paris, 7 Bull.*: 75–100.

Davis, D. D. 1938. Notes on the anatomy of the tree shrew *Dendrogale*. *Publications Field Mus. nat. Hist. Zool.* 20: 383–404.

----- 1958. Mammals of the Kelabit plateau, northern Sarawak. *Fieldiana, Zool.* 39: 119–147.

----- 1962. Mammals of the lowland rain-forest of North Borneo. *Bulletin natn. Mus. St. Singapore* 131: 1–129, 23pls.

----- 1964. The giant panda, a morphological study . . . *Fieldiana Zool. Mem.* 3: 339pp.

Davis, J. H. *et al.* 1979. The chromosomes of the African spotnecked otter, *Hydrictis maculicollis* . . . *Chromosome Inf. Serv.* 26: 11–13.

Davison, G. W. H. 1979. Some notes on Savi's pigmy shrew. *Malayan Nat. J.* 32: 227–231.

----- 1984. New records of Peninsular Malayan and Thai shrews. *Malayan Nat. J.* 36: 211–215.

Davison, G. W. H. *et al.* 1982. The Sunda shrew, *Crocidura monticola*: a correction and first records from Peninsular Malaysia. *Malayan Nat. J.* 36: 133–134.

Dawson, M. R. 1971. Fossil mammals of Java 1. Notes on Quaternary Leporidae . . . *Proceedings K. ned. Akad. Wet.* B, 74: 27–32.

Day, F. 1863. *Land of the Permauls.* Madras, x, 577, xxxvii pp.

DeBlase, A. F. 1980. The bats of Iran . . . *Fieldiana Zool.* 4:xvii, 424pp.

Dehne, J. F. A. 1841. Micromys agilis, *Kleinmaus, ein neues Säugethiere der Fauna von Dresden* . . . Hoflossnitz, Dresden, 10pp, 1pl.

De Jong, N. & Bergmans, W. 1981. A revision of the fruit bats of the genus *Dobsonia* . . . from Sulawesi . . . *Zoologische Abh. st. Mus. Tierk. Dresden* 37: 209–224.

Delacour, J. 1951. La systematique des gibbons Indochinois. *Mammalia* 15: 118–123.

Delson, E. 1980. Fossil macaques . . . Pp. 10–30 in Lindburg, D. G. (*q. v.*).

Dene, H. *et al.* 1980. Molecular evidence for the affinities of Tupaiidae. Pp. 269–291 in Luckett, W. P. (*q. v.*).

Deniger, K. 1910. Uber Babirusa. *Bericht naturf. Ges. Freiburg I. BR.* 18: 1–22, pls 1–3.

Denys, C. 1990. The oldest *Acomys* . . . from the lower Pliocene of South Africa . . . *Palaeontographica* A, 210: 79–91.

Deraniyagala, P. E. P. 1939a. Some aspects of the Asiatic elephant. *Journal Ceylon Brch r. Asiat. Soc.* 34 (1938): 126–162, pl. 1.

----- 1939b. Some fossil animals from Ceylon, part II. *Journal Ceylon Brch r. Asiat. Soc.* 34(1938): 231–239.

----- 1944. Some mammals of the extinct Ratnapura fauna of Ceylon. *Spolia zeylan.* 24: 19–56, 4pls.

----- 1945a. Some fossil animals from Ceylon, part IV. *Journal Ceylon Brch. r. Asiat. Soc.* 36: 80.

----- 1945b. Some aspects of the prehistory of Ceylon, part 2. *Spolia zeylan.* 24: 121–155, 2pls.

----- 1946. Some mammals of the extinct Ratnapura fauna of Ceylon (part II). *Spolia zeylan.* 24: 161–171, 3pls.

----- 1947. Some fossil animals from Ceylon, part V. *Journal Ceylon Brch. r. Asiat. Soc.* 37: 221–230, 1 pl.

----- 1948. Some mystacetid whales from Ceylon. *Spolia zeylan.* 25,2: 61–63.

----- 1949. *Some vertebrate animals of Ceylon. Vol. 1.* Colombo & Oxford: National Museum of Ceylon, Pictorial Series, ii, 119pp.

----- 1950. The elephant of Asia. *Proceedings 5th ann. Session Ceylon Assoc. Sci.*

----- 1951a. *Elephas maximus*, the elephant of Ceylon (parts 1 and 2). *Spolia zeylan.* 26: 21–68, 17pls; 161–176, 8pls.

----- 1951b. Some mammals of the extinct Ratnapura fauna of Ceylon (part III). *Spolia zeylan.* 26: 117–124.

----- 1952. Some Sinhala combative, field and aquatic sports and games. *Spolia zeylan.* 26: 179–215, 2 pls.

----- 1953a. The wild buffalo of Ceylon, a new subspecies. *Spolia zeylan.* 27: 103–105.

----- 1953b. Some aspects of the prehistory of Ceylon, part 3. *Spolia zeylan.* 27: 125–131, pls 1–13.

----- 1955a. A new subspecies of jungle cat from south India. *Spolia zeylan.* 27: 291.

----- 1955b. A new race of leaf monkey from Ceylon. *Spolia zeylan.* 27: 293–294.

----- 1956. A new subspecies of rusty-spotted cat from Ceylon. *Spolia zeylan.* 28: 113–114.

----- 1957. The races of the Stone Age and of the Ferrolithic of Ceylon. *Journal Ceylon Brch r. Asiat. Soc.* 5: 1–23, pls 1-xiii.

----- 1958a. *Podihik* gen. nov. A new shrew. *Administration Rep. Director nat. Mus. Ceylon* (1957), pt 4, Ed. Sci. Art (E): 5.

----- 1958b. *The Pleistocene of Ceylon.* Ceylon Nat. Mus. nat. Hist. Ser. :ix, 164pp, 58pls.

----- 1960a. Some mammals of the extinct Ratnapura fauna of Ceylon. Part iv. *Spolia zeylan.* 29: 1–14, 3pls.

----- 1960b. Some southern temperate zone . . . whales that enter the Ceylon area. *Spolia zeylan* 29: 79–85.

----- 1963a. Some mammals of the extinct Ratnapura fauna of Ceylon, part V. *Spolia zeylan.* 30: 5–25, pls 1–5.

----- 1963b. Mass mortality of the new subspecies of little piked whale *Balaenoptera acutorostrata thalmaha* . . . *Spolia zeylan.* 30: 79–84, pl. 1.

----- 1963c. *Ceylon Today* 12,3: 13.

----- 1965a. Some aspects of the fauna of Ceylon. *Journal Ceylon Br. r. Asiat. Soc.* 9: 165–219.

----- 1965b. A new subspecies of macaque, *Macaca sinica longicaudata* from Ceylon. *Spolia zeylan.* 30: 261–262, pls 1,2.

Desmarest, A. G. 1804. Tableaux méthodiques des mammifères. Pp. 5–38 in Tableaux methodique d'histoire naturelles, 238pp. in *Nouveau dictionnaire d'histoire naturelle* . . . 24, Paris.

----- 1816. Chat. Pp. 73–123 in *Nouveau dictionnaire d'histoire naturelle* 6.

----- 1817. Civette, vol. 7: 159–171; Dauphin, *Delphinus*, vol. 9: 146–179; Ecureuil, vol. 10: 94–111, in *Nouveau dictionnaire d'histoire naturelle* . . . Paris.

----- 1818. Moufette, vol. 21: 512–520; Polatouche, *Pteromys*, vol. 25: 401–407; *Phalanger*, vol. 25: 470–477, in *Nouveau dictionnaire d'histoire naturelle* . . . Paris.

----- 1819. Rat, vol. 29: 40–71; Tapir, vol. 32: 451–458, in *Nouveau dictionnaire d'histoire naturelle.*

----- 1820. *Mammalogie ou description des espèce de mammifères. In: Encyclopédie méthodique* . . . part 1. Paris: Agasse.

----- 1822. *Ibid*, part 2: 277–555.

----- 1823. Mangouste, pp. 54–63 in *Dictionnaire des sciences naturelle* 29. Strasbourge & Paris.

Desmoulins, L. A. 1822,1824,1825. *Dictionnaire classique d'histoire naturelle* vols 2,5, 7. Paris.

Deuve, J. & Deuve, M. 1963. Contribution à la connaissance des mammifères de Laos. *Bulletin Soc. r. Sci. nat. Laos* 8: 49–62; 9: 35–46.

De Vis, C. W. 1905. Bats. *Annals Queensl. Mus.* 6: 36–40.

Dhaliwal, S. S. 1962. Studies on body measurements and skeletal variation of two taxa of *Rattus rattus* in Malaya. *Journal Mammal.* 43: 248–260.

Diard, P. M. 1820. [Report of a meeting on Mar. 10th] *Asiatic J. month. Reg.* 10: 472–478 ? (*n. v.*).

Didier, R. & Petter, F. 1960. L'os pénien de *Jaculus blanfordi*. *Mammalia* 24: 171–176.

Dieterlen, F. 1986. *Kurzfassungen der Vortrege und Posterdemonstration, Deutsche Gesellschaft für Säugetierkunde Hauptversammlung, Stuttgart* . . . Hamburg & Berlin, Parey.

----- 1989. Revocation of name, p. 65 in Ansell, W. F. H. *African mammals 1938–1988*. Zennor: Trendrine Press, 77pp.

Dinnerstein, E. & Mehta, J. N. 1989. The clouded leopard in Nepal. *Oryx* 23: 199–201.

Divyabhanusinh 1987. Notes on a sighting of a caracal (*Felis caracal*) at the Sariska National Park. *Journal Bombay nat. Hist. Soc.* 84: 201.

Dobroruka, L. J. 1967. On the nomenclature and taxonomy of the lesser mouse-deer of Java. *Mammalia* 31: 456–458.

Dobson, G. E. 1871a. On some new species of Malayan bats . . . *Proceedings Asiat. Soc. Bengal* : 105–106.

----- 1871b. Description of four new species of Malayan bats . . . *Journal Asiat. Soc. Bengal* 40: 260–267, 2pls.

----- 1871c. On a new genus and species of Rhinolophidae . . . *Journal Asiat. Soc. Bengal* 40: 455–461,1pl.

----- 1871d. Notes on nine species of Indian and Indo-Chinese Vespertilionidae . . . *Proceedings Asiat. Soc. Bengal* : 210–215.

----- 1871e. On a new species of *Vespertilio*. *Journal Asiat. Soc. Bengal* 40,2: 186–188.

----- 1872a. Brief descriptions of five new species of rhinolophine bats. *Journal Asiat. Soc. Bengal* 41: 336–338.

----- 1872b. Notes on the Asiatic species of *Taphozous* Geoff. *Proceedings Asiat. Soc. Bengal* : 151–154.

----- 1872c. Pp. 220–223 in Stoliczka, F. 1872 (*q. v.*).

----- 1872d. Notes on some bats . . . in the northwestern Himalaya. *Proceedings Asiat. Soc. Bengal* : 208–210.

----- 1873a. Notes on the Pteropi of India . . . *Proceedings Asiat. Soc. Bengal* : 147–148.

----- 1873b. On the Pteropidae of India . . . *Journal Asiat. Soc. Bengal* 42: 194–205, pl. xiv.

----- 1873c. Description of a remarkable new species of *Molossus* (*Nyctinomus*) . . . *Proceedings Asiat. Soc. Bengal* : 22–23.

----- 1873d. Description of a new species of *Vespertilio* from the northwestern Himalaya. *Journal Asiat. Soc. Bengal* 42,2: 205–206.

----- 1873e. On the genera *Murina* and *Harpyiocephalus* of Gray. *Proceedings Asiat. Soc. Bengal* : 107–110.

----- 1874a. On the Asiatic species of Molossi. *Journal Asiat. Soc. Bengal* 43,2: 142–144.

----- 1874b. List of Chiroptera inhabiting the Khasia Hills . . . *Journal Asiat. Soc. Bengal* 43,2: 234–236.

----- 1874c. Descriptions of new species of Chiroptera from India and Yunan. *Journal Asiat. Soc. Bengal* 43,2: 237–238.

----- 1874d. *Catalogue of the Chiroptera in the Indian Museum, Calcutta* (*n. v.*).

----- 1875a. Description of a new species of *Taphozous* from Labuan. *Annals Mag. nat. Hist.* 16: 232.

----- 1875b. Conspectus of the suborders, families and genera of Chiroptera . . . *Annals Mag. nat. Hist.* 16: 345–357.

----- 1875c. On the genus *Scotophilus* . . . *Proceedings zool. Soc. Lond.* : 368–373.

----- 1875d. Postscript to the list of Chiroptera inhabiting the Khasia Hills. *Proceedings Asiat. Soc. Bengal* : 85.

----- 1875e. Descriptions of new or little-known species of bats. *Proceedings zool. Soc. Lond.* : 470–474.

----- 1875f. Description of new species of Vespertilionidae. *Annals Mag. nat. Hist.* 16: 260–262.

----- 1876a. *Monograph of the Asiatic Chiroptera and catalogue of the species of bats in the collection of the Indian Museum, Calcutta*. London.

----- 1876b. A monograph of the genus *Taphozous* Geoff. *Proceedings zool. Soc. Lond.* (1875): 546–556.

----- 1877. Notes on a collection of Chiroptera from India and Burma . . . *Journal Asiat. Soc. Bengal* 46,2: 310–313.

----- 1878. *Catalogue of the Chiroptera in the collection of the British Museum*. London: British Museum.

----- 1879. Notes on recent additions to the collection of Chiroptera in the Muséum d'Histoire Naturelle at Paris . . . *Proceedings zool. Soc. Lond.* (1878); 873–880.

----- 1880. Description of a new species of *Cynopterus* (*Ptenochirus*) from Sarawak. *Annals Mag. nat. Hist.* 6: 163–164.

----- 1887. Description of new species of Soricidae . . . *Annali Mus. civ. Stor. nat. Giacomo Doria* 4: 564–657.

----- 1888. Description of two new species of Indian Soricidae. *Annals Mag. nat. Hist.* 1: 427–429.

----- 1890a. *Monograph of the Insectivora*, part 3: pls 23–28. London: Gurney & Jackson.

----- 1890b. Descriptions of new species of *Crocidura*. *Annals Mag. nat. Hist.* 6: 494–497.

----- 1890c. Description of a new species of *Crocidura* from the Amur Region. *Annals Mag. nat. Hist.* 5: 31–33.

Dolgov, V. A. 1974. Diagnosis of *Crocidura suaveolens* and *C. leucodon*. *Zoologicheskii Zh.* 53: 912–918.

Dollman, G. 1909. A new species of *Presbytis* . . . *Annals Mag. nat. Hist.* 4: 204–205.

----- 1912. *Rhinopithecus avunculus* sp. n. A new snub-nosed monkey. *Abstracts Proc. zool. Soc. Lond.* : 18; *Proceedings zool. Soc. Lond.* : 503–504.

----- 1930. On mammals obtained by Mr Shaw Mayer in New Guinea. *Proceedings zool. Soc. Lond.* : 429–435, 4pls.

Domico, T. 1988. *Bears of the world*. New York, Oxford: Facts on File, 189pp.

Dorst, J. 1947. Une nouvelle chauve-souris de l'Indochine française. *Bulletin Mus. nat. Hist. Paris* 19: 436–437.

----- 1954. Précisions sur un Rhinolophidé rare du Tonkin. *Mammalia* 18: 212–215.

D'Souza, F. 1974. A preliminary field report on the lesser tree shrew *Tupaia minor*. Pp. 167–182 in Martin, R. D. *et al.* (eds), *Prosimian biology*. London: Duckworth.

Dubois, E. 1892. Palaeontologische anderzoekingen op Java. *Verslagen Meinw. Batavia* : 10–14.

----- 1894. Pithecanthropus erectus, *eines menschenähnliche Uebergangsform aus Java*. Batavia, 39pp.

----- 1907. Eenige van Nederlandschen kant verkregen uitkomsten met betrekking tot de kennis der Kendeng-fauna . . . *Tijdschrift K. ned. aardrijksk. Genoot.* 24: 449–458.

----- 1908. Das geologische Alter der Kendeng- oder Trinil-Fauna. *Tijdschrift K. ned. aardrijksk. Genoot.* 25: 1235–1270.

----- 1921. The proto-Australian fossil man of Wadjak, Java. *Proceedings K. Ned. Akad. Wet.* 23: 1013–1051, 2pls.

Dufresne, C. 1797. Sur une nouvelle espèce de singe. *Bulletin Soc. philom. Paris* 1,7: 49.

Dukelski, N. M. 1928. Zwei neue Nagetiere aus dem Ussuri-Gebiet. *Zoologischer Anz.* 77: 39–45.

----- 1930. Zur Kentnis der Säugetierfauna Westsibiriens. *Zoologischer Anz.* 88: 75–84.

Duméril, A. M. C. 1806. *Zoologie analytique* . . . Paris, xxxii, 344pp.

Duncan, J. F. 1976. Karyotypes of four rats . . . from Sulawesi. *Cytologia* 41: 481–486.

Duncan, J. F. & Van Peenen, P. F. D. 1971. Karyotypes of ten rats . . . from southeast Asia. *Caryologia* 24: 331–346.

Duncan, J. F. *et al.* 1974. Karyotypes of five taxa of *Rattus* . . . from Indonesia. *Cytologia* 39: 295–302.

Durden, L. A. *et al.* 1990. Some ectoparasites of bats from Seram Island, Indonesia. *Entomological News* 101: 48–56.

Dutta, A. K. 1976. Occurrence of fossil lion and spotted hyena from Pleistocene deposits of Susunia, Bankura district, West Bengal. *Journal geol. Soc. India* 17: 386–391.

Duvernoy, -- 1842. Notices pur servir a la monographie du genre musaraigne *Sorex* Cuv. *Magasin Zool. Paris* 5: 48pp., pls 38–54.

Eaton, R. G. (ed.) 1973. *The World's cats. Vol. 1. Ecology and conservation*. Oregon: Wiley, viii, 349pp.

Eisenberg, J. F. & McKay, G. M. 1970. An annotated checklist of the recent mammals of Ceylon with keys to the species. *Ceylon J. Sci.,biol. Sci.* 8: 69–99.

Elbert, J. 1911. *Die Sunda-Expedition des Vereins* . . . vol. 1. Frankfurt am Main: Minjon.

Elera, C. de 1915. *Contribucion a la faune Filipina*. Manila (*n. v.*).

Ellerman, J. R. 1940,1941,1949. *The families and genera of living rodents, vols. 1,2, 3*. London: British Museum (Nat. Hist.), 689,690,191pp.

----- 1947a. A key to the Rodentia inhabiting India, Ceylon and Burma . . . *Journal Mammal.* 28: 357–387.

----- 1947b. Notes on some Asiatic rodents in the British Museum. *Proceedings zool. Soc. Lond.* 117: 259–271.

----- 1947c. Further notes on two little-known Indian murine genera . . . *Annals Mag. nat. Hist.* 13: 204–208.

----- 1948. Key to the rodents of South-West Asia . . . *Proceedings zool. Soc. Lond.* 118: 765–816.

----- 1961. *The fauna of India. Mammalia* (*2nd ed.,vol. 3 Rodentia*). Dehli, 884pp. (2 vols).

Ellerman, J. R. & Morrison-Scott, T. C. S. 1951. *Checklist of Palaearctic and Indian mammals* . . . London: British Museum (Natural History) (2nd ed., 1966).
----- 1952. Checklist of Palaearctic and Indian mammals—new name. *Journal Mammal.* 33: 399.
----- 1955. *Supplement to Chasen (1940) A handlist of Malaysian mammals* . . . London: British Museum (Natural History), 66pp.
Elliot, D. G. 1882–83. *A monograph of the Felidae* . . . London.
----- 1896. On sundry collections of mammals received by the Field Columbian Museum . . . *Publications Field Mus. zool. Ser.* 1: 67–82, pls 6–13.
----- 1897. Remarks upon two species of deer of the genus *Cervus*. *Publications Field Mus. zool.* 1,7:[159].
----- 1906. Description of an apparently new species of monkey of the genus *Presbytis* from Sumatra . . . *Proceedings biol. Soc. Wash.* 19: 49–50.
----- 1909. Descriptions of apparently new species and subspecies of monkeys . . . *Annals Mag. nat. Hist.* 4: 244–274.
----- 1910a. The genus *Presbytis* . . . *Bulletin Am. Mus. nat. Hist.* 28: 151–154.
----- 1910b. Descriptions of some new species of monkeys of the genera *Pithecus* and *Pygathrix* . . . *Proceedings US nat. Mus.* 38: 343–352.
----- 1913. A review of the Primates. *Monographs Am. Mus. nat. Hist.* 1: cxxvi, 317, xxxviii pp, 32pls; 2: xviii, 382, xxvi pp, 39pls; 3: xiv, 262, clxvii pp, 39pls.
Elliot, O. 1971. Bibliography of the tree shrews 1780–1969. *Primates* 12: 323–414.
Elliot, O. S. *et al.* 1969. Karyological study of *Tupaia* from Thailand. *Journal Heredity* 60: 97–100, 153–157.
Elliot, W. 1839. A catalogue of the species of Mammalia found in the southern Mahratta country . . . *Madras J. Litt. Sci.* 10: 92–108, 205–233, 1pl.
Ellis, W. 1860. In Gray, J. E., Early notice of the *Tupaia* found in Pulo Condore. *Annals Mag. nat. Hist.* 5: 71.
Epstein, H. 1969. *Domestic animals of China*. Farnham Royal: Commonwealth Agric. Bur., xviii, 166pp, 205pls.
----- 1984. Ass, mule and onager. Pp. 174–184 in Mason, I. L. 1984a (*q. v.*).
Epstein, H. & Bichard, M. 1984. Pig. Pp. 145–162 in Mason, I. L. 1984a (*q. v.*).
Epstein, H. & Mason, I. L. 1984. Cattle. Pp. 6–27 in Mason, I. L. 1984a (*q. v.*).
Erxleben, J. C. P. 1777. *Systema regni animales . . . Classis I. Mammalia*. Lipsia: xlviii, 636pp.
Eschricht, D. F. 1849. Undersögelser over hvaldyrene. *Kongelige Danske Vidensk. Selsk. Skr.* 5: 85–138.
Eschscholtz, J. F. von 1821. Beschreibung einer neuen Affengattung, *Presbytis mitrata*. In Kotzebue, O. von *Entdeckungs-Reise in die Süd-See* . . . Weimar: Hofman.
----- 1831. *Zoologischer Atlas . . . um die Welt*. Berlin.
Evans, P. G. H. 1989. *The natural history of whales and dolphins*. London: Christopher Helm, 343pp.
Everts, W. 1968. Beitrag zur Systematik der Sonnendachse. *Zeitschrift Säugetierk.* 33: 1–19.
Eydoux, F. & Gervais, P. 1836a. *Voyage autour du monde . . . sur La Favorite . . . Zoologie*. Paris.
----- 1836b. Voyage autour du monde de la corvette Favorite . . . *Magazin zool. Paris* 6: 7–8.
Eydoux, F. & Souleyet, F. A. 1841–42. *Voyage autour du monde . . . La Favorite . . . Zoologie*. Paris, pp. 1–106 (1841), 107–328 (1842).

Fa, J. E. 1989. The genus *Macaca*: a review of taxonomy and evolution. *Mammal Rev.* 17: 45–81.
Falconer, H. 1859. *Descriptive catalogue of the fossil remains of Vertebrata of the Asiatic Society of Bengal*. Calcutta, 261pp.
----- 1868. *Palaeontological memoirs and notes . . . Vol. 1. Fauna antiqua sivalensis*. London: Hardwicke, 590 pp.
Falconer, H. & Cautley, P. T. 1836a. Note on the fossil hippopotamus of the Siwalik Hills. *Asiatic Res.* 19: 39–53.
----- 1836b. Note on the fossil camel of the Siwalik Hills. *Asiatic Res.* 19: 115–134, pls xx, xxi.
----- 1836c. Note on the *Felis cristata*, a new fossil tiger, from the Siwalik Hills. *Asiatic Res.* 19: 135–142.
----- 1843. On some fossil remains of . . . giraffe, from the Siwalik Hills . . . *Proceedings geol. Soc.* 4: 235–249, 2pls.
----- 1846–49. *Fauna antiqua sivalensis, being the fossil zoology of the Sewalik Hills . . . Illustrations*.
Fan, N. & Shi, Y. 1982. A revision of the zokors of subgenus *Eospalax*. *Acta theriol. sin.* 2: 183–197.
Farid Ahsan, M. D. 1989. Occurrence of the crab-eating mongoose (*Herpestes urva*) in Sylhet, Bangladesh. *Bangladesh J. Zool.* 17: 87–88.
Feiler, A. 1977. Uber die intraspecifische Variation des *Phalanger ursinus* . . . *Zoologische Abh. st. Mus. Tierk. Dresden* 34: 187–197.
----- 1978a. Bemerkungen über *Phalanger* der "*Orientalis*- Gruppe" . . . *Zoologische Abh. st. Mus. Tierk. Dresden* 34: 385–395.
----- 1978b. Uber artliche Abgrenzung . . . bei *Phalanger maculatus*. *Zoologische Abh. st. Mus. Tierk. Dresden* 35: 1–30.
----- 1978c. Zur morphologischen Characteristik des *Phalanger celebensis* . . . *Zoologische Abh. st. Mus. Tierk. Dresden* 35: 161–168.
----- 1980. *Taphozous saccolaimus* Temminck, 1841 auf Sulawesi . . . *Zoologische Abh. st. Mus. Tierk. Dresden* 36: 225–228.
----- 1981. Weitere Nachweise von *Hipposideros inexpectatus* . . . *Faunistische Abh. st. Mus. Tierk. Dresden* 8: 143–145.
----- 1990. On the mammals of the Sangihe and Talaud Islands . . . *Zoologische Abh. st. Mus. Tierk. Dresden* 46: 75–94.
Feldhammer, G. A. 1980. *Cervus nippon*. *Mammalian Species* 128: 7pp.
Felten, H. 1960. Festlegung eines Lectotypus von *Pteropus modiglianii* Thomas . . . *Senckenbergiana biol.* 41: 265–266.
----- 1971. Eine neue Art der Fledermaus-Gattung *Eptesicus*. *Senckenbergiana biol.* 52: 371–376.
----- 1977. In Felten, H. *et al.*, Zur Kleinsäugerfauna West- Anatoliens. IIIa. *Senckenbergiana biol.* 58: 1–44.
Feng, Z. & Zheng, C. 1985. Studies on the pikas (genus *Ochotona*) of China . . . *Acta theriol. sin.* 5: 269–285.
Feng, Z-j. *et al.* 1986. [*The mammals of Xizang*.] Beijing: Science Press, 421pp.
Findley, J. S. 1972. Phenetic relationships among bats of the genus *Myotis*. *Systematic Zool.* 21: 31–52.
Fischer, G. 1803. *Das National Museum der Naturgeschichte zu Paris* . . . vol. 2. Frankfurt am Main.
----- 1804. *Anatomie der Maki, und der ihnen verwandten Thiere*. Frankfurt: Andreäische Buchhdl., xxiv, 194pp.
----- 1813. *Zoognosia tabulis synopticis illustrata* vol. 2. Moscow: Vsevolozsky, xxiii, 605pp.
----- 1814. Ibid.,vol. 3, 732pp.
----- 1817. Adversaria zoologica. *Memoires Soc. Nat. Moscou* 5: 368–428.
Fischer, J. B. 1829. *Synopsis mammalium*. Stuttgart, xlii, 527pp.
Fitzgibbon, J. 1966. On a new species of *Salpingotus* . . . from north-western Baluchistan. *Mammalia* 30: 431–440.
Fitzinger, L. J. (F. J.) 1855. *Wissenschaftliche-populäre Naturgeschichte der Säugethiere*, 1. Wien.
----- 1860. *Ibid*, 6.
----- 1861. Die Ausbeute der österreichischen Naturforscher an Säugethieren . . . Novara. *Sitzungsberichte Akad. Wiss. Wien* 42(1860): 383–416.
----- 1866. In Heuglin, T. von, Systematische Ubersicht der Säugethiere Nordost-Afrika's . . . *Sitzungsberichte Akad. Wiss. Wien* 54,1: 537–611.
----- 1867. Die naturliche Familie der Igel (Erinacei) . . . *Sitzungsberichte Akad. Wiss. Wien* 56,1: 844–890.
----- 1868a. Kritische Untersuchungen über die . . . Soriciden . . . *Sitzungsberichte Akad. Wiss. Wien* 57: 121–180.
----- 1868b, 1869a. Revision der zur natürlichen Familie der Katzen (Feles) gehörigen Formen. *Sitzungsberichte Akad. Wiss. Wien* 58: 421–519; 60: 173–262.
----- 1869b. Die Gattungen der Familie der Antilopen . . . *Sitzungsberichte Akad. Wiss. Wien.* 59,1: 128–182.
----- 1870. Kritische Durchsicht der Ordnung der Flatterthiere. *Sitzungsberichte Akad. Wiss. Wien* 60(1869): 823–890; 61: 123–198; 62: 13–144, 211–317, 353–438, 527–579.
----- 1872. Die natürliche Familie der Schuppenthiere. *Sitzungsberichte Akad. Wiss. Wien* 65,1: 9–83.
----- 1874. Die Gattungen der Familie der Hirsche (*Cervi*) . . . *Sitzungsberichte Akad. Wiss. Wien* 68,1: 332–362.
----- 1875. Kritische Untersuchungen über die Arten der natürlichen Familie der Hirsche (*Cervi*) II. *Sitzungsberichte Akad. Wiss. Wien* 70(1874): 296–298.
Flannery, T. *et al.* 1987. The phylogenetic relationships of living phalangerids . . . Pp. 477–506 in Archer, M. (*q. v.*).
Fleay, D. 1947. *Gliders of the gum trees*. Melbourne: Bread and Cheese Club, 113pp.
Fleetwood, J. D. L. 1963. Notes on mammals new to Kenya Colony. *Durban Mus. Novit.* 6: 203–211.
Fleming, J. 1822. *The philosophy of zoology* vol. 2. Edinburgh: Constable, lii, 618pp.
Flerov, C. 1929. Preliminary note on the diagnostic characters of the genus *Moschus* . . . *Doklady Akad. Nauk SSSR* 1928A: 515–519.
Flerov, K. K. 1952. *Fauna of the USSR. Mammals vol. 1 (2), Musk deer and deer*. Moscow & Leningrad. (English translation: Jerusalem, 1960).
Flower, W. H. 1865. Notes on the skeletons of whales . . . *Proceedings zool. Soc. Lond.* (1864): 384–420.
----- 1884. On the characters and divisions of the family Delphinidae. *Proceedings zool. Soc. Lond.* (1883): 466–519.
Flower, W. H. & Lydekker, R. 1891. *An introduction to the study of mammals* . . . London, 763pp.
Flux, J. E. C. 1990. The Sumatran rabbit *Nesolagus netscheri*. Pp. 137–139 in Chapman, J. A. & Flux, J. E. C. (*q. v.*).
Flynn, J. J. *et al.* 1988. Phylogeny of the Carnivora. Pp. 73–116 in Benton, M. J., 1988b (*q. v.*).
Flynn, L. J. 1982a. Systematic revision of the Siwalik Rhizomyidae. *Geobios, Lyon* 15: 327–389.

----- 1982b. A revision of fossil rhizomyid rodents from northern India. *Geobios, Lyon* 15: 583–588.

----- 1983. Sur l'age de la faune de vertebrès du Bassin de Bamian, Afghanistan. *Compte r. Acad. Sci. Paris* 297: 411–414.

Flynn, L. J. *et al.* 1990. Additional fossil rodents from the Siwalik Beds of India. *Proceedings K. ned. Akad. Wet.* 93: 7–20.

Flynn, R. W. & Tajuddin Abdullah, M. 1984. Distribution and status of the Sumatran rhinoceros in Peninsular Malaysia. *Biological Conserv.* 28: 253–273.

Focillon, A. 1850. Du genre pangolin (*Manis* Linn.) . . . sec. 2. *Revue Mag. Zool.* 2: 513–534, pls 10,11.

Fooden, J. 1964. Rhesus and crab-eating macaques: intergradation in Thailand. *Science NY* 143: 363–365.

----- 1969. Taxonomy and evolution of the monkeys of Celebes. *Bibliotheca primatol.* 10: 148pp.

----- 1975. Taxonomy and evolution of liontail and pigtail macaques. *Fieldiana Zool.* 67: 169pp.

----- 1976. Provisional classification and key to living species of macaques . . . *Folia primatol.* 25: 225–236.

----- 1979. Taxonomy and evolution of the *sinica* group of macaques: 1. Species and subspecies accounts of *Macaca sinica*. *Primates* 20: 109–140.

----- 1980. Classification and distribution of living macaques. Pp. 1–8 in Lindburg, D. G. (*q. v.*).

----- 1981. Taxonomy and evolution of the *sinica* group of macaques. 2. Species and subspecies accounts of the Indian bonnet macaque, *Macaca radiata*. *Fieldiana Zool.* 9: 52pp.

----- 1982. Taxonomy and evolution of the *sinica* group of macaques. 3. Species and subspecies accounts of *Macaca assamensis*. *Fieldiana Zool.* 10: 52pp.

----- 1983. Taxonomy and evolution of the *sinica* group of macaques. 4. Species account of *Macaca thibetana*. *Fieldiana Zool.* 17: 20pp.

----- 1986. Taxonomy and evolution of the *sinica* group of macaques. 5. Overview of natural history. *Fieldiana Zool.* 29: 22pp.

----- 1987. Type locality of *Hylobates concolor leucogenys*. *American J. Primatol.* 12: 107–110.

----- 1988. Taxonomy and evolution of the *sinica* group of macaques. 6. Interspecific comparison and synthesis. *Fieldiana Zool.* 45: 1–44.

----- 1990. The bear macaques, *Macaca arctoides* . . . *Journal Human Evol.* 19: 607–686.

----- 1991a. Systematic review of Philippine macaques . . . *Fieldiana Zool.* 64: 44pp.

----- 1991b. Eastern limit of distribution of the slow loris, *Nycticebus coucang*. *International J. Primatol.* 12: 287–290.

Fooden, J. & Feiler, A. 1988. *Pygathrix nemaeus* in Hainan? New evidence, no resolution. *International J. Primatol.* 9: 275–279.

Fooden, J. & Lanyon, S. M. 1989. Blood-protein allele frequencies and phylogenetic relationships in *Macaca*: a review. *American J. Primatol.* 17: 209–241.

Fooden, J. *et al.* 1981. Redefinition of rhesus macaque-bonnet macaque boundary in peninsular India . . . *Journal Bombay nat. Hist. Soc.* 78: 463–474.

----- 1985. The stumptail macaques of China. *American J. Primatol.* 8: 11–30.

Foote, R. B. 1874. Fauna of the Indian fluviatile deposits. *Rhinoceros deccanensis*, a new species . . . *Memoirs geol. Surv. India* 10,1, 1: 17pp.

Fox, M. W. (ed.) 1975. *The wild canids* . . . New York etc.: Van Norstrand Reinhold, 508pp.

Francis, C. M. 1989. Notes on fruit bats . . . from Malaysia and Brunei. *Canadian J. Zool.* 67: 2878–2882.

Francis, C. M. & Hill, J. E. 1986. A review of the Bornean *Pipistrellus* . . . *Mammalia* 50: 43–55.

Fraser, F. C. 1957. A new Sarawak Dolphin. *Sarawak Mus. J.* 7 (1956): 478–503.

Fraser, L. 1849. *Zoologia typica* . . . London.

Fredga, K. 1972. Comparative chromosome studies in mongooses, 1. *Hereditas* 71: 1–74, 24pls.

Frick, F. 1969. Die Hohenstufenverteilung der nepalischen Säugetiere. *Säugetierkundliche Mitt.* 17: 161–173.

Frisch, J. L. 1775. *Das Natur-System der Vierfüssigen Thiere*. Golgau. Rejected for nomenclatorial purposes: Opinion 258 (1954).

Fritze, A. 1894. Die Fauna der Liu-Kiu-Insel Okinawa. *Zoologische Jb. Syst.* 7: 852–926, 12pls.

Fry, T. B. 1931. Proposed classification of the smaller Indian field (or jungle) mice. *Journal Bombay nat. Hist. Soc.* 34: 916–921.

Furley, C. W. *et al.* 1988. Systematics and chromosomes of the Indian gazelle, *Gazella bennetti* . . . *Zeitschrift Säugetierk.* 53: 48–54.

Gadi, I. K. & Sharma, T. 1983. Cytogenetic relationships in *Rattus, Cremnomys, Millardia, Nesokia* and *Bandicota* . . . *Genetica* 61: 21–40.

Gaisler, J. 1970. The bats (Chiroptera) collected in Afghanistan. *Prirodovedne Pr. Cesk. Akad. Ved.* 4,6: 1–56, 8pls.

----- 1971. Systematic review . . . of the bats hitherto recorded in Afghanistan. *Zoologické Listy* 20: 97–110.

Gambell, R. 1985. Sei whale—*Balaenoptera borealis* and fin whale —*Balaenoptera physalus*. Pp. 155–192 in Ridgeway, S. H. & Harrison, R. (*q. v.*).

Gao, Y. 1985. Classification and distribution of the musk deer (*Moschus*) in China. Pp. 113–116 in Kawamichi, T. (*q. v.*).

Gao, Y. *et al.* 1987. *Fauna Sinica, Mammalia, vol. 8, Carnivora*. Beijing: Science Press, v, 377pp, 10pls, in Chinese.

Garg, G. S. & Sharma, T. 1972. Chromosomes of the flying squirrel *Hylopetes a. alboniger* . . . *Indian Biologist* 4: 45–49.

Gaur, R. 1987a. New and additional Suidae . . . from the Upper Siwalik subgroup . . . *Revista ital. Paleont. Stratigr.* 93: 127–143.

----- 1987b. *Environment and ecology of early man in northwest India* . . . Dehli: B. R. Publishing Co., 252pp (*n. v*).

Gaur, R. & Chopra, S. R. K. 1984. On a new subspecies of *Equus* from Pinjor Formation of Upper Siwaliks . . . *Journal palaeont. Soc. India* 29: 19–25.

Gee, E. P. 1956. A new species of langur in Assam. *Journal Bombay nat. Hist. Soc.* 53: 252–254.

Gemmeke, H. & Niethammer, J. 1982. Zur Charakterisierung der Waldmaus (*Apodemus*) Nepals. *Zeitschrift Säugetierk.* 47: 33–38.

----- 1984. Zur Taxonomie der Gattung *Rattus* . . . *Zeitschrift Säugetierk.* 49: 104–116.

Geoffroy (-Saint Hilaire), E. 1796a. Mammifères. Mémoire sur les rapports naturels des Makis Lemur . . . *Magazin encycl.* (2) 1: 20–50, 1pl.

----- 1796b. Extrait d'un mémoire sur le *Myrmecophaga capensis* Gme. *Bulletin Sci. Soc. philom. Paris* 1: 102–103.

----- 1803. *Catalogue des mammifères du Muséum National d'Histoire Naturelle*. Paris. Has been considered unavailable for nomenclatural purposes but reasons for accepting it were given by Hill (1980b).

----- 1804. Note sur un nouveau genre de mammifère, de l'ordre des rongeurs, sous le nom d'hydromys. *Bulletin Soc. Philomath. Paris* 3(93): 253–254.

----- 1806. Mémoire sur le genre et les espèces de vespertilion. *Annales Mus. Hist. nat. Paris* 8: 187–205, 3pls.

----- 1810a. Description des rousettes et des cephalotes . . . *Annales Mus. Hist. nat. Paris* 15: 86–108, 4pls.

----- 1810b. Sur les phyllostomes et les mégadermes . . . *Annales Mus. Hist. nat. Paris* 15: 157–198, 3pls.

----- 1812. Tableau des quadrumanes . . . *Annales Mus. Hist. nat. Paris* 19: 85–122; 156–170.

----- 1813a. De l'organisation et de la détermination des Nyctères . . . *Annales Mus. Hist. nat. Paris* 20: 11–20, 1pl.

----- 1813b. Sur un genre de chauve-souris sous le nom de Rhinolophes. *Annales Mus. Hist. nat. Paris* 20: 254–266, 2pls.

----- 1818. *Description de l'Egypte* . . . *Description des mammifères*. 2. Paris.

----- [1828]. *Cours de l'histoire naturelle des mammifères* . . . Paris.

Geoffroy (-Saint-Hilaire), E. & Cuvier, G. 1795. Mémoire sur une nouvelle division des mammifères . . . *Magazin encycl.* 2: 164–190.

----- 1819–1842. *Histoire naturelle des mammifères* . . . Paris. (Many names usually attributed to this are invalid since they were not allocated to a genus.)

Geoffroy (-Saint Hilaire), I. 1811. Memoire sur les espèces des genres musaraigne et mygale. *Annales Mus. Hist. nat. Paris* 17: 169–194.

----- 1826. Lièvre *Lepus*. Pp. 378–387 in Bory de Saint Vincent (ed.) *Dictionnaire classique d'histoire naturelles* . . . vol. 9.

----- 1827. Memoire sur quelques espèces nouvelles ou peu connues de genre musaraigne. *Memoires Mus. Hist. nat. Paris* 15: 117–144, pl. 4.

----- 1828. *Dictionnaire classique d'histoire naturelles* . . . vol. 14.

----- 1831–34. Mammifères, in Bélanger, C., *Voyage aux Indes-Orientales* . . . Paris: Bertrand. For dates see Sherborn & Woodward (1901).

----- 1836. In Gervais, P. *Résumé des leçons de mammalogie* . . . Paris: Museum d'Histoire Naturelle.

----- 1838. Notes sur des rongeurs épinaux . . . *Annales Sci. nat., Zool.* 10: 122–127.

----- 1843a. Description des mammifères . . . *Archives Mus. Hist. nat. Paris* 2: 485–592, 6pls.

----- 1843b. Sur les singes de l'ancien monde . . . *Compte r. hebd. Acad. Sci. Paris* 15 (1842): 716–720.

----- 1844. IV Mammifères, in Jacquemont, V. *Voyage dans l'Inde*. Paris.

----- 1850. Note sur plusieurs espèces nouvelles de mammifères de l'ordre des Primates. *Compte r. hebd. Acad. Sci. Paris* 31: 873–876.

----- 1851. *Catalogue méthodique de la collection des mammifères*. Part 1. Paris: Museum d'Histoire Naturelle.

George, -- 1869. Études zoologiques sur les hémiones . . . *Annales Sci. nat.,Zool.* 12: 5–48.

George, G. 1987. Characterisation of the living species of cuscus . . . Pp. 507–526 in Archer, M. (ed.) (*q. v.*).

Gerbe, Z. 1880. Espèce nouvelle de vespertilionien de Chine. *Bulletin Soc. zool. France* 5: 70–71.

Gervais, P. (F. L.) 1836. *Dictionaire pittoresque Hist. nat.* 4: 617.

----- 1841. In Eydoux, F. & Souleyet, F. A., Zoologie, in *Voyage autour du monde . . . sur la Bonite, vol. 1*. Paris.

----- 1850. Mémoire sur la famille des cétacés ziphioides. *Annales Sci. nat., Zool.* 14: 5–17.

Gervais, P. (F. L.) 1854–5. *Histoire naturelle des mammifères* . . . Paris, 344pp.

----- 1859. *Zoologie et paléontologie françaises* . . ., 2nd ed. Paris.

----- 1870. Mémoire sur les formes cérébrales propres aux carnivores . . . *Nouvelles Arch. Mus. Hist. nat. Paris* 6: 103–162.

----- 1883. Sur une nouvelle espèce du genre Mégaptère . . . *Compte r. Acad. Sci. Paris* 97: 1566–1569.

Gheyara, F. D. 1956. Rats in human dietary. *Journal Bombay nat. Hist. Soc.* 53: 689–690.

Ghosal, D. K. 1973. Occurrence of *Nesokia indica* . . . in Nepal. *Labdev* (B) 11: 85.

Ghose, R. K. 1964. A new rat of the genus *Rattus* from Darjeeling District . . . *Proceedings zool. Soc. Calcutta* 17: 193–197.

----- 1965. A new species of mongoose . . . from West Bengal, India. *Proceedings zool. Soc. Calcutta* 18: 173–178.

----- 1967. Notes on the taxonomy and distribution of the Pachmarhi hare, *Lepus nigricollis mahadeva* . . . *Journal Bombay nat. Hist. Soc.* 64: 355–358.

----- 1976a. Further new records of mammals from Rajasthan, India. *Journal zool. Soc. India* 26: 149–150.

----- 1976b. A record of the shrew, *Suncus murinus griffithi* from Darjeeling . . . *Journal zool. Soc. India* 26: 151–152.

Ghose, R. K. & Saha, S. S. 1981. Taxonomic review of Hodgson's giant flying squirrel, *Petaurista magnificus* . . . *Journal Bombay nat. Hist. Soc.* 78: 93–102.

Ghose, R. K. *et al.* 1972. A new Pleistocene buffalo from Tiruchirapalli district, Tamil Nadu, India. *Journal zool. Soc. India* 24: 97–102.

Ghosh, A. K. & Biswas, B. 1976. A note on the ecology of golden langurs: *Presbytis geei* Khajuria. *Journal Bombay nat. Hist. Soc.* 72: 524–527.

Ghosh, A. K. & Chowdhury, M. 1986. Record of a plumbeous dolphin (*Sousa plumbea* Cuvier) from Duduya River, north Bengal, India. *Record zool. Surv. India* 83: 49–52.

Ghosh, M. K. 1989. New locality records for *Myotis montivagus peytoni* . . . *Journal Bombay nat. Hist. Soc.* 86: 93–94.

Giebel, C. G. 1855. *Odontographie* . . . Leipzig: Abel.

----- 1863. Neue Spitzratte, *Gymnura alba*, von Borneo. *Zeitschrift ges. naturw. Halle* 22: 277–290, pls 1–2.

Giglioli, E. H. 1870. *Note intorno alla distribuzione della fauna vertebtrata nell'oceano* . . . Firenze.

Gill, T. 1865. On two species of Delphinidae from California . . . *Proceedings Acad. nat. Sci. Philad.* [17]: 177–178.

----- 1871. The sperm whales, giant and pygmy. *American Nat.* 4: 725–743.

----- 1872. Arrangement of the families of mammals. *Smithsonian misc. Collns* 11(1):vi, 98pp.

----- 1875. Synopsis of insectivorous mammals. *Bulletin US geol. geogr. Surv. Territ.* 2(2): 91–120.

Ginsburg, L. & Mein, P. 1987. *Tarsius thailandica* nov. sp., première Tarsiidae . . . fossile d'Asie. *Compte r. hebd. Acad. Sci.* Paris 304: 1213–1215.

Gistel, J. von 1848. *Naturgeschichte des Thierreichs* . . . . Stuttgart, 216pp, 32pls.

Gittelman, J. L. (ed.) 1989. *Carnivore behaviour, ecology and evolution.* Ithaca: Cornell U. P.

Glass, G. E. & Todd, N. B. 1977. Quasi-continuous variation of the second upper premolars in *Felis bengalensis* . . . *Zeitschrift Säugetierk.* 42: 36–44.

Glatson, A. R. (ed.) 1989. *Red panda biology.* Hague: SPB Academic Publishing, 187pp.

Gloger, C. W. L. 1841. *Gemeinnütziges Hand- und Hilfsbuch der Naturgeschichte* . . . *Band 1.* Breslau, xxxiv, 495 pp.

Glover, R. 1956. Notes on the sika deer. *Journal Mammal.* 37: 99–105.

Gmelin, S. G. 1770. De *Capra saiga* et *Erinaceo aurito. Novi Comment. Acad. Sci. imp. Petropol.* 14: 512–524.

----- 1788. *C. a. Linné* . . . *systema naturae*. 13th ed. vol. 1. Lugdun, 232pp.

Goldfüss, G. A. 1809. *Vergleichende Naturbeschreibung Säugethiere.* Erlangen.

----- 1820. *Handbuch der Zoologie* 2. Nürnberg.

Goldman, D. *et al.* 1989. Molecular genetic-distance estimates among the Ursidae . . . *Evolution* 43: 282–295.

Goldman, E. A. 1913. A new generic name for the Asiatic tapir. *Proceedings biol. Soc. Wash.* 26: 65–66.

Gong, Z-d. *et al.* 1989. New record of murid rodent from China— Brahma rat, *Niviventer brahma. Endemic Diseases Bull.* 4: 28–29.

Goodwin, R. E. 1979. The bats of Timor . . . *Bulletin Am. Mus. nat. Hist.* 163: 73–122.

Gould, J. 1850. On new species of Mammalia and birds from Australia. *Proceedings zool. Soc. Lond.* (1849): 109–112.

----- 1854. *The mammals of Australia.* 3. London.

Gray, A. P. 1972. *Mammalian hybrids: a check-list with bibliography.* Slough: Commonwealth Agricultural Bureaux, 262pp.

Gray, J. E. 1821. On the natural arrangement of vertebrose animals. *London Med. Reposit.* 15: 296–310.

----- 1824. A revision of the family Equidae. *Zoological J.* 1: 241–248, pl. ix.

----- 1825. Outine of an attempt at the disposition of the Mammalia into tribes and families . . . *Annals Phil.* 10: 337–344.

----- 1827a. A synopsis of the species of the class Mammalia. Vol. 5 in Griffith, *The animal kingdom* London: Whittaker, 391pp.

----- 1827b. Description of the skulls of two apparently undescribed species of dolphins . . . *Philosophical Mag.* 2: 375–376.

----- 1828–1830. *Spicilegia zoologica* London. Part 1:pp 1–8, pls 1–6 (1828); part 2:pp. 9–12, pls 7–11 (1830).

----- 1830–1835. *Illustrations of Indian zoology chiefly selected from the collection of Major-General Hardwicke, FRS.* London.

----- 1831a. *Zoological miscellany.* London.

----- 1831b. [Exhibition of Mammalia] *Proceedings zool. Soc. Lond.* : 94–96.

----- 1832. [Specimens and drawings of numerous animals referable to the genus *Paradoxurus*] *Proceedings zool. Soc. Lond.* : 63–68.

----- 1834. Characters of a new species of bat (*Rhinolophus*) . . . *Proceedings zool. Soc. Lond.* 52–53.

----- 1836. On the genus *Moschus* of Linnaeus . . . *Proceedings zool. Soc. Lond.* 4: 63–66.

----- 1837a. Description of some new or little known Mammalia . . . *Magazine nat. Hist.* 1: 577–587.

----- 1837b. [A series of Mammalia selected from the collection of the Society . . .] *Proceedings zool. Soc. Lond.* (1836): 87–88.

----- 1837d. [The horn of a deer supposed to come from India . . .] *Proceedings zool. Soc. Lond.* 5: 45.

----- 1838a. A revision of the genera of bats . . . *Magazine Zool. Bot.* 2: 483–505.

----- 1838b. [Arrangement of the Sorices.] *Proceedings zool. Soc. Lond.* (1837): 123–126.

----- 1838c. [Meeting of] June 27th, 1837. *Proceedings zool. Soc. Lond.* : 67–69.

----- 1842. Description of some new genera and fifty unrecorded species of Mammalia. *Annals Mag. nat. Hist.* 10: 255–267.

----- 1843a. *List of the specimens of Mammalia in the . . . British Museum.* London: British Museum, 216pp.

----- 1843b. Descriptions of some new genera and species of Mammalia . . . *Annals Mag. nat. Hist.* 11: 117–119.

----- 1844. Mammalia. In Hinds, R. B. *The zoology of the voyage of H. M. S. Sulphur* . . . Vol. 1. London.

----- 1846a. On the British Cetacea. *Annals Mag. nat. Hist.* 17: 82–85.

----- 1846b. *Catalogue of the specimens and drawings of Mammalia and birds of Nepal and Tibet* . . . London.

----- 1846c. On the cetaceous animals. Pp. 15–53, 37pls, in Richardson, J. & Gray, J. E. (eds) *The zoology of the voyage of HMS Erebus and Terror.* London: Longmans etc.

----- 1846d. New species of Mammalia. *Annals Mag. nat. Hist.* 18: 211–212.

----- 1847a. Characters of six new genera of bats . . . *Proceedings zool. Soc. Lond.* : 14–16.

----- [1847]b. *Catalogue of the specimens of Mammalia and birds of Nepal and Thibet presented by B. H. Hodgson* . . . London: British Museum. (Dated 1846, published Jan. 1847.)

----- 1847c. On the porcupines of the older or eastern continent. *Proceedings zool. Soc. Lond.* 15: 97–104.

----- 1847d. *List of the osteological specimens in the collection of the British Museum*. London: British Museum, 147pp.

----- 1848a (4 Mar.). *Literary Gazette* 1624: 167 (*n. v*).

----- 1848b (1 Aug.). Description of a new genus of insectivorous Mammalia, or Talpidae [sic], from Borneo. *Proceedings zool. Soc. Lond.* : 23–24.

----- 1850a. *Gleanings from the menagerie and aviary at Knowlesley Hall.* Knowlesley, 76pp, 59pls.

----- 1850b. *Catalogue of the specimens of Mammalia in the . . . British Museum, part 1.* London: British Museum, 153pp.

----- 1851. Notice of two Viverridae from Ceylon . . . *Proceedings zool. Soc. Lond.* : 131, pls 30,31.

----- 1852. *Catalogue of the specimens of Mammalia in the collection of the British Museum. Part III. Ungulata furcipeda.* London: British Museum, 286pp, 27pls.

----- 1853. Observations on some rare Indian animals. *Proceedings zool. soc. Lond.* : 190–192, pls 35–38.

----- 1854a. On the painted pig of the Cameroons (*Potamochoerus penicillatus*). *Proceedings zool. Soc. Lond.* (1852): 129–132.

----- 1854b. On a new species of *Rhinoceros. Proceedings zool. Soc. Lond.* : 250–251.

----- 1856. [Report of Zoological Society.] *Athenaeum* 1517: 1437.

----- 1857. On a new species of squirrel (*Sciurus macrotis*) from Borneo. *Proceedings zool. Soc. Lond.* (1856): 341, pl. xlvi.

----- 1858. Observations on the genus *Cuscus* . . . *Proceedings zool. Soc. Lond.* : 100–105, pls lxi, lxii.

----- 1860a. Note on the female of *Cuscus ornatus. Proceedings zool. Soc. Lond.* (1860, 1): 374, pl. 74.

----- 1860b. On some new species of Mammalia . . . from Cambojia. *Annals Mag. nat. Hist.* 6: 217–218.

----- 1860c. Description of a new species of squirrel (*Sciurus siamensis*). . . *Annals Mag. nat. Hist.* 5: 500–501.

----- 1861. List of Mammalia . . . in Cambodia. *Proceedings zool. Soc. Lond.* : 135–140.

----- 1862a. Additional observations on the genus *Cuscus. Proceedings zool. Soc. Lond.* (1861): 314–321.

----- 1862b. Notice of a new 'wild goat' (*Capricornis Swinhoei*) from the island of Formosa. *Annals Mag. nat. Hist.* 10: 320.

----- 1863a. Description of some new species of Mammalia. *Proceedings zool. Soc. Lond.* (1862): 261–263.

----- 1863b. *Catalogue of the specimens of Mammalia and birds of Nepal and Thibet presented by B. H. Hodgson . . . , 2nd edition.* London: British Museum.

----- 1864a. On the Cetacea which have been observed in the seas surrounding the British isles. *Proceedings zool. Soc. Lond.* : 195–248.

----- 1864b. A revision of the genera and species of ursine animals (Ursidae) . . . *Proceedings zool. Soc. Lond.* : 677–709.

----- 1865a. Notice of a species of *Tupaia* from Borneo . . . *Proceedings zool. Soc. Lond.* : 322.

----- 1865b. Revision of the genera and species of entomophagous Edentata . . . *Proceedings zool. Soc. Lond.* : 359–386, pls xvii-xix.

----- 1865c. Revision of the genera and species of Mustelidae . . . *Proceedings zool. Soc. Lond.* : 100–154, pl. 7.

----- 1865d. A revision of the genera and species of viverrine animals . . . *Proceedings zool. Soc. Lond.* (1864): 502–579.

----- 1865e. Notice of a new genus of delphinoid whales from the Cape of Good Hope . . . *Proceedings zool. Soc. Lond.* : 522–529.

----- 1865f. Notice of a new species of Australian sperm whale. *Proceedings zool. Soc. Lond.* : 439–442.

----- 1866a. A revision of pteropine bats (Pteropidae) . . . *Proceedings zool. Soc. Lond.* : 62–67.

----- 1866b. Notes on some Mammalia from Port Albany . . . *Proceedings zool. Soc. Lond.* : 219–221.

----- 1866c. A revision of the genera of Rhinolophidae . . . *Proceedings zool. Soc. Lond.* : 81–83.

----- 1866d. Synopsis of the genera of Vespertilionidae and Noctilionidae. *Annals Mag. nat. Hist.* 17: 89–93.

----- 1866e. On the species of porcupine in the gardens of the Society and in the British Museum. *Proceedings zool. Soc. Lond.* : 306–311, pl. 31.

----- 1866f. Notice of an ape (*Macaca inornatus*) . . . *Proceedings zool. Soc. Lond.* : 202–203, 1pl.

----- 1866g. Short account of part of a skeleton of a finner whale . . . from the coast of Formosa. *Proceedings zool. Soc. Lond.* (1865): 725–728.

----- 1866h. *Catalogue of seals and whales in the British Museum.* 2nd ed. London: British Museum, 402pp.

----- 1866i. Notes on the skulls of dolphins, or bottlenose whales, in the British Museum. *Proceedings zool. Soc. Lond.* : 211–216.

----- 1867a. Synopsis of the Asiatic squirrels (Sciuridae) in the British Museum . . . *Annals Mag. nat. Hist.* 20: 270–286.

----- 1867b. Notes on the variegated or yellow-tailed rats of Australia. *Proceedings zool. Soc. Lond.* : 597–600.

----- 1867c. Notes on the skulls of hares (Leporidae) and picas (Lagomyidae) . . . *Annals Mag. nat. Hist.* 20: 219–225.

----- 1867d. Notice of *Lutronectes whitleyi*, an otter from Japan. *Proceedings zool. Soc. Lond.* : 180–182.

----- 1867e. Notes on certain species of cats in the collection of the British Museum. *Proceedings zool. Soc. Lond.* : 394–405, pls 24,25.

----- 1867f. Notes on the skulls of the cats (Felidae). *Proceedings zool. Soc. Lond.* : 258–277.

----- 1868a. Synopsis of the species of pigs (Suidae) in the British Museum. *Proceedings zool. Soc. Lond.* : 17–49.

----- 1868b. Notice of a badger from China (*Meles chinensis*) . . . *Proceedings zool. Soc. Lond.* : 206–209.

----- 1868c. Notice of *Macacus lasiotis*, a new species of ape from China . . . *Proceedings zool. Soc. Lond.* : 60–61, pl. 6.

----- 1868d. *Synopsis of the species of whales and dolphins in the . . . British Museum.* London: Quaritch, 53pp, 37pls.

----- 1868e. Observation on the preserved specimens and skeletons of the Rhinocerotidae . . . *Proceedings zool. Soc. Lond.* (1867): 1003–1032.

----- 1869. *Catalogue of carnivorous, pachydermatous and edentate Mammalia in the British Museum.* London: British Museum.

----- 1870. Notes on the skulls of the genus *Orca* . . . *Proceedings zool. Soc. Lond.* : 70–77.

----- 1871a. The Chinese long-tailed goat antelope (*Urotragus caudatus*). *Annals Mag. nat. Hist.* 8: 371–372.

----- 1871b. *Supplement to the catalogue of seals and whales in the British Museum.* London: British Museum, 103pp.

----- [1871]c. *Catalogue of monkeys, lemurs and fruit-eating bats in the collections of the British Museum.* London: British Museum, viii, 137pp. (Dated 1870).

----- 1872. *Catalogue of ruminant Mammalia . . . in the British Museum.* London: British Museum, 102pp, 4pls.

----- 1873a. Notes on the rats. *Annals Mag. nat. Hist.* 12: 416–419.

----- 1873b. *Handlist of the edentate, thick-skinned and ruminant mammals in the British Museum.* London: British Museum, 176pp, 41pls.

----- 1873c. On the dentition of rhinoceroses . . . *Annals Mag. nat. Hist.* 11: 356–361, pl. xi.

----- 1873d. Observations on pigs . . . and their skulls . . . *Annals Mag. nat. Hist.* 11: 431–441.

----- 1874a. Description of a new species of cat (*Felis badia*) from Sarawak. *Proceedings zool. Soc. Lond.* : 322–323, pl. xlix.

----- 1874b. List of the species of feline animals (Felidae). *Annals Mag. nat. Hist.* 14: 351–356.

----- 1875. *Feresa attenuata. Journal Mus. Godeffroy* 4,8: 52 (p. 184 of volume).

Gregori, J. & Petrov, B. 1976. Scientific results of the Yugoslav 1972 Himalaya Expedition. Mammalia. *Razprave Slov. Akad. Znan. Umet.* 19: 1–20.

Griffith, E. 1821. *General and particular descriptions of the vertebrated animals . . .* London, 143pp.

Griffiths, T. A. *et al.* 1991. The systematic relationship of *Emballonura nigrescens* . . . *American Mus. Novit.* 2996.

Grigson, C. 1989. Complex cattle. *Asian wild Cattle spec. Group Newsl.* 2: 9–10.

Grine, F. E. *et al.* (eds) 1988. *Primate phylogeny*. Academic Press.

Gropp, A. *et al.* 1972. Robertsonian chromosomal variation in the long-tailed tree mouse (*Vandeleuria*). *Zeitschrift zool. Syst. EvolForsch.* 10: 210–214.

----- 1973. Chromosomal findings in the spiny mice of Thailand (genus *Mus*). *Zeitschrift Säugetierk.* 38: 159–168.

Groves, C. P. 1965. Description of a new subspecies of rhinoceros from Borneo . . . *Säugetierkundliche Mitt.* 13: 128–131.

----- 1967a. On the rhinoceroses of south-east Asia. *Säugetierkundliche Mitt.* 15: 221–237.

----- 1967b. Geographical variation in the hoolock or white-handed gibbon (*Hylobates hoolock* Harlan, 1834). *Folia primatol.* 7: 276–283.

----- 1968. A new subspecies of white-handed gibbon from northern Thailand . . . *Proceedings biol. Soc. Wash.* 81: 625–628.

----- 1969a. Systematics of the anoa. *Beaufortia* 17: 1–12.

----- 1969b. On the smaller gazelles of the genus *Gazella* . . . *Zeitschrift Säugetierk.* 34: 38–60.

----- 1971a. Species characters in rhinoceros horns. *Zeitschrift Säugetierk.* 36: 238–252.

----- 1971b. Systematics of the genus *Nycticebus*. Pp. 44–53 in Biegert, J. & Leutenegger, W. (eds) *Proceedings 3rd int. Congr. Primat., Zurich* 1970, *vol. 1. Taxonomy* . . . . Basel etc. : Karger.

----- 1971c. *Pongo pygmaeus. Mammalian Species* 4: 6pp.

----- 1972. Systematics and phylogeny of gibbons. Pp. 1–89 in Rumbaugh, D. M. (ed.) *Gibbon and siamang, vol. 1.* Basel: Karger.

----- 1974a. *Horses, asses and zebras in the wild.* Newton Abbot: David & Charles, 192pp.

----- 1974b. A note on the systematic position of the muntjac . . . *Zeitschrift Säugetierk.* 39: 369–372.

----- 1976. The origin of the mammalian fauna of Sulawesi (Celebes). *Zeitschrift Säugetierk.* 41: 201–216.

----- 1980a. Notes on the systematics of *Babyrousa* . . . *Zoologische Meded. Leiden* 55: 29–46.

----- 1980b. Speciation in *Macaca*: the view from Sulawesi. Pp. 84–124 in Lindburg, D. G. (*q. v.*).

----- 1981a. Ancestors for the pigs: taxonomy and phylogeny of the genus *Sus. Technical Bull., Dept. Prehist., Res. Sch. Pacific Stud., Aust. nat. Univ.* 3: 96pp.

----- 1981b. Systematic relationships in the Bovini . . . *Zeitschrift zool. Syst. EvolForsch.* 19: 264–278.

----- 1982a. Cranial and dental characteristics in the systematics of the Old World Felidae. *Carnivore* 5: 28–39.

----- 1982b. A note on geographical variation in the Indian blackbuck (*Antilope cervicapra*). *Record zool. Surv. India* 76: 125–138.

----- 1982c. Geographical variation in the barasingha or swamp deer (*Cervus duvauceli*). *Journal Bombay nat. Hist. Soc.* 79: 620–629.

----- 1983. Phylogeny of the living species of rhinoceros. *Zeitschrift zool. Syst. EvolForsch.* 21: 293–313.

----- 1984. A new look at the taxonomy and phylogeny of the gibbons. Pp. 542–561 in Preuschoft, H. *et al.* (eds) *The lesser apes. Evolutionary and behavioural biology*. Edinburgh: Edinb. Univ. Press.

----- 1985a. Plio-Pleistocene mammals in island southeast Asia. *Modern Quatern. Res. SE Asia* 9: 43–54.

----- 1985b. An introduction to the gazelles. *Chinkara* 1: 4–16.

----- 1986. The taxonomy, distribution and adaptations of Recent equids. Pp. 11–65 in Meadow, R. H. & Uerpmann, H-P. (eds) *Equids in the ancient world.* Wiesbaden: Reichert.

----- 1987. On the cuscuses . . . of the *Phalanger orientalis* group. Pp. 569–579 in Archer, M. (*q. v.*).

----- 1988. A catalogue of the genus *Gazella*. Pp. 193–198 in Dixon, A. & Jones, D. (eds) *Conservation and biology of desert antelopes.* London: Christopher Helm.

----- 1989. *A theory of human and primate evolution.* Oxford: Clarendon Press, 388pp.

Groves, C. P. & Feng, Z. 1986. The status of the musk-deer from Anhui Province, China. *Acta theriol. sin.* 6: 101–106.

Groves, C. P. & Grubb, P. 1982. The species of muntjac (genus *Muntiacus*) in Borneo . . . *Zoologische Meded. Leiden* 56: 203–214.

----- 1985. Reclassification of the serows and gorals . . . Pp. 45–50 in Lovari, S. (ed.) *The biology and management of mountain ungulates*. Beckenham: Croom Helm.

----- 1987. Relationships of living Cervidae. Pp. 21–59 in Wemmer, C. M. (ed.) *Biology and management of the Cervidae*. Washington: Smithsonian Institute.

Groves, C. P. & Guerin, C. 1980. Le *Rhinoceros sondaicus annamiticus* . . . d'Indochine . . . *Geobios, Lyon* 13: 199–208.

Groves, C. P. & Holthius, L. B. 1985. The nomenclature of the orang utan. *Zoologische Meded. Leiden* 59: 411–417.

Groves, C. P. & Kurt, F. 1972. *Dicerorhinus sumatrensis*. *Mammalian Species* 21: 6pp.

Groves, C. P. & Wang, W-x. 1990. The gibbons of the subgenus *Nomascus* . . . *Zoological Res.* 11: 147–154.

Groves, C. P. & Willoughby, D. P. 1981. Studies on the taxonomy and phylogeny of the genus *Equus*. 1. Subgeneric classification of the Recent species. *Mammalia* 45: 321–354.

Grubb, P. 1977. Notes on a rare deer, *Muntiacus feai*. *Annali Mus. Civ. Stor. nat. Genova* 81: 202–207.

----- 1982. The systematics of Sino-Himalayan musk deer (*Moschus*). *Säugetierkundliche Mitt.* 30: 127–135.

Grubb, P. & Groves, C. P. 1983. Notes on the taxonomy of the deer of the Philippines. *Zoologischer Anz.* 210: 119–144.

Gruber, U. F. 1969. Tiergeographische . . . Untersuchungen an kleinen Säugetieren in Ost-Nepal. *Khumbu Himal* 3: 197–312.

Gu, Y-m. 1980. A Pliocene macaque tooth from Hubei. *Vertebrata palasiat.* 18: 324–326.

Guedman, D. *et al.* 1989. Molecular genetic-distance estimates among the Ursidae . . . *Evolution* 43: 282–293.

Gueldenstaedt, A. I. 1770. Peregusna, nova mustelae species. *Novi Comment. Acad. Sci. Imp. Petropol.* 14,1: 441–455.

----- 1776. Chaus, animal feli affine descriptum. *Novi Comment. Acad. Sci. Imp. Petropol.* 20: 483–500, pls 14,15.

----- 1780. *Antilope subgutturosa* descripta. *Acta Acad. Scient. Imp. Petropol.* 1778,1: 251–274.

Günther, A. 1873. Description of three new species of flying squirrels . . . *Proceedings zool. Soc. Lond.* : 413–414, pls 37–38.

----- 1876. Remarks on some Indian and . . . Bornean mammals. *Proceedings zool. Soc. Lond.* : 424–428.

----- 1877. Report on some of the additions to the collections of Mammalia in the British Museum. *Proceedings zool. Soc. Lond.* (1876): 735–751.

----- 1879a. List of the mammals, reptiles and batrachians sent by Mr Everett from the Philippine Islands. *Proceedings zool. Soc. Lond.* : 74–79.

----- 1979b. Description of a new species of porcupine from the Philippine islands. *Annals Mag. nat. Hist.* 4: 106–107.

Guo, Z. *et al.* 1978. A new subspecies of sika deer from Sichuan. *Acta zool. sin.* 24: 187–192.

Gupta, V. J. & Sahni, A. 1981. *Theropithecus delsoni* a new cercopithecine species from the Upper Siwaliks of India. *Bulletin Indian geol. Assn* 14: 69–71.

Gureev, A. A. 1971. [*Shrew fauna (Soricidae) of the world*] Leningrad: Nauka, 253pp (in Russian).

----- 1979. [*Fauna SSSR*/ *mammals*] IV (2) *Insectivora*. Leningrad: Nauka, 502pp.

Gyldenstolpe, N. 1917. Zoological results of the Swedish zoological expeditions to Siam . . . V Mammals II. *Kungliga svenska VetenskAkad. Handl.* 57,2: 59pp, 6pls.

----- 1920. On a collection of mammals made in eastern and central Borneo . . . *Kungliga svenska VetenskAkad. Handl.* 60,6 (1919): 62pp, 6pls.

Haimoff, E. H. *et al.* 1984. A phylogeny and classification of gibbons . . . Pp. 614–632 in Preuschaft, H. *et al.* (eds). *The lesser apes; evolutionary and behavioural biology*. Edinburgh: Edinburgh Univ. Press.

Haldeman, S. S. 1841. [Written communication] *Proceedings Acad. nat. Sci. Philad.* 1: 127.

Hall, D. G. E. 1981. *A history of South-East Asia*. 4th ed. London: Macmillan, xiii, 1070pp.

Hall, E. R. 1981. *The mammals of North America*, vol. 1. New York etc. : Wiley, xv, 600,90pp.

Haltenorth, T. 1963. Klassifikation der Säugetiere: Artiodactyla. *Handbuch der Zoologie* (8) 32: 1–167.

Hamilton-Smith, E. 1984. Comments upon Maeda and vagrant *Miniopterus*. *Australian Bat Res. News* no. 20.

Han, D. & Xu, C. 1985. Pleistocene mammalian faunas of China. Pp. 267–289 in Wu, T. & Wilson-Orlando, J. W. *Palaeoanthropology and palaeolithic archaeology in the Peoples' Republic of China*. Academic Press.

Han, D. *et al.* 1975. Quaternary mammalian fossils from Bijianshan, Liuzhou, Guangxi. *Vertebrata palasiat.* 13: 150–156.

Hanak, V. & Gaisler, J. 1969. Notes on the taxonomy and ecology of *Myotis longipes* (Dobson, 1873). *Zoologické Listy* 18: 195–206.

Handley, C. O. 1966. A synopsis of the genus *Kogia* (pygmy sperm whales). Pp. 62–69 in Norris, K. S. (ed.) *Whales, dolphins and porpoises*. Berkeley & Los Angeles: Univ. California Press.

----- 1980. Inconsistencies in formation of family-group names in Chiroptera. *Proceedings 5th int. Bat Res. Conf.* : 9–13.

Harada, M. & Kobayashi, T. 1980. Studies on the small mammal fauna of Sabah . . . *Contributions biol. Lab. Kyoto Univ.* 26: 83–95.

Harada, M. & Takada, S. 1985. Karyotypes of two species of Insectivora from Taiwan . . . *Experientia* 41: 510–511.

Hardjasasmita, H. S. 1987. Taxonomy and phylogeny of the Suidae in Indonesia. *Scripta geol.* 85: 1–65, pls 1–13.

Hardwicke, T. 1804. Description of a large species of rat . . . *Transactions Linn. Soc. Lond.* 7: 306–308, pl. 7.

----- 1807. Description of a species of jerboa, found in . . . Hindustan . . . *Transactions Linn. Soc. Lond.* 8: 279–281, pl. 7.

----- 1821. Descriptions of the wild dog of Sumatra . . . *Transactions Linn. Soc. Lond.* 13: 235–238.

----- 1825a. Descriptions of two species of antelope from India. *Transactions Linn. Soc. Lond.* 14: 518–524.

----- 1825b. Description of a new species of tailed bat (*Taphozous*) found in Calcutta. *Transactions Linn. Soc. Lond.* 14: 525–526.

----- 1827. On the *Bos gour* [*sic*] of India. *Journal Zool.* 3: 231–233.

Harlan, R. 1826. Description of an hermaphrodite orang outan . . . *Journal Acad. nat. Sci. Philad.* 5: 229–236, pls ix, x.

----- 1831. Description of a new species of orang . . . *Transactions Am. phil. Soc.* 4,1: 52–59, pl. 2.

Harper, F. 1940. The nomenclature and type localities of certain Old World mammals. *Journal Mammal.* 21: 191–203; 322–332.

----- 1945. *Extinct and vanishing mammals of the Old World*. New York, xv, 850pp.

Harris, W. P. 1931. A new squirrel from China. *Occasional Pap. Mus. Zool. Univ. Mich.* 228: 2pp.

----- 1951. A substitute name for *Petaurista petaurista rufipes* Sody. *Journal Mammal.* 32: 234.

Harrison, D. L. 1972. *The mammals of Arabia. Vol. 3: Lagomorpha, Rodentia*. London: Benn, pp. 385–670.

----- 1974. A new form of the Indian rock rat, *Cremnomys cutchicus* . . . *Mammalia* 38: 681–687.

Harrison, J. L. 1958. *Chimarrogale hantu*, a new water shrew from the Malay Peninsula . . . *Annals Mag. nat. Hist.* 1: 282–290.

----- 1961. Ecology of the forms of *Rattus rattus* in the Malay Peninsula. *Proceedings 9th Pacific Sci. Congr.* 19: 19–24.

----- 1966. *An introduction to the mammals of Singapore and Malaya*. Singapore: Malayan Nature Soc., Singapore Branch, 340pp.

Harrisson, T. 1960. South China Sea dolphins. *Malayan Nat. J.* 14: 87–89.

----- 1970. The prehistory of Borneo. *Asian Perspectives* 13: 17–45.

Hartert, E. 1898. On the birds of Lomblen, Pontar, and Alor. *Novitates zool.* 5: 455–465.

Hartl, G. B. *et al.* 1988. On the biochemical systematics of the Bovini. *Biochemical Syst. Ecol.* 16: 575–579.

Hassinger, J. D. 1970. Shrews of the *Crocidura zarudnyi-pergrisea* group . . . *Fieldiana Zool.* 58: 5–8.

----- 1973. A survey of the mammals of Afghanistan . . . *Fieldiana Zool.* 60: 1–195.

Hatori, J. 1915. [On the brown musk shrew of Formosa.] *Taiwan Igaku Zasshi*, Jan. no.

Hayasaki, K. *et al.* 1988. Phylogenetic relationships among Japanese, rhesus, Formosan and crab-eating monkeys . . . *Molecular Biol. Evol.* 5: 270–281.

Hayman, R. W. 1946. A new genus of fruit-bat and a new squirrel, from Celebes. *Annals Mag. nat. Hist.* 12: 569–578.

----- 1957. A new race of the Indian smooth-coated otter from Iraq. *Annals Mag. nat. Hist.* 9: 710–712.

----- 1961. The red goral of the North-east Frontier Region. *Proceedings zool. Soc. Lond.* 136: 317–324.

Hayman, R. W. & Hill, J. E. 1971. Chiroptera, in Meester, J. & Setzer, H. W. (eds) *The mammals of Africa. An identification manual*. Washington: Smithsonian Institution Press.

Heaney, L. R. 1978. Island area and body size of insular mammals . . . *Evolution, Lawrence* 32: 29–44.

----- 1979. A new species of tree squirrel (*Sundasciurus*) from Palawan Island, Philippines. *Proceedings biol. Soc. Wash.* 92: 280–286.

----- 1985. Systematics of oriental pygmy squirrels of the genera *Exilisciurus* and *Nannosciurus*. *Miscellaneous Publ. Mus. Zool. Univ. Mich.* 170: 58pp.

----- 1991. An analysis of patterns of distribution and species richness among Philippine fruit bats (Pteropodidae). *Bulletin Am. Mus. nat. Hist.* 206: 145–167.

Heaney, L. R. & Alcala, A. C. 1986. Flat-headed bats (Mammalia, *Tylonycteris*) from the Philippine Islands. *Silliman J.* 33: 117–123.

Heaney, L. R. & Heideman, P. D. 1987. Philippine fruit bats: endangered and extinct. *Bats* 5: 3–5,7.

Heaney, L. R. & Morgan, G. S. 1982. A new species of gymnure, *Podogymnura* . . . *Proceedings biol. Soc. Wash.* 95: 13–26.

Heaney, L. R. & Peterson, R. L. 1984. A new species of tube-nosed fruit bat (*Nyctimene*) from Negros Island, Philippines . . . *Occasional Pap. Mus. Zool. Univ. Michigan* 708: 16pp.

Heaney, L. R. & Rabor, D. S. 1982. Mammals of Dinagat and Siargao Islands, Philippines. *Occasional Pap. Mus. Zool. Univ. Michigan* 699: 30pp.

Heaney, L. R. *et al.* 1987. An annotated checklist of the taxonomic and conservation status of land mammals in the Philippines. *Silliman J.* 34: 32–66.

Heberer, G. 1965. *Menschliche Abstammungslehre* ... Stuttgart: Fischer, viii, 481pp.

Heilprin, A. 1887. *The geographical and geological distribution of animals*. London: Kegan Paul, 435pp.

Heim de Balsac, H. & Heim de Balsac, M-H. 1956. *Suncus murinus* (L) a La Reunion et en Nouvelle-Guinee ... *Naturaliste malagache* 8: 143–147.

Heissig, K. 1972. Paläontologische und geologische Untersuchungen im Tertiär von Pakistan. 5. Rhinocerotidae ... *Abhandlungen bayer. Akad. Wiss.* 152: 18–103, 25pls.

Heller, K. G. & Volleth, M. 1984. Taxonomic position of '*Pipistrellus societatis*' ... *Zeitschrift zool. Syst. EvolForsch.* 22: 65–77.

----- 1989. Fledermäuse aus Malaysia. 1. Beobachtungen zur Biologie, Morphologie und Taxonomien ... *Senckenbergiana Biol.* 69 (1988): 243–276.

Heltne, P. G. (ed.) 1985. *The lion-tailed macaque: status and conservation.* (Monographs in Primatology 7). New York: Lus, 412pp.

Helvoort, B. E. van *et al.* 1985. A leopard skin and skull (*Panthera pardus* L.) from Kangean Island, Indonesia. *Zeitschrift Säugetierk.* 50: 182–184.

Hemmer, H. 1967. Fossilbelege zur Verbreitung und Art-geschichte des Löwen, *Panthera leo* ... *Säugetierkundliche Mitt.* 15: 289–300.

----- 1971. Fossil mammals of Java, 3. Zur Kenntnis der Evolution javanischer Kleinkatzen. *Proceedings K. ned. Akad. Wet. B* 74: 365–375, 2pls.

----- 1974a. Studien zur Systematik und Biologie der Sandkatze *Felis margarita. Zeitschrift Köln. Zoo* 17: 11–20.

----- 1974b. *Felis margarita scheffeli*, eine neue Sandkatzen-Unterart aus der Nushki-Wüste, Pakistan. *Senckenbergiana biol.* 55: 29–34.

----- 1978. The evolutionary systematics of living Felidae ... *Carnivore* 1: 71–79.

Hemmer, H. & Koenigswald, G. H. R. von 1964. Fossile Nebelparden (*Neofelis*) aus dem Pleistozän Sudchinas und Javas. *Proceedings Akad. Wet. B* 67: 1–16, 1pl.

Hemprich, F. G. & Ehrenberg, C. G. 1833. *Symbolae physicae* ... Mammalia, 2. Berlin.

Henry, A. 1890. Notes on two mountain-antelopes of central China. *Proceedings zool. Soc. Lond.* : 93–94.

Heptner, V. G. 1937. Notes on Gerbollidae [*sic*] ... *Bulletin Soc. Nat. Moscou* 46: 189–190.

----- 1960. Uber das Java-Nashorn auf New-Guinea. *Zeitschrift Säugetierk.* 25: 128–129.

----- 1971. Systematic status of the Amur wildcat and some other East-Asian cats referred to *Felis bengalensis. Zoologicheskii Zh.* 50: 1720–1727.

----- 1975. Materials to the morphology and systematics of the tridactyl jerboas (genus *Jaculus*) ... *Byulleten Mosk. Obshch. Ispyt. Prir. Otd. Biol.* 80,3: 5–15.

Heptner, V. G. *et al.* 1961. *Mammals of the Soviet Union. Vol. 1. Artiodactyla and Perissodactyla.* Moscow. (English translation 1988, New Dehli: Amerind Publ. Co., 1147pp.)

Hermann, J. 1804. *Observationes zoologicae quibus novae complures aliaeque animalium species* ... Argentorate [= Strasbourg] & Paris.

Herrington, S. J. 1986. *Phylogenetic relationships of the wild cats of the world.* Ph. D. thesis, University of Kansas.

----- 1987. Subspecies and the conservation of *Panthera tigris*: preserving genetic heterogeneity. Pp. 51–61 in Tilson, R. L. & Seal, U. S. (*q. v.*).

Hershkovitz, P. 1966. Catalog of living whales. *Bulletin US natn. Mus.* 246: 259pp.

Heude, P. M. 1882. Note sur quelques cerfs de Chine. *Bulletin Soc. philomath. Paris* 6: 183–189.

----- 1884. [Letter] *Compte r. hebd. Séanc. Acad. Sci. Paris* 98: 1017.

----- 1885. Catalogue des cerfs tachetés ... *Memoires Hist. nat. Emp. Chin.* 1,3, 2: 1–12.

----- 1888. Etude sur les ruminants de l'Asie Orientale. *Mémoires Hist. nat. Emp. Chin.* 2: 1–64, pls 0-xix.

----- 1892. Etude odontologique; Etudes sur les suilliens, chapitres I, II. *Mémoires Hist. nat. Emp. Chin.* 2: 65–115, pls xixA- xxixC.

----- 1894. Etude odontologique ... etc. *Mémoires Hist. nat. Emp. Chin.* 2 (3,4): 117–247, pls xixB-xl.

----- 1896. Etude odontologique ... etc. *Mémoires Hist. nat. Emp. Chin.* 3 (1): 1–52.

----- 1897. Essai sur l'armiture frontale des ruminants ... etc. *Mémoires Hist. nat. Emp. Chin.* 3 (2–4): 53–198, pls ii-v, xv-xxxix.

----- 1898. Capricornes de Moupin, etc. *Memoires Hist. nat. Emp. Chin.* 4 (1,2): 1–111, pls i-xxii.

----- 1899. Etude sur les suilliens, chapitre iv, etc. *Mémoires Hist. nat. Emp. Chin.* 4 (3,4): 113–211, pls xxiii-xliii.

----- 1901. Notes sur quelques ursidés ... etc. *Mémoires Hist. nat. Emp. Chin.* 5(1): 1–43, pls i-xi.

Heyning, J. E. 1989a. Cuvier's beaked whale—*Ziphius cavirostris* G Cuvier, 1823. Pp. 289–308 in Ridgeway, S. H. & Harrison, R. (*q. v.*).

----- 1989b. Comparative facial anatomy of beaked whales ... *Contributions Sci. nat. Hist. Mus. Los Angeles Co.* 405: 64pp.

Heyning, J. E. & Dahlheim, M. E. 1988. *Orcinus orca. Mammalian Species* 304: 9pp.

Hill, J. E. 1958a. Some observations on the fauna of the Maldive Islands. Part II. Mammals. *Journal Bombay nat. Hist. Soc.* 55: 3–10.

----- 1958b. The bat *Rhinolophus creaghi. Sarawak Mus. J.* 11: 470–475.

----- 1959a. A north Bornean pygmy squirrel *Glyphotes simus* Thomas. *Bulletin Br. Mus. nat. Hist.* (*Zool*) 5: 257–266.

----- 1959b. Bats in limestone caves of Sarawak. *Annals Mag. nat. Hist.* 2: 85–91.

----- 1960. The Robinson collection of Malaysian mammals. *Bulletin Raffles Mus.* 29: 1–112.

----- 1961a. Fruit-bats from the Federation of Malaya. *Proceedings zool. Soc. Lond.* 136: 629–642.

----- 1961b. Indo-Australian bats of the genus *Tadarida. Mammalia* 25: 29–56.

----- 1962a. Notes on some insectivores and bats from Upper Burma. *Proceedings zool. Soc. Lond.* 139: 119–137.

----- 1962b. Notes on flying squirrels of the genera *Pteromyscus, Hylopetes* and *Petinomys. Annals Mag. nat. Hist.* 4(1961): 721–738.

----- 1963a. Occurrence of the European free-tailed bat ... in India. *Journal Bombay nat. Hist. Soc.* 60: 723–725.

----- 1963b. A revision of the genus *Hipposideros. Bulletin Br. Mus. nat. Hist.* (*Zool.*) 11: 1–129.

----- 1963c. Some notes on the *Pipistrellus ceylonicus* group ... *Mammalia* 27: 532–536.

----- 1964. Notes on some tube-nosed bats, genus *Murina* ... *Federation Mus. J.* 8 (1963): 48–59, 4pls.

----- 1965. Asiatic bats of the genera *Kerivoula* and *Phoniscus* (Vespertilionidae) ... *Mammalia* 29: 524–556.

----- 1966. A review of the genus *Philetor* ... *Bulletin Br. Mus. nat. Hist.* (*Zool.*) 14: 373–387.

----- 1967. The bats of the Andaman and Nicobar Islands. *Journal Bombay nat. Hist. Soc.* 64: 1–9.

----- 1969. The generic status of *Glischropus rosseti* ... *Mammalia* 33: 133–139.

----- 1971a. Bats from the Solomon Islands. *Journal nat. Hist.* 5: 573–581.

----- 1971b. A note on *Pteropus* ... from the Andaman Islands. *Journal Bombay nat. Hist. Soc.* 68: 1–8.

----- 1971c. The status of *Vespertilio brachypterus* Temminck, 1840 ... *Zoologische Meded. Leiden* 45: 139–146.

----- 1972a. New records of Malayan bats, with taxonomic notes and the description of a new *Pipistrellus. Bulletin Br. Mus. nat. Hist.* (*Zool.*) 23: 21–42.

----- 1972b. A note on *Rhinolophus rex* ... and *Rhinomegaphyllus paradoxolophus* ... *Mammalia* 36: 428–434.

----- 1974a. A new family, genus and species of bat ... from Thailand. *Bulletin Br. Mus. nat. Hist.* (*Zool.*) 27: 301–336.

----- 1974b. New records of bats from southeastern Asia ... *Bulletin Br. Mus. nat. Hist.* (*Zool.*) 27: 127–138.

----- 1974c. A review of *Scotoecus* ... *Bulletin Br. Mus. nat. Hist.* (*Zool.*) 27: 167–188.

----- 1975. Systematic section. Pp. 8–38 in CTNRC staff, Hill, J. E. & McNeely, J. A. The bats and bat's parasites of Thailand. *Report FE-516–1, US Army Res. Dev. Group, Far East.*

----- 1976. Bats referred to *Hesperoptenus* ... *Bulletin Br. Mus. nat. Hist.* (*Zool.*) 30: 1–28, 4pls.

----- 1977. A review of the Rhinopomatidae ... *Bulletin Br. Mus. nat. Hist.* (*Zool.*) 32: 29–43.

----- 1980a. The mammals of Sri Lanka. *Spolia Zeylanica* 35: 203–211.

----- 1980b. The status of *Vespertilio borbonicus* E Geoffroy, 1803 ... *Zoologische Meded.* 55: 287–295.

----- 1982. A review of the leaf-nosed bats *Rhinonycteris, Cloeotis* and *Triaenops* ... *Bonner zool. Beitr.* 33: 165–186.

----- 1983. Bats ... from Indo-Australia. *Bulletin Br. Mus. nat. Hist.* (*Zool.*) 45: 103–208.

----- 1986. A note on *Rhinolophus pearsonii* ... and *Rhinolophus yunanensis* ... *Journal Bombay nat. Hist. Soc.* 83: 12–18.

----- 1988. A record of *Rhinolophus arcuatus* ... from Sulawesi. *Mammalia* 52: 588–589.

----- 1991. Bats ... from the Togian Islands, Sulawesi, Indonesia. *Bulletin Am. Mus. nat. Hist.* 206: 168–175.

Hill, J. E. & Beckon, W. N. 1978. A new species of *Pteralopex* from the Fiji Islands. *Bulletin Br. Mus. nat. Hist.* (*Zool.*) 34: 65–82.

Hill, J. E. & Boeadi 1978. A new species of *Megaerops* from Java. *Mammalia* 42: 427–434.

Hill, J. E. & Francis, C. M. 1984. New bats ... from Borneo and Malaya. *Bulletin Br. Mus. nat. Hist.* (*Zool.*) 47: 305–329.

Hill, J. E. & Harrison, D. L. 1987. The baculum in the Vespertilionidae ... with a ... synopsis of *Pipistrellus* and *Eptesicus* ... *Bulletin Br. Mus. nat. Hist.* (*Zool.*) 52: 225–305.

Hill, J. E. & Pratt, T. K. 1981. A record of *Nyctophilus timoriensis* ... from New Guinea. *Mammalia* 45: 264–266.

Hill, J. E. & Rozendaal, F. G. 1989. Records of bats ... from Wallacea. *Zoologische Meded. Leiden* 63: 97–122.

Hill, J. E. & Schlitter, D. A. 1982. A record of *Rhinolophus arcuatus* from New Guinea ... *Annals Carnegie Mus.* 51: 455–464.

Hill, J. E. & Smith, J. D. 1984,1985,1988. *Bats. A natural history*. London: British Museum (Nat. Hist.).

Hill, J. E. & Smith, S. E. 1981. *Craseonycteris thonglongyai. Mammalian Species* 160: 4pp.

Hill, J. E. & Thonglongya, K. 1972. Bats from Thailand and Cambodia. *Bulletin Br. Mus. nat. Hist.* (*Zool.*) 22: 171–196.

Hill, J. E. & Topál, G. 1973. The affinities of *Pipistrellus ridleyi* ... and *Glischropus rosseti* ... *Bulletin Br. Mus. nat. Hist.* (*Zool.*) 24: 447–454.

----- 1990. Records of Marshall's horseshoe bat, *Rhinolophus marshalli* ... from Vietnam. *Mammalia* 54: 490–491.

Hill, J. E. & Yenbutra, S. 1984. A new species of the *Hipposideros bicolor* group ... from Thailand. *Bulletin Br. Mus. nat. Hist.* (*Zool.*) 47: 77–82.

Hill, J. E. & Yoshiyuki, M. 1980. A new species of *Rhinolophus* ... from Iriomote Island ... *Bulletin natn. Sci. Mus. Tokyo* 6: 179–189.

Hill, J. E. & Zubaid, A. 1989. The Dyak leaf-nosed bat, *Hipposideros dyacorum* ... in Peninsular Malaysia. *Mammalia* 53: 307–308.

Hill, J. E. *et al.* 1985. *Hipposideros lekaguli*, a new leaf-nosed bat recorded in Peninsular Malaysia. *Malayan Nat. J.* 39: 147–148.

----- 1986. The taxonomy of leaf-nosed bats of the *Hipposideros bicolor* group ... *Mammalia* 50: 535–540.

----- 1990. Records of Theobald's tomb bat *Taphozous theobaldi* from Borneo and Sulawesi. *Mammalia* 54: 314–315.

Hill, W. C. O. 1933. A monograph on the genus *Loris* ... *Spolia zeylan.* 18: 89–132, pls vii-xi.

----- 1939. A new leaf-monkey hybrid ... *Spolia zeylan.* 21: 265–275, pls 32,33.

----- 1942a. The slender loris of the Horton Plains, Ceylon ... *Journal Bombay nat. Hist. Soc.* 43: 73–78, 1pl.

----- 1942b. The highland macaque of Ceylon. *Journal Bombay nat. Hist. Soc.* 43: 402–406.

Hill, W. C. O. & Phillips, W. W. A. 1932. A new race of slender loris from the highlands of Ceylon. *Spolia zeylan.* 17: 109–122, pls xxv, xxvi.

Hilzheimer, M. 1905a. Uber einige Tigerschädel aus der Strassburger zoologischer Sammlung. *Zoologischer Anz.* 28: 594–599.

----- 1905b. Neue chinesische Säugetiere. *Zoologischer Anz.* 29: 297–299.

----- 1906a. *Papio mundamensis, Felis deliensis, Canis reissii* und andere neue Säugetiere. *Zoologischer Anz.* 30: 109–118.

----- 1906b. Eine kleine Sendung chinesischer Säugetiere. *Abhandlungen Ber. Mus. Nat. -u. Heimatk.* 1: 165–184.

----- 1910. Neue tibetanische Säugetiere. *Zoologischer Anz.* 35: 309–311.

Hinton, H. E. & Dunn, A. M. S. 1967. *Mongooses/ their natural history and behaviour*. Edinburgh & London: Oliver & Boyd, vii, 44pp.

Hinton, M. A. C. 1918. Report on the house rats of India, Burma, and Ceylon [part 1]. *Journal Bombay nat. Hist. Soc.* 26: 59–88.

----- 1919. Ditto, parts 2 and 3. *Journal Bombay nat. Hist. Soc.* 26: 384–416; 716–725.

----- 1922. Note on *Soriculus nigrescens* and its subspecies. *Journal Bombay nat. Hist. Soc.* 28: 1052–1055.

----- 1923a. On the voles collected by Mr G. Forrest in Yunnan ... *Annals Mag. nat. Hist.* 11: 145–162.

----- 1923b. The nomenclature and subspecies of the purple-faced langur ... *Annals Mag. nat. Hist.* 11: 506–515.

----- 1923c. On the capped langurs (*Pithecus pileatus* Blyth and its allies). *Annals Mag. nat. Hist.* 29: 77–83.

----- 1925. Two new nose-leaf bats from Buru. *Annals Mag. nat. Hist.* 16: 255–257.

----- 1926. *Monograph of the Microtinae*. London: British Museum (Nat. Hist.).

----- 1933. Diagnoses of new genera and species of rodents from Indian Tertiary deposits. *Annals Mag. nat. Hist.* 12: 620–622.

Ho, C. K. 1988. Human origins in Asia? *Human Evol.* 3: 357–365.

Ho, H. J. 1936. On some small mammals from South China. *Contributions biol. Lab. Sci. Soc. China, Zool.* 12,4: 75–87.

Hodgson, B. H. 1831a. Some account of a new species of *Felis. Gleanings Sci.* 3: 177–178.

----- 1831b. Contributions in natural history. *Gleanings Sci.* 3: 320–324, pl. xxi.

----- 1832. [Various Mammmalia and birds, collected in Nepal by B. H. Hodgson. . .] *Proceedings zool. Soc. Lond.* : 10–16.

----- 1833a. On the ratwa deer of Nepal. *Asiatic Res.* 18,2: 139–146, 1pl.

----- 1833b. Description of the wild dog of the Himalaya. *Asiatic Res.* 18,2: 221–237, 3pls.

----- 1833c. The wild goat and the wild sheep of Nepal. *Asiatic Res.* 18,2: 129–138, 3pls.

----- 1834. On the Mammalia of Nepal. *Proceedings zool. Soc. Lond.* : 95–99.

----- 1835a. Synopsis of the Vespertilionidae of Nipal. *Journal Asiat. Soc. Bengal* 4: 699–701.

----- 1835b. Description of the little musteline animals, denominated Kathiah Nyul ... *Journal Asiat. Soc. Bengal* 4: 702–704.

----- 1835c. Specific name and characters of a new species of *Cervus* ... *Journal Asiat. Soc. Bengal* 4: 648–649, pl. 53.

----- 1836a. Description of three new species of *Paradoxurus* ... *Asiatic Res.* 19: 72–86.

----- 1836b. Synoptical description of sundry new animals ... *Journal Asiat. Soc. Bengal* 5: 231–238.

----- 1836c. Ilustrations of Nipalese zoology. . . *Journal Asiat. Soc. Bengal* 4: 521–525.

----- 1836d. Postscript to the account of the wild goat in Nipal. *Journal Asiat. Soc. Bengal* 4: 710.

----- 1836e. Indication of a new genus of the Carnivora ... *Asiatic Res.* 19: 60–68, pl. viii.

----- 1837a. [Description of the Gauri Gau of the Nipal forest] *Journal Asiat. Soc. Bengal* 6: 499.

----- 1837b. On a new genus of the plantigrades. *Journal Asiat. Soc. Bengal* 6: 560–564.

----- 1837c. On the Bibos, Gauri Gau or Gaurika Gau of the Indian forests. *Journal Asiat. Soc. Bengal* 6: 745–750, pl. xxxix.

----- 1838. [Classified catalogue of Nepalese Mammalia] *Annals Mag. nat. Hist.* 1: 152–154.

----- 1839a. On three new species of musk (*Moschus*) ... *Journal Asiat. Soc. Bengal* 8: 202–203.

----- 1839b. Summary description of four new species of otter. *Journal Asiat. Soc. Bengal* 8: 319–320.

----- 1840. On the common hare of the Gangetic Provinces and of the Sub-Himalayas ... *Journal Asiat. Soc. Bengal* 9: 1183–1186.

----- 1841a. New species of *Rhizomys* discovered in Nepal. *Calcutta J. nat. Hist.* 2: 60–61.

----- 1841b. Classified catalogue of mammals of Nepal ... *Calcutta J. nat. Hist.* 2: 212–221.

----- 1841c. On the civet of the continent of India, *Vivera orientalis. Calcutta J. nat. Hist.* 2: 47–56.

----- 1841d. On a new species of *Prionodon, P. Pardicotor nobis. Calcutta J. nat. Hist.* 2: 57–60, pl. 1.

----- 1841e. Three new species of monkey ... *Journal Asiat. Soc. Bengal* 9 (1840): 1211–1213.

----- 1841f. Note on the *Cervus Elephas* (?) of the Sal forest of Nepal ... *Journal Asiat. Soc. Bengal* 10: 721–724.

----- [1842.] Classified catalogue of mammals of Nepal. *Journal Asiat. Soc. Bengal* 10: 907–916.

----- 1843a. *Journal Asiat. Soc. Bengal* 12: 414.

----- 1843b. On a new species of *Cervus, Cervus dimorphé. Journal Asiat. Soc. Bengal* 12: 897, 1pl.

----- 1844a. Summary description of two new species of flying squirrel. *Journal Asiat. Soc. Bengal* 13: 67–68, 1pl.

----- 1844b. Classified catalogue of the mammals of Nepal ... *Calcutta J. nat. Hist.* 4: 284–294.

----- 1845. On the rats, mice, and shrews of the central region of Nepal. *Annals Mag. nat. Hist.* 15: 266–270.

----- 1847a. The slaty blue Megaderme, *Megaderma schistacea* N. S. *Journal Asiat. Soc. Bengal* 16,2: 889–894, 1pl.

----- 1847b. On a new species of porcupine. *Journal Asiat. Soc. Bengal* 16: 771–774.

----- 1847c. On the cat-toed subplantigrades of the sub- Himalayas. *Journal Asiat. Soc. Bengal* 16: 1113–1129.

----- 1847d. Observations on the manners and structure of *Prionodon Pardicolor. Calcutta J. nat. Hist.* 8: 40–45, pl. 1.

----- 1847e. On a new form of the hog kind or Suidae. *Journal Asiat. Soc. Bengal* 16: 423–428, pls xii, xiii.

----- 1847f. On the four-horned antelopes of India. *Calcutta J. nat. Hist.* 8: 87–94, pl. iv.

----- 1847g. On the various genera of the ruminants. *Journal Asiat. Soc. Bengal* 16: 685–711.

----- 1847h. On a new species of *Plecotus. Journal Asiat. Soc. Bengal* 16: 894–896.

----- 1849. On the physical geography of the Himalaya. *Journal Asiat. Soc. Bengal* 18: 761–788.

----- 1850. On the takin of the eastern Himalaya ... *Journal Asiat. Soc. Bengal* 19: 65–75, pls 1–3.

----- 1858. Description of a new species of Himalayan mole *Talpa macrura. Journal Asiat. Soc. Bengal* 27: 176.

Hoffmann, B. 1887. Uber Säugethiere aus dem Ostindischen Archipel. *Abhandlungen Ber. K. zool. anthrop-eth. Mus. Dresden* 3: 29pp, 1pl.

Hoffmann, R. S. 1984. A review of the shrew-moles (genus *Uropsilus*) ... *Journal mammal. Soc. Japan* 10: 69–80.

----- 1986. A review of the genus *Soriculus* ... *Journal Bombay nat. Hist. Soc.* 82: 459–481.

----- 1987. A review of ... Chinese red-toothed shrews ... *Acta theriol. sin.* 7: 100–139.

Hohmann, G. 1988. Analysis of loud calls provides new evidence for hybridization between two Asian leaf monkeys (*Presbytis johnii, P. entellus*). *Folia primatol.* 51: 209–213.

Hollister, N. 1911a. Two new species of *Epimys* from Luzon. *Proceedings biol. Soc. Wash.* 24: 89–90.

----- 1911b. Description of a new Philippine flying squirrel. *Proceedings biol. Soc. Wash.* 24: 185–186.

----- 1912. Five new mammals from Asia. *Proceedings biol. Soc. Wash.* 25: 181–184.

----- 1913a. Review of the Philippine land mammals in the United Stated National Museum. *Proceedings US natn. Mus.* 46: 299–341, pls 27–29.

----- 1913b. Two new Philippine fruit bats. *Proceedings biol. Soc. Wash.* 26: 111–112.

----- 1913c. Two new bats of the genus *Taphozous. Proceedings biol. Soc. Wash.* 26: 157–158.

Holmes, J. R. S. 1970. Himalayan tahr, *Hemitragus jemlahicus* . . . in Bhutan. *Journal Bombay nat. Hist. Soc.* 67: 106.

Holt, S. J. 1988. Indian Ocean whale sanctuary—background and update. Pp. 45–60 in Silas, E. G. (*q. v.*).

Holthuis, L. B. 1987. The scientific name of the sperm whale. *Marine Mammal Sci.* 3: 87–88.

Honacki, J. H. *et al.* (eds) 1982. *Mammal species of the world.* Lawrence, Kansas: Allen Press, ix, 694pp.

Honda, T. *et al.* 1977. An unusual sex chromosome constitution found in the Amami spinous country rat, *Tokudaia osimensis osimensis. Japanese J. Genetics* 52: 247–249.

Hong, Z-f. 1981. A new subspecies of reed vole from Fujian— *Microtus fortis fujianensis* . . . *Acta zootaxon. sin.* 6: 444–445.

Hooijer, D. A. 1942. On the nomenclature of some fossil hippopotami. *Archives néerl. Zool.* 6: 279–282.

----- 1946. Prehistoric and fossil rhinoceroses from the Malay Archipelago and India. *Zoologische Meded. Leiden* 26: 1–138, pls 1- 10.

----- 1947. On fossil and prehistoric remains of *Tapirus* from Java . . . *Zoologische Meded. Leiden* 27: 253–299, 2pls.

----- 1948a. A new race of the leaf-monkey *Presbytis aygula* (L.) from Deli, northeastern Sumatra. *Proceedings K. ned. Akad. Wet.* 51: 234–237.

----- 1948b. Pleistocene vertebrates from Celebes III . . . *Proceedings K. ned. Akad. Wet.* 51: 1322–1330.

----- 1948c. Prehistoric teeth of Man and of the orang-utan from central Sumatra. *Zoologische Meded. Leiden* 29: 175–301, 9pls.

----- 1949a. Observations on a calvarium of *Equus sivalensis* . . . *Archives néerl. Zool.* 8: 243–266, 1pl.

----- 1949b. Pleistocene vertebrates from Celebes IV . . . *Zoologische Meded* 30: 205–226, 2pls.

----- 1950a. Man and other mammals from Toalian sites in southwestern Celebes. *Verhandelingen K. ned. Akad. Wet.* 46: 1–164.

----- 1950b. The fossil Hippopotamidae of Asia . . . *Zoologische Verh. Leiden* 8: 1–124, 22pls.

----- 1951. Two new deer from the Pleistocene of Wanhsien, Szechwan, China. *American Mus. Novit.* 1495: 18pp.

----- 1952. Some remarks on the subspecies of *Phalanger ursinus* . . . *Zoologische Meded.* 31: 233–246.

----- 1954a. Pleistocene vertebrates from Celebes, VIII . . . *Zoologische Verh. Leiden* 24: 1–46, pls 1–6.

----- 1954b. A pygmy *Stegodon* from the Middle Pleistocene of eastern Java. *Zoologische Meded. Leiden* 33: 91–102, pl. 19.

----- 1957. Three new giant prehistoric rats from Flores, Lesser Sunda Islands. *Zoologische Meded. Leiden* 35: 299–314, 2pls.

----- 1958. Fossil Bovidae from the Malay Archipelago and the Punjab. *Zoologische Verh. Leiden* 38: 1–112, 9pls.

----- 1960a. The giant extinct pangolin (*Manis palaeojavanica* Dubois) from Niah. *Sarawak Mus. J.* 9: 350–355.

----- 1960b. Quaternary gibbons from the Malay Archipelago. *Zoologische Verh. Leiden* 46: 1–42, 2pls.

----- 1962. Quaternary langurs and macaques from the Malay Archipelago. *Zoologische Verh. Leiden* 55: 1–62, 2pls.

----- 1964. New records of mammals from the middle Pleistocene of Sangiran, central Java. *Zoologische Meded. Leiden* 40: 73–88.

----- 1974. *Elephas celebensis* (Hooijer) from the Pleistocene of Java. *Zoologische Meded. Leiden* 48: 85–93.

Hooper, E. T. 1941. The type specimen of the water buffalo, *Anoa mindorensis* Steere. *Occasional Pap. Mus. Zool. Univ. Michigan* 443: 4pp.

Hopwood, A. T. 1925. A new species of buffalo from the Pleistocene of China. *Annals Mag. nat. Hist.* 16: 238–239.

----- 1936. The former distribution of the caballine and zebrine horses in Europe and Asia. *Proceedings zool. Soc. Lond.* : 897–912, 2pls.

Horácek, I & Hanák, V. 1985–86. Generic status of *Pipistrellus savii* . . . *Myotis* 23–24: 11–16.

Horikawa, Y. 1928. *Transactions nat. Hist. Soc. Formosa* 18: 339–342.

----- 1929. Material for the study of Formosan mammals, pt 1. *Transactions nat. Hist. Soc. Formosa* 19: 75–82.

Horsfield, T. 1821–24. *Zoological researches in Java* . . . London: Kingbury, Parbury & Allen.

----- 1825a. Description of the Rimou-dakau of the inhabitants of Sumatra, a new species of *Felis* . . . *Zoological J. Lond.* 1: 542–554, pl. 21.

----- 1825b. Description of the *Helarctos euryspilus* . . . *Zoological J. Lond.* 2: 221–234.

----- 1828. Notice of a new species of *Mustela*, found in India. *Zoological J. Lond.* 4: 238–240, pl. 8.

----- 1830. Catalogue of zoological specimens. Mammalia. Pp. 633–647 in Raffles [Lady Sophia] *Memoirs of Sir Stamford Raffles.* London.

----- 1831. Observations on two species of bats, from Madras . . . *Proceedings zool. Soc. Lond.* : 113–114.

----- 1840. List of Mammalia and birds collected in Assam . . . *Proceedings zool. Soc. Lond.* (1839): 146–167.

----- 1849. Brief notice of several Mammalia and birds discovered by B. H. Hodgson . . . *Annals Mag. nat. Hist.* 3: 202–203.

----- 1851. *A catalogue of the Mammalia in the museum of the East India Company.* London, 212pp.

----- 1855. Brief notices of . . . Mammalia . . . in Nepal. *Annals Mag. nat. Hist.* 16: 101–114.

----- 1856. Catalogue of a collection of Mammalia from Nepal . . . *Proceedings zool. Soc. Lond.* : 393–406, pls 47–50.

Horsfield, T. & Vigors, N. A. 1829. Observations on some of the Mammalia contained in the collection of the Zoological Society. *Zoological J. Lond.* 4: 380–384, pls 13,39 (suppl).

Horst, M. D. & Raadt, O. L. E. de 1918. De identiteit van *Mus diardii* Jentink. *Zoologische Meded. Leiden* 4: 67–69.

Hose, C. 1893a. *A descriptive account of the mammals of Borneo.* London.

----- 1893b. Description of a new deer from Mount Dulit, eastern Sarawak. *Annals Mag. nat. Hist.* 12: 206.

----- 1900. A new flying-squirrel from Borneo. *Annals Mag. nat. Hist.* 5: 214–216.

Houzé, E. 1896. *Revue Univ. Bruxelles* : 1 (*n.v.*).

Howell, A. B. 1926a. Three new mammals from China. *Proceedings biol. Soc. Wash.* 39: 137–140.

----- 1926b. A new name for *Felis* (*Catopuma*) *melli* Matschie . . . *Proceedings biol. Soc. Wash.* 39: 143.

----- 1927a. Five new Chinese squirrels. *Journal Wash. Acad. Sci.* 17: 80–84.

----- 1927b. Two new Chinese rats. *Proceedings biol. Soc. Wash.* 40: 43–46.

----- 1928a. A new white-toothed shrew from Fukien, China. *Journal Mammal.* 9: 60.

----- 1928b. New Chinese mammals. *Proceedings biol. Soc. Wash.* 41: 41–44.

Howell, A. H. 1915. Description of a new genus and seven new races of flying squirrels. *Proceedings biol. Soc. Wash.* 28: 109–114.

Hsu, L. & Wu, J. 1981. A new subspecies of *M. flavigula* from Hainan Island. *Acta theriol. sin.* 1: 145–148.

Hsu, T. C. & Johnson, M. L. 1963. Karyotypes of two mammals from Malaya. *American Nat.* 97: 127–129.

Hu, C-c. 1973. Ape-man teeth from Yuanmou, Yunnan. *Acta geol. sin.* 1: 65–71.

Hu, C. & Qi, T. 1978. Gongwangling Pleistocene mammalian fauna of Lantian, Shaanxi. *Palaeontologia sin.* C, 21: 1–64.

Hu, Z-z. & Wang, Y-z. (eds) 1984. [*Sichuan fauna economica*, vol. 2] (In Chinese).

Huang, T. 1966. A new species of a whale tympanic bone from Taiwan, China. *Proceedings palaeont. Soc. Japan* 61: 183–187, 1pl.

Huang, W. 1982. A fossil hominid and mammalian fauna found in Hexian County, Anhui. *INQUA* 11: 109.

----- 1985. The demarcation line between the Palaearctic and Oriental Regions in eastern China. Pp. 133–141 in Kawamichi, T. (*q. v.*).

Huang, W. *et al.* 1982. Preliminary study on the fossil hominid skull and fauna of Hexian, Anhui. *Vertebrata palasiat.* 20: 248–256.

----- 1983. A fossil *Ailuropoda* and *Capricornis* from Guiliu County, Kwangsi. *Vertebrata palasiat.* 21: 151–159.

Hubrecht, A. A. W. 1891. A new mammal from Sumatra. *Notes Leyden Mus.* 13: 241–242.

Huet, J. 1887. Note sur une espèce nouvelle de mammifère du genre *Mydaus* . . . *Naturaliste* 9 (13): 149–151.

----- 1888. Sur deux espèces nouvelles de sangliers. *Naturaliste* 2,10: 5–7.

Hume, A. O. 1875. Notes on two apparently undescribed species of goat from northern India . . . *Proceedings Asiat. Soc. Bengal* (1874): 240–241.

----- 1877. On a supposed new sheep from the central hills of Kelat. *Journal Asiat. Soc. Bengal* 46,2: 327–328.

Husar, S. L. 1978. *Dugong dugon. Mammalian Species* 88: 7pp.

Husson, A. M. 1955a. Note on the cotypes of *Vespertilio harpia* Temminck, 1840 . . . *Zoologische Meded. Leiden* 33: 121–125.

----- 1955b. Notes on the mammals collected by the Swedish New Guinea Expedition 1948–1949. *Nova Guinea* 6: 283–306.

----- 1963. On *Blarina pyrrhonota* and *Echimys macrourus* . . . *Studies on the Fauna of Suriname and other Guyanas* 5: 34–41.

Husson, A. M. & Holthuis, L. B. 1953. The dates of publication of

'Verhandelingen over de natuurlijke Geschiedenis des Nederlandsche Overzeesche Bezittingen'. *Zoologische Meded. Leiden* 34: 17–24.

Hutchison, J. H. 1974. Notes on the type specimens of European Miocene Talpidae and a tentative classification of Old World Tertiary Talpidae. *Geobios, Lyon* 7: 211–256.

Hutterer, R. 1979. Ergebnisse der Bhutan-Expedition 1972 . . . *Bonner zool. Beitr.* 30: 1–6.

Hutton, T. 1842. The wild sheep of Afghanistan . . . *Calcutta J. nat. Hist.* 2: 514–542, pls xix, xx.

----- 1872. On the bats of the north-western Himalayas . . . *Proceedings zool. Soc. Lond.* : 690–714.

Ibáñez, C. & Valverde, J. A. 1985. Taxonomic status of *Eptesicus platyops* . . . *Zeitschrift Säugetierk.* 50: 241–242.

Ichihara, T. 1966. The pygmy blue whale, *Balaenoptera musculus brevicauda*, a new subspecies from the Antarctic. Pp. 79–113 in Norris, K. S. (ed.) *Whales, dolphins and porpoises.* Berkeley: Univ. California Press.

Illiger, C. 1811. *Prodromus systematis mammalium* . . . Berlin: Salfeld, xvii, 301pp.

----- 1815. Ueberblick der Säugethiere . . . *Abhandlungen preuss. Akad. Wiss.* (1804–11): 39–159.

Imaizumi, Y. 1960. *Coloured illustrations of the mammals of Japan.* Osaka: Hoikusha Publ. Co., 196pp.

----- 1961. Taxonomic status of *Crocidura dsinezumi orii. Journal mammal. Soc. Japan* 2: 17–22.

----- 1967. A new genus and species of cat from Iriomote, Ryukyu Islands. *Journal mammal. Soc. Japan* 3: 75–106.

----- 1970. *The handbook of Japanese land mammals, vol. 1.* Tokyo: Shin-Shicho-Sha co., 350pp.

----- 1974. Notes on *Crocidura orii* Kuroda . . . *Journal mammal. Soc. Japan* 6: 87.

Imaizumi, Y. & Yoshiyuki, M. 1965. Taxonomic studies on *Tadarida insignis* from Japan. *Journal mammal. Soc. Japan* 2: 105–108.

----- 1968. A new species of insectivorous bat of the genus *Nyctalus* from Japan. *Bulletin natn. Sci. Mus. Tokyo* 11: 127–134.

Ingles, J. M. *et al.* 1980. The first record of a rare murine rodent *Diomys* . . . from Nepal. *Bulletin Br. Mus. nat. Hist. (Zool.)* 39: 205–211.

Iredale, T. & Troughton, E. le G. 1934. A checklist of the mammals recorded from Australia. *Memoirs Aust. Mus.* 6:xi, 122pp.

IUCN 1981–1986. *Identification Manual, vol. 1: Mammalia.* Gland: IUCN, looseleaf.

Jacobi, A. 1923. Zoologische Ergebnisse . . . nach Szetschwan. 1. Mammalia. *Abhandlungen Ber. Mus. Tierk. Völkerk. Dresden* 16: 1–22.

Jacobs, L. L. 1977. A new genus of murid rodent from the Miocene of Pakistan and comments on the origin of the Muridae. *Paleobios* 25: 11pp.

----- 1978. Fossil rodents . . . from Neogene Siwalik deposits, Pakistan. *Bulletin Mus. nth Arizona* 52:xi, 103pp.

----- 1980. Siwalik fossil tree shrews. Pp. 205–216 in Luckett, W. P. (*q. v.*).

Jacobs, L. L. *et al.* 1989. Neogene rodents of southern Asia. Pp. 157–177 in Black, C. C. & Dawson, M. R. (eds) *Papers on fossil rodents in honour of Albert Elmer Woods.* Nat. Hist. Mus. Los Angeles Co., Science Series 33.

Jacquinot, H. & Pucheran, -- 1853. Zoologie. Mammifères et oiseaux. Pp. 5–166 in Zoologie, vol. 3 (vol. 2 in error on title page) in Dumont-d'Urville, J. S. C., *Voyage au Pôle Sud . . . sur les corvettes l'Astrolabe et la Zélée* . . . Paris.

Jaeger, J-J. 1988. Rodent phylogeny: new data and old problems. Pp. 177–199 in Benton, M. J. 1988b (*q. v.*).

James, D. B. 1984. Capture of a false killer whale *Pseudorca crassidens* at Port Blair, Andamans. *Marine Fish. Inf. Serv.* 55: 17.

Jameson, E. W. & Jones, G. S. 1977. The Soricidae of Taiwan. *Proceedings biol. Soc. Wash.* 90: 459–482.

Janczewski, D. N. *et al.* 1990. Molecular genetic divergence of orangutan (*Pongo pygmaeus*) subspecies . . . *Journal Hered.* 81: 375–387.

Janis, C. 1984. Tragulids as living fossils. Pp. 87–94 in Eldridge, N. & Stanley, S. M. (eds) *Living fossils.* New York: Springer.

Jardine, W. 1834. *Naturalist's library. Mammalia, vol. II. The Felinae.* Edinburgh, 276pp.

Jenkins, P. D. 1976. Variation in Eurasian shrews of the genus *Crocidura* . . . *Bulletin Br. Mus. nat. Hist. (Zool.)* 30: 271–309.

----- 1982. A discussion of Malayan and Indonesian shrews of the genus *Crocidura* . . . *Zoologische Meded.* 56: 267–279.

----- 1987. *Catalogue of primates in the British Museum (Natural History) . . . Part IV. Suborder Stripsirrhini* . . . London: British Museum (Nat. Hist.), 189pp.

----- 1990. *Ibid. Part V. The apes, superfamily Hominoidea.* London: Natural History Museum Publications, 137pp.

Jenkins, P. D. & Groves, C. 1992. *Bulletin zool. Nom.* 49: 58–61.

Jenkins, P. D. & Hill, J. E. 1981. The status of *Hipposideros galeritus* . . . *Bulletin Br. Mus. nat. Hist. (Zool.)* 41: 279–294.

----- 1982. Mammals from Siberut, Mentawai Islands. *Mammalia* 46: 219–224.

Jentink, F. A. 1879a. On various species of *Mus*, collected . . . in Celebes. *Notes Leyden Mus.* 1: 7–13.

----- 1879b. On three new squirrels. *Notes Leyden Mus.* 1: 36–42.

----- 1879c. On a new genus and species of bat from Celebes. *Notes Leyden Mus.* 1: 117–119.

----- 1879d. On a new species of *Echimys. Notes Leyden Mus.* 1: 97–98.

----- 1880a. A Celebian mouse renamed. *Notes Leyden Mus.* 2: 11–12.

----- 1880b. On some hitherto undescribed species of *Mus* in the Leyden Museum. *Notes Leyden Mus.* 2: 13–19.

----- 1883. A list of species of mammals from West-Sumatra and North-Celebes . . . *Notes Leyden Mus.* 5: 170–181.

----- 1887. Catalogue ostéologique des mammifères, Tome 10, in Schlegel, H. G. & Jentink, F. A., *Muséum d'Histoire Naturelle des Pays-Bas. Revue méthodique* . . . Leiden.

----- 1888a. Catalogue systematique des mammifères, Tome 12 in Schlegel, H. G. & Jentink, F. A. *Muséum d'Histoire Naturelle des Pays-Bas. Revue méthodique* . . . Leiden.

----- 1888b. On the shrews from the Malayan Archipelago. *Notes Leyden Mus.* 10: 161–167.

----- 1888c. On a collection of mammals from East-Sumatra. *Notes Leyden Mus.* 11: 17–30.

----- 1889a. On a new genus and a new species in the macroglossine-group of bats. *Notes Leyden Mus.* 11: 209–212.

----- 1889b. On a new shrew from the Indian Archipelago. *Notes Leyden Mus.* 11: 213–214.

----- 1890a. Mammalia from the Malay Archipelago, in Weber, M. C. W., *Zoologische Ergebnisse einer Reise in Niederländische Ost-Indien* . . . Leiden, 460pp, 25pls.

----- 1890b. On a new flying squirrel from Deli, Sumatra. *Notes Leyden Mus.* 12: 145–148.

----- 1890c. On a collection of mammals from Billiton. *Notes Leyden Mus.* 12: 149–154.

----- 1891. On *Lepus netscheri* Schlegel . . . *Notes Leyden Mus.* 13: 213–222.

----- 1892. On a new species of rat from the island of Flores. Pp. 78–83 in Weber, M. C. W., *Zoologische Ergebnisse einer Reise in Niederländisch Ost-Indien* . . . Vol. 3. Leiden.

----- 1895. On two mammals from the Calamanianes. *Notes Leyden Mus.* 17: 41–48.

----- 1897. The dog of Tengger. *Notes Leyden Mus.* 18: 217–220.

----- 1902. Revision of the genera *Macroglossus* and *Syconycteris. Notes Leyden Mus.* 23: 131–142.

----- 1903. A new Bornean *Herpestes. Notes Leyden Mus.* 23: 223–228.

----- 1905. *Sus* studies in the Leyden Museum. *Notes Leyden Mus.* 26: 155–195, pls 2–14.

----- 1907. On a new *Taphozous* from Java. *Notes Leyden Mus.* 29: 65–67.

----- 1910a. Two undescribed shrews from Mount Pangerango, Java. *Notes Leyden Mus.* 32: 197–198.

----- 1910b. On a new mouse from Java. *Notes Leyden Mus.* 33: 69.

----- 1910c. *Chrysopteron bartelsii, novum genus et nova species,* from Java. *Notes Leyden Mus.* 32: 73–77.

Jerdon, T. C. 1867. *The mammals of India* . . . Rorkee: 319pp. (Reprinted, London, 1874, 335pp.)

Ji, H. 1985. Geographical history of genus *Elaphurus. Vertebrata palasiat.* 23: 214–222.

Jiang, N-r. *et al.* 1989. Strates cenozoiques tardives et fossiles humains dans le Bassin de Yuanmou, Yunnan, China. *Anthropologie (Paris)* 92 (1988): 939–944.

Jiang, X. *et al.* 1991. Taxonomic revision and distribution of subspecies of rhesus monkey (*Macaca mulatta*) in China. *Zoological Res.* 12: 241–247.

Johnsingh, A. J. T. 1985. Distribution and status of dhole *Cuon alpinus* Pallas, 1811 in south Asia. *Mammalia* 49: 203–208.

Johnson, D. H. 1946a. The spiny rats of the Riu Kiu Islands. *Proceedings biol. Soc. Wash.* 59: 169–172.

----- 1946b. A new Philippine rat allied to "*Bullimus*" *bagobus* Mearns. *Journal Wash. Acad. Sci.* 36: 317–320.

----- 1952. The status of *Nycticebus coucang brachycephalus* Sody. *Journal Mammal.* 33: 115–116.

----- 1962. Two new murine rodents. *Proceedings biol. Soc. Wash.* 75: 317–319.

Johnson, D. H. *et al.* 1980. Mammals from Nepal. *Journal Bombay nat. Hist. Soc.* 77: 56–63.

Jones, G. S. & Jones, D. B. 1976. *A bibliography of the land mammals of Southeast Asia.* Special Pub. Dept Entom. Bernice P Bishop Mus., Honolulu, 238pp.

Jones, G. S. & Maa, T. C. 1976. *Scotophilus kuhlii* from Sulawesi. *Journal mammal. Soc. Japan* 7: 35–38.

Jones G. S. & Mumford, R. E. 1971. *Chimarrogale* from Taiwan. *Journal Mammal.* 52: 228–231.

Jones, G. S. & Tseng, C-s. 1968. A melano *Callosciurus flavimanus* from . . . Taiwan. *Formosan Sci.* 23: 12–13.

Jones, G. S. *et al.* 1971. A key to the mammals of Taiwan. *Chinese J. Microbiol.* 4: 267–278.

Jones, J. K. & Packard, R. L. 1956. *Feresa intermedia* (Gray) preoccupied. *Proceedings biol. Soc. Wash.* 69: 167–168.

Jotterand-Bellomo, M. & Schauenberg, P. 1988. Les chromosomes du rat de Cuming, *Phloeomys cumingi* . . . *Genetica* 76: 181–190.

Jouanin, C. 1986. Une espèce inattendue pour la faune française: un écureuil asiatique acclimaté sur le Cap d'Antibe. *Revue Ecol. Terre Vie* 41: 107–109.

Jouffroy, F. K. 1959. Un crane subfossile de macaque du Pléistocène du Viet-Nam. *Bulletin Mus. Hist. nat. Paris* 31: 209–216.

Jourdan, [E.] 1837a. *Echo Monde Sav.* (4) 275: 156.

----- 1837b. Mémoire sur deux mammifères nouveaux de l'Inde . . . *Compte r. hebd. Acad. Sci. Paris* 5: 442–447.

----- 1837c. Mémoire sur quelques mammifères nouveaux. *Compte r. hebd. Acad. Sci. Paris* 5: 521–524.

Kaneko, Y. 1987. Skull and dental characters and skull measurements of *Microtus kikuchii* . . . *Journal mammal. Soc. Japan* 12: 31–39.

Kao, Y. *et al.* 1962. Mammals of the Hsi-Shuan-Pan-Na area in southern Yunnan. *Acta zool. sinica* 14: 180–196.

Kastschenko, N. 1907. *Chodsigoa* subgen. nov. . . . *Ezhegodnik zool. Muz.* 10(1905): 251–254.

----- 1925. *Zapiski fiz. -mat. Vidd. vseukr. Akad. Nauk* 1,4: 21 (*n. v.*).

Kastschenko, N. F. & Akimov, M. P. 1917. *Rhinolophus bocharicus* sp. nov. *Ezhegodnik. zool. Muz.* 22: 221–223.

Kasuya, T. 1972. Some information on the growth of the Ganges dolphin . . . *Scientific Rep. Whales Res. Inst. Tokyo* 24: 87–108.

Kasuya, T. & Nishiwaki, M. 1971. First record of *Mesoplodon densirostris* from Formosa. *Scientific Rep. Whales Res. Inst. Tokyo* 23: 129–137, 5pls.

Kaup, [J. J.] 1828. Ueber Hyaena, Uromastix, Basiliscuo . . . *Isis, Jena* 21:cols 1144–1150.

----- 1829. *Skizzirte Entwickelungs-Geschichte und natürliche System der Europäischen Thierwelt* . . . *Theil 1* Darmstadt & Leipzig: 203pp.

Kawamichi, T. (ed.) 1985. *Contemporary mammalogy in China and Japan.* Osaka: Mammalogical Society of Japan, 194pp.

Kawamura, Y. 1989. Quaternary rodent faunas in the Japanese islands (part 2). *Memoirs Fac. Sci. Kyoto Univ., Ser. Geol. Mineral.* 54: 1–235, pls 1–26.

Kelaart, E. F. 1850a. List of Mammalia observed or collected in Ceylon. *Journal Ceylon Brch r. Asiat. Soc.* 2: 203–207 (316–320 in reprint of 1890).

----- 1850b. Description of new species and varieties of mammals found in Ceylon. *Journal Ceylon Brch r. Asiat. Soc.* 2: 208–215 (321–328 in reprint of 1890).

----- 1850c. Description of additional mammals. *Journal Ceylon Brch r. Asiat. Soc.* 2: 216–217 (329–330 in reprint of 1890).

----- 1851. Catalogue of the Mammalia of Ceylon. *Annals Mag. nat. Hist.* 8: 339–340.

----- 1852,1853. *Prodromus faunae zeylanicae* . . . Colombo: published by author, 197,66pp.

Keller, A. 1988. Comparative study on the different hair structures of . . . *Crateromys schadenbergi* . . . and *Phloeomys cumingi. Revue. suisse Zool.* 95: 137–144.

Kellogg, R. 1944. A new macaque from an island off the east coast of Borneo. *Proceedings biol. Soc. Wash.* 57: 75–76.

----- 1945a. Two rats from Morotai Island. *Proceedings biol. Soc. Wash.* 58: 65–68.

----- 1945b. Two new Philippine rodents. *Proceedings biol. Soc. Wash.* 58: 121–124.

Kelm, H. 1939. Ein Pustelschwein von Sumatra. *Zoologischer Anz.* 125: 219–224.

Kerr, R. 1792. *The animal kingdom* . . . *Class 1 Mammalia.* London: Murrey, 400pp. On Official List of works approved as available for use in zoological nomenclature: Opinion 890 (1969).

Keyserling, A. & Blasius, J. H. 1839. Uebersicht der Gattungs- und Artcharaktere der europäischer Fledermäuse. *Archiv Naturgesch.* 5: 293–331.

----- 1840. Nachträgliche Bemerkungen zur Uebersicht der Gattungs- und Artcharaktere der europäischer Fledermäuse. *Archiv Naturgesch.* 6: 1–12.

Khajuria, H. 1953. Taxonomic status of the Celebes ashy-black monkey . . . *Record Indian Mus.* 50: 301–306.

----- 1956. A new langur . . . from Goalpara District, Assam. *Annals Mag. nat. Hist.* 9: 86–88.

----- 1970. A new leaf-nosed bat from central India. *Mammalia* 34: 622–627.

----- 1980. Taxonomical and ecological studies on the bats of Jabalpur Dist . . . II. *Miscellaneous Pub. occ. Pap. Rec. zool. Surv. India* 19: 69pp, 8pls.

----- 1981. On habits of Blanford's rat, *Rattus blanfordi* Thomas. *Bulletin zool. Surv. India* 3: 25–26.

Khan, M. A. R. 1984. *Wildlife of Bangladesh: a checklist.* Dhaka Univ., 173pp.

----- 1985. *Mammals of Bangladesh| a field guide.* Dhaka, 92pp.

Khan, M. K. bin M. 1989. *Asian rhinos. An action plan for their conservation.* Gland: IUCN, 23pp.

Khauna, N. D. *et al.* 1990. Population trends and distribution of camel population in India. *Indian J. Anim. Sci.* 60: 331–337.

Kinnear, N. B. 1920. The past and present distribution of the lion in south eastern Asia [error for south western]. *Journal Bombay nat. Hist. Soc.* 27: 33–39.

----- 1952. The history of Indian mammalogy . . . *Journal Bombay nat. Hist. Soc.* 50: 766–778.

Kirsch, J. A. W. 1968. Prodromus of the comparative seriology of marsupials. *Nature, Lond.* 217: 418–420.

Kishida, K. 1924a. On the formosan Chiroptera. *Zoological Mag.,Tokyo* 36: 30–49.

----- 1924b. Additions to the Japanese Chiroptera. *Zoological Mag.,Tokyo* 36: 127–139.

----- 1926. [Nezumi (rats and mice).] *Dobutsu Kyôzai no Konponteki Kenkyû* : 139–150 (*n. v.*).

----- 1929. A new fruit bat from N. Loo-choo. *Lansania* 1: 125–128.

----- 1932. *Lansania* 4 (40): 153 (*n. v.*).

Kitchener, D. J., Aplin, K. P. & Boeadi 1991. A new species of *Rattus* from Gunung Mutis, South West Timor Island, Indonesia. *Records West Austr. Mus.* 15: 445–461.

Kitchener, D. J. & Foley, S. 1985. Notes on a collection of bats . . . from Bali . . . *Records West Austr. Mus.* 12: 223–232.

Kitchener, D. J., How, R. A. & Maharadatunkamsi 1991a. A new species of *Nyctophilus* from Lembata Island . . . *Records West Austr. Mus.* 15: 97–107.

----- 1991b. *Paulamys sp. cf P. naso* (Musser, 1981) . . . from Flores Island. *Records West Austr. Mus.* 15: 171–189.

----- 1991c. A new species of *Rattus* from the mountains of West Flores, Indonesia. *Records West Austr. Mus.* 15: 611–626.

Kitchener, D. J. & Maharadatunkamsi 1991. Description of a new species of *Cynopterus* . . . from Nusa Tenggara, Indonesia. *Records West Austr. Mus.* 15: 307–363.

Kitchener, D. J. *et al.* 1986. Revision of Australian-Papuan *Pipistrellus* . . . *Records West Austr. Mus.* 12: 435–495.

----- 1990. Wild mammals of Lombok Island . . . Systematics and natural history. *Records West Austr. Mus., Suppl.* 33: 129pp.

Klingener, D. 1984. Gliroid and dipodoid rodents. Pp. 381–388 in Anderson, S. & Jones, J. K. (*q. v.*).

Klingener, D. & Creighton, G. K. 1984. On small bats of the genus *Pteropus* from the Philippines. *Proceedings biol. Soc. Wash.* 97: 395–403.

Kloss, C. B. 1908. New mammals from the Malay Peninsula. *Journal fed. Malay St. Mus.* 2: 143–147.

----- 1910. On mammals and birds from the lowlands of Pahang. *Journal fed. Malay St. Mus.* 4: 144–166.

----- 1911. Diagnosis of new mammals from the Trenganu Archipelago . . . *Annals Mag. nat. Hist* 7: 115–119.

----- 1915a. On two squirrels from the inner gulf of Siam. *Journal nat. Hist. Soc. Siam* 1: 157–162.

----- 1915b. On two new rats from the inner Gulf of Siam. *Journal nat. Hist. Soc. Siam* 1: 221–224.

----- 1916a. On a collection of mammals from Siam. *Journal nat. Hist. Soc. Siam* 2: 1–32.

----- 1916b. On a collection of mammals from the coast and islands of South-east Asia. *Proceedings zool. Soc. Lond.* : 25–75.

----- 1916c. On a new race of flying squirrel from Siam. *Journal nat. Hist. Soc. Siam* 2: 33–36.

----- 1916d. Diagnoses of four new squirrels from Siam. *Journal nat. Hist. Soc. Siam* 2: 178.

----- 1916e. On two rodents new to the fauna of the Malay Peninsula . . . *Journal fed. Malay St. Mus.* 6: 249–252.

----- 1916f. Notes on the *hypomelanus* fruit-bats of the Straits of Malacca . . . *Journal fed. Malay St. Mus.* 6: 245–252.

----- 1916g. A new mouse-deer from upper Siam. *Journal nat. Hist. Soc. Siam* 2: 88–89.

----- 1917a. On five new mammals from Siam. *Journal nat. Hist. Soc. Siam* 2: 282–287.

----- 1917b. On two new pygmy shrews from the Malay Peninsula. *Journal fed. Malay St. Mus.* 7: 127–128.

----- 1917c. On a new murine genus and species from Siam. *Journal nat. Hist. Soc. Siam* 2: 279–281.

----- 1917d. Notes on the type specimens of some Burmese and Himalayan rats. *Records Indian Mus.* 13: 5–10.

----- 1917e. On the mongooses of the Malay Peninsula. *Journal fed. Malay St. Mus.* 7: 123–125.

----- 1917f. On a new mongoose from Siam. *Journal nat. Hist. Soc. Siam* 2: 215–217.

----- 1917g. Description of a new macaque from Siam. *Journal nat. Hist. Soc. Siam* 2: 247–249.

----- 1918a. Notes on some hares in the Indian Museum . . . *Records India Mus. Calcutta* 15: 89–96.

----- 1918b. Notes on Malayan and other mouse-deer. *Journal fed. Malay St. Mus.* 7: 245–256.

----- 1919a. Three new mammals from Siam. *Journal nat. Hist. Soc. Siam* 3: 73–77.

----- 1919b. On mammals collected in Siam. *Journal nat. Hist. Soc. Siam* 3: 333–407.

----- 1919c. New and other white-toothed rats from Siam. *Journal nat. Hist. Soc. Siam* 3: 79–82.

Kloss, C. B. 1920a. Two new *Leggada* mice from Siam. *Journal nat. Hist. Soc. Siam* 4: 59–63.

----- 1920b. Two new races of *Sciurus finlaysoni*. *Journal nat. Hist. Soc. Siam* 4: 103–104.

----- 1920c. A new giant squirrel from Pulo Condore. *Journal nat. Hist. Soc. Siam* 4: 71–72.

----- 1921a. Seven new Malayan mammals. *Journal fed. Malay. St. Mus.* 10: 229–234, pl. 3.

----- 1921b. A new squirrel from north Sarawak. *Journal Straits Brch r. Asiat. Soc.* 83: 151–152.

----- 1921c. On a small collection of mammals from Cambodia. *Journal nat. Hist. Soc. Siam* 4: 99–102.

----- 1921d. Some rats and mice of the Malay Archipelago. *Treubia* 2: 115–124, pls 2,3.

----- 1926a. A new squirrel from Tenggol Island . . . *Journal Malay Brch r. Asiat. Soc.* 4: 259–261.

----- 1926b. Mammals from Pulo Condore . . . *Journal nat. Hist. Soc. Siam, nat. Hist. Suppl.* 6: 357–359.

----- 1926c. A new race of monkey from Annam. *Annals Mag. nat. Hist.* 18: 214.

----- 1928. A new race of barking deer from Annam. *Annals Mag. nat. Hist.* 1: 399–400.

----- 1929. Some remarks on the gibbons . . . *Proceedings zool. Soc. Lond.* : 113–127.

----- 1931a. A contribution to the zoology of Mangalum Island, North-west Borneo. *Bulletin Raffles Mus.* 5: 87–89.

----- 1931b. A new sub-species of Malaysian rat. *Bulletin Raffles Mus.* 5: 105–107.

----- 1932. Notes on the Malayan races of *Ratufa affinis*. *Bulletin Raffles Mus.* 7: 1–2.

Knottnerus-Meyer, T. 1933. Ein neuer Schlankaffe aus Siam. *Zoologische Gart. Lpz.* 6: 259.

Kobayashi, T. *et al.* 1980. Studies of the small mammal fauna of Sabah, East Malaysia. 1. Order Chiroptera and genus *Tupaia* (Primates). *Contributions Biol. Lab. Kyoto Univ.* 26: 67–82.

Kock, D. 1969a. *Dyacopterus spadiceus* . . . auf dem Philippinen . . . *Senckenbergiana biol.* 50: 1–7.

----- 1969b. Eine neue Gattung und Art cynopteriner Flughunde von Mindanao, Philippinen . . . *Senckenbergiana biol.* 50: 319–327.

----- 1969c. Eine bemerkenswerte neue Gattung und Art Flughunde von Luzon, Philippinen . . . *Senckenbergiana biol.* 50: 329–338.

----- 1969d. Die Fledermaus-Fauna des Sudan . . . *Abhandlungen senckenb. naturforsch. Ges.* 521: 1–238.

----- 1974. Eine neue *Suncus*-Art von Flores . . . *Senckenbergiana biol.* 55: 197–203.

----- 1980. Zwei Fledermäuse neu für Pakistan. *Senckenbergiana biol.* 61: 1–9.

----- 1981. *Philetor brachypterus* auf New Britannien und den Philippinen . . . *Senckenbergiana biol.* 61: 313–319.

Kock, D. & Posamentier, H. 1983. *Cannomys badius* . . . in Bangladesh. *Zeitschrift Säugetierk.* 48: 314–316.

----- 1984. *Vulpes bengalensis* . . . in Bangladesh. *Zeitschrift Säugetierk.* 49: 374–376.

----- 1986. *Tupaia glis* . . . in Bangladesh . . . *Zeitschrift Säugetierk.* 51: 146–152.

Koenigswald, G. H. R. von 1933. Beitrage zur Kenntnis der fossilen Wirbeltiere Javas. 1. *Wetenschappelijke Meded. Dienst Mijnb. Ned. Indië* 23: 127pp, 28pls.

----- 1934. Zur Stratigraphie des javanischen Pleistocän . . . *De Ingenieur Ned. Indië* 1(iv), 11: 185–201, 4pls.

----- 1935. Eine fossile Säugetierfauna mit *Simia* aus Südchina. *Proceedings k. ned. Akad. Wet.* 38: 872–879.

----- 1936. Erste Mitteilung über einen fossilen Hominiden aus den Altpleistocän Ostjavas. *Proceedings k. ned. Akad. Wet.* 39: 1000–1009, 1pl.

----- 1940. Neue *Pithecanthropus*-Funde 1936–1938. *Wetenschappelijke Meded. Dienst Mijnb. Ned. Indië* 28: 1–132,14 pls.

----- 1952. *Gigantopithecus blacki* von Koenigswald, a giant fossil hominoid from the Pleistocene of southern China. *Anthropological Pap. Am. Mus. nat. Hist.* 43: 293–325, pls 48,49.

----- 1954. Status of *Szechuanopithecus* from the Pleistocene of China. *Nature, Lond.* 173: 643–644.

----- 1956. Fossil mammals from the Philippines. *Proceedings 4th Far-eastern Prehistoric Congress* 1,2: 333–369.

----- 1963. Fossil pygmy Suidae from Java and China. *Proceedings k. ned. Akad. Wet.* 66, C4: 192–197, 1pl.

Koffler, B. R. 1972. *Meriones crassus*. *Mammalian Species* 9: 4pp.

Kohlbrugge, J. H. F. [1896a]. Zoogdieren van Zuidoost Borneo. *Natuurkundig Tijdschr. Ned. Indië* 55: 176–200.

----- 1896b. Zoogdieren van den Tengger. *Natuurkundig Tijdschr. Ned. Indië* 55: 261–298.

Kohler-Rollefson, I. U. 1991. *Camelus dromedarius*. *Mammalian Species* 375: 8pp.

Kolenati, F. A. 1856. Europa's Chiroptera. 1. *Allgemeine dt. naturh. Ztg.* 2: 121–133.

----- 1858. Eine neue österreichische Fledermaus. *Sitzungsberichte Akad. Wiss. Wien* 29: 250–256.

----- 1863. Beiträge zur Kenntniss der Phthirio-Myiaren. *Horae Soc. ent. Ross.* 2,2: 9–109.

Koop, B. F. *et al.* 1989. A molecular view of primate phylogeny . . . *Molecular Biol. Evol.* 6: 580–613.

Koopman, K. F. 1972. *Eudiscopus denticulus*. *Mammalian Species* 19: 2pp.

----- 1973. Systematics of Indo-Australian pipistrelles. *Periodicum biol.* 75: 113–116.

----- 1975. Bats of the Sudan. *Bulletin Am. Mus. nat. Hist.* 154: 353–444.

----- 1982–85. A synopsis of the families of bats. Parts i-vii, errata. *Bat Res. News* 23: 15–16,26–27; 24: 1–2,17–18,36–38; 25: 15–17,29; 26: 5.

----- 1982. Bats from Eastern Papua and the Eastern Papuan Islands. *American Mus. Novit.* 2747: 34pp.

----- 1983. A significant range extension for *Philetor* . . . *Journal Mammal.* 64: 525–526.

----- 1984a. Bats. Pp. 145–186 in Anderson, S. & Jones, J. K. (*q. v.*).

----- 1984b. Taxonomic and distributional notes on tropical Australian bats. *American Mus. Novit.* 2778: 48pp.

----- 1989. Distributional patterns of Indo-Malayan bats . . . *American Mus. Novit.* 2942: 19pp.

Koopman, K. F. & Danforth, T. N. 1989. A record of the tube-nosed bat (*Murina florium*) from western New Guinea . . . *American Mus. Novit.* 2934: 5pp.

Kopstein, F. 1931. *Rattus concolor otteni, nomen novum*. *Geneesk Tijdschr. ned. - Indië* 71: 3–5.

Kral, B. & Zima, J. 1980. Karyosystematics of the family Felidae. *Gazella* 2/3: 45–53.

Krantz, G. S. 1975. An explanation for the diastema of Javan *erectus* skull IV. Pp. 361–372 in Tuttle, R. H. (*q. v.*).

Kratochvil, J. 1982. The karyotype and system of the family Felidae . . . *Folia zool.* 31: 289–304.

Krefft, G. 1867. Notes on the mammals and birds of Cape York . . . *Proceedings zool. Soc. Lond.* : 317–319.

Kretzoi, M. 1947. New names for mammals. *Annales hist. nat. Mus. natn. hung.* 40: 285–287.

Krumbiegel, I. 1978. Die Kurzschwanz-Stumpfnase, *Simias concolor*. *Säugetierkundliche Mitt.* 26: 59–75.

Kuhl, H. 1819. Die deutschen Fledermause. *Annalen wetterauisch. Ges.* 4: 11–49,185–215, 2pls.

Kumar Sharma, I. 1980. Habitats, feeding, breeding and reaction to man of the desert cat, *Felis libyca* (Gray) in the Indian Desert. *Journal Bombay nat. Hist. Soc.* 76: 498–499.

Kundin, W. D. *et al.* 1970. *Suncus* as a potential reservoir of leptospirosis . . . *SE Asian J. trop. Med. pub. Hlth* 1: 270–274.

Kunkel d'Herculais, J. 1884. Le gibon du Tonkin. *Science Nat. Paris* 2: 86–90.

Kuroda, N. 1920. On two species of Muridae obtained in the central mountains of Formosa. *Zoological Mag.,Tokyo* 32: 36–43.

----- 1921. On three new mammals from Japan. *Journal Mammal.* 2: 208–211.

----- 1924. *On new mammals from the Riu Kiu Islands* . . . Tokyo: published by author.

----- 1925. Description of a new species of the genus *Mus* from Formosa. *Dobutsugaku Zasshi* 37: 16.

----- 1933a. A revision of the genus *Pteropus* found in the islands of the Riu Kiu chain, Japan. *Journal Mammal.* 14: 312–316.

----- 1933b. Mammals collected in Java, Bali and Lombok. *Journal Mammal.* 14: 339–442.

----- 1935. Formosan mammals preserved in the collection of Marquis Yamashina. *Journal Mammal.* 16: 277–291.

----- 1938. *A list of the Japanese mammals*. Tokyo: published by author, 111,122pp.

----- 1940. *A monograph of the Japanese mammals* . . . Tokyo & Osaka: Sanseido (in Japanese).

----- 1943. The occurrence of the spinous rat in Okinawa Island. *Bulletin biogeogr. Soc. Japan* 13: 49–64, pl. 2.

----- 1952a. Mammalogical history of Formosa . . . *Quarterly J. Taiwan Mus.* 5: 267–304.

----- 1952b. Description of three new forms of *Rattus* from Hokkaido and south China. *Journal mammal. Soc. Japan* 1: 1–4.

Kurten, B. 1956. The status and affinities of *Hyaena sinensis* Owen and *Hyaena ultima* Matsumoto. *American Mus. Novit.* 1764: 48pp.

----- 1957. *Percrocuta* Kretzoi . . . a group of Neogene hyenas. *Acta zool. Cracov.* 2: 375–404.

Kurup, G. U. 1966. Mammals of Assam and adjoining areas 1. *Proceedings zool. Soc. Calcutta* 19: 1–21.

----- 1968. Mammals of Assam and adjoining areas. 2. A distributional list. *Proceedings zool. Soc. Calcutta* 21: 79–99.

----- 1989. Rediscovery of the small Travancore flying squirrel. *Oryx* 23: 2–3.

Lacépède, B. G. E. [= La Cepède] 1799. *Tableaux des divisions, sous-divisions, ordres et genres des mammifères*. Paris: 18pp. [Also published in *Memoires Inst. nat. Sci. Arts* 3: 489–502, 1802.]

----- 1800. Nouveau table méthodique de la classe des mammifères. *Séances des Ecoles normales* . . . 8: 66–86.

----- 1804. *Histoire naturelle des cétacées.* Paris: Plassan, xxxiii, 329pp, 16pls.

Lair, R. C. 1988. The number and distribution of domesticated elephants in Thailand. *Natural Hist. Bull. Siam Soc.* 36: 143–160.

Lakhotia, S. C. *et al.* 1973. Studies on rodent chromosomes VI. Coexistence of *Rattus rattus* with 38 and 42 chromosomes in South Western India. *Cytologia* 38: 403–410.

Lal, J. P. 1976. Occurrence of the little Indian field mouse *Mus booduga* . . . in Jammu and Kashmir. *Newsletter zool. Surv. India* 2: 153–154.

----- 1984. Taxonomic status of the Indian subspecies of the Kelaart's pipistrelle, *Pipistrellus ceylonicus* Kelaart . . . *Bulletin zool. Surv. India* 6: 159–161.

Lal Mohan, R. S. 1988. Observations on the by-catch of dolphins in the gill nets off Calicut Coast, India. Pp. 78–83 in Silas, E. G. (*q. v.*).

Lambert, A. B. 1804. Description of *Bos frontalis,* a new species from India. *Transactions Linn. Soc. Lond.* 7: 57–59, pl. 4 (and pp. 302–305).

Lampe, E. 1900. Catalog des Säugetier-Sammlung des Naturhistorischen Museums zu Wiesbaden. *Jahrbuch Nassau. Ver. Naturk.* 53, Suppl. : 1–39.

Lang, J. & Lavocat, R. 1968. Première découverte d'une faune de vertébrés dans le tertiare d'Afghanistan. *Compte r. Acad. Sci. Paris* 266D: 79–82.

Langham, N. P. E. 1982. The ecology of the common tree shrew, *Tupaia glis* in peninsular Malaya. *Journal Zool. Lond.* 197: 323–344.

Langham, N. P. E. & Lam Y. M. 1977. *Mus caroli* Bonhote, 1902, a new mammal of peninsular Malaysia. *Malayan Nat. J.* 30: 39–44.

Lataste, F. 1881. Mammifères nouveaux d'Algerie (suite). *Naturaliste* 1: 506–508.

----- 1882. Mammifères nouveaux d'Algerie (suite). *Naturaliste* 2: 126–127.

----- 1884. Analyse avec annotations critiques [footnote], in Poliakoff, Revue systematique des campagnols de Sibérie. *Annali Mus. Civ. Stor. nat. Giocomo Doria* 20: 253–301.

Latreille, P. A. [1801]. *Histoire naturelle des singes* . . . 2. Paris.

Laurie, A. 1982. Behavioural ecology of the greater one-horned rhinoceros (*Rhinoceros unicornis*). *Journal Zool.,Lond.* 196: 307–341.

Laurie, A. & Seidensticker, J. 1977. Behavioural ecology of the sloth bear (*Melursus ursinus*). *Journal Zool., Lond.* 182: 187–204.

Laurie, E. M. O. & Hill, J. E. 1954. *List of land mammals of New Guinea, Celebes and adjacent islands 1758–1952.* London: British Museum (Natural History), 175pp, 3pls.

----- 1957. Renaming of *Miniopterus australis minor* Laurie & Hill. *Journal Mammal.* 38: 128.

Laurie, W. A. *et al.* 1983. *Rhinoceros unicornis. Mammalian Species* 211: 6pp.

Lawrence, B. 1939. Collections from the Philippine Islands. Mammals. *Bulletin Mus. comp. Zool. Harv.* 86: 28–73.

Lawrence, M. A. 1982. Western Chinese arvicolines . . . *American Mus. Novit.* 2745: 19pp.

Lawrence, M. A. & Musser, G. G. 1990. Mammal holotypes in the American Museum of Natural History: the lectotype of *Prionailurus bengalensis alleni* Sody (1949). *American Mus. Novit.* 2973: 11pp.

Laxmann, E. 1769. *Sibirische Briefe* . . . Göttingen & Gotha, 104pp.

Lay, D. M. 1983. Taxonomy of the genus *Gerbillus* . . . *Zeitschrift Säugetierk.* 48: 329–354.

Lay, D. M. & Nadler, C. F. 1975. A study of *Gerbillus* east of the Euphrates River. *Mammalia* 39: 423–445.

Leach, W. E. 1814. *The zoological miscellany* . . . *vol. 1.* London: 144pp.

----- 1816. *Systematic catalogue of the specimens of the indigenous Mammalia* . . . London: British Museum.

----- 1821. The characters of three new genera of bats without foliaceous appendages to the nose (pp. 69–72); The characters of seven genera of bats with foliaceous appendages to the nose (pp. 73–82). *Transactions Linn. Soc. Lond.* 13: 69–82.

Leatherwood, S. & Reeves, R. R. 1983. *The Sierra Club handbook of whales and dolphins.* San Francisco: Sierra Club Books, 302pp.

----- (eds) 1990. *The bottlenose dolphin.* San Diego etc. : Academic Press, 653pp.

----- 1991. *Marine mammal research and conservation in Sri Lanka 1985–1986*, 2nd ed. Nairobi: UN Environment Programme, Marine Mammal Tech. Rep. 1, 138pp.

Lebeck, H. J. 1801. *Delphinus gangeticus. Neue Schrift. Ges. naturf. Fr. Berlin* 3: 280.

Lee, P-F. *et al.* 1986. Ecological studies on two sympatric *Petaurista* species in Taiwan. *Bulletin Inst. Zool. Acad. Sinica* 25: 113–124.

Legendre, S. *et al.* 1988. Miocene fossil vertebrates from the . . . Phitsanulok Basin, Thailand. *Journal Vertebr. Paleontol.* 8: 278–289.

Lehmann, E. von 1955. Die Säugetiere aus Fukien . . . *Bonner zool. Beitr.* 6: 147–170.

Lei, Ciyu 1985. Study on the Mid-Miocene apes discovered in Jiangsu, China. *Acta geol. sin.* 59: 17–24.

Lekagul, B. & McNeely, J. A. 1977. *Mammals of Thailand.* Bangkok: Assoc. Conservation Wildlife, li, 758pp. (Reprinted without change, except for revised preface, 1988.)

Lesson, R. P. 1826. Zoologie. Vol. 1, part 1, in Duperrey, L. I. (ed.), *Voyage autour du Monde* . . . *sur la corvette La Coquille.* Paris: Bertrand, 360pp.

----- 1827a. *Manuel de mammalogie* . . . Paris, 441pp.

----- 1827b. Description d'un orang-outang . . . *Bulletin Sci. nat. Geol.* 13: 111–113.

----- 1828. *Complément des oevres de Buffon. Histoire naturelle. 1. Cétacés.* Paris.

----- 1830a. *Histoire naturelle* . . . *des mammifères et des oiseaux découverte depuis 1788* . . . *IV* Paris.

----- 1830b-1832. *Centurie zoologique* . . . Paris: Levrault.

----- 1835. *Illustrations de zoologie* . . . Paris: Bertrand, 60 pls, each with unpaginated text.

----- 1838. *Compléments de Buffon, 2nd ed. 1. Races humaines et mammifères.* Paris.

----- 1840. *Species des mammifères bimanes et quadrimanes* . . . Paris & London: Baillière, xiv, 292pp.

----- 1842. *Nouveau tableau de règne animal, mammifères.* Paris: Bertrand.

Lesueur, C. A. & Petit, N. 1807. In Peron, M. F. *Voyage de découverte aux Terres Australes.* 3. Atlas. Paris.

Lever, C. 1985. *Naturalized mammals of the world.* London: Longman, 487pp.

Lewis, G. E. 1939. Siwalik fossil *Mastomys. American J. Sci.* 237: 341–344.

Lewis, W. 1834. Description of a gibbon. *Boston J. nat. Hist. Soc.* 1: 32–40.

Leyhausen, P. 1973. Verhaltensstudien an Katzen. *Zeitschrift Tierpsych.* Suppl. 2, 3rd ed., 232pp.

Li, B. & Chen, F. 1989. A taxonomic study and new subspecies of the subgenus *Eospalax*, genus *Myospalax. Acta zool. sin.* 35: 90–95.

Li, S-s. & Wang, Y-x. 1981. A new subspecies of Pallas's squirrel. *Zoological Res.* 2: 71–76.

Li, Z. -x. & Ma, S. -l. 1980. A revision of the white-headed langur. *Acta zootax. sin.* 5: 440–442.

Liais, E. 1872. *Climats, géologie, faune et géographie botanique au Brésil.* Paris, viii, 640pp.

Lichtenstein, M. H. C. 1814. *Gesellschaft Natur-Freunde Berlin Mag.* 6: 152–173.

----- 1823a. *Verzeichniss der Doubletten des zoologischen Museums der* . . . *Universität.* Berlin.

----- 1823b. Vorrede. In Eversmann, E. F., *Reise von Orenburg nach Buchara* . . . Berlin, viii, 150pp, 2pls.

----- 1825. Uber die Springmäuse oder die Arten der Gattung *Dipus. Abhandlungen preuss. Akad. Wiss.* : 133–161, 9pls.

Lidicker, W. Z. 1968. A phylogeny of New Guinea rodent genera based on phallic morphology. *Journal Mammal.* 49: 609–643.

Lilljeborg, W. 1866. *Systematisk öfversigt af de gnagande däggdjuren, Glires.* Upsala: Kongl. Akad., 59pp, 2pls.

----- 1874. *Sveriges och Norges ryggradsdjur, 1: däggdjuren.* Uppsala: Schultz.

Lim B. L. 1966. Land molluscs as food of Malayan rodents and insectivores. *Journal Zool.,Lond.* 148: 554–560.

----- 1967. Notes on the food habits of *Ptilocercus lowii* . . . *Journal Zool.,Lond.* 152: 375–379.

Lim B. L. & Muul, I. 1975. Notes on a rare species of arboreal rat, *Pithecheir parvus* Kloss. *Malayan Nat. J.* 28: 181–185.

Lin, Y-p. & Zhang, X-y. 1978. Mammalian fossils and palaeoliths from Lijiang Basin, Yunnan. *Professional Pap. Stratigr. Palaeont.* 7: 80–85.

Lindburg, D. G. (ed.). 1980. *The macaques—studies in ecology, behaviour and evolution.* New York: Van Nostrand Reinhold.

Lindsay, H. M. 1926. Further examination of Singhalese species of *Funambulus* . . . *Journal Bombay nat. Hist. Soc.* 31: 239–240.

----- 1928. A note on *Viverra civettina. Journal Bombay nat. Hist. Soc.* 33: 146–148.

----- 1929a. Indian shrews. *Journal Bombay nat. Hist. Soc.* 33: 326–340.

----- 1929b. A new flying squirrel from Nepal. *Journal Bombay nat. Hist. Soc.* 33: 565–569.

Link, H. F. 1795. *Beiträge zur Naturgeschichte. Band 1, Stück 2. Ueber die Lebenskräfte in naturhistorisher Rücksicht und die Klassifikation der Säugethiere.* Rostock & Leipzig, 126pp.

Linnaeus, C. 1758. *Systema naturae* . . . vol. 1,10th ed. Holmia: 824pp.

----- 1760. Anthropomorpha, in *Amoenitates academicae* 6: 63–76.

----- 1761. *Fauna Suecica* . . . 2nd ed. Stockholm, 578pp, 2pls.

----- 1766. *Systema naturae*, 12th ed., 532pp.

----- 1771. *Mantissa plantarum altera.* Holmia.

Liu, C-S. *et al.* 1988. On classification of *Apodemus agrarius ningpoensis* and *A. a. pallidior. Abstacts Symp. Asian-Pacific Mamm., Beijing, July 1988.*

Liu, J. & Zhen, S. 1981. Quaternary mammalian fauna from N Anhui. *Scientia sin.* 24: 216–227.

Liu, J. *et al.* 1982. Some new species of mammals in the Middle Pleistocene from Chao Xian, Anhui . . . *Memoirs Beijing nat. Hist. Mus.* 19: 1–29.

Liu, T-s. *et al.* 1978. Pliocene vertebrates of Lantian, Shensi. *Professional Pap. Stratigr. Palaeont.* 7: 149–200.

Ljungh, S. I. 1801. *Sciurus Plantani*, en foga kand Ikorn. *Kungliga svenska VetenskAkad. Handl.* 22: 99–103, pl. 1.

Loche, -- 1858. Description d'une nouvelle espèce de chat. *Revue Mag. Zool.* 10: 49–50, pl. 1.

Long, C. A. 1978. A listing of recent badgers of the world . . . *Report Fauna Flora Wisconsin* 14: 1–6.

Long, C. A. 1981. Provisional classification and evolution of the badgers. *Proceedings Wldwide Furbearer Conf.* 1: 55–85.

Long, C. A. & Killingley, C. A. 1983. *The badgers of the world.* Springfield, Illinois: Charles & Thomas, 444pp.

Lönnberg, E. 1923. Notes on *Arctonyx*. *Annals Mag. nat. Hist.* 11: 322–326.

----- 1924. On a new fossil porcupine from Honan . . . *Palaeontologia sin.* ser. c, 1,3: 15pp, 1pl.

----- 1925. Notes on a paradoxure from central Borneo. *Nyt Mag. Naturvid.* 62: 60–62.

----- 1926. Some remarks on mole-rats of the genus *Myospalax* from China. *Arkiv Zool.* 18A, 21: 11pp.

Lönnberg, E. & Mjöberg, E. 1925a. Mammalia from Mount Murud and the Kelabit Country. *Annals Mag. nat. Hist.* 16: 508–513, pl. 27.

----- 1925b. Mammalia from Mount Dulit and the Baram District. *Annals Mag. nat. Hist.* 16: 513–516.

Lowenstein, J. M. *et al.* 1982. Piltdown jaw confirmed as orang. *Nature, Lond.* 299: 294.

Lu, C-k. *et al.* 1965. On the mammals from the Lin-Tsang area, west Yunnan. *Acta zootax. sin.* 2: 279–295.

Lu, H-g. & Sheng, H-l. 1984. Status of the black muntjac, *Muntiacus crinifrons*, in eastern China. *Mammal Rev.* 14: 29–36.

Luckett, W. P. (ed.) 1980a. *Comparative biology and evolutionary relationships of tree shrews.* New York & London: Plenum Press, xv, 314pp.

----- 1980b. The suggested evolutionary relationships and classification of tree shrews. Pp. 3–31 in Luckett (1980a).

Luckett, W. P. & Hartenberger, J. L. 1985. *Evolutionary relationships among rodents . . .* New York & London: Plenum Press, 721pp.

Ludeking, E. W. A. 1862. Natuur- an geneeskundige topographische schets der Assistant-residentie Agam . . . *Geneesk Tijdschr. Ned- Indie* 4: 1–153.

Lue, K. Y. 1987. A preliminary study on the ecology of Formosan serow *Capricornis crispus swinhoei*. Pp. 125–133 in Soma, H. (*q. v.*).

Luo, Z. 1988. *The Chinese hare.* China Forestry Publishing House, 186pp, 3 colour pls, in Chinese.

Lütken, C. F. 1889. Bidrag til kundkab om de tre pelagiske tandhval-slaegter *Steno, Delphinus* og *Prodelphinus*. *Kongelige Danske Vidensk. Selsk. Skr.* 5: 1–61, pl. 1.

Lydekker, R. 1876. Molar teeth and other remains of Mammmalia. *Memoirs geol. Surv. India* 10,1: 19–87 (also paginated 1–69).

----- 1877. Notices of new and other Vertebrata from Indian Tertiary and Secondary rocks. *Records geol. Surv. India* 10: 30–43.

----- 1878a. Notices of Siwalik mammals. *Records geol. Surv. India* 11: 64–104.

----- 1878b. Crania of ruminants from the Indian Tertiaries. *Memoirs geol. Surv. India, Palaeont. Indica* 10,1: 88–171 (also paginated 1–84).

----- 1880. Preface [to Indian Tertiary and post-Tertiary Vertebrata, vol. 1] *Memoirs geol. Surv. India, Palaeont. Indica* 10,1 :vii-xix.

----- 1881. Siwalik Rhinocerotidae. *Memoirs geol. Surv. India, Palaeont. Indica* 10,2: 1–62.

----- 1882. Note on some Siwalik and Jamna mammals. *Records geol. Surv. India* 15: 28–33.

----- 1884a. Indian Tertiary and post-Tertiary Vertebrata. Siwalik and Narboda Carnivora. *Memoirs geol. Surv. India, Palaeont. Indica* 10,2: 178–363, pls 1–40.

----- 1884b. Indian Tertiary and post-Tertiary Vertebrata, parts 1–5. *Memoirs geol. Surv. India, Palaeont. Indica* 10,3:i-xxiv, 1–154 (each part also paginated separately).

----- 1885a. Note on a second species of Siwalik camel . . . *Records geol. Surv. India* 18: 78–79.

----- 1885b. A revision of the antelopes of the Siwaliks. *Geological Mag.* 2: 169–171.

----- 1886a. Preliminary note on the Mammalia of the Karnaul Caves. *Records geol. Surv. India* 19: 120–122.

----- 1886b. The fauna of the Karnul Caves. *Memoirs geol. Surv. India, Palaeont. Indica* 4: 23–58, pls 7–10.

----- 1886c. Siwalik Mammalia—Supplement 1. *Memoirs geol. Surv. India, Palaeont Indica* 10,4, 1: 1–21, pls 1–6.

----- [1886d] *Catalogue of the fossil Mammalia in the British Museum (Natural History), Part 2.* London: British Museum (Natural History), 324pp.

----- 1896. *A geographical history of mammals.* Cambridge UP, xii, 400pp.

----- 1898a. *Deer of all lands.* London: Rowland Ward, 330pp.

----- 1898b. On the geographical races of the banting. *Proceedings zool. Soc. Lond.* : 276–278, pl. xxv.

----- 1898c. *Wild oxen, sheep and goats of all lands . . .* London: Rowland Ward, 318pp.

----- 1900. An undescribed type of rusine deer. *Annals Mag. nat. Hist.* 6: 204–205.

----- 1901a. An instance of adaptation among the deer. *Nature, Lond.* 64: 257.

----- 1901b. Notice of an apparently new estuarine dolphin from Borneo. *Proceedings zool. Soc. Lond.* : 88–91, pl. 8.

----- 1903a. The Burmese gaur, or pyoung. *Zoologist* 7: 264–266.

----- 1903b. On two dolphins from Madras. *Journal Bombay nat. Hist. Soc.* 15: 408–414, pls C, D.

----- 1903c. Notes on the Trivandrum cetaceans. *Journal Bombay nat. Hist. Soc.* 15: 40–41, pl. B.

----- 1904a. Large muntjac antlers. *Field* 104: 780.

----- 1904b. The Ichang tufted deer. *Proceedings zool. Soc. Lond.* (1904,2): 166–169.

----- 1904c. On two lorises. *Proceedings zool. Soc. Lond.* (1904,2): 345–346, pl. xxiii.

----- 1905a. The Ceylon chital. *Field* 105: 947.

----- 1905b. The Bonin Island sambar. *Annals Mag. nat. Hist.* 15: 391–2.

----- 1905c. The gorals of India and Burma. *Zoologist* 9: 81–84.

----- 1905d. The wild cattle of the Malay Peninsula. *Field* 105: 151.

----- 1905e. A new (?) anoa. *Field* 106: 378.

----- 1907a. *Game animals of India . . .* London: Rowland Ward, 409pp.

----- 1907b. The Bhutan takin. *Field* 110: 887.

----- 1908a. A Siamese hog-deer. *Field* 111: 583.

----- 1908b. A grey takin. *Field* 111: 790.

----- 1909a. On the skull of a black bear from eastern Tibet . . . *Proceedings zool. Soc. Lond.* : 607–610.

----- 1909b. [Photograph of a spotted bull tsaine or bantin . . . shot . . . in Siam.] *Proceedings zool. Soc. Lond.* : 668–669.

----- 1909c. On an Indian dolphin and porpoise. *Proceedings zool. Soc. Lond.* (1908): 802–806, pls xliv, xlv.

----- 1910a. The gazelles of Seistan. *Nature, Lond.* 83: 201–202.

----- 1910b, 1911. On a wapiti and a muntjac. *Abstracts Proc. zool. Soc. Lond.* (1910): 38; *Proceedings zool. Soc. Lond.* (1910): 987–991.

----- 1912. The Bornean bantin. *Proceedings zool. Soc. Lond.* : 902–906.

----- 1913a. *Catalogue of the . . . Indian big game bequeathed by A. O. Hume . . .* London: British Museum (Natural History), 45pp.

----- 1913b. *Catalogue of the ungulate mammals in the British Museum (Natural History).* Vol. 1. London: British Museum (Nat. Hist.), xvii, 249pp.

----- 1914a. [The Malay race of the Indian elephant.] *Abstracts Proc. zool. Soc. Lond.* (130): 20.

----- 1914b. *Catalogue of the ungulate mammals in the British Museum (Natural History).* Vol. 3. London: British Museum (Nat. Hist.), xv, 283pp.

----- 1915. *Ibid*, Vol. 4, xxii, 438pp.

----- 1916. *Ibid*, Vol. 5, xlv, 207pp.

Lyon, M. W. 1904. Classification of the hares and their allies. *Smithsonian misc. Collns* 45: 321–447.

----- 1906a. Mammals of Banka, Mendanau and Billiton . . . *Proceedings US natn. Mus.* 31: 575–612.

----- 1906b. The pygmy squirrels of the *Nannosciurus melanotis* group. *Proceedings biol. Soc. Wash.* 19: 51–56.

----- 1906c. Notes on the slow lemurs. *Proceedings US natn. Mus.* 31: 527–538, pl. xiii.

----- 1907a. Mammals of Batam Island . . . *Proceedings US natn. Mus.* 31: 653–657.

----- 1907b. A new flying squirrel from the island of Terutau. *Proceedings biol. Soc. Wash.* 20: 17–18.

----- 1907c. Description of a new squirrel of the *Sciurus prevostii* group . . . *Smithsonian misc. Collns* 48: 275–276.

----- 1907d. Mammals collected in western Borneo . . . *Proceedings US natn. Mus.* 33: 547–571.

----- 1907e. Notes on the porcupines of the Malay peninsula and archipelago. *Proceedings US natn. Mus.* 32: 575–594, pls 54–57.

----- 1907f. The squirrels of the *Sciurus vittatus* group in Sumatra. *Smithsonian misc. Collns* 48: 277–283.

----- 1907g. Remarks on the giant squirrels of Sumatra . . . *Proceedings US natn. Mus.* 32: 439–445.

----- 1907h. Notes on some squirrels of the *Sciurus hippurus* group. *Smithsonian misc. Collns* 50: 24–29.

----- 1907i. Notes on a small collection of mammals from the province of Kan-Su, China. *Smithsonian misc. Collns* 50: 133–137.

----- 1908a. On a collection of mammals from the Batu Islands. *Annals Mag. nat. Hist.* 1: 137–140.

----- 1908b. Mammals collected in eastern Sumatra . . . *Proceedings US natn. Mus.* 34: 619–679, pls 52–54.

----- 1909a. Remarks on the insectivores of the genus *Gymnura*. *Proceedings US natn. Mus.* 36: 449–456.

----- 1909b. Additional notes on mammals of the Rhio-linga Archipelago . . . *Proceedings US natn. Mus.* 36: 479–491, pl. 39.

----- 1909c. A new squirrel from Direction Island, South China Sea. *Proceedings US natn. Mus.* 36: 509–510.

----- 1911a. Mammals collected by Dr W. L. Abbott on Borneo . . . *Proceedings US natn. Mus.* 40: 53–146, pls 1–7.

----- 1911b. *Sciurus poliopus* Lyon changed to *Sciurus conipus*. *Proceedings biol. Soc. Wash.* 24: 98.

----- 1911c. Decriptions of four new treeshrews. *Proceedings biol. Soc. Wash.* 24: 167–170.

----- 1913. Treeshrews: an account of the mammalian family Tupaiidae. *Proceedings US natn. Mus.* 45: 1–186.

----- 1916a. Mammals collected . . . on islands off the western coast of Sumatra. *Proceedings US natn. Mus.* 52: 437–462.

----- 1916b. Two new mammals from Sumatra. *Proceedings biol. Soc. Wash.* 29: 209–211.

Ma, S. & Wang, Y. 1986. The taxonomy and distribution of the gibbons of southern China . . . *Zoological Res.* 7: 393–410.

Ma, S., Wang, Y-x. & Groves, C. P. 1988. Taxonomic notes on the subspecies of the Indian muntjac (*Muntiacus muntjak*) in Yunnan, China. *Acta theriol. sin.* 8: 95–104.

Ma, S., Wang, Y-x. & Poirier, F. E. 1988. Taxonomy, distribution and status of gibbons (*Hylobates*) in southern China . . . *Primates* 29: 277–286.

Ma, S., Wang, Y-x. & Xu, L. 1986. Taxonomic and phylogenetic studies on the genus *Muntiacus*. *Acta theriol. sin.* 6: 191–209.

Ma, S. *et al.* 1989. Taxonomy and distribution of the Francois' langur (*Presbytis francoisi*). *Primates* 30: 233–240.

----- 1990. A new species of the genus *Muntiacus* from Yunnan, China. *Zoological Res.* 11: 47–53.

Ma, Y. 1964. A new species of hedgehog from Shansi Province, *Hemiechinus sylvaticus* sp. n. *Acta zootaxon. sin.* 1: 31–36.

Mackenzie, J. M. D. 1951. A note on some natural barriers in Burma. *Journal Bombay nat. Hist. Soc.* 51: 189–199.

MacKinnon, J. R. & Stuart, S. N. (eds) 1989. *The kouprey/ an action plan for its conservation.* Gland, Switzerland: IUCN, 19pp.

MacKinnon, K. S. 1978. Stratification and feeding differences among Malayan squirrels. *Malayan Nat. J.* 330: 593–608.

Maeda, K. 1980. Review on the classification of little tube-nosed bats. *Mammalia* 44: 531–551.

----- 1982. Studies on the classification of *Miniopterus* . . . *Mammalian Sci.*, Suppl. 1: 1–176.

----- 1983. Geographic and sexual variation . . . in *Miniopterus macrodens* . . . *Journal mammal. Soc. Japan* 9: 291–301.

Major, C. I. F. 1897. On *Sus verrucosus* . . . from the Eastern Archipelago. *Annals Mag. nat. Hist.* 19: 521–542.

----- 1899. On fossil and Recent Lagomorpha. *Transactions Linn. Soc. Lond. Zool.* 7: 433–530.

Mandal, A. K. 1984. A note on the distribution of the little Indian field mouse, *Mus booduga* . . . *Bulletin zool. Surv. India* 6: 301–303.

----- 1986. Report on the occurrence of the brown spiny mouse, *Mus platythrix* . . . in West Bengal, India. *Journal Bombay nat. Hist. Soc.* 82: 643–644.

Mandal, A. K. & Das, P. K. 1969. Taxonomic notes on . . . *Anourosorex squamipes* from India. *Journal Bombay nat. Hist. Soc.* 66: 608–612.

Mandal, A. K. & Ghosh, M. K. 1984. Report on . . . *Mus cervicolor* . . . on the Andaman and Nicobar Islands. *Journal Bombay nat. Hist. Soc.* 81: 465–466.

Mandal, A. K. & Ghosh, S. 1981. A new subspecies of the metad, *Millardia meltada* from West Bengal. *Bulletin zool. Surv. India* 4: 235–238.

Mani, M. S. 1974. *Ecology and biogeography in India.* The Hague: Junk (Monographiae biol., no. 23), 785pp.

Manouvrier, -- 1896. *Bulletin Mém. Soc. Anthrop. Paris* 7: 396.

Markvong, A. 1977. Cytogenetic studies on genus *Mus* . . . with special reference to Thailand. *Dissertation Abst. int.* (B) 38: 1048.

Marsden, W. 1811. *The history of Sumatra* . . . London, viii, 479pp, 19pls.

Marsh, H. *et al.* 1989. Irrawaddy dolphin—*Orcaella brevirostris*. Pp. 101–118 in Ridgeway, S. H. & Harrison, R. (*q. v.*).

Marshall, J. 1976, 1977a. Family Muridae (rats and mice). Pp. 397–487 in Lekagul & McNeely, 1977 (*q. v.*). (Distributed separately in 1976—see Musser, 1981b: 237.)

----- 1977b. A synopsis of the Asian species of *Mus*. *Bulletin Am. Mus. nat. Hist.* 158: 173–220.

----- 1979. Classification of genus *Mus*. Pp. 212–220 in Altman, P. C. & Katz, D. D. (eds), *Inbred and genetically defined strains of laboratory animals. Part 1. Mouse and rat.* Bethesda: Fed. Am. Soc. exp. Biol.

----- 1986. Systematics of the genus *Mus*. *Current Topics Microbiol. Immun.* 127: 12–18.

Marshall, J. & Sugardjito, J. 1986. Gibbon systematics. Pp. 137–185 in Swindler, D. R. & Erwin, J. (eds), *Comparative primate biology, vol. 1. Systematics, evolution and anatomy.* New York: Liss.

Marshall, L. G. *et al.* 1990. Phylogenetic relationships of the families of marsupials. Pp. 431–505 (chapter 10) in Genoways, H. H. (ed.) *Current mammalogy* vol. 2. New York & London: Plenum Press.

Marshall, P. 1967. *Wild mammals of Hong Kong.* Hong Kong: Oxford UP, 64pp.

Martens, J. & Niethammer, J. 1972. Die Waldmäuse (*Apodemus*) Nepals. *Zeitschrift Säugetierk.* 37: 144–154.

Martin, A. R. *et al.* 1990. *Whales and dolphins.* London, New York: Salamander Books, 192pp.

Martin, E. B. *et al.* 1987. Conservation crisis—the rhinoceros in India. *Oryx* 21: 212–218.

Martin, K. 1886. Fossile Säugetierreste von Java und Japan. *Sammlung geol. Reichsmus. Leiden* 1,4: 25–68, pls ii-ix.

Martin, W. [C. L.] 1837. [A new species of the genus *Felis*.] *Proceedings zool. Soc. Lond.* (1836): 107–108.

----- 1838. Monograph of the genus *Semnopithecus*. *Magazine nat. Hist.* 2: 320–326; 434–441.

----- 1841. *A natural history of quadrupeds and other mammiferous animals* . . . London: Whitehead.

Mason, G. E. 1908. On the fruit bats of the genus *Pteropus* inhabiting the Andaman and Nicobar Archipelago . . . *Records Indian Mus.* 2: 159–166.

Mason, I. L. (ed.) 1984a. *Evolution of domesticated animals.* London: Longman, 452pp.

----- 1984b. Camels. Pp. 106–115 in Mason, I. L. (above).

----- 1984c. Goat. Pp. 85–99 in Mason, I. L. (above).

Matschie, P. 1893. Uber zwei von Schreber beschreibene Affen . . . *Sitzungsberichte Ges. naturf. Freunde Berl.* : 60–62.

----- 1898. Uber Säugethiere von der Philippinen. *Sitzungsberichte Ges. naturf. Freunde Berl.* : 38–43.

----- 1899. *Die Fledermäuse des Berliner Museums für Naturkunde. 1. Die Megachiroptera* . . . Berlin.

----- 1900. Den Schakel des Menam-Gebietes in Siam. *Sitzungsberichte Ges. naturf. Freunde Berlin* : 144–145.

----- 1901a. Die Säugetiere . . . auf Halmahera . . . *Abhandlungen Senckenb. naturforsch. Ges.* 25: 247–296, taf. xi-xiii.

----- 1901b. Rumänische Säugethiere. *Sitzungsberichte Ges. naturf. Freunde Berl.* : 220–228.

----- [1907]. Mammalia. Pp. 134–244 in *Wissenschaftliche Ergebnisse der Expedition Filchner nach China und Tibet.* Berlin: Mittler. (Dated 1908 but issued 1907.)

----- 1912a. Die achtzenhnte deutsche Geweihausstellung zu Berlin 1912. *Deutsche Jäger-Zeitung* 58/59: 1–88.

----- 1912b. Zur Kenntnis der südchinesischen Kurzschwanz-Makaken. *Sitzungsberichte Ges. naturf. Freunde Berl.* : 305–310.

----- 1912c. Über einige Rassen des Steppenluchses *Felis* (*Caracal*) *caracal*. *Sitzungsberichte Ges. naturf. Freunde Berl.* : 55–67.

----- 1916. Die Verbreit der Beuteltiere auf Neu Guinea . . . *Mitteilungen zool. Mus. Berlin* 8: 256–309.

Matsumoto, H. 1915. On some fossil mammals from Sze-Chuan, China. *Science Rep. Tohoku Univ. Geol.* 3,1: 1–28, pls 1–10.

----- 1926. On some new fossil cervicorns from Kazusa and Liukiu. *Science Rep. Tohoku Univ.,Geol.* 10,2: 17–25, pl. xi.

Matthew, W. D. 1929. Critical observations upon Siwalik mammals. *Bulletin Am. Mus. nat. Hist.* 56: 437–560.

Matthew, W. D. & Granger, W. 1923. New fossil mammals from the Pliocene of Sze-Chuan, China. *Bulletin Am. Mus. nat. Hist.* 48: 563–598.

Mayer, -- 1856. Zur Anatomie des Orang-Utang . . . *Archiv Naturgesch.* 22: 281–304.

Mayr, E. 1944. Wallace's line in the light of recent zoogeographic studies. *Quarterly Rev. Biol.* 19: 1–14.

----- 1986. Uncertainty in science: is the giant panda a bear or a raccoon? *Nature, Lond.* 323: 769–771.

Mazak, V. 1963. Asiatische Wildesel im Pragen zoologischer Garten. *Zeitschrift Säugetierk.* 28: 278–293.

----- 1968. Nouvelle sous-espèce de tigre provenent de l'Asie du Sud-Est. *Mammalia* 32: 104–112.

----- 1981. *Panthera tigris. Mammalian Species* 152: 8pp.

McAllan, I. A. W. & Bruce, M. D. 1989. Some problems in vertebrate nomenclature. 1. Mammals. *Bollettino Mus. reg. Scienze nat.* 7: 443–460.

McBee, K. *et al.* 1986. Standard karyology of nine species of vespertilionid bats . . . from Thailand. *Annals Carnegie Mus.* 55: 95–116.

McCann, C. 1978. Animal colouration in response to environment with special reference to *Callosciurus sladeni* . . . . Unpublished typescript in British Museum (Natural History), 75pp.

McClelland [= M'Clelland], J. 1838. On the genus *Hexaprotodon* of Dr Falconer and Captain Cautley. *Journal Asiat. Soc. Bengal* 7: 1038–1047, pl. lix.

----- 1839. List of Mammalia and birds collected in Assam. *Proceedings zool. Soc. Lond.* 7: 146–167.

----- 1842a. Editor's note [following 'Further notice of a nondescript species of deer . . . . ' by P. Eld.] *Calcutta J. nat. Hist.* 2: 417, pl. xii.

----- 1842b. On collections. *Calcutta J. nat. Hist.* 2: 456–459.

----- 1843. Description of the Sungnai, *Cervus* (*Rusa*) *frontalis*. *Calcutta J. nat. Hist.* 3: 401–409, pls xiii, xiv.

McFarlane, D. A. & Blood, B. R. 1986. Taxonomic notes on a collection of Rhinolophidae from northern Thailand . . . *Zeitschrift Säugetierk.* 51: 219–223.

McKay, G. M. 1984. Ecology and biogeography of mammals. Pp. 413–429 in Fernando, C. H. (ed.), *Ecology and biogeography in Sri Lanka.* The Hague: Junk (Monographiae biol. vol. 57).

McKenna, M. C. 1962. *Eupetaurus* and the living petauristine sciurids. *American Mus. Novit.* 2104: 38pp.

Mead, J. G. 1989. Bottlenose whales . . . ; Beaked whales of the genus *Mesoplodon*. Pp. 321–348; 349–430 in Ridegeway, S. H. & Harrison, R. (*q. v.*).

Mead, J. I. 1989. *Nemorhaedus goral. Mammalian Species* 335: 5pp.

Mearns, E. A. 1905. Descriptions of new ... mammals from the Philippine Islands. *Proceedings US natn. Mus.* 28: 425–460.

Mech, L. D. 1974. *Canis lupus. Mammalian Species* 37: 6pp.

Medina, F. I. S. & Leonard, A. 1977. Observations sur les chromosomes de *Suncus luzoniensis* Peters. *Acta zool. Path. antverp.* 69: 183–188.

Medway, Lord 1960. The Malay tapir in late Quaternary Borneo. *Sarawak Mus. J.* 9: 356–360.

----- 1961. The status of *Tupaia splendidula* ... *Treubia* 25: 269–272.

----- 1963. Niah Cave animal bone. VI—new records. *Sarawak Mus. J.* 11: 188–191.

----- 1965a. Comments on the status of *Rattus inas* ... *Federation Mus. J.* 9: 95–101.

----- 1965b, 1977a. *Mammals of Borneo* ... Malay Branch r. Asiatic Soc., 193pp. (Also issued as vol. 36, part 3 of *Journal Malayan Brch r. Asiatic Soc.* with same pagination.) 2nd edition, 1977.

----- 1969,1978,1983. *The wild mammals of Malaya* ... Kuala Lumpur: Oxford Univ. Press, 127pp. 1978: 2nd ed. ; 1983: 2nd ed. reprinted with corrections.

----- 1970. The monkeys of Sundaland ... Pp. 515–553 in Napier, J. R. & Napier, P. H. (eds) *Old World monkeys: evolution, systematics and behaviour*. London: Academic Press.

----- 1977b. The ancient domestic dogs of Malaysia. *Journal Malay. Br. r. Asiat. Soc.* 50: 14–27.

Medway, Lord & Lim, B. L. 1966. The specific relation of *Rattus tiomanicus* Miller. *Bulletin natn. Mus. Singapore* 34: 33–38.

Medway, Lord & Yong, H. S. 1976. Problems in the systematics of the rats (Muridae) of Peninsular Malaysia. *Malaysian J. Sci.* 4: 43–53, pls 1–4.

Mees, G. F. 1957. Over het belang van Temminck's 'Discours preliminaire' voor de zoologische nomenclatur. *Zoologische Meded. Leiden* 35: 205–227.

Meester, J. A. J. *et al.* 1986. Classification of southern African mammals. *Monographs Transvaal Mus.* 5: 359pp.

Mein, P. & Tupinier, Y. 1977. Formule dentaire et position systématique du minioptère ... *Mammalia* 41: 207–211.

Mein, P. *et al.* 1990. Nouveaux rongeurs du Miocene de Li (Thailande). *Compte r. Acad. Sci.* Ser. II, 310: 861–865.

Mell, R. 1922. Beitrage zur Fauna sinica. *Archiv Naturgesch.* 88,10: 1–159.

Melville, D. S. 1984. Notes on some small mammals from Chiang Mai Province ... *Natural Hist. Bull. Siam Soc.* 31 (1983): 157–162.

Menu, H. 1985. Morphotypes dentaires actuels et fossiles des chiroptères vespertilionines. 1e partie ... *Palaeovertebrata* 15: 71–128.

----- 1987. Morphotypes dentaires actuels et fossiles des chiroptères ... *Palaeovertebrata* 17: 77–150, 5pls.

Menzies, J. I. & Pernetta, J. C. 1986. A taxonomic revision of cuscuses allied to *Phalanger orientalis* ... *Journal Zool. Lond.* (B) 1: 551–618.

Merriam, C. H. 1897. The generic names *Ictis, Arctogale* and *Arctogalidea*. *Science N. Y.* 5: 302.

Mertens, R. 1929. Zwei neue Palmenroller von den Kleinen Sunda-Inseln. *Zoologischer Anz.* 84: 333–336.

----- 1936. Die Säugetiere der Inseln Bali, Lombok, Sumbawa und Flores. *Zoologische Jahrb. Jena* (*Syst.*) 68: 273–324, 2pls.

Meyen, F. J. F. 1833. Beiträge zur Zoologie ... Säugethiere. *Nova Acta Acad. Caesar. Leop. -Carol.* 16: 549–610, 6pls.

Meyer, A. B. 1890. Description of a new squirrel from the Philippine Islands. *Proceedings zool. Soc. Lond.* : 599–601.

----- 1894/95. Eine neue *Tarsier* Art. *Abhandlungen Ber. K. zool. anthrop. -ethn. Mus. Dresden* 1: 1–12.

----- 1897. Säugethiere vom Celebes- und Philippinen- Archipel. *Abhandlungen Ber. K. zool. anthrop. -ethn. Mus. Dresden* 6,6: 36pp, 15pls.

----- 1898. Ueber zwei Eichhörnchenarten von Celebes. *Abhandlungen Ber. K. zool. anthrop. -ethn. Mus. Dresden* 7,4: 3pp.

----- 1899. Säugetiere vom Celebes- und Philippinen-Archipel II. *Abhandlungen Ber. K. zool. anthrop. -ethn. Mus. Dresden* 7,7:viii, 55pp, 11pls.

Meyer, F. A. A. 1793. *Systematische-summarische Uebersicht der neuersten zoologischen Entdeckungen* ... Leipzig, 198pp.

----- 1794. Ueber de la Metheries schwarzen Panther. *Zoologische Annalen Weimar* 1: 394–396.

Meyer, J. N. E. von 1826. *Beiträge zur Anatomie des Tigers* .... Wien: 62pp, 1pl.

Michaelis, B. 1969. Die Schleichkatzen (Viverriden) Afrikas. *Säugetierkundliche Mitt.* 20: 1–110.

Mickleburgh, S. P. *et al.* (in prep). Old-World fruit bats. An action plan for the family Pteropididae. IUCN.

Miller, C. 1779. An account of the island of Sumatra. *Philosophical Trans. r. Soc.* 68 (1778): 160–179.

Miller, G. S. 1896. Genera and subgenera of voles and lemmings. *North American Fauna* 12: 78pp, 3pls.

----- 1897. Description of a new vole from Kashmir. *Proceedings biol. Soc. Wash.* 11: 141.

----- 1898. List of bats collected by Dr W. L. Abbott in Siam. *Proceedings Acad. nat. Sci. Philad.* : 316–325.

----- 1899. The voles collected by Dr W. L. Abbott in central Asia. *Proceedings Acad. nat. Sci. Philad.* : 281–298, pls 12–13.

----- 1900a. Mammals collected ... on islands in the South China Sea. *Proceedings Wash. Acad. Sci.* 2: 203–246.

----- 1900b. Descriptions of two new squirrels from Trong, Lower Siam. *Proceedings Wash. Acad. Sci.* 2: 79–81.

----- 1900c. The giant squirrels of Burmah and the Malay Peninsula. *Proceedings Wash. Acad. Sci.* 2: 69–77.

----- 1900d. Seven new rats collected by Dr W. L. Abbott in Siam. *Proceedings biol. Soc. Wash.* 13: 137–150.

----- 1900e. Mammals collected by Dr W. L. Abbott on Pulo Lankawi and the Batang Islands. *Proceedings biol. Soc. Wash.* 13: 187–193.

----- 1900f. A new mouse deer from Lower Siam. *Proceedings biol. Soc. Wash.* 13: 185–186.

----- 1901a. Mammals collected by Dr W. L. Abbott on the Natuna Islands. *Proceedings Wash. Acad. Sci.* 3: 111–138.

----- 1901b. Description of three new Asiatic shrews. *Proceedings biol. Soc. Wash.* 14: 157–159.

----- 1901c. A new squirrel from Borneo. *Proceedings biol. Soc. Wash.* 14: 33–34.

----- 1901d. The subgenus *Rhinosciurus* of Trouessart. *Proceedings biol. Soc. Wash.* 14: 23.

----- 1901e. A new name for *Mus obscurus* Miller. *Proceedings biol. Soc. Wash.* 14: 178.

----- 1902a. Mammals collected in ... Sumatra. *Proceedings Acad. nat. Sci. Philad.* 54: 143–159.

----- 1902b. The mammals of the Andaman and Nicobar Islands. *Proceedings US natn. Mus.* 24: 751–795.

----- 1902c. Two new tropical Old World bats. *Proceedings biol. Soc. Wash.* 15: 245–246.

----- 1902d. A new pig from Sumatra. *Proceedings biol. Soc. Wash.* 15: 51–52.

----- 1902e. Two new Malayan mouse deer. *Proceedings biol. Soc. Wash.* 15: 173–175.

----- 1903a. Seventy new Malaysian mammals. *Smithsonian misc. Collns* 45: 1–73.

----- 1903b. Mammals collected ... northwest Sumatra. *Proceedings US natn. Mus.* 26: 437–484, pls xviii-xix.

----- 1903c. A new squirrel from lower Siam. *Proceedings biol. Soc. Wash.* 16: 147–148.

----- 1903d. A new name for *Mus atratus* Miller. *Proceedings biol. Soc. Wash.* 16: 50.

----- 1903e. Descriptions of eleven new Malayan mouse deer. *Proceedings biol. Soc. Wash.* 16: 31–44.

----- 1905. A new genus of bats from Sumatra. *Proceedings biol. Soc. Wash.* 18: 229–230.

----- 1906a. Mammals collected ... in the Karimata Islands ... *Proceedings US natn. Mus.* 31: 55–66.

----- 1906b. The mammals collected ... in the Rhio-Linga Archipelago. *Proceedings US natn. Mus.* 31: 247–286.

----- 1906c. Twelve new genera of bats. *Proceedings biol. Soc. Wash.* 19: 83–86.

----- 1906d. Seven new Malayan bats. *Proceedings biol. Soc. Wash.* 19: 61–65.

----- 1906e. The mammals of Engano Island, West Sumatra. *Proceedings US natn. Mus.* 30: 819–825.

----- 1906f. A new name for *Rhinolophus minutus* Miller. *Proceedings biol. Soc. Wash.* 19: 41.

----- 1906g. Two new carnivores from the Malay Peninsula. *Proceedings biol. Soc. Wash.* 19: 25–27.

----- 1906h. Notes on Malayan pigs. *Proceedings US natn. Mus.* 30: 737–758, pls 39–64.

----- 1906i. The monkeys of the *Macaca nemestrina* group. *Proceedings US natn. Mus.* 29: 555–563, pls xiii-xx.

----- 1907. The families and genera of bats. *Bulletin US natn. Mus.* 57:xxvii, 282pp, 14pls.

----- 1910. Description of ... mammals from the Philippine Islands. *Proceedings US natn. Mus.* 38: 391–404.

----- 1911a. Two new shrews from Kashmir. *Proceedings biol. Soc. Wash.* 24: 241–242.

----- 1911b. Description of six new mammals from the Malay Archipelago. *Proceedings biol. Soc. Wash.* 24: 25–28.

----- 1911c. A new mouse-deer from the Rhio-Lingga Archipelago. *Proceedings biol. Soc. Wash.* 24: 165–166.

----- 1912. *Catalogue of the mammals of Western Europe*. London: BM(NH), 1019pp.

----- 1913a. A new shrew from Baltistan. *Proceedings biol. Soc. Wash.* 26: 113–114.

----- 1913b. Fifty-one new Malayan mammals. *Smithsonian misc. Collns* 61,21: 30pp.

----- 1913c. A new pteropine bat from Luzon. *Proceedings biol. Soc. Wash.* 26: 73–74.

----- 1913d. Some overlooked names of Sicilian mammals. *Proceedings biol. Soc. Wash.* 26: 80–81.

----- 1913e. Two new murine rodents from Baltistan. *Proceedings biol. Soc. Wash.* 26: 197–198.

----- 1915a. A new squirrel from northeastern China. *Proceedings biol. Soc. Wash.* 28: 115–116.

----- 1915b. The jaw of the Piltdown Man. *Smithsonian misc. Collns* 65,12: 31pp, 5pls.

----- 1918. A new river-dolphin from China. *Smithsonian misc. Collns* 68,9: 12pp, 13pls.

----- 1928. A new bat of the genus *Coelops*. *Proceedings biol. Soc. Wash.* 41: 85–86.

----- 1931. Two new mammals from Celebes. *Journal Mammal.* 12: 412–413.

----- 1933. The classification of the gibbons. *Journal Mammal.* 14: 158–159.

----- 1934a. The langurs of the *Presbytis femoralis* group. *Journal Mammal.* 15: 124–137.

----- 1934b. A new monkey from Dutch Northeast Borneo. *Proceedings biol. Soc. Wash.* 47: 15–17.

----- 1940a. A new mole from Annam. *Journal Mammal.* 21: 203–204.

----- 1940b. Notes on some moles from southeastern Asia. *Journal Mammal.* 21: 442–444.

----- 1942. Zoological results of the George Vanderbilt Sumatran Expedition. V. Mammals . . . *Proceedings Acad. nat. Sci. Philad.* 94: 107–165, pls 3–6.

Miller, G. S. & Allen, G. M. 1928. The American bats of the genera *Myotis* and *Pizonyx*. *Bulletin US natn. Mus.* 144:viii, 218pp.

Miller, G. S. & Hollister, N. 1921a. Description of sixteen new murine rodents from Celebes. *Proceedings biol. Soc. Wash.* 34: 66–76.

----- 1921b. Twenty new mammals collected . . . in Celebes. *Proceedings biol. Soc. Wash.* 34: 93–104.

----- 1922. A new phalanger from Celebes. *Proceedings biol. Soc. Wash.* 35: 115–116.

Miller, S. D. & Everett, D. D. 1986. *Cats of the world: biology, conservation and management*. Washington DC & Kingsville: Nat. Wildl. Fed. etc, xii, 501pp.

Milne-Edwards, A. 1864. Recherches anatomique . . . sur la famille des chevrotains. *Annales Sci. nat. Zool.* 2: 49–167, pls 2–12.

----- 1866. Notes sur le Mi-lou . . . *Annales Sci. nat.,Zool.* 5: 380–382.

----- 1867a. Observations sur quelques mammifères du nord de la Chine. *Annales Sci. nat.,Zool.* (5)7: 375–377; 8: 374–376.

----- 1867b. Description de quelques espèces nouvelles d'écureuils . . . *Revue Mag. Zool.* 19: 193–197; 225–232.

----- 1867c. Note sur une nouvelle espèce du genre *Nycticebus* . . . *Nouvelles Archs Mus. Hist. nat. Paris* 3, Bull. : 9–13.

----- 1870a. Note sur quelques mammifères du Thibet oriental. *Compte r. Acad. Sci. Paris* 70: 341–342.

----- 1870b. Note sur quelques mammifères du Thibet oriental. *Annales Sci. nat. Zool.* 13,10: 1p.

----- 1871a. Etude pour servir à l'histoire de la faune mammologique de la Chine. Pp. 67–229 in *Recherches pour servir à l'histoire des mammifères*. Paris: Masson.

----- 1871b. Note sur une nouvelle espèce de semnopithèque . . . *Nouvelles Archs Mus. Hist. nat. Paris* 6, Bull. : 7–9, pl. 1.

----- 1872a. [Description of mammals—footnotes] Pp. 91–93 In David, A., Rapport adressé à mm les professeurs-administrateurs du Museum d'Histoire Naturelle. *Nouvelles Archs Mus. Hist. nat. Paris* 7: 73–100.

----- 1872b, 1874. Memoire sur la faune mammalogique du Tibet oriental. Pp. 231–304 (1872), 305–386 (1874) in *Recherches pour servir à l'histoire des mammifères*. Paris: Masson.

----- 1877. Sur quelques mammifères et crustacés nouveaux. *Bulletin Soc. philom. Paris* 13 (1876): 8–10.

----- 1892. Observations sur les mammifères du Thibet. *Revue gén. Sci. pur. appl.* 3: 670–672.

----- 1897a. In Trouessart, E-L., Descriptions de deux espèces nouvelles de singes . . . *Naturaliste* 11: 9–10.

----- 1897b. Note sur une nouvelle espèce du genre Rhinopithèque . . . *Bulletin Mus. Hist. nat. Paris* 3: 156–160.

Minasian, S. M. *et al.* 1984. *The world's whales: the complete illustrated guide*. Washington: Smithsonian Books, 224pp.

Miranda-Ribeiro, A. de 1936. Notas cetologicas . . . *Boletim Mus. nac. Rio de Janeiro* 12: 3–23 (Portuguese);25–48 (English), 2pls.

Mishra, A. C. & Dhanda, V. 1975. Review of the genus *Millardia* . . . *Journal Mammal.* 56: 76–80.

Misonne, X. 1969. African and Indo-Australian Muridae . . . *Annales Mus. r. Afr. cent.,Sci. zool.* 172: 219pp, 27pls.

----- 1990. New record for Iran: *Golunda elliotti* . . . *Mammalia* 54: 494.

Mitchell, R. 1975. A checklist of Nepalese mammals (excluding bats). *Säugetierkundliche Mitt.* 23: 152–157.

----- 1979. The sciurid rodents . . . of Nepal. *Journal Asian Ecology* 1: 21–28.

Mitchell, R. M. & Derksen, D. V. 1976. Additional new mammal records from Nepal. *Mammalia* 40: 55–61.

Mivart, St G. 1867. Notes on the osteology of the Insectivora. *Journal Anat. Physiol.* 1: 281–312.

----- 1890. *A monograph of the Canidae*. London: Porter & Dulau, 216pp.

Miyakazi, N. & Wada, S. 1978. Fraser's dolphin in the western north Pacific. *Scientific Rep. Whales Res. Inst.* 30: 231–244.

Mohr, E. 1964. Zur Nomenklatur und Systematik der Quastenstachler, Gattung *Atherurus* . . . *Zeitschrift Säugetierk.* 29: 93–116.

Montagu, G. 1808. An account of the larger and lesser species of horse-shoe bats . . . *Transactions Linn. Soc. Lond.* 9: 162–173.

----- 1821. Description of a specimen of *Delphinus* . . . *Memoirs Wernerian nat. Hist. Soc.* 3: 75–82, pl. 3.

Moore, J. C. 1956. A new subspecies of an Oriental squirrel, *Dremomys lokriah*. *American Mus. Novit.* 1816: 2pp.

----- 1958a. New striped tree squirrels from Burma and Thailand. *American Mus. Novit.* 1879: 6pp.

----- 1958b. New genera of East Indian squirrels. *American Mus. Novit.* 1914: 5pp.

----- 1958c. A new species and a redefinition of the squirrel genus *Prosciurillus* of Celebes. *American Mus. Novit.* 1890: 5pp.

----- 1959a. Relationships among living squirrels of the Sciurinae. *Bulletin Am. Mus. nat. Hist.* 118: 153–206.

----- 1959b. A new subgenus and new species of squirrel from Borneo. *American Mus. Novit.* 1944: 10pp.

----- 1960. Squirrel geography of the Indian subregion. *Systematic Zool.* 9: 1–17.

----- 1968. Relationships among the living genera of beaked whales. *Fieldiana Zool.* 53: 209–298.

Moore, J. C. & Gilmore, R. M. 1965. A beaked whale new to the western hemisphere. *Nature, Lond.* 205: 1239–1240.

Moore, J. C. & Tate, G. H. H. 1959. On the identity of the tree squirrel *Macroxus chinensis* Gray, 1867. *American Mus. Novit.* 1970: 7pp.

----- 1965. A study of the diurnal squirrels, Sciurinae, of the Indian and Indo-Chinese subregions. *Fieldiana Zool.* 48: 1–351.

Moriwaki, K. *et al.* 1986. Genetic features of major geographical isolates of *Mus musculus*. Pp. 55–61 in Potter, M. *et al.* (*q. v.*)

Morselli, -- 1901. *Il precursore dell'Uomo* (*n. v.*).

Moses, S. T. 1948. Notes on two whales stranded in Baroda State . . . *Proceedings Indian Sci. Congr.* 34: 188.

Motley, J. & Dillwyn, L. L. 1855. *Contribution to the natural history of Labuan* . . . London: van Voorst, 62pp, 12pls.

Moulton, J. C. 1921. Occurrence of the Malayan badger or teledu in Borneo. *Journal Straits Brch r. Asiat. Soc.* 83: 142.

Mueller, P. L. S. 1776. *Des Ritters C. von Linné . . . vollständiges Natursystem, Supplement*. Nürnberg.

Müller, F. 1911. Beiträge zur Kenntnis der Stachelschweine Asiens . . . *Sitzungsberichte Ges. naturf. Freunde Berl.* : 110–130.

Müller, S. 1838. Over eenige nieuwe zoogdieren van Borneo. *Tijdschrift Natuurl. Gesch. Physiol.* 5: 134–150.

----- 1839–1840. Over de zoogdieren van den Indischen Archipel. Pp. 1–8, pls 1–3,17 (1839); pp. 9–57, table, pls 4–6,8–12 (1840). In Temminck, C. J., *Verhandelingen natuurlijke geschiedenis Nederlandische overzeesche Bezittingen*. Leiden.

Müller, S. & Schlegel, H. 1840–41. Monographisch oversigt van het Geslacht *Semnopithecus*. Pl. 10 (1840), pp. 57–84 (1841) in Temminck, C. J., as above.

----- 1841,1844. Over de tot heden bekunde eekhorens (*Sciurus*) van den Indischen Archipel. Pls 13–15 (1841), pp. 85–102 (1844) in Temminck, C. J., as above.

----- 1843,1845a. Over de wilde zwijnen van den Indischen Archipel. Pls 18–31 (1843), pp. 169–182 (1845) in Temminck, C. J., as above.

----- 1845b. Over de herten van den Indischen Archipel. Pp. 209–228, pls 42–45, in Temminck, C. J., as above.

Munthe, J. 1980. Rodents of the Miocene Daud Khel local fauna, Mianwali District, Pakistan. Part 1. *Contributions Biol. Geol. Milwaukee publ. Mus.* 34: 36pp.

Munthe, J. & West, R. M. 1980. Insectivora of the Miocene Daud Khel local fauna . . . Pakistan. *Contributions Biol. Geol. Milwaukee publ. Mus.* 38: 17pp.

Murray, A. 1866. *The geographical distribution of mammals*. London: Day, xvi, 420pp.

Murray, J. A. 1884a. Additions to the present knowledge of the vertebrate zoology of Persia. *Annals Mag. nat. Hist.* 14: 97–106.

----- 1884b. *The vertebrate zoology of Sind* . . . London & Bombay, xvi, 424pp, 13pls.

----- 1884c. A contribution to the knowledge of the marine fauna of Kurrachee. *Annals Mag. nat. Hist.* 13: 348–352.

----- 1885. Description of a new species of *Mus* from Sind. *Proceedings zool. Soc. Lond.* : 809–810.

----- 1886. Description of a new *Gerbillus* from Sind. *Annals Mag. nat. Hist.* 17: 246–248.

Musser, G. G. 1969a. Notes on the taxonomic status of *Rattus aspinatus* Tate and Archbold and *Mus callitrichus* Jentink . . . *American Mus. Novit.* 2365: 9pp.

----- 1969b. A new genus and species of murid rodent from Celebes. *American Mus. Novit.* 2384: 41pp.

----- 1969c. Taxonomic notes on *Rattus dollmani* and *Rattus hellwaldi* . . . of Celebes. *American Mus. Novit.* 2386: 24pp.

Musser, G. G. 1970a. Reidentification and reallocation of *Mus callitrichus* . . . *American Mus. Novit.* 2440: 35pp.

----- 1970b. The taxonomic identity of *Mus bocourti* A. Milne-Edwards . . . *Mammalia* 34: 484–490.

----- 1970c. *Rattus masaretes*: a synonym of *Rattus rattus moluccarius*. *Journal Mammal.* 51: 606–609.

----- 1970d. Species-limits of *Rattus brahma* . . . *American Mus. Novit.* 2406: 27pp.

----- 1971a. The taxonomic status of *Rattus dammermani* and *Rattus toxi* . . . *Beaufortia* 18: 205–216.

----- 1971b. The taxonomic association of *Mus faberi* Jentink with *Rattus xanthurus* . . . *Zoologische Meded.* 45: 107–118.

----- 1971c. Taxonomic status of *Rattus tatei* and *Rattus frosti*. *American Mus. Novit.* 2454: 19pp.

----- 1971d. The taxonomic status of *Rattus tondanus* Sody . . . *Zoologische Meded.* 45: 147–157.

----- 1971e. The identities and allocations of *Taeromys paraxanthus* and *T. tatei* . . . *Zoologische Meded.* 45: 127–138.

----- 1972a. The species of *Hapalomys* . . . *American Mus. Novit.* 2503: 27pp.

----- 1972b. Identities of taxa associated with *Rattus rattus* . . . of Sumba Islands, Indonesia. *Journal Mammal.* 53: 861–865.

----- 1973a. Notes on additional specimens of *Rattus brahma*. *Journal Mammal.* 54: 267–270.

----- 1973b. Species-limits of *Rattus cremoriventer* and *Rattus langbianus* . . . *American Mus. Novit.* 2525: 65pp.

----- 1973c. Zoological significance of the ricefield rat, *Rattus argentiventer*, on Celebes . . . *American Mus. Novit.* 2511: 30pp.

----- 1977a. *Epimys benguetensis*, a composite, and one zoogeographic view of rat and mouse faunas in the Philippines and Celebes. *American Mus. Novit.* 2624: 15pp.

----- 1977b. Notes on the Philippine rat, *Limnomys* . . . *American Mus. Novit.* 2636: 14pp.

----- 1979. The species of *Chiropodomys*. *Bulletin Am. Mus. nat. Hist.* 162: 377–445.

----- 1981a. The giant rat of Flores and its relatives east of Borneo and Bali. *Bulletin Am. Mus. nat. Hist.* 169: 67–176.

----- 1981b. Notes on the systematics of Indo-Malayan murid rodents . . . *Bulletin Am. Mus. nat. Hist.* 168: 225–334.

----- 1981c. A new genus of arboreal rat from west Java, Indonesia. *Zoologische Verh. Leiden* 189: 35pp, 4pls.

----- 1982a. A new genus of arboreal rat from Luzon Island in the Philippines. *American Mus. Novit* 2730: 23pp.

----- 1982b. The Trinil rats. *Modern quatern. Res. SE Asia* 7: 65–85.

----- 1982c. The definition of *Apomys* . . . *American Mus. Novit.* 2746: 43pp.

----- 1982d. *Crunomys* and the small-bodied shrew rats native to the Philippine Islands and Sulawesi. *Bulletin Am. Mus. nat. Hist.* 174: 1–95.

----- 1984. Identities of subfossil rats from caves in southwestern Sulawesi. *Modern quatern. Res. SE Asia* 8: 61–94.

----- 1986. Sundaic *Rattus*: definitions of *Rattus baluensis* and *Rattus korinchi*. *American Mus. Novit.* 2862: 24pp.

----- 1987a. The mammals of Sulawesi. Pp. 73–93 in Whitmore, T. C. (*q. v.*).

----- 1987b. The occurrence of *Hadromys* . . . in Early Pleistocene Siwalik strata in northern Pakistan. *American Mus. Novit.* 2883: 36pp.

----- 1991. Sulawesi rodents: descriptions of new species of *Bunomys* and *Maxomys* . . . *American Mus. Novit.* 3001: 41pp.

Musser, G. G. & Boeadi 1980. A new genus of murid rodent from the Komodo Islands . . . *Journal Mammal.* 61: 395–413.

Musser, G. G. & Califia, , D. 1982. Identities of rats from Pulau Maratua . . . *American Mus. Novit.* 2726: 30pp.

Musser, G. G. & Chiu, S. 1979. Notes on taxonomy of *Rattus andersoni* and *R. excelsior* . . . *Journal Mammal.* 60: 581–592.

Musser, G. G. & Dagosto, M. 1987. The identity of *Tarsius pumilus*. *American Mus. Novit.* 2867: 53pp.

Musser, G. G. & Freeman, P. W. 1981. A new species of *Rhynchomys* from the Philippines. *Journal Mammal.* 62: 154–159.

Musser, G. G. & Gordon, L. K. 1981. A new species of *Crateromys* from the Philippines. *Journal Mammal.* 62: 513–525.

Musser, G. G. & Heaney, L. R. 1985. Philippine *Rattus*: a new species from the Sulu Archipelago. *American Mus. Novit.* 2818: 32pp.

Musser, G. G. & Holden, M. E. 1991. Sulawesi rodents . . . morphological and geographical boundaries of species in the *Rattus hoffmanni* group . . . *Bulletin Am. Mus. nat. Hist.* 206: 322–413.

Musser, G. G. & Newcomb, C. 1983. Malaysian murids and the giant rat of Sumatra. *Bulletin Am. Mus. nat. Hist.* 174: 329–598.

----- 1985. Definition of Indochinese *Rattus losea* and a new species from Vietnam. *American Mus. Novit.* 2814: 32pp.

Musser, G. G., Gordon, L. K. & Sommer, H. 1982. Species-limits in the Philippine murid *Chrotomys*. *Journal Mammal.* 63: 514–521.

Musser, G. G., Koopman, K. F. & Califia, D. 1982. The Sulawesian *Pteropus arquatus* and *P. argentatus* are *Acerodon celebensis*. *Journal Mammal.* 63: 319–328.

Musser, G. G. *et al.* 1979. Definition and contents of the Sundaic genus *Maxomys* . . . *Journal Mammal.* 60: 592–606.

----- 1985. Philippine rats: a new species of *Crateromys* . . . *American Mus. Novit.* 2821: 25pp.

----- 1986. *Paulamys*, a replacement name for *Floresomys* Musser, 1981 . . . *American Mus. Novit.* 2850: 10pp.

Muul, I. 1980. An additional species of squirrel, *Petinomys vordermanni*, in Thailand. *Natural Hist. Bull. Siam Soc.* 28: 135–136.

----- 1989. Rodents of conservation concern in the Southeast Asian region. *Occasional Pap. IUCN Species Survival Comm.* 4:51–52.

Muul, I. & Lim, B. L. 1971. New locality records for some mammals of West Malaysia. *Journal Mammal.* 52: 430–437.

----- 1978. Comparative morphology, food habits, and ecology of some Malaysian arboreal rodents. Pp. 361–368 in *The ecology of arboreal folivores*. Washington DC: Smithsonian Institute Press.

Muul, I. & Thonglongya, K. 1971. Taxonomic status of *Petinomys morrisi* (Carter). *Journal Mammal.* 52:362–369.

Nabhitabhata, J. *et al.* 1982. Ein Zwerg unter den Säugetieren— die Fledermaus *Craseonycteris thonglongyai* aus Thailand. *Natur Mus.,Frankf.* 112: 81–86.

Nachtrieb, H. F. 1892. A new Lemur (Menagensis). *Zoologischer Anz.* 15: 147–148.

Nadler, C. F. & Hoffmann, R. S. 1970. Chromosomes of some Asian and South American squirrels . . . *Experientia* 26: 1383–1386.

Nadler, C. F. *et al.* 1975. Chromosomes of three species of Asian tree squirrels, *Callosciurus* . . . *Experientia* 31: 166–167.

Nagorsen, D. 1985. *Kogia simus*. *Mammalian Species* 239: 6pp.

Naik, S. N. 1978. Origin and domestication of zebu cattle (*Bos indicus*). *Journal hum. Evol.* 7: 23–30.

Namiya, M. 1909. [On small mammals of Okinawa and Amami-Oshima Islands.] *Zoological Mag. Tokyo* 21: 452–457 (in Japanese).

Napier, J. R. & Napier, P. H. 1985. *Natural history of the primates*. London: British Museum (Nat. Hist.).

Napier, P. H. 1981,1985. *Catalogue of primates in the British Museum (Natural History)* . . . Parts 2 and 3. London: British Museum (Nat. Hist.), 203, 111pp.

Nash, W. G. & O'Brien, S. J. 1987. A comparative chromosome banding analysis of the Ursidae . . . *Cytogenetics Cell Genet.* 45: 206–212.

Nath, B. 1952. On a new mouse from the Khasi Hills, Assam (India). *Journal zool. Soc. India* 4: 85–88.

Nath, S. 1987. On a collection of long-eared bats of the genera *Otonycteris* and *Plecotus* . . . *Journal Bombay nat. Hist. Soc.* 84: 425.

Neas, J. E. & Hoffmann, R. S. 1987. *Budorcas taxicolor*. *Mammalian Species* 277: 7pp.

Nehring, A. 1885. Uber eine neue Art von Wildschweinen . . . aus Südost-Borneo. *Zoologischer Anz.* 8: 347–353.

----- 1886. Uber zwei Schädel des *Sus longirostris* Nehring von Borneo und Java. *Sitzungsberichte Ges. naturf. Freunde Berl.* : 80–85.

----- 1889. Ueber *Sus celebensis* und Verwandte. *Abhandlungen Ber. K. zool. anthrop-ethn. Mus. Dresden* (1888–89), 2: 34pp, 2pls.

----- 1890. Uber Säugethiere der Philippinen . . . *Sitzungsberichte Ges. naturf. Freunde Berl.* : 101–108.

----- 1894. Säugethiere von den Philippinen . . . *Sitzungsberichte Ges. naturf. Freunde Berl.* : 179–193.

----- 1902. *Nesokia gracilis*, n. sp., von der Insel Ceylon. *Sitzungsberichte Ges. naturf. Freunde Berl.* : 116–121.

Neitzel, H. 1982. *Karyotypen Evolution und deren Bedeutung für den Speciationprozess der Cerviden* . . . Berlin: Freien Universität, inaugural dissertation.

Nellis, D. W. 1989. *Herpestes auropunctatus*. *Mammalian Species* 342: 6pp.

Neuhauser, H. N. 1970. First positive record of *Pipistrellus savii* from India. *Journal Bombay nat. Hist. Soc.* 67: 319–320.

Nevill, H. 1887. Species of *Herpestes* in Ceylon. *Taprobanian* 1: 61–63.

Nevo, E. 1985. Genetic differentiation and speciation in spiny mice, *Acomys*. *Acta zool. fenn.* 170: 131–136.

Niemitz, C. (ed.) 1984. *Biology of tarsiers*. New York: Fischer, xx, 357pp.

Niemitz, C. *et al.* 1991. *Tarsius dianae*: a new primate species from Central Sulawesi (Indonesia). *Folia primatol.* 56: 105–116.

Niethammer, J. 1973. Zur Kenntnis der Igel (Erinaceidae) Afghanistans. *Zeitschrift Säugetierk.* 38: 271–276.

----- 1975. Zur Taxonomie und Ausbreitungsgeschichte der Hausratte (*Rattus rattus*). *Zoologischer Anz.* 194: 405–415.

----- 1977. Versuch der Rekonstruktion der phylogenetischen Beziehungen zwischen einiger zentralasiatischen Muriden. *Bonner zool. Beitr.* 28: 236–248.

Niethammer, J. & Martens, J. 1975. Die Gattungen *Rattus* und *Maxomys* in Afghanistan und Nepal. *Zeitschrift Säugethierk.* 42: 325–355.

Nishiwaki, M. & Kamiya, T. 1958. A beaked whale *Mesoplodon* stranded at Oiso Beach, Japan. *Scientific Rep. Whales Res. Inst. Tokyo* 13: 53–83, 17pls.

Nishiwaki, M. & Marsh, H. 1985. Dugong—*Dugong dugon*. Pp. 1–32 in Ridgeway, S. H. & Harrison, R. (*q. v.*).

Nishiwaki, M. & Norris, K. S. 1966. A new genus, *Peponocephala*, for the odontocete cetacean species *Electra electra*. *Scientific Rep. Whales Res. Inst. Tokyo* 20: 95–100, 5pls.

[Nordmann, A. de] 1840. Catalogue raisonné des mammifères de la faune Pontique. Pp. 9–65 in Demidoff, A de, *Voyage dans la Russie méridionale* . . . Paris: Bourdin.

Novacek, M. J. *et al.* 1988. The major groups of eutherian mammals. In Benton, M. J. 1988a (*q. v.*).

Obaba, I. 1967. *Mustela sibirica itatsi*, introduced to Iriomote- jima, the Ryukyu Islands. *Journal mammal. Soc. Japan.* 3: 127–128.

O'Brien, S. J. *et al.* 1985. A molecular solution to the riddle of the giant panda's phylogeny. *Nature, Lond.* 317: 140–144.

----- 1986. The cheetah in genetic peril. *Scientific American* 254 (5): 68–76.

Oda, S. *et al.* 1988. Domestication of . . . *Suncus murinus*. Abstracts, Symp. Asian-Pacific Mammal., Beijing 1988.

Oey [= Oei], H. P. 1951. A new species of bat from Cambodge, *Glischropus rosseti sp. nov. Beaufortia* 1,8: 1–6.

----- 1960. Notes on bats from Bali, Lombok and Sumba. *Hemera Zoa* 67: 23–32.

Oey, H. P. *et al.* 1958. Some remarks on *Hipposideros speoris* and *Hipposideros larvatus* . . . *Beaufortia* 6: 215–241.

Ogilby, W. 1829. Description of a new species of Paradoxure . . . *Zoological J. Lond.* 4: 300–304.

----- 1831. [A new species of Indian deer . . .] *Proceedings zool. Soc. Lond.* 1: 136.

----- 1837a. [The skin of a fox from the Himalayan Mountains] *Proceedings zool. Soc. Lond.* 4(1836): 103.

----- 1837b. [Generic distinction among the Ruminantia.] *Proceedings zool. Soc. Lond.* (1836): 131–139.

----- 1838a. [On a new gibbon (*Hylobates chromandus*) . . .] *Proceedings zool. Soc. Lond.* (1837): 68–69.

----- 1838b. [Skins of two species of a new genus *Kemas*.] *Proceedings zool. Soc. Lond.* (1837): 81.

----- 1839a. [A new species of muntjac deer . . .] *Proceedings zool. Soc. Lond.* (1838): 105.

----- 1839b. [A new species of monkey.] *Proceedings zool. Soc. Lond.* 7: 31.

----- 1840a. Memoir on the mammalogy of the Himalayas. *Madras J. Lit.* 12: 139–171.

----- 1840b. On a new species of gibbon (*Hylobates leucogenys*). *Proceedings zool. Soc. Lond.* : 20–21.

----- 1840c. Memoir on the mammalogy of the Himalayas. Pp. lvi–lxxix in Royle, J. F. (*q. v.*).

----- 1841. [A new species of monkey.] *Proceedings zool. Soc. Lond.* 8 (1840): 56.

Ognev, S. I. 1914. *Izvestya imp. Obshch. Lyub. Estest. Zool.* 2: 102.

----- 1921. Contribution à la classification des mammifères de la Russie. *Ezhegodnik zool. Muz.* 22: 311–350.

----- 1924. *The fauna of the terrestrial vertebrates of the government of Voronesh.* Moscow, 254pp.

----- 1926. A new genus and species of cat from the Transcaspian region. *Ezhegodnik zool. Muz.* 27: 356–362.

----- 1928. *Mammals of Eastern Europe and northern Asia, vol. 1, Insectivora and Chiroptera.* Moscow (English translation Jerusalem, 1962).

----- 1931. *Ibid*, vol. 2. Moscow (English Translation Jerusalem, 1962).

----- 1951. [On the hedgehogs (Erinaceidae) of the Far East.] *Byull. Mosk. Obschch. Ispyt. Prir. Otd. biol.* 56: 8–14.

Ohtaisha, N. & Gao, Y. 1990. A review of the distribution of all species of deer . . . in China. *Mammal Rev.* 20: 125–144.

Oken, L. 1816. *Lehrbuch der Naturgeschichte. 3. Zoologie. Abt. 2. Fleischtiere.* Leipzig, Jena.

Oliver, W. L. R. 1977. The continuing decline of the pygmy hog and hispid hare. *Dodo* 14: 80–85.

----- [1980] The pigmy hog. The biology and conservation of the pigmy hog *Sus (Porcula) salvanius* and the hispid hare *Caprolagus hispidus*. *Special sci. Rep. Jersey Wildl. Pres. Trust* 1: 80pp, 14pls.

Oliver, W. L. R. *et al.* 1991. The Philippine spotted deer conservation project. *Oryx* 25: 199–205.

Olsen, Ø. 1912. *Tidens Tegn.* Nov. 12,1912 (Norwegian newspaper.)

----- 1913. On the external characters and biology of Bryde's whale (*Balaenoptera brydei*) . . . *Proceedings zool. Soc. Lond.* : 1073–1090, pls cix-cxiii.

Oppenheimer, F. C. & Oppenheimer, J. R. 1975. Certain behavioural features in the pariah dog (*Canis familiaris*) in West Bengal. *Applied Animal Ethol.* 2: 81–92.

Oppenoorth, W. F. F. 1932. *Homo* (*Javanthropus*) *soloensis*, een Plistoceene mensch van Java. *Wetenschappelijke Meded. Dienst Mijnb. Ned. Oost-Indië* 20: 49–63, 6pls.

Osbeck, P. 1765. *Reise nach Ostindien und China.* Rostock, xxiv, 552pp, 13pls.

Osborn, H. F. 1929. New Eurasiatic and American proboscideans. *American Mus. Novit.* 393: 23pp.

----- 1936,1942. *Proboscidea. A monograph of the . . . elephants of the world.* New York: American Mus. nat. Hist., vol. 1 (1936): 1–802, pls 1–12; vol. 2 (1942): 805–1675, pls 13–33.

Osgood, W. H. 1918. A new flying squirrel from the Philippine Islands. *Proceedings biol. Soc. Wash.* 31: 1–2.

----- 1932. Mammals of the Kelley-Roosevelts and Delacour Asiatic expeditions. *Publications Field Mus. nat. Hist. Zool.* 18: 193–339.

Otsuka, H. & Lin, C-c. 1984. Fossil *Rhinoceros* from the T'ouk oushan Group in Taiwan. *Journal Taiwan Mus.* 37: 1–35.

Otsuka, H. & Shikama, T. 1978. Fossil cervids from the Taokoushan group in Taiwan. *Reports Fac. Sci. Kagoshima Univ.* 2: 27–59, 6pls.

Otten, L. 1917. Over de biologie van *Mus concolor*. *Mededelingen burg. geneesk. Dienst. Ned. -Indië* 6: 82–119.

Otto, A. W. 1835. Uber die *Viverra hermaphrodita* Pallas's . . . *Nova Acta physico-Med.* 17: 1089–1102.

Ouwens, P. A. 1910. Contribution à la connaissance des mammifères de Célèbes. *Bulletin Dep. Agric. Indes Néerl.* 38: 7pp, 1pl.

Owen, D. et al. 1987. *Proyek Kelelawar. A study of bats and invertebrates in an archipelago's caves. Final report of the Oxford University Expedition to the Togian Islands, Sulawesi, Indonesia* . . . 84pp.

Owen, R. 1837. [On a new orang (*Simia morio*).] *Proceedings zool. Soc. Lond.* 4 (1836): 91–96.

----- 1843. Report on the British fossil Mammalia. *Report Br. Assoc. Adv. Sci.* 12: 54–74.

----- 1846. *A history of British fossil mammals and birds.* London: Van Voorst, 560pp.

----- 1853. *Descriptive catalogue of the osteological series* . . . London: Royal College of Surgeons, xvi, 914pp.

----- 1866. On some Indian Cetacea collected by Walter Elliot, Esq. *Transactions zool. Soc. Lond.* 6,1: 17–47, pls 3–14.

----- 1870. On fossil remains of mammals found in China. *Quarterly Jl geol. Soc. Lond.* 26: 417–434, pls 28,29.

Owen, R. D. & Qumsiyeh, M. B. 1987. The subspecies problem in the trident leaf-nosed bat, *Asellia tridens* . . . *Zeitschrift Säugetierk.* 52: 329–337.

Pallas, P. S. 1766. *Miscellanea zoologica* . . . Hagae Comitum, 224pp.

----- 1767. *Spicilegia zoologica* . . . *animalium species.* Berlin.

----- 1769. De ossibus Siberiae fossilibus cranis praesertim rhinocerotum . . . *Novi Comment. Acad. scient. Imp. Petrop.* 13: 436–477.

----- 1771,1773. *Reise durch verschiedene Provinzen des Russischen Reichs.* St Petersburg.

----- 1775. *Equus hemionus*, Mongolis Dshikketaei dictus. *Novi Comment. Acad. scient. Imp Petrop.* 19 (1774): 394–417, pl. vii.

----- 1777. *Spicilegia zoologica* . . . fasc. 12. Berlin, 71pp.

----- 1778,1779a. *Novae species quadrupedum e glirum ordine*. . . Erlangae: Walther, pp. 1–69 (1778), 70–388 (1779).

----- 1779b. Moschi historia naturalis. *Spicilegia zoologica* fasc. 13: 45pp.

----- 1783. Galeopithecus volans, camellii. Descriptus. *Acta Acad. scient. Imp. Petrop.* 1 (1780): 208–222.

----- 1785. Sorices aliquot illustrati. *Acta Acad. scient. Imp. Petrop.* 2(1781): 314–348.

----- 1811. *Zoographia Rosso-Asiatica* . . . vol. 1. Petropoli, 568pp.

Palmer, T. S. 1898. Random notes on the nomenclature of the Chiroptera. *Proceedings biol. Soc. Wash.* 12: 109–114.

----- 1899a. The family name of the dormouse. *Science* 10: 412–413.

----- 1899b. Notes on three genera of dolphins. *Proceedings biol. Soc. Wash.* 13: 23–24.

----- 1904. *Index generum mammalium; a list of the genera and families of mammals.* North American Fauna, no. 23.

Paradiso, J. L. 1971. A new subspecies of *Cynopterus sphinx* . . . from Serasen Island, Indonesia. *Proceedings biol. Soc. Wash.* 84: 293–300.

Pasteur, N. *et al.* 1982. Generic differentiation in Indonesian and French rats of the subgenus *Rattus*. *Biochemical Syst. Evol.* 10: 191–196.

Patnaik, S. K. & Acharjyo, L. N. 1990. White tiger in India—its past and present. *Tigerpaper* 17,1: 8–10.

Patterson, B. 1979. Pholidota and Tubulidentata. In Maglio, V. J. & Cooke, H. B. S. (eds), *Evolution of African mammals.* Harvard.

Pavlinov, I. Ya. 1980a. Taxonomic status of *Calomyscus* . . . *Zoologicheskii Zh.* 59: 312–316.

----- 1980b. Superspecies groupings Cardiocraniinae Satunin . . . *Vestnik Zool.* 1980 (2): 47–51.

----- 1982. Phylogeny and classification of the subfamily Gerbillinae. *Byulleten Mosk. Obshch. Ispyt. Prir. Biol.* 87,2: 19–31.

Pavlinov, I. Ya. *et al.* 1990. [Gerbils of the world fauna.] Moscow: Nauka, 363pp. In Russian.

Payne, J. B. 1979a. Synecology of Malayan tree squirrels with special reference to the genus *Ratufa*. PhD thesis, Cambridge, 406pp.

----- 1979b. Abundance of diurnal squirrels at the Kuala Lompat post of the Krau Game reserve, Pahang, peninsular Malaysia. Pp. 37–52 in Marshall, A. G. (ed.) *The abundance of animals in Malesian rain forests.* Univ. Hull Dept. Geog. Misc. Ser. no. 22.

Payne, J. *et al.* 1985. *A field guide to the mammals of Borneo.* Sabah Soc./World Wildlife Fund, Malaysia, 332pp.

Peale, T. R. 1848. Mammalogy and ornithology. *US Exploring Exped.* vol. 8. Philadelphia.

Pearson, J. 1832. [A stuffed specimen of a species of *Felis* . . .] *Journal Asiat. Soc. Bengal* 1: 75.

----- 1836. On the *Canis Vulpes montana*, or hill fox. *Journal Asiat. Soc. Bengal* 5: 313–314.

-----1841. Catalogue of Mammalia in the museum of the Asiatic Society. *Journal Asiat. Soc. Bengal* 10: 660–663.

----- 1956. Quaternary mammalian fossils from Hsintsai, southeastern part of Honan. *Acta palaeont. sin.* 4: 77–99, pl. 1.

Pelzeln, A. von & Kohl, F. F. 1886. Ueber eine Sendung von Säugethieren und Vögeln aus Ceylon. *Verhandelungen zool. - bot. Ges. Wien* 35: 525–528.

Pen, H-s. *et al.* 1962. Report on mammals from southwestern Szechuan. . . *Acta zool. sin.* 14 (suppl.): 105–132, 1pl.

Peng, H. & Wang, Y. 1981. New mammals from the Gaoligong Mountains. *Acta theriol. sin.* 1: 167–176.

Pennant, T. 1769. *Indian Zoology.* London: 14pp, 12pls.

Penny, M. 1987. *Rhinos/ endangered species.* London: Christopher Helm, 116pp.

Perrin, W. F. 1990. Subspecies of *Stenella longirostris* . . . *Proceedings biol. Soc. Wash.* 103: 453–463.

Perrin, W. F. *et al.* 1987. Revision of the spotted dolphins, *Stenella* spp. *Marine Mammal Sci.* 3: 99–170.

Perry, G. 1811. *Arcana; or the museum of natural history* . . . London [346pp, 84pls, un-numbered].

Peters, W. 1859. Neue Beiträge zur Kenntniss der Chiropteren. *Monatsberichte K. preuss. Akad. Wiss.* : 222–225.

----- 1860. Uber einige merkwürdige Nagethiere (*Spalacomys indicus*. . .) *Physikalische Math. Abh. K. Akad. Wiss. Berlin* : 139–156, taf. 1,2.

----- 1861. Über die von Hrn. F. Jagor bisher auf Malacca, Borneo, Java und den Philippinen gesammelten Säugethiere . . . *Monatsberichte K. preuss. Akad. Wiss.* : 706–712.

----- 1862a. In Hensel, R., Berichte über die Leistungen in der Naturgeschichte der Säugethiere . . . *Archiv Naturgesch.* (2): 83–160.

----- 1862b. Mitteilung über einem neuen Flederhund, *Pteropus scapulatus* . . . *Monatsberichte K. preuss. Akad. Wiss.* : 574–577.

----- 1865a. Vorlage von Abbildungen zu einer Monographie der Chiropteren . . . *Montasberichte K. preuss. Akad. Wiss.* : 256–258.

----- 1865b. Uber einige weniger bekannte Flederthiere . . . *Monatsberichte K. preuss. Akad. Wiss.* : 641–648.

----- 1865c. Über die zu den *Vampyris* gehörigen Flederthiere . . . *Monatsberichte K. preuss Akad. Wiss.* : 503–524.

----- 1866a. On some Mammalia . . . Burmah. *Proceedings zool. Soc. Lond.* : 426–429, pl. xxxv.

----- 1866b. Notes on a collection of mice . . . *Proceedings zool. Soc. Lond.* 21: 558–559.

----- 1866c. Uber einige neue oder weniger bekannte Flederthiere. *Monatsberichte K. preuss. Akad. Wiss.* : 16–25.

----- 1866d. Mittheilung über neue. . . Flederthiere . . . *Monatsberichte K. preuss. Akad. Wiss.* : 392–411.

----- 1867a. [Herpetologische Notizen] *Monatsberichte K. preuss. Akad. Wiss.* : 13, footnote.

----- 1867b. Uber die Flederhunde, Pteropi . . . *Monatsberichte K. preuss. Akad. Wiss.* : 319–333.

----- 1867c. Ueber eine neue Gattung von Nagern, *Uromys*, aus Nordaustralien. *Monatsberichte K. Preuss. Akad. Wiss.* : 343–345.

----- 1867d. Uber Flederthiere (*Pteropus gouldii* . . .) *Monatsberichte K. preuss. Akad. Wiss.* : 703–712.

----- 1867e. On some mammals collected . . . at Moulmein, Burmah. *Proceedings zool. Soc. Lond.* (1866): 426–430, 1 pl.

----- 1868. Mitteilung über eine neue Nagergattung, *Chiropodomys penicillatus* . . . *Monatsberichte K. preuss. Akad. Wiss.* 448–460, pls 1,2.

----- 1869a. Bemerkungen über neue oder weniger bekannte Flederthiere . . . *Monatsberichte K. preuss. Akad. Wiss.* : 391–406.

----- 1869b. Mittheilung über die von dem Hon. Marquis Giacomo Doria in Sarawak und Borneo gesammelten Flederthiere. *Monatsberichte K. preuss. Akad. Wiss.* (1868): 626–627.

----- 1870. Uber neue Arten von Spitzmäusen . . . *Monatsberichte K. preuss. Akad. Wiss.* 584–596.

----- 1871a. Uber die Gattungen und Arten der Hufeisennasen, Rhinolophi. *Monatsberichte K. preuss. Akad. Wiss.* 301–332.

----- 1871b. In Swinhoe, R. (*q. v.*).

----- 1871c. On some bats collected by Mr F. Day in Burma. *Proceedings zool. Soc. Lond.* : 513–514.

----- 1872a. Uber die Arten der Chiropterengattung *Megaderma*. *Monatsberichte K. preuss. Akad. Wiss.* : 192–196.

----- 1872b. Mitteilung über neue Flederthiere . . . *Monatsberichte K. preuss Akad. Wiss.* : 256–264.

----- 1872c. Uber den *Vespertilio calcaratus* Prinz zu Wied . . . *Monatsberichte K. preuss. Akad. Wiss.* : 701–706.

----- 1873. Uber einige zu der Gattung *Cynopterus* gehörige Arten der Flederhunde . . . *Monatsberichte K. preuss. Akad. Wiss.* : 485–488.

----- 1874. Diagnosi di tre nuovi mammiferi della Nuava Guinea ad isole Key. *Annali Mus. civ. Stor. nat. Giacomo Doria* 6: 303.

----- 1881. Uber die Chiropterengattung *Mormopterus* . . . *Monatsberichte K. preuss Akad. Wiss.* : 482–485.

----- 1906. *Chiroptera of Berlin Museum. Plates.* Berlin.

Peters, W. & Doria, G. 1875. Diagnosi di alcune nuove specie di marsupiali . . . *Annali Mus. civ. Stor. nat. Giacomo Doria* 7: 541–544.

----- 1881. Enumerazione dei mammiferi. . . nella Nova Guinea . . . *Annali Mus. civ. Stor. nat. Giacomo Doria* 16: 665–707, 14 pls.

Peterson, B. V. *et al.* 1990. Some ectoparasites of bats from Halmahera Island, Indonesia. *Entomological News* 101: 39–47.

Peterson, R. L. 1968. Notes on an unusual specimen of *Scotophilus* from Vietnam. *Canadian J. Zool.* 46: 1079–1081.

----- 1981. Systematic variation in the *tristis* group of the bent-winged bats of the genus *Miniopterus* . . . *Canadian J. Zool.* 59: 828–843.

Peterson, R. L. & Fenton, M. B. 1970. Variation in the bats of the genus *Harpionycteris* . . . *Life Sci. occ. Pap. r. Ont. Mus.* 17: 15pp.

Petit, D. & Dutrillaux, B. 1985. Phylogenie chromosomique de sept espèces de sciurines. *Annales Genet.* 28: 13–18.

Petocz, R. 1989. Status of the tamaraw. *Asian wild Cattle spec. Group Newsl.* 2: 1–4.

Petter, F. 1961. Eléments d'une révision des lièvres européens et asiatiques du sous-genre *Lepus. Zeitschrift Säugetierk.* 26: 30–40.

----- 1963. Un nouvel insectivore du Nord de l'Assam: *Anourosorex squamipes schmidi* nov. subsp. *Mammalia* 27: 444–445.

Pettigrew, J. D. 1991. Wings or brain? Convergent evolution in the origin of bats. / A fruitful, wrong hypothesis? Response to Baker, Novacek and Simmons. *Systematic Zool.* 40: 199–216; 231–239.

Pfeffer, P. 1957. Notes sur le peuplement mammmalien der Îles de Flores, Komodo et Rintja. *Mammalia* 21: 405–419.

Phillips, C. J. 1969. Review of central Asian voles of the genus *Hyperacrius. Journal Mammal.* 50: 457–474.

Phillips, W. W. A. 1926. Guide to the mammals of Ceylon. IV. Monkeys. *Spolia zeylan.* B, 13: 169–280, 4pls.

----- 1927. A new *Pithecus* monkey from Ceylon . . . *Spolia zeylan.* B, 14: 57–59.

----- 1928. Ceylon shrews. *Spolia zeylan.* 14: 295–331, pls 19–20.

----- 1929a. New and rare Ceylon shrews. *Spolia zeylan.* 15: 113–118.

----- 1929b. Two new rodents from the Highlands of Ceylon. *Spolia zeylan.* 15: 165–168.

----- 1931. The giant squirrels (*Ratufa*) of Ceylon. *Spolia zeylan.* 16: 209–216.

----- 1932a. Additions to the fauna of Ceylon. No. 3. A new pigmy shrew . . . *Spolia zeylan.* 17: 123–126, pl. 27.

----- 1932b. Additions to the fauna of Ceylon. *Spolia zeylan.* 16: 323–327.

----- 1936. The mole-rats (*Gunomys*) of Ceylon. *Spolia zeylan.* 20: 93–97.

----- 1980,1981,1984. *Manual of the mammals of Sri Lanka* 2nd ed. Sri Lanka: Wildlife and Nature Protection Society, 390pp.

Pickford, M. 1976. An Upper Miocene pholidote from Pakistan. *Pakistan J. Zool.* 8: 21–24.

Pilgrim, G. E. 1910. Notices of new mammalian genera and species from the Tertiaries of India. *Records geol. Surv. India* 40: 63–71.

----- 1913. Correlation of the Siwaliks with mammal horizons of Europe. *Records geol. Surv. India* 43: 264–326, pls 26–28.

----- 1926. The fossil Suidae of India. *Memoirs geol. Surv. India* 8,4: 1–65, pls 1-xx.

----- 1932. The fossil Carnivora of India. *Memoirs geol. Surv. India Palaeont. indica* 18: 1–232.

----- 1937. Siwalik antelopes and oxen in the American Museum of Natural History. *Bulletin Amer. Mus. nat. Hist.* 72: 729–874.

----- 1939. The fossil Bovidae of India. *Palaeontologica Indica* 26,1 :iii, 356pp, 8pls.

Pillai, R. S. 1973. Record of the slow loris, *Nycticebus coucang* from Meghalaya, India. *Journal Assam sci. Soc.* 16: 205–207.

Pilleri, G. & Gihr, M. 1972. Contribution to the knowledge of the cetaceans of Pakistan . . . *Investigations Cetacea* 4: 107–162, 41pls.

----- 1975. On the taxonomy and ecology of the finless black porpoise, *Neophocaena* . . . *Mammalia* 39: 657–673, 8pls.

----- 1976. Record of *Pseudorca crassidens* off Karachi. *Investigations Cetacea* 7: 205–207.

----- 1977. Neotype for *Platanista indi* Blyth, 1859. *Investigations Cetacea* 8: 77–81.

Pilleri, G. *et al.* 1982. Speciation in the Platanistoidea . . . *Investigations Cetacea* 14: 15–46.

Pinel, -- 1792. Recherches sur une nouvelles méthode de classification des quadrupèdes . . . *Actes Soc. Hist. nat. Paris* 1: 50–60.

Pocock, R. I. 1908. Notes upon some species and geographical races of serows (*Capricornis*). . . *Abstracts Proc. zool. Soc. Lond.* : 12; *Proceedings zool. Soc. Lond.* : 173–202.

----- 1914. Description of a new species of goral . . . *Journal Bombay nat. Hist. Soc.* 23: 32–33.

----- 1917a. A new genus of Ursidae. *Annals Mag. nat. Hist.* 20: 128–130

----- 1917b. The classification of existing Felidae. *Annals Mag. nat. Hist.* 20: 329–350.

----- 1921. On the external characters and classification of the Mustelidae. *Proceedings zool. Soc. Lond.* : 803–837.

----- 1923a. On the external characters of *Elaphurus* . . . *Proceedings zool. Soc. Lond.* : 181–207.

----- 1923b. The classification of the Sciuridae. *Proceedings zool. Soc. Lond.* : 209–246.

----- 1924a. The external characters of the pangolins, Manidae. *Proceedings zool. Soc. Lond.* : 707–723.

----- 1924b. A new genus of monkeys. *Abstracts Proc. zool. Soc. Lond.* : 17; *Proceedings zool. Soc. Lond.* : 330–331.

----- 1928. The langurs, or leaf monkeys, of British India. *Journal Bombay nat. Hist. Soc.* 32: 472–504; 660–677, 4pls.

----- 1929. Tigers. *Journal Bombay nat. Hist. Soc.* 33: 505–541.

----- 1930. The panthers and ounces of Asia. *Journal Bombay nat. Hist. Soc.* 34: 64–82; 307–336, pls 1–13.

----- 1931a. The long-tailed macaque monkeys . . . of southern India and Ceylon. *Journal Bombay nat. Hist. Soc.* 35: 276–287.

----- 1931b. The pig-tailed macaques (*Macaca nemestrina*). *Journal Bombay nat. Hist. Soc.* 35: 297–311.

----- 1932a. The black and brown bears of Europe and Asia. Part 2. *Journal Bombay nat. Hist. Soc.* 30: 101–138.

----- 1932b. The marbled cat (*Pardofelis marmorata*) and some other Oriental species . . . *Proceedings zool. Soc. Lond.* : 741–766.

----- 1932c. The rhesus macaques (*Macaca mulatta*). *Journal Bombay nat. Hist. Soc.* 35: 530–551.

----- 1933a. The civet cats of Asia. *Journal Bombay nat. Hist. Soc.* 36: 423–449; 629–656.

----- 1933b. The rarer genera of Oriental Viverridae. *Proceedings zool. Soc. Lond.* : 969–1035, pls 1–3.

----- 1933c. The palm civets or 'Toddy cats' of the genera *Paradoxurus* and *Paguma* . . . *Journal Bombay nat. Hist. Soc.* 36: 854–877.

----- 1934a. The palm civets or 'Toddy cats'. . . *Journal Bombay nat. Hist. Soc.* 37: 172–192; 314–346.

----- 1934b. The geographical races of *Paradoxurus* and *Paguma* . . . *Proceedings zool. Soc. Lond.* : 613–683, pls 1,2.

----- 1935. The monkeys of the genus *Pithecus* . . . found to the east of the Bay of Bengal. *Proceedings zool. Soc. Lond.* (1934): 895–961, 2pls.

----- 1936a. The Asiatic wild dog or dhole (*Cuon javanicus*). *Proceedings zool. Soc. Lond.* : 33–55.

----- 1936b. The Javan race of the Oriental yellow-throated marten, *Lamprogale flavigula*. *Annals Mag. nat. Hist.* 17: 403–404.

----- 1936c. The oriental yellow-throated marten (*Lamprogale*). *Proceedings zool. Soc. Lond.* : 531–553.

----- 1937. The mongooses of British India . . . *Journal Bombay nat. Hist. Soc.* 39: 211–245.

----- 1939, 1941. *The fauna of British India: Mammalia. Primates and Carnivora.* London.

----- 1940a. The hog-badgers (*Arctonyx*) of British India. *Journal Bombay nat. Hist. Soc.* 41: 461–469.

----- 1940b. Notes on some British Indian otters . . . *Journal Bombay nat. Hist. Soc.* 41 : 514–517.

----- 1943a. Notes on the Asiatic elephant (*Elephas maximus*). *Annals Mag. nat. Hist.* 10: 273–280.

----- 1943b. The larger deer of British India. Part II. *Journal Bombay nat. Hist. Soc.* 43: 554–572.

----- 1947. Two new local races of the Asiatic wild ass. *Journal Bombay nat. Hist. Soc.* 47: 143–144.

----- 1951. *Catalogue of the genus* Felis. London.

Poduschka, W. & Poduschka, C. 1985. Contribution to the study of the genus *Podogymnura* . . . *Sitzungsberichte ost. Akad. Wiss.* 194: 1–22.

Pohl, L. 1911. Eine Höhenvarietät von *Siamanga syndactylus* Desm. *Zoologischer Anz.* 38: 51–53.

Pohle, H. 1920. Die Unterfamilie der Lutrinae . . . *Archiv Naturgesch.* 85, A, 9: 246pp.

----- 1944. *Hemitragus jemlahicus schaeferi* sp. n. *Zoologischer Anz.* 144: 184–191.

----- 1950. Zur Säugetier-geographie der Kleinen Sundainseln. *Kleinchmidt Syllegomena Biologica, Leipzig* : 319–346.

Pohlig, H. 1884. *Sitzungsberichte neiderrhein. Ges. Nat. u. Heilk.* : 47–61 (*n. v.*).

Poirier, F. E. 1986. A preliminary study of the Taiwan macaque (*Macaca cyclopis*). *Zoological Res.* 7: 411–421.

Pomel, A. 1848. Etudes sur les carnassiers insectivores. *Archives Sci. phys. nat. Genève* 9: 159–165, 244–251.

Potter, M. *et al.* (eds) 1986. *The wild mouse in immunology*. Berlin etc. : Springer-Verlag, 395pp. (*Current Topics in Microbiology and Immunology* no. 127).

Pousargues, E. de 1896. Sur la faune mammalogiques du Setchuan. *Bulletin Mus. Hist. nat. Paris* 2: 11–16.

----- 1897. Sur un gibbon d'espèce nouvelle provenant du Haut-Tonkin. *Bulletin Mus. Hist. nat. Paris* 2 (1896): 367–369.

----- 1898. Note préliminaire sur un nouveau semnopithèque . . . *Bulletin Mus. Hist. nat. Paris* 4: 319–321.

Pradhan, M. S. *et al.* 1989. Proposal of an additional species in the genus *Bandicota* . . . from India. *Mammalia* 53: 369–376.

Prakash, I. 1972. Extension of the range of *Suncus stoliczkanus* in the Rajasthan desert. *Journal Bombay nat. Hist. Soc.* 69: 643–644.

----- 1977. Replacement of sympatric rodent species in the Indian Desert. *Proceedings All-India Rodent Seminar, 1975* : 29–31.

Prakash, I & Kumari, S. 1978. Morphological variation in *Meriones hurrianae* . . . in the Indian desert. *Säugetierkundliche Mitt.* 26: 281–284.

Prasad, M. R. N. 1954. Food of the Indian gerbille *Tatera indica cuvieri* (Waterhouse). *Journal Bombay nat. Hist. Soc.* 52: 321–325.

Prashad, B. 1936. Mammalia. Pp. 15–59, 7 pls in Animal remains from Harappa. *Memoirs archaeol. Surv. India* 51.

Prater, S. H. 1948,1965,1971. *The book of Indian animals*, eds 1–3. Bombay: Bombay nat. Hist. Soc.

Preble, E. A. 1899. Revision of the jumping mice of the genus *Zapus*. *North Amer. Fauna* 15: 42pp, 1pl.

Pritchard, J. S. 1989. Ilin Island cloud rat extinct? *Oryx* 23: 126.

Prothero, D. R. & Schoch, R. M. (eds) 1989. *The evolution of perissodactyls*. Oxford UP, 320pp.

Prothero, D. R. *et al.* 1986. The phylogeny of the Rhinocerotidae . . . *Zoological J. Linn. Soc.* 87: 341–366.

Prouty, L. A. *et al.* 1983. *Bunopithecus*: a genus-level taxon for the hoolock gibbon. *American J. Primatol.* 5: 83–87.

Pucheran, 1855. Notes mammalogiques. *Revue Mag. Zool.* 7: 392–394.

Pycraft, W. P. 1925. Diagnoses of four species and one subspecies of the genus *Homo*. *Man* 25: 162–164.

Qiu, Z. 1980. *Viverra peii*, a new species from the 'cap' travertine of Zhoukoudien. *Vertebrata palasiat.* 18: 304–313.

----- 1987. Fossil tupaiid from the hominoid locality of Lufeng, Yunnan. *Vertebrata palasiat.* 24: 308–319.

----- 1989. Fossil platacanthomyids from the hominid locality of Lufeng, Yunnan. *Vertebrata palasiat.* 27: 268–283.

Qiu, Z. & Liu, Y. 1986. The Aragonian vertebrate fauna of Xiacaowan, Jiangsu. 5. Sciuridae. *Vertebrata palasiat.* 24: 195–209.

Quan, G. & Wang, W. 1984. Karyotype of pangolins (*Manis pentadactyla aurita*). *Acta theriol. sin.* 4: 89–92.

Quan, G. *et al.* 1987. A new record of primate species in China. *Acta theriol. sin.* 7: 158.

Qumsiyeh, M. B. & Jones, J. K. 1986. *Rhinopoma hardwickii* and *Rhinopoma muscatellum*. *Mammalian Species* 263: 5pp.

Quoy, J. R. C. & Gaimard, J. P. 1830. *Voyage de . . . l'Astrolabe . . . Zoologie*. Vol. 1, Atlas, L, 268pp.

Rabinowitz, A. 1988. The clouded leopard in Taiwan. *Oryx* 22: 46–47.

Rabor, D. S. 1939. *Sciuropterus mindanensis* sp. nov., a new species of flying squirrel from Mindanao. *Philippine J. Sci.* 69: 389–393.

----- 1952. Two new mammals from Negros Island, Philippines. *Chicago Acad. Sci. nat. Hist. Misc.* 96: 7pp.

----- 1977. *Philippine birds and mammals*. Quezon City: Univ. Philippine Press, 284pp.

----- 1986. Philippine mammals. Pp. 107–211 in *Guide to Philippine flora and fauna* 11. Manila.

Radtke, M. & Niethammer, J. 1985. Zur Stellung der Pestratte (*Nesokia indica*) im System der Murinae. *Säugetierkundliche Mitt.* 32: 13–16.

Raffles, T. S. 1821. Descriptive catalogue of a zoological collection made . . . in Sumatra. *Transactions Linn. Soc. Lond.* 13: 239–340.

Rafinesque, C. S. 1814. *Précis de décourtes et travaux somiologiques* . . . Palerme, 55pp.

----- 1815. *Analyse de la nature*. Palerme, 224pp.

----- 1820. *Annals of nature* . . . Lexington, Kentucky.

----- 1821. *Annales Sci. Phys. Brux.* 7: 214 (*n. v.*).

----- 1832. Description of a new otter, *Lutra concolor* from Assam in Asia. *Atlantic J.* 1: 62.

Raghavan, P. 1990. Fossil rodent assemblages from the basal Pinjor Formation . . . *Research Bull. Panjab Univ.* 40: 245–268.

Rahmani, A. R. 1990. Distribution, density, group size and conservation of the Indian gazelle . . . *Biological Conserv.* 51: 177–189.

Rajagopalan, P. K. 1968. Notes on the Malabar spiny dormouse, *Platacanthomys lasiurus* . . . *Journal Bombay nat. Hist. Soc.* 65: 214–215.

Ramachandran, K. K. 1989. Endangered grizzled giant squirrel habitat. *Journal Bombay nat. Hist. Soc.* 86: 94–95.

Rana, B. D. 1986. Ecological distribution of *Rattus meltada* in India. *Journal Bombay nat. Hist. Soc.* 82: 573–580.

Rana, T. 1989. Record of palm civet from Rawalpindi and Mansehra District. *Pakistan J. Zool.* 21: 390–392.

Ranjitsinh, M. K. 1985. A possible sighting of Blanford's fox in Kutch. *Journal Bombay nat. Hist. Soc.* 82: 395–396.

----- 1989. *The Indian blackbuck.* Dehradun: Natraj, 156pp.

Rao, S. K. & Aswathanarayana, N. V. 1978. On the occurence of Horsfield's shrew, *Crocidura horsfieldi* (Tomes) in peninsular India. *Journal Bombay nat. Hist. Soc.* 75: 473.

Ratajszczak, R. & Cox, R. 1991. Back-striped weasel in Vietnam. *Mustelid and viverrid Conservation* 4: 17.

Raven, H. C. 1935. Wallace's line and the distribution of Indo-Australian mammals. *Bulletin Am. Mus. nat. Hist.* 68: 179–293.

Reeves, R. R. & Brownell, R. L. 1989. Susu—*Platanista gangetica* . . . and *Platanista minor* . . . Pp. 69–100 in Ridgeway, S. H. & Harrison, R. (*q. v.*).

Reichenbach, H. G. L. 1862,1863. Affen der alten Welt. Lief. 14 in *Central-Atlas für zoologishe Gärten* . . . Dresden & Leipzig.

Reid, J. 1837. On a new species of monkey (*Semnopithecus obscurus*). *Proceedings zool. Soc. Lond.* : 14.

Reinhardt, J. 1862. [En for den Danske fauna ny delphinart.] *Oversigt K. Danske Videnskabernes Selskabs Forhandlingen* : 103–152.

Repenning, C. A. 1967. Subfamilies and genera of the Soricidae. *Professional Pap. US geol. Surv.* 565: 74pp.

Revilliod, P. 1911. Über einige Säugeteire von Celebes. *Zoologischer Anz.* 37: 513–517.

----- 1914. Les mammifères de la Nouvelle-Calédonie. . . *Nova Caledonia* 1: 341–365, 1pl.

Reza Khan, M. A. 1980. Range extension of crabeating mongoose in Bangladesh. *Journal Bombay nat. Hist. Soc.* 76 (1979): 153–154.

Rice, C. G. 1988. Habitat, population dynamics and conservation of the Nilgiri tahr . . . *Biological Conserv.* 44: 137–156.

Rice, D. W. 1977. A list of the marine mammals of the world, 3rd ed. *NOAA tech. Rep. NMFS SSRF* 711: 15pp.

----- 1989. Sperm whale—*Physeter macrocephalus* . . . Pp. 177–234 in Ridgeway, S. H. & Harrison, R. (*q. v.*).

Rickert, E. A. & Heaney, L. R. 1991. A new species of *Chrotomys* . . . from Luzon Island, Philippines. *Proceedings biol. Soc. Wash.* 104: 387–398.

Ridgeway, S. H. & Harrison, R. (eds) 1985,1989. *Handbook of marine mammals. Vol. 3. The sirenians and baleen whales; Vol. 4. River dolphins and the larger toothed whales.* London etc. : Academic Press, 362,442pp.

Rieger, I. 1981. *Hyaena hyaena. Mammalian Species* 150: 5pp.

Riffel, M. 1991. An update on the Javan ferret-badger *Melogale orientalis* . . . *Mustelid viverrid Newsl.* 5: 2–3.

Rishi, K. K. & Puri, U. 1978. Mitotic, meiotic and G-banded chromosomes of *Mus saxicola sadhu* . . . *Caryologia* 31: 103–107.

----- 1984. Chromosomes of *Rattus cutchicus cutchicus* and its systematic position. *Folia biol. Krakow* 32: 209–212.

Robbins, L. W. & Sarich, V. M. 1988. Evolutionary relationships in the family Emballonuridae . . . *Journal Mammal.* 69: 1–13.

Roberts, M. S. & Gittelman, J. L. 1984. *Ailurus fulgens. Mammalian Species* 222: 8pp.

Roberts, T. J. 1977. *The mammals of Pakistan.* London: Ernest Benn, 361pp.

Robin, H. A. 1881a. Diagnoses de deux chiroptères nouveaux . . . *Bulletin Soc. philomath. Paris* (7)5: 90–91.

----- 1881b. Recherches anatomiques sur les mammifères de l'ordre des chiroptères. *Annales Sci. nat. Zool.* 12,2: 1–180, 9pls.

Robinson, H. C. 1912. On new mammals from the islands of the Johore Archipelago . . . *Annals Mag. nat. Hist.* 10: 589–595.

----- 1916a. A note on *Callosciurus finlaysonii* . . . *Journal fed. Malay St. Mus.* 7: 35–36.

----- 1916b. A collection of mammals and birds from Pulau Panjang . . . *Journal fed. Malay St. Mus.* 7: 59–72.

----- 1916c. On a new race of *Callosciurus vittatus* from Singapore Island. *Journal fed. Malay St. Mus.* 7: 73–74.

----- 1917a. On two new subspecies of squirrel from southern India. *Records Indian Mus.* 13: 41–42.

----- 1917b. On three new races of Malayan mammals. *Journal fed. Malay St. Mus.* 7: 101–105.

----- 1921. Two new Indo-Malayan rats. *Annals Mag. nat. Hist.* 7: 234–235.

Robinson, H. C. & Kloss, C. B. 1911a. On six new mammals from the Malay Peninsula . . . *Journal fed. Malay St. Mus.* 4: 169–174.

----- 1911b. On new animals from the Malay Peninsula . . . *Journal fed. Malay St. Mus.* 4: 241–246.

----- 1914a. On new mammals . . . from the Malay Peninsula. *Annals Mag. nat. Hist.* 13: 223–234.

----- 1914b. On a collection of mammals from the Siamese province of Bandon. *Journal fed. Malay St. Mus.* 5: 111–127.

----- 1915. *Aeromys*, a new genus of flying squirrel. *Journal fed. Malay St. Mus.* 6: 23.

----- 1916. Preliminary diagnosis of some new. . . mammals and birds obtained in Korinchi, West Sumatra . . . *Journal Straits Brch Asiat. Soc.* 73: 269–278.

----- 1917. On the southern Malayan race of the white-whiskered palm-civet. *Journal fed. Malay St. Mus.* 7: 243–244.

----- 1918a. Results of an expedition to Korinchi Peak, Sumatra. *Journal fed. Malay St. Mus.* 8,2: 1–80; 311–319.

----- 1918b. Notes on the genus *Petaurista* . . . *Journal fed. Malay St. Mus.* 7: 223–225.

----- 1919a. On a collection of mammals from . . Sumatra. *Journal fed. Malay States Mus.* 7: 257–291.

----- 1919b. On five new mammals from Java. *Annals Mag. nat. Hist.* 4: 374–378.

----- 1919c. On mammals, chiefly from the Ophir District, West Sumatra. . . *Journal fed. Malay St. Mus.* 7: 299–323.

----- 1920. Notes on Viverridae. *Record Indian Mus.* 19,4: 175–179.

----- 1922. New mammals from French Indo-China and Siam. *Annals Mag. nat. Hist.* 9: 87–99.

----- 1927. A new rat from the mountains of East Java. *Journal fed. Malay St. Mus.* 13: 209.

Robinson, H. C. & Thomas, O. 1917. A new mink-like *Mustela* from Java. *Annals Mag. nat. Hist.* 20: 261–262.

Robinson, H. C. & Wroughton, R C. 1911a. Notes on Indo-Malayan squirrels. *Journal fed. Malay St. Mus.* 4: 166–168.

----- 1911b. On five new sub-species of oriental squirrels. *Journal fed. Malay St. Mus.* 4: 233–235.

----- 1916. On a new race of *Callosciurus atrodorsalis* (Gray) from north Siam. *Journal fed. Malay St. Mus.* 7: 91.

Robinson, M. 1990. Project Barito Ulu Chiroptera survey . . . *Bat Res. News* 31: 21–23.

Robinson, R. 1982. Evolution of the domestic cat. *Carnivore* 5: 4–13.

Rode, P. 1941. Catalogue des types de mammifères . . . chiroptères. *Bulletin Mus. nat. Hist. Paris* 13: 227–252.

Roguin, L. de 1988. Notes sur quelques mammifères du Baluchistan iranien. *Revue suisse Zool.* 95: 595–606.

Rolleston, G. 1877. On the domestic pig of prehistoric times in Britain . . . *Transactions Linn. Soc. Lond.* 1: 251–286, pls 41–43.

Rollinson, D. H. L. 1984. Bali cattle. Pp. 28–34 in Mason, I. L. (*q. v.*).

Romer, J. D. 1963. First record of the grey shrew, *Crocidura a. attenuata* Milne-Edwards, in Hong Kong. *Memoirs Hong Kong nat. Hist. Soc.* 6: 13.

----- 1974. Annotated checklist with keys to the bats of Hong Kong. *Memoirs Hong Kong nat. Hist. Soc.* 9: 1–6.

Rookmaaker, L. C. 1977a. The identity of the one-horned rhinoceros in Berlin 1874–1884. *International Zoo News* 24: 15.

----- 1977b. The distibution and status of the rhinoceros, *Dicerorhinus sumatrensis*, in Borneo . . . *Bijdragen Dierk.* 47: 197–204.

----- 1983. *Bibliography of the rhinoceroses* . . . Rotterdam: Balkema, 312pp.

----- 1984. The taxonomic history of the recent forms of Sumatran rhinoceros . . . *Journal Malay. Brch R. Asiat. Soc.* 57: 12–25.

----- 1988. The Indo-Chinese rhinoceros. *Oryx* 22: 241.

Rookmaaker, L. C. & Bergmans, W. 1981. Taxonomy and geography of *Rousettus amplexicaudatus* . . . *Beaufortia* 31: 1–29.

Roonwal, M. L. *et al.* 1984. Geographical boundary between the northern and southern tail styles in the common south Asian langur, *Presbytis entellus. Journal zool. Soc. India* 36: 15–26.

Roonwal, M. M. (ed.) 1987. Recent advances in rodentology in India. *Record zool. Surv. India, misc. Publ.* 105: 126pp.

Roonwal, M. V. 1948. Three new Muridae . . . from Assam. *Proceedings nat. Inst. Sci. India* 14: 385–387.

Rörig, A. 1896. *Geweihsammlung Kön. Landw. Hochsch. Berlin* (*n. v.*).

Rosenberg, H. von 1878. *Der Malayische Archipel.* Leipzig.

Ross, G. J. B. 1984. The smaller cetaceans of the south east coast of southern Africa. *Annals Cape Prov. Mus.,nat. Hist.* 15: 173–410.

Ross, G. J. B. & Cockroft, V. G. 1990. Comments on Australian bottlenose dolphins and the taxonomic status of *Tursiops aduncus* . . . Pp. 101–128 in Leatherwood, S. & Reeves, R. R. (*q. v.*).

Roux, J. 1910. Beitrag zur Kenntnis der *Sciurus*-Arten von Celebes. *Zoologischer Anz.* 35: 515–520.

Roxburgh, W. 1801. An account of a new species of *Delphinus* . . . *Asiatick Researches* [Calcutta ed.] 7: 170–174.

Royle, J. F. 1833–1840. *Illustrations of the botany . . . of the Himalayan Mountains* . . . London, lxxxii, 472pp, 100pls.

Rozendaal, F. G. 1984. Notes on macroglossine bats from Sulawesi. *Zoologische Meded.* 58: 187–212.

Ruedi, M. *et al.* 1990. The *Crocidura fuliginosa* species complex in Peninsular Malaysia . . . *Biochemical Syst. Ecol.* 18: 573–581.

Rümmler, H. 1938. Die Systematik und Verbreitung der Muriden Neuguineas. *Mitteilungen zool. Mus. Berlin* 23: 1–297.

Rumpler, Y. *et al.* 1987. Chromosomal phylogeny of three Lorisidae . . . *Folia Primatol.* 48: 216–220.

Rütimeyer, L. 1865. Beiträge zu einen palaeontologischen Geschichten der Wiederkauer . . . *Verhandlungen naturf. Ges. Basel* 4: 299–354.

----- 1867. Versuch einen natürlichen Geschichte des Rindes . . . 2. *Neue Denkschr. allg. schweiz. Ges. ges. Naturw.* 22,2, 3: 175pp, 4pls.

----- 1878. Die Rinder der Tertiär-Epoche . . . 2. *Abhandlungen schweiz. paläont. Ges.* 5: 73–208, 6pls.

Rutty, J. 1772. *An essay towards a natural history of the county of Dublin* . . . vol. 1. Dublin.

Ryan, R. M. 1965. Taxonomic status of the vespertilionid genera *Kerivoula* and *Phoniscus. Journal Mammal.* 46: 517–518.

Ryder, M. L. 1984. Sheep. Pp. 63–85 in Mason, I. L. (*q. v.*).

Ryley, K. V. 1913. Scientific results from the Mammal Survey V. *Journal Bombay nat. Hist. Soc.* 22: 434–443.

----- 1914. Scientific results from the Mammal Survey VI. *Journal Bombay nat. Hist. Soc.* 22: 658–664.

Saha, S. S. 1977. A new subspecies of the flying squirrel, *Petaurista nobilis* (Gray), from Bhutan. *Proceedings zool. Soc. Calcutta* 28 (1975): 27–29.

----- 1980. Notes on some mammals recently collected from Andaman and Nicobar Islands. *Records zool. Surv. India* 77: 119–126.

----- 1981. A new genus and a new species of flying squirrel . . . from northern India. *Bulletin zool. Surv. India* 4: 331–336.

----- 1984. Occurrence of the tail-less fruit bat *Megaerops ecaudatus* . . . in Arunchal Pradesh . . . *Bulletin zool. Surv. India* 6: 343–344.

----- 1985. Mammalia. In Fauna of Namdapha, Arunchal Pradesh. A proposed biosphere reserve. *Records zool. Surv. India* 82: 321–330.

Sanborn, C. C. 1931. Bats from Polynesia, Melanesia, and Malaysia. *Publications Field Mus. nat. Hist. Zool.* 18: 7–29.

----- 1939. Eight new bats of the genus *Rhinolophus. Publications Field Mus. nat. Hist. Zool.* 24: 37–43.

----- 1950. New Philippine fruit bats. *Proceedings biol. Soc. Wash.* 63: 189–190.

----- 1952a. Philippine zoological expedition 1946–1947, mammals. *Fieldiana Zool.* 33: 89–158.

----- 1952b. The mammals of the Rush-Watkins Zoological Expedition to Siam. *Natural Hist. Bull. Siam Soc.* 15: 1–20.

----- 1954. Mammals from Mindanao . . . *Videnskabelige Meddr dansk naturh. Foren.* 115 (1953): 283–288.

Santiapillai, A. & Santiapillai, C. 1990. Status, distribution and conservation of the sloth bear (*Melursus ursinus*) in Sri Lanka. *Tigerpaper* 17: 13–15.

Santiapillai, C. & Ashby, K. R. 1988. The clouded leopard in Sumatra. *Oryx* 22: 44–45.

Santiapillai, C. & Jackson, P. 1990. *The Asian elephant. An action plan for its conservation.* Gland: IUCN, 80pp.

Santiapillai, C. & Ramono, W. S. 1990. The status and conservation of the Malayan tapir in Sumatra, Indonesia. *Tigerpaper* 17,4: 6–11.

Santiapillai, C. & Suprahan, H. 1985. On the status of the leopard cat (*Felis bengalensis*) in Sumatra. *Tigerpaper* 12,3: 8–13.

Sarker, S. U. & Sarker, N. J. 1984. Mammals of Bangladesh—their status, distribution and habitat. *Tigerpaper* 11,1: 8–13.

Sartono, S. 1975. Implications arising from *Pithecanthropus* VIII. Pp. 327–360 in Tuttle, R. H. (*q. v.*).

----- 1976. *Abstracts Union int. Sci. prehist. protohist., Nice.* (*n. v.*).

Sarwar, M. 1977. Taxonomy and distribution of the Siwalik Proboscidea. *Bulletin Dept Zool. Univ. Panjab* 10: 1–172.

----- 1979. A new subspecies of *Elephas hysudricus* from the Upper Siwaliks . . . *Pakistan J. Zool* 11: 329–335.

----- 1988. A new progressive bunodont suid from the Tatrotian of the Upper Siwaliks. . . *Pakistan J. Zool.* 20: 401–408.

Satunin, K. 1901. Zwei neue Säugethiere aus Transkaukasien. *Zoologischer Anz.* 24: 361–464.

----- 1903. Neue Nagetiere aus Centralasien. *Ezhegodnik zool. Muz.* 7: 547–589.

----- 1904. Neue Katzenarten aus Central-Asien. *Ezhegodnik zool. Muz.* 9: 524–535.

----- 1907. Über neue und wenig bekannte Igel . . . *Ezhegodnik zool. Muz.* 11(1906): 167–190.

----- 1909. Beitrage zur Kenntnis der Säugetierfauna Kaukasiens . . . *Mitteilungen Kaukas Mus.* 4: 269–300.

----- 1911. [On the systematics of the family Mustelidae] *Izvestia kavkaz. Muz.* 5: 243–280.

Satya Prakash, K. C. & Aswathanarayana, N. V. 1973. Diversity of the karyotype in the long-tailed tree mouse, *Vandeleuria oleracea. Current Sci.* 42: 755–756.

----- 1976. Chromosome complexity in the Indian long-tailed tree mouse. *Journal Heredity* 67: 249–250.

Savage, D. E. & Russell, D. E. 1983. *Mammalian paleofaunas of the world.* Reading, Mass. : Addison-Wesley, xvii, 432pp.

Savi, P. 1822. *Nuovo Giorn. Lett. Sci. Pisa* 1: 60 (*n. v.*).

----- 1825. Descrizione del *Dinops cestonii* . . . *Nuovo Giorn. Lett. Sci. Pisa* 10: 229–236.

Schaller, G. 1973. Observations of Himalayan tahr (*Hemitragus jemlahicus*). *Journal Bombay nat. Hist. Soc.* 70: 1–24.

----- 1977. *Mountain monarchs.* Chicago: Univ. Chicago Press, 425pp.

Schaller, G. B. *et al.* 1985. *The giant pandas of Wolong.* Chicago & London: Univ. Chicago Press, 320pp.

----- 1990. Javan rhinoceros in Vietnam. *Oryx* 24: 77–80.

Schaub, S. 1953. Remarks on the distribution and classification of the "Histricomorpha". *Verhandlunden naturf. Ges. Basel* 64: 389–400.

Schevill, W. 1987. Reply to 'The scientific name of the sperm whale'. *Marine Mammal Sci.* 3: 89–90.

Schinz, H. R. 1821. *Das Thierreiche . . . vol. 1* Stuttgart, Tubingen.

----- 1825. *Ibid*, vol. 5.

----- 1844,1845. *Systematisches Verzeichniss aller bis jetzt bekannten Säugethiere oder Synopsis Mammalium.* Solothurn. Vol. 1 (1844), 587pp; vol. 1, Nachträge (1845a), 48pp; vol. 2 (1845b), 574pp.

Schlegel, H. 1841. *Abhandlungen aus dem Gebiete der Zoologie . . . Cetaceen.* Leiden, 44pp, 6pls.

----- 1857. *Handleiding tot de beoefening der dierkunde.* Leiden: Koninklijke Akad. Zee- en Landmagt.

----- 1863. Notice sur les écureuils à ventre rouge . . . *Nederlandsch Tijdschr. Dierk.* 1: 24–30.

----- 1866. Contribution à la faune de Madagascar . . . *Nederlandisch Tijdschr. Dierk.* 3: 73–89.

----- 1876. Les singes: Simiae. *Monographe Mus. Hist. nat. Pays-Bas* 40: 365pp.

----- 1879. *Paradoxurus musschenbroekii. Notes Leyden Mus.* 1: 43.

----- 1880. On an anomalous species of hare discovered in the isle of Sumatra: *Lepus netscheri. Notes Leyden Mus.* 2: 59–65.

Schlegel, H. & Müller, S. 1845. Over de op de Oost-Indische Eilanden levende sorten van het Geschlacht *Hylogale* (pp. 159–168); Over de ossen van den Indischen Archipel (pp. 195–208), in Temminck, C. J. *Verhandelingen natuurlijke geschiedenis Nederlandische overzeesche Bezittingen.* Leiden.

Schlitter, D. A. & Setzer, H. W. 1973. New rodents . . . from Iran and Pakistan. *Proceedings biol. Soc. Wash.* 86: 163–173.

Schlitter, D. A. & Thonglongya, K. 1971. *Rattus turkestanicus* . . . the valid name for *Rattus rattoides* . . . *Proceedings biol. Soc. Wash.* 84: 171–174.

Schlitter, D. A. *et al.* 1983. Taxonomic review of Temminck's trident bat, *Aselliscus tricuspidatus* . . . *Annals Carnegie Mus.* 52: 337–358.

Schlosser, M. 1903. Die fossilen Säugethiere Chinas . . . *Abhandlungen bayer. Akad. Wiss* 22: 1–221, 14pls.

----- 1924. Fossil primates from China. *Palaeontologia sin.* C, 1,2: 14pp, 1pl.

Schneider, J. G. 1800. In Schreber, J. C. D. *Die Säugethiere.* Leipzig.

Schrank, F. von P von 1798. *Fauna Boica* . . . 1,1. Nürnberg.

Schreber, J. C. D. von 1774–1785. *Die Säugethiere in Abbildungen nach der Natur* . . . Erlangen: 1112pp, 347pls.

----- 1799. *Die Säugethiere in Abbildungen* . . . Supplement.

Schreiber, A. *et al.* 1989. *Weasels, civets, mongooses and their relatives/ An action plan for the conservation of mustelids and viverrids.* Gland, Switzerland: IUCN, 100pp.

Schrenk, L. von 1859. *Reisen im Amur-Lande* vol. 1. St Petersburg.

Schultz, A. H. 1932. The generic position of *Symphalangus klossii. Journal Mammal.* 13: 368–369.

Schultz, W. 1970. Einige Bemerkungen zum . . . *Harpionycteris whiteheadi* . . . *Zeitschrift Säugetierk.* 35: 81–89.

Schulze, E. 1897. Mammalia Europaea. *Helios, Berlin* 14: 75–100.

Schütt, G. 1972. Fossil mammals of Java 4. Zur Kenntnis der Pleistozänen Hyänen Javas. *Proceedings K. ned. Akad. Wet.* B, 75: 261–288, 2pls.

----- 1973. Fossil mammals of Java. 5. Pleistozäne Caniden . . . *Proceedings K. ned. Akad. Wet.* B, 76: 446–470.

Schwartz, J. H. 1986. Primate systematics and a classification of the order. Pp. 1–41 in Swindler, D. R. & Erwin, J. 1986. *Comparative primate biology, vol. 1. Systematics, evolution and anatomy.* New York: Liss, 820pp.

----- (ed.) 1989. *Orang-utan biology.* Oxford Univ. Press, 390pp.

Schwartz, J. H. & Tattershall, I. 1985. Evolutionary relationships of living lemurs and lorises . . . *Anthropological Pap. Am. Mus. nat. Hist.* 60: 1–100.

Schwarz, E. 1910a. Two new oriental Viverridae. *Annals Mag. nat. Hist.* 6: 230–232.

----- 1910b. Notes on some palm-civets. *Annals Mag. nat. Hist.* 5: 422–424.

----- 1911. Seven new Asiatic mammals. *Annals Mag. nat. Hist.* 7: 634–640.

----- 1912a. On a new palm-civet from Timor. *Annals Mag. nat. Hist.* 10: 131–132.

----- 1912b. Notes on Malay tigers . . . *Annals Mag. nat. Hist.* 10: 324–326.

----- 1913a. On a kangaroo and a new palm-civet in the British Museum. *Annals Mag. nat. Hist.* 12: 288–289.

----- 1913b. Two new mammals from the Malay Archipelago. *Annals Mag. nat. Hist.* 11: 296–298.

----- 1914. Säugetiere von Timor. Pp. 113–161 in Haniel, C. B. *Zoologie von Timor. Lief. 2.* Stuttgart.

----- 1939. A new porcupine (*Thecurus*) from Borneo. *Journal Mammal.* 20: 246–248.

----- 1947. Colour mutants of the Malay short-tailed mongoose, *Herpestes brachyurus* Gray. *Proceedings zool. Soc. Lond.* : 79–80.

Schwarz, E. 1948. Revision of the old world moles of the genus *Talpa*. *Proceedings zool. Soc. Lond.* 118: 36–48.

Schwarz, E. & Schwarz, H. K. 1943. The wild and commensal stocks of the house mouse, *Mus musculus*. *Journal Mammal.* 24: 59–72.

----- 1967. A monograph of the *Rattus rattus* group. *Annales Esc. nac. Cienc. biol. Mex.* 14: 79–178.

Sclater, P. L. 1858. On the general geographical distribution of the members of the class Aves. *Journal Linn. Soc. Lond.* (*Zool.*) 2: 130–145.

----- 1862. Note on the deer of Formosa. *Proceedings zool. Soc. Lond.* : 150–152, pls 16,17.

----- 1870a. [The arrangement of the family Cervidae.] *Proceedings zool. Soc. Lond.* : 114–116.

----- 1870b. [Meeting of June 9, 1870.] *Proceedings zool. Soc. Lond.* : 380–381, pl. 28.

----- 1871. Notes on rare or little-known animals now or lately in the Society's garden. *Proceedings zool. Soc. Lond.* : 221–237, pls 14–17.

----- 1872. [Report on additions to the Society's menagerie . . .] *Proceedings zool. Soc. Lond.* : 493–496, pls 23–26.

----- 1873. Additions to the menagerie. *Proceedings zool. Soc. Lond.* : 193–194, 1 pl.

----- 1875. On several rare or little-known mammals now or lately in the Society's collection. *Proceedings zool. Soc. Lond.* : 417–423, pls 47–51.

----- 1876. On the rhinoceroses now or lately living in the Society's menagerie. *Transactions zool. Soc. Lond.* 9: 650, pls xcv–xcix.

----- 1877. [Additions to the Society's menagerie] *Proceedings zool. Soc. Lond.* : 679–683.

----- 1883. *List of vertebrated animals living in the garden of the Zoological Society of London, 8th ed.* . London: xvi, 682pp.

----- 1885. [Report on additions made to the Society's menagerie . . .] *Proceedings zool. Soc. Lond.* : 1–2, pl. 1.

----- 1888. The "Tamarou" of the Philippine Islands. *Nature, Lond.* 38: 363–364.

----- 1898. [Report on the additions to the Society's menagerie] *Proceedings zool. Soc. Lond.* : 1–3. pl. 1.

Scopoli, G. A. 1786. *Deliciae florae et faunae Insubricae* . . . Part 2. Ticini.

Scott, R. W. 1969. Notes on the interbreeding between domestic and wild buffalo in Upper Assam. *Journal Bombay nat. Hist. Soc.* 66: 159–166.

Scrivenor, J. B. *et al.* 1943. A discussion of the biogeographic division of the Indo-Australian Archipelago. *Proceedings Linn. Soc. Lond.* 154: 120–165.

Scudder, S. H. 1882. Nomenclator zoologicus. *Bulletin US natn. Mus.* 19: 1–340.

Scully, J. 1881a. On some mammals from Kandahar. *Annals Mag. nat. Hist.* 8: 222–229.

----- 1881b. On the mammals of Gilgit. *Proceedings zool. Soc. Lond.* : 197–209.

Seabra, A. F. de 1898. Noticia sobre una nove especie de genero "Cynonycteris" . . . *Jornal Sci. math. phys. nat.* 5: 157–162.

Seba, A. 1734. *Locupletissimi rerum naturalium* . . . Amsterdam.

Seebeck, J. H. 1989. Sciuridae. Pp. 932–978 in Walton, D. W. & Richardson, B. J. *Fauna of Australia, vol. 1, B. Mammalia*. Canberra: Aust. Govt Pub. Service.

Seitz 1892. *Mitteilungen Dtsch. Naturk. Ostasiens* 5: 364.

Selenka, E. 1896. Die Rassen und der Zahnwechsel des Orang-Utan. *Sitzungsberichte k. pruss. Akad. Wiss.* 16: 381–392.

Sélys Longchamps, M. E. de 1839. *Etude de micromammalogie* . . . Paris, 165pp, 3pls.

Sen, S. 1983. Rongeurs et lagomorphes du Gisement pliocene de Pul-e Charkhi, Bassin de Kabul, Afghanistan. *Bulletin Mus. natn. Hist. nat. Paris* (C) 5: 33–74.

Sen, S. & Sharma, T. 1983. Role of constitutive heterochromatin in evolutionary divergence . . . in *Mus musculus, M. booduga* and *M. dunni*. *Evolution, Lawrence* 37: 628–636.

Sen, S. *et al.* 1979. Découverte de rongeurs 'africans' dans le Pliocène d'Afghanistan . . . *Bulletin Mus. natn. Hist. nat. Paris* (C), 1: 65–75.

Seth, P. K. & Saxena, M. B. 1977. Cytological mechanisms in the evolution of mammalian chromosomes. *Nucleus, Calcutta* 20: 42–52.

Severtzov, N. 1858. Notice sur la classification multisériale des carnivores . . . *Revue Mag. Zool.* 10: 385–393.

----- 1873. [Vertical and horizontal distribution of Turkestan animals.] *Izvestiya imp. Obschch. Lyub. estest. Antrop. Etnogr. imp. mosk. Univ.* 8,2: 157pp, 10pls.

Severzoff, N. 1876. The mammals of Turkestan. *Annals Mag. nat. Hist.* 18: 40–57; 168–174; 208–225; 325–336; 377–388. (Translation of Severtzov, 1873 by F. C. Craemers.)

Shahi, S. P. 1982. Status of the grey wolf (*Canis lupus pallipes* Sykes) in India . . . *Journal Bombay nat. Hist. Soc.* 79: 493–502.

Shamel, H. H. 1930. Mammals collected on the island of Koh Tau off the east coast of the Malay peninsula. *Journal Mammal.* 11: 71–74.

----- 1940a. Three new mammals from Asia. *Journal Mammal.* 21: 76–78.

----- 1940b. The insectivorous bats collected by H. C. Raven in Celebes. *Journal Mammal.* 21: 352–354.

----- 1942. A collection of bats from Thailand. *Journal Mammal.* 23: 317–328.

----- 1944. A new *Myotis* from Formosa. *Journal Mammal.* 25: 191–192.

Sharma, T. & Rajiva Raman 1972. Odd diploid number in both sexes and a unique multiple sex chromosome system of a rodent, *Vandeleuria o. oleracea* . . . *Cytogenetics* 11: 247–258.

Shaw, G. 1793. *Museum Leverianum*, vol. 1 (2). London.

----- 1800,1801. *General zoology . . . vol. I, vol. II, part 1. Mammalia*. London: 553, 226pp.

Shaw, G. & Nodder, F. P. 1791. *The naturalist's miscellany* vol. 2. London: Nodder.

Shaw, T-h. & Wong, S. 1959. A new insectivore from Hainan, *Acta zool. sin.* 11: 422–428.

Shaw, T-h. *et al.* 1966. A survey of the mammals of Hainan Island, China. *Acta zootaxon. sin.* 3: 260–276.

Shelford, R. 1901. On a new bat from Borneo. *Annals Mag. nat. Hist.* 8: 113–114.

Sherborn, C. D. & Woodward, B. B. 1961. Notes on the dates of publication of the natural history portions of some French voyages . . . *Annals Mag. nat. Hist.* 7: 388–392.

Shi, L. & Chen, Y. 1989. The karyotype analysis of Yunnan mouse deer (*Tragulus javanicus williamsoni*). *Acta zool. sin.* 35: 41–44.

Shi, M. *et al.* 1981. Pliocene mammals. . . from Zheotung, Yunnan. *Memoirs Beijing nat. Hist. Mus.* 11: 155pp, 2pls.

Shih, C. M. 1930a. Preliminary report on the mammals from Yaoshan, Kwangsi . . . *Bulletin Dept. Biol. Sun Yatsen Univ. Canton* 4: 10pp.

----- 1930b. Note on the mammals from south-western Hunan. *Bulletin Dept. Biol. Sun Yatsen Univ. Canton* 9: 4pp.

----- 1930c. Notes on a small collection of mammals from Yaoshan, North River, Kwangtang. *Bulletin Dept Biol. Sun Yatsen Univ. Canton* 8: 2pp.

----- 1931. Further notes on mammals from Yaoshan, North River, Kwangtung. *Bulletin Dept. Biol. Sun Yatsen Univ. Canton* 12: 8pp.

Shikama, T. 1937. Fossil cervifauna of Syatin near Tainan, southwestern Taiwan (Formosa). *Science Rep. Tohoku Univ.,Geol.* 19: 75–85, pl. xvi.

Shoshani, J. & Eisenberg, J. F. 1982. *Elephas maximus*. *Mammalian Species* 182: 8pp.

Shoshani, J. *et al.* 1985. Protein and anatomical evidence of the phylogenetic position of *Mammuthus primigenius* . . . *Acta zool. fenn.* 170: 237–240.

Shou, Z-h. 1964. [*Handbook of economic animals of China, Mammals*.] Peking, 554pp, 72pls.

Siddiqi, M. S. 1969. *Fauna of Pakistan*. Karachi: Agricultural Research Council.

Siebold, P. T. von. 1824. *Spicilegia faunae Japoniae* . . . Batavia.

Sigé. B. 1974. Données nouvelles sur le genre *Stehlinia* . . . *Palaeovertebrata* 6: 253–272.

Sigurdsson, J. B. & Yang, C. M. 1990. Marine mammals of Singapore. Pp. 25–37 in *Essays in zoology*. Singapore: National University.

Silas, E. G. (ed.) 1988. *Proceedings of the symposium on endangered marine animals* . . . Cochin, India: Marine Biol. Assoc. India, 506pp.

Silva, P. H. D. H. de 1988. Cetaceans . . . recorded off Sri Lanka, India . . . *Journal Bombay nat. Hist. Soc.* 84: 505–525.

Simoons, F. J. 1984. Gayal or mithan. Pp. 34–39 in Mason, I. L. (*q. v.*).

Simpson, G. G. 1945. The principles of classification and a classification of the mammals. *Bulletin Am. Mus. nat. Hist.* 85:xvi, 350pp.

Sinha, Y. P. 1969. A new pipistrelle bat . . . from Burma. *Proceedings zool. Soc. Calcutta* 22: 83–86.

----- 1970. Taxonomic notes on some Indian bats. *Mammalia* 34: 81–92.

----- 1973. Taxonomic studies on the Indian horseshoe bats of the genus *Rhinolophus*. *Mammalia* 37: 603–630.

----- 1977. The oriental bats of genus *Megaderma* . . . *Geobios, Jodhpur* 4: 9–12.

----- 1981. New record of . . . *Nesokia indica* from Bihar. *Bulletin zool. Surv. India* 3: 267–268.

----- 1990. Occurrence of Dobson's long-tongued fruit bat *Eonycteris spelaea* . . . in Meghalaya. *Journal Bombay nat. Hist. Soc.* 87: 134.

Sinha, Y. P. & Chakraborty, S. 1971. Taxonomic status of the vespertilionid bat, *Nycticejus emarginatus* Dobson. *Proceedings zool. Soc. Calcutta* 24: 53–59.

Sjarmidi, A. & Gerard, J. F. 1988. Autour de la systématique et la distribution des suidés. *Monitore zool. ital.* 22: 415–448.

Slack, J. H. 1867. Mammalogical notices. *Proceedings Acad. nat. Sci. Philad.* 19: 34–38.

Smee, W. 1834. [Meeting of Dec. 10, 1833] *Proceedings zool. Soc. Lond.* (1833): 140.

Smielowski, J. M. & Praduman, P. R. 1988. The Indian wild ass— wild and captive populations. *Oryx* 22: 85–88.

Smiet, A. C. *et al.* 1978. Geographical distribution and variation of *Bandicota bengalensis* in Pakistan. *Pakistan J. Zool.* 10: 43–47.

Smith, A. 1840. *Illustrations of the zoology of South Africa . . . vol. 1. Mammalia*. London.

Smith, C. H. 1827. The Carnassiers (vol. 2); Order VII—Ruminantia (Pecora, Lin.). (vol. 4: 1–428; vol. 5: 296–376) in Griffith, E. *et al., Cuvier's animal kingdom. Synopsis of the species of the class Mammalia*. London: Whittaker.

----- 1839. *The natural history of dogs* . . . (Naturalist's Library, Mammalia, vol. 9). Edinburgh: Lizars, 267pp. 31pls.

----- 1842. *An introduction to the Mammalia* . . . (Naturalist's Library, vol. 13 (35). Edinburgh: Lizars, 313pp, 30pls.

Smith, J. D. & Hood, C. S. 1983. A new species of tube-nosed fruit bat (*Nyctimene*) from the Bismark Archipelago . . . *Occasional Pap. Mus. Texas Tech Univ.* 81: 1–14.

Smith, M. J. 1973. *Petaurus breviceps. Mammalian Species* 30: 5pp.

Smit-van Dort, M. 1989. Skin, skull and skeleton characters of mouse deer (Mammalia, Tragulidae) . . . *Bulletin zool. Mus. Univ. Amst.* 12: 89–96.

Sobti, R. C. & Gill, S. S. 1984. Chromosomes and protein polymorphism in certain *Rattus* species from northern India. *Research Bull. Panjab Univ. Sci.* 35: 203–210.

Sody, H. J. V. 1928. Twee nieuwe subspecies van *Pachyotis temmincki* Horsf. *Natuurkundig Tidjschr. ned.-Indië* 88: 86–91.

----- 1929a. Over *Callosciurus notatus* . . . *Natuurkundig Tijdschr. ned.-Indië* 88 (1928): 325–332.

----- 1929b. Naamlijst van de zoogdieren van Java. *Natuurkundig Tijdschr. ned.-Indië* 89: 160–166.

----- 1930a. On some new . . . mammals from Java, Borneo and Celebes. *Natuurkundig Tijdschr. ned.-Indië* 90: 258–273.

----- 1930b. Oversicht van de ratten van Java . . . *Zoologische Meded. Leiden* 13: 100–140.

----- 1930c. Over *Rattus rattus roquei* Sody . . . *Zoologische Meded. Leiden* 13: 94–99.

----- 1930d. *Pithecus aygula fredericae* n. subsp. *Tropische Natuur.* 19: 68.

----- 1931a. Two new races of *Rattus lepturus* from Java. *Natuurkundig Tijdschr. Ned. -Indië* 91: 212–215.

----- 1931b. Six new mammals from Sumatra, Java, Bali and Borneo. *Natuurkundig Tijdschr. Ned. -Indië* 91: 349–360.

----- 1932a. Six new Indo-Malayan rats. *Natuurhistorisch Maandblad* 21: 157–160.

----- 1932b. Four new mammals from Bali and Soemba. *Natuurkundig Tijdsch. Ned.-Indië* 92: 334–340.

----- 1933a. Ten new mammals from the Dutch East Indies. *Annals Mag. nat. Hist.* 12: 430–442.

----- 1933b. On the mammals of Bali . . . *Natuurkundig Tijdschr. Ned. -Indië* 93: 56–95.

----- 1934. New rats from Java and New Guinea. *Natuurkundig Tijdschr. Ned. -Indië* 94: 170–177.

----- 1936a. Seventeen new generic, specific and subspecific names for Dutch East Indian mammals. *Natuurkundig Tijdschr. Ned.-Indië* 96: 42–55.

----- 1936b. A new flying squirrel from Sumatra. *Natuurkundig Tijdschr. Ned. -Indië* 96: 146–147.

----- 1936c. A new race of *Callosciurus notatus* from north Sumatra. *Natuurkundig Tijdschr. ned.-Indië* 96: 217–218.

----- 1937a. Notes on some mammals from Sumatra . . . *Temminckia* 2: 211–220.

----- 1937b. On the mammals of Banka. *Temminckia* 2: 221–250.

----- 1940a. On the mammals of Enggano. *Treubia* 17: 381–401.

----- 1940b. Notes on *Rattus rattus jalorensis* and *R. r. roquei* . . . *Treubia* 17: 403–405.

----- 1941a. On a collection of rats from the Indo-Malayan and Indo-Australian regions . . . *Treubia* 18: 255–325.

----- 1941b. A new race of *Sus verrucosus* from Madoera Island. *Treubia* 18: 393–394.

----- 1946. Review of: Prehistoric and fossil rhinoceroses of the Malay Archipelago and India, by D. A. Hooijer. *Natuurwetenschappelijk Tijdschr. Ned. Indië* 102: 151.

----- 1949a. On a collection of Sciuridae . . . *Treubia* 20: 57–120.

----- 1949b. Notes on some Primates, Carnivora and the babirusa from the Indo-Malayan and Indo-Australian Regions. *Treubia* 20: 121–190.

----- 1959. Das javanische Nashorn *Rhinoceros sondaicus. Zeitschrift Säugetierk.* 24: 109–240.

----- 1989. Land mammals of Java. Pp. 137–164 (published posthumously) in Becking, J. H. (*q. v.*).

Soma, H. (ed.) 1987. *The biology and management of* Capricornis *and related mountain antelopes.* London: Croom Helm, xi, 391pp.

Sonakia, A. 1984. The skull-cap of early man and associated mammalian fauna from Narmada Valley . . . *Record geol. Surv. India* 113,6: 159–172.

Sondaar, P. Y. 1984. Faunal evolution and the mammalian biostratigraphy of Java. *Courier ForschInst. Senckenberg* 69: 219–235.

Sondaar, P. Y. *et al.* 1989. The insular proboscideans of S. E. Asia. *Abstracts 5th ITC, Rome*: 161–162.

Song, S. 1985. A new subspecies of *Cricetulus triton* from Shaanxi, China. *Acta theriol. sin.* 5: 137–139.

Soota, T. D. & Chaturvedi, Y. 1980. New locality record of *Pipistrellus camortae* Miller from Car Nicobar . . . *Records zool. Surv. India* 77: 83–87.

Sowerby, A. de C. 1933. Horns of a new deer and other relics from the waste of Yin, Honan, China. *China J.* 19: 141–144, 1pl.

----- 1945. A new species of shrew from the Shanghai area. *Musée Heude Notes Mammal.* 3: 2pp.

Sparrman, A. 1778. Beskrifning pa *Sciurus bicolor*, et nytt species Ikorn, fran Java. *Handlingar Götheb. Wet. Saml. Wet. afd.* 1: 70–71.

----- 1780. Lufspringare gazellen . . . *Kungliga svenska VetenskAkad. Handl.* 1: 275–281.

Srivastava, J. P. & Verma, B. C. 1972. New fossil forms of Rhinocerotidae and Suidae from Pinjar beds . . . *Journal palaeont. Soc. India* 15: 76–82, 3pls.

Stanyon, R. *et al.* 1987. Banded karyotypes of the 44-chromosome gibbons. *Folia primatol.* 48: 54–64.

Stebbings, R. E. & Griffith, F. 1984,1986. *Distribution and status of bats in Europe.* Abbots Ripton: Institute of Terrestrial Ecology.

Steere, J. B. 1889. [Letter to the Secretary.] *Proceedings zool. Soc. Lond.* (1888): 413–415.

----- 1890. *A list of the birds and mammals collected by the Steere Expedition to the Philippines* . . . Ann Arbor, Michigan, 30pp.

Stehlin, H. G. & Graziosi, P. 1935. Ricerche sugli asinidi fossili d'Europa. *Mémoires soc. Paléont. suisse* 56: 1–73, pls 1–9.

Stein, G. 1933. Weitere Mitteilungen zur Systematik papuanischer Säuger. *Zeitschrift Säugetierk.* 8: 87–95.

Stewart, B. S. & Leatherwood, S. 1985. Minke whale—*Balaenoptera acutorostrata.* Pp. 91–136 in Ridgeway, S. H. & Harrison, R. (*q. v.*).

Stogov, I. I. 1985. On two little studied species of white-toothed shrews . . . *Zoologicheskii Zh.* 64: 264–268.

Stoliczka, F. 1872. Notice of the mammals and birds inhabiting Kachh. *Journal Asiatic Soc. Bengal* 41,2: 211–258.

Stone, W. 1900. Description of a new rabbit from the Liu Kiu Islands and a new flying squirrel from Borneo. *Proceedings Acad. Sci. Philad.* : 460–463.

Stone, W. & Rehn, J. A. G. 1902. A collection of mammals from Sumatra . . . *Proceedings Acad. nat. Sci. Philad.* 54: 127–142.

Storr, G. C. C. 1780. *Prodromus methodi mammalium* . . . Tubingen, 43pp.

Strand, E. 1926. Miscellanea nomenclatorica zoologica et palaeontologica. *Archiv Naturgesch.* 92, A, 8: 30–75.

Strasser, E. & Delson, E. 1988. Cladistic analysis of cercopithecoid relationships. *Journal Hum. Evol.* 16: 81–99.

Strelkov, P. P. 1986. The Gobi bat *Eptesicus gobiensis* . . . *Zoologicheskii Zh.* 65: 1103–1108.

Stremme, H. 1911a. Die Säugetierfauna der *Pithecanthropus*-Schichten. *Zentralblatt Miner. Geol. Paläont.* : 54–60; 83–89.

----- 1911b. Die Säugetiere mit Ausnahme der Proboscidien. Pp. 82–150 in Selenka, M. L. & Blankenhorn, M., *Die* Pithecanthropus- *Schichten auf Java.* Leipzig.

Stroganov, S. V. 1941. Insectivore mammals of the fauna of the USSR. *Doklady Akad. Nauk SSSR* 33: 270–272.

----- 1948. [Systematics of moles, Talpidae.] *Trudy zool. Inst. Leningr.* 8,2: 1–239.

Stroganova, A. S. 1957. On the Sciuridae-fauna in the southern part of Yunnan. *Zoologicheskii Zh.* 36: 1761–1769.

Stuart, C. T. 1984. The distribution and status of *Felis caracal* Schreber, 1776. *Säugetierkundliche Mitt.* 31: 197–203.

Subba, P. B. 1989. The status and conservation of the golden langur (*Presbytis geei* Khajuria, 1956) in the Manas National Park, Bhutan. *Tigerpaper* 16,4: 16–18.

Sugardjito, J. *et al.* 1989. Population survey of macaques in northern Sulawesi. *American J. Primatol.* 18: 285–302.

Sukumar, R. 1989. *The Asian elephant: ecology and management.* Cambridge: Cambridge Univ. Press, 251pp.

Sundevall, C. J. 1842. Ofversigt of slogtet *Erinaceus. Kungliga svenska VetenskAkad. Handl.* (1841): 215–239.

----- 1843a. Om Professor J Hedenborgs insamlingen af däggdjur i Nordöstra Africa och Arabien. *Kungliga svenska VetenskAkad. Handl.* (1842): 189–244.

----- 1843b. Ofversigt af slägtet *Manis. Kungliga svenska VetenskAkad. Handl.* (1842): 245–282.

----- 1846. Methodisk öfversigt af Idislande djuren, Linnés Pecora. *Kungliga svenska VetenskAkad. Handl.* (1844): 121–210.

Supriatna, J. *et al.* 1990. Secondary intergradation between *Macaca maurus* and *Macaca tonkeana* in Sulawesi Selatan. *American J. Primatol.* 20: 236.

Swainson, W. 1835. *On the natural history and classification of quadrupeds.* London: Longman etc., 397pp.

Swinhoe, R. 1863. On the mammals of the island of Formosa. *Proceedings zool. Soc. Lond.* (1862): 347–365.

----- 1864a. On a new rat from Formosa. *Proceedings zool. Soc. Lond.* : 185–187.

----- 1864b. [Letter from Mr R Swinhoe] *Proceedings zool. Soc.* Lond. : 378–383.

----- 1866a. On a new species of beech-marten from Formosa. *Annals Mag. nat. Hist.* 18: 286.

----- 1866b. [Letter.] *Proceedings zool. Soc. Lond.* : 556.

----- 1870a. Zoological notes of a journey from Canton to Peking and Kalgan. *Proceedings zool. Soc. Lond.* : 427–451.

----- 1870b. On the mammals of Hainan. *Proceedings zool. Soc. Lond.* : 224–239.

----- 1870c. On a new deer from China. *Proceedings zool. Soc. Lond.* : 89–92, pls 6,7.

----- 1870d. [Report of Zoological Society.] *Athenaeum* 2208: 264.

----- 1871. Catalogue of the mammals of China . . . *Proceedings zool. Soc. Lond.* (1870): 615–653.

----- 1873a. On Chinese deer . . . *Proceedings zool. Soc. Lond.* : 572–576.

----- 1873b. Notes on Chinese Mammalia observed near Ningpo. *Proceedings zool. Soc. Lond.* (1872): 813–818.

----- 1874. On a small, tufted, hornless deer from the mountains near Ningpo. *Proceedings zool. Soc. Lond.* : 452–454, pl. 59.

Sykes, W. H. 1831. [Skins of . . . Mammalia obtained in Dukhun.] *Proceedings zool. Soc. Lond.* 1: 99–105.

Sylvestre, J-P. 1985. On the intergeneric hybrids in cetaceans. *Aquatic Mamm.* 11: 101–108.

Szalay, F. S. & Delson, E. 1979. *Evolutionary history of the primates.* New York: Academic Press, xiv, 580pp.

Takashima, H. 1954. Considerations on the change of animal life in Japan. II. *Miscellaneous Rep. Yamachima Inst. Orn. Zool.* 4: 146–155.

Tan, B. 1989. Conservation and economic importance of the mustelids and viverrids in China. *Mustelid and viverrid conservation (IUCN)* 1: 5–6.

T'an, P-c. 1957. Rare catches by Chinese animal collectors. *Zoo Life* 12: 61–63.

Tate, G. H. H. 1936. Some Muridae of the Indo-Australian region. *Bulletin Am. Mus. nat. Hist.* 72: 501–728.

----- 1940. Notes on the types of certain early described species of monotremes . . . from the Indo-Australian Region. *American Mus. Novit.* 1061: 10pp.

----- 1941a. Remarks on some Old World leaf-nosed bats. *American Mus. Novit.* 1140: 11pp.

----- 1941b. Molossid bats of the Archbold collections. *American Mus. Novit.* 1142: 4pp.

----- 1941c. A review of the genus *Hipposideros* . . . *Bulletin Am. Mus. nat. Hist.* 78: 353–393.

----- 1941d. Notes on Oriental *Taphozous* and allies. *American Mus. Novit.* 1141: 5pp.

----- 1941e. Review of *Myotis* of Eurasia. *Bulletin Am. Mus. nat. Hist.* 78: 537–565.

----- 1941f. Notes on vespertilionid bats. *Bulletin Am. Mus. nat. Hist.* 78: 567–597.

----- 1942a. Pteropodidae (Chiroptera) of the Archbold collections. *Bulletin Am. Mus. nat. Hist.* 86: 331–347.

----- 1942b. Review of the vespertilionine bats . . . *Bulletin Am. Mus. nat. Hist.* 80: 221–297.

----- 1943. Further notes on the *Rhinolophus philippinensis* group (Chiroptera). *American Mus. Novit.* 1219: 5pp.

----- 1944. *A list of the mammals of the Japanese War area. Part 3. Lesser Sunda Islands, Moluccas, Celebes.* New York: American Museum of Natural History.

----- 1945a. The marsupial genus *Phalanger*. *American Mus. Novit.* 1283: 41pp.

----- 1945b. Notes on the squirrel-like and mouse-like possums. *American Mus. Novit.* 1305: 12pp.

----- 1946. Geographical distribution of the bats in the Australasian archipelago. *American Mus. Novit.* 1323: 21pp.

----- 1948. Studies in the Peramelidae . . . *Bulletin Am. Mus. nat. Hist.* 92: 313–346.

----- 1951a. *Harpionycteris*, a genus of rare fruit bats. *American Mus. Novit.* 1522: 9pp.

----- 1951b. The rodents of Australia and New Guinea. *Bulletin Am. Mus. nat. Hist.* 97: 183–430.

----- 1952. A new *Rhinolophus* from Queensland . . . *American Mus. Novit.* 1578: 3pp.

----- 1954. A new squirrel from Burma. *American Mus. Novit.* 1676: 2pp.

Tate, G. H. H. & Archbold, R. 1935a. Twelve apparently new forms of *Rattus* from the Indo-Australian region. *American Mus. Novit.* 802: 10pp.

----- 1935b. Twelve apparently new forms of Muridae . . . from the Indo-Malayan Region. *American Mus. Novit.* 803: 9pp.

----- 1936. A new race of *Hyosciurus*. *American Mus. Novit.* 846: 1p.

----- 1939a. A revision of the genus *Emballonura* . . . *American Mus. Novit.* 1035: 14pp.

----- 1939b. Oriental *Rhinolophus* . . . *American Mus. Novit.* 1036: 12pp.

Taylor, E. H. 1934. Philippine land mammals. *Monographs Philipp. Bureau Sci.* 30: 548pp, 25pls.

Taylor, J. 1891. Description of a new species of palm-civet (*Paradoxurus*) found in Orissa. *Journal Bombay nat. Hist. Soc.* 6: 429–432, pl.

Taylor, J. M. *et al.* 1982. A revision of the genus *Rattus* . . . in the New Guinea region. *Bulletin Am. Mus. nat. Hist.* 173: 177–336.

Tehsin, R. H. 1981. Occurrence of the large brown flying squirrel and mouse deer near Udaipur, Rajasthan. *Journal Bombay nat. Hist. Soc.* 77: 498.

Teilhard de Chardin, P. 1940. The fossils from Locality 18 near Peking. *Palaeontologica sinica* 9: 100pp, 3pls.

Teilhard de Chardin, P. & Leroy, P. 1942. Chinese fossil mammals. A complete bibliography . . . *Publications Inst. Géobiol. Pekin* 8: 142pp.

----- 1945a. Les félidés de Chine. *Publications Inst. Géobiol. Pekin* 11: 58pp.

----- 1945b. Les mustélidés de Chine. *Publications Inst. Géobiol. Pekin* 12: 56pp.

Teilhard de Chardin, P. & Trassaert, M. 1937. The Pliocene Camelidae, Giraffidae and Cervidae of South Eastern Shansi. *Palaeont. sin.* C, 1: 56pp, 6pls.

Temminck, C. J. 1820. Sur le genre Hyène . . . *Annales gén. Sci. phys. Brux.* 3: 46–57, pl. 35.

----- 1824–1827. *Monographies de mammalogie* . . . Paris: Dufour.

----- 1832. Monographie over een nieuw geslacht van knaagdier, onder den naam van *Nyctocleptes*. *Bijdragen Natuurk. Wet. Amst.* 7: 1–8, pl. 1.

----- 1834. Over een geslachk der vleugelhandige zoogdieren . . . *Tijdschrift Natuurl. Gesch. Physiol.* 1: 1–30, 1pl.

----- 1835,1837,1840, 1841a. *Monographies de mammalogie* . . . tome 2. Leiden & Paris, 392pp, pls xxvi-lxx.

----- [1836.] *Coup d'oeil sur la faune des Iles de la Sonde et de l'empire du Japon.* Leiden, 30pp.

----- 1838. Over de geslachten *Taphozous, Emballonura, Urocryptus* en *Diclidurus*. *Tijdschrift Natuurl. Gesch. Physiol.* 5: 3–34, 1pl.

----- 1839. Over de kennis en de verbreiding de zoogdieren van Japan. *Tijdschrift Natuurl. Gesch. Physiol.* 5: 273–293.

----- 1841b. Nouveau genre de chauve-souris. *Echo du Monde Savant* 654: 452.

----- 1843–1844. *Aperçu general et specifique sur les mammifères.* In Siebold, P. T. von, *Fauna Japonica* . . . Amsterdam: Muller. (Pp. 1–24, pls 1–10: 1843; pp. 25–59, pls 11–20 and marine mammals pp. 1–26, pls 21–30: 1844.)

----- 1846–1849. *Coup d'oeil général sur les possessions Néerlandaises dans l'Inde Archipélagique.* Vol. 1 (1846); vol. 2 (1847); vol. 5 (1849). Leiden.

----- 1853. *Equisses zoologiques sur la côte de Guiné. 1. Mammifères.* Leiden: Brill, xvi, 256pp.

Templeton, W. E. 1848. In Blyth, E., Report of Curator . . . *Journal Asiat. Soc. Bengal* 17: 247–255.

Thenius, E. 1948. Bemerkungen über fossile Ursiden . . . *Anzeiger öst. Akad. Wiss. Math. -Naturwiss. Klasse* 84 (1947): 11–12.

Thomas, O. 1880. Description of a new bat from Java, of the genus *Kerivoula*. *Annals Mag. nat. Hist.* 5: 472–473.

----- 1881. Description of a new species of *Mus* from southern India. *Annals Mag. nat. Hist.* 7: 24.

----- 1882. Description of a new species of rat from China. *Proceedings zool. Soc. Lond.* : 587–588, pl. 44.

----- 1886a. Diagnoses of three new Oriental mammals. *Annals Mag. nat. Hist.* 17: 84.

----- 1886b. On the mammals presented by Allan O. Hume . . . *Proceedings zool. Soc. Lond.* : 54–79, pls 5,6.

----- 1887a. Description of two new squirrels from North Borneo. *Annals Mag. nat. Hist.* 20: 127–129.

----- 1887b. Description of a new rat from North Borneo. *Annals Mag. nat. Hist.* 20: 269–270.

----- 1887c. Report on a zoological collection . . . Christmas Island, Indian Ocean. *Proceedings zool. Soc. Lond.* : 511–514, pls 41–42.

----- 1888a. *Catalogue of the Marsupialia* . . . London: British Museum (Nat. Hist.), xiii, 401pp.

----- 1888b. Diagnoses of four new mammals from the Malayan region. *Annals Mag. nat. Hist.* 2: 407–409.

----- 1888c. *Eupetaurus*, a new form of flying squirrel from Kashmir. *Journal Asiat. Soc. Bengal* 57,2: 256–260.

----- 1889a. On the mammals of Christmas Island. *Proceedings zool. Soc. Lond.* (1888): 532–534.

----- 1889b. Description of a new Bornean monkey belonging to the genus *Semnopithecus*. *Proceedings zool. Soc. Lond.* : 159–190, pl. 16.

----- 1889c. On the mammals of Mount Kina Balu, North Borneo. *Proceedings zool. Soc. Lond.* : 228–236, pl. 24.

----- 1890a. Diagnosis of a new *Cynopterus* from Borneo. *Annals Mag. nat. Hist.* 5: 235–236.

----- 1890b. Description of a new squirrel from Borneo. *Annals Mag. nat. Hist.* 6: 171–172.

----- 1891a. Diagnoses of three new mammals . . . in Burma. *Annali Mus. civ. Stor. nat. Giacomo Doria* 10: 884.

----- 1891b. Descriptions of three new bats . . . *Annals Mag. nat. Hist.* 7: 527–580.

----- 1891c. Description of a new vole from China. *Annals Mag. nat. Hist.* 8: 117–118.

----- 1892a. On some new Mammalia from the East Indian Archipelago. *Annals Mag. nat. Hist.* 9: 250–254.

----- 1892b. Description of two new Bornean squirrels. *Annals Mag. nat. Hist.* 10: 214–216.

----- 1892c. On the Mammalia collected by Signor Leonardo Fea in Burma and Tenasserim. *Annali Mus. civ. Stor. Giacomo Doria* 10: 913–949.

----- 1892d. Diagnosis of a new subspecies of hare from Corea. *Annals Mag. nat. Hist.* 9: 146–147.

----- 1892e. Description of a remarkable new *Semnopithecus* from Sarawak. *Annals Mag. nat. Hist.* 10: 474–477.

----- 1892f. Note on the gibbon of the island of Hainan . . . *Annals Mag. nat. Hist.* 9: 145–146.

----- 1893a. On some new Bornean Mammalia. *Annals Mag. nat. Hist.* 11: 341–347.

----- 1893b. Description of a new Bornean *Tupaia*. *Annals Mag. nat. Hist.* 12: 53–54.

----- 1893c. Description of a new *Sciuropterus* from the Philippines. *Annals. Mag. nat. Hist.* 12: 30–31.

----- 1893d. Descriptions of two new North-Bornean mammals. *Annals Mag. nat. Hist.* 12: 230–232, pl. 7.

----- 1893e. Description of a new species of *Sminthus* from Kashmir. *Annals Mag. nat. Hist.* 11: 184–186.

----- 1893f. Description of a new monkey of the genus *Semnopithecus* from northern Borneo. *Proceedings zool. Soc. Lond.* (1892): 582–583, pl. 41.
----- 1894a. On the Palawan representative of *Tupaia ferruginea*. *Annals Mag. nat. Hist.* 13: 367.
----- 1894b. On two new Chinese rodents. *Annals Mag. nat. Hist.* 13: 363–365.
----- 1894c. A preliminary revision of the Bornean species of the genus *Mus*. *Annals Mag. nat. Hist.* 14: 449–460.
----- 1894d. On some mammals from Engano Island, west of Sumatra. *Annali Mus. civ. Stor. nat. Giacomo Doria* 14: 105–110.
----- 1894e. Descriptions of two new bats of the genus *Kerivoula*. *Annals Mag. nat. Hist.* 14: 460–462.
----- 1894f. Description of a new species of *Vespertilio* from China. *Annals Mag. nat. Hist.* 14: 300–301.
----- 1895a. Description of a new species of *Helictis* from Borneo. *Annals Mag. nat. Hist.* 15: 331–332.
----- 1895b. Revised determinations of three of the Natuna rodents. *Novitates zool.* 2: 26–28.
----- 1895c. Description of a new Chinese squirrel. *Annals Mag. nat. Hist.* 16: 242.
----- 1895d. On some mammals ... in Sipora, Mentawei Islands. *Annali Mus. civ. Stor. nat. Giacomo Doria* 14: 660–672.
----- 1895e. On a special mountain race of the plantain squirrel from Mount Dulit, E Sarawak. *Annals Mag. nat. Hist.* 15: 529–531.
----- 1895f. Preliminary diagnoses of new mammals from northern Luzon ... *Annals Mag. nat. Hist.* 16: 160–164.
----- 1896a. On mammals from Celebes, Borneo and the Philippines. *Annals Mag. nat. Hist.* 18: 241–250.
----- 1896b. On mammals collected ... on Woodlark Island ... *Novitates zool.* 3: 526–529.
----- 1897a. [Mammals obtained ... in the Philippine Islands] *Minutes Proc. zool. Soc. Lond.* 15 June: 1.
----- 1897b. Description of a new rat from China. *Annals Mag. nat. Hist.* 20: 113–114.
----- 1897c. On the genera of rodents ... *Proceedings zool. Soc. Lond.* (1896): 1012–1028.
----- 1897d. On some bats obtained in the Surat and Thana Districts ... *Journal Bombay nat. Hist. Soc.* 11: 274–276.
----- 1898a. Descriptions of two new cuscuses (*Phalanger*). *Novitates zool.* 5: 433–434.
----- 1898b. Description of new Bornean and Sumatran mammals. *Annals Mag. nat. Hist.* 2: 245–251.
----- 1898c. On the mammals obtained by Mr John Whitehead ... in the Philippines. *Transactions zool. Soc. Lond.* 14: 377–412.
----- 1898d. Description of a new bat from North Borneo. *Annals Mag. nat. Hist.* 1: 243.
----- 1898e. Description of a new bat from Selangore. *Annals Mag. nat. Hist.* 1: 360–362.
----- 1898f. Descriptions of three new mammals from the East Indian Archipelago ... *Novitates zool.* 5: 1–4.
----- 1899. On mammals collected ... at Kuatun, N. W. Fokien, China. *Proceedings zool. Soc. Lond.* (1898): 769–775.
----- 1900a. The red flying squirrel of the Natuna Islands. *Novitates zool.* 7: 592.
----- 1900b. Descriptions of two new Sciuropteri ... in Borneo. *Annals Mag. nat. Hist.* 5: 275–276.
----- 1900c. A new bat from the Key Islands. *Annals Mag. nat. Hist.* 5: 145.
----- 1901a. [Exhibition of a pair of antlers ...] *Minutes, Proc. zool. Soc. Lond.* : 54.
----- 1901b. A new scotophiline bat from British East Africa ... *Annals Mag. nat. Hist.* 7: 263–265.
----- 1902a. A new *Hipposideros* from Borneo. *Annals Mag. nat. Hist.* 9: 271–272.
----- 1902b. On two new mammals from China. *Annals Mag. nat. Hist.* 10: 163–166.
----- 1902c. A new vole from the lower Yang-tse-kiang. *Annals Mag. nat. Hist.* 10: 166–169.
----- 1902d. On the panda of Sze-chuen. *Annals Mag. nat. Hist.* 10: 251–252.
----- 1902e. On the species of *Mydaus* found in Borneo ... *Annals Mag. nat. Hist.* 9: 442–444.
----- 1902f. On a new genus of vespertilionine bat from New Guinea. *Annals Mag. nat. Hist.* 9: 220–222.
----- 1903a. On the species of the genus *Rhinopoma*. *Annals Mag. nat. Hist.* 11: 496–499.
----- 1903b. Exhibition of a skin and description of a new species of monkey ... *Proceedings zool. Soc. Lond.* : 224–225, pl. 21.
----- 1904a. In Schneider, G., Ergebnisse zoologischer Forschungsreisen in Sumatra. *Zoologischer Anz.* 27: 722–724.
----- 1904b. New bats and rodents from West Africa, the Malay Peninsula and Papuasia. *Annals Mag. nat. Hist.* 14: 196–202.
----- 1904c. Three new bats, African and Asiatic. *Annals Mag. nat. Hist.* 13: 384–388.
----- 1905a. On a remarkable new squirrel from Borneo. *Annals Mag. nat. Hist.* 16: 314–315.
----- 1905b. [On a collection of mammals obtained ... in Persia and Armenia] *Abstracts Proc. zool. Soc. Lond.* 24: 22–23.
----- 1905c. A new genus and two new species of bats. *Annals Mag. nat. Hist.* 16: 572–576.
----- 1905d. On some Australasian mammals. *Annals Mag. nat. Hist.* 16: 422–428.
----- 1906a. On a second species of *Lenothrix* from the Liu Kiu Islands. *Annals Mag. nat. Hist.* 17: 88–89.
----- 1906b. On a collection of mammals from Persia and Armenia. *Proceedings zool. Soc. Lond.* (1905): 519–527, pl. 16.
----- 1906c. List of mammals obtained by Mr M. P. Anderson in Japan. *Proceedings zool. Soc. Lond.* (1905): 331–363, pl. 9.
----- 1906d. New Asiatic mammals of the genera *Kerivoula, Eliomys* and *Lepus*. *Annals Mag. nat. Hist.* 7: 423–426.
----- 1907a. The Duke of Bedford's zoological exploration in Eastern Asia. V. Second list of mammals from Korea. *Proceedings zool. Soc. Lond.* : 462–466.
----- 1907b. A new flying-squirrel from Formosa. *Annals Mag. nat. Hist.* 20: 522–523.
----- 1907c. The Duke of Bedford's zoological exploration in eastern Asia. II. List of small mammals from Korea and Quelpart. *Proceedings zool. Soc. Lond.* (1906): 868–865.
----- 1907d. A subdivision of the old genus *Nesokia* ... *Annals Mag. nat. Hist.* 20: 202–207.
----- 1907e. [On mammals collected in Mindanao ...] *Abstracts Proc. zool. Soc. Lond.* (39): 5.
----- 1907f. On mammals obtained ... in the Philippine Islands. *Proceedings zool. Soc. Lond.* 140–142.
----- 1907g. On a new dormouse from Asia Minor, with remarks on the subgenus "*Dryomys*". *Annals Mag. nat. Hist.* 20: 406–407.
----- 1908a. *Abstracts Proc. zool. Soc. Lond.* 1908: 44–45.
----- 1908b. The nomenclature of the flying-lemurs. *Annals Mag. nat. Hist.* 1: 252–255.
----- 1908c. The genera and subgenera of the *Sciuropterus* group. *Annals Mag. nat. Hist.* 1: 1–8.
----- 1908d. On the large flying-squirrels referred to *Petaurista nitida* Desm. *Annals Mag. nat. Hist.* 1: 250–252.
----- 1908e. On mammals from the Malay Peninsula and islands. *Annals Mag. nat. Hist.* 2: 301–306.
----- 1908f. On the generic position of the groups of squirrels typified by "*Sciurus*" *berdmorei* and *pernyi* ... *Journal Bombay nat. Hist. Soc.* 18: 244–249.
----- 1908g. The squirrels described as *Sciurus steerii* from Balabac and Palawan. *Annals Mag. nat. Hist.* 2: 498.
----- 1908h. The Duke of Bedford's zoological exploration in eastern Asia. VI. List of mammals from the Shantung Province, north China. *Proceedings zool. Soc. Lond.* : 5–10.
----- 1908i. New Asiatic *Apodemus, Evotomys* and *Lepus*. *Annals Mag. nat. Hist.* 1: 447–450.
----- 1908j. List of mammals from the provinces of Chih-li and Shan-si, northern China. *Proceedings zool. Soc. Lond.* : 635–646, pl. 32.
----- 1908k. On the mammals collected ... on Tioman and Aor Islands. *Journal fed. Malay St. Mus.* 2: 101–106.
----- 1908l. New bats and rodents in the British Museum collection. *Annals Mag. nat. Hist.* 2: 370–375.
----- 1909a. The Duke of Bedford's zoological exploration in Eastern Asia. XI. On mammals from the provinces of Shan-si and Shen-si, northern China. *Proceedings zool. Soc. Lond.* (1908): 963–983.
----- 1909b. New species of *Paradoxurus* ... *Annals Mag. nat. Hist.* 3: 374–377.
----- 1909c. A new gibbon from Annam. *Annals Mag. nat. Hist.* 4: 112–113.
----- 1910a. Three new Asiatic mammals. *Annals Mag. nat. Hist.* 5: 534–536.
----- 1910b Two new mammals from the Malay Peninsula. *Annals Mag. nat. Hist.* 5: 424–426.
----- 1910c. New mammals from the East Indian Archipelago. *Annals Mag. nat. Hist.* 5: 383–387.
----- 1910d. New genera of Australasian Muridae. *Annals Mag. nat. Hist.* 6: 506–508.
----- 1910e. [Exhibition of ... mammals from China ...] *Abstracts Proc. zool. Soc. Lond.* 83: 25–26.
----- 1911a. [Mammals collected in ... western China ...] *Abstracts Proc. zool. Soc. Lond.* 90: 3–5.
----- 1911b. The Duke of Bedford's exploration of Eastern Asia. XIII. On mammals from the provinces of Kan-su and Sze-chwan, Western China. *Proceedings zool. Soc. Lond.* : 158–180.
----- 1911c. [Mammals in ... W China ...] *Abstracts Proc. zool. Soc. Lond.* 100: 48–50.
----- 1911d. A new murine genus and species from Sind ... *Journal Bombay nat. Hist. Soc.* 20: 996–999a.
----- 1911e. New Asiatic Muridae. *Annals Mag. nat. Hist.* 7: 205–209.
----- 1911f. On mammals from southern Shen-si, central China. *Proceedings zool. Soc. Lond.* : 687–695, pl. 21.

Thomas, O. 1911g. [On mammals collected in southern Shen-si . . .] *Abstracts Proc. zool. Soc. Lond.* 95: 26–27.

----- 1911h. New rodents from Sze-chuan . . . *Annals Mag. nat. Hist.* 8: 727–729.

----- 1911i. Three new rodents from Kan-su. *Annals Mag. nat. Hist.* 8: 720–723.

----- 1912a. On a collection of small mammals from the Tsin-ling mountains, central China . . . *Annals Mag. nat. Hist.* 10: 395–403.

----- 1912b. The Duke of Bedford's zoological exploration of Eastern Asia. XV. On mammals from the provinces of Szechuan and Yunnan, western China. *Proceedings zool. Soc. Lond.* : 127–141.

----- 1912c. On insectivores and rodents collected by Mr F. Kingdon Ward in N. W. Yunnan. *Annals Mag. nat. Hist.* 9: 513–519.

----- 1912d. New species of *Crocidura* and *Petaurista* from Yunnan. *Annals Mag. nat. Hist.* 9: 686–688.

----- 1912e. [Exhibition of the skin and skull of a viverrine carnivore from Tonkin.] *Proceedings zool. Soc. Lond.* : 17–18.

----- 1912f. Two new genera and a new species of viverrine Carnivora. *Proceedings zool. Soc. Lond.* : 498–503.

----- 1913a. Four new shrews. *Annals Mag. nat. Hist.* 11: 214–218.

----- 1913b. On some mammals from Bali and Mysol . . . *Annals Mag. nat. Hist.* 11: 504–507.

----- 1913c. A new shrew from the Andaman Islands. *Annals Mag. nat. Hist.* 11: 468–469.

----- 1913d. Some new Ferae from Asia and Africa. *Annals Mag. nat. Hist.* 12: 88–92.

----- 1913e. On a remarkable new free-tailed bat from southern Bombay. *Journal Bombay nat. Hist. Soc.* 22: 87–91, 1pl.

----- 1914a. The tree-shrews of the *Tupaia belangeri-chinensis* group. *Annals Mag. nat. Hist.* 13: 243–245.

----- 1914b. A new *Soriculus* from the Mishmi Hills. *Journal Bombay nat. Hist. Soc.* 22: 683.

----- 1914c. On small mammals collected in Tibet and the Mishmi Hills . . . *Journal Bombay nat. Hist. Soc.* 23: 230–233.

----- 1914d. Scientific results from the mammal survey, VII. *Journal Bombay nat. Hist. Soc.* 23: 23–31.

----- 1914e. Scientific results from the Mammal Survey, VIII. *Journal Bombay nat. Hist. Soc.* 23: 197–205.

----- 1914f. Note on the genus *Leggada*. *Journal Bombay nat. Hist. Soc.* 22: 682.

----- 1914g. New Asiatic and Australian bats and a new bandicoot. *Annals Mag. nat. Hist.* 13: 439–444.

----- 1914h. Second list of small mammals from western Yunnan . . . *Annals Mag. nat. Hist.* 14: 472–475.

----- 1914i. On mammals from Manus Island . . . *Annals Mag. nat. Hist.* 13: 434–439.

----- 1915a. A new shrew of the genus *Blarinella* from Upper Burma. *Annals Mag. nat. Hist.* 15: 335–336.

----- 1915b. The penis-bone, or "baculum", as a guide to the classification of certain squirrels. *Annals Mag. nat. Hist.* 15: 383–387.

----- 1915c. Scientific results from the Mammal Survey, IX. *Journal Bombay nat. Hist. Soc.* 23: 413–416.

----- 1915d. A second species of *Coelomys* from Ceylon. *Journal Bombay nat. Hist. Soc.* 24: 49–50.

----- 1915e. On some specimens of *Vandeleuria* from Bengal, Bihar and Orissa. *Journal Bombay nat. Hist. Soc.* 24: 54–55.

----- 1915f. List of mammals . . . collected on the upper Congo . . . *Annals Mag. nat. Hist.* 16: 465–481.

----- 1915g. On some pteropine bats from Vulcan and Dampier . . . New Guinea. *Annals Mag. nat. Hist.* 15: 387–389.

----- 1915h. Notes on *Emballonura*, with descriptions of new species. *Annals Mag. nat. Hist.* 15: 137–140.

----- 1915i. Notes on the Asiatic bamboo rats (*Rhizomys*, etc.). *Annals Mag. nat. Hist.* 16: 56–61.

----- 1915j. Further notes on Asiatic bamboo-rats. *Annals Mag. nat. Hist.* 16: 313–317.

----- 1915k. Scientific results from the Mammal Survey, X. *Journal Bombay nat. Hist. Soc.* 23: 607–613a.

----- 1915l. On bats of the genera *Nyctalus, Tylonycteris*, and *Pipistrellus*. *Annals Mag. nat. Hist.* 15: 225–232.

----- 1915m. Two new species of *Leuconoe*. *Annals Mag. nat. Hist.* 15: 170–172.

----- 1915n. Notes on *Taphozous* and *Saccolaimus*. *Journal Bombay nat. Hist. Soc.* 24: 57–63.

----- 1915o. On pipistrelles of the genera *Pipistrellus* and *Scotozous*. *Journal Bombay nat. Hist. Soc.* 24: 29–36.

----- 1915p. A special genus for the Himalayan bat known as *Murina grisea*. *Annals Mag. nat. Hist.* 16: 309–310.

----- 1916a. A new flying squirrel from Tenasserim. *Journal Bombay nat. Hist. Soc.* 24: 422–423.

----- 1916b. The races of *Dremomys pernyi*. *Annals Mag. nat. Hist.* 17: 391–394.

----- 1916c. Scientific results from the Mammal Survey, XIII. *Journal Bombay nat. Hist. Soc.* 24: 404–430.

----- 1916d. Scientific results from the Mammal Survey, XIV. *Journal Bombay nat. Hist. Soc.* 24: 639–644.

----- 1916e. A new rat from Tenasserim. *Annals Mag. nat. Hist.* 17: 425.

----- 1916f. Notes on the genus *Cremnomys*. *Journal Bombay nat. Hist. Soc.* 24: 224–243, 1pl.

----- 1916g. A new bamboo rat from Perak. *Annals Mag. nat. Hist.* 18: 445–446.

----- 1916h. The porcupines of Tenasserim and southern Siam. *Annals Mag. nat. Hist.* 17: 136–139.

----- 1916i. A new binturong from Siam. *Annals Mag. nat. Hist.* 17: 270–271.

----- 1916j. List of Microchiroptera . . . in the collections of the Federated Malay States Museums. *Journal fed. Malay St. Mus.* 7: 1–6.

----- 1916k. On *Rattus* as a generic name . . . *Annals Mag. nat. Hist.* 18: 70–72.

----- 1917a. Scientific results from the mammal survey, XVI. *Journal Bombay nat. Hist. Soc.* 25: 199–206.

----- 1917b. Two new rats of the *Rattus confucianus* group. *Annals Mag. nat. Hist.* 20: 198–200.

----- 1918a. Four new species of hedgehogs. *Annals Mag. nat. Hist.* 1: 230–233.

----- 1918b. A selection of lectotypes of Indian mammals . . . *Journal Bombay nat. Hist. Soc.* 25: 368–372.

----- 1918c. The nomenclature of the geographical forms of the Panolia deer (*Rucervus eldi*) and its relatives. *Journal Bombay nat. Hist. Soc.* 25: 363–367.

----- 1919a. A synopsis of the groups of true mice found within the Indian Empire. *Journal Bombay nat. Hist. Soc.* 26: 417–420.

----- 1919b. Notes on the genus *Cheliones*. *Journal Bombay nat. Hist. Soc.* 26: 726–727.

----- 1919c. Some new mammals from Mesopotamia. *Journal Bombay nat. Hist. Soc.* 26: 745–749.

----- 1919d. Notes on gerbils referred to the genus *Meriones*. *Annals Mag. nat. Hist.* 3: 263–273.

----- 1920a. On mammals from Ceram. *Annals Mag. nat. Hist.* 6: 422–431.

----- 1920b. Four new squirrels of the genus *Tamiops*. *Annals Mag. nat. Hist.* 5: 304–308.

----- 1920c. A new bat of the genus *Rhinopoma* from S. E. Persia. *Journal Bombay nat. Hist. Soc.* 27: 25.

----- 1920d. Scientific results from the Mammal Survey, XXI. *Journal Bombay nat. Hist. Soc.* 26: 933–940.

----- 1920e. Some notes on *Babirussa*. *Annals Mag. nat. Hist.* 5: 185–188.

----- 1920f. Two new asiatic bats of the genera *Tadarida* and *Dyacopterus*. *Annals Mag. nat. Hist.* 5: 283–285.

----- 1920g. A new bat of the genus *Myotis* from Sikkim. *Journal Bombay nat. Hist. Soc.* 27: 248–250.

----- 1921a. On small mammals from the Kachin Province, northern Burma. *Journal Bombay nat. Hist. Soc.* 27: 499–505.

----- 1921b. A new monkey and a new squirrel from the Middle Mekong . . . *Annals Mag. nat. Hist.* 7: 181–183.

----- 1921c. Scientific results from the Mammal Survey, XXV. *Journal Bombay nat. Hist. Soc.* 27: 596–599.

----- 1921d. On a new genus and species of shrew, and some new Muridae from the East-Indian Archipelago. *Annals Mag. nat. Hist.* 7: 243–249.

----- 1921e. On a collection of rats and shrews from the Dutch East Indian islands. *Treubia* 2: 109–114.

----- 1921f. Two new rats from Assam. *Journal Bombay nat. Hist. Soc.* 28: 26–27.

----- 1921g. The *Arctonyx* of Annam. *Annals Mag. nat. Hist.* 7: 524.

----- 1921h. The masked civets (*Paguma*) of western China. *Annals Mag. nat. Hist.* 8: 617–618.

----- 1921i. The mungooses of the *Herpestes smithii* group. *Journal Bombay nat. Hist. Soc.* 28: 23–26.

----- 1921j. The geographical races of *Herpestes brachyurus* Gray. *Annals Mag. nat. Hist.* 8: 134–136.

----- 1921k. Two new species of slow-loris. *Annals Mag. nat. Hist.* 8: 627–628.

----- 1921l. The geographical races of *Scotomanes ornatus*. *Journal Bombay nat. Hist. Soc.* 27: 772–773.

----- 1922a. On mammals from the Yunnan Highlands . . . *Annals Mag. nat. Hist* 10: 391–406.

----- 1922b. Scientific results of the mammal survey, XXXII. *Journal Bombay nat. Hist. Soc.* 28: 428–432.

----- 1922c. Two new rodents from the Mergui Archipelago. *Journal Bombay nat. Hist. Soc.* 28: 1067–1968.

----- 1922d. New mammals from New Guinea . . . *Annals Mag. nat. Hist.* 9: 261–265.

----- 1922e. On some new forms of *Ochotona*. *Annals Mag. nat. Hist.* 9: 187–193.

----- 1922f. Some notes on ferret-badgers. *Annals Mag. nat. Hist.* 9: 193–196.

----- 1922g. The generic classification of the *Taphozous* group. *Annals Mag. nat. Hist.* 9: 266–267.

----- 1923a. On some small mammals . . . from the East Indian Archipelago. *Annals Mag. nat. Hist.* 11: 250–255.

----- 1923b. On the tree-shrews from the Mergui Archipelago. *Journal Bombay nat. Hist. Soc.* 29: 84.

----- 1923c. On mammals from the Li-kiang Range, Yunnan .. *Annals Mag. nat. Hist.* 11: 655–663.

----- 1923d. Geographical races of *Petaurista alborufus*. *Annals Mag. nat. Hist.* 12: 171–172.

----- 1923e. On some squirrels from the Mergui Archipelago. *Journal Bombay nat. Hist. Soc.* 29: 376–377.

----- 1923f. [Exhibition of the skull of a pygmy fruit-bat from Sumatra.] *Abstracts Proc. zool. Soc. Lond.* 235: 14.

----- 1923g. Exhibition of the skull of a pygmy fruit-bat from Sumatra. *Proceedings zool. Soc. Lond.* : 178.

----- 1923h. A new mouse from Madura, S India. *Journal Bombay nat. Hist. Soc.* 29: 87.

----- 1923i. On the large squirrels of the *Ratufa gigantea* group. *Journal Bombay nat. Hist. Soc.* 29: 85–86.

----- 1923j. The distribution and geographical races of the Gulandi bush-rats (*Golunda ellioti*). *Journal Bombay nat. Hist. Soc.* 29: 373–375.

----- 1923k. A new *Uromys* from the Kei Islands. *Treubia* 3: 442.

----- 1923l. On some mammals from Simalur Island . . . *Annals Mag. nat. Hist.* 12: 591–593.

----- 1923m. The mouse deer (*Tragulus*) of the Mergui Archipelago. *Journal Bombay nat. Hist. Soc.* 29: 84–85.

----- 1923n. On the forms contained in the genus *Harpiocephalus*. *Journal Bombay nat. Hist. Soc.* 29: 88–89.

----- 1924a. A new genus and species of shrew from Ceylon. *Spolia zeylan.* 13,1: 93–95.

----- 1924b. On some Ceylon mammals. *Annals Mag. nat. Hist.* 13: 239–243.

----- 1924c. On a new field mouse from Nepal . . . *Journal Bombay nat. Hist. Soc.* 29: 888–889.

----- 1925. The mammals obtained... on the Sladen-Godman Expedition to Tonkin. *Proceedings zool. Soc. Lond.* : 495–506.

----- 1926a. Two new flying squirrels from the Mergui Archipelago. *Journal Bombay nat. Hist. Soc.* 31: 22.

----- 1926b. Two new subspecies of *Callosciurus quinquestriatus*. *Annals Mag. nat. Hist.* 17: 639–641.

----- 1926c. A new *Myotis* from Ladak. *Annals Mag. nat. Hist.* 17: 609–610.

----- 1927. The Delacour Exploration of French Indo-China— mammals. *Proceedings zool. Soc. Lond.* : 41–58.

----- 1928a. A remarkably coloured squirrel from North Siam. *Annals Mag. nat. Hist.* 2: 100–101.

----- 1928b. The south-Tenasserim race of *Paguma leucomystax*, Gray. *Annals Mag. nat. Hist.* 2: 101–102.

----- 1928c. The Delacour Exploration of French Indo-China— Mammals. II. On mammals collected during the winter of 1926–27. *Proceedings zool. Soc. Lond.* (Part 1): 139–150.

----- 1929a. A new mole from western Siam. *Annals Mag. nat. Hist.* 3: 206–207.

----- 1929b. The Delacour Exploration of French Indo-China— mammals. III . . . *Proceedings zool. Soc. Lond.* (1928): 831–841.

----- 1929c. A new bandicoot rat from western Siam. *Annals Mag. nat. Hist.* 3: 205.

Thomas, O. & Doria, G. 1886. Note intorno ad alcune Chirotteri. *Annali Mus. civ. Stor. nat. Giacomo Doria* 4: 201–207.

----- 1889. Diagnosi di una nuova specie del genero *Cervulus* . . . *Annali Mus. civ. Stor. nat. Giacomo Doria* 7: 92.

Thomas, O. & Hartert, E. 1894. First glimpses of the zoology of the Natuna Islands. III. List of the first collection of mammals. *Novitates zool.* 1: 652–660.

----- 1895. On a second collection of mammals from the Natuna Islands. *Novitates zool.* 2: 489–492.

Thomas, O. & Robinson, H. C. 1921. Some new insular squirrels . . . from Siam. *Annals Mag. nat. Hist.* 7: 118–123.

Thomas, O. & Ryley, K. V. 1913. A new species of *Leggada* from Mysore. *Journal Bombay nat. Hist. Soc.* 22: 242.

Thomas, O. & Wroughton, R. C. 1909a. Diagnoses of new mammals . . . Malayan Peninsula and Rhio Archipelago. *Annals Mag. nat. Hist.* 3: 439–441.

----- 1909b. Diagnosis of new mammals . . . in the islands of the Straits of Malacca. *Annals Mag. nat. Hist.* 4: 534–536.

----- 1909c. On a collection of mammals from western Java . . . *Abstracts Proc. zool. Soc. Lond.* : 18–19.

----- 1909d. On a collection of mammals from western Java . . . *Proceedings zool. Soc. Lond.* : 371–392.

----- 1915a. The giant squirrels of Ceylon. *Journal Bombay nat. Hist. Soc.* 24: 34–37.

----- 1915b. The Singhalese species of *Funambulus*. *Journal Bombay nat. Hist. Soc.* 24: 37–41.

----- 1916a. A new flying squirrel from the Chin Hills. *Journal Bombay nat. Hist. Soc.* 24: 423–424.

----- 1916b. On the squirrels obtained . . . on the Chindwin River, upper Burma . . . *Journal Bombay nat. Hist. Soc.* 24: 224–239.

----- 1916c. The squirrels of the *Tomeutes lokroides* and *mearsi* groups. *Journal Bombay nat. Hist. Soc.* 24: 419–422.

Thonglongya, K. 1972. A new genus and species of fruit bat from S India. . . *Journal Bombay nat. Hist. Soc.* 69: 151–158.

----- 1973a. First record of *Rhinolophus paradoxolophus* . . . from Thailand . . . *Mammalia* 37: 587–597.

----- 1973b. Note on the validity of *Scotophilus castaneus* . . . *Natural Hist. Bull. Siam Soc.* 25: 161–166.

Thonglongya, K. & Hill, J. E. 1974. A new species of *Hipposideros*. *Mammalia* 38: 285–294.

Thorington, R. W. 1984. Flying squirrels are monophyletic. *Science* 225: 1048–1050.

Thornton, I. W. B. (ed.) 1986. 1985 zoological expedition to the Krakataus . . . *Miscellaneous Ser. Dept Zool. La Trobe Univ.* 2: 1–63.

Thunberg, C. P. 1820. Beskrifning och Teckning på ett nytt species, *Hyaena brunnea*. *Kungliga svenska VetenskAkad. Handl.* : 59–65.

Tickell, S. R. 1842. On the Oral, or Singhoom flying squirrel, *Pteromys oral*. *Calcutta J. nat. Hist.* 2: 401–408.

Tidemann, C. R. *et al.* 1990. Recolonisation of the Krakatau Islands . . . by bats. *Philosophical Trans. r. Soc. Lond.* B, 328: 123–130.

Tiedemann, F. 1808. *Zoologie. Zu seinen Vorlesungen entworfen. Vol. 1.* Landshut & Heidelberg.

----- 1820. *Abhandlung über das vermeintliche bärenartige Faulthiere.* Heidelberg, 11pp, 1pl.

Tikader, B. K. & Das, A. K. 1985. *Glimpses of animal life of Andaman and Nicobar Islands.* Calcutta: Zoological Survey of India, 170pp.

Tilson, R. C. & Seal, U. S. (eds) 1987. *Tigers of the world* . . . Park Ridge, New Jersey: Noyes, 510pp.

Tinker, S. W. 1988. *Whales of the world.* Leiden etc. : Brill, 310pp.

Tiwari, K. K. *et al.* 1971. Notes on a collection of small mammals from Western Ghats . . . *Journal Bombay nat. Hist. Soc.* 68: 378–384.

Todd, N. B. 1966. Metrical and non-metrical variation in the skulls of Gir lions. *Journal Bombay nat. Hist. Soc.* 62: 507–520.

Tokuda, M. 1937. The alpine Muridae of Formosa . . . *Botany Zool. Tokyo* 5: 1115–1122.

----- 1941. A revised monograph of the Japanese and Manchou-Korean Muridae. *Biogeographia* 4: 1–155, pls 1–10.

Tokuda, M. & Kano, T. 1936. A bat and a new shrew from Koto-sho (Botel-Tobago). *Annotnes zool. jap.* 15: 427–432.

Tomes, R. F. 1856. Pp. 22–23 in Blyth, E., Memoir on the Indian species of shrews. *Annals Mag. nat. Hist.* 17: 11–28.

----- 1857. Descriptions of four undescribed species of bats. *Proceedings zool. Soc. Lond.* : 50–54.

----- 1858a. On the characters of four species of bats inhabiting Europe and Asia . . . *Proceedings zool. Soc. Lond.* : 78–90, 1pl.

----- 1858b. A monograph of the genus *Kerivoula*. *Proceedings zool. Soc. Lond.* : 322–338, 1pl.

----- 1858c. A monograph of the genus *Nyctophilus*. *Proceedings zool. Soc. Lond.* : 25–37.

----- 1858d. A monograph of the genus *Miniopterus*. *Proceedings zool. Soc. Lond.* : 115–128, 1 pl.

----- 1859a. Notice of five species of bats . . . in Labuan . . . *Proceedings zool. Soc. Lond.* (1858): 536–540.

----- 1859b. Description of six hitherto undescribed species of bats. *Proceedings zool. Soc. Lond.* : 68–79.

Topal, G. 1970a. On the systematic status of *Pipistrellus annectans* . . . *Annales hist. -nat. Mus. natn. Hung. Zool.* 62: 373–379.

----- 1970b. The first record of *Ia io* Thomas, 1902 in Vietnam and India . . . *Opuscula Zool. Budapest* 10: 341–347.

----- 1975. Bacula of some Old World leaf-nosed bats . . . *Vertebrata hung.* 16: 21–53, 9pls.

Trouessart, E-L. 1878. Catalogue des mammifères vivant et fossiles. Chiroptera. *Revue Mag. Zool.* (3)6: 201–254.

----- 1879a. *Ibid.* Insectivora. *Revue Mag. Zool.* (3)7: 219–285.

----- 1879b. Révision synoptique du genre Semnopithèque (*Semnopithecus*). *Revue Mag. Zool.* (3)7: 48–59.

----- 1880a. Description d'une espèce nouvelle de musaraigne . . . *Naturaliste* 1: 330.

----- 1880b. Revision du genre écureuil (*Sciurus*). *Naturaliste* 1: 290–293.

----- 1880c. Catalogue des mammifères vivants et fossiles. Ordre des rongeurs. *Bulletin Soc. Etud. scient. Angers* 10: 58–104.

----- 1881. *Ibid* : 105–217.

----- 1895. Sur les variations de pelage du *Putorius auriventer*. *Bulletin Mus. Hist. nat. Paris* 1: 235–236.

----- 1897–1905. *Catalogus mammalium* . . . Berlin: Friedlander.

----- 1909. *Neotetracus sinensis*, a new insectivore of the family Erinaceidae. *Annals Mag. nat. Hist.* 4: 389–391.

----- 1911. On a new species of *Semnopithecus* . . . from Tonkin. *Annals Mag. nat. Hist.* 8: 271–272, 1pl.

Troughton, E. le G. 1929. A new genus and species of bat (Kerivoulinae) from the Solomons . . . *Records Austr. Mus.* 17: 85–99.

Troughton, E. le G. 1941,1943. *Furred animals of Australia*. Sydney.

True, F. W. 1894a. Notes on mammals of Baltistan and the Vale of Kashmir . . . *Proceedings US natn. Mus.* 17: 1–16.

----- 1894b. On the rodents of the genus *Sminthus* in Kashmir. *Proceedings US natn. Mus.* 17: 341–343.

Trumler, E. 1961. Entwurf einer Systematik der rezenten Equiden . . . *Säugetierkundliche Mitt.* 9: 109–125.

Tuttle, R. H. (ed.) 1975. *Paleoanthropology: morphology and paleoecology*. The Hague: Mouton, xvi, 454pp, 12pls.

Tytler, R. C. 1854. Miscellaneous notes on the fauna of Dacca . . . *Annals Mag. nat. Hist.* 14: 168–177.

----- 1864. Description of a new species of *Paradoxurus* from the Andaman Islands. *Journal Bombay nat. Hist. Soc.* 33: 188.

U Tun Yin 1967. *Wild animals of Burma*. Rangoon: Rangoon Gazette, 301pp.

Udvardy, M. D. F. 1975. A classification of the biogeographical provinces of the world. *IUCN occ. Pap.* 18: 48pp.

Urbain, A. 1937. Le kou prey ou boeuf gris Cambodgien. *Bulletin Soc. zool. France* 62: 305–307.

Vahl, M. 1797. Beskrivelse paa tre nye Arter Flagermuse. *Skrifter Naturh. -Selsk. Kiobenhavn* 4: 121–138.

Valenciennes, A. 1856. Sur une nouvelle espèce de panthère . . . *Compte r. Acad. Sci. Paris* 42: 1035–1039.

van Bemmel, A. C. V. 1949a. Revision of the rusine deer in the Indo-Australian Archipelago. *Treubia* 20: 191–262.

----- 1949b. Notes on Indo-Australian mammals (1–2). *Treubia* 20: 375–380.

----- 1952. Contribution to the knowledge of the genera *Muntiacus* and *Arctogalidea* . . . *Beaufortia* 16: 20pp.

van Bemmelen, R. W. 1949. *The geology of Indonesia*. Vol. 1. The Hague.

van Beneden, J. & Gervais, P. 1880. *Ostéographie des cétacées*. Paris, viii, 634pp.

van Bree, P. J. H. 1968. Einige Bemerkungen über den Glatotter *Lutrogale perspicillata* . . . *Zoologische Gart. Lpz.* 35: 30–36.

----- 1971. *Delphinus tropicalis*, a new name for *Delphinus longirostris* G Cuvier, 1829. *Mammalia* 35: 345–346.

----- 1972. On the validity of the subspecies *Cephalorhynchus hectori bicolor* Oliver, 1946. *Investigations Cetacea* 4: 182–186.

van Bree, P. J. H. & Boeadi 1978. Notes on the Indonesian mountain weasel, *Mustela lutreolina* . . . *Zeitschrift Säugetierk.* 43: 166–171.

van Bree, P. J. H. & Glallagher, M. D. 1978. On the taxonomic status of *Delphinus tropicalis* van Bree, 1971. *Beaufortia* 28: 1–8.

van der Maarel, F. H. 1932. Contribution to the knowledge of the fossil mammalian fauna of Java. *Wetenschappelijke Meded. Dienst Mijnb. Ned. Oost-Indië* 15: 208pp, 20pls.

van der Zon, A. P. M. 1979. *Mammals of Indonesia (draft version)*. Bogor: UNDP/FAO National Park Development Project, 152pp.

van Dort, M. 1988. Note on the skull size in the two sympatric mouse deer species, *Tragulus javanicus* and *T. napu* . . . *Zeitschrift Säugetierk.* 53: 124–125.

Van Gelder, R. G. 1978. A review of canid classification. *American Mus. Novit.* 2646: 10pp.

van Mourik, S. & Schurig, V. 1985. Hybridization between sambar (*Cervus* (*Rusa*) *unicolor*) and rusa (*Cervus* (*Rusa*) *timorensis*) deer. *Zoologischer Anz.* 214: 177–184.

Van Peenen, P. F. D. *et al.* 1969. *Preliminary identification manual for mammals of South Vietnam*. Washington: Smithsonian Institution.

----- 1974. Observations on *Rattus bartlesii*. *Treubia* 28: 83–117.

van Strien, N. J. 1975. *Dicerorhinus sumatrensis* (Fischer), the Sumatran or two-horned Asiatic rhinoceros. *Medelingen ned. Comm. int. NatBescherm.* 22: 82pp.

----- 1986. *Abbreviated checklist of the mammals of the Australasian Archipelago*. Bogor: School of Environmental Conservation Management, 91pp.

Van Valen, L. 1967. New Palaeocene insectivores and insectivore classification. *Bulletin Am. Mus. nat. Hist.* 135: 219–284.

----- 1979. The evolution of bats. *Evolution Theory* 4: 103–121.

van Weers, D. J. 1976. Notes on southeast Asian porpupines. I. On the taxonomy of the genus *Trichys*. *Beaufortia* 25: 15–31.

----- 1977. Notes on south-east Asian porcupines. . . II. On the taxonomy of the genus *Atherurus* . . . *Beaufortia* 26: 205–230.

----- 1978. Notes on southeast Asian porcupines. III. On the taxonomy of the subgenus *Thecurus* . . . *Beaufortia* 28: 17–33.

----- 1979. Notes on Southeast Asian porcupines . . . 4. On the taxonomy of the subgenus *Acanthion* . . . *Beaufortia* 29: 215–272.

----- 1983. Specific distinction in Old World porcupines. *Zoologische Garten* 53: 226–232.

----- 1985. *Hystrix gigantea*, a new fossil porcupine species from Java. . . *Senckenbergiana leth.* 66: 111–119.

van Wurmb, F. 1787. De kahau, een lang-startige aap. *Verhandlingen batav. genoot. Kunst. Wet.* 3: 345–354.

Van Zyll de Jong, C. G. 1987. A phylogenetic study of the Lutrinae. . . *Canadian J. Zool.* 65: 2536–2544.

Venkataramanujam, K. *et al.* 1988. On a sei whale, *Balaenoptera borealis* stranded at Tuticorin, east coast of India. *Investigations Cetacea* 21: 247–249.

Verma, K. 1965. Notes on the biology and anatomy of the Indian tree-shrew, *Anathana wroughtoni*. *Mammalia* 29: 289–330.

Vigors, N. A. & Horsfield, T. 1827. Description of two species of the genus *Felis*. *Zoological J.* 3: 449–451, pl. 12.

----- 1828. Observations on some of the Mammalia contained in the museum of the Zoological Society. *Zoological J.* 4: 105–113, pl. iv.

Vinogradov, B. S. 1922. [Notes on a peculiar new genus and species of jerboa from Hara-Hoto.] Pp. 540–545 in Kozlov, P. K. [*Mongolia and Amdo and the dead city of Hara-Hoto*]. In Russian.

Vogel, P. & Besançon, F. 1979. On the systematic position of the genera *Nectogale* and *Chimarrogale* . . . *Revue suisse Zool.* 86: 335–338.

Vorontzov, N. N. & Potapova, E. G. 1979. Taxonomy of the genus *Calomyscus*. . . *Zoologicheskii Zh.* 58: 1391–1397.

Vorontsov, N. N. & Shenbrot, G. I. 1984. A systematic review of the genus *Salpingotus* . . . *Zoologicheskii Zh.* 63: 731–744.

Vorontzov, N. N. *et al.* 1979. Systematics of the genus *Calomyscus* . . . *Zoologicheskii Zh.* 58: 1213–1224.

Vos, J. de *et al.* 1987. A historical review of the Siwalik Hyaenidae . . . *Proceedings K. ned. Akad. Wet.* B, 90: 333–369.

Wagler, J. 1830. *Natürliches System der Amphibien* . . . München etc., 354pp.

----- 1832. Mittheilungen über einige merkwurdige Thiere. *Isis, Jena* : 275–281.

Wagner, J. A. 1835. *Die Säugethiere in Abbildungen nach der Natur* . . . 6. Leipzig.

----- 1839a. *Ibid*, Suppl. 1. Leipzig: Weigel, 551pp, pls ib-lxiib.

----- 1839b. Beschreibung einiger neuer oder wenig bekannter Säugethiere. *Gelehrte Anz. Akad. Wiss. München* 9(184): 433–440.

----- 1841a. *Die Säugethiere in Abbildungen nach der Natur* . . . Suppl. 2, Raubthiere. Leipzig: Weigel.

----- 1841b. Gruppirung der Gattungen der Nager. *Gelehrte Anz. Akad. Wiss. München* 12: 409–414, 417–422, 425–430.

----- 1841c. Gruppirung der Gattungen der Nager in natürlichen Familien . . . *Archiv Naturgesch.* 7,1: 111–138.

----- 1843,1844a. *Die Säugethiere in Abbildungen nach der Natur. Suppl. 4*. Leipzig.

----- 1844b. Beiträge zur Säugethier-Fauna von Kaschmir. In Huegel, C. A. A. von., *Kaschmir und das Reich der Siek*. IV (2). Stuttgart.

----- 1846. *Die Säugethiere in Abbildungen nach der Natur* . . . 7. Leipzig, vii, 427pp.

----- 1853,1855. *Die Säugethiere in Abbildungen nach der Natur* Suppl. 5. Leipzig: Wiegel, pp. 1–336 (1853), 337–810 (1855).

Wall, W. S. 1851. History and description of the skeleton of a new sperm whale . . . *Memoirs Austr. Mus.* 1: 66pp, 2pls. (Reprinted 1867.)

Wallace, A. R. 1869. *The Malay Archipelago*. London: Macmillan.

----- 1876. *The geographical distribution of animals*. New York: Harper, 503,553pp.

Wallin, L. 1969. The Japanese bat fauna. A comparative study of chorology . . . *Zoologiska Bijdr. Upps.* 37: 223–440.

Wang, F. 1985. Preliminary study on the ecology of *Trogopterus xanthipes*. *Acta theriol. sin.* 5: 103–110.

Wang, J-k. 1978. A description of Sichiaoshan fossil elephant's skull in Kwangtung Province. *Vertebrata palasiat.* 16: 123–128.

Wang, K. M. 1931. Die Hohlen-ablangerungen und Fauna in der Drachen-Maul-Hohle von Kiangsen, Chekiang. *Contributions natn. Res. Inst. Geol. Shanghai* 1: 41–64, 3pls.

Wang, Q. *et al.* 1982. On a new subspecies of Siberian musk-deer, *Moschus moschiferus anhuiensis* . . . *Acta theriol. sin.* 2: 133–137.

Wang, S. & Zheng, C. 1981. On the subspecies of the Chinese sulphur-bellied rat—*Rattus niviventer* . . . *Sinozoologia* 1: 1–8.

Wang, S. *et al.* 1962. On the mammals from southwestern Kwangsi, China. *Acta zool. Sin.* 14: 555–570, 2pls.

Wang, T-k. 1974. [On the taxonomic status of species, geological distribution and evutionary history of *Ailuropoda*.] *Acta zool. sin.* 20: 191–201.

Wang, Y. 1982. New subspecies of the pipistrels . . . from Yunnan, China. *Zoological Res.* 3 (Suppl.): 343–348.

----- 1985. A new genus and species of Gliridae—*Chaetocauda sichuanensis* gen. et sp. nov. *Acta theriol. sin.* 5: 67–73.

Wang, Y. *et al.* 1980. A new species of Murinae—*Vernaya foramena* sp. nov. *Acta zool. sin.* 26: 393–397.

Wang, Y-c. *et al.* 1966. Notes on some small mammals from Szechuan Province . . . *Acta zootax. sin.* 3: 85–89.

Wang, Y-x. 1987. Taxonomic research on Burma-Chinese tree shrew, *Tupaia belangeri* . . . *Zoological Res.* 8: 213–230.

Wang, Y-x. & Li, C. 1982. A new subspecies of shrew-hedgehog, *Neotetracus sinensis* . . . from Yunnan, China. *Zoological Res.* 3: 427–430.

Wang, Y-x. & Li, Z-x. 1980. New records of Chinese mammals of Insectivora. *Zoological Res.* 1: 563–564.

Wang, Y-x. & Yang, G-h. 1986. Pp. 134–151 in [Data collection 1951–1986.] Control and Research Institute of Epidemic Diseases of Yunnan Province. (In Chinese.)

Wang, Y-x. & Xu, L-h. 1981. A new subspecies of palm-civet . . . from Hainan, China. *Acta zootaxon. sin.* 6: 446–448.

Wang, Y-x. *et al.* 1980. Karyotypes and evolution of three species of Chinese squirrels . . . *Zoological Res.* 1: 501–521.

----- 1985. Small mammals and their vertical distribution in Jawu Mountain, Northwest Yunnan. Pp. 20–27 in Kawamichi, T. (*q. v.*).

Wang, Y., Luo, Z. & Feng, Z. 1985. Taxonomic revision of Yunnan hare, *Lepus comus* . . . *Zoological Res.* 6: 101–109.

Wang, Z. & Du, R. 1988. [*Cytogenetics of Chinese deer.*] Beijing: Science Press, 163pp.

Wang, Z-y. & Wang, S. 1962. The discovery of a flying fox (*Pteropus giganteus* Brünnich) from Chin-Hai Province, northwestern China. *Acta zool. sin.* 14: 494.

Ward, O. G. & Wurster-Hill, D. H. 1990. *Nyctereutes procyonoides. Mammalian Species* 358: 5pp.

Ward, R. 1907. *Records of big game*, 5th ed. London: Rowland Ward, 526pp.

Waterhouse, G. R. 1837. Observations of the palm squirrel (*Sciurus palmarum* of authors). *Magazine nat. Hist.* 1: 496–499.

----- 1838a. [Report of meeting on] September 26th, 1837. *Proceedings zool. Soc. Lond.* (1837): 87–90.

----- 1838b. [A new species . . of *Gerbillus*, and a new *Herpestes*] *Proceedings zool. Soc. Lond.* : 55–56.

----- 1838c. *Catalogue of the Mammalia in the museum of the Zoological Society of London, 2nd ed.* London: Taylor, 68pp.

----- 1838d. [A new species of squirrel . . .] *Proceedings zool. Soc. Lond.* 4: 19–20.

----- 1838e. [Report of Zoological Society.] *Athenaeum* 580: 880. (*n.v.-fide* McAllan & Bruce, 1989.)

----- 1839a. [A new species of squirrel . . . from the Philippine Islands.] *Proceedings zool. Soc. Lond.* : 117–118.

----- 1839b. Observations on certain modifications observed in the dentition of the flying opposums (*Petaurus*). *Proceedings zool. Soc. Lond.* 6: 149–154.

----- 1839c. [On a new species of rodent . . .] *Proceedings zool. Soc. Lond.* : 107–108.

----- 1839d. [On the flying lemurs (*Galeopithecus*)] *Proceedings zool. Soc. Lond.* (1838): 119–120.

----- 1841. [A species of hare from Little Thibet.] *Proceedings zool. Soc. Lond.* : 7–8.

----- 1842. Observations on the Rodentia. *Annals Mag. nat. Hist.* 10: 197–203; 344–347.

----- 1843a. [Proceedings of Zoological Society meeting, October 11, 1842.] *Annals Mag. nat. Hist.* 12: 130–135.

----- 1843b. On various species of bats . . . in the Philippine Is. *Proceedings zool. Soc. Lond.*: 66–69.

----- 1845. Descriptions of bats collected in the Philippine Islands . . . *Proceedings zool. Soc. Lond.* : 3–10.

----- 1850. Description of a new species of *Tupaia* discovered in the continent of India . . . *Proceedings zool. Soc. Lond.* (1849): 106–108.

Watson, L. 1981. *Sea guide to whales of the world.* London: Hutchinson, 302pp.

Wayne, R. K. *et al.* 1989. Molecular and biochemical evolution within the Carnivora. Pp. 465–494 in Gittelman, J. L. (*q. v.*).

Weber, M. 1902. *Der Indo-australische Archipel und die Geschichte seinen Tierwelt.* Jena, 46pp.

Wegner, A. M. R. 1950. The cause of the enormous proportions and the extraordinary shape of some buffalo horns from the Island of Sumba. *Hemera Zoa Buitenzorg* 57: 708–710.

Weidenreich, F. 1932. *Zeitschrift Anat. Entw. Gesch.* 99: 212.

----- 1944. Giant early man from Java and South China. *Science* 99: 479–482.

Weigel, I. 1961. Das Fellmuster der wildlebenden Katzenarten and der Hauskatze . . . *Säugetierkundliche Mitt.* 9: 1–129.

----- 1969. Systematische Übersicht über die Insektenfresser und Nager Nepals . . . *Khumbu Himal* 3: 149–196.

Weitzel, V. & Groves, C. P. 1985. The nomenclature and taxonomy of the colobine monkeys of Java. *International J. Primatol.* 6: 399–409.

Weitzel, V. *et al.* 1988. A catalogue of primates in the Singapore Zoological Reference Collection . . . *Raffles Bull. Zool.* 36: 1–166.

Wells, D. R. 1989. Notes on the distribution and taxonomy of peninsular Malaysian mongooses (*Herpestes*). *Natural Hist. Bull. Siam Soc.* 37: 87–97.

Wells, D. R. & Francis, C. M. 1988. Crab-eating mongoose, *Herpestes urva*, a mammal new to Peninsular Malaysia. *Malayan Nat. J.* 42: 37–41.

Wemmer, C. & Wathing, D. 1986. Ecology and status of the Sulawesi palm civet *Macrogalidia musschenbroeckii. Biological Conserv.* 35: 1–17.

Wenzel, E. & Haltenorth, T. 1972. System der Schleichkatzen (Viverridae). *Säugetierkundliche Mitt.* 20: 110–127.

West, J. D. *et al.* 1978. *Mus musculus x Mus caroli* hybrids: mouse mules. *Journal Hered.* 69: 321–326.

Wharton, C. H. 1950. Notes on the Philippine tree shrew *Urogale everetti* Thomas. *Journal Mammal.* 31: 352–354.

Whitmore, T. C. (ed.) 1987. *Biogeographical evolution of the Malay Archipelago.* Oxford: Clarendon Press, ix, 147pp.

Whitten, A. J. 1987. The *Presbytis* of Sumatra. *Primate Conserv.* 8: 46–47.

Whitten, A. J. *et al.* 1987a. *The ecology of Sulawesi.* Yogyakarta: Gadjah Mada Univ. Press, 777pp.

----- 1987b. *The ecology of Sumatra.* 2nd ed. Yogyakarta: Gadjah Mada Univ. Press, 583pp.

Whitten, J. E. J. 1981. Ecological separation of three diurnal squirrels in tropical rainforest on Siberut Island, Indonesia. *Journal Zool. Lond.* 193: 405–420.

Willemsen, G. F. 1986. *Lutrogale palaeoleptonyx* (Dubois, 1908), a fossil otter from Java . . . *Proceedings K. ned. Akad. Wet.* (*Palaeont. Geol. Phys. Chem.*) 89: 195–200.

Wilson, C. & Wilson, W. 1976. Behavioural and morphological variation among primate populations in Sumatra. *Yearbook phys. Anthrop.* 20: 207–233.

Wilson, P. 1985. Maeda's *Miniopterus* taxonomy. *Macroderma* 1: 29–36.

Wilson, W. C. & Wilson, C. C. 1975. Species-specific vocalizations and the determination of phylogenetic affinities of the *Presbytis aygula-melalophus* group in Sumatra. Pp. 459–463 in Kondo, S. *et al. Contemporary primatology.* Basel: Karger.

Winge, H. 1887. *Jordfundne og nulevende Gnavere* (*Rodentia*) *fra Brasilien . . . vol. 1*(*3*). Copenhagen: Museo Lundii, 178pp, 8pls.

Winking, H. *et al.* 1979. Karyotype and sex chromosomes in *Vandeleuria oleracea. Zeitschrift Säugetierk.* 44: 195–201.

Winn, H. E. & Reichley, N. E. 1985. Humpback whale—*Megaptera novaeangliae.* Pp. 241–274 in Ridgeway, S. H. & Harrison, R. (*q. v.*).

Wirth, R. 1989. Mustelid and viverrid wealth in Sikkim. *Mustelid and viverrid Conservation* (*IUCN*) 1: 10.

Wirth, R. & Groves, C. 1988. A new deer in China. *Species* 11: 15.

Winton, W. E. de 1899. On Chinese mammals . . . *Proceedings zool. Soc. Lond.* : 572–578, pls 31,32.

Wolfheim, J. H. 1983. *Primates of the world: distribution, abundance and conservation.* Chur (Switzerland): Harwood, 831pp.

Woo, J-k. 1965. Preliminary report on a skull of *Sinanthropus lantianensis* . . . *Scientia sin.* 14: 1032–1035.

Woodward, A. S. 1915. On the skull of an extinct mammal related to *Aeluropus* . . . *Proceedings zool. Soc. Lond.* : 425–428, pl. 1.

World Conservation Monitoring Centre 1990. *1990 IUCN red list of threatened animals.* Gland, Switzerland & Cambridge, UK, 228pp.

Wozencraft, W. C. 1989a. The phylogeny of the Recent Carnivora. Pp. 495–535 in Gittelman, J. L. (*q. v.*).

----- 1989b. Appendix: classification of the Recent Carnivora. Pp. 560–593 in Gittelman, J. L. (*q. v.*).

Wroughton, R. C. 1899. Some Konkan bats. *Journal Bombay nat. Hist. Soc.* 12: 716–725.

----- 1905. "The" common striped palm squirrel. *Journal Bombay nat. Hist. Soc.* 16: 406–413, 1pl.

----- 1906. Notes on the genus *Tatera* . . . *Annals Mag. nat. Hist.* 17: 474–499.

----- 1907. Notes on some rats of the *Mus mettada* group. *Journal Bombay nat. Hist. Soc.* 17: 997–1000.

----- 1908a. A new squirrel from Burmah. *Annals Mag. nat. Hist.* 11: 491.

----- 1908b. On the forms of squirrel hitherto classed under *Sc. finlaysoni*, Hors. *Annals Mag. nat. Hist.* 11: 393–401.

----- 1908c. On some Indian forms of the genus *Micromys. Journal Bombay nat. Hist. Soc.* 18: 280–283.

----- 1908d. Notes on the classification of the bandicoots. *Journal Bombay nat. Hist. Soc.* 18: 736–752.

----- 1910. On a local form of the Chinese toddy-cat taken in north Burma . . . *Journal Bombay nat. Hist. Soc.* 19: 793–794.

----- 1911a. Oriental flying squirrels of the "Pteromys" group. *Journal Bombay nat. Hist. Soc.* 20: 1012–1023.

----- 1911b. On a small collection of rodents from lower Sind. *Journal Bombay nat. Hist. Soc.* 20: 1000–1001a.

----- 1912a. Some new Indian rodents. *Journal Bombay nat. Hist. Soc.* 21: 338–342.

----- 1912b. Some new Indian mammals. *Journal Bombay nat. Hist. Soc.* 21: 767–773.

----- 1915a. The Ceylon hare. *Journal Bombay nat. Hist. Soc.* 24: 41–42.

----- 1915b. The genus *Epimys* in Ceylon. *Journal Bombay nat. Hist. Soc.* 24: 46–49.

----- 1915c. The Burmese civets. *Journal Bombay nat. Hist. Soc.* 24: 63–64.

----- 1915d. The common Indian mongoose. *Journal Bombay nat. Hist. Soc.* 24: 50–54.

----- 1915e. A new monkey from the Chindwin River. *Journal Bombay nat. Hist. Soc.* 24: 55–57.

----- 1916a. New rodents from Sikkim. *Journal Bombay nat. Hist. Soc.* 24: 425–430.

----- 1916b. The squirrels of the *Funambulus palmarum-tristriatus* group in the Peninsula. *Journal Bombay nat. Hist. Soc.* 24: 644–649.

----- 1916c. The Indian jackals. *Journal Bombay nat. Hist. Soc.* 24: 649–653.

----- 1916d. Bombay Natural History Society's Mammal Survey. . . Report no. 26. Darjiling District. *Journal Bombay nat. Hist. Soc.* 24: 773–782.

-----1916e. The langurs of Assam. *Journal Bombay nat. Hist. Soc.* 24: 653–655.

----- 1917. Scientific results from the Mammal Survey xv. *Journal Bombay nat. Hist. Soc.* 25: 40–51.

----- 1918a. A new Indian hare. *Journal Bombay nat. Hist. Soc.* 25: 361–362.

Wroughton, R. C. 1918b. Summary of the results from the Indian mammal survey. Parts 1 & 2. *Journal Bombay nat. Hist. Soc.* 25: 547–598; 26: 19–58.

----- 1919a. Summary of the results from the Indian mammal survey. Parts 3 & 4. *Journal Bombay nat. Hist. Soc.* 26: 338–379; 776–802.

----- 1919b. On the genus *Tadarida* . . . *Journal Bombay nat. Hist. Soc.* 26: 731–733.

----- 1921a. Bombay Natural History Society's mammal survey of India . . . Report no 35. Prome. *Journal Bombay nat. Hist. Soc.* 27: 553–554.

----- 1921b. Scientific results from the Mammal Survey, XXVI. *Journal Bombay nat. Hist. Soc.* 27: 599–601.

----- 1921c. On the *erythraeus* group of squirrels. *Journal Bombay nat. Hist. Soc.* 27: 773–777.

----- 1921d. A renaming of *Mungos mungo ellioti* Wroughton. *Journal Bombay nat. Hist. Soc.* 28: 23.

Wroughton, R. C. & Davidson, W. M. 1919. Two new forms of the "*Funambulus tristriatus*" group. *Journal Bombay nat. Hist. Soc.* 26: 728–730.

Wroughton, R. C. & Ryley, K. V. 1913. Scientific results from the Mammal Survey, III. *Journal Bombay nat. Hist. Soc.* 22: 13–21.

Wu, A. *et al.* 1989. The karyotype of *Eothenomys miletus miletus*. *Acta theriol. sin.* 9: 231.

Wu, D. 1982. On subspecific differentiation of brown rat (*Rattus norvegicus* Berkenhaut) in China. *Acta theriol. sin.* 2: 107–112.

----- 1988. A preliminary study of genus *Rattus* in China. *Abstracts Symp. Asian Pacific Mammal.* Beijing.

Wu, D. & Deng, X. 1984. A new species of tree mice from Yunnan, China. *Acta theriol. sin.* 4: 207–212.

Wu, D. & Wang, G. 1984. A new subspecies of *Typhlomys cinereus* Milne-Edwards from Yunnan, China. *Acta theriol. sin.* 4: 213–215.

Wu, J. 1986. Study of system and distribution of Chinese takin (*Budorcas taxicolor* . . .). *Zoological Res.* 7: 167–175.

Wu, J. & Li, G. 1982. A report on the mammals of Ankang Region, Shanxi Province. *Zoological Res.* 3: 59–68. [= Shaanxi]

Wu, M. & Zhang, S. 1983. New discovery of the palaeoliths in Guizhou Province. *Acta anthropol. sin.* 2: 320–330.

Wu, X. 1981. A well-preserved cranium of an archaic type of early *Homo sapiens* from Dali, China. *Scientia sin.* 24: 530–541.

Wurster-Hill, D. H. & Atkin, N. B. 1972. Muntjac chromosomes: a new karyotype for *Muntiacus muntjak*. *Experientia* 28: 972–973.

Wurster-Hill, D. H. & Seidel, B. 1985. The G-banded chromosomes of Roosevelts' muntjac, *Muntiacus rooseveltorum*. *Cytogenetics Cell Genet.* 39: 75–76.

Wurster-Hill, D. H. *et al.* 1987. Banded chromosome study of the Iriomote cat. *Journal Hered.* 78: 105–107.

Wüst, E. 1922. Beiträge zur Kenntnis der diluvialen Nashörner Europas. *Zentralblatt Miner. Geol. Paläont.* : 641–656; 680–688.

Xia, W. 1984. A study on Chinese *Apodemus* . . . *Acta theriol. sin.* 4: 93–98.

Xu, L. & Liu, Z. 1985. Mammalian fauna of Hainan Island. Pp. 146–151 in Kawamichi, T. (*q. v.*).

Xu, L. & Yu, S. 1985. A new subspecies of Edwards' rat from Hainan Island, China. *Acta theriol. sin.* 5: 131–135.

Xu, L. *et al.* 1983. [*Birds and mammals of Hainan Island.*] Beijing: Science Press.

Xu, W. & Chen, F. 1989. Three new subspecies of *Callosciurus erythraeus* (Pallas). *Acta theriol. sin.* 9: 289–302.

Xu, Y. & Wei, F. 1980. [The discovery of new deer fossils in Long Dong, Fuyang Gaoshan, Zhejang.] *Vertebrata palasiat.* 18: 176–177.

Xue, X. 1981. An early Pleistocene mammalian fauna and its stratigraphy of the river Yon, Weinan, Shensi. *Vertebrata palasiat.* 19: 35–44.

Yang, A. & Fang, L. 1988. Phallic morphology of 13 species of the family Muridae from China . . . *Acta theriol. sin.* 8: 275–287.

Yang, C. M. *et al.* 1990. Wild mammals of Singapore. Pp. 1–23 in *Essays in zoology* . . . Singapore: National University.

Yang, D-h. 1988. Black snub-nosed monkeys in China. *Oryx* 22: 41–43.

Yang, G. & Wang, Y-x. 1987. A new subspecies of *Hadromys humei* from Yunnan, China. *Acta theriol. sin.* 7: 46–50.

Yang, G-r. & Wu, D-l. 1979. Two new records of Chinese rodents. *Acta zootaxon. sin.* 4: 192–193.

Yates, T. L. 1982. Historical zoogeography and evolution of North American moles . . . *Abstracts 3rd int. theriol. Congr.* : 264.

Yenbutra, S. & Felten, H. 1983. A new species of the fruit bat genus *Megaerops* from SE Asia . . . *Senckenbergiana biol.* 64: 1–11.

----- 1986. Bat species and their distribution in Thailand . . . *Courier ForschInst. Senckenberg* 87: 9–45.

Yochem, P. K. & Leatherwood, S. 1985. Blue whale—*Balaenoptera musculus*. Pp. 193–240 in Ridgeway, S. H. & Harrison, R. (*q. v.*).

Yonekawa, H. *et al.* 1986. A hybrid origin of Japanese mice *Mus musculus molossinus*. *Current Topics Microbiol. Immunol.* 127: 62–67.

Yong, H-s. 1969. Karyotypes of Malayan *Rattus* . . . *Chromosoma* 27: 245–267.

----- 1970. A Malayan view of *Rattus edwardsi* and *R. sabanus*. *Zoological J. Linn. Soc.* 49: 359–370.

----- 1971. Rat from Pulau Tenggol, Trengganu. *Malay Nat. J.* 24: 87–89.

----- 1972. The systematic status of Malayan *Rattus rajah* and *Rattus surifer*. *Bulletin Br. Mus. nat. Hist. Zool.* 23: 155–163, pl. 1.

----- 1973. Chromosomes of the pencil-tailed tree-mouse *Chiropodomys gliroides* . . . *Malayan Nat. J.* 26: 159–162.

Yong, H-s. & Dhaliwal, S. S. 1972. Supernumary (B-) chromosomes in the Malayan house rat, *Rattus rattus diardii*. *Chromosoma* 36: 256–262.

----- 1976. Variations in the karyotype of the red giant squirrel *Petaurista petaurista* . . . *Malaysian J. Sci.* 4: 9–12.

Yong, H-s. *et al.* 1975. Karyotypes of four species of *Callosciurus* . . . from peninsular Malaysia. *Malaysian J. Sci.* 3: 1–5.

----- 1982. Karyotypes of *Hapalomys* and *Pithecheir* . . . from Peninsular Malaya. *Cytologia* 47: 535–538.

Yongzu, Z. & Zhao, K. 1978. On the zoogeographical regions of China. *Acta zool. sin.* 24: 196–202.

Yoon, M. H. *et al.* 1990. Taxonomic validity of scientific names in Japanese *Vespertilio* species . . . *Journal mammal. Soc. Japan* 14: 119–128.

Yoshiyuki, M. 1971. On the external and cranial characters of *Aonyx cinerea*. *Journal mammal. Soc. Japan* 5: 117–119, 1 pl.

----- 1979. A new species of the genus *Ptenochirus* . . . from the Philippine Islands. *Bulletin natn. Sci. Mus. Tokyo* (A), Zool. 5: 75–81.

----- 1989. *A systematic study of the Japanese Chiroptera.* Tokyo: National Science Museum.

----- 1990. Notes on Thai mammals 2. Bats of the *pusillus* and *philippinensis* groups of the genus *Rhinolophus* . . . *Bulletin natn. Sci. Mus. Tokyo* A, 16: 21–40.

Yoshiyuki, M. *et al.* 1989. Taxonomic analysis of two rare bats from the Amami Islands . . . *Memoirs natn. Sci. Mus. Tokyo* 22: 215–225.

Yosida, T. H. 1980. *Cytogenetics of the black rat* . . . Tokyo: Univ. Tokyo Press; Baltimore: Univ. Park Press, 256pp.

----- 1982. Karyotype of hybrid male between the domestic cat and the leopard cat. *Report natn. Inst. Genet. Misima* 32 (1981): 66–67.

----- 1985. The evolution and geographical differentiation of the house shrew karyotypes. *Acta zool. fenn.* 170: 31–34.

You, Y-z. & Qi, G-q. 1973. New materials of Pleistocene mammals in Yuanmu, Yunnan. *Vertebrata Palasiat.* 11: 66–85, 4 pls.

Young, C-c. 1927. Fossile Nagetiere aus Nord-China. *Palaeontologia sin.* C5,3: 82pp, 3pls.

----- 1934. On the Insectivora, Chiroptera, Rodentia and Primates other than *Sinanthropus* from Locality 1 at Choukoutien. *Palaeontologia sin.* C8,3: 139pp, 10pls.

----- 1937. New finds of fossil *Bubalus* in China. *Bulletin geol. Soc. China* 15: 505–516, pl. 1.

----- 1947. Notes on a Pleistocene microfauna from Loping, Kiangsi. *Bulletin geol. Soc. China* 27: 163–170.

----- 1948. *Budorcas*, a new element in the protohistoric Anyang fauna of China. *American J. Sci.* 246: 157–164.

Young, C-c. & Liu, P-t. 1951. On the mammalian fauna at Koloshan, near Chungking, Szechuan. *Bulletin geol. Soc. China* 30: 43–90.

Yu, Y. & Chen, F. G. 1988. A taxonomic study on the noctule (*Nyctalus*). *Abstracts Symp. Asian-Pacific Mammal.*, Beijing 1988.

Zagorodnyuk, I. V. 1990. Karyotype variability and systematics of the Arvicolini . . . *Vestnik Zool.* 1990(2): 26–37.

Zahn, W. 1942. Die Riesen-, Streifen-, und Spitznasen-hornchen der orientalischen Region. *Zeitschrift Säugetierk.* 16: 1–182.

Zdansky, O. 1924. Jungtertiäre Carnivoren Chinas. *Palaeontologia sin.* C2 (1): 145pp, 33pls.

Zelebor, J. 1869. Säugethiere. 42pp, 3pls in Wüllerstorf-Urbair, B. von (ed.) *Reise der . . . Novara . . . Zoologische Teil, 1.* Wien.

Zhang, M. 1985. Preliminary study of the Holocene (Neolithic site) mammalian fauna of South China. Pp. 163–167 in Kawamichi, T. (*q. v*).

Zhang, Y. & Zhao, K. 1978. On the zoogeographical regions of China. *Acta zool. sin.* 24: 196–202.

Zhang, Z. & Zhao, M. 1984. A new subspecies of the sulphur-bellied rat from Jilin . . . *Acta zool. sin.* 30: 99–102.

Zhao, Z. 1980. Fossil mammals from Jiaoshan Wuming, Guangxi. *Vertebrata palasiat.* 18: 299–303.

Zhen, X-y. 1987. A survey of the bats (Chiroptera) of Fujian Province. *Wuyi Sci. J.* 7: 237–242.

Zheng, C. & Wang, S. 1985. On the insectivore fauna of Qinghai-Xizang. . . Plateau. *Acta theriol. sin.* 5: 35–40.

Zheng, S. 1983. Micromammals from the Hexian Man locality. *Vertebrata palasiat.* 21: 230–240.

----- 1985. Remains of the genus *Anourosorex* . . . from Pleistocene of Guizhou district. *Vertebrata palasiat.* 23(1): 39–51.

----- 1988. Preliminary studies of fossil *Hapalomys, Chiropodomys, Vernaya* and *Vandeleuria* . . . from southwest China. *Abstracts Symp. Asian-Pacific Mamm., Beijing, 1988.*

Zheng, Y. & Xu, L. 1983. Subspecific study of the ferret-badger (*Melogale moschata*) in China. *Acta theriol. sin.* 3: 165–171.

Zhou, K. *et al.* 1979. The osteology and the systematic position of the baiji, *Lipotes vexillifer*. *Acta zool. sin.* 25: 58–74.

Zhuge, Y. 1982. On the geographical distribution and the mammalian fauna of Zhejiang Province. *Acta theriol. sin.* 2: 157–166.

Zimmermann, E. A. W. 1777. *Specimen zoologiae geographicae, quadrupedum . . .* Lugduni Batavorum: xxiv, 685pp.

----- 1780. *Geographische Geschichte des Menschen . . .* Leipzig.

Zimmermann, K. 1962. Die Untergattungen der Gattung *Apodemus*. *Bonner zool. Beitr.* 13: 198–208.

Zubaid, A. 1988. The second record of a trident horseshoe bat, *Aselliscus stoliczkanus . . .* from Peninsular Malaysia. *Malayan Nat. J.* 42: 29–30.

Zubaid, A. & Davison, G. W. H. 1988. *Rhinolophus marshalli*, a new horseshoe bat recorded in Peninsular Malaysia. *Sains Malaysiana* 17: 367–368.

Zukovsky, L. 1927. Bemerkungen über die rassenweise Verschiedenheit der Hirschziegenantilope. *Illustrierte Tier- Menschenwelt* 2: 124–127.

----- 1928. Weitere Mitteilungen über die Systematik der Hirschziegenantilope. *Illustrierte Tier- Menchenwelt* 3: 60–62.

----- 1964. Weitere Mitteilungen über Persische Panther. *Zoologische Gart. Leipz.* 28: 151–182, 17pls.

Index of species- and genus-group names

The names used in the text for valid Recent genera and species are printed here in bold. Species-group names are followed first by the current generic name, then by the original generic name if different.